Canine and Feline Skin Cytology

Francesco Albanese

Canine and Feline Skin Cytology

Second Edition

Springer

Francesco Albanese
Naples, Italy

ISBN 978-3-031-95712-3 ISBN 978-3-031-95713-0 (eBook)
https://doi.org/10.1007/978-3-031-95713-0

This Springer imprint is published by the registered company Springer Nature Switzerland AG
The registered company address is: Gewerbestrasse 11, 6330 Cham, Switzerland

Unfortunately you are no longer with us in this world that you have lived with joy and respect, as only the right people can do. You have been an exceptional person and, to me, a brother rather than a colleague.
I dedicate this book to you wholeheartedly, the credit for which is also yours for all you have done for me during our years of university studies. Without your patience and help, I would probably never have been able to graduate.
Dear Leo, my friend, I miss you so much.

Acknowledgments

Well, I have come to write the second edition of the Skin Cytology book, and this is indicative that colleagues appreciated the first edition and this fills me with pride.

More than 5 years have passed and it is time to update the text and enrich it with new images of cases that I have been lucky enough to see in recent years, both as a clinician and as a lover of both cytological and histological skin pathology.

The thanks are always numerous and the list of people I have to thank is always very long, but for this second edition I would like to thank mainly five people, for me very special, who enrich my life both private and scientific.

First of all, I want to thank Luisa Vera Muscatello, an incredible pathologist at the University of Veterinary Medicine of Bologna (Italy), with whom I share many scientific ideas regarding general pathology on a daily basis and who always keeps alive in my mind, the flame of knowledge, thank you Luisa, you are a continuous stimulus.

I want to thank two other colleagues, my "skin cytologist brothers," with whom I share skin cytology cases on a daily basis, namely, Fabiano Necci and Amanda Prolo.

A big thank you to Francesca Abramo, a great dermatopathologist, and a sister more than a colleague.

Thanks to Chiara Caporali, colleague and wife, who supported me and, why not, put up with me every day and with whom I share so many interesting clinical cases of canine and feline dermatology as well the life with our two children Veronica and Emilio.

Thanks to all of you guys, thanks for everything, sincerely.

I would also like to thank the biologists of both histopathology and cytology departments of the CDVet, the analysis laboratory where I work, who supported me in the last year in the clippings and in the coloring of the slides from which many of the photos in this book were taken; in strict alphabetical order I thank Justyna Barbara Koczwara, Giulia Manfredi, Flavia Menichini, Camilla Paradisi, Roberta Proietto, and Eleonora Scorsi.

A sincere thanks goes also to all the colleagues who have sent me cytological preparations in these more than 30 years of profession, allowing me to gain a great cytological experience, and to all those who have provided me with the clinical photos that I was missing, allowing me to enrich the book iconographically. Thanks to Tommaso Natalia for the schematic drawings of the skin lesions used in the Chap. 1.

Lastly, thanks to Oscar Tronconi of TiEsseLab of Milan (Italy), who provided a 20 MP digital camera and a microscope, NeXcope NE 920-D, with which almost all of the cytological and histopathological photographs published in this second edition were taken.

Competing Interests The author has no competing interests to declare that are relevant to the content of this manuscript.

Contents

About the Author

DVM Francesco Albanese graduated in Veterinary Medicine at the University of Naples "Federico II" in 1993. Since 1994 he has been focusing his interest in veterinary dermatology by attending many congresses and courses in Italy and abroad. From 1996 to 1998 he attended the advanced course in dermatology at the European School for Advanced Veterinary Studies (ESAVS). He is a full member of both the European (ESVD) and the Italian (SIDEV) Society of Veterinary Dermatology for which he held the position of treasurer and advisor for several years. He is an invited speaker in many national and international courses, seminars, and meetings on canine and feline skin cytology. He has published in both national and international journals. He has authored some books on feline and canine dermatology and skin cytology. He currently works as freelancer on referred dermatological cases. He is a consultant of dermatology and cytology for the private veterinary laboratory "CDVet" in Rome.

Skin Lesions Morphology and Techniques of Sampling and Preparation of Cytological Specimens

1

1.1 Introduction

The lesions that can be observed on the skin of dogs and cats are numerous and characterized by a wide polymorphism, linked to the disease that caused them. Papules, pustules, collarettes, scales, erosions, ulcers, plaques, and nodules are frequently observed on the skin of dogs and cats affected by dermatological diseases and are macroscopically easy to recognize by visual examination. Some of these lesions may be present at the same time, representing different stages of the same disease or clinical signs of different coexisting diseases.

Cytology is one of the fastest, cheapest, and simplest diagnostic tools that doctors can perform in their daily practice and is essential to obtain an immediate diagnosis or to decide on the choice of further diagnostic tests to be performed; as well as for other tests, a good quality of the specimens is mandatory for a right interpretation. Even if more than one sampling technique can be performed to collect cells from the same lesion, there is always the most appropriate technique that offers more guarantees of obtaining a "representative" sample. This chapter discusses the sampling methods in canine and feline skin cytology in the light of clinical presentation; therefore, starting from clinical lesions, the reader is provided with fundamental information for the correct preparation of slides. The execution of a correct cytological sampling cannot disregard three basic steps:

(a) *The choice of the lesion to be sampled*
(b) *The choice of the most appropriate sampling method for that specific lesion*
(c) *A proper preparation of the slide*

1.2 Sampling Techniques

The most commonly used sampling techniques include *imprint smears*, *fine-needle biopsy* (FNB) or *fine-needle aspiration biopsy* (FNAB), and *scraping.*

1.2.1 Imprint Smears

The *imprint smear* is the only method we can perform when cells must be sampled from flattened and superficial exudative lesions. With this method, it is possible to collect cells simply by gently placing a slide on the skin surface. As the cells adhere to the slide, it is sufficient to gently place it on the lesion without exerting excessive pressure which could damage the cell morphology. Nevertheless, there are some precautions that differ according to the lesion that must be investigated, which are discussed below.

1.2.1.1 Cytological Sampling from *Papules*

Papules are small, solid, raised primary lesions, usually erythematous, smaller than 1 cm (usually a few millimeters) (Figs. 1.1 and 1.2). Histopathologically, papules are composed of inflammatory or, rarely, neoplastic cells, located in the superficial dermis and usually cause a slight hyperplasia of the overlying epidermis. Intra-epidermal exocytosis of inflammatory cells or small ulceration of the epidermidis, the latter often located on the top of the papule, cause oozing of a small amount of exudate that, once dehydrated, gives rise to a small hematic crust (Fig. 1.3). These *crusted papules* are commonly observed in cats with a peculiar clinical presentation named "miliary dermatitis" while in dogs are commonly observed in case of sarcoptic mange (Fig. 1.4).

Papules with intact surfaces are too small for FNB and cells cannot be collected using the imprint technique; therefore, only the ulcerated lesions, or those covered by crusts, permit the collection of a few cells using the imprint technique. The term *nodular papule* is used to describe a raised lesion that is round to oval in shape, of intermediate size between a papule and a nodule; in fact, they are less than 1 cm but do not appear large enough to be defined as nodules (Fig. 1.5). These lesions, unlike the papules, provide more rich specimens since a higher number of cells can be collected through FNB.

F. Albanese, *Canine and Feline Skin Cytology*, https://doi.org/10.1007/978-3-031-95713-0_1

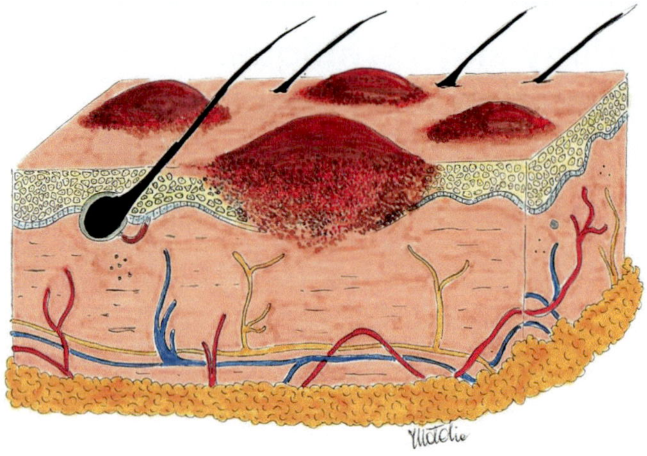

Fig. 1.1 Schematic representation of a papule

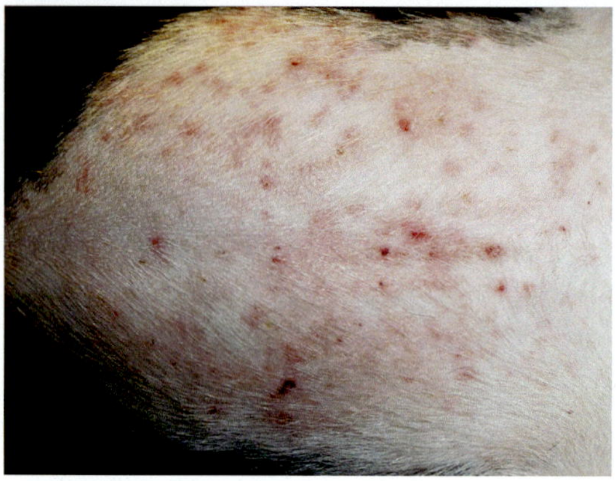

Fig. 1.4 Multiple crusted papules on the skin of a dog affected by sarcoptic mange

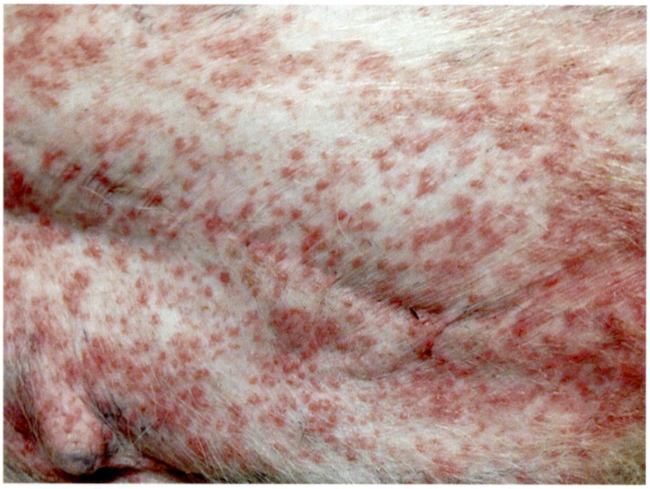

Fig. 1.2 Multiple erythematous papules on the abdomen of a dog with superficial pyoderma

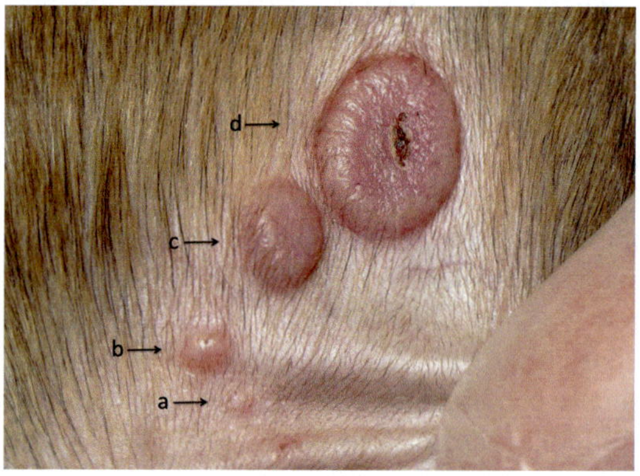

Fig. 1.5 Nomenclature based on size: (**a**) papule (few mm in size); (**b**) papular nodule (less than 1 cm); (**c**) nodule (1 cm or more); and (**d**) plaque (1 cm or more that grows more in width than in height). (Courtesy of DVM F. Leone—Italy)

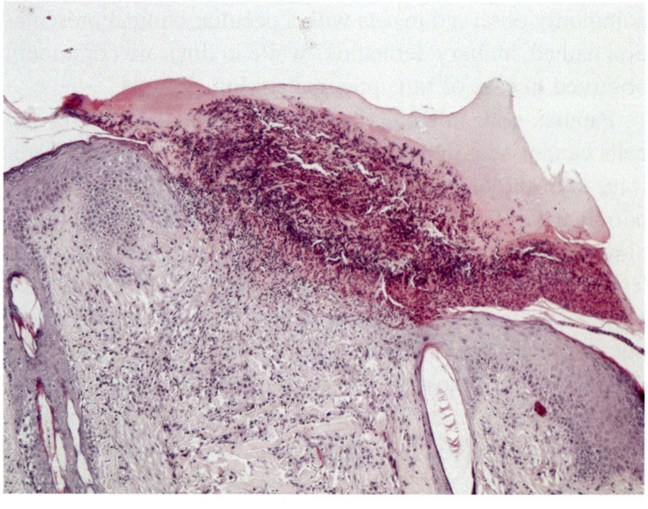

Fig. 1.3 Histology of a crusted papule: note the crust covering the ulcer

The author prefers to treat the cytological aspects of *papular nodular* lesions separately from papules, as he believes that the former are more characteristic of some specific skin diseases and, for this reason, deserve a more detailed clinical differentiation and cytological description.

Sampling Technique

Usually, *papules* do not release many cells, but, in some cases, even a few cells may be useful for their interpretation.

When the epidermidis is intact, the collection of cells is not possible using the imprint technique, while, when the overlying epidermis is ulcerated or when leukocytic exocytosis through the epidermis occurs, it is possible to collect some cells by gently placing a slide on its top. In order to collect cells from papules covered by crusts, the latter must be removed before placing the slide on the wet material (Fig. 1.6). In the case that papules are not so small, it is instead possible to try collecting cells via FNB using a small needle (23 G), as discussed below (Fig. 1.7).

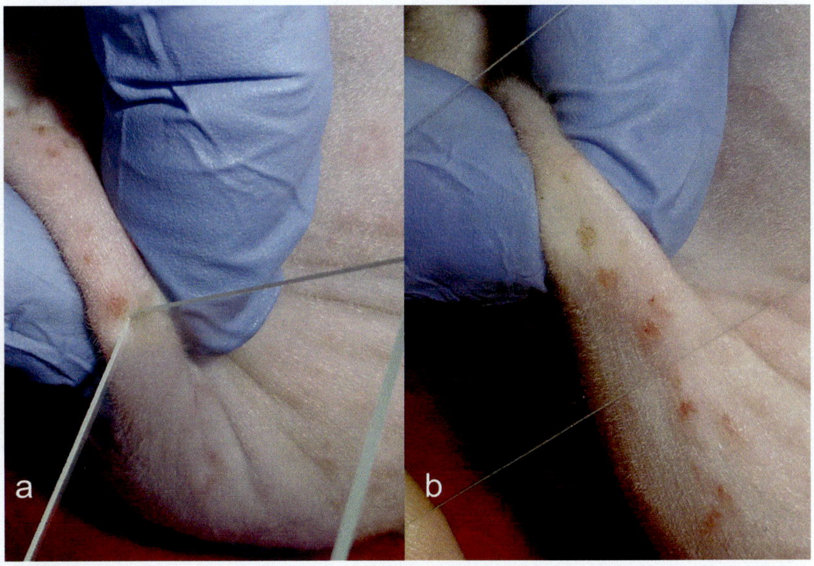

Fig. 1.6 (**a**) Removal of a crust with a slide and (**b**) placing the slide onto the exposed wet material

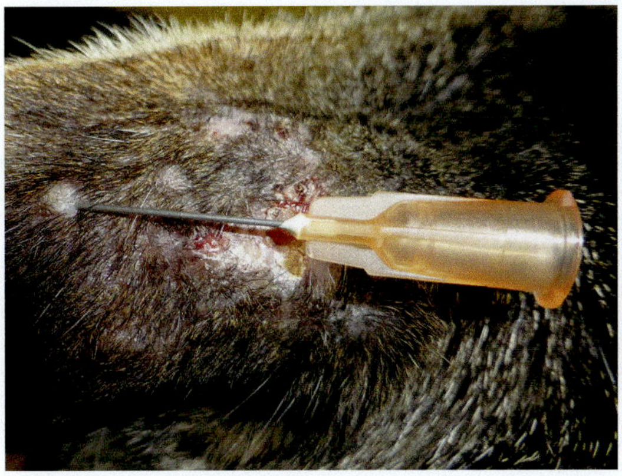

Fig. 1.7 FNB: a small needle of 23 G, not connected to a syringe, can be inserted into a small nodular papule

1.2.1.2 Cytological Sampling from *Pustules*

Pustules are raised, soft, yellowish, intra-epidermal or intra-follicular lesions. The yellowish color of the pustules is closely related to their content of granulocytes, neutrophils or eosinophils or both, which are the main cells that compose the purulent exudate (Figs. 1.8 and 1.9).

By medical definition, pus is an inflammatory exudate composed of karyolytic neutrophils, necrotic debris, and bacteria; according to this definition, the term pus should only be used to describe the content of pustules in case of bacterial infection (pyoderma). As also different *sterile* pustular diseases are recognized in canine and feline dermatology, the term "pus" should be referred only to describe its granulocytic composition and should not imply a bacterial infection. This nomenclature will be used in this book.

In addition to granulocytes, pustules may contain other types of cells (keratinocytes) or microorganisms (bacteria), and parasites (*Demodex* spp.); their cytological finding provides important diagnostic clues.

Clinically, pustules can be divided into *follicular* and *non-follicular. Follicular pustules* are located in the follicular lumen and are clinically recognizable by their small size and, when still present, by the presence of one hair that emerges from its center. Since dog hair follicles are compounds, each follicle is composed of a central primary hair surrounded by 3–15 smaller secondary hairs, and all hairs emerge from one pore. Therefore, one should not be surprised if more hairs emerge from the center of a small follicular pustule. Follicular pustules are very small, and they are from roundish to oval in shape and of few millimeters in size (Fig. 1.10). The term *mural follicular pustule* is used in histopathology to describe a pustule located in the thickness of the follicular wall, as is sometimes observed in pemphigus foliaceus; in these cases, the pus is not free in the follicular lumen, and therefore the pus does not escape from the follicular ostium in the form of a pustule (Fig. 1.11).

Therefore, when the reader reads in this book the term "follicular pustule," the latter must always be interpreted from a clinical point of view and refers to lesions characterized by the presence of pus in the follicular lumen. Follicular pustules are always indicative of pyoderma (bacterial folliculitis); in fact, although even in the course of demodicosis it is possible to observe follicular pustules, it should be highlighted that they have a bacterial origin too and therefore, it can be concluded that, in any case of a clinically visible follicular pustule, we are certain that a bacterial infection is present.

In *non-follicular pustules*, since follicles are not involved, pus is not located in the follicular lumen, but it is contained in intra-epidermal or intra-corneal layers (Fig. 1.12). They can be of variable size, ranging from a few millimeters to over 2 cm, and can be round or irregular in shape. When the non-follicular pustules are very large, it is clinically possible to see many hairs emerging from several parts of them; in

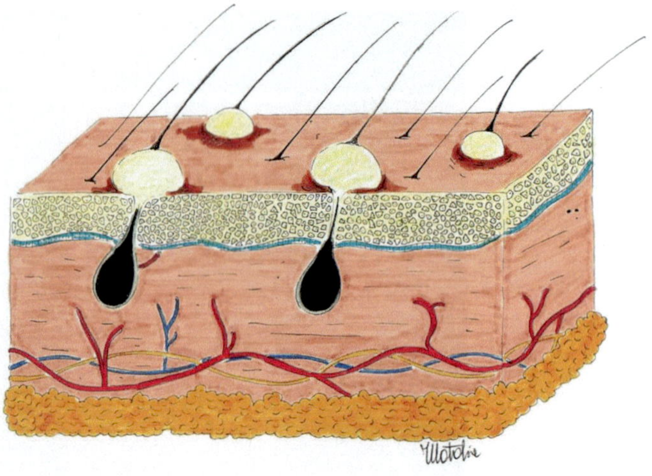

Fig. 1.8 Schematic drawing of a "follicular" pustule

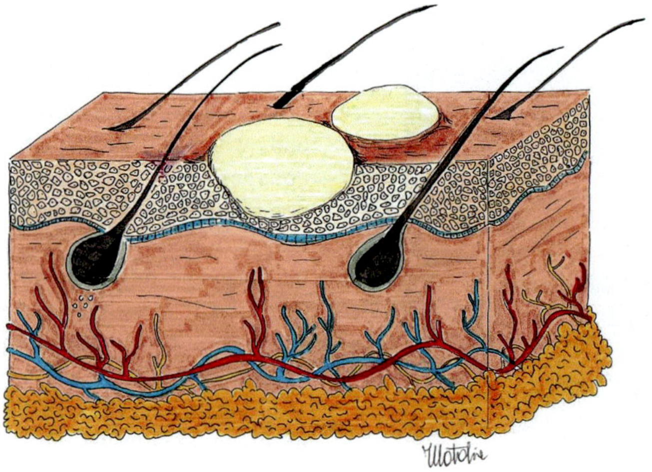

Fig. 1.9 Schematic drawing of a "non-follicular" pustule

Fig. 1.10 Follicular pustule in a dog with demodicosis: (**a**) histology: note how the pustule is located on the center of the follicle and (**b**) clinical aspect of a follicular pustule: a hair shaft emerges from the center of the pustule

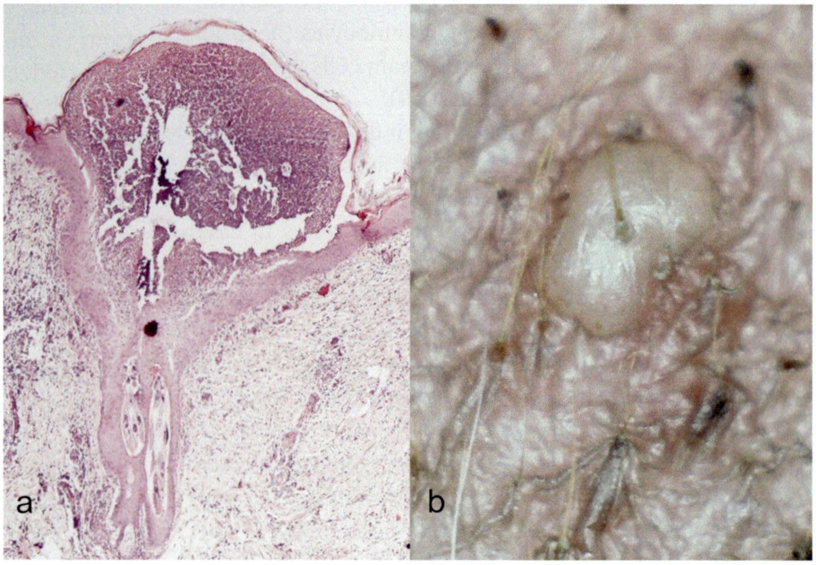

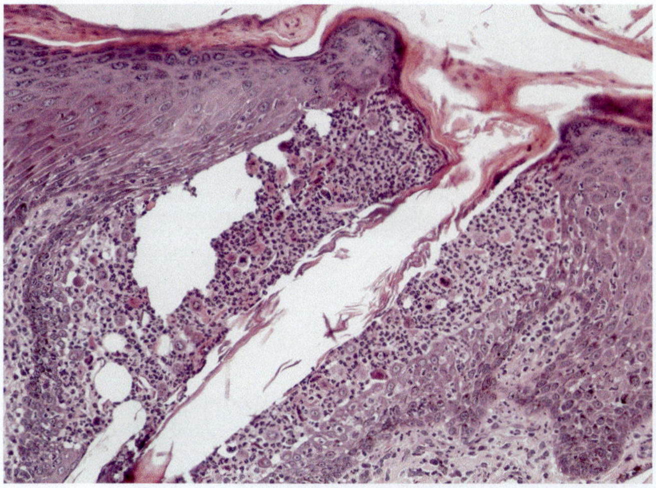

Fig. 1.11 Pemphigus foliaceus: histology of two intramural acantholytic follicular pustules. Note how the pustules are confined in the follicular wall and, as they do not emerge from the center of the follicle, they are not macroscopically evident; in fact, they can be only detected via histological examination

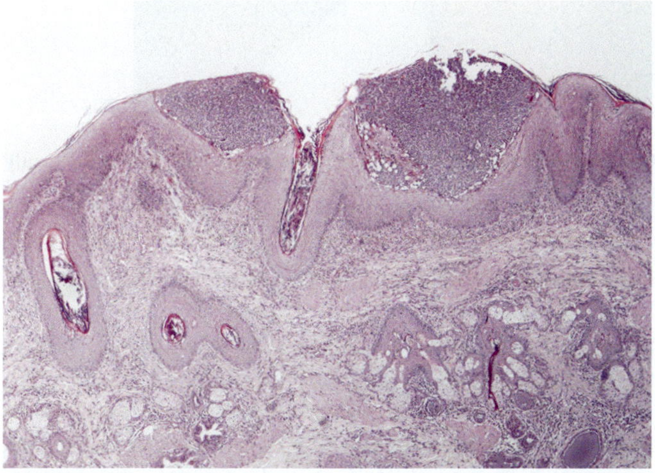

Fig. 1.12 Histology of two non-follicular pustules: note how the pustules do not involve the follicle

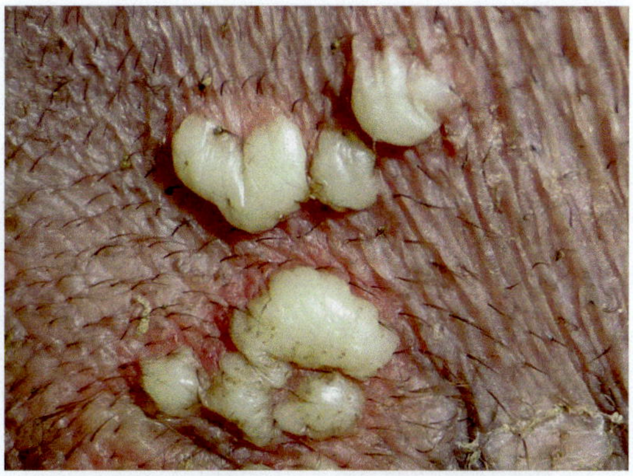

Fig. 1.13 Large and non-follicular pustules of irregular shapes in a dog with juvenile impetigo

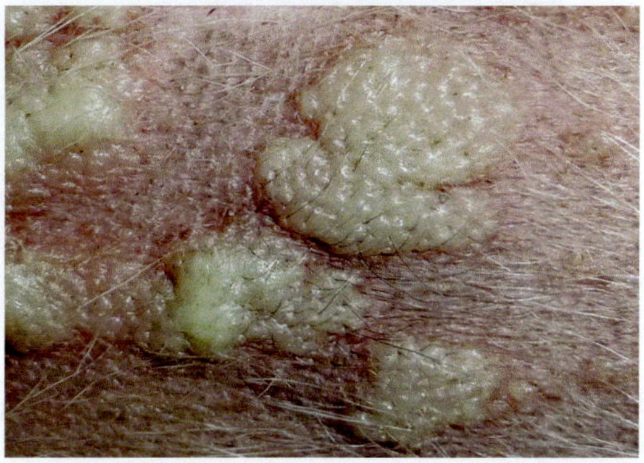

Fig. 1.14 Non-follicular pustules: note as many hairs emerge from the pustules since the latter is very large

these cases, since the follicles are not involved in the pathological process, the hairs do not necessarily fall and they come out of the pustule. In practice, hair shafts emerge from the pustule only because the latter is extended and occupies a large area in which many not-infected follicular units are anatomically present (Figs. 1.13 and 1.14).

The clinical aspect and localization of pustules do not have only a descriptive meaning, but it is very useful when clinicians must interpret them along with the cytological results.

Sampling Technique

Pustules are one of the most important lesions in canine skin cytology, as they usually represent the direct action of a disease, both infectious (bacteria) and sterile. In the case of large pustules, cells can be collected using the imprint smear technique; this method provides the breakage of the base of the pustule with a fine needle and lifts the keratinocytes that form its "roof." The pus is exposed, and a sample can be taken by gently placing a slide on it. This method is a variant of the technique conceived by Dr. Arnault Tzanck, a French dermatologist of human medicine, who was a pioneer of diagnostic cytology in the early twentieth century, and for this reason, the technique is worldwide known as the *Tzanck test* (Fig. 1.15).

When pustules are too small, as follicular ones, the *Tzanck test* is impossible to perform; in these cases, the shorter edge of the slide must be slightly pushed on the lateral base of the pustule, allowing the pus to be transferred onto the slide (Fig. 1.16).

Once collected, the material must be gently smeared onto another slide and left to dry to air. Pustules tend to rapidly dehydrate and transform into yellowish crusts or epidermal collarettes. In these cases, it is possible to obtain cells by collecting the exudate present below the crusts or after lifting the edges of the collarettes, as discussed below.

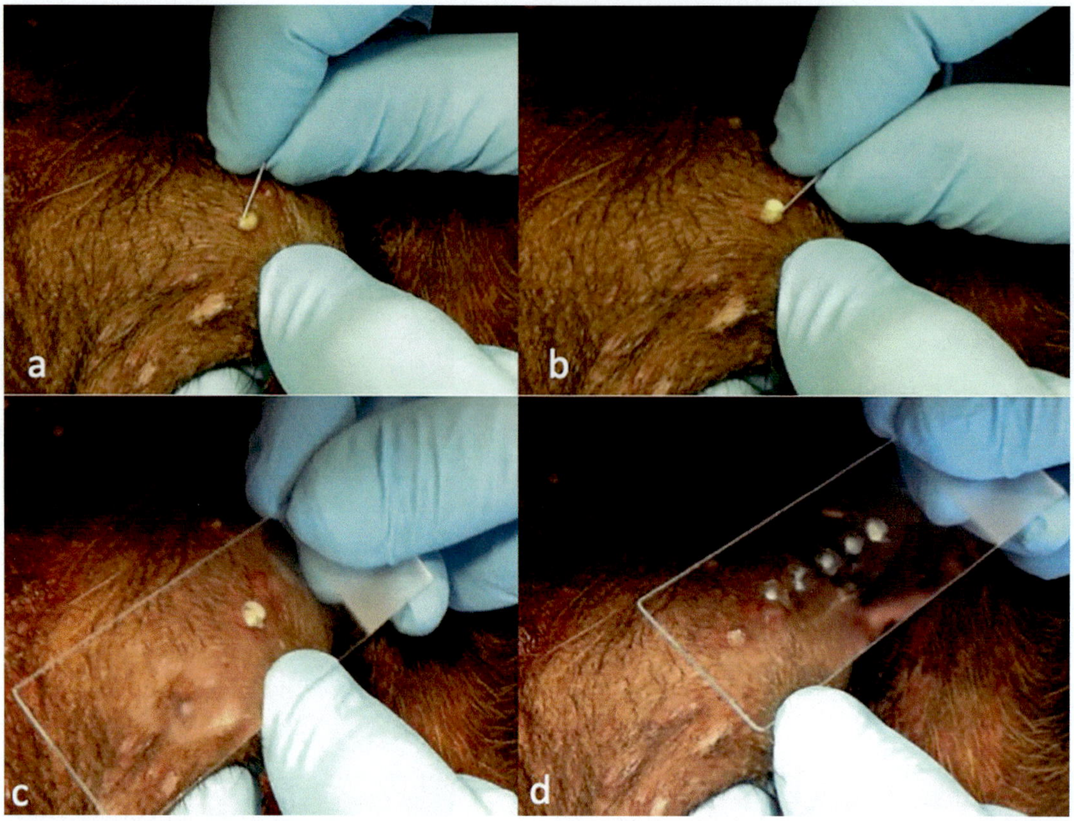

Fig. 1.15 Tzanck test: (**a** and **b**) rupture of the base of a large pustule with a needle and (**c** and **d**) the pus is transferred onto the slide by gently placing a slide (impression technique)

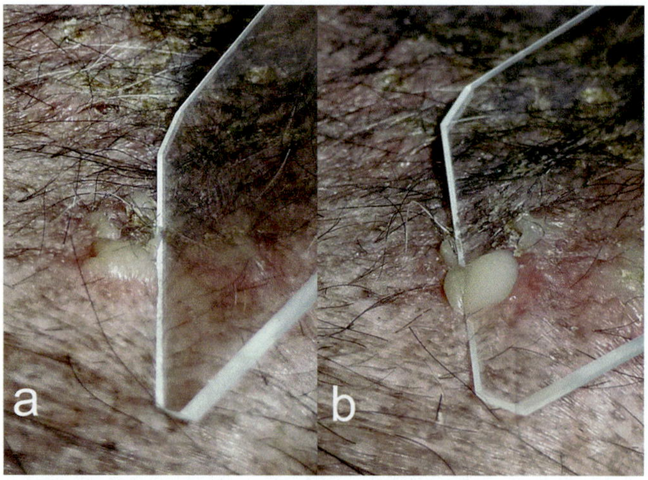

Fig. 1.16 Sampling technique for a small pustule: (**a**) lateral pressure is placed on a small pustule with a slide and (**b**) the pus is transferred onto the slide

1.2.1.3 Cytological Sampling from Intact Skin Covered with "Dry" or "Oily/Waxy"

Scales

Scales are aggregates of corneocytes that represent the final product of the physiological keratinization process of the epidermidis and are continuously released into the environment, both singly and in large groups (Fig. 1.17). In healthy skin in which the epidermidis renewal is regular, scales are not clinically visible. When scales are clearly visible to the naked eye, the clinical dermatological pattern is described as *exfoliative dermatitis*, which is observed in many both canine and feline skin diseases. Scaling diseases recognize both *primary* and *secondary* causes; the formers are usually linked to genetic or idiopathic disorders, and in these cases, cytology has no diagnostic interest. Scales are also very common in secondary keratinization defects, caused by microorganisms (dermatophytosis, superficial pyoderma, *Malassezia* overgrowth), parasites (demodicosis, Cheyletiella spp. infestation), protozoa (leishmaniasis), autoimmune diseases (pemphigus foliaceus), and rarely to skin neoplasia (epitheliotropic lymphoma); in all these cases, cytology can provide diagnostic results or can direct clinicians in selecting additional diagnostic tests.

Sampling Technique

Placing a slide onto an oily/waxy skin allows keratinocytes and microorganisms present on their surface to be collected. In patients with dry scaly dermatitis, the imprint technique does not permit the cells to adhere to the slide; in these cases, sampling requires the use of a piece of transparent acetate tape. This method is very useful when searching for yeasts

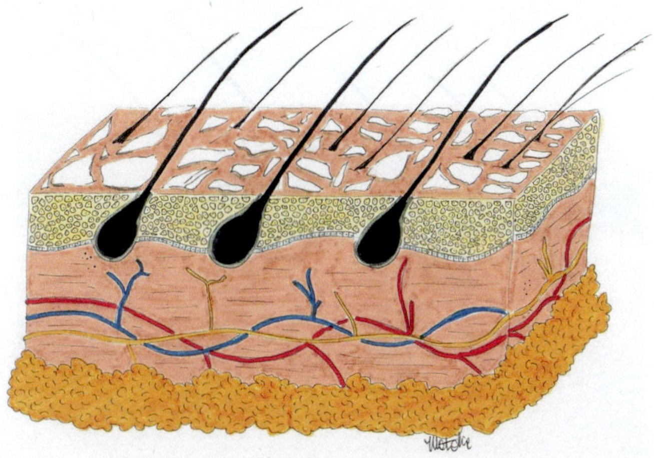

Fig. 1.17 Schematic representation of scales

Fig. 1.18 Sampling technique from exfoliative dermatitis: (**a**) a piece of transparent tape is placed directly on the scaly skin and (**b**) note as many scales remain stuck to the adhesive part of the acetate tape

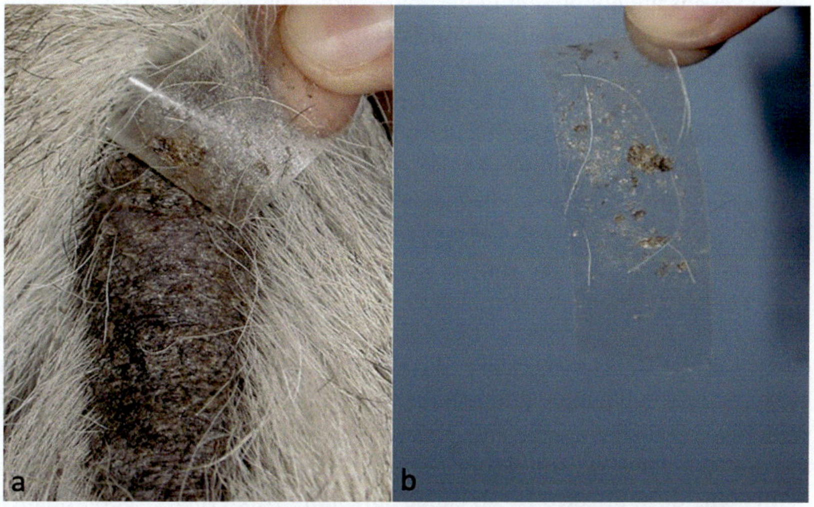

and dermatophytes, but it can be used to collect any type of cells or parasites present on the skin surface (Fig. 1.18).

The use of transparent tape is fundamental when cells must be collected from areas with deep folds, such as interdigital areas or lips; in fact, these body areas cannot be easily sampled by placing a slide (Fig. 1.19).

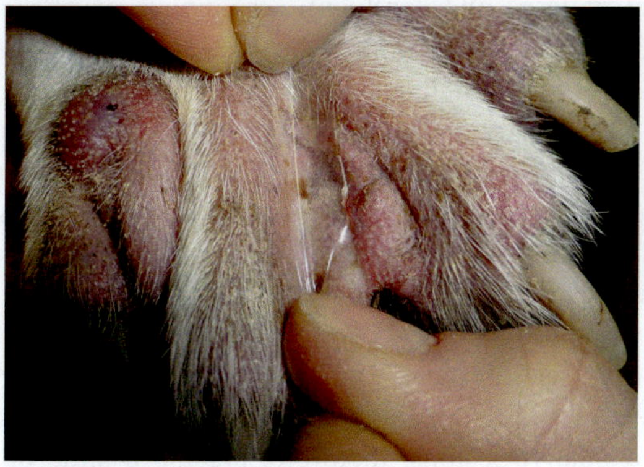

Fig. 1.19 Transparent acetate tape: a piece of tape is placed on the interdigital skin

1.2.1.4 Cytological Sampling from *Erosions*

Erosions are superficial lesions characterized by the loss of one or more layers of the epidermis, without involving the basement membrane and so, the dermis (Fig. 1.20). Determining the integrity of the basement membrane is not always clinically possible, particularly when the erosion is deep and its base is composed only of the basal keratinocytes. Since

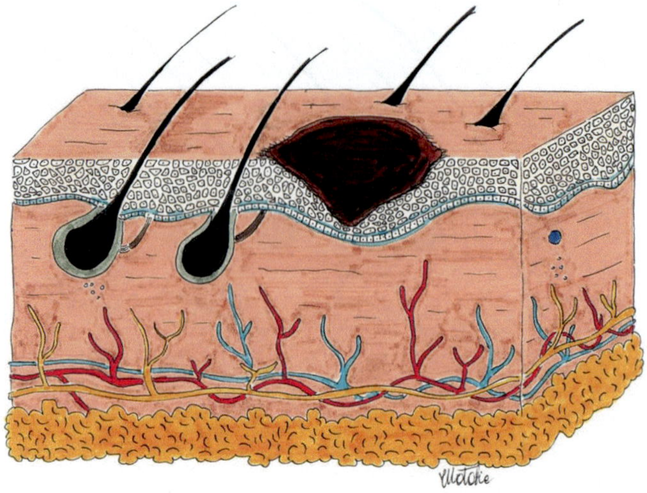

Fig. 1.20 Schematic representation of an erosion. Note how the lesion is located in the epidermis with no dermal involvement

blood vessels are not present in the epidermidis, the cytological examination can confirm or rule out an ulcer, based on the presence or not of red blood cells, respectively.

Epidermal collarettes are secondary lesions and are the most frequent erosions observed in canine dermatology; they are very useful for cytological examination. Collarettes are roundish erosion, with an erythematous or, in chronic cases, hyperpigmented central area and are peripherally bordered by yellowish crusts which is another clinical indicator of the presence of granulocytes (dehydrated and broken pustules). Collarettes are very common in dogs due to the high frequency of superficial pyoderma in this species and are usually observed in combination with papules and pustules. Rarely, epidermal collarettes are the only type of detectable lesion on the skin and, thus, the only lesion that can be sampled.

In some dogs affected by so-called *spreading pyoderma*, the intra-corneal diffusion of the infection leads to the development of large and confluent, serpiginous or plyciclic in shape, *collarette-like* lesions, which give a clinical presentation similar to exfoliative dermatitis.

Less frequently, collarettes are observed in other pustular sterile diseases such as pemphigus foliaceus, in which cytology sampled under the peripheral crusts could be very useful in highlighting segmented neutrophils along with acantholytic keratinocytes.

Sampling Technique

Because of their superficial location, erosions are flat and often exudative lesions; therefore, placing a slide on their surface can provide useful information. As for all lesions that are exposed to the external environment, cytological specimens obtained via impression smear technique must always be interpreted with caution. Clinicians must be aware of the frequent possibility of bacterial contamination of specimens coming from both the environment and oral cavity in case of licking such as *Simonsiella* spp. (*Conchiformibius*), rod-shaped and filamentous bacteria.

The sampling of cells from epidermal collarettes with dry central skin is performed by lifting the crusts that border the lesion and placing a slide on the exposed fresh pus (Fig. 1.21).

1.2.1.5 Cytological Sampling from *Ulcers*

Ulcers are deep lesions that represent an excavation in the skin with a loss of all epidermal layers and part of the dermis (Fig. 1.22). Based on this definition, it is clear as it is normal to find red blood cells in cytological specimens collected from ulcerated lesions (Fig. 1.23). Samples are usually of poor quality and constantly hemocontaminated; therefore, only in rare cases, diagnostic specimens are obtained. Most ulcerative diseases, such as cutaneous lupus, vesiculobullous sub-epidermal diseases, erythema multiforme, and so on, can be diagnosed only by histopathology; nevertheless, there are some ulcerative diseases such as leishmaniasis or other infectious diseases such as sporotrichosis or pythiosis, for which cytology may provide an immediate diagnosis.

Sampling Technique

The method of sampling differs according to the depth of the ulcer. In ulcers with deep bottom and raised edges, the imprint smear is not indicated because the slide cannot reach the bottom of the ulcer; in these cases, cells can be sampled by scraping (Fig. 1.24). In less deep ulcers with loss of only superficial portions of the dermis, the cells may be collected by imprint smear technique.

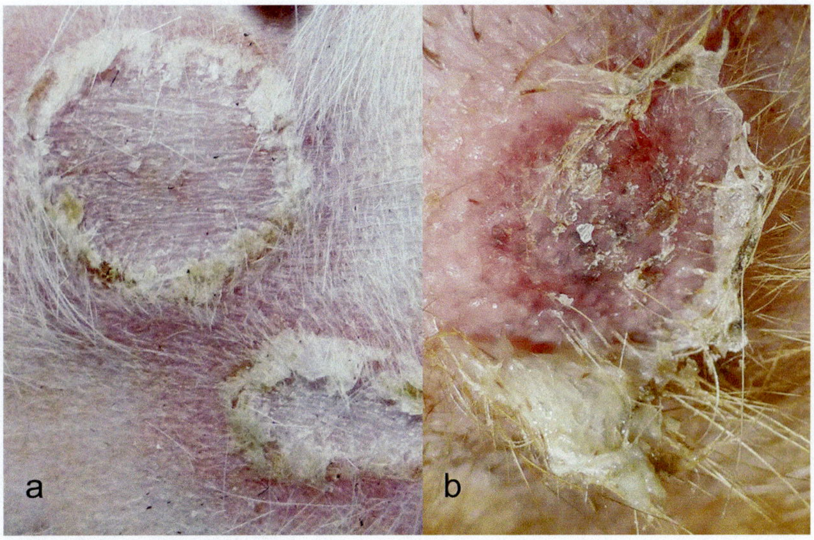

Fig. 1.21 Sampling technique from epidermal collarettes: (**a**) epidermal collarettes with a dry central area and (**b**) purulent exudate is present under the peripheral crusts

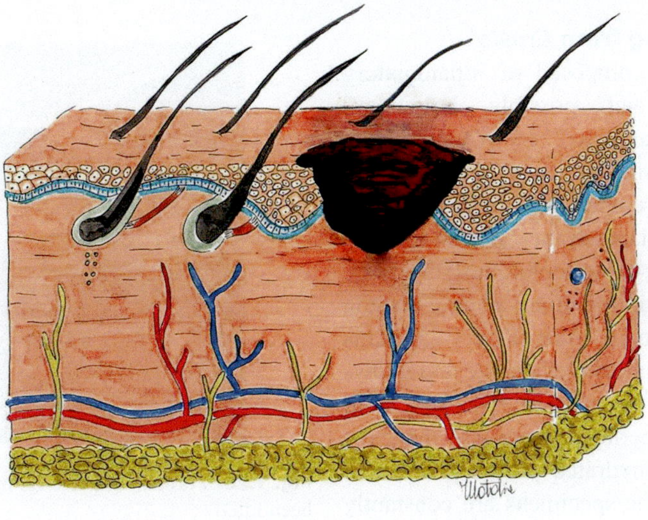

Fig. 1.22 Schematic drawing of an ulcer. Note how the lesion is deep and involves the dermis

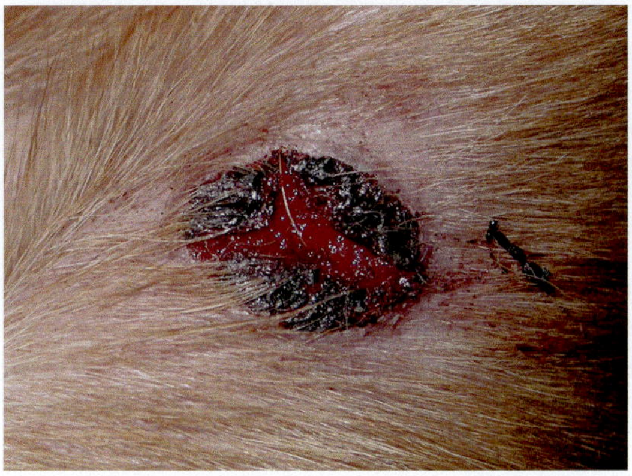

Fig. 1.23 Round ulcer, peripherally covered with hematic crusts

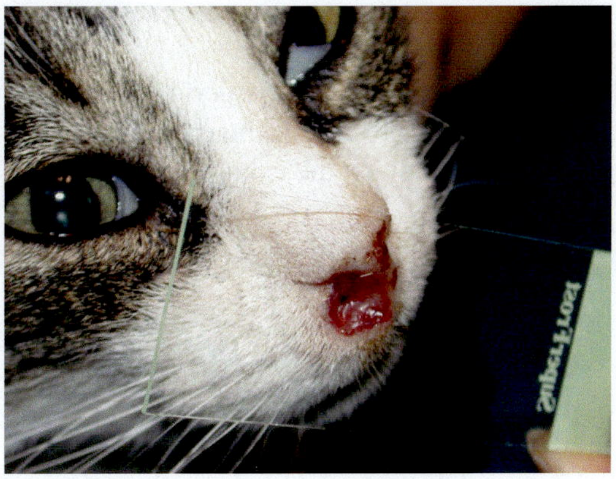

Fig. 1.24 In the case of deep ulcers, placing a slide does not permit the cells to be collected

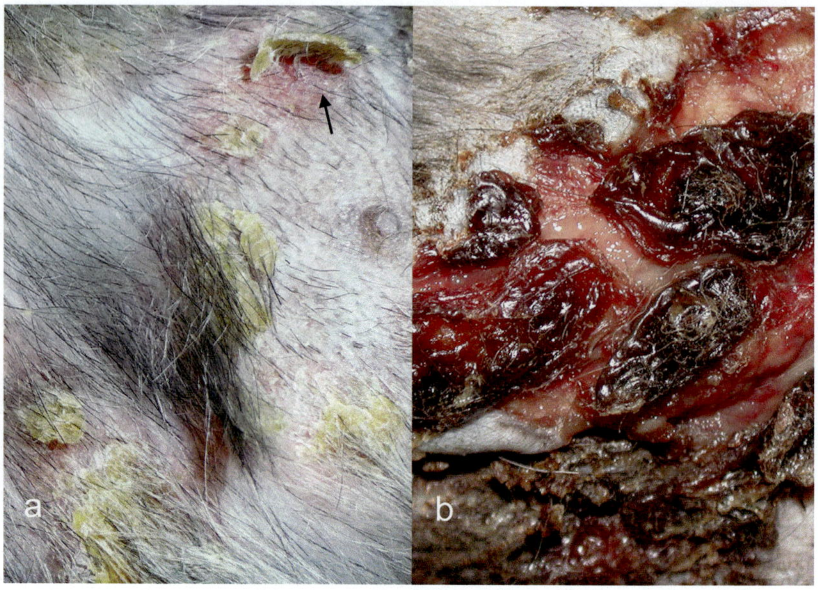

Fig. 1.25 Macroscopic aspects of crusts: (**a**) yellowish superficial crust secondary to pustule dehydration; note as under the crust, a small amount of fresh exudate is still present (arrow; **b**) dark-red hematic crusts secondary to a deep ulcerative lesion

1.2.1.6 Cytological Sampling from *Crusts*

Crusts are secondary lesions, composed of inflammatory cells, keratinocytes, debris, and, possibly, hair shaft fragments, blood, and microorganisms.

Clinically, the crusts are classified as *superficial* and *deep*. The former are confined to the epidermis, and therefore, as this portion of skin is not vascularized, they are not hematic and characterized by a yellowish-brown color; deep crusts are instead secondary to an ulcer and therefore, as they contain blood, appear red or red-brown in color (Fig. 1.25). This different appearance is easily clinically evaluable and allows the clinician to obtain useful information regarding their etiology. Cytologically, the superficial crusts are the most interesting as they represent dehydrated pustules, whereas, in the case of hematic crusts, the specimens are constantly hemocontaminated and can provide less useful results. As discussed for collarettes, also in superficial crusts, it is possible, once lifting the crust, to sample the pus by imprint technique.

Sampling Technique

Sampling of cells from the crust is performed following its removal and by gently pressing its inner surface a slide on or placing the slide directly on the exposed pus (Fig. 1.26).

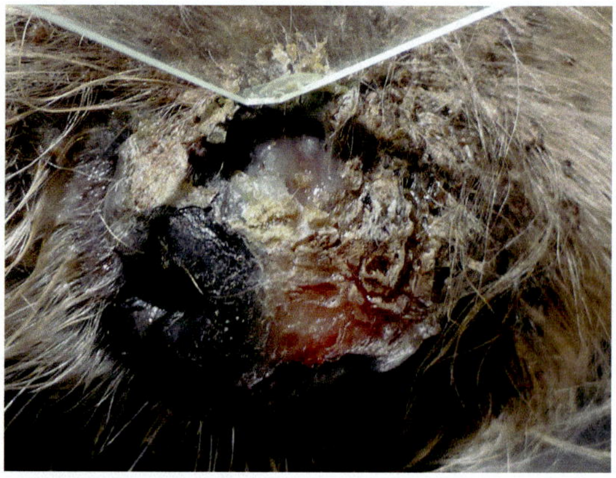

Fig. 1.26 Purulent exudate on the exposed dermis after the crust has been lifted

1.3 FNB and FNAB

1.3.1 Cytological Sampling from Nodules and Plaques

By definition, *nodules* are raised lesions, dermal or subcutaneous in location, greater than 1 cm in size and with a tendency to grow in height rather than in width (Figs. 1.27 and 1.28). The *plaques* are also raised lesions larger than 1 cm, but they tend to extend further in width than in height, giving the lesion a flattened appearance (Fig. 1.28). Large nodules are defined *masses*. Nodules and plaques may be composed of inflammatory (infectious or sterile diseases) or neoplastic cells or can be secondary to dermal deposits of fat (xanthogranuloma) or mineral salts (calcinosis cutis) (Fig. 1.29).

All nodules and plaques can be investigated using FNB, which is the best method when small nodules must be sampled. When cells are few and samples are not representative of the lesion, or when it is necessary to sample a large cutaneous or subcutaneous mass, it is better to perform a forced aspiration with needles of larger gauges (FNAB).

Sampling Technique

In cases of plaques or nodules, the FNB or the FNAB are the most appropriate techniques (Fig. 1.30).

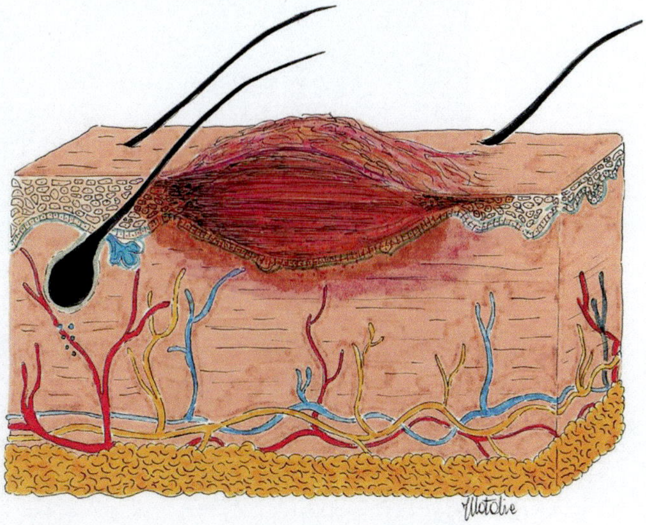

Fig. 1.27 Schematic representation of a dermal nodule

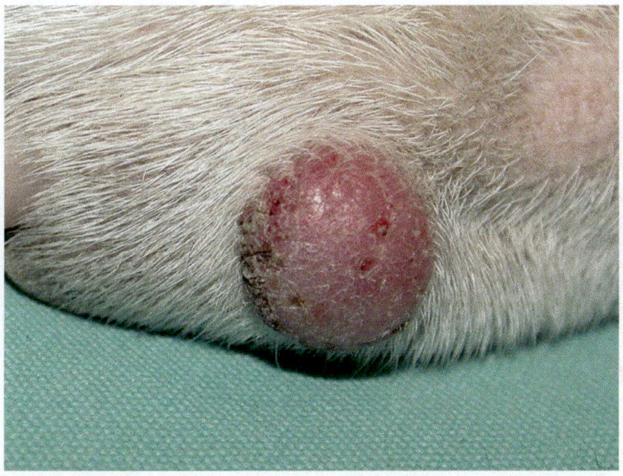

Fig. 1.28 Skin nodule: round and alopecic benign cutaneous histiocytoma in a dog

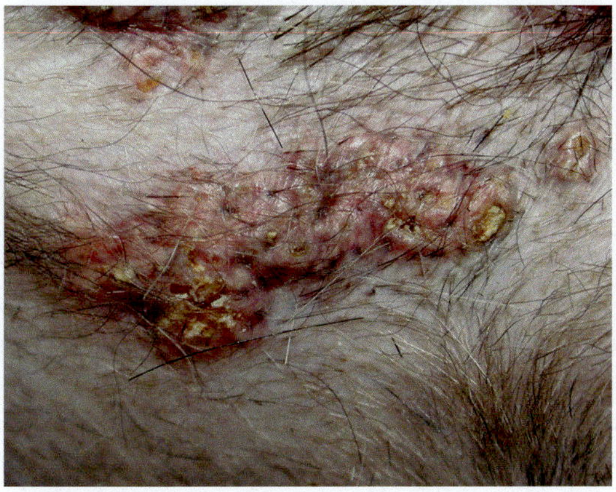

Fig. 1.29 Plaque in a dog affected by *calcinosis cutis*: note, as the raised lesion is wider than the height

The FNB technique is based on the simple insertion of a needle into a nodule, and in most cases, this is enough for collecting a high number of cells. A needle with a small gauge (23–25 G) is usually used for this purpose. The author routinely uses small needles as they reduce the possibility of hemocontamination. The needle must be inserted into different areas of the lesion taking care to perform continuous and repeated *rotational* and *back-and-forth* movements (Fig. 1.31). These measures are fundamental because many nodules and plaques may have a mixed composition; for example, the same nodule can be morphologically heterogeneous in its composition or can contain necrotic areas or a single or multiple areas containing fluid. As many nodules and plaques are ulcerated, the impression smear technique can provide contaminated samples. In these cases, FNB must be always preferred as it permits cells to be collected from deeper and not-inflamed or contaminated areas, reducing the possibility of sampling only inflammatory non-diagnostic cells.

FNAB can be also used in case of flattened lesions; in fact, in some non-nodular lesions, such as in the case of slightly raised skin lesions or in case of exfoliative dermatitis with intact skin, it is possible to try to collect the cells located in the superficial dermis, just below the epidermis.

In some cases of canine leishmaniasis, when other techniques have not proved diagnostic, for example, when no protozoa have been observed with lymph node cytology or via imprint from ulcerative lesions, it is possible to use the intradermal FNAB technique: a 23 gauge needle, connected to a 5-mL syringe, is inserted into the sub-epidermal dermis and aspiration movements are performed. In lucky cases, a variable number of amastigotes, usually few, can be collected

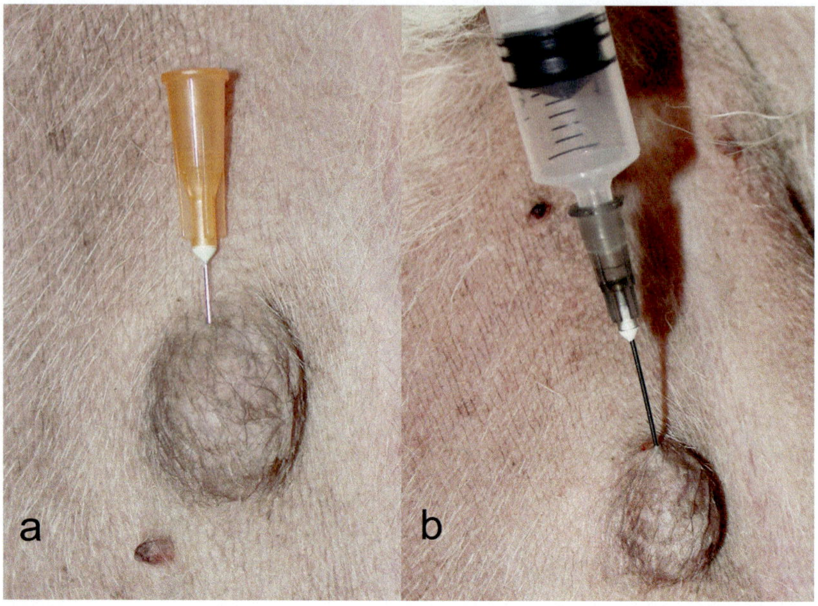

Fig. 1.30 FNB techniques: (**a**) FNB and (**b**) FNAB

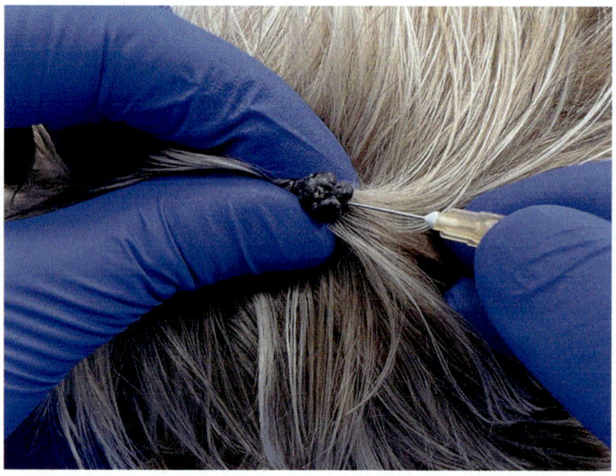

Fig. 1.31 FNB: a small needle is inserted into a nodule

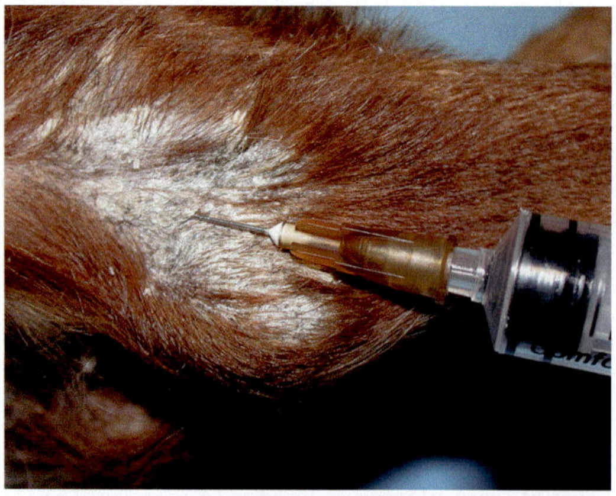

Fig. 1.32 Intradermal FNAB in a dog with leishmaniasis

together with a few inflammatory cells (Figs. 1.32, 1.33, and 1.34). The author also uses this technique in cases of suspected exfoliative form epitheliotrophic lymphoma or cutaneous lymphocytosis. The detection of many lymphoid cells is not diagnostic but, when clinical lesions are compatible, reinforces the suspicion of these two diseases and orients clinicians to perform a histological sampling.

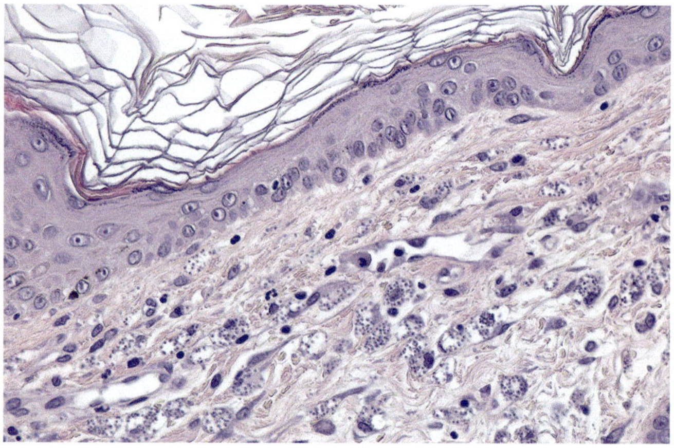

Fig. 1.33 Histology: all the small blue dots are amastigotes of leishmania located in the cytoplasm of macrophages located at the upper dermis. An intradermal needle aspiration biopsy can collect them

In the presence of large masses, especially those that release few cells such as lipomas, FNB without aspiration may not be sufficient, also if longer needles are used. In these cases, it is advisable to use longer and larger caliber needles, connected to 2.5- or 5-mL syringes; the technique needs to perform aspiration maneuvers while keeping pulled the plunger of the syringe and directing the needle into different parts of the mass in order to collect a sample as representative as possible (Fig. 1.35). However, in any nodular lesion that does not yield cells with FNB, it is advisable to attempt to sample cells using forced aspiration (FNAB).

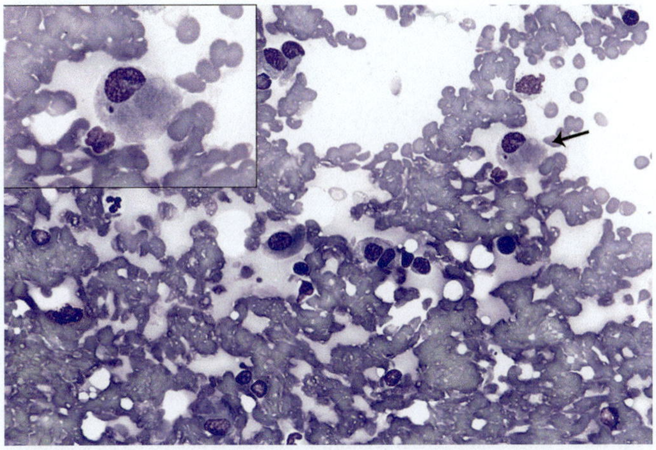

Fig. 1.34 Intradermal FNAB in a dog with leishmaniasis: few inflammatory cells are collected. Note the presence of amastigotes of leishmania in the cytoplasm of a macrophage (*arrow—inset*)

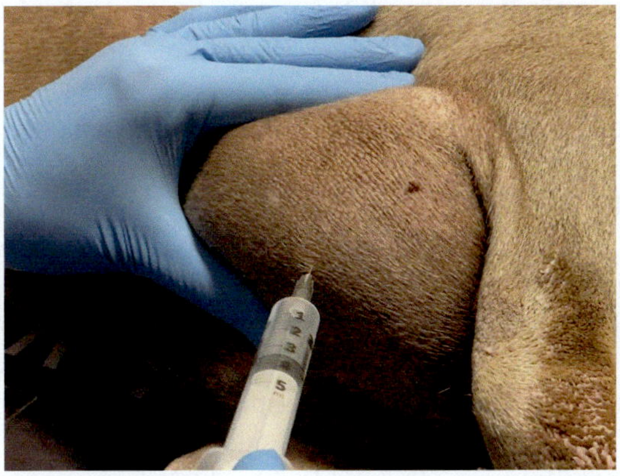

Fig. 1.35 FNAB of a large lipoma

1.4 Scraping

1.4.1 Cytological Sampling from Ulcers

As mentioned, samples collected from ulcers using the impression smear technique can produce non-diagnostic slides, often composed only of blood, debris, and environmental or oral contaminant bacteria (Fig. 1.36).

Scraping is a traumatic method that allows numerous cells to be collected. The traumatic detachment of cells from the original tissue has the disadvantage of causing bleeding, and so hemocontaminated specimens are common. Scraping is, therefore, the most suitable technique for deep ulcers because it enables the collection of cells from areas unreachable with other less traumatic methods. A typical example is squamous cell carcinoma in which the neoplastic keratinocytes pass the basal membrane and infiltrate the dermis, causing ulcerative lesions. In these cases, FNB is not the best method of sampling, whereas scraping provides more possibilities for obtaining diagnostic cells.

Some veterinarians use the *brushing technique*, which is widely used in endoscopy, which can be considered a less traumatic variant of scraping. It finds application especially for collecting cells from superficial lesions which cannot be easily sampled by scraping with a blade, such as on the eyelid conjunctiva or both oral and nasal mucosa.

Sampling Technique

As ulcers are typically covered with a hematic crust containing inflammatory–necrotic cells or debris, sampling by scraping the upper material could be not diagnostic. Therefore, the first scrapes could be carried out to remove

Fig. 1.36 Sampling technique of an ulcer: (**a**) impression smear technique used on the ulcerated nose of a cat with squamous cell carcinoma and (**b**) blood and many contaminant bacteria dispersed and arranged in large groups

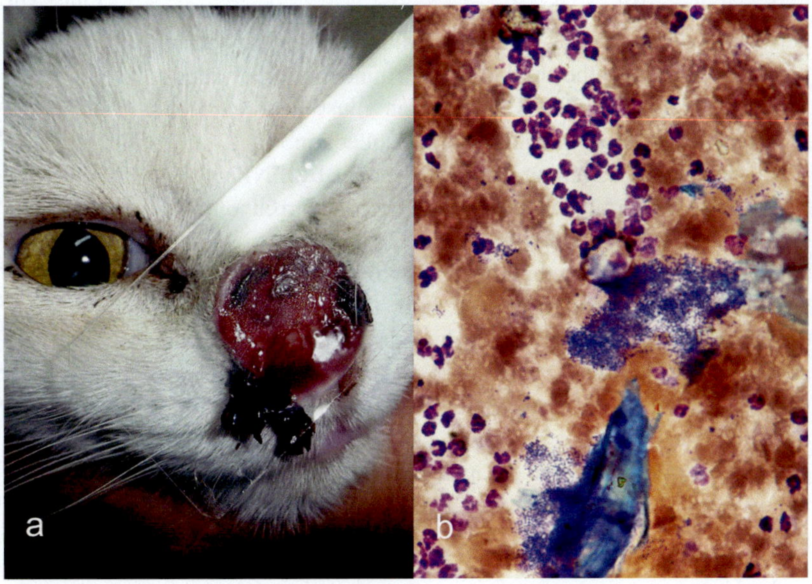

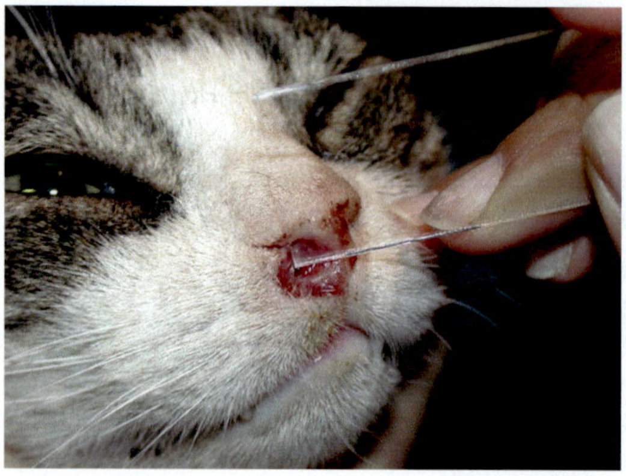

Fig. 1.37 Cells collected via scraping of an ulcerative lesion

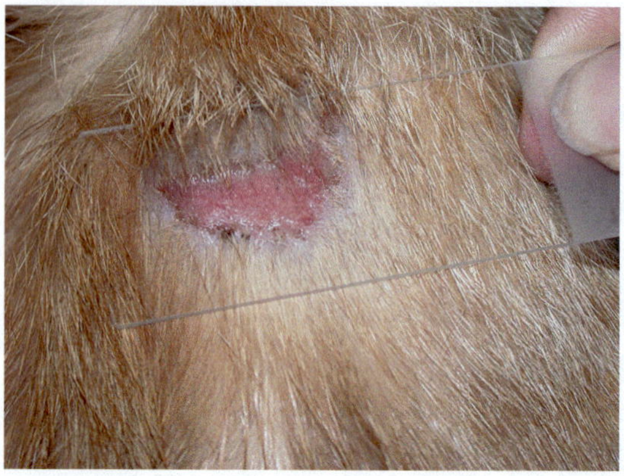

Fig. 1.38 Impression smear technique: a slide is placed on an exudative lesion

the superficial debris, and after these, other scrapes must be gently performed to collect more representative specimens.

It is possible to use any sharp instrument, such as a surgical blade or the edge of a glass slide; whichever tool is used, it is critical that multiple delicate scrapings must be repeated to gradually reach the bottom of the ulcer or the inner part of the edges with minimal bleeding (Fig. 1.37).

Regarding the brushing technique, cells are collected by gently brushing the affected area and deposited on the slide by gently rolling the brush over the latter in a rotary motion.

1.5 Preparation of Slides

The reading of a cytological specimen requires the slides to be correctly prepared. As already mentioned, once the lesion and the sampling technique have been chosen, the success of good preparation also depends on the manual skill of the operator who, with experience, is able to prepare slides of high quality. When the impression smear technique is performed on wet and exudative skin, it is sufficient to gently place the slide onto the surface to determine the direct transfer of cells (Fig. 1.38). The clinician must take care to minimize or avoid cell damage during collection.

In the case of sampling by needle, either by FNB or by FNAB, the material collected must be transferred onto a slide. The needle must be connected to a syringe in which a variable amount of air has been previously aspirate. The amount of air to be aspirated into the syringe before connecting the needle must be evaluated based on the amount of material present in the lumen of the needle and its consistency. To avoid exerting too much pressure on the cells, the use of a 1-mL syringe is usually sufficient, and only if a smaller amount of material escapes from the needle, is possible to use a 2.5- or 5-mL syringe which ensures better pressure.

This step is very delicate and important to obtaining good-quality slides. The collected material should not be sprayed too energetically onto a slide but must be gently deposited to approximately 1 cm from the frosted area of the slide and smeared (Figs. 1.39 and 1.40).

In this way, after smeared, all the cells are contained in the central part of the slide and disposed of in a single layer (Fig. 1.41).

At the same time, when collecting a large amount of material, it is mandatory not to deposit all of it on a single slide. Distributing only a small amount of material onto more slides has two advantages: (a) obtaining a specimen distributed in a single layer that permits the routinely used rapid dyes to better penetrate and stain the cells (e.g., Romanowsky-type dye) and (b) obtaining more slides, some of which can be used for special cytochemical or immunocytochemical stainings.

Once sufficient material has been deposited on the slide, this must be gently smeared using a second slide. The two slides, maintaining a constant pressure and speed, are gently slid in opposite directions. It will be down to the experience of the operator to evaluate, case by case, the pressure to be exerted on the two slides to obtain a high-quality sample. On well-prepared slides, as aforementioned, the material must be contained in the central area without reaching the edges, forming a sort of ellipse (Fig. 1.42).

In the case of fluids, such as that collected from apocrine cystic lesions, the slides may stick together, making it difficult for one slide to slip onto the other. This phenomenon can cause marked cell damage, which can be avoided by preparing slides of liquid consistency with the same method as that used for blood smears.

Conversely, when the material has a very dense consistency such as, for example, in the course of follicular cystic

Fig. 1.39 Cytological material sprayed and not smeared on a slide

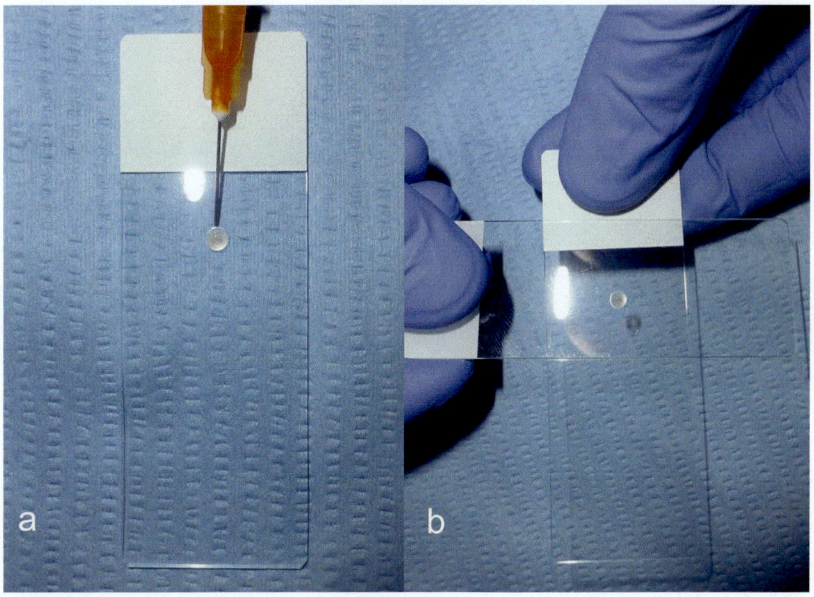

Fig. 1.40 Slide preparation technique: (**a**) the collected material is gently transferred onto the slide and (**b**) smear technique: another slide is placed on the collected material and slipped onto it

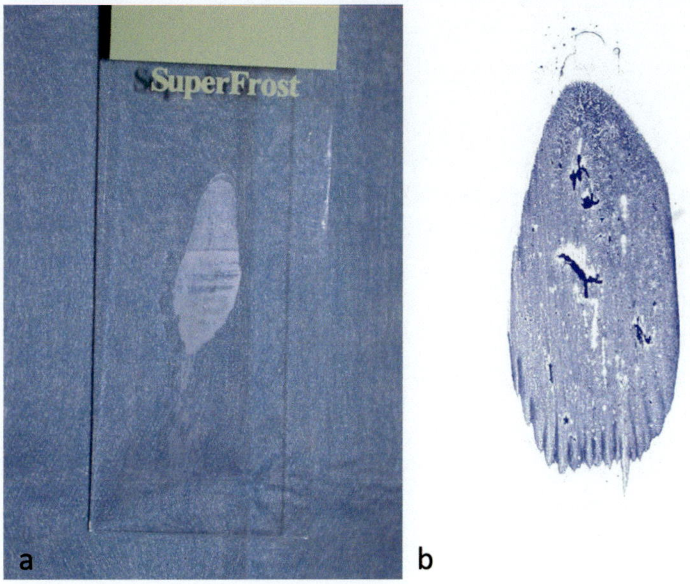

Fig. 1.41 Smear technique: (**a**) ellipsoidal shape of the smeared material is indicative of a well-prepared slide and (**b**) good-quality preparation: the oval shape of the smear ensures that all cells are arranged in a monolayer

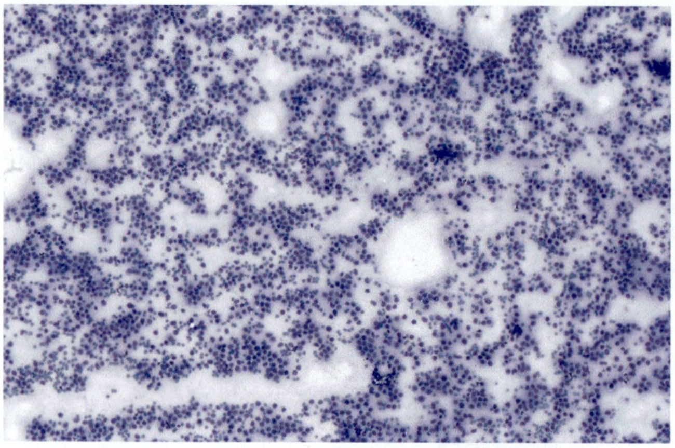

Fig. 1.42 Good-quality slide: microscopic evaluation of as cells are well preserved since arranged in a single layer and perfectly stained

lesions with keratin content or in keratinizing follicular tumors, the pressure must be greater in order to obtain a cell monolayer.

Material collected by *scraping* must be transferred onto a slide. In general, because, with this method, a large number of cells are collected and rapid stains fail to penetrate specimens that are too thick, it is necessary to exert an adequate pressure on the slides. With experience, clinicians acquire the skill of producing excellent slides.

2 Staining Techniques of Cytological Specimens: Routine and "Special" Cytochemical Dyes

2.1 Introduction

The slide, once prepared, must be correctly stained in order not to invalidate the work done up for it, that is, the choice of the lesion, the sampling method, and the preparation of the sample; therefore, the staining of the slide is the last but no less important procedure, before proceeding with the observation of the sample at the microscope.

Romanowsky-type dye is the stain usually used by clinicians in their daily practice, and it is adequate for satisfactorily staining all components of cells (nuclei, nucleoli, and cytoplasm) and for highlighting most microorganisms; in some cases, since the dye cannot enter the cell wall of some bacteria (e.g., mycobacteria) or some fungi, special cytochemical stains must be used.

Cytochemical dyes are not only useful for highlighting microorganisms, but they can be also important in many other situations, such as in the staining of both intra- and extracellular substances such as amyloid and lipids, or to demonstrate the Calcium composition of mineralized tissue, allowing to orient or confirm a diagnosis. A limitation for labs regarding the use of these additional stains is the lack of non-stained slides that can be used for this purpose. However, some articles have been published, which demonstrate how, starting from already colored slides, it is possible by means of bleaching techniques, to decolorize them and make them available for further cytochemical dyes. Recolored slides with cytochemical stains provide satisfactory results and allow us to avoid resampling new slides from the skin lesion [1].

2.2 Routine Staining of Slides

2.2.1 Romanowsky Stains (Wright's and Giemsa)

Romanowsky-type dye, originally composed of methylene blue and eosin, is the most used polychromatic rapid stain in veterinarians' practice; it is a panchromatic stain that allows the adequate display of nuclear characteristics and is excellent at highlighting cytoplasmic details, such as granules and microorganisms.

The dye takes its name from the Russian physician Dimitri Leonidovich Romanowsky (1861–1921), who was the first to use it. Although it was conceived for blood and bone marrow smears, over time and with the expansion of cytopathology, it has found wide use for staining any type of cytological sample.

Stains such as Giemsa, Wright, May–Grünwald, and Leishman are all related to or derived from the Romanowsky-type stains.

The technique is very easy and quick but requires some precautions that the veterinarian must consider in order to avoid altering the final result. As discussed in Chap. 1, since specimens that are too thick do not permit the Romanowsky-type dye to properly penetrate into the cells, it is mandatory that cells must be distributed in a thin mono-layer during the slide preparation; furthermore, slides must be air-dried before to be stained, as the dye does not appropriately color the wet specimens.

The staining method comprises three very fast steps: a first dip in an alcoholic fixative (methyl alcohol) and two consecutive passages in two different stains; the first staining is Eosin Y (Tetrabromofluorescein), an acid dye, while the second is Azure B (N,N,N-trimethylamine), a basic dye that is a metabolite of methylene blue. The technique is very simple and fast; there is no predetermined number of dips or seconds to be met for each dip, as much depends on how the dyes are used and how often they are filtered or renewed. In general, the number of dips for each solution depends on how long it takes to permeate the cells. Three to four dips of 1–2 s each may be sufficient for good staining; in time, the operator acquires the expertise to perform efficient staining (Fig. 2.1).

With this dye, nuclei take on a *purple color*, whereas cytoplasm stains *azure-blue*. As aforementioned, the Romanowsky-type stain permits to obtaining good-quality

F. Albanese, *Canine and Feline Skin Cytology*, https://doi.org/10.1007/978-3-031-95713-0_2

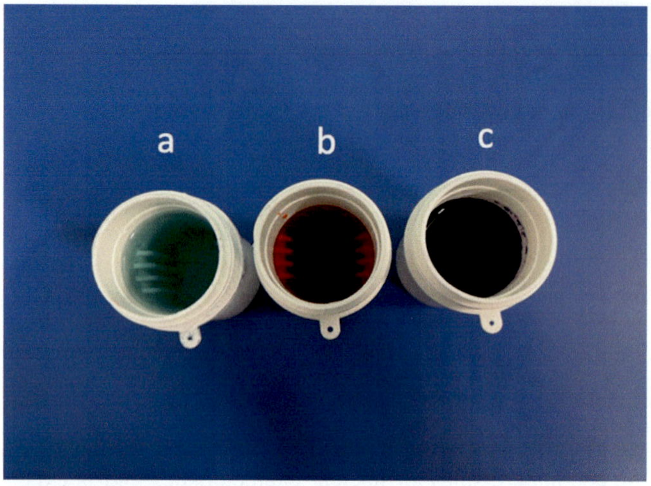

Fig. 2.1 Rapid *Romanowsky-type staining*: (**a**) fixative; (**b**) first dye (Eosin Y); and (**c**) second dye (Azure B)

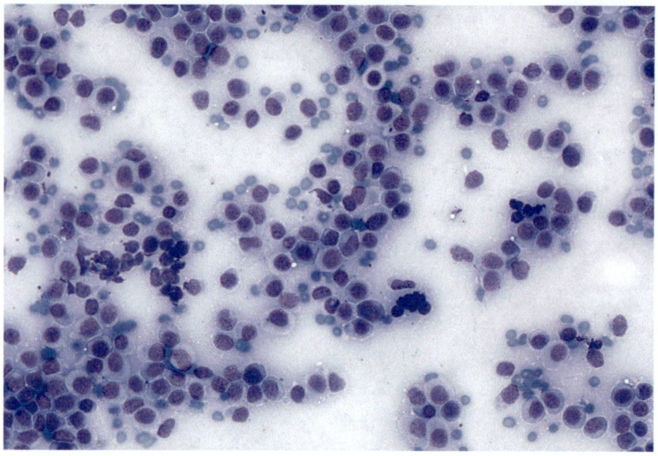

Fig. 2.2 Romanowsky-type staining: single layer of cells. Nuclei are stained violet and the cytoplasm is pale blue (round cell tumor). Note that the chromatin pattern is also recognizable

morphological findings of the nuclei and an excellent quality of cytoplasmic details and most infectious agents; for this reason, this is the routine staining used from vets in all the private clinics (Figs. 2.2, 2.3, 2.4, and 2.5). To obtain better detail of nuclei and chromatin patterns, other stains such as *Papanicolau* dye would be more suitable. Unfortunately, for this stain, a more complex organization of the laboratory inside the clinic is required, and the presence of a fume hood is suggested. The author believes that the Romanowsky-type stain is more than sufficient to obtain qualitative slides for evaluating skin cytology.

Since most clinicians use external laboratories for reading cytological preparations, it is emphasized that it is essential to avoid storing *unstained* cytology slides near formalin-preserved histological specimens; in fact, even if the container with formalin is hermetically sealed, its vapors can escape from the package and fix the cells of cytological specimens. As a result, the dye will not be able to penetrate and stain the cells which will appear only faintly colored, poorly detailed, and with a *faded* appearance (Fig. 2.6). For this reason, it's very important not to send, in the same shipping box, unstained slides for cytology along histological samples.

The Romanowsky-type dye is also used when transparent acetate tape must be stained. Some clinicians use to dip the tape into the three solutions as do for slides, but, since most tapes used in practice can curl or become opaque once immersed in the alcoholic fixative, this makes it impossible to have good details of cells. To avoid this problem, it is possible to use only the blue dye: a few drops of the latter are deposited on a slide to which the lower surface of the adhesive tape, on which the cells are attached, is placed. Before looking at the slide under the microscope, the excess dye, which would make it difficult to view the cellular details, is removed with a piece of absorbent paper through gentle pressure on the tape (Figs. 2.7, 2.8, 2.9, and 2.10). Vital stains such as *lactophenol blue* or *crystal violet* can be used with the same result.

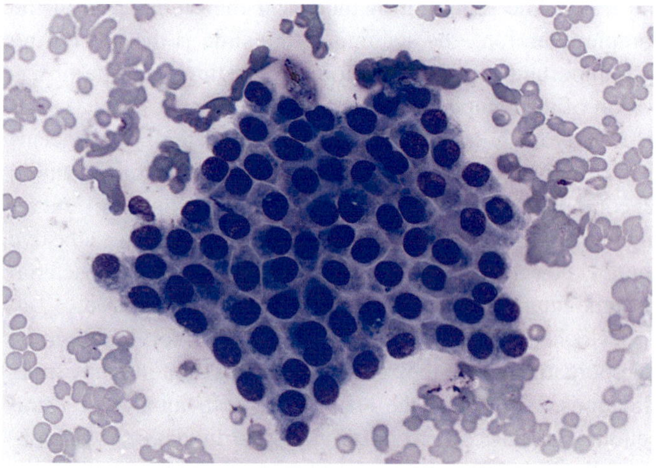

Fig. 2.3 Romanowsky-type staining: the outline of cells and the presence of a scarce amount of intracytoplasmic pale blue secretion are clearly detailed

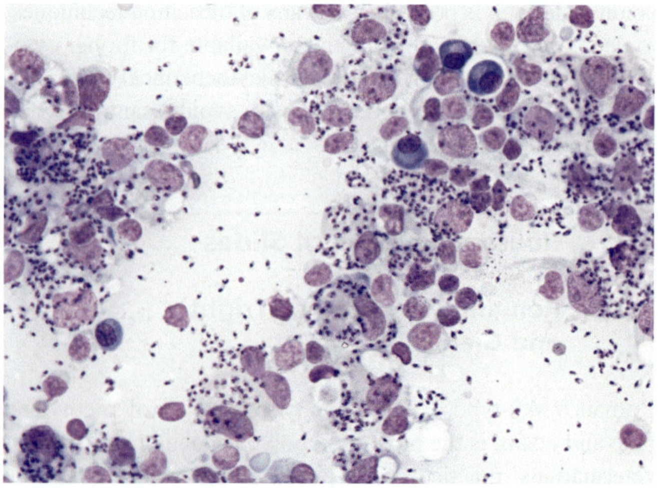

Fig. 2.4 Romanowsky-type staining: the morphology of amastigotes of *leishmania infantum* (cytoplasm, nucleus, and kinetoplast) is perfectly evident with this dye

2.3 Cytochemical "Special" Stains

In human and veterinary histopathology, the use of histochemical stains is very common. The visualization of some bacteria or fungi is sometimes very difficult on specimens stained with the routine dye for histology (hematoxylin-eosin (H-E)), and for this reason, many pathologists take advantage of other special dyes, especially when they have the suspicion that infectious agents are present [2].

With the increase in the knowledge and use of cytology in the routine of many veterinary practices, the use of cytochemical stains has also become very common for the staining of cytological slides. Although many staining methods are not yet standardized in veterinary cytology, and not many kits dedicated to cytological specimens are available on the market, histopathological staining kits are commonly used for staining cytological slides with good results.

The most common or useful cytochemical special stains are discussed below.

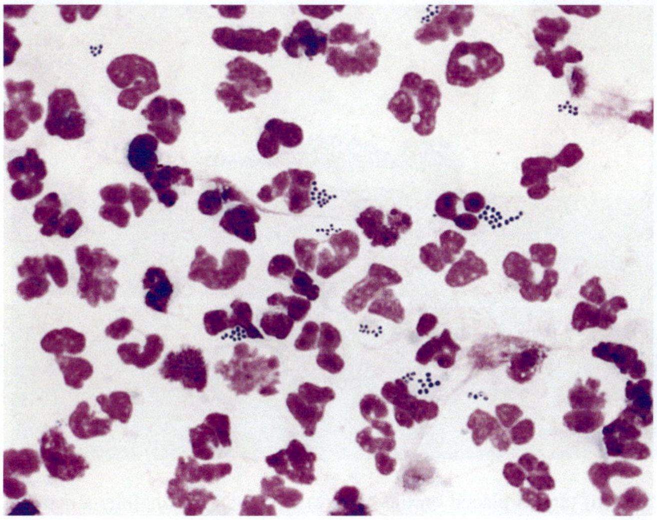

Fig. 2.5 Bacteria stains dark blue with Romanowsky-type staining and are easily recognizable

2.3.1 Periodic Acid–Schiff

Periodic acid–Schiff (P.A.S.) is a dye commonly used in histology to mark molecules that contain a high percentage of carbohydrates, such as glycogen, mucin, and those that compose the fungal wall [3].

In veterinary skin histopathology, it is most commonly used to highlight fungi. This stain demonstrates the presence of polysaccharides (simple and mucopolysaccharides) in the fungal wall which will take on a *red-magenta* color after the transformation of some pre-existing chemical groups into aldehydes: this chemical transformation occurs through the action of periodic acid, while the red coloring-magenta is induced by the Schiff reagent (Fig. 2.11).

In veterinary dermatology, dermatophytic infections are very frequent, while deep fungal diseases caused by environmental saprophytic fungi that penetrate the skin through infected foreign bodies or cutaneous involvement in the course of systemic mycoses such as *Cryptococcosis*, are much less common. In the latter diseases, however, the fungi are usually very voluminous and numerous, and Romanowsky-type staining is sufficient to highlight them (Figs. 2.12, 2.13, and 2.14). For this reason, P.A.S. dye is not so important on cytological specimens coming from deep fungal infection.

Fig. 2.6 Slides stained with Romanowsky-type dye: (**a**) cells are not stained because of contact with the formalin vapors and (**b**) the same sample is not exposed to formalin fumes

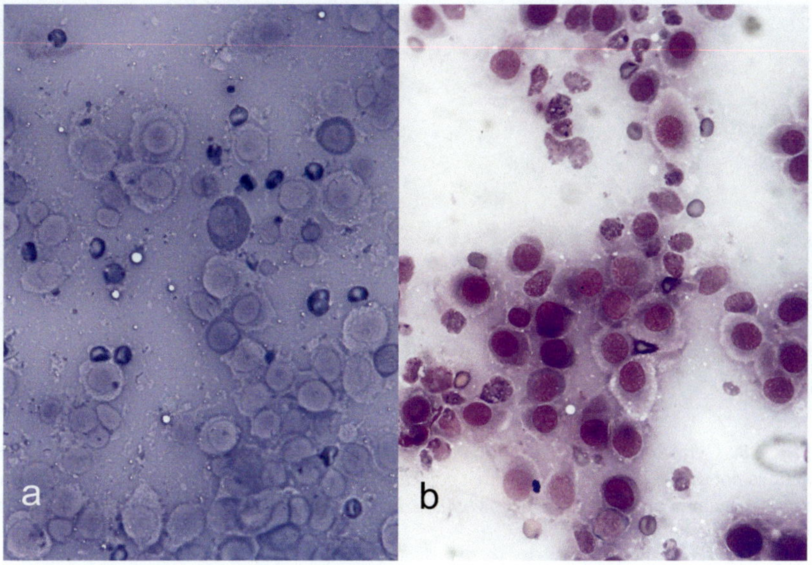

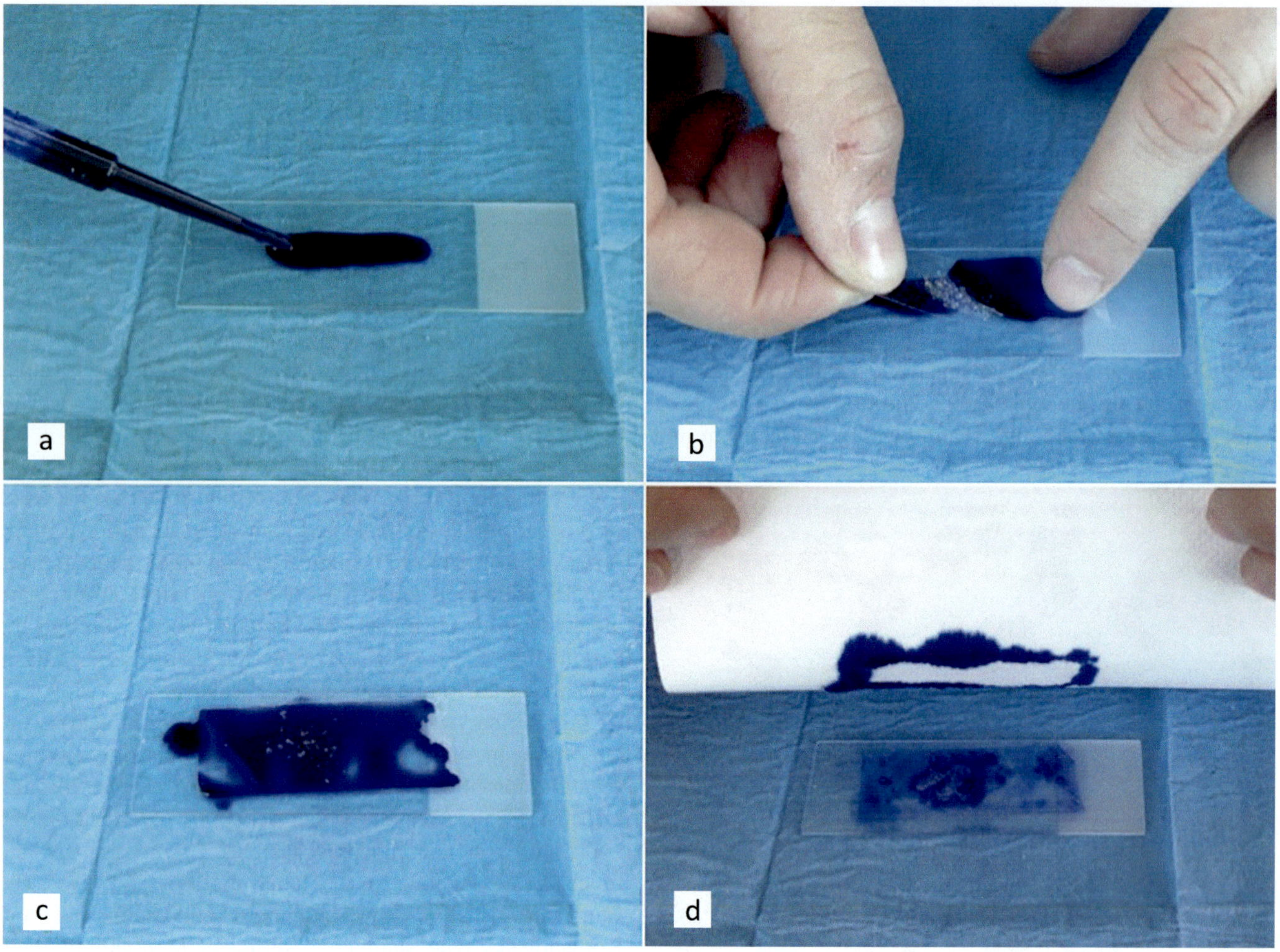

Fig. 2.7 Staining technique of the transparent tape: (**a**) blue dye on the slide; (**b**) tape is placed on the dye; (**c**) too much dye inhibits a detailed visualization of cells; and (**d**) excess of dye is removed before reading the slide at the microscope

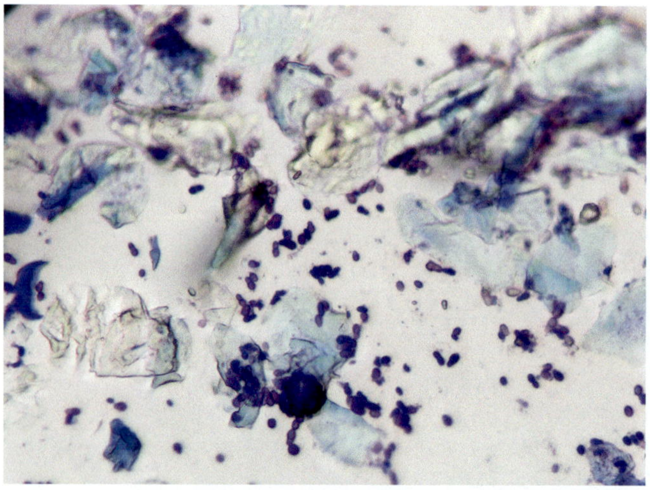

Fig. 2.8 Cytology with tape sampling method: despite the focus of the slide is not on the same plane, note that cells are adequately stained, and many peanut-shaped and budding *Malassezia* spp. yeasts are clearly recognizable

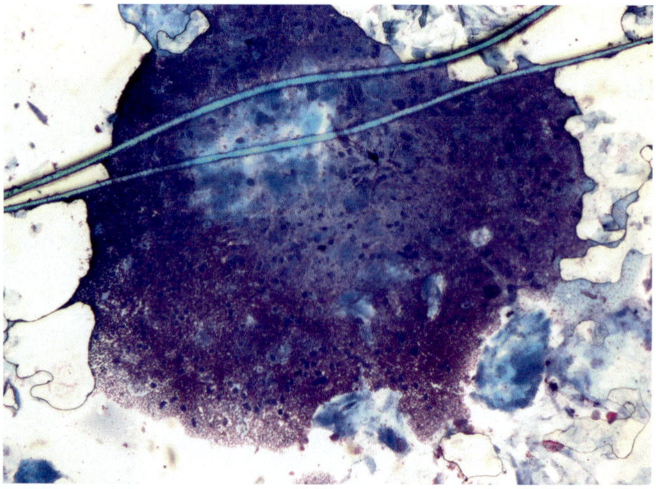

Fig. 2.9 Cytology with adhesive tape sampling method: neutrophils along with many acantholytic keratinocytes are perfectly stained and interpretable (pemphigus foliaceus)

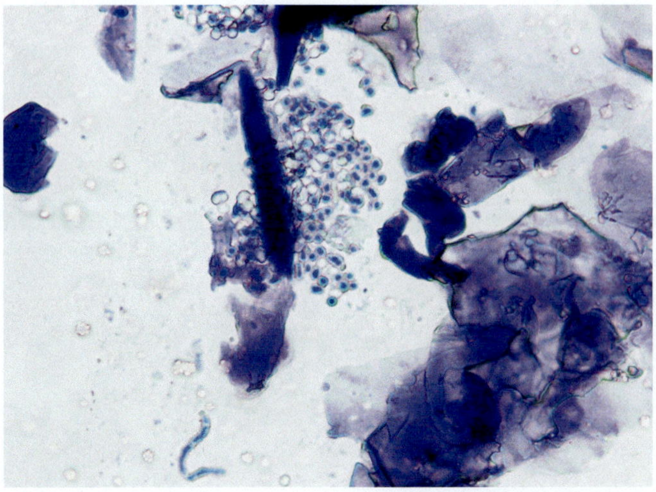

Fig. 2.10 Cytology with adhesive tape sampling method: many small arthroconidia of *Microsporum canis* are collected from the surface of the skin of a dog and stained only with the "blue dye"

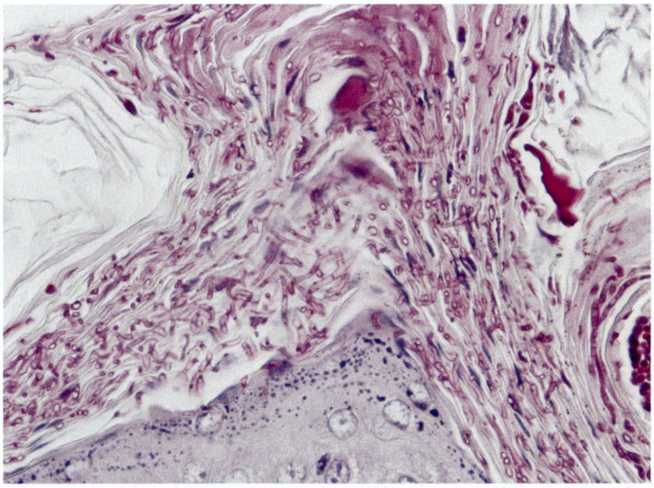

Fig. 2.11 Histology P.A.S. staining: many dermatophyte hyphae are perfectly highlighted with a magenta-red color

There are some situations where the use of P.A.S. on cytological samples is very useful and allows an immediate diagnosis, as happens in some cases when fungal bodies are small (i.e., dermatophytic Kerion in dogs) or hidden in amorphous and thick material (i.e., dermatophytic pseudomycetoma in cats).

In the course of dermatophytic Kerion, when the fungal elements are few and not visible in the context of a highly cellular mixed inflammatory population, the PAS may be able to highlight the few arthroconidia present (Fig. 2.15). Unfortunately, however, it must be emphasized that it is not always so easy to visualize the fungal bodies, since the PAS dye is not selective towards fungi, but tends to color

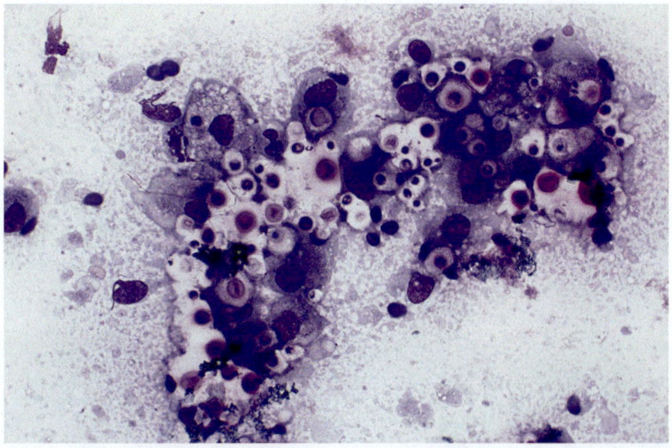

Fig. 2.12 Romanowsky-type dye: many *Cryptococcus* spp. are easily recognizable as round fungal bodies with a large clear achromatic halo

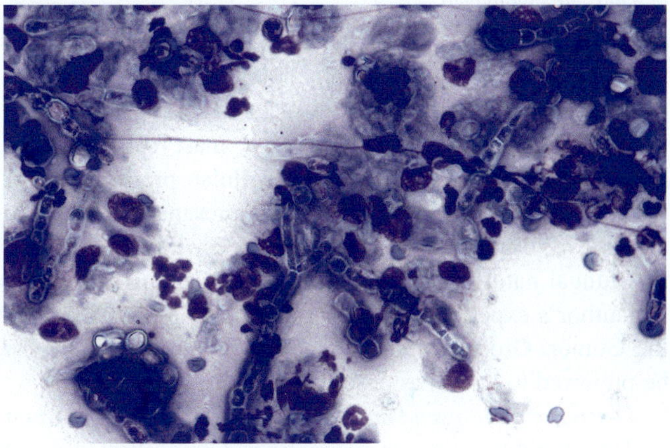

Fig. 2.13 Romanowsky-type dye: long septate hyphae are easily detected without special stains

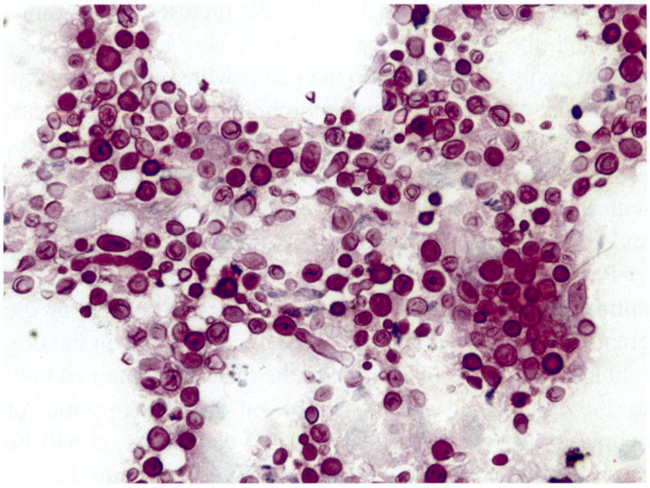

Fig. 2.14 P.A.S. staining: fungal bodies of *Cryptococcus* spp. stained red magenta

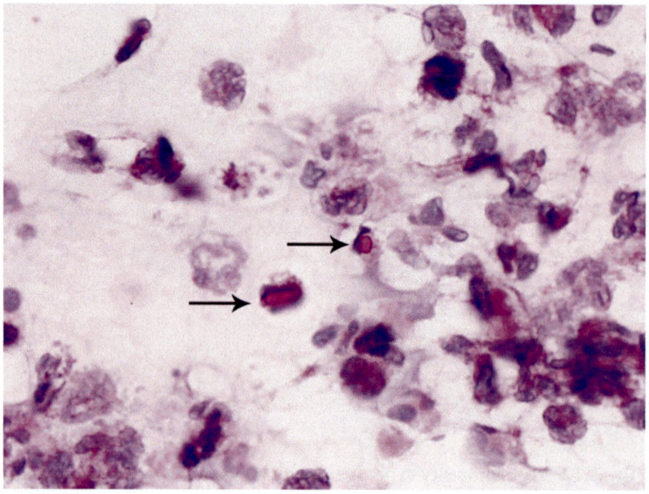

Fig. 2.15 P.A.S. staining: fungal oval and elongated arthroconidia of *M. canis* are highlighted

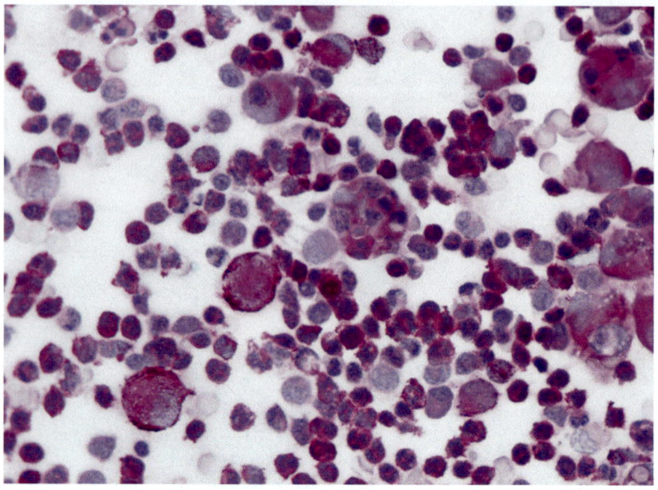

Fig. 2.16 P.A.S. staining: note that the stain colors a lot of non-fungal material into the cytoplasm of inflammatory cells

substances containing carbohydrates such as material phagocytized by macrophages, material dispersed on the bottom, granules of eosinophils and mast cells, and so on. All this means that, in very cellular preparations in which spores are not detected, even if present, the latter are difficult to visualize and can be confused with material of a non-fungal nature (Figs. 2.16 and 2.17). For this reason, in the author's experience, for searching dermatophytic spores, the Gomori-Grocott Methenamine Silver (GMS) stain must be preferred to P.A.S. as described below.

Dermatophytic pseudomycetoma is a rare clinical variant of dermatophytosis in cats, characterized by multiple dermal and hypodermal granulomas composed of many fungal bodies surrounded by a deposit of antigen–antibody complexes produced from the host (Splendore-Hoeppli phenomenon). These histological aspects are clinically represented as grains. Although the fungal bodies are always very numerous, they are not easily visualized into the grains with a Romanowsky-type stain, with which this amorphous material takes on a dark blue color. In this case, when a cytological suspicion of pseudomycetoma is present, it is possible to stain some slides with P.A.S., which allows you to visualize the fungal bodies and confirm the diagnostic suspicion (Figs. 2.18 and 2.19).

P.A.S stain can also find use in the course of non-infectious diseases, such as to mark the content of the granules of a rare variant of basal follicular tumor in the dog (granular cell trichoblastoma), or the material contained both in the cytoplasm and dispersed on the background of neoplasms such as adnexal clear cell carcinoma as will be discussed in the chapter dedicated to skin neoplasms.

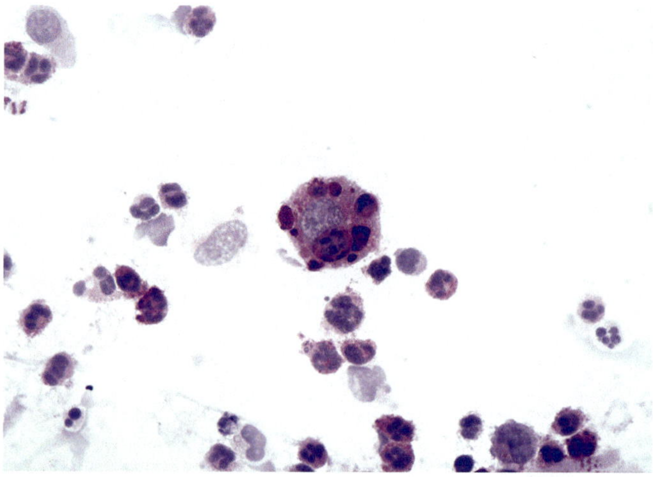

Fig. 2.17 P.A.S. staining: multiple P.A.S.-positive roundish formations in the cytoplasm of macrophages could be confused with fungal bodies

2.3.2 Gomori-Grocott Methenamine Silver

GMS is a dye that is commonly used in histology to better visualize fungal organisms. In most fungi, the cell wall is composed of chitin, a polymer of N-acetylglucosamine, which can be linked to polymers of D-glucose, D-mannose, proteins, and lipids [3].

With the use of GMS dye, the aldehydic groups that are produced by the oxidation of mucopolysaccharides present in the fungal wall reduce the silver chloride to metallic silver, permitting their visualization.

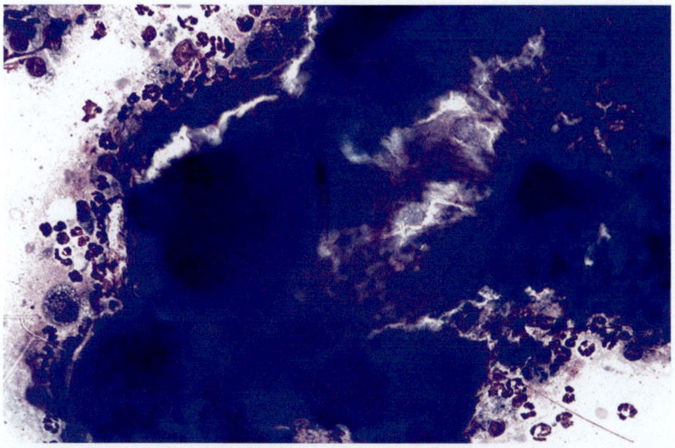

Fig. 2.18 Romanowsky-type dye: grains are cytologically represented by large mass of amorphous dark blue acellular material surrounded by a pyogranulomatous inflammation

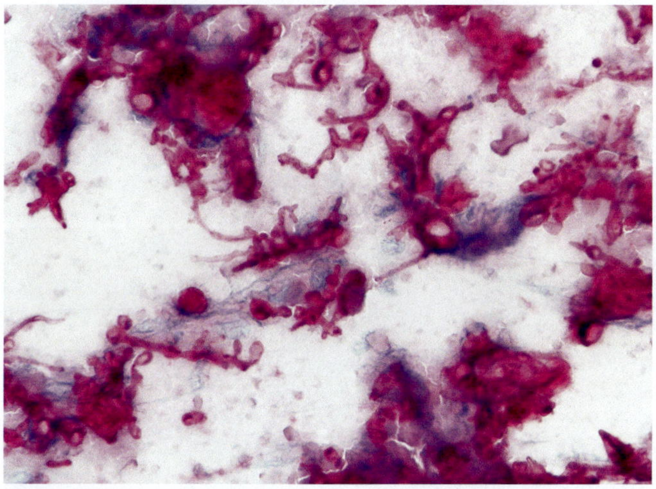

Fig. 2.19 P.A.S. staining: many fungal bodies, represented by septate and distorted branching hyphae along with large roundish bodies, are easily highlighted with P.A.S. stain

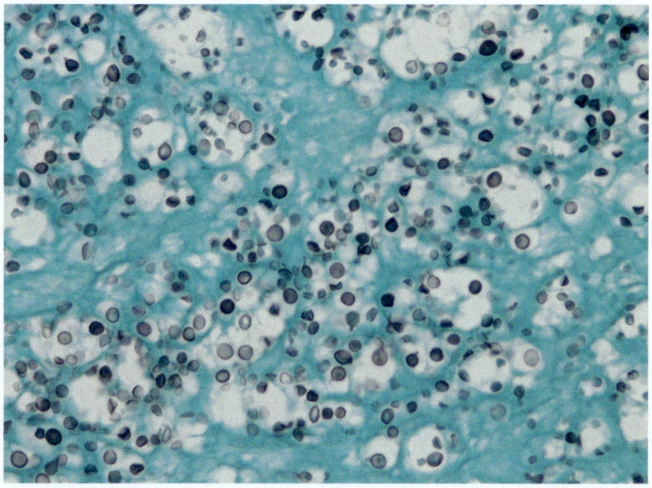

Fig. 2.20 *GMS* staining: many fungal bodies of *Cryptococcus* spp. are easily detected because of their black color against a green background

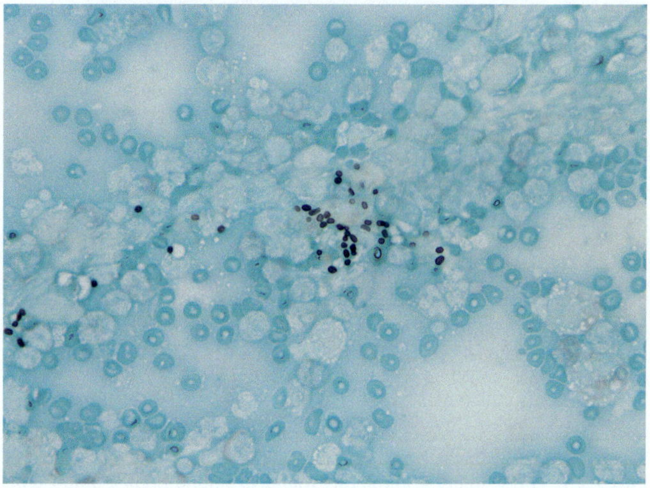

Fig. 2.21 *GMS staining*: small, round, black-stained arthroconidia of *M. canis* are highlighted since the green background

As discussed for P.A.S., in the case of fungal skin diseases caused by numerous and large in-size fungi, the GMS stain is not fundamental to detect them. Nevertheless, according to the author, it is better than P.A.S. dye, when we have to find smaller fungal structures such as dermatophytic spores; indeed, since with GMS dye, fungi assume a black color and the inflammatory cells and the background stain pale green, the arthroconidia can be easily highlighted (Figs. 2.20, 2.21, and 2.22).

Also in the case of feline dermatophytic pseudomycetoma, GMS dye is very useful for highlighting fungal bodies that are contained but not recognizable in the basophilic granular material (grains), when the latter are stained with Romanowsky-type dye (Fig. 2.23).

2.3.3 Ziehl–Neelsen

Ziehl-Neelsen (Z-N) staining is used in histopathology to color mycobacteria since they are usually not detectable with routine histopathological staining H-E [4, 5]. Acid resistance is a peculiar characteristic of the mycobacteria consisting of the ability of these micro-organisms, once stained with carbol fuchsin, to maintain their red color, even when subjected to energetic bleaching treatment. The acid resistance is due to the characteristic composition of the cell wall of mycobacteria, which is very rich in lipids; mycobacteria have, in fact, a complex wall that does not allow the entry of the routinely used dyes.

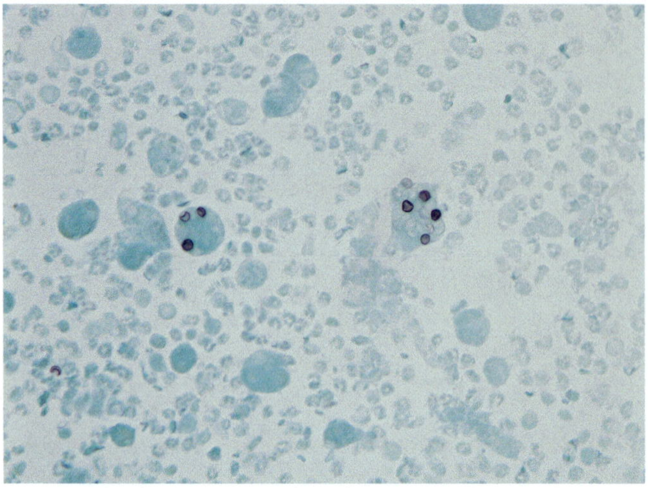

Fig. 2.22 *GMS* staining: small, round, black-stained arthroconidia are easily detectable in the cytoplasm of two macrophages

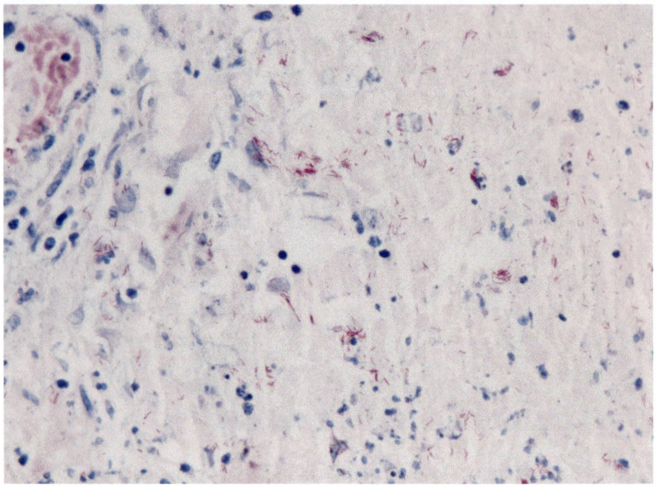

Fig. 2.24 Z-N staining—histology: many bright red bacilli (mycobacteria) in a skin nodule of a cat affected by feline leprosy

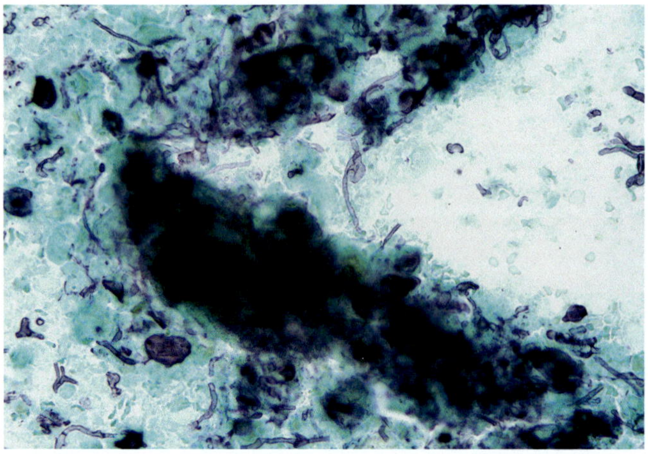

Fig. 2.23 *GMS*: many voluminous roundish fungal bodies along with numerous septate hyphae are clearly recognizable on a green background

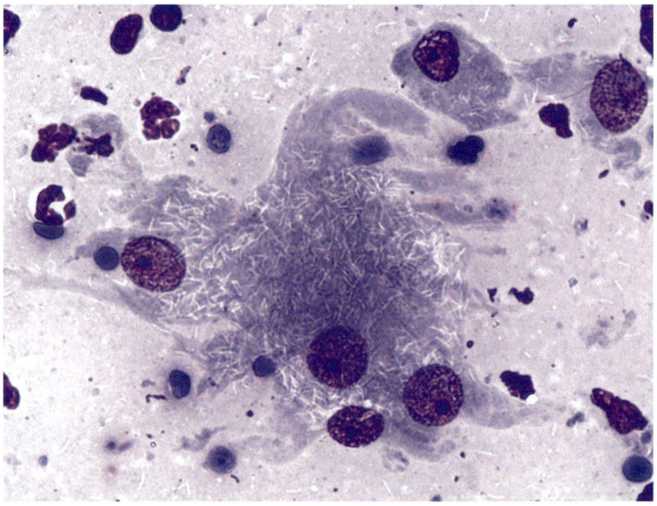

Fig. 2.25 Romanowsky-type stain cytology: many rod-shaped achromatic areas, representing unstained mycobacteria, are clearly evident in the cytoplasm of macrophages

With Z-N staining, the pigment (carbol-fuchsin red) is "forced" to enter into bacteria by a combination of heat and chemicals, making the mycobacteria clearly visible as *bright red bacilli* whereas the nuclei take on a pale blue color.

In skin cytology, on slides stained with Romanowsky-type dye, the presence of mycobacteria is strongly suspected when many achromatic rod-shaped areas are present in the cytoplasm of macrophages. To confirm the presence of bacteria, a Z-N dye is necessary (Figs. 2.24, 2.25, 2.26, and 2.27).

In case of abdominal mycobacterial panniculitis observed in cats and caused by facultative non-obliged mycobacteria, it is not so easy to microscopically detect the microorganisms; in fact, since their tropism for fat, they are not usually phagocytized by macrophages but live into the fat and for this reason, special histochemical stains such as Z-N or Fite Faraco, are needed to highlight them into the adipose tissue.

It should be noted that also *Nocardia* spp., a bacterium non-belonging to the mycobacteria family, could be partially positive for Z-N staining. Fortunately, the morphology of *Nocardia* is usually very different from *Mycobacteria*, but when only a few microorganisms positive to Z-N stains are present, we must be aware of this particular tinctorial behavior of *Nocardia* spp. In any case, also to evaluate the zoonotic risk, further diagnostic tests such as culture or, better, PCR tests are mandatory for their differentiation and typifying.

2.3.4 Oil Red O

Oil Red O (ORO) staining is used to highlight *lipid substances*, which are soluble in non-polar solvents and therefore insoluble in water. This dye belongs to the category of lysochromes, a group of fat-soluble substances that have

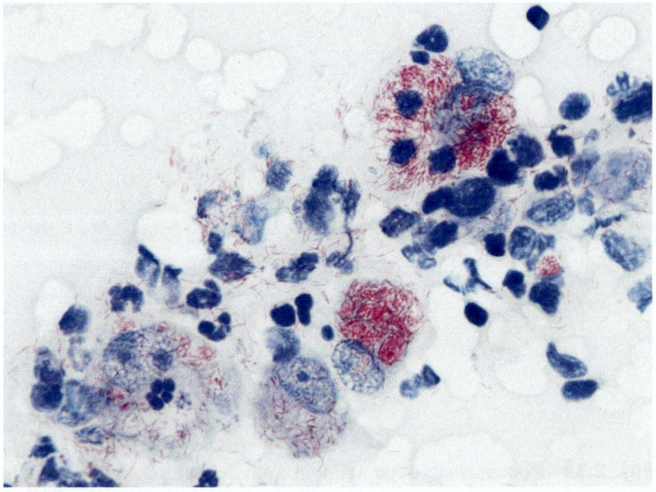

Fig. 2.26 Z-N staining—cytology: many red-stained bacilli confirm the presence of mycobacteria on cytological specimens

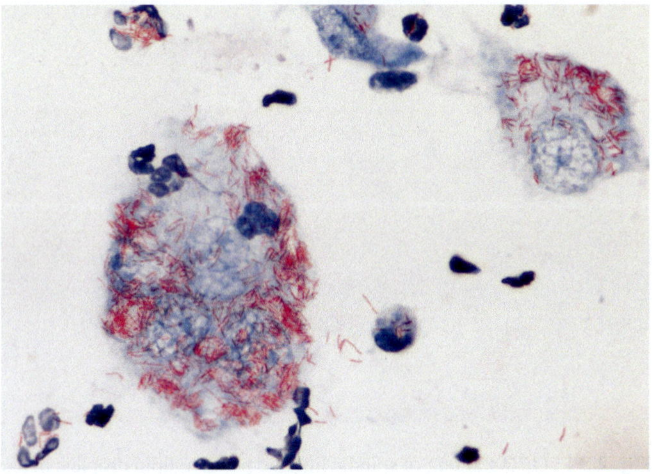

Fig. 2.27 Rod-shaped mycobacteria in the cytoplasm of macrophages are highlighted with Z-N staining

the ability to dissolve in the lipid molecules, transmitting their color.

The ORO dye is fat-soluble and therefore selectively binds to the lipid structures of the cells. Staining can only be done on fresh samples, as alcohol fixation removes most of the lipids; for this reason, this staining cannot be done on formalin-fixed histopathological specimens and frozen fresh tissue is necessary. In cytology instead, ORO can be easily used, but it is mandatory that slides have not been previously immersed in the alcoholic solution. With this staining, lipids will stain bright *red-orange*, while the nuclei will stain blue (hematoxylin).

All the cells containing fat are positive to this staining, such as adipocytes or sebocytes; anyway, when those cells are well differentiated and easily recognizable with routinely Romanowsky-type stains, ORO dye is useless (Figs. 2.28, 2.29, and 2.30).

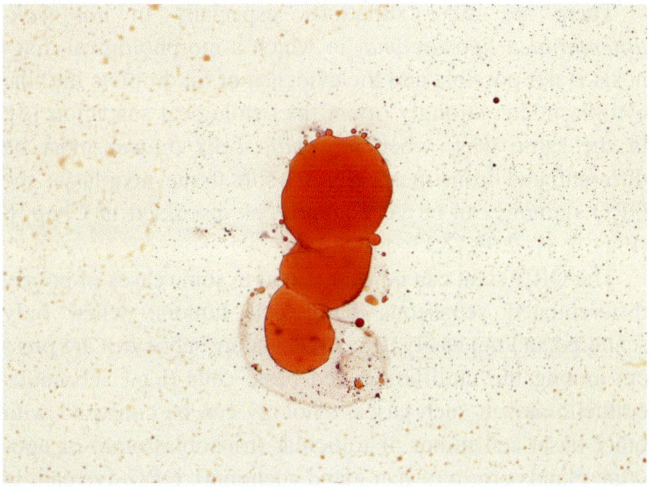

Fig. 2.28 *ORO* staining: positive brightly red-stained lipid droplets

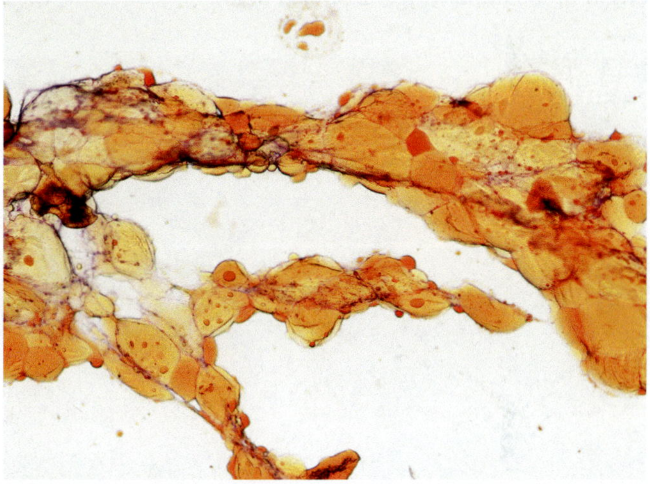

Fig. 2.29 *ORO* staining of lipoma: positive lipid droplets in the cytoplasm of adipocytes

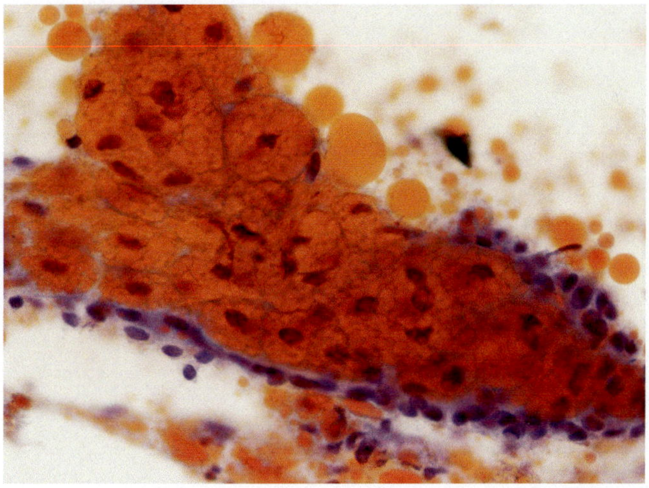

Fig. 2.30 *ORO* staining of Sebaceous glands: positive lipid droplets in the cytoplasm of mature sebocytes. Note the non-lipidized basal reserve cells at the periphery of the cluster

There are other situations, especially in *not well-differentiated liposarcoma*, in which a morphological diagnosis is not possible and the evidence of fat droplets into the cytoplasm can strongly orient the cytological suspicion [6]; in the same way, when a *spindle cells lipoma* must be differentiated from mesenchymal soft tissue neoplasia, the ORO staining can be very helpful, as discussed in Chap. 8 (Figs. 2.31, 2.32, 2.33, and 2.34).

The ORO stain can be very useful in some cases of poorly differentiated sebaceous basal cell tumors, when only scattered and not easily detectable mature sebocytes are present among the undifferentiated basal cells (e.g., sebaceous epithelioma); in such cases, cytology can be confused with other basal cell tumor of follicular (trichoblastoma) or apocrine glands origin (sweat gland adenoma). *ORO* dye can, in fact, highlight the rare mature sebocytes between the immature basal cells (reserve cells), confirming the sebaceous origin of the tumor (Figs. 2.35 and 2.36).

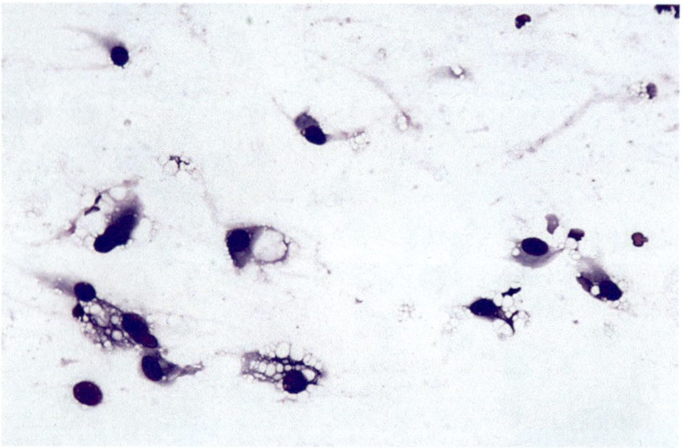

Fig. 2.33 Romanowsky-type stain of spindle lipoma: many spindle cells, with scattered achromatic roundish intracytoplasmic areas. Differentiate between a spindle cell lipoma and other Soft tissue neoplasia, it is impossible with routine staining

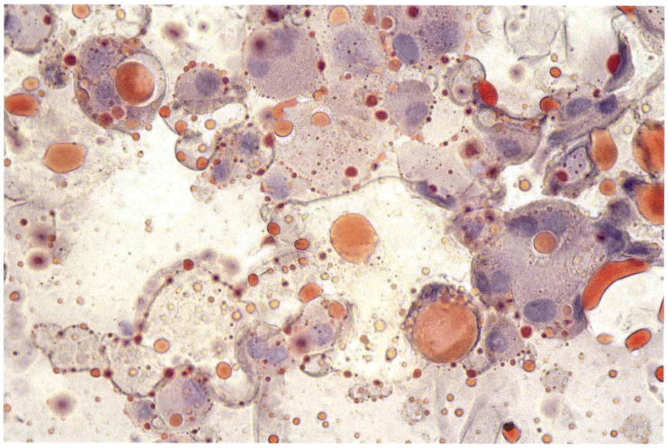

Fig. 2.31 *ORO* staining of liposarcoma: positive brightly red-stained lipid drops into the cytoplasm of malignant adipocytes

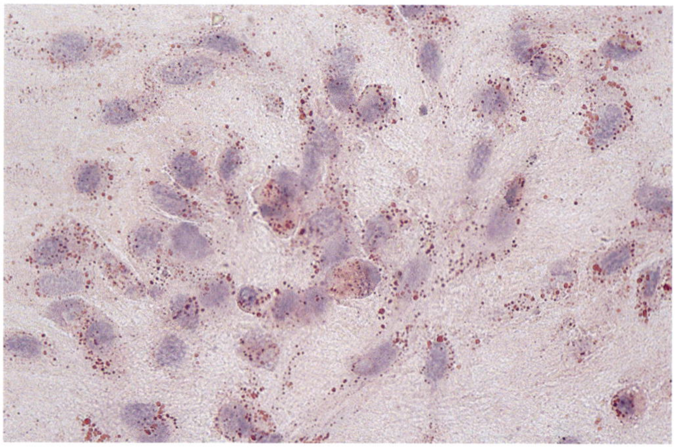

Fig. 2.34 *ORO* staining of spindle lipoma: positive lipid droplets in the cytoplasm of spindle-shaped adipocytes

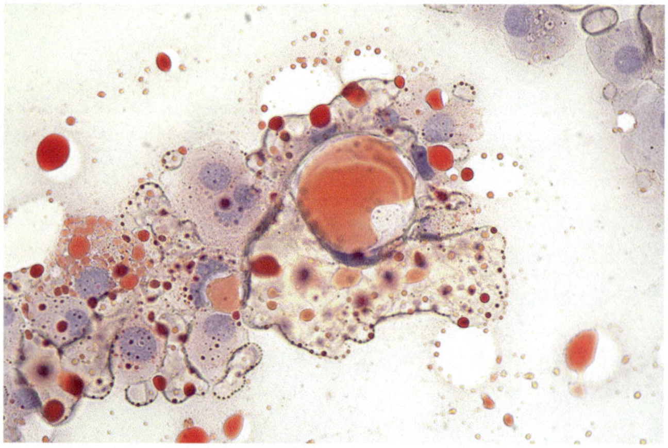

Fig. 2.32 *ORO* staining of liposarcoma: note how the positive lipid material pushes to the cell periphery and deforms the nucleus of the neoplastic adipocyte

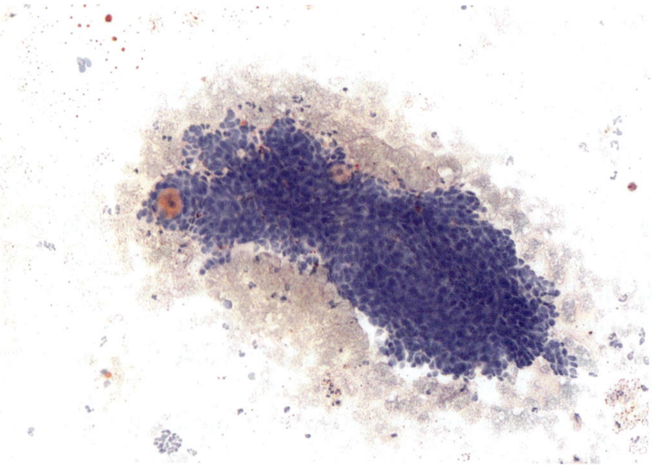

Fig. 2.35 *ORO* staining of sebaceous epithelioma: a cluster of immature basal cells with only two mature sebocytes

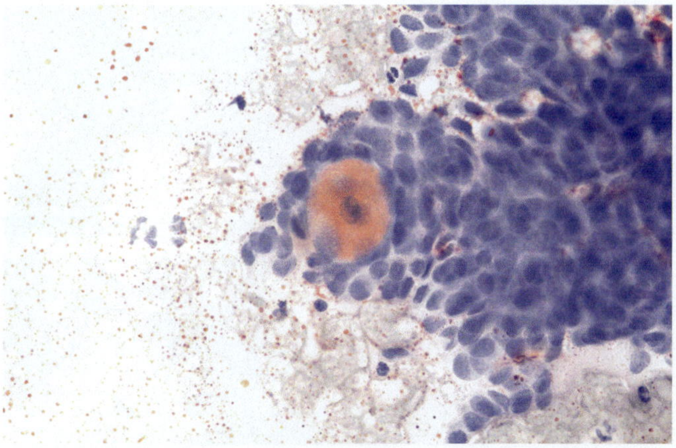

Fig. 2.36 *ORO* staining of sebaceous epithelioma: closeup of Fig. 2.35 in which intracytoplasmic lipid droplets are well recognizable after staining

Finally, ORO stain is very useful to confirm the presence of lipid in the vacuoles of macrophages and multinucleated giant macrophagic cells when a xanthogranuloma is suspected (Figs. 2.37 and 2.38).

2.3.5 Von Kossa

Von Kossa staining is used to highlight calcium ions in histological sections [7]. A *substitution reaction* is at the basis of this stain. This method is not specific for calcium itself, but tissues are treated with a silver nitrate solution and the silver is deposited by replacing the calcium reduced by the strong light, and so can be visualized as metallic silver (black in color). In specimens, the areas in which calcium salts were present are colored *black*, while nuclei stain *red* (Figs. 2.39, 2.40, 2.41, and 2.42).

Such coloration may be useful for confirming, in cytological specimens, the presence of calcium salts in skin lesions from dogs with cutaneous mineralization (calcinosis cutis), especially due to both spontaneous and iatrogenic hypercortisolism (Figs. 2.43 and 2.44).

When from a nodule, a chalk-like material is collected with cytology, *calcinosis circumscripta* should be suspected. Sometimes, the material collected is granular and amorphous, and it is not cytologically recognizable as a mineral; staining the slide with Von Kossa can confirm its calcium salt composition.

2.3.6 Toluidine Blue

Metachromasia is the physical-chemical phenomenon whereby, by coloring a histological sample, some elements of the tissue take on a color different from that which is

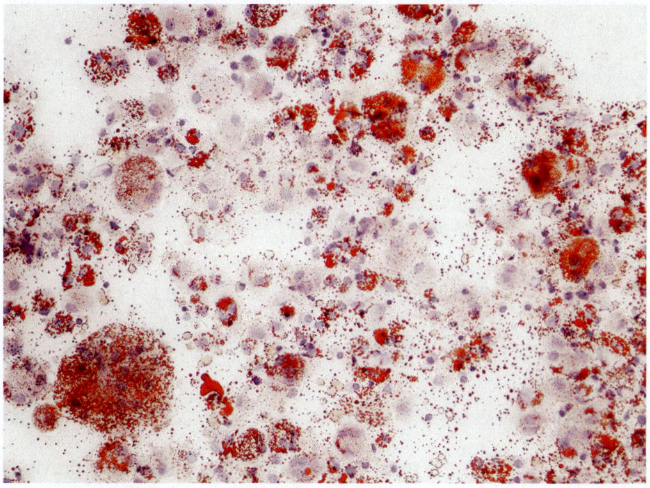

Fig. 2.37 *ORO* staining of xanthogranuloma: many red-stained droplets of lipid are clearly highlighted after staining

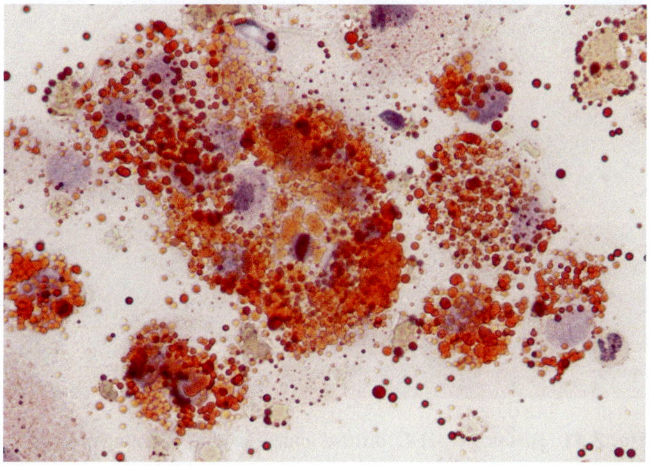

Fig. 2.38 *ORO* staining of xanthogranuloma: closeup of positive lipid droplets

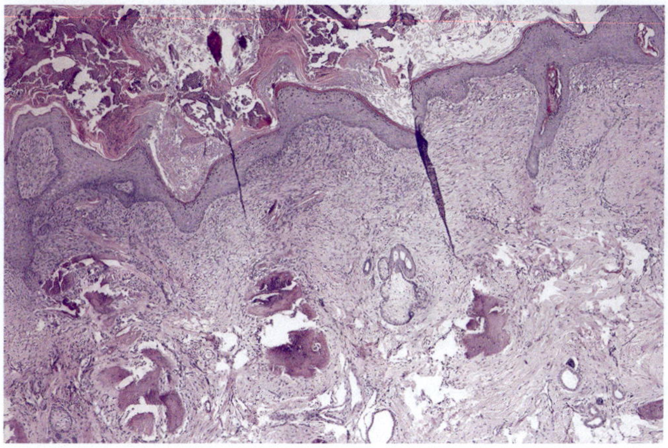

Fig. 2.39 Histology—H-E stain: many foci of amorphous deeply pinkish mineralized material are present in the dermis and in the stratum corneum

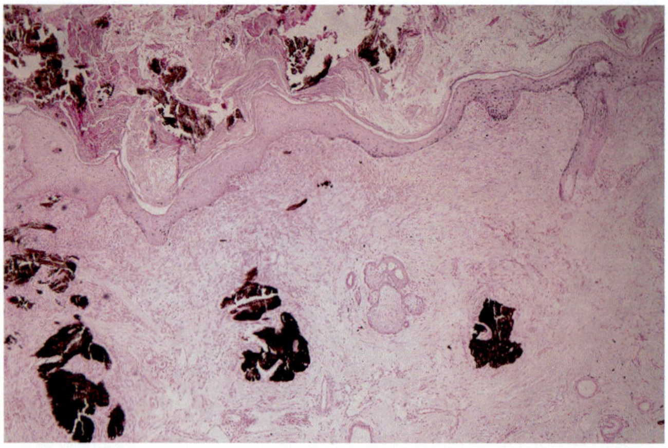

Fig. 2.40 Histology—Von Kossa stain: the same case as the previous image, in which the black areas demonstrate that foci represent mineralized areas composed of calcium salts

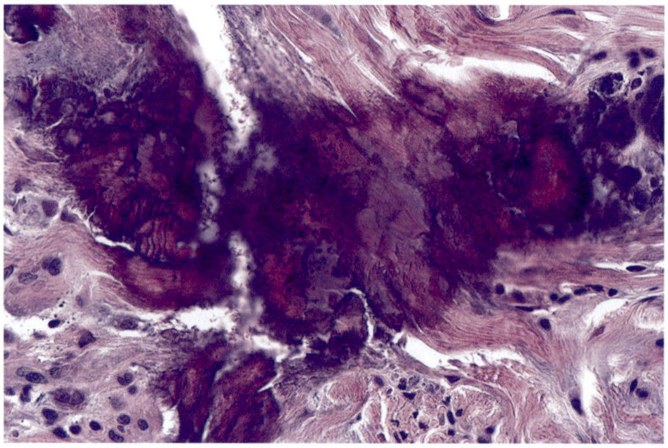

Fig. 2.41 Histology—H-E: diffuse mineralization of collagen fibers

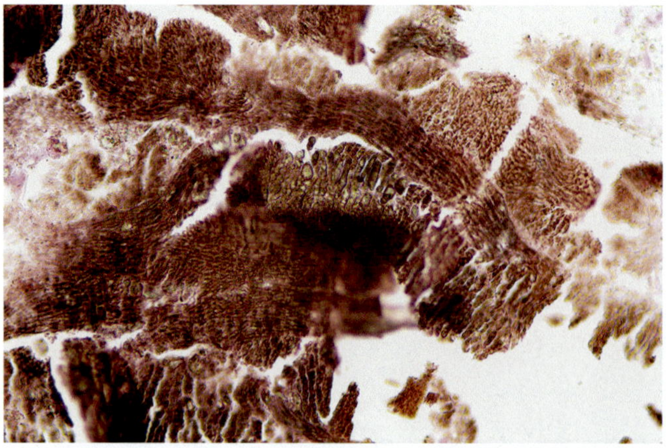

Fig. 2.42 Histology—Von Kossa: the area where *calcium* salts were present stains black

expected to take on with the dye used. Metachromasia is a property of some basic dyes such as those of aniline (thiazide, toluidine blue (TB), and gentian violet) to change the original

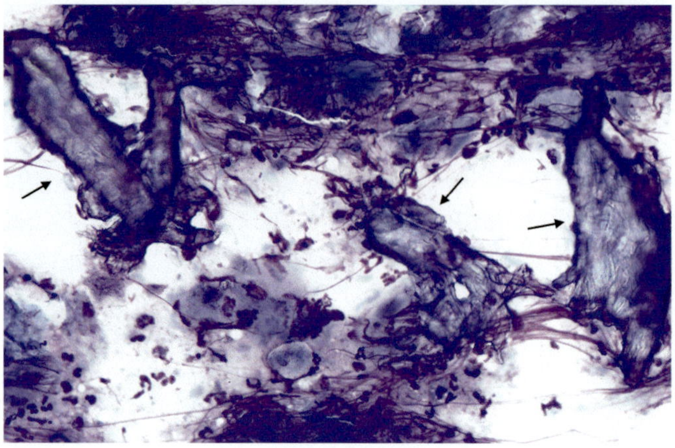

Fig. 2.43 Cytology of calcium salts: pyogranulomatous inflammation and amorphous mineral material (arrow) in a dog with *calcinosis cutis*

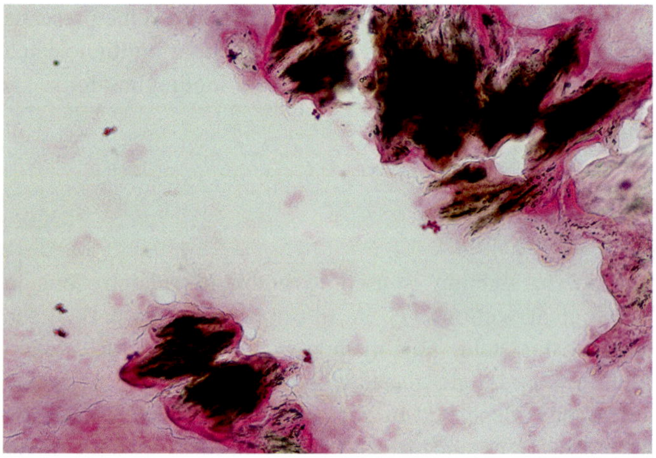

Fig. 2.44 Cytology of Von Kossa stain: the same specimen of the previous image in which the area where *calcium* salts were present stains black. The inflammatory cells stain pale pink and are barely perceptible around *calcium* salts

color when they come into contact with certain polyanionic substances (called metachromatic).

Usually, the granules of mast cells are easily stained with the routine Romanowsky-type dyes. In rare cases of mast cell tumors, in both dogs and cats, the cytoplasmic granules are present but not stained.

TB dye is an alkaline stain used for the metachromatic staining of acidic substances. Due to the high acidity of the heparin and histamine present in the granules of mast cells, the alkaline staining undergoes a color change. This property (*metachromasia*) is typical of mast cells. With TB, secretory granules of mast cells stain red-violet [8].

In practice, it is utilized in the attempt to highlight granules in poor-granulated mast cells in which the low number of granules, or their absence in specimens stained with the Romanowsky-type dye, can be the cause of missing a diagnosis of MCT (Figs. 2.45 and 2.46).

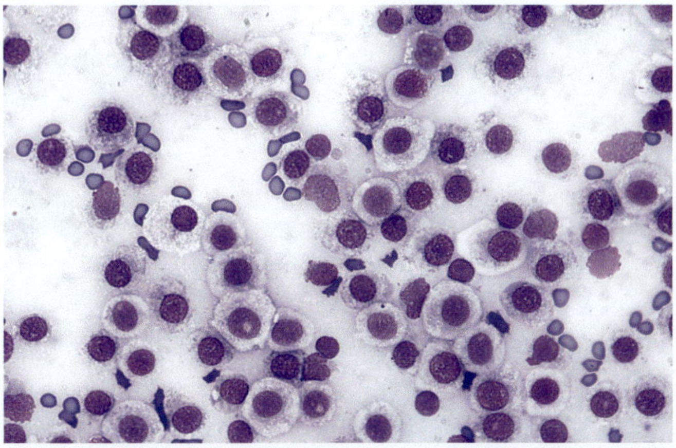

Fig. 2.45 Romanowsky-type dye: mast cell tumor in cat. Many punctate vacuole-like achromatic areas are visible in the cytoplasm of mast cells. No granules are viewable

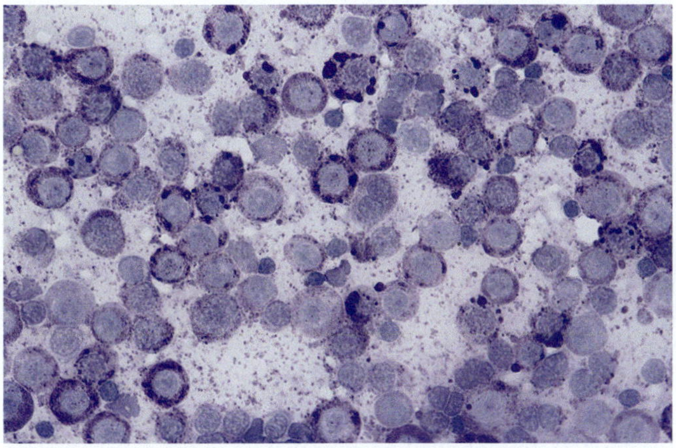

Fig. 2.46 TB: same specimen of the previous image after staining with TB. Notice that many metachromatic intracytoplasmic granules are highlighted

2.3.7 Congo Red

Amyloid is a homogeneous structure consisting of protein fibrils; amyloidosis results from the abnormal folding of proteins, which become insoluble, aggregate, and deposit as fibrils in extracellular tissues.

In reality, this substance can have a very different composition and pathogenesis, even if the microscopic aspect is superimposable. In familiar forms, such as in Abyssinian cats and Shar-Pei dogs, and in those associated with some inflammatory processes, amyloidosis is caused by a lack of degradation of a derivative of the acute phase serum amyloid protein A. In the forms associated with lymphoid malignancies such as plasma cell tumors, it is instead derived from an accumulation of pathological immunoglobulin light chains.

Congo red is an anionic dye and is able to deposit in amyloid fibrils; the link between amyloid and Congo Red is not clear whether it is due to the protein or polysaccharide components of the amyloid itself [9].

Congo red stained tissue appears *orange-red* with a light microscope. Under polarized light, however, amyloid deposits appear as double refractive images of bright green on a dark background; this phenomenon seems to be due to the ability of the amyloid fibrils to form crossed β sheets. Since the amyloid fibril protein of the amyloid light chain type is produced from free Ig light chains secreted by a monoclonal population of plasma cells, this dye can be used in cutaneous samples, to highlight the amyloid produced in some plasma cell tumors or in rare cases of nodular cutaneous amyloidosis.

With Romanowsky-type staining, the amyloid cytologically appears as an amorphous fibrillar to globular, bright pink material that cannot be easily differentiated from other amorphous acellular substances, such as extracellular matrix or amorphous collagen. In samples very rich in amyloid, the latter can be observed as a very thick and deep blue amorphous material and not as the classic pinkish fibrillar material that we are used to observe on slides; staining these slides with Congo red allows us to highlight how, the intense blue colored amorphous material, is simply a thick amount of amyloid (Figs. 2.47, 2.48, 2.49, and 2.50).

It should be remembered that other substances, such as collagen, could also be colored with Congo red, which, however, does not assume the same behavior as amyloid when observed under polarized light. Since, in practice, the microscopes supplied in veterinary clinics are not equipped with polarized light, a mild Congo Red positivity must be interpreted with caution, especially if on the slides, some collagen fibers are also present along with amorphous pinkish material.

2.3.8 Alkaline Phosphatase

Bone alkaline phosphatase is the specific bone isoform of the alkaline phosphatase; it is a glycoprotein that is located on the surface of the osteoblasts and reflects the biosynthetic activity of these cells. The demonstration of the reactivity of the phosphatase alkaline by cancer cells on stained slides is useful for differentiating osteosarcoma from other types of skin neoplasia, which can be cytologically similar. On the

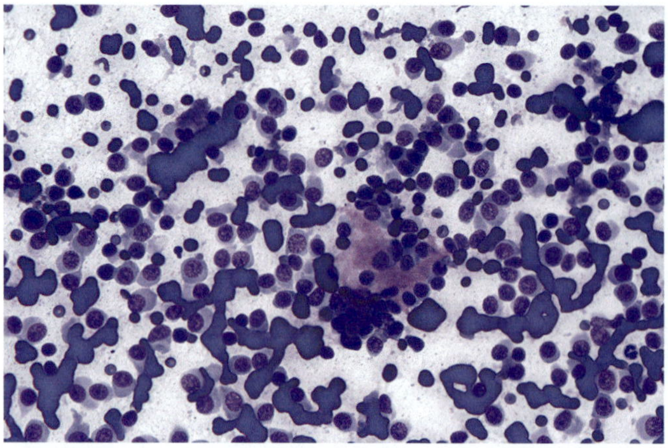

Fig. 2.47 Romanowsky-type stain: amyloids can be detected as pinkish amorphous material between neoplastic plasma cells

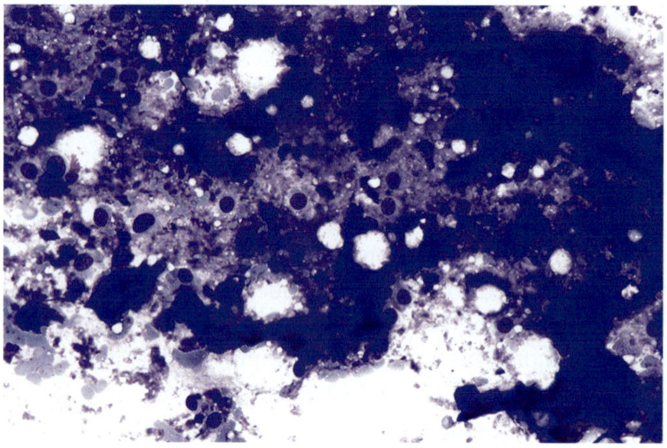

Fig. 2.48 Romanowsky-type stain: in cases where the amount of amyloid is very abundant, the latter can be represented by amorphous material from deep violet to blue

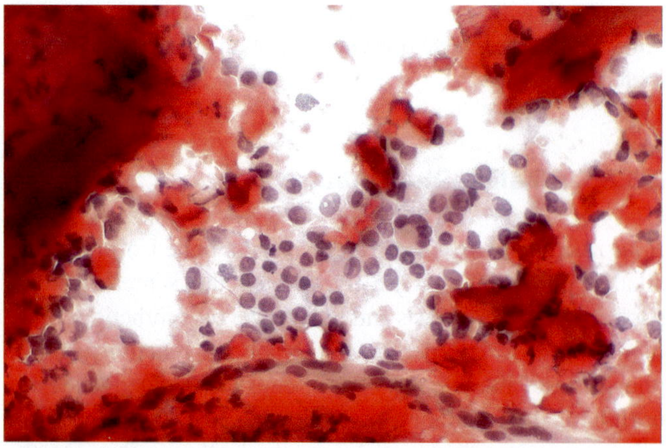

Fig. 2.49 Congo red staining: the same specimen of the previous image. Many plasma cells are intermingled in large amounts of bright red-stained amyloid

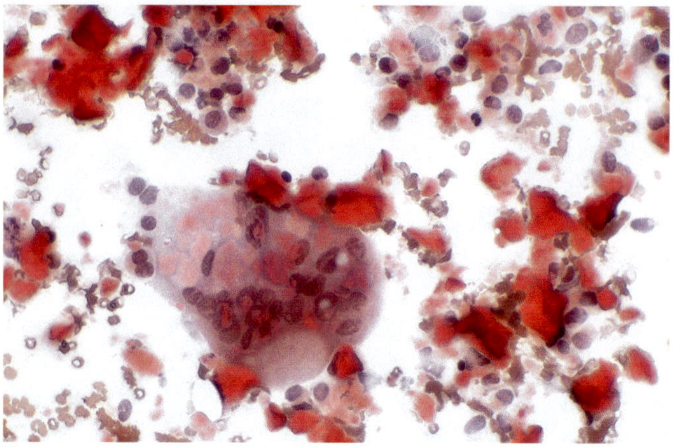

Fig. 2.50 Congo red staining: the same specimen of the previous image. The multinucleated giant cell at the center of the picture is a macrophage filled with amyloid

market, there is a ready-to-use kit, which allows highlighting the presence of alkaline phosphatase in neoplastic osteoblasts on cytological preparations previously stained with the Romanowsky-type routine dyes. Although both primary and metastatic cutaneous osteosarcomas are very rare in dogs and cats, the use of a quick test that can differentiate neoplastic osteoblasts from other morphologically similar cells such as those observed in other sarcomas or in some plasma cell tumors can be of great help in cytological diagnostic, as discussed in Chap. 8.

The 5-bromo chromogen, 4-cloror, 3-indolil phosphate/nitroblual tetrazium (BCIP/NBT) is a very sensitive substrate for the detection of alkaline phosphatase in the cells; alkaline phosphatase hydrolyzes the BCIP, which is oxidized by the atmospheric oxygen, producing an insoluble product and a change of color (blue) on the reaction site (cytoplasmic expression). In the presence of NBT (P-nitrobluar tetrazolium chloride), after a reduction reaction, an insoluble dark blue cytoplasmic precipitate is formed. On the slides already colorful with Romanowsky-type stain, the cytoplasm changes color to dark gray-brown (Figs. 2.51, 2.52, 2.53, and 2.54). The intensity of the color varies according to the pH and the type of dye previously used; therefore, not all cases of osteosarcoma will have the same intensity of color of cytoplasm. It should be remembered that this reaction does not differentiate reactive osteoblasts from neoplastic ones, and so cell morphology and clinical presentation are always of considerable importance in order to avoid a misdiagnosis. In a study performed on stained slides indicative of osteosarcoma, the sensitivity and specificity of the expression of alkaline phosphatase detected using the BCIP/NBT substrate were, 88% and 94%, respectively [10, 11].

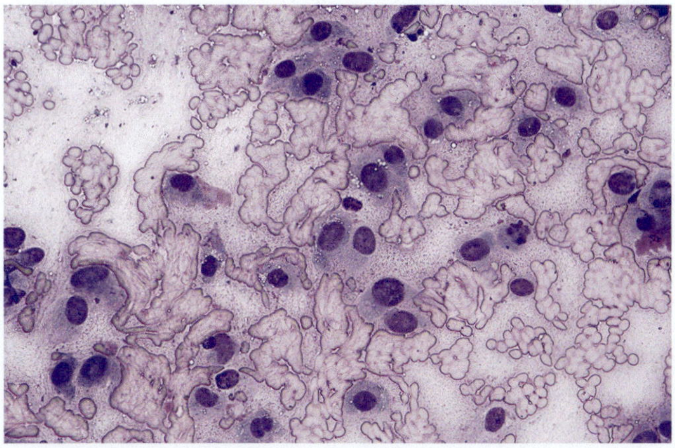

Fig. 2.51 Romanowsky-type stain: spindle-shaped neoplastic osteoblasts: note that it is very difficult to differentiate them from cells of mesenchymal soft tissue sarcomas

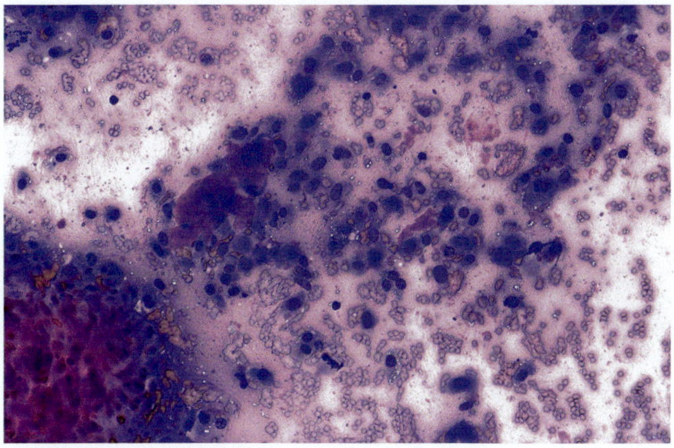

Fig. 2.52 Romanowsky-type stain: many neoplastic osteoblasts are embedded in a pinkish extracellular material representing bone matrix

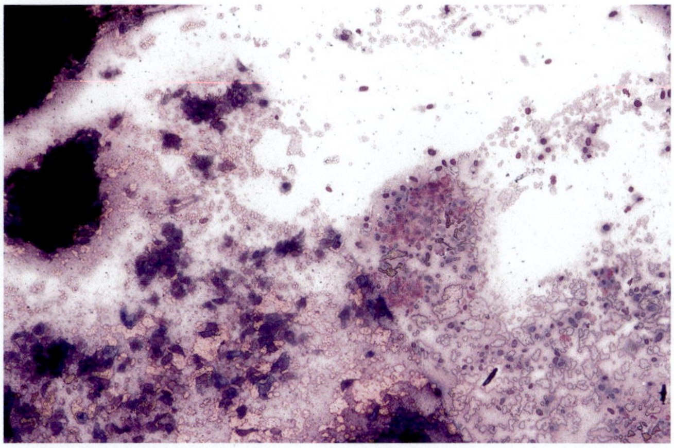

Fig. 2.53 The oblique line marks the transition between the two stains: on the right, the standard Romanowsky-type stain, on the left, it can be the same slide treated with BCIP/NBT and which can be clearly recognizable as the cytoplasm is strongly positive for the presence of alkaline phosphatase (dark cytoplasm)

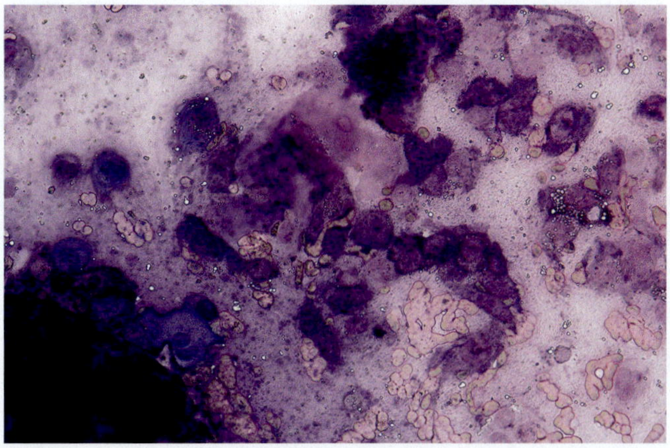

Fig. 2.54 BCIP/NBT substrate: close up of the Fig. 2.53; note as cells positive to alkaline phosphatase show dark-black granular cytoplasm

References

1. Marcos R, Santos M, Santos N et al (2009) Use of destained cytology slides for the application of routine special stain. Vet Clin Pathol 38(1):94–102
2. Riva MA, Manzoni M, Isimbaldi G et al (2014) Histochemistry: historical development and current use in pathology. Biotech Histochem 89(2):81–90
3. Shalin SC, Ferringer T, Cassarino DS (2020) PAS and GMS utility in dermatopathology: review of the current medical literature. J Cutan Pathol 47(11):1096–1102
4. Van Deun A, Hossain MA, Gumusboga M et al (2008) Ziehl-Neelsen staining: theory and practice. Int J Tuberc Lung Dis 12(1): 108–110
5. Selvakumar N, Sekar MG, Rahman et al (2005) Comparison of variants of carbol-fuchsin solution in Ziehl-Neelsen for detection of acid-fast bacilli. Int J Tuberc Lung Dis 9(2):226–229
6. Masserdotti C, Bonfanti U, De Lorenzi D, Ottolini N (2006) Use of oil red O stain in the cytologic diagnosis of canine liposarcoma. Vet Clin Pathol 35(1):37–41
7. Schneider MR (2021) Von Kossa and his staining technique. Histochem Cell Biol 156(6):523–526. Epub 2021 Nov 20
8. Ribatti D (2018) The staining of mast cells: a historical overview. Int Arch Allergy Immunol 176(1):55–60
9. Rowland PH, Valentine BA, Stebbins KE et al (1991) Cutaneous plasmacytomas with amyloid in six dogs. Vet Pathol. 28(2):125–130
10. Barger A, Graca R, Bailey K et al (2005) Use of alkaline phosphatase staining to differentiate canine osteosarcoma from other Vimentin-positive tumors. Vet Pathol 42:161–165
11. Ryseff JK, Bohn AA (2012) Detection of alkaline phosphatase in canine cells previously stained with Wright–Giemsa and its utility in differentiating osteosarcoma from other mesenchymal tumors. Vet Clin Pathol 41:391–395

3 Morphology and Function of the Cells that Compose the Skin

3.1 Introduction

To better interpret the cytological specimens collected from skin lesions, it is mandatory to know the morphology of the cells that compose the epidermidis, the dermis, the adnexa, and the subcutaneous tissue, along with cells that normally reside in the dermis.

Anatomically, the skin is composed of three main different parts: the *epidermidis*, the *dermis* with the *follicular* and *glandular adnexa*, and the *hypodermis*, also known as *subcutis* or *adipose panniculus*.

In the skin, there are also both articulated *vascular* and *nervous* networks. The vascularization of the skin is guaranteed by a dense network of arteries, capillaries, and veins, stratified in both horizontal and parallel arrangements.

The skin is a richly innervated sensory organ in which innervation develops in a segmental way thanks to *nerve fibers* coming from *sensory neurons* whose cell bodies are located in the spinal nerve ganglia.

3.2 The Cells of the Epidermidis

More than 90% of the *epidermidis* consists of epithelial nucleated cells that mature to become anucleate and are completely keratinized through the physiological *keratinization* process. The remaining intraepidermal cells comprise melanocytes and intraepidermal dendritic *Langerhans cells* (antigen-presenting cells (APCs)), which are observed on cytological specimens coming from their neoplastic transformation, respectively, in *melanocytomas/melanomas* and in *benign cutaneous histiocytoma/Langerhans cell histiocytosis*. In the basal layer, there are also *Merkel cells*, oval-shaped receptors that function as tactile sensors that are responsible for recording the pressure exerted on the skin; they are mainly grouped in *Tylotrich pads*. These cells are cytologically visible only in the very rare cases of Merkel cell tumor (carcinoma).

As epidermal cells are mainly composed of keratin, they are named *keratinocytes*, which compose the four layers of the epidermidis: *basale*, *spinosum*, *granulosum*, and *corneum*; since the stratum corneum is composed of fully keratinized cells, the latter are more commonly named *corneocytes* (Fig. 3.1).

Under physiological conditions, and in animals with no skin lesions, corneocytes are the only cells that can be collected on the skin surface, since they are the result of the physiological process of keratinization.

3.2.1 Stratum Basale

The cells of the basal layer are the smaller among the keratinocytes and are also those less frequently observed in non-neoplastic cutaneous specimens. They are cuboidal to polygonal in shape in histological section but appear roundish when observed on cytological specimens; they measure 8–10 μm and have round and central nuclei, sometimes with a single evident nucleolus, and a high nucleus–cytoplasm ratio (N/C ratio) (Fig. 3.2). The cytoplasm is scarce and deep blue in color. Since basal cells are nonkeratinized, they can confuse with other non-epidermal cells; a cytological characteristic of some basal keratinocytes that can orient the cytologist to recognize them as epidermal cells when they are not arranged in clusters is the presence of a thin clear halo around the nucleus (Fig. 3.3). As aforementioned, basal keratinocytes are not commonly observed in inflammatory specimens, while they are frequently observed in neoplastic skin cytology since the basal cell tumors, which can originate from the basal cells of epidermidis or follicular and glandular adnexa, are extremely common in dogs and cats, as discussed in Chap. 8.

F. Albanese, *Canine and Feline Skin Cytology*, https://doi.org/10.1007/978-3-031-95713-0_3

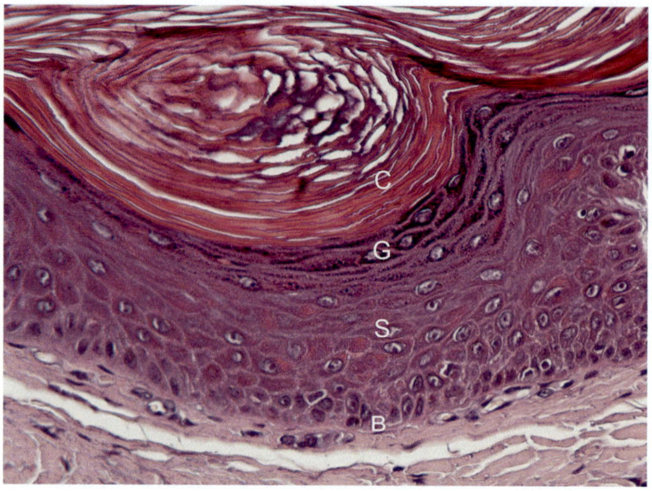

Fig. 3.1 Histology of the epidermidis: hyperplastic epidermidis in which the different epidermal layers are well recognizable: (***b***) stratum basale, (***s***) stratum spinosum, (***g***) stratum granulosum, and (***c***) stratum corneum

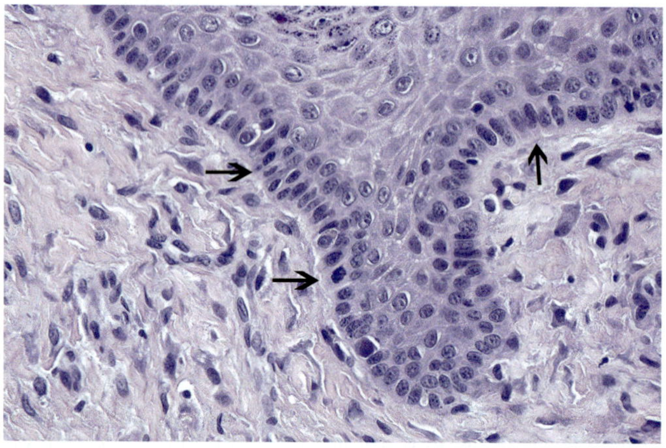

Fig. 3.2 Histology of the epidermidis: cells of the basal layer are arranged in palisade and have smaller and hyperchromatic nuclei (arrows)

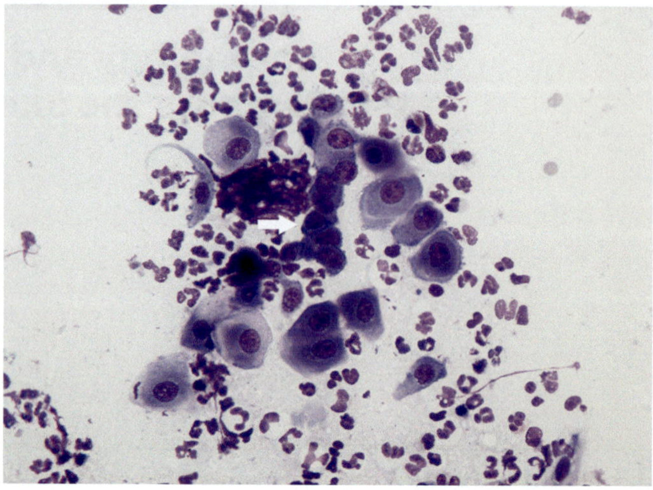

Fig. 3.3 Cytology of the stratum basale: a cluster of basal keratinocytes organized in a short row (white arrow). Note the barely perceptible achromatic perinuclear halo and the small size compared with larger keratinocytes of the spinous layer

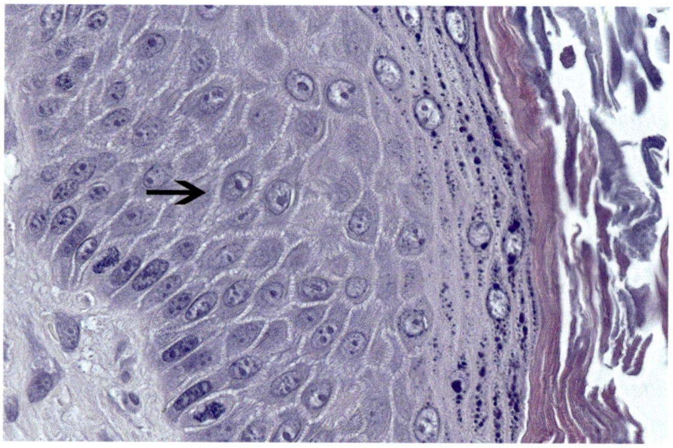

Fig. 3.4 Histology of the epidermidis: keratinocytes of the spinous layer are the most numerous nucleated cells of epidermidis; linear peripheral spines are clearly recognizable in the intercellular spaces (arrow)

3.2.2 Stratum Spinosum

The cells of the *spinosum layer* are larger than basal ones and are more frequently observed on cytological specimens as they are the most numerous nucleated cells of the epidermidis. The term "spinous" derives from their appearance on histopathological preparations following the fixation processes of skin tissues, and the numerous intercellular connections that mimic spines become clearly evident (Fig. 3.4). *Keratinocytes* of this layer are very different in size, depending on the degree of keratinization; they are usually large and polygonal cells with angled borders and round to oval nuclei, usually central, with a lower N/C ratio if compared with basal cells. The cytoplasm is large and of varying intensity of colors, which range from pale to dark blue. They can be observed both as a single cell and as clusters of cells arranged as the tiles of a pavement (Figs. 3.5 and 3.6). Spinous keratinocytes can be observed in any diseases of the skin that cause epidermal hyperplasia or when ulcerative lesions are sampled. Spinous keratinocytes are commonly observed in pemphigus foliaceus. In this autoimmune pustular disease, keratinocytes are very different from those described above; in fact, the acantholytic keratinocyte assumes that a predominantly rounded shape, sometimes oval or with short cytoplasmic tails, has a central nucleus and a non-keratinized cytoplasm, of an intense blue color (Fig. 3.7).

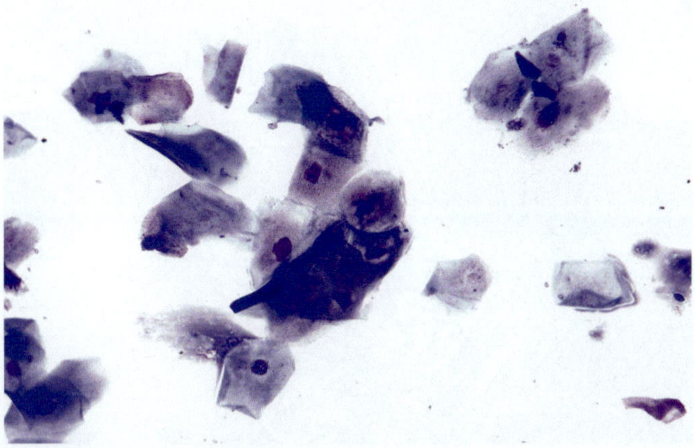

Fig. 3.5 Many large, polygonal, spinous cells with angled borders and central round to oval nucleus along corneocytes

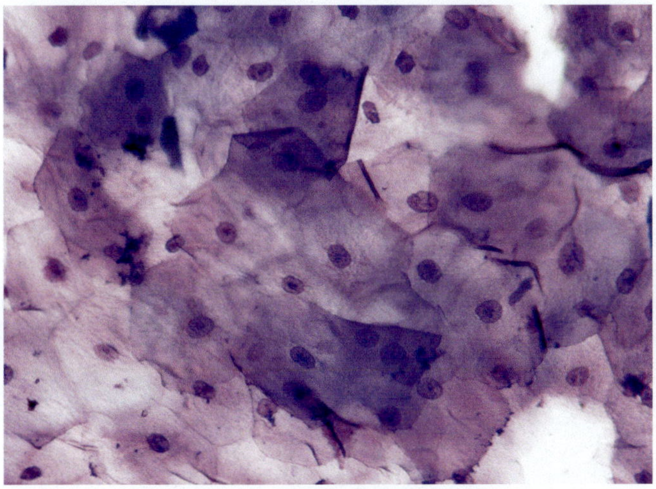

Fig. 3.6 Large cluster of spinous keratinocytes with squamous cytoplasm arranged as tiles of a pavement

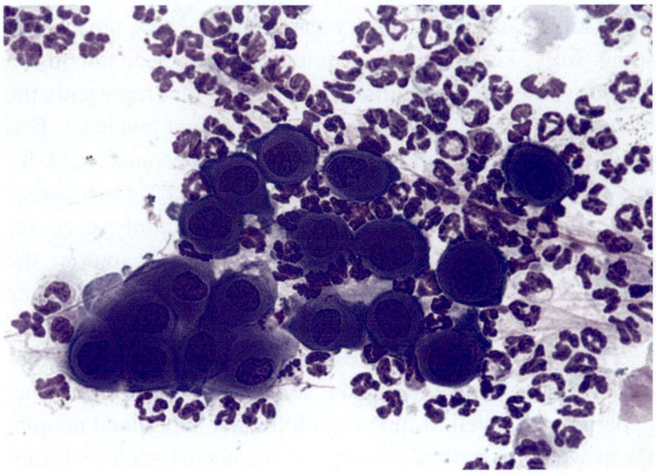

Fig. 3.7 Pemfigus foliaceus: medium-sized spinous acantholytic keratinocytes with roundish to oval silhouette and not keratinizing deep blue cytoplasm

3.2.3 Stratum Granulosum

The cells of the *granulosum layer* are so called because they contain intracytoplasmic *keratohyaline granules* of different sizes and round in shape. Anyway, from a structural point of view, the so-called keratohyalin granules present in the cytoplasm are not real granules; indeed; they lack a membrane, are described as insoluble aggregates, and are the morphological equivalent of the structural protein profilaggrin, which is the precursor of filaggrin that is synthesized in the stratum granulosum. The "granules" are roundish to oval, different in shape, and take on a characteristic dark blue color in histopathological specimens (Fig. 3.8).

On cytology, they, instead, assume a violet color when stained with the Romanowsky-type dye; the pink color makes them easily recognizable and cytologically distinguishable from any other type of granule, pigment, or microorganism, which can be detected on the surface of keratinocytes. The size of the cells is variable, but they are usually medium-large with a polygonal shape and a low N/C ratio (Figs. 3.9 and 3.10). They can be observed in the same skin disorders described for spinous keratinocytes and, since hypergranulosis is one of the typical histopathological features of papillomavirus infection, the keratinocytes of the granulosum layer could be observed in specimens collected via scraping technique from the cut surface of a cutaneous viral papilloma.

3.2.4 Stratum Corneum

The normal epidermal keratinization process results in the formation of the *stratum corneum*, which is composed of keratinocytes that have lost the nucleus to become completely keratinized. For this reason, they are commonly

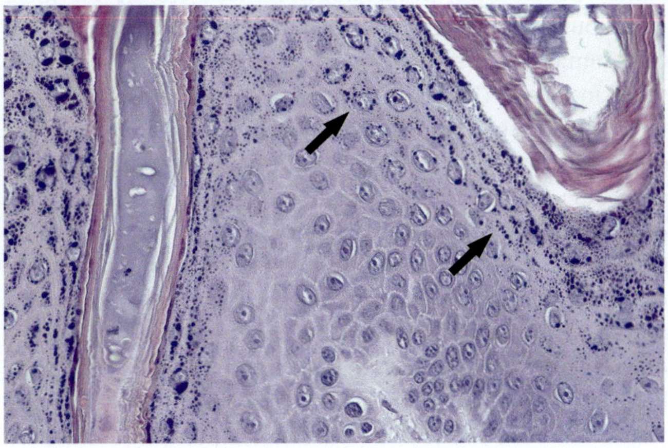

Fig. 3.8 Histology of the epidermidis: keratinocytes of granulosum layer are histologically characterized by the presence of many intracytoplasmic keratohyalin "granules" which stain dark blue (arrows)

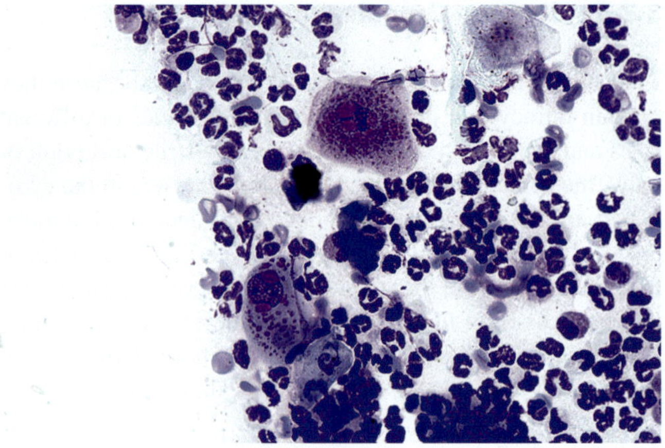

Fig. 3.9 Keratinocytes of the granulosum stratum: large polygonal keratinocytes with cytoplasm filled with roundish and violet keratohyalin "granules"

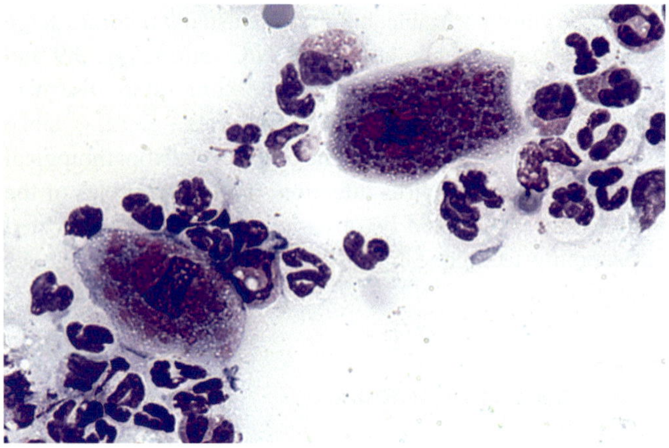

Fig. 3.10 Cytology of the stratum granulosum: high magnification in which from small to large polygonal keratinocytes with cytoplasm filled with roundish and violet keratohyalin granules are easily recognizable. Many neutrophils and eosinophils are also present

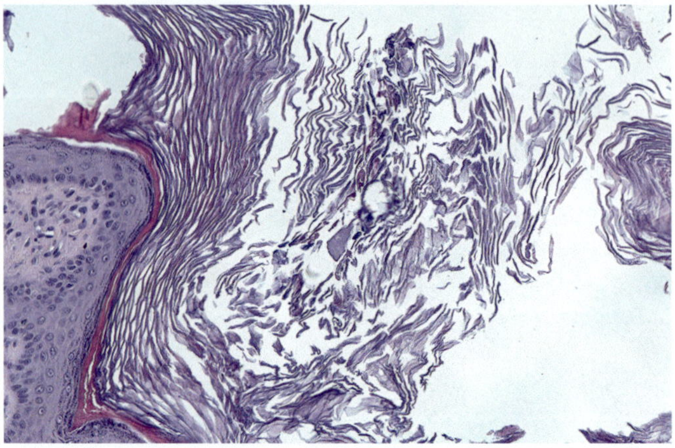

Fig. 3.11 Histology of the epidermidis: corneocytes of the *stratum corneum* represent the final process of keratinization of epidermidis. The space between fully keratinizated cells is in nature occupied by lipid intercellular material

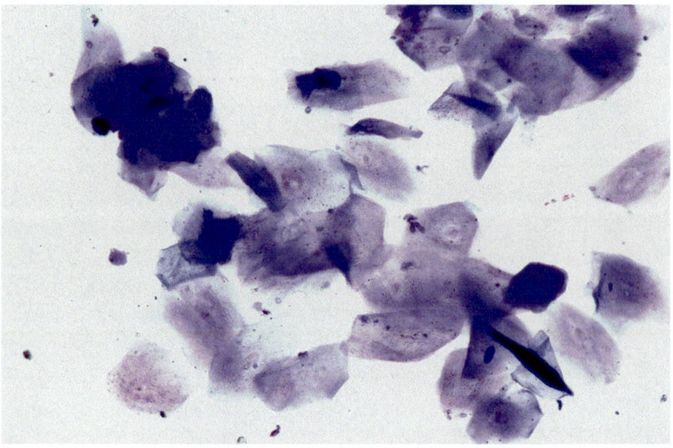

Fig. 3.12 Cytology of the stratum corneum: many anucleate cornified keratinocytes (corneocytes): note the melanin pigment on the surface of many of them

named *corneocytes*. In histology, corneocytes are the most numerous cells that are embedded in a lipid matrix composed of ceramides and sphingosines; all together, they compose the skin barrier (Fig. 3.11). In cytology, the morphology of the corneocytes is very variable and varies from large polygonal, flattened, and anucleate cells with angular edges, to deep blue-colored, rolled keratinocytes that give cells a *lanceolate* or *papyrus-like* appearance (Figs. 3.12 and 3.13). Color differences in corneocytes range from completely unstained cells to cells that are colored in various shades of purple and pale or deep blue. These differences depend not only on the amount of keratin present in the individual cells but also on the failure to dissolve lipids, part of the film that forms the skin barrier and that envelopes the surface of keratinocytes. The excessive presence of lipids can, in fact, reduce or inhibit the fixing of cells and so the good penetration of stains.

In some cases, along with normal corneocytes, are evident some fully keratinized keratinocytes which maintain on cytology, a central roundish circular area that represents the membrane profile of the previously present nucleus. This aspect gives the nuclei a shadow-like appearance, and for this reason, these corneocytes are also named "ghost cells." Since, in skin cytology, this term is commonly used for defining keratinocytes that produce the hair's matrix, the author prefers to use for these superficial corneocytes the term "false ghost cells" or "superficial ghost cells."

These false ghost cells are commonly observed in cases of orthokeratotic hyperkeratosis associated with epidermal hyperplasia as well as in benign follicular superficial neoplasia in which a marked hyperkeratosis occurs such as infundibular keratinizing acanthoma (IKA) or trichofolliculoma (Figs. 3.14 and 3.15).

Fig. 3.13 Cytology of the stratum corneum: many lanceolate and deeply blue stained corneocytes

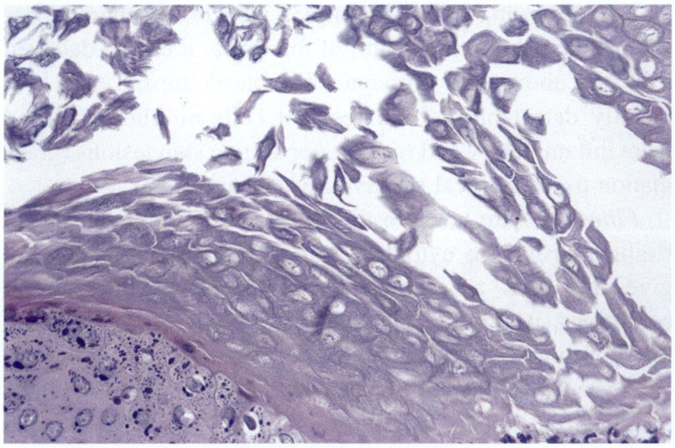

Fig. 3.14 Histopathology. Epidermal hyperplasia with orthokeratotic hyperkeratosis: notice that corneocytes can show a "ghost" appearance before becoming the conventional anucleate corneocytes

Chapter 8 will exhaustively discuss the different cytological aspects of "true" metrical ghost cells and their importance in helping the recognition of follicular neoplasia such as Trichoepitelioma and Pilomatricoma and how do not confuse them with the "false" ghost cells that can observed in case of hyperkeratosis and in other superficial follicular neoplasia such as IKA.

Under normal conditions, only *corneocytes* (anucleated keratinocytes) can be collected from the surface of the skin; cells can be sampled only through the transparent adhesive tape technique. The same cytology is sampled from the surface of many exfoliative skin diseases in which there is an orthokeratotic hyperkeratosis (Fig. 3.16). Instead, if with adhesive transparent acetate tape, many nucleated keratinocytes are sampled from superficial desquamative or crusty lesions, an altered keratinization process must be suspected (Fig. 3.17). These cytological findings can be observed in some *primary keratinization defects* in which a *parakeratotic hyperkeratosis* occurs as in the case of superficial necrolytic dermatitis (hepato-cutaneous syndrome), zinc-responsive or generic food dermatosis, and idiopathic nasal parakeratosis of the Labrador retriever (Figs. 3.18 and 3.19). It must be pointed out that a small number of nucleated keratinocytes collected from the skin surface, along with anucleated superficial corneocytes, is a common finding in many chronic inflammatory canine and feline dermatitis (e.g., canine atopic dermatitis) or in some skin neoplasia (i.e., Bowenoid in situ squamous cell carcinoma in cats) and their presence must not be interpreted as linked to primary keratinization defects; in these cases, the clinical presentation can guide the clinician in the suspicion of a specific disease in which parakeratosis is a histological hallmark.

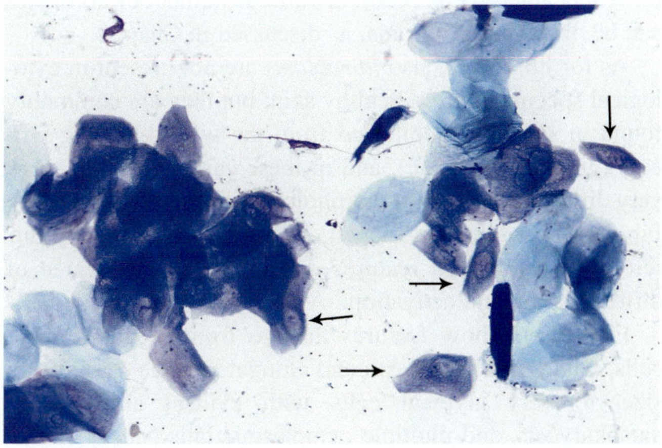

Fig. 3.15 Cytology of the same sample of the previous image. Some large "false ghost keratinocytes" show the roundish central halo of the old nucleus (arrows). Conventional fully cornified corneocytes are also present

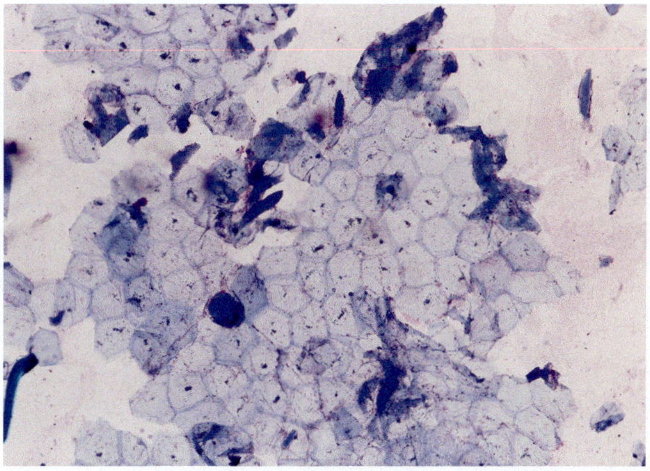

Fig. 3.16 Cytology of the normal skin surface: large sheet of corneocytes collected using a piece of transparent acetate tape

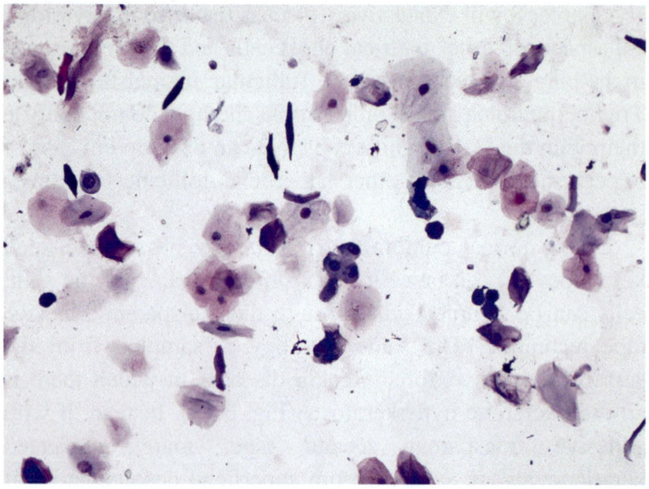

Fig. 3.17 Cytology collected from the nose of a Labrador retriever affected by idiopathic nasal parakeratosis. Note the presence of nuclei in all the keratinocytes sampled. The different sizes and shapes of nucleated superficial keratinocytes must alert cytologists for an altered keratinization process

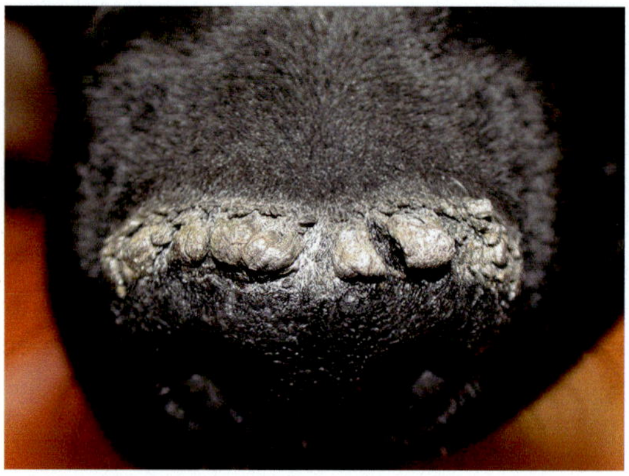

Fig. 3.18 Adherent scales on the nose of a black Labrador retriever with idiopathic nasal parakeratosis

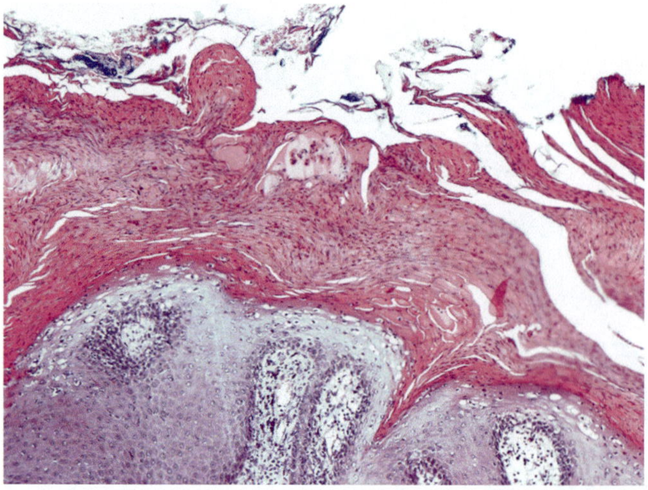

Fig. 3.19 Histopathology: a thick parakeratotic corneum layer with many serum lakes is a typical microscopic alteration of biopsy specimens from Labrador with nasal parakeratosis

3.3 The Cells of the Dermis and Adnexa

The *dermis* is the intermediate portion of the skin located between the *epidermidis* and the *adipose tissue* (hypodermis or adipose panniculus). Anatomically, it is the most complex part of the skin because it is made up of the *connective tissue*, *vascular* and *nervous* networks, *piloerector muscles*, and the adnexa, the latter represented by *hair follicles* and both *sebaceous* and *sweat glands*.

The connective tissue is composed of *fibers*, *ground substances,* and *cells*. The fibers are both insoluble such as *collagen* and *elastin* and soluble such as *proteoglycans* and *hyaluronans* (hyaluronic acid). The extracellular matrix is produced by fibroblasts, and is composed of fibrous protein and amorphous ground substances; it is composed of fibrillar, cable-like striated structures that also represent the myxoid ground substance observed in some mesenchymal soft tissue neoplasia (glycosaminoglycans or mucopolysaccharides).

Without going into the complex world of connective tissue, which is beyond the aim of this book, it can be deduced from the above that fibrocytes/fibroblasts are the most frequently detectable dermal resident cells on slides coming from inflammatory and reactive/reparative skin lesions (granulation tissue or focal reactive fibroplasia).

Fibrocytes are inactive *fibroblasts*, with elliptical nuclei, slightly basophilic cytoplasm, and opposing thin tails that give the cells an elongated spindle shape. In normal conditions, it is quite impossible to collect fibrocytes from the not-injured dermis, since cells are few and strictly connected to collagen fibers. In the case of very lucky slides coming from hyperplastic lesions of collagen such as some skin tag (fibroepithelial polips), few fibrocytes can be collected, sometimes along with collagen fibers, the latter represented by thick and amorphous eosinophilic material which sometimes can be more thin and slender with a vermiform appearance (Fig. 3.20). A variable number of fibrocytes can be observed in fibroma, as discussed in Chap. 8.

As for fibrocytes, also *fibroblasts* are not present in cytological specimens from healthy skin, but they are commonly found in specimens collected from the wound-healing process (granulation tissue), and in these cases, they can show very different shapes and morphologies (Figs. 3.21 and 3.22). Fibroblasts are spindle-shaped cells that can show very different morphological features depending on the degree of differentiation and activation.

Fibroblasts show features that go from regular spindle cells, with tailed cytoplasm and elongated nuclei of uniform size, to very atypical cells with evident anisocytosis, anisokaryosis, and multiple cytoplasmic tails (*starry* or *tentacular* appearance); the nuclei are oval, plump, or spindle in shape, and the cytoplasm is variably basophilic and sometimes with small achromatic vacuolar-like areas (Figs. 3.23, 3.24, and 3.25). In some cases, fibroblasts are released in groups (pseudo-aggregates), held together by the

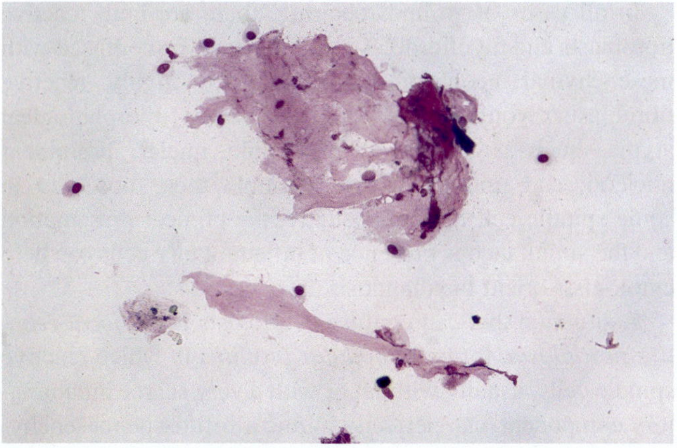

Fig. 3.20 Cytology: pinkish in color collagen fibers along with a few fibrocytes collected from a skin tag

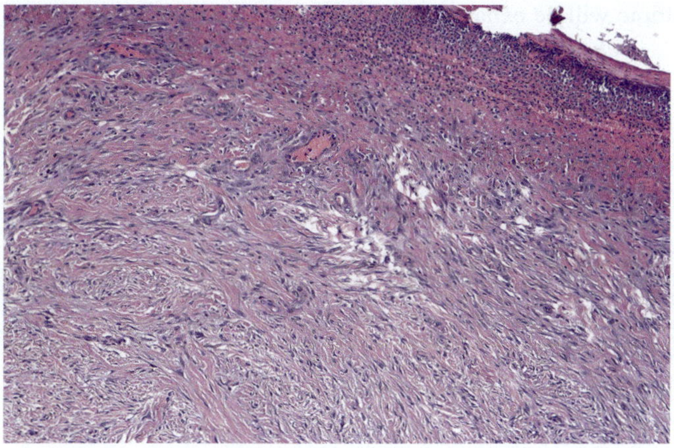

Fig. 3.21 Histopathology: reparative granulation tissue in an inflammatory process; many activate spindloid fibroblasts/myofibroblasts beneath an ulcerated lesion

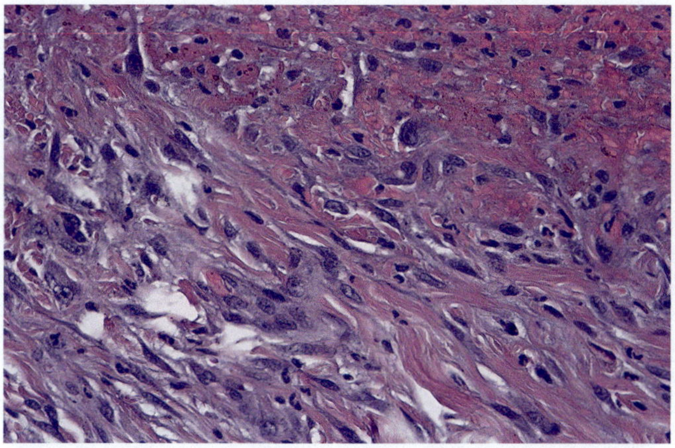

Fig. 3.22 Histopathology: closeup of the lesion 3.21 in which, embedded in a collagenous stroma, many reactive fibroblasts and inflammatory cells are recognizable

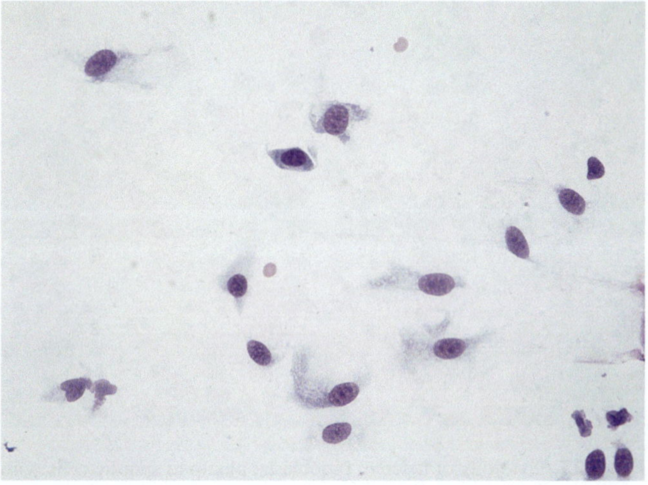

Fig. 3.23 Cytology of skin tag: uniform in size spindle cells, with oval nuclei, on a clear background

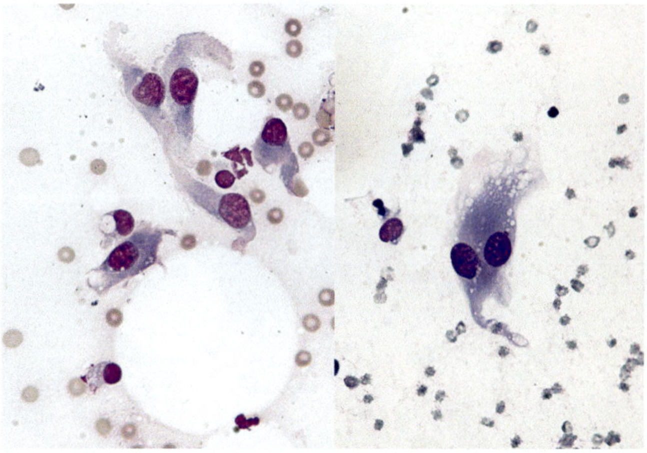

Fig. 3.24 Cytology of a few reactive fibroblasts from granulation tissue: spindle cells with elongated nuclei and tailed cytoplasm with ill-defined margins

extracellular matrix produced by themselves; the latter is cytologically recognizable as a pinkish, amorphous, and fibrillar material between the cells (Figs. 3.26 and 3.27).

Another cell present in the reparative tissue is the *myofibroblast* which morphologically resembles fibroblast and from which cannot be cytologically differentiated. In practice, the myofibroblasts are fibroblasts with contractile activity since they contain, into the cytoplasm, alfa-smooth muscle actin; like fibroblasts, these cells are involved in tissue repairing (wound contraction), but they can also present in scary lesions (fibrosis) or in the stroma of tumors. When spindle cells are collected from granulation tissue or other reactive or neoplastic lesions, it is not possible to differentiate myofibroblasts from fibroblasts with both cytology and histology and immunohistochemistry markers must be used.

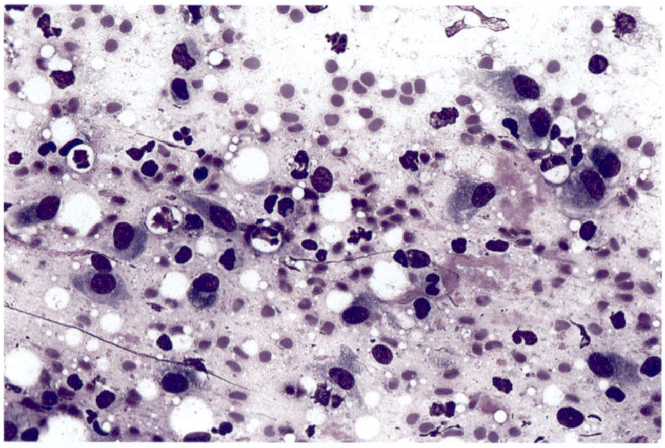

Fig. 3.25 Cytology of reactive fibroblasts: plump to spindle cells with morphological atypia collected from chronic panniculitis with severe fibroplasia

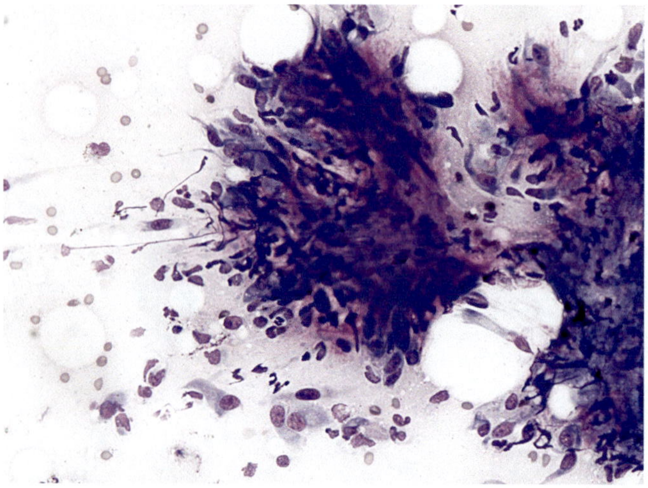

Fig. 3.26 Cytology of the extracellular matrix: bright eosinophilic material that holds together many reactive fibroblasts. This material represents the ground substance produced by fibroblasts

In all cases of wound repairing, there are both reactive fibroblasts and myofibroblasts that must not be confused with mesenchymal neoplastic cells; in fact, highly reactive fibroblasts/myofibroblasts may show so morphological atypia, such as anisocytosis, double nuclei, prominent nucleoli, and mitosis, which resemble those observed in some spindle cell tumors. Usually, the clinical presentation and the simultaneous presence of inflammatory cells can help cytologists orient the diagnosis.

A situation that can confuse cytologists is the *focal reactive fibroplasia*, formerly *nodular fasciitis*, in which reactive spindle cells, usually without or with a very scarce inflammatory component, can perfectly mimic a soft tissue mesenchymal neoplasia; for this reason, a cytological diagnosis of soft tissue neoplasia must be done with caution, especially if the history of the lesion is not detailed, specimens are of scarce cellularity, no inflammatory cells are present, and no severe cytological atypia is observed (Figs. 3.28 and 3.29). This topic will be exhaustively discussed in Chap. 8.

Dendritic APCs originate, like macrophages, from a common precursor present in the bone marrow; their job is to process antigens and travel to secondary lymphoid organs, where they communicate with lymphocytes to orchestrate adaptive immune responses. In *inflammatory processes,* the APCs are involved, but they are not distinguishable from non-activated macrophages on cytological specimens.

Reactive proliferation or neoplastic transformation of dermal APCs are included in the group of the so-called *histiocytic diseases/disorders*, such as benign cutaneous histiocytoma, Langerhans cells histiocytosis, cutaneous and systemic reactive canine histiocytosis, and histiocytic sarcomas in dogs, which are discussed in Chap. 8; in cats, feline progressive histiocytosis is the most common histiocytic dendritic disorders of the skin.

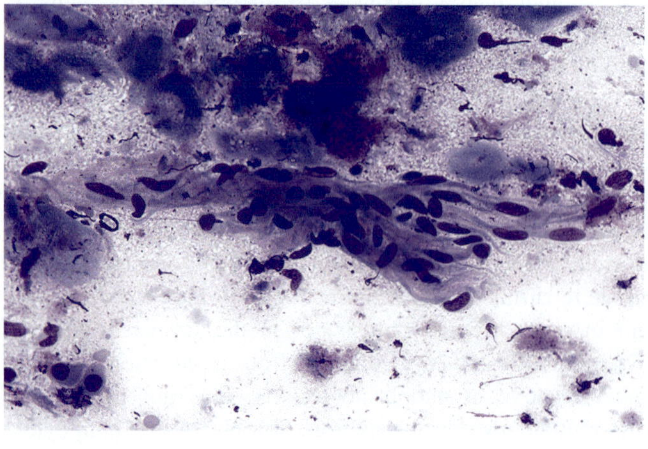

Fig. 3.27 Cytology of reactive fibroblasts organized in bundles in the context of a granulation tissue

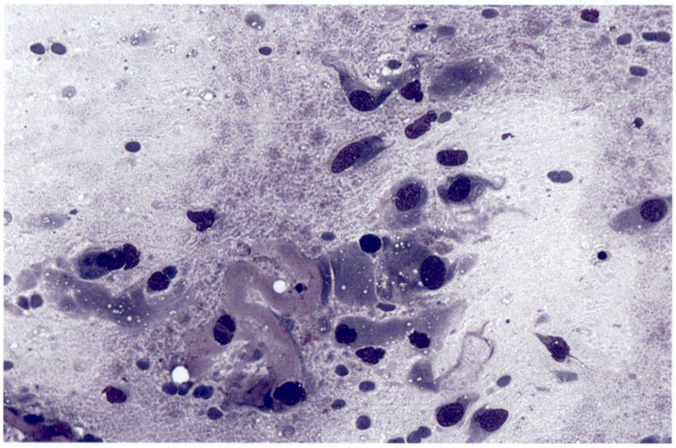

Fig. 3.28 Cytology of *focal reactive fibroplasia*: on a myxoid background, many fibroblast/myofibroblasts with moderate atypia are present. A single collagen fiber is also recognizable

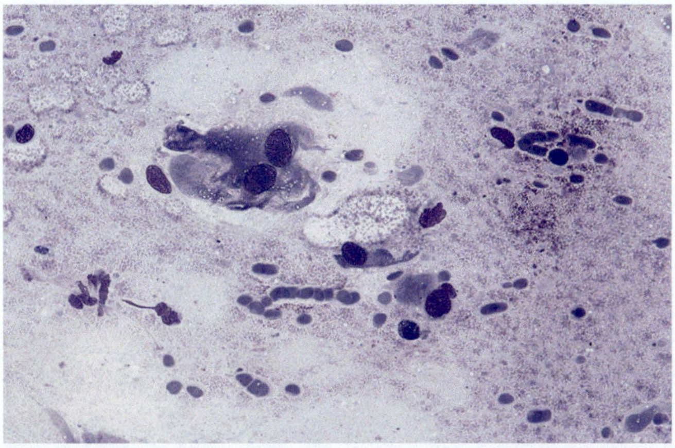

Fig. 3.29 Cytology of *focal reactive fibroplasia*: high magnification of spindle and starry-like shaped fibroblasts/myofibroblasts

Finally, *melanocytes* are physiologically present around the vessels and in the hair follicles, particularly in dogs and cats with black-pigmented coats; these cells are not visible in cytology until their neoplastic transformation occurs (melanocytoma or melanoma).

In many inflammatory processes especially those characterized with granulation tissue, it is possible to find *blood capillaries*, represented by linear and sometimes branched structures, composed of endotheliocytes with elongated nuclei and indistinct cytoplasm that delimitate a thin central lumen in which some red blood cells (RBCs) may be still evident (Figs. 3.30, 3.31, and 3.32). Blood vessels are also observed in many skin neoplasia such as perivascular tumors, hepatoid gland neoplasia, neoplasia of adipose tissue, and so on.

Collagen fibers are protein components of connective tissue produced by fibroblasts, which are cytologically recognizable as fibrillar linear or branched eosinophilic structures (Fig. 3.33). Their discovery is uncommon in preparations sampled from inflammatory lesions, while they are usually very numerous in most well-differentiated mast cell tumors and in some soft tissue sarcomas such as fibrosarcoma and keloidal fibroma/fibrosarcoma (Fig. 3.34). In some cases of collagen benign tumors such as collagenous nevi or hamartomas, skin tags, or fibro-adnexal dysplasia, a variable number of collagen fibers can be collected through cytology. In these groups of benign diseases of collagen, multiple forced movements of aspiration with the plugger of the syringe (fine-needle aspiration biopsy (FNAB)) are strongly recommended if some fibers of collagen want to be collected.

The dermal *adnexa* are represented by *hair follicles* and both *sebaceous* and *apocrine sweat glands*.

The follicles are a continuation of the external epidermidis, which direct downward to form a sac bordered by keratinocytes specialized in the production of different types of keratins comprising those that give origin to hairs.

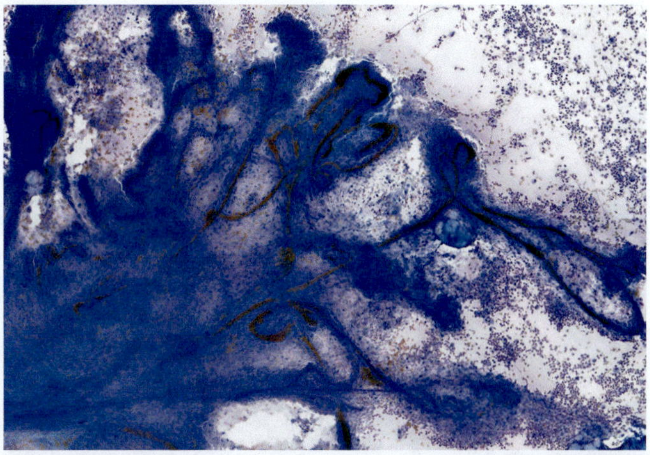

Fig. 3.30 Linear and convoluted blood vessels in the context of an inflammatory process: the orange color is linked to the presence of intravascular RBCs

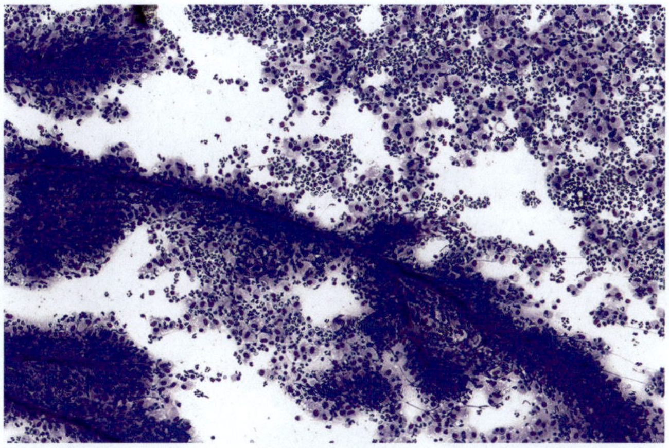

Fig. 3.31 Cytology of vessels: multiple anatomizing linear vessels are surrounded by many inflammatory cells

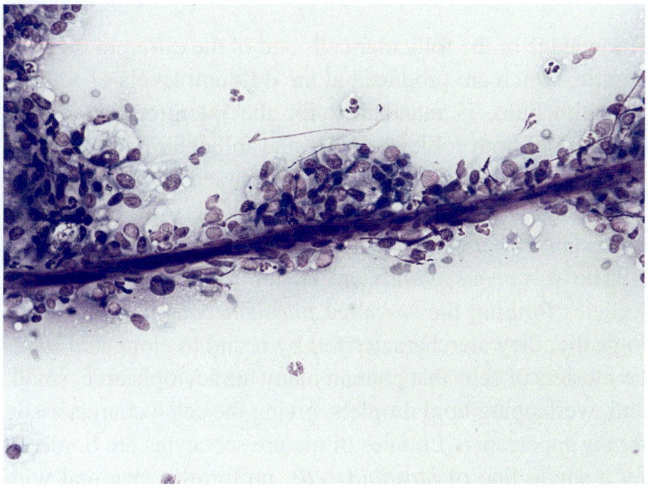

Fig. 3.32 The flattened and slender nuclei composing vessels represent endotheliocytes

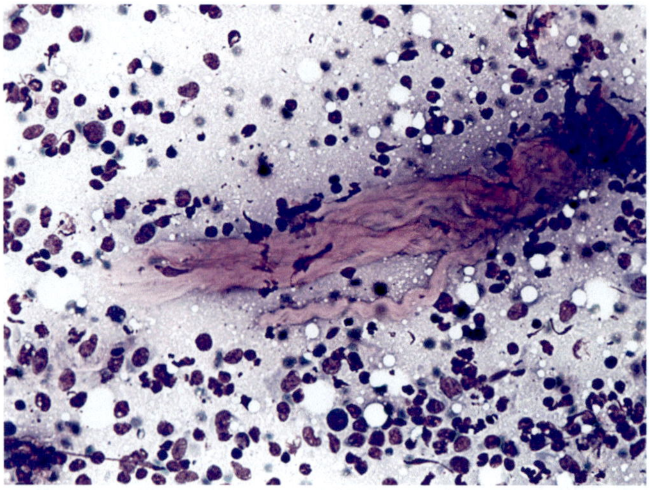

Fig. 3.33 Cytology of collagen fibers: collagen fibers are cytologically recognizable as linear and wavy eosinophilic filamentous structures

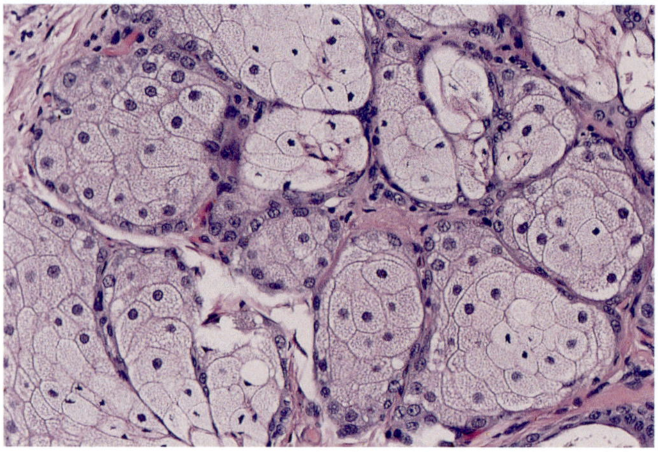

Fig. 3.35 Histology of normal sebaceous gland in cats: lobules of sebaceous cells in which more small and regular in size lipid droplets are evident

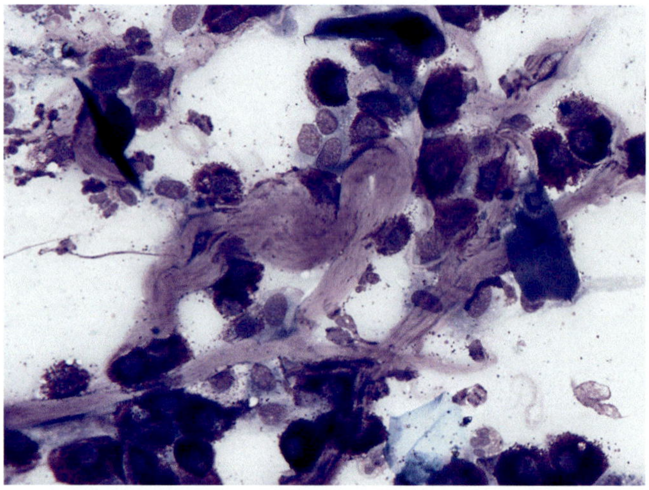

Fig. 3.34 Cytology of mast cell tumor: multiple and long eosinophilic collagen fibers are intermingled with neoplastic mastocytes

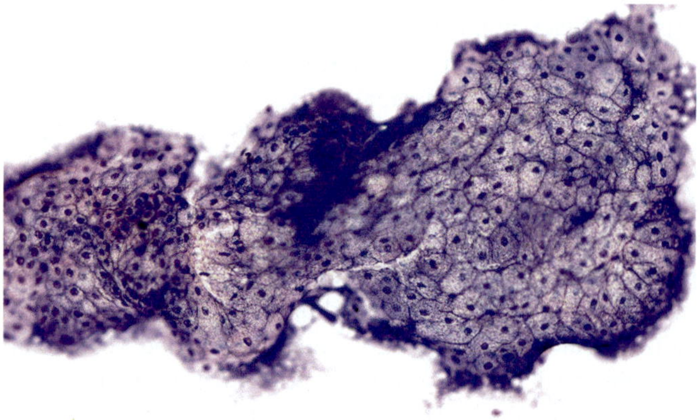

Fig. 3.36 Cytology of normal sebaceous gland: clusters of sebaceous cells in which cytoplasms are filled with lipid droplets

Knowledge of the follicular cells and of the different types of keratin, which are produced at the different levels of follicular epithelium, is mandatory for the interpretation of the cytology of both follicular cysts and follicular neoplasms. A more detailed description will be discussed in Chap. 8.

During inflammatory processes, clusters of both *sebaceous* and *sweat glands* can be collected.

The *sebaceous glands* are anatomically associated with follicles forming the so-called *pilosebaceous unit.* Morphologically, they are characterized by round to elongated lobular clusters of cells that contain many intracytoplasmic, small, and overlapping lipid droplets, giving the cell a characteristic *foamy* appearance. Lobules of mature sebocytes are bordered by a single line of *basaloid cells*, uniform in size and with scant basophilic cytoplasm, representing immature *germinative cells* (reserve cells), which will mature into sebocytes (Figs. 3.35 and 3.36). In cats, the morphology of the mature sebocytes is similar to those of dogs, the difference being that the size of the cytoplasmic lipid droplets is smaller and more uniform in feline sebaceous glands. The secretory duct, named *pilosebaceous canal*, is short, lined with squamous epithelium, and ends directly in the follicular lumen, in which sebum is released (Fig. 3.37). Sebaceous glands have a *holocrine* way of secretion, in which the plasma membrane dissolves and the intracytoplasmic sebum is released into the lumen. In a few words, the *sebum* represents the product of destroyed cytoplasms of mature sebocytes (Figs. 3.38 and 3.39). In cytological samples from sebaceous glands, debris of dead sebocytes and a variable number of keratinocytes, the latter composing the inner part of the ductal epithelium, can be observed on the sebaceous lipid background (sebum) (Figs. 3.40 and 3.41). Clusters of well-differentiated sebocytes are commonly observed in dogs with hyperplasia and neoplasia of the sebaceous glands as we will discuss in Chap. 8. As above mentioned, in some cases, a few clusters of sebocytes can be accidentally collected during the

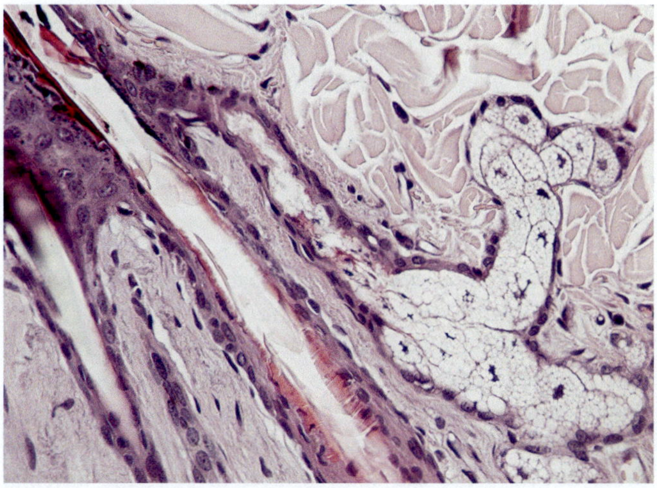

Fig. 3.37 Histology of sebaceous gland: the short sebaceous duct containing sebum starts from the gland and ends in the follicular lumen

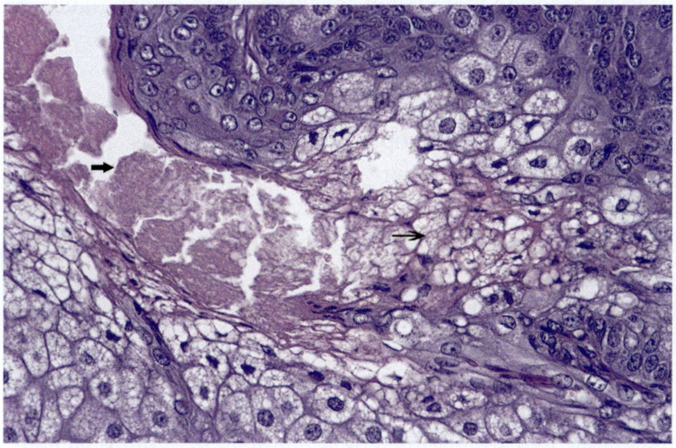

Fig. 3.38 Histology of sebaceous gland: olocrin way of secretion. Note that the cytoplasm of sebocytes undergoes spontaneous destruction (thin arrow) giving origin to sebum (short arrow) that is released into the duct

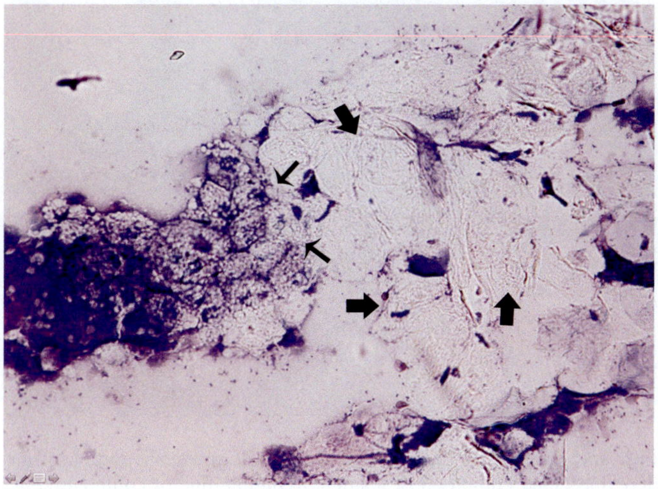

Fig. 3.39 Cytology of normal sebaceous way of secretion: the cytoplasm of sebocytes dissolves (thin arrows) and gives origin to the sebum (thick arrows)

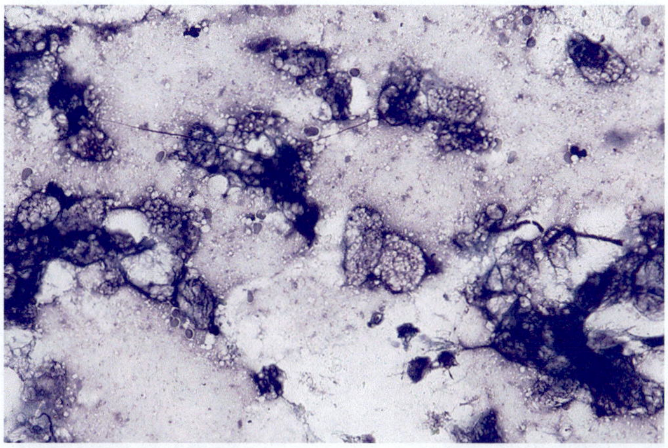

Fig. 3.40 Cytology of the normal sebaceous way of secretion: multiple groups of destroyed cytoplasms of sebocytes transform themselves in sebum

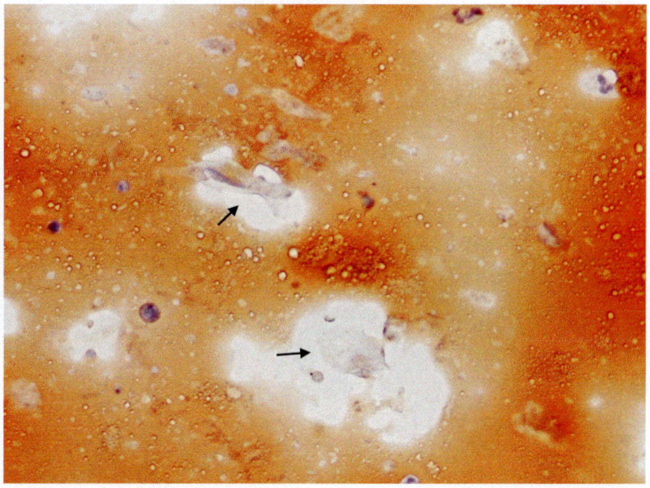

Fig. 3.41 Cytology of the sebum: with oil red O stain, the sebum is colored *red-orange*. Note the presence of some corneocytes that come from the inner surface of the sebaceous ducts

sampling of inflammatory lesions; in these cases, they must be interpreted as *innocent bystanders*, as they are normally residents of the dermis (Fig. 3.42). When multiple large clusters of mature sebocytes are observed along with inflammatory cells, primary sebaceous lesion with secondary inflammation must be considered (sebaceous hyperplasia or adenoma); especially if a nodular lesion is present (Figs. 3.43 and 3.44).

In the canine skin, some *modified* or *specialized* sebaceous glands are also physiologically present in specific body areas; since their microscopic morphology resembles hepatocytes, they are defined as *hepatoid glands* and have a secretory duct that opens into the follicular lumen. These glands are commonly affected by hyperplastic and neoplastic changes, and even if they are mostly anatomically located in the perianal area and on the tail gland, they can be detected at other body sites such as the preputial area or exceptionally, on the back, trunk,

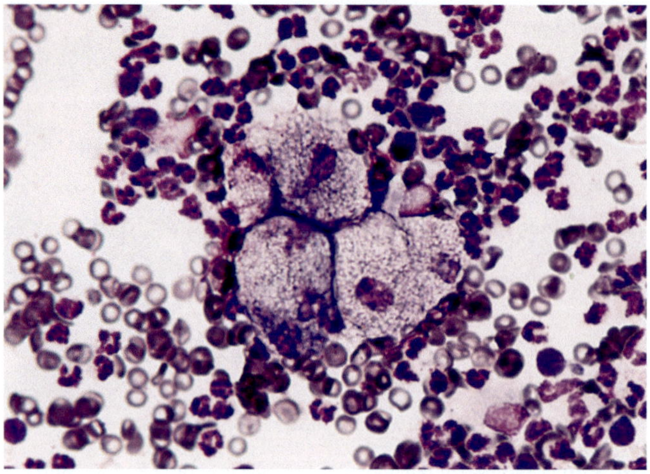

Fig. 3.42 Cytology of the sebaceous glands: a small cluster of mature sebocytes is immersed in an inflammatory population of cells

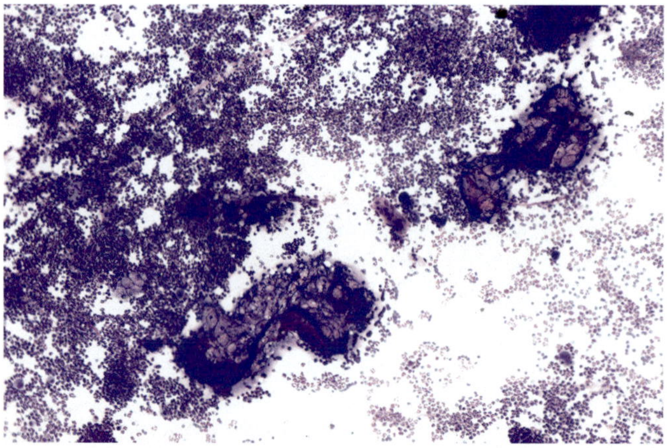

Fig. 3.43 Cytology of inflamed sebaceous adenoma. Some large clusters are embedded in a mixed inflammation

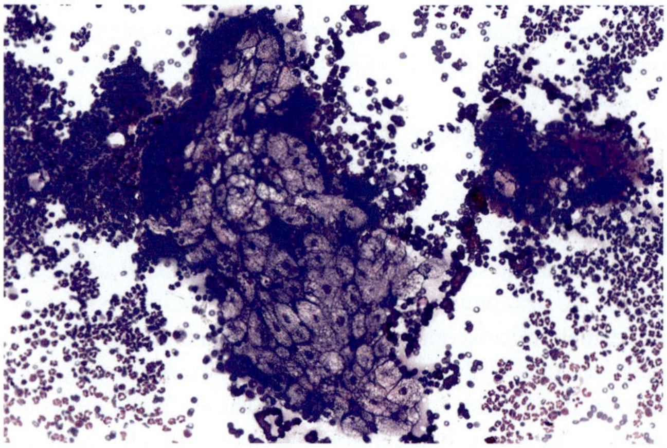

Fig. 3.44 Cytology of inflamed sebaceous adenoma. Large mature benign sebocytes in cluster

and neck. Classical hepatoid glands seem to be not present in cats around the anus. The "tail gland" of cats is composed only of sebaceous glands that, in the case of hyperplasia, a large amount of sebum is present on the surface of the skin, originating the clinical disorder named "stud tail." The cytological features of the hepatoid glands hyperplasia and neoplasia are exhaustively discussed in Chap. 8.

A second typology of adnexal glands, present all over the dermis, is the *sweat glands*, which can be morphologically divided into *epitrichials* and *atrichials*.

The epitrichial glands are characterized by the fact that their secretory ducts open into the follicular lumen; these glands have an *apocrine* way of secretion. The *atrichial* glands are both of *eccrine* and *merocrine* nature and release sweat directly onto the skin surface; in pets, the atrichial glands are exclusively present in the skin of the paw pads. As epitrichial glands account for almost all of the sweat glands in the skin, the term *apocrine glands* is commonly used to describe them when speaking about sweat glands. In dogs and cats, these glands have no thermoregulatory function as in human beings (Figs. 3.45 and 3.46).

On slides, apocrine glands are observed in the case of benign or malignant neoplasms of sweat glands. Nevertheless, a variable number of well-differentiated apocrine aggregates are commonly observed in specimens coming from non-neoplastic disorders, such as adnexal hamartomas or inflammatory skin diseases causing glandular hyperplasia as occurs in acral lick granulomas or in inflamed calluses. Cytologically, the apocrine glands collected from inflammatory or reactive lesions are well differentiated and appear as clusters from cohesive polygonal cells with a honeycomb or a pavement-like arrangement or as cuboidal to cylindrical cells, uniform in size, arranged in short palisades (Figs. 3.47, 3.48, and 3.49). Nuclei are uniform in size and shape, round,

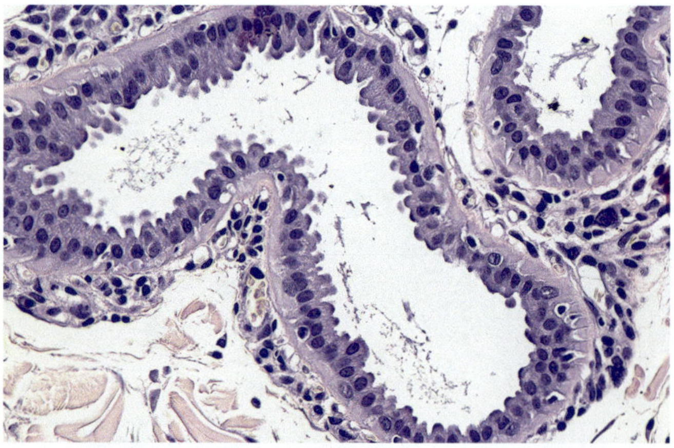

Fig. 3.45 Histology of the sweat glands: the secretory portion of the apocrine gland composed of a single to double line of columnar cells. Note the secretion through *cytoplasmic decapitation* (blebs) and the scant blue reticular secretion present in the lumen

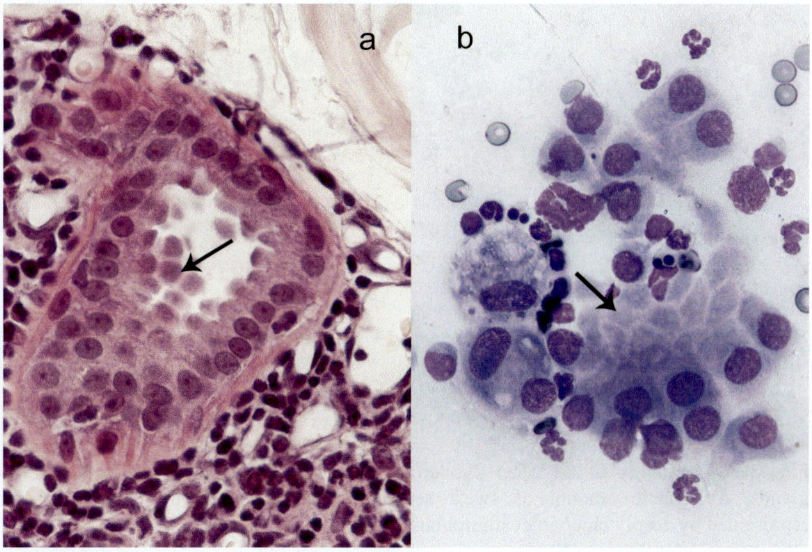

Fig. 3.46 Histology and cytology of the sweat glands: (**a**) blebs of cytoplasm are released in the ductal lumen (*arrow*) and (**b**) cytological features are identical to those shown in the previous histological image (*arrow*)

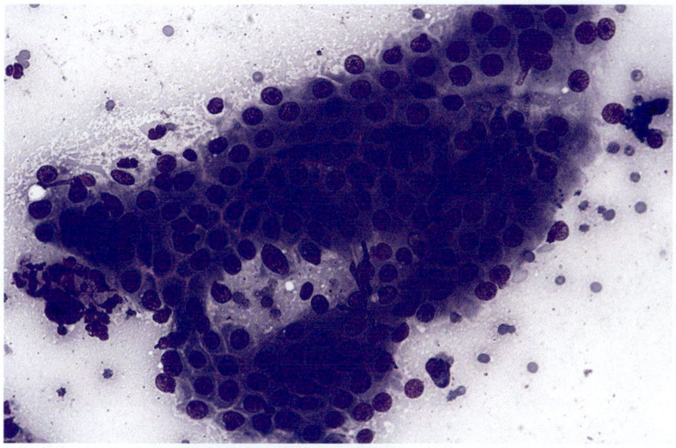

Fig. 3.47 Cytology of the sweat glands: a large sheet of polygonal in shape apocrine gland cells arranged in a honeycomb-like disposition

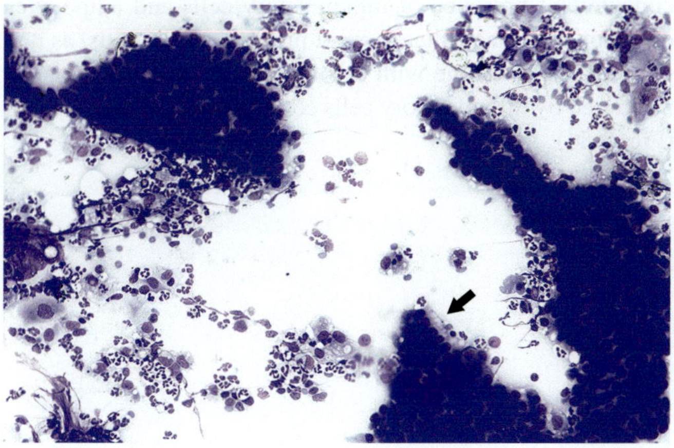

Fig. 3.48 Cytology of the sweat glands: two large clusters of sweat glands embedded in an inflammatory background. The arrow highlights a typical palisade arrangement

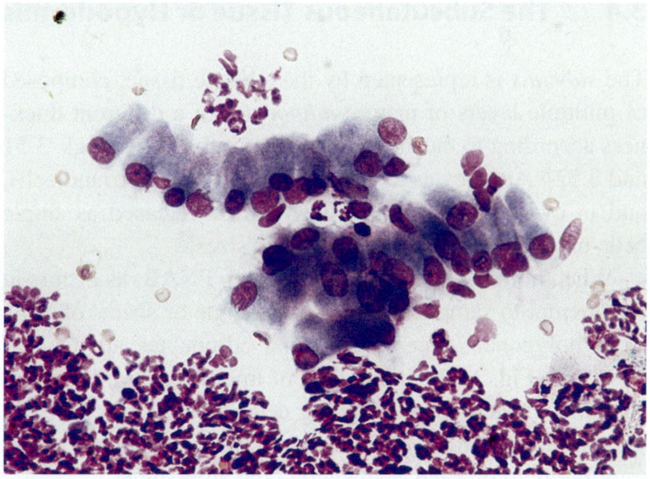

Fig. 3.49 Sweat gland cytology: unmistakable palisade arrangement of cylindrical apocrine cells sampled from an inflammatory lesion

central, or basal in location, and the cytoplasm is small to moderate in size, blue in color, and sometimes containing coarse dark blue secretory material representing sweat. Cytoplasmic blebs on the apical surface of the cylindrical cells, representing the typical apocrine secretion via cytoplasmic decapitation of the upper part of the cytoplasm of the apocrine cells, are also detectable. Along with clusters of glands, a variable amount of sweat secretion can be observed, represented by deeply blue amorphous material, of variable size and with smooth borders and surfaces (Fig. 3.50).

Ceruminous glands are specialized apocrine glands located in the external ear canal, while *anal sac glands* consist of apocrine glands that cover the inner surface of the anal sacs from which very aggressive malignant tumors can originate (anal sac adenocarcinoma). The cytological features of both are discussed with neoplastic lesions (Chap. 8).

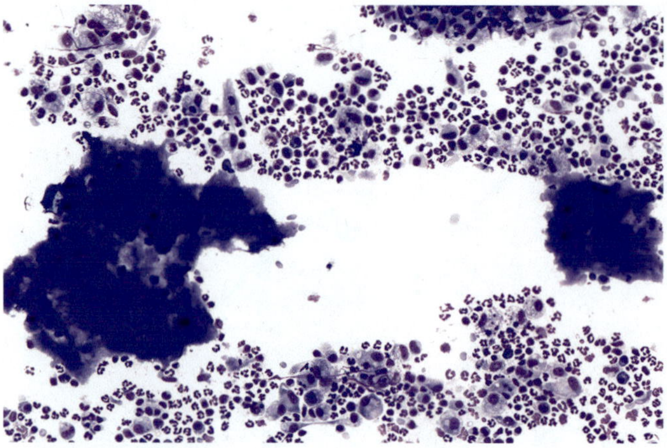

Fig. 3.50 Cytology of the deep pyoderma: along with clusters of glands, a variable amount of sweat secretion can be observed, represented by deeply blue/violet amorphous and homogeneous material

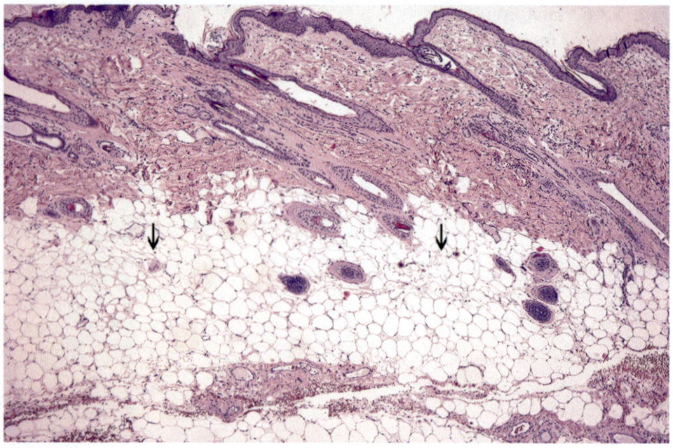

Fig. 3.51 Histopathology adipose tissue: large achromatic adipocytes compose the thick subcutis (arrows)

3.4 The Subcutaneous Tissue or Hypodermis

The *subcutis* is represented by the adipose tissue, composed of multiple layers of mature *adipocytes* of a different thickness according to the anatomical area of the body (Figs. 3.51 and 3.52). Adipocytes are large from polygon to round cells, and in cytological samples, they can be released as single cells or as groups of cells of variable sizes.

When normal fat is collected through FNAB, as occurs in the attempt to sample a deep lesion, single or sheets of large from polygonal to perfectly round adipocytes are usually observed (Fig. 3.53). In the case of lipoma, voluminous and roundish clusters of overlapping cells are usually observed, with optically empty cytoplasm, in which single or large multiple lipid droplets characterize the adipocytes.

Although the overlapping of adipocytes and their arrangement around hematic capillaries are cytological features more suggestive of lipoma compared with normal fat, a differentiation between normal and lipomatous fat is not possible with cytology. The nuclei, when visible, are small, mainly hyperchromatic and oval in shape, of pyknotic aspect, and often located at the periphery of the cells. Some adipocytes can occasionally be collected from inflammatory lesions; the concomitant presence of inflammatory cells and adipocytes does not authorize a diagnosis of primary panniculitis, as it is not possible to define with cytology if the fat is secondarily infiltrated by inflammatory cells coming from a dermal lesion (Fig. 3.54).

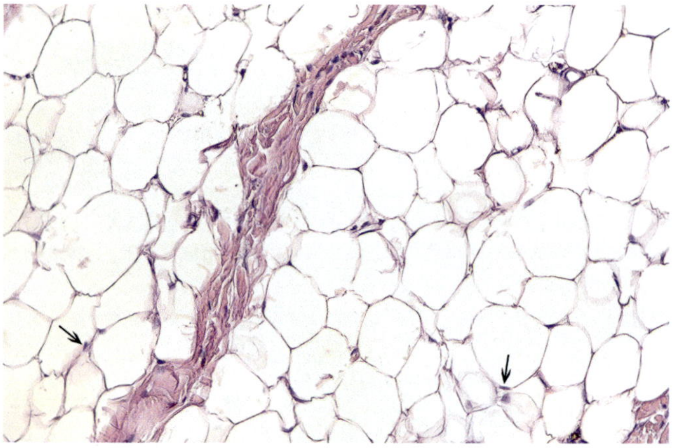

Fig. 3.52 Histopathology adipose tissue: closeup of mature norma adipocytes in which nuclei are rarely recognizable at the periphery of the cells (arrows)

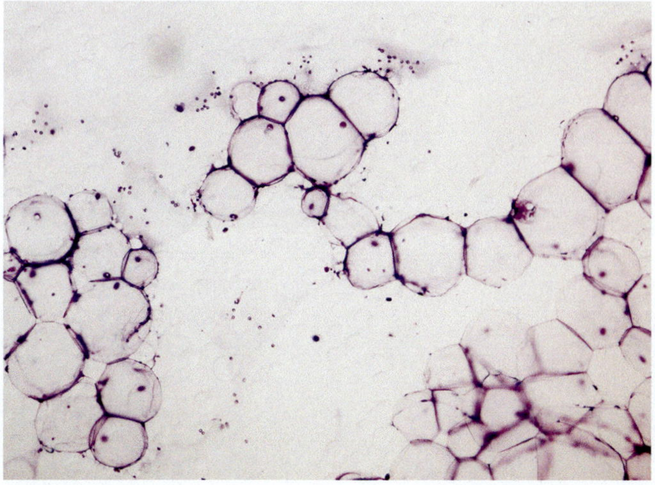

Fig. 3.53 Cytology of normal fat: polygonal to round large adipocytes with a single small oval nucleus

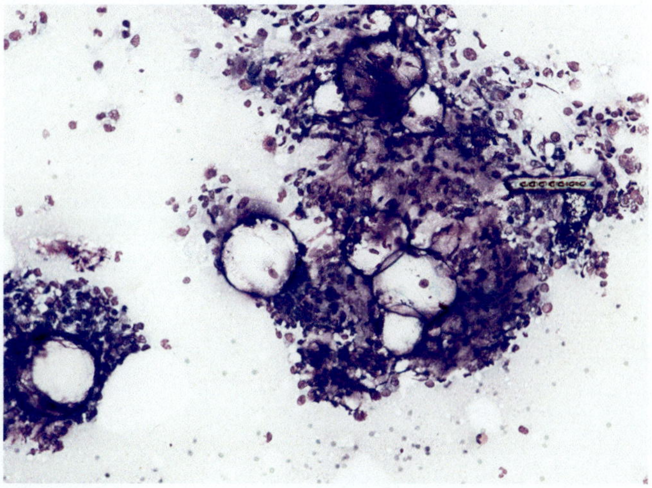

Fig. 3.54 Cytology of the deep pyoderma: single or small clusters of adipocytes are sampled along with many neutrophils and macrophages

The Inflammatory Cells of the Skin and Related Cytological Patterns

4

Inflammatory and non-neoplastic skin diseases are a vast and very important chapter in dog and cat dermatology. The typology of the cells that make up the inflammatory infiltrate is varied, and their percentage determines the different *inflammatory patterns* that can be observed on the cytological slides. To better interpret the cytological features of inflammatory processes causing skin lesions, it is mandatory to know the morphology of the cells that, produced by the bone marrow, reach the skin via the bloodstream.

4.1 The Cells of Inflammation

Cells of hematic origin and those residing in the dermis that can commonly be detected in a cutaneous inflammatory process are described below.

4.1.1 Red Blood Cells

Red blood cells (RBCs) are anucleate round cells, measuring 6–7 μm, that are pale pink to deep violet in color and are commonly present in samples from deep skin lesions. In inflamed and reparative tissues, marked neo-vascularization is often present, and in ulcerated lesions, bleeding is normally observed; moreover, because some sampling techniques such as scraping or FNB/FNAB can traumatically cause the breakage of dermal vessels, it is easy to imagine why blood contamination is commonly observed on cytological slides.

Their presence in skin cytology is usually of scarce pathological meaning but can help to define the depth of the lesions when we will not differentiate between an erosion, in which no RBCs are present (the epidermidis is not vascularized), from an ulcer, in which the epidermal layers are not present and blood is collected; furthermore, RBCs can be also used as a *cytological meter* to compare the size of other cells (Fig. 4.1).

RBCs may be a n important indication when after repeated sampling attempts from the same lesion, slides exclusively composed of blood are obtained; in these cases, especially if the cone of the needle immediately fills with blood as inserted into a nodule, a vascular lesion composed of only blood such as hemangioma or hemangiosarcoma must be suspected.

4.1.2 Neutrophils

Neutrophils are the most frequently observed cells in inflammatory skin lesions. Neutrophils leave the bloodstream and, together with other leukocytes, migrate into the dermis in response to chemotactic molecules released from the same cells of the skin or from microorganisms such as bacteria and fungi, parasites, foreign bodies, and so on. Neutrophils are cells of the acute phase of inflammation, but in all chronic and persistent inflammatory processes, they can be still present. Morphologically, they are characterized by hyperchromatic and multi-lobulated nuclei with granular chromatin; the number and morphology of the lobes are strictly dependent on the age of the cells and the nuclear injuries caused by infectious agents. Healthy neutrophils are composed of three to five nuclear lobes, connected to each other by very thin chromatin bridges that are not always clearly detectable with cytology. Unlike eosinophils and basophils, the cytoplasm of neutrophils is apparently free of granules; the latter are not detectable with routine staining, and the cytoplasm, rarely visible, takes on a scant gray–blue. In rare cases, the cell membrane is barely perceptible (Figs. 4.2 and 4.3).

The morphology of the nuclei can provide the cytopathologist with useful information about the possible underlying pathogenesis of their alterations.

Healthy multilobed neutrophils, with well-preserved morphology, are defined as *segmented*, whereas those with fewer lobes showing degenerative changes are defined as *degener-*

F. Albanese, *Canine and Feline Skin Cytology*, https://doi.org/10.1007/978-3-031-95713-0_4

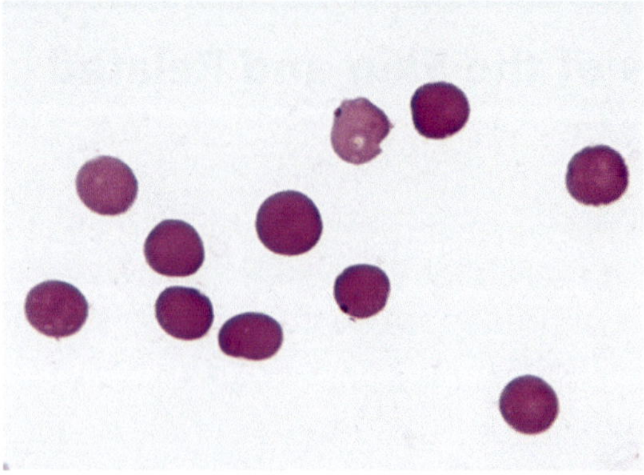

Fig. 4.1 Cytology of RBCs: small round anucleate cells cytologically characterize the RBCs

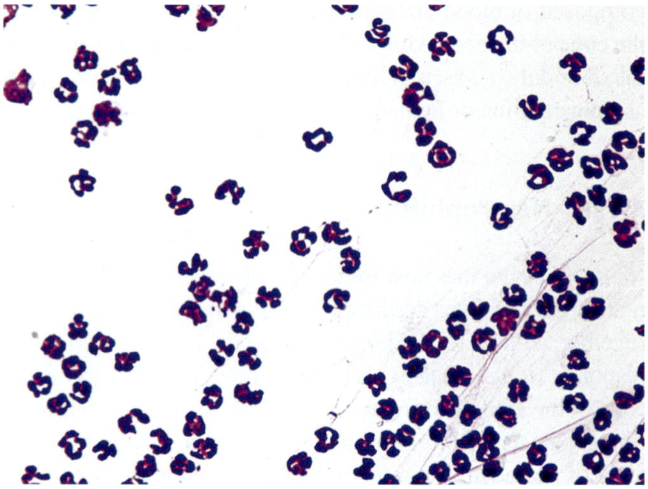

Fig. 4.2 Cytology of neutrophils: polymorph nucleated cells, with no evident granules and cytoplasm. Nuclear lobes are hyperchromatic and well preserved

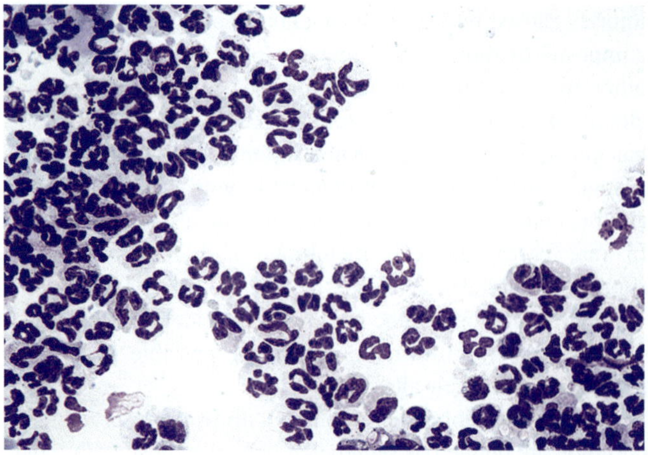

Fig. 4.3 Cytology of neutrophils: cytoplasm and cellular membrane are fairly evident in some cells. Nuclear lobes are well preserved

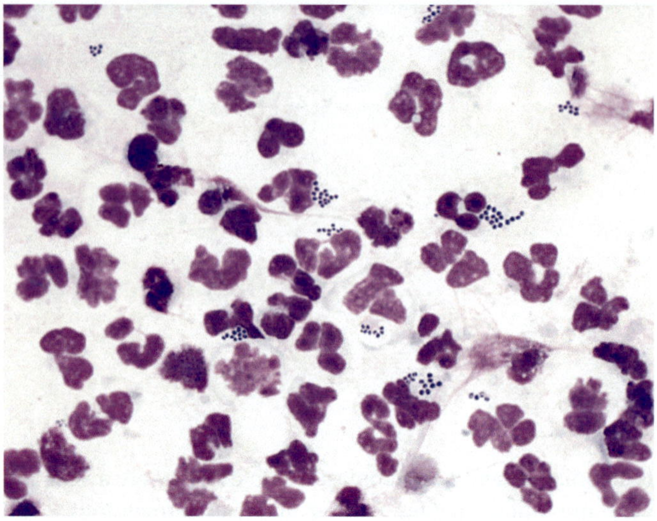

Fig. 4.4 Cytology of karyolytic neutrophils: pale and swollen nuclei characterize the nuclear degeneration termed *karyolysis*. The cause of these nucleic changes is the bacterial toxins of *Staphylococcus*

ate. The three main morphological nuclear degenerative alterations are *karyolysis*, *karyopyknosis* (*pyknosis*), and *karyorrhexis*.

Karyolysis is the most common form of nuclear degeneration as it is indicative of severe cell damage usually caused by bacterial toxins (e.g., *Staphylococcus* spp.).

Karyolytic nuclei undergo complete dissolution of the chromatin because of the enzymatic degradation by endonucleases; nuclei appear swollen and hypochromatic (pale), with a reduction in the number of the lobes, due to water penetration into the cells through damaged cell membrane (Figs. 4.4 and 4.5). As the high frequency of pyoderma in dogs, in any cases in which marked nuclear degenerative features are observed in neutrophils, cytological searching for *Staphylococcus pseudintermedius*, the main pathogen of skin infections, must be constantly performed by cytologists. Less frequently, karyolitic aspects and nuclear streaks are also detected in not-infectious diseases; indeed, in chronic lesions, the aging of the cells makes the nuclei degenerated and more easily damaged (Fig. 4.6).

Many long streaks of nucleic material, linked to the rupture of severely degenerated nuclei, are often present on the slides coming from pyoderma; although excessive pressure during the sampling could be the cause of these nuclear streaks, the latter are commonly observed on specimens from pyoderma since the karyolytic nuclei are less resistant and more prone to the rupture (Figs. 4.7 and 4.8). When the nuclear streaks are very thin and long, beginners must be careful not to confuse them with filamentous bacteria; the color of the streaks is essential to avoid confusion since the nucleic material is pink anyway and not blue like that of bacterial bodies.

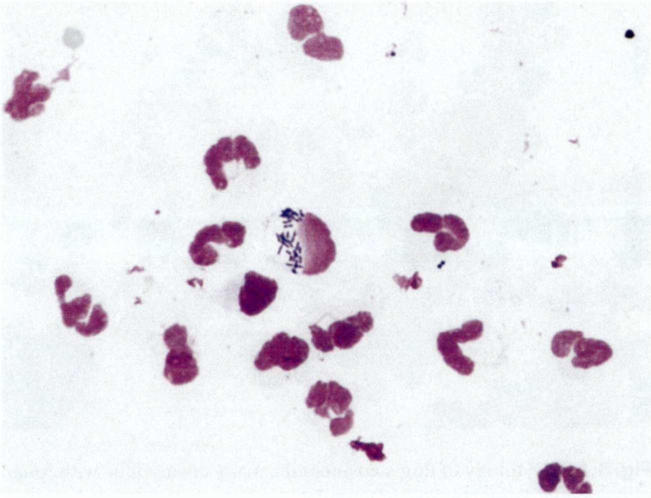

Fig. 4.5 Cytology of karyolytic neutrophils: pale and swollen fragments of nuclei characterize the nuclear degeneration termed *karyolysis*. Cocci and rod-shaped bacteria justify the severe degenerative nuclear changes

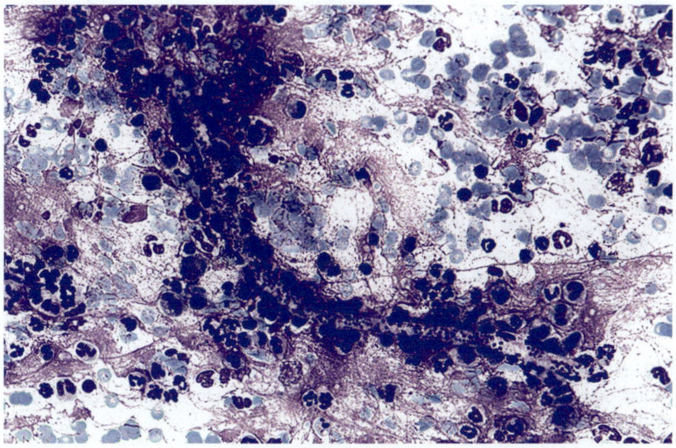

Fig. 4.6 Cytology of karyolytic neutrophils: many steaks of necrotic nucleic material, along with karyolytic neutrophils, are observed in a sample collected from a sterile subcutaneous nodule caused by *Dirofilaria repens*. Note the larva and the lack of bacterial infection

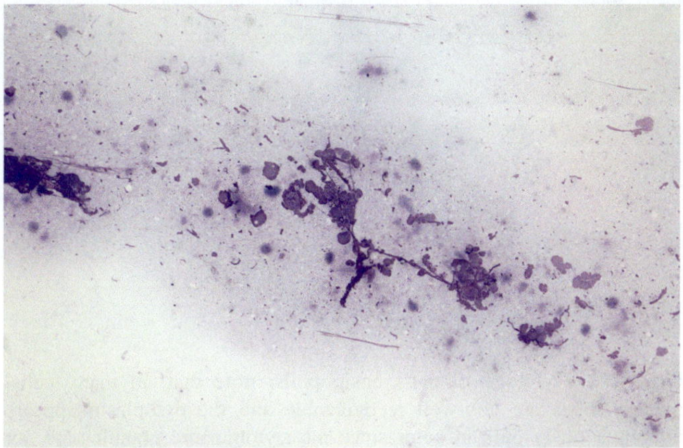

Fig. 4.7 Cytology of karyolytic neutrophils: short linear nuclear streaks representing severely degenerated nuclei of neutrophils

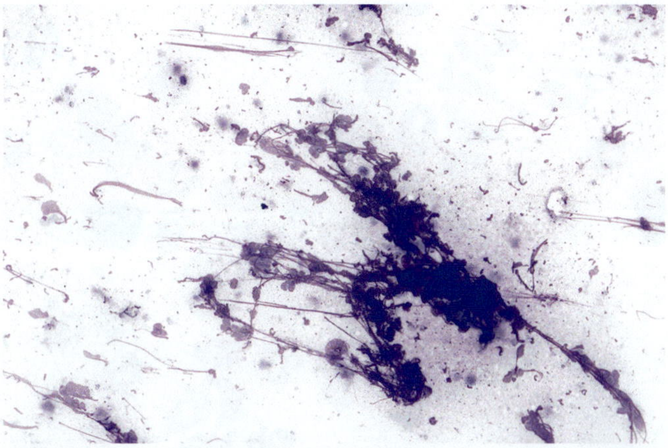

Fig. 4.8 Long linear streaks of karyolytic nucleic material during the slide preparation

Pyknosis is a less severe change of nuclear material. Unlike karyolysis, it is not usually linked to bacterial toxins. Pyknosis is the result of chromatin condensation, but the nuclear morphology cannot be cytologically differentiated from *apoptotic* cells. The nuclear size progressively reduces as the chromatin condenses into both single and multiple roundish dark blue nuclear fragments (Figs. 4.9 and 4.10).

Karyorrhexis is the destructive fragmentation of the nucleus of a dying cell into many small punctiform to dusty and hyperchromatic chromatin material of different sizes, much smaller than those observed in pyknosis. It is usually preceded by pyknosis and can occur as a result of apoptosis (programmed cell death), senescence, or necrosis.

Pyknosis and karyorrhexis, but not karyolysis, are therefore most frequently seen in apoptotic cells.

Cytologically, it is not possible to define if pyknosis and karyorrhexis are secondary to apoptosis or may be part of the cytomorphological changes in the case of non-programmed necrosis. Fortunately, their presence has no diagnostic significance.

The neutrophils are cells with phagocytic activity, and therefore intracytoplasmic microorganisms such as bacteria, fungi, and protozoa are commonly observed in many infectious diseases.

4.1.3 Eosinophils

Eosinophils are observed in many cutaneous inflammatory processes, especially in cats. Their presence is common in both acute and chronic lesions, even if the exact mechanisms by which cell infiltrates are plentiful in eosinophils, for example, in *feline eosinophilic skin diseases*, are not yet fully clear. Formerly, the presence of eosinophils was ideally associated with parasitic or hypersensitivity diseases, but this is not the rule; in fact, parasitic skin diseases in dogs such as demodicosis and dirofilariasis do not usually stimulate a relevant eosinophilic inflammation, or better still, eosinophils

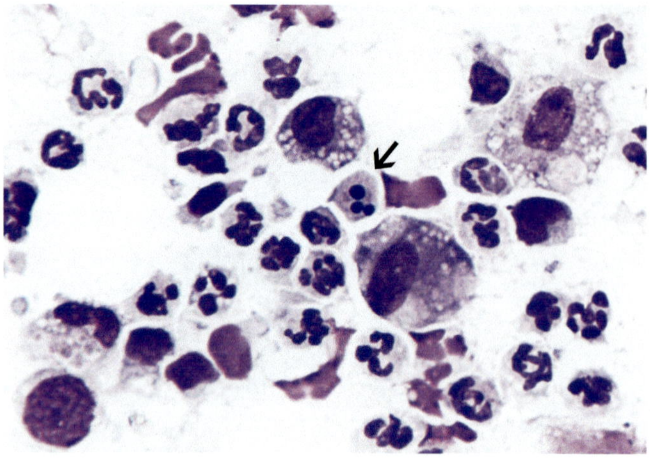

Fig. 4.9 Neutrophil: cytology of nuclear pyknosis: condensed chromatin causes the roundish and deeply blue nuclear material that characterizes the pyknosis

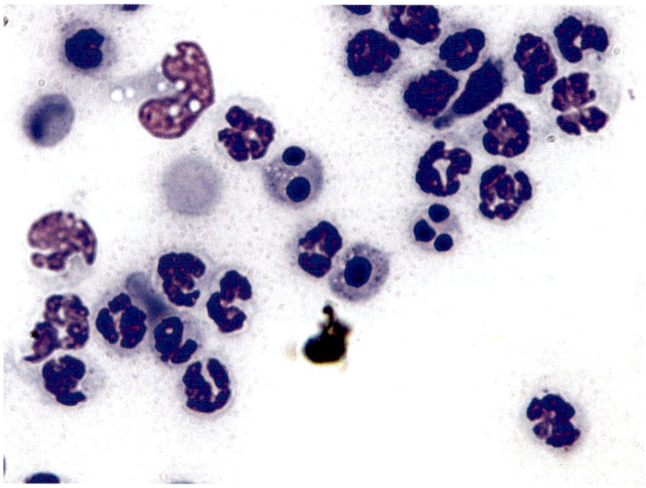

Fig. 4.10 Cytology of nuclear pyknosis: both single and multiple pyknosis of neutrophils

are usually not present during these parasitic infestation. Moreover, many eosinophils are present in other diseases such as *feline herpesvirus infection* or in autoimmune diseases such as *pemphigus foliaceus* both in cats and dogs; these are two classical examples of how eosinophils must not be exclusively associated with parasitic or allergic diseases.

Eosinophils are morphologically similar to neutrophils but are few larger, with less evident nuclear lobes and with the presence of orange-pinkish cytoplasmic granules, which are *round* in dogs and *rod-shaped* in cats (Figs. 4.11 and 4.12). In cats, lobes of eosinophils are usually less visible than in dogs, and, at low magnification, the granules, as they are very small and thin, are not always clearly recognizable as single structures; for this reason, in most cases, eosinophils are recognized for their diffuse orange-pinkish color of the cytoplasm (Figs. 4.13 and 4.14). During the preparation of slides,

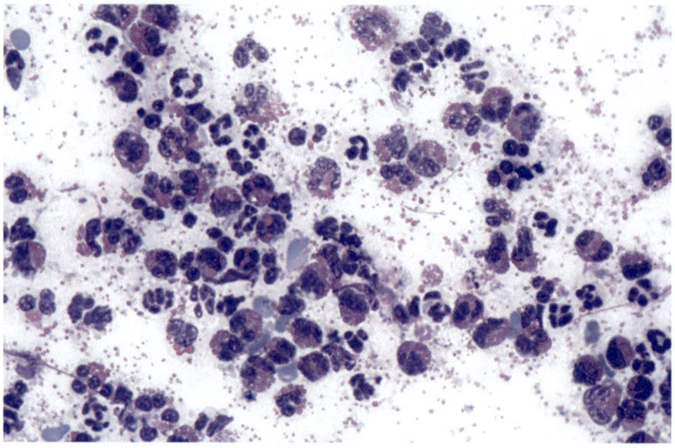

Fig. 4.11 Cytology of dog's eosinophils: many eosinophils with round pink–orange intracytoplasmic granules

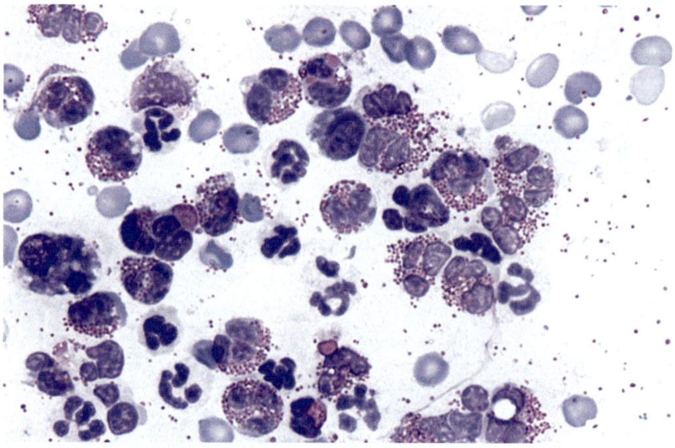

Fig. 4.12 Cytology of dog's eosinophils: note the variable in size of round-shaped pink–orange granules that characterize the eosinophils of dogs

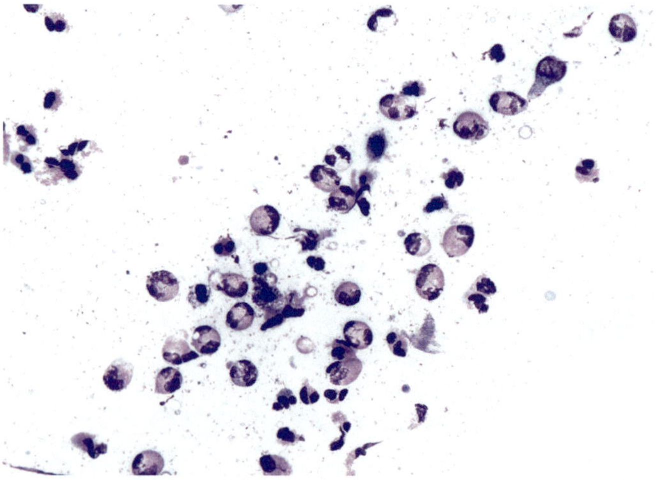

Fig. 4.13 Cytology of cat's eosinophils: note that, in many cells, nuclear lobes are non-well recognizable and the cytoplasm appears homogeneously pink in color since intracytoplasmic granules are not clearly evident

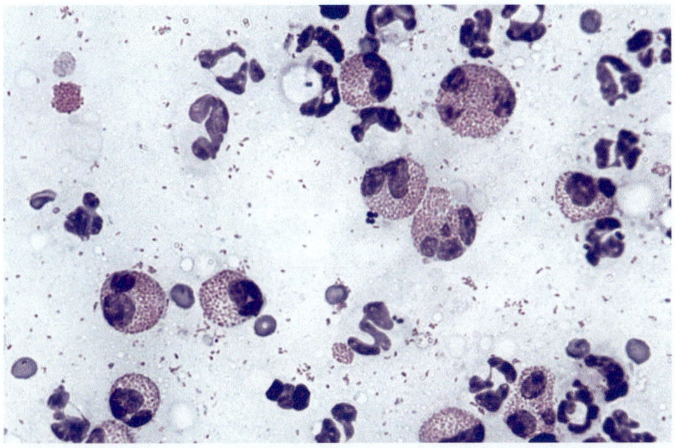

Fig. 4.14 Cytology of cat's eosinophils: to higher magnification, the rod-shaped pink–orange granules that characterize the eosinophils of cats are better recognizable, both into the cytoplasm and free on the background of the slide. Most nuclei appear as single and not multi-lobated

cells can break, and the spreading of eosinophilic granules in the background is a common finding. An inexperienced cytologist must not confuse rod-shaped granules of a cat's eosinophils with bacteria; indeed, as mentioned, with rapid Romanowsky-type dye, the bacteria are always stained dark blue and never pink–orange.

4.1.4 Basophils

Basophils are rarely observed in skin cytology. They appear slightly larger than eosinophils and are characterized by a multi-lobed nucleus and a cytoplasm full of fine pale blue-gray from round to slightly rod-shaped granules (Figs. 4.15 and 4.16). In cats affected by eosinophilic diseases, basophils are sometimes found in large numbers; in the author's experience, it occurs mostly in samples coming from papules in cats affected by mosquito bite hypersensitivity.

4.1.5 Lymphocytes

Lymphocytes are cells that play an important role in inflammatory processes, particularly in those that are immune-mediated or when there is a chronic antigenic stimulus, such as *granulomas*. Usually, a variable number of lymphocytes are observed in any chronic inflammatory process as occurs in chronic deep pyoderma in dogs or in such immune-mediated/autoimmune diseases as lupus, mucocutaneous pyoderma, erythema multiforme group, insect or tick bite local reaction, and so on. In many of these diseases, they are accompanied by a variable number of plasma cells.

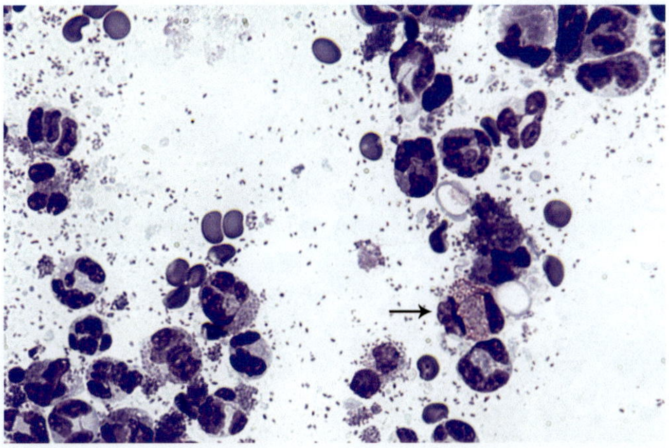

Fig. 4.15 Cytology of basophils: pale blue/gray granules in the cytoplasm of many basophils. Note the difference in the color of granules compared with those orange of an eosinophil (*arrow*)

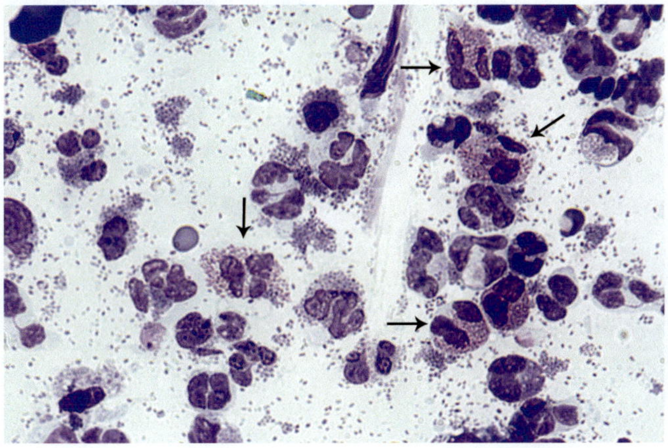

Fig. 4.16 Cytology of basophils: eosinophils are differentiable from basophils from the size and shape of intracytoplasmic granules. The arrows highlight eosinophils. Many granules of basophils are present in the background

Lymphocytes are discrete round cells, slightly larger than RBCs, with round, sometimes indented nuclei that occupy almost the entire cell and with chromatin distributed in hyperchromatic areas (heterochromatin) interspersed with lighter areas (euchromatin); the cytoplasm is scarce, located at the periphery of the nucleus or confined to a small peripheral area that assumes an intense basophilic color (Figs. 4.17 and 4.18).

Along with mature small lymphocytes, it is common to see a variable number of immature lymphoid cells (Fig. 4.19). Lymphocytes are commonly observed, as infiltrating cells, also in many skin neoplasia, for example, can be frequently observed in slides coming from canine benign histiocytomas in the regression phase; in these cases, their number could be higher than histiocytes, and a lymphocytic inflammation can be misdiagnosed.

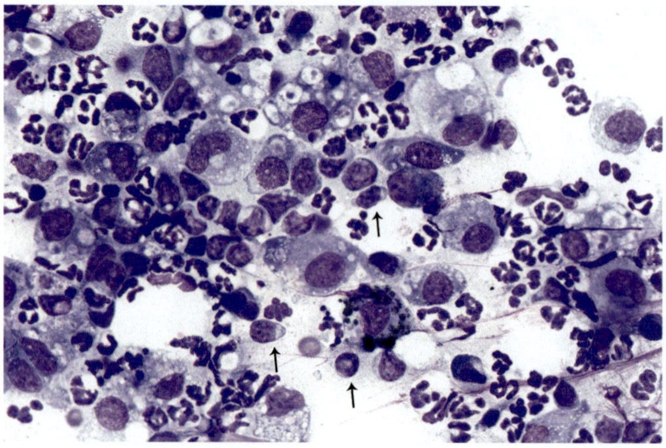

Fig. 4.17 Cytology of lymphocytes: many both mature and immature lymphoid cells are observed in chronic inflammatory processes (arrows)

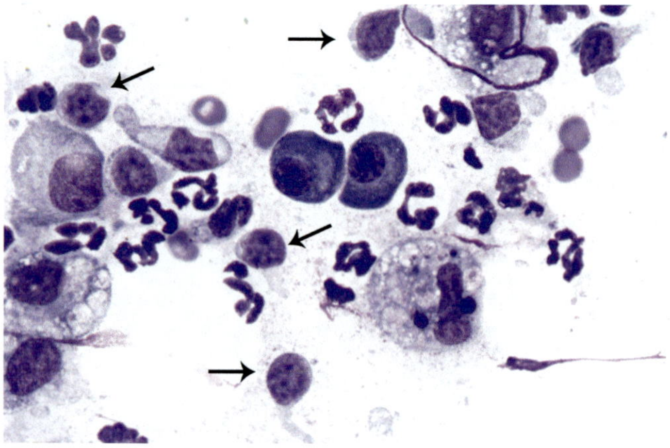

Fig. 4.18 Cytology of lymphocytes: lymphocytes are easily recognizable as round cells with scarce deep blue cytoplasm and coarse chromatin (arrows). Two plasma cells are recognizable at the center

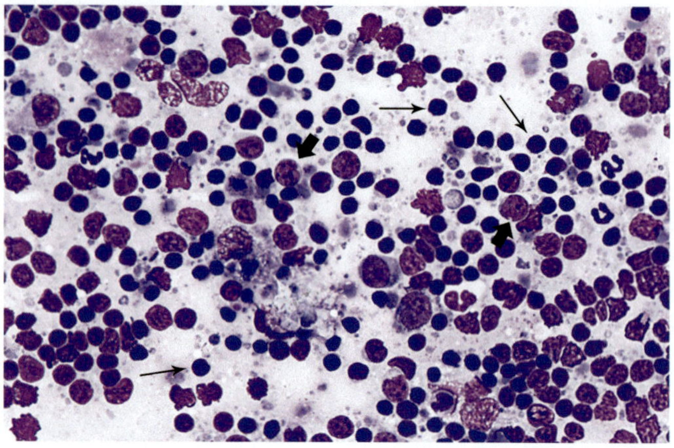

Fig. 4.19 Cytology of lymphocytes: along with the most cellular mature lymphocytic population of cells (thin arrows), some immature lymphoid cells are present (large arrows)

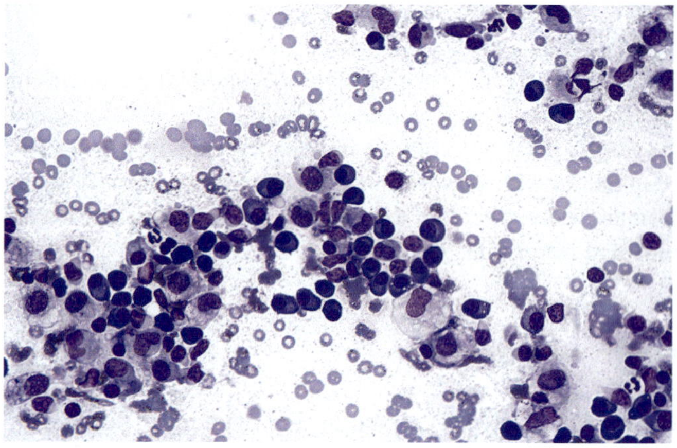

Fig. 4.20 Cytology of plasma cells: many plasma cells along with macrophages. The former are recognizable by the smaller size, deep blue cytoplasm and eccentric nuclei

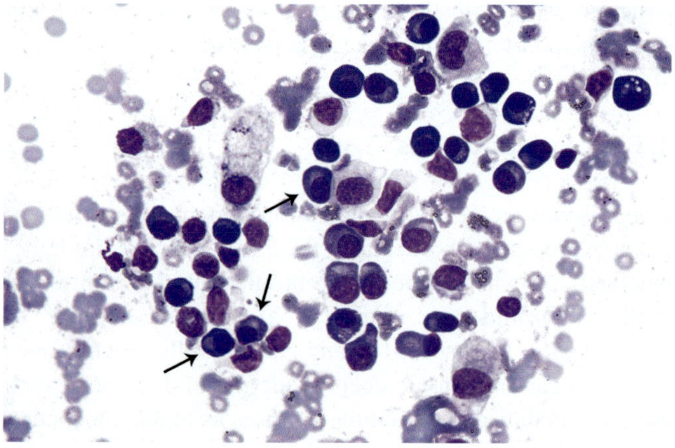

Fig. 4.21 Cytology of plasma cells: the Golgi apparatus is easily recognizable in most plasma cells

4.1.6 Plasma Cells

Plasma cells come from B-lymphocytes and are specialized in the production of antibodies.

The number of plasma cells is usually few in inflammatory processes, a part of some specific diseases in which a strong antigenic stimulus is present as discussed in Chap. 6. Cytologically, plasma cells have mostly an oval silhouette, with round eccentric nuclei and a deep blue-in-color cytoplasm, which is separated from the nucleus by a characteristic arc-shaped achromatic area, the *Golgi apparatus* (Figs. 4.20 and 4.21). Some activated plasma cells, the *Mott cells*, contain numerous intracytoplasmic roundish achromatic or pale blue overlapped areas, the so-called *Russell bodies*, which give the cytoplasm a *foamy* appearance; these roundish areas represent products of immunoglobulin synthesis (Fig. 4.22).

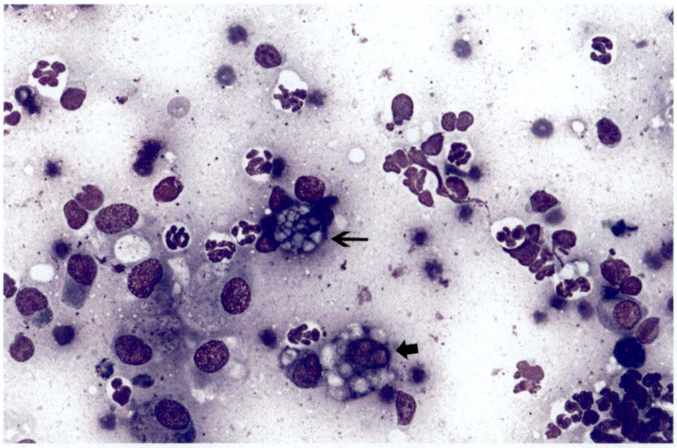

Fig. 4.22 Cytology of plasma cells: a Mott cell, filled with Russell bodies (thin arrow), must not be confused with activated macrophages with many intracytoplasmic vacuoles (large arrow)

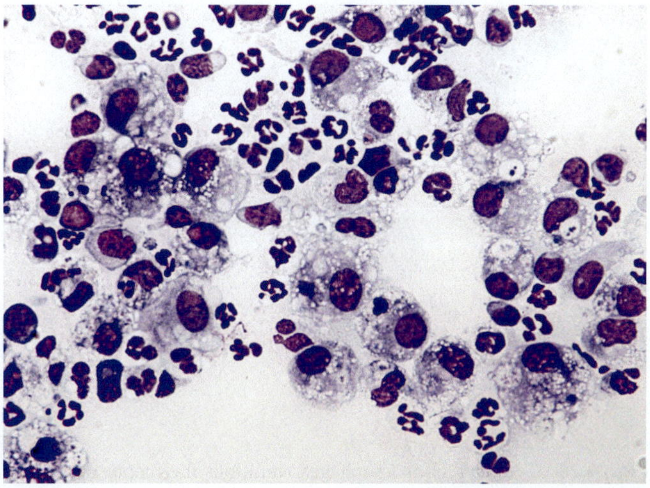

Fig. 4.23 Cytology of activated macrophages: many macrophages with different size intracytoplasmic achromatic vacuoles

4.1.7 Macrophages

Macrophages are histiocytic cells that represent the transformation in the tissue of circulating monocytes, which, in turn, migrate into the dermis, to carry out their main function: *Phagocytosis*. Macrophages are the key cells of chronic inflammatory processes affecting the skin as well as other organs and apparatus. The macrophages have an extremely variable morphology depending on the stage of activation: they are usually roundish to oval, with a very high variability in size, with round to oval nuclei, centrally or eccentrically located, very often folded and reniform and with regular chromatin and inconspicuous nucleoli. The cytoplasm may be scarce to very abundant, from pale gray to dark blue and, when activated, filled with vacuoles of different sizes (Fig. 4.23). In some cases, cytoplasm is not vacuolated, and the cytological aspect of macrophages cannot so immediately recognizable.

As the main activity of the macrophages is phagocytosis, the cytoplasm can contain cellular debris, fragments of leukocytes, lipids, melanin pigment (melanophages), small foreign bodies, different infectious agents such as protozoa, fungi, mycobacteria, and so on (Figs. 4.24, 4.25, 4.26, 4.27, and 4.28). In the case of hematoma or chronic hemorrhagic lesions, erythrophagocytosis is usually observed, and the cytoplasm of macrophages can contain products of hemoglobin degradation such as *hemosiderin* and *hematoidin*; the first appears as a dark blue material of granular aspect and the latter as shiny yellow-orange rhomboid crystals (Fig. 4.29).

Since macrophages are not able to eliminate large microorganisms or foreign bodies, they can transform themselves into larger cells, with round to oval and often indented nuclei and blue in color and not-vacuolated cytoplasm. This particular morphology resembles those of some epithelial cells, and therefore, these macrophages are called *epithelioid cells* or *epithelioid macrophages*. Since epithelioid

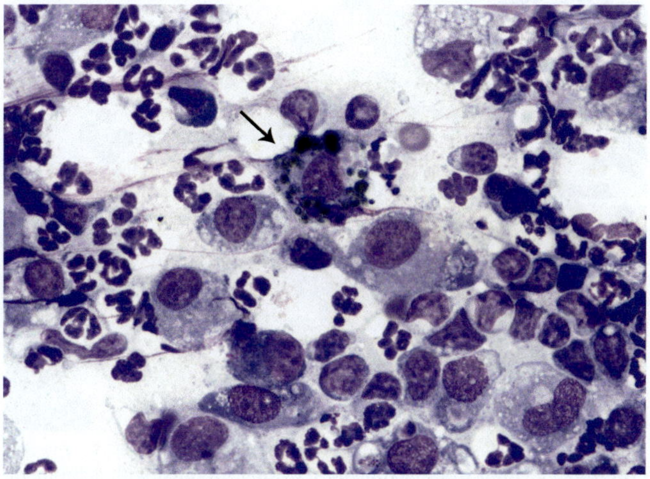

Fig. 4.24 Cytology of macrophages: macrophages with cytoplasm filled with clumps of melanin pigment take the name of "melanophages" (arrow)

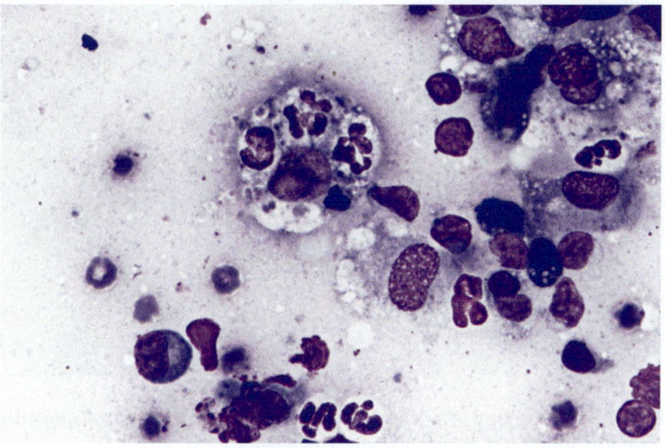

Fig. 4.25 Cytology of macrophages: macrophages with cytoplasm filled with karyolytic and pyknotic neutrophils (leukophagocytosis)

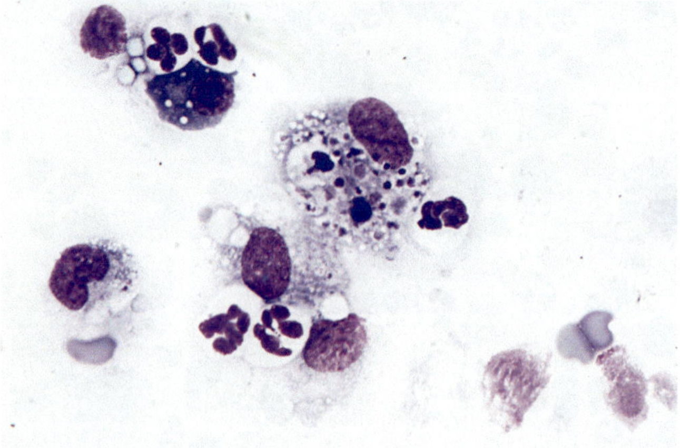

Fig. 4.26 Cytology of macrophages: multiple fragments of nucleic debris must not be confused with bacteria or protozoa

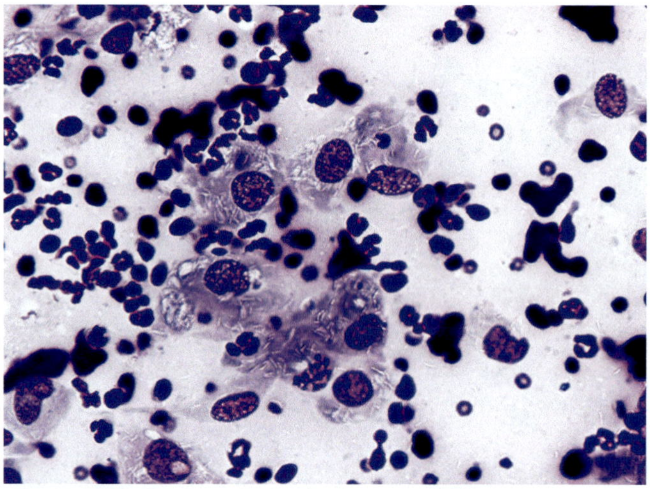

Fig. 4.27 Cytology of macrophages: cytoplasm filled with achromatic linear areas representing unstained *Mycobacterium* spp.

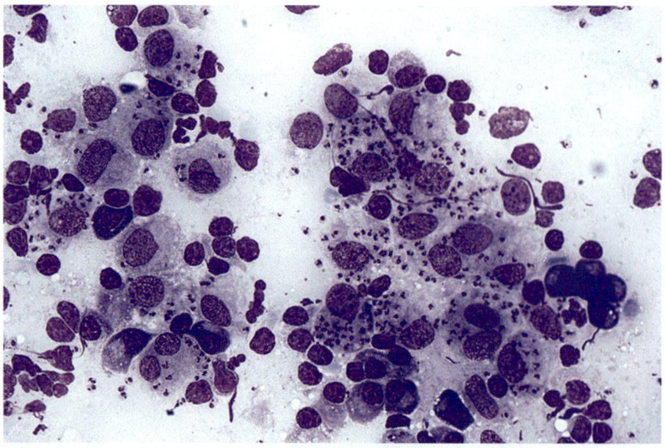

Fig. 4.28 Cytology of macrophages: many amastigotes of *Leishmania infantum* into the cytoplasm of macrophages

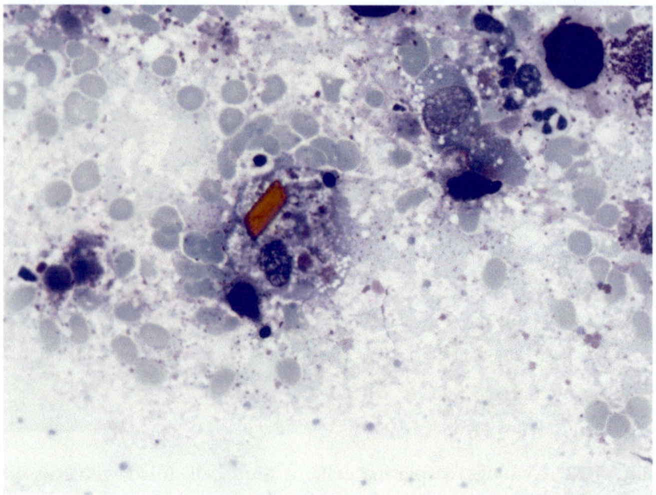

Fig. 4.29 Cytology of macrophages: a single hematoidin crystal is clearly recognizable as an orange polygonal structure

macrophages are the "cells of granuloma" they are very commonly observed on slides taken from granulomatous lesions, and since in granuloma the epithelioid cells are tightly packed in large numbers, this arrangement can create doubts and these cells can be confused with epithelial cells (Fig. 4.30). As with all histiocytic cells, also epithelioid macrophages can assume a spindled shape, which cannot be confused with spindle mesenchymal cells such as reactive fibroblasts (Fig. 4.31). Epithelioid macrophages tend to be arranged both individually and in groups, especially around foreign bodies (e.g., corneocytes), forming epithelial-like arrangements, as is commonly observed in the case of furunculosis (Fig. 4.32).

Some macrophages can transform into *multinucleated giant cells*. The mechanism underlying the formation of these cells, also called *foreign body* cells, is not fully understood, but it is believed that they form following the fusion of several macrophages and not following a splitting of the nucleus of a single macrophage. The giant cells are very large and are characterized by a multiple number of nuclei ranging from 3 to over 50 (Figs. 4.33 and 4.34). The presence of these reactive histiocytic giant cells is needed in all cases in which very large foreign bodies (vegetal, keratin, calcium salts, lipid, etc.), particular microorganisms (mycobacteria and protozoa), or very large infectious agent (fungi) must be eliminated; anyway, their presence is observed, along epithelioid cells, when a granulomatous lesion occurs (Figs. 4.35, 4.36, 4.37, and 4.38).

The nuclei of giant cells can be haphazardly distributed in the cytoplasm or arranged at the periphery in a *horseshoe shape*. The former giant cells are named *foreign body giant cells* and are the most frequently observed in canine and feline skin cytology, especially when an exogenous material

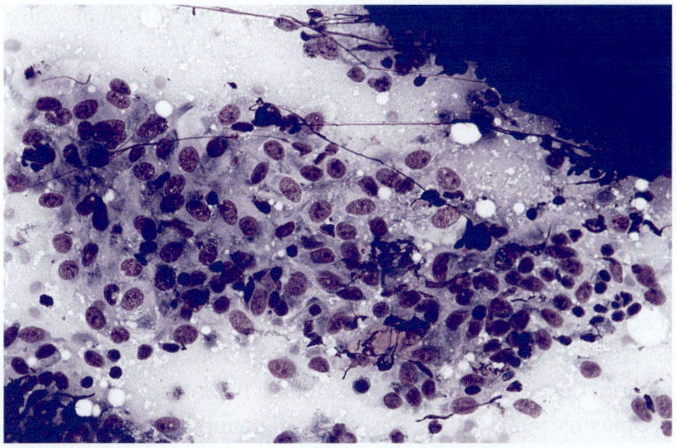

Fig. 4.30 Cytology of epithelioid macrophages: many epithelioid cells, arranged in a large group

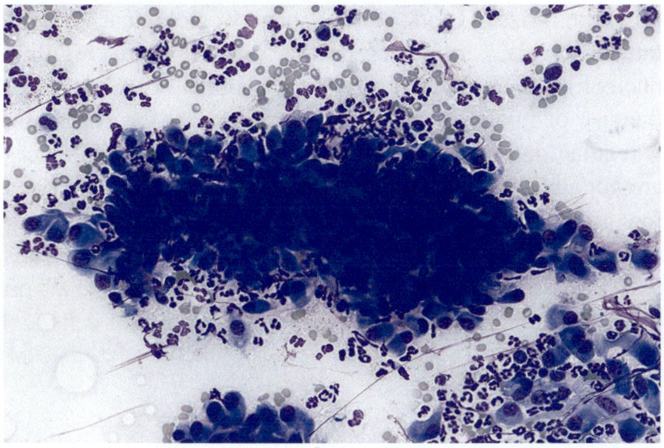

Fig. 4.31 Cytology of epithelioid macrophages: many epithelioid cells with spindled shape

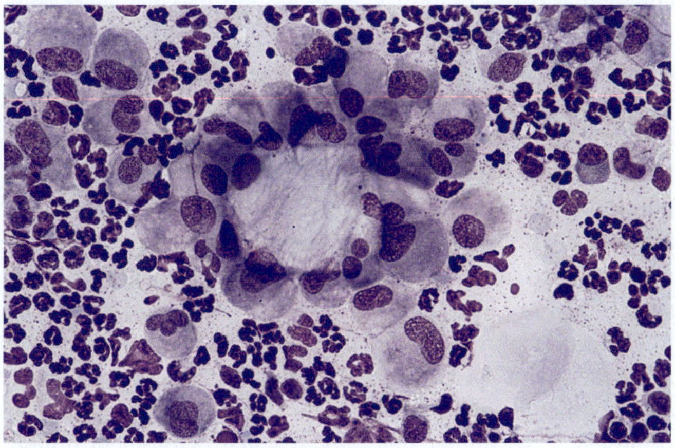

Fig. 4.32 Cytology of epithelioid macrophages: epithelioid macrophages arranged around corneocytes: note that their morphology resembles epithelial cells

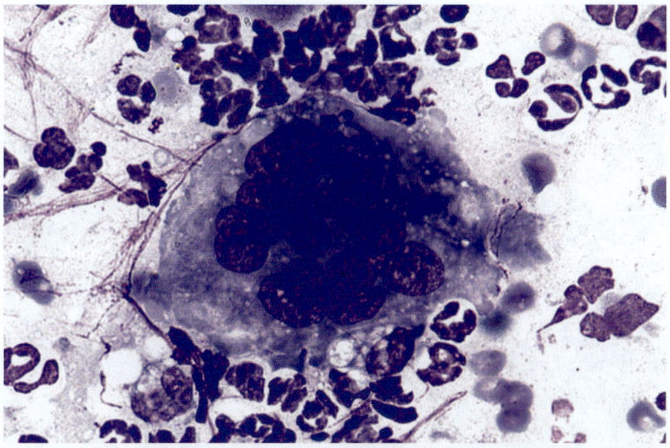

Fig. 4.33 Cytology of histiocytic giant cells: single multinucleate macrophage. Nuclei are uniform in size and overlapped

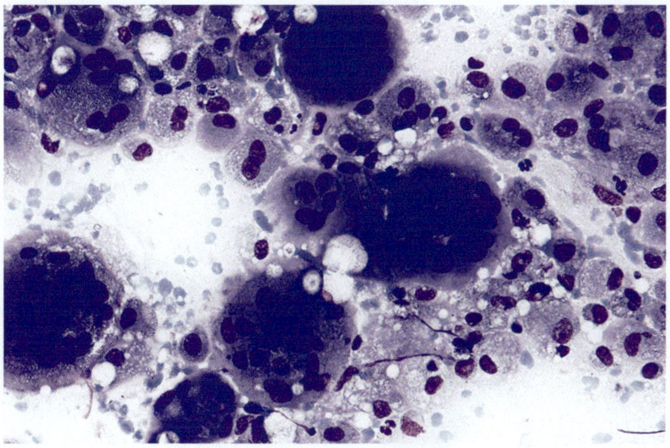

Fig. 4.34 Cytology of histiocytic giant cells: multiple foreign bodies giant cells in a granulomatous inflammatory

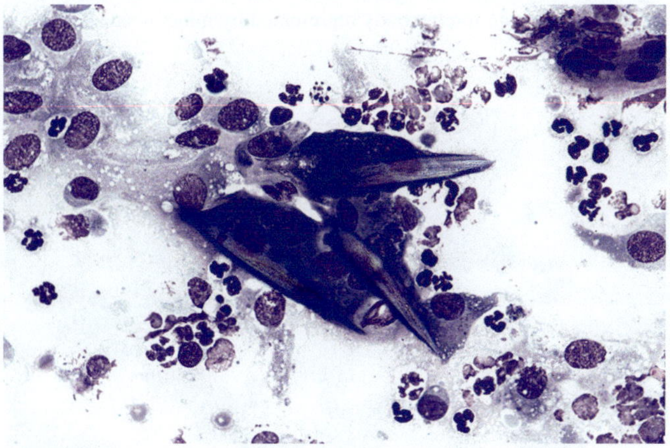

Fig. 4.35 Cytology of giant cells: multinucleated macrophages in the attempt to eliminate a large foreign body represented by a cluster of corneocytes

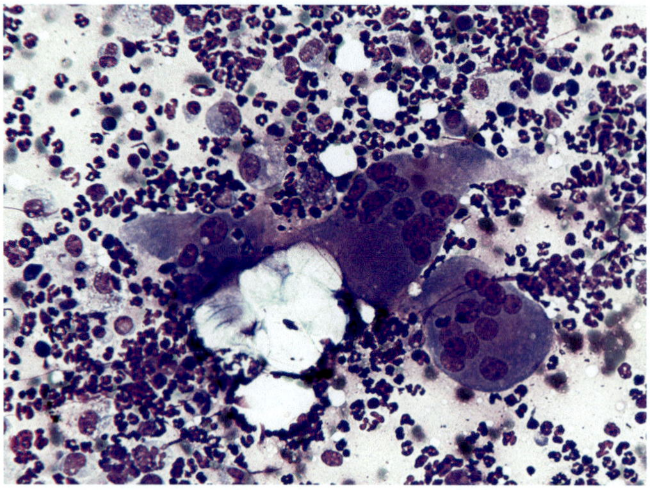

Fig. 4.36 Cytology of giant cells: three multinucleated giant cells in the attempt to eliminate a group of corneocytes

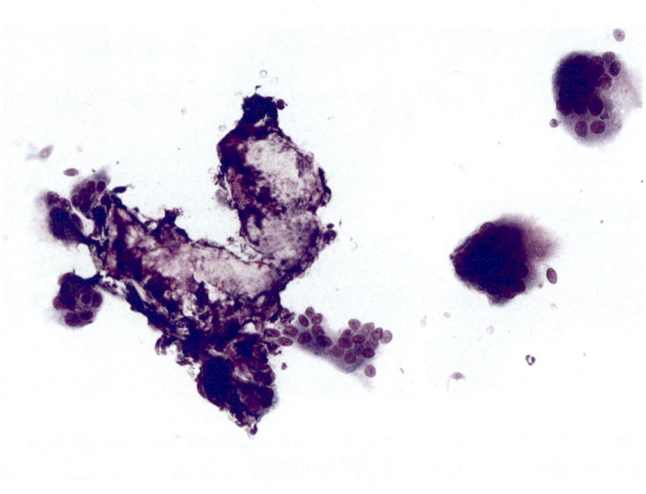

Fig. 4.37 Cytology of giant cells: many multinucleated cells surrounding a large foreign body represented by mineral salts

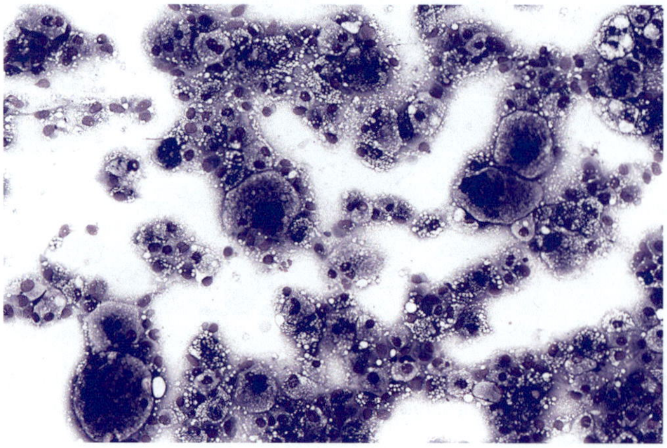

Fig. 4.38 Cytology of giant cells: many multinucleated macrophages filled with lipid droplets

is present (e.g., keratin). The second type of giant cells, named *Langhans cells*, is rarer and seems more commonly observed in human beings with tubercular mycobacterial infection or in some deep fungal diseases. According to the author, this morphological difference is not useful in canine and feline cutaneous cytology for two reasons: the first because cutaneous tuberculosis is extremely rare in dogs, and also since the so-called Langhans cells can be occasionally observed in non-tuberculous granulomatous reactions also.

As discussed in Chap. 8, many multinucleated reactive histiocytic cells are also stimulated from some malignant neoplasms such as in the case of anaplastic pleomorphic sarcomas of soft tissue, especially in cats. A cytological feature that helps to differentiate the malignant multinucleated cells from multinucleated reactive histiocytic giant cells is that, in the former, the nuclei, since malignant, are of different sizes and can show cytological atypia (anisokaryosis, irregular profile, macro and multiple nucleoli), whereas, in histiocytic inflammatory giant cells, the size of nuclei is usually uniform in size, the chromatin is regular, and nucleoli are inconspicuous. These differences are not always respected, as it is rarely possible to observe a few giant cells with different-sized nuclei in case of inflammatory processes; therefore, the interpretation of such cells must be assessed based on the clinical signs and, above all, on the cytological features of coexisting population of cells.

4.1.8 Mast Cells

Mast cells originate from the bone marrow precursors that migrate through the bloodstream and reach the dermis, where they turn into *mast cells* and become resident cells of the connective tissue mainly located around vessels. Mast cells are mainly engaged in hypersensitivity diseases, especially in cats. Morphologically, they are discrete round cells, with central round nuclei, often completely hidden by the presence of numerous intracytoplasmic metachromatic granules that, with the rapid Romanowsky-type dyes, take on a distinctive purple color (Fig. 4.39); this phenomenon is linked to the high acidity of the heparin contained in these granules which, when stained with alkaline dyes, cause a change in color (purple and not blue as expected). The number of mast cells in canine inflammatory diseases is usually poor and usually associated with hypersensitivity diseases; the same occurs in cats in which the number is usually higher than in dogs and in some cases, so higher that creates some cytological doubts that cells can belong to a mast cell tumor (MCT). Urticaria pigmentosa-like is a benign mast cell hyperplasia reported in Sphinx and Devon rex cats; sampling from papular lesions can sometimes release a significant amount of mast cells (Figs. 4.40, 4.41, and 4.42). Mast cells are commonly

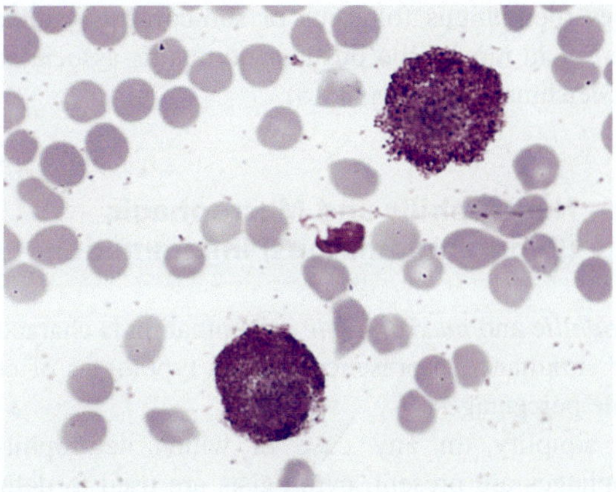

Fig. 4.39 Cytology of mast cells: numerous intracytoplasmic purple granules characterize two well-differentiated mast cells

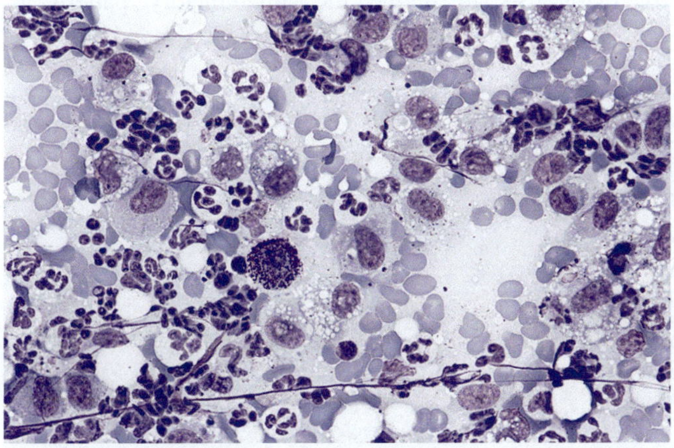

Fig. 4.40 Cytology of mast cells: a single mast cell is recognizable in the contest of a pyogranulomatous inflammation

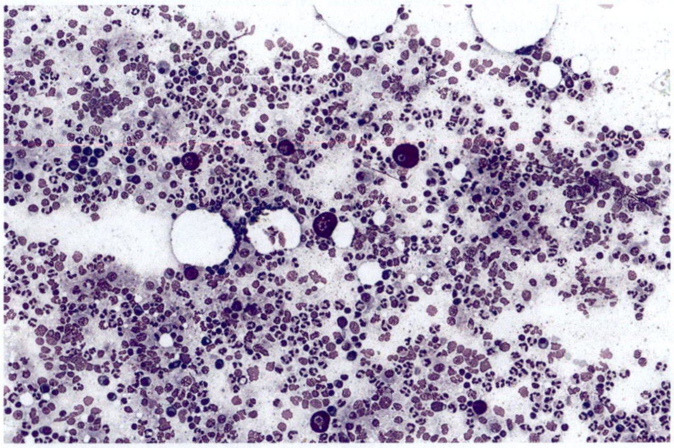

Fig. 4.41 Cytology of mast cells: many well-differentiated mast cells in mixed chronic inflammation

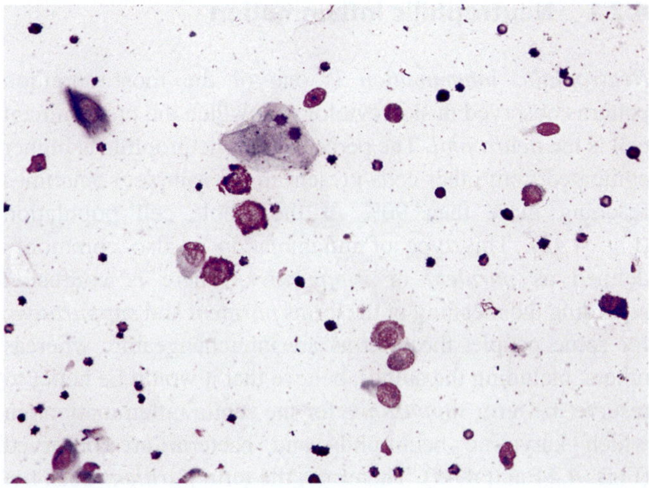

Fig. 4.42 Cytology of urticarial pigmentosa-like in cat: many well-differentiated mast cells are evident

observed in neoplastic cytology since the MCT is the most common skin tumor in dogs and it is also quite frequent in cats.

4.2 The Inflammatory Patterns

The typology of the aforementioned inflammatory cells and their percentage on slides define the so-called *inflammatory patterns*. The latter, evaluated in the light of clinical signs, often allows the clinicians to reach a diagnosis or, at least, to guide them in the choice of which other diagnostic investigations to be performed. A brief description of the cutaneous cytological inflammatory patterns is discussed below, while the inflammatory diseases of the skin in which these patterns can be observed are exhaustively discussed in Chap. 6.

The inflammatory patterns of the skin can be classified as follows:

1. *Neutrophilic.*
2. *Neutrophilic and macrophagic* (commonly named "*Pyogranulomatous*").
3. *Granulomatous.*
4. *Eosinophilic.*
5. *Lympho-plasmacellular.*
6. *Lympho-histiocytic.*
7. *Lympho-histio-plasmacellular.*

4.2.1 Neutrophilic Inflammation

Neutrophilic inflammation is one of the most common patterns observed in skin cytology, in which the predominant cell is the *neutrophil*. The percentage of neutrophils is higher compared with other cells present in the sample, sometimes reaching more than 90% of the whole cell population (Fig. 4.43). This type of inflammation is also commonly defined as *purulent* or *suppurative*. There is confusion regarding the meaning of the terms *purulent* and *suppurative*; for some people, these terms are interchangeable, whereas others, including the author, believe that it would be better to reserve the term *suppurative* for the septic inflammations in which karyolitic neutrophils and bacteria are observed (Figs. 4.44 and 4.45). Therefore, the term *purulent* does not necessarily indicate a bacterial etiology, as a high presence of neutrophils (pus) can also be found in "sterile" processes such as some pustular immune-mediated/autoimmune diseases (pemphigus foliaceus or sub-corneal pustular dermatitis) or in rare sterile pustular dermatitis associated with canine leishmaniasis (Fig. 4.46).

4.2.2 Neutrophilic and Macrophagic (Pyogranulomatous) Inflammation

Neutrophilic and *macrophagic* inflammation is characterized by the simultaneous presence of both typologies of cells in variable percentages.

To simplify, in any case in which neutrophils and macrophages are present, cytologists are used to define the pattern as *pyogranulomatous*.

From a pathological point of view, it must be stressed that the presence of neutrophils and macrophages does not authorize cytologists to define the inflammatory pattern as

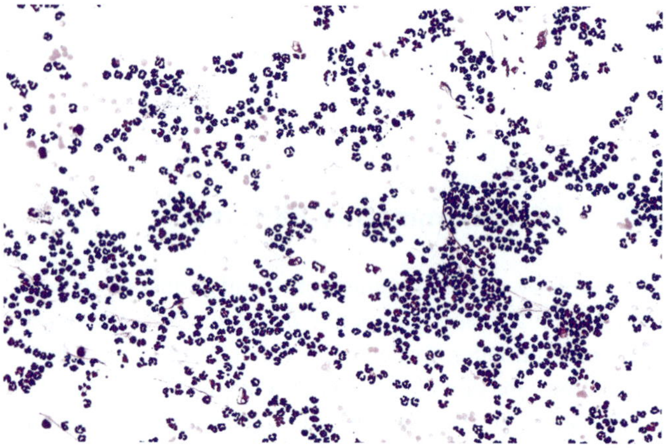

Fig. 4.43 Cytology of neutrophilic inflammation: the totality of the cells is represented by segmented neutrophils

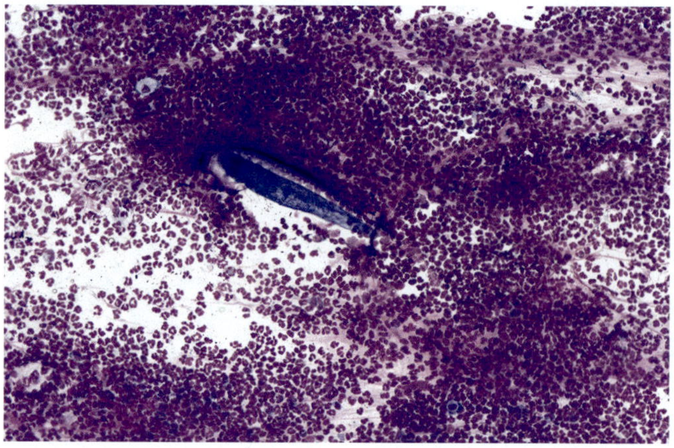

Fig. 4.45 Cytology of neutrophilic *suppurative* inflammation: a pure purulent septic inflammatory pattern is also observed in the course of demodectic folliculitis

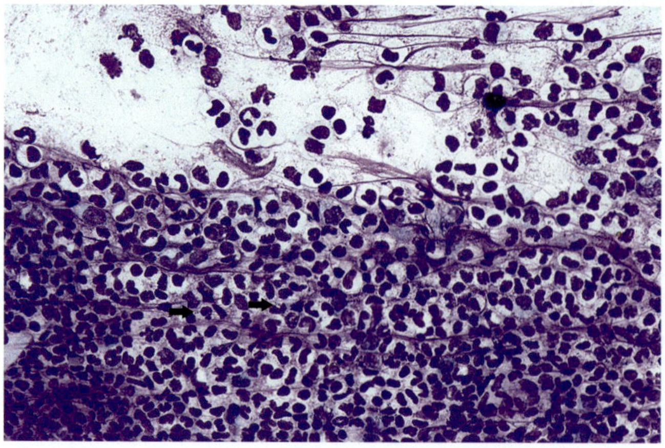

Fig. 4.44 Cytology of neutrophilic *suppurative* inflammation: severely karyolytic nuclei in the course of staphylococcal pyoderma; few cocci are recognizable (arrows)

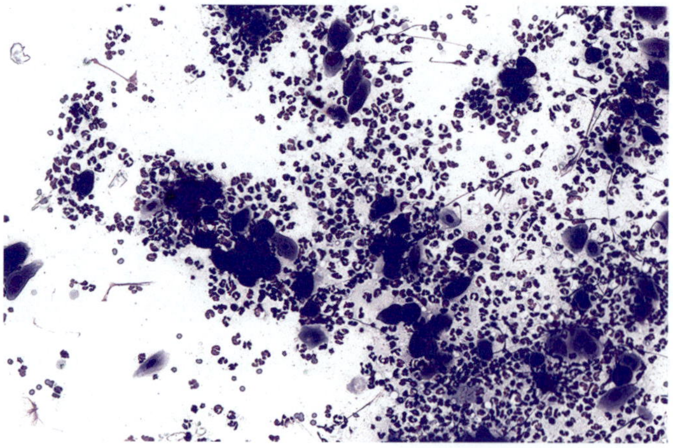

Fig. 4.46 Cytology of neutrophilic *sterile* inflammation in the course of pemphigus foliaceus: many *segmented* neutrophils with numerous acantholytic keratinocytes

pyogranulomatous because such cells can be distributed diffusely in the tissues and not necessarily organized as a granuloma; so, regarding the correct cytological terminology, when epithelioid macrophages (the cells of granuloma) or when foreign bodies, both endogenous or exogenous (vegetal, hair shafts, keratinocytes, calcium salts, etc.) are not detected on slides, the cytological pattern should be defined as *neutrophilic and macrophagic* and not *pyogranulomatous* (Fig. 4.47). At the same time, the fact of not finding in the context of a mixed neutrophilic and macrophage inflammation, cytological findings indicative of the presence of a granuloma as afore-discussed, does not rule out a granulomatous lesion; ultimately, in these cases, only histological examination allows the inflammatory pattern to be defined (Fig. 4.48). The recognition of cytological findings indicative of a granulomatous lesion may not only have a purely didactic importance but also help to cytologist to orient clinical suspicion, especially if the person reading the slide is the same who is visiting the animal.

Pyogranulomatous inflammatory pattern is characterized by a variable percentage of neutrophils, macrophages, epithelioid macrophages, and giant cells and can be due to many different diseases. The infectious and parasitic causes include, among others, bacteria (deep staphylococcal pyoderma and mycobacteriosis), fungi (dermatophytosis, phaeohyphomycosis, and cryptococcosis), protozoa (leishmaniasis), and parasites (demodicosis); nevertheless, many sterile diseases, such as cellulitis or panniculitis (juvenile or adult-onset cellulitis and sterile nodular panniculitis), reactions to endogenous foreign bodies (follicular keratin or hair shaft fragments), accumulation of fat (xanthogranuloma), or mineral salts deposit in the dermis (cutaneous calcinosis), are able to evoke pyogranulomatous inflammations.

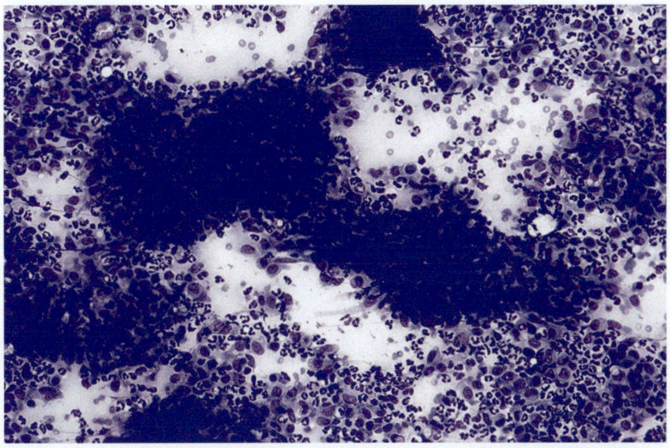

Fig. 4.47 Cytology of *pyogranulomatous* inflammation (canine juvenile cellulitis): many neutrophils and macrophages; the epithelioid macrophages arranged in a large group represent a granulomatous formation allowing the inflammatory pattern to be defined as pyogranulomatous

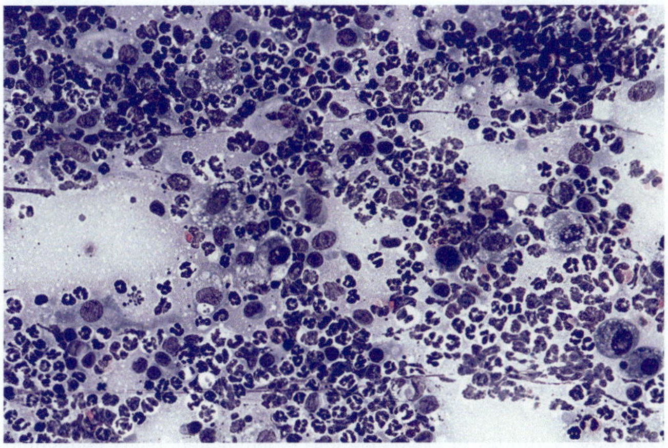

Fig. 4.48 Cytology of *pyogranulomatous* inflammation: segmented neutrophils and vacuolated macrophages collected from a dermatophytic kerion are singularly spread; although kerion is a granulomatous lesion due to the spreading of keratin into the dermis, no evidence of granuloma is present in this picture

4.2.3 Granulomatous Inflammation

Some inflammatory processes, both on an infectious and sterile basis, can mainly determine the recall of histiocytes; in these cases, epithelioid cells and multinucleated giant macrophages can be, almost, the unique type of cells observed on slides.

In granuloma in which neutrophils are absent or very few, the term *naked granuloma* is used, and the inflammatory pattern is defined as *granulomatous* (Figs. 4.49 and 4.50). A rare disease that is cytologically characterized by a pure granulomatous inflammation is canine reactive cutaneous sarcoidosis (Chap. 6).

4.2.4 Eosinophilic Inflammation

Eosinophils are present, in a variable number, in many inflammatory processes, usually in association with other cells such as neutrophils, macrophages, lymphocytes, and, especially in cats, mast cells. When the number of eosinophils is so high as to exceed that of other cells, the inflammatory process is defined as *eosinophilic inflammation.*

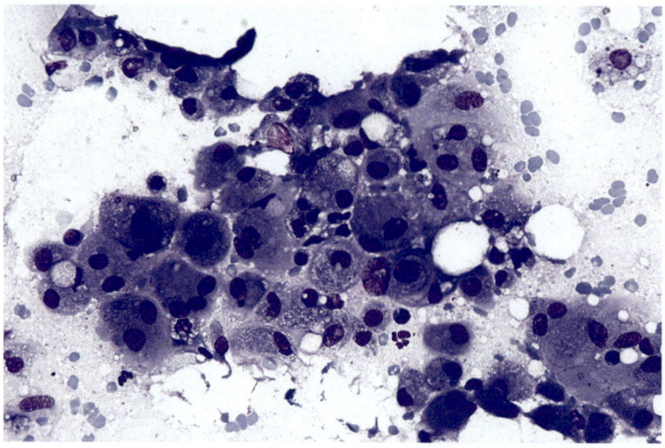

Fig. 4.49 Cytology of *granulomatous* inflammation: many macrophages, epithelioid macrophages, and few giant multinucleated macrophages sampled from a "naked granuloma"; note that neutrophils are lacking

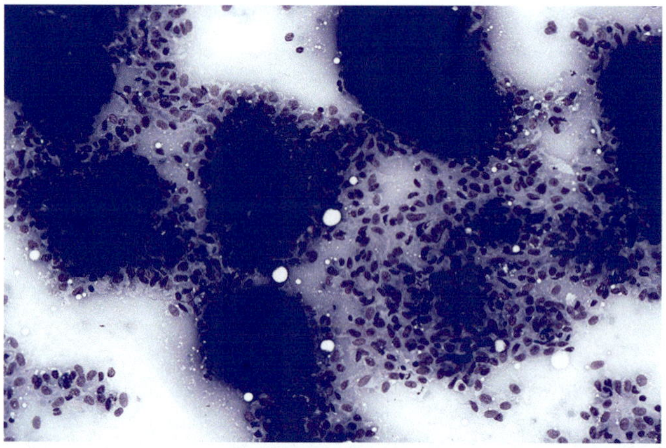

Fig. 4.50 Cytology of *granulomatous* inflammation: many epithelioid cells arranged in overlapped groups representing granulomas in a dog with cutaneous sarcoidosis (sarcoid reaction)

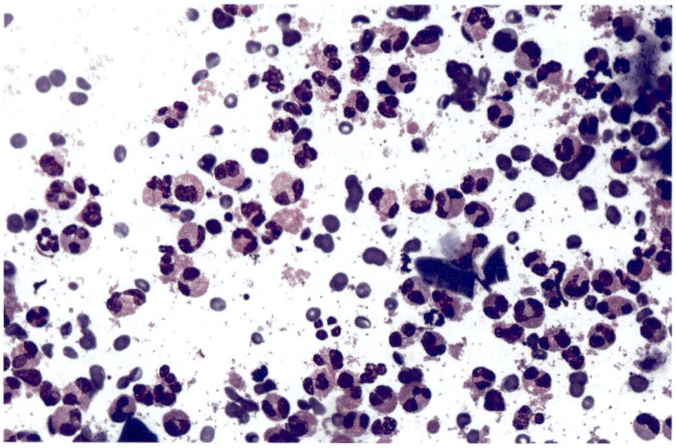

Fig. 4.51 Cytology of eosinophilic inflammation in cats: many eosinophils with rod-shaped granules sampled from a feline eosinophilic plaque

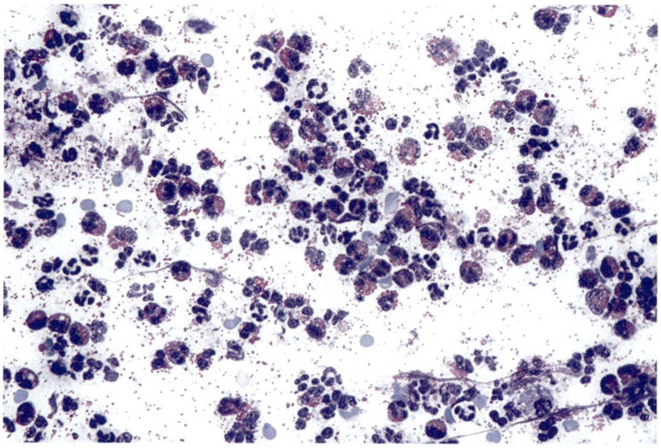

Fig. 4.52 Cytology of eosinophilic inflammation in dogs: many eosinophils collected from a facial eosinophilic furunculosis

Eosinophilic inflammation is very common in cats, considering the frequency of *feline eosinophilic diseases* (e.g., eosinophilic granuloma complex, miliary dermatitis, mosquito bite hypersensitivity), whereas it is much less common in dogs (e.g., oral eosinophilic granuloma and facial eosinophilic furunculosis) (Figs. 4.51 and 4.52).

As aforementioned, it must be stressed, that not only parasitic or allergic diseases are able to recruit eosinophils in the tissues; in fact, other diseases such as *pemphigus foliaceus* as well as skin localization of *feline herpesvirus* can be frequently characterized by a high amount of eosinophils. When a high number of macrophages are present along with eosinophils, the term *eosinophilic-granulomatous pattern* can be used.

Finally, it should be highlighted that eosinophils frequently infiltrate canine MCTs; their presence is of great help in suspecting an MCT when neoplastic cells do not present the characteristic purple metachromatic granules, as happens in some undifferentiated tumors or in the cases in which the granules are normally present, but they do not stain with the rapid Romanowsky-type dyes, as discussed in Chap. 8.

4.2.5 Lympho-Plasmacellular Inflammation

Skin diseases characterized by pure *lymphocytic* and *plasma cell inflammation* are less common in dogs and cats. Their presence is usually only indicative of a chronic, persistent, non-specific antigenic stimulus.

This inflammatory pattern characterizes the feline *plasma cell pododermatitis*, an immune-mediated disease that affects cats and is characterized by dermal inflammation rich in lymphocytes and especially in plasma cells (Fig. 4.53). In dogs, severe *lympho-plasma cellular* inflammation can also be observed in case of chronic deep pyoderma as well as in not-specific chronic inflammatory processes (Fig. 4.54). Less frequently, in not secondarily complicated cases, a prevalent number of lymphocytes and plasma cells can also be observed in specimens collected by imprint from ulcerated lesions located at the mucocutaneous junctions such as in mucocutaneous pyoderma or some autoimmune skin diseases (e.g., lupus complex) or protozoal disease (leishmaniasis), in which a marked sub-epidermal lympho-plasma cellular infiltrate is present.

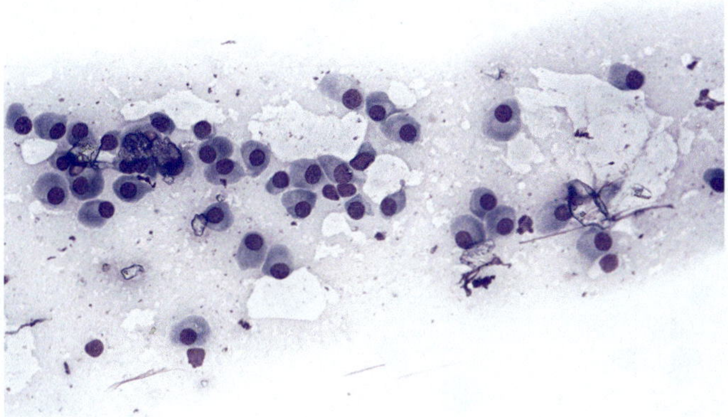

Fig. 4.53 Cytology of lympho-plasma cellular inflammation in cats: many plasma cells and some lymphoid cells from a feline plasma cell pododermatitis

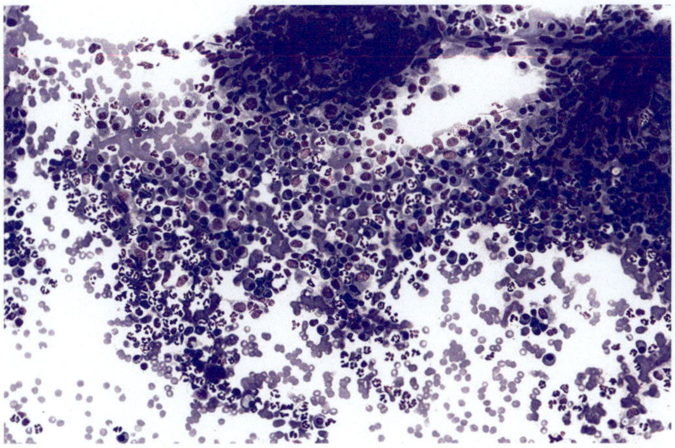

Fig. 4.54 Cytology of lympho-plasma cellular inflammation in a dog with chronic pyoderma: many plasma cells along with few lymphocytes; the former are recognizable also at low magnification thanks to the smaller size, deep blue cytoplasm and eccentric nuclei

4.2.6 Lympho-Histiocytic Inflammation

The inflammatory *lympho-histiocytic* pattern is composed of many lymphocytes and histiocytes. This cytological pattern can be observed in the case of rare diseases such as *sterile granulomatous syndrome* (SGS), in which the neutrophilic percentage is very low or totally absent (Fig. 4.55). Even if in *canine reactive histiocytosis* (CRH) histiocytes do not belong to the macrophagic lineage but to the dendritic antigen-presenting cells family, the cytological pattern is lympho-histiocytic as well (Fig. 4.56); it must be stressed that it is not possible to differentiate with cytology, the morphology of reactive dendritic histiocytes of CRH from not-activated macrophages or epithelioid cells, as it will be discussed in Chap. 6.

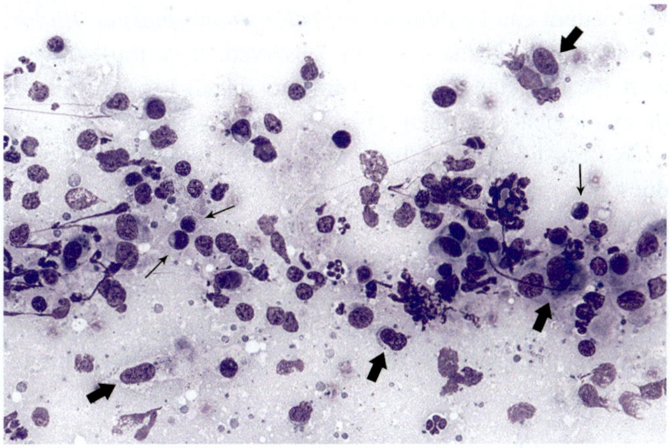

Fig. 4.55 Cytology of lympho-histiocytic inflammation: many lymphoid cells (thin arrows) and histiocitic macrophages (large arrows) characterize the cytology of SGS

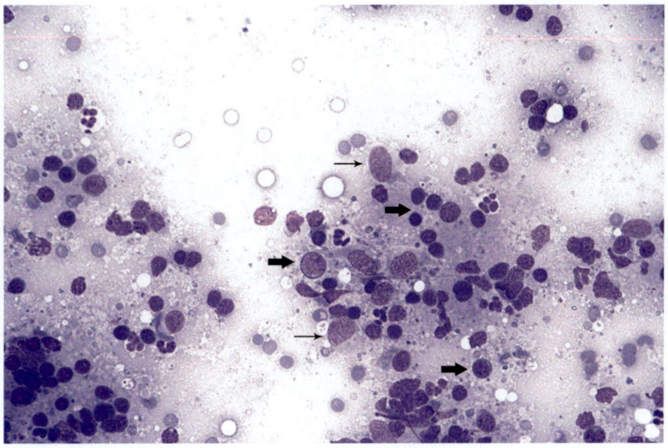

Fig. 4.56 Cytology of lympho-histiocytic inflammation: many lymphoid cells (large arrows) and dendritic histiocytes (thin arrows) characterize the cytology of CRH

A lympho-histiocytic cytological pattern can also be observed during the regression phase of benign canine histiocytoma; indeed, in the advanced stage of neoplasia, the number of infiltrating lymphoid cells is greatly superior to histiocytes and if a history of a nodule present of several weeks with the typical clinical aspect of histiocytoma is lacking, cytology is difficult to differentiate from an aspecific lymphocytic inflammatory lesion.

4.2.7 Lympho-Histio-Plasmacellular Inflammation

The coexistence of numerous plasma cells along with lymphocytes and histiocytes is a very common cytological finding of canine leishmaniasis, especially in the papular variant of the disease (Chap. 6). In these cases, the inflammatory pattern can be defined as *histio-lympho-plasmacellular*. When many plasma cells are observed, they must always alert the clinician and direct him to search for amastigotes on cytological specimens (Figs. 4.57 and 4.58).

We must be aware that a scarce number of plasma cells does not rule out leishmania; therefore, also in case of a lympho-histiocytic pattern, the protozoa must be always searched, especially if a strong clinical suspicion is present; at the same time, many plasma cells does not mean leishmaniasis, since they can also be observed in chronic aspecific inflammatory lesions in which a persistent antigenic stimulus is present.

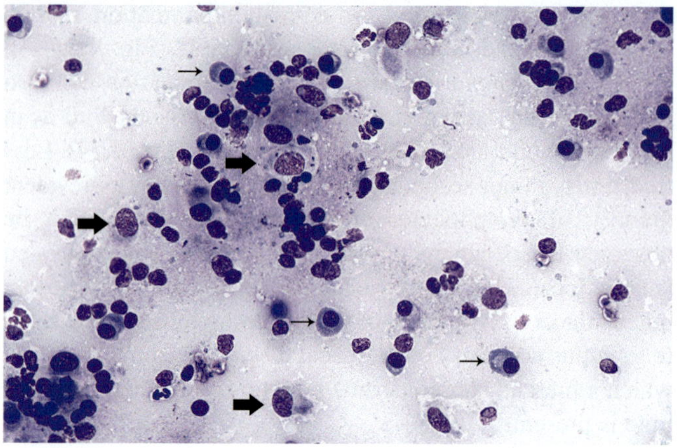

Fig. 4.57 Cytology of lympho-histio-plasmacellular inflammation: many plasma cells (*thin arrows*) and histiocytes macrophages (*large-short arrows*); note that three amastigotes of *leishmania* spp. are into the cytoplasm of a macrophage (*arrow* at the center of the image)

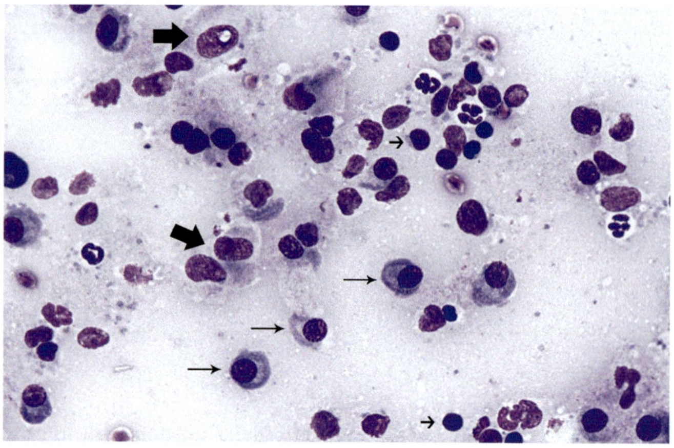

Fig. 4.58 Cytology of lympho-histio-plasmacellular inflammation: many lymphoid cells (*short arrows*), macrophages (*large arrows*), and plasma cells (*thin-long arrows*) collected from a skin lesion in a dog with leishmaniasis

The "Background" in Skin Cytology

5

The term *background* is slangily used to define the material that characterizes the bottom of the slides; it can be composed of only acellular material, or from the latter along with cells. Knowledge of the cytological features of these backgrounds is of considerable help to the cytologist in interpreting slides under the microscope and sometimes could add cytological details necessary to reach the diagnosis.

The backgrounds of cytological interest in skin cytology are as follows:

- *Hematic*
- *Necrotic*
- *Apocrine*
- *Lipid*
- *Pigmented (melanin)*
- *Myxoid*
- *Tigroid*

The above-mentioned backgrounds may also be present simultaneously on the same sample, as occurs, for example, with hematic and necrotic ones.

5.1 Hematic Background

Because of the rupture of the blood vessels that frequently occurs during the sampling maneuvers, the presence of blood on the bottom of the slides is a very common finding in skin cytology; this finding occurs in any specimen collected via fine-needle biopsy and fine-needle aspiration biopsy.

Furthermore, the presence of red blood cells is still considered normal when sampling an ulcerated lesion which, since it involves the dermis, releases blood. For this reason, *hemocontaminated* slides are a very common evenience, and, in almost all cases, they have no diagnostic or prognostic meaning.

Although blood samples are often due to an incorrect sampling method, there are two situations in which the blood background could guide the diagnosis: when the slide is composed of only red blood cells or when the latter are associated with spindle cells. In the first case, especially if the needle cone fills with dark blood at the time of insertion into the nodule, a neoplasm of vascular origin such as an angioma or a hemangiosarcoma or, more rarely, an angiomatosis must be strongly suspected; in fact, especially in the most frequent histotype of hemangioma, the cavernous one, composed of large vascular lacunae filled with blood and lined by a single flattened layer of endotheliocytes, the possibility to collect endotheliocytes, is quite impossible and slides are usually composed of only blood (Fig. 5.1).

A second possibility is when all the slides show a diffuse blood background with a variable number of spindle cells; also in this case, a vascular lesion of endothelial origin or belonging to the group of perivascular wall tumors (PWTs) must be suspected (Fig. 5.2).

Therefore, when a hematic background is present, it must be always cytologically interpreted in light of the clinical appearance of the lesions sampled. A marked hematic background can also be observed in non-neoplastic lesions, such as hematomas, or in lesions where hemorrhage has occurred. In these cases, along with a blood background, macrophages are detected phagocytosing granular debris, from blue to gray in color (*hemosiderin)* or bright orange rhomboid structures (*hematoidin* crystals) (Figs. 5.3 and 5.4).

F. Albanese, *Canine and Feline Skin Cytology*, https://doi.org/10.1007/978-3-031-95713-0_5

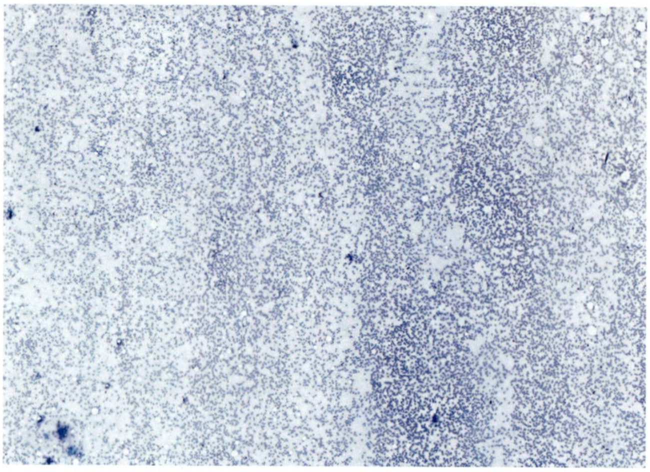

Fig. 5.1 Hematic background in the course of a sampling of a hemangioma: the slide is composed of only blood cells

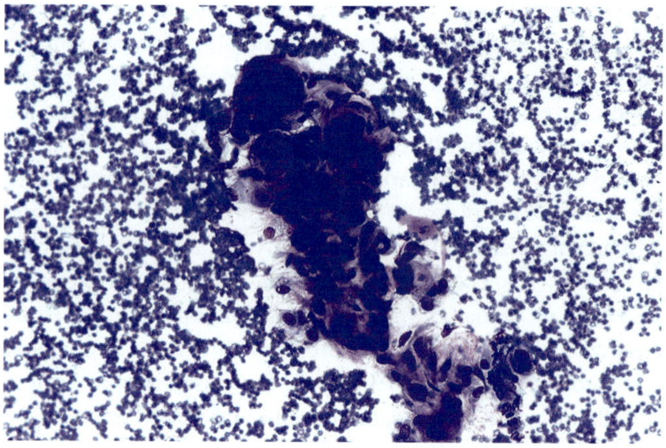

Fig. 5.2 Hematic background in the course of a sampling of a perivascular wall tumor recognizable by the whorled arrangements of neoplastic spindle cells

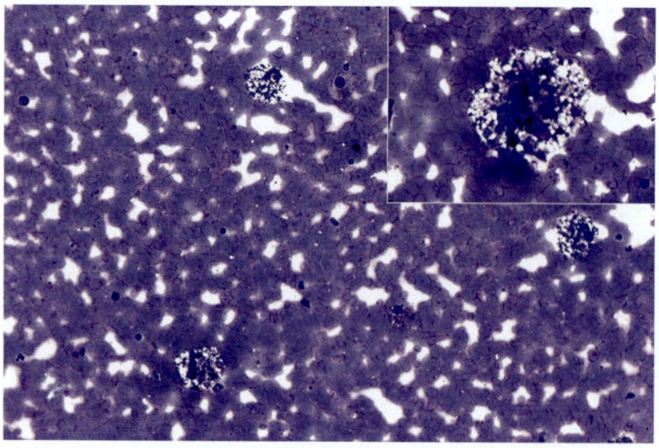

Fig. 5.3 Cytology of a hemorrhagic lesion: hematic background with many macrophages filled with hemosiderin

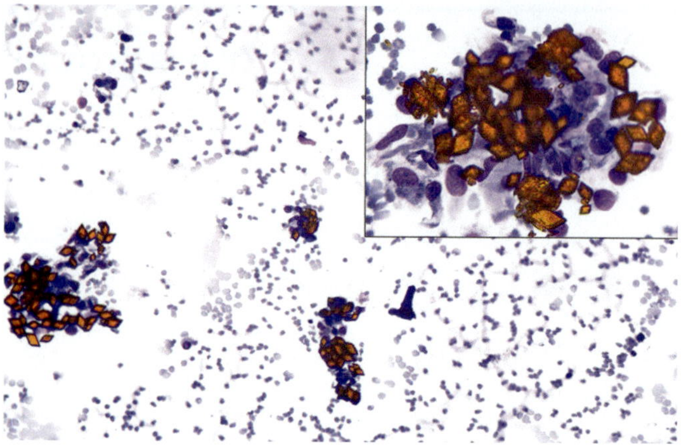

Fig. 5.4 Cytology of hemorrhagic lesion: many hematoidin crystals are observed in the cytoplasm of macrophages

5.2 Necrotic Background

The term *necrosis* refers to the premature death of cells whose causes are multiple and include hypoxia, trauma, infection, vascular damages, granulomas and neoplastic diseases, and so on.

From a pathologic point of view, different types of cell necrosis are possible; indeed, the cell death can be not-programmed (necrosis) or regulated by signaling pathways which result in defined biochemical and functional outcomes (apoptosis). *Apoptosis* is the first and the most studied programmed cell death, but other non-apoptotic regulated cell death pathways including pyroptosis, ferroptosis, and necroptosis have been discovered; seven patterns of necrosis are recognized in medicine, which discussion is out of the aim of this book, in which only the massive not-programmed necrosis that is possible to observe as background in slides, how it is cytologically recognizable, and its eventual diagnostic meaning will be discussed.

In cytology, necrosis is recognized as an amorphous, granular, debris, deep blue in color material in which fragments of nuclear material or naked nuclei are embedded; sometimes, in association with these cytological features, inflammatory cells such as neutrophils and macrophages reach the area of necrosis in an attempt to clear it by the process of phagocytosis (Fig. 5.5).

Usually, necrosis in medicine tends to almost always have a negative connotation and is also used, together with others, as a negative prognostic parameter during the grading of some groups of tumors such as, for example, soft tissue sarcomas; anyway, it is observed in many other malignant tumors such as in histiocytic sarcoma, lymphoma, apocrine glands carcinoma, skin metastasis of internal malignancies, and so on [1, 2] (Figs. 5.6, 5.7, and 5.8).

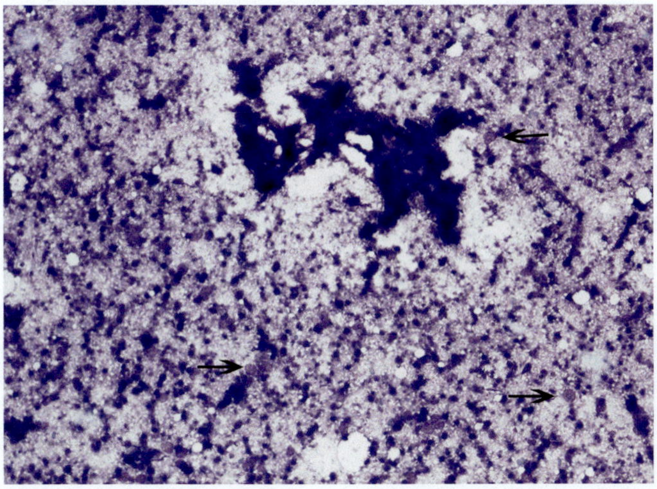

Fig. 5.5 Cytology of the necrotic background: amorphous acellular material, deep blue, represents necrosis. Scattered naked nuclei are barely noticeable (*arrows*)

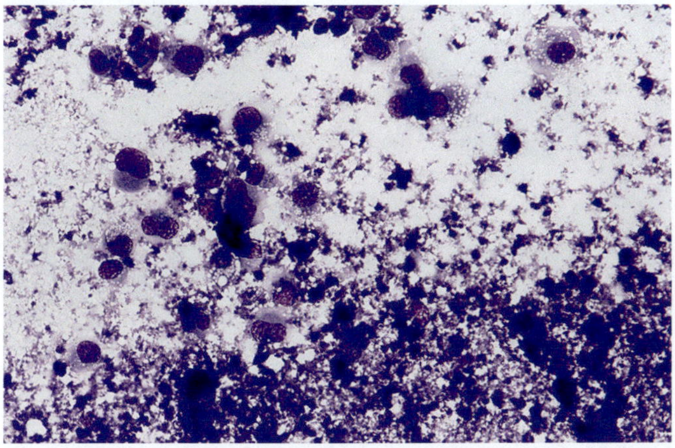

Fig. 5.6 Cytology of necrotic background: histiocytic malignant cells (histiocytic sarcoma) on a necrotic background

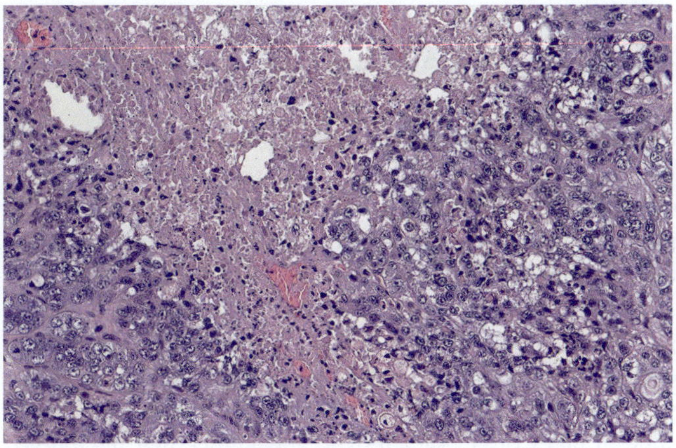

Fig. 5.7 Histology of necrotic background in an apocrine carcinoma: eosinophilic amorphous material in which scattered dusty necrotic fragments of destroyed nuclei represent necrosis. Note the apocrine malignant epithelial cells at the periphery of the pictures

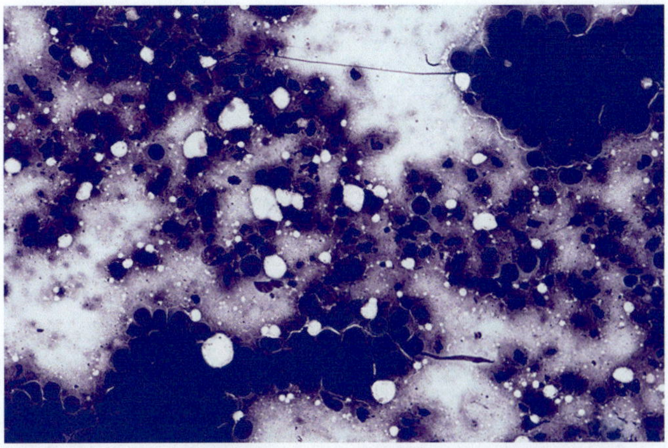

Fig. 5.8 Cytology of necrotic background in an apocrine carcinoma: necrotic background along with two large clusters of malignant apocrine glandular neoplastic cells

However, the necrotic background is not always synonymous with malignancy; in fact, it can commonly be observed in the course of both non-neoplastic lesions (necrotic foci in feline eosinophilic granulomas, vasculitis, post-injection panniculitis) and benign neoplasms as occurs during the sampling of the cystic areas of apocrine or follicular neoplasia (e.g., solid-cystic apocrine adenoma or cystic trichoblastoma) or in the regression phase of benign cutaneous histiocytomas in dogs (Figs. 5.9, 5.10, 5.11, 5.12, 5.13, and 5.14).

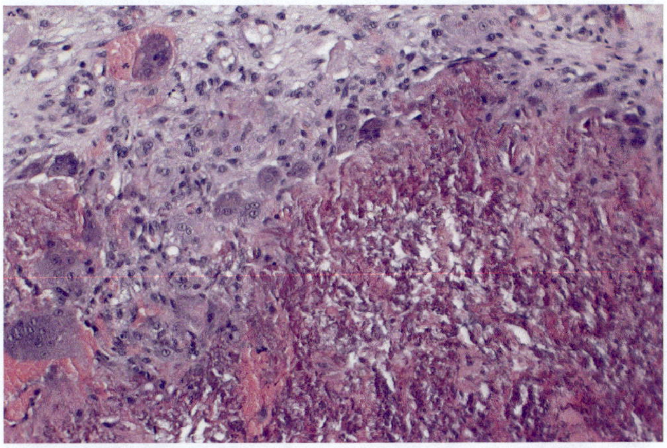

Fig. 5.9 Histology of necrotic foci in an eosinophilic granuloma: large area of pinkish in color necrotic background surrounded by many inflammatory cells; note the multinucleated macrophages

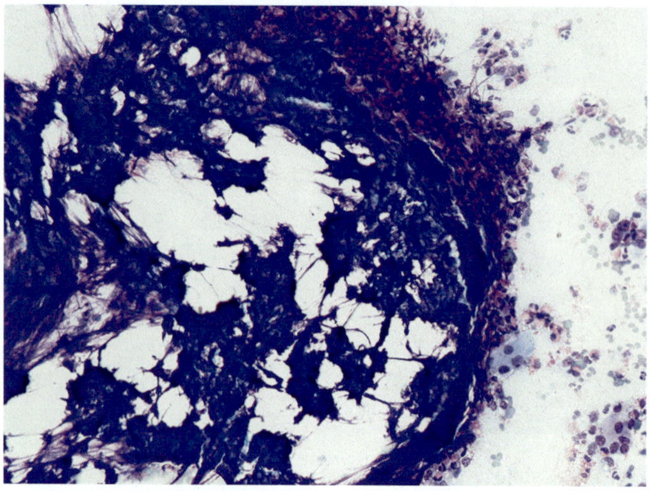

Fig. 5.10 Cytology of necrotic material in an eosinophilic granuloma: note the numerous eosinophils surrounding the pinkish-in-color necrotic background

Fig. 5.11 Histology of necrotic material in post-injection panniculitis: note the large amount of eosinophilic necrosis surrounded by the inflammatory infiltrate typically arranged peripherally

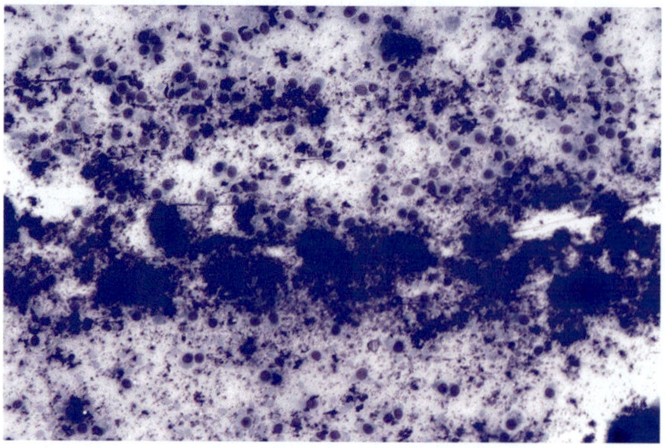

Fig. 5.12 Cytology of necrotic background in a benign histiocytoma: many round-shaped dendritic histiocytes immersed in a necrotic background

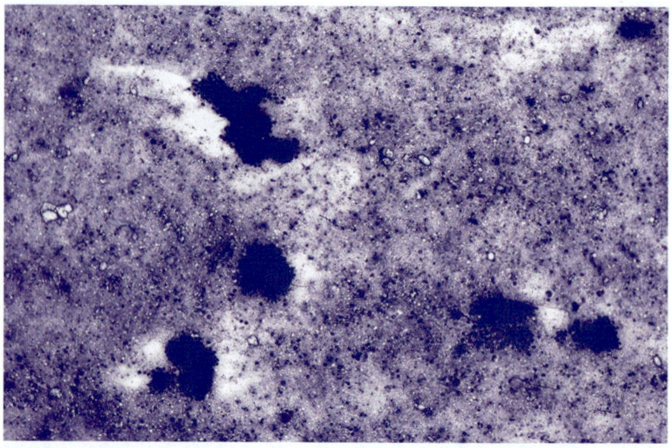

Fig. 5.13 Cytology of necrotic background in a cystic trichoblastoma: massive presence of necrotic amorphous basophilic material in which some clusters of basal follicular cells are present

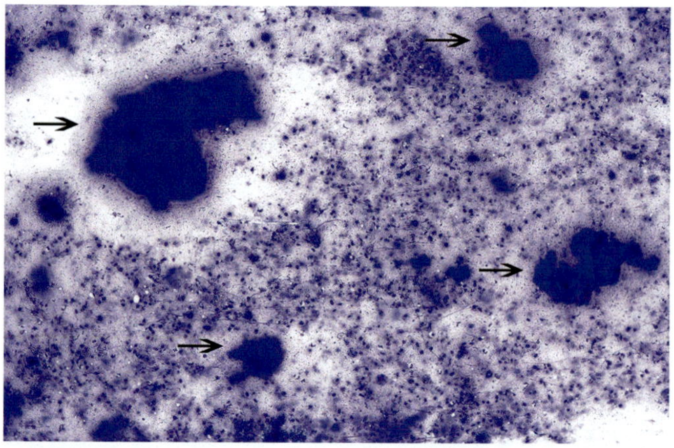

Fig. 5.14 Cytology of necrotic background in an apocrine solid-cystic adenoma: massive presence of necrotic amorphous basophilic material. Apocrine cells are not present in this image. The arrows indicate the sweat secretion

5.3 Apocrine Background

In many cystic or neoplastic cystic lesions of apocrine nature, for example, benign neoplasms such as cystadenoma, it is common to take a transparent liquid with the needle which, placed on the slide, is absolutely transparent after the fixation and staining processes.

During both reactive and neoplastic lesions of the sweat glands, a hypersecretion activity of apocrine glands can occur, and therefore, tubules are more dilated and filled with stagnant sweat. When the secretion is observed on the slide, it takes on a characteristic appearance that the cytologist must recognize and differentiate from other amorphous and acellular materials such as necrosis and amorphous isthmic (trichilemmal) keratin.

The term *apocrine background* is used by the author to define the presence of variable quantities of sweat on the bottom of the slide, rarely as the only material released by the lesion. The sweat glands are mainly apocrine in nature since the eccrine ones are located only on the pads. The type of secretion that occurs through the decapitation of the apical portion of the cytoplasm causes the material to pour into the lumen of the tubules and then into the follicle through a duct. Sweat is cytologically characterized by usually well-defined, acellular, homogeneous, and apparently "shiny and smooth" basophilic material that must not be confused with necrosis; the latter is, in fact, more amorphous, grainy, irregular, and with scattered necrotic nuclear material (Fig. 5.15).

A cytological feature that can help differentiate necrosis from sweat is that, in the latter, linear slits with clear margins can be observed in the context of the aforementioned homogeneous and compact basophilic material; furthermore, since the more compact consistency of sweat, slits can separate pieces of sweat from the background (Figs. 5.16 and 5.17). In practice, when a background completely represented by sweat is observed, even in the absence of cells, a lesion of apocrine glandular origin (sweat glands) must be considered.

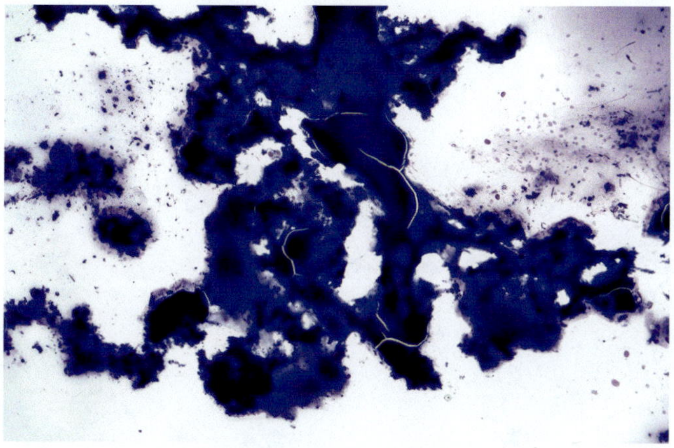

Fig. 5.15 Cytology of basophilic amorphous background coming from an apocrine gland: the sweat must not confused with necrosis; the former is more uniform, homogeneous, and with smooth and shiny aspect compared with the granulosus and dendritic aspect of necrosis

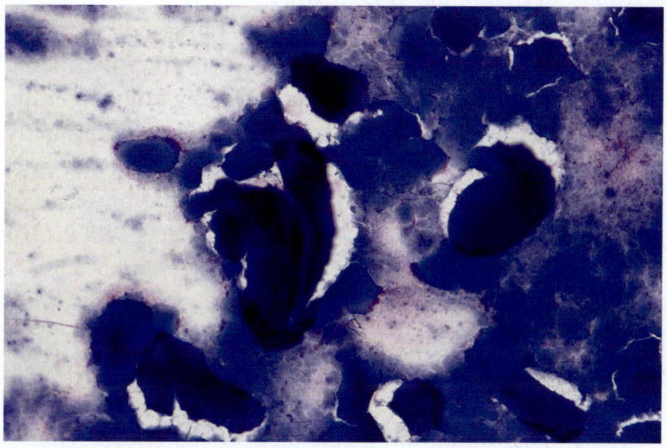

Fig. 5.16 Cytology of basophilic amorphous background coming from an apocrine gland: the sweat must not be confused with necrosis; the former is more uniform, homogeneous, and with smooth and shiny aspect compared with the granulosus and dendritic aspect of necrosis. Note the slits that separate the secretion from the background

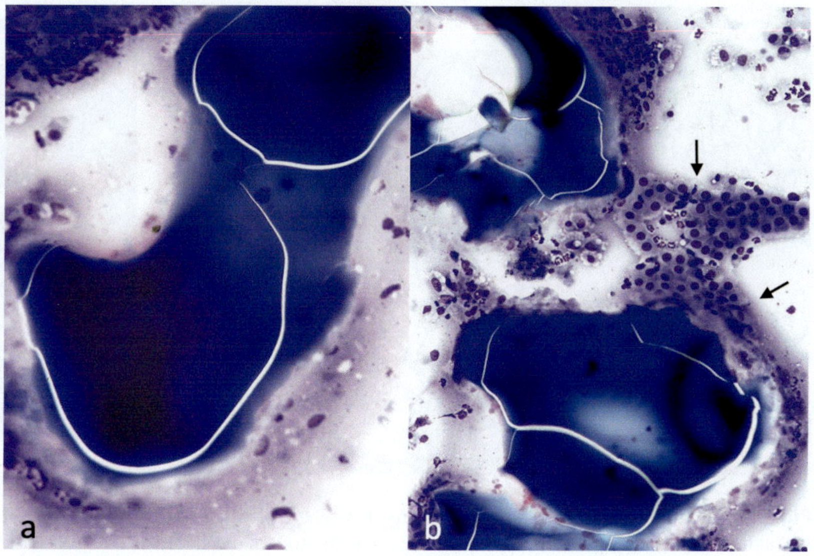

Fig. 5.17 Apocrine background: (**a**) shiny and smooth in appearance sweat glands secretion. Note the sharp slits; (**b**) slitted sweat glands secretion, surrounded by large clusters of well-differentiated apocrine glands. Inflammatory cells are also recognizable

5.4 Lipid Background

The presence of some drops of lipids on slides obtained from deep skin lesions is to be considered an absolutely normal finding in skin cytology; therefore, their presence in scarce amounts does not allow cytologists to interpret the lesion as surely originated or exclusively localized in the panniculus; in fact, in some cases, the fat can be accidentally sampled and representing only an "innocent bystander."

It is different when the quantity of "fatty" material is very abundant and occupies most part, if not all, of the bottom of the slide; in this case, the hypothesis of accidental sampling of fat is very unlikely, and a connection between the lesion and the panniculus must be suspected and searched. The typical aspect of the lipid background is the presence of numerous achromatic and roundish areas representing various sizes of lipid drops (Fig. 5.18).

The lipid background can be observed in both non-neoplastic and neoplastic lesions; therefore, its finding is not indicative of a specific disease. Among non-neoplastic lesions, a disease in which the lipid background is constantly observed is *panniculitis*, the inflammation of fat; in this case, the bottom of the slide is characterized by the presence of numerous fat drops, of various sizes, associated with a variable amount of inflammatory cells (Figs. 5.19 and 5.20).

Being liposarcoma a tumor that originates from adipocytes, slides are obviously characterized by the presence of a lipid background, also because the intracytoplasmic fat tends to disperse on the background following the rupture of some cells during the preparation of the slides. This is especially true in cases of well-differentiated liposarcomas, while obviously a lipid background is not present, or can be very few, in poorly lipogenic liposarcomas; for this reason, adipose cells neoplasia must not be ruled out if a lipid background is not clearly evident (Fig. 5.21).

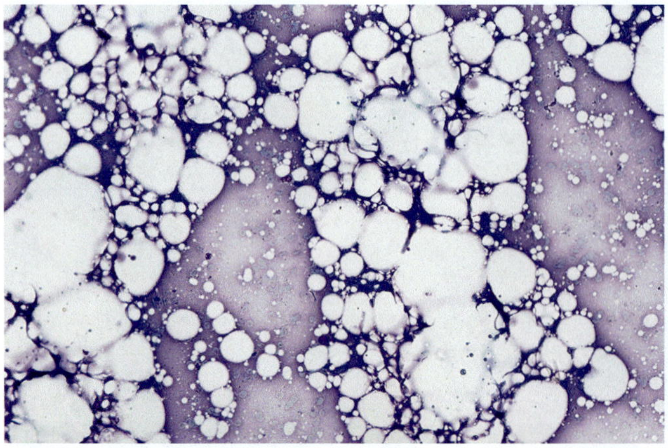

Fig. 5.18 Cytology of lipid background: many different in size, roundish, and achromatic droplets of fat characterize all the bottom of the slide

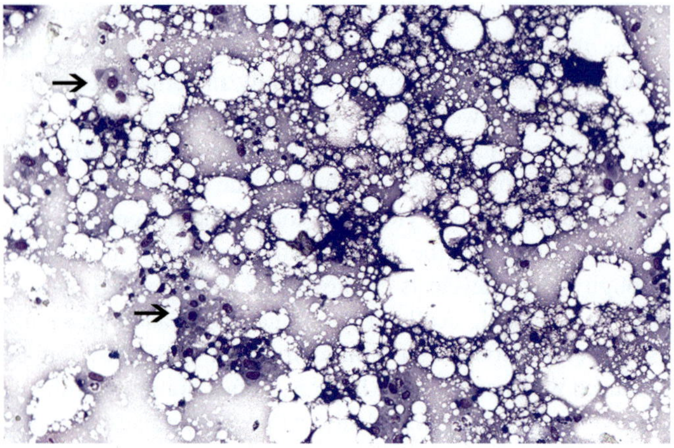

Fig. 5.19 Cytology of lipid background coming from panniculitis: many droplets of lipid material in which some inflammatory cells are recognizable (*arrows*)

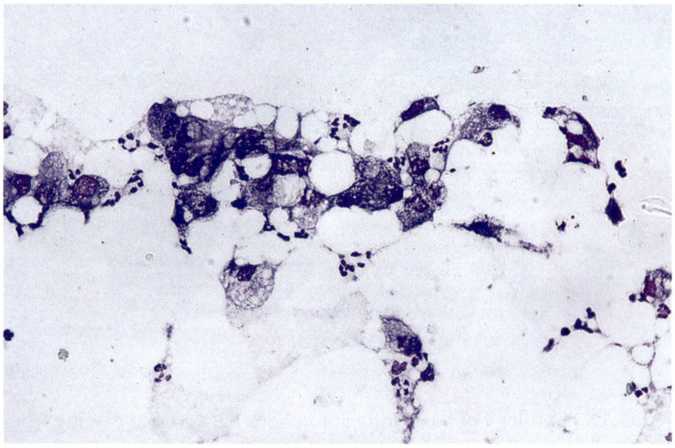

Fig. 5.20 Cytology of lipid background coming from panniculitis: at high magnification, lipid droplets both on the bottom and into the cytoplasm of macrophages are evident (lipo-phagocytosis)

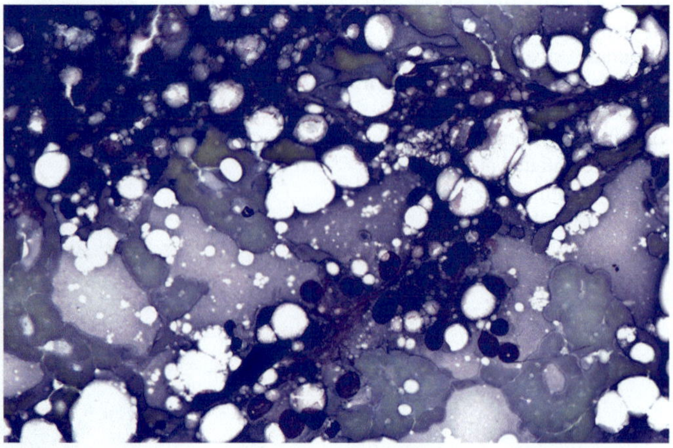

Fig. 5.21 Cytology of lipid background coming from well-differentiated liposarcoma: lipid droplets both on the bottom and into the cytoplasm of neoplastic cells are evident

The term lipid should not be restricted to the material present exclusively in the adipose tissue, since the fat contained in other skin structures is also composed of lipid material. A classic example is *sebum*, which cytologically presents itself as an amorphous material with a greasy appearance and which can be observed during both hyperplastic and neoplastic lesions of the sebaceous glands. As already discussed in Chap. 3, the sebaceous glands have a type of holocrine secretion in which the complete dissolution of the cytoplasm occurs, rich in drops of sebum that are dispersed on the bottom giving a lipid appearance; this, unlike the fat of the adipocytes, does not appear in the form of large rounded achromatic drops but takes on a more homogeneous, amorphous, and often thicker as to not always be squashed in a mono-layer on slides. In almost all cases, sebum takes no color when stained with routinely Romanowsky-type stainings but can assume a slightly barely perceptible yellowish aspect. Sebum can be better recognized in non-mounted slides since the mounting glue can inhibit a good definition of it (Figs. 5.22 and 5.23).

The quantity of sebum is not usually high, except in cases where benign hyperplastic or adenomatous lesions of the sebaceous glands are characterized by the presence of large "cystic" intraglandular areas or have markedly dilated sebaceous ducts filled with sebum. In these cases, the amount of sebum present on the bottom of the slides can be notable. The presence of small groups of mature sebocytes, or of cytoplasmic residues containing micro-droplets of sebum, allows us to define the sebaceous origin of the lipid background (Fig. 5.24).

Fig. 5.22 Cytology of lipid sebaceous background coming from a sebaceous adenoma with ducts filled with sebum

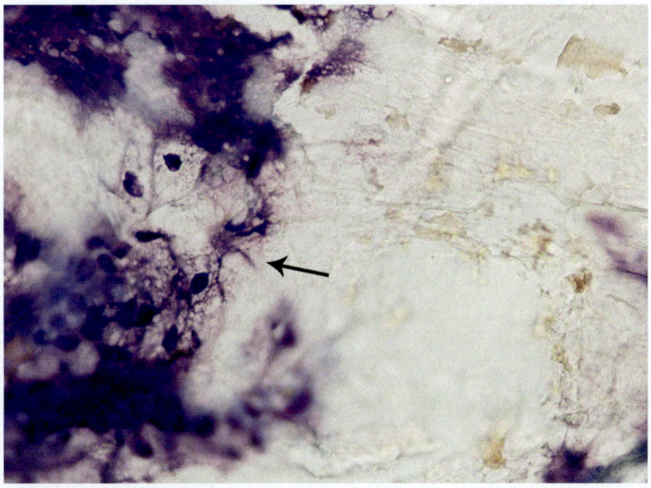

Fig. 5.23 Cytology of holocrine secretion: the arrow highlights the dissolution of mature sebocytic cells in sebum

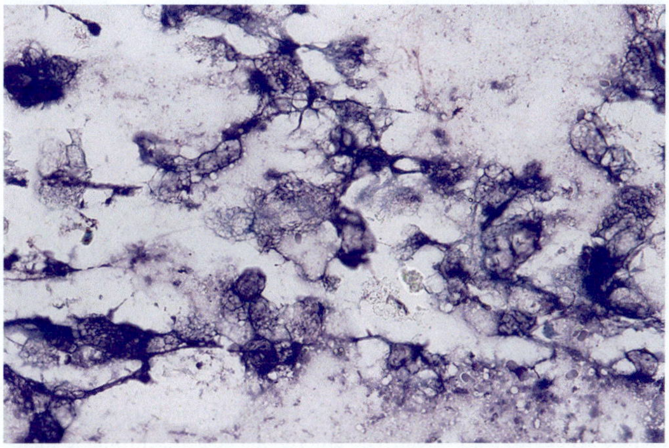

Fig. 5.24 Cytology of sebum: the cytoplasm dissolved by the sebocytes is yet recognizable by its foamy appearance due to the microdroplets

The latter can be of great help in cases where basal tumors of the sebaceous glands, sebaceous epitheliomas, are not well differentiated and therefore do not show mature sebocytes on cytological examination, but only basal cells; the presence of a lipid background associated with basal cells would therefore allow us to suspect a sebaceous epithelioma, avoiding an incorrect diagnosis of trichoblastoma or other basal epithelial neoplasia. In these cases, an oil red O stain can help define the sebaceous nature of the epithelial cells, as discussed in Chap. 2 (Figs. 5.25, 5.26, and 5.27).

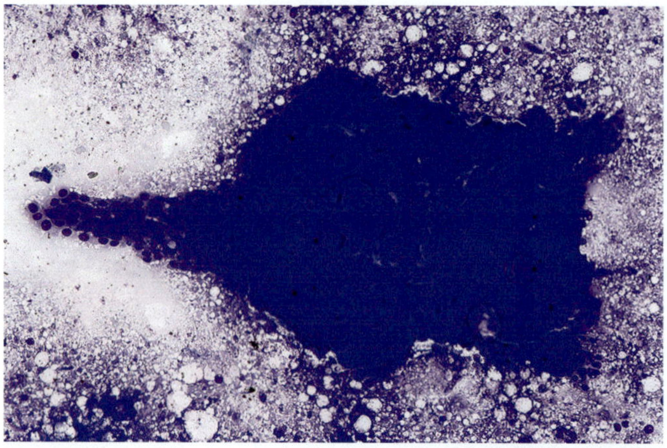

Fig. 5.25 Cytology of sebaceous epithelioma on lipid background: note that mature sebocytes are not clearly present among the basal cells

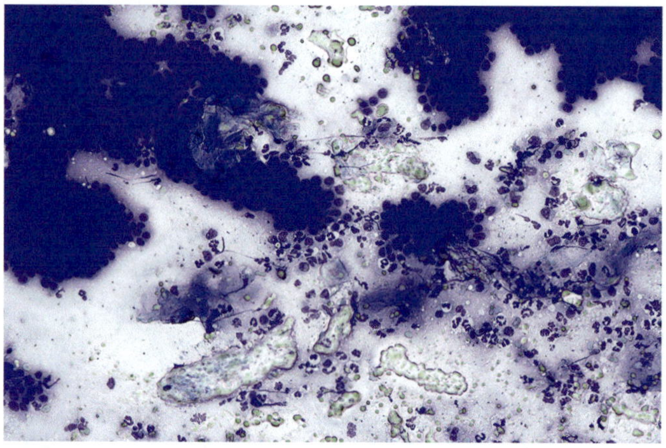

Fig. 5.26 Cytology of sebaceous epithelioma on lipid background: no mature sebocytes are present. The amorphous coarse achromatic material is sebum and its recognition can orient the diagnosis

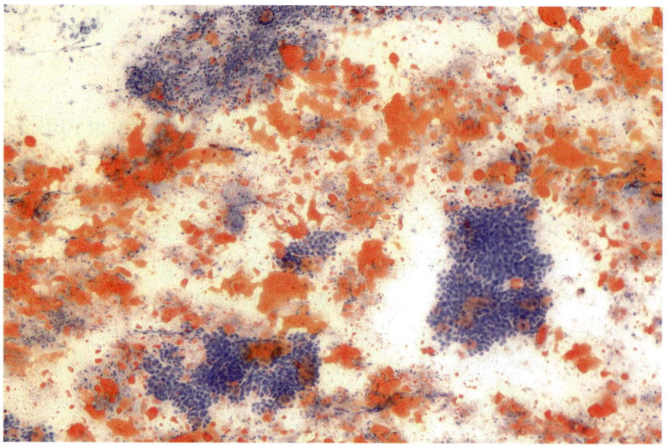

Fig. 5.27 Oil red O staining: lipid background is highlighted by the specific for lipid material cytochemical stain

5.5 Pigmented (Melanin) Background

The presence of abundant melanin pigment is a very common finding in highly pigmented melanocytic tumors. In the latter, the quantity of cytoplasmic pigment is so high that it marks the entire bottom of the slide black, preventing correct visualization of the neoplastic melanocytes. At small magnification, it is, in fact, recognizable as a background composed of black pigment with a dusty appearance, sometimes with shades of dark green, and which often tends to aggregate and give rise to short linear arrangements of a worm-like appearance (Figs. 5.28 and 5.29).

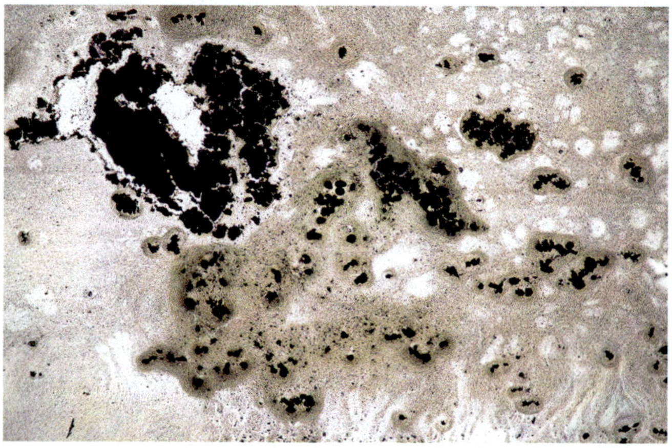

Fig. 5.28 Cytology of melanin background: heavily pigmented melanocytic tumor. Note the linear arrangements of melanin

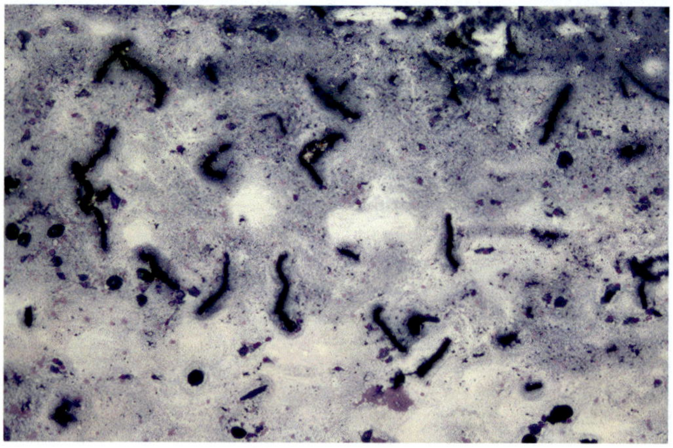

Fig. 5.29 Cytology of melanin background: heavily pigmented neoplastic melanocytes in which nuclear details are only barely valuable. Note the dusty melanin both free and arranged in linear configurations on the bottom

Although in almost all cases in which a melanin background is observed, we are faced with a melanocytic neoplasm, there are some cases in which both follicular and basal epithelial neoplasms of the epidermis can be strongly pigmented. The amount of pigment can be so high that the sampled material appears inky black when smeared onto the slide; this macroscopic characteristic must absolutely not lead to the formulation of a misdiagnosis of melanocytic neoplasia without observing the preparations under the microscope. Furthermore, given the marked presence of melanin, it can sometimes be difficult to cytologically differentiate the aforementioned neoplasias, and, also for skilled cytologists, an accurate cytological observation is necessary for defining diagnosis. This situation must be kept in mind by beginners (Figs. 5.30, 5.31, and 5.32).

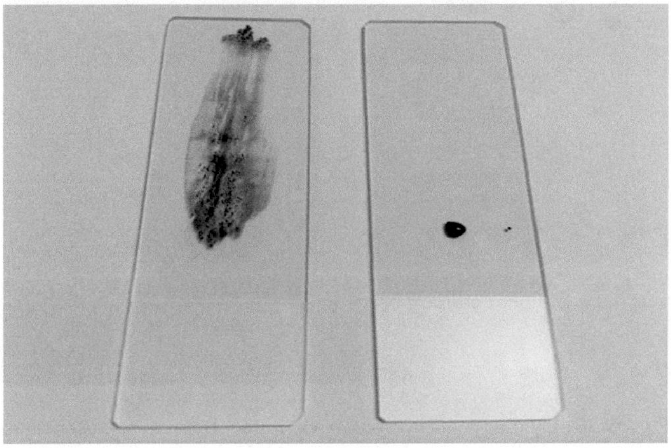

Fig. 5.30 Cytology of melanin background: grossly aspect of strongly pigmented material collected from a hyperpigmented trichoblastoma

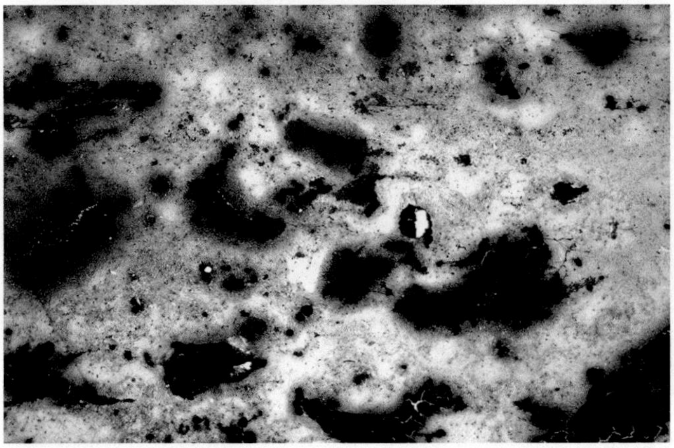

Fig. 5.31 Cytology of melanin background: at low magnification, the high amount of black pigmented bottom does not permit a clear definition of cell morphology

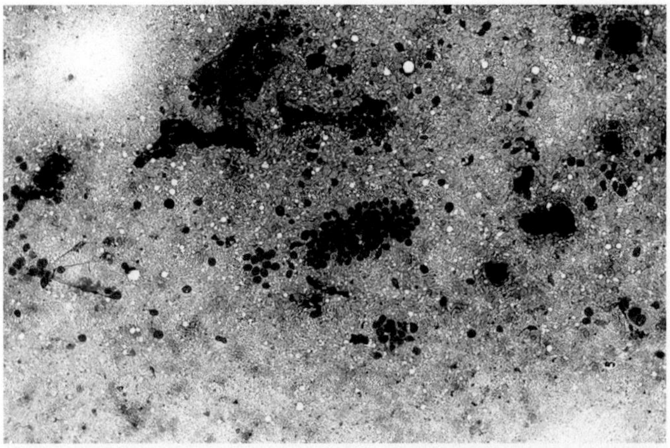

Fig. 5.32 Cytology of melanin background: at high magnification, a few clusters of basal cells, some of them also singularly released, are better recognizable

5.6 Myxoid Background

The myxoid substance is secreted by the neoplastic mesenchymal spindle cells and consists of a gelatinous mucopolysaccharide matrix composed of sulfated and non-sulfated glycosaminoglycans. Macroscopically, this material appears as a dense, viscous, stringy substance that comes out during cytological sampling, and since it is usually produced by neoplastic cells belonging to the group of mesenchymal spindle cell tumors, its visualization is a useful alarm bell in suspecting these neoplasia (Fig. 5.33).

When mucopolysaccharides are produced in low quantities, it is not visible macroscopically during the sampling, but it is possible to observe them on slides as pinkish-violet material, dispersed among the neoplastic cells (Fig. 5.34).

In some soft tissue neoplasia, however, large quantities of myxoid material can be produced, such as to affect the entire bottom of the slides and give the so-called *myxoid background*. On cytological slides, the myxoid material, being viscous, determines a characteristic alignment of the neoplastic cells, and of the red blood cells if present, to form long linear chains (*windrowing appearance*) (Fig. 5.35).

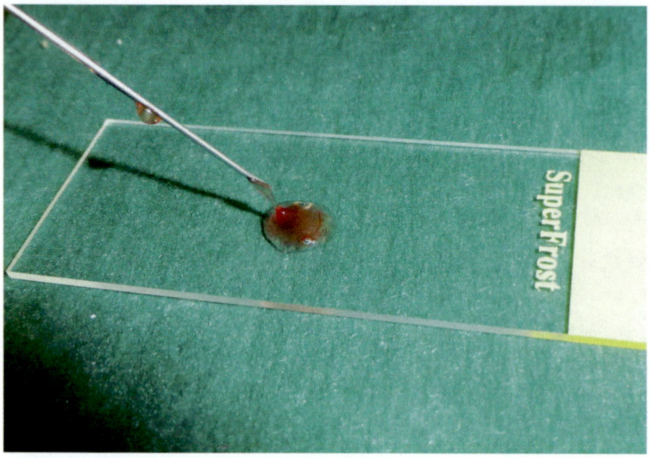

Fig. 5.33 Grossly appearance of myxoid material collected from a spindle cell neoplasia of the skin: note the filamentous and viscous aspect

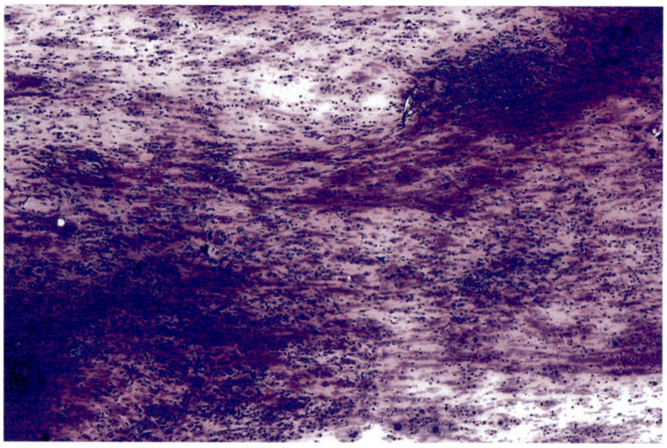

Fig. 5.34 Cytology of myxoid background: deeply violet in color mucopolysaccharides represent myxoid background

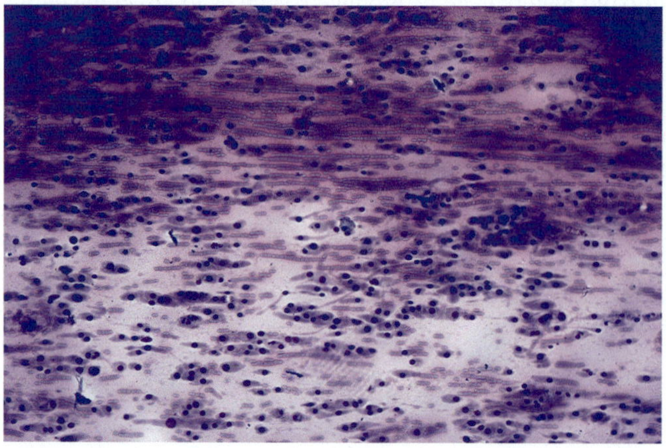

Fig. 5.35 Cytology of myxoid background from a myxosarcoma: pinkish to violet myxoid matrix in which neoplastic cells are arranged in a linear disposition (windrowing appearance)

The finding of myxoid material is not in itself diagnostic of a specific mesenchymal neoplasm since different types of neoplasms belonging to this group can produce myxoid matrix; usually, myxoma and myxosarcoma (myxoid fibrosarcoma) as for their definition produce myxoid matrix that is an integral component of the tumor, but, sometimes, other neoplasms can occasionally also do so such as PWTs, NSTs, spindle cell lipoma, and any other type of sarcoma as well. The myxoid material is cytologically interpretable as an amorphous, acellular, from pinkish to violet until slightly blue material. With alcian blue staining, it can be highlighted by taking a typical azul color (Figs. 5.36, 5.37, and 5.38).

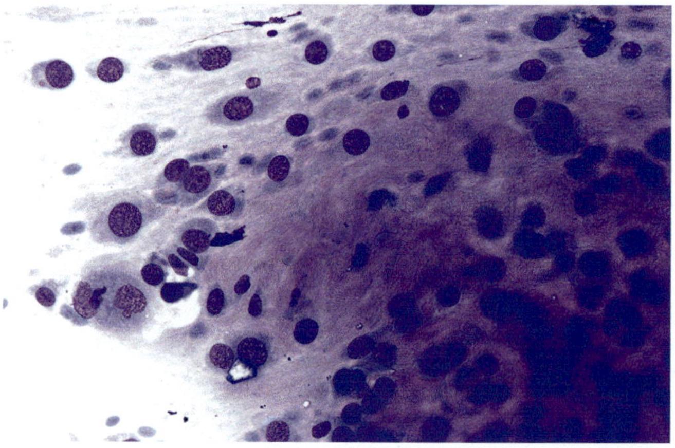

Fig. 5.36 Cytology of myxoid background from a myxosarcoma: malignant spindle cells embedded in a high amount of myxomatous material

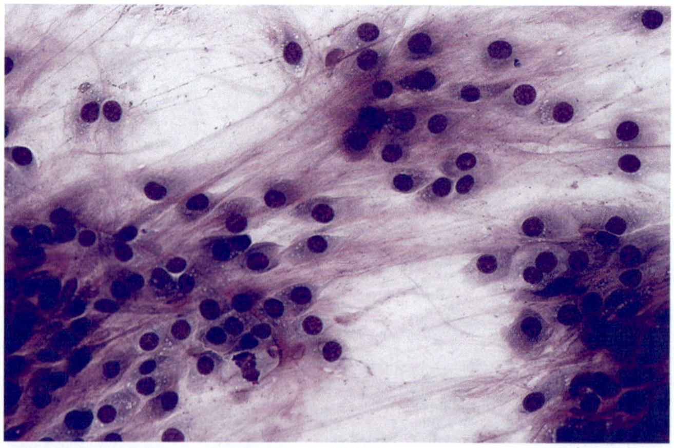

Fig. 5.37 Cytology of myxoid background from a PWT: spindle cells produced much amount of myxoid matrix

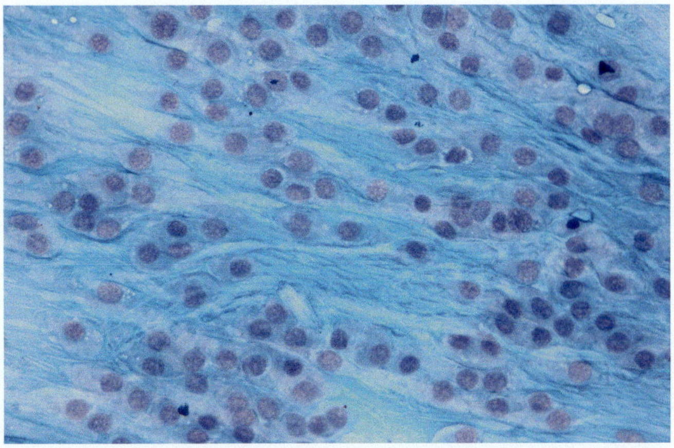

Fig. 5.38 Cytology of myxoid background from the same sample of Fig. 5.37: with alcian blue dye, the myxoid background takes on an azul color

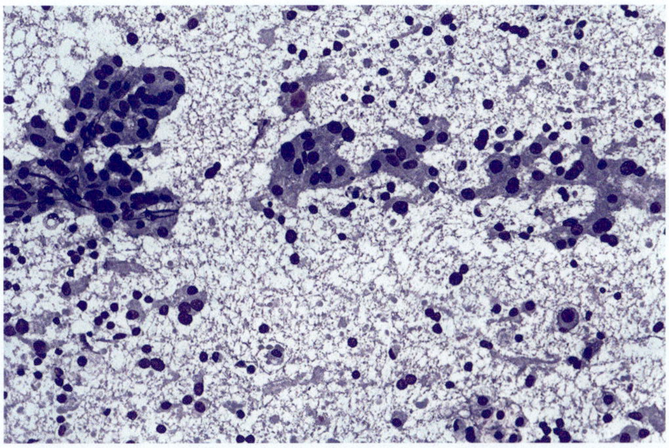

Fig. 5.39 Cytology of tigroid background from a clear cell adnexal carcinoma: a violet, reticulate to lacy in appearance material, characterizes this particular background

5.7 Tigroid Background

The background term "tigroid" is used in human cytology to describe a particular reticulated, granular, and sluggish appearance that the background material can assume [4]. This term was initially used in 1976 by Dr. Paul Lopes-Cardozo, one of the pioneers of the cytopathological science, to describe the background seen in preparations obtained from seminomas, cells with glycogen-rich cytoplasm that tend to rupture during preparation, releasing the fragmented cytoplasmic contents onto the slide; glycogen seems the common denominator, since, as demonstrated by electron microscopy, most tumors showing tigroid background contain large amounts of glycogen. In practice, the background is the content of the cytoplasm of broken cells.

In canine and feline skin cytology, this background is extremely rare and is observed almost exclusively in a rare carcinomatous neoplasm of dogs characterized by "clear" cells on histopathological samples, the so-called *Clear cell adnexal carcinoma.*

The background, visible only if colored with Romanowsky-type dyes, takes on a purplish color and is interposed between the cells with woven material with a reticular and lacy appearance which recalls the stripes present on the tiger's coat (tigroid appearance) or the appearance of the coat of the Astrakhan sheep breed (Astrakhan lambskin appearance). Since the clear cells of neoplasms have, as seminoma cells, a marked cytoplasmic fragility, slides show many "bare" nuclei (Figs. 5.39 and 5.40). As the cytoplasm material is rich in glycogen, the tigroid background is strongly positive for P.A.S. staining, taking a magenta color (Fig. 5.41).

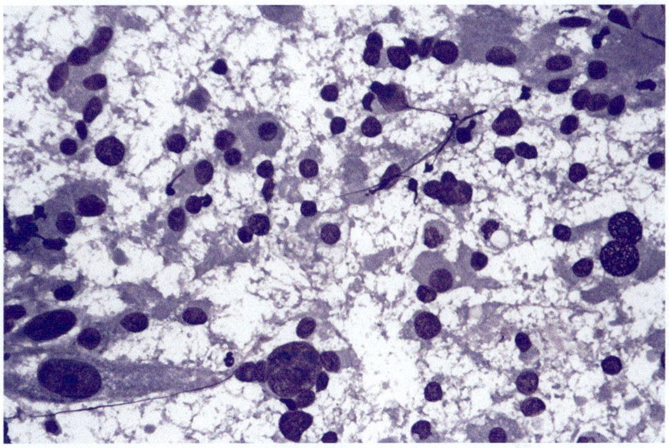

Fig. 5.40 Cytology of tigroid background from a clear cell adnexal carcinoma: many atypical neoplastic cells embedded in a tigroid background

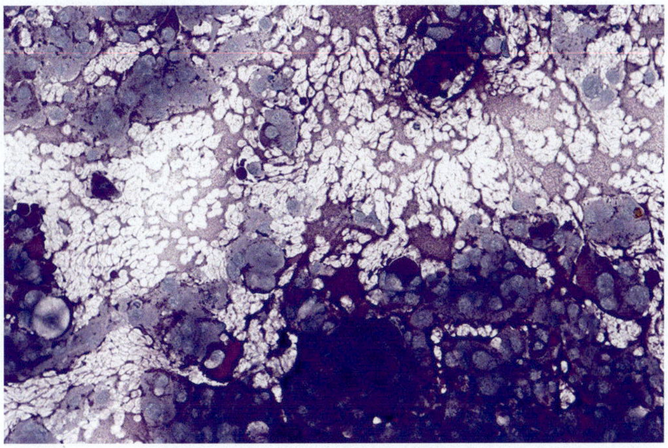

Fig. 5.41 Cytology of tigroid background from a clear cell adnexal carcinoma: P.A.S. staining highlights positive material both into the cytoplasm of neoplastic cells and on the bottom of the slides

Further indication regarding the cytological features of clear cell adnexal carcinoma in dogs is reported in Chap. 8.

References

1. Dennis MM, McSporran KD, Bacon NJ, Schulman FY, Foster RA, Powers BE (2011) Prognostic factors for cutaneous and subcutaneous soft tissue sarcomas in dogs. Vet Pathol 48(1):73–84
2. Seo YJ, Shin H, Lee HW et al (2021) Causes of necrotic features in fine-needle aspirates from cervical lymph nodes. J Pathol Transl Med 55(1):60–67. https://doi.org/10.4132/jptm.2020.09.28. Epub 2020 Nov 27. Necrosis
3. Wong NL, DI F. (2007) Fine needle aspiration cytology of myxoid lesions of soft tissues: a study of 24 cases. Zhonghua Bing Li Xue Za Zhi 36(9):619–623. Chinese
4. Jiménez-Heffernan JA, Rodríguez-García AM, Cima L et al (2022) A comprehensive review of the "tigroid" background cytological concept: what, when, where and why? Pathologica 114(2):121–127

6 Cytology of Canine and Feline Non-Neoplastic Skin Diseases

6.1 Introduction

Unlike cutaneous neoplasms, which manifest as nodular lesions in most cases, non-neoplastic inflammatory skin diseases are characterized by a wide clinical polymorphism.

Although the following classification is not useful for diagnostic purposes, skin lesions are usually classified as *primary*, which represent the direct damage of the disease on the skin, and *secondary*, that are the evolution of lesions over time or the outcome of self-trauma. Skin lesions that are observed on a dermatological patient are numerous and morphologically very different, and although the macroscopic appearance makes them easily recognizable, cytology is the fastest technique that allows us to interpret them. Unfortunately, not always their nature can be established through cytology, and other diagnostic techniques, such as histopathology, are necessary in some cases. In this chapter, only the skin lesions in which a cytological investigation may be diagnostic or useful for diagnosis are discussed. The author believes that, to better interpret the cutaneous inflammatory cytology, it is mandatory to be able to distinguish the clinical appearance of the lesions from which the cells have been sampled, and for this reason, he prefers to discuss the cytological findings starting from clinical signs and not from the cytological pattern. This more practical approach allows veterinarians, even those less confident with dermatology, to interpret cytological results in an easier and more rational way, and to gain from them all the information needed for their interpretation.

6.2 Papules

As mentioned in Chap. 1, it is not easy to obtain a diagnosis through cytological examination from *papules*, as they usually result in poor cellular specimens.

Papules with an intact surface are too small for fine-needle biopsy (FNB) and cells cannot be collected by the imprint technique; therefore, only crusted papules or those in which exudate can discharge from its top permit the collection of a few cells, which, even if they are not diagnostic, may guide the clinician in the choice of further diagnostic tests.

6.2.1 Papular Diseases in Dogs

6.2.1.1 Pyoderma

In dogs, *papules* are frequently observed during *superficial staphylococcal pyoderma* [1, 2]. In this disease, papular lesions are very often associated with pustules and epidermal collarettes, which help the clinician to clinically suspect the disease (Fig. 6.1). Pustules must always be preferred to papules for sampling cells, but when papules are the only lesions available, an attempt to collect cells via the imprint technique must be done.

Cytological Findings

Slides from bacterial papules are usually poor cellular and hemocontaminated and show few from segmented to karyolitic neutrophils. In more lucky slides, a variable number of cocci phagocytized by neutrophils can be detected. Since superficial crusts that cover the papules usually entrap large groups of coccoid bacteria, their presence must not be interpreted as pathogenic, but simply indicative of environmental contamination, especially if their number is more numerous than inflammatory cells and if they are organized in large extracellular groups (Figs. 6.2, 6.3, and 6.4).

6.2.1.2 Scabies and Fleabite Allergic Dermatitis

Papules are very common in both parasitic and hypersensitivity diseases in dogs. Scabies and fleabite allergic dermatitis (FAD) are characterized by papules and small crusted papules. In these cases, cytology does not permit making a

F. Albanese, *Canine and Feline Skin Cytology*, https://doi.org/10.1007/978-3-031-95713-0_6

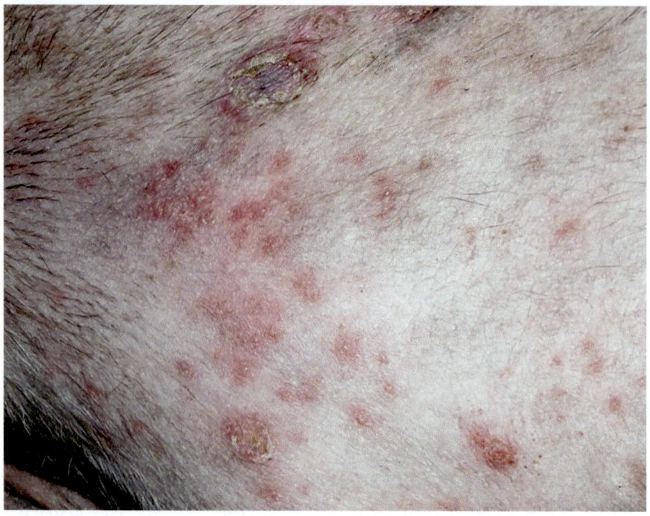

Fig. 6.1 Papules, pustules, and epidermal collarettes on the abdomen of a dog with superficial pyoderma

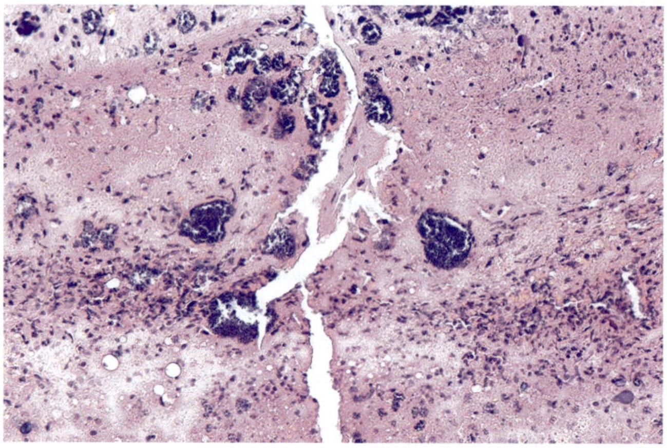

Fig. 6.2 Histology of superficial crust: many groups of contaminant bacteria are present in the contest of sero-hematic crusts

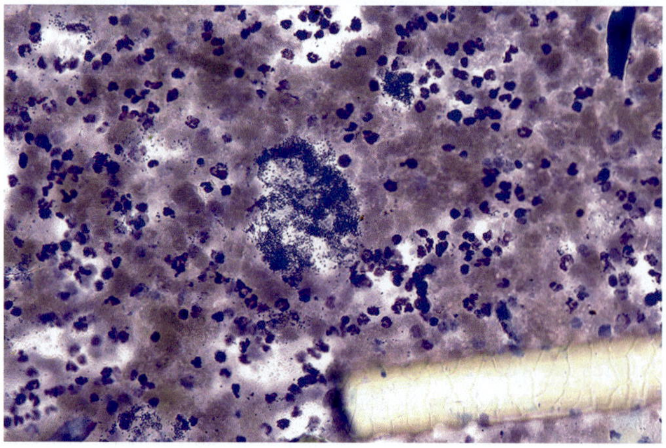

Fig. 6.3 Cytology of contaminant bacteria: hyperchromatic neutrophils and many bacteria, most of them singular or in extracellular large groups. A hair shaft fragment is also present

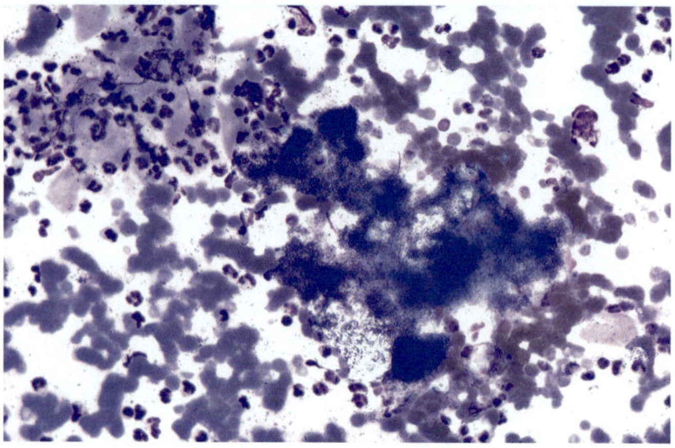

Fig. 6.4 Cytology of contaminant bacteria: many extracellular, both singular and in groups, must not be considered as pathogens

diagnosis; in fact, also if many eosinophils are detected on slides, the latter is a cytological finding common to all hypersensitivity diseases. The history of severe pruritus and the localization of the lesions on the body can help the clinician to suspect the two diseases. To make a diagnosis, multiple skin scrapings must be performed for scabies, while the detection of flea or their stools on the skin confirms a FAD.

6.2.2 Papular Diseases in Cats

Papular lesions in cats are observed in numerous diseases, and when they are spread on wide areas of the skin, the clinical pattern is commonly named *miliary dermatitis*; this name originates since the papular lesions have the same size as millet seeds.

Miliary dermatitis is not pathognomonic of disease but represents an aspecific, but very common feline skin reaction, whose cause includes a large group of diseases such as allergic, parasitic, idiopathic, and neoplastic. All the non-neoplastic diseases characterized by miliary dermatitis share the presence of a dermal infiltrate mainly composed of eosinophils and mast cells. As for dogs, cytology of crusted papules very rarely permits reaching a diagnosis but may be helpful in confirming a clinical suspicion or addressing the clinicians in the selection of further tests to perform.

6.2.2.1 Hypersensitivity Diseases

The most common causes of miliary dermatitis in cats include *feline atopic syndrome*, *fleabite dermatitis,* and *food allergy* [3–8].

Skin lesions could be identical and, for this reason, do not allow clinical differentiation between them (Fig. 6.5).

Another feline skin disease characterized by papules, and less frequently by papular–nodular lesions, is *mosquito bite hypersensitivity* (MBH). The cause of this disease is an

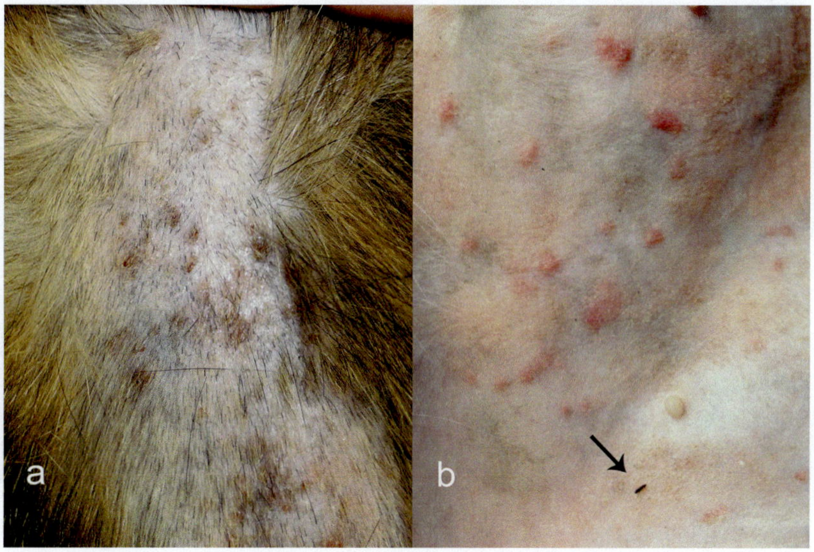

Fig. 6.5 Papules: (**a**) crusted papules on the back of a cat with miliary dermatitis and (**b**) erythematous papules in a cat with fleabite dermatitis: note the adult flea (*arrow*)

immune-mediated type I hypersensitivity reaction against antigens contained in the saliva of mosquitos. Clinically, it is characterized by crusted papules on the head and especially on the outer surface of the pinna, the bridge of the nose, and the nose and muzzle (Figs. 6.6 and 6.7) [6]. Less frequently, lips, eyelids, and paws (especially interdigital areas) are affected. Anecdotally, it seems that black-coated domestic shorthaired (DSH) cats are more predisposed to becoming hypersensitive, although the cause of this preference has never been investigated or scientifically demonstrated. Skin lesions are very pruritic and tend to ulcerate and change aspects due to self-trauma.

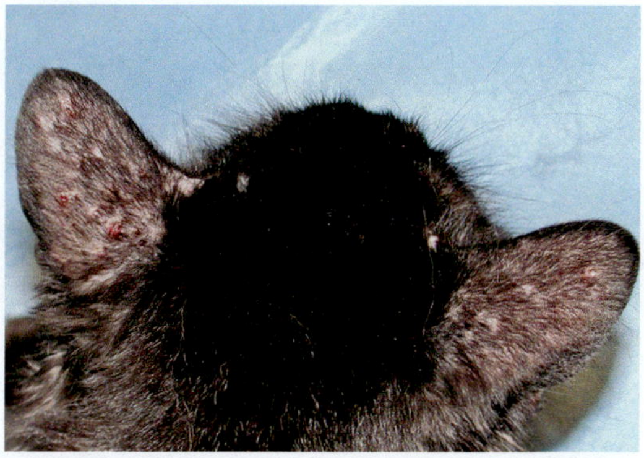

Fig. 6.6 Multiple ulcerated and crusted papules on the pinna of a black-coated cat with MBH

Cytological Findings

In all hypersensitivity diseases grossly presented with *miliary dermatitis*, cytological specimens show a variable number of eosinophils. In some cases, in small crusted-papules, the slides will be composed only of blood, with an acellular basophilic proteic background, few eosinophils intermingled with neutrophils, and a variable number of macrophages can also be observed (Figs. 6.8 and 6.9). In some ulcerated papules or when the latter are larger (papular–nodular in size), it is possible to obtain highly cellular specimens, rich in eosinophils, and a variable amount of macrophages (Figs. 6.10, 6.11, and 6.12).

In some cats, it is possible to observe a large number of *basophils*, distinguishable from eosinophils by their larger size, and especially for characteristic intracytoplasmic basophilic granules, which stain pale blue with Romanowsky-

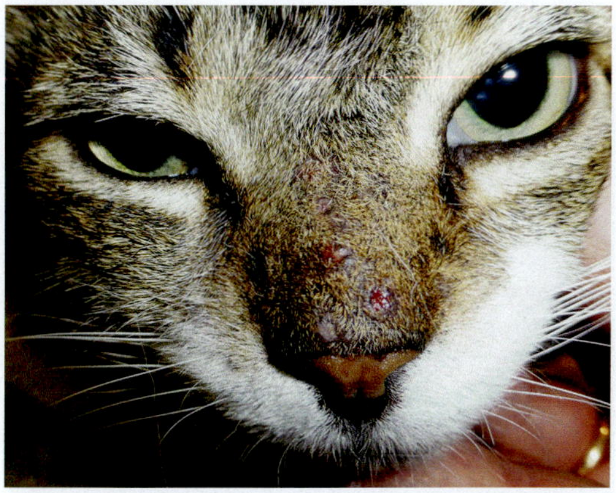

Fig. 6.7 Crusted papules on the nose of a cat with MBH

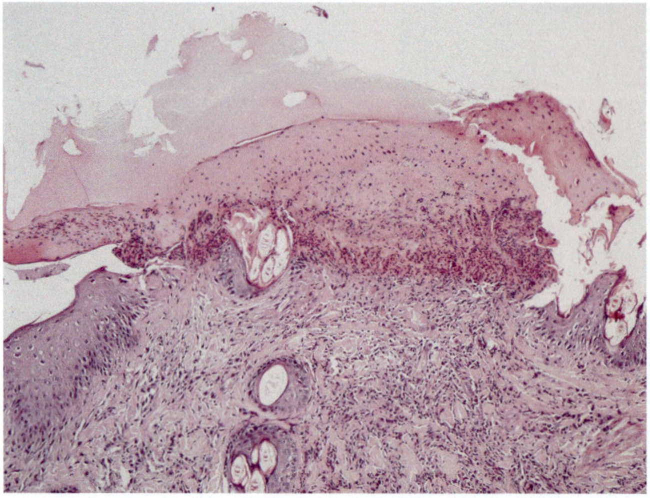

Fig. 6.8 Histopathology of a crusted papule: sero-neutrophilic crusts covering a small erythematous papule

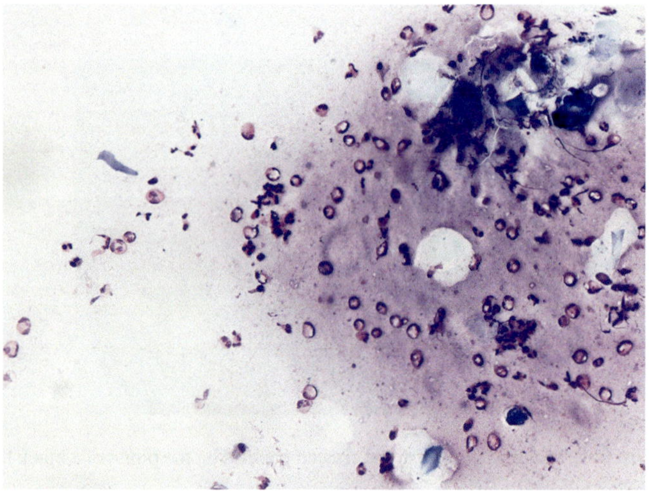

Fig. 6.9 Cytology of miliary dermatitis: many eosinophils and few corneocytes on a proteinaceous background

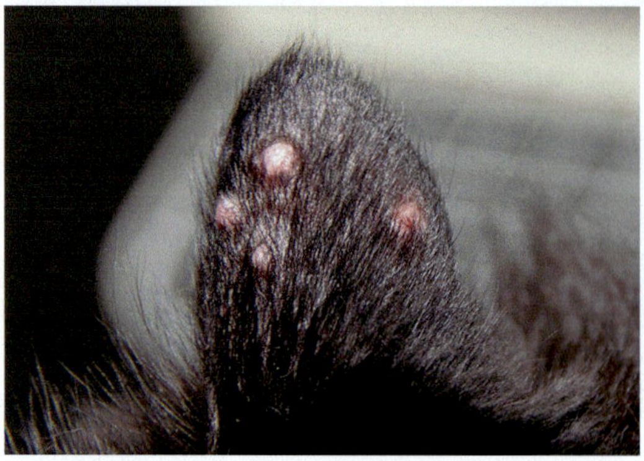

Fig. 6.10 Large papular lesions allow higher cellular samples to be collected

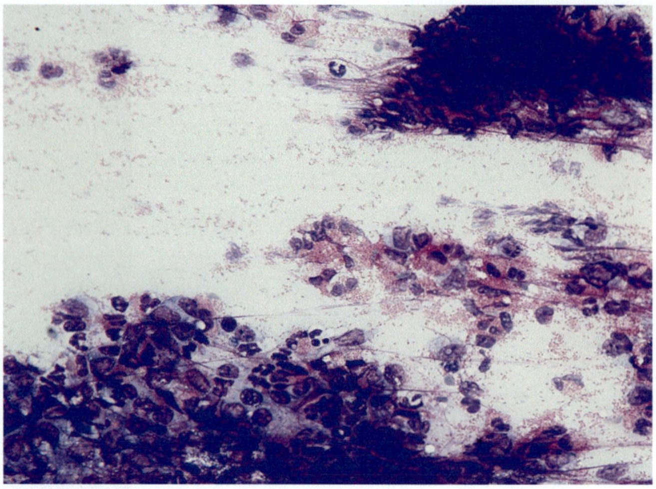

Fig. 6.11 Cytology of MBH: many eosinophils and some macrophages

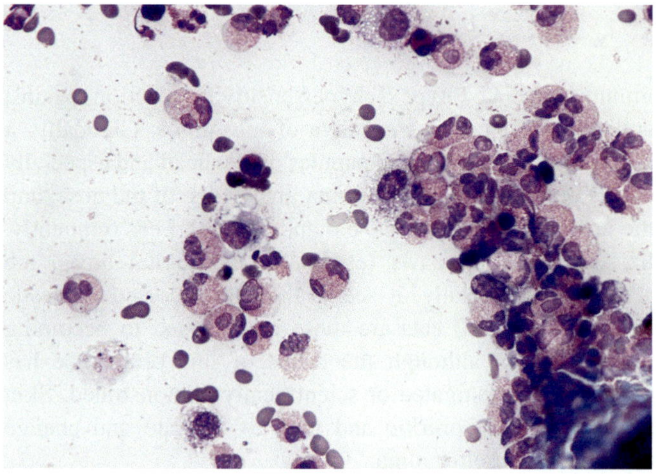

Fig. 6.12 Cytology of MBH: many eosinophils and few vacuolated macrophages

type dye (Figs. 6.13 and 6.14). In the author's experience, basophils are frequently observed in lesions due to MBH compared with other feline eosinophilic inflammatory diseases. Maybe, as demonstrated in men affected by MBH, basophils activated by mosquito bites are involved in the initiation and development of severe skin reactions [9].

6.2.3 Benign Papular Mastocytic Hyperplasia

In human medicine, *Urticaria pigmentosa* is a form of cutaneous mastocytosis. An idiopathic papular disease, named *benign papular mastocytic hyperplasia*, also known as *urticaria pigmentosa-like*, has been reported in cats [12, 13]. This disease has been reported in Sphynx and

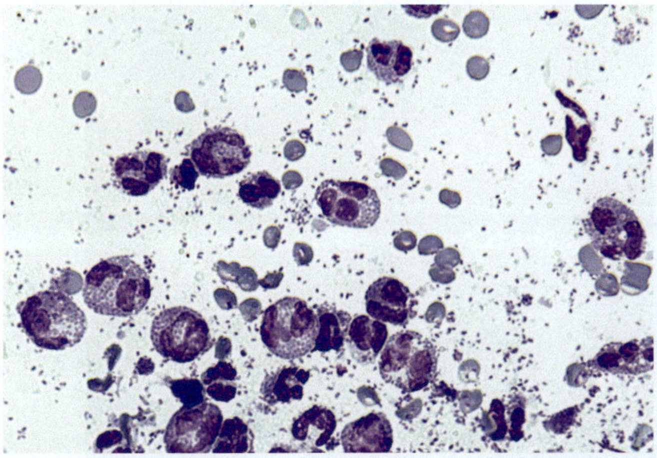

Fig. 6.13 Many basophils are recognizable for the gray to pale blue, oval to roundish granules

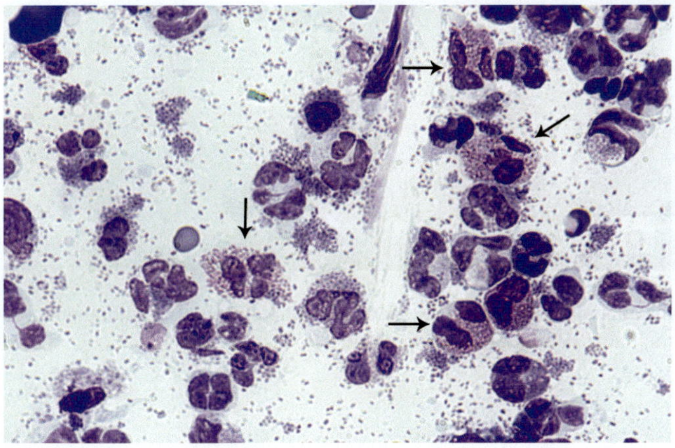

Fig. 6.14 Many basophils and eosinophils; the latter are easily differentiable from the orange color of the granules (*arrows*)

Devon rex cats as a clinical form of mastocytosis exclusively localized to the skin. The affected cats develop many crusted papules (miliary dermatitis appearance), which spread all over the body; in other cats, a linear configuration is quite common on the ventral region of the abdomen. In some patients, skin lesions result in small-flattened plaques with thickened skin (Fig. 6.15). The cause of this disease is not known, and as it is observed almost exclusively in the two above-mentioned feline breeds, it is assumed that there may be a genetic cause. The healing of the lesions following immunomodulatory therapy and the non-specificity of the histopathological pattern make it impossible to differentiate from a feline hypersensitivity disease, which must always be ruled out. Therefore, the clinical appearance of this feline dermatological disorder is varied, and in a recent article on 13 cats, a subclassification of the lesions has been proposed on the basis of the clinical aspects observed on the current human consensus of urticaria pigmentosa. The three clinical forms include: (1) *polymorphic maculopapular cutaneous mastocytosis* characterized by spontaneously healing and clinically by large papular lesions and wheals, observed in various areas of the body such as the head, shoulders, ventral neck and axilla; (2) *monomorphic maculopapular cutaneous mastocytosis* with poorer prognosis, in which erythematous dermatitis is observed, with small maculopapular lesions often associated with crusts; and (3) *pigmented maculopapular cutaneous mastocytosis*, more similar to the "urticaria pigmentosa" in man, which manifests as more chronic dermatitis characterized by lichenification and hyperpigmentation. Histopathology confirms the presence of mast cells and eosinophils without allowing differentiation between the various forms [11].

In rare cases, miliary dermatitis may be secondary to dermatophytic hypersensitivity. Some Devon rex cats affected by dermatophytosis and with clinical and histopathological lesions resembling urticaria pigmentosa-like disease have been reported [10].

Cytological Findings

Histologically, the lesions are characterized by slight superficial to deep perivascular dermatitis with a significant number of mast cells and few eosinophils (Fig. 6.16). Cytology obtained using the imprint technique, after removal of the crust that covers the top of crusted papules, is hemodiluted and usually with few neutrophils. Rarely, in more successful slides, specimens can show some well-differentiated mast cells and some eosinophils, which do not authorize the diagnosis of *benign papular mastocytic hyperplasia*, but reinforce the suspicion of the disease, which must be always confirmed with histopathology (Figs. 6.17 and 6.18).

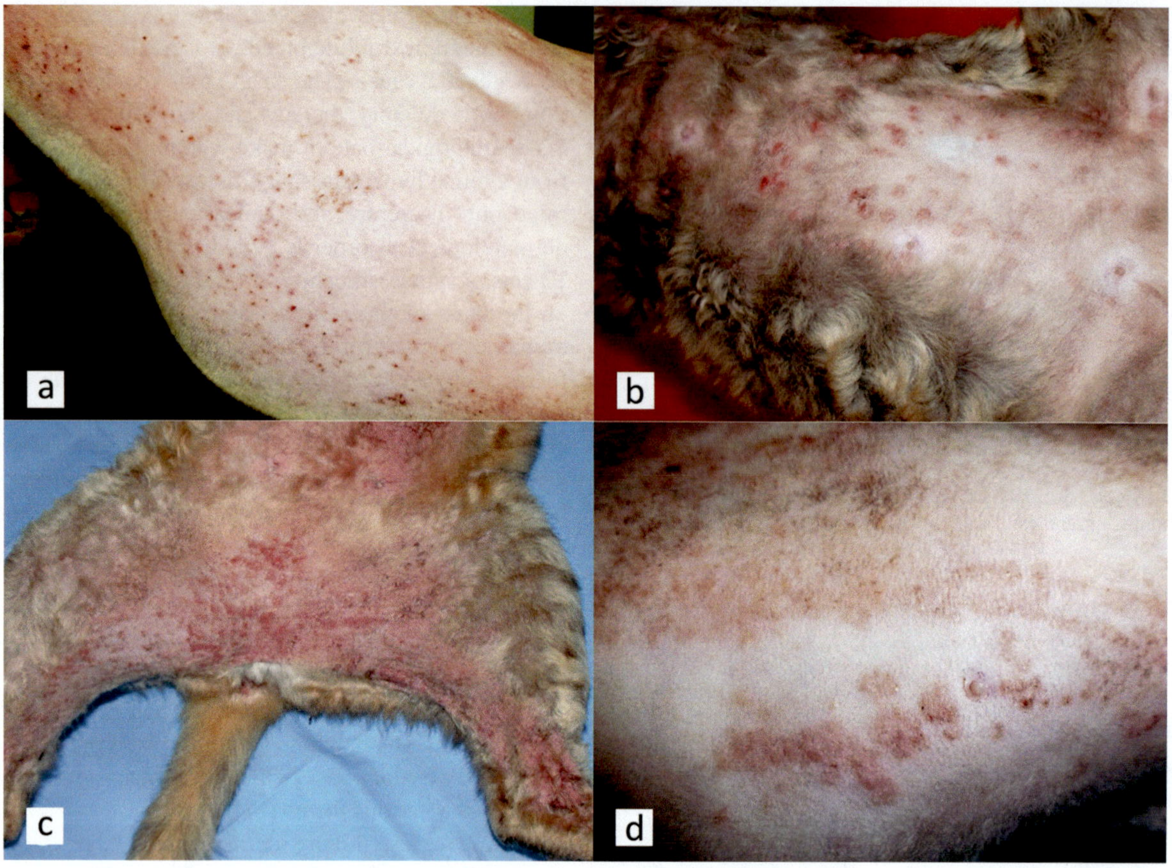

Fig. 6.15 Benign papular mastocytic hyperplasia: (**a**) miliary dermatitis on the abdomen of a Sphynx cat; (**b**) multiple papules on the abdomen of a Devon rex; (**c**) multiple papules and small plaques on the abdomen of a Devon rex; and (**d**) multiple papules on abdomen: note the linear configuration of the lesions

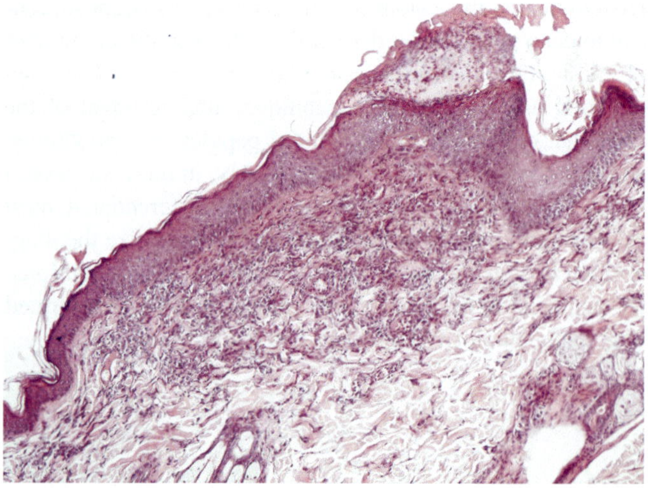

Fig. 6.16 Histopathology of benign papular mastocytic hyperplasia: superficial dermal infiltration of mast cells and eosinophils with mild hyperplasia of the epidermidis

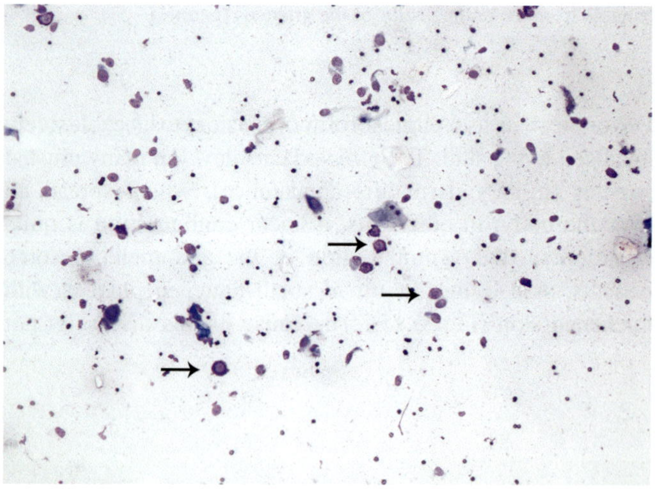

Fig. 6.17 Cytology of benign papular mastocytic hyperplasia: well-differentiated mast cells collected from papules (arrows)

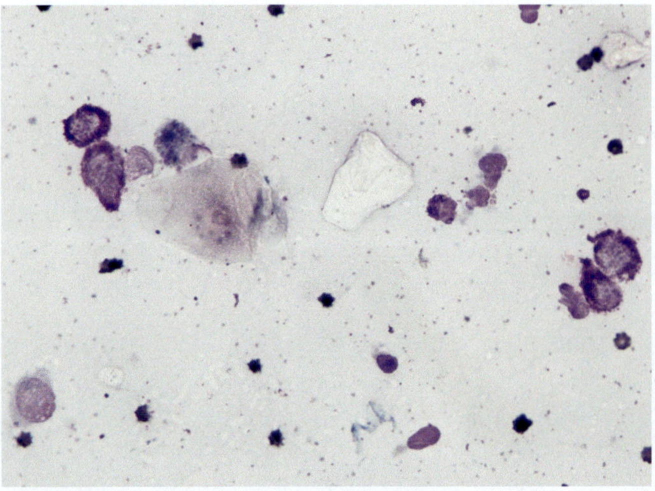

Fig. 6.18 Cytology of benign papular mastocytic hyperplasia: many well differentiated mast cells are clearly evident

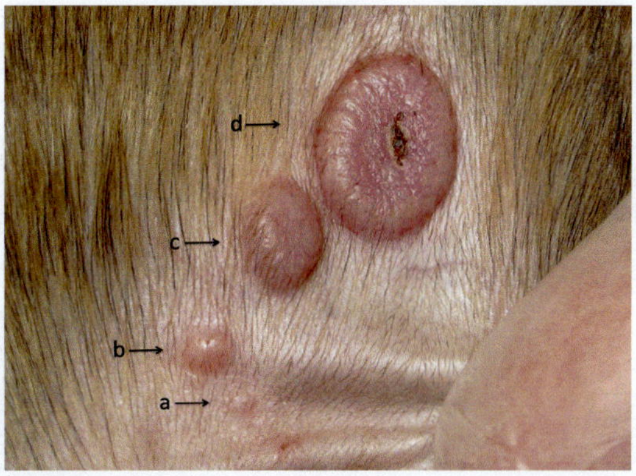

Fig. 6.19 Nomenclature based on size: (**a**) papule (few millimeters in size); (**b**) papular-nodule (less than 1 cm); (**c**) nodule (1 cm or more); and (**d**) plaque (1 cm or more that grows more in width than in height). (Courtesy of DVM F. Leone—Italy)

6.3 Nodular Papules

As described in Chap. 1, *nodular papules* are roundish lesions, of intermediate size between a papule and a nodule (<1 cm but no >1 cm) but do not appear large enough to be defined as nodules (Fig. 6.19). These types of lesions provide more diagnostic specimens if compared with papules because their size permits more cells to be collected.

6.3.1 Deep Pyoderma (Furunculosis)

Deep pyoderma is a common cause of papular–nodular lesions in dogs, with the latter that are the clinical expression of furunculosis. Histologically, the rupture of the hair follicles following bacterial staphylococcal folliculitis creates small dermal pyogranulomas focused on keratin material; the latter include corneocytes and hair shaft fragments, both coming from follicular lumen after the breakage of its wall, and secondarily spreading into the dermis (Figs. 6.20, 6.21, and 6.22) [16] *Deep pyoderma* can manifest with many different clinical presentations, depending on the chronicity of the infection and the body site involved; when inflammation is located in a single follicle or involves a few near follicles, it is clinically characterized by papular–nodular lesions. This type of lesion is visible until the individual dermal granulomas merge to give rise to widespread inflammation of the dermis and cause a change in the clinical morphology (plaques, draining tracts, etc.). So, in the early stage of furunculosis, lesions are well-defined papular nodules that are mostly detectable on the glabrous areas such as the muzzle, chins, legs, and between the interdigital areas, especially in shorthaired breeds (Fig. 6.23) [15]. Many

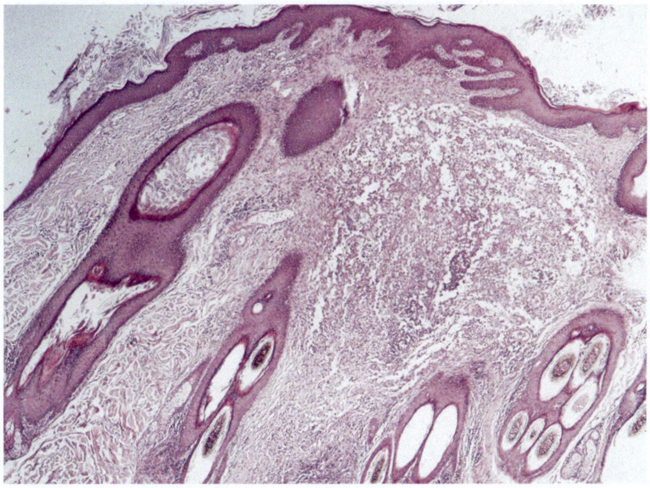

Fig. 6.20 Histopathology of deep pyoderma: large nodular pyogranulomatous inflammation replacing normal pilo-sebaceous units in a dog with deep pyoderma

nodular papules fistulate on the skin surface, where a purulent exudate is discharged; therefore, a large number of cells can be collected for cytopathological examination via FNB of intact lesions or via imprint smears on the exudate released from draining tracts.

Cytological Findings

Samples obtained by FNB are certainly of better quality and are usually composed of many inflammatory cells, mainly represented by numerous both karyolitic and segmented neutrophils and high amounts of macrophages. The latter are of variable size and morphology, from few activated to large and often vacuolated cytoplasm, and can show phagocytosis of leukocytes, cellular debris, amorphous material, and, sometimes, melanin pigment (melanophages) in animals with pigmented skin or with post-inflammatory

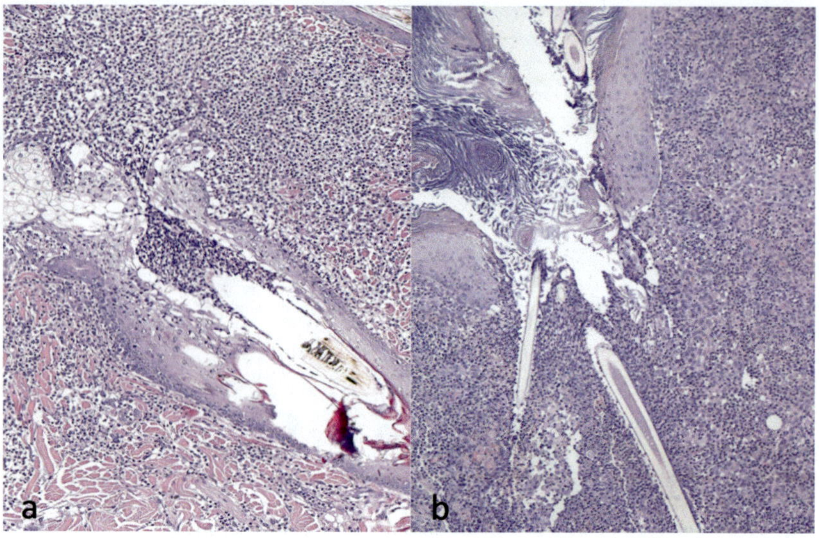

Fig. 6.21 Histopathology of deep pyoderma: (**a**) neutrophilic folliculitis with the early phase of follicular wall rupture and (**b**) hair shafts spread into the dermis following the disrupted follicular wall

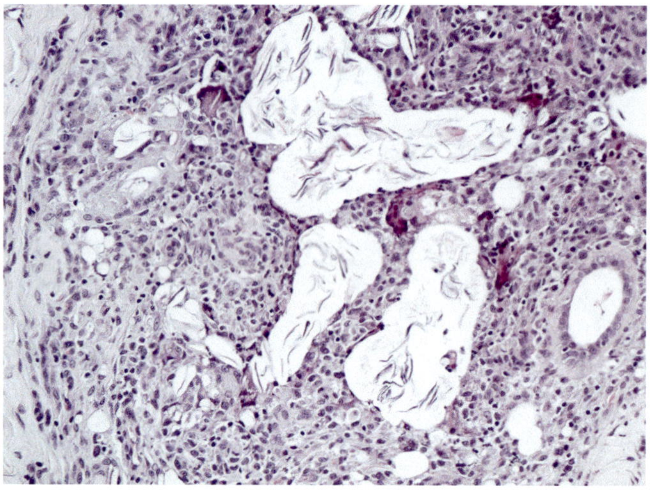

Fig. 6.22 Histopathology of deep pyoderma: granulomatous inflammation around lamellar corneocytes

hyperpigmentation (Fig. 6.24). These features are common to many other inflammatory skin diseases. In dogs with furunculosis secondary to bacterial folliculitis, keratin is spread in the dermis as lamellar corneocytes or as hair shaft fragments; because keratin is strongly antigenic and irritating, it attracts many inflammatory cells, but, as keratinocytes are very large cells, normal macrophages are not able to eliminate them. For this reason, macrophages tend to transform into so-called *epithelioid macrophages*, larger cells with wide cytoplasm that resemble epithelial cells (hence the term "epithelioid") and also give origin to *histiocytic multinucleate giant cells,* as discussed in Chap. 4. On slides, both cells are variably numerous and often arranged around single or rafts of corneocytes or hairs (Figs. 6.25, 6.26, 6.27, and 6.28).

We must be aware that these cytological features are not pathognomonic of deep pyoderma, since many other granulomatous sterile diseases can show identical cytological pictures.

For this reason, to make a cytological diagnosis of deep pyoderma, intracytoplasmic bacteria must be detected. Unfortunately, in most cases of deep pyoderma, bacteria are not easily observed, especially when the deep infection follows a folliculitis; a justification for this phenomenon could be that since the keratin acts as a true foreign body, the following severe pyogranulomatous reaction "dilutes" the number of bacteria, so they result less detectable (Figs. 6.29 and 6.30). This cytological aspect is very important since the failure to see bacteria in samples collected from deep pyogranulomatous lesions does not authorize us to rule out a pyoderma. In chronic lesions, cytological samples are enriched with a variable amount of lymphocytes, plasma cells, as well as fibroblasts due to a chronic irritating stimulation and a strong tissue-repairing activity (Figs. 6.31 and 6.32).

Similar clinical, cytological, and histopathological findings are observed in dogs with furunculosis secondary to demodicosis; indeed, in this parasitic disease, the pathogenetic mechanism of furunculosis is the same as deep pyoderma, and although cytological examination is not the best technique for diagnosing demodicosis, deep skin scrapings and trichoscopy are usually more performing, it is sometimes possible to collect through FNB from papular–nodular

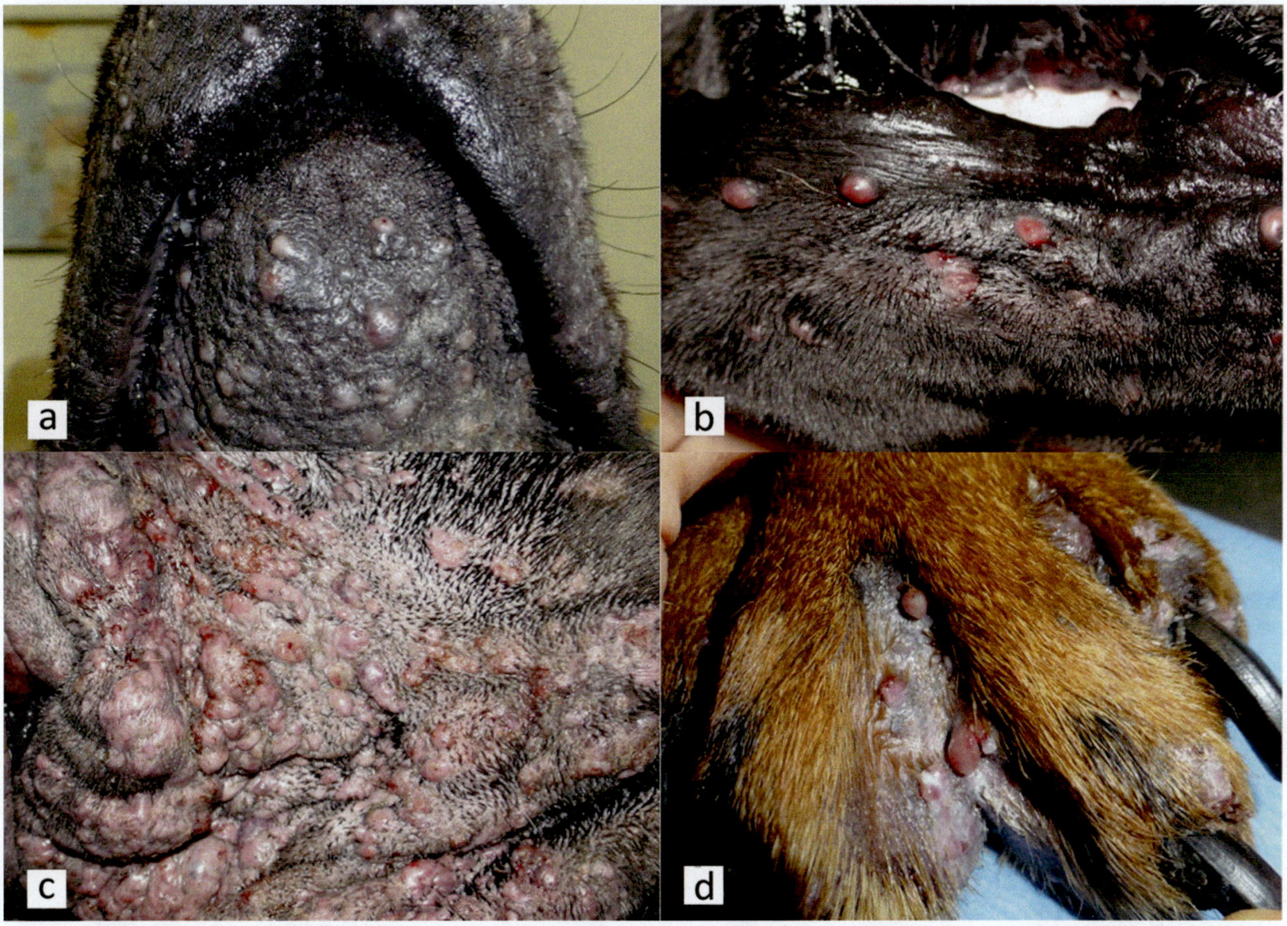

Fig. 6.23 Clinical aspects of deep pyoderma: (**a**) multiple nodular papules on the muzzle of a Doberman; (**b**) small alopecic nodule on the chin of a dog; (**c**) diffuse and confluent nodular papules on the muzzle and lips of a Neapolitan mastiff; and (**d**) papular–nodular lesions in the interdigital areas

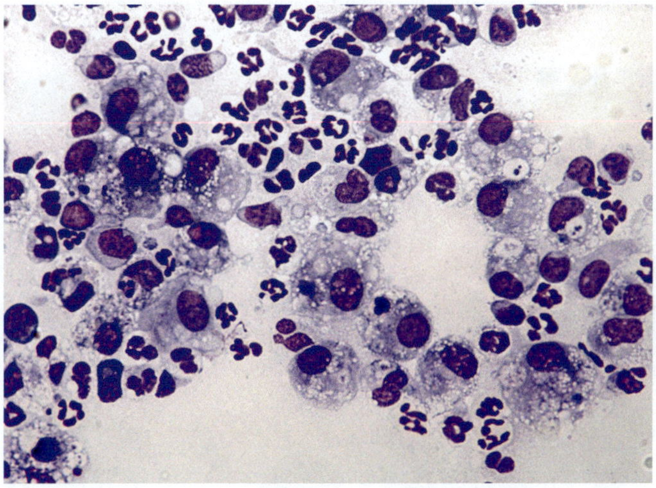

Fig. 6.24 Cytology of deep pyoderma: segmented neutrophils and many vacuolated macrophages showing leukophagocytosis

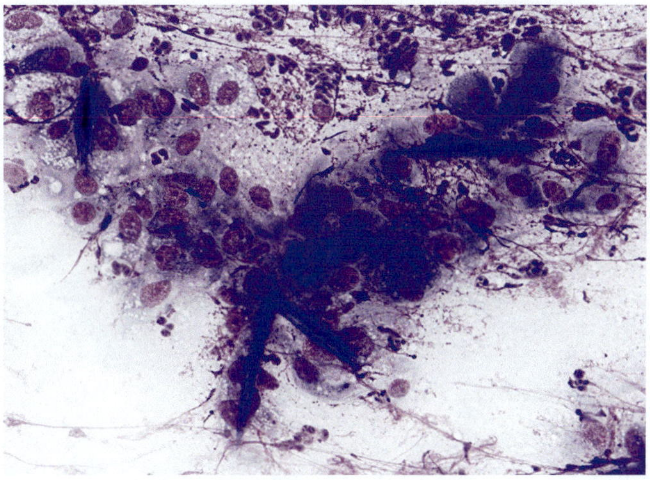

Fig. 6.25 Cytology of deep pyoderma: macrophages and many epithelioid macrophages surrounding numerous hair shaft fragments

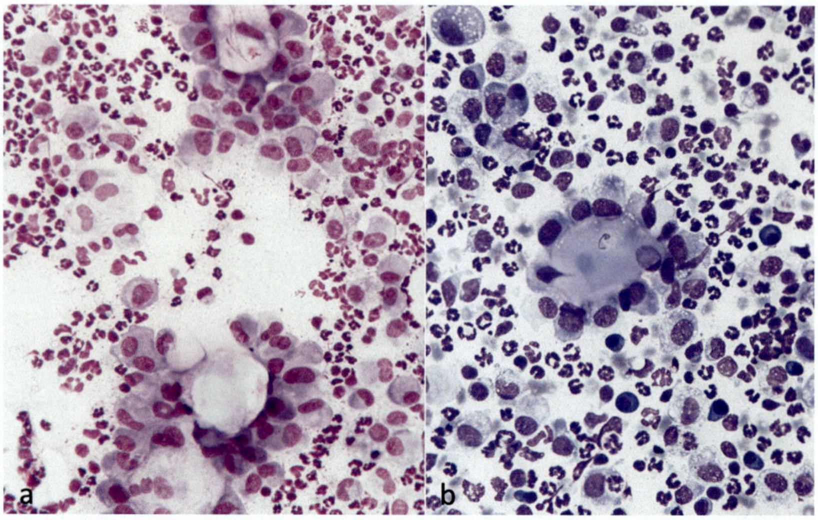

Fig. 6.26 Cytology of deep pyoderma: (**a**) many epithelioid macrophages try to eliminate large groups of corneocytes and (**b**) macrophages and epithelioid cells around corneocytes

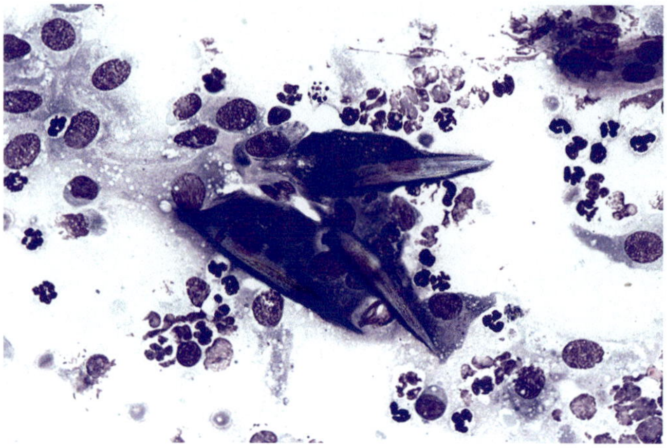

Fig. 6.27 Cytology of deep pyoderma: hair shaft fragments attacked by multinucleated macrophages

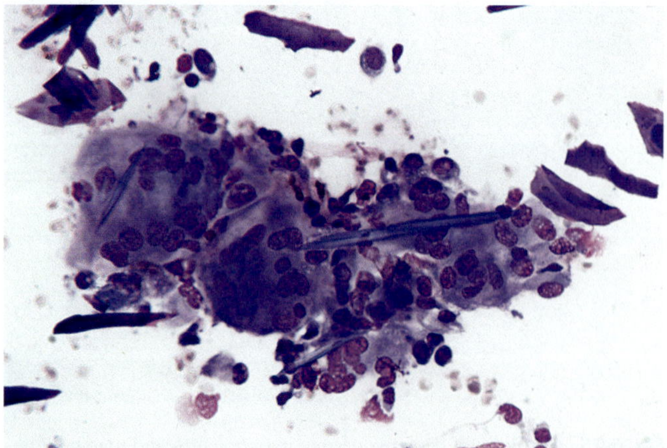

Fig. 6.28 Cytology of deep pyoderma: many giant cells in phagocytosis of hair shaft fragments

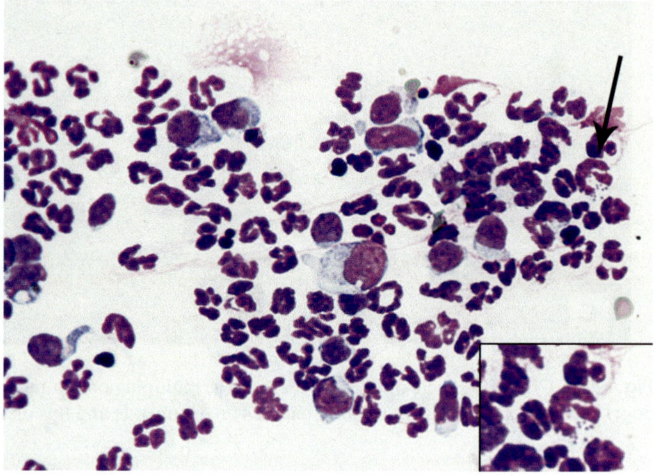

Fig. 6.29 Cytology of deep pyoderma: few macrophages and many karyolytic neutrophils with intracytoplasmic cocci (*arrow* and *inset*)

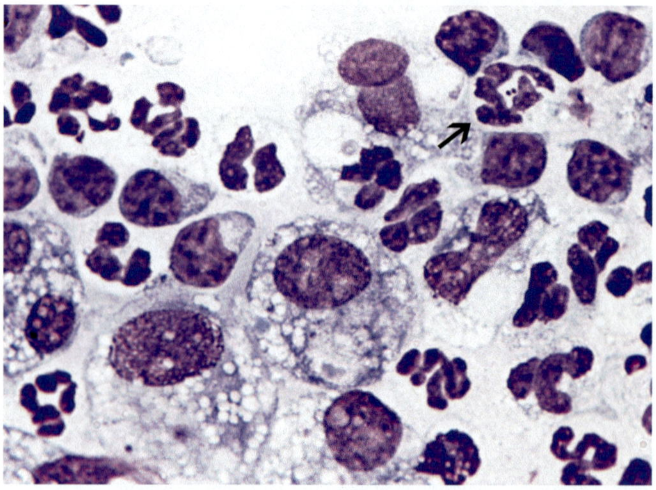

Fig. 6.30 Cytology of deep pyoderma: two *Staphylococcus pseudintermedius* into the cytoplasm of a neutrophil confirm a bacterial cause (arrow)

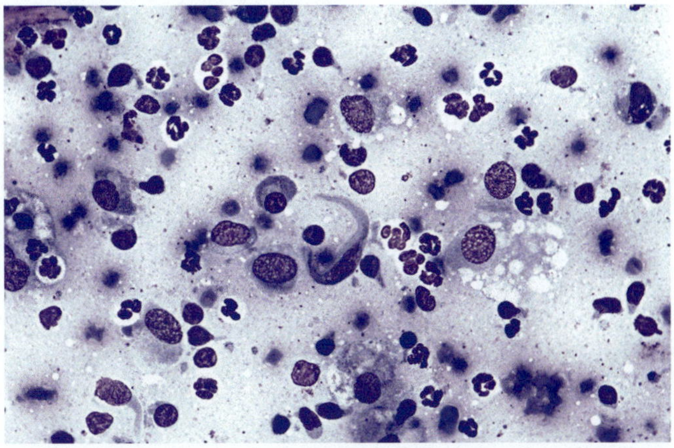

Fig. 6.31 Cytology of chronic deep pyoderma: single fibroblast along with macrophages, lymphocytes, and plasma cells

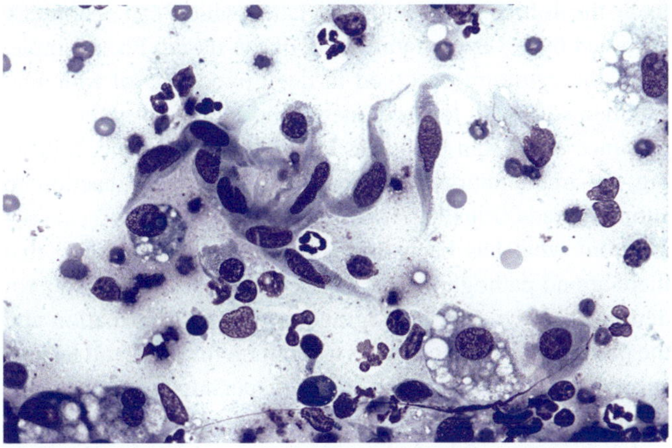

Fig. 6.32 Cytology of chronic deep pyoderma: many reactive fibroblasts are easily recognizable by their typical spindle-shape

lesions, few *Demodex canis* mites or their eggs along with a typical pyogranulomatous inflammation [14–18] (Figs. 6.33, 6.34, 6.35, 6.36, 6.37, and 6.38). The overgrowth of mites into the follicular lumen causes a secondary bacterial folliculitis, so the visualization of phagocytized cocci from neutrophils must be expected in cytological slides sampled from papular lesions of demodectic dogs (Fig. 6.39).

As aforementioned, the same cytological features are evident in many other non-infectious follicular diseases (sterile follicular conditions) in which the infundibular portion of the follicle is dilated and filled with lamellar keratin (comedones or small follicular cysts): for example, in miniature Schnauzer with comedo syndrome, in dogs with milia due to hypercortisolism or in dogs affected by solar dermatitis/

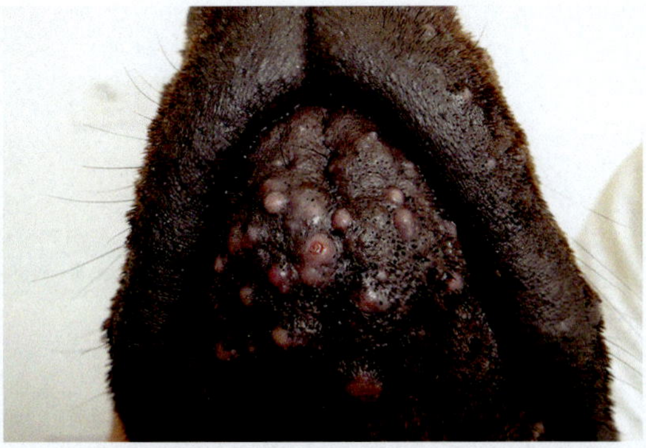

Fig. 6.33 Papular–nodular lesions with follicular plugs and comedones on the muzzle of a Dobermann with demodicosis

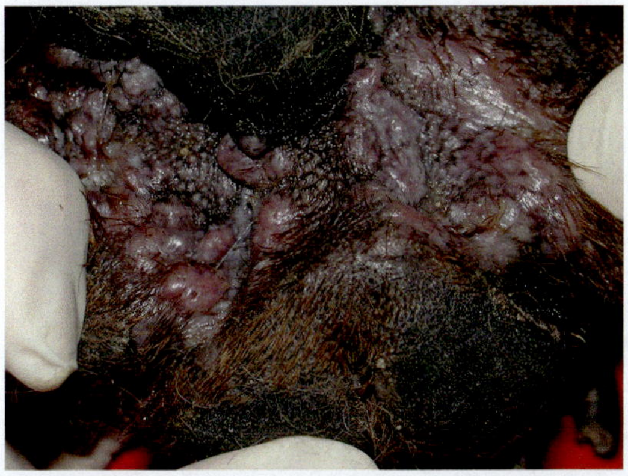

Fig. 6.34 Multiple confluent nodular papules on the interdigital area of a dog with demodicosis

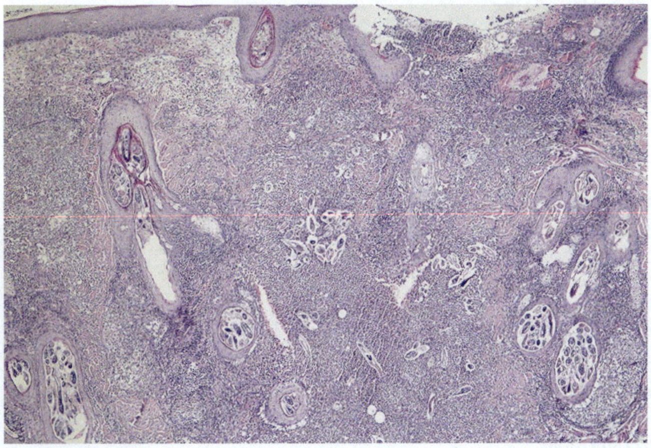

Fig. 6.35 Histopathology of demodicosis: many *Demodex canis* both intrafollicular and spread into the dermis causing severe pyogranulomatous inflammation

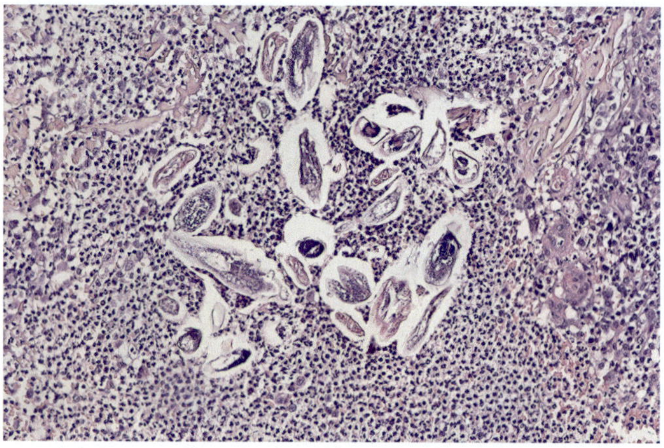

Fig. 6.36 Histopathology of demodicosis: closeup of the previous image. Many *Demodex canis* both intrafollicular and spread into the dermis, with the latter evoking severe pyogranulomatous inflammation

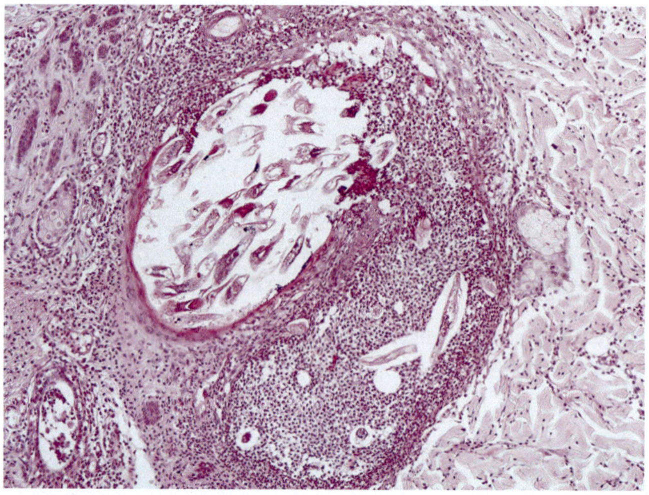

Fig. 6.37 Histopathology of demodicosis: note that many mites spread out from a broken follicle

actinic keratosis in which actinic comedones develop (Figs. 6.40 and 6.41). In practice, in any case in which the rupture of the wall of the follicles occurs and lamellar keratin spreads into the dermis. An important difference with cytological features of deep pyoderma could be that in comedones or micro-cystic lesions, since the breakage of the follicular wall occurs when the lumen is extremely dilated and filled with keratin lamellae, the latter are often found more numerous and organized in very large groups on slides, compared to those a follicle with folliculitis could produce, deep pyoderma; in these situations, cytological features resemble more those sampled from a small and inflamed infundibular cyst of the follicle (Figs. 6.42 and 6.43).

Finally, pictures of pyogranulomatous inflammation with epithelioid and giant cells, although with a scarce number of anucleate fully keratinized keratinocytes (corneocytes), can also be observed in a lesion that does not originate primarily from the follicles but with the latter which can secondary involved by severe inflammation of the tissue. For example, the same cytology observed in deep pyoderma of follicular origin, also if with fewer corneocytes, can be observed in the presence of foreign bodies as happens, for example, following the implantation of a vegetal fragment or reaction to suture threads (Figs. 6.44, 6.45, and 6.46).

From what has been discussed, we can summarize that there are no pathognomonic cytological aspects of deep pyoderma since the severe pyogranulomatous inflammation and the presence of corneocytes is a common finding in all lesions secondary to follicular rupture and dispersion of horny lamellae in the dermis. The clinical picture, the breed, the location and distribution of the lesions, and, in the absence of bacteria in the cytological examination, the bacterial culture of exudate will allow the diagnosis of deep pyoderma to be formulated.

Fig. 6.38 Cytology of papular demodicosis: (**a**) an adult and a larva of *Demodex* canis immersed into a pyogranulomatous inflammation and (**b**) at higher magnification, a pyogranulomatous inflammation with karyolitic neutrophils and many macrophages are clearly recognizable

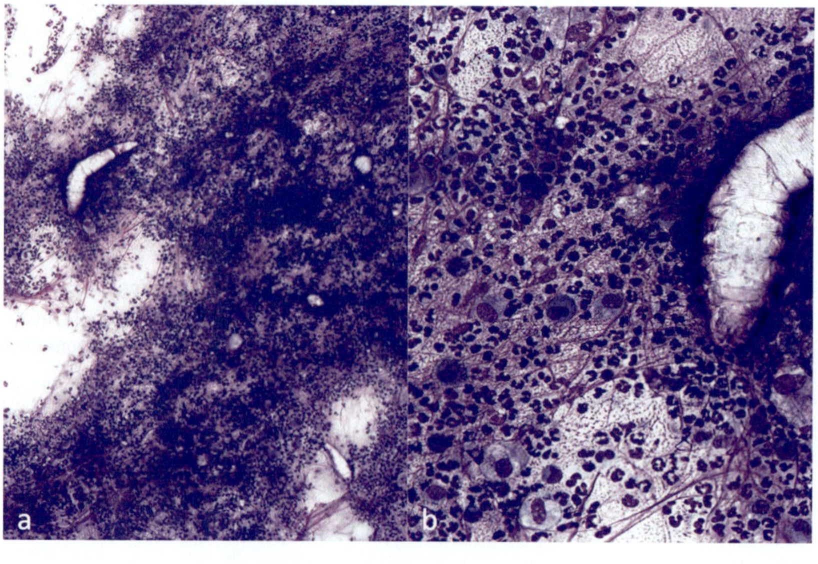

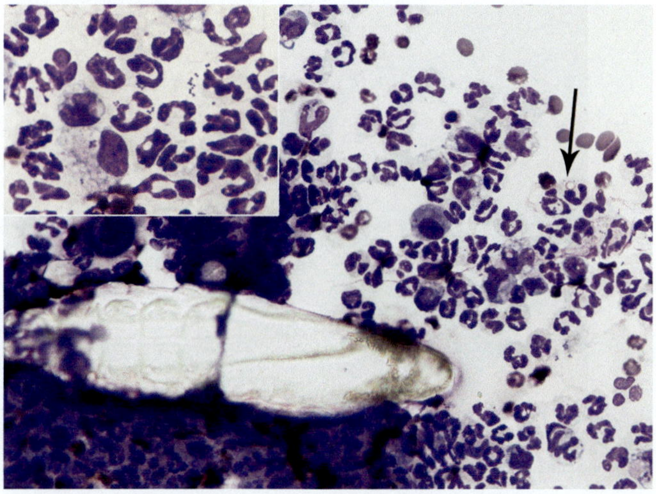

Fig. 6.39 Cytology of demodicosis: a *Demodex canis* mite surrounded by macrophages and neutrophils, the latter with many intracytoplasmic cocci (*arrow*); bacteria are well recognizable in the cytoplasm of neutrophils and into a macrophage (*inset*)

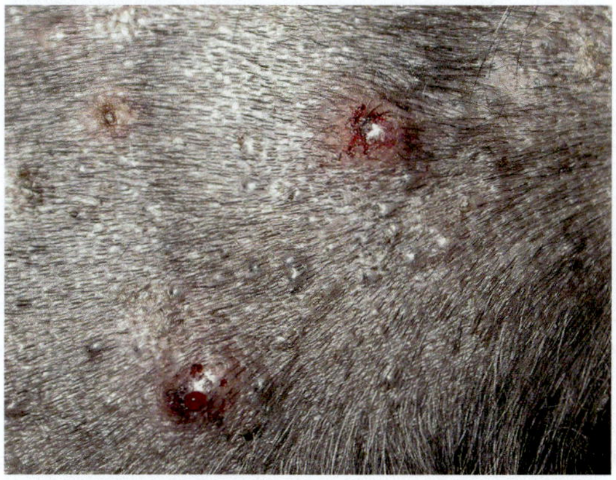

Fig. 6.40 Comedones an papular–nodular lesions on the trunk of a miniature Schnauzer affected by Comedo syndrome

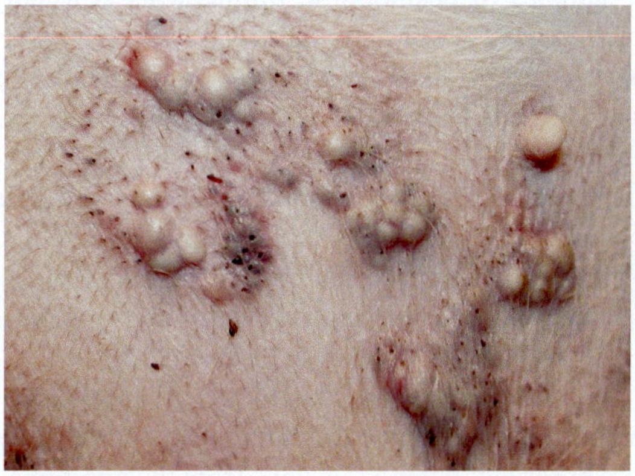

Fig. 6.41 Numerous milia (small infundibular cysts) and few comedones on the ventral chest of a dog with spontaneous hypercortisolism

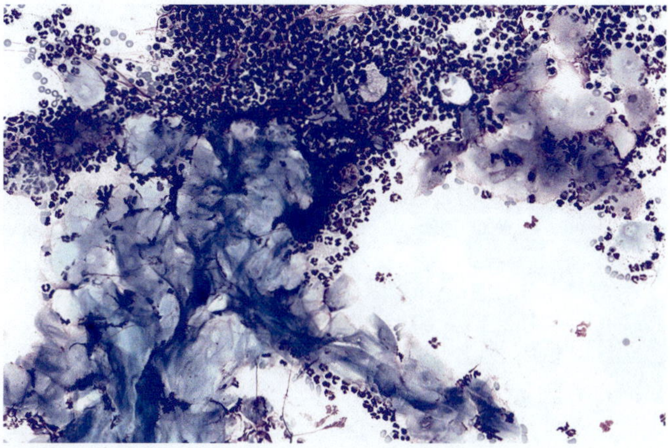

Fig. 6.42 Large amount of groups of corneocytes immersed in a pyogranulomatous inflammation

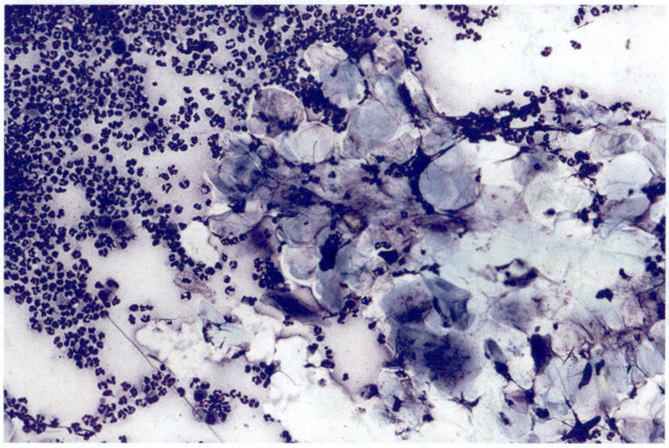

Fig. 6.43 Large amount of groups of corneocytes immersed in a pyogranulomatous inflammation

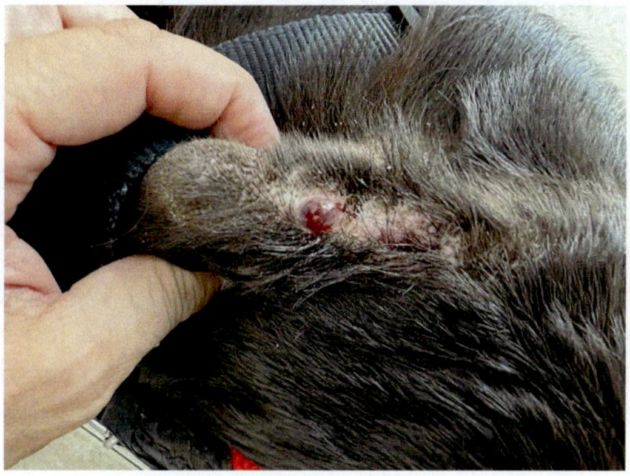

Fig. 6.44 Papular–nodular lesion following suture threads reaction

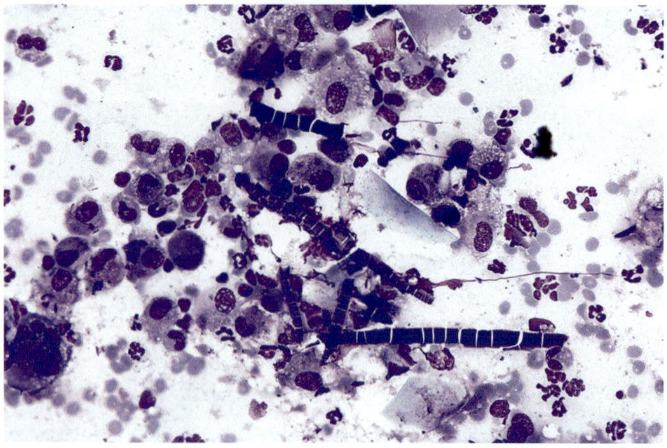

Fig. 6.45 Cytology of suture threads reaction: pyogranulomatous inflammation with few corneocytes. Note the presence of segmented deeply violet structure representative of suture threat

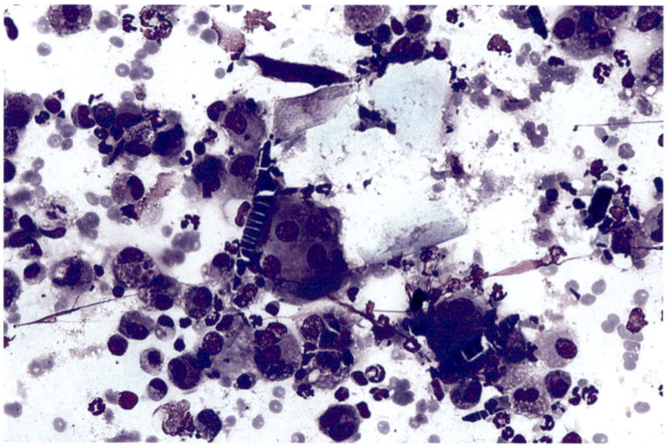

Fig. 6.46 Cytology of suture threads reaction: pyogranulomatous inflammation with few corneocytes and multinucleated giants macrophages in the attempt to eliminate the large foreign body (suture threat)

6.3.2 Facial Eosinophilic Furunculosis

Eosinophilic furunculosis is a hypersensitivity disease of dogs, characterized by an acute onset of papular–nodular lesions, mostly located on the face, lesions on the pinna, and, less frequently, on the neck, abdomen, and limbs, can also be associated. Early lesions are represented by nodular papules, mainly spread over the bridge of the nose and muzzle (Figs. 6.47 and 6.48) [19, 20, 22]. In a few hours, nodular papules tend to grow and merge, giving rise to plaques and nodules, which tend to ulcerate and become infected (Figs. 6.49 and 6.50). Eosinophilic furunculosis is usually very pruritic, and this is the reason why self-trauma can rapidly change the initial papular–nodular appearance (Fig. 6.51).

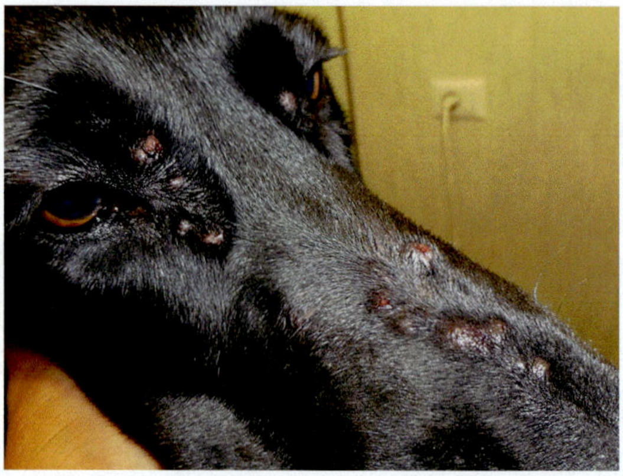

Fig. 6.47 Many small erythematous nodular papules on the bridge of the nose and eyelids of a dog with eosinophilic furunculosis

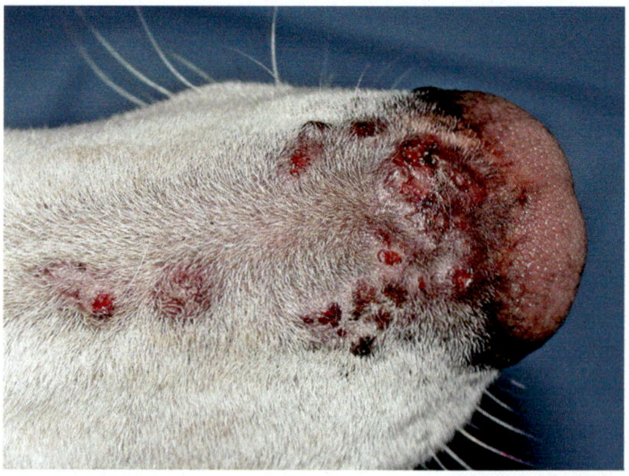

Fig. 6.48 Merging papular nodular lesions on the nose of a dog with eosinophilic furunculosis

Cytological Findings

Although histologically the target of the inflammatory process is the follicle, the number of lamellar keratin, epithelioid macrophages, and macrophagic giant cells observed in cytological samples of eosinophilic furunculosis is usually low compared to what is normally observed in deep pyoderma or other sterile furunculosis, in which an intrafollicular inflammation and keratosis is more commonly present (Figs. 6.52 and 6.53).

Samples of early and uninfected lesions are therefore characterized by a high number of eosinophils and macrophages and a variable number of neutrophils [21]. Epithelioid macrophages and multinucleated cells are present in variable quantities (Figs. 6.54, 6.55, and 6.56).

In the case of secondarily infected lesions, along with eosinophils, karyolitic neutrophils with intracytoplasmic cocci are observed. After a few days from the onset of clinical lesions, eosinophils can be very few or totally absent on cytological specimens, and lymphocytes and plasma cells increase in number.

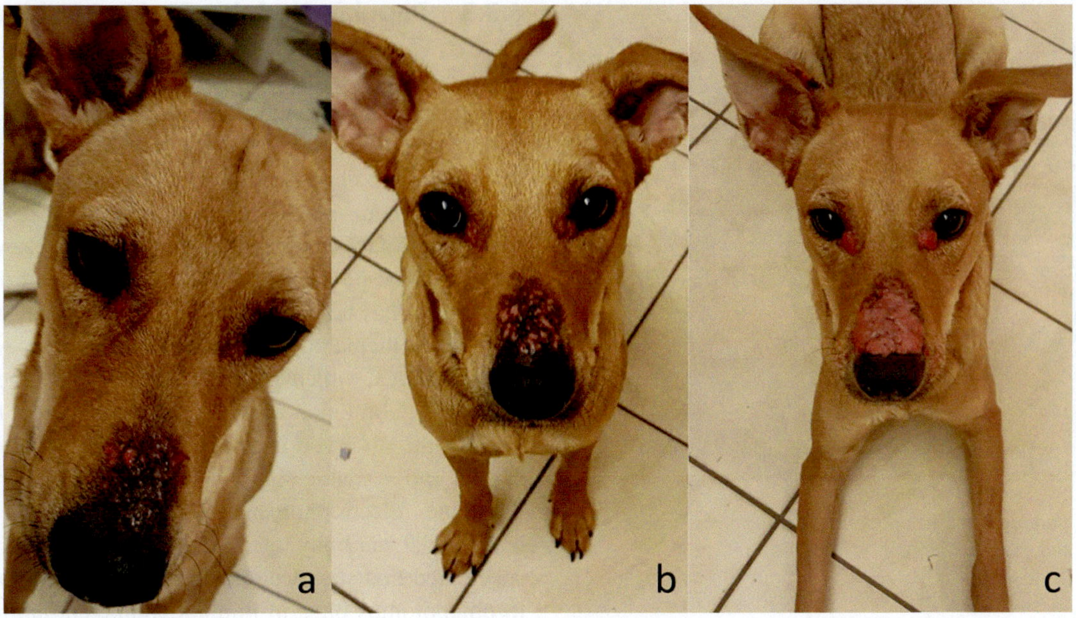

Fig. 6.49 Modification of morphology during the time course of eosinophilic furunculosis: (**a**) Day 1: onset of small papules; (**b**) Day 4: papules become small erythematous plaque; and (**c**) Day 7: large and merging ulcerated plaques (Courtesy of DVM P. Rosi-Italy)

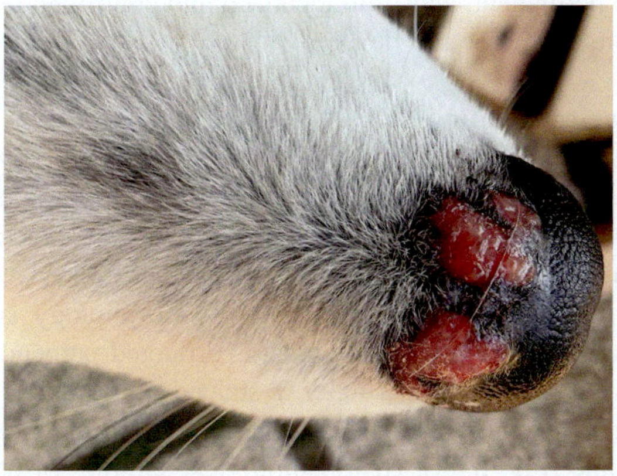

Fig. 6.50 Ulcerated plaques on the bridge of the nose of a dog with eosinophilic furunculosis

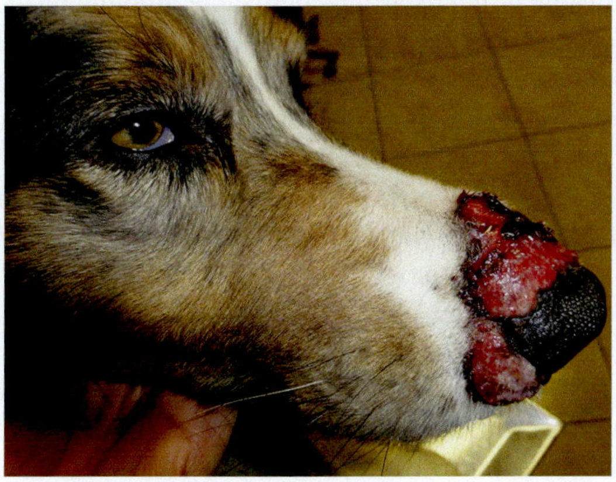

Fig. 6.51 Ulcerative lesions secondary to self-trauma in an Australian shepherd with eosinophilic furunculosis. Note that the nasal planum is spared since follicles are not present in that area

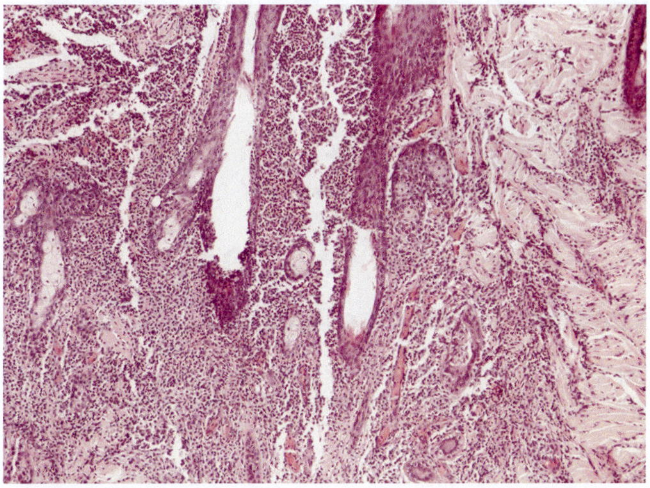

Fig. 6.52 Histopathology of eosinophilic furunculosis: note how the target of the inflammation is the wall of the follicles, which undergoes a total destruction

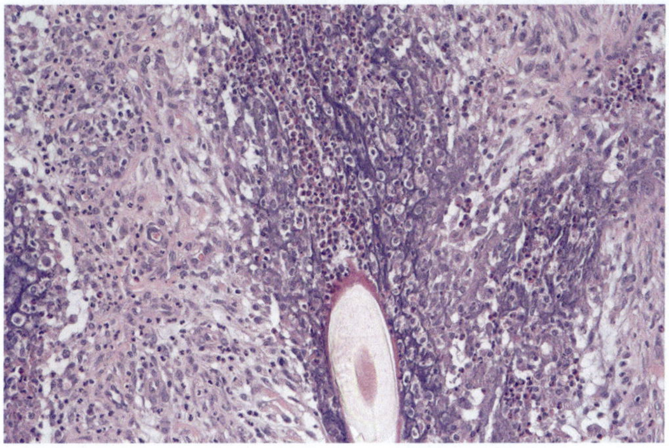

Fig. 6.53 Histopathology of eosinophilic furunculosis: note that the main typology of intramural cells is eosinophils

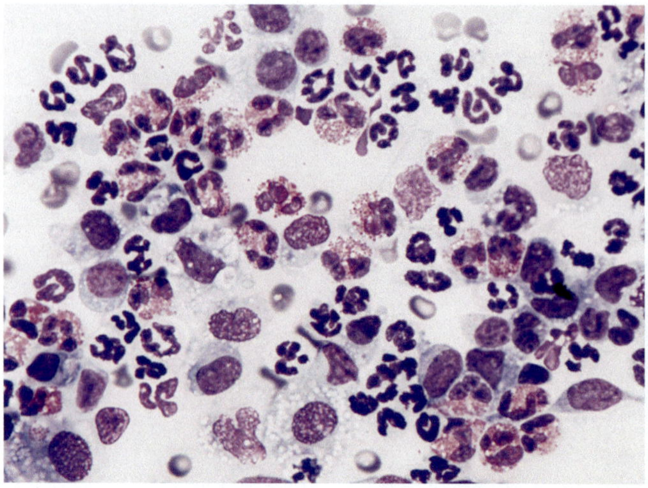

Fig. 6.54 Cytology of eosinophilic furunculosis: neutrophils, macrophages, and many eosinophils

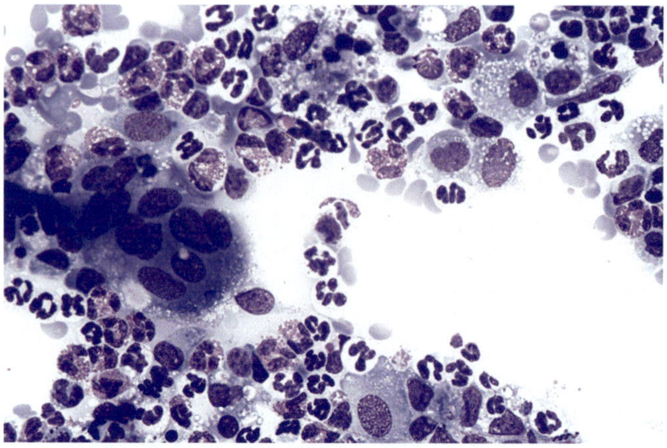

Fig. 6.55 Cytology of eosinophilic furunculosis: many eosinophils and macrophages. A multinucleated macrophage is also recognizable

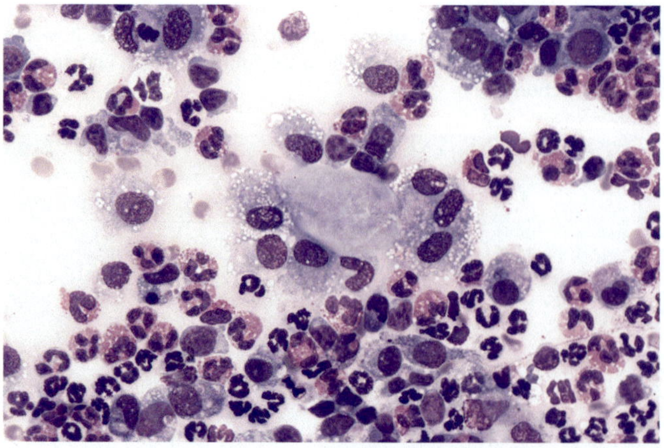

Fig. 6.56 Cytology of eosinophilic furunculosis: eosinophils and macrophages, the latter mainly localized around corneocytes coming from the broken follicles

6.3.3 Papular–Nodular Canine Leishmaniasis

Canine leishmaniasis (CanL) is a protozoan disease characterized by a wide clinical polymorphism, depending on the type of immune response developed by affected dogs. In the clinical classification of CanL, a form characterized only by *papular–nodular* lesions, mainly located on hairless areas of the face, muzzle, bridge of the nose, eyelids, inner surface of the pinna, and the glabrous areas of the abdomen and thighs, has been reported [22–26]. Lesions are alopecic, slightly erythematous, and very often with a small central ulcerated area, which gives the lesions a characteristic, but not pathognomonic, *umbilicated* appearance (Fig. 6.57). Lesions are observed in young dogs with no other dermatological and systemic signs, with normal blood values, unremarkable electrophoresis, and negative or low anti-leishmanial antibody titers. For these aspects, it has been speculated that such lesions might represent a local reaction to phlebotomus bites in immunocompetent animals; in fact, in most of these patients, spontaneous regression of the lesions occurs [24, 27, 28].

Nevertheless, the immune response in these patients could not be permanent, and from localized and controlled lesions, a systemic disease can develop.

Cytological Findings

Leishmania infantum amastigotes are recognizable by their characteristic oval shape, measuring approximately 1–2 × 2–5 μm, containing a nucleus and a characteristic rod-shaped hyperchromatic structure, that represents the *kinetoplast*. Leishmania's amastigotes morphology is easy to recognize when the kinetoplast is arranged perpendicular to the nucleus, giving a letter T-shaped appearance; when overlapped with a nucleus, the kinetoplast is not easily recognizable (Fig. 6.58). Histopathological samples of papular leishmaniasis reveal a nodular to diffuse lympho-histio-plasmacellular dermatitis; in most cases, a variable number of protozoa can be detected and in case they are very few and not viewable or not clearly recognizable, immunohistochemical staining for *Leishmania* can be performed on paraffined material (Figs. 6.59, 6.60, and 6.61) [24, 28].

In most cases, since clinical lesions are small, cytological slides are of scarce cellularity, and only scattered macrophages, lymphocytes, and plasma cells can be collected with a variable amount of histiocytes represented by bare nuclei. In lucky cytological specimens, a mixed population of inflammatory cells, mainly composed of histiocytes with medium/large pale blue and not vacuolated cytoplasm, along with a high number of both small and intermediate lymphocytes and, a variable amount, usually high, of plasma cells is constantly observed (lympho-histio-plasmacellular inflammatory pattern); the visualization of amastigotes in

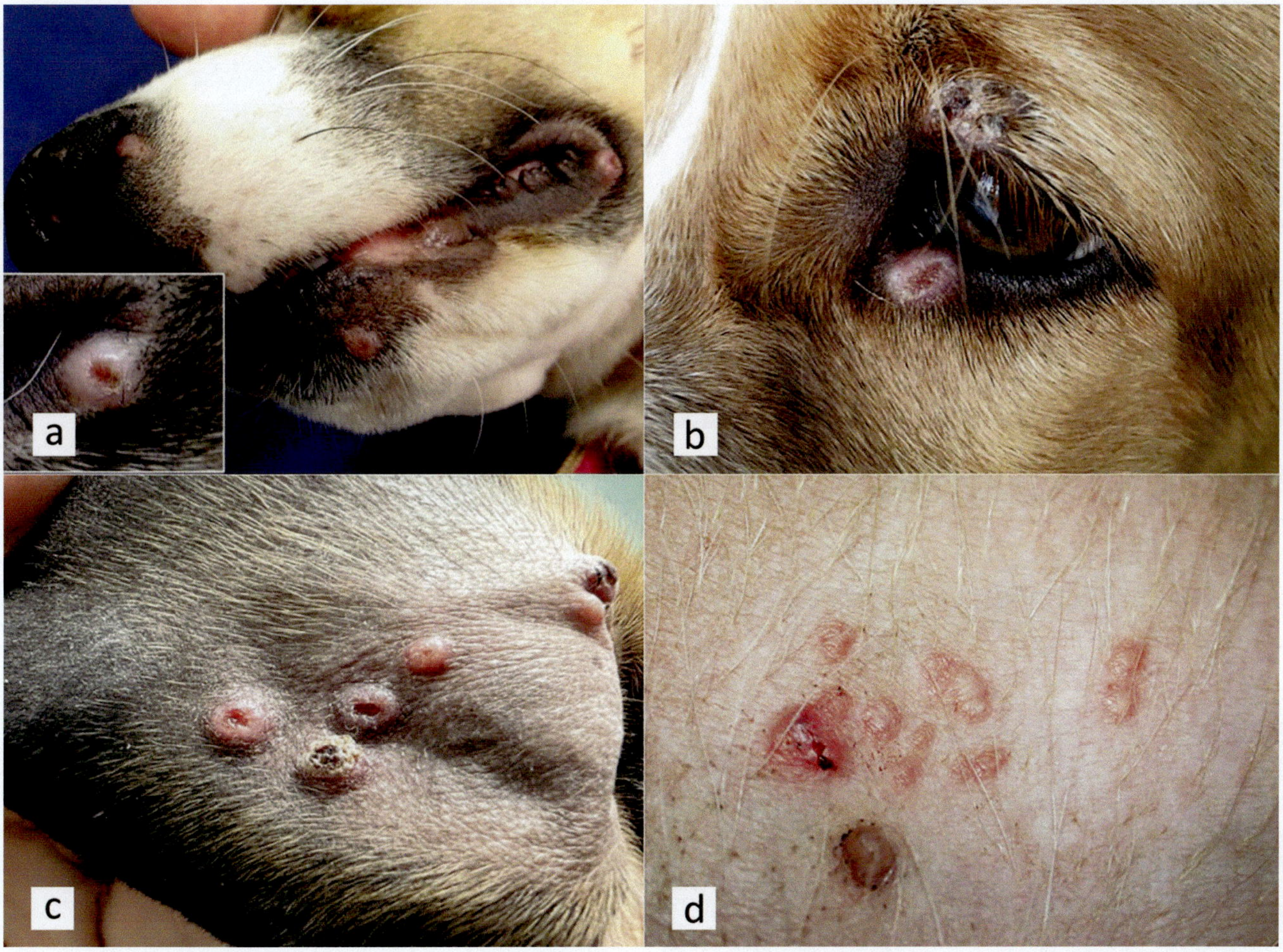

Fig. 6.57 Papular leishmaniasis: (**a**) multiple papular–nodular lesions on the nose and lips of a young dog. Inset: typical umbilicated shape of a nodular papule; (**b**) two umbilicated nodular papules on the eyelids; (**c**) multiple characteristic papular lesions on the inner surface of the pinna; and (**d**) multiple nodular papules, some of them merged, on the abdomen of a Dogue de Bordeaux

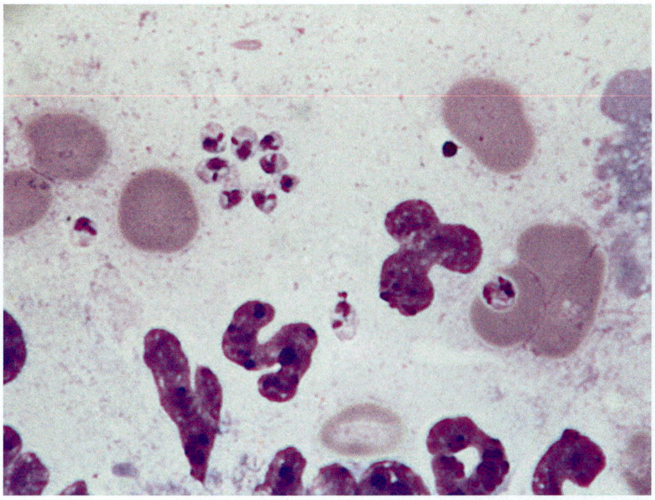

Fig. 6.58 Cytology of leishmaniasis: amastigotes of *Leishmania* spp. in which kinetoplast is easily recognizable as a rod-shaped structure

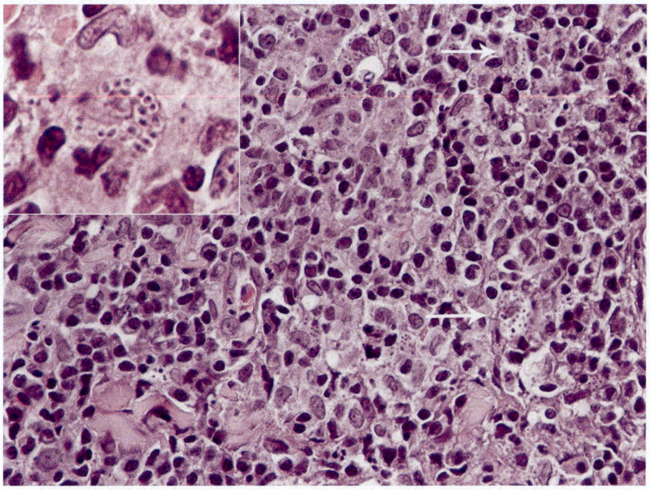

Fig. 6.59 Histopathology of leishmaniasis: diffuse dermatitis composed mostly of macrophages, lymphocytes, and plasma cells. Some macrophages are filled with amastigotes (*arrows* and *inset*)

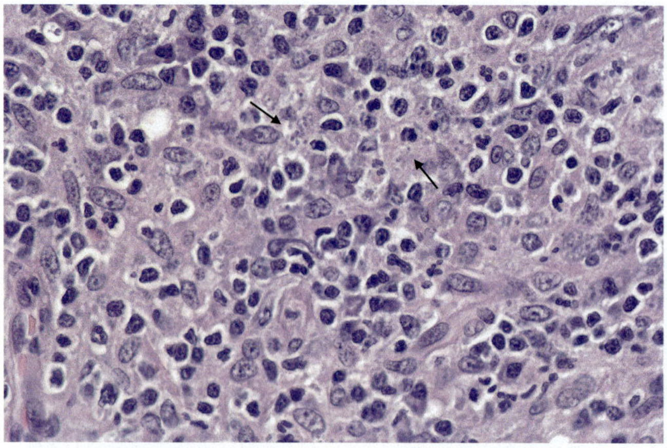

Fig. 6.60 Histopathology of leishmaniasis: diffuse lympho-histio-plasma cells dermatitis; few suspected amastigotes are detectable (arrows)

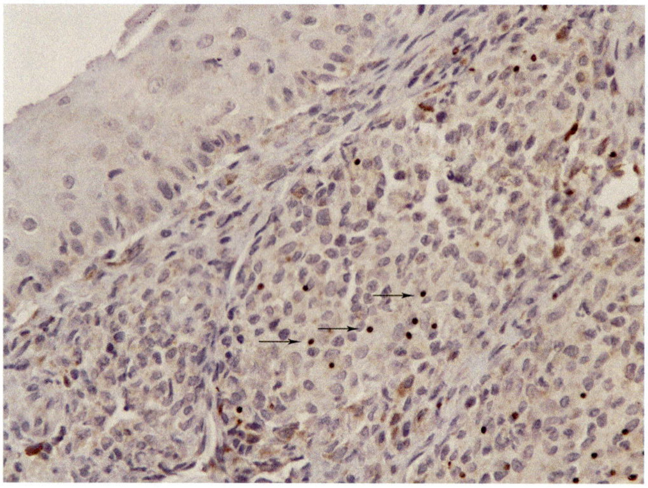

Fig. 6.61 Histopathology of leishmaniasis: immunohistochemical staining is able to highlight amastigotes of *leishmania* as dark orange in color roundish to oval bodies (arrows)

the cytoplasm of macrophages or free on the background confirms the clinical suspicion (Figs. 6.62, 6.63, 6.64, 6.65, 6.66, and 6.67). In the literature, there are discrepancies about the number of protozoa that can be detected with cytology; some authors report more frequently a low number of parasites, whereas, in many cases, the number of amastigotes detected is higher [24, 28]. When they are few, amastigotes must be carefully searched on slides.

In the author's experience, also if protozoa are few, in most cases, the latter are detected on slides but, when they are not present and histopathology cannot be performed, amastigotes could be searched through immunocytochemistry; in lucky slides, protozoa can be indeed highlighted as brown orange bodies (Fig. 6.68).

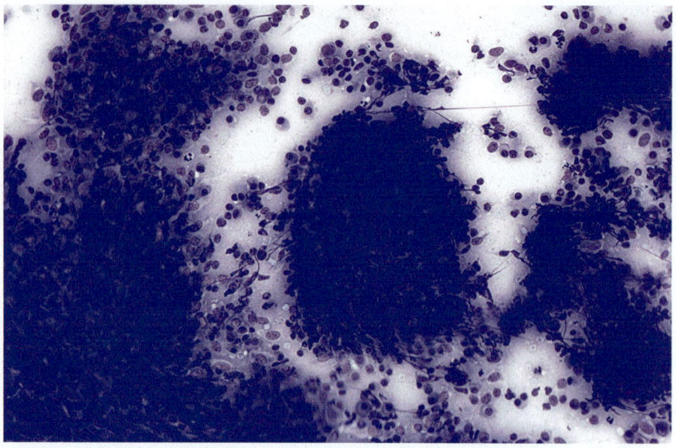

Fig. 6.62 Cytology of papular leishmaniasis: highly cellular specimens composed with a lympho-histio-plasma cells inflammatory pattern

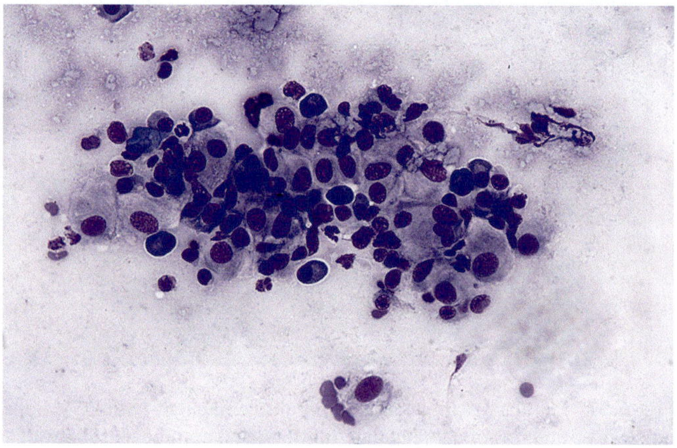

Fig. 6.63 Cytology of papular leishmaniasis: macrophages, lymphoid cells, and plasma cells are recognizable. No protozoa are present

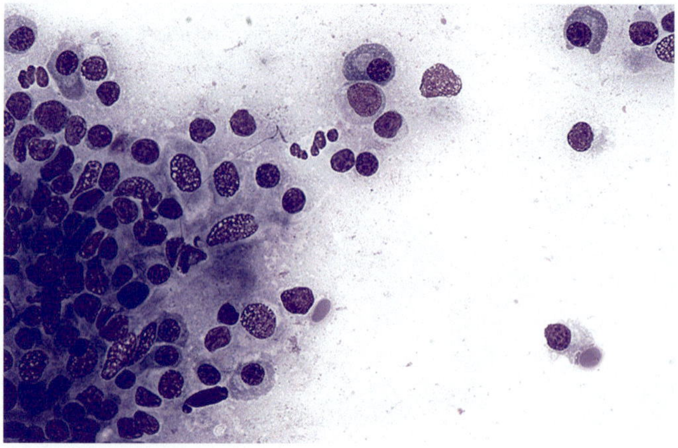

Fig. 6.64 Cytology of papular leishmaniasis: at high magnification, the morphology of inflammatory cells is well recognizable. No protozoa are present in this picture

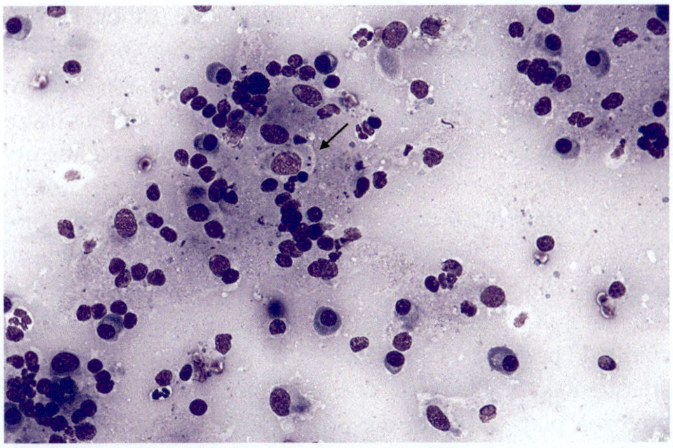

Fig. 6.65 Cytology of papular leishmaniasis: at low magnification, three suspected amastigotes are present in the cytoplasm of a macrophage (arrow)

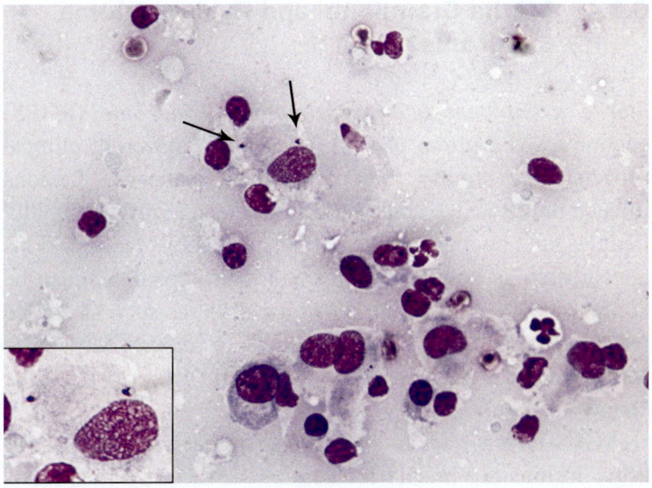

Fig. 6.66 Cytology leishmaniasis: two protozoa are evident in the cytoplasm of macrophages (arrows, inset)

Fig. 6.67 Cytology leishmaniasis: many amastigotes are evident in the cytoplasm of macrophages (arrow, inset). Note that the typical morphology of amastigotes is clearly recognizable

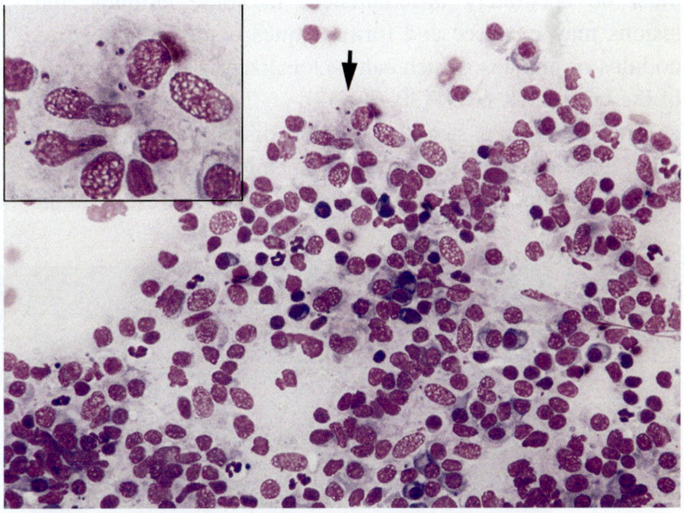

Fig. 6.68 Immunocytochemistry: (**a**) two amastigotes are highlighted as brown oval in shape bodies (arrows), and (**b**) at higher magnification, on a background in which poorly stained inflammatory cells are present, the typical morphology of amastigotes is clearly recognizable

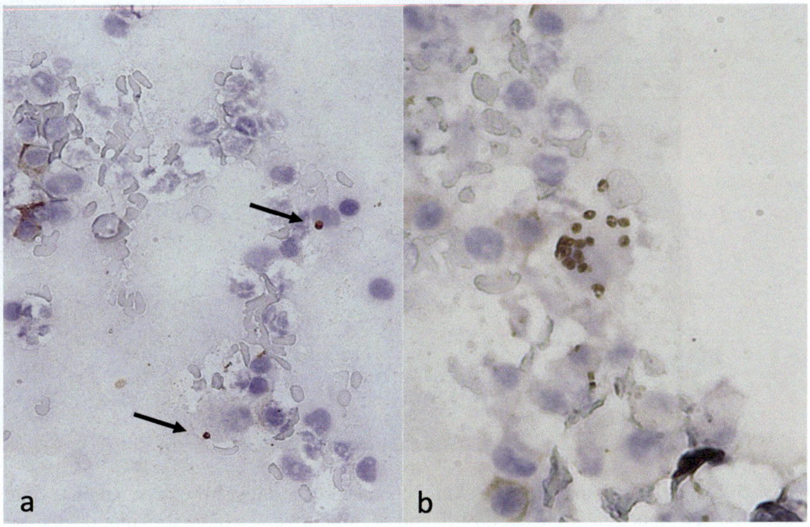

6.3.4 Sterile Granuloma and Pyogranuloma Syndrome

Sterile granuloma and pyogranuloma syndrome (SGPS) encompasses a group of skin lesions with different clinical aspects and similar histopathological patterns. Lesions include multifocal to coalescent alopecia and scaling, papular nodules, and nodules or plaques: since in some cases, lesions appear only as nodular papules and their appearance can be confused with other papular–nodular diseases such as canine reactive histiocytosis (CRH) or the above-discussed papular form of CanL, this disease is discussed in this part of the chapter.

SGPS is very rare in dogs and extremely rare in cats. In dogs, lesions are mainly located on the face, eyelids, nose, bridge of the nose, and pinna, and with regard to clinical aspect, size, and location, they are very similar to those observed in papular–nodular leishmaniasis from which it must be accurately differentiated. In some animals, the lesions may coalesce and form plaques or directly arise as nodules or plaques, which can be localized in different areas of the body (Fig. 6.69) [30, 31, 33].

The SGPS is a sterile idiopathic disease and, although some authors have supposed that an infectious antigen can cause a dysfunction of the immune system that results in an aberrant lympho-histiocytic response, its real pathogenesis is still unknown; all studies aimed at searching for bacterial agents that could act as triggers of granulomatous lesions have never given positive results. Furthermore, some histological features and the good response to immune modulatory drugs suggest the sterile nature of the lesions [29, 30, 31, 33]. In a study of 46 dogs with a histomorphological diagnosis of SGPS, 21 were positive for *Leishmania* according to polymer chain reaction (PCR) and immunohistochemical staining [28]. The meaning of this positivity needs to be clarified because the detection of DNA fragments of amastigotes in the skin does not necessarily imply that protozoa play a role in the development of the lesions. In another study, the skin microbiota of 30 cases of sterile granulomatous and pyogranulomatous dermatitis/panniculitis was investigated. Microbial DNA was amplified using primers targeting the bacterial 16S rRNA V1-V3 and fungal ITS-1 regions. The results showed that the type of bacteria mainly detected in these cases were normal inhabitants of the canine skin, suggesting that their presence must be interpreted as secondary contaminants and therefore confirming that these clinical forms of granulomatous lesions are really sterile [32].

Nevertheless, the term sterile should only be attributed to cases where granulomatous/pyogranulomatous lesions with

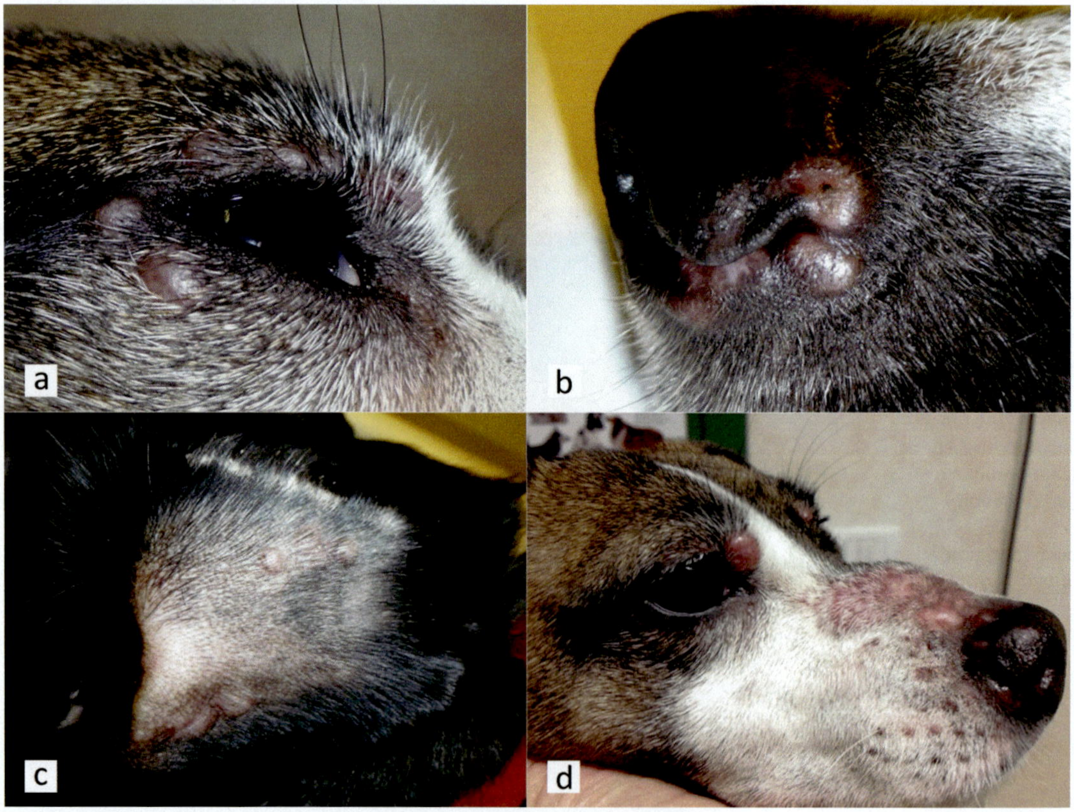

Fig. 6.69 SGPS: (**a**) papular nodules on the eyelids of a dog; (**b**) papular nodules on the inner surface of the pinna; (**c**) papular nodules on the mucocutaneous junction of the left nostril; and (**d**) nodules and a large irregular plaque on the bridge of the nose of a Jack Russell terrier with SGPS

typical clinical and histopathological findings and in which any microorganisms have been excluded by special stains, PCR, immunohistochemistry, and bacterial culture.

Cytological Findings

Histopathological features of SGS or SGPS are characterized by *perifollicular granulomas,* sometimes with a typical *elongated* or *sausage-like* shape (Figs. 6.70, 6.71, and 6.72).

As granulomas are small and located around the isthmic area of follicles, the sampling of a high number of cells is not possible; indeed, in these cases, it is very difficult to center the small granulomas with the needle and only a few histiocytes, often represented with bare nuclei, along with lymphocytes can be collected. However, when granulomas merge and give origin to larger dermal granulomas, transforming the dermatitis from nodular to diffuse, it is easier to obtain slides of high cellularity. Despite the term pyogranuloma, the number of neutrophils is usually scarce and usually located in the center of the granulomas.

Slides are therefore frequently characterized by a lympho-histiocytic inflammation, represented by a variable number of macrophages and epithelioid macrophages, with oval to kidney-shaped nuclei and moderate to large, slightly gray–blue cytoplasm. As mentioned, many histiocytes are represented by bare nuclei. In association, an important number of lymphocytes are also constantly present, while only occasional plasma cells are observed.

From a cytological point of view, the macrophages of SGS are not differentiable from dendritic histiocytes observed in CRH in which a cytologically similar lympho-histiocytic pattern is observed. Furthermore, in specimens in which amastigotes are not detected, it is difficult to cytologically differentiate an SGS from the papular form of CanL; the high number of plasma cells usually observed in the protozoal disease helps to orient the diagnosis (Figs. 6.73, 6.74, and 6.75).

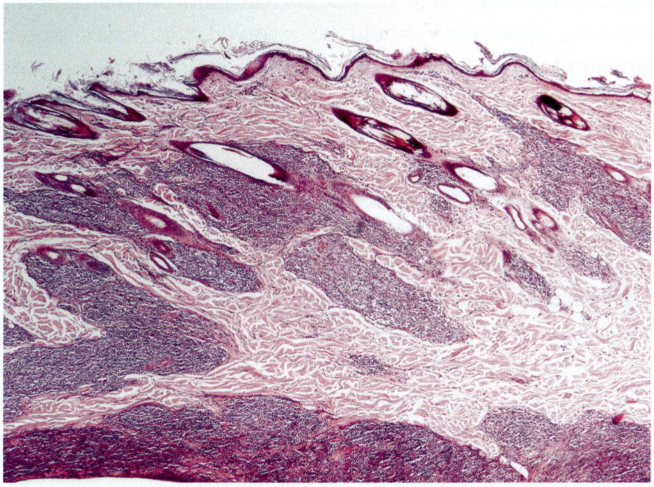

Fig. 6.70 Histopathology of SGPS: multiple elongated perifollicular nodules

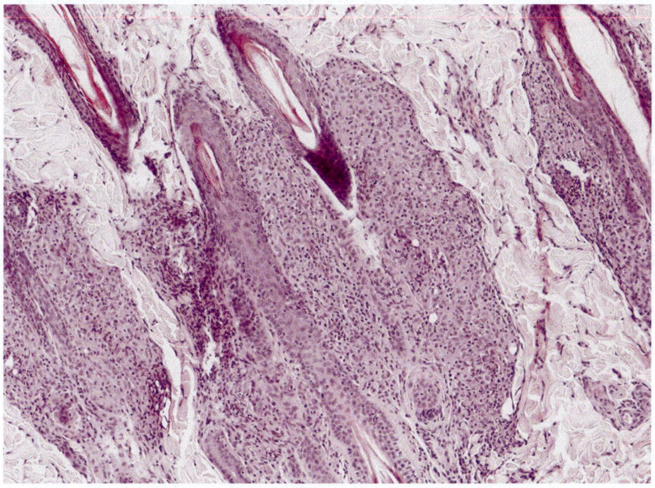

Fig. 6.71 Histopathology of SGPS: at high magnification, the lympho-histiocytic infiltrate is better recognizable in perifollicular granulomas

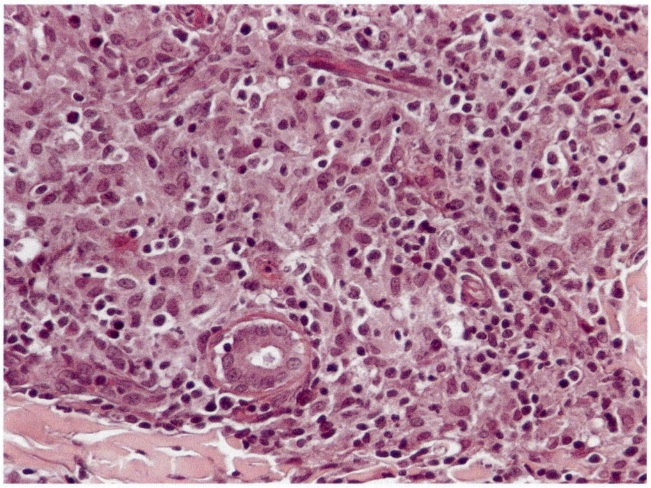

Fig. 6.72 Histopathology of SGPS: at high magnification, it is possible to recognize many histiocytes and lymphocytes

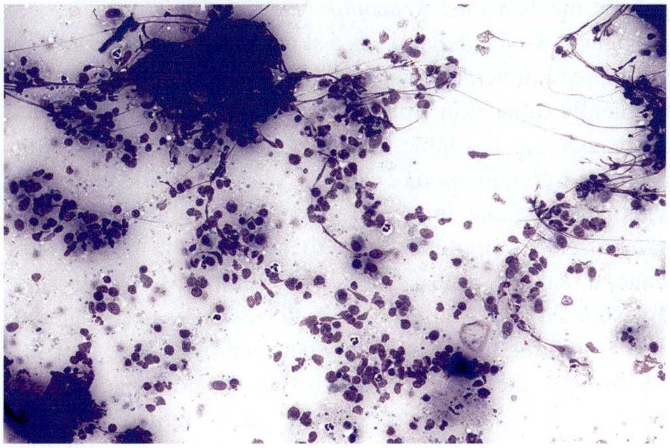

Fig. 6.73 Cytology of SGPS: very cellular sample in which a lympho-histiocytic inflammatory pattern is recognizable

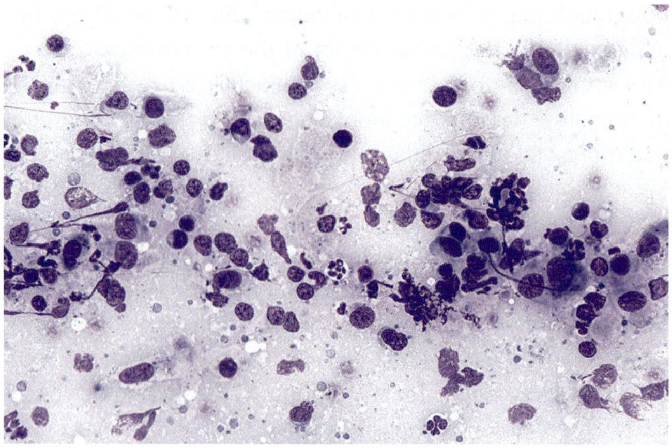

Fig. 6.74 Cytology of SGPS: histiocytes and lymphocytes along many bare nuclei. Note the presence of lympho-glandular bodies indicative of the rupture of the cytoplasm

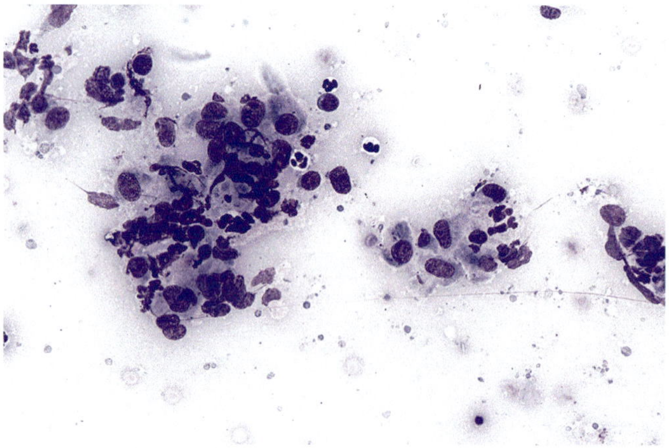

Fig. 6.75 Cytology of SGPS: histiocytes and rare lymphocytes

6.3.5 Calcinosis Cutis

Skin tissue mineralization can occur through different pathogenetic mechanisms: *dystrophic, metastatic, idiopathic,* and *iatrogenic*. *Calcinosis cutis* is a skin disease characterized by widespread mineralization mainly of collagen fibers and less frequently other skin structures. The mineral salts deposited in the skin are insoluble inorganic compounds mainly composed of a combination of calcium, phosphate, and carbonate salts. Because the former are the most common, the term *calcinosis* is commonly used to describe the presence of mineral salts, even if their composition is usually not investigated; in a study in dogs with *calcinosis cutis*, calcium apatite was the most commonly detected mineral [34, 36].

In dogs, the most frequent cause of skin calcium salt deposits is spontaneous or iatrogenic hypercortisolism [37]. The reason why, in this disease, calcium salts are deposited in the dermis and epidermidis is not clear and the

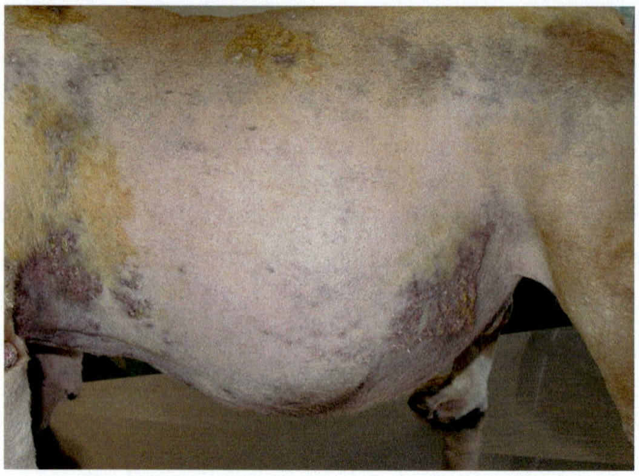

Fig. 6.76 Alopecia, pendulous abdomen, and diffuse calcinosis in an English bulldog with spontaneous hypercortisolism

definition of dystrophic calcinosis used by many authors to describe calcium salt deposits in the course of hypercortisolism could be not correct until proven; indeed, the term dystrophic should be reserved with cases where structural damage of the tissue, which may lead to intra- or extracellular deposits of minerals, has been microscopically identified (Fig. 6.76).

Clinical lesions morphology varies greatly and can modify depending on the chronicity of the lesions, and as skin ulceration and secondary inflammation/infection occur, lesions range from small papular–nodular lesions, often alopecic, erythematous, and crusty, which tend to merge and form different in size plaques, usually irregular in shape, with a pinkish smooth surface or covered by crusts [37, 38]. Papular and nodular lesions are better visible in the hairless areas of the axilla, groin, abdomen, and inner surface of thighs (Figs. 6.77 and 6.78). A pathognomonic aspect of cutaneous calcinosis is flattened plaques with multiple and disposed regularly small holes on the surface, which give the surface a morphology that resembles "parmesan cheese crisps (Figs. 6.79 and 6.80). In some dogs, especially some apparently predisposed breeds to the cutaneous deposit of mineral salts such as English Bulldog, Rottweiler, and Boxer, calcinosis could be more severe and spread, frequently observed over the dorsal part of the neck and trunk; in most severe cases represented with a large, thick with irregular edges and usually ulcerated, crusty, infected and pruritic plaque that can extend from the head to the back (Fig. 6.81). As the organism tries to eliminate calcium salts as a foreign body through the epidermidis, chronic lesions are often ulcerated and covered with crusts and pus as a result of a secondary bacterial infection (Fig. 6.82).

Cytological Findings

Cytological features may be different depending on the presence of ulceration and complications secondary to it. In not ulcerated papular lesions, histopathology is characterized by

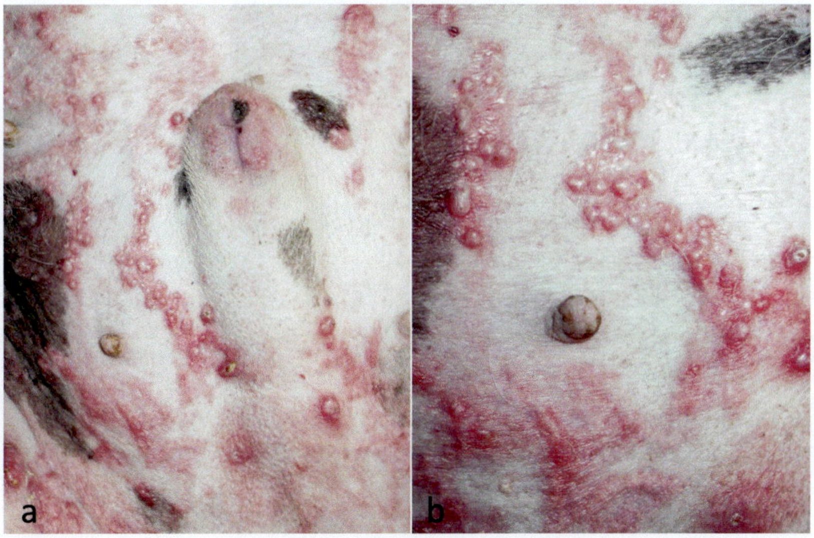

Fig. 6.77 Calcinosis cutis: (**a**) multiple erythematous nodular papules on the abdomen of a dog with calcinosis cutis secondary to iatrogenic hypercortisolism and (**b**) closeup of the previous image

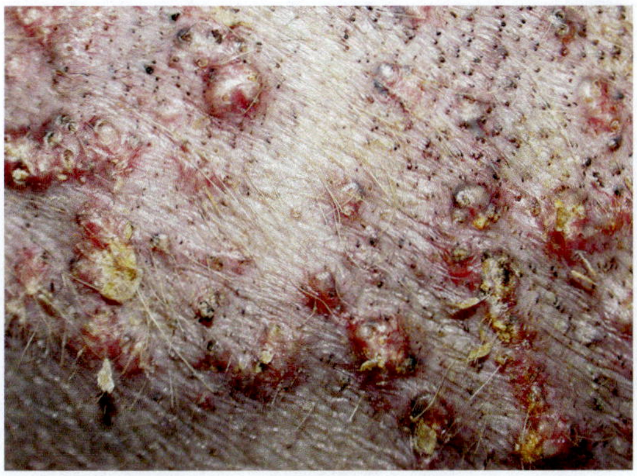

Fig. 6.78 Small papule covered with scales in a dog receiving for a long time corticosteroids; note the presence of numerous comedones

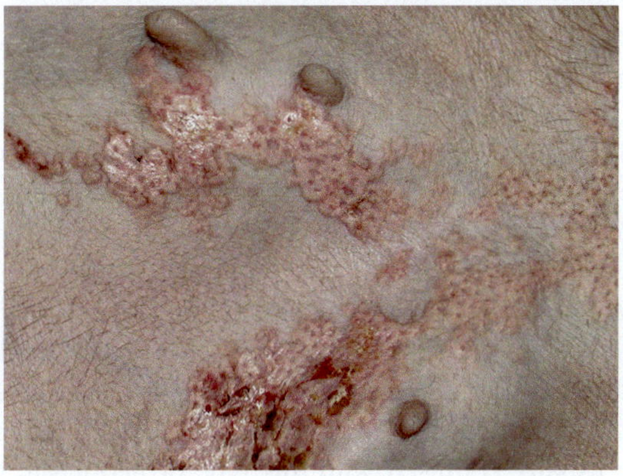

Fig. 6.80 *Calcinosis cutis:* Closeup of a lesion with "parmesan cheese crisps" appearance

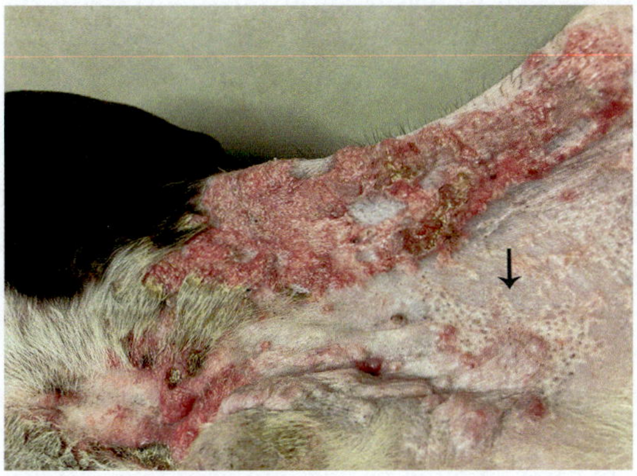

Fig. 6.79 Large erythematous plaques in a dog treated with a long course of corticosteroids; yellowish and flattened irregular in shape mineralized plaques with a "parmesan cheese crisps" appearance are also recognizable (arrow)

Fig. 6.81 Diffuse large crusty plaques over the dorsal neck and trunk in a Boxer with spontaneous pituitary-dependent hyperadrenocorticism

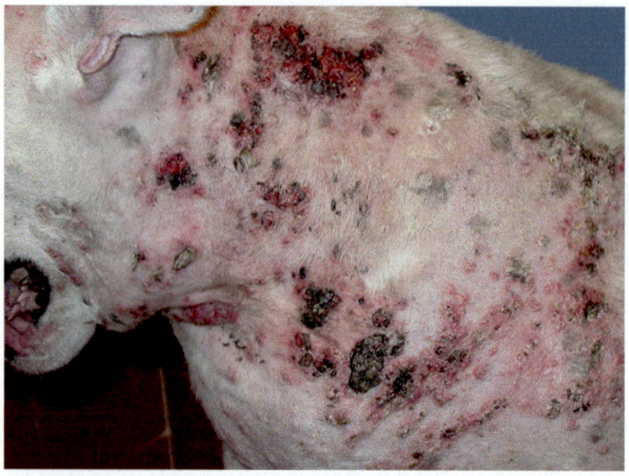

Fig. 6.82 *Calcinosis cutis*: multiple nodular papules, plaques, and hematic crusts in a bull terrier with iatrogenic hypercortisolism

multifocal dermal deposits of mineral salts, surrounded by an inflammatory response composed of epithelioid macrophages and multinucleated giant cells (Figs. 6.83 and 6.84).

In these lesions, cytology shows the same findings described for histopathology, with many epithelioid macrophages and large multinucleated histiocytes surrounding variable amounts of calcium salts that behave like a foreign body. A severe fibroblastic reaction, cytologically represented by spindle cells without clear aspects of atipia, is also commonly observed. To an inexpert cytologist, especially when mineral salts are not evident, the simultaneous presence of both spindle cells and large multinucleated cells could be misinterpreted as a pleomorphic soft tissue sarcoma with giant cells; fortunately, the clinical aspects of calcinosis cannot be usually confused with this type of neoplasia, but in case calcinosis is represented by a single plaque, care must be done in interpreting cytological findings (Figs. 6.85, 6.86, and 6.87).

In advanced stages of *calcinosis cutis* with ulceration of the skin, cytology findings change, and a pyogranulomatous inflammatory pattern can be observed enriching neutrophils and vacuolated macrophages along with lymphoid cells and plasma cells. A secondary bacterial infection can occur (Figs. 6.88 and 6.89). In some cases, slides are few cellular, hemocontaminated, and with only scattered multinucleated cells; when this cytological feature is observed, in the presence of a compatible clinical pattern, a cutaneous calcinosis must be suspected.

In cytology, the morphology of calcium salts can be very different, ranging from amorphous granular material, which can be achromatic or violet or bluish in color, to transparent rhomboidal with irregular borders or needle-like in appearance.

In cases in which mineral salts are few, they may be hardly detectable in the contest of hypercellular inflammatory specimens or they could be confused with lamellar keratin; so to highlight them, it is possible to stain some slides with Von Kossa dye, which colors in black the area where calcium salts are present, as discussed in Chap. 2 (Figs. 6.90, 6.91, and 6.92).

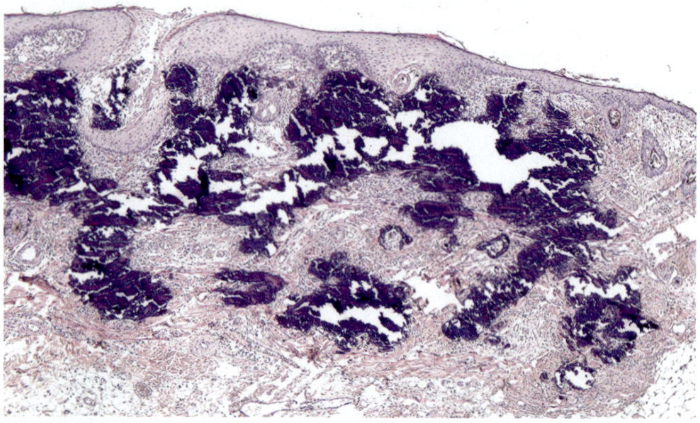

Fig. 6.83 Histopathology of *calcinosis cutis*: multifocal mineralization of the dermis; calcium salts are represented by deep violet amorphous material

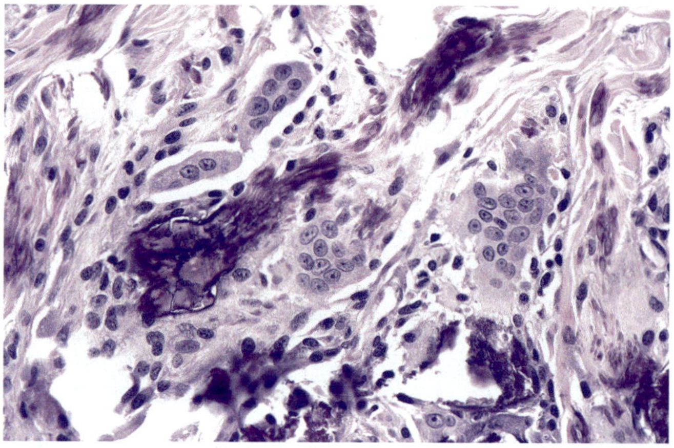

Fig. 6.84 Histopathology of *calcinosis cutis*: multinucleated giant cells and reactive fibroblasts surrounding mineral salts

When a needle is introduced into a mineralized skin lesion, it is possible to perceive a crunching sensation, which helps to support the suspicion of the presence of mineral salts in the tissue. In chronic cases, calcium salts can transform into bone tissue (osseous metaplasia) and lesions turn from *Calcinosis cutis* to *Osteoma cutis*; in these cases, the suspicion may be generated when needles cannot penetrate the firm nodular papules or plaques or when osteoblasts can be collected via FNAB using a large needle [35, 37] (Figs. 6.93, 6.94, 6.95, and 6.96).

In rare cases, if large needles are used (e.g., 19 gauge), a few fragments of bony spicules along with some osteoblasts can be collected through FNAB; with this clinical presentation and history, a cytological diagnosis of *Osteoma cutis* could be cytologically suspected.

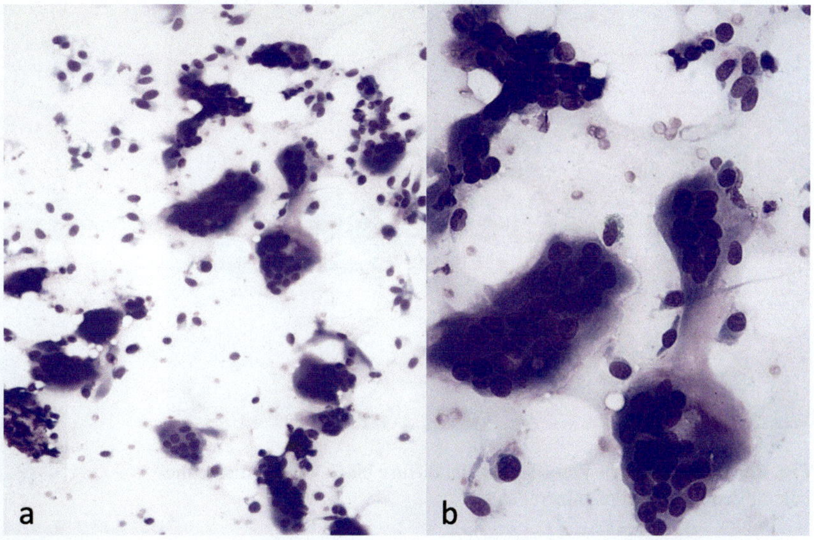

Fig. 6.85 Cytology of *calcinosis cutis*: (**a**) many spindle fibroblasts and giant cells and (**b**) spindle fibroblasts and giant cells. Mineral salts are not evident in this picture

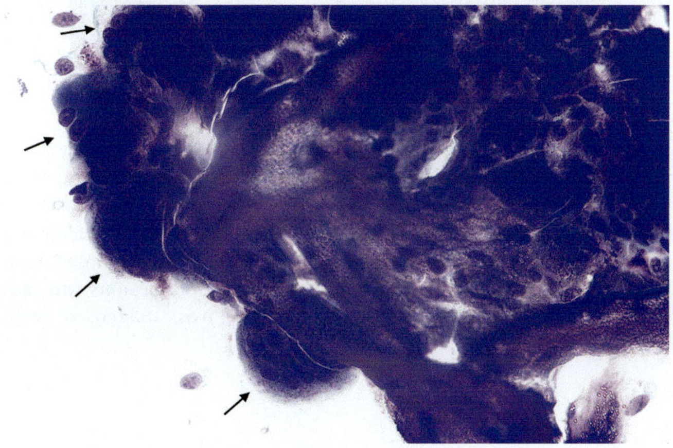

Fig. 6.86 Cytology of *calcinosis cutis*: many multinucleated giant cells and fibroblasts surrounding mineral salts (arrows). Calcium salts are recognizable as amorphous, voluminous, and grainy in appearance material

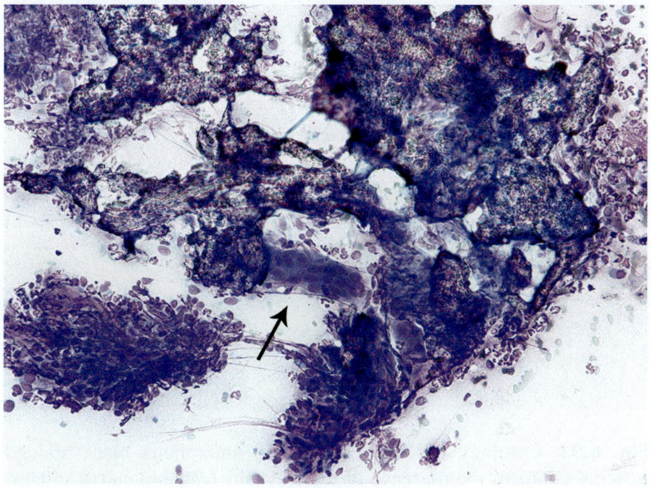

Fig. 6.88 Cytology of *calcinosis cutis*: pyogranulomatous inflammation surrounding mineral salts. Note the giant cell (arrow)

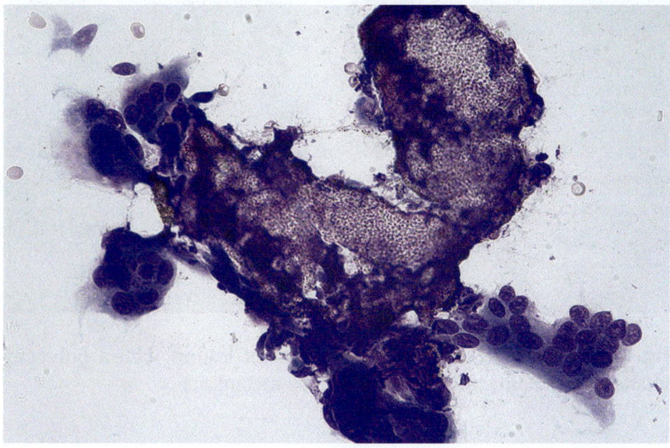

Fig. 6.87 Cytology of *calcinosis cutis*: many giant cells attacking calcium salts

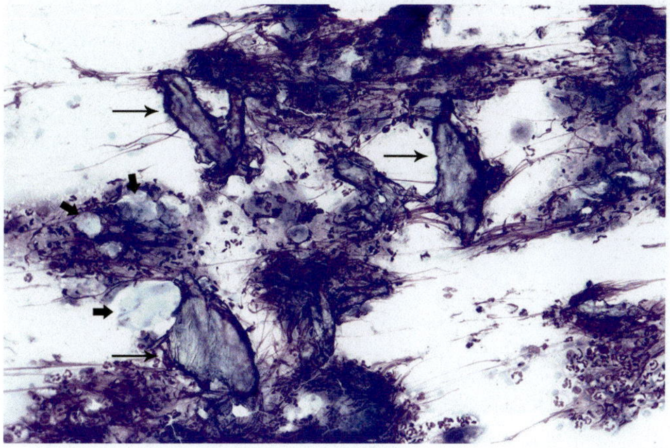

Fig. 6.89 Cytology of *calcinosis cutis*: many amorphous calcium salts (thin arrows) along with secondary pyogranulomatous inflammation. Calcium salts must be not confused with corneocytes (short arrow)

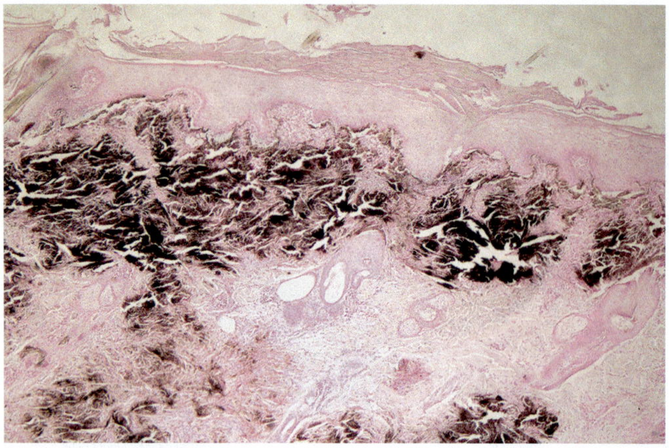

Fig. 6.90 Histology of *Von Kossa* stain: diffuse black areas indicate the calcium composition of mineral salts

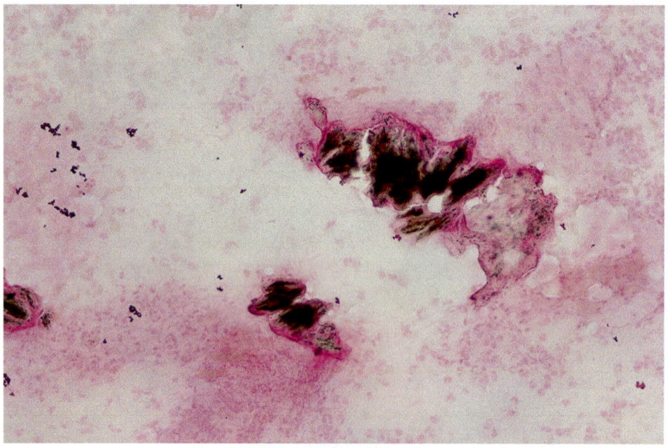

Fig. 6.91 Cytology of *Von Kossa* stain: amorphous black stained material confirms the presence of mineral salts (calcium salts); inflammatory cells are poorly recognizable in pink color

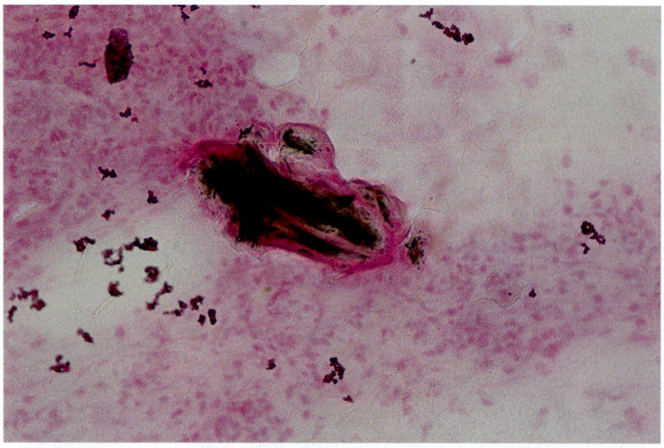

Fig. 6.92 Cytology of *Von Kossa* stain: amorphous calcium salts and secondary pyogranulomatous inflammation

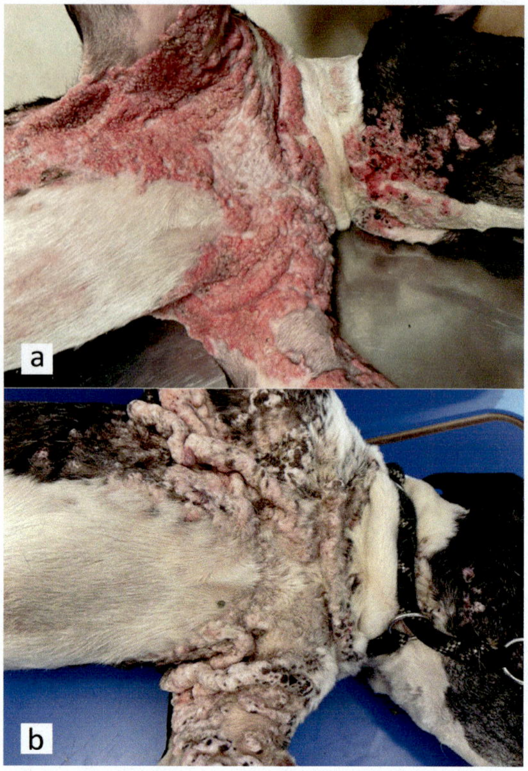

Fig. 6.93 Calcinosis cutis: (**a**) day 0: diffuse erythematous calcinosis cutis with plaques and nodules; *calcinosis cutis* was diagnosed with histopathology and (**b**) the same dog at day 300: lesions are less erythematous and thicker; *Osteoma cutis* was diagnosed with histopathology

Fig. 6.94 *Osteoma cutis*: nodular confluent lesions with a polypoid appearance in a dog with diffuse osseous metaplasia

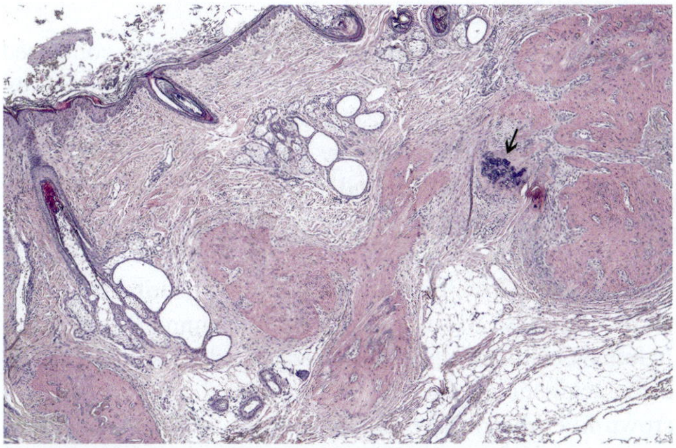

Fig. 6.95 Histopathology of *osteoma cutis*: osseous metaplasia secondary to chronic deposit of *calcium* salts in the dermis. Note that a focus of cartilaginous metaplasia is also present (arrow)

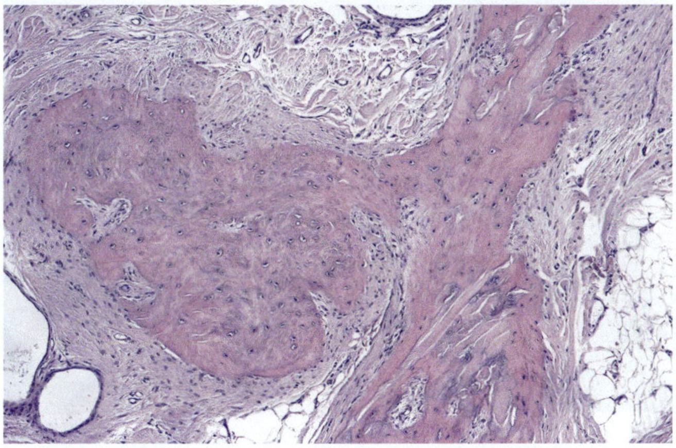

Fig. 6.96 Histopathology of *osteoma cutis*: bone tissue surrounded with a fibroblastic reaction

6.3.6 Xanthogranuloma/Xanthomatosis

Cutaneous *Xanthomatosis* is a rare skin disease of cats and dogs in which, because of dyslipoproteinemia, an accumulation of cholesterol, triglycerides, and phospholipids into the dermis occurs. In cats, xanthomas are mostly due to diabetes mellitus, both spontaneous and iatrogenic, the latter usually secondary to chronic therapy with corticosteroids, mainly administered to manage pruritic diseases (especially allergic dermatitis). Less commonly, feline xanthomatosis develops in cats fed with a high-fat diet, in young patients suffering from idiopathic familial hyperlipidemia or, exceptionally, in patients with pancreatitis [40, 44, 47].

In dogs, this disease is less common and usually secondary to an excessively fatty diet or to diabetes mellitus [39]. In both species, xanthomatosis can be idiopathic; indeed, in some cases, no underlying diseases are detected and no alteration of the lipid metabolism is evident in blood tests.

In any case, lesions ranging from papular lesions to small plaque or nodules represent skin lesions; lesions located in the dermis, alopecic, pale pink, or yellow-orange in color in not-pigmented skin [41]. Single nodular lesions are also reported in both normolipemic dogs and cats [42]; in dogs, the scrotum seems to be the most frequently involved area. In the case of the singular lesion, the term *xanthogranuloma* is more correct since xanthomatosis should refer to multiple skin lesions associated with systemic disease. The cause of the development of single xanthogranuloma in the absence of dyslipoproteinemia is not known; local trauma has been hypothesized as a possible cause (Figs. 6.97, 6.98, 6.99, 6.100, and 6.101).

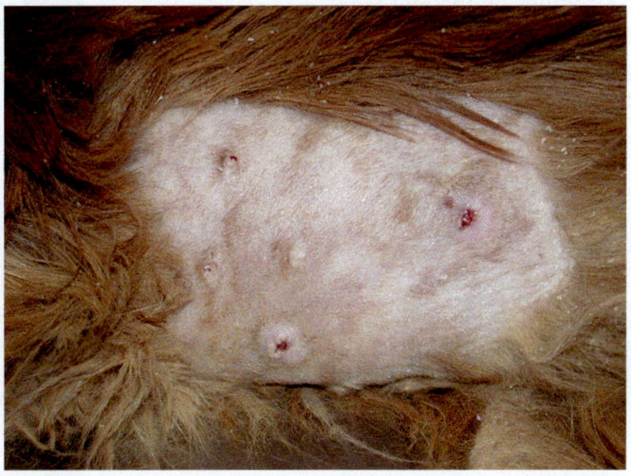

Fig. 6.97 Multiple ulcerated xanthomas on the chest of a cat with diabetes mellitus

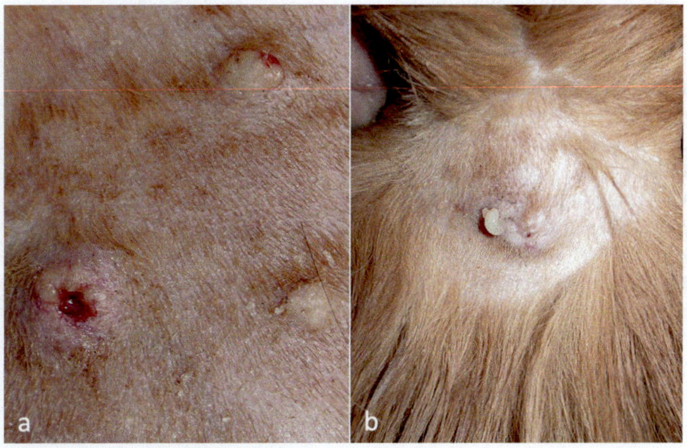

Fig. 6.98 Cutaneous xanthomatosis: (**a**) intact and ulcerated papules of the same cat of the previous image and (**b**) whitish and greasy material discharge from a draining tract of a diabetic cat with xanthomatosis

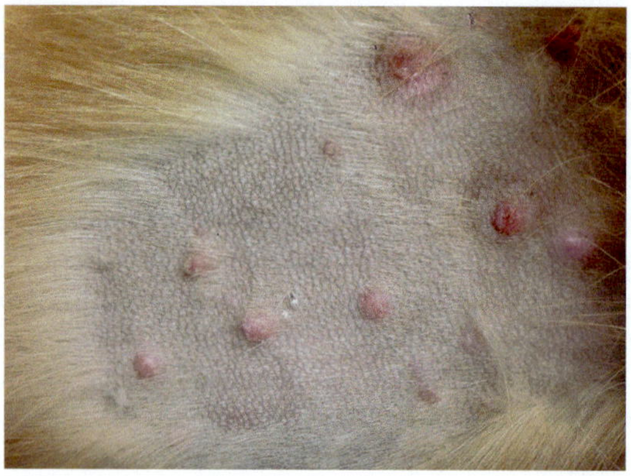

Fig. 6.99 Multiple papular nodules on the chest of a dog with idiopathic xanthomatosis

Cytological Findings

Xanthogranulomatous nodules are histologically composed of a dermal infiltrate of histiocytic cells represented by macrophages with cytoplasm rich in overlapped vacuoles, containing lipids that give the cell a characteristic *foamy appearance* (Figs. 6.102 and 6.103) [43]. Neutrophils are few or absent while a variable amount of lymphocytes is usually present. A variable number of multinucleated giant macrophages, usually with haphazardly distributed nuclei and with diffuse foamy cytoplasm, are commonly observed. In human medicine, the multinucleated cells commonly show a characteristic appearance; in fact, nuclei are arranged in a crown-like disposition at the center of the cytoplasm, surrounding a central homogeneous and eosinophilic in

Fig. 6.100 Idiopathic familial hyperlipidemia: (**a**) multiple and ulcerated nodules on the rear legs of a kitten affected by idiopathic hyperlipidemia and (**b**) closeup of an interdigital papule

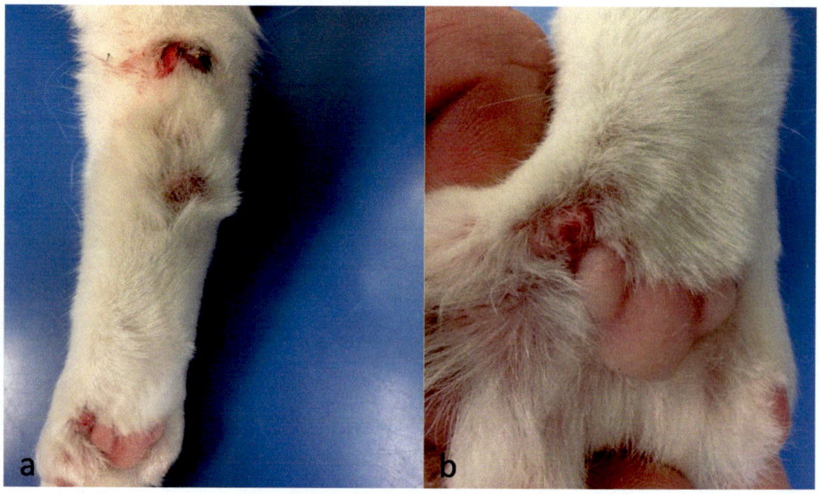

Fig. 6.101 Xanthogranuloma: (**a**) small xanthogranuloma on the nose of a cat; (**b**) dermoscopic view of a xanthogranuloma in a dog: the yellowish color is better evaluated with magnification. (Courtesy of DVM F. Necci-Italy); (**c**) yellowish in color xanthogranuloma on the scrotum of a dog. (Courtesy DVM C. Caruso-Italy); and (**d**) xanthogranuloma on the scrotum of a dog. (Courtesy DVM G. Calzavara—Italy)

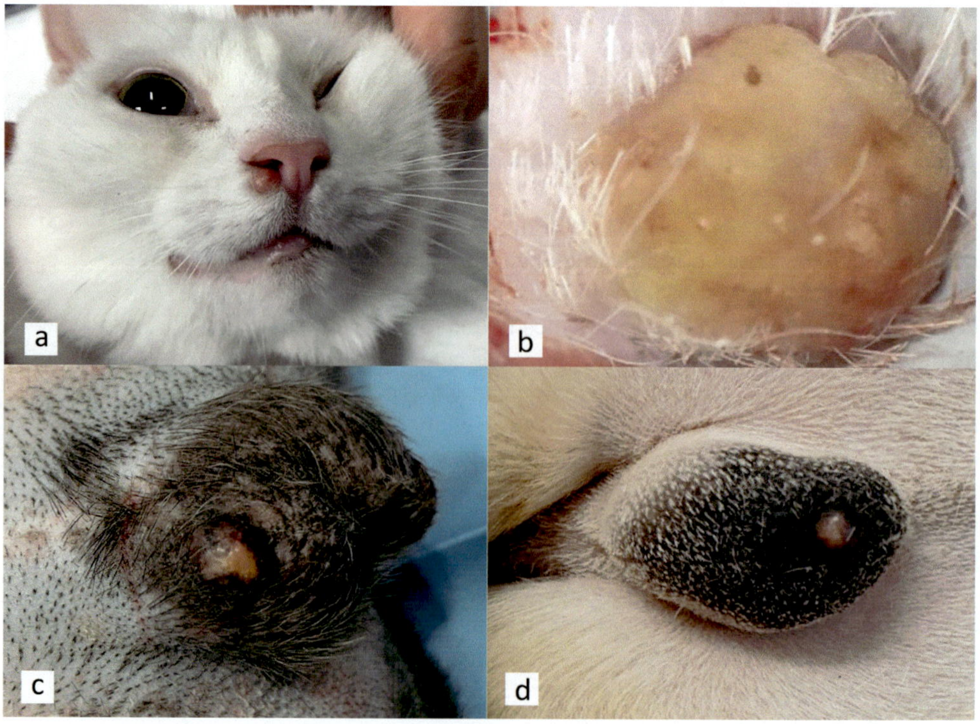

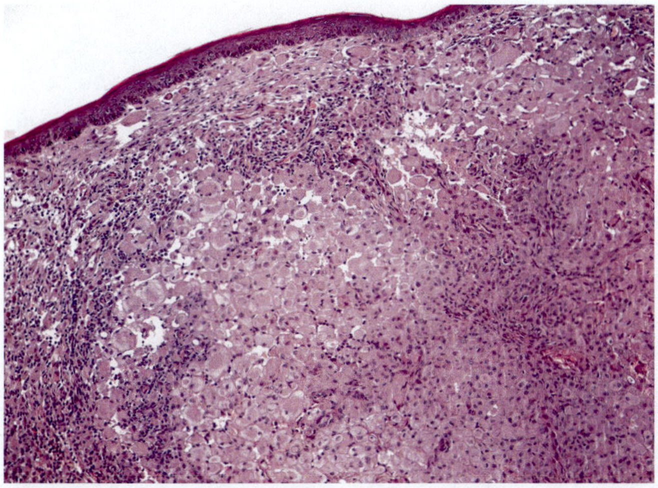

Fig. 6.102 Histopathology of xanthogranuloma: many large macrophages infiltrating the dermis

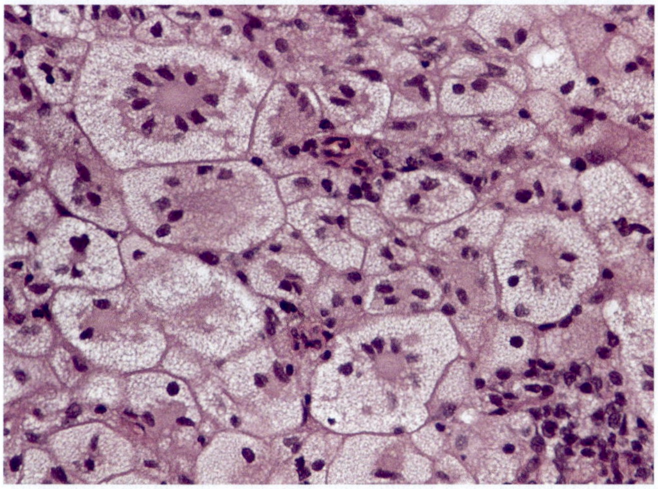

Fig. 6.103 Histopathology of xanthogranuloma: Touton cells are foamy multinucleated macrophages with nuclei arranged in a crown-like disposition at the center of the cytoplasm, a homogeneous and pinkish in color central area, and a highly lipidized peripheral cytoplasm

color area, while the peripheral cytoplasm is diffusely lipidized. These cells with this peculiar morphology, known as *Touton cells*, are not commonly observed in xanthogranuloma of dogs and cats in which nuclei are, as above mentioned, mostly haphazardly distributed (Figs. 6.104, 6.105, and 6.106). To better show the lipid content of the vacuoles, the *Oil Red O* dye can be used: lipids show a characteristic bright red color (Fig. 6.107). For beginners, especially when single lesions are investigated, the large vacuolated macrophages and the giant cells can be confused with neoplastic sebaceous cells since the hyper-lipidized cytoplasm is a physiological cytological feature of sebocytes. Immunohistochemical marker IBA-1 can help to define the histiocytic origin of cells (Fig. 6.108).

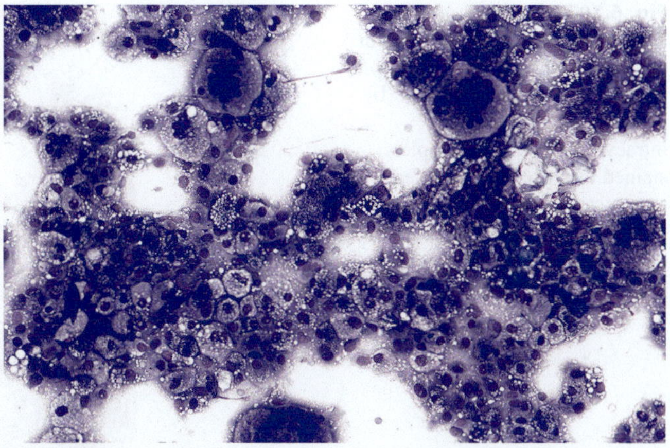

Fig. 6.104 Cytology of xanthogranuloma: many lipidized macrophages and multinucleated macrophages with haphazardly distributed nuclei characterize this cytology

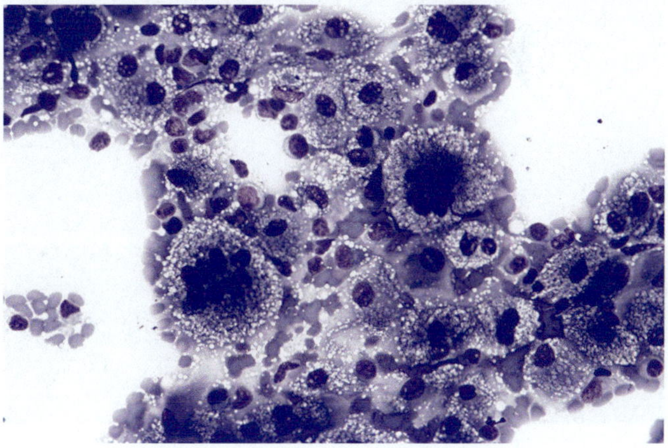

Fig. 6.105 Cytology of xanthogranuloma: many lipidized macrophages and multinucleated macrophages with crown-like arrangement (Touton cells) are present along lymphoid cells

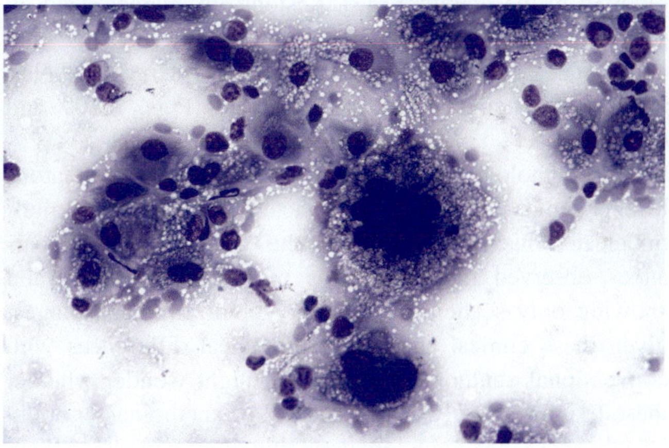

Fig. 6.106 Cytology of xanthogranuloma: a closeup of Touton cells in which a dark central area and a vacuolated cytoplasm are, respectively, at the center and at the periphery of the nuclei; note that the latter show a crown-like arrangement

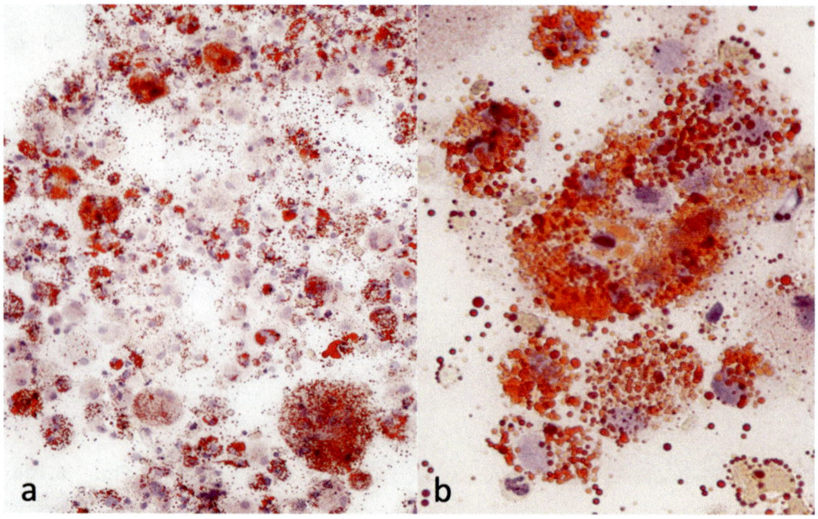

Fig. 6.107 Cytology of xanthogranuloma. Oil red O stain: (**a**) many macrophages show lipid intracytoplasmic content and (**b**) at higher magnification, intracytoplasmic lipid droplets stained brightly red can be easily recognizable

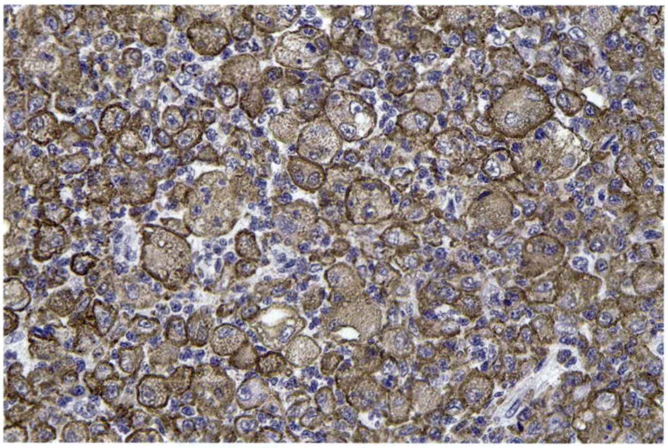

Fig. 6.108 Immunohistochemistry: strong and diffuse positivity to marker IBA-1 confirms the histiocytic origin of both mono- and multinucleated macrophages. Most of the unstained cells are lymphocytes

A *preauricular localization of xanthomas* in cats is reported in the literature [43]. Lesions are characterized by plaques and nodules localized in the preauricular and periorbital area, with a tendency to merge, develop intralesional hemorrhages, and ulcerate (Figs. 6.109 and 6.110).

Histopathology of these lesions shows granulomatous lesions composed of many compacted both mono- and multinucleated macrophages lacking the typical "foamy appearance" observed in other forms of xanthogranuloma and showing only scattered and poorly lipidized macrophages; given these clinical and histopathological differences with conventional xanthogranuloma, one might wonder whether these lesions are truly a clinical form of xanthomas or could be a distinct feline granulomatous skin disease in which there is a scare amount of intra-macrophagic lipids, as occurs in many other granulomatous diseases (Figs. 6.111, 6.112, 6.113, and 6.114).

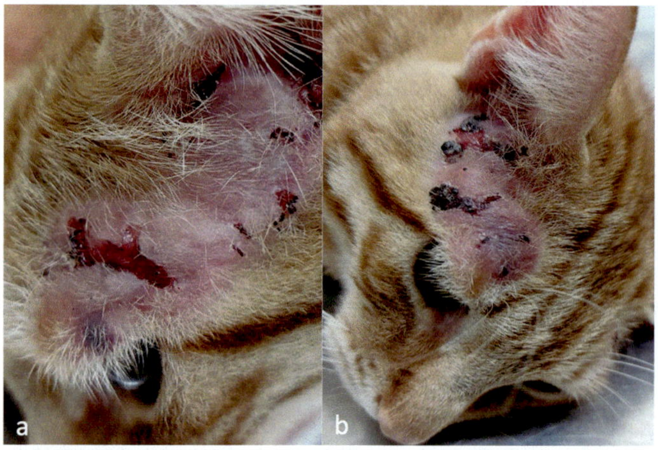

Fig. 6.109 Preauricular xanthogranuloma: (**a**) large plaque with ulcers and crusts on the right preauricular area: note the purple color of the lesion indicatives of intralesional hemorrhages, and (**b**) contralateral lesions. (Courtesy of DVM S. Gattuso—Italy)

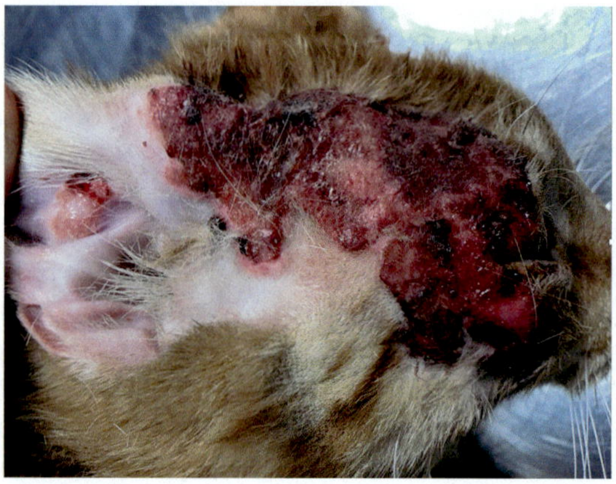

Fig. 6.110 Severely ulcerated merged plaques with raised edges in a cat with "so-named" preauricular xanthogranuloma

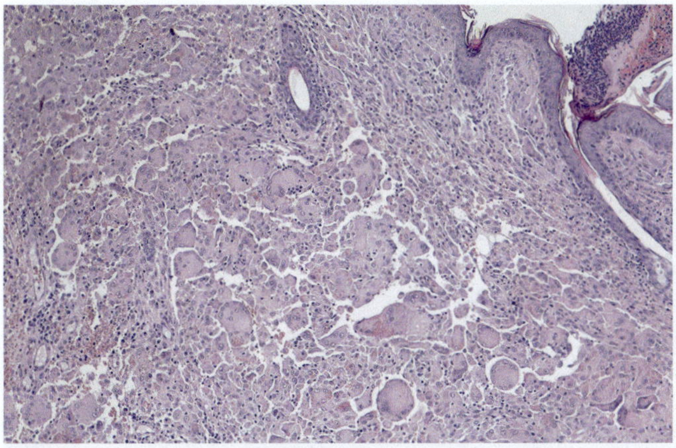

Fig. 6.111 Histopathology: diffuse granulomatous inflammation with many giant macrophages lacking lipidized cytoplasm

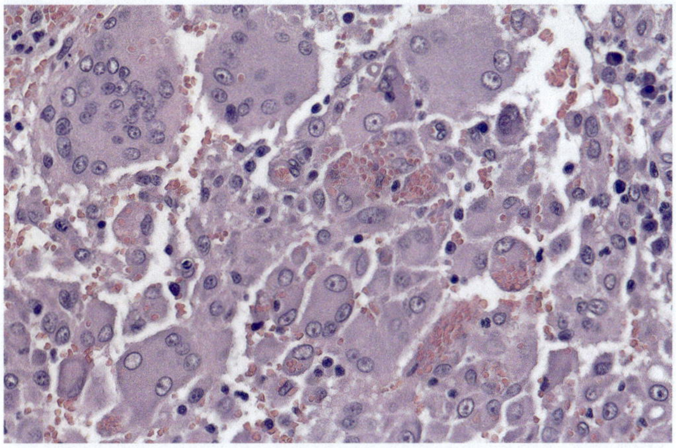

Fig. 6.112 Histopathology: at high magnification, are clearly evident epithelioid and multinucleated macrophages; note the absence of intracytoplasmic lipid and marked erythrophagocytosis, the latter may be indicative of intralesional hemorrhagic episodes

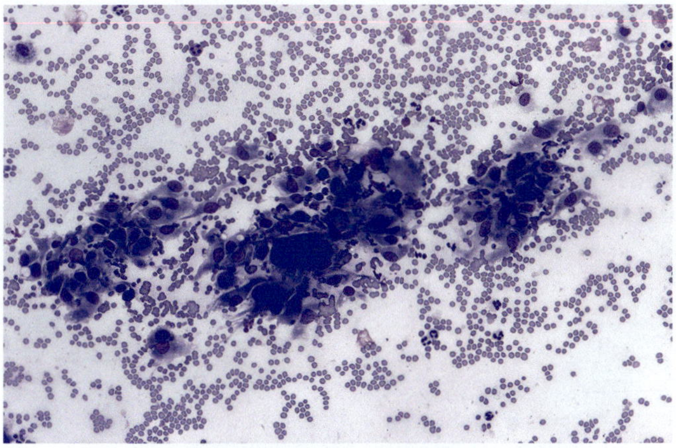

Fig. 6.113 Cytology: many round to spindled epithelioid macrophages along with a single multinucleated macrophage are present. Note the absence of intracytoplasmic lipid

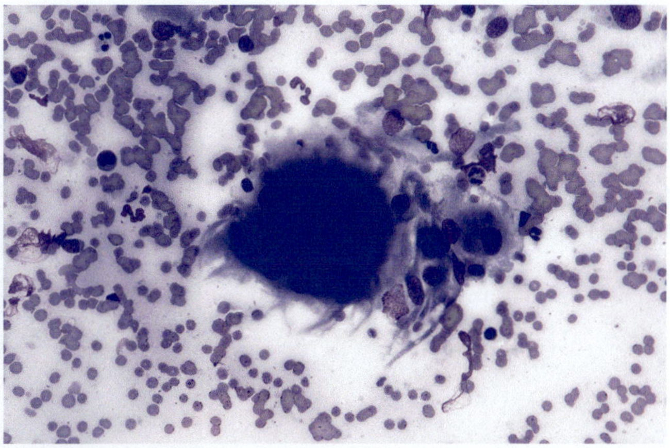

Fig. 6.114 Cytology: multinucleated giant cells and some macrophages. Note the absence of intracytoplasmic lipid

6.3.7 Sarcoidosis/Sarcoid Reaction

In human medicine, sarcoidosis is a granulomatous, chronic, immune-mediated disease whose pathogenesis is not yet fully understood. It is usually a multisystem disease involving internal organs such as lungs, kidneys, and spleen, associated with lesions in lymph nodes, joints, muscles, eyes, and so on. About 30% of cases also involve the skin. When the skin is the only affected body site, lesions are defined as "Sarcoid reaction" and not sarcoidosis [48].

In dogs, it is an extremely rare disease, only four cases have been published, and, in none of them, an involvement of the internal organs or metabolic alterations has been reported; therefore, to date, it would be more correct to speak of a *Sarcoid reaction* in canine patients. Since the lack of a large number of reported cases in veterinary dermatology in dogs, the lesions are not clinically well cataloged, and from what is published, they are mostly represented by not-pruritic small papules, papular–nodular and plaque-like lesions on the trunk, neck, face, and pinna [45–47] (Fig. 6.115).

Cytological Findings

Histopathology of *sarcoid reaction* shows many confluent and round-in-shape granulomas, with a linear sub-epidermal plaque-like way of growth. These granulomas are composed of epithelioid macrophages intermingled with a few lymphocytes and surrounded by fibroblasts. Neutrophils are not usually present (naked granulomas) (Figs. 6.116, 6.117, and 6.118).

Cytologically, in good-quality slides, due to the particular structure of granulomas, histiocytes are released in large and roundish cohesive groups of overlapped epithelioid cells, which, at low magnification, show an organization resembling that they have in origin (Fig. 6.119). Histiocytes are easily recognizable at the periphery of groups and have oval

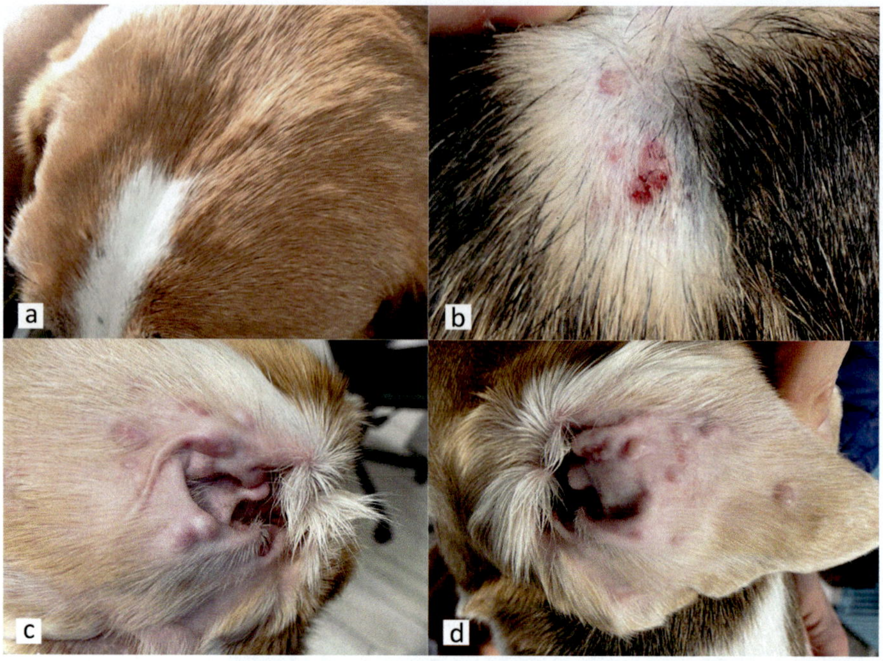

Fig. 6.115 Sarcoid reaction: (**a**) small papules on the haired skin of the head; (**b**) small papules on the skin of the right shoulder; (**c**) papular nodules on the inner surface of the right pinna; and (**d**) left pinna in a Beagle with cutaneous sarcoidosis. (Courtesy of DVM C. Bonfoco—Italy)

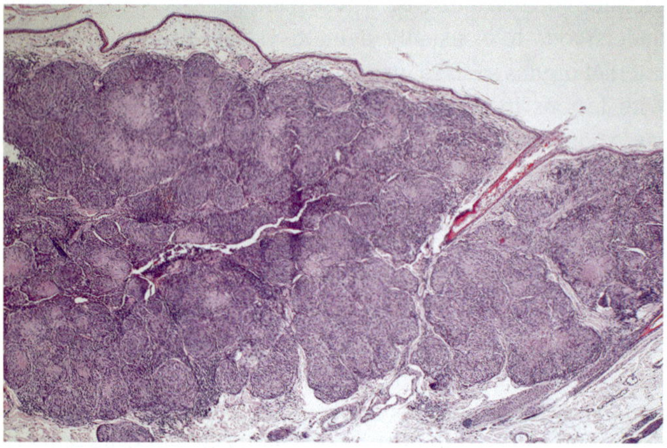

Fig. 6.116 Sarcoid reaction histopathology: granulomatous lesions in the upper part of the dermis with a lateral way of growth

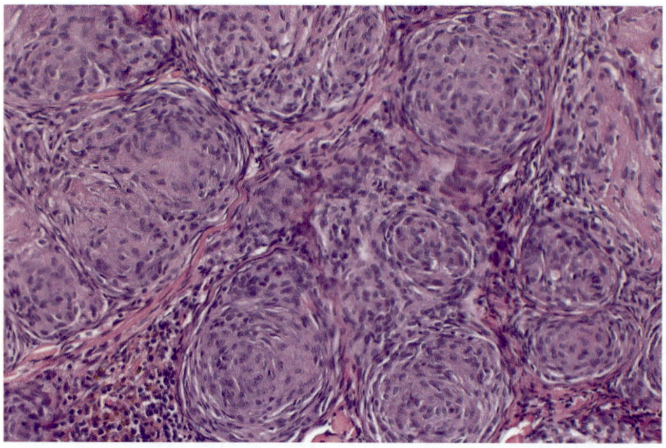

Fig. 6.117 Sarcoid reaction histopathology: multiple roundish granulomas separated by a thin stroma rich of lymphocytes

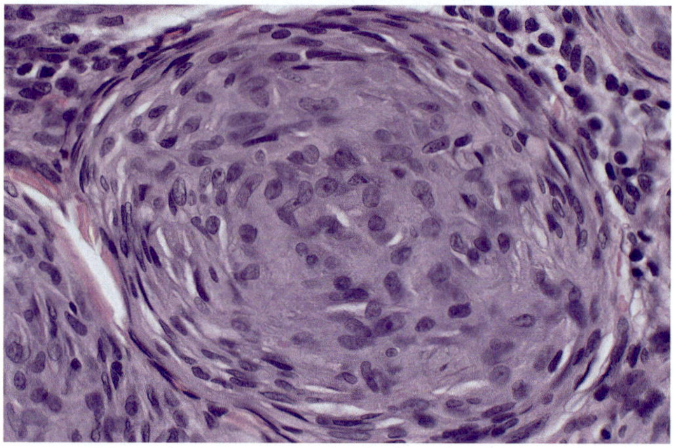

Fig. 6.118 Sarcoid reaction histopathology: high magnification of a sarcoid granuloma, in which epithelioid macrophages surrounded by thin spindle fibroblasts are recognizable

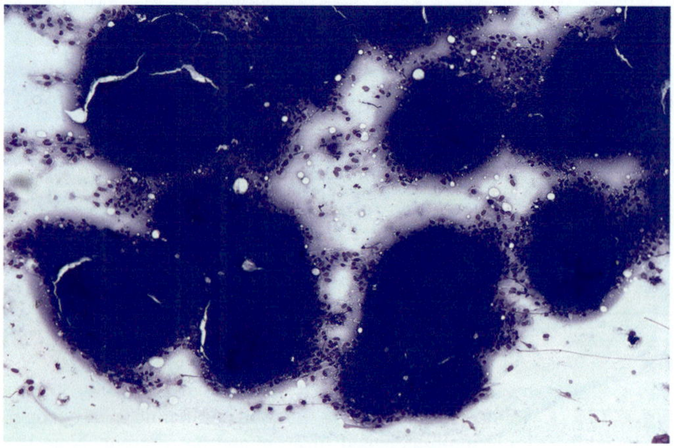

Fig. 6.119 Sarcoid reaction cytology: many roundish groups of cells are collected

to reniform nuclei and medium-large pale blue/gray cytoplasm. There is a variable amount of small lymphocytes and scarce spindle cells of a fibroblastic/myofibroblastic nature, which tend, in histopathological specimens, to surround the granulomas (Figs. 6.120 and 6.121).

The cytological features of this very rare skin disorder are not differentiable from other granulomatous lesions such as the sterile granuloma syndrome, canine reactive cutaneous histiocytosis, and CanL, also if in the latter a high number of plasma cells is expected; furthermore, since clinical presentations of sarcoid reaction can be very similar to the aforementioned diseases, histopathology is needed for final diagnosis.

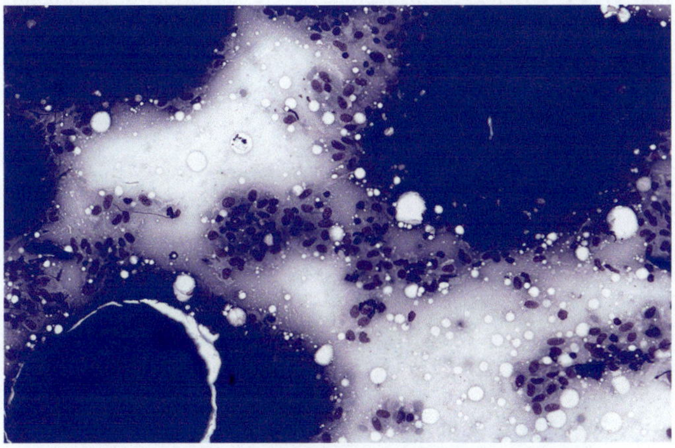

Fig. 6.120 Sarcoid reaction cytology: epithelioid macrophages are recognizable at the periphery of granuloma

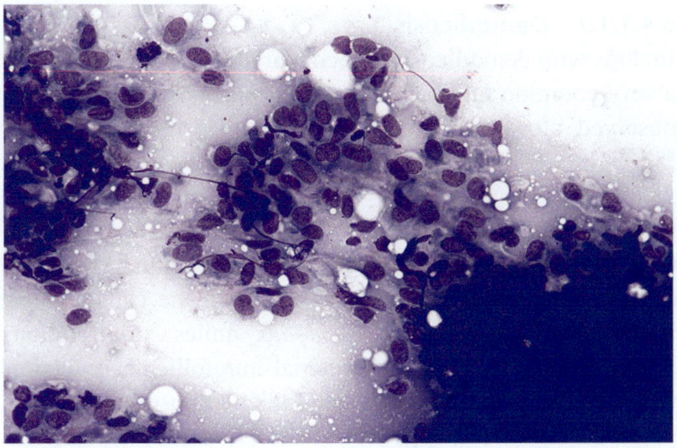

Fig. 6.121 Sarcoid reaction cytology: epithelioid macrophages show from oval to reniform nuclei, fine chromatin, inconspicuous nucleoli, and pale gray/blue cytoplasm

6.4 Pustules

Specimens collected from pustules are always useful and very often provide an immediate diagnosis. The exact anatomical localization of the pustules, follicular or non-follicular, is very important to determine the underlying cause.

6.4.1 Pustular Diseases in Dogs

6.4.1.1 Diseases Characterized by Follicular Pustules

Follicular pustules indicate an infection localized in the follicular lumen, and, when observed, an intrafollicular disease such as bacterial folliculitis or demodicosis must be suspected.

6.4.1.1.1 Bacterial Folliculitis

The most common cause of follicular pustules in dogs is *bacterial folliculitis*. Pyoderma is a very common disease in dogs, and the predisposing causes are numerous and of different origins. Microorganisms, parasites, allergic diseases, keratinization disorders, and endocrine disorders can frequently cause an injury to the skin barrier, allowing resident bacteria such as *Staphylococcus pseudintermedius* to cause infection [51].

Clinically, pustules are mainly follicular, but because the latter are delicate and transient, they rapidly dehydrate and form crusts or epidermal collarettes (Fig. 6.122) [49–51].

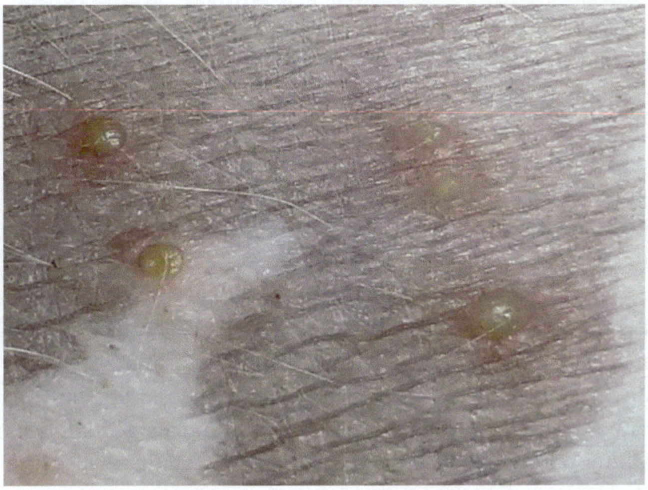

Fig. 6.122 Multiple follicular pustules: note that hairs emerge from the center of the pustules

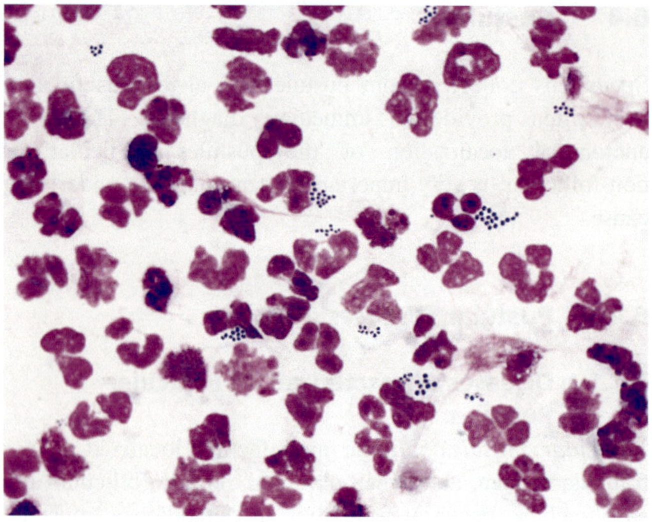

Fig. 6.123 Cytology of pyoderma: karyolitic neutrophils with many intracytoplasmic cocci

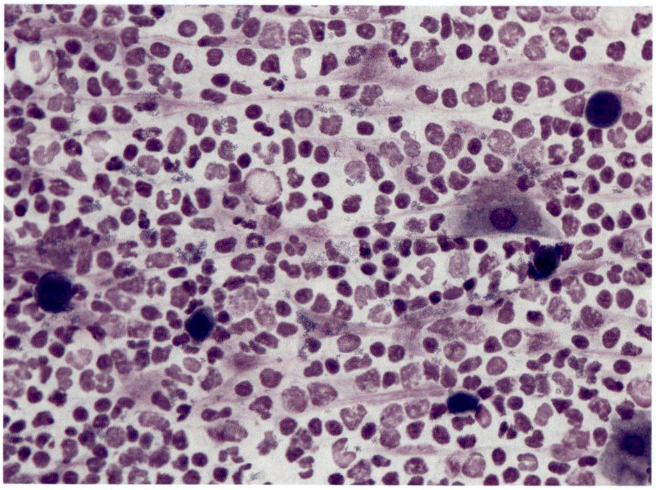

Fig. 6.124 Cytology of pyoderma: severe karyolysis of neutrophils with many cocci and AKs

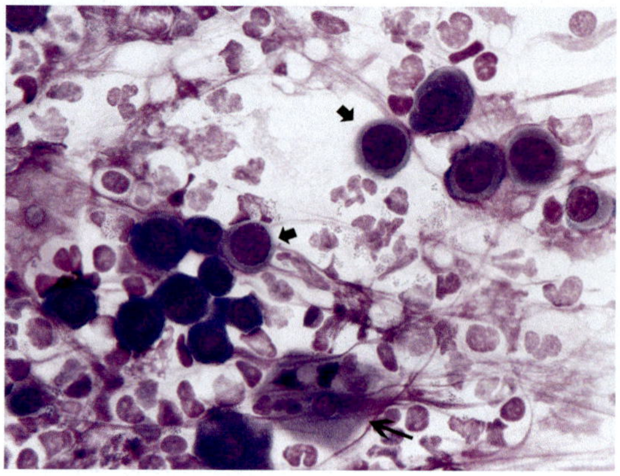

Fig. 6.125 Cytology of pyoderma: severe karyolysis of neutrophils with many cocci and both squamous not AKs (long arrow) and true AKs (short arrows)

Cytological Findings

Slides from an intact pustule contain a large number of neutrophils characterized by nuclear degeneration represented by swollen and hypochromatic nuclei with a loss of the classical multilobed silhouette, due to severe cell membrane damage that permits water to penetrate into the cell. As mentioned in Chap. 4, this common type of nuclear degeneration is known as *karyolysis*.

In the cytoplasm of neutrophils, a variable number of coccoid bacteria are usually present, arranged singly, in pairs, or in tetrads. Many bacteria are extracellular, but this is linked to the rupture of severely degenerated neutrophils that inevitably occur during the preparation of the slides (Fig. 6.123).

Karyolitic neutrophils have "sick" nuclei and once broken during slide preparation, the nucleic material forms streaks of variable length. Together with neutrophils, it is common to find a few of both lamellar anucleate keratin (corneocytes) and squamous nucleated large keratinocytes belonging to the granulosum or spinous layers, which are the cells that compose the roof of the pustule and that usually exfoliate into the pus.

The latter must not be confused with acantholytic keratinocytes (AKs), roundish and hyperbasophilic epidermal cells more characteristic of pemphigus foliaceus (PF); anyway, it must be aware that not only auto-antibodies but either bacterial proteolytic enzymes can induce acantholysis acting on cell junctions and splitting the keratinocytes from themselves. For this reason, in some pustular pyoderma, a variable number of AKs, usually less than in PF, can be observed (Figs. 6.124 and 6.125) [49–51].

6.4.1.1.2 Demodicosis

In dogs with demodicosis, a secondary bacterial folliculitis is a very common finding, so follicular pustules are very often observed along with comedones (Figs. 6.126 and 6.127) [52–54].

Cytological Findings

Cytology specimens from pustules are similar to those described in pyoderma because, although the cause of the lesions is intrafollicular *Demodex* mites proliferation, pustules are caused by the bacterial intrafollicular infection that secondarily occurs; so, when follicular pustules are present in a demodectic dog, they are surely caused by bacteria (Figs. 6.128 and 6.129). In addition to karyolitic neutrophils that phagocytize cocci, a variable number of adult and/or immature mites and their eggs are observed (Figs. 6.130, 6.131, and 6.132). As mentioned for pyoderma, it is also possible to observe a few AKs.

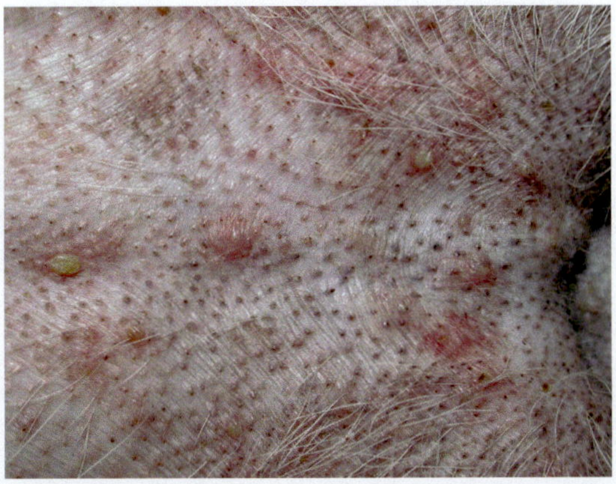

Fig. 6.126 Follicular pustules and comedones in a puppy with demodicosis

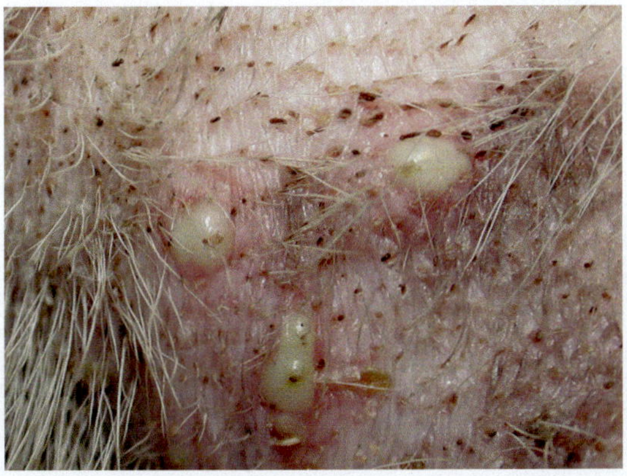

Fig. 6.127 Follicular pustules: at high magnification, a hair that emerges from the center of the pustule can be easily recognizable

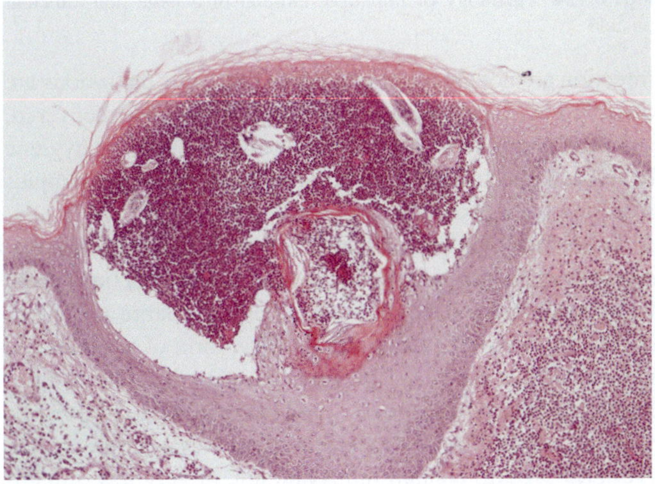

Fig. 6.128 Histology of pustular demodicosis: follicular pustule containing many *Demodex* mites

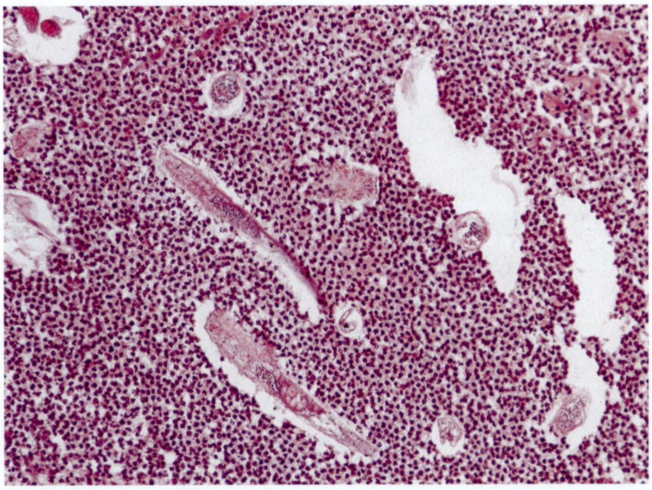

Fig. 6.129 Histology of pustular demodicosis: at high magnification, mites are clearly recognizable

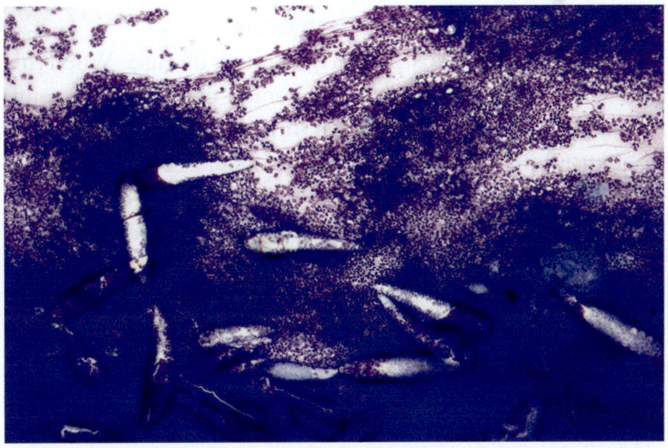

Fig. 6.130 Cytology of pustular demodicosis: a large number of *Demodex canis* are collected from a follicular pustule

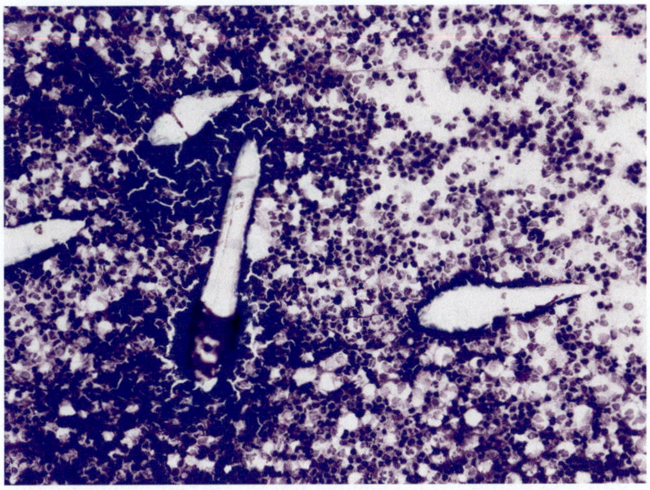

Fig. 6.131 Cytology of pustular demodicosis: many *Demodex canis* immersed in a strongly karyolitic neutrophilic exudate

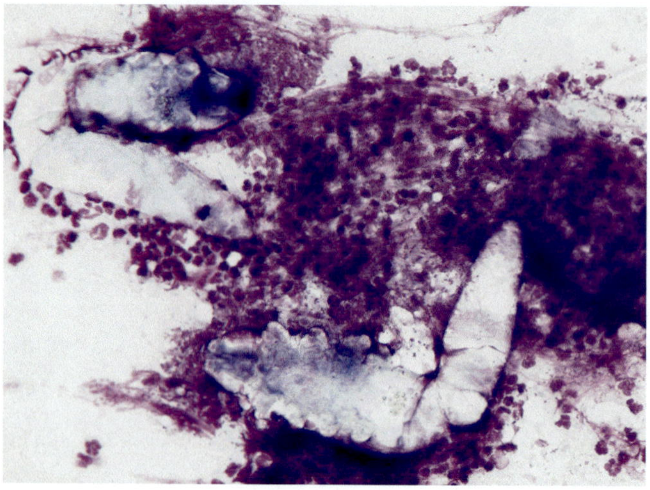

Fig. 6.132 Cytology of pustular demodicosis: two adult *Demodex canis* and karyolitic neutrophils, which phagocytize many cocci

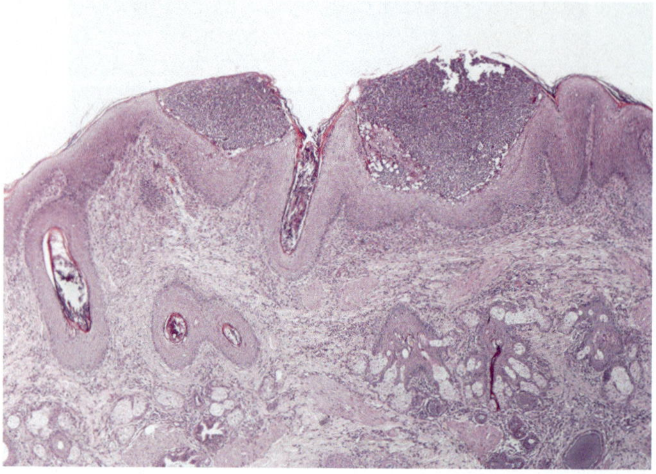

Fig. 6.133 Histology of impetigo: two large non-follicular pustules. Note that follicles are not involved

6.4.1.1.3 Pustular Dermatophytosis

In dogs, dermatophytosis is not a common disease. When this occurs, the infection involves the keratin of stratum corneum and hairs. The development of a follicular pustule secondary to neutrophil recruitment into the follicular lumen is rarely observed in histopathological samples; when it occurs, the amount of pus is not so high to determine the clinical visualization of a follicular pustule.

Therefore, we can state that, in the presence of clinically visible follicular pustules, it is very unlikely, that the cause could be a dermatophyte. Non-follicular pustules are instead reported in rare forms of dermatophytosis as discussed below.

6.4.1.2 Diseases Characterized by Non-Follicular Pustules

Unlike follicular pustules that are always linked to bacterial infections, different diseases, both infectious and sterile, can cause non-follicular pustules. As mentioned, non-follicular pustules vary in size according to the cause and their extension into the epidermal layers.

6.4.1.2.1 Pustular Non-Follicular Staphylococcal Infection (Impetigo)

In *superficial pyoderma,* it is possible to observe intraepidermal subcorneal pustules that are not localized in the follicular lumen (Figs. 6.133 and 6.134) [56]. *Impetigo* is a form of surface pyoderma mostly reported in puppies and young dogs, mainly hunting breeds, clinically characterized by small to large-sized *non-follicular pustules*, localized on the ventral part of the body, especially on the abdomen, groin, and axilla [57]. Pustules are of varying shapes, both small and round, and with a tendency to merge and form very large pustules of irregular silhouette. The causes of bacterial infection are not always ascertained, but, since the lesions are observed in young animals and do not recur once recovered, stressful events or local trauma associated with poor hygiene have been speculated to be possible causes. Rarely, puppies suffering from distemper can develop impetigo.

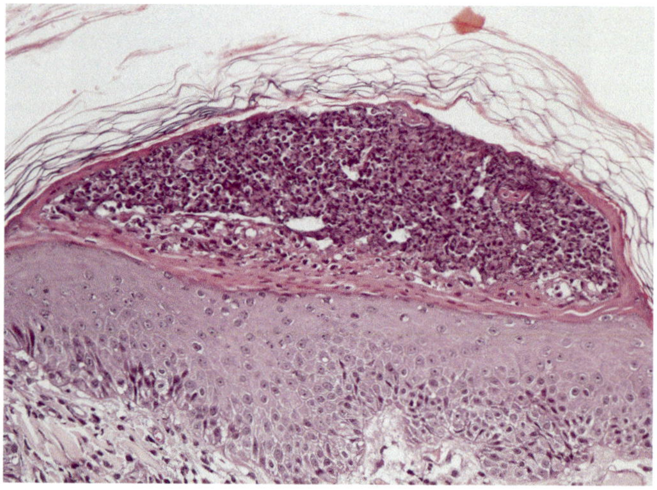

Fig. 6.134 Histology of impetigo: closeup of a large non-follicular pustule

In adults or old dogs affected by several immunosuppressive systemic diseases such as spontaneous or iatrogenic hypercortisolism, it is common to observe large and often coalescent non-follicular pustules. This form of pyoderma is named *bullous impetigo;* the latter is a term that derives from human medicine, and from a medical etymological point of view, it would be an incorrect term because the bullae (vesicles >1 cm) contain fluids and not, or very few, inflammatory cells; however, this term is commonly used in dermatological clinical practice to describe very large pustules (Fig. 6.135). In this non-follicular pustular pyoderma, exotoxins produced by *Staphylococcus pseudintermedius* bind to the extracellular portion of the desmoglein-1 and, in association with the development of proteolytic enzymes, are the cause of the keratinocytes splitting and of the consequent development of pustules [55, 58, 59].

Cytological Findings

The cytology of an impetiginous pustule is very similar to that described for a follicular pustule. In many cases of impetigo, especially in the bullous variant, the karyolitic nuclear changes are so severe as to make it difficult, for an inexperienced cytologist, to interpret those cells as severely degenerated neutrophils. Another characteristic is the high number of bacteria detected compared with those observed in typical follicular pustules. This difference could be explained because *bullous impetigo* is usually secondary to severe and systemic immunosuppressive diseases that can impair the phagocytic ability of neutrophils, allowing them to proliferate massively (Figs. 6.136 and 6.137).

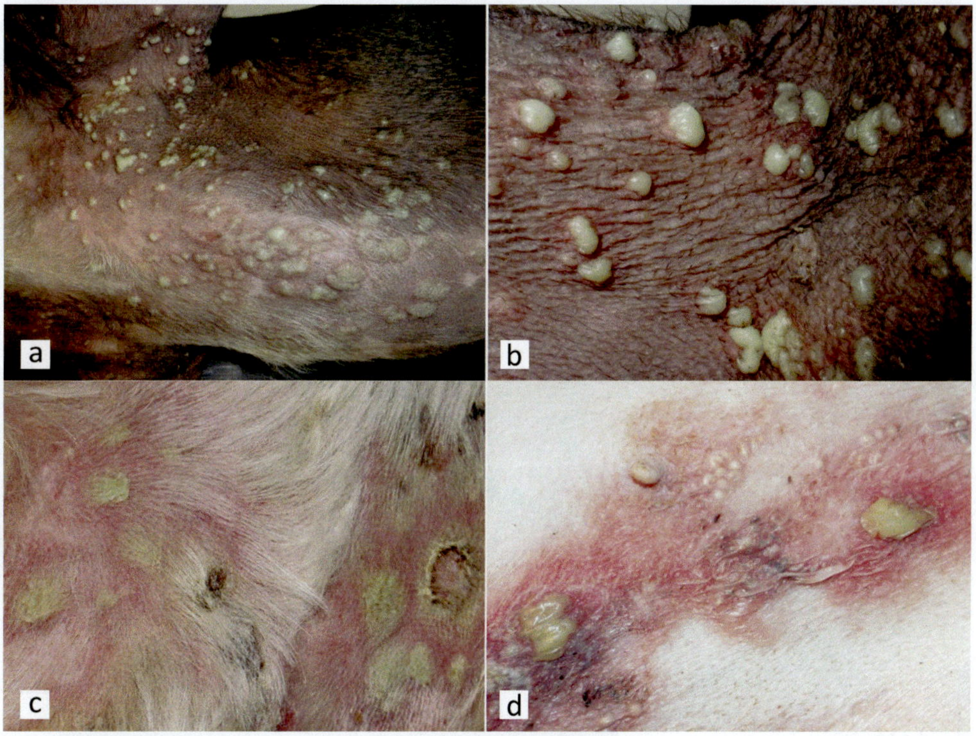

Fig. 6.135 Impetigo: (**a**) multiple large non-follicular pustules on the sternum and chest of a dog with juvenile impetigo; (**b**) closeup of the previous lesions; (**c**) bullous impetigo in a dog treated with chronic systemic corticosteroid therapy; and (**d**) bullous impetigo in an old dog with spontaneous hypercortisolism. Note the simultaneous presence of comedones and skin atrophy

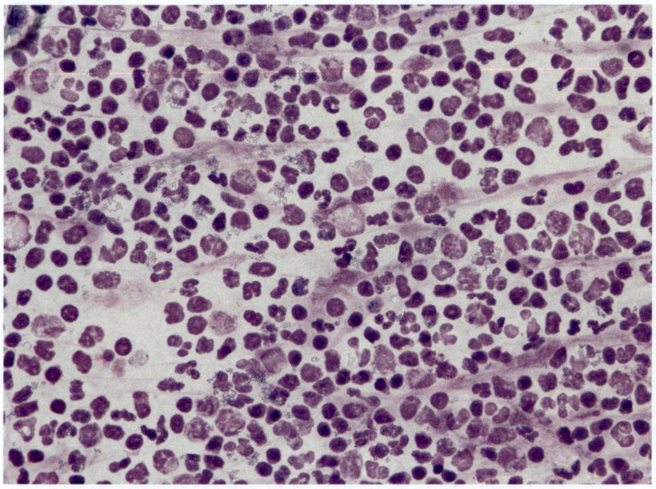

Fig. 6.136 Cytology of bullous impetigo: many karyolytic neutrophils characterized by non-lobulated and hypochromatic and swollen nuclei. Many cocci are immersed in the pus

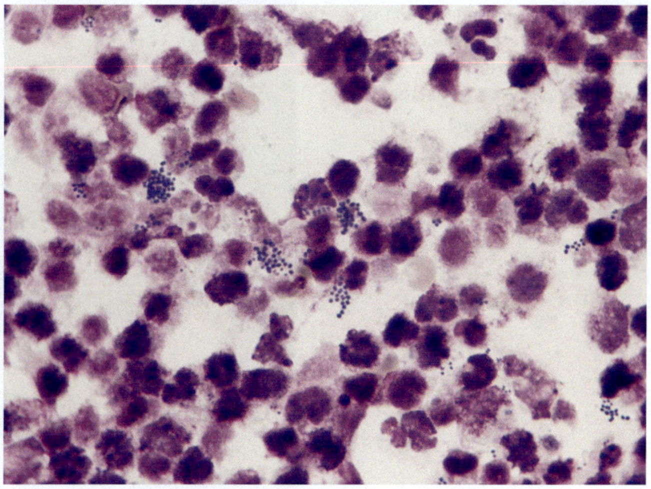

Fig. 6.137 Cytology of bullous impetigo: note how many coccoid bacteria are present. Most of the nuclei of neutrophils are faded and poorly recognizable

6.4.1.2.2 Pemphigus Foliaceus

PF is the most common sterile disease characterized by non-follicular pustules. PF is an autoimmune disease histologically characterized by intra-epidermal pustules, which can be localized sub-corneal or in the spinous or granular, layers [60–62]. The pathogenic mechanism of the pustular formation is *acantholysis* during which the splitting of keratinocytes following the breakage of the intercellular junctions (*desmosomes*) occurs. Auto-antibodies of the PF target the extracellular component of some cadherins (adhesion molecules) that make up *desmosomes* (desmocollin-1 and to a lesser extent desmoglein-1 and plakoglobin) [62, 63]. Following the attack to desmosomes by autoantibodies, neutrophils that fill the intra-epidermal cleft created by acantholysis are recruited and give origin to the typical lesions of PF, the pustules.

PF is a disease that is clinically characterized by large *non-follicular pustules* (Fig. 6.138). Since pustules rapidly dehydrate, they are not always present on the skin of dogs affected with PF but can be detectable as yellowish crusts, mainly located on the skin of the abdomen, face, pinna, muzzle, and bridge of the nose. The skin of the nose often shows an "inflammatory" depigmentation with loss of the typical cobblestone appearance and development of yellowish crusts over the dorsal surface. Pads, scrotum, and mucocutaneous junctions are other sites where pustular or crusty lesions are often observed, and, in generalized cases, lesions can be spread all over the body, sometimes associated with fever and malaise (Fig. 6.139) [60–62].

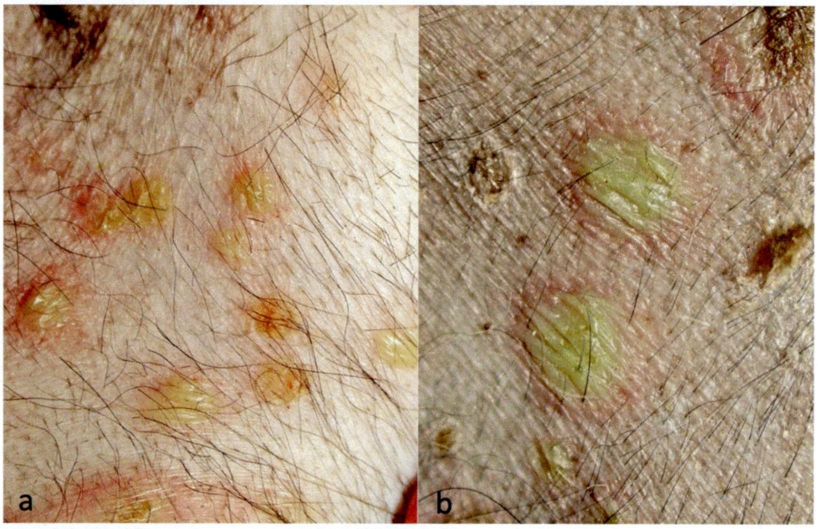

Fig. 6.138 Non-follicular pustules in PF: (**a**) large non-follicular pustules in a dog affected by PF and (**b**) closeup of two large non-follicular pustules

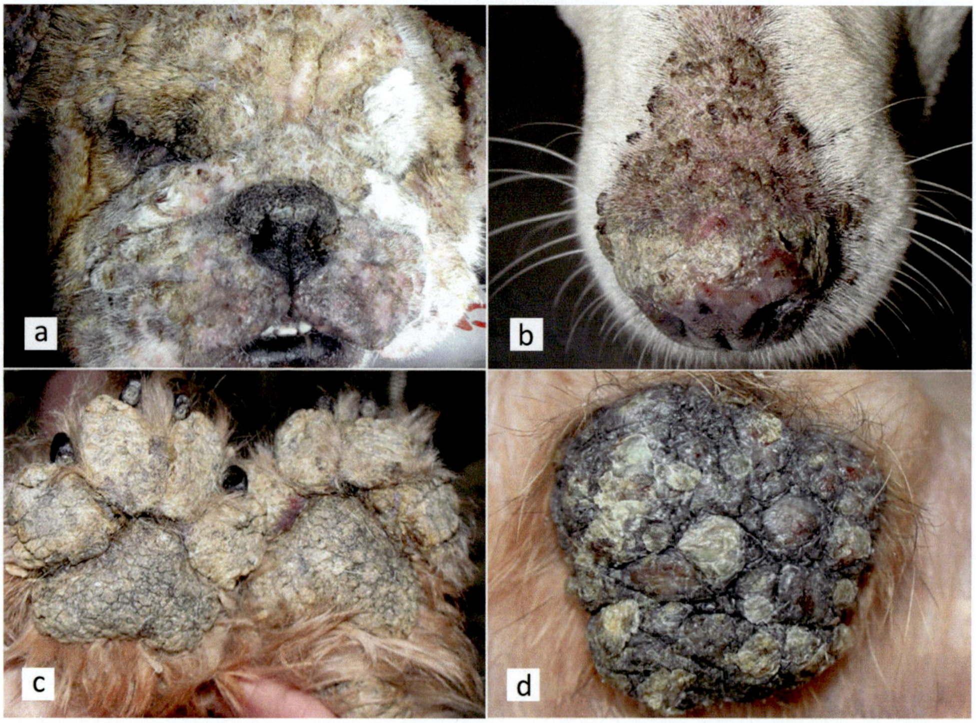

Fig. 6.139 PF in dogs: (**a**) yellowish crusts spread all over the face of an English bulldog; (**b**) inflammatory depigmentation of the nose and yellowish crusts; (**c**) diffuse crusts on the pads; and (**d**) yellowish crusts on the scrotum

Cytological Findings

As aforementioned, the histopathology of lesions in PF is characterized by intra-epidermal pustules that usually contain many acantholytic epithelial cells (Figs. 6.140, 6.141, and 6.142).

Because PF is a sterile disease, cytological samples obtained by intact pustules are typically characterized by many non-degenerated neutrophils with *segmented* nuclei in which nuclear lobes are usually well preserved. In intact and very large pustules that have arisen for several days, neutrophils can manifest morphological alterations due to cellular aging, such as poorly evident nuclear lobes and hyperchromatic nuclei. Together with neutrophils, a large number of nucleated keratinocytes are observed. These cells are the typical cells of PF, linked to the process of acantholysis that characterizes the pathogenesis of the pustule's formation; for this reason, these epidermal cells are named *acantholytic cells* or AKs.

AKs come from both the spinous and granular layers, depending on which part of the epidermis the acantholysis occurs; they have mostly a typical round silhouette but, since they are epithelial cells, can show polygonal or oval and less frequently spindle shape. This different morphology of AKs must be taken into account when the classic round-shaped epithelial cells are few compared to those with less classical morphology because they could confuse inexperienced cytologists. The cytoplasm is often so dark (deep blue) that it can mask the nuclear profile when viewing the slides at low magnification. The nuclei are round and central, with a single nucleolus often evident. AKs can be released into the pustules, both singly and in small clusters (Figs. 6.143, 6.144, 6.145, and 6.146).

Although typical, AKs are not pathognomonic of PF as it is possible to observe them in other infectious (bacteria and dermatophytes) and sterile pustular diseases as well. In any case, when many AKs are collected from an animal with

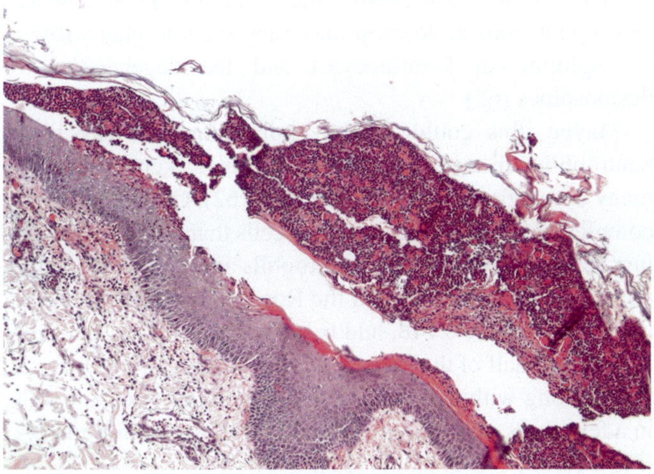

Fig. 6.140 Histology of PF in a dog: large non-follicular pustule; the pinkish more pale areas in the pus are AKs

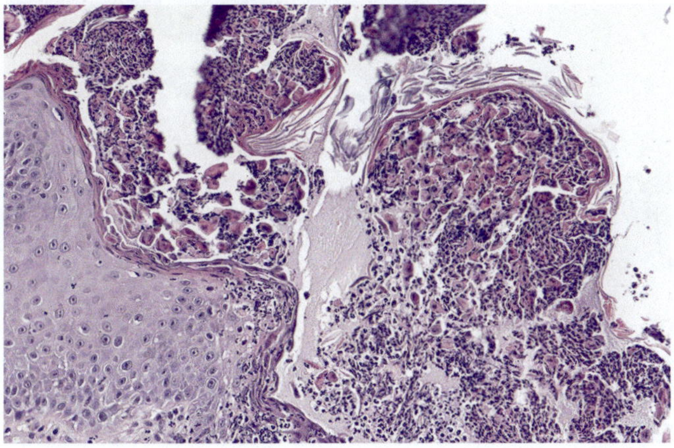

Fig. 6.141 Histology of PF in a cat: large subcorneal non-follicular neutrophilic pustule filled with many AKs

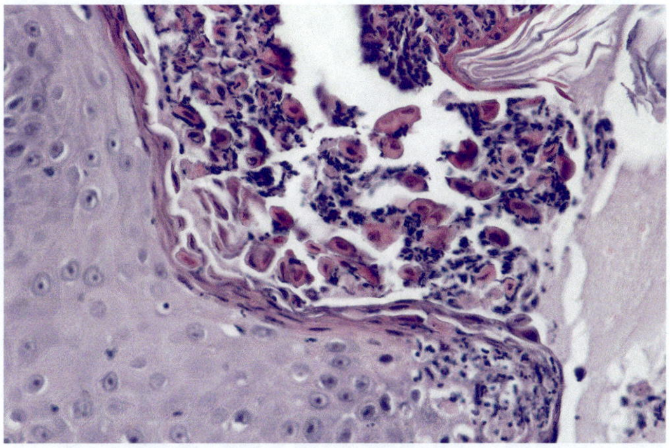

Fig. 6.142 Histology of PF in a cat: at high magnification, acantholytic cells are well recognizable

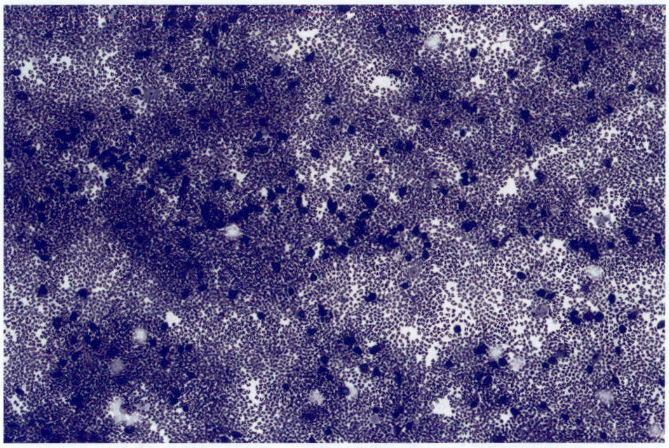

Fig. 6.143 Cytology of PF: at low magnification, acantholytic cells are not clearly recognizable, but they can be interpreted as roundish cells with very dark cytoplasm

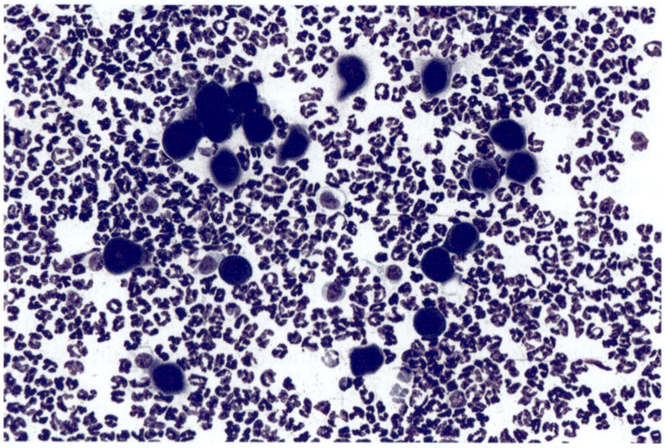

Fig. 6.144 Cytology of PF: many well-preserved segmented neutrophils, no bacteria, and both single and small clusters of AKs. Note that, in many AKs, the nuclei are not clearly visible due to deep blue cytoplasm

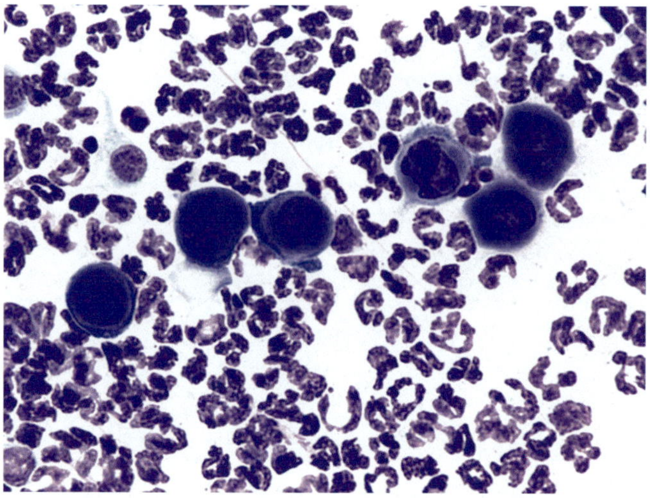

Fig. 6.145 Cytology of PF: both roundish and polygonal AKs

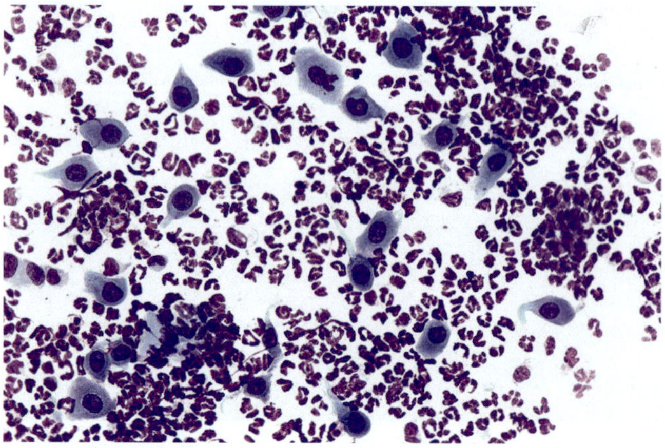

Fig. 6.146 Cytology of PF: neutrophils and many AKs; the latter show a more large and pale blue cytoplasm along with a spindled morphology

typical clinical lesions, they are highly indicative of PF and histology must be performed to confirm the disease.

In many slides, it is possible to observe many single AKs surrounded by a single or double line of granulocytes arranged as *flower's petals* that resemble a *rosette-like* arrangement; this cytological finding is also not pathognomonic of PF, but, when rosette-like arrangements are very numerous, the autoimmune disease should be suspected. Exceptionally, many of these arrangements can be observed in the case of pustular pyoderma, so the interpretation must always be done in the light of clinical presentation.

Rosette-like arrangement in PF could be explained by a work on four dogs affected by PF in which electron microscopy of pustules revealed that at the points where neutrophils contact AKs, half-desmosomes of AKs were observed in the invaginations of neutrophils. Furthermore, seems that neutrophil granules are secreted to the surface of AKs and contribute to degenerating the half-desmosome. Finally, neutrophils were also observed within the epidermis adjacent to the pustule. These findings speculate that, in the area where neutrophils contact desmosomes, they seem to play a role in the splitting of keratinocytes and the degeneration of desmosomes [65].

Maybe, this could be the explanation for why many neutrophils are arranged around the single keratinocytes in many cytological specimens from PF [62]. Often neutrophils completely surround the epithelial cells that are barely visible just below them; when the neutrophils are not overlapped on the AKs, the morphology of the Rosettes is clearly recognizable (Figs. 6.147, 6.148, and 6.149).

In about half of the dogs with PF, cytology of pustules can show, along with neutrophils, a high number of eosinophils; in a few cases, they are more numerous than neutrophils [64] (Figs. 6.150, 6.151, and 6.152).

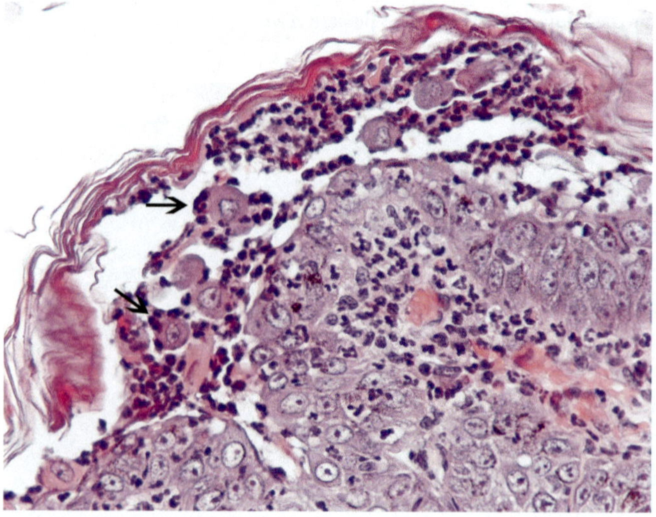

Fig. 6.147 Histology of PF: rosette-like arrangement; neutrophils are arranged around acantholytic cells (*arrows*)

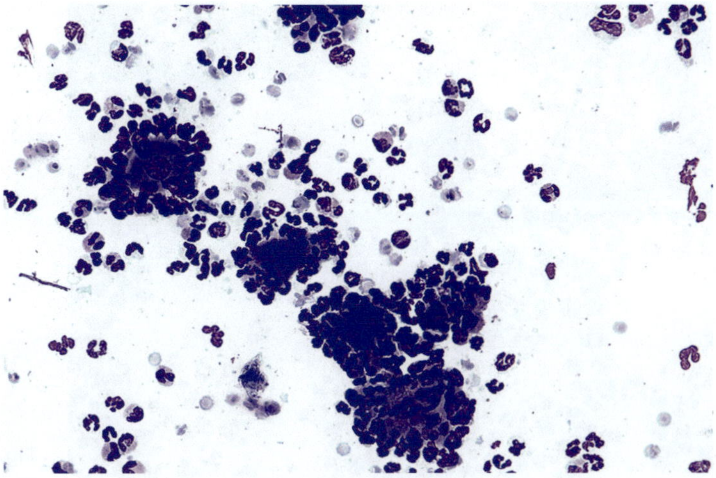

Fig. 6.148 Cytology of PF: many neutrophils surround single acantholytic cells obscuring most of them

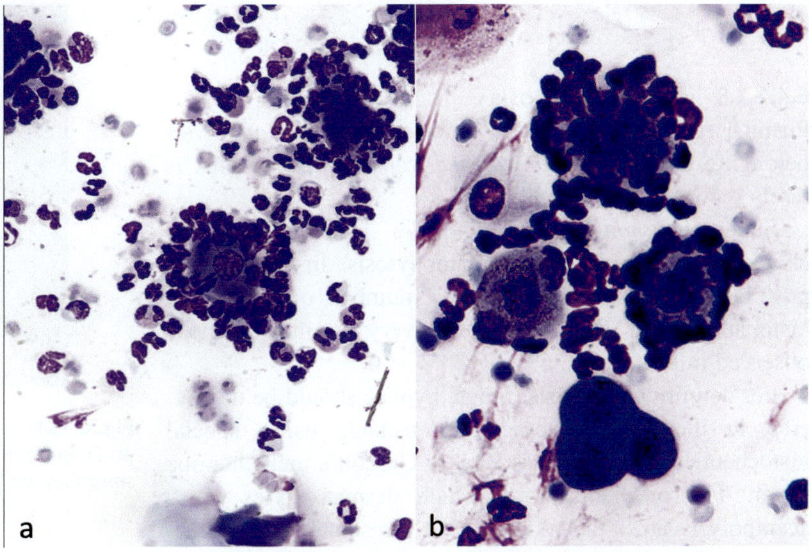

Fig. 6.149 Cytology of PF: (**a**) typical features of rosette-like arrangement. Note the presence of eosinophils; and (**b**) rosette-like arrangement in three cells. A small cluster of AKs is also present

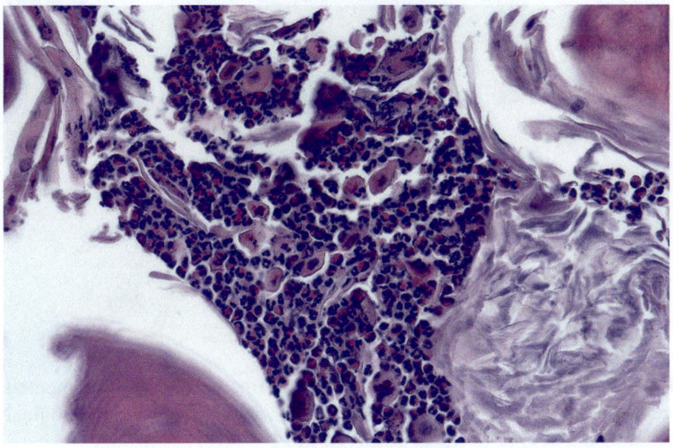

Fig. 6.150 Histology of PF: acantholytic pustule with a high percentage of eosinophils (more numerous than neutrophils)

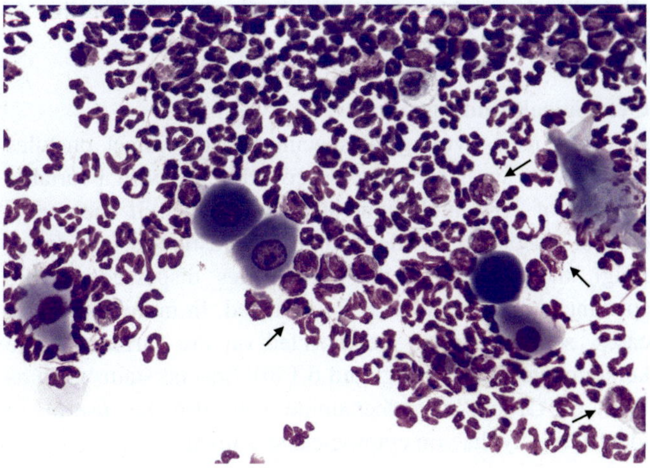

Fig. 6.151 Cytology PF: neutrophils, acantholytic cells, and some eosinophils (arrows)

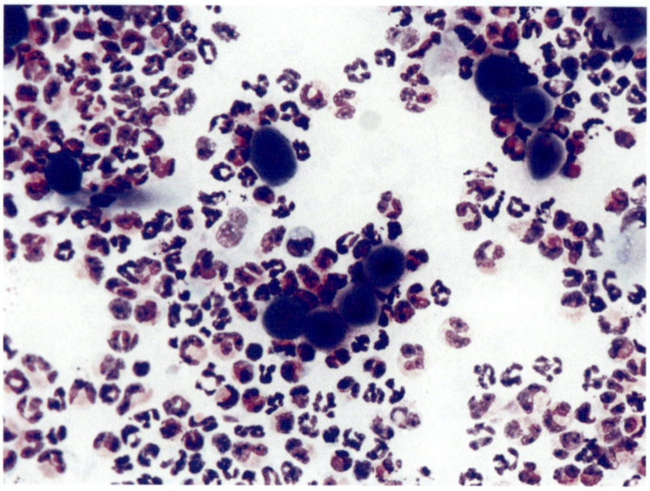

Fig. 6.152 Cytology PF: note that eosinophils are more numerous than neutrophils

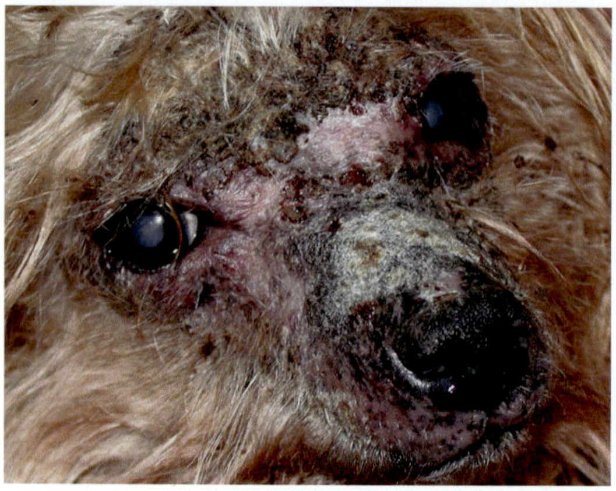

Fig. 6.153 Pustular dermatophytosis: yellowish crusts on the face and on the bridge of the nose. Note that clinical presentation is similar to PF. (Courtesy of DVM. C. Caporali—Italy)

6.4.1.2.3 Dermatophytosis

Pustular dermatitis in dogs affected by *dermatophytes* belonging to the *Trichophyton mentagrophytes* complex is rarely reported [68, 70, 71].

Superficial *non-follicular pustules* histopathologically characterize this form of dermatophytosis. In some dogs, pustules may contain a variable number of AKs and histopathological findings can be very similar to those observed in PF [66, 67, 69].

For definitive diagnosis, fungal hyphae should be discovered in the stratum corneum, detectable using special histochemical stains such as PAS or Grocott's methenamine silver. The mechanisms by which dermatophytes cause acantholysis are still unknown, but, presumably, it occurs because of proteases secreted by the fungi [63, 70].

Superficial pustules and yellowish crusts characterize the clinical lesions with a facial distribution resembling that of PF (Figs. 6.153 and 6.154).

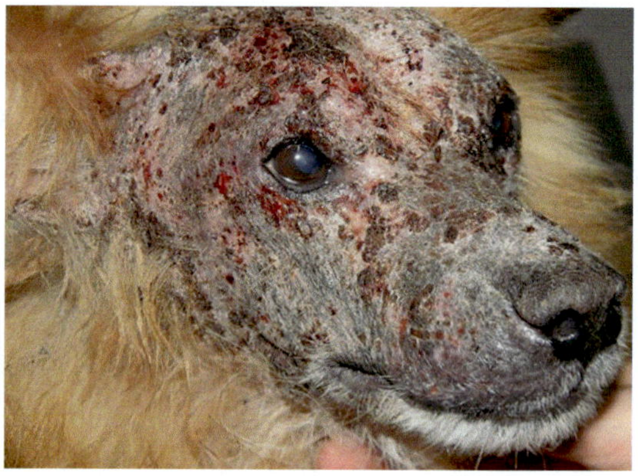

Fig. 6.154 *T. mentagrophytes* complex pustular dermatophytosis: alopecia and crusty lesions on face; intact pustules are not more visible since they are transformed with time in brown dehydrated crusts

Cytological Findings

As mentioned, cytological specimens from intact pustules may be very similar to those described in PF, with a variable number of AKs and segmented neutrophils. If there is a clinical suspicion of dermatophytic pustulosis, to detect fungi, the surface of the corneocytes dispersed on the specimens must be accurately evaluated. In more successful cases, septate hyphae are detected on the surface of the keratinocytes (Figs. 6.155 and 6.156). Special stains such as PAS or Grocott's methenamine can also be useful in identifying hyphae on cytological specimens.

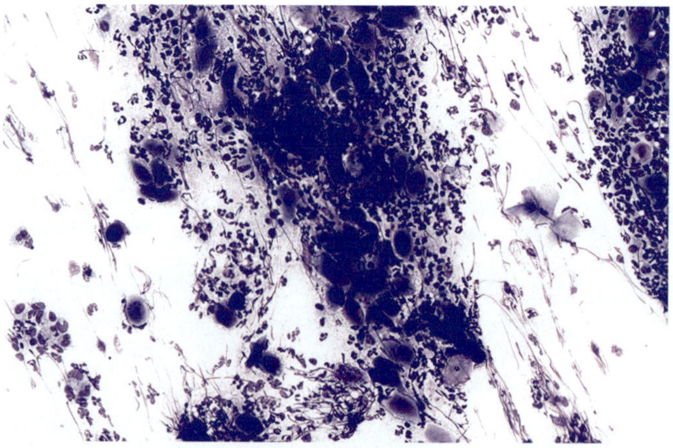

Fig. 6.155 Cytology of pustular dermatophytosis: segmented neutrophils and many acantholytic cells. Note that the cytological features are similar to those observed in PF

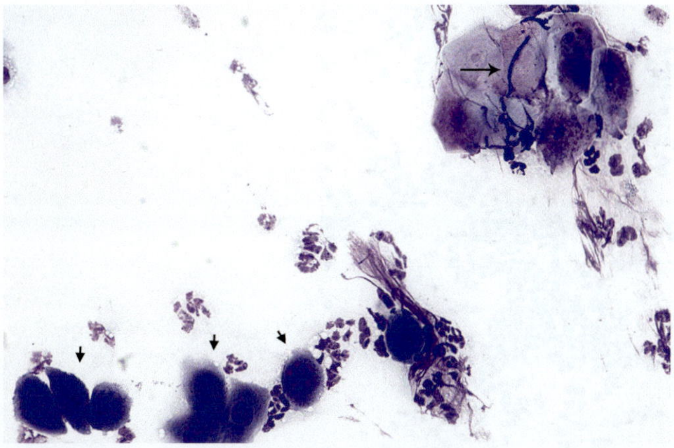

Fig. 6.156 Cytology of pustular dermatophytosis: along with AKs (short arrows) some hyphae of *T. mentagrophytes* are detected on the surface of the corneocytes (long arrow)

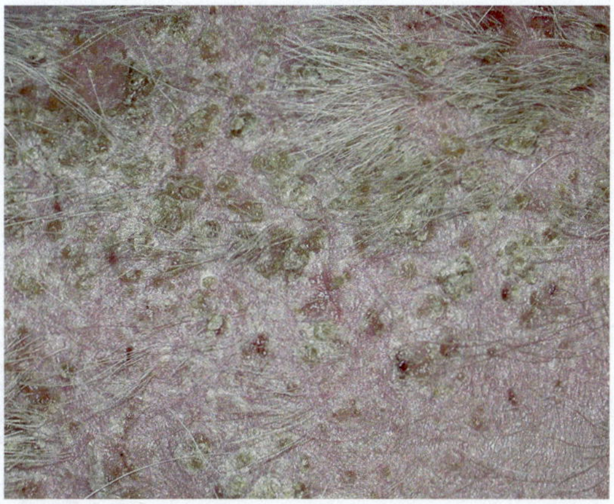

Fig. 6.158 Closeup of the same lesions of the previous image

6.4.1.2.4 Idiopathic and Drug-Related Sterile Pustular Diseases

Apart from PF, other immune-mediated sterile pustular diseases, such as *subcorneal pustular dermatosis* with *IgA deposits*, superficial acantholytic pustulosis due to *drug reaction*, and *idiopathic pustular dermatitis*, are very rarely reported in dogs. Given the limited literature on these forms of pustular dermatitis, their clinical aspects and localization of the lesions are not well cataloged; clinically, they are characterized by non-follicular pustules spread all over the body, which can mimic a PF (Figs. 6.157 and 6.158) [72–78].

Cytological Findings

Histology shows multiple superficial non-follicular neutrophilic pustules. Cytological slides are not diagnostic, as they are composed of segmented neutrophils and no or a few acantholytic cells (Figs. 6.159, 6.160, 6.161, and 6.162).

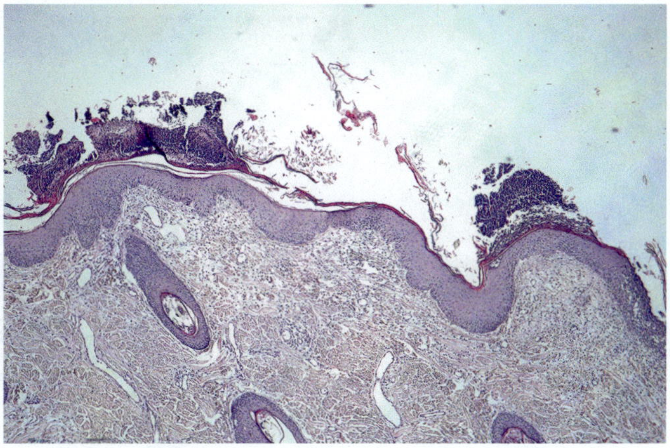

Fig. 6.159 Histology of sterile subcorneal pustulosis: two non-follicular pustules

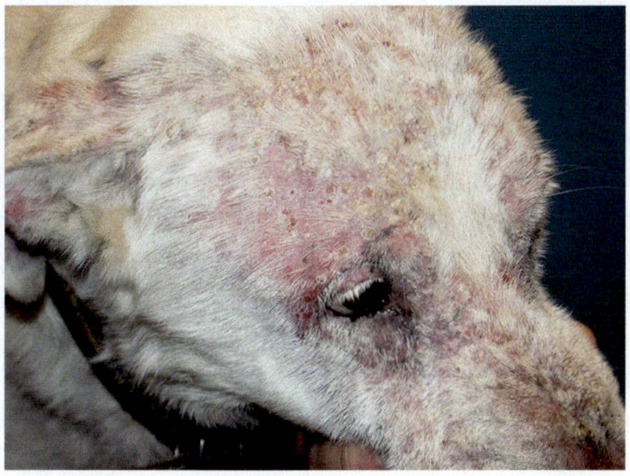

Fig. 6.157 Yellowish crusts secondary to pustule dehydration in a dog with sterile neutrophilic subcorneal pustulosis

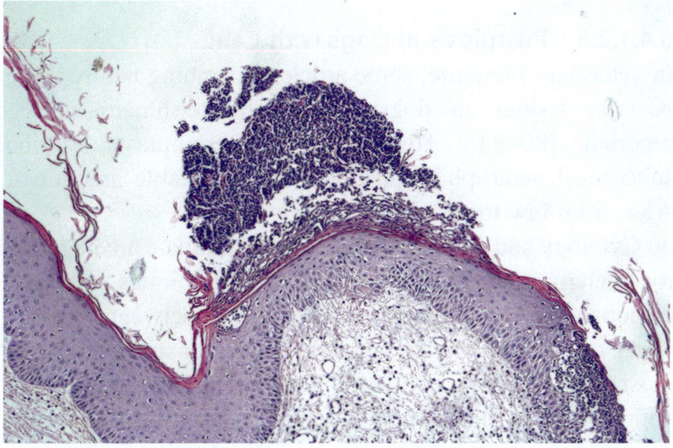

Fig. 6.160 Histology of sterile subcorneal pustulosis: no AKs are evident to high magnification

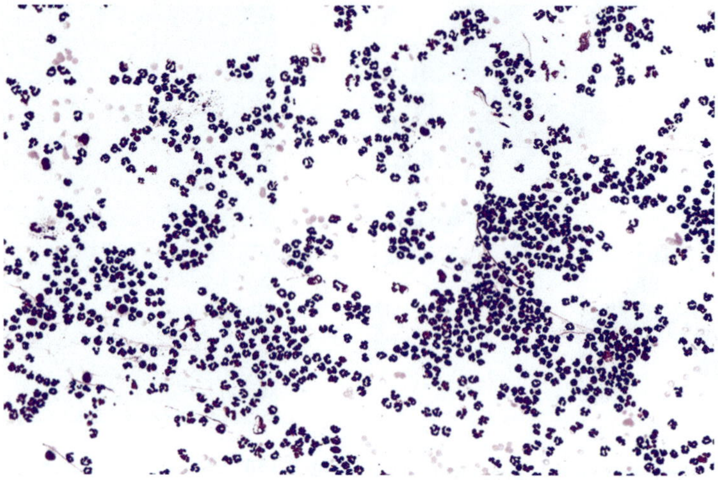

Fig. 6.161 Cytology of sterile subcorneal pustulosis: many well-preserved (segmented) neutrophils and no acantholytic cells

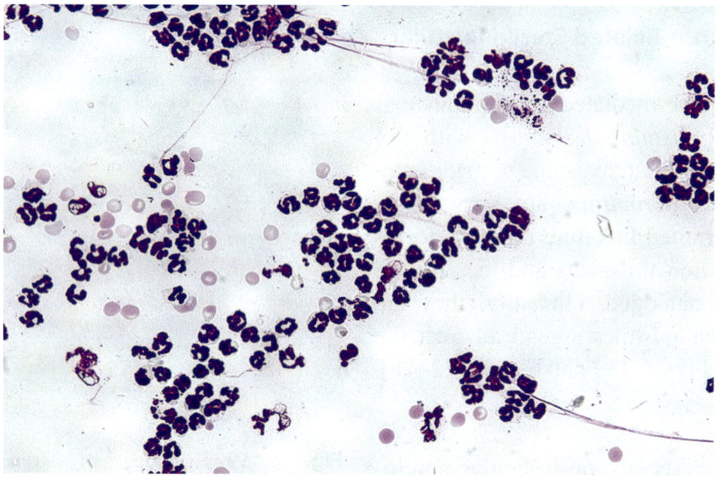

Fig. 6.162 Cytology of sterile subcorneal pustulosis: at high magnification, the nuclear lobes of neutrophils are well recognizable and no bacteria are detectable

6.4.1.2.5 Pustulosis in Dogs with CanL

In veterinary literature, some articles describing neutrophilic pustular lesions in dogs affected by leishmaniasis are reported [80–83]. Histopathological samples describe subcorneal neutrophilic pustules with a variable amount of AKs, from few to no one [79].

Cytology and cultures of the purulent exudate are negative for bacteria, and all dogs treated with antibiotics were not responsive to therapy. The relationship between clinical signs and leishmania is not easy to demonstrate since, in most dogs, pustular dermatitis does not improve following anti-protozoal therapy and systemic immunosuppressive therapy is necessary. This aspect indicates that lesions are due to an immune-mediated mechanism, probably triggered by leishmaniasis. This therapeutical behavior has been also confirmed in another case series of dogs with CanL [80, 81].

As CanL is endemic in many countries, it is not possible to exclude the possibility that two different diseases might affect animals simultaneously; for example, if many AKs are present in the pustules, a PF could be simultaneously present along with CanL. A criticism of this possibility is that sterile pustular diseases are very rare in dogs, and it seems strange that most dogs with this clinical presentation are simultaneously affected by leishmaniasis. However, to date, it has not been possible to demonstrate the real role of Leishmania in the development of pustular eruptions. Pustules have an aspect and distribution similar to those described for PF, and as occurs in the latter, systemic symptoms such as fever and malaise are sometimes associated probably because, also in these dogs with CanL, circulating antibodies, which cause pustular eruption, are produced (Fig. 6.163).

Cytological Findings

Cytological and histopathological findings are not indicative of PF, and in most cases, AKs are occasional or totally absent. When they are present, they are less numerous than in PF. Histopathologically, pustules are non-follicular, and in some dogs, intra-macrophagic amastigotes can be detected in the subepidermal dermis (Figs. 6.164, 6.165, and 6.166). In some dogs, to detect a few amastigotes in the dermis, immunohistochemical staining must be performed (Fig. 6.167). Cytology of pustules is composed of segmented neutrophils, with few or no AKs (Figs. 6.168 and 6.169).

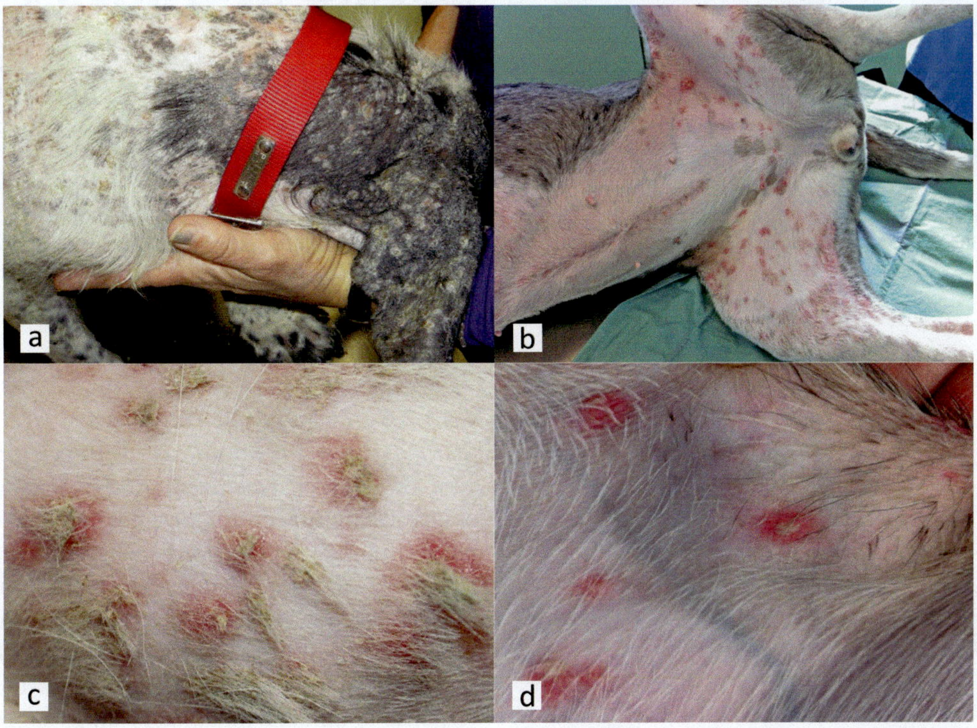

Fig. 6.163 Pustulosis associated with Leishmaniasis: (**a**) diffuse yellowish crusty lesions indicative of dehydrated crusts in a dog affected with CanL; (**b**) large non-follicular pustules on the abdomen of a dog with CanL; (**c**) Large crusty pustules on the sternal area in a dog with CanL; and (**d**) closeup of large non-follicular pustules with an erythematous halo in a dog with CanL

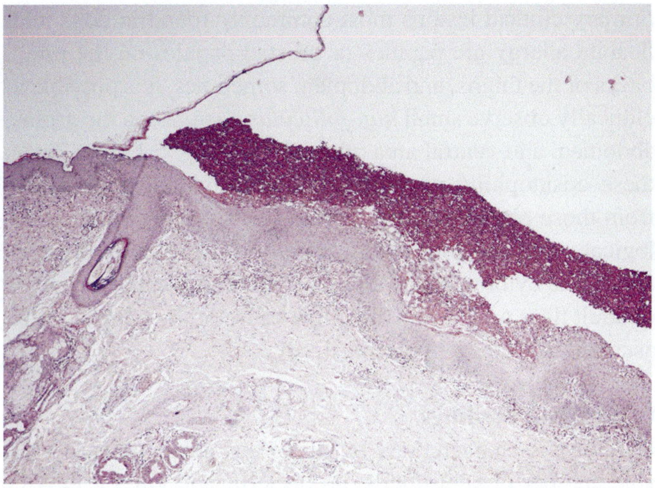

Fig. 6.164 Histology of sterile pustules in a leishmaniotic dog: very large subcorneal non-follicular pustule

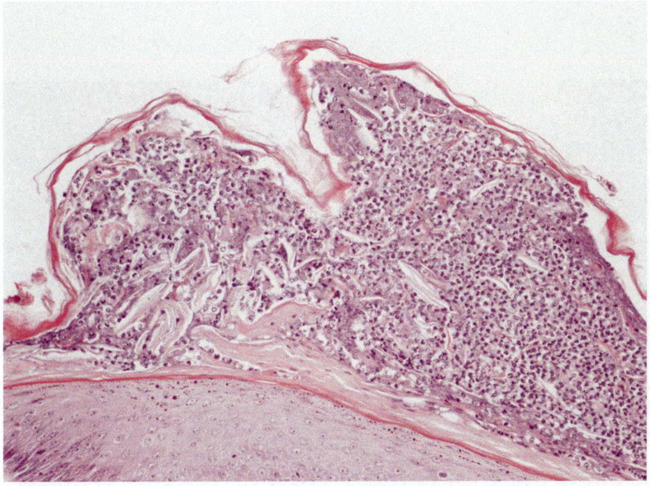

Fig. 6.165 Histology of a sterile pustule in a leishmania: note that the morphology of well-preserved neutrophils is yet recognizable at high magnification. No AKs are present

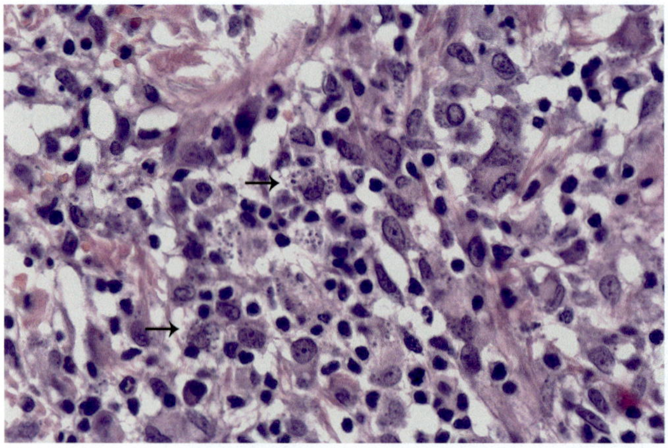

Fig. 6.166 Histology of leishmaniasis: many amastigotes of *Leishmania infantum* in the cytoplasm of macrophages are easily detectable (arrows)

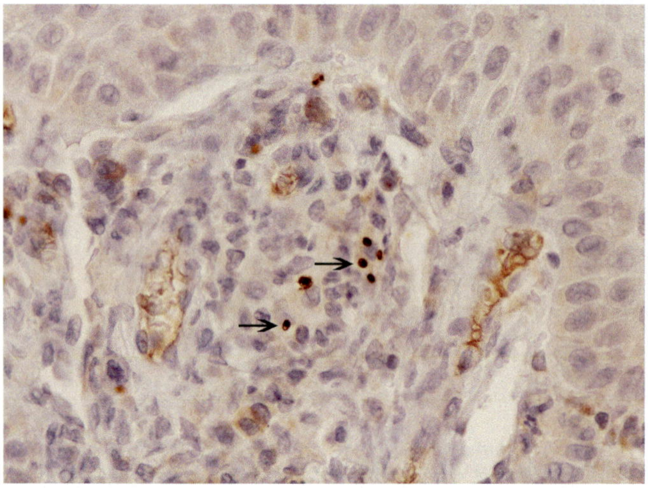

Fig. 6.167 Immunohistochemistry: amastigotes of *Leishmania infantum* are highlighted by immunohistochemical staining as brown oval to roundish bodies (*arrows*)

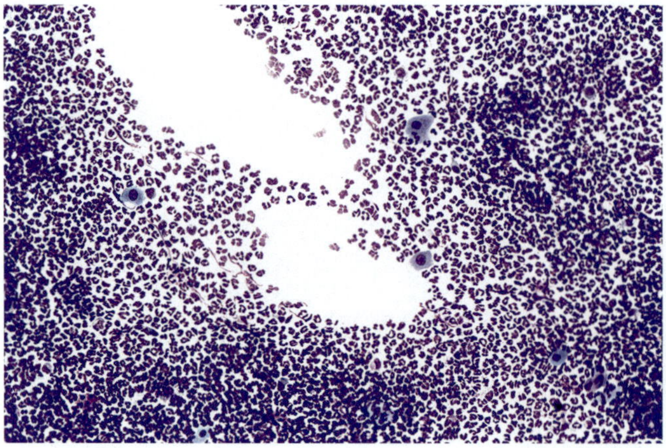

Fig. 6.168 Cytology of a sterile pustule in a dog with leishmaniasis: many segmented neutrophils and few acantholytic cells

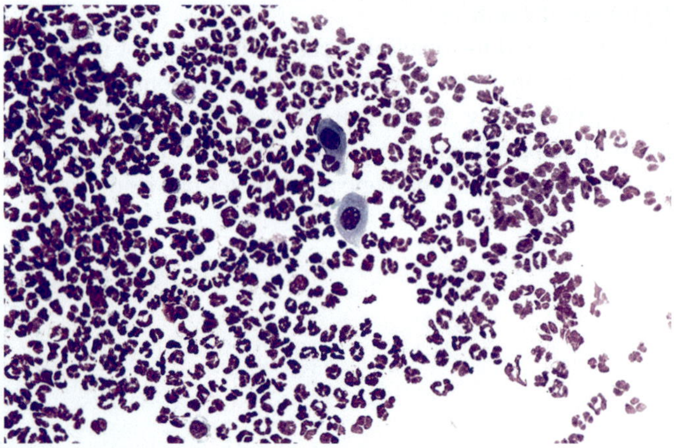

Fig. 6.169 Cytology of a sterile pustule in a dog with leishmaniasis: many neutrophils and few acantholytic cells. Note that nuclei of neutrophils are well-segmented and without features of nuclear degeneration

6.4.1.2.6 Eosinophilic Pustulosis

Sterile eosinophilic pustulosis (SEP) is a rare idiopathic disease of dogs characterized by the development of intraepidermal or subcorneal pustules, predominantly localized on the trunk and associated with systemic symptoms such as fever, malaise, lymphadenopathy, and eosinophilia [84, 85, 88]. No clinical differences between the pustules of this disease with those of PF or other sterile pustulosis are reported. The cause of skin lesions is unknown, but the response to immune modulatory drugs suggests an immune-mediated pathogenesis.

Intra-epidermal microscopic *eosinophilic pustules* are a common finding in histopathological samples from animals affected by *FAD* and *atopic dermatitis* [85, 86] (Figs. 6.170 and 6.171). In these diseases, pustules are intraepidermal and very small to be not clinically appreciable. Although the primary clinical lesions most commonly found in dogs with fleabite allergy are papules or crusted papules on the rump, backs of the thighs, and abdomen, sometimes, it is possible to clinically observe small *non-follicular pustules* on the groins, abdomen, and ventral area of the thighs (Fig. 6.172) [87]. As these eosinophilic pustules are clinically indistinguishable from those observed in superficial canine pyoderma, a cytological investigation could rapidly differentiate between them. The cytological findings and the concomitant presence of adult fleas on the skin allow the cause of the pustules to be ascribed to fleabite hypersensitivity.

Cytological Findings

Cytology is characterized by numerous eosinophils, a variable number of neutrophils, no bacteria, and rare or no AKs (Figs. 6.173 and 6.174).

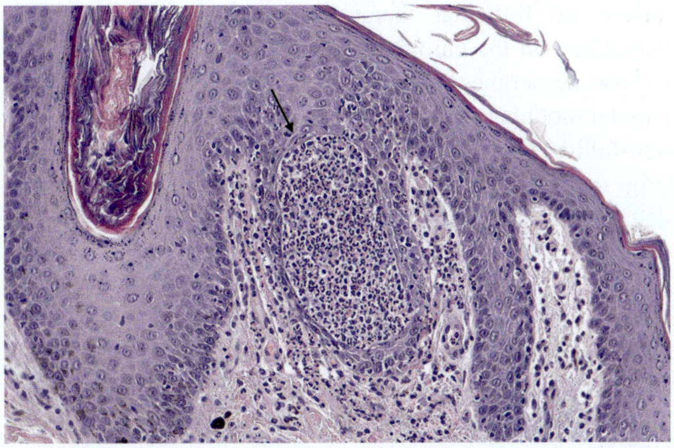

Fig. 6.170 Histopathology eosinophilic pustule in an atopic dog: epidermal hyperplasia with intraepidermal pustule. Note that it is so small that cannot be visualized macroscopically (arrow)

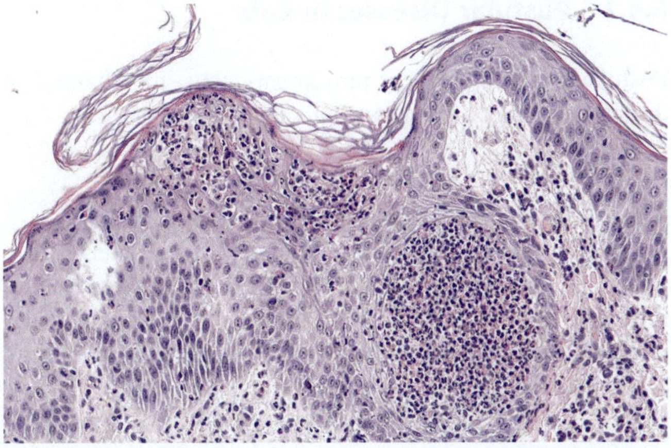

Fig. 6.171 Histopathology of SEP: large both deeply intraepidermal and superficial pure eosinophilic pustules. The latter is superficial and can be clinically visualized

Fig. 6.172 Eosinophilic pustules: (**a**) multiple small and non-follicular pustules on the inner surface of the legs in a dog with fleabite dermatitis and (**b**) closeup of the previous image

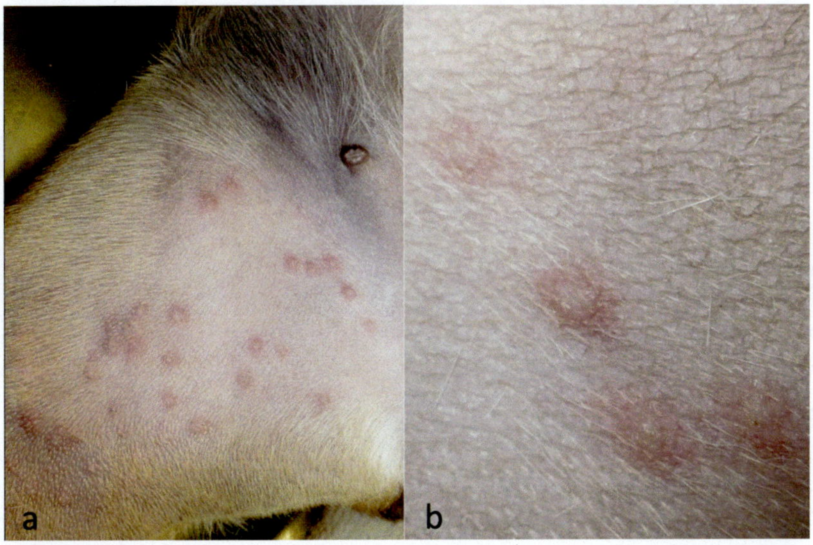

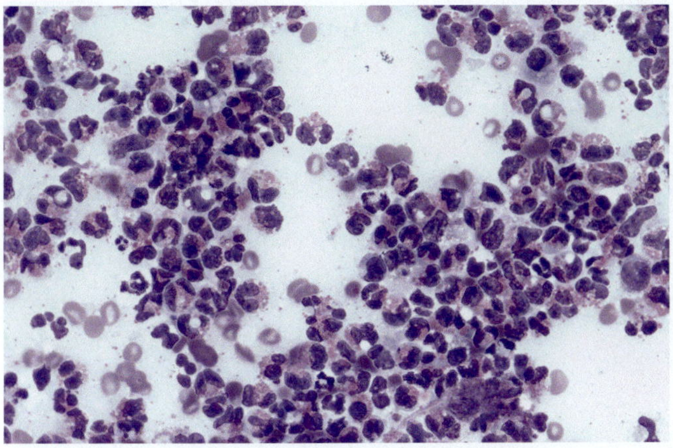

Fig. 6.173 Cytology of eosinophilic pustule: pure eosinophilic inflammation

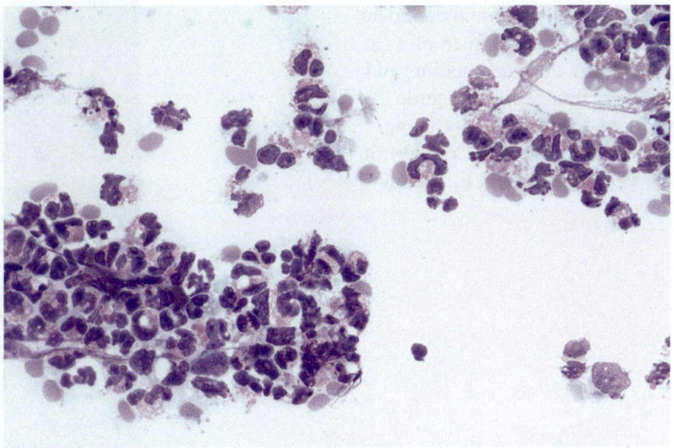

Fig. 6.174 Cytology of eosinophilic pustule: the round pinkish granules of eosinophils are well recognizable

6.4.2 Pustular Diseases in Cats

Pustular diseases are very rare in cats; thus, the chances of detecting intact pustules on the skin of cats are extremely low.

6.4.2.1 Bacterial Infection

Apart from *abscesses* and other uncommon bacterial skin infections (mycobacteriosis), pyoderma in cats is usually related to the complication of self-trauma in patients with pruritic diseases, and the pathogenic role of bacteria in these cases seems to be not determining. However, clinical presentations in cat pyoderma are very different from those seen in dogs.

An uncommon form of localized pyoderma that exceptionally shows *pustules* is observed in *feline acne*. Feline acne is considered an idiopathic disease associated with a defect of follicular keratinization. Skin lesions vary depending on the chronicity and the development of furunculosis; sometimes, together with comedones, papules, and papular–nodular lesions due to furunculosis, a few non-follicular impetiginous pustules can be observed (Fig. 6.175). Exceptionally, non-follicular pustules may be detected in cats with systemic diseases such as endocrine disorders like diabetes or in the case of sterile immune drug-related diseases such as drug reaction (Fig. 6.176).

Cytological Findings

As in dogs, the cytology of bacterial pustules in cats consists of karyolitic neutrophils with intracytoplasmic cocci, while, in sterile disease, only well-preserved neutrophils, sometimes with occasional keratinocytes, are present (Figs. 6.177 and 6.178).

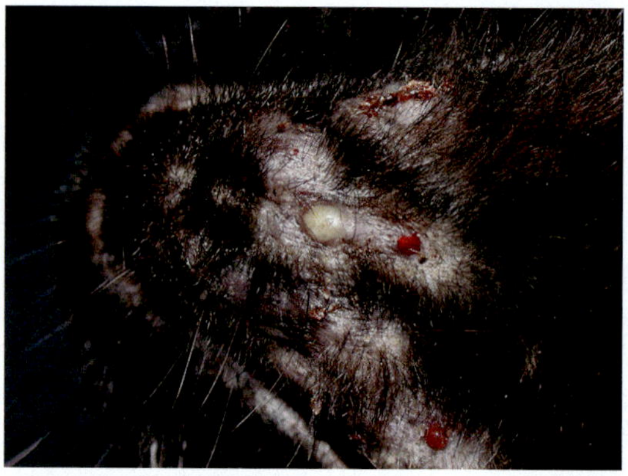

Fig. 6.175 Large non-follicular pustule on the chin of a cat with feline acne

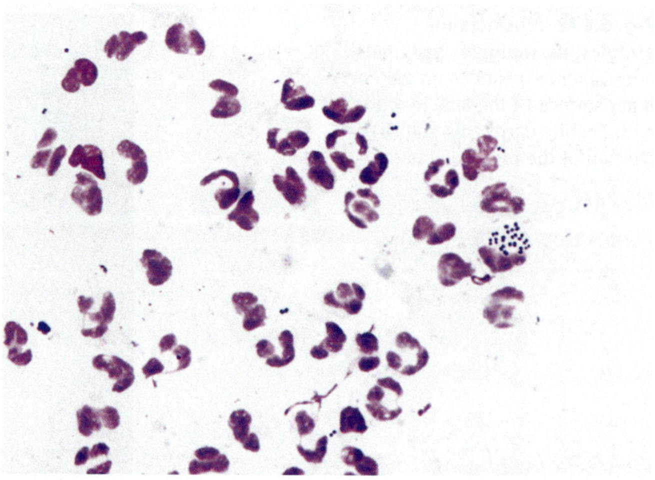

Fig. 6.177 Cytology of pyoderma: karyolitic neutrophils that phagocytize many cocci

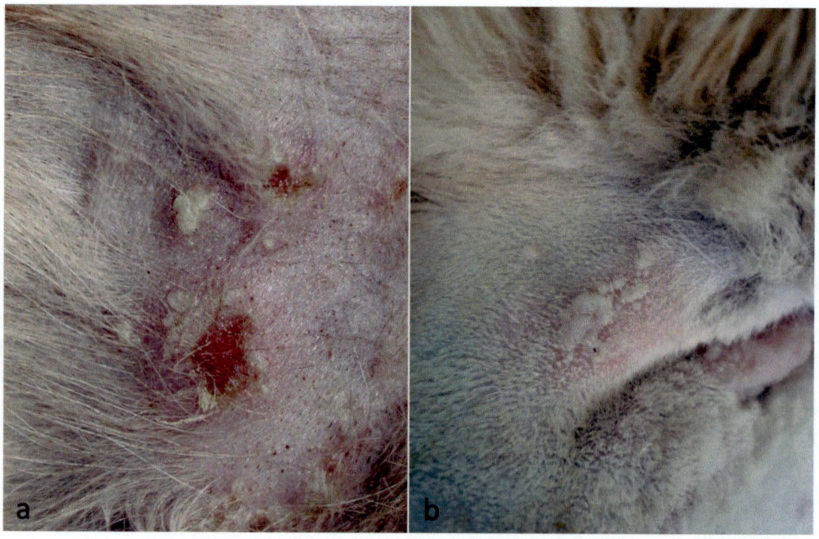

Fig. 6.176 Pustules in cats: (**a**) large non-follicular and confluent pustules on the abdomen of a cat affected by diabetes mellitus and (**b**) large non-follicular sterile pustules secondary to systemic antibiotic treatment. (Courtesy of DVM L. Bellina and F. Bertini—Italy)

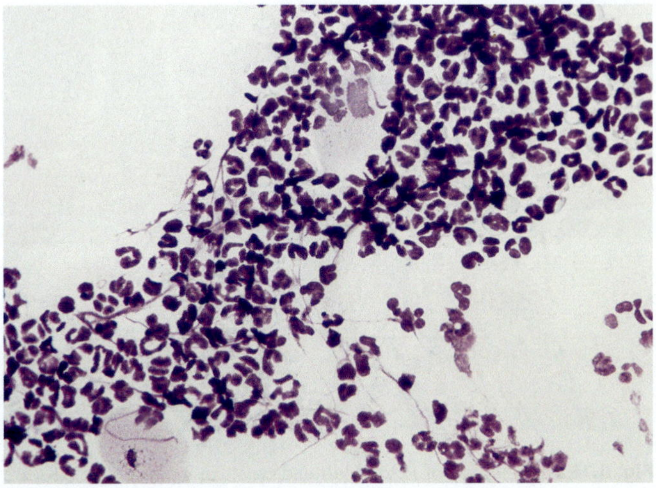

Fig. 6.178 Cytology of sterile pustule: segmented neutrophils with no both karyolitic features and bacteria

6.4.2.2 Pemphigus Foliaceus

Even in cats, PF is a pustular disease. Unlike the dog, the skin of the cats is thinner and the possibility of finding intact pustules is very rare. Pustules rapidly dehydrate and give rise to yellowish crusts. Pustules or crusts located on the pinna, face, nose, and around the nipples are highly indicative of feline PF. In association, multidigital paronychia with the nail fold covered by dry pus with a characteristic *cheesy* appearance is commonly observed in feline patients (Fig. 6.179) [89–91].

Cytological Findings

Histopathology of feline PF is similar to those for PF in dogs, with a population of segmented neutrophils and a variable but usually high number of acantholytic cells/keratinocytes (AKs) [89]. In most cats, it is very hard to find intact pustules; thus, cytological sampling must be performed from the inner surface of the crusts and histopathological diagnosis of PF is

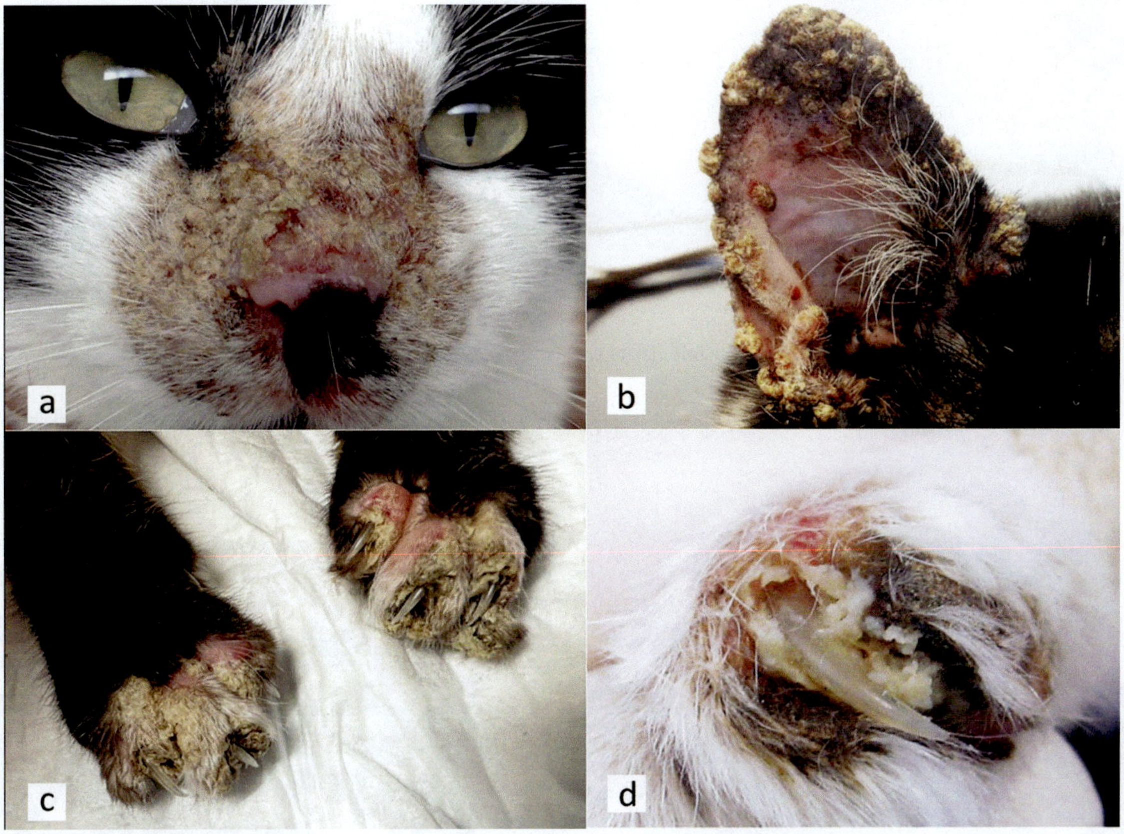

Fig. 6.179 PF in cats: (**a**) yellowish crusts around the nose of a cat; (**b**) diffuse yellowish crusts from dehydrated pustules on the pinna; (**c**) alopecia, swelling, and large yellowish crusts on the feet; and (**d**) paronychia with dried pus with a *cheesy* appearance on the nail fold

often based on the detection of AKs in the crusts (Figs. 6.180, 6.181, and 6.182). Cytological examination shows similar features to those described for dogs. It must be kept in mind that the cytological evaluation of the dry and dense pus present in the nail folds of cats with PF is rarely diagnostic; in fact, slides are usually characterized by numerous severely karyolitic neutrophils, along with many bacteria due to an unavoidable secondary infection that occurs in that area. Furthermore, AKs are few to absent in most slides collected from this area, while both many squamous superficial nucleated keratinocytes and corneocytes are numerous; if lesions are present in other body areas, the latter must be preferred for sampling (Figs. 6.183, 6.184, and 6.185).

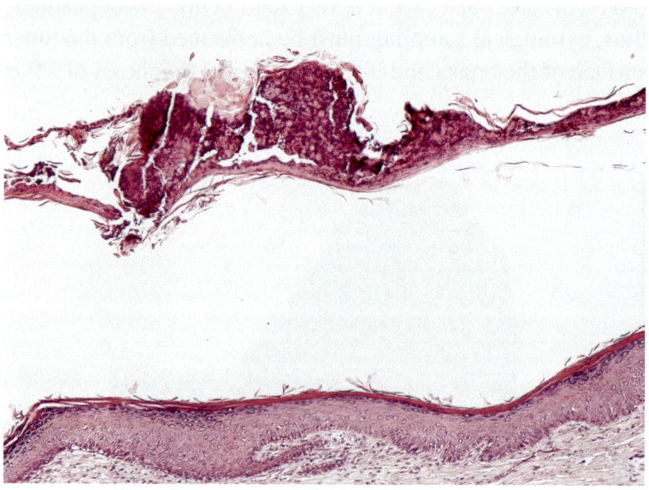

Fig. 6.180 Histology of PF: a crust of acantholytic pustule detached from the epidermidis

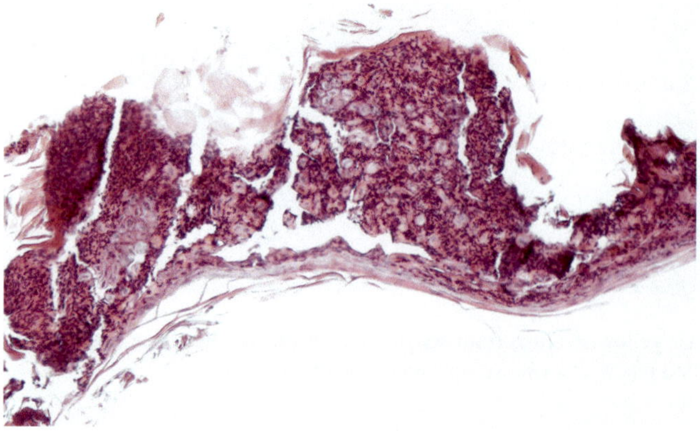

Fig. 6.181 Histology of PF: at high magnification, many AKs can be recognizable

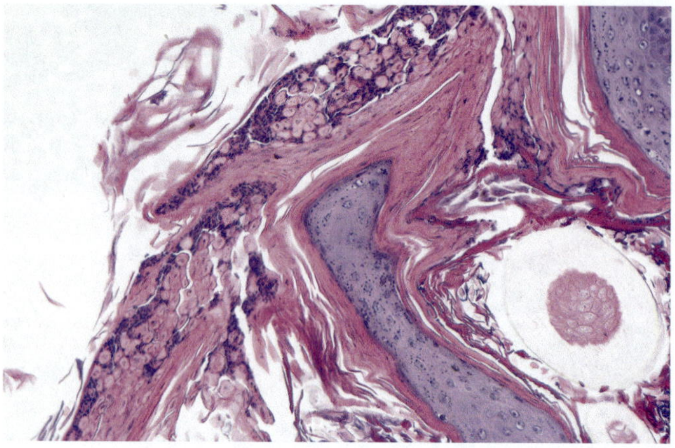

Fig. 6.182 Histology of PF: dehydrated crusts in which many AKs are evident along with coated neutrophils

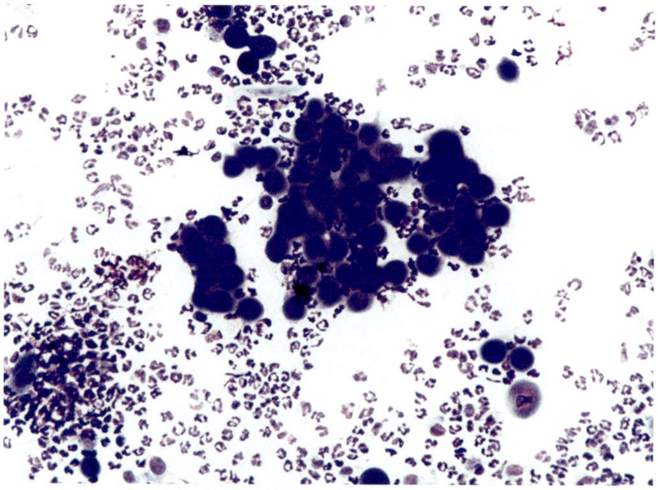

Fig. 6.183 Cytology of PF: many roundish and dark blue cytoplasm acantholytic cells are mostly released in large clusters

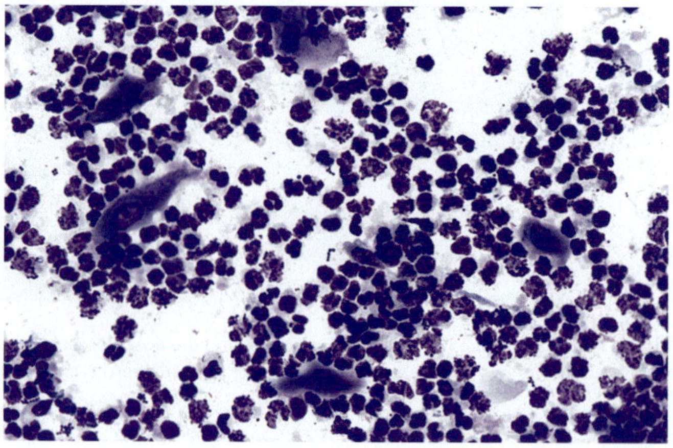

Fig. 6.184 Cytology of PF collected from nail fold: many karyolitic keratinocytes with few spindle-shaped epithelial cells. Note how many coccoid bacteria are present

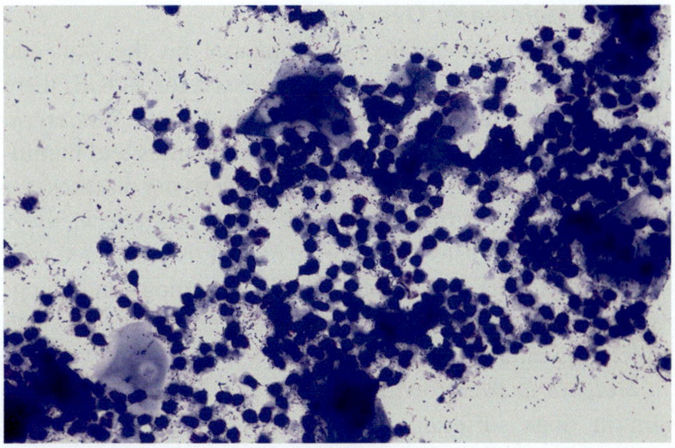

Fig. 6.185 Cytology of PF collected from nail fold: many severely kariolytic nuclei of neutrophils along with corneocytes. Bacteria are both intracellular and free on the bottom of the slide

6.5 Scales

In many dogs and cats, collection using the impression smear technique from the skin with an intact epidermal surface covered in dry or oily/waxy scales can provide useful diagnostic results.

6.5.1 Scaling Diseases in Dogs and Cats

6.5.1.1 Bacterial Overgrowth Syndrome

Bacterial overgrowth syndrome (BOGS) is a dermatological disorder observed in dogs characterized by colonization of bacteria on the skin surface in the absence of true neutrophilic inflammation [92, 93]. An underlying disease usually affects dogs with BOGS, most frequently *atopic dermatitis*, as it causes injury to the cutaneous barrier and consequently impairs the defensive mechanisms favoring bacterial overgrowth and toxin overproduction. From the above, to resolve BOGS, the underlying predisposing disease must be diagnosed and treated. Skin lesions are characterized by erythema and pruritus, with skin usually covered in a shiny, sticky film and frequently associated with chronic lesions such as lichenification and hyperpigmentation, especially on the groin and axilla (Figs. 6.186 and 6.187).

Cytological Findings

Cytology is characterized by a large number of corneocytes with many coccoid or/and rod-shaped bacteria on their surface, but with no inflammatory cells. Anecdotally, although the number of bacteria is usually much higher, it is believed that to diagnose BOGS, more than five bacteria per immersion field (100×) must be detected (Figs. 6.188 and 6.189).

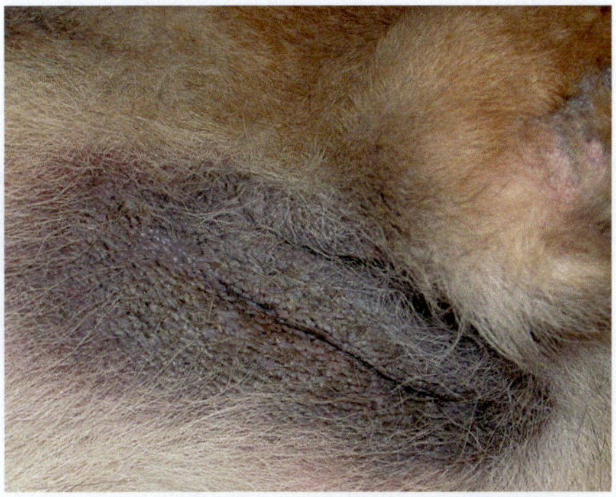

Fig. 6.186 Lichenification and skin hyperpigmentation of the axilla in a dog with BOGS

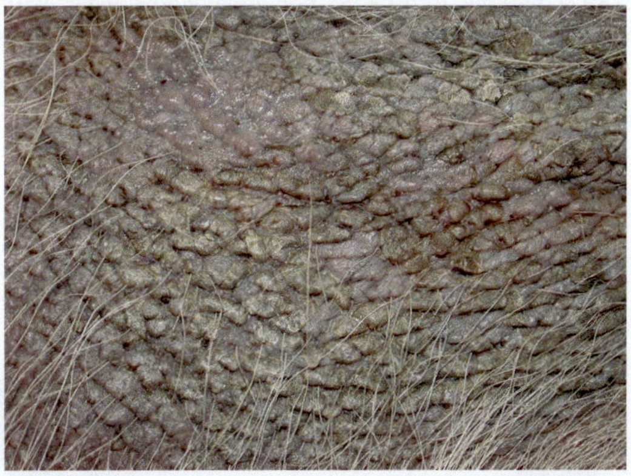

Fig. 6.187 Lichenification with humid and shiny surface in a dog with BOGS

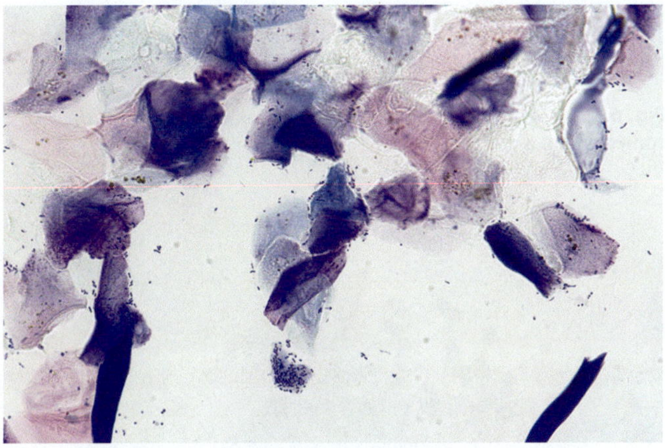

Fig. 6.188 Cytology of BOGS: polygonal corneocytes with many cocci and rod-shaped bacteria on their surface. Note the absence of neutrophils

In some cases, *Simonsiella* bacteria can be observed along with BOG; since these bacteria are residents of the oral mucosa, their presence is indicative that dogs are licking the pruritic areas (Figs. 6.190 and 6.191).

Fig. 6.189 Cytology of BOGS: polygonal corneocytes with many cocci and rod-shaped bacteria on their surface. Note the absence of neutrophils

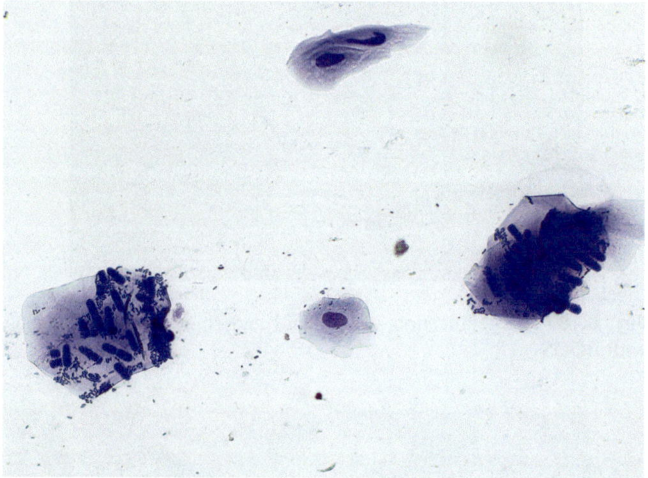

Fig. 6.190 Many *Simonsiella* spp. bacteria adhered to the surface of a corneocyte in a dog with BOGS

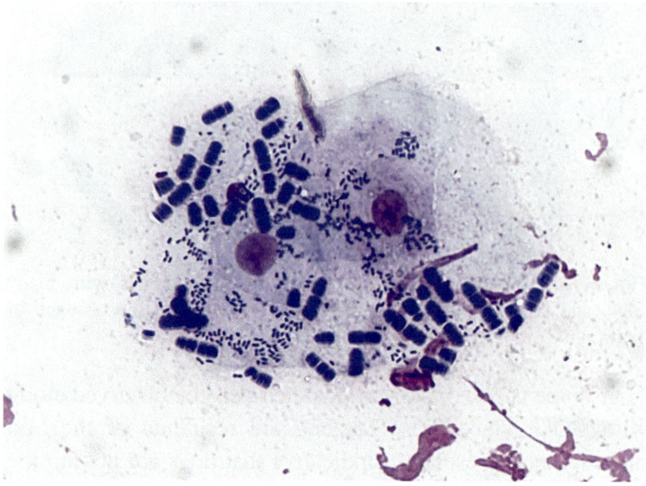

Fig. 6.191 Many *Simonsiella* spp. bacteria along with coccoid and rod-shaped bacteria. Note the parallel disposition of rod-shaped bacteria

6.5.1.2 *Malassezia* Overgrowth

Malassezia pachydermatis is commensal on the skin and mucous membranes of dogs. As discussed for BOGS, there are factors that promote pathogenicity and allow *Malassezia* overgrowth (MOGS), such as high temperatures and humidity, excessive secretion of lipids, and underlying diseases, for example, atopic dermatitis [96, 99–103]. *Malassezia overgrowth* usually causes very pruritic dermatitis associated with scaling, erythema, and a bad smell [95]. Typology and topography of lesions are the same as described for BOGS, with which it is often associated.

Malassezia overgrowth/dermatitis is a frequent complication in dogs suffering from allergic diseases (especially atopic dermatitis), even though any disease that affects the skin surface ecosystem can cause excessive multiplication of yeasts. In dogs, clinical lesions are characterized by erythema, greasy scaling, often with a waxy appearance, orange in color, and with rancid smell, usually associated with lichenification and hyperpigmentation (Figs. 6.192 and 6.193). Such findings are mainly evident in more humid body areas such as the interdigital area, the nail fold, and the ventral surface of the neck, axilla, and groin, but in some individuals, they may be present throughout the body.

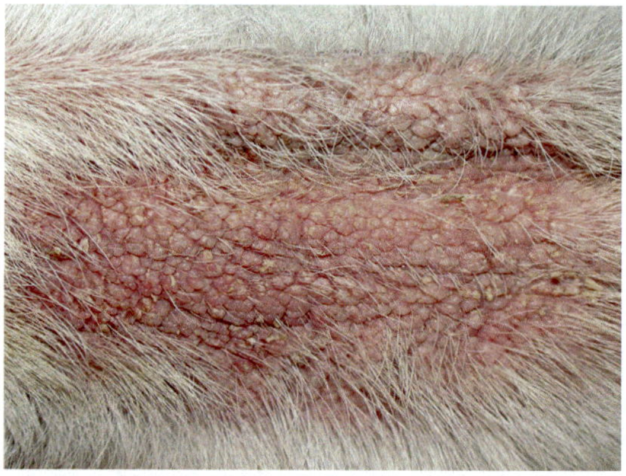

Fig. 6.192 Atopic dermatitis: alopecia, lichenification, and waxy scales in a dog with *Malassezia* overgrowth

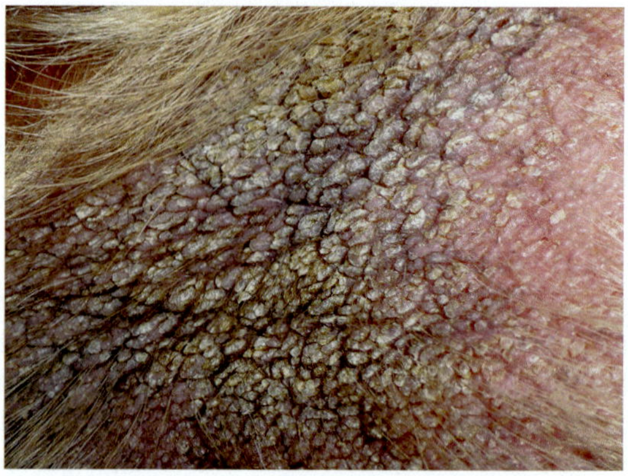

Fig. 6.193 Alopecia, erythema, lichenification, and waxy scales in a dog with *Malassezia* dermatitis due to atopic dermatitis

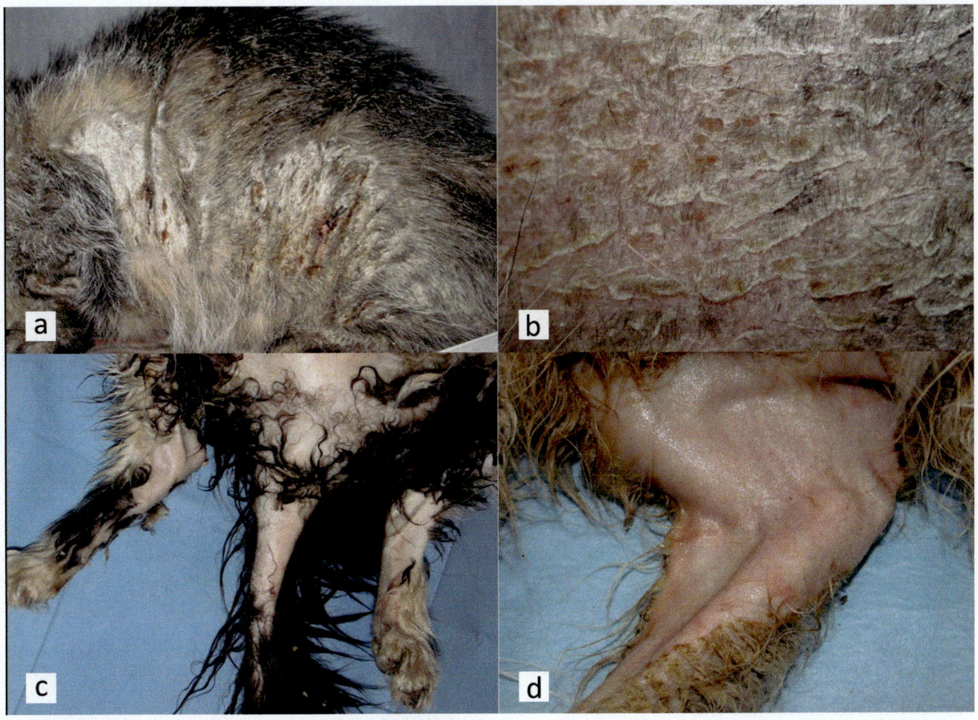

Fig. 6.194 Feline Paraneoplastic syndromes: (**a**) alopecia and diffuse scaling dermatitis in a cat with thymoma; (**b**) closeup of the skin of the same cat; (**c**) shiny alopecia in a cat with paraneoplastic alopecia due to hepatocellular carcinoma: MOGS was diagnosed with skin surface cytology; and (**d**) shiny alopecia in a cat with paraneoplastic alopecia due to an endocrine pancreatic carcinoma. MOGS was diagnosed by imprint technique

Frequently, whether associated with other skin lesions or not, dogs present ceruminous otitis with a yeast overgrowth characterized by yellowish and smelly secretion.

In feline patients, *Malassezia overgrowth* is rarer than in dogs [94, 107]. Usually, yeasts are observed in cats affected by hypersensitivity or immune-suppressive diseases, such as paraneoplastic alopecia associated with internal organs neoplasia and exfoliative dermatitis associated or not with thymoma [97, 98, 101, 104] (Fig. 6.194).

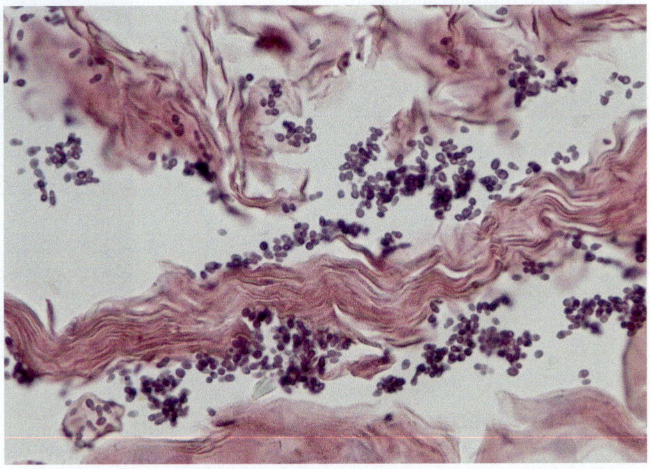

Fig. 6.195 Histology of *Malassezia* overgrowth: many yeasts in the stratum corneum

Cytological Findings

As mentioned, *Malassezia* spp. is a commensal microorganism that lives on the skin surface, and when overgrowth, a high number of yeasts are usually sampled. *Malassezia* has an oval shape, measuring 3–8 μm, that assumes a characteristic bilobate appearance during the budding phase that gives yeasts an *American peanut* appearance (Fig. 6.195). The number of yeasts present in cytopathological preparations is variable and dependent on the type of lesions, the sampling method, and the breed. It has been shown that, in some feline breeds such as the Devon rex or in canine breeds such as the Basset hound, the number of yeasts on the skin is physiologically higher than in other breeds [97]. Anecdotally, it is believed that when diagnosing a *Malassezia* overgrowth, more than two yeasts per immersion field (100×) need to be found. Although the number of yeasts usually found is greater than that considered relevant to diagnosing a *Malassezia* overgrowth, in some dogs with chronic lichenification, it is possible to obtain negative samples. This occurrence is because severe epidermal hyperplasia may not allow the sampling of yeasts *hidden* between the epidermal folds. To avoid this misdiagnosis, it is advisable to carry out repeated sampling from the same area to try to collect more deeply located yeasts. In a severe infection, the number of yeasts is so high that many *Malassezia* spp. can be easily collected with a piece of transparent acetate tape (Figs. 6.196 and 6.197).

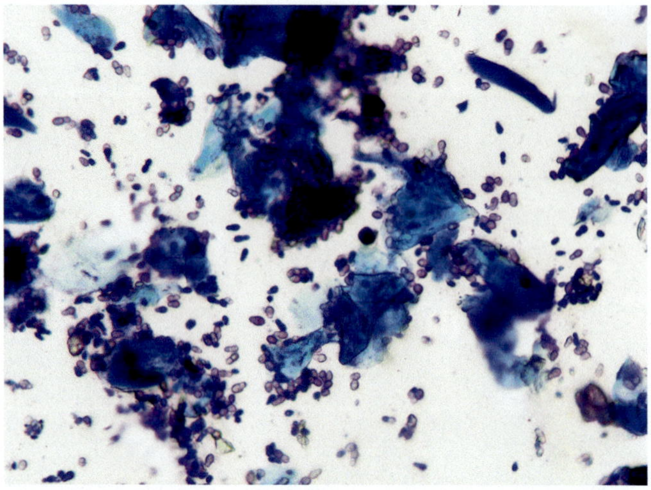

Fig. 6.196 Cytology of *Malassezia* dermatitis: numerous *Malassezia pachidermatis* collected from the skin surface with an acetate transparent tape

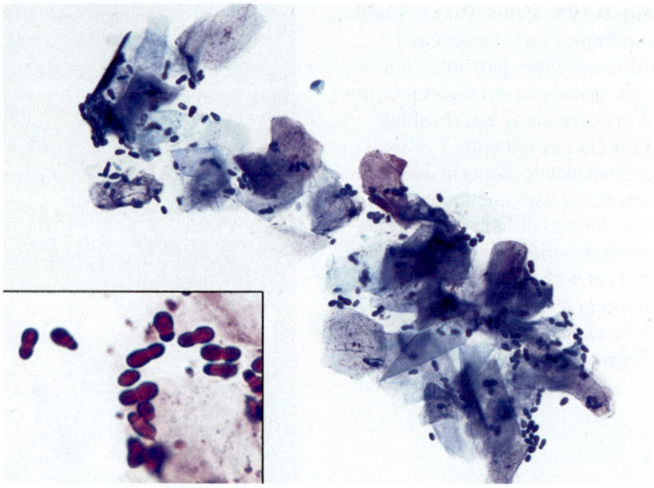

Fig. 6.197 Cytology of *Malassezia* overgrowth: many yeasts adhering to the surface of the corneocytes. *Inset*: at high magnification, the characteristic peanut shape of yeasts is easily recognizable

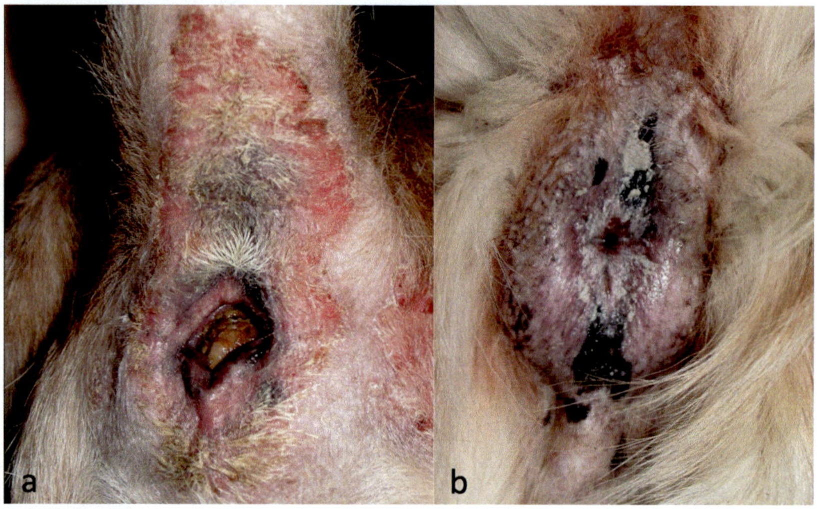

Fig. 6.198 Candidiasis: (**a**) perianal erythema in a dog with hepatocutaneous syndrome and (**b**) erythema and whitish patina on the anal mucocutaneous junction in an immune depressed dog with cutaneous candidiasis

6.5.1.3 Cutaneous Candidiasis

Candida albicans is commensal mycelial yeast that colonizes the mucous membranes of dogs and cats. In addition to *C. albicans*, other yeasts such as *Candida parapsilosis* and *Candida guilliermondii* have been reported to cause dermatitis in dogs [105, 107–109]. These yeasts can occasionally cause dermatitis, but when it happens, they are usually opportunistic infections due to an underlying disease that impairs the immune system, such as metabolic diseases (superficial necrolytic dermatitis/hepatocutaneous syndrome), hormonal disorders (hypercortisolism and diabetes mellitus), visceral neoplasia, or chronic treatment with immune-suppressive drugs. Usually, the more humid body areas such as mucocutaneous areas, interdigital, and skin fold areas are more frequently affected, and sometimes, a smelly whitish film is observed on the skin surface (Figs. 6.198 and 6.199) [105, 106, 109, 110].

Cytological Findings

Like *Malassezia*, *Candida* may also be easily sampled using the imprint technique with a slide or a piece of adhesive transparent tape.

Candida albicans yeasts live indeed on the superficial layers of the epidermidis, and for this reason, they are usually observed together with a variable number of keratinocytes (Fig. 6.200). *Candida* is a dimorphic fungus and a normal inhabitant of the mucous membranes. Unlike *Malassezia*, it is mycelial yeast and can be observed in the form of both roundish and oval spores, measuring 3–6 µm, and sometimes arranged in pseudo-mycelia or in chains or as branched hyphae (bipolar or multipolar narrow-based budding). Spores stain blue with the Romanowsky-type stain and have a thin, transparent, peri-cellular halo (Figs. 6.201 and 6.202).

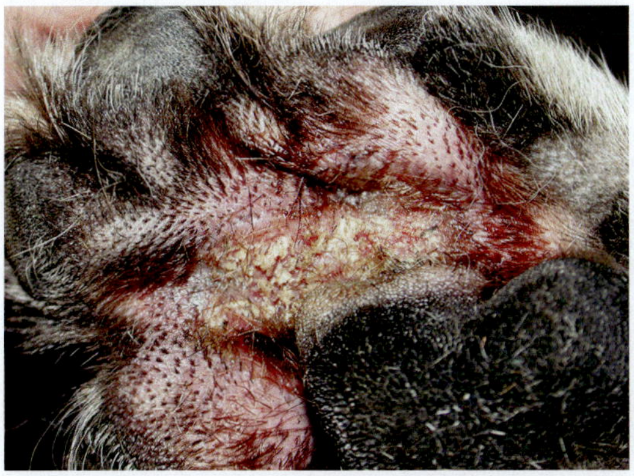

Fig. 6.199 Erythema and whitish patina in the interdigital area of a dog with pododermatitis due to *Candida albicans*

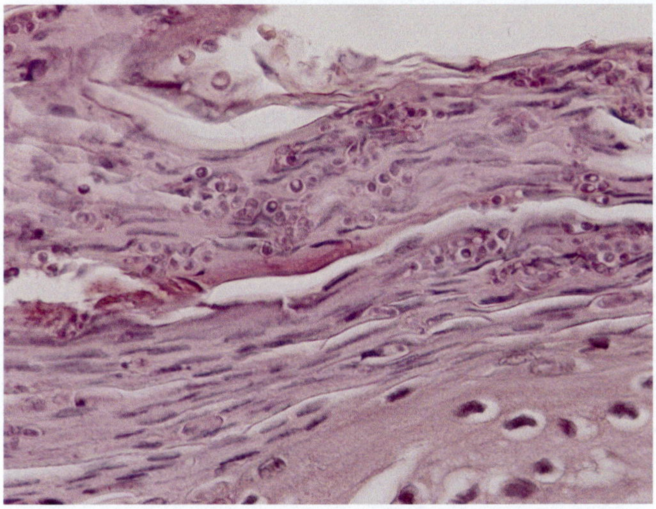

Fig. 6.200 Histology of cutaneous *candidiasis*: many round to oval yeasts in the stratum corneum

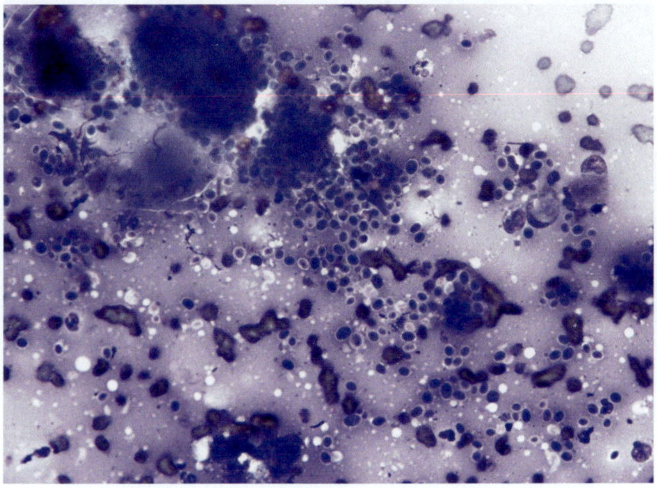

Fig. 6.201 Cytology of cutaneous *candidiasis*: many oval-shaped yeasts with a clear peripheral halo, which adhere to corneocytes

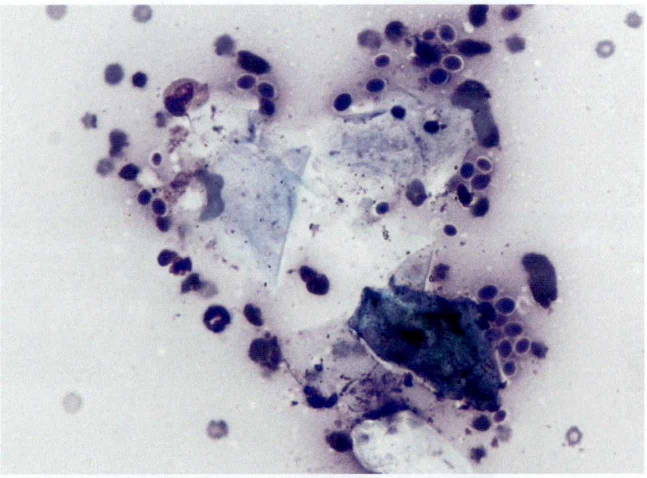

Fig. 6.202 Cytology of cutaneous *candidiasis*: many oval yeasts with an achromatic peripheral halo

6.5.1.4 Dermatophytosis

Dermatophytosis is a fungal disease caused by keratinophilic fungi, mostly represented by *Microsporum canis* and *Microsporum gypseum*, the latter renamed *Nannizzia gypsea* in honor of Arturo Nannizzi an Italian mycologist, or by fungi belonging to the *Trichophyton mentagrophytes* complex [112, 114, 117]. As dermatophytes infect keratin, one of the most common clinical signs is exfoliative dermatitis associated or not with alopecia, erythema, and papules (Figs. 6.203 and 6.204). For this reason, it is not so uncommon to detect arthroconidia and/or hyphae on corneocytes' surfaces, especially in cats [113].

In the case of infections caused by *Microsporum persicolor*, as this fungus is corneophylic, which means that it does not infect hairs but only the keratin of keratinocytes, the sampling of scales is the only way to try to find fungi through cytology [111].

Cytological Findings

The infection of the stratum corneum permits the collection of a variable amount of fungal elements using the acetate tape test. With this technique also, small fragments of infected hair shafts that emerge from the follicular ostium and cannot be visible macroscopically can be collected (Figs. 6.205 and 6.206). The quantity of fungi is strictly linked to the severity of the infestation (Figs. 6.207 and 6.208).

Spores (arthroconidia) are very small, rounded, or oval and are cytologically characterized by a pericellular clear halo, whereas hyphae are linear branched filaments, with multiple segmentations that give fungi a *bamboo pole* appearance. Hyphae are of several different lengths, whereas

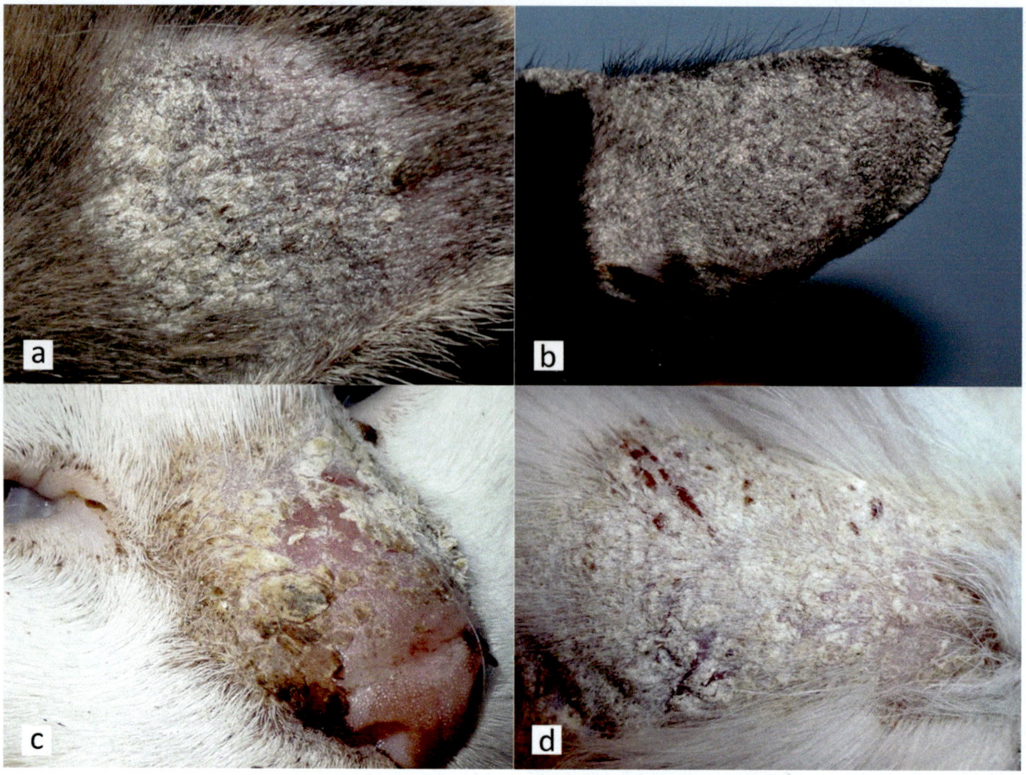

Fig. 6.203 Feline dermatophytosis: (**a**) focal alopecia with dermatophytosis: note the typical fine scaly aspect that resembles "cigarette ash"; (**b**) alopecia and scales on a pinna infected by *M. canis*; (**c**) yellowish thick scales on the nose of a cat with dermatophytosis; and (**d**) diffuse exfoliative dermatitis in the same cat of the previous picture

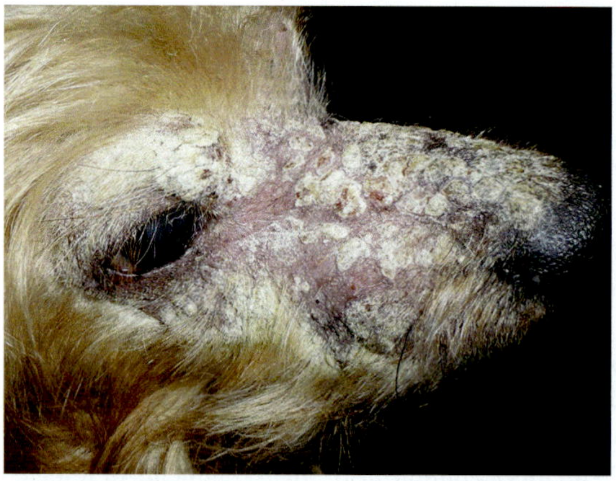

Fig. 6.204 Yellowish scales on the face of a Yorkshire terrier with generalized dermatophytosis

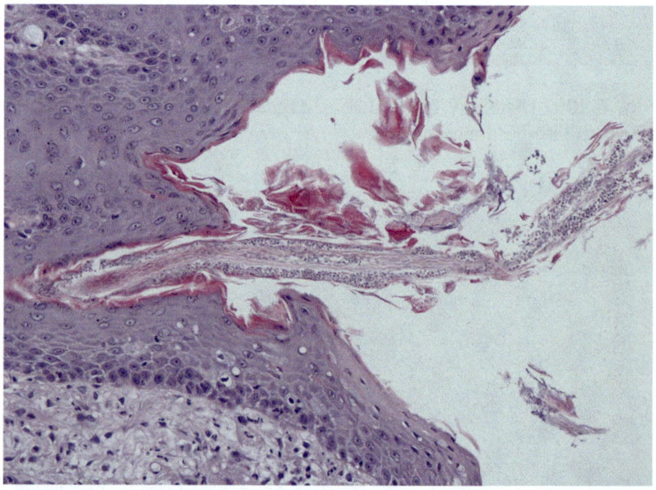

Fig. 6.205 Histology of dermatophytosis (HE staining): a fragment of infected hair emerges for a few millimeters from the follicle

arthroconidia vary in size from 2 to 5 μm, smaller for *M. canis*, larger for *N. gypsea (M. gypseum)* and *T. mentagrophytes*, and assume a blue color with rapid stains, magenta with PAS and black with Grocott's methenamine staining (Figs. 6.209, 6.210, and 6.211).

It should be remembered that any structure found on cytological slides resembling *macroconidia*, should not be confused with pathogenic fungi, as the macroconidia of dermatophytes grow exclusively on cultures and not on the skin (Fig. 6.212).

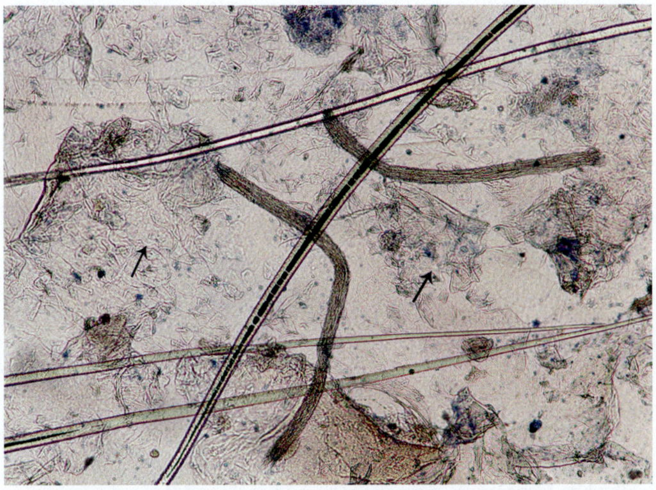

Fig. 6.206 Trichoscopy acetate tape technique: two swollen and deconstructed and infected hair fragments are recognizable. Note that many non-stained worm-like, segmented hyphae are present on the scales (*arrows*)

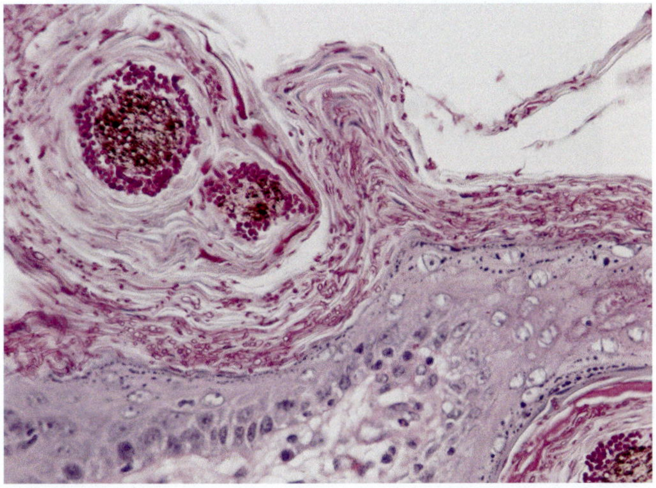

Fig. 6.207 Histology of dermatophytosis (PAS staining): many arthroconidia surround hair shafts and slender segmented hyphae are present in the stratum corneum

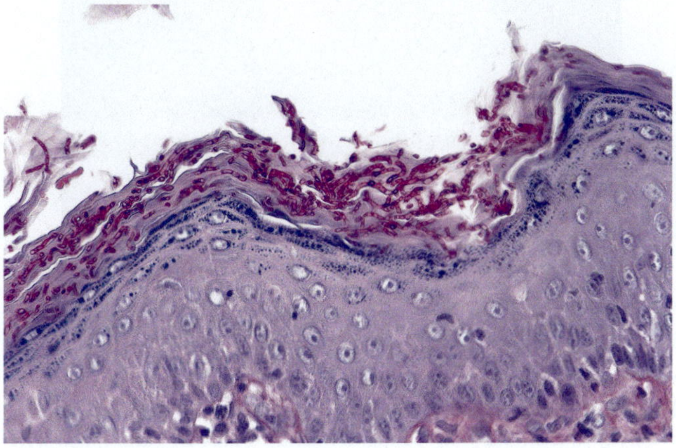

Fig. 6.208 Histology of dermatophytosis (PAS staining): numerous hyphae in the stratum corneum

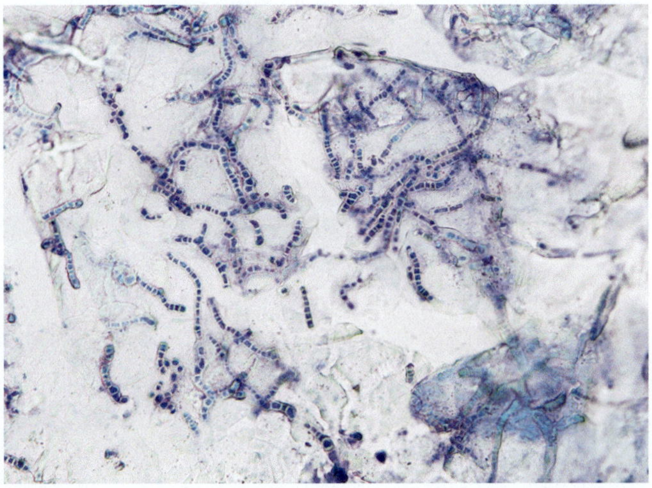

Fig. 6.209 Imprint technique cytology: numerous elongated and septate hyphae on the surface of corneocytes

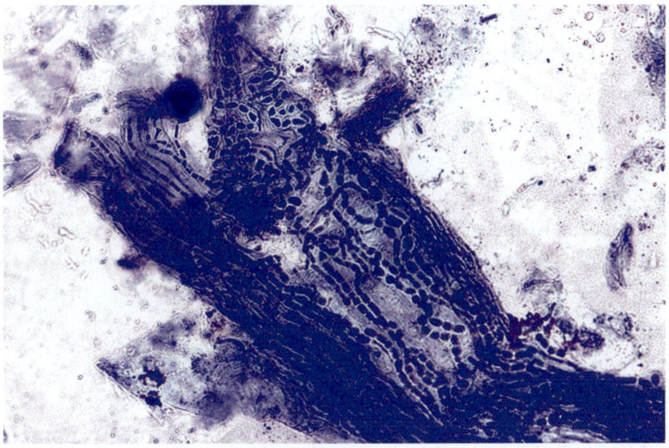

Fig. 6.210 Cytology of dermatophytosis: septate hyphae are dividing in arthrospores

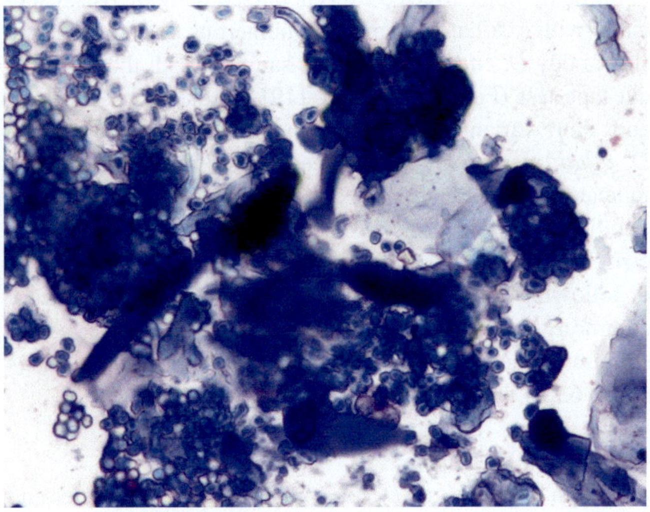

Fig. 6.211 Cytology of dermatophytosis: numerous arthrospores on the surface of the corneocytes

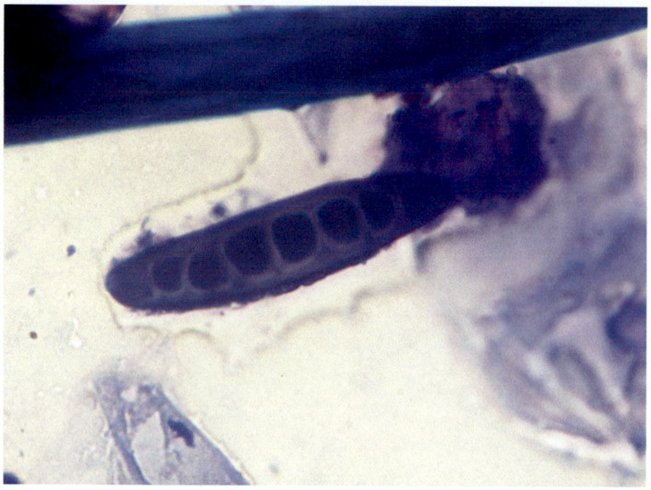

Fig. 6.212 Cytology: a non-dermatophytic fungal body that looks like pathogenic dermatophytic macroconidia

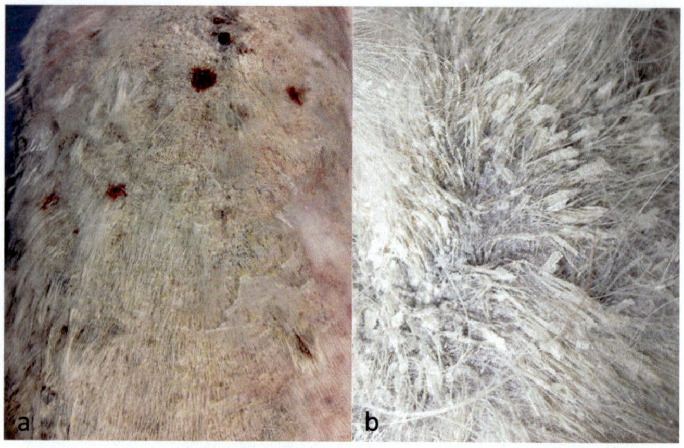

Fig. 6.213 Demodicosis in dogs: (**a**) scaly dermatitis in a West Highland white terrier with generalized demodicosis and (**b**) scaly dermatitis with follicular cysts is a mixbreed with many *D. cornei* (short body)

6.5.1.5 Demodicosis and Other Superficial Ectoparasites

Demodex mites are normal commensals of the skin of dogs and cats. Genetic predisposition and imbalances of the immune system predispose animals to develop an uncontrolled proliferation of mites. Diagnosis of demodicosis can be made using various techniques; among them, the deep skin scraping technique is the more sensitive [117, 118]. In dogs and cats, different *Demodex* spp. are recognized, showing morphological differences. In both species, mites with long and short bodies are distinguished; in dogs, a study has demonstrated that the short-body mite formerly named *Demodex cornei* is genotypically strictly related to *Demodex canis* and seems to be only a morphological variant of the latter [120].

In many dogs affected by generalized demodicosis, especially with exfoliative dermatitis, numerous both normal and short-body *D. canis* mites can be sampled with the transparent tape test (Fig. 6.213) [116–119]. Unlike dogs, the long and short variants of feline *Demodex*, named *D. cati* and *D. gatoi,* respectively, have been recently demonstrated to indicate two different types of mites [115]. In cats affected by exfoliative demodicosis, *D. gatoi* is frequently observed, especially in patients with an immune-suppressive disorder or following corticosteroid therapy. In these cases, it is very common to collect mites on the surface of the skin, as the stratum corneum seems to be the habitat of these mites (Figs. 6.214 and 6.215).

Cytological Findings

Although normal-sized *D. canis* mites live in the follicles and sebaceous glands, they mate on the skin surface, after which the females go into the follicular lumen in which they lay

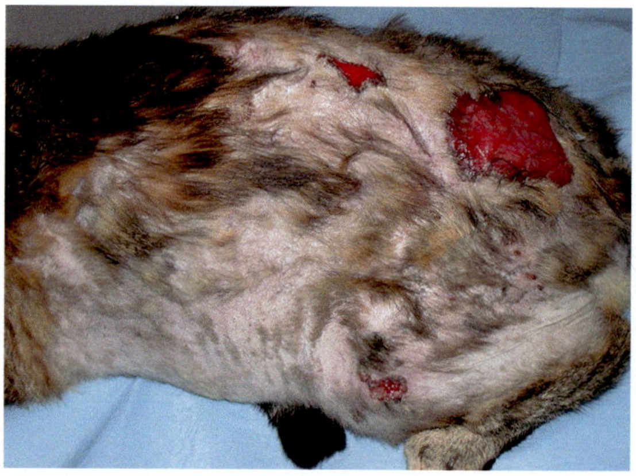

Fig. 6.214 Skin fragility syndrome due to iatrogenic administration of steroids with secondary scaly dermatitis caused by *Demodex gatoi* overgrowth

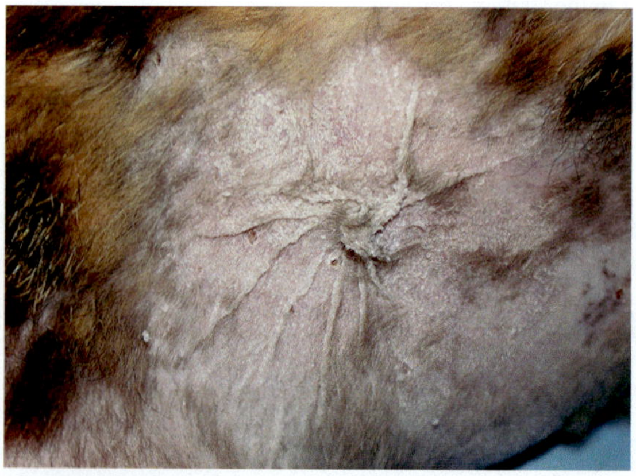

Fig. 6.215 Same cat of Fig. 6.270: note the atrophic skin, with loss of elasticity and diffuse fine scaling dermatitis. Many *D. gatoi* mites were collected on skin surface

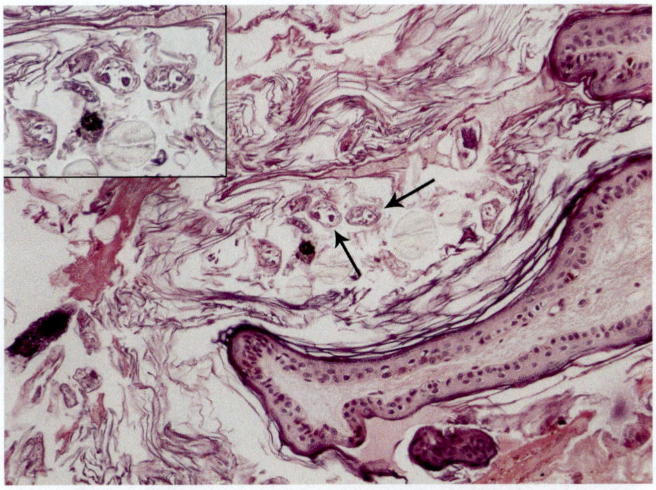

Fig. 6.216 Histology of demodicosis: many *Demodex canis* in the stratum corneum (*arrows, inset*)

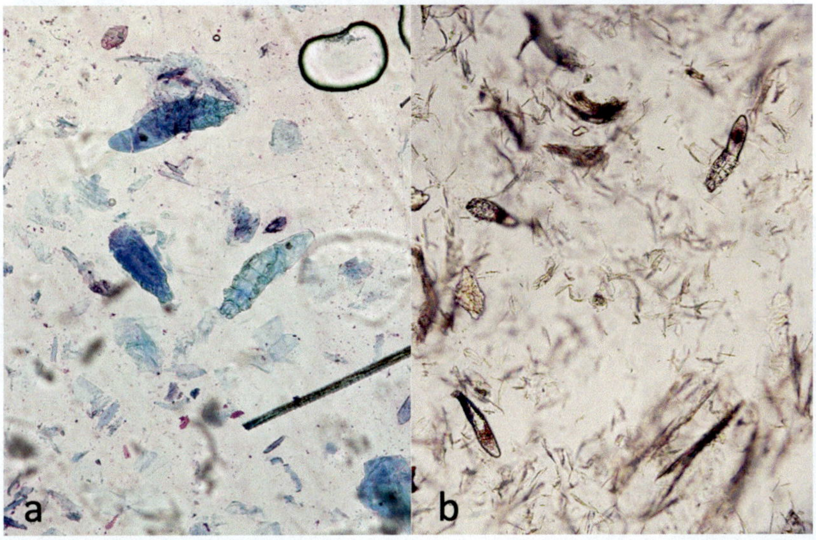

Fig. 6.217 Cytology of demodicosis: (**a**) adults of short bodies *Demodex canis* mites collected by the imprint technique with transparent adhesive tape and (**b**) adults of short bodies *Demodex gatoi*

eggs; for this reason, in many cases of generalized demodicosis, it is also possible to collect them by simply placing a piece of transparent tape on the skin surface (Figs. 6.216 and 6.217). In the case of secondary bacterial infection, karyolitic neutrophils with phagocytized cocci could also be observed (Fig. 6.218).

In the rare cases of canine *Crustous sarcoptic mange* slangy named *Norwegian scabies*, unlike classic scabies infections, the number of mites is very high and they can easily be collected via the acetate tape test. This finding is normal in cats affected by notoedric mange, in which the chance of collecting mites via the scotch test is very high because *Notoedres cati* mites are always numerous on the skin (Figs. 6.219 and 6.220).

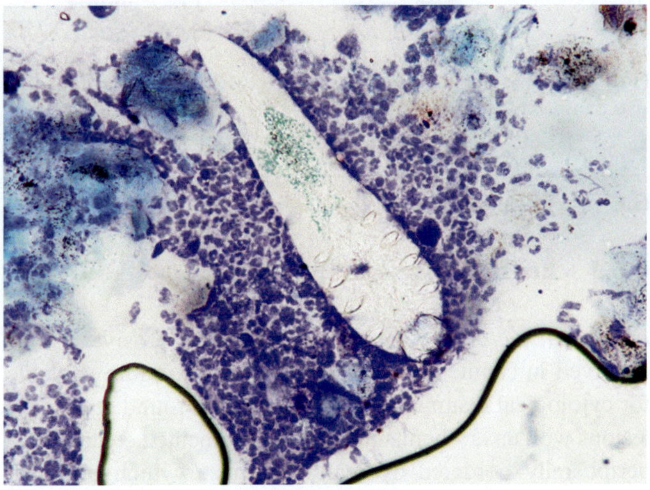

Fig. 6.218 Cytology of demodicosis: an adult *Demodex canis* immersed in a neutrophilic inflammatory background

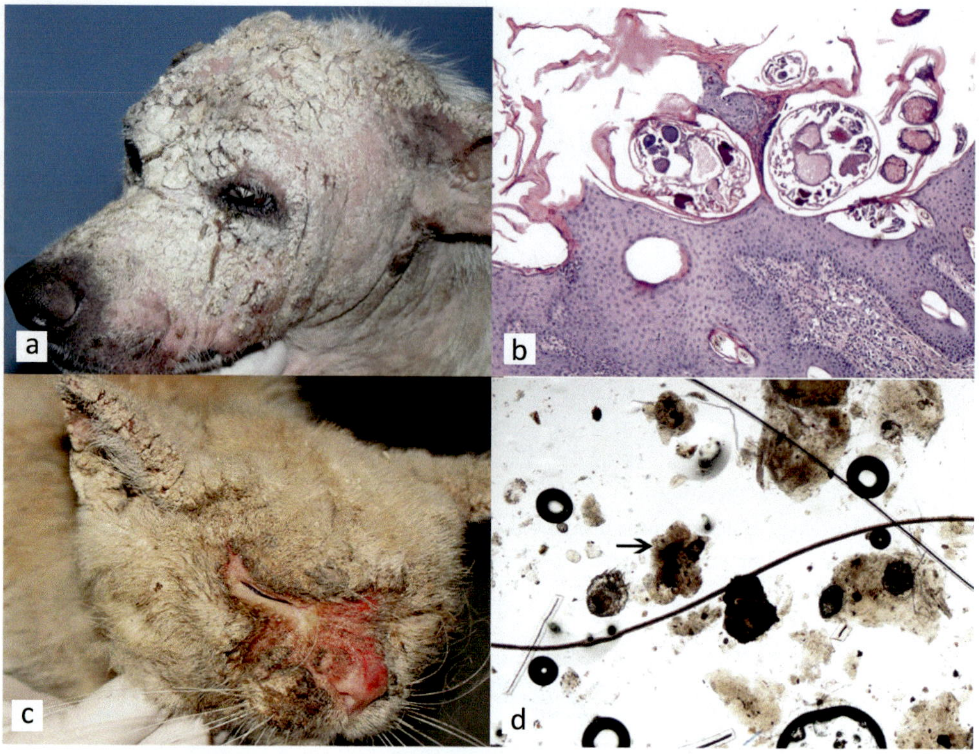

Fig. 6.219 Mange: (**a**) Norwegian scabies in a mixbreed dog. Very thick scales spread all over the head; (**b**) histology of scabies: two adult mites of *Sarcoptes scabiei* var. *canis* on the skin surface; (**c**) notoedric mange: severe scaling and crusty lesions on the head of a cat; and (**d**) many mites of *Sarcoptes scabiei* var. canis are easily collected with transparent tape. Eggs are present in keratin material (arrow)

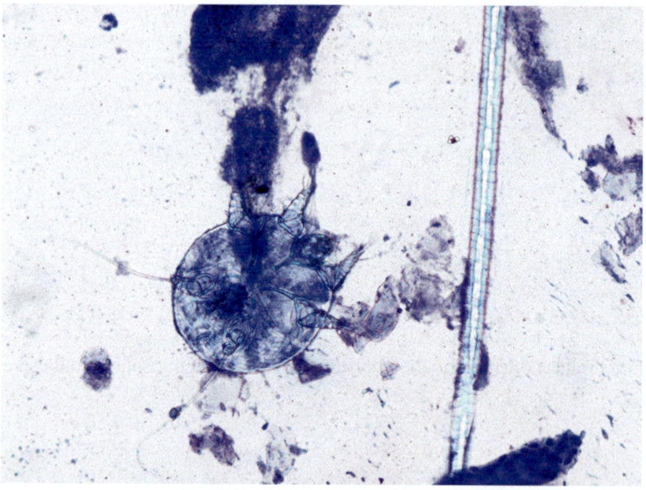

Fig. 6.220 Cytology: an adult *Notoedres cati* sampled placing adhesive transparent tape on the scales

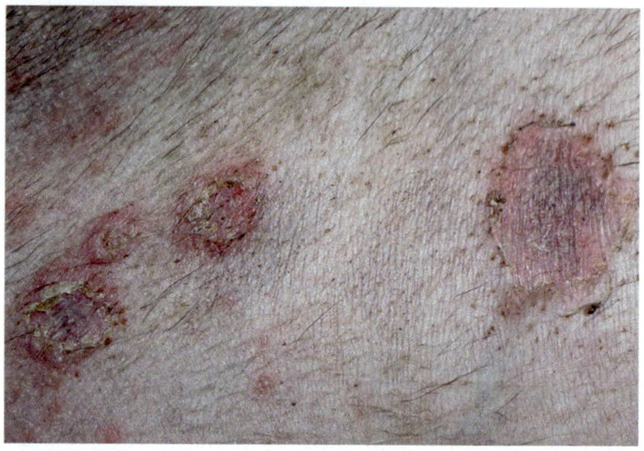

Fig. 6.221 Epidermal collarettes in a dog with superficial pyoderma. Typical erythematous roundish erosion bordered by scales and hyperpigmented central area

6.6 Erosions

6.6.1 Erosive Diseases in Dogs

Epidermal collarettes are the most frequent erosive lesions observed in canine dermatology and are also more valuable for cytological examination. Collarettes are round secondary lesions with an erythematous or hyperpigmented central area, peripherally bordered by scales or crusts, which originate following the breakage of a pustule [121–123] (Fig. 6.221).

Usually, in canine superficial pyoderma, collarettes are present in association with papules and intact pustules, but, sometimes, they are the only lesions present. Large and confluent lesions that resemble collarettes can be observed in *spreading pyoderma*, an exfoliative variant of superficial staphylococcal pyoderma (Figs. 6.222 and 6.223).

Although pyoderma is the most frequent cause of *collarettes*, they can develop in all cases of *pustular dermatosis*. In dogs affected by PF, during the rest phase of pustular eruption, skin lesions may only be represented by yellowish crusts and epidermal collarettes, and the clinical lesions may mimic those observed in superficial and spreading pyoderma

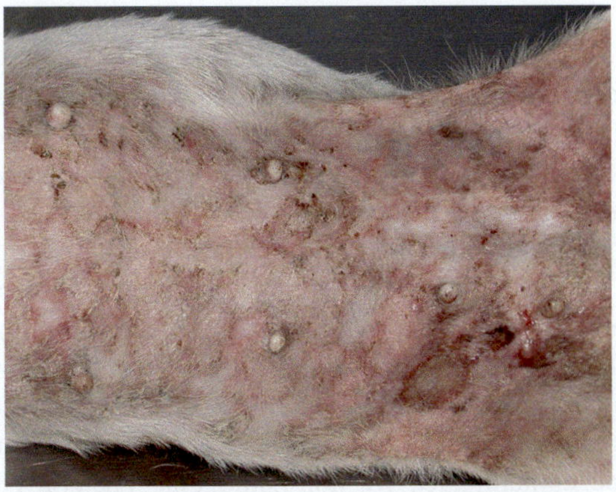

Fig. 6.222 Multiple and coalescent erosive and scaly lesions that mimic epidermal collarettes in a dog with superficial spreading pyoderma

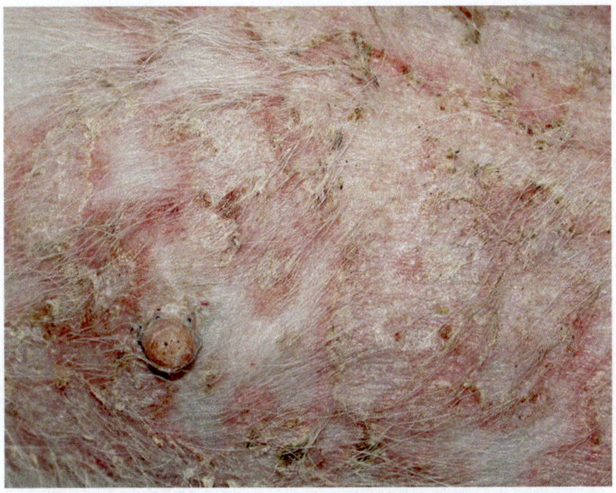

Fig. 6.223 Closeup of the lesions of the previous image

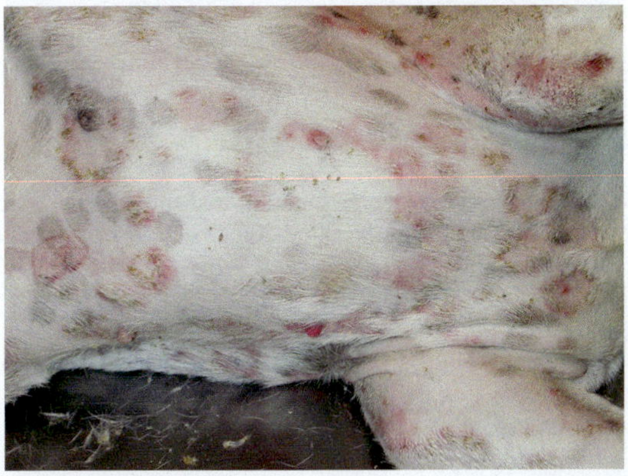

Fig. 6.224 Multiple and diffuse epidermal collarettes in an English bulldog with PF

(Figs. 6.224 and 6.225). In this case, cytology from collarettes can provide important indications for the diagnosis of autoimmune disease when segmented neutrophils, no bacteria, and many acantholytic cells are observed.

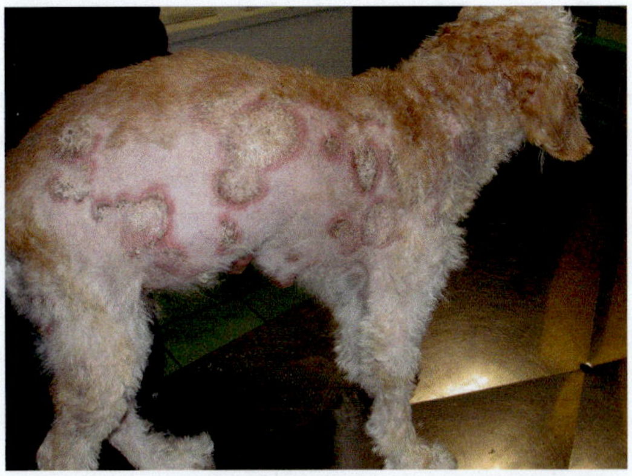

Fig. 6.225 Very large and coalescent epidermal collarettes in a dog with PF

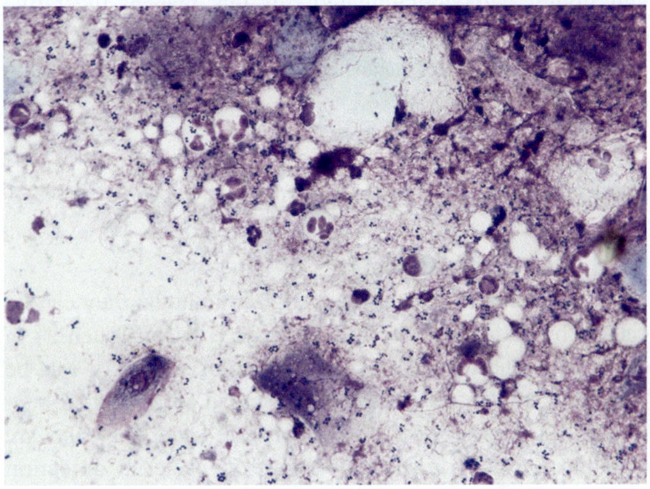

Fig. 6.226 Cytology collarettes: karyolitic neutrophils with many cocci both intracytoplasmic and on the bottom of the slide

Cytological Findings

Cytology from collarettes in dogs with pyoderma shows the same cytology as that observed in specimens from intact pustules.

Specimens are less cellular and composed of more corneocytes or nucleated keratinocytes, the latter from the spinous and granular layers, together with many entire or broken karyolitic neutrophils (nuclear streaks) and cocci, both phagocytized and free on the slide's background (Fig. 6.226).

As mentioned, collarettes can be detected in all pustular diseases, regardless of the cause.

In PF, the cytological detection of segmented neutrophils and AKs can orient the clinicians to perform a skin biopsy.

When sampling erosions, it should be remembered that the visualization of bacteria should be interpreted with more caution, as erosions are exposed to environmental and oral contaminant microorganisms. The aspect of lesions, the history, and the morphology of bacteria orient the clinicians toward a correct interpretation of the detected bacteria.

6.6.2 Erosive Diseases in Cats

Epidermal collarettes are extremely rare in cats since pustules are very rare lesions in this species. In pustular diseases such as PF, rather than collarettes, yellowish crusts indicative of the dehydration of pustules are more frequently observed. Round lesions with scaly borders similar to epidermal collarettes are often observed in cats affected by *dermatophytosis*, but, in these cases, they are not secondary to pustules.

6.7 Ulcer

As mentioned in the chapter regarding the diagnostic techniques, samples from *ulcers* are usually of poor quality and highly hemocontaminated; for this reason, ulcerative diseases rarely supply useful specimens. Nevertheless, there are some ulcerative diseases for which cytology may provide an immediate diagnosis.

6.7.1 Ulcerative Diseases in Dogs

6.7.1.1 Deep Pyoderma

Ulcers can characterize *deep pyoderma* in dogs; they can be caused by self-trauma or following deep lesions such as in cases of bacterial cellulitis or furunculosis. Ulcers can be localized as in infected calluses or spread all over the body. In all these cases, ulcerative lesions are secondary and are usually not the direct expression of bacterial infection [126, 127]. In dogs, the so-called *mucocutaneous pyoderma* is characterized by depigmentation and skin ulceration located at the mucocutaneous junctions of the lips, eyelids, nares, and, less frequently, the anal and genital junctions (Figs. 6.227 and 6.228) [124, 130]. These particular variants of pyoderma are clinically indistinguishable from some autoimmune-mediated diseases such as mucocutaneous lupus, and in many cases, the histopathological findings are not sufficient for a definitive diagnosis either [128, 130]. Another form of deep pyoderma, characterized by ulcerative lesions, is the so-called *idiopathic recurrent pyoderma* of the German shepherd dog. The pathogenesis of this form of pyoderma is unclear and the role of bacteria in the development of the lesions is debated, because many forms are speculated to be idiopathic. Furthermore, many dogs recover without the use of antibacterial drugs, and dramatic improvement with therapies based on immune modulators (e.g., cyclosporine) suggests that an immune-mediated pathogenesis may be on a genetic base (Fig. 6.229) [125, 127, 129, 131]. As with erosions, ulcers may be contaminated by non-pathogenic bacteria coming from the environment or, in cases of excessive licking, transferred onto the skin from the oral cavity. In such cases, clinicians must carefully interpret the presence of bacteria based on the clinical lesions from which cells are sampled.

Cytological Findings

Samples from *ulcerative pyoderma* are composed of blood, amorphous debris, and a high number of neutrophils and macrophages, with a variable number of lymphocytes and plasma cells. Keratinocytes from different epidermal layers are usually present.

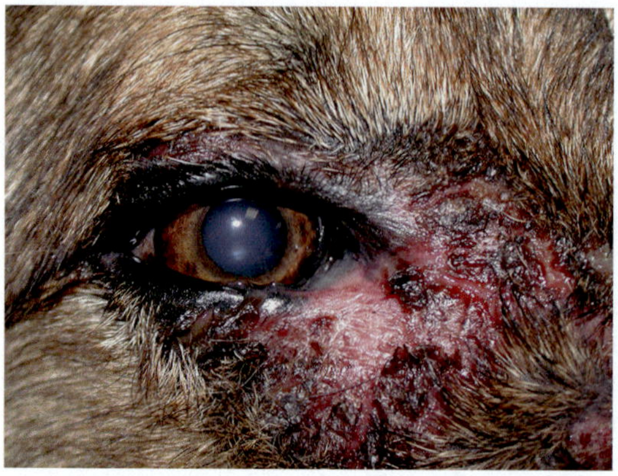

Fig. 6.227 Ulcerative blepharitis in a German shepherd with mucocutaneous pyoderma

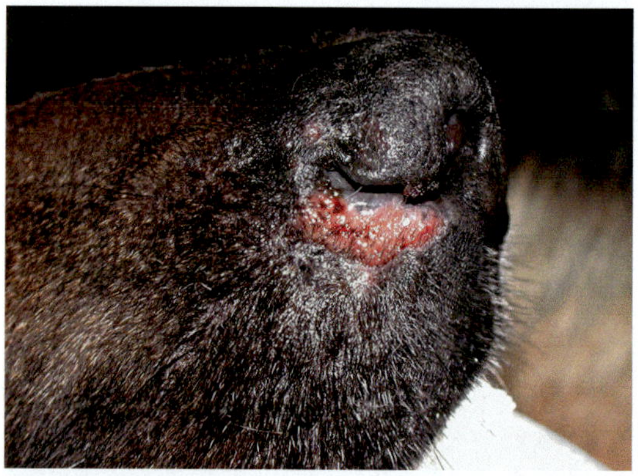

Fig. 6.228 Inflammatory depigmentation and ulcer of the mucocutaneous junction of a nostril of the same dog of the previous image

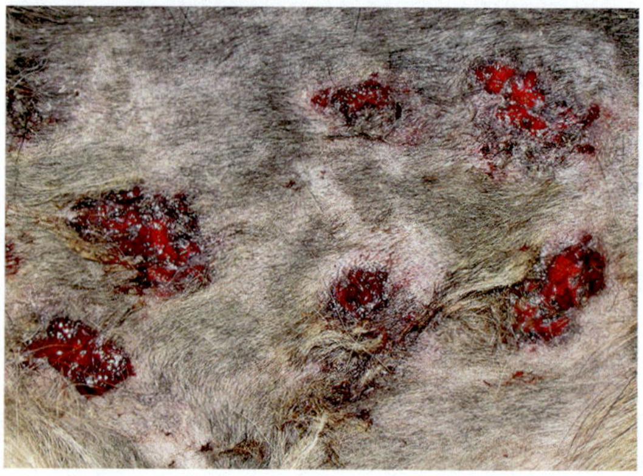

Fig. 6.229 Multiple bleeding ulcers on the trunk of a German shepherd with *idiopathic deep pyoderma*

As mentioned, bacteria are the hallmark for a cytological diagnosis of pyoderma, but their presence must be interpreted with caution; large numbers of extracellular bacteria arranged in clusters and spread between the inflammatory cells must be interpreted as contaminants, whereas few intracytoplasmic bacteria have a definite pathological significance. It must be remembered that, as in deep pyoderma, especially when characterized by ulcerative lesions, the failure to visualize bacteria is a very frequent occurrence and does not allow the infection to be excluded (Fig. 6.230).

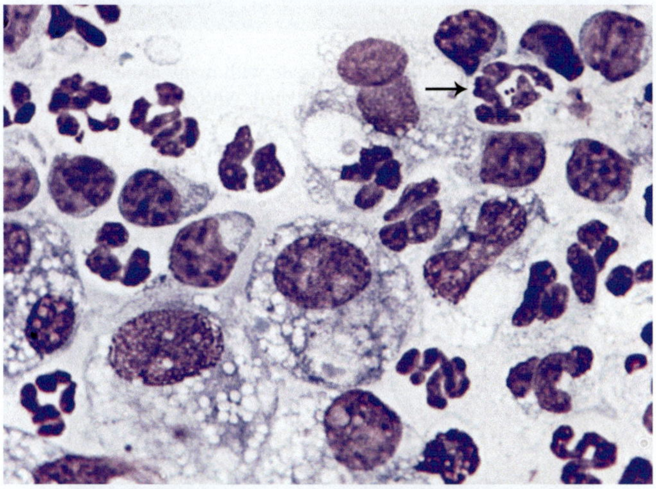

Fig. 6.230 Cytology of deep pyoderma: few coccoid bacteria are present in the cytoplasm of a neutrophil (arrow)

6.7.1.2 Leishmaniasis

CanL is a disease clinically characterized by a large polymorphism, and ulcers are very commonly observed in sick animals [132, 133] In endemic areas, the presence of bleeding ulcers on the tip of the tail and pinna is very common in affected dogs; lesions on these body sites are usually interpreted as secondary to vasculitis, although, in most cases, skin biopsies are not performed or if carried out, do not always exhibit the histopathological findings of vasculitis. Chronic ulcers that tend not to heal, usually with raised edges and commonly seen on the pressure points such as the hook, elbow, and carpus, or on the mucous membranes, can sometimes provide diagnostic samples (Fig. 6.231).

Cytological Findings

As in other ulcerative diseases, cytological specimens from ulcerative lesions provide hemocontaminated samples together with many inflammatory cells. In samples from non-bleeding ulcers, granulomatous/pyogranulomatous inflammation with a variable number of amastigotes, from low to very high, can be detected. Protozoa are usually localized in the cytoplasm of macrophages or are free on the background of the slides following cell rupture during the preparation of the specimens (Figs. 6.232 and 6.233).

Fig. 6.231 Leishmaniasis ulcerative lesions**:** (**a**) ulcerative blepharitis; (**b**) severe ulcer and hematic crusts on the pinna ulcers; and (**c**) deep ulcer on the paw pad; (**d**) ulceration of the preputial mucosa

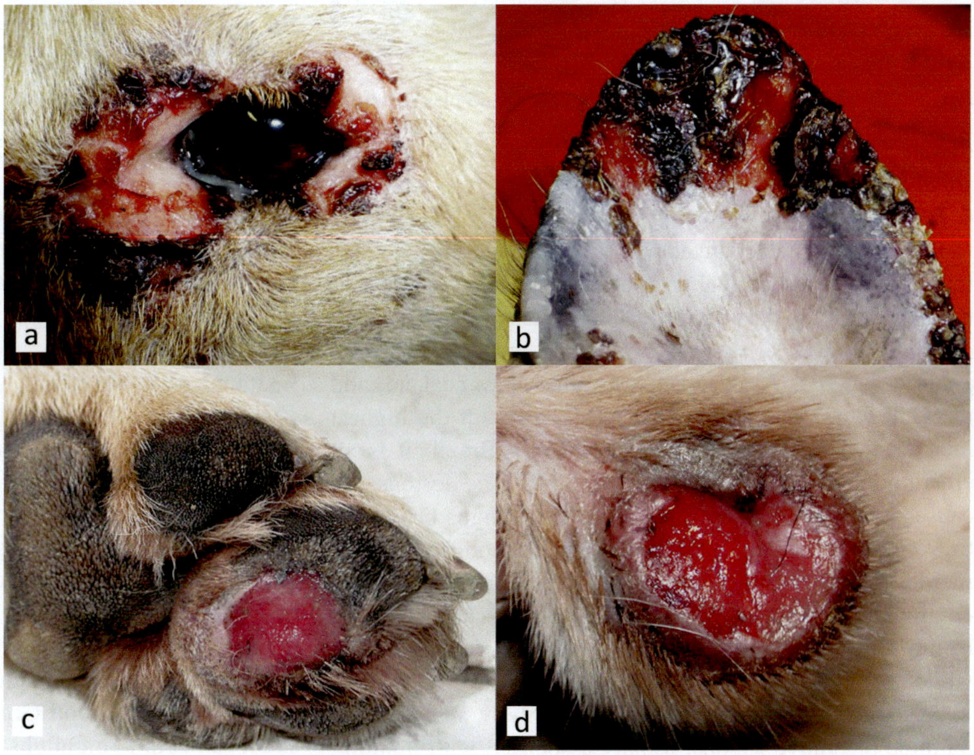

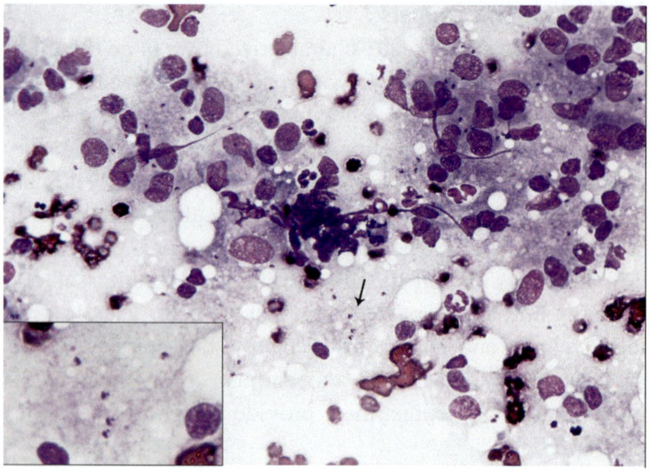

Fig. 6.232 Cytology of leishmaniasis: lympho-histiocytic inflammation. The typical morphology of amastigotes is clearly recognizable (*arrow—inset*)

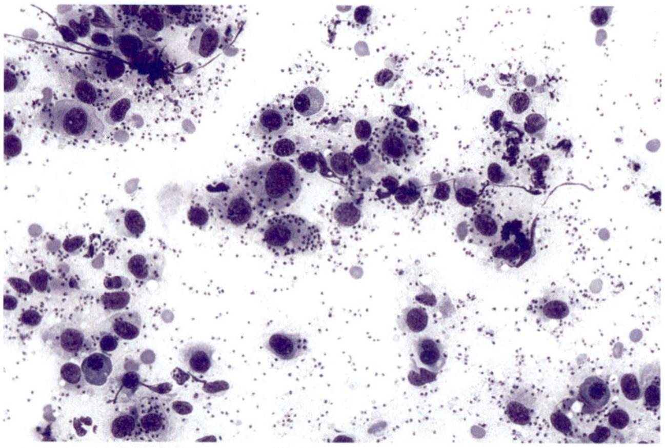

Fig. 6.233 Cytology of leishmaniasis: lympho-histio-plasmacellular inflammation. Many amastigotes both intracytoplasmic and free on the background of the slide

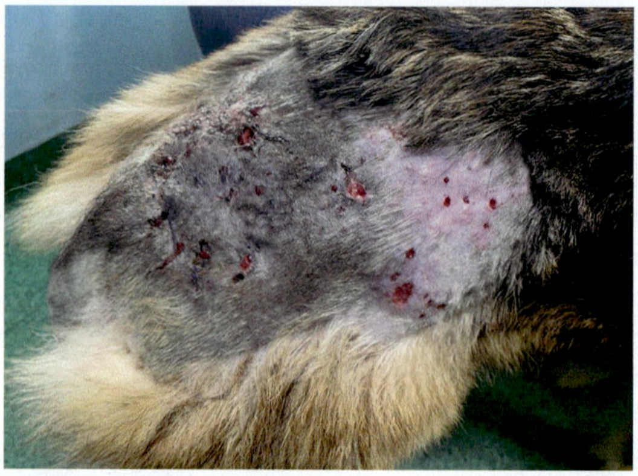

Fig. 6.234 Diffuse swelling of the tissue with many draining tracts on the back of a German shepherd affected with cutaneous *Pythiosis* (Courtesy of DVM F. Necci-Italy)

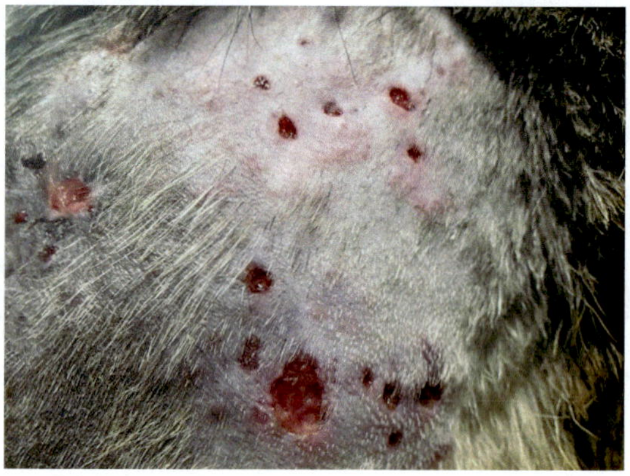

Fig. 6.235 Close up of the lesions of the same dog in Fig. 6.234 (Courtesy of DVM F. Necci-Italy)

6.7.1.3 Pythiosis

Pythiosis is a rare deep mycotic infection that can affect dogs mainly present in tropical and sub-tropical areas, which primarily affects the gastrointestinal apparatus [137]. This fungal infection is a disease present in many worldwide countries such as Asia, Japan, Australia, New Zealand, and the Gulf Coast states of the United States with the highest incidence in South America, especially in Brazil; it is an emerging disease in Europe [136, 138]. The lesions are caused by *Pythium insidiosum*, an aquatic fungus belonging to *Oomycetes* family. The infective stage is thought to be the motile, flagellate zoospore, which is produced in aquatic environments (e.g., lakes), and that probably causes disease by invading damaged skin or intestinal mucosa, the latter while dogs drink stagnant water.

In some dogs, also the skin can be involved, and single or multiple ulcerative lesions, usually associated with draining tracts, are observed [134–136, 138] (Figs. 6.234 and 6.235). Final diagnosis is not so easy to perform with cytology since the clinical–pathological features are not differentiable from zygomycosis or lagenidiosis and skin biopsy, fungal culture, or, better, PCR test are usually necessary to typify the fungus.

Cytological Findings

Pythium insidiosum causes a pyogranulomatous inflammation, with a variable number of multinucleated macrophages (multinucleated histiocytes) (Fig. 6.236). Fungal elements are usually easily recognizable on cytological specimens since they are represented by long, broad, usually non-segmented, with non-parallel wall and with branched hyphae [137] (Figs. 6.237, 6.238, and 6.239). As with other fungal bodies, in many cases, the routine dyes are unable to stain the hyphae which are visible as slender achromatic linear formations. PAS, or better Grocott-Gomori methenamine silver (GMS) stain, can help to highlight fungi (Fig. 6.240).

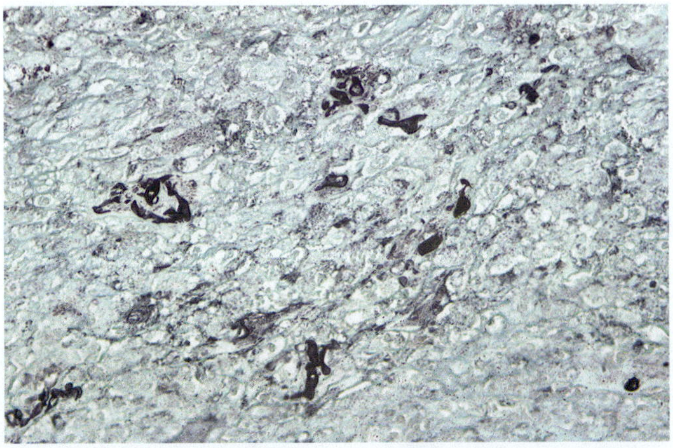

Fig. 6.236 Histology of *Pythiosis* (GMS): hyphae with nonparallel walls and irregular branching typical of fungi belonging to oomycetes

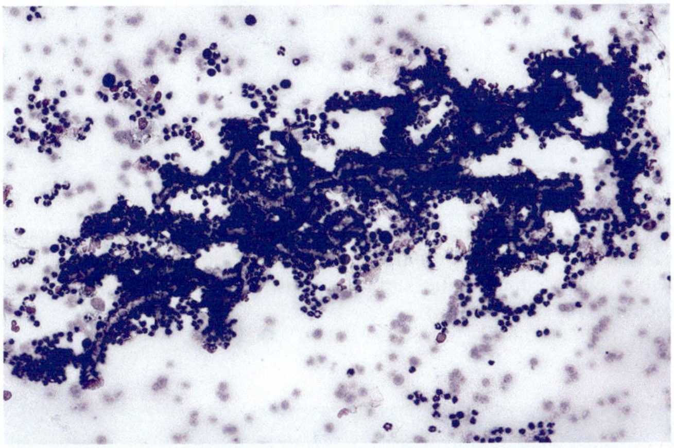

Fig. 6.237 Cytology of *Pythiosis*: inflammatory cells surround achromatic long and branched not segmented hyphae

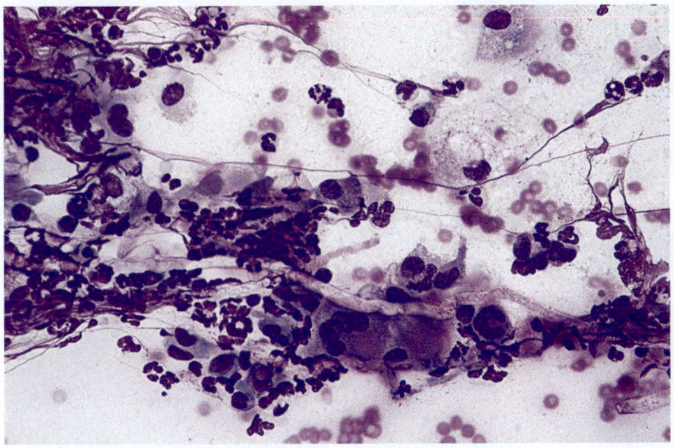

Fig. 6.238 Cytology of *Pythiosis*: achromatic not segmented hyphae in a pyogranulomatous inflammation. Note the giant macrophage in attempting to eliminate the fungus

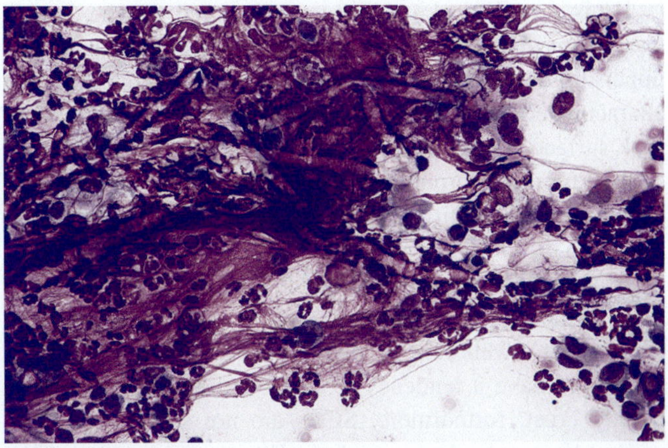

Fig. 6.239 Cytology of *Pythiosis*: the lack of septa into the hyphae is well visible to higher magnification

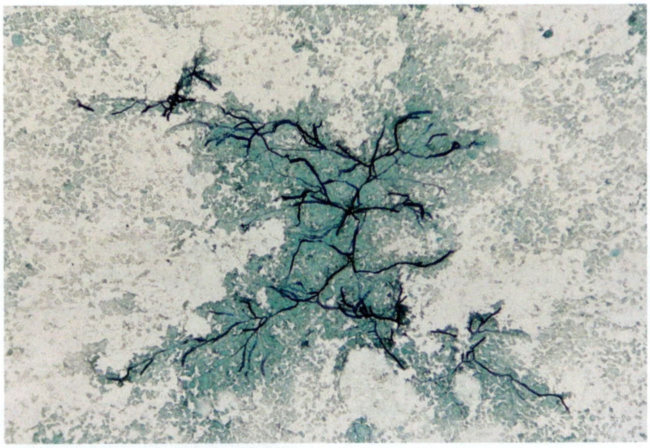

Fig. 6.240 Cytology of *Pythiosis* (GMS): the length and morphology of hyphae are highlighted with cytochemical special staining

6.7.2 Ulcerative Diseases in Cats

Ulcerative dermatitis is very common in cats, and most cases are observed following self-trauma in pruritic allergic cats, with lesions mainly located on the head and neck. Other feline spontaneous inflammatory ulcerative lesions can be investigated through cytology.

6.7.2.1 Feline Indolent Ulcer

Feline IU is a characteristic lesion localized to the lip and sometimes extending to the filter of the nose [139–141]. These are grouped into a so-called *eosinophilic granuloma complex* (EGC), a recognized entity in cats along with *eosinophilic granuloma* (EG) and *eosinophilic plaque* (EP). The latter, although sometimes manifested as ulcers, are more frequently nodular and, for this reason, are discussed in detail with nodular lesions.

IU is a typical skin ulceration observed on the lips of cats. Clinically, they can be single or symmetrical on the upper lips; the lesions are not painful to palpation and characterized by a smooth, non-bleeding, dry, yellowish-pink surface.

The typical morphology and localization of IU do not usually require confirmatory cytology. Some authors include as a main differential squamous cell carcinoma (SCC), which is usually ulcerative in cats. Unlike IU, carcinomas do not have a smooth and dry surface with yellowish foci, and the edges are not well defined as in IU but are more irregular in shape and have a tendency to bleed and be covered with a bloody crust; furthermore, SCCs are not symmetrical and rarely affect only the upper lips. The suspicion of an SCC may arise only when a severe SCC starting from the nasal planum involves the filter and upper lips; in these cases, cytological or histopathological investigations can be performed to differentiate the two diseases (Fig. 6.241).

Cytological Findings

Although included in the EGC, the inflammatory infiltrate of IUs is not always characterized by eosinophils as might be expected. The histopathology of many IUs is characterized by a few or no eosinophils, and this may also depend on the chronicity of the lesions [140]. Another problem is linked to the characteristic IU surface, which, being dry and firm, could not provide any cells if sampled using the imprint technique. Finally, the mucocutaneous location inevitably leads to specimens contaminated by bacteria from the normal resident flora of the oral cavity. In the early phases of the disease, a sampling made by gentle surface scarification can help to remove the first superficial and contaminated layers and to collect a significant number of eosinophils as occurs in other eosinophilic lesions. Contamination bacteria are anyway almost always present in these slides (Figs. 6.242 and 6.243).

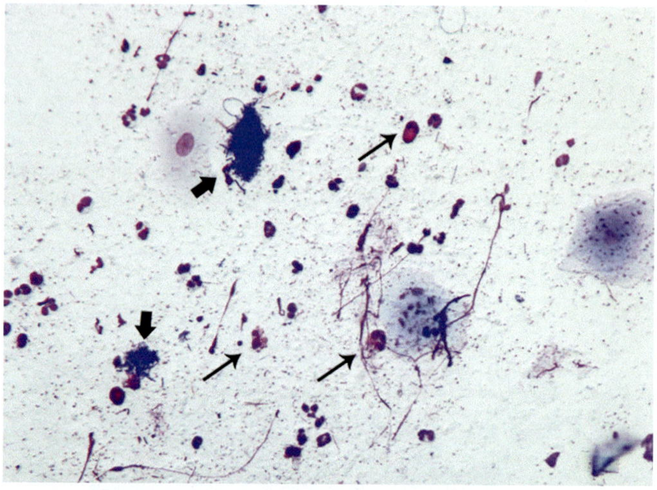

Fig. 6.242 Cytology of an IU: eosinophils, nuclear stripes, aggregates of coccoid bacteria from oral cavity contamination (short arrows), and debris collected from the surface of the ulcer. Few eosinophils are recognizable (thin arrows)

Fig. 6.241 Feline IU: (**a**) typical symmetrical and bilateral non-bleeding ulcer; (**b**) dry and yellowish surface on the upper lips of a cat with an IU; (**c**) IU on the center of the upper lips with dry non-bleeding surface showing yellowish foci; and (**d**) large IU: the well-defined edges and yellowish foci help to orient a clinical suspicion of IU more than an SCC

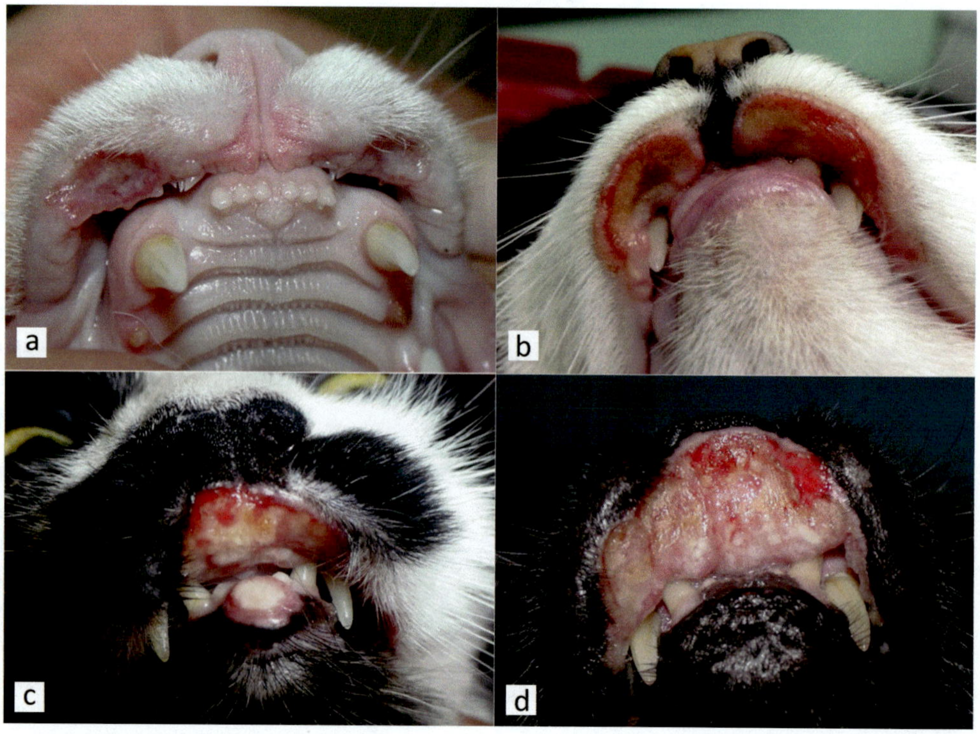

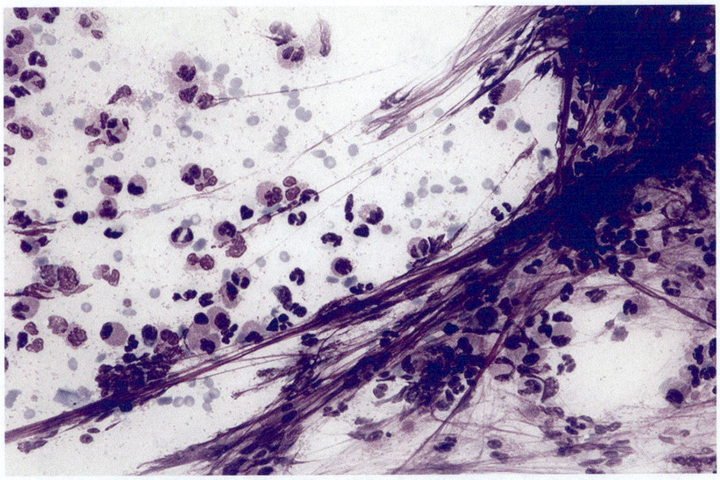

Fig. 6.243 Cytology of an IU: many eosinophils are collected along with neutrophils

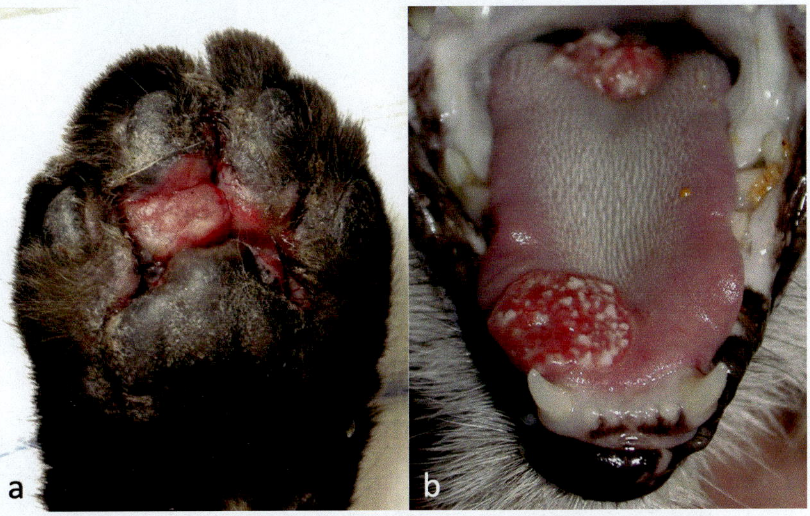

Fig. 6.244 EG: (**a**) interdigital ulceration in a cat with EG and (**b**) ulcerated plaques on the tongue of a cat

6.7.2.2 Eosinophilic Granuloma

As mentioned, the EG is a polymorphous lesion that ranges from papules and nodules to ulcers [142–146].

Ulcers are usually due to the strong dermal eosinophilic infiltrate, which leads to epidermal ulceration. Clinically, a typical ulcerative granulomatous lesion on the paws and in the interdigital spaces is observed in affected cats. Ulcerated lesions belonging to the feline EG are localized in the oral cavity, where single or multiple ulcerated nodules and plaques can be detected (Fig. 6.244).

Cytological Findings

Cytology from interdigital EG lesions is not easy to perform and, like every ulcer, rarely provides specimens of high cellularity and good quality. Secondary pyoderma, together with constant licking, can further complicate the specimens. In some cases, diagnostic slides could be obtained via FNB that permit eosinophils to be collected directly from the dermis avoiding superficial contaminants. A predominant population of eosinophils with many macrophages and a variable amount of epithelioid macrophages and giant macrophages can be collected; the amount of neutrophils depends on the sampling technique used and on secondary infections (Fig. 6.245).

Since the histopathology of CGE has peculiar characteristics and, in many cases, necrotic foci and flame figures are present, these cytological aspects will be discussed in nodular lesions of CGE.

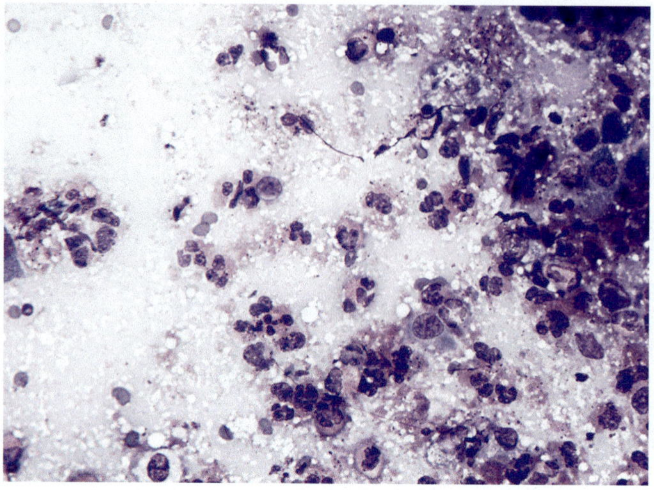

Fig. 6.245 Cytology of EG: many eosinophils and some macrophages and lymphocytes are also recognizable

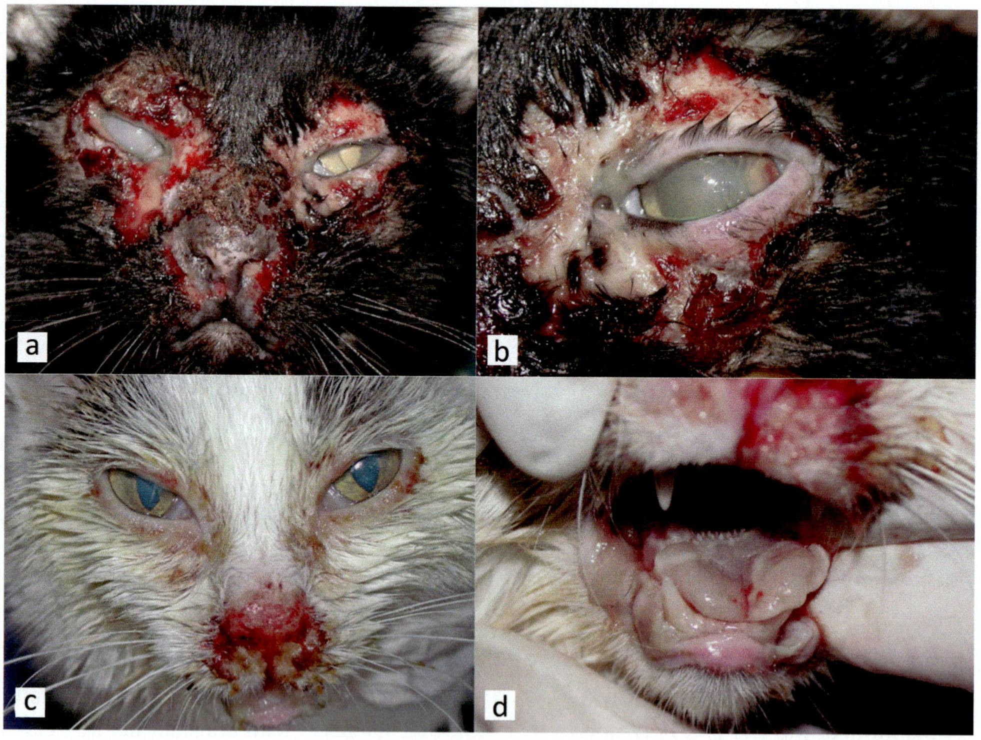

Fig. 6.246 Feline herpesvirus: (**a**) severe ulcerative facial dermatitis and corneal edema in a cat with herpesvirus infection that has received chronic steroid therapy; (**b**) closeup of the same cat of the previous image: ulcers, crusts, and severe ocular lesions; (**c**) blepharitis and ulcerative lesions on the nose and muzzle; and (**d**) severe ulcerative lingual lesions with detachment of the mucosal epithelium

6.7.2.3 Herpesvirus

Type I herpesvirus is the etiological agent of *feline rhinotracheitis*, causing the well-known upper respiratory and ocular symptoms [149, 151, 152].

The viral tropism for keratinocytes rarely causes ulcerative lesions and is usually confined to the facial skin with involvement of the nose, muzzle, and eyelids. In rare cases, the virus can be the cause of severe and extensive *necrotizing facial dermatitis*. The oral cavity is often involved with vesicular/bullous and ulcerative lesions (Fig. 6.246) [147, 148, 150]. Histopathologically, the herpesvirus lesions are characterized by necrotic dermatitis involving the epidermis and the follicle wall (mural folliculitis) in association with strong eosinophilic and neutrophilic diffuse dermal inflammation [148] The diagnosis of herpesvirus is very difficult but can be formulated if viral inclusion bodies in the nuclei of keratinocytes are detected by histology. Unfortunately, it is hard to find those viral inclusions in histological specimens, and it is almost impossible to observe them through cytology (Fig. 6.247).

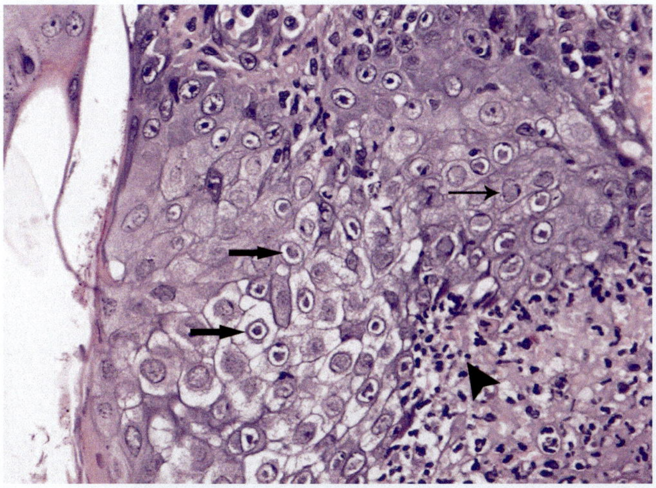

Fig. 6.247 Histology of herpesvirus: intranuclear viral inclusion (arrows), some of them amphophilic (thin arrow); note the eosinophilic infiltrate in the superficial dermis (*head of arrow*). (Courtesy of Prof. DVM F. Abramo—Italy)

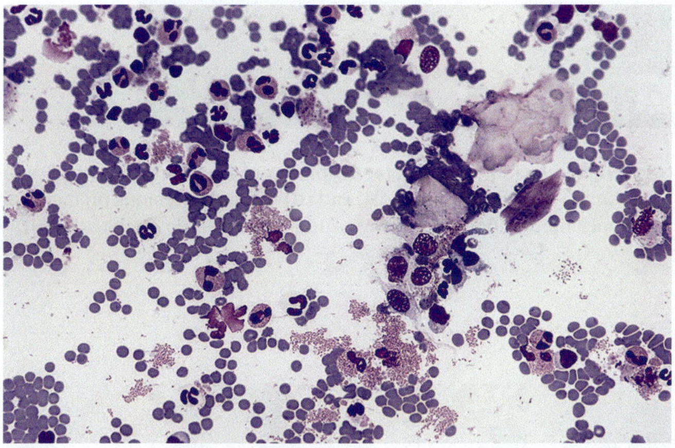

Fig. 6.248 Cytology of herpesvirus: many eosinophils and macrophages collected using the impression smear technique from the inner surface of a facial crust

Cytological Findings

Slides obtained through impression smear from ulcers or from the lower surface of the crusts are often hemocontaminated and characterized by an aspecific mixed exudate with a varying amount of neutrophils and macrophages. In more lucky slides, a large number of eosinophils can be found since the herpes virus typically evokes an eosinophilic inflammation. Since facial lesions also characterize allergic cats, care must be taken to make a diagnosis of allergic diseases based only on eosinophils detection on slides; in fact, a treatment with systemic steroids, often used by practitioners for the management of many allergic diseases, may have serious consequences in cats

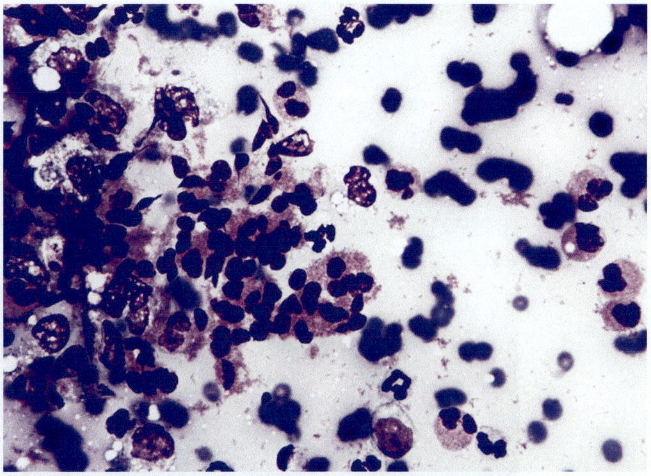

Fig. 6.249 Cytology of herpesvirus: at high magnification, the eosinophilic inflammation is clearly visible

with herpesvirus infection (Figs. 6.248 and 6.249). Ulcers covered with hematic crusts, the lack of pruritus, and a history of previous upper respiratory symptoms and conjunctivitis can direct clinicians to suspect a viral infection and perform a skin biopsy.

6.7.2.4 Sporotrichosis

Sporothrix schenckii is a ubiquitous dimorphic fungus that causes *sporotrichosis*, a cutaneous and subcutaneous fungal disease of cats and dogs. The disease is observed mainly in certain geographical areas, such as the United States, and in some areas of South America, such as Brazil [153–159]. Since the disease is mainly found in cats, it has been hypothesized that *Sporothrix* spp. may penetrate the dermis through bites and scratches; furthermore, since the skin lesions mainly affect the extremities, such as the head and paws, this hypothesis could be further supported.

Clinically, three different forms of the diseases are recognized: a *localized* or *fixed cutaneous form* that seems more frequent in dogs, a *cutaneous lymphatic* form, and a *disseminated form*, the latter usually associated with the lymphatic form [153]. In cats, the cutaneous lymphatic form is most common, with skin lesions presenting with nodules and ulcers, and as the lymphatic vessel involvement creates mostly ulcerative lesions, the author prefers to discuss this fungal disease in the group of ulcerative lesions (Fig. 6.250).

Cytological Findings

Cytological specimens show many macrophages filled with fungal bodies; the latter usually seem to be less numerous in samples collected from dogs (Fig. 6.251). The fungal

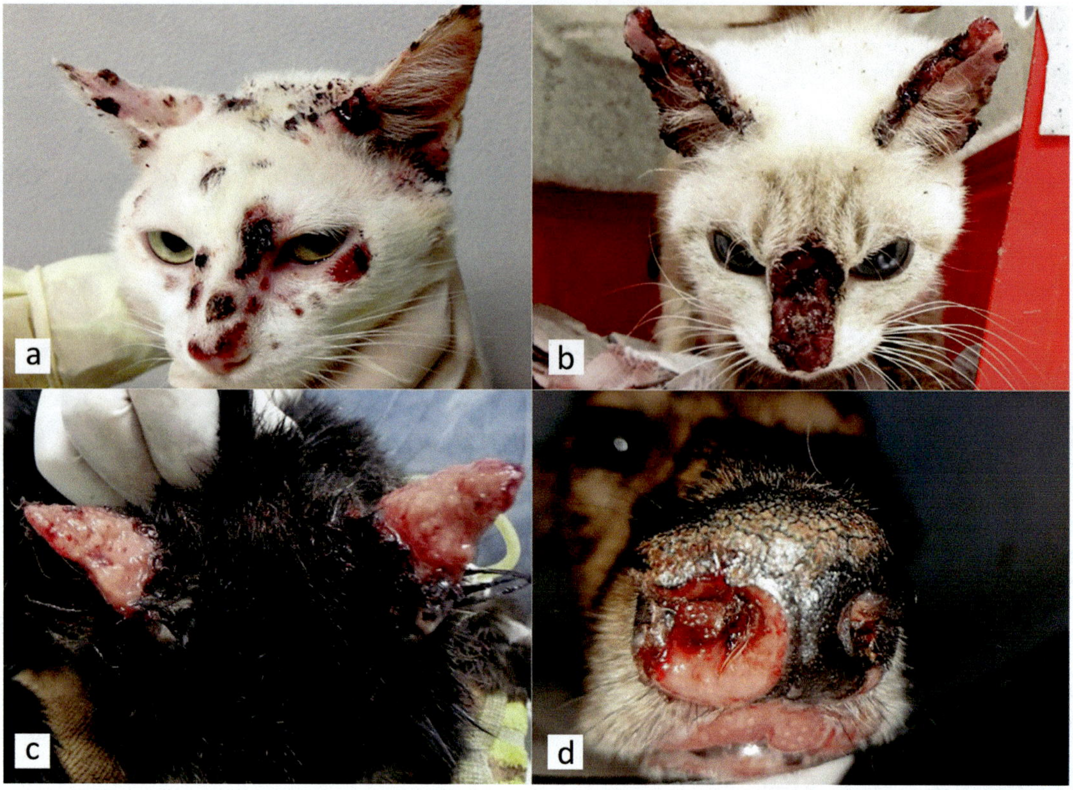

Fig. 6.250 Sporotrichosis: (**a**) multiple ulcerative lesions on the face and on the pinna. (Courtesy of DVM A. Vieira Pereira—Brazil); (**b**) ulcerative dermatitis on the pinna and on the bridge of the nose in a cat affected by sporotrichosis. (Courtesy of Dr. A. Vieira Pereira—Brazil); (**c**) bilateral ulcerative lesions on the pinnas of a cat. (Courtesy of DVM Luís Américo de Borba—Brazil); and (**d**) ulcerative lesions on the nose and upper lip of a dog. (Courtesy of Luís Américo de Borba—Brazil)

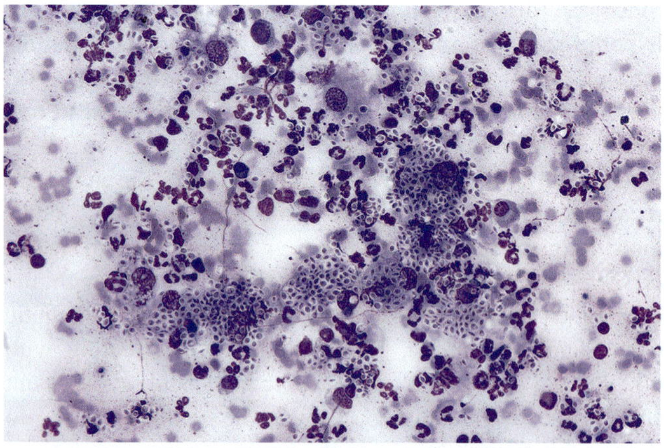

Fig. 6.251 Cytology of sporotrichosis: many *Sporothrix schenkii* in the cytoplasm of macrophages

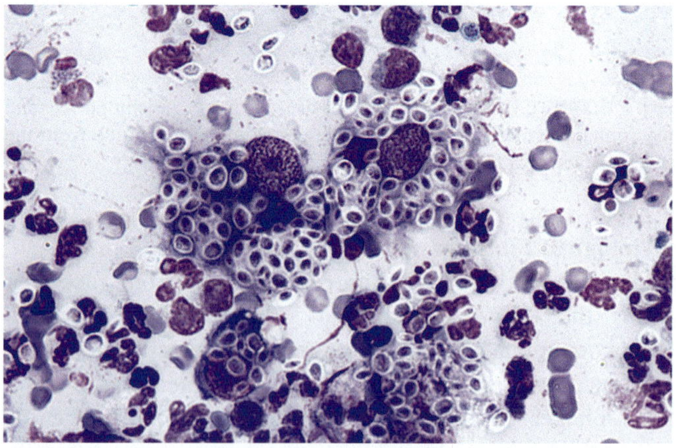

Fig. 6.252 Cytology of sporotrichosis: many oval-shaped *Sporothrix schenkii* in the cytoplasm of macrophages

bodies can show double morphology, ranging from *oval* to a more characteristic *cigar* shape measuring 3–10-µm long and 2–4-µm diameter with a thin, peripheral, clear halo (Figs. 6.252 and 6.253). As mentioned, *Sporothrix* is easily observed, but in dogs and rarely in some cats, PAS or Grocott's methenamine staining can help to detect fungal bodies immersed in the inflammatory cells (Figs. 6.254 and 6.255).

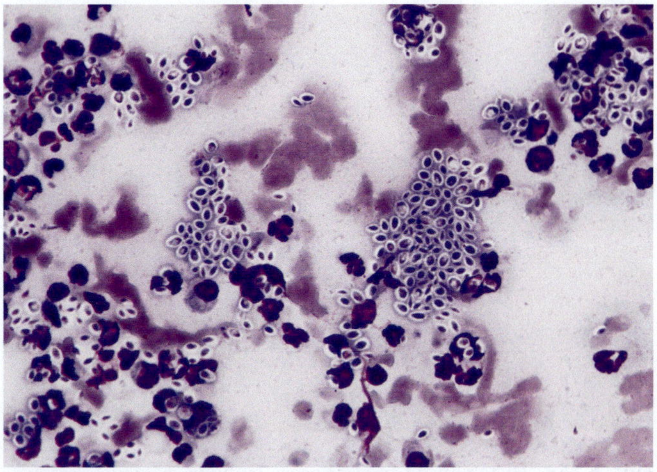

Fig. 6.253 Cytology of sporotrichosis: many cigar-shaped *Sporothrix schenkii*

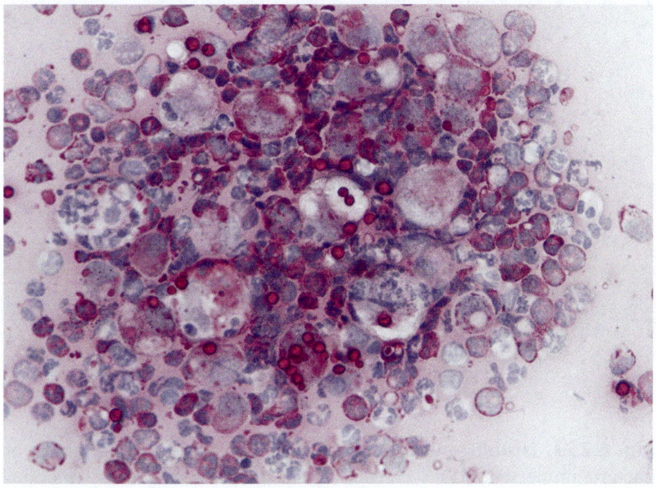

Fig. 6.254 Cytology of sporotrichosis (PAS staining): well-defined PAS-positive fungal bodies represented by roundish red color microorganisms are visible despite PAS stains also the content of inflammatory cells

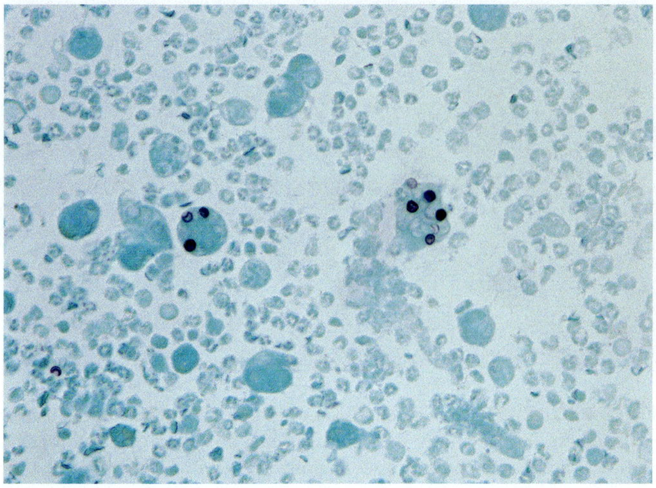

Fig. 6.255 Cytology of sporotrichosis (Grocott's methenamine silver staining): fungal bodies stain black on a green background

6.8 Plaques and Nodules

In pets, many skin lesions grow as *nodules* or *plaques* and include numerous both *neoplastic* and *non-neoplastic* (inflammatory) skin diseases.

Unfortunately, skin nodules are clinically not often differentiable from each other; for this reason, to simplify the discussion about those of non-neoplastic origin, they have been divided into two groups, based on the presence or absence of infectious agents. Two groups are as follows: *infectious* and *parasitic diseases*, in which a causative microorganism or a parasite is the cause of the lesion, and *non-infectious (sterile) diseases*, in which the cause is not linked to an infectious agent or a parasite.

6.8.1 Infectious and Parasitic Causes

6.8.1.1 Bacteria

Bacterial Nodular Diseases in Dogs and Cats

Excluding papular–nodular deep pyoderma caused by *Staphylococcal* spp. *folliculitis*, the development of nodules in dogs with pyoderma is due to the penetration of bacteria into the dermis or subcutis.

6.8.1.1.1 Nocardiosis and Actinomycosis

These nodular forms of deep pyoderma are caused by the deep penetration into the dermis and adipose tissue, of Gram-positive filamentous bacteria belonging to the *Actinomycetaceae* family.

Nocardia spp. are Gram-positive, partially alcohol–acid-resistant bacteria among which the *N. asteroides* complex includes *N. brasiliensis* and *N. otitidiscaviarum,* whereas, in cats, *N. nova* is most frequently involved [163–166].

Actinomycosis is caused by *Actinomyces,* Gram-positive and acid-fast negative bacteria, mostly represented by *A. viscosus*, *A. hordeovulneris,* and *A. odontolyticus* [160].

The list of these groups of bacteria expands continuously with some change in their names. Although sharing the filamentous morphology, the two bacteria have completely different habitats; *Nocardia* lives as a saprophyte in the soil and penetrates the skin mostly via deep infected wounds, whereas *Actinomyces* is a commensal of the oral cavity and intestines of dogs and cats and penetrates the skin with bites, but also via infected vegetal foreign bodies [160–162].

The skin lesions are characterized by single or multiple nodules or plaques that often tend to ulcerate or develop draining tracts, which discharge purulent and smelling exudate that may contain visible *white* or *brown grains* similar to those observed in true mycetoma (Fig. 6.256). In cats with actinomycosis, lesions are seen more frequently on the extremities, face, and neck, which are the sites that are most prone to bites. In some cases, lesions are represented as swelling and poorly defined nodules, with panniculitis,

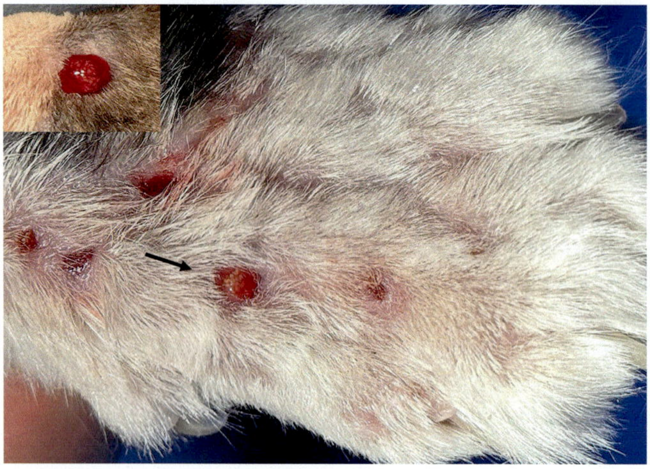

Fig. 6.256 Cat with cutaneous nocardiosis: draining tracts oozing purulent exudate in a in which many grains are clearly visible (arrow and inset) (Courtesy of DVM E. Konarzewska-Poland)

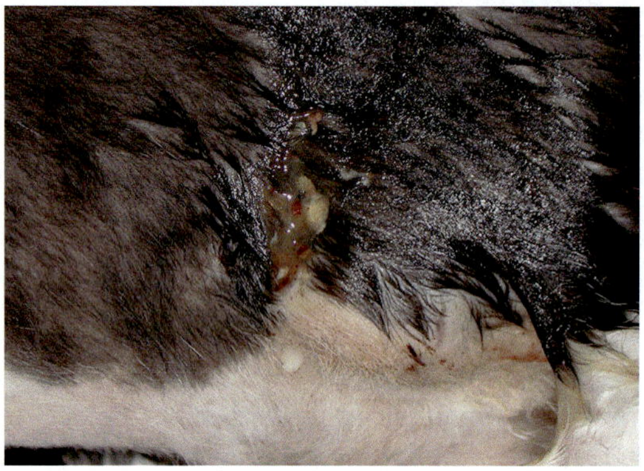

Fig. 6.257 Nocardiosis: draining tracts discharging grainy purulent exudate in a cat

especially on the abdominal area; clinical findings resemble those observed in cats with opportunistic mycobacterial infections [167] (Fig. 6.257).

Cytological Findings

Histologically, multiple granulomatous foci composed of bacteria surrounded by *Splendore–Hopplei* (S-H) reaction and inflammatory epithelioid and multinucleated macrophages (Figs. 6.258 and 6.259). Cytologically, *filamentous* microorganisms are easy to recognize because of their characteristic silhouette. At low magnification, when present, *grains* are composed of extracellular clusters of bacteria and are represented as deep blue or reddish amorphous material immersed in neutrophilic and macrophagic inflammation. Neutrophils are karyolitic and macrophages are vacuolated and show leukophagocytosis. At high magnification, elongated, slender, filamentous, often branched, linear, or beaded bacteria are recognizable (Figs. 6.260, 6.261, 6.262, and 6.263). It is not possible to typify bacteria through cytology,

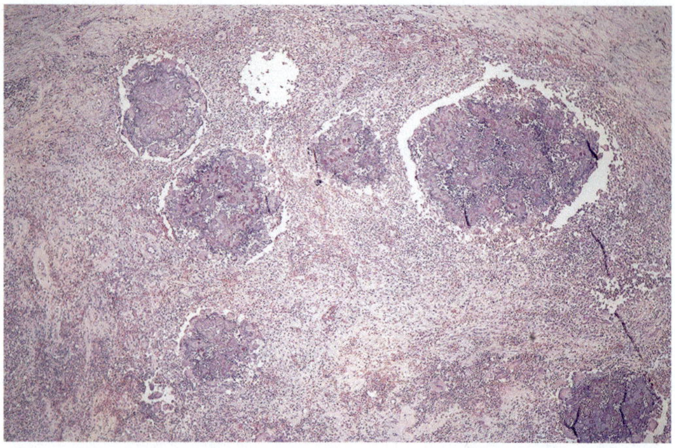

Fig. 6.258 Histology of nocardiosis: multiple bacterial granules immersed in a pyogranulomatous inflammation

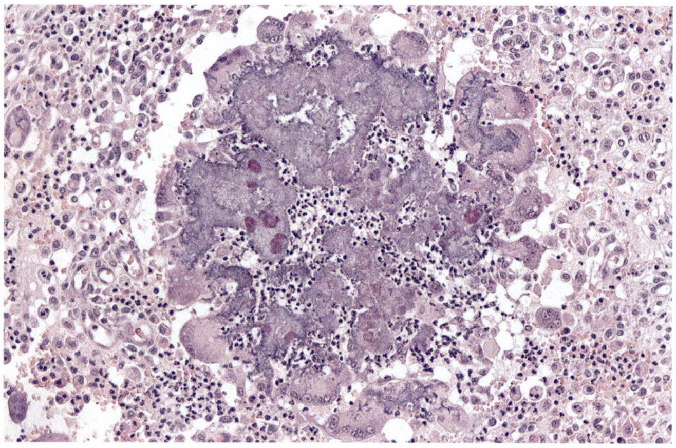

Fig. 6.259 Histology of nocardiosis: at high magnification, groups of bacteria immersed in an S-H reaction and surrounded by many multinucleated macrophages

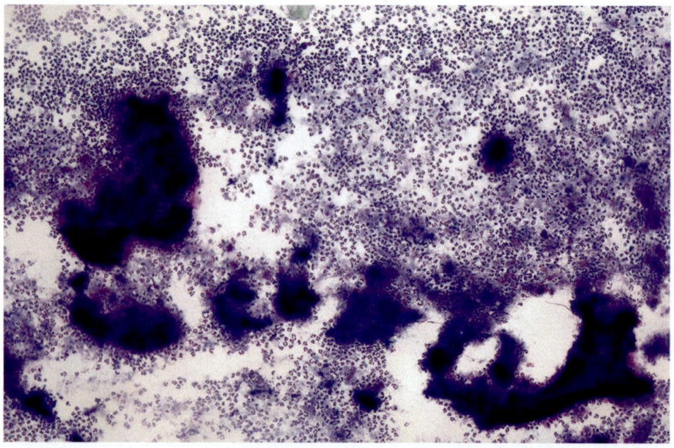

Fig. 6.260 Cytology of nocardiosis: multiple irregular shape clusters of bacteria, representing the same grains observed in histological samples, are recognizable at low magnification

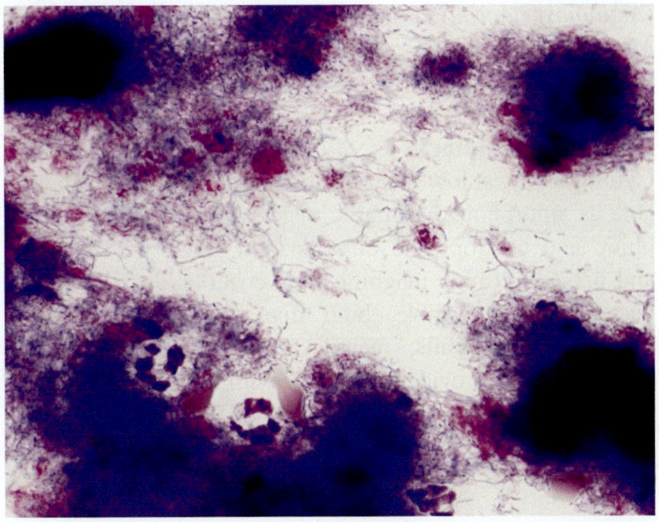

Fig. 6.261 Cytology of nocardiosis: at high magnification, many slender and filamentous *Nocardia asteroides* bacteria are recognizable

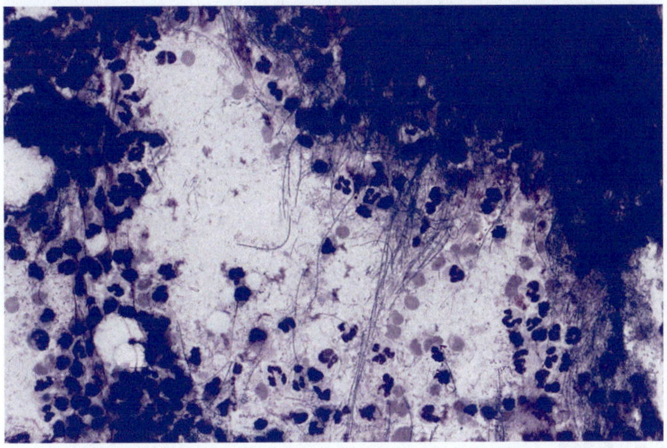

Fig. 6.262 Many long filamentous bacteria belonging to *Actinomycetaceae* family along with karyolitic neutrophils

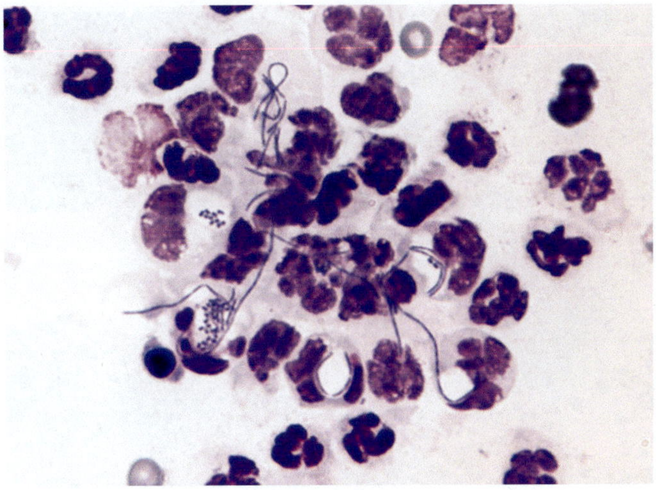

Fig. 6.263 Many cocci, some rods, and filamentous bacteria belonging to *Actinomycetaceae* family phagocytized by neutrophils

with which we can only define the shape; for a final typing, an anaerobic culture or molecular tests such as PCR are mandatory.

6.8.1.1.2 Botryomycosis

In some cases, deep nodular pyoderma can be linked to dermal or hypodermal localization of *non-branching* and *non-filamentous* bacteria such as *cocci* and *rod-shaped* bacteria.

The deep infections not linked to filamentous bacteria are called *Botryomycosis*. As non-filamentous bacteria do not usually form grains, as normally some fungi do them (e.g., *true mycetomas*), this form of deep granular bacterial infection is also defined by the confusing term of *bacterial pseudomycetoma* [167–169].

The most frequently isolated bacteria include *P. aeruginosa, E. coli, Proteus* spp., *P. acne, S. marcescens, P. multocida,* and so on, which penetrate the skin through contaminated foreign bodies and tend to arrange themselves in large clusters surrounded and isolated by the inflammatory cells.

Clinically, single or multiple nodules, swellings, and draining tracts can be observed, and, in rare cases, lesions are represented by abdominal panniculitis resembling those observed in cats with opportunistic mycobacterial infections (Figs. 6.264 and 6.265). Bacteria, together with a peripheral host reaction, comprise the *grains*, which are clinically recognizable as small granular formations immersed in purulent exudate (mycetoma-like). Histopathological specimens reveal nodular-to-diffuse pyogranulomatous dermatitis and panniculitis with many grains that resemble a *bunch of grapes* (Figs. 6.266 and 6.267). At the periphery of the grains, an eosinophilic material composed of immunoglobulin and fibrin, named the S-H *reaction*, is observed. This reaction is not pathognomonic of botryomycosis as is always observed in the true mycetomas. It is sometimes detected in deep pyoderma because of filamentous bacteria

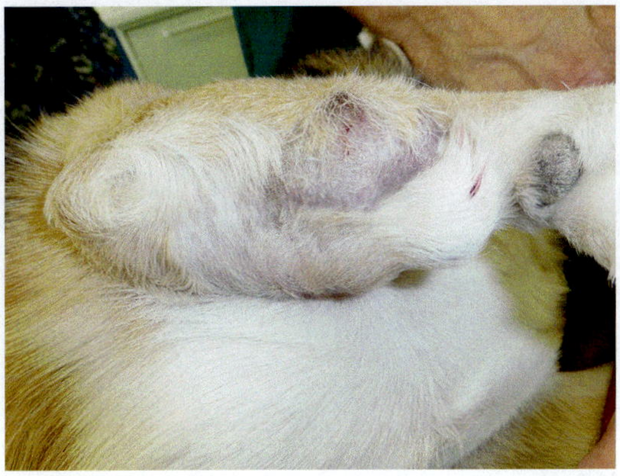

Fig. 6.264 Subcutaneous nodule on the leg of a dog with *Botryomycosis*

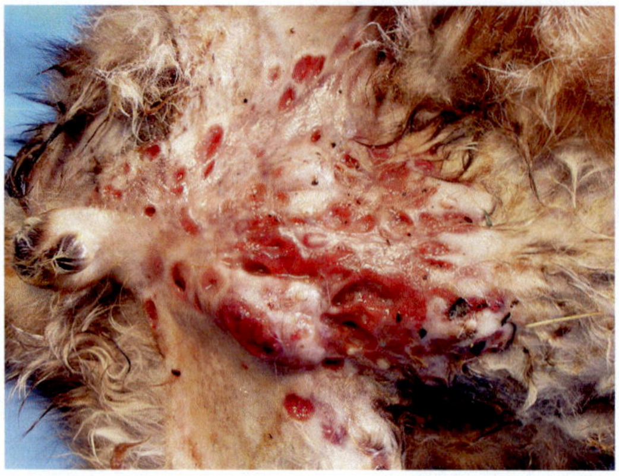

Fig. 6.265 Multiple subcutaneous nodules and draining tracts on the abdomen of a cat infected with *Rhodococcus equi.* (Courtesy of DVM C. Caporali—Italy)

(*Actinomycetaceae*) and in rare cases of deep localization of *Microsporum canis* (dermatophytic pseudomycetoma (DP)) [167].

Cytological Findings

Cytologically, at low magnification, the *grains* are recognizable as amorphous basophilic material immersed in neutrophilic and macrophagic inflammation (Figs. 6.268, 6.269, and 6.270). At high magnification, cocci and/or rod-shaped bacteria are detectable, both free and phagocytized by karyolitic neutrophils; many vacuolated leukophagocytic macrophages are evident, together with variable amounts of lymphocytes (Fig. 6.271).

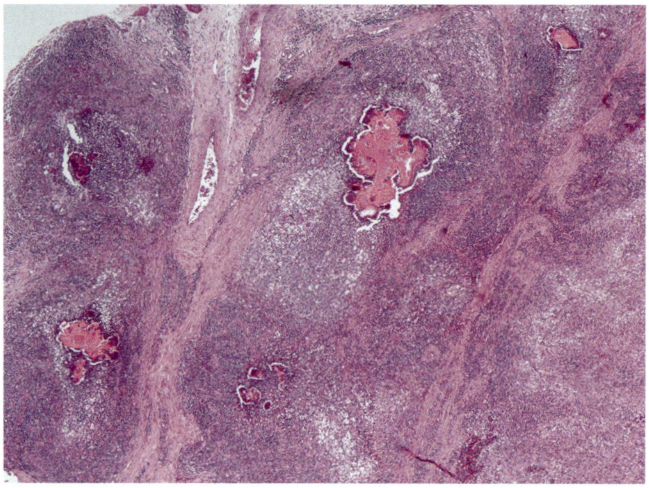

Fig. 6.266 Histology of *Botryomycosis:* many groups of bacteria representing grains

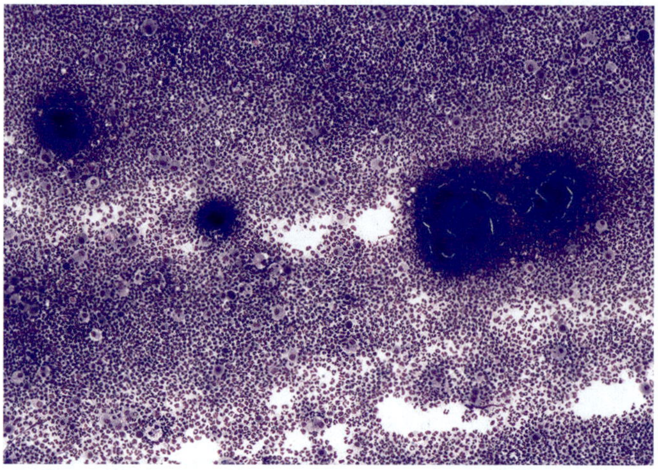

Fig. 6.268 Cytology of *botryomycosis*: three clusters of bacteria (grains) immersed in pyogranulomatous inflammation

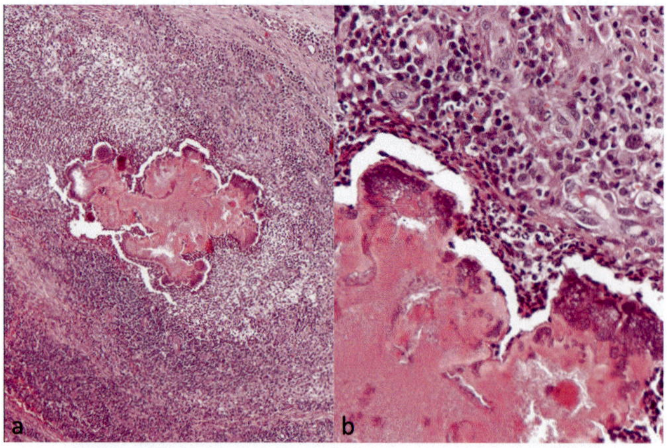

Fig. 6.267 Histology of *Botryomycosis:* (**a**) bunch of grapes appearance of a grain and (**b**) at higher magnification, pyogranulomatous inflammation surrounding the cluster of bacteria is clearly appreciable

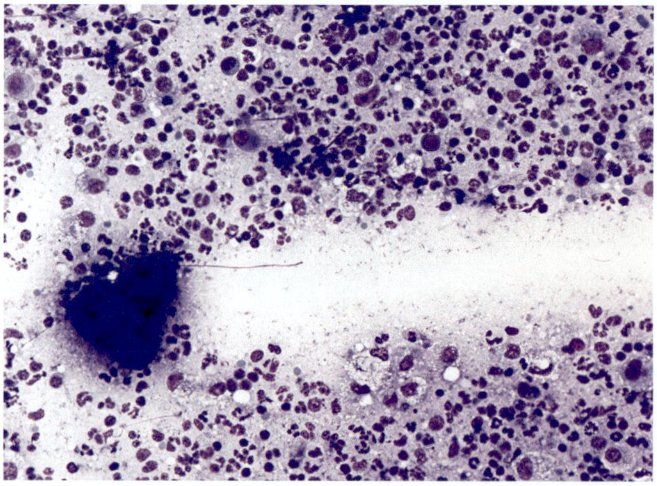

Fig. 6.269 Cytology of botryomycosis: a grain is smeared on the slides leading to an achromatic trail

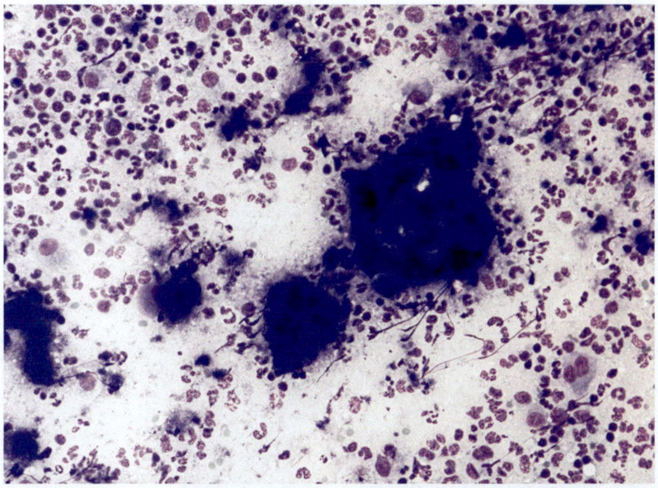

Fig. 6.270 Cytology of *botryomycosis*: at medium magnification, the morphology of bacteria is not evident

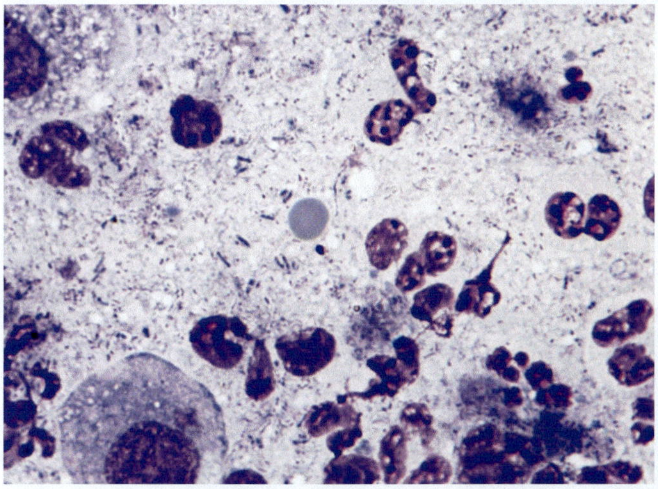

Fig. 6.271 Cytology *botryomycosis:* at high magnification, cocci and rod-shaped bacteria are evident, free, or phagocytized by neutrophils

6.8.1.1.3 Mycobacteriosis

Mycobacteria are a group of rod-shaped, alcohol–acid-resistant bacteria that can be classified into two main groups, namely: *obliged* and *facultative* pathogens, the latter also known by the terms *opportunistic* or *atypical mycobacteria* [172–175, 179].

These microorganisms are causative of three different syndromes: (a) *Tuberculosis* (*M. tuberculosis* complex which comprises *M. tuberculosis (mostly dog), M. bovis, M. microti, M. africanum, M. pinnipedii, M. caprae, and M. canetti*); (b) the *feline leprosy syndrome* (fastidious/leproid) (*M. lepraemurium, M. visibile, M. leprae,* and *M. lepraefelis*); and (c) facultative pathogenic opportunistic saprophytes (non-tuberculous mycobacteriosis) which represent a group of ubiquitous *mycobacteria* (*M. fortuitum, M. smegmatis, M. genavense, M. avium,* etc.) that live in soil and water and, through traumatic wounds or infected foreign bodies, can cause infections in immunodeficient cats and, less frequently, in dogs.

Infections caused by the so-called *Mycobacterium avium complex* (MAC) also belong to the latter group and are also frequently reported [172, 176, 177, 178].

Clinical lesions are characterized by single or multiple nodules, often ulcerated and with draining tracts [179].

In cats affected by leprosy, *Mycobacteria* is inoculated subcutaneously via bites from infected rats. Depending on the modality of transmission, lesions are characterized by single or multiple nodules of different sizes, often ulcerated, particularly located at the extremities and face (Fig. 6.272).

In dogs, cutaneous mycobacteriosis is extremely rare and a self-limiting nodular form of infection named *canine leproid syndrome* has been documented, particularly in Australia and Brazil, in which single or multiple nodules on the ears and head are observed. Few cases have been reported in Europe [171]. Boxer dogs and other shorthair breeds are over-represented and the constant localization of the lesions suggests that mycobacteria could be transmitted through insect bites [180] (Fig. 6.273).

In cats, characteristic abdominal panniculitis caused by opportunistic mycobacteria with lipid tropism is well-recognized. Lesions range from poorly defined nodules to fistulas, which drain a purulent exudate mixed with fat that, in chronic untreated cases, leads to the destruction of all the adipose abdominal tissue (Figs. 6.274 and 6.275).

Cytological Findings

Feline leprosy can be histologically divided into two forms: *lepromatous* and *tuberculoid*. In the former, cytological specimens are composed of neutrophils and many macrophages, epithelioid macrophages, and giant cells filled with intracytoplasmic mycobacteria; in the tuberculoid form, cytology seems instead to be characterized by fewer giant cells and mycobacteria and a higher number of lymphoid cells [170–173] (Fig. 6.276).

Cytological findings in the course of mycobacterial infections differ depending on the causative bacteria. In some cases, they are very few and, when a clinical suspicion is present, they must be carefully searched on slides. In leprosy, slides show pyogranulomatous inflammation with many intracellular and extracellular, achromatic linear rod-shaped areas representing unstained bacteria (negative images) (Figs. 6.277, 6.278, and 6.279).

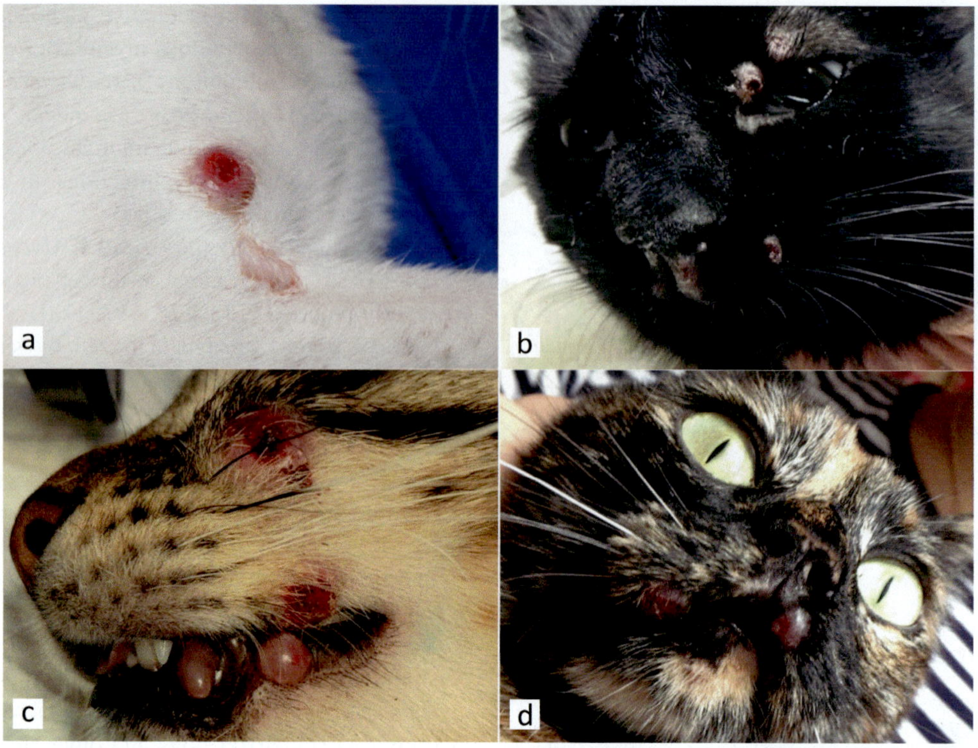

Fig. 6.272 Feline mycobacteriosis: (**a**) ulcerated nodule on the foreleg of a cat with feline leprosy. (Courtesy of DVM V. Imburgia—Italy); (**b**) multiple erythematous papular nodular lesions on the face of a cat affected by mycobacteria belongings to MAC. (Courtesy of Vet Center Anacaprese—Italy); (**c**) multiple erythematous papular nodular lesions on the muzzle and lip. (Courtesy of DVM L. Gulotta—Italy); (**d**) multiple erythematous papular nodular lesions on the muzzle and lip. (Courtesy of DVM D. Guarneri—Italy)

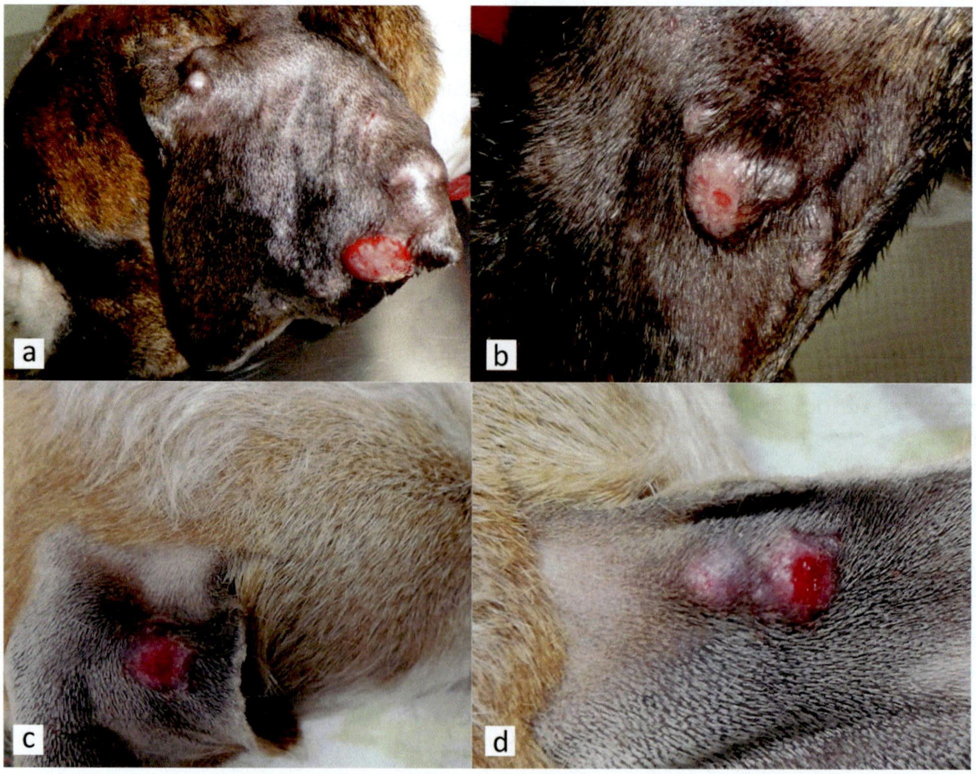

Fig. 6.273 Canine leproid syndrome: (**a**) multiple ulcerated nodules on the pinna of a Boxer with canine leproid syndrome. (Courtesy of DVM L. Cotta—Brazil); (**b**) nodules of different sizes on the pinna of the same dog of the previous image; (**c**) large and ulcerated nodule on the pinna of a mixbreed. (Courtesy of DVM D. Perreca—Italy); and (**d**) nodules on the pinna of the same dog of the previous image

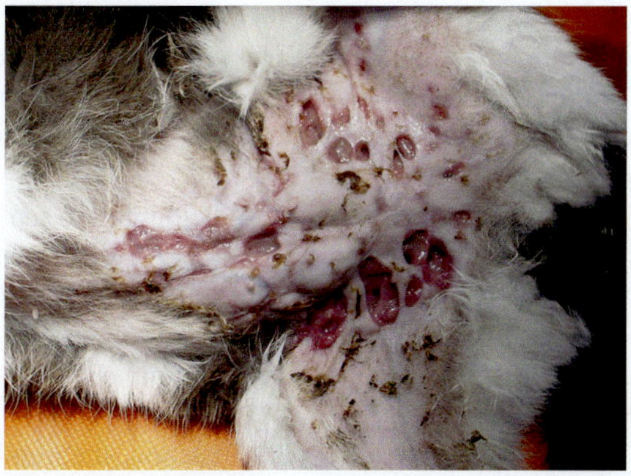

Fig. 6.274 Multiple ulcers and draining tracts in a cat with panniculitis caused by opportunistic *mycobacteria.* (Courtesy of DVM S. Fanfoni—Italy)

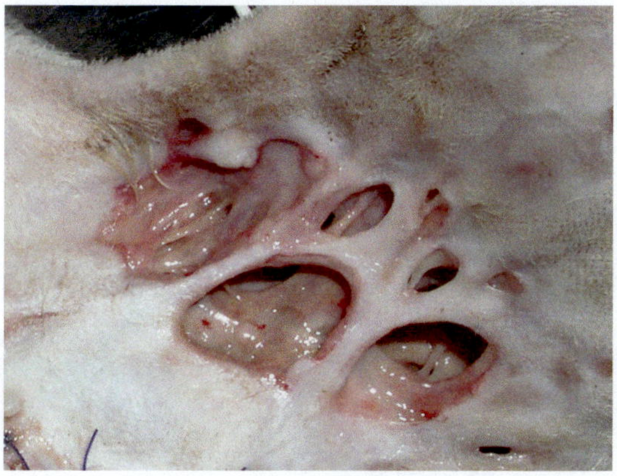

Fig. 6.275 Closeup of the lesions of the same cat of the previous image. (Courtesy of DVM S. Fanfoni—Italy)

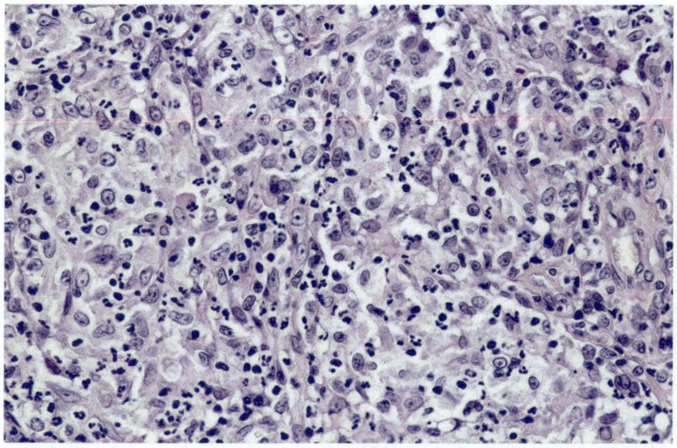

Fig. 6.276 Histology of feline leprosy: histiocytes, lymphocytes, and few neutrophils. With routine HE staining, mycobacteria are not visible

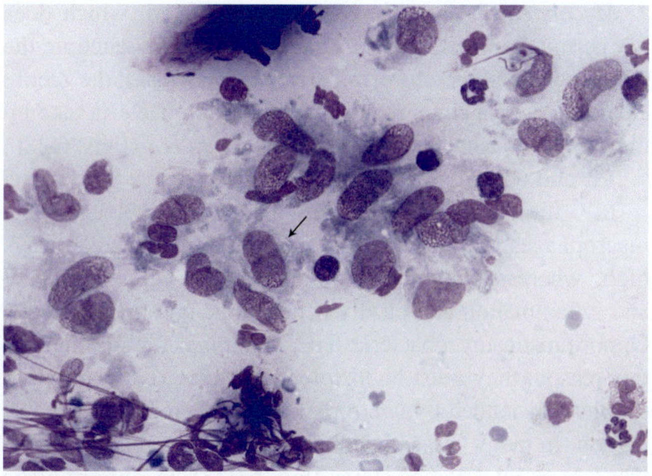

Fig. 6.277 Cytology of feline leprosy: only a few achromatic linear images are faintly visible in the cytoplasm of epithelioid macrophage (*arrow*)

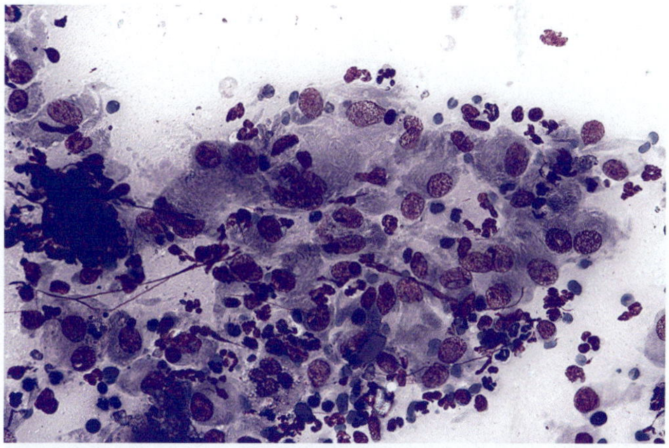

Fig. 6.278 Cytology of mycobacteriosis: many achromatic linear images are faintly visible both intracellular and on the background of the slides; the latter are better recognizable than the former since they are highlighted by the proteinaceous background

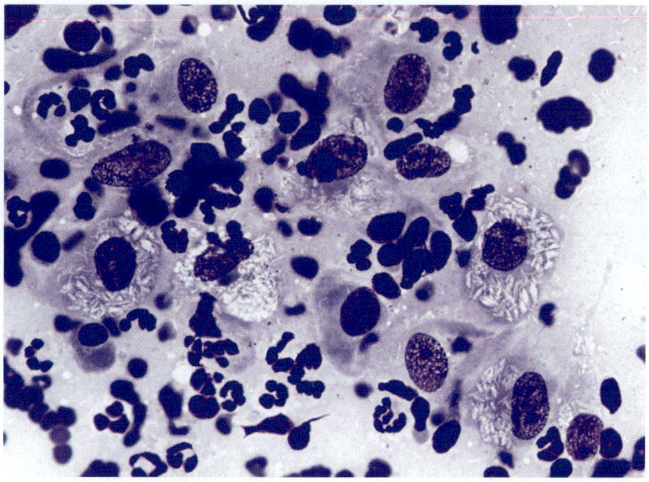

Fig. 6.279 Cytology of feline leprosy: many macrophages filled with linear achromatic spaces in the cytoplasm. The larger achromatic intracytoplasmic areas mean that more bacteria are grouped

Mycobacteria are covered by mycolic acid, which does not permit standard Romanowsky-type dyes to penetrate the microorganism; therefore, to detect *mycobacteria*, the Ziehl–Neelsen (Z-N) stain is required, with which bacteria take up the carbol fuchsin and typically stain *bright red* (Figs. 6.280, 6.281, and 6.282).

In the atypical (opportunistic) mycobacteriosis, macrophages are present in varying numbers, but always high, whereas the number of giant cells seems irrelevant and inflammation is typically neutrophilic and macrophagic. Opportunistic mycobacteria have a tropism for fat, and for this reason, they must be histologically checked in the lipid droplet surrounded by neutrophilic and macrophagic inflammation. In many cases, mycobacteria are viewable only with the Z-N or Fite–Faraco stains (Fig. 6.283). In *opportunistic mycobacteria panniculitis*, cytology is similar to that observed in other forms of fat inflammation, and it is also very difficult to see bacteria with Z-N staining. Specimens are composed of segmented neutrophils together with normal or foamy lipophagocytic macrophages and may also contain some lymphocytes and plasma cells. With special stains, a few mycobacteria can be found in the fat, but not in the macrophages (Fig. 6.284).

Infection caused by mycobacteria belonging to the MAC seems histologically characterized by granulomatous/pyogranulomatous inflammation with a variable number of giant cells and, in some variants, with an impressive fibroblastic reaction that can cytologically mimic a soft-tissue sarcoma with giant cells.

In MAC with a poor fibroblastic reaction, the number of microorganisms is nevertheless high. It must be stressed that it is not possible to typify mycobacteria through cytology, for which culture examinations (often difficult to perform), molecular biology tests (PCR), and interferon-gamma release (assay) are mandatory.

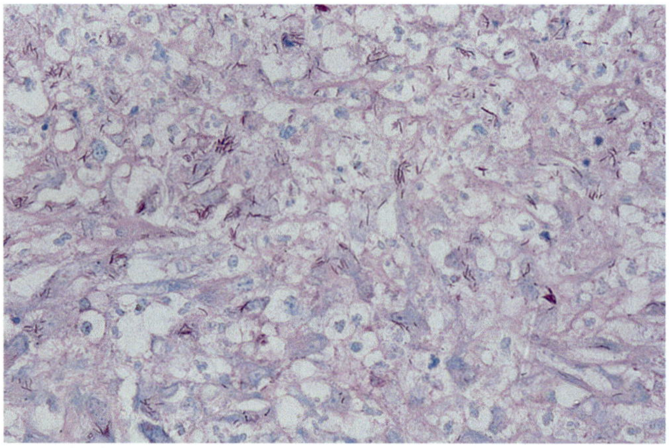

Fig. 6.280 Histology of mycobacteriosis: many bacillary in shape positive mycobacteria are evident using Z-N stain

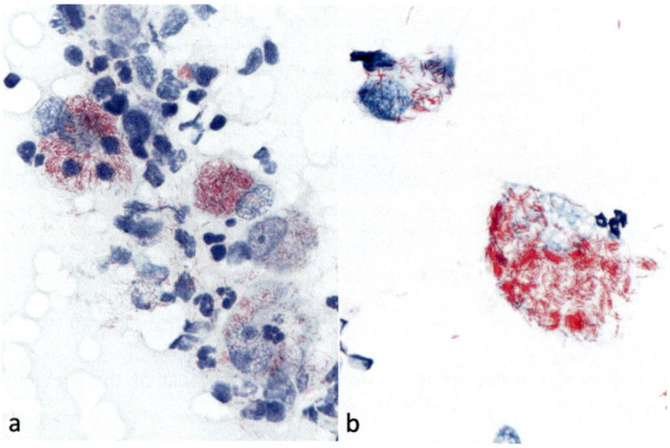

Fig. 6.282 Cytology of mycobacteriosis: Ziehl-Neelsen dye: (**a**) pyogranulomatous inflammation with many positive mycobacteria and (**b**) closeup in which red bacilli are clearly recognizable

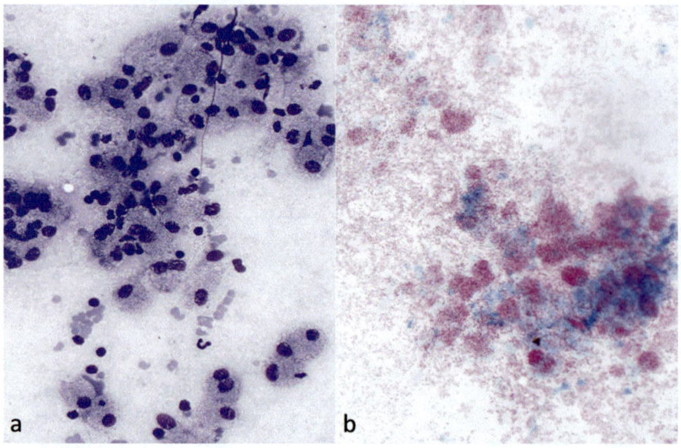

Fig. 6.281 Cytology of feline leprosy: (**a**) Romanowsky-type dye: countless mycobacteria are suspected as achromatic linear spaces recognizable both in the cytoplasm and on the bottom of the slide and (**b**) bacillary in shape positive mycobacteria are evident using Z-N stain

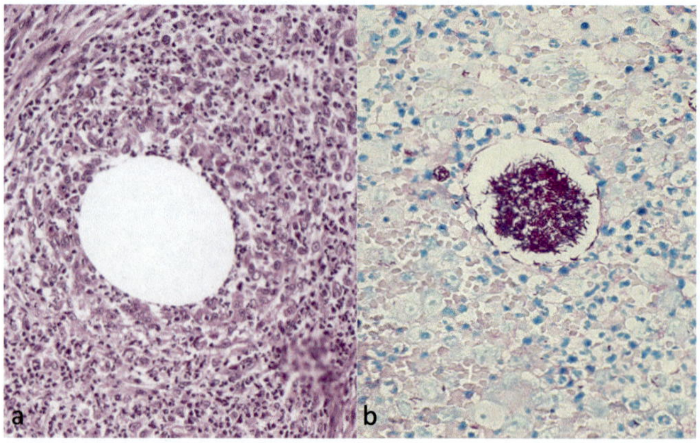

Fig. 6.283 Histology of mycobacteriosis: (**a**) inflammatory cells surround lipid material in which no bacteria are viewable and (**b**) with Z-N stain, a group of mycobacteria can be highlighted. (Courtesy of DVM L. Pazzini—Italy)

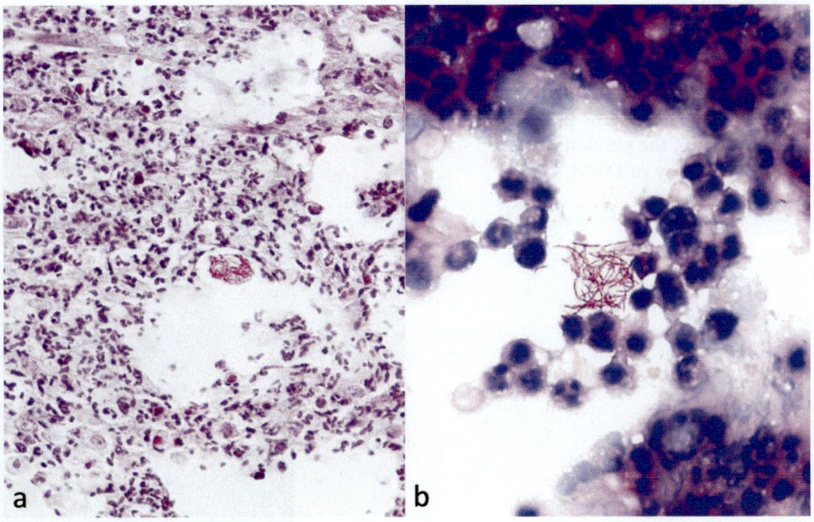

Fig. 6.284 Histology of opportunistic mycobacteriosis: (**a**) note that Z-N-positive bacteria are well recognizable in the lipid tissue and not intracellular and (**b**) many Z-N-positive bacteria are present in lipid material, but not in the cytoplasm of inflammatory cells. (Courtesy of DVM W. Bertazzolo—Italy)

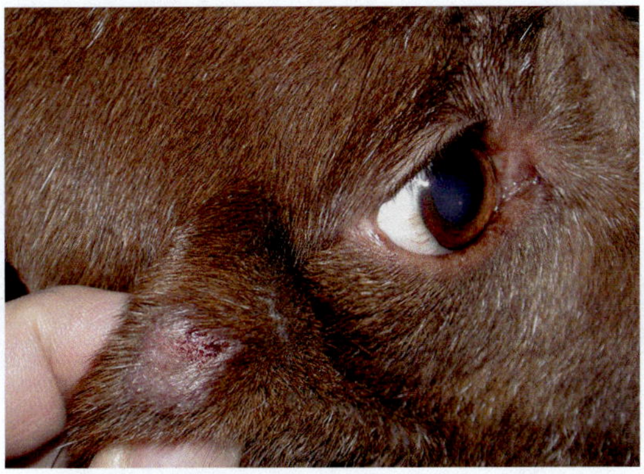

Fig. 6.285 Single alopecic nodule on the face of a dog with systemic leishmaniasis

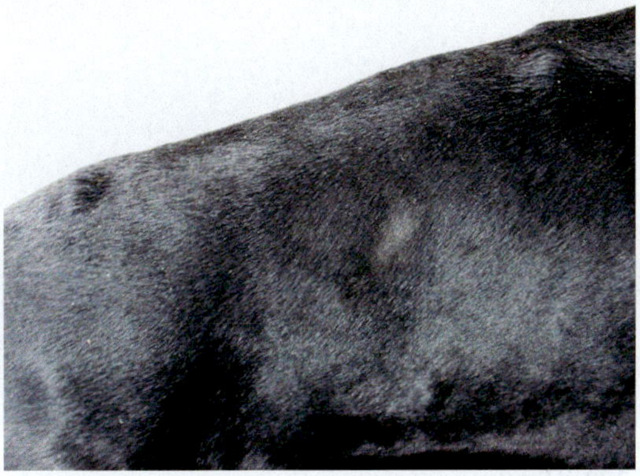

Fig. 6.286 Multiple subcutaneous nodules in a Dobermann with systemic leishmaniasis

6.8.1.2 Protozoa

6.8.1.2.1 Leishmaniasis

As already mentioned, skin lesions in CanL are very polymorph, and nodular lesions can be reported also as unique lesions. Apart from the papular–nodular form described above, which is considered a local reaction to sandfly bites in young immunocompetent animals, dogs suffering from systemic leishmaniasis can develop both single and multiple cutaneous/subcutaneous nodules (Figs. 6.285, 6.286, and 6.287) [185]. *Leishmaniasis* in cats is rare, but even though the typology of skin lesions has not yet been characterized, in a recent systematic review, where only cases in which amastigotes were observed in cytology or histopathology have been included, nodules and ulcerative lesions were the most common clinical lesions (Fig. 6.288) [181, 183, 184, 185].

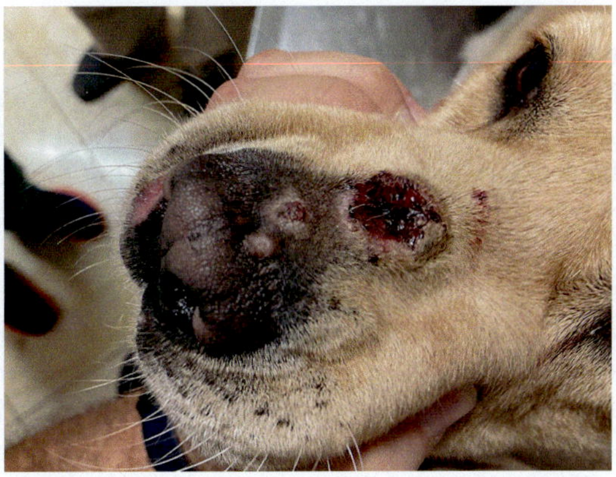

Fig. 6.287 Ulcerated plaque and nodules on the bridge of the nose in a dog affected with CanL

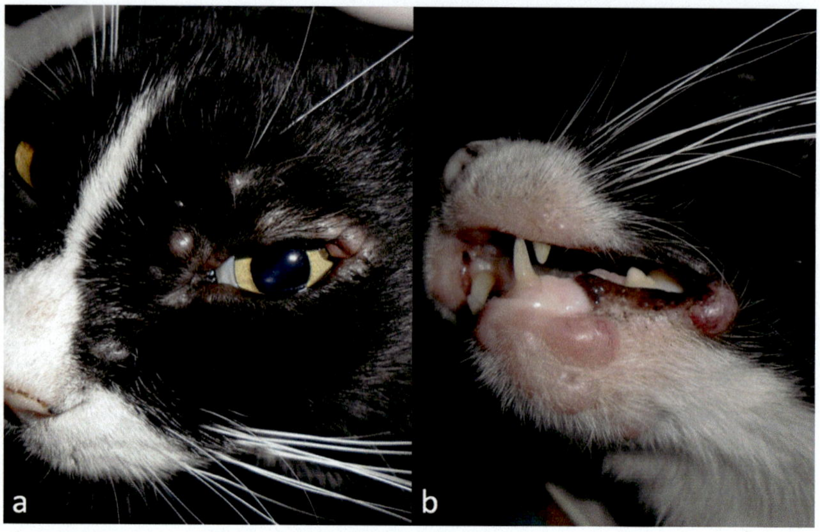

Fig. 6.288 Feline leishmaniasis: (**a**) multiple cutaneous nodules on the eyelids of a cat with leishmaniasis and (**b**) multiple cutaneous nodules on the lips of the same cat of the previous image. (Courtesy of DVM I. Fileccia—Italy)

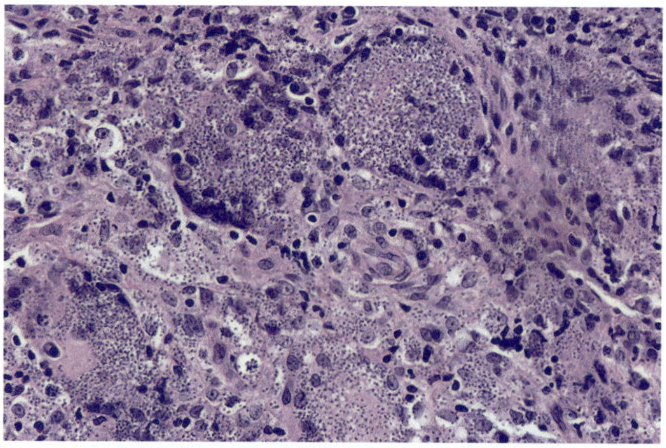

Fig. 6.289 Histology of feline leishmaniasis: pyogranulomatous inflammation with many amastigotes of *Leishmania infantum* in the cytoplasm of both macrophages and giant cells

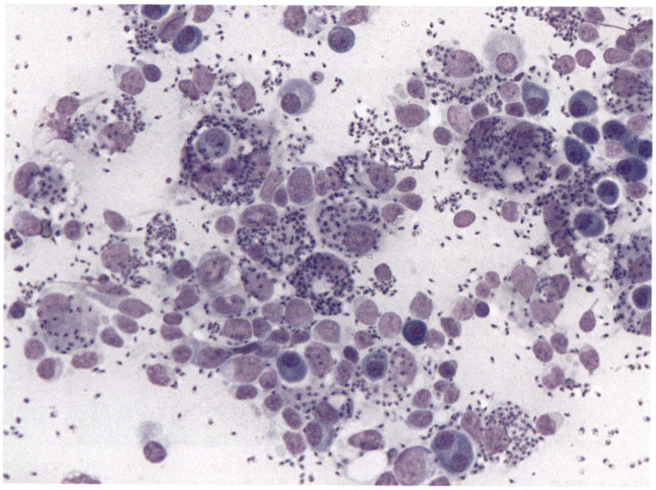

Fig. 6.290 Cytology of feline leishmaniasis: many amastigotes of *Leishmania infantum*, both free and in the cytoplasm of macrophages and giant cells. Note the high number of plasma cells

Cytological Findings

Unlike some papular lesions above discussed, in sick animals with systemic disease, nodular lesions are usually rich in amastigotes. They can be localized both in the dermis and involve the panniculus. Inflammatory cells are represented by macrophages and a variable number of epithelioid macrophages and giant cells; lymphocytes, and especially plasma cells, are usually numerous. The number of segmented neutrophils is variable but usually few in not ulcerated lesions [181, 182]. Amastigotes are visible in the cytoplasm of both histiocytes and neutrophils and scattered on the slide background following the breakage of some of the cells during slide preparation (Figs. 6.289, 6.290, and 6.291).

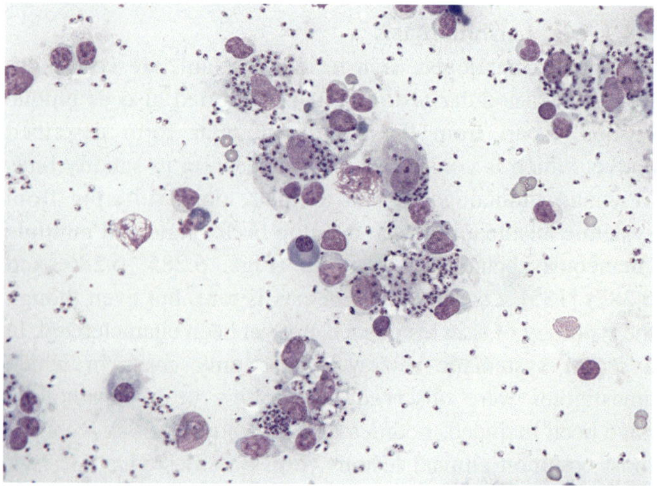

Fig. 6.291 Cytology of feline leishmaniasis: many amastigotes of *Leishmania infantum*. The typical T-shape of amastigotes is well recognizable

6.8.1.2.2 Toxoplasmosis and Neosporosis

Toxoplasma gondii is a protozoa that infects all warm-blooded animals. Cats are the definitive hosts, whereas both felids and non-felids serve as intermediate hosts. Transmission of *Toxoplasma* can occur via ingestion of oocysts from feline feces or ingested feeding of infected intermediate animals. Systemic toxoplasmosis occurs in kittens or in immune-compromised adults such as those affected by feline immunodeficiency virus (FIV) and/or feline leukemia virus. Rarely, a cutaneous localization of the *T. gondii* tachyzoites has been reported in patients with systemic protozoal disease; skin lesions consisted of multiple cutaneous and subcutaneous nodules [186, 189, 191, 195, 200]. Rarely, cutaneous localization of *Toxoplasma* has been reported in dogs [192]. *Neospora caninum* infection with cutaneous involvement has been also reported in dogs [187–190, 193, 194, 197, 199, 201]. In most of these cases, the lesions were secondary to immune-suppressive therapy with corticosteroids [196, 198]. The skin lesions were mostly characterized by multiple nodules and draining tracts (Fig. 6.292).

Cytological Findings

Histologically, skin lesions of both diseases are characterized by a pyogranulomatous dermatitis rich with protozoa. Cytologically, the inflammatory cells can phagocyte a variable amount, but usually high, of tachyzoites. When tachyzoites are intracytoplasmic, they are contained in a pseudo-cyst called *parasitophorous vacuole*; the latter is a structure derived from the plasma membrane of the host and that acts as a defense of protozoa, a niche in which protozoa can replicate [187]. In this phase of replication, microorganisms are many and crowded between themselves, and since the typical crescent-shaped silhouette of microorganisms is not always well recognizable, they can be confused with amastigotes of *Leishmania* spp. (Figs. 6.293 and 6.294). The characteristic crescent-shaped silhouette of these protozoa is more recognizable when parasites are free on the background of the slides. *Tachyzoites* measure 6–7 × 2–3 μm, with oval nuclei and basophilic cytoplasm, and are easily identifiable when their crescent-shaped silhouette is clearly evident (Figs. 6.295, 6.296, and 6.297). *T. gondii* and *N. caninum* cannot be differentiated using cytology and should undergo serology, immunohistochemistry, electron microscopy, or PCR for identification.

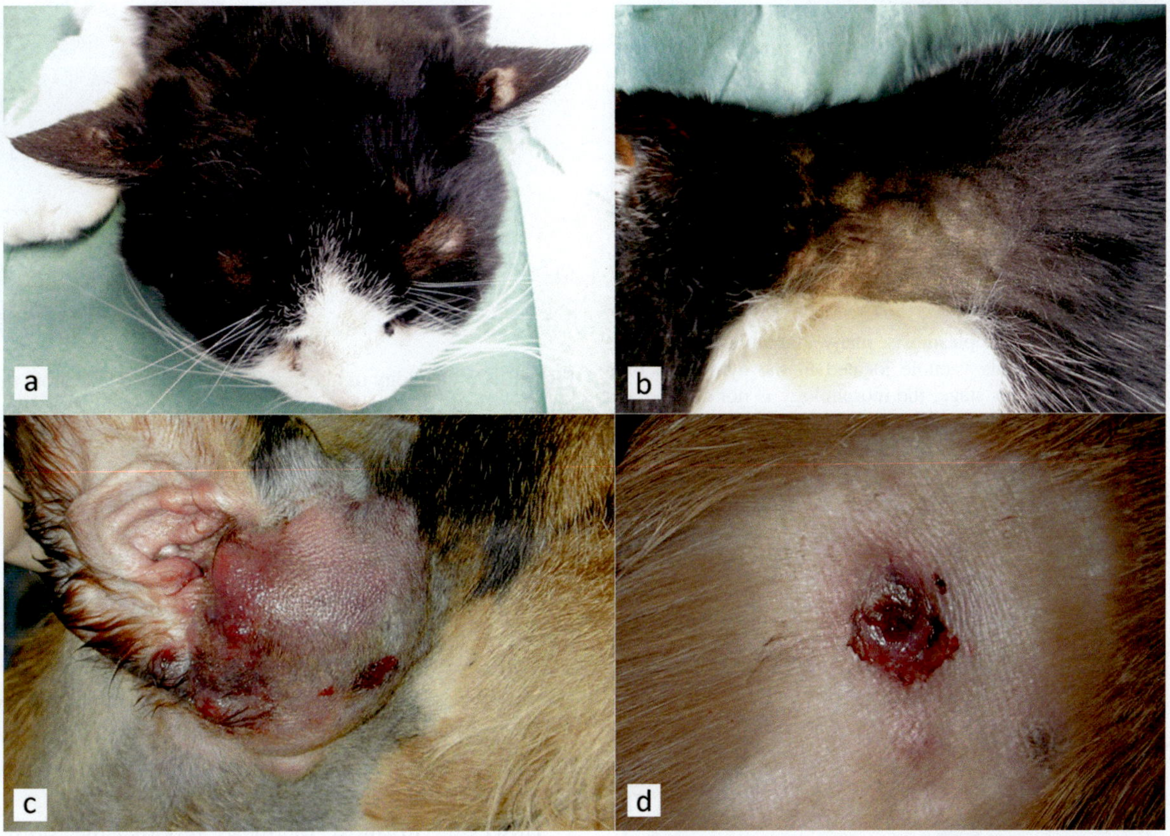

Fig. 6.292 Cutaneous toxoplasmosis and neosporosis: (**a**) multiple subcutaneous nodules on the head and pinna of a cat with toxoplasmosis. (Courtesy of DVM. C. Caporali—Italy); (**b**) alopecic nodules on the neck of the same cat of the previous image; (**c**) large nodule at the base of the ear of a dog with neosporosis. (Courtesy of DVM G. Zanna—Italy); and (**d**) ulcerated nodule in the same dog of the previous image

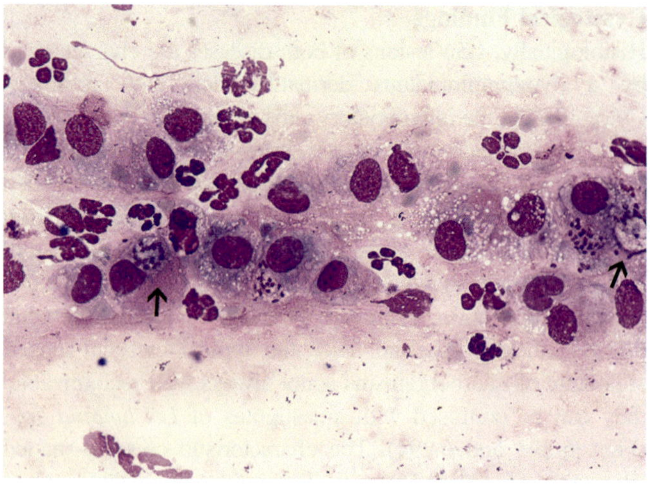

Fig. 6.293 Cytology of toxoplasmosis: pyogranulomatous inflammation. Note the presence of many protozoa in the cytoplasm of macrophages. A parasitophorous vacuole is yet recognizable (arrows)

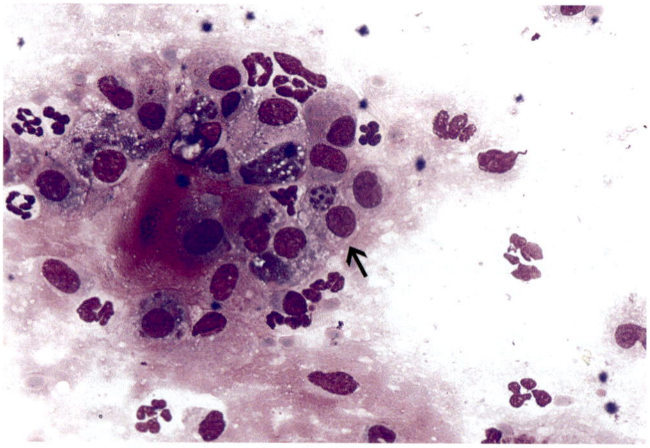

Fig. 6.294 Cytology of toxoplasmosis: many immature protozoa are content in a parasitophorous vacuole located in the cytoplasm of a macrophage (*arrow*). In this stage, the morphology is not defined and microorganisms can be confused with amastigotes of *Leishmania* spp.

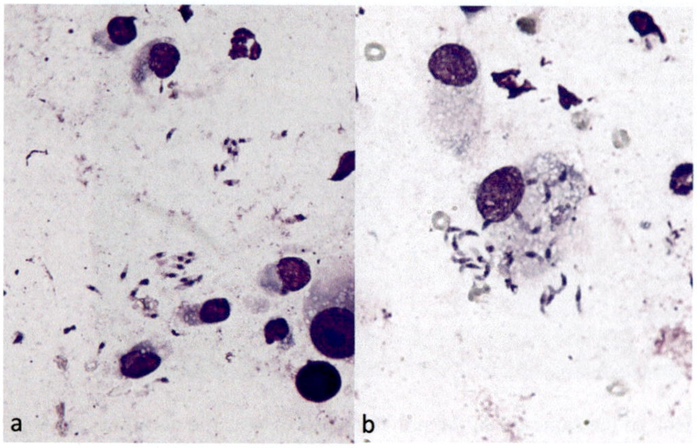

Fig. 6.295 Cytology of toxoplasmosis: (**a**) many crescent-shaped *tachyzoites* of *T. gondii* are better recognizable when free on the bottom of the slides and (**b**) a macrophage with intracytoplasmic *Toxoplasma protozoa*

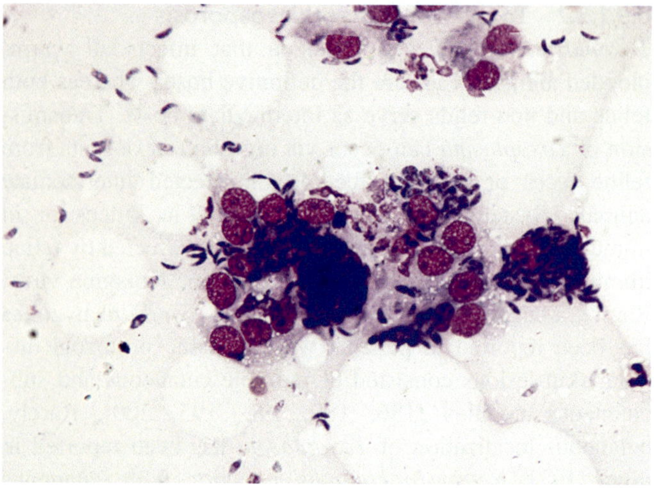

Fig. 6.296 Cytology of neosporosis: pyogranulomatous inflammation. Mono- and multi-nucleated macrophages are engulfed with protozoa

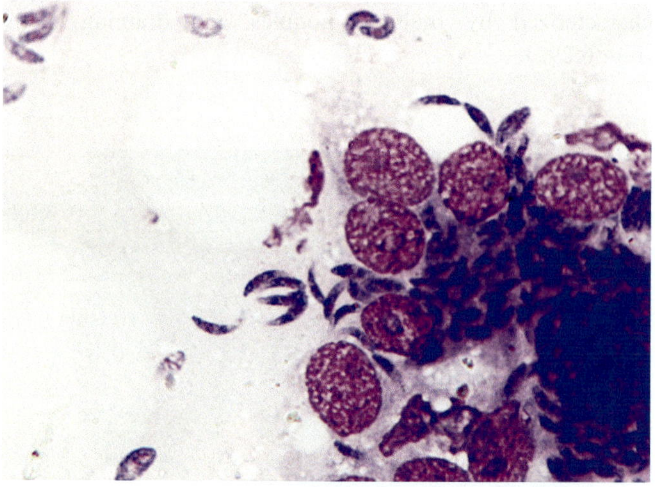

Fig. 6.297 Cytology of neosporosis: at high magnification, the typical crescent shape of *Neospora caninum* protozoa is clearly evident

6.8.1.3 Worms

6.8.1.3.1 Dirofilariasis

Adult worms of *Dirofilaria repens* live in the subcutaneous tissues of dogs and cats, where the mature females release microfilariae (Larva 1) into the blood. Although *D. repens* seems to be responsible for a subclinical disease in dogs, a plethora of non-specific cutaneous findings (e.g., erythema, papules, nodules, alopecia, and pruritus) have been reported [204, 206].

However, in most of these reports, the presence of *D. repens* in skin lesions is not always ascertained; thus, the role of parasites in the development of lesions has only been presumed [206]. In a study of 16 dogs, the only lesions observed were single or multiple subcutaneous nodules containing adult worms and/or microfilariae [202]. Nodules of cutaneous dirofilariasis may vary in size and are often barely perceptible on palpation as they are localized in the panniculus, the normal habitat of adult worms (Figs. 6.298, 6.299, and 6.300). During *dirofilaria repens* infestation, skin lesions are present in only a few cases, since adult worms live in the subcutis without necessarily giving origin to nodules or other skin lesions. Spontaneous regression of nodular lesions is reported.

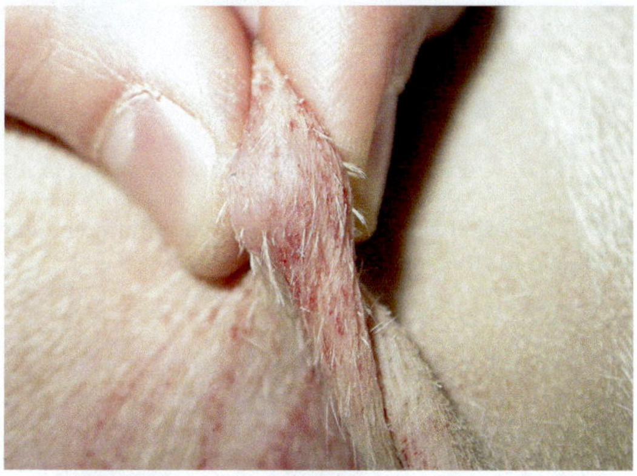

Fig. 6.298 Cutaneous dirofilariasis: small subcutaneous nodule

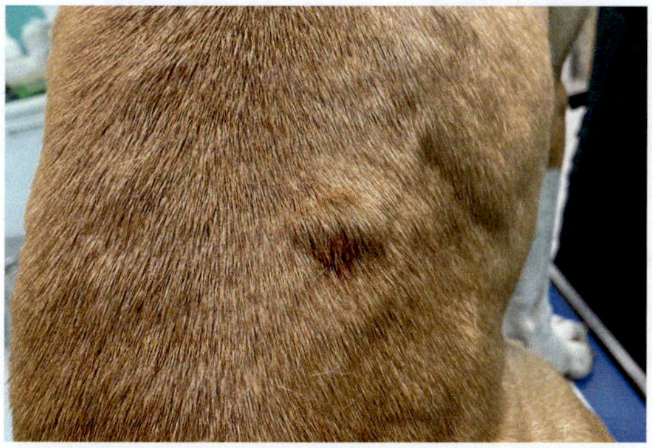

Fig. 6.299 Subcutaneous nodule caused by *Dirofilaria repens*

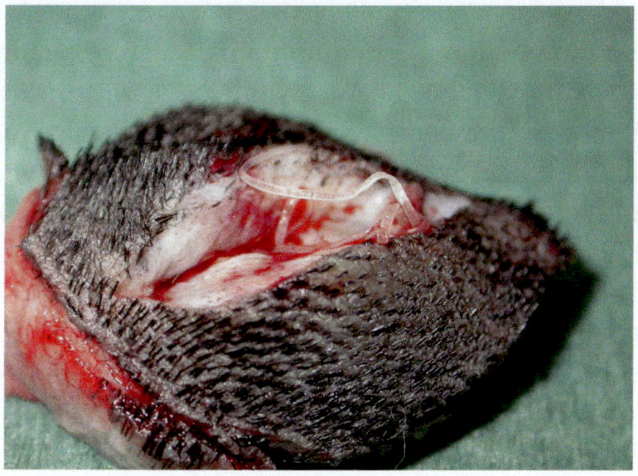

Fig. 6.300 Sectioned nodule after surgery: note the long adult *Dirofilaria repens* worm

Cytological Findings

Cytological findings vary depending upon whether the nodule presents adult worms or if the latter are not yet present; furthermore, a secondary inflammatory reaction can occur in many cases, especially if microfilariae were spread into the dermis.

In the latter cases, slides are characterized by a variable amount of microfilariae immersed in an exudate composed of segmented to degenerated neutrophils along with activated macrophages; few eosinophils can also be present (Fig. 6.301). In some cases, few eosinophils are observed. When the adult worms entirely occupy the nodules and a secondary inflammatory reaction is only composed of reactive fibrosis, the body parts of some adult worms are inevitably collected via FNB. In these cases, small pieces of uterus containing numerous eggs and immature larvae can be observed on the slides (early stage of ova, fertilized ova, shelled mature morulae, coiled or stretched microfilariae, etc.) [203] (Figs. 6.302, 6.303, 6.304, 6.305, and 6.306).

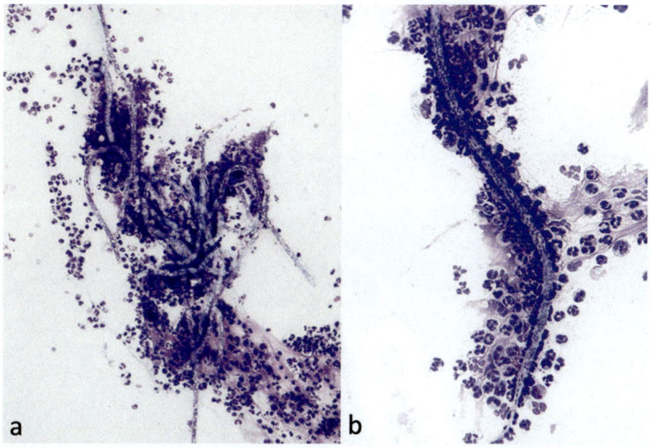

Fig. 6.301 Cytology of cutaneous dirofilariasis: (**a**) many microfilariae immersed in a pyogranulomatous inflammation and (**b**) immature larva of *Dirofilaria repens* surrounded by inflammatory cells

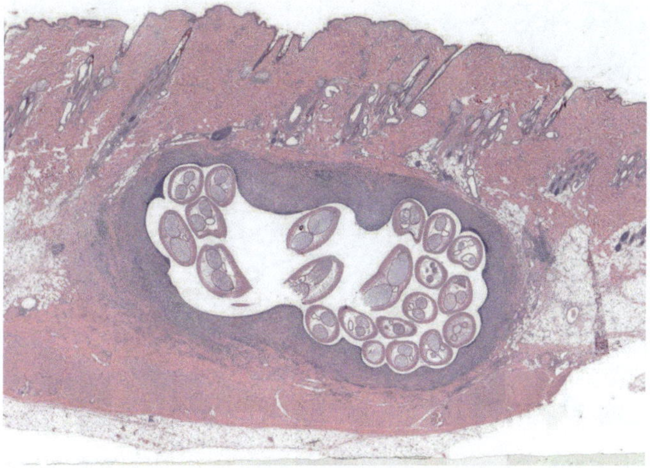

Fig. 6.302 Histology of dirofilariasis: female adult worm of *Dirofilaria repens* in the deep dermis, surrounded by mild inflammation

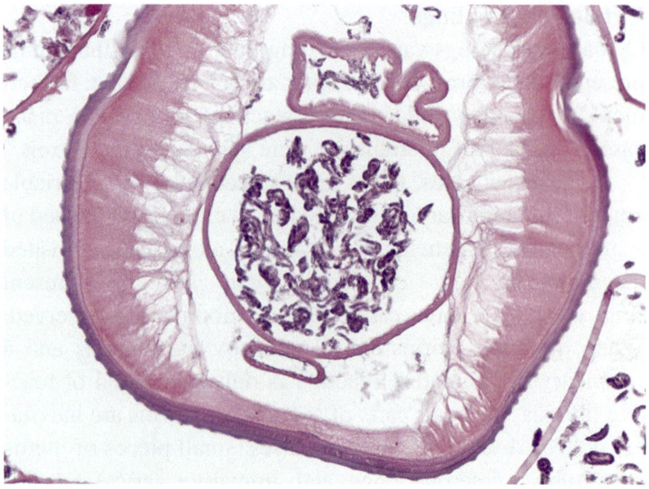

Fig. 6.303 Histology of dirofilariasis: section of a female adult worm in which the uterus, filled with immature microfilariae, is evident

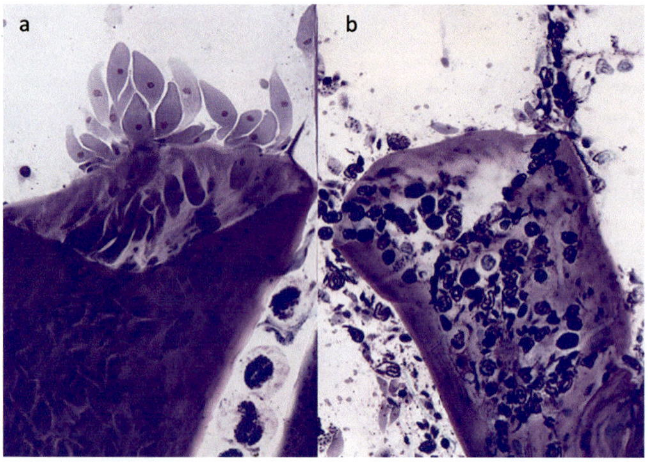

Fig. 6.304 Cytology of dirofilariasis. Uterus with different stages of eggs and larvae: (**a**) many intrauterine plump and spindle-shaped immature eggs and (**b**) many intrauterine immature larvae

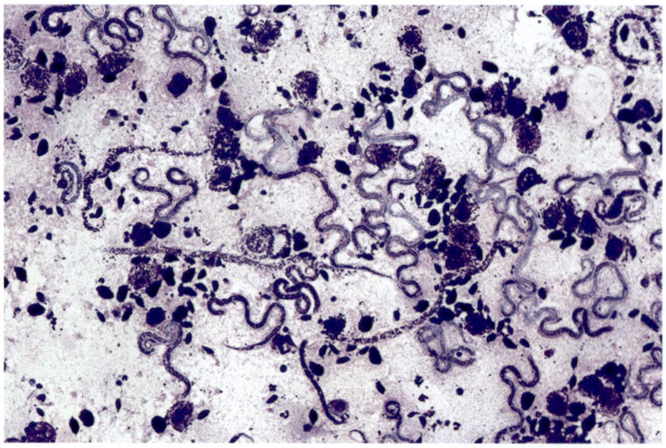

Fig. 6.305 Cytology of dirofilariasis: many larvae along with immature morulae

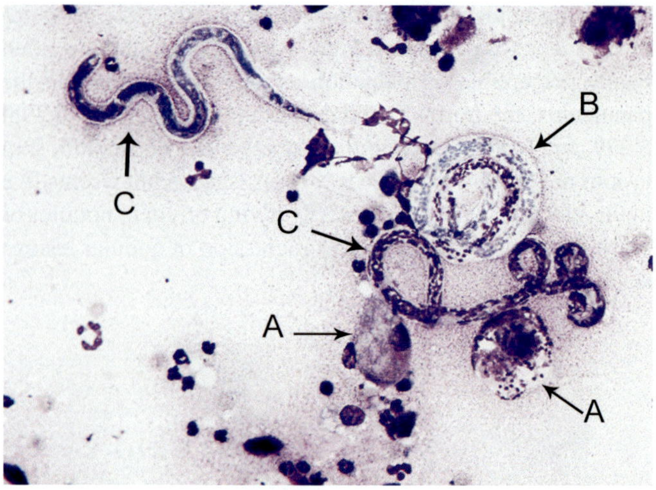

Fig. 6.306 Cytology of dirofilariasis: (**a**) ova at different stages; (**b**) coiled larva; and (**c**) free larvae at different stages

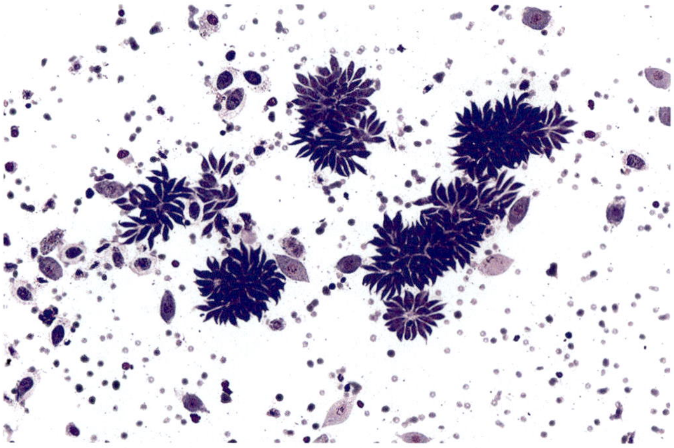

Fig. 6.307 Cytology of dirofilariasis: at low magnification, many immature ova with a "fireworks" arrangement could be confused with spindled neoplastic cells coming from a soft tissue sarcoma

When microfilariae are present, it is easy to diagnose nodular dirofilariasis; instead, in rare cases, only early-stage eggs and no larvae are collected. Immature eggs can show different morphology depending on the stage of maturation. A particular spindle shape with an arrangement that resembles the "fireworks" or the "flower's petals," can be observed; if these cytological features are not known, they cannot be recognized as immature eggs and the diagnosis of cutaneous dirofilariasis will be missed [205]. Cells of an adult worm's body can also be observed (Figs. 6.307 and 6.308).

Filarial nodule undergoes spontaneous remission with chronicity, and adult worms degenerate and undergo autolysis and calcification. Clinicians should be aware that the sampling of microfilariae from a cutaneous nodule that yields a hemocontaminated specimen with no inflammatory cells must not be interpreted as a cutaneous localization of worms;

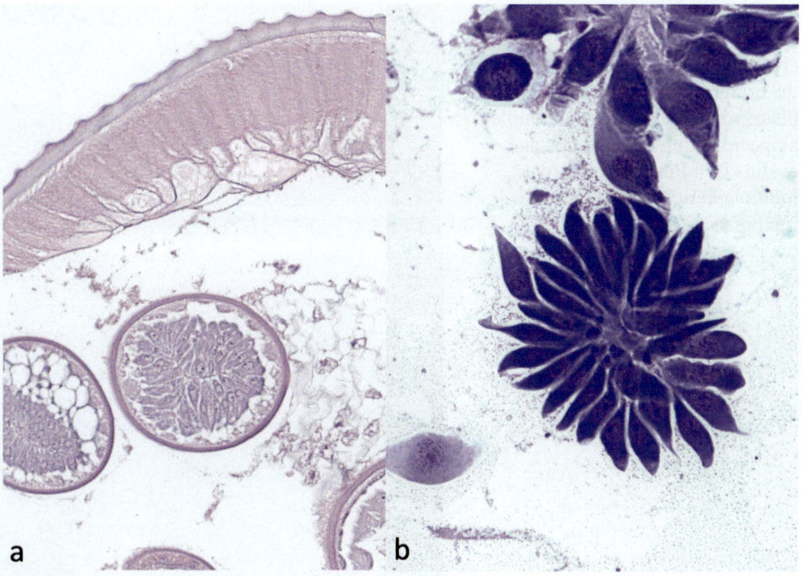

Fig. 6.308 Dirofilariasis: (**a**) histology: spindled immature ova are easily recognizable in the uterus of an adult worm and (**b**) cytology: high magnification of early stage of ova with spindle shape and flowers' petal-like disposition

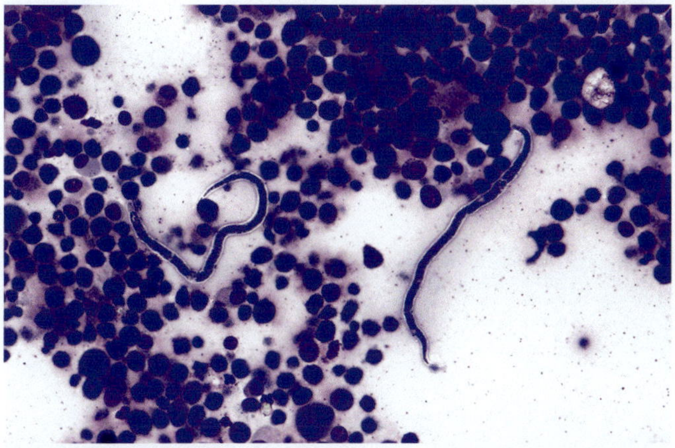

Fig. 6.309 *D. repens* larvae accidentally collected during the sampling of a subcutaneous mast cell tumor. The larvae come from the bloodstream (microfilaremia)

indeed, it may only indicate the presence of microfilaremia. This event can be verified in any case in which cytology is performed from a skin nodule, and a hemocontaminated sample with some microfilariae coming from the bloodstream is accidentally collected; this condition occurs mostly in case of vascular neoplasia (hemangioma) but can be observed in any lesion (Fig. 6.309).

6.8.1.4 Fungi

6.8.1.4.1 Dermatophytic Kerion

Dermatophytic kerion is a pyogranulomatous nodular skin lesion due to the spread into the dermis of dermatophytes, which infect the hairs; the breakage of the follicular wall is the cause of the spreading of keratin and fungal elements into the dermis and the secondary development of a pyogranulomatous inflammation [207–212]. Since usually dermatophytes do not cause nodular lesions in dogs, it is speculated that Kerion represents the outcome of an abnormal response of the immune system against dermatophytes which, in most cases, infect keratin without causing a neutrophilic folliculitis; anyway, in Kerion, the rupture of the follicular wall and secondary granulomatous inflammation occurs. *Kerion* is usually observed as a single plaque/nodule, located mostly on the face, especially on the bridge of the nose, muzzle, and lips, but it can occur in any body site and, less frequently, multiple nodules can be present and spread all over the body (Fig. 6.310). In the early stage, kerion is characterized by roundish raised lesions, usually button-shaped, covered in hairs or crusted exudate; in a few days, it becomes alopecic, with smooth to irregular surfaces on which small draining tracts discharging a purulent exudate. Some lesions can be morphologically very similar to skin neoplasia, especially to benign cutaneous histiocytoma (Fig. 6.311). Although some textbooks anecdotally report that kerion is more often caused by *M. gypseum* (*Nannizzia gypsea*) and *T. mentagrophytes* infection [211], in some publications, *M. canis* was the most frequently isolated dermatophyte [207, 208, 211]. The species of causative fungus is strictly linked to the environment in which the dogs live, with most cases of *M. canis* if the animal lives in an urban area and most cases of the other dermatophytes, if the dog lives in a rural area.

Cytological Findings

The histopathology of a dermatophytic kerion is characterized by multiple pyogranuloma that, as above written, are secondary to the breakage of the follicular wall and following the spreading of infected hair shafts into the dermis

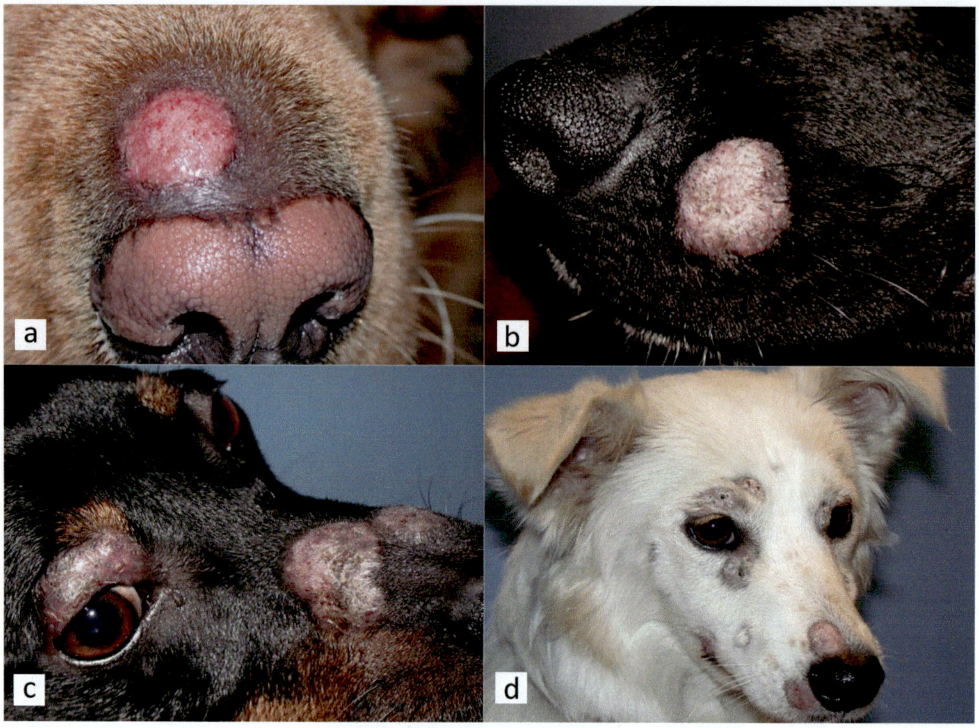

Fig. 6.310 Dermatophytic kerion: (**a**) alopecic and erythematous roundish plaque on the bridge of the nose; (**b**) alopecic plaque on the muzzle; (**c**) multiple alopecic nodules on the lips and eyelids of a Pinscher; and (**d**) multiple kerion on the face of a mixbreed

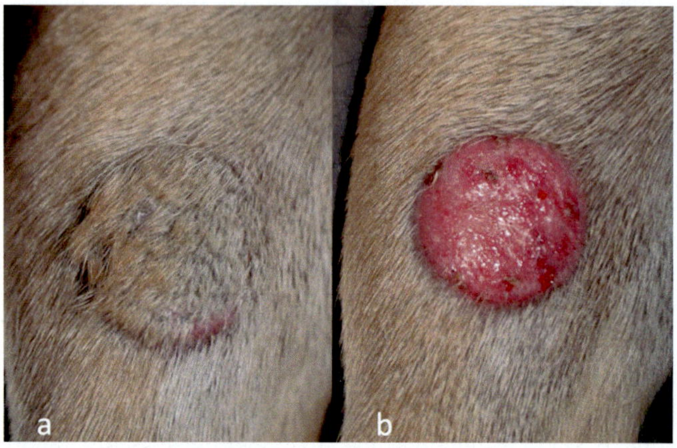

Fig. 6.311 Dermatophytic kerion: (**a**) plaque covered by hairs and (**b**) the same lesion after the removal of the hairs. Note that the morphology can mimic a canine benign histiocytoma

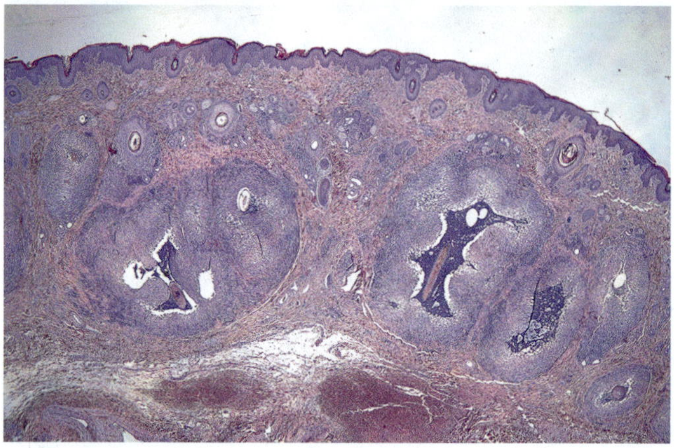

Fig. 6.312 Histology of a dermatophytic kerion: multiple nodular granulomas centered on fragments of infected hairs

(Figs. 6.312 and 6.313) [209]. In cytology, kerion shows, similarly to histology, a mixed pyogranulomatous inflammatory pattern, composed of segmented neutrophils, many variably vacuolated macrophages, epithelioid macrophages, and only occasional giant cells. In almost all cases, a variable number of eosinophils and lymphocytes is present, while plasma cells are less numerous (Fig. 6.314).

In more successful specimens, it is possible to find, free or in the cytoplasm of neutrophils and macrophages, a variable number of fungal arthroconidia, which allow a rapid diagnosis of dermatophytosis. Round, oval to elongated basophilic bodies measuring 2–5 μm, surrounded by a thin clear halo, characterize the dermatophytic arthroconidia (Figs. 6.315 and 6.316). Hyphae are less commonly detected. It is not easy to define with cytology to which dermatophyte the spores belong, but it is well known that smaller arthroconidia belong to *M. canis*, while larger spores are more indicative of *M. gypseum* or belong to *T. mentagrophytes* complex (Fig. 6.317).

Fig. 6.313 Histology of a dermatophytic kerion: (**a**) longitudinal section of infected hair. Long hyphae in the medulla and cortex and many spores at the periphery are clearly recognizable, and (**b**) two infected hairs transversally sectioned in which positive arthroconidia are well recognizable around the hair shafts

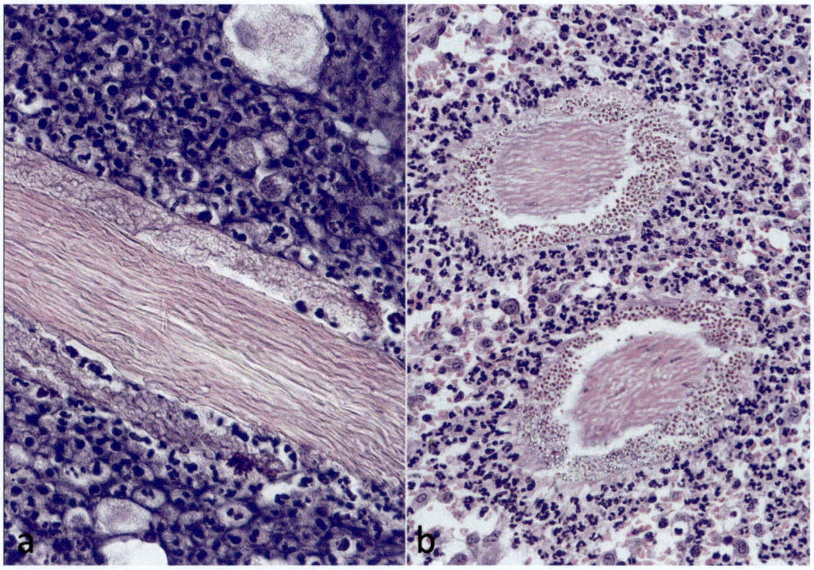

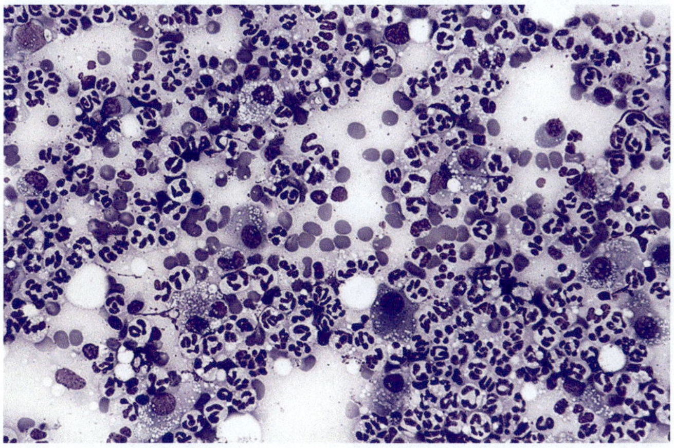

Fig. 6.314 Cytology of a dermatophytic kerion: a mixed flogosis with many well-differentiated neutrophils and vacuolated macrophages characterize most of the cytology slides of dermatophytic kerion

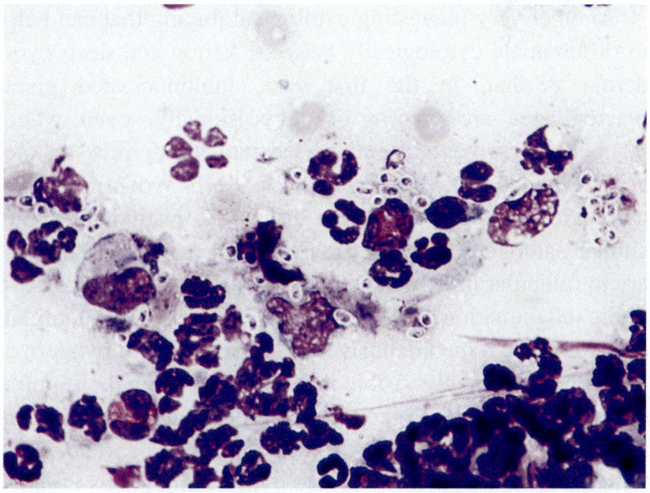

Fig. 6.316 Cytology of dermatophytic kerion: at high magnification, spores are recognizable as round to oval bodies surrounded by a clear halo

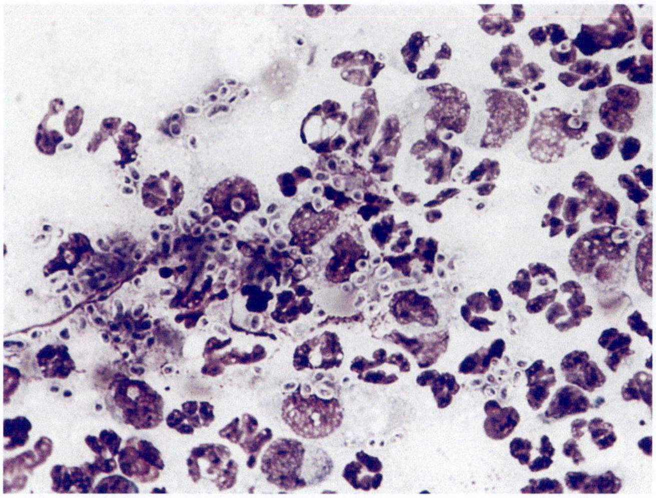

Fig. 6.315 Cytology of dermatophytic kerion: neutrophils, macrophages, and many arthroconidia of *M. canis*

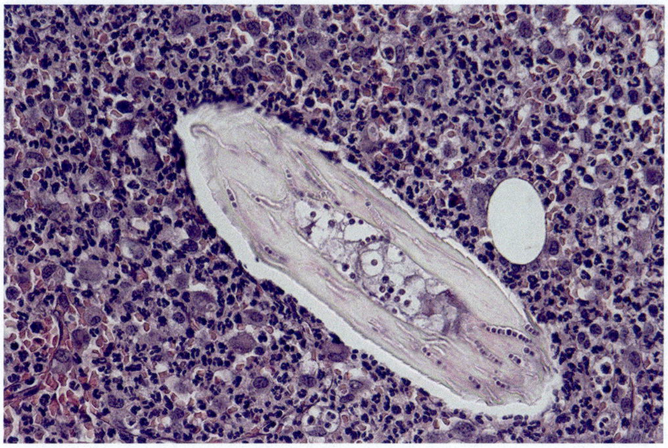

Fig. 6.317 Histology of dermatophytic kerion: many hyphae and very large spores belonging to *T. mentagrophytes*

In slides sampled from some kerion, severely altered hair shafts are very often recognizable; the latter are completely deconstructed and affected by a trichorrhexis that involves the entire hair shaft and that presents with multiple longitudinal fractures that give it an appearance reminiscent of "glass wool fibers." This cytological aspect is not pathognomonic of dermatophytic kerion but is very suggestive of it; once completely fractured, the individual hair fragments disperse in the pyogranulomatous inflammation.

Even in deep pyoderma secondary to folliculitis, fragments of hair shafts can be observed in cytology, but, usually, they are single and small fragments and not large parts of hairs with diffuse trichorrhexis.

In pyoderma, on the other hand, compared to kerion, corneocytes of intrafollicular origin are observed much more frequently. Therefore, in the presence of a clinically compatible lesion, when many fragmented hairs are present on slides, it is always advisable to search for dermatophytes (Figs. 6.318, 6.319, 6.320, and 6.321).

Another very interesting cytological finding that can help to differentiate cytologically between kerion and deep pyoderma is that, in the first case, multinucleated giant macrophages are present only occasionally, even when many hair fragments are dispersed into the dermis; on the contrary, they are very common in deep pyoderma. This cytological feature could seem difficult to justify given the similar pathogenesis between the two diseases, both secondary to follicular infection; a justification could be that neutrophilic inflammation present in pyoderma stimulates a luminal orthokeratotic hyperkeratosis a secondary event to pyoderma that is not commonly visible in kerion. So, when the rupture of the follicular wall occurs, neutrophils and bacteria are spread into the dermis along with numerous corneocytes, with the latter that evoke a severe pyogranulomatous inflammation with many giant cells. In practice, very few

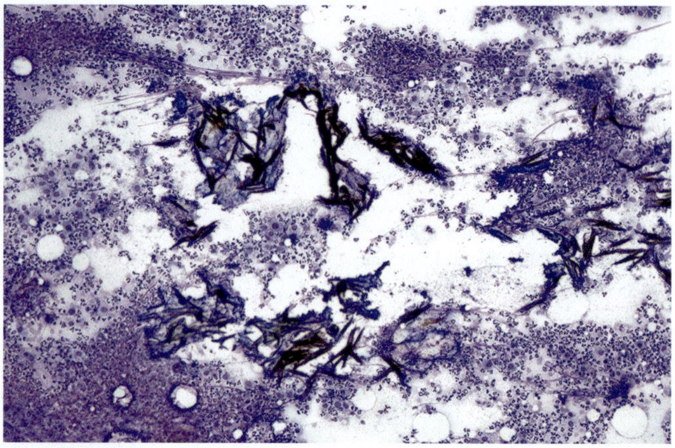

Fig. 6.319 Cytology of dermatophytic kerion: at low magnification, many pigmented fragments of hair hafts are evident

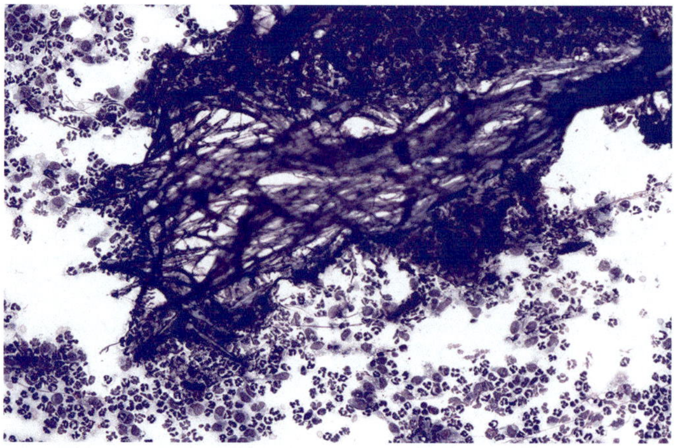

Fig. 6.320 Cytology of dermatophytic kerion: longitudinal splitting of the infected and destroyed hair shaft keratin gives an aspect that resembles glass wool fibers

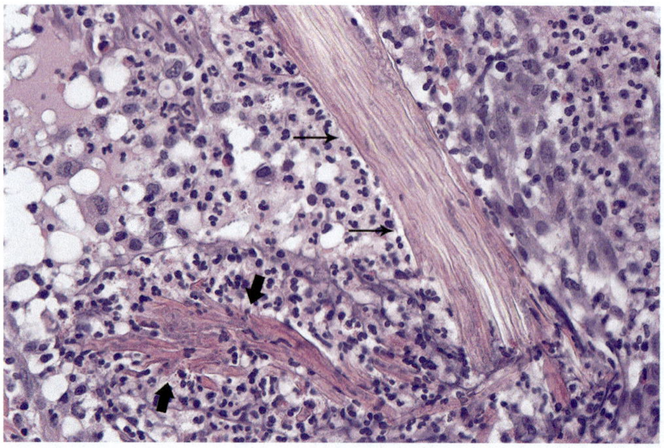

Fig. 6.318 Histology of dermatophytic kerion: trichorrhexis of an infected hair (*short arrow*). Note that the hairs are completely deconstructed and plentiful of both hyphae and spores (*thin arrows*)

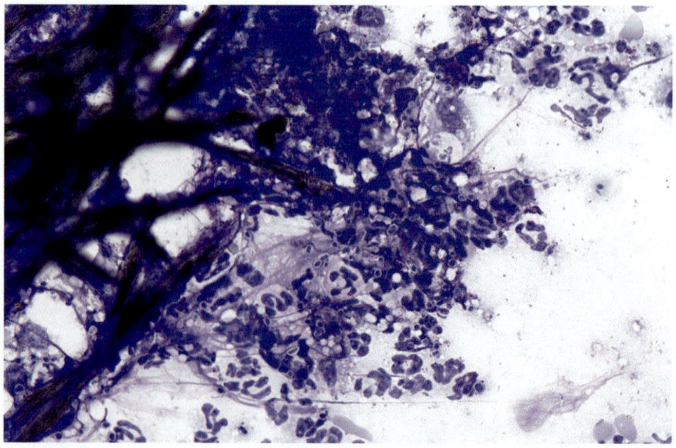

Fig. 6.321 Cytology of dermatophytic kerion: at high magnification, spores are recognizable as round to oval bodies surrounded by a clear halo around the hair shaft fragments

corneocytes may be present in kerion cytology, but not as numerous as in deep pyoderma.

When fungal elements are very few and not easily detectable with routine Romanowsky-type stains, slides can be stained with special stains such as PAS or GMS, which color fungal bodies *magenta* or *black,* respectively. Unfortunately, as discussed in Chap. 2, PAS staining colors not only fungal bodies but also intracytoplasmic material phagocytized by macrophages or spread in the background, the granules of eosinophils and mast cells, and so on; therefore, only in few cases, PAS stain is a useful stain for kerion, and especially in highly cellular inflammation, GMS must be preferred. With the latter, the fungal elements are highlighted in black on a diffuse green background (Figs. 6.322, 6.323, and 6.324).

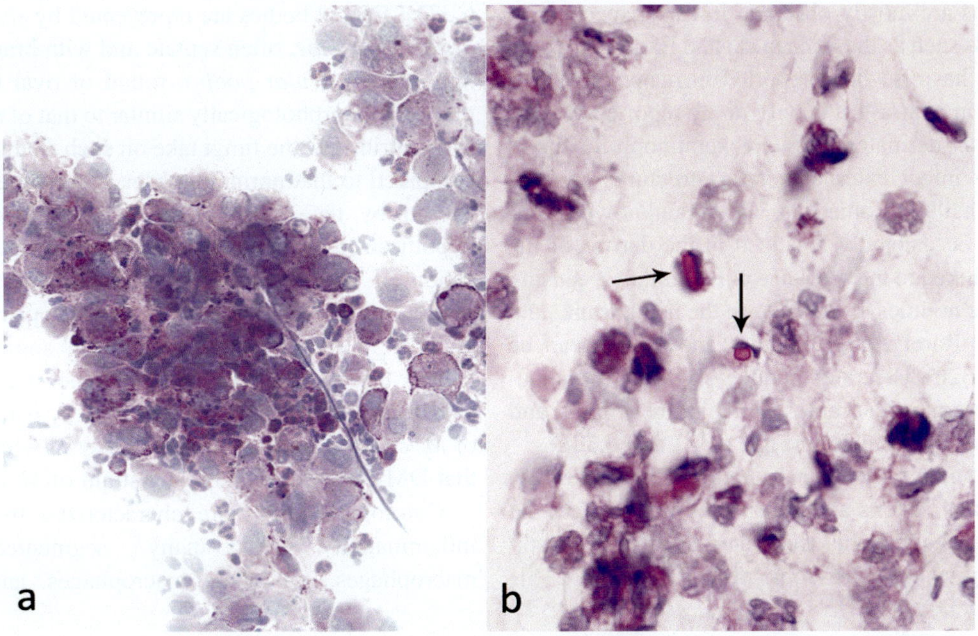

Fig. 6.322 Cytology of dermatophytic kerion (PAS stain): (**a**) amorphous acellular PAS-positive material in the cytoplasm of inflammatory cells can be misdiagnosed as fungal bodies and (**b**) the magenta-stained oval bodies are arthroconidia of dermatophytes (*arrows*)

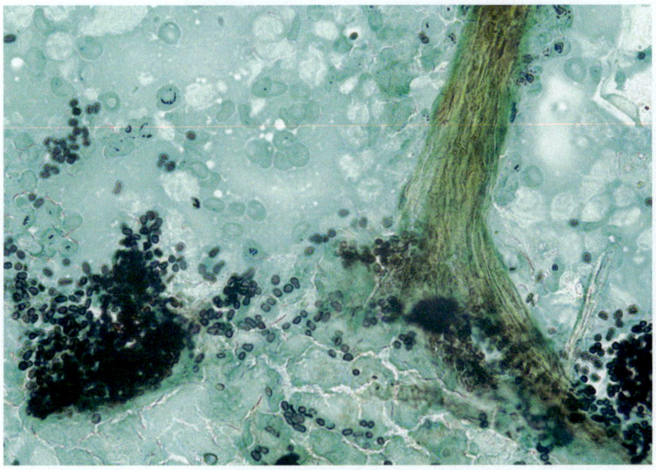

Fig. 6.323 Cytology of dermatophytic kerion (GMS): many arthroconidia stained black on a green background

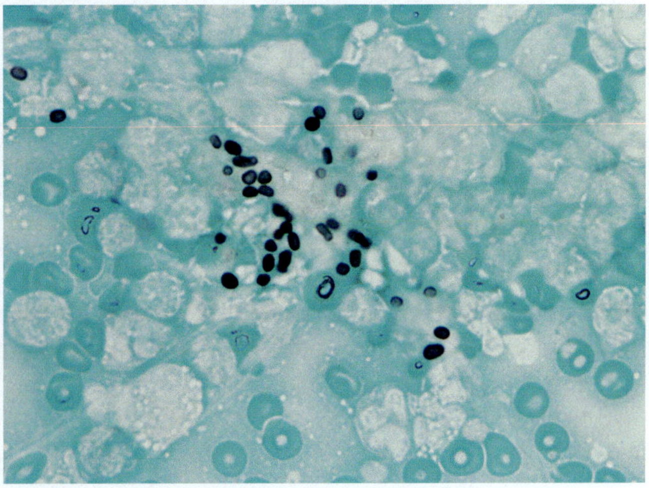

Fig. 6.324 GMS stain: arthroconidia stained black on a green background in which inflammatory cells are recognizable, which confirms the diagnosis of dermatophytic kerion

6.8.1.4.2 Dermatophytic Pseudomycetoma

DP is an uncommon clinical presentation of *Microsporum canis* deep infection. DP is characterized by single or multiple nodules/plaques, both dermal and hypodermal in location. In nature, only some fungi can normally produce grains in infected tissue (*true mycetomas*), so, when grains are observed in the course of dermatophytic infection, they must be considered unusual; for this reason, the infection is defined as DP. To date, *M. canis* is the only species isolated from DP. Lesions are mostly observed in cats, especially in longhaired breeds such as the Persian cat and Exotic purebred cats, although it has also been reported in many DSH cats [214, 216, 217]; in contrast, it is only rarely reported in dogs [213]. Usually, as dermatophytes are keratinophilic fungi, they exclusively infect external keratin structures such as scales, hair, and nails; in some cats, with an unknown mechanism, *M. canis* occasionally localized in the dermis or the panniculus, not linked to the breakage of the follicles, causing single or multiple nodules, especially on the head, trunk, and tail (Fig. 6.325). Rarely, large nodules and masses can be observed (Fig. 6.326). Ulcerated nodules show many characteristic yellowish grains that represent a mixture of fungi and pyogranulomatous inflammation (Figs. 6.327 and 6.328).

Furthermore, some cases of intra-abdominal localization of *M. canis* causing lesions as DP have been published; the implantation of fungi in the abdominal cavity is mostly secondary to surgery or laparoscopic maneuvers [214, 216].

Cytological Findings

Histologically, lesions are characterized by a multifocal nodular dermal/hypodermal pyogranulomatous inflammation around many aggregates of fungi immersed in a characteristic antigenic reaction known as Splendore-Hoeppli (S-H) phenomenon (Fig. 6.329) [215]. Nevertheless, their large size, fungal bodies are slightly visible since they are embedded with the S-H reaction; in fact, they can better be highlighted when special histochemical stains are used (Figs. 6.330 and 6.331). Fungal bodies are represented by *short*, *swollen,* and *distorted hyphae*, often septate and with branches, and many large and *vesicular bodies*, round or oval in shape, with a thick wall morphologically similar to that of *chlamydospores*. Why during PD the fungi take on such a strange morphology compared to that normally observed in superficial infections caused by the same fungus is not known; it could be hypothesized that this unusual dermal or hypodermal localization of dermatophytes, uncommon for a keratinophilic fungus, is the cause for which dermatophytes develop this atypical morphology or, maybe that some unknown host factors can play a role. On the other hand, the fact that pseudomycetoma can be caused only by particular genotypes of *M. canis* was ruled out by a study in which it was shown that DM can be caused by any strain of *M. canis* [217].

Cytological slides are characterized by granulomatous inflammation with many segmented neutrophils, macrophages, epithelioid macrophages, and, constantly, a

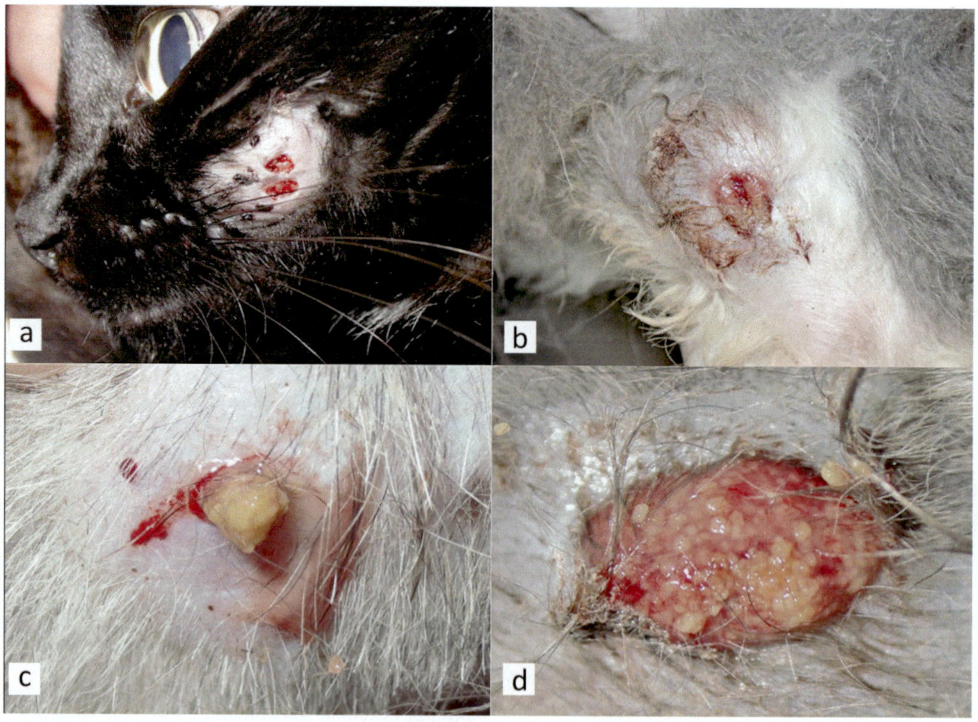

Fig. 6.325 DP: (**a**) ulcerated plaque on the face of DSH cat; (**b**) ulcerated nodule from which granular material oozes; (**c**) yellowish material containing grains; and (**d**) at high magnification, many yellowish grains are better visualized

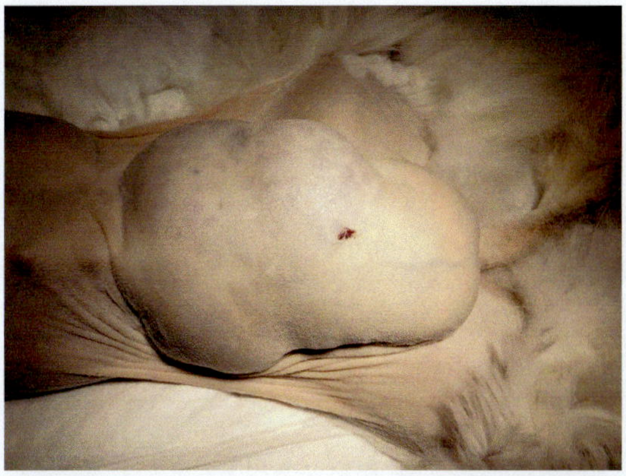

Fig. 6.326 DP: large mass on the abdominal area mimicking a mammary tumor. (Courtesy of Ferraravet clinic—Italy)

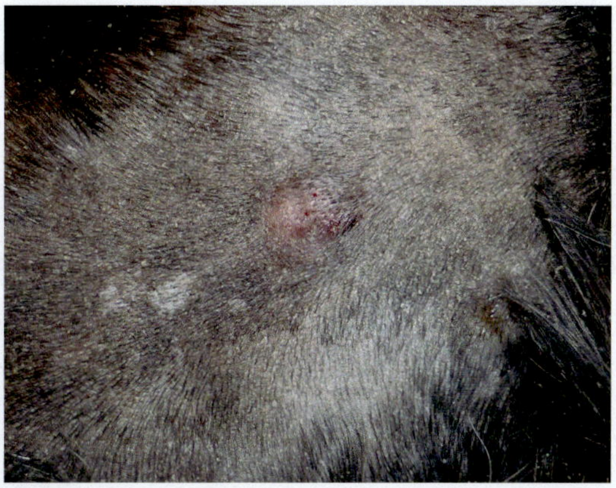

Fig. 6.328 Small nodular lesions in a dog with DP

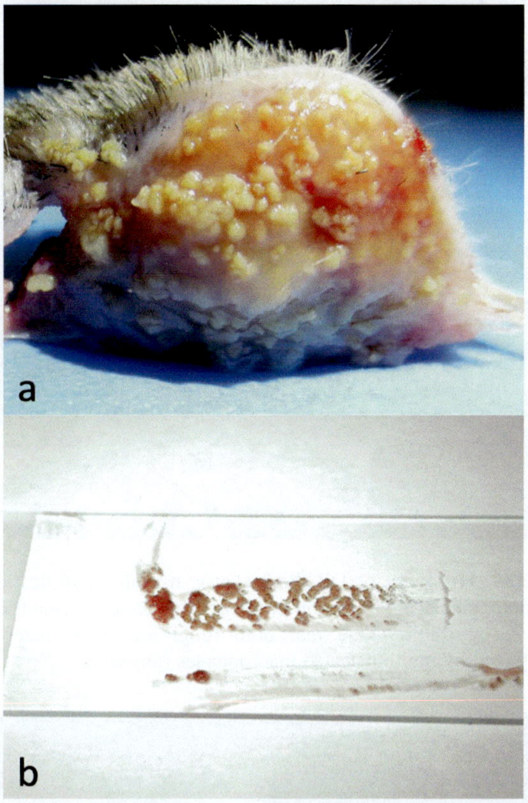

Fig. 6.327 DP: (**a**) ulcerated nodule oozing a yellowish grainy exudate and (**b**) grains squashed on a slide. (Courtesy of DVM F. Necci—Italy)

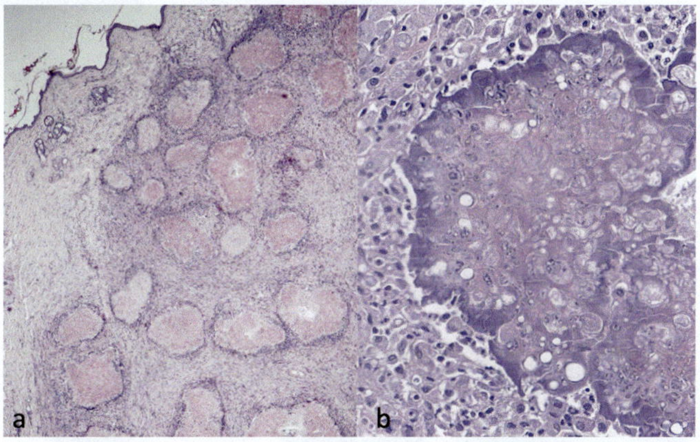

Fig. 6.329 Histology of DP (HE staining): (**a**) multiple granulomatous nodules in the dermis and (**b**) high magnification of a grain. Note that not well-defined fungal bodies, mostly roundish and large, are immersed in a pinkish-violet amorphous material (S-H reaction). Granulomatous inflammation surrounds the grain

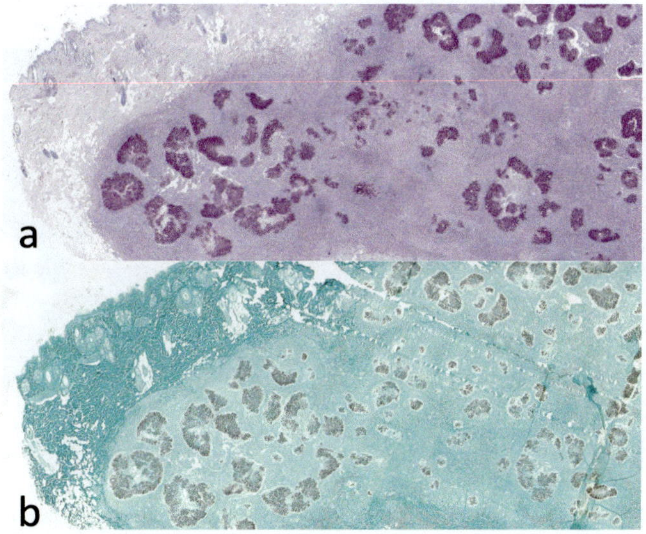

Fig. 6.330 Histology of DP: (**a**) PAS stain and (**b**) GMS stain. With special stains, the presence of fungi is better identifiable

large number of giant cells arranged around grains; the latter are composed of an amorphous, from granular to coarse deep blue material that represents the S-H reaction observed on histopathology and that contains many fungal bodies (Figs. 6.332, 6.333, and 6.334). The amorphous blue material must be carefully scan to detect fungi, as the latter tend to be only partially stained using the standard rapid Romanowsky-type dye; fungal bodies are better detectable at the periphery of grains (Figs. 6.335 and 6.336). In case the clinical

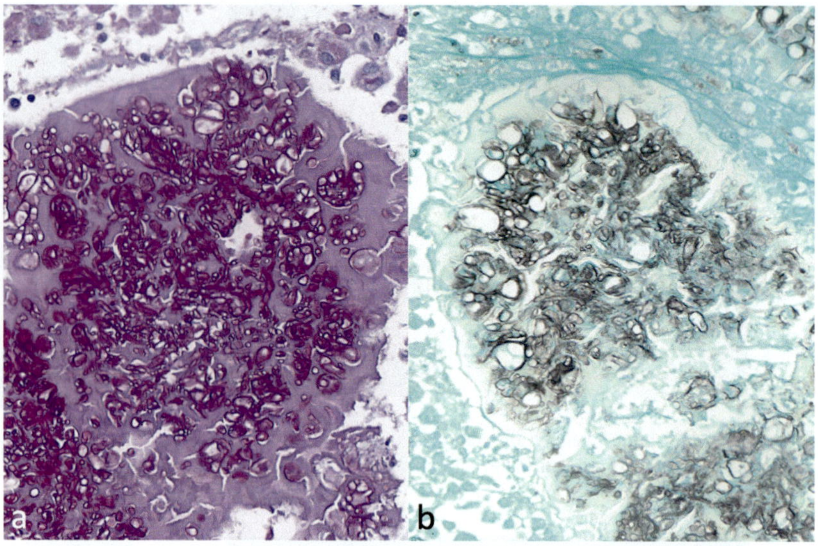

Fig. 6.331 Histology of DP: (**a**) PAS stain: fungi are strongly positive. Note that they are immersed in an amorphous slightly violet material (S-H reaction) and (**b**) (GMS stain): black fungal bodies are highlighted on a green background

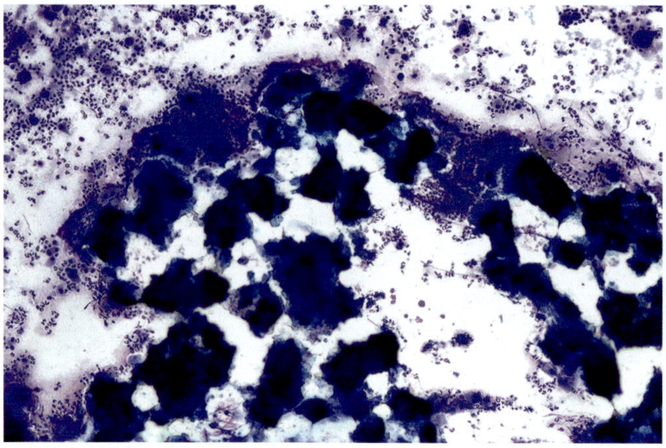

Fig. 6.332 Cytology of DP: multiple grains are surrounded by pyogranulomatous inflammation

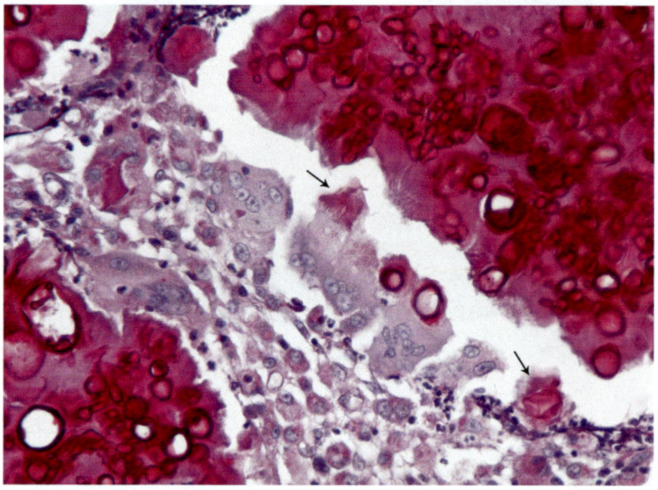

Fig. 6.334 Histology of DP (PAS stain): at high magnification, the pyogranulomatous inflammation composed of numerous giant cells is clearly recognizable. Note that both fungi and S-H material are phagocytized by multinucleated macrophages (arrows)

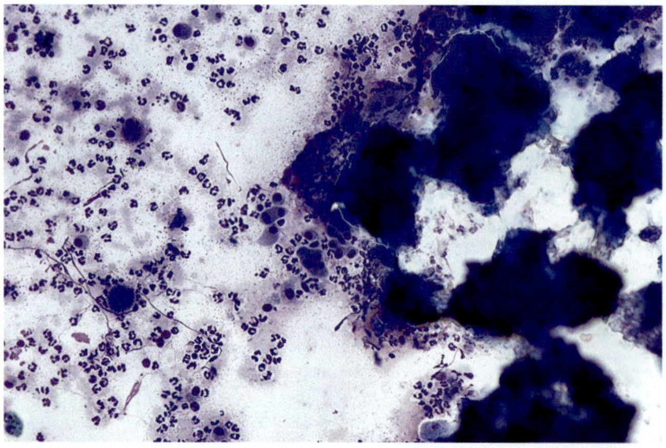

Fig. 6.333 Cytology of DP: amorphous deeply blue material grain is surrounded by a pyogranulomatous inflammation

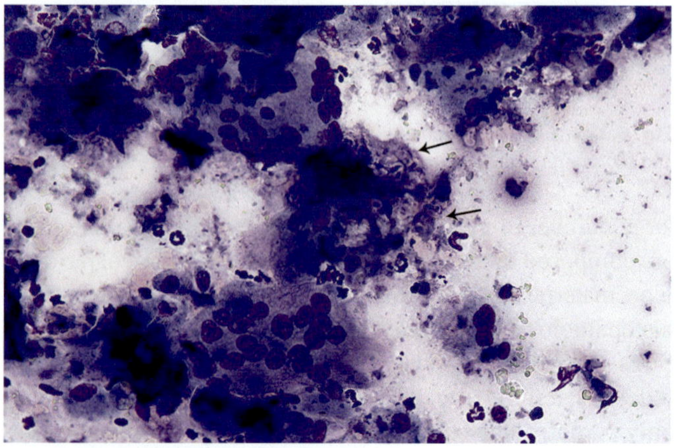

Fig. 6.335 Cytology of DP: macrophages and giant cells attacking the amorphous deep blue material in which some fungal bodies are fairly recognizable (arrows)

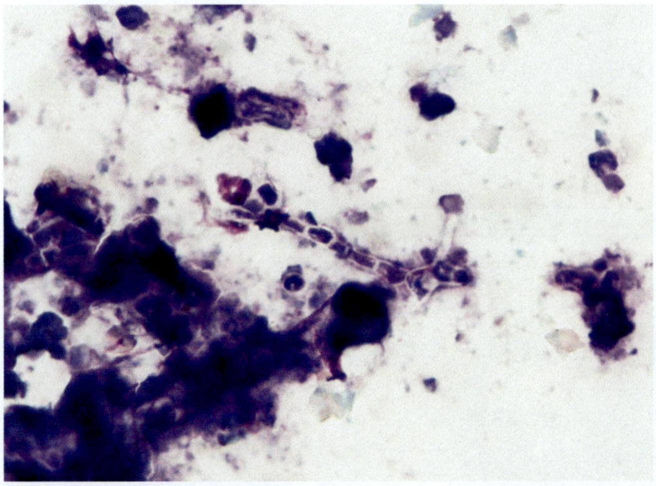

Fig. 6.336 Cytology of DP: septate hyphae are visible at the periphery of a grain

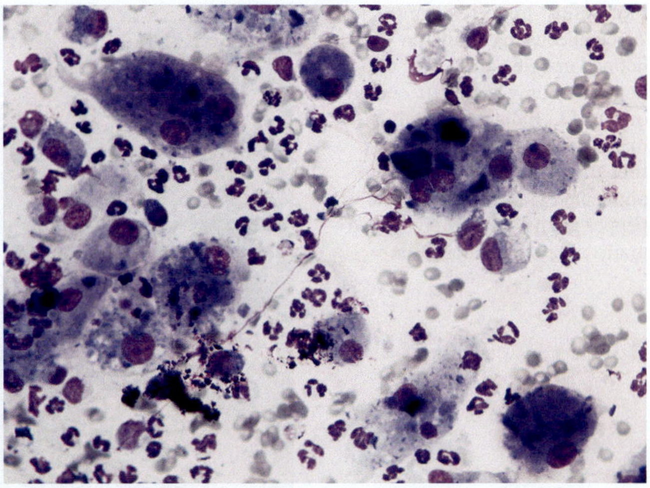

Fig. 6.337 Cytology of DP: macrophages and giant cells that phagocyte deeply basophilic acellular material representing S-H

Fig. 6.338 Cytology of DP: (**a**) (PAS stain): septate hyphae at large round fungal bodies stained magenta and (**b**) (GMS stain): septated hyphae and large roundish fungal bodies stained black

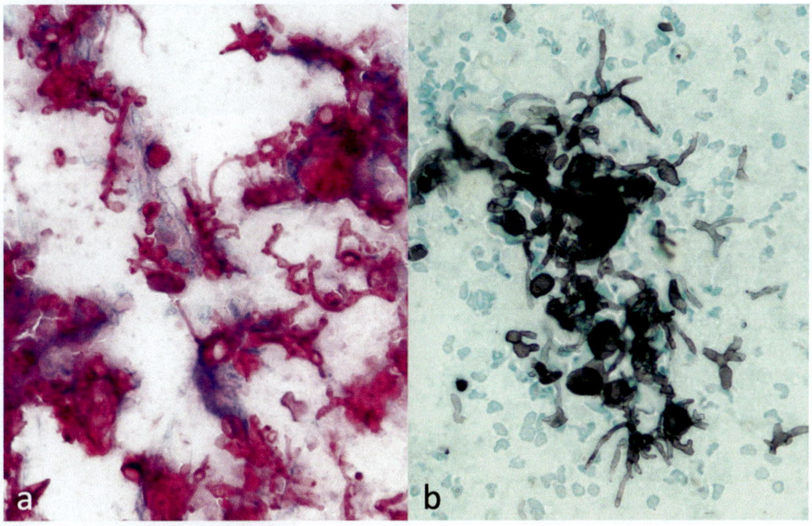

presentation is strongly suggestive of DP and fungal elements are not visible, the presence of this dark-blue stained material into the cytoplasm of inflammatory cells, representing pieces of S-H reaction, must alert cytologists to use cytochemical stains such as PAS or Grocott's methenamine silver (Figs. 6.337 and 6.338).

6.8.1.4.3 Eumycetoma

Eumycetoma or *true mycetoma* is a deep fungal disease characterized by single or, less frequently, multiple dermal or subcutaneous nodules in which the causative fungi, for their nature, form *grains* into the tissues.

These fungal infections, mainly observed in temperate, tropical, and subtropical climates, are uncommon in human medicine and only exceptionally reported in dogs. Eumycotic mycetomas can be classified as "black grains" when caused by pigmented hyphae (*Curvularia lunata*, *Curvularia geniculata*, *Cladophialophora bantiana,* and less commonly *Phialophora oxyspora* and *Madurella* spp.) or "white grains" when the hyphae are non-pigmented (*Pseudallescheria (Scedosporium) boydii* and *Penicillium (Thermomyces) dupontii*) [221]. True mycetoma can occur secondarily to a wound infection or can be due to the direct implantation of a foreign body contaminated with environmental fungi. In animals, there are only a few reports [220, 222]. A case of deep subcutaneous infection by *Madurella pseudomycetomatis* has been recently reported in a Great Dane with a large subcutaneous mass on the hindlimb in which many small black color grains characterized the lesion [218] (Fig. 6.339).

Cytological Findings

Histopathology is characterized by dermal to subcutaneous multifocal aggregates of pyogranulomatous inflammation

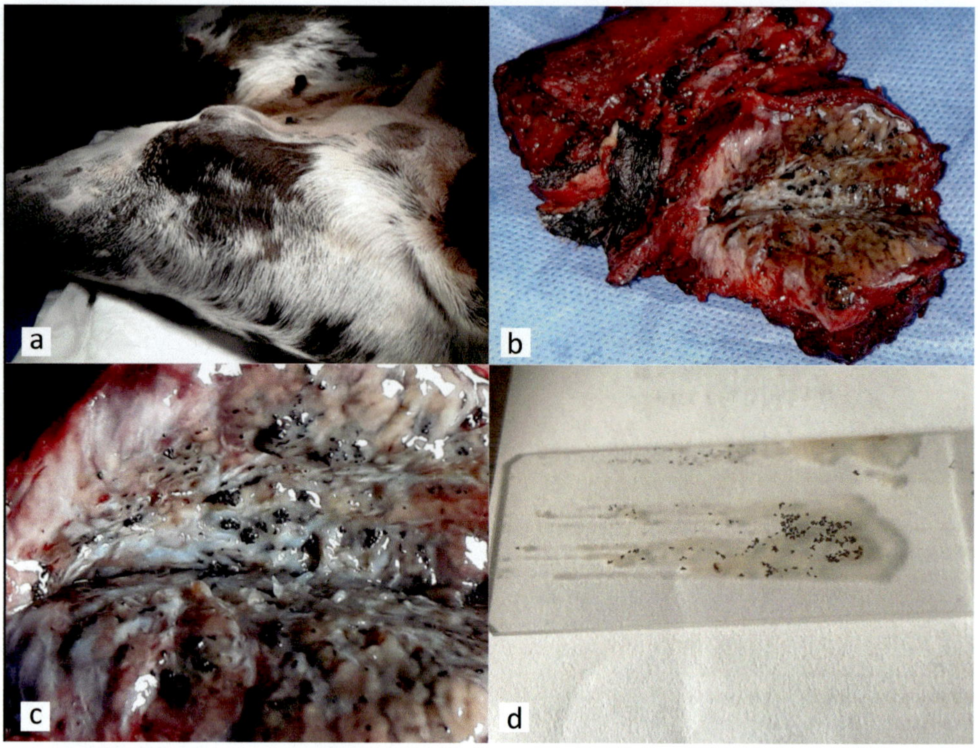

Fig. 6.339 Eumycetoma: (**a**) subcutaneous mass with no well-defined margins, on the leg of a dog with Eumycetoma. (Courtesy of DVM L. Ballantani—Italy); (**b**) many small pointed pigmented fungal grains; (**c**) closeup of the pigmented grains that look like pepper grains; and (**d**) pigmented grains after slide preparation

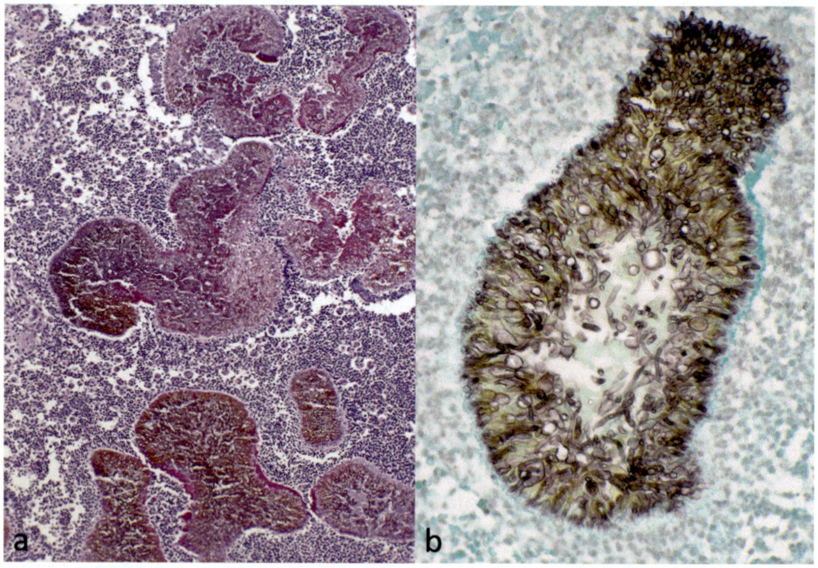

Fig. 6.340 Histopathology of black grains eumycetoma: (**a**) (H-E stain) multiple grains containing pigmented fungi and (**b**) GMS stain highlights the morphology of fungi represented from both hyphae and chlamydoconidia

surrounding, pigmented or not pigmented, fungal hyphae, frequently septate, with irregular dichotomous branching and with frequent formation of terminal cystic dilatations 5–20 μm in diameter termed *chlamydoconidia* [219]. As in other infectious grainy diseases, the grains are histologically represented by fungi surrounded by a marked pyogranulomatous inflammation and by an S-H reaction which hold together the fungal elements, the latter better highlighted with PAS or Grocott' methenamine silver staining (Fig. 6.340).

Cytology mirrors what is reported with histology. When slides are observed at low magnification, multiple pigmented foci of apparently acellular material are detected. At high magnification, the latter are composed of amorphous deep blue, sometimes containing pigmented areas, which hide many fungal bodies. These foci represent grains, which are

Fig. 6.341 Cytology of black grains eumycetoma: a large amount of pigmented material (grains) immersed in a pyogranulomatous inflammation

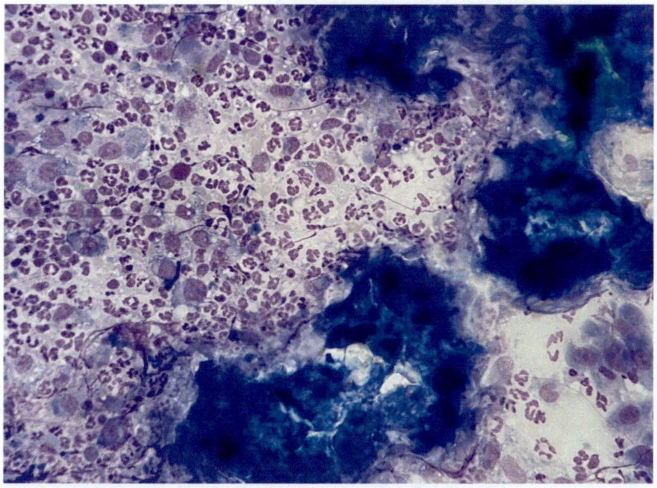

Fig. 6.342 Cytology of black grains eumycetoma: high magnification of amorphous dark and pigmented material (grain) along with inflammatory cells

surrounded by inflammatory cells mostly represented by neutrophils, activated macrophages, and epithelioid macrophages along with many multinucleate giant cells and a variable amount of lymphocytes. In some cases, fungal bodies can be observed with routine Romanowsky-type staining at the periphery of grains (Figs. 6.341, 6.342, and 6.343).

As mentioned for DP, in cases in which fungi are hidden by amorphous material that composes grains and are not cytologically detectable with Romanowsky-type dye, special cytochemical dyes such as PAS and Grocott's methenamine silver can be used to highlight them.

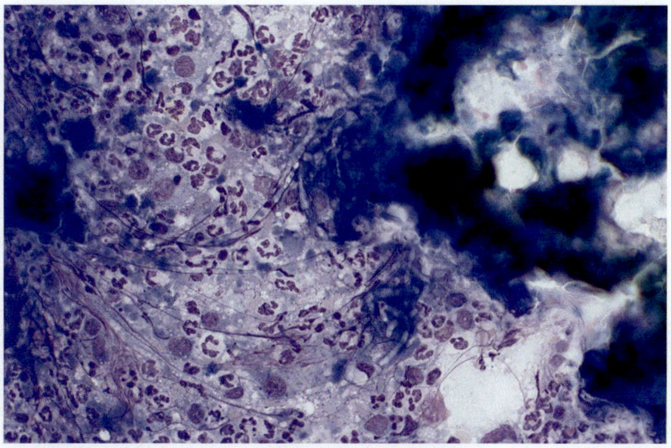

Fig. 6.343 Cytology of black grains eumycetoma: pigmented grains with many fungal hyphae at their periphery

6.8.1.4.4 Phaeohyphomycosis

Phaeohyphomycosis comprises a group of deep mycoses affecting both cats and dogs, in which fungi are localized in the dermis and/or panniculus. This group of infectious agents includes many different saprophytic, free-living environmental molds that can cause nodular lesions following implantation of contaminated foreign bodies [226–228]. Fungal isolates belong to different groups, and the list of microorganisms that cause *phaeohyphomycosis* is increasing based on the literature; among these are reported *Alternaria alternata*, *Bipolar* spp., *Curvularis* spp., *Phialophora verrucosa*, *Fonsecaea*, and many others [223–225].

Fungi included under the umbrella of *phaeohyphomycosis* are dematiaceous (dark) fungi, as they give rise, when cultured, to pigmented fungal colonies. When fungi do not produce pigmented colonies on cultures, they are included in the group of *hyalohyphomycosis* (Fig. 6.344). Some molds can also be dark in nature, and in these cases, nodules on animals can grossly show black pigmentation. Clinically, lesions are mostly represented as single nodules, often ulcerated and with draining tracts from which a purulent material discharges; they are usually located on the extremities and on the nose, both areas in contact with ground, and rarely diffuse via lymphatic into adjacent areas. The skin that covers the lesion can be hyperpigmented or not, depending on the implanted fungus (Fig. 6.345). Less frequently, in immune-compromised animals, lesions can be multiple, and disseminated cases in internal organs are also occasionally reported [229].

Cytological Findings

Since these deep mycotic skin diseases are not characterized by grain formation, histopathology shows a diffuse dermal to hypodermal granulomatous inflammation, in which fungal

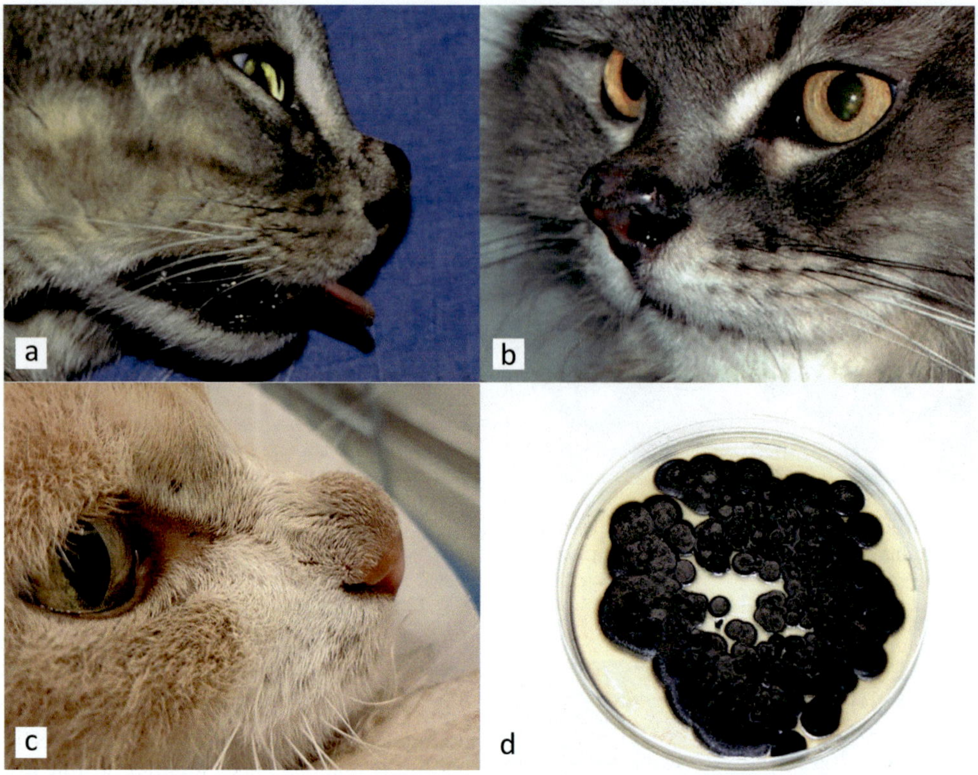

Fig. 6.344 Phaeohyphomycosis: (**a**) pigmented and ulcerated nodule on the nose of a cat. (Courtesy of DVM D.N. Carlotti—France); (**b**) pigmented neoformation on the nose of a cat affected by phaeohyphomycosis. (Courtesy of DVM F. Bastelli—Italy); (**c**) not pigmented swelling on the nose of a cat affected by phaeohyphomycosis due to *Aspergillus* spp. (Courtesy of DVM P. Mascherpa—Italy); and (**d**) pigmented colonies of *Phialophora verrucosa.* (Courtesy of Prof. DVM A. Peano—Italy)

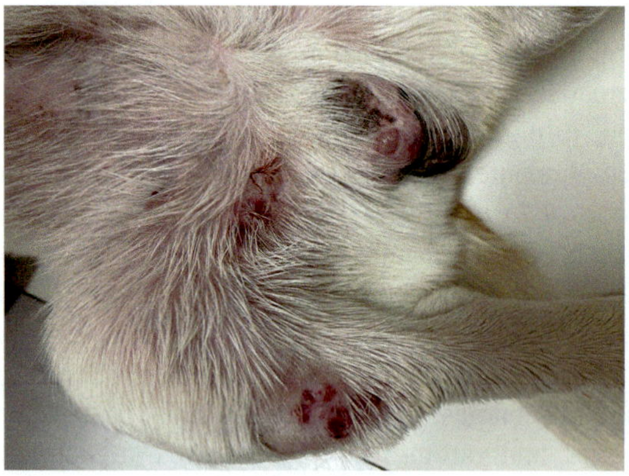

Fig. 6.345 Phaeohyphomycosis in dogs: multiple nodules, sometimes ulcerated, on the skin of an immune-compromised dog affected by cutaneous *Aspergillus flavus*

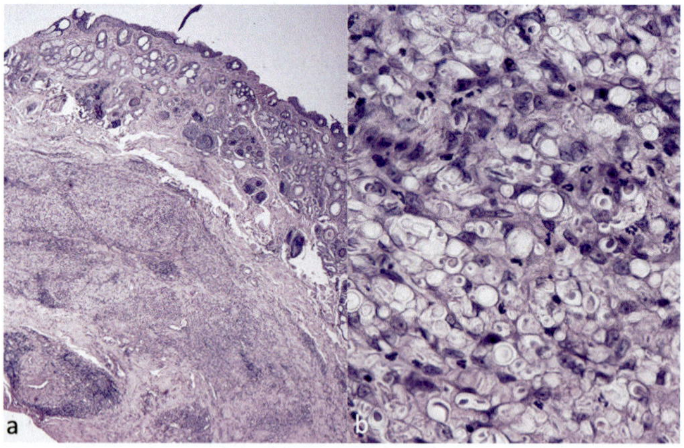

Fig. 6.346 Histology of phaeohyphomycosis: (**a**) multiple subcutaneous and confluent nodules and (**b**) at high magnification, many fungi are recognizable as oval to roundish achromatic and capsulated bodies along with short septate hyphae

bodies are usually numerous; as for all fungal diseases, especially when microorganisms are few, histochemical staining, such as PAS or Grocott's methenamine silver, can better mark the fungal bodies (Figs. 6.346 and 6.347). Cytological slides show a variable number of neutrophils, from segmented to moderately degenerate, accompanied by many macrophages, epithelioid macrophages, and a variable number of multinucleated giant cells; lymphocytes and plasma cells can also be found in variable amounts. The morphology of fungi is variable depending on the family to which they belong; in fact, they can be represented by short to long septate and branched hyphae, of different lengths, together

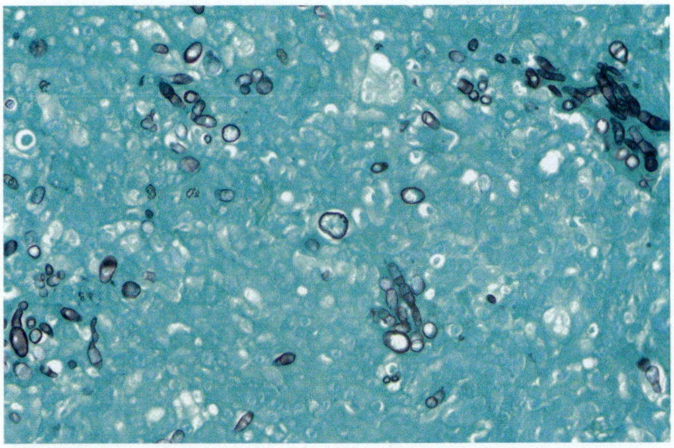

Fig. 6.347 Histology of phaeohyphomycosis: black hyphae and chlamydospores-like fungal bodies are better visible once stained with Grocott's methenamine silver dye

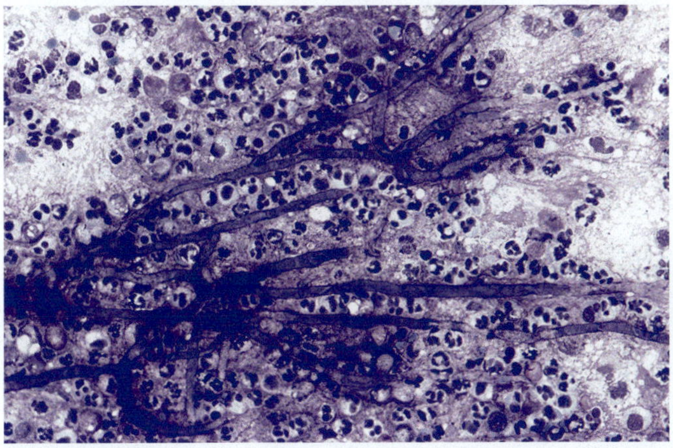

Fig. 6.348 Cytology of phaeohyphomycosis due to *Aspergillus flavus*: many elongated, septate, and branched hyphae along with inflammatory cells

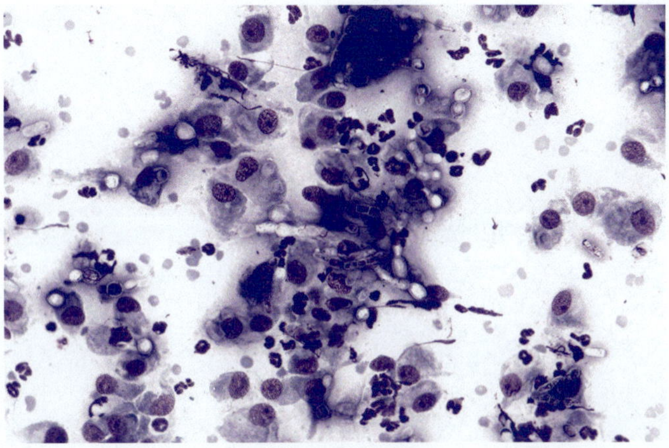

Fig. 6.349 Cytology of phaeohyphomycosis: a morphologically mixed population of fungal bodies, both small and roundish bodies along with septate hyphae immersed in a pyogranulomatous inflammation

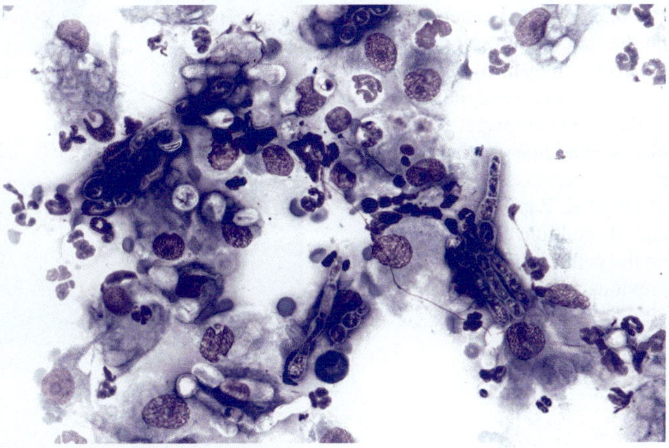

Fig. 6.350 Cytology of phaeohyphomycosis: septate hyphae and *chlamydospores*-like bodies immersed in a pyogranulomatous inflammation

with yeast-like roundish bodies of different sizes and with vesicular appearance, with a distorted profile and a thick wall, that look similar to *chlamydospores*. In other cases, fungi are smaller in size and represented by small, oval-shaped fungal bodies or short distorted hyphae. A thin achromatic halo can be clearly recognizable around many fungal bodies (Figs. 6.348, 6.349, 6.350, and 6.351). In cases in which fungi are pigmented in nature, dark fungal bodies can also be detected on cytological specimens (Figs. 6.352 and 6.353). When nodules are strongly pigmented, a lot of dark black material can be grossly visible on the unstained slides after squashing material on it. In cytological specimens, fungi are usually numerous, but sometimes, they are few; in these cases, cytochemical stains such as PAS and Grocott's methenamine silver permit their detection.

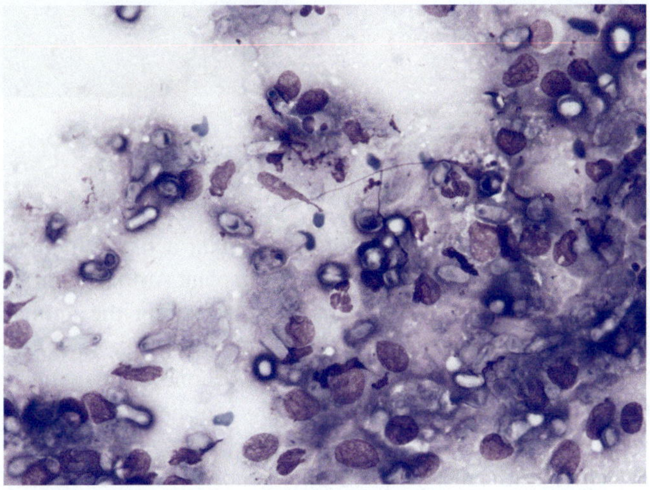

Fig. 6.351 Cytology of phaeohyphomycosis: oval to roundish fungal bodies. Note the thin halo that surrounds many fungi

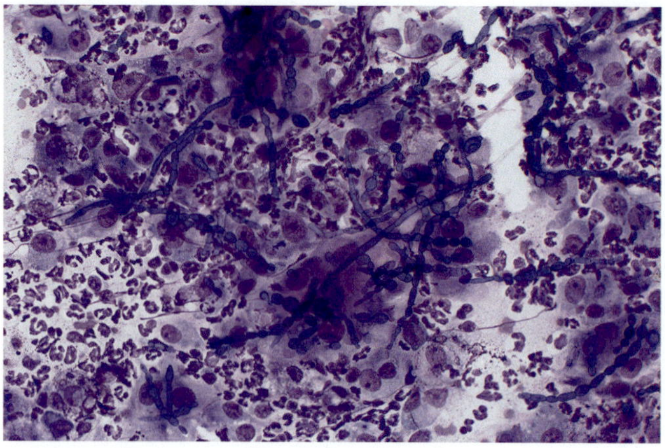

Fig. 6.352 Cytology of phaeohyphomycosis: many long, septate, and branched heavily hyperpigmented hyphae

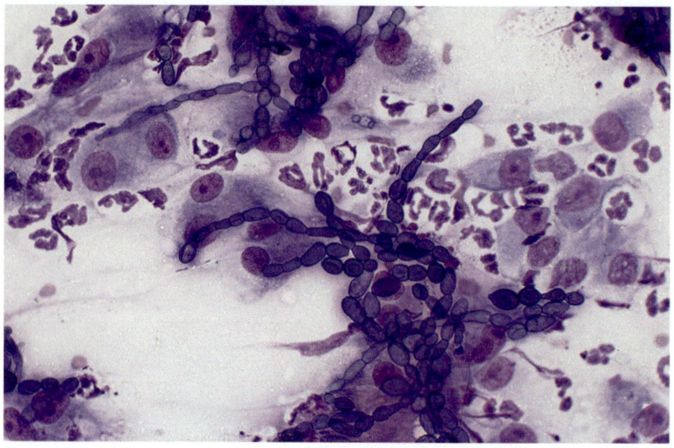

Fig. 6.353 Cytology of phaeohyphomycosis: many long, septate, and branched strongly hyperpigmented hyphae

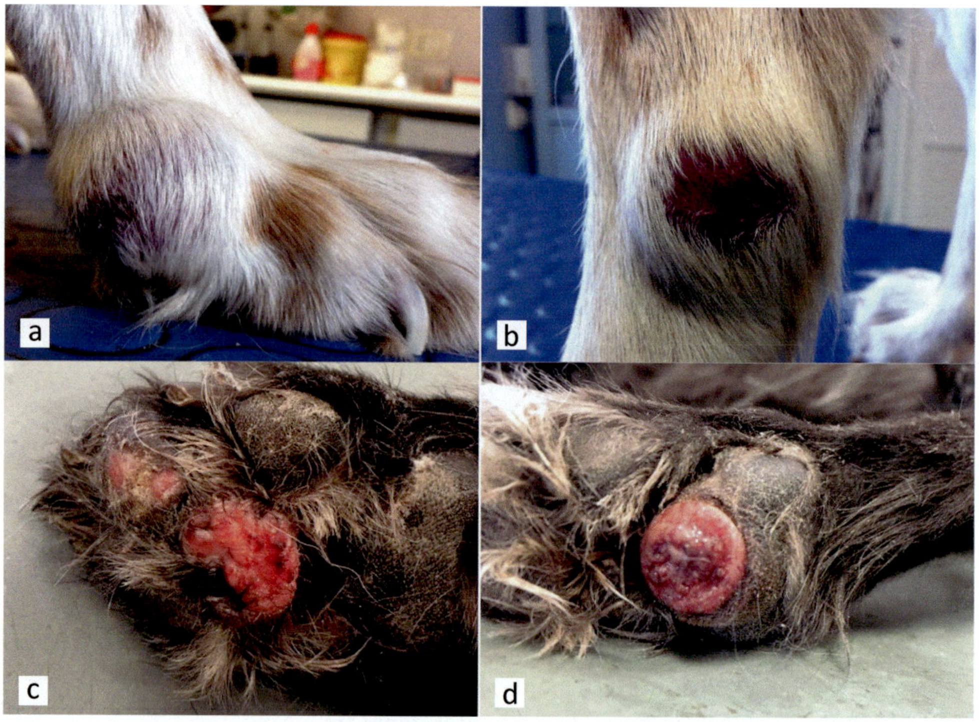

Fig. 6.354 Cutaneous *Protothecosis*: (**a**) intact nodule on the foot. (Courtesy of DVM G. Noris—Italy); (**b**) ulcerated nodules on the leg of the same dog of the previous image; (**c**) ulcerated nodules on the foot of an immune-depressed dog. (Courtesy of DVM K. Zappaterra—Italy); and (**d**) multiple ulcerated nodules on the pads of the same dog of the previous image

6.8.1.4.5 Protothecosis

Prototheca spp. is a ubiquitous unicellular achlorophyllous alga that may be routinely isolated from plant surfaces, soil or water, and may occur as transient flora within the animal gastrointestinal tract. The actual *Prototheca* taxonomy is very complex and, based on the mitochondrially-encoded CYTB gene marker, includes many species: *Prototheca wickerhamii*, *P. bovis* (formerly *P. zopfii* genotype 2), *P. ciferrii* (formerly *P. zopfii* genotype 1), *P. cutis, P. miyajii, and P. blaschkeae* [230–235].

Traumatic implantation or contamination of open wounds has been associated with skin infections, which are characterized by nodules, plaques, and ulcerative lesions [237] (Fig. 6.354). Dogs with disseminated protothecosis, most commonly, have clinical signs consistent with distal gastrointestinal tract imbalances, central nervous system, and eye infections. In these cases, via of entrance of alga into the organism is thought to be the gastrointestinal tract following the ingestion of contaminated water or food. Immunosuppressive diseases predispose the dissemination of algae. Some dogs develop simultaneous lesions on the skin and other organs [230]. The skin involvement in cats is absolutely rare [232, 233].

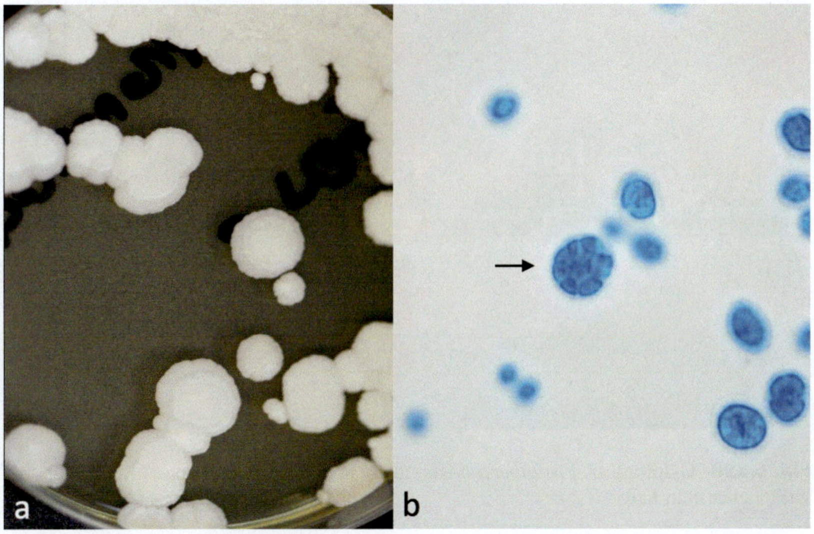

Fig. 6.355 Colony of *Prototheca bovis* (ex *P. zopfii*) on Prototheca isolation medium. (**a**) colonies are creamy and white with irregular and jagged edges. (Courtesy of DVM P. Danesi—Italy) and (**b**) spores and sporangia (morula-like) stained with lactophenol blue (arrow). (Courtesy of DVM P. Danesi-Italy)

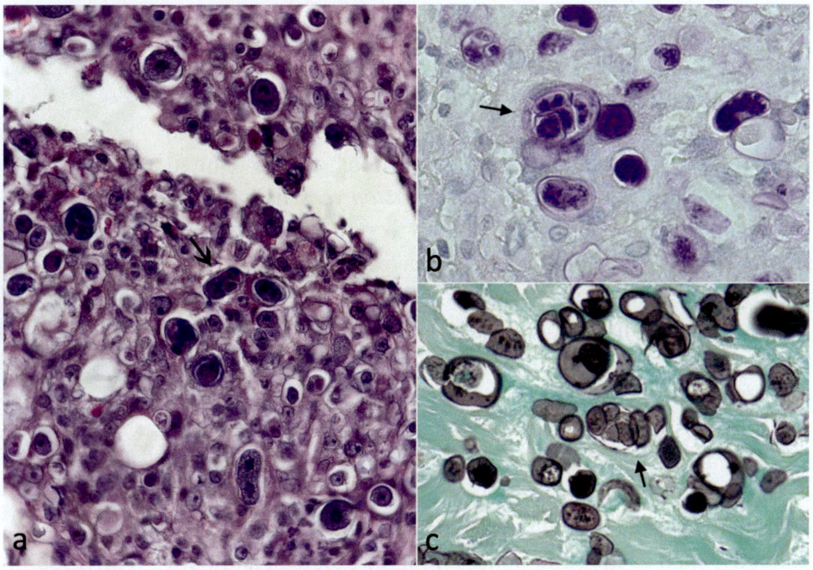

Fig. 6.356 Histology of *Prototheca bovis*: (**a**) (H-E stain): many Prototheca bodies and many sporangia (arrow); (**b**) (PAS stain): algal bodies with clear halo and sporangia morphology (arrow); and (**c**) (GMS stain): very large algal bodies and one sporangia (arrow) are highlighted as black bodies on a green background

Cytological Findings

Prototheca undergoes asexual reproduction, so *Sporangia*, the mother cells, through a process of splitting of their cytoplasm, give origin to endospores; the sporangia break under the pressure of the growing spores, and the latter spread outside the cell. When spores become numerous, Sporangia assume a morphology that resembles a *Morula* or takes a cartwheel-like appearance (Figs. 6.355 and 6.356). On slides, many encapsulated, roundish or oval algal bodies with a peripheral thin clear halo are usually observed along with a pyogranulomatous inflammation; they are detectable both in the cytoplasm of macrophages and free extracellular on the background. Sporangia are also evident in variable numbers (Figs. 6.357 and 6.358).

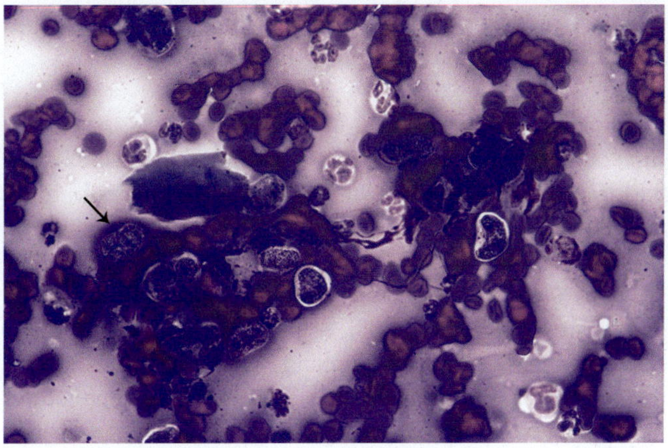

Fig. 6.357 Cytology of *Prototheca bovis*: many spores and a sporangia (arrow)

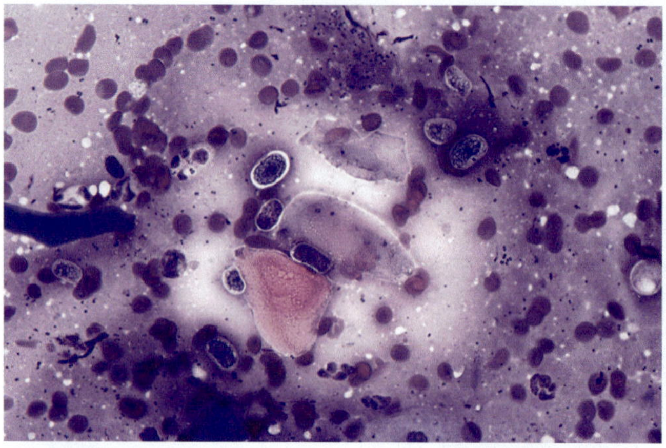

Fig. 6.358 Cytology of *Prototheca bovis*: many spores oval in shape with a clear thin halo

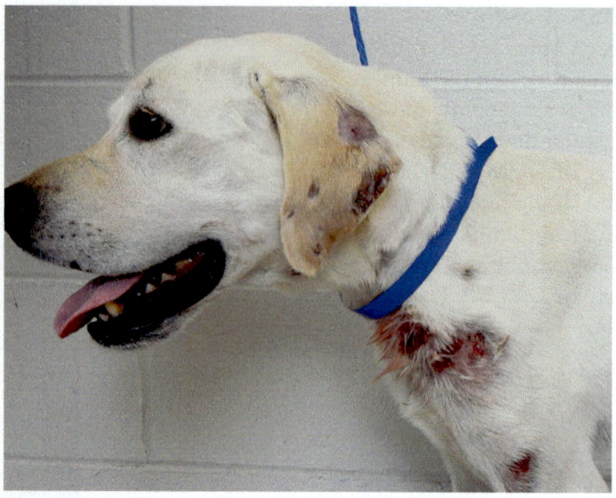

Fig. 6.359 Multiple ulcerated nodules, with draining tracts in a Labrador affected by Blastomycosis and with skin involvement. (Courtesy DVM E. Layne—United States—UW-Madison SVM Teaching Materials)

6.8.1.4.6 Blastomycosis

Blastomycosis is a human and canine systemic disease caused by *Blastomyces dermatitidis*, a non-pathogenic, saprophytic fungus diffusely present in all of Africa as well as in the southcentral and southeastern portions of the United States. The disease is rare in cats. *Blastomyces* lives in the environment especially in the soil, in very humid areas and rich in decaying material, such as leaves and other organic material. The fungus grows as a mold in the environment that produces tiny fungal spores, and when soil or organic material is disturbed by the animal while digging, the tiny spores can be released into the air and inhaled by the animals. For this reason, sporting or hunting dogs are at an increased risk. Once in the lungs, body temperature allows the spores to develop into yeast, where they can either remain localized over there or spread through the bloodstream to other parts of the body, such as the skin, bones and joints, organs, and both central (brain) and peripheral (spinal cord) nervous system. The skin lesions appear as single or multiple nodules that rapidly tend to fistulate and usually ooze a purulent-hemorrhagic material [236–239] (Fig. 6.359).

Cytological Findings

Blastomyces is a roundish large yeast, which measures from 5 to 20 μm in diameter and has a refractile, deeply basophilic, and thick cell wall, that divides by broad-based budding compared with the narrow-based budding of *Cryptococcus* sp. In cytological specimens, yeasts are usually found in the cytoplasm of macrophages although extracellular microorganisms are also commonly detected; other inflammatory cells associated are neutrophils, epithelioid macrophages, multinucleated giant cells, and lymphocytes.

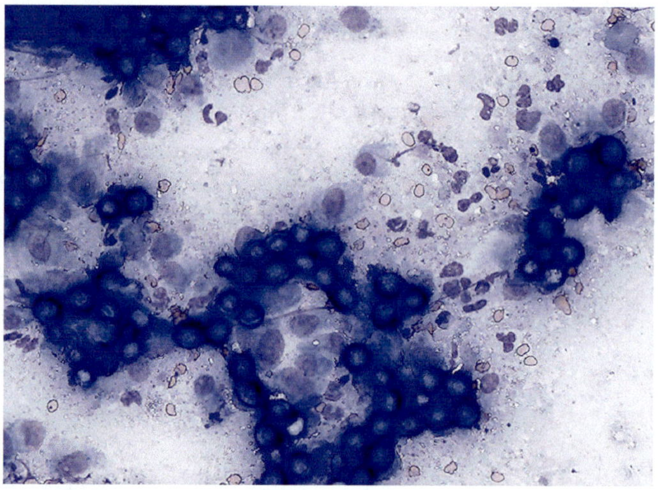

Fig. 6.360 Multiple large round yeasts, with thick cell walls along with a pyogranulomatous inflammation. (Courtesy of DVM M. Dell'Orco—Italy)

In very thick samples, PAS and Grocott's methenamine silver staining can help to highlight the yeasts (Figs. 6.360 and 6.361).

Histoplasmosis

Histoplasmosis is a systemic mycosis caused by *Histoplasma capsulatum*. In endemic areas, such as the United States, although its distribution is worldwide, it is the most common systemic fungal disease in dogs and the second most common fungal infection in cats [240, 242]. The fungi enter the body by inhalation and less frequently by ingestion and can be

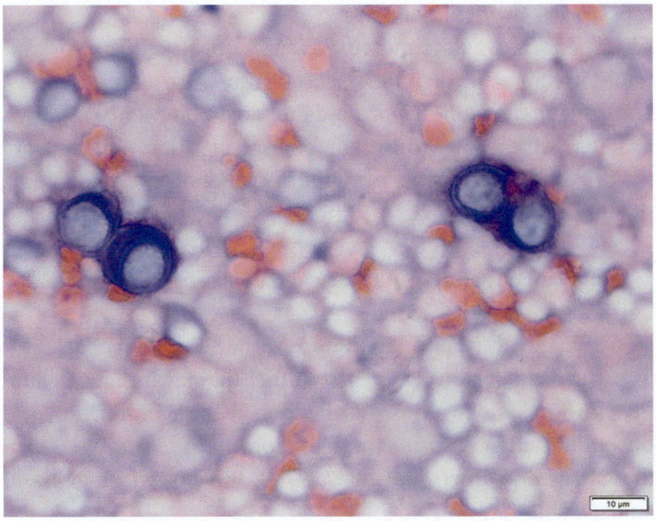

Fig. 6.361 *Blastomyces dermatitidis* yeasts: note the tick cell wall. (Courtesy of DVM S. Torres—United States)

disseminated via the bloodstream or lymphatic system predominantly to the gastrointestinal tract and, less frequently, to the skin, bones, nervous system, and eyes. As histoplasmosis is a systemic disease, clinical signs are often nonspecific, and affected animals often show weight loss, loss of appetite, and lethargy [241, 243] (Fig. 6.362).

Cytological Findings

Histoplasma capsulatum is an obligate intracellular microorganism that can be seen within macrophages and neutrophils as well as extracellular. Macrophages are very often filled with organisms. The yeasts are small, with an average diameter of 2–5 µm with a single nucleus purple in color and pale basophilic cytoplasm surrounded by a thin and clear halo (Figs. 6.363, 6.364, and 6.365).

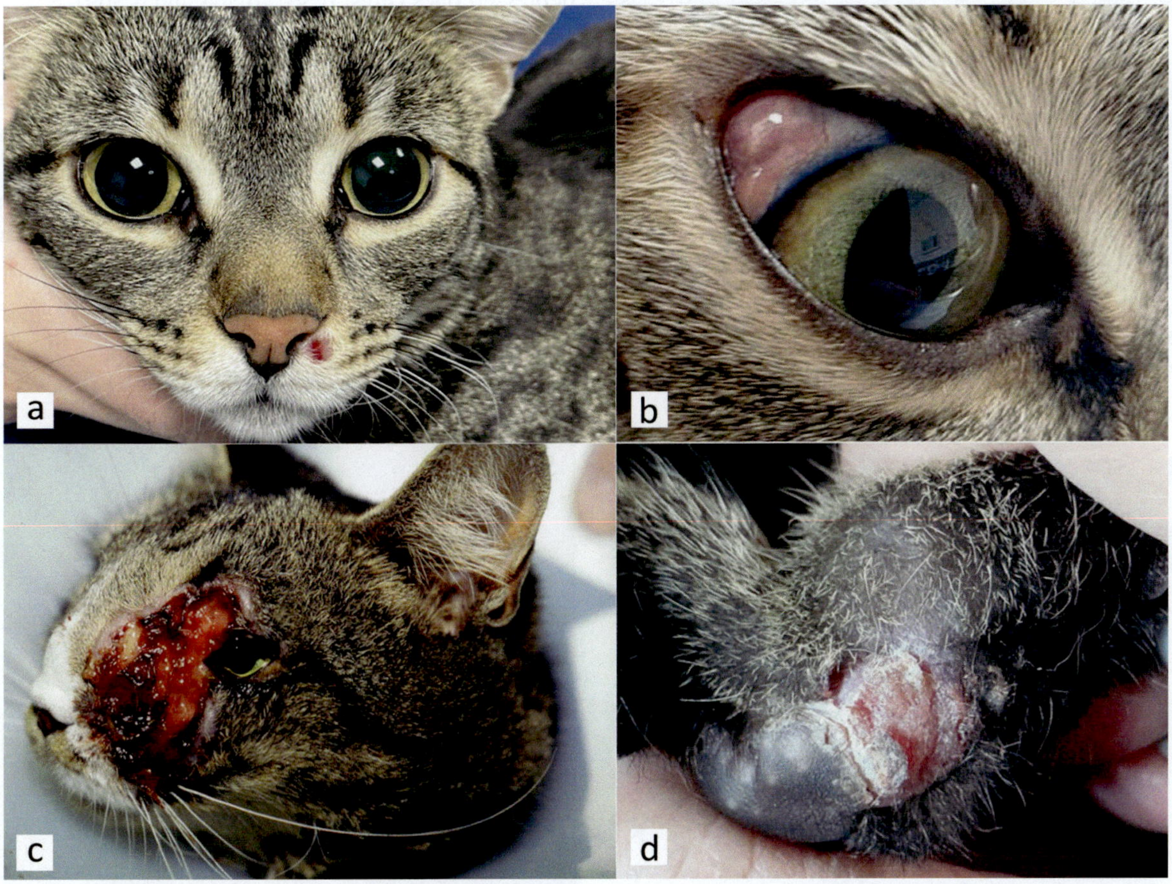

Fig. 6.362 Histoplasmosis: (**a**) small nodule on the muzzle of a cat with Histoplasmosis. (Courtesy of DVM I. Remy—Belgium); (**b**) nodule on the conjunctiva of the same cat of the previous image. (Courtesy of DVM I. Remy—Belgium); (**c**) large and ulcerated mass on the face of a cat with. (Courtesy of DVM M. Rosenbaum—United States); and (**d**) erythematous and ulcerate lesion on the paw of a cat with histoplasmosis. (Courtesy of Dermawet—Poland)

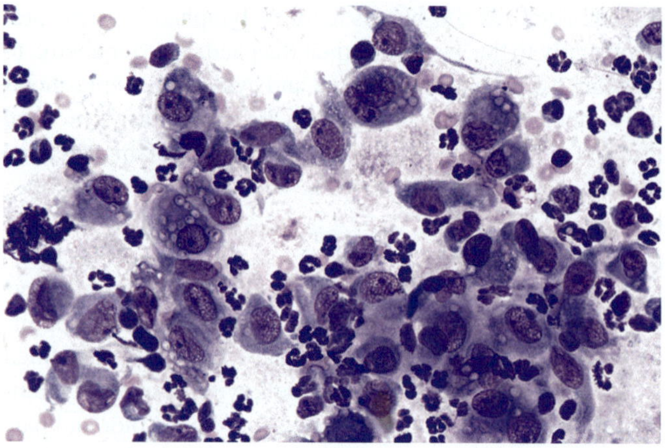

Fig. 6.363 Histoplasmosis cytology: many intracytoplasmic fungi are visible

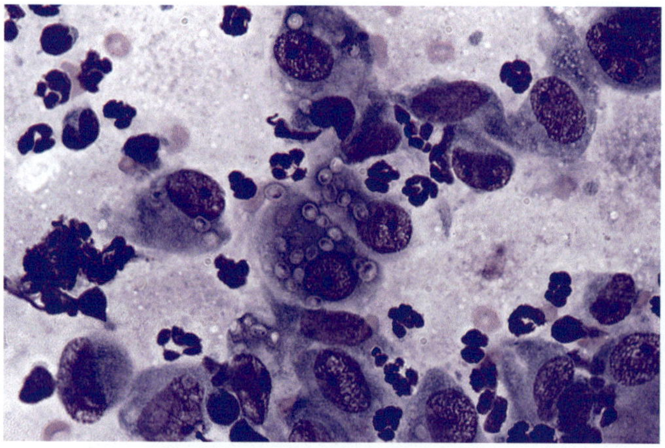

Fig. 6.364 Histoplasmosis cytology: at high magnification, the fungal oval silhouette is well recognizable

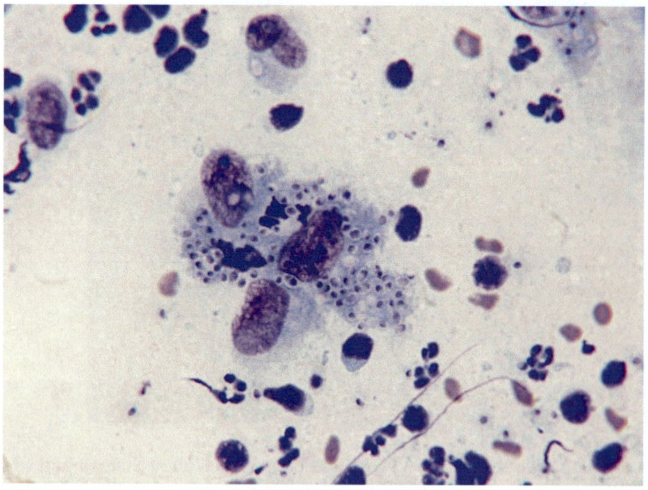

Fig. 6.365 *Histoplasma capsulatum* cytology: many fungal bodies into the cytoplasm of macrophages (Courtesy of R. Gomes Olinda – Brazil)

6.8.1.4.7 Cryptococcosis

Cryptococcosis is the most frequently observed form of systemic mycosis in cats; the disease is rarely observed in dogs. *C. neoformans* and *C. gattii* are the most commonly reported species, which have different biological reservoirs that are mainly the pigeon stools for the former and the eucalyptus plants for the *C. gattii.*

As *cryptococcosis* is a systemic disease, many affected patients can develop respiratory and neurological signs [244–250].

Since the entry into the body is through the upper respiratory tract, the lesions often also affect the nasal mucosa and, in some cases, swelling of the dorsal area of the nose is observed which deforms the facial profile, precisely because the infection affects the airways of the nostrils and nasal sinuses (Fig. 6.366).

In case of skin involvement, papular nodules or nodules and plaques that could be single or multiple, often ulcerated and located mainly on the face, head, nose, and extremities, are the most common lesions. In some dogs and cats, lesions can be spread all over the body, especially in immune-depressed ones (Figs. 6.367, 6.368 and 6.369).

Cytological Findings

In histopathological specimens, fungal bodies are very numerous and usually associated with an inflammation composed mainly of macrophages and giant cells and of few neutrophils. Fungal bodies are always very numerous and thus easily detectable in the cytoplasm of the inflammatory cells [245] (Figs. 6.370 and 6.371).

Fig. 6.366 Swelling of the nose and facial profile alteration in cats with cryptococcosis

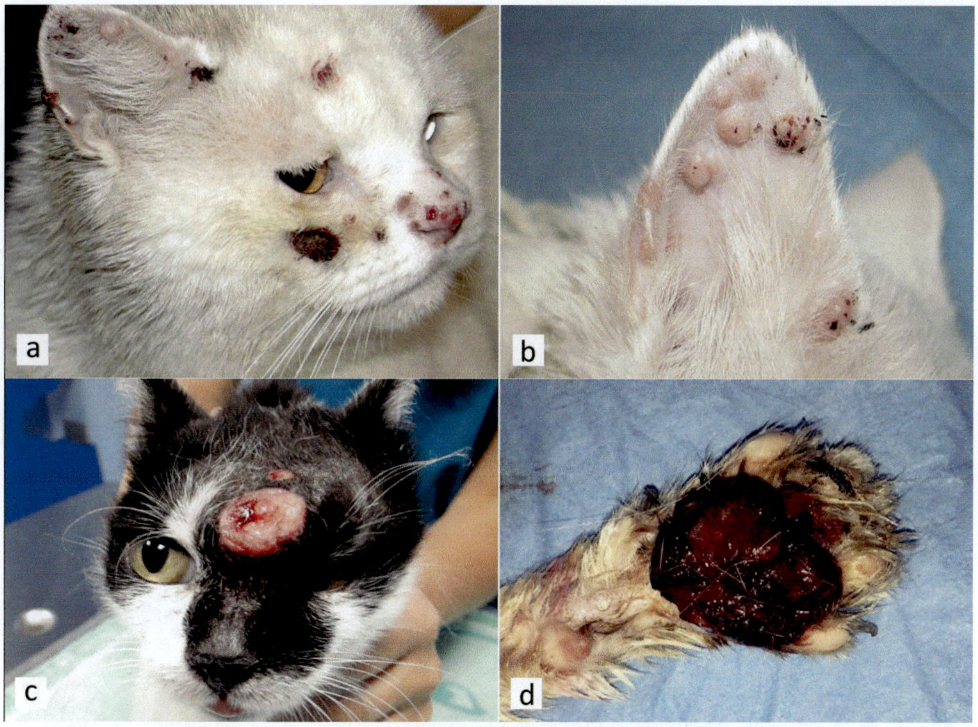

Fig. 6.367 Cryptococcosis: (**a**) multiple nodules, some of them ulcerated on the face, pinna, and nose; (**b**) multiple nodules on the pinna; (**c**) single ulcerated plaque on the head of a cat. (Courtesy of DVM C. Penna—Italy); and (**d**) large ulcerated mass on the paw in cats with cryptococcosis

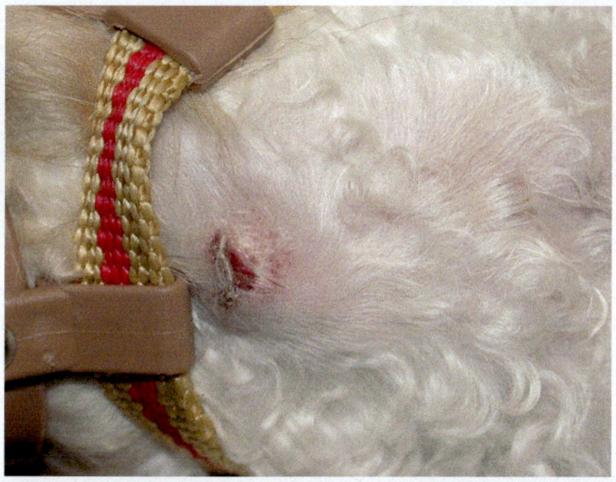

Fig. 6.368 Canine cryptococcosis: single ulcerated nodule on the neck of a Cotton de Tulear with systemic cryptococcosis

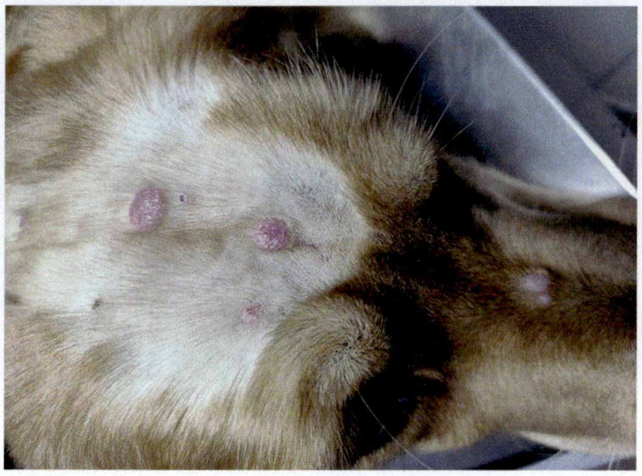

Fig. 6.369 Canine cryptococcosis: multiple nodules on the head in a dog in which an underlying cause was not found. (Courtesy of DVM C. Penna—Italy)

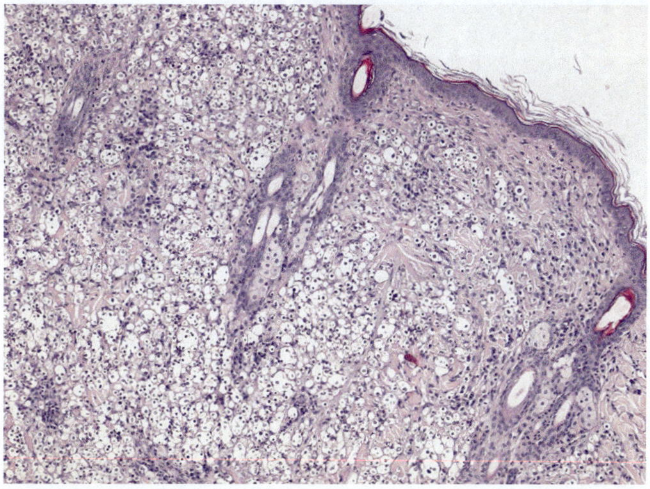

Fig. 6.370 Histology of cryptococcosis: diffuse granulomatous dermatitis with many fungal bodies

Cryptococcus neoformans is a dimorphic fungus that assumes in the tissues a characteristic yeast-like morphology, round or oval, measuring 4–20 μm in diameter, with a round, oval, or elliptical nucleus and a characteristic wide mucopolysaccharide capsule that does not stain with the common rapid Romanowsky-type dyes. This large achromatic pericellular area of variable size is a hallmark of cryptococcosis, together with typical narrow-based unipolar budding,

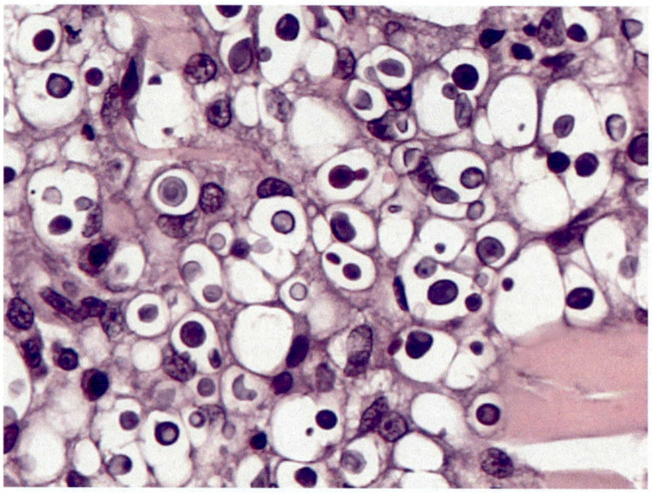

Fig. 6.371 Histology of cryptococcosis: the large achromatic capsule and the narrow-based budding characterize the *Cryptococcus* fungi

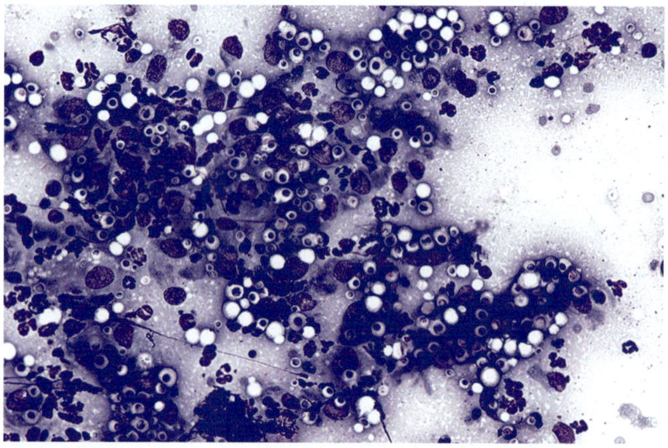

Fig. 6.372 Cytology of cryptococcosis: pyogranulomatous inflammation with many fungal bodies with large achromatic halo

which provides a characteristic *teardrop-like* appearance. Consecutive unipolar budding formations may give rise to more structures similar to short hyphae (pseudo-hyphae) (Figs. 6.372 and 6.373).

The typical narrow-based budding and large achromatic capsule can be highlighted when special stains for fungi are used, such as PAS and Grocott's methenamine silver stains; furthermore, putting some India ink drops on a wet smeared specimen (not air dried) permits highlighting on a black background unstained capsules as a large and clear halo (Figs. 6.374 and 6.375).

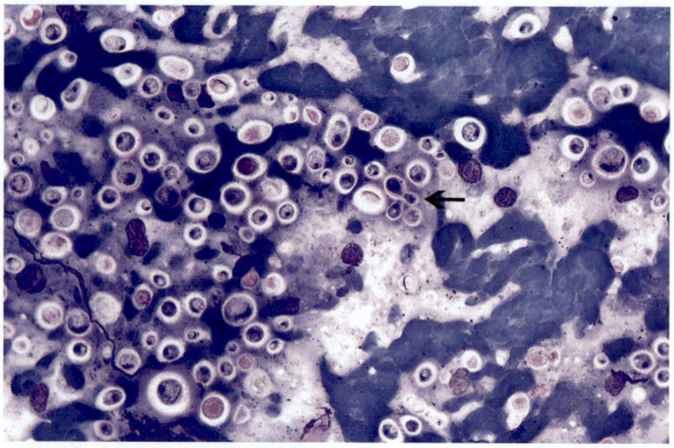

Fig. 6.373 Cytology of cryptococcosis: typical narrow-based budding (arrow)

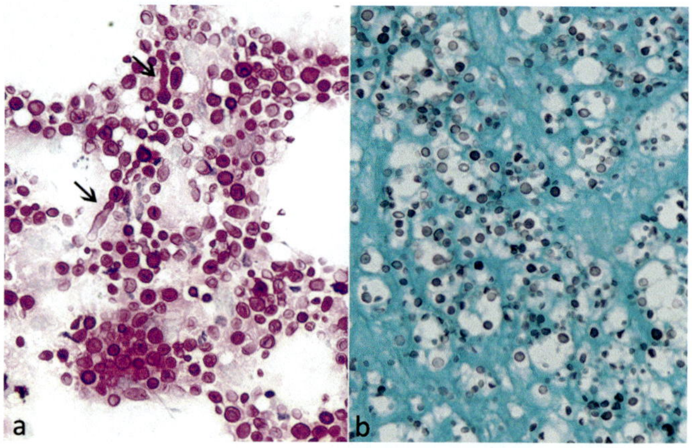

Fig. 6.374 Cytology of cryptococcosis: (**a**) (PAS stain): narrow-based unipolar budding and pseudo-hyphae (arrows) are clearly recognizable and (**b**) (GMS stain) multiple fungal bodies stain black on a green background

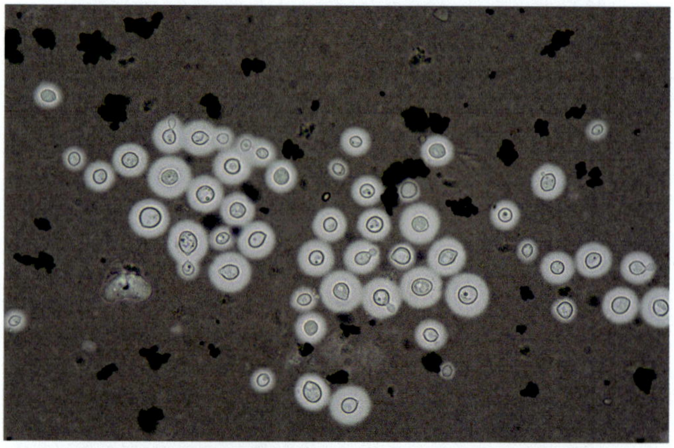

Fig. 6.375 Cytology of cryptococcosis (India ink): with India ink, the capsule of *Cryptococcus* spp. is highlighted as an achromatic peripheral area. Note also that the narrow-based budding is very easily recognizable

6.8.2 Sterile Diseases

This group of diseases includes several skin disorders of different natures, which share the characteristic of not being caused by an infectious microorganism or a parasite. For some diseases belonging to this group, the pathogenesis is clear, whereas it is completely unknown for others.

Hypersensitivity diseases, immune-mediated disorders, metabolic diseases, skin deposits or accumulation of lipids or mineral salts, and traumatic lesions may be the cause of the development of single or multiple nodular lesions observed in dogs and cats.

6.8.2.1 Cutaneous Mineralization

Skin mineralization is characterized by deposits of *calcium salts* in the dermis, epidermis, and rarely in the subcutaneous tissue [252–254].

In addition to *calcinosis cutis,* discussed in papular–nodular lesions and almost exclusively observed in dogs with spontaneous or iatrogenic hypercortisolism, there are other forms of mineralization characterized by single or multiple skin nodules not linked to corticosteroids.

Metastatic calcinosis due to chronic kidney disease (CKD) has been reported in cats, but rarely in dogs [251]. Affected animals develop hypercalcemia by altering the Ca:P ratio, especially when the *Ca:P solubility product* exceeds a value of 70. Lesions are small, firm, irregular, and often ulcerated nodules localized to the paws and between the toes (Fig. 6.376) [251].

Dystrophic calcinosis follows traumatic causes in which the *calcium* deposit occurs in damaged tissue as is thought should happen in *calcinosis circumscripta*. It is commonly observed as a single nodule or mass, especially in young dogs at the pressure points of body areas, such as digits or elbow; probably, the affected areas are the most predisposed to local trauma [253, 255, 256]. German shepherd and Rottweiler dogs seem to be over-represented, but lesions can occasionally develop in any breed (Fig. 6.377). Finally, in *idiopathic cutaneous calcinosis*, skin nodules are reported in patients with no history of trauma and no hypercalcemia.

Due to the nodule composition, a characteristic crunching sensation can be felt by the operator following the passage of the needle into the mineralized tissue during the sampling, especially if calcium salts are well-organized. Furthermore, another macroscopic indicator of the mineral composition of a nodule is the typical whitish and *chalk-like* aspect of the collected material when smeared onto a slide; it can be more fluid or grainy in consistency (Figs. 6.378 and 6.379).

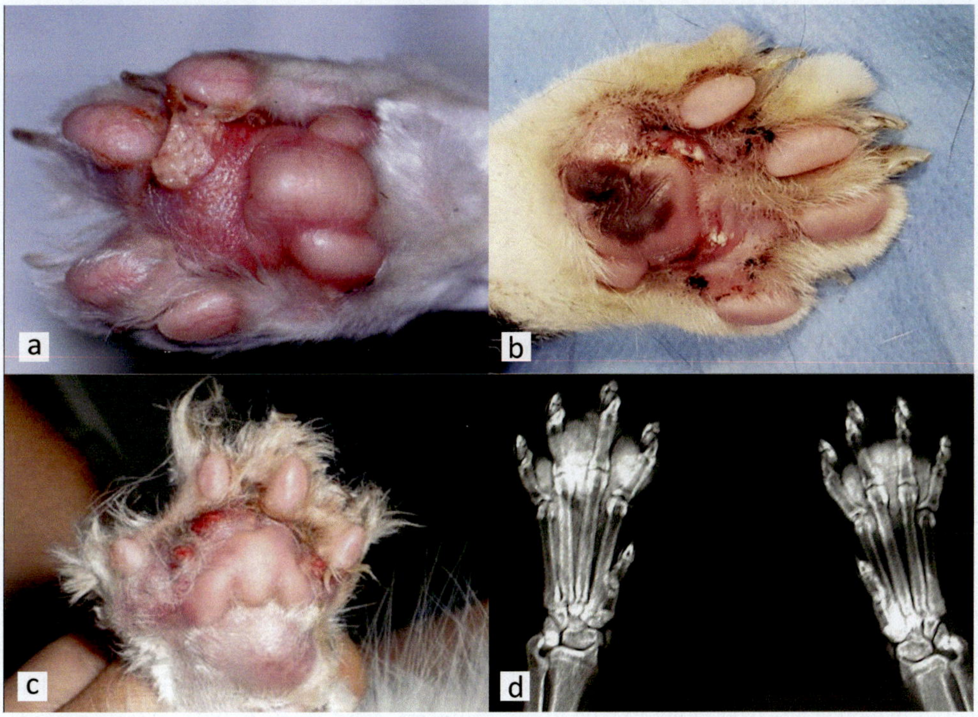

Fig. 6.376 Metastatic mineralization cat: (**a**) interdigital swelling with a calcified nodule in a cat with CKD. (Courtesy of DVM W. Bertazzolo—Italy); (**b**) whitish chalky-like material goes out from ulcerated nodules on the pads; (**c**) swollen of the finger and interdigital spaces infiltrated with mineral salts; and (**d**) radiography of the paws in which multiple calcified interdigital areas are observed. (Courtesy of DVM E. Antoniazzi—Italy)

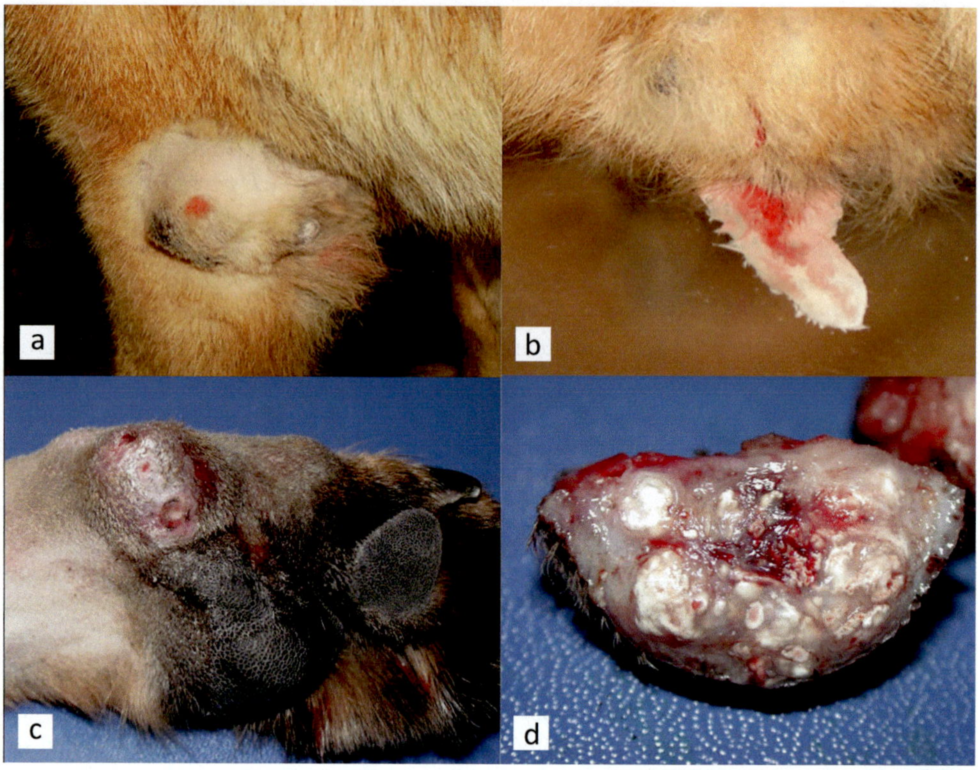

Fig. 6.377 *Calcinosis circumscripta* in dogs: (**a**) large mass on the elbow of a German shepherd dog; (**b**) chalky-like material, mixed to blood, oozes after FNB of the same lesion of the previous image; (**c**) single nodule of *Calcinosis circumscripta* on the pressure point of the paw of a young German shepherd dog; and (**d**) multiple with foci of calcification are recognizable at the section of the surgically removed nodule

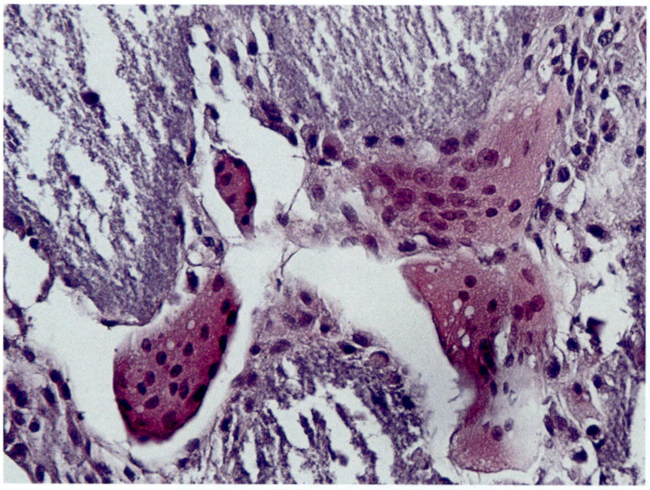

Fig. 6.378 Histology of *Calcinosis circumscripta*: multinucleated giant cells surrounding large foci of mineralization

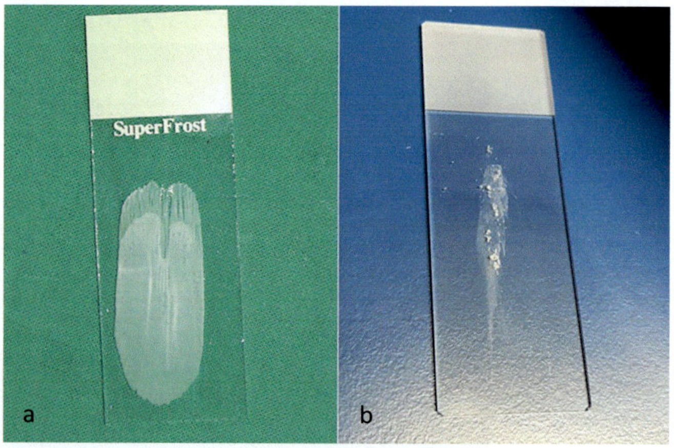

Fig. 6.379 Chalk-like aspect of the material collected from a mineralized nodule: (**a**) fluid appearance and (**b**) grainy appearance

Cytological Findings

Calcium salts are cytologically characterized by amorphous, punctiform, grainy, from irregularly geometric or rhomboidal to needle-like material, which stains pale violet to pale gray-blue, commonly intermingled with transparent mineral materials that do not take color (Figs. 6.380 and 6.381). *Calcium* salts act as a true foreign body; thus, slides from *Calcinosis circumscripta,* as for other calcified lesions (*Calcinosis cutis*), evoke a strong pyogranulomatous inflammation composed of macrophages, epithelioid macrophages, and many multinucleated giant cells that try to eliminate mineral salts (Fig. 6.382). As discussed in Chap. 2, the *calcium* composition of mineral salts can be identified cytologically using Von Kossa staining, which colors black the areas in which *calcium* salts are present (Fig. 6.383).

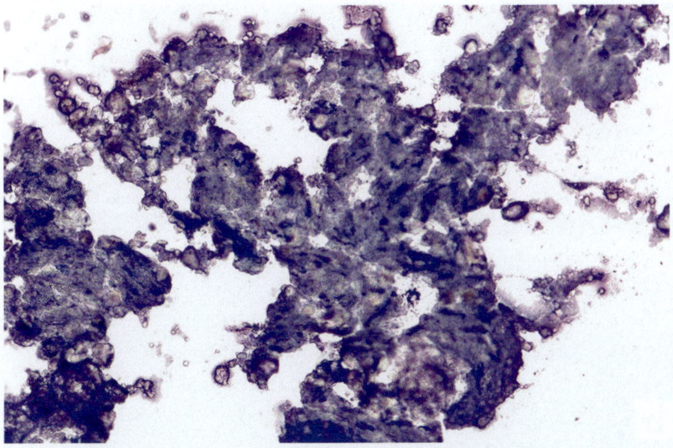

Fig. 6.380 Cytology of *Calcinosis circumscripta*: amorphous, acellular, and granular material representing *calcium* salts

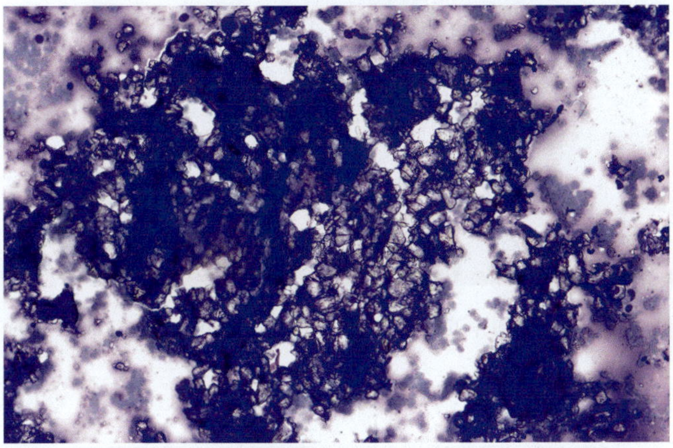

Fig. 6.381 Cytology of *Calcinosis circumscripta*: irregular geometric, acellular, and granular material representing *calcium* salts

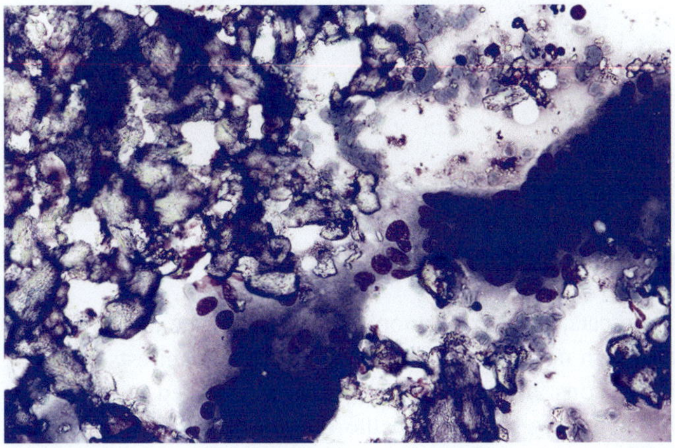

Fig. 6.382 Cytology *Calcinosis circumscripta*: two multinucleated macrophages (giant cells) surrounding amorphous, transparent *calcium* salts

Fig. 6.383 Cytology of *Calcinosis circumscripta* (Von Kossa stain): all the amorphous mineral salts take a black color with Von Kossa staining

6.8.2.2 EG and EP

In cats, EG, EP, and IU are well-recognized skin lesions grouped into the so-called EGC. As these lesions show marked differences in clinical aspects and histopathological findings, the definition of CGE is widely considered obsolete and many dermatologists prefer to use the more generic term *feline eosinophilic dermatitis* [258, 266]. The only aspect shared by these lesions is the presence of an eosinophilic inflammatory pattern [259, 261, 264, 265].

EG clinically manifests with different lesions, but in most cases, the inflammatory cells lead to nodular lesions. EG is very common in cats and rare in dogs.

In cats, a single nodule on the chin, or more precisely, in the area between the lower lip and chin, can develop; when this lesion is large enough, it gives the cat a defined *pouty-face* appearance A linear, cord-like nodular lesion distributed along the rear profile of one or both thighs is well known in cats; due to its configuration, this form of EG is named *linear* EG. The surface of linear granuloma is often covered with scales and crusts, but on palpating the lesion, the presence of small nodules is noticeable and in some cases also grossly evident.

Finally, it is not uncommon to observe nodules and ulcerated plaques in the mouth, especially on the hard palate and tongue. The cause of most EGs is hypersensitivity disease, although idiopathic causes have been reported, especially in young cats, probably due to a temporary genetic alteration. All the lesions belonging to the EGC can be observed simultaneously on the same cat.

EG in dogs is very rare and observed mainly as ulcerated oral nodules and plaques, especially in the Siberian husky and Cavalier King Charles [262, 263] Occasionally, oral EG has been reported in other breeds (Fig. 6.384).

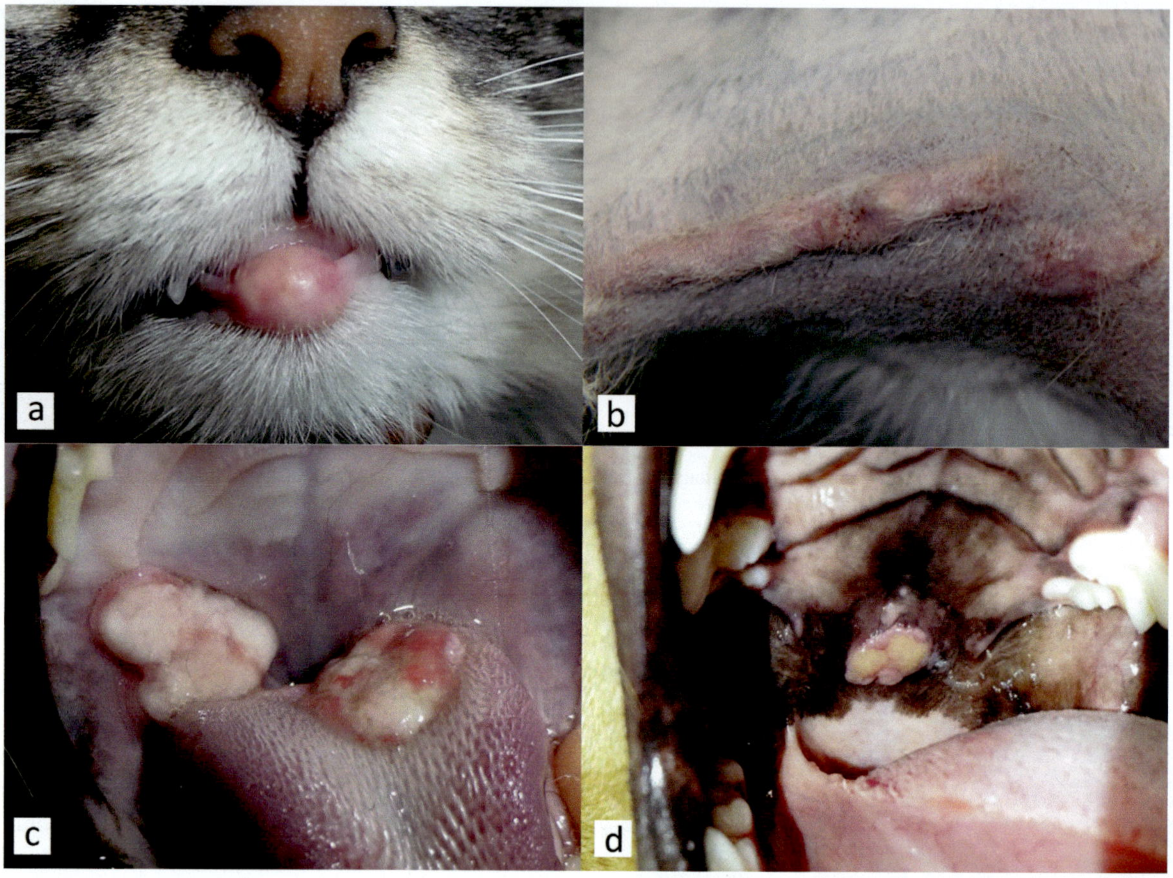

Fig. 6.384 EG in cats: (**a**) alopecic nodule on the chin; (**b**) cord-like configurations on the rear profile of the thigh characterize *linear* EG; (**c**) ulcerated plaques on the tongue and palate of a cat with EG; and (**d**) plaque with typical yellowish surface on the hard palate of a Cavalier king with oral EG

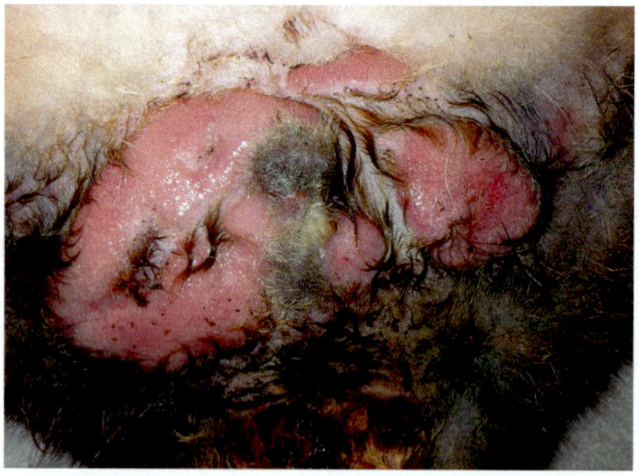

Fig. 6.385 Large and exudative EP on the abdominal skin of an allergic cat

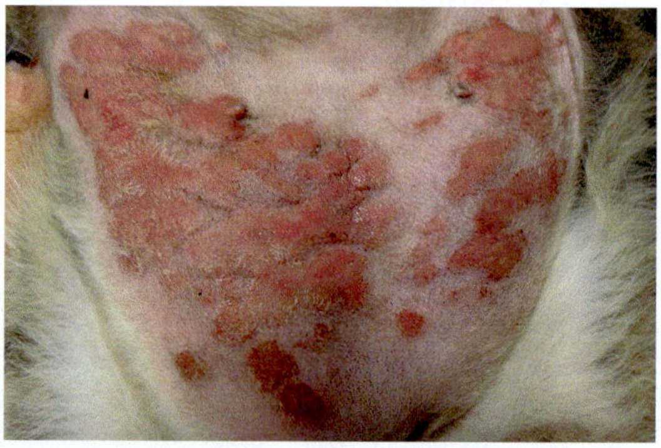

Fig. 6.386 Multiple confluent plaques on the abdomen of a cat with EP

The EP is a common eosinophilic lesion observed in cats. EP is clinically characterized by an erythematous and exudative, single or coalescent, usually pruritic plaque, mainly localized on the abdomen and medial thighs (Figs. 6.385 and 6.386) [259]. As for EG, hypersensitivity diseases are the most common causes of EPs as well [266].

In cats, another eosinophilic lesion can be observed that, although not included in the CGE group, manifests as nodular lesions and is characterized by marked eosinophilic inflammatory infiltrate in the dermis; this clinical entity is defined as *feline perforating dermatitis*.

Perforating dermatoses are a group of rare but well-known diseases in human medicine that include Kyrle's disease, elastosis perforans serpiginosa, perforating folliculitis, and reactive perforating collagenosis (RPC). The latter presents

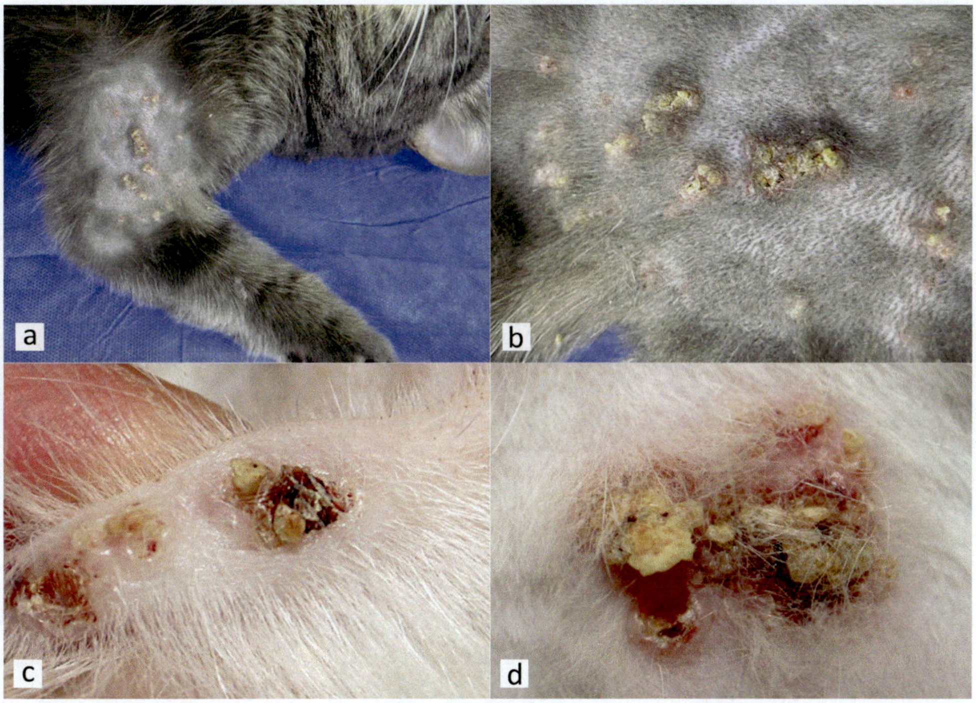

Fig. 6.387 Feline perforating dermatitis: (**a**) multiple small nodules on the axilla; (**b**) closeup of the same cat of the previous image; (**c**) at high magnification, a thick material protruding from a cup-shaped nodular lesion is better viewable; and (**d**) lifting the crust, an ulcer can be observed

with an inherited familiar form, typical of childhood, and an acquired form, mostly occurring in adult patients with underlying diseases such as diabetes mellitus and chronic renal failure [260].

Perforating dermatitis is uncommon in cats and skin lesions have a characteristic morphology, represented by an umbilicated central area from which an amorphous yellow crusty material, tightly adherent to the underlying skin, has extruded; bleeding results if the crust is removed. No spontaneous regression has been reported in cats [257].

Very few cases have been reported in cats, but the pathogenesis is not clear. No cases have been associated with the internal disease as occurs in humans; it might also be possible that perforating dermatitis in cats is a pattern reaction, similar to although much less common than miliary dermatitis or EGC. However, the classification of perforating dermatosis is still controversial and diagnostic criteria are not well defined for all entities (Fig. 6.387).

Cytological Findings

Histopathological findings of EG in cats show diffuse eosinophilic dermatitis in which few to many *flame figures* are evident; the latter are areas in which collagen fibers are trapped in the amorphous eosinophilic material, consisting of the fragmentation and degranulation of eosinophils [257, 258, 260]. Specimens collected from EGs show a very large number of eosinophils and a variable amount of mast cells and macrophages; neutrophils are very few in non-ulcerated lesions. Exceptionally when the needle collects flame figures, some collagen fibers, immersed on a deep blu amorphous material can be observed (Figs. 6.388, 6.389, 6.390, and 6.391).

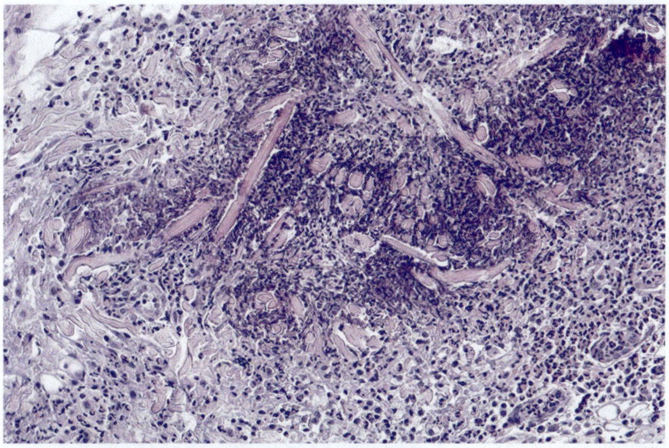

Fig. 6.388 Histology of EG: many *flame figures* immersed in eosinophilic inflammation

Histologically, large areas of *necrotic foci,* composed of large amorphous granular eosinophilic to lightly basophilic material, surrounded by fibroblasts, epithelioid cells, and numerous giant cells, are also observed in histopathological specimens. Unlike flame figures, in necrotic foci, collagen fibers are not usually observed. The same aspects are detectable in cytological specimens where many eosinophils and histiocytic giant cells are arranged around deep blue amorphous and acellular material [265] (Figs. 6.392, 6.393, 6.394, 6.395, and 6.396).

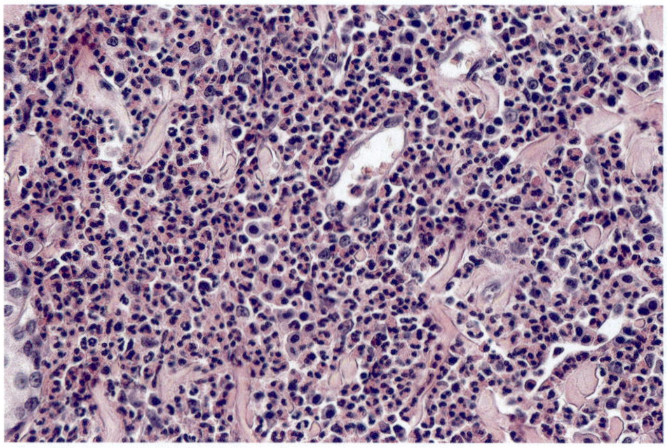

Fig. 6.389 Histology of EG: diffuse eosinophilic infiltrate along with few mast cells and macrophages

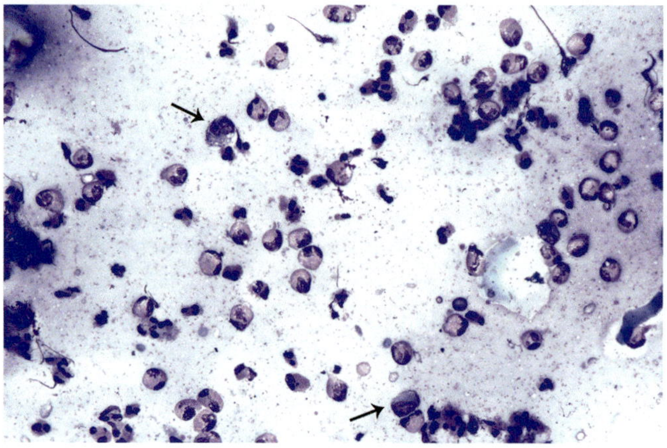

Fig. 6.390 Cytology of EG: many eosinophils and rare macrophages (*arrows*)

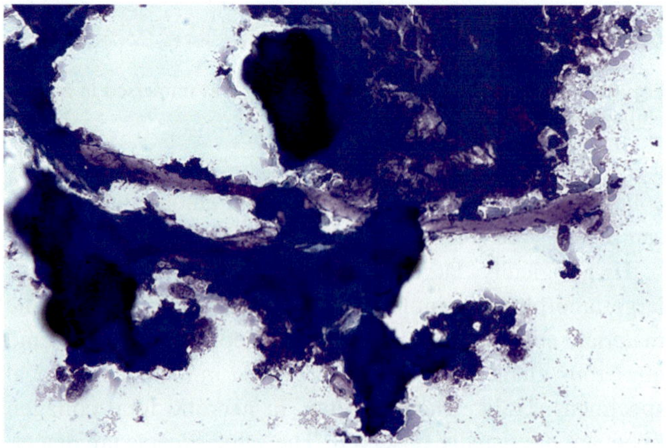

Fig. 6.391 Cytology of *flame figures*: many eosinophils and rare macrophages around basophilic necrotic material that entraps collagen fibers

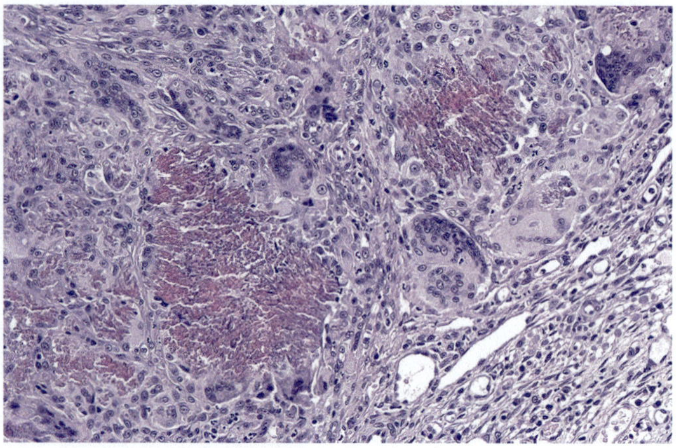

Fig. 6.392 Histology of EG (*necrotic foci*): many giant cells surrounding many areas of amorphous eosinophilic material, consisting of fragmentation and degranulation of eosinophils

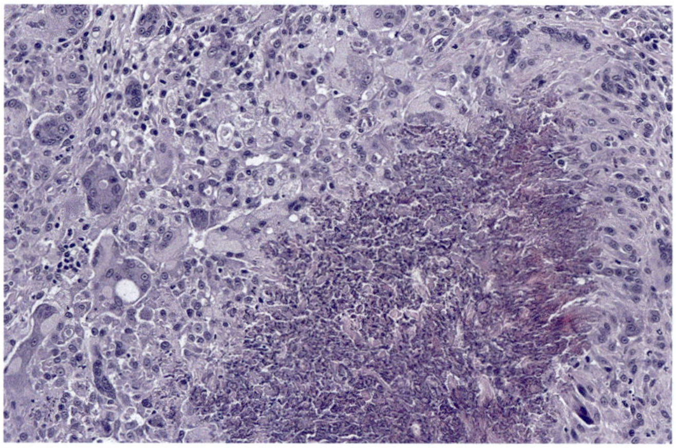

Fig. 6.393 Histology of EG (*necrotic foci*): at high magnification, the material composing the necrotic foci is better viewable. Note that collagen fibers are not present and the high number of giant cells

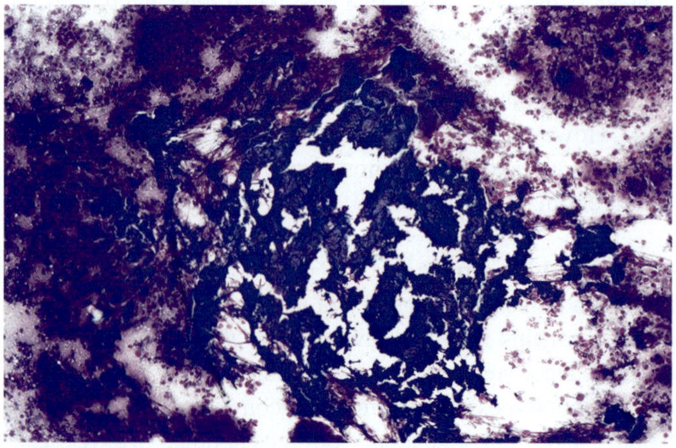

Fig. 6.394 Cytology of EG (*necrotic foci*): the eosinophilic material observed in histology is cytologically recognizable as deep basophilic amorphous material surrounded by an eosinophilic-granulomatous inflammatory pattern

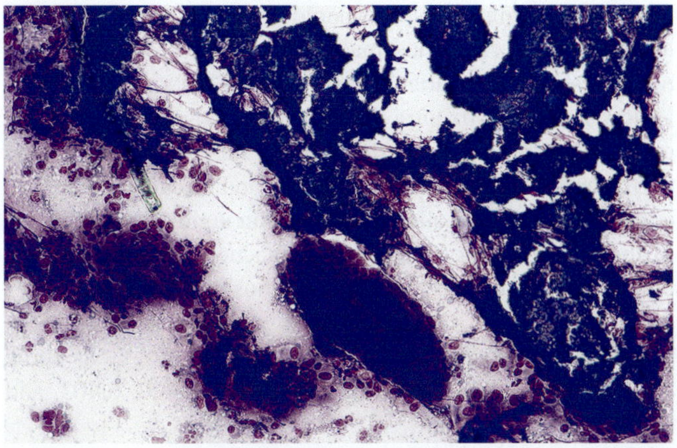

Fig. 6.395 Cytology of EG (*necrotic foci*): large giant macrophages around the necrotic material representing necrotic foci

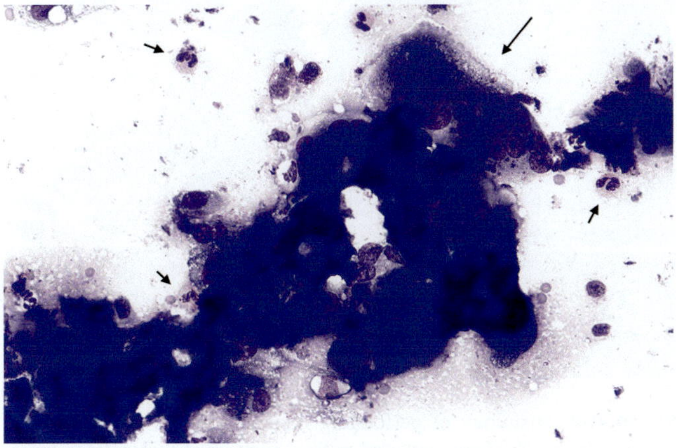

Fig. 6.396 Cytology of EG (*necrotic foci*): at high magnification, giant macrophages (long arrow) attacking necrotic material are better detectable. Note eosinophils (short arrows)

In dogs, as in cats, EG shows the typical eosinophilic inflammatory pattern with a variable number of macrophages, epithelioid macrophages, and giant cells, to confirm the granulomatous infiltrate into the dermis (Fig. 6.397).

EP is histologically characterized by a diffuse dermal eosinophilic infiltrate with a smaller amount of histiocytes, mast cells, and neutrophils, with the latter increased beneath the ulcers [258, 266] (Fig. 6.398). The epidermis is frequently ulcerated and covered in a crust filled with blood, debris, and many aggregates of contaminant bacteria (Fig. 6.399). Cytology obtained via FNB from uncomplicated plaques is composed almost completely of eosinophils and a variable number of mast cells and macrophages (Fig. 6.400). In association with eosinophils, in specimens sampled using the impression smear technique, neutrophils with intracytoplasmic cocci or rod-shaped bacteria also dispersed on the background of the slide are commonly seen; in some

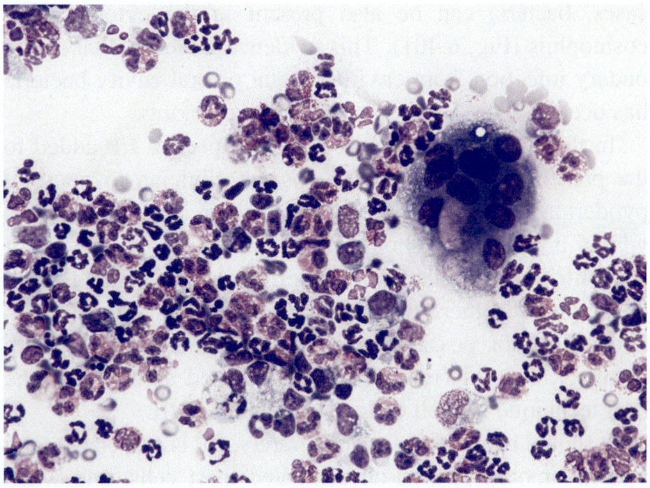

Fig. 6.397 Cytology of eosinophilic furunculosis in dogs: a large multinucleated macrophage along with eosinophils, neutrophils, and macrophages

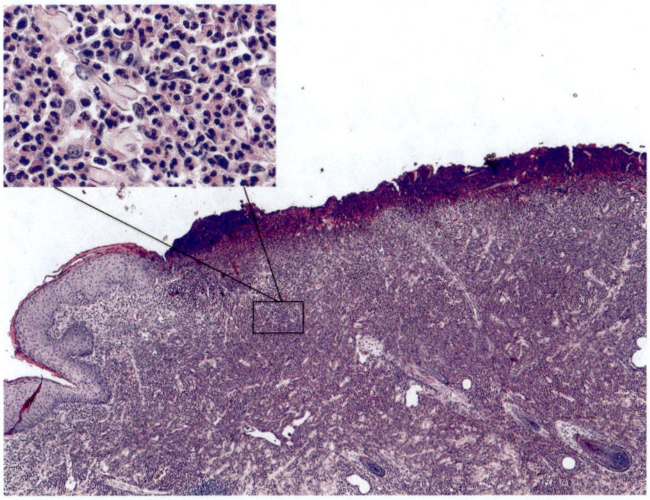

Fig. 6.398 Histology of EP: Large ulcerative lesion with many eosinophils infiltrating the dermis (inset)

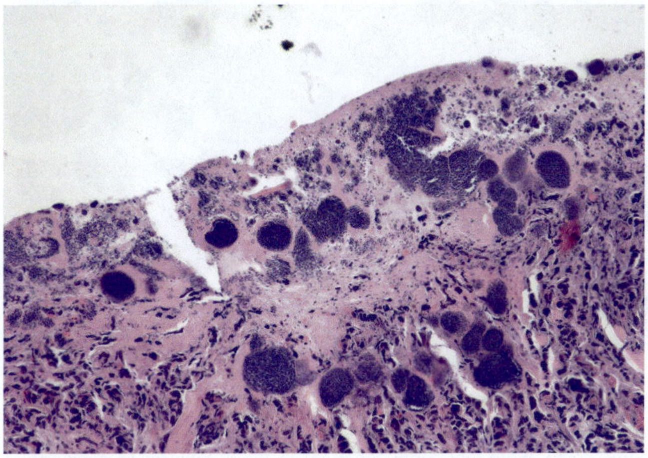

Fig. 6.399 Histology of EP: ulcerated epidermis covered by a crust, which entraps blood, debris, and many aggregates of contaminant bacteria

cases, bacteria can be also present in the cytoplasm of eosinophils (Fig. 6.401). This evidence indicates that a secondary infection from environmental or oral cavity bacteria has occurred, the latter due to excessive licking.

In these cases, the typical clinical aspect of EP added to the presence of eosinophils helps the clinician to interpret pyoderma as secondary. However, there are situations in which a specimen should be exclusively composed of karyolitic neutrophils with bacterial phagocytosis without eosinophils. In these cases, even if there is a clinical suspicion, it is not possible to make a diagnosis of EP with cytology. For this reason, in every raised skin lesion, the best technique for cell collection remains FNB.

It should be remembered that, rarely, EP is associated with a large amount of well-differentiated mast cells, and when these cells are numerous, we should be careful not to formulate a diagnosis of cutaneous mast cells that, in some cats, can notoriously recall a variable number of infiltrating eosinophils.

Regarding *perforating dermatitis*, cytology results depend on the sampling method used; when a superficial sampling is performed, a large amount of keratin and a scarce number of eosinophils can be collected. To obtain significant cytology, lifting the crust that is tenaciously attached to the dermis and placing a slide on the exposed area must collect cells. Slides are not always very cellular and are often very hemocontaminated. In more lucky cases, cytology can show an important number of eosinophils (Figs. 6.402, 6.403, and 6.404).

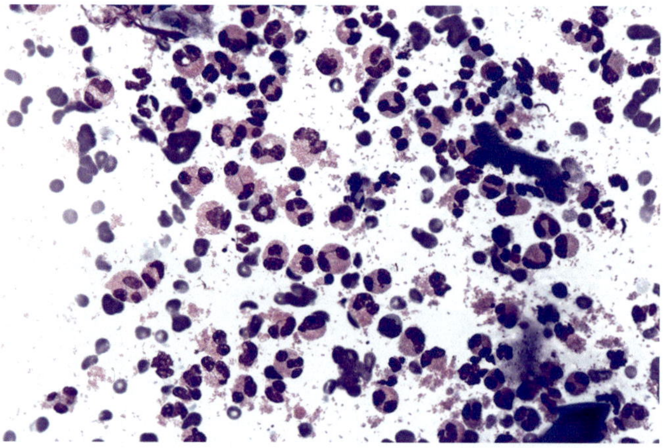

Fig. 6.400 Cytology of EP: pure eosinophilic inflammation collected through FNB

Fig. 6.402 Histology of perforating dermatitis: large necrotic lesion covered with cup-shape, covered with a thick crust that entraps many necrotic eosinophils, corneocytes, and collagen fibers. At the base of necrotic material, a diffuse eosinophilic inflammation is present (*inset*)

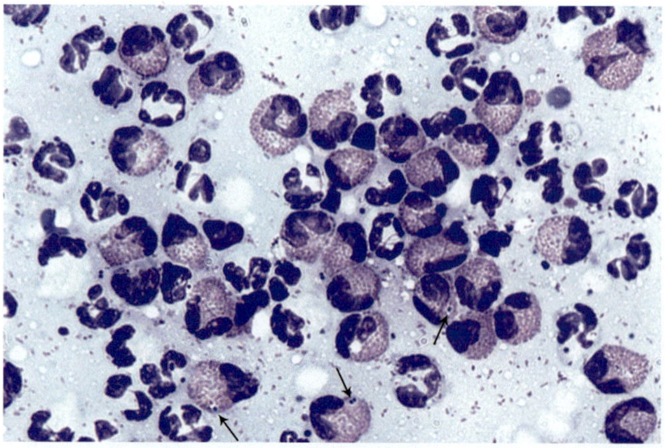

Fig. 6.401 Cytology of EP (imprint technique): mixed inflammation composed of eosinophils and some neutrophils, which phagocytize some cocci. Note that some cocci are in the cytoplasm of eosinophils (arrows)

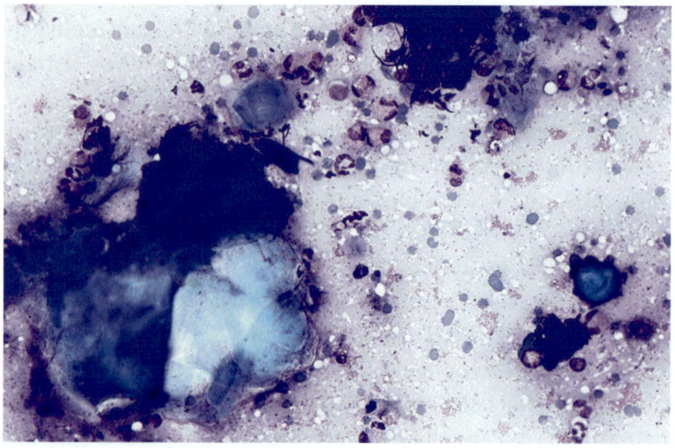

Fig. 6.403 Cytology of perforating dermatitis: corneocytes with only few eosinophils. A lot of granules of eosinophils are spread on the background

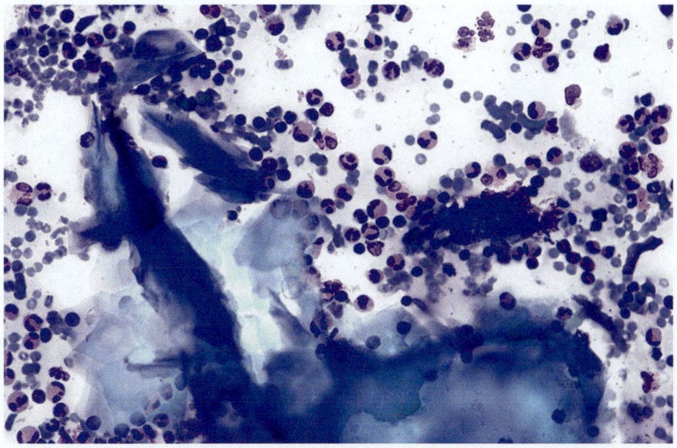

Fig. 6.404 Cytology of perforating dermatitis: many eosinophils and a large group of corneocytes are present

6.8.2.3 Panniculitis

Adipose tissue inflammation is named *panniculitis*. Panniculitis is a multifactorial inflammatory condition of the subcutaneous fat; in dogs and cats, causes are many and include infectious agents (e.g., *Mycobacteria* and *Leishmania infantum*), trauma (post-injecting and foreign bodies), metabolic (pancreatitis), drugs, vaccines, immune-mediated (rheumatoid arthritis and lymphoplasmacytic colitis), and sterile idiopathic diseases [266–268, 270–273, 276].

Clinical signs are single or multiple nodules based on the pathogenesis, localized into the fat, usually painful to palpation, and fistulous draining tracts or ulcerated surfaces. Single nodules are most commonly observed following drug injections such as antibiotics or vaccines, whereas, in the case of multifocal panniculitis, *sterile nodular panniculitis* should be considered. In this uncommon immune-mediated disease, multiple painful and subcutaneous nodules develop on the trunk; lesions tend to ulcerate rapidly and drain a lipid material mixed with blood. In severe cases, the recovered lesions result in permanent scars. Associated with skin lesions, systemic symptoms such as malaise and poor appetite due to hyperthermia are always present [274] (Fig. 6.405).

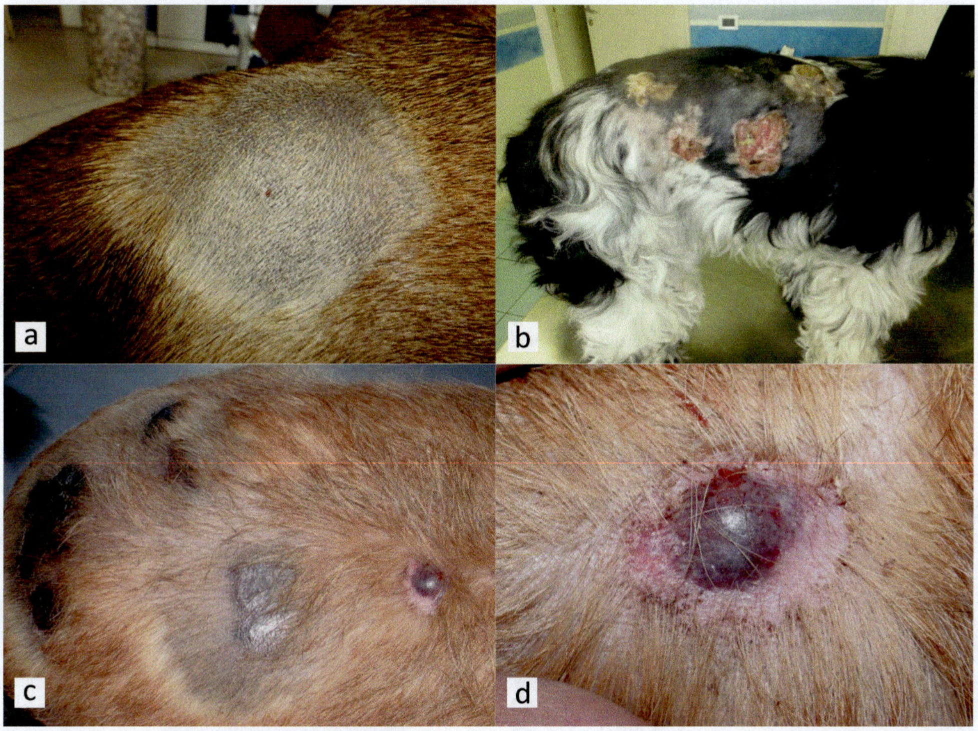

Fig. 6.405 Panniculitis in dogs: (**a**) iatrogenic subcutaneous nodule in a dog (panniculitis) due to an antibiotic injection; (**b**) multiple ulcers with granulation tissue, resulting from fistulization of nodules in a dog with idiopathic sterile immune-mediated panniculitis; (**c**) multiple subcutaneous nodules, some of them ulcerated, in a dog with sterile panniculitis. Note the scar in the recovered areas; and (**d**) closeup of a nodule of panniculitis on the dorsum of the same dog of the previous image

In cats, characteristic panniculitis located at the injection site of rabies vaccine has been well-documented [269].

Cytological Findings

Histopathology of panniculitis depends on the disease from which has been caused. Inflammatory cells are located in the adipose tissue and are mostly represented by macrophages and neutrophils (Fig. 6.406).

Cytologically, at low magnification, it is clearly recognizable as a lipid background; the latter is represented by many lipid droplets, that can be recognized as from round to irregular in shape and different in size achromatic areas. Immersed in this lipid background, many inflammatory cells, usually represented by neutrophils and macrophages, are detected; in chronic cases, lymphocytes and plasma cells can be numerous, as well as fibroblasts in the attempt from the body to repair the damaged tissues [269]. Usually, in most cases, a cytological hallmark is the presence of lipophagocytosis that is cytologically represented by macrophages with cytoplasm filled with micro-droplets of lipid that give cells a *foamy appearance* (Figs. 6.407 and 6.408).

In some cases, necrosis of fat can be observed in the adipose cells, represented by brightly yellowish to orange in color amorphous material. The latter can be phagocytized from the macrophages (Figs. 6.409 and 6.410).

As the lesion becomes chronic, healing processes occur, and a proliferation of reactive fibroblasts and myofibroblasts occurs. These reactive spindle cells may assume "atypical" morphological aspects that mimic the alterations observed in mesenchymal spindle cell neoplasms of soft tissues (Figs. 6.411 and 6.412).

In the so-named *post-injection panniculitis*, the cytological features can be very different from the above-described panniculitis; in fact, histopathology is characterized by a

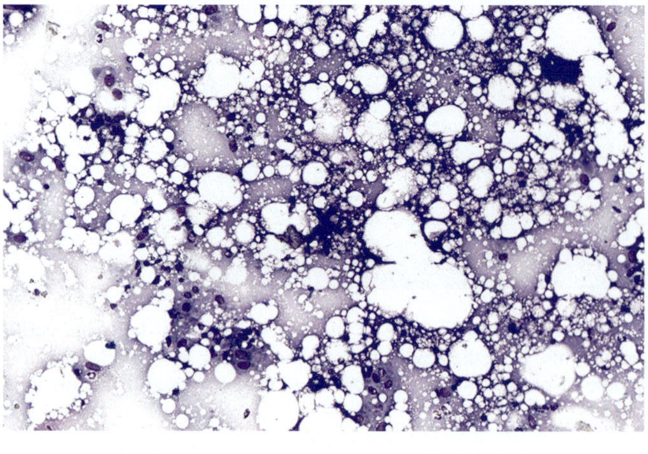

Fig. 6.406 Cytology of panniculitis: on a protein and basophilic background, many lipid droplets of different sizes along with a few inflammatory cells

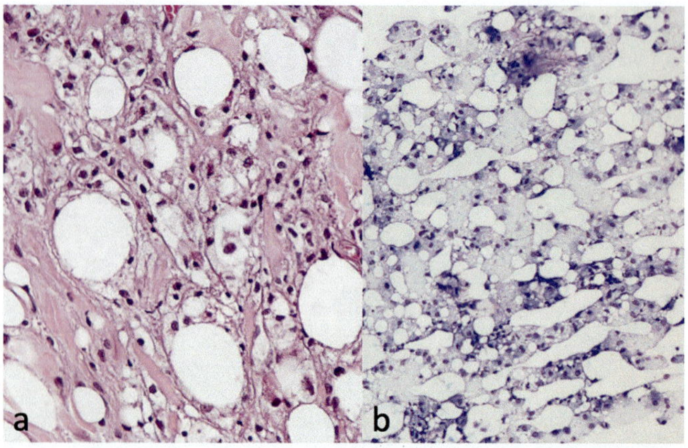

Fig. 6.407 Panniculitis: (**a**) histology: globules of fat along with a macrophagic inflammation. Note the vacuolar appearance of the cytoplasm of histiocytic cells (lipophagocytosis); and (**b**) cytology: on a lipid background many lipid macrophages are recognizable (lipophagocytosis)

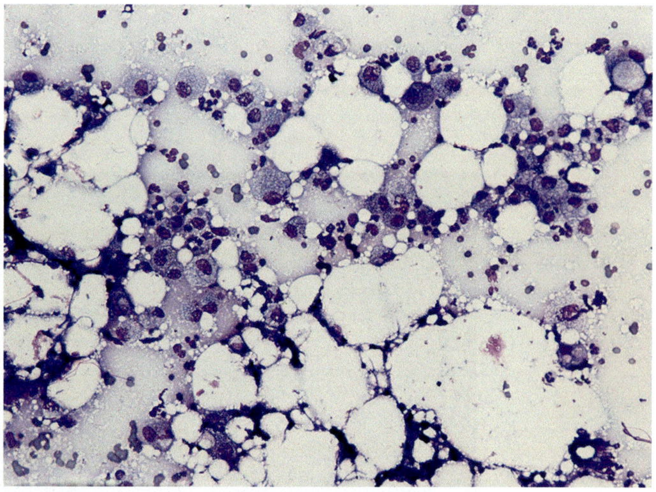

Fig. 6.408 Cytology of panniculitis: globules of fat along with a macrophagic inflammation. Note the vacuolar appearance of the cytoplasm of histiocytic cells (lipophagocytosis)

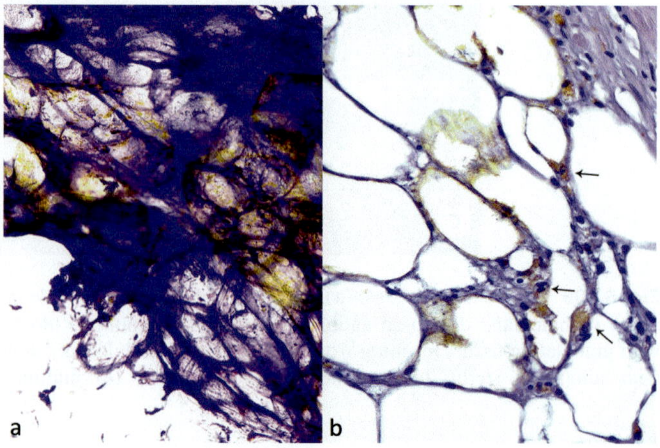

Fig. 6.409 Panniculitis: (**a**) cytology: necrotic adipocytes with a large amount of orange in color necrotic fat and (**b**) histology of panniculitis: necrotic fat is represented by orange amorphous material. Note that macrophages phagocytize it (arrow)

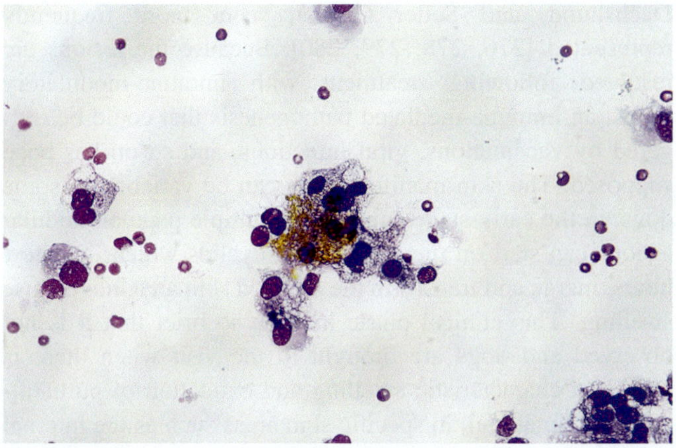

Fig. 6.410 Cytology of panniculitis: foamy macrophages filled in lipid and in phagocytosis of necrotic fat (orange amorphous intracytoplasmic material)

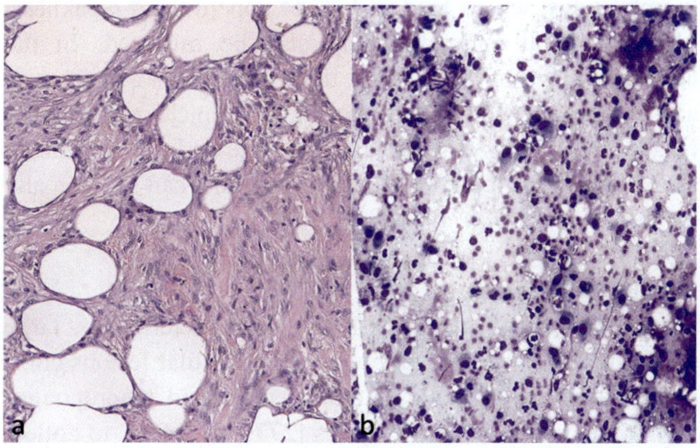

Fig. 6.411 Fibrous panniculitis: (**a**) histology of panniculitis with severe fibroplasia: many spindle reactive fibro/myofibroblasts are recognizable in the adipose tissue, and (**b**) on a lipid and hematic background, a mixed inflammation with many reactive fibro/myofibroblasts is observed

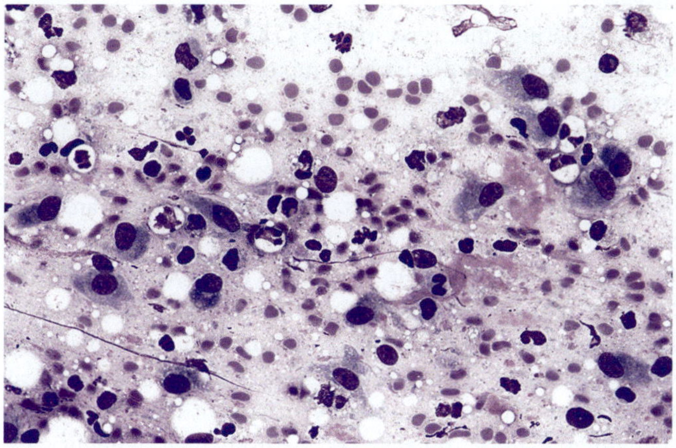

Fig. 6.412 Cytology of fibrous panniculitis: high magnification in which many reactive fibro/myofibroblasts are present

marked central necrotic area, surrounded by many inflammatory cells that delimit peripherally the lesion like a pseudocapsule.

Cytology is characterized by a variable amount of basophilic amorphous necrotic material along with inflammatory cells, mainly represented by lymphocytes, histiocytes, and plasma cells. If the needle has been inserted into the center of the large necrotic area, the lipid material that characterizes the cytological findings of all the types of panniculitis cannot be observed and only necrosis can be observed (Figs. 6.413 and 6.414).

In cats, a similar situation can be secondary to rabies vaccination; a high amount of necrotic debris, surrounded by strong granulomatous inflammation rich in giant cells, is characteristically observed in cytology. In these cases, the history of a previous anti-rabies vaccination can orientate the diagnosis (Fig. 6.415).

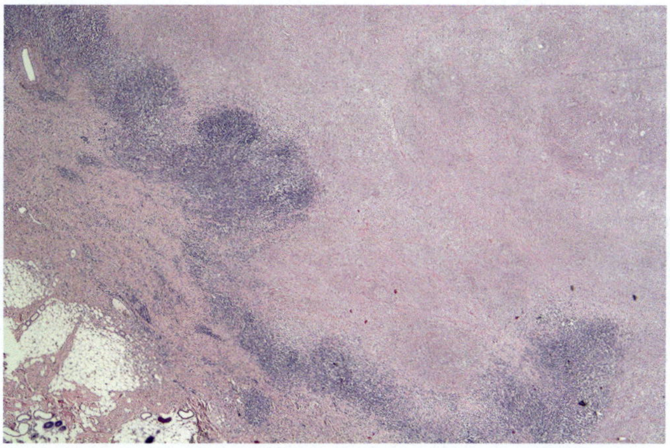

Fig. 6.413 Histopathology post-injection panniculitis: a peripheral line of inflammatory cells (dark blue) delimits a central necrotic area (pink)

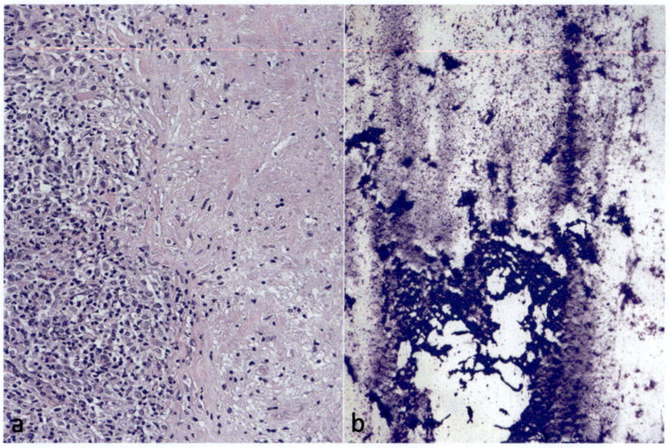

Fig. 6.414 Post-injection panniculitis: (**a**) histopathology: a peripheral lympho-histiocytic inflammation delimits a necrotic eosinophilic amorphous material and (**b**) the slides are composed of only necrotic background with only scattered bare nuclei of inflammatory cells

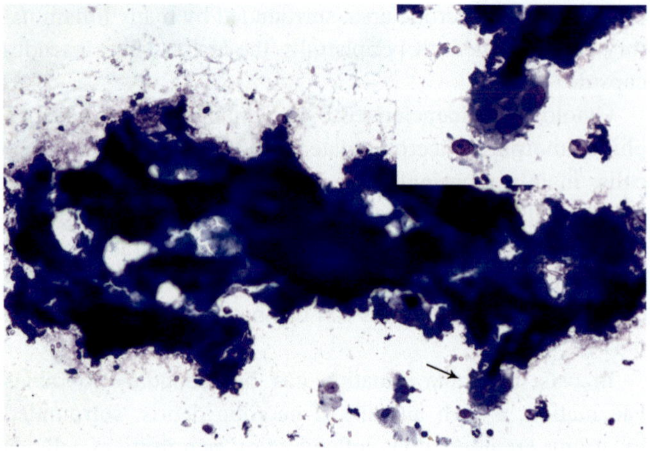

Fig. 6.415 Cytology post-rabies vaccination panniculitis: necrotic amorphous material surrounded by a granulomatous reaction represented mostly by giant cells (arrow, inset)

Since many infectious agents can cause panniculitis, such as leishmania, opportunistic mycobacteria, fungi, and so on, if they are not detected with cytology, before defining the panniculitis as sterile, microbiological culture, special histochemical stains, and molecular biology tests must be performed. Finally, in animals with diffuse panniculitis, a careful check for metabolic disorders or neoplasia must be carried out (pancreatitis or pancreatic carcinoma).

6.9 Other Sterile Diseases

In this group, two diseases have been included that, because of the typology of clinical lesions, are difficult to include in the groups dealt with above. Lesions are not typically only nodular and can change the clinical aspect over time.

6.9.1 Sterile Granulomatous Dermatitis and Panniculitis and Lymphadenitis (Juvenile Cellulitis)

Juvenile cellulitis, as the name given to the disease defines, is an immune-mediated, sterile disease of unknown cause that mostly affects puppies from a few weeks of age. A recent retrospective study on 90 adult dogs demonstrated that these dogs showed, from a clinical, pathological, and behavioral point of view, the same lesions observed in puppies with juvenile cellulitis [275]. In light of this report, the term juvenile cellulitis should be reevaluated in favor of a more generic definition such as *sterile granulomatous dermatitis and lymphadenitis*. Certain breeds such as Cocker spaniel, Dachshund, and Setter Gordon seem more frequently represented [276, 278, 279, 280]. Because the lesions are resolved following treatment with immune-modulatory drugs, an immune-mediated pathogenesis that could be triggered by vaccinations, viral infections, and so on has been supposed. The skin manifestations can be variable; in some dogs, in the early stage, small and multiple papular–nodular lesions can start on the muzzle and eyelids which, in a few hours, merge and transform the affected skin area into diffuse swelling. This clinical phase may be so brief that it is not observed and dogs are brought to the visit when there is already a characteristic swelling and exudation of purulent-appearing material, in specific skin areas such as the internal surface of the pinnae, the eyelids, the muzzle, and the bridge of the nose; these lesions tend to rapidly fistulate covering the skin with purulent exudate (Fig. 6.416). In this stage of the disease, skin lesions are usually associated with general malaise, fever, joint pain, and enlarged and severely painful lymph nodes, which lead to loss of appetite and weakness. Mucocutaneous junctions may also be involved. In the advanced stage of the disease, deep ulcerations and secondary pyoderma can modify the initial clinical presentation (Fig. 6.417). Some dogs develop multiple nodules of variable size, mainly distributed on the trunk but that can grow in any cutaneous areas (Fig. 6.418).

Cytological Findings

Histopathology of juvenile cellulitis is characterized by large perifollicular sterile pyogranulomatous nodular lesions composed of macrophages and many epithelioid macrophages with a central core of neutrophils [277]; therefore, to collect cells that are as representative as possible of the disease, a needle must be inserted into the dermis with the attempt to collect cells directly from granuloma, avoiding so the superficial bacterial contamination (Figs. 6.419 and 6.420). Slides obtained from dogs with juvenile cellulitis may vary according to the sampling method and the stage of the disease.

Considering the type of skin lesions represented by draining tracts that discharge a mixed hematic and purulent exudate that undergoes bacterial contamination, sampling cells by imprint technique is not the ideal way to sampling.

In good-quality specimens, cytological findings are very suggestive and if evaluated in the light of the clinical signs, the diagnosis of juvenile cellulitis could be made. Well-segmented neutrophils, macrophages with variably vacuolated cytoplasm, and many epithelioid macrophages, organized singularly or in large groups, compose the typical pyogranulomatous inflammatory pattern of this disease. As the inflammatory process is perifollicular without the

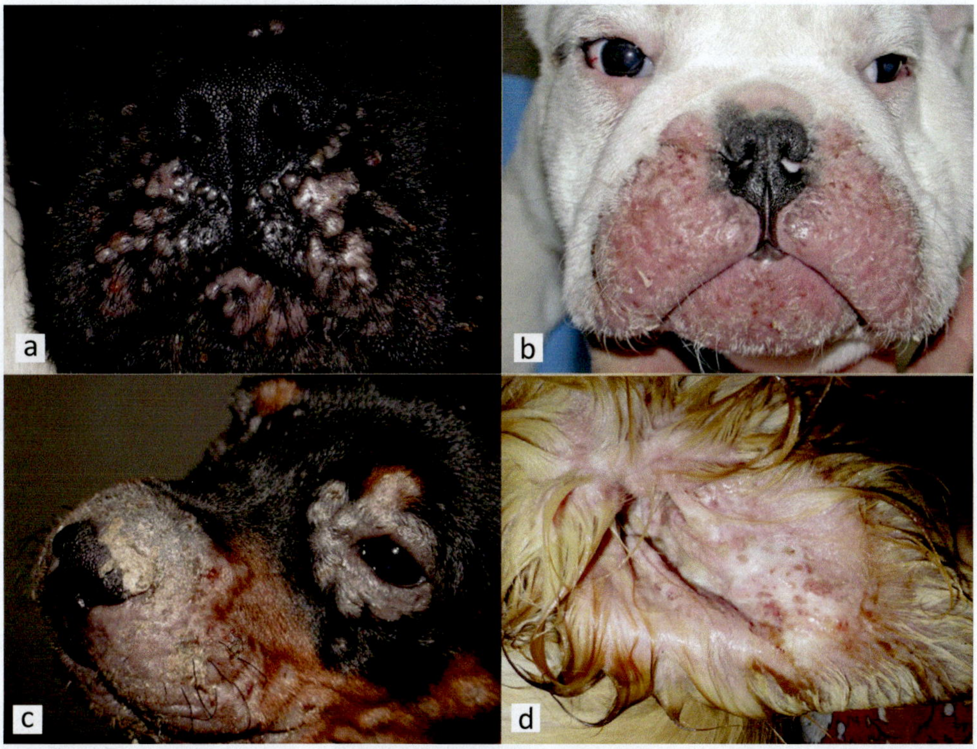

Fig. 6.416 Sterile granulomatous dermatitis and lymphadenitis (Juvenile cellulitis): (**a**) early stage: multiple confluent papular–nodular lesions on the muzzle of a puppy; (**b**) swelling of the muzzle and eyelids in a young English Bulldog; (**c**) blepharitis and swelling of the muzzle; and (**d**) purulent-like exudate on the inner surface of the pinna

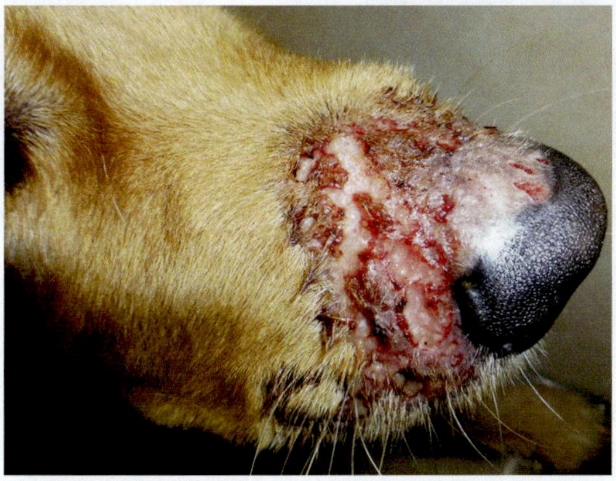

Fig. 6.417 Sterile granulomatous dermatitis and lymphadenitis: ulcerative dermatitis in a young Golden Retriever with chronic disease

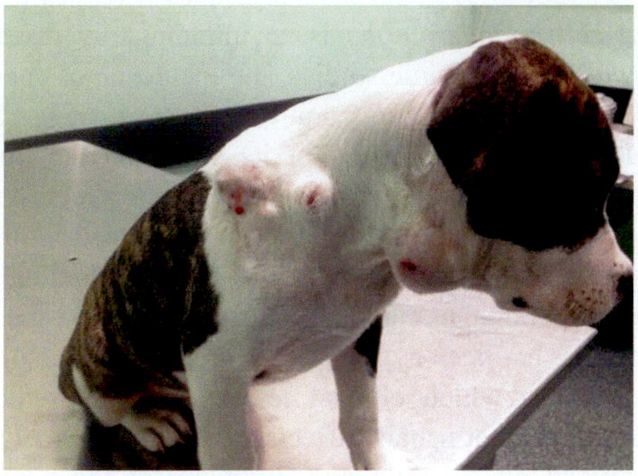

Fig. 6.418 Multiple cutaneous nodules in a young Pitt bull with juvenile cellulitis. (Courtesy of DVM A. Rauseo—Italy)

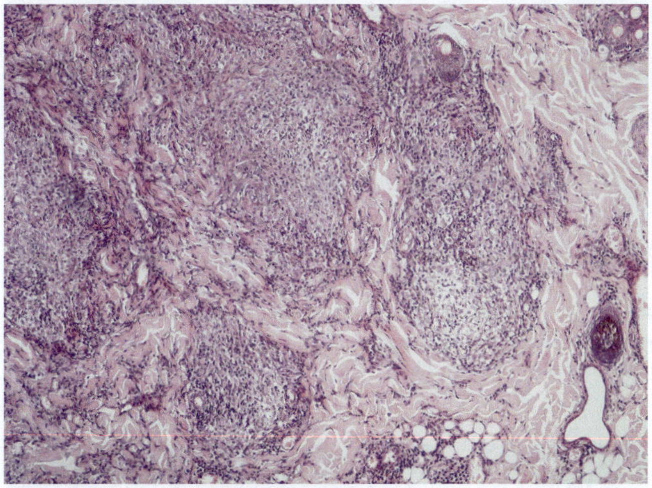

Fig. 6.419 Histology of sterile granulomatous dermatitis and panniculitis: multiple perifollicular granulomas

involvement of the follicles, the possibility of finding giant cells is very low; when the latter are clearly present, secondary furunculosis must be suspected (Figs. 6.421, 6.422, and 6.423). The cytology of nodular skin lesions, especially those present for a long time, differs from that described above; in fact, the number of neutrophils increases, many of which can show degenerative change linked to age, the macrophages are more vacuolated, and epithelioid cells are less numerous (Fig. 6.424).

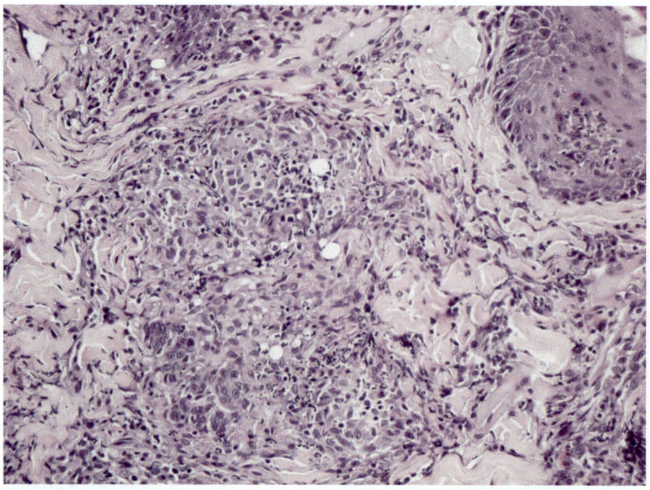

Fig. 6.420 Histology of sterile granulomatous dermatitis and panniculitis: at high magnification, nodules are composed mostly of histiocytes and of a variable number of neutrophils

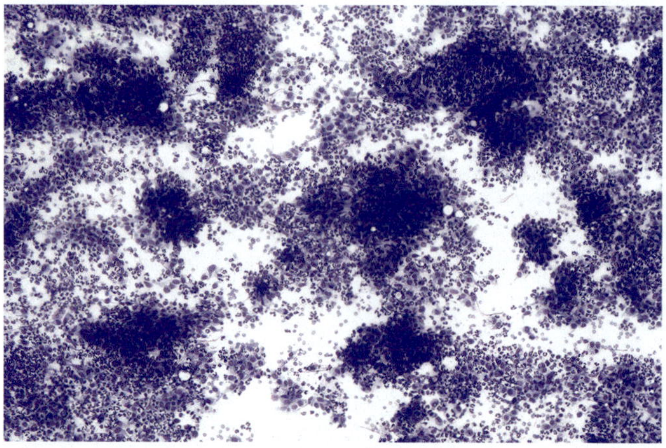

Fig. 6.421 Cytology of sterile granulomatous dermatitis and panniculitis: pyogranulomatous inflammation in which many granulomas are recognizable

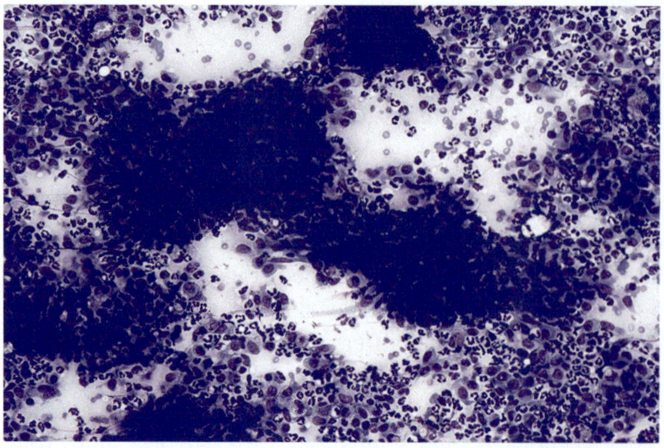

Fig. 6.422 Cytology of sterile granulomatous dermatitis and panniculitis: the large groups of overlapped epithelioid macrophages represent granulomas. Note that neutrophils are well segmented

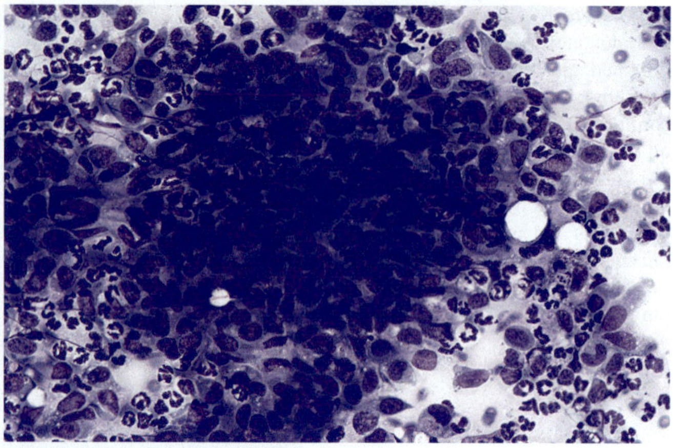

Fig. 6.423 Cytology of sterile granulomatous dermatitis and panniculitis: cytological representation of a granuloma. Many overlapped epithelioid macrophages are easily detectable at this magnification

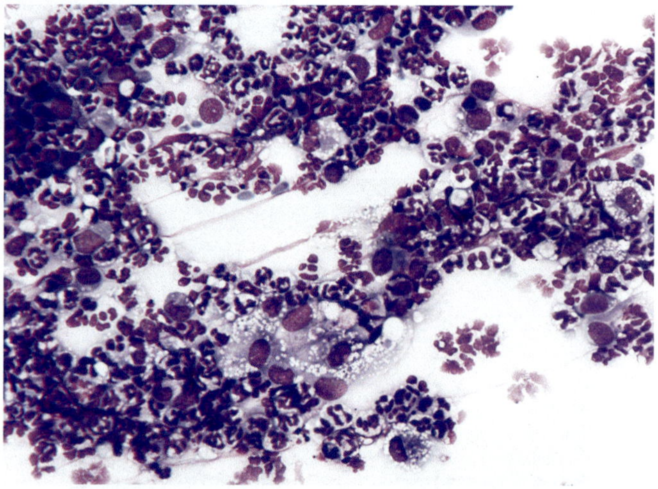

Fig. 6.424 Cytology of sterile granulomatous dermatitis and panniculitis: segmented neutrophils and vacuolated macrophages

6.9.2 Plasma Cell Pododermatitis

Plasma cell pododermatitis is an inflammatory disease of unknown cause. In some studies, the incidence of concomitant FIV infection ranged from 44% to 62%, suggesting a possible action of this virus or its effect on the immune system; the link with retrovirus infection was speculated but never demonstrated [282]. Since the widespread infiltration of plasma cells, the hypergammaglobulinemia, and the good response to immune modulatory drugs, immune-mediated pathogenesis has been hypnotized. This disease affects exclusively cats, in which clinical lesions are represented by pododermatitis that usually affects multiple to all feet and is characterized by swelling and erythema of the paw pads

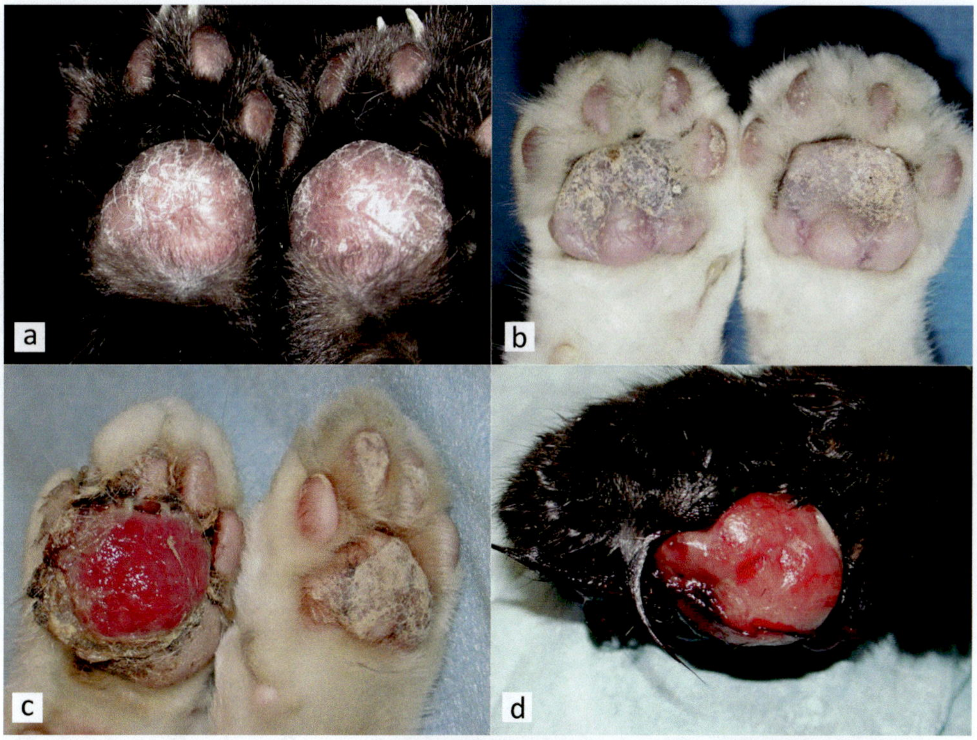

Fig. 6.425 Plasma cell pododermatitis: (**a**) swelling and erythema of the pads; (**b**) soft pads, with loss of elasticity and a "deflated ball" appearance; (**c**) lesions that mainly affect a single pad; and (**d**) severe ulceration with the presence of a bleeding lesion with a fleshy appearance

[283, 286–288]. In the early stage, paws are swollen and painful; the cat can show lameness; with chronicity, the skin of the pads could ulcerate, and an erythematous and hemorrhagic *fleshy* in appearance tissue can protrude from it. In the advanced stage, the consistency of the pads is pasty, appears soft, and apparently emptied of their contents with the permanence of a pit to pressure (deflated balloon appearance) (Fig. 6.425). In some cats, swelling of the dorsum of the nose is also present and characterized histopathologically by a diffuse plasmacellular infiltrate in the dermis; this particular type of presentation can be associated with pododermatitis or presented as only a clinical sign [281, 284] (Fig. 6.426).

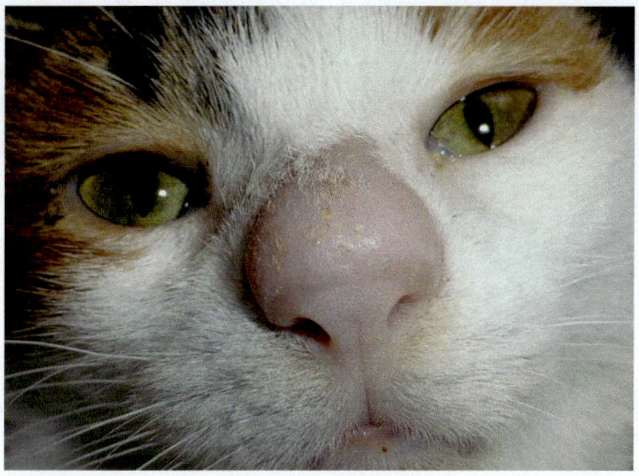

Fig. 6.426 Swelling of the nose in a cat with concomitant plasma cell pododermatitis

Cytological Findings

Cytology obtained from non-ulcerated lesions should be the most representative and reflect the typical plasma cell inflammation that characterizes the histopathological samples [285] (Figs. 6.427 and 6.428). Unfortunately, slides are almost always poorly cellular and only in very successful specimens; mature lymphocytes and a variable amount of plasma cells can be collected (Figs. 6.429, 6.430, and 6.431). In some cases, *Mott cells*, representing hyper-activated plasma cells, filled with large and irregular vacuoles named *Russell bodies* (immunoglobulin aggregates), are also detected. Finally, specimens from ulcerated lesions can be infected and rich with neutrophils, macrophages, and a scarce number of eosinophils.

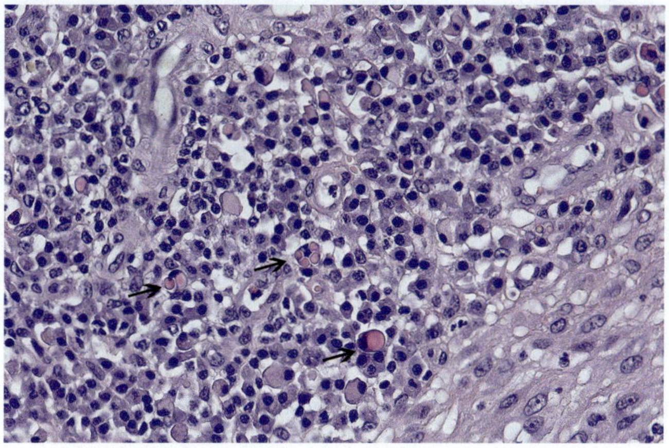

Fig. 6.427 Histology of plasma cell pododermatitis: diffuse plasma cells inflammation. Note how many Mott cells are present (*arrows*)

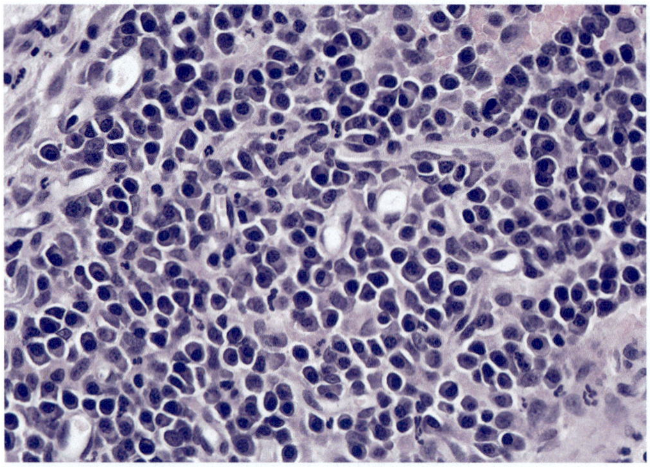

Fig. 6.428 Histology of plasma cell pododermatitis: the typical silhouette of plasma cells is clearly recognizable

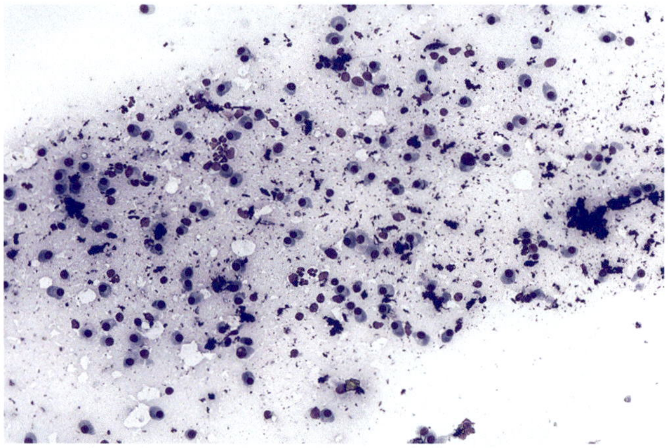

Fig. 6.429 Cytology of plasma cell pododermatitis: many plasma cells

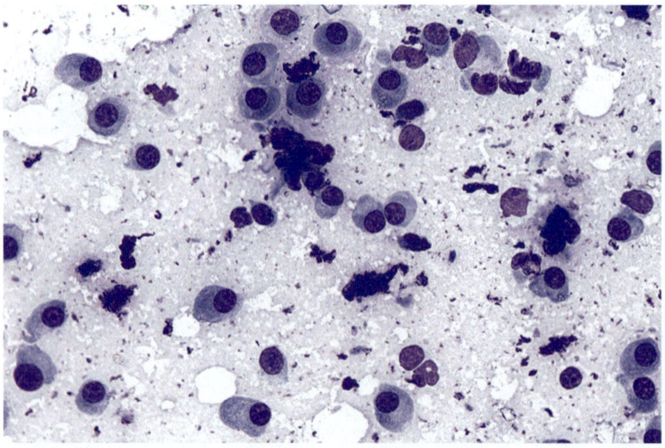

Fig. 6.430 Cytology plasma cell pododermatitis: at high magnification, well-differentiated plasma cells are easily recognizable

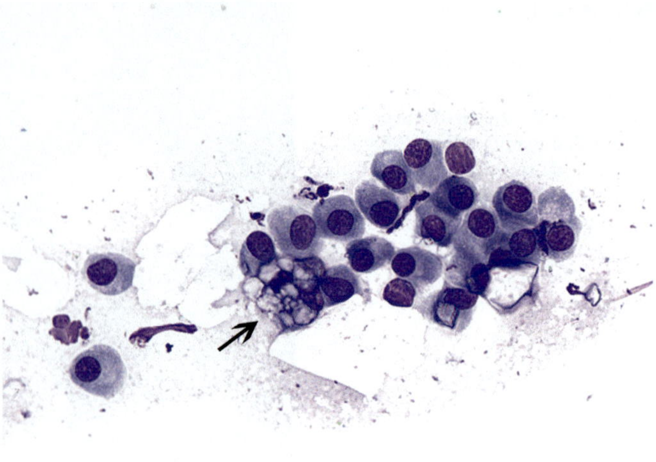

Fig. 6.431 Cytology plasma cell pododermatitis: two Mott cells filled with Russel bodies are recognizable (arrow)

References

Papular Diseases in Dogs

Pyoderma

1. Ihrke PJ (1996) Bacterial skin disease in the dog: a guide to canine pyoderma. Veterinary Learning Systems, Trenton, pp 1–97
2. Miller WH, Griffin CE, Campbell KL (2013) Superficial bacterial fulliculitis. In: Muller & Kirk's- small animal dermatology, 7th edn. Elsevier, St. Louis, pp 194–195

Papular Diseases in Cats

Hypersensitivity Diseases

3. Favrot C, Steffan J, Seewald W et al (2011) Establishment of diagnostic criteria for feline nonflea-induced hypersensitivity dermatitis. Vet Dermatol 23:45–51
4. Hobi S, Linek M, Marignac G (2011) Clinical characteristics and causes of pruritus in cats: a multicentre study on feline hypersensitivity-associated dermatoses. Vet Dermatol 22(5): 406–413
5. Miller WH, Griffin CE, Campbell KL (2013) Feline mosquito hypersensitivity. In: Muller & Kirk's- small animal dermatology, 7th edn. Elsevier, St. Louis, pp 412–414
6. Nagata M, Ishida T (1997) Cutaneous reactivity to mosquito bites and its antigens in cats. Vet Dermatol 8(1):19–26
7. Ravens PA, Xu BJ, Vogelnest LJ (2014) Feline atopic dermatitis: a retrospective study of 45 cases (2001–2012). Vet Dermatol 25:95–104
8. Santoro D, Pucheu-Haston CM, Prost C, Mueller RS, Jackson H (2021) Clinical signs and diagnosis of feline atopic syndrome: detailed guidelines for a correct diagnosis. Vet Dermatol 32:26–e6
9. Sakakibara Y, Wada T, Muraoka M et al (2015) Basophil activation by mosquito extracts in patients with hypersensitivity to mosquito bites. Cancer Sci 106(8):965–971

Benign Papular Mastocytic Hyperplasia

10. Colombo S, Scarampella F, Ordeix L et al (2012) Dermatophytosis and papular eosinophilic/mastocytic dermatitis (urticaria pigmentosa-like dermatitis) in three Devon Rex cats. J Feline Med Surg 14(7):498–502
11. Ngo J, Morren MA, Bodemer C, Heimann M, Fontaine J (2018) Feline maculopapular cutaneous mastocytosis: a retrospective study of 13 cases and proposal for a new classification. JMFS:1–11
12. Noli C, Colombo S, Abramo F et al (2004) Papular eosinophilic/mastocytic dermatitis (feline urticaria pigmentosa) in Devon Rex cats: a distinct disease entity or a histopathological reaction pattern? Vet Dermatol 15(4):253–259
13. Vitale CB, Ihrke PJ, Olivry T et al (1996) Feline urticaria pigmentosa in three related Sphinx cats. Vet Dermatol 7(4): 227–233

Papular–Nodular Diseases in Dogs

Deep Pyoderma (Furunculosis)

14. Gross TL, Ihrke PJ, Walder EJ et al (2005) Pustular and nodular diseases with adnexal destruction. In: Skin diseases of the dog and cat: clinical and histopathological diagnosis, 2nd edn. Blackwell Science, Ames, pp 442–449
15. Miller WH, Griffin CE, Campbell KL (2013) Deep folliculitis, furunculosis and cellulitis. In: Muller & Kirk's- small animal dermatology, 7th edn. Elsevier, St. Louis, pp 198–203 and 305–314
16. Mueller RS (2004) Treatment protocols for demodicosis: an evidence-based review. Vet Dermatol 15(2):75–89
17. Mueller RS, Rosenkrantz W, Bensignor E et al (2020) Diagnosis and treatment of demodicosis in dogs and cats: clinical consensus guidelines of the World Association for Veterinary Dermatology. Vet Dermatol 31(1):5–27
18. Mueller RS, Bensignor E, Ferrer L et al (2012) Treatment of demodicosis in dogs: 2011 clinical practice guideline. Vet Dermatol 23(2):86–96, e20–e21

Facial Eosinophilic Furunculosis

19. Curtis CF, Bond R, Blunden AS et al (1995) Canine eosinophilic folliculitis and furunculosis in three cases. J Small Anim Pract 36(3):119–123
20. Gross TL (1993) Canine eosinophilic furunculosis of the face. In: Ihrke PJ et al (eds) Advances in veterinary dermatology, vol II. Pergamon Press, New York, p 239
21. Gross TL, Ihrke PJ, Walder EJ et al (2005) Eosinophilic furunculosis of the face. In: Skin diseases of the dog and cat: clinical and histopathological diagnosis, 2nd edn. Blackwell Science, Ames, pp 450–453
22. Guaguère E et al (1996) Furonculose éosinophilique chez le chien. Prat Méd Chir Anim Comp 31:413

Papular-Nodular Canine Leishmaniasis

23. Lombardo G, Pennisi MG, Lupo T et al (2014) Papular dermatitis due to *Leishmania infantum* infection in seventeen dogs: diagnostic features, extent of the infection and treatment outcome. Parasit Vectors 7:120
24. Ordeix L, Solano-Gallego L, Fondevila D et al (2005) Papular dermatitis due to *Leishmania* spp. infection in dogs with parasite-specific cellular immune responses. Vet Dermatol 16:187–191
25. Ordeix L, Dalmau A, Osso M et al (2017) Histological and parasitological distinctive findings in clinically-lesioned and normal-looking skin of dogs with different clinical stages of leishmaniosis. Parasit Vectors 10(1):121
26. Solano-Gallego L, Koutinas A, Miro G et al (2009) Directions for the diagnosis, clinical staging, treatment and prevention of canine leishmaniosis. Vet Parasitol 165:1–18
27. Solano-Gallego L, Miro G, Koutinas A et al (2011) LeishVet guidelines for the practical management of canine leishmaniosis. Parasit Vectors 4:86

Sterile Granuloma and Pyogranuloma Syndrome

28. Cornegliani L, Fondevilla D, Vercelli A et al (2005) PCR technique detection of Leishmania spp. but not Mycobacterium spp. in canine cutaneous 'sterile' pyogranuloma/granuloma syndrome. Vet Dermatol 16(4):233–238
29. Gross TL, Ihrke PJ, Walder EJ et al (2005) Sterile granuloma and pyogranuloma syndrome. In: Skin diseases of the dog and cat: clinical and histopathological diagnosis, 2nd edn. Blackwell Science, Ames, pp 320–323
30. Miller WH, Griffin CE, Campbell KL (2013) Sterile granuloma/pyogranuloma syndrome. In: Muller & Kirk's- small animal dermatology, 7th edn. Elsevier, St. Louis, pp 704–706
31. Panich R, Scott DW, Miller WH (1991) Canine cutaneous sterile pyogranuloma/granuloma syndrome: a retrospective analysis of 29 cases (1976–1988). J Am Anim Hosp Assoc 27:519–528
32. Rosa FB, Older CE, Meason-Smith C et al (2018) Analysis of bacterial and fungal nucleic acid in canine sterile granulomatous and pyogranulomatous dermatitis and panniculitis. Vet Pathol 55(1):124–132
33. Scott DW, Buerger RG, Miller WH Jr (1990) Idiopathic sterile granulomatous and pyogranulomatous dermatitis in cats. Vet Dermatol 1(3):129–137

Calcinosis Cutis

34. Doerr KA, Outerbridge CA, White DS et al (2013) Calcinosis cutis in dogs: histopathological and clinical analysis of 46 cases. Vet Dermatol 24:355–e79
35. Frazier KS, Hullinger GA, Liggett AD et al (1998) Multiple cutaneous metaplastic ossification associated with iatrogenic hyperglucocorticoidism. J Vet Diagn Invest 10:303–307
36. Gross TL, Ihrke PJ, Walder EJ et al (2005) Calcinosis cutis. In: Skin diseases of the dog and cat: clinical and histopathological diagnosis, 2nd edn. Blackwell Science, Ames, pp 373–378
37. Miller WH, Griffin CE, Campbell KL (2013) Calcinosis cutis. In: Muller & Kirk's- small animal dermatology, 7th edn. Elsevier, St. Louis, p 680–682, 831–832
38. White SD, Kass PH, Shiraki R et al (2013) Calcinosis cutis in dogs: histopathological and clinical analysis of 46 cases. Vet Dermatol 24(3):355–361, e78–e79

Xanthogranuloma/Xanthomatosis

39. Banajee KH, Orandle MS, Ratterree W et al (2011) Idiopathic solitary cutaneous xanthoma in a dog. Vet Clin Path 40(1):95–98
40. Chanut F, Colle MA, Deschamps JY et al (2005) Systemic xanthomatosis associated with hyperchylomicronemia in a cat. J Vet Med A Physiol Pathol Clin Med 52(6):272–274
41. Gross TL, Ihrke PJ, Walder EJ et al (2005) Cutaneous xanthoma. In: Skin diseases of the dog and cat: clinical and histopathologic diagnosis. Blackwell Publishing, Oxford, pp 330–333
42. Ravens PA, Vogelnest LJ, Piripi SA (2013) Unique presentation of normolipaemic cutaneous xanthoma in a cat. Aust Vet J 91(11): 460–463
43. Russell EB, Courtman NF (2019) Unique cytologic and histologic features of a suspected cutaneous xanthoma in a dog. Vet Clin Pathol 00:1–5
44. Vitale CB, Ihrke P, Gross TL (1998) Diet-induced alterations in lipid metabolism and associated cutaneous xanthoma formation in 5 cats. In: Kwochka KW, Willemse T, von Tscharner C (eds) Advances in veterinary dermatology, vol 3. Butterworth-Heinemann, Oxford, pp 243–249

Sarcoidosis/ Sarcoid Reaction

45. Iyori K, Kinoshita M, Imai A et al (2019) Clinical and histopathological presentation and successful treatment with ciclosporin for canine sarcoidosis: a case report. Vet Dermatol 30(5):438–e136
46. Noe M, Rosenbach M (2017) Cutaneous sarcoidosis. Curr Opin Pulm Med 23(5):482–486
47. Scott DW, Noxon JO (1990) Sterile sarcoidal granulo- matous skin disease. Canine Pract 15:11–18
48. Sève P, Pacheco Y, Durupt F et al (2021) Sarcoidosis: a clinical overview from symptoms to diagnosis. Cells 10(4):766

Pustular Diseases in Dogs

Staphylococcal Folliculitis

49. Gross TL, Ihrke PJ, Walder EJ et al (2005) Sterile granuloma and pyogranuloma syndrome. In: Skin diseases of the dog and cat: clinical and histopathological diagnosis, 2nd edn. Blackwell Science, Ames, pp 406–410
50. Ihrke P (1996) Bacterial skin disease in the dog: a guide to canine pyoderma. Veterinary Learning Systems, Trenton, pp 1–97
51. Miller WH, Griffin CE, Campbell KL (2013) Superficial bacterial fulliculitis. In: Muller & Kirk's- small animal dermatology, 7th edn. Elsevier, St. Louis, pp 194–195

Demodicosis

52. Gross TL, Ihrke PJ, Walder EJ et al (2005) Pustular and nodular diseases with adnexal destruction. In: Skin diseases of the dog and cat: clinical and histopathological diagnosis, 2nd edn. Blackwell Science, Ames, pp 442–449
53. Miller WH, Griffin CE, Campbell KL (2013) Demodicosis. In: Muller & Kirk's- small animal dermatology, 7th edn. Elsevier, St. Louis, pp 305–314
54. Mueller RS, Bensignor E, Ferrer L et al (2011) Treatment of demodicosis in dogs: clinical practice guideline. Vet Dermatol 23(2):86–96, e20–e21

Pustular Non-follicular Staphylococcal Infection (Impetigo)

55. Amagai M (2009) The molecular logic of pemphigus and impetigo: the desmoglein story. Vet Dermatol 20(5–6):308–312
56. Gross TL, Ihrke PJ, Walder EJ et al (2005) Pustular diseases of epidermis. In: Skin diseases of the dog and cat: clinical and histopathological diagnosis, 2nd edn. Blackwell Science, Ames, pp 4–9
57. Ihrke P (1996) Bacterial skin disease in the dog: a guide to canine pyoderma. Veterinary Learning Systems, Trenton, pp 1–97
58. Nishifuji K, Sugai M, Amagai M (2008) Staphylococcal exfoliative toxins: "molecular scissors" of bacteria that attack the cutaneous defense barrier in mammals. J Dermatol Sci 49:21–31
59. Terauchi R, Sato H, Hasegawa T et al (2003) Isolation of exfoliative toxin from Staphylococcus intermedius and its local toxicity in dogs. Vet Microbiol 94:19–29

Pemphigus Foliaceus

60. Gross TL, Ihrke PJ, Walder EJ et al (2005) Pemphigus foliaceous. In: Skin diseases of the dog and cat: clinical and histopathologic diagnosis. Blackwell Publishing, Oxford, pp 13–19
61. Miller WH, Griffin CE, Campbell KL (2013) Canine pemfigus foliaceus. In: Muller & Kirk's- small animal dermatology, 7th edn. Elsevier, St. Louis, pp 441–445
62. Olivry T (2006) A review of autoimmune skin diseases in domestic animals: I – superficial pemphigus. Vet Dermatol 17:291–305
63. Olivry T, Linder KE (2009) Dermatoses affecting desmosomes in animals: a mechanistic review of acantholytic blistering skin diseases. Vet Dermatol 20(5–6):313–326
64. Vaughan DF, Hodgin EC et al (2010) Clinical and histopathological features of pemphigus foliaceus with and without eosinophilic infiltrates: a retrospective evaluation of 40 dogs. Vet Dermatol 21(2):166–174
65. Yabuzoe A, Nishifuji K, Sekiguchi M et al (2008) Neutrophils contact to plasma membrane of keratinocytes including desmosomal structures in canine pemphigus foliaceus. J Vet Med Sci 70(8):807–812

Dermatophytosis

66. Fairley RA (2001) The histopathological lesions of *Trichophyton mentagrophytes* var. *erinacei* infection in dogs. Vet Dermatol 12: 119–122
67. Gross TL, Ihrke PJ, Walder EJ et al (2005) Superficial pustular dermatophytosis. In: Skin diseases of the dog and cat: clinical and histopathologic diagnosis. Blackwell Publishing, Oxford, pp 11–13
68. Parker WM, Yager JA (1997) Trichophyton dermatophytosis – a disease easily confused with pemphigus erythematosus. Can Vet J 38:502–505
69. Peters J, Scott DW, Erb HN et al (2007) Comparative analysis of canine dermatophytosis and superficial pemphigus for the

prevalence of dermatophytes and acantholytic keratinocytes: a histopathological and clinical retrospective study. Vet Dermatol 18:234–240

70. Poisson L, Olivry T, Lemons C (1998) Subcorneal neutrophilic acantholytic pustular dermatitis: an unusual manifestation of dermatophytosis resembling canine pemphigus foliaceus. In: Kwochka KW, Willemse T, von Tscharner C (eds) Advances in veterinary dermatology, vol 3. Butterworth-Heinemann, Oxford, pp 456–457
71. Scott DW (1994) Marked acantholysis associated with dermatophytosis due to Trichophyton equinum in 2 horses. Vet Dermatol 5:105–110

Idiopathic and Drug Related Sterile Pustular Diseases

72. Bizikova P, Linder KE, Olivry T (2014) Fipronil–amitraz–*S*-methoprene-triggered pemphigus foliaceus in 21 dogs: clinical, histological and immunological characteristics. Vet Dermatol 25(2):103–111
73. Bizikova P, Moriello KA, Linder KA et al (2015) Dinotefuran/pyriproxyfen/permethrin pemphigus-like drug reaction in three dogs. Vet Dermatol 26(3):206–208, e45–e46
74. Gross TL, Ihrke PJ, Walder EJ et al (2005) Subcorneal pustular dermatosis. In: Skin diseases of the dog and cat: clinical and histopathologic diagnosis. Blackwell Publishing, Oxford, pp 19–25
75. Kalaher KM, Scott DW (1990) Subcorneal pustular dermatosis in dogs and in human being: comparative aspects. J Am Acad Dermatol 22:1023–1028
76. Miller WH, Griffin CE, Campbell KL (2013) Subcorneal pustular dermatosis. In: Muller & Kirk's- small animal dermatology, 7th edn. Elsevier, St. Louis, pp 712–713
77. Oberkirchner U, Linder KE, Dunston SM et al (2011) Metaflumizone–amitraz (Promeris)-associated pustular acantholytic dermatitis in 22 dogs: evidence suggests contact drug-triggered pemphigus foliaceus. Vet Dermatol 22:436–448
78. White SD, Carlotti DN, Pin D et al (2002) Putative drug-related pemphigus foliaceus in four dogs. Vet Dermatol 13:195–202

Pustular Pemphigus-Like in Dogs with Canine Leishmaniasis

79. Bardagì M (2012) Canine leishmaniasis: the challenge of histopathological diagnosis. In: Proceedings of the International Society of Veterinary Dermatopathology (ISVD). Precongress Day of the 7th World Congress of Veterinary Dermatology. Vancouver, pp 22–30
80. Bardagí M, Monaco M, Fondevila D (2020) Sterile or nonantibiotic-responsive pustular dermatitis and canine leishmaniosis: a 14 case series description and a statistical association study on 2420 cases. Vet Dermatol 31(3):197-e41
81. Colombo S, Abramo F, Borio S, Albanese F et al (2016) Pustular dermatitis in dogs affected by leishmaniosis: 22 cases. Vet Dermatol 27(1):9–e4
82. Ferrer L, Rabanal R, Fondevila D et al (1988) Skin lesions in canine leishmaniasis. J Small Anim Pract 29:381–388
83. Saridomichelakis MN, Koutinas AF (2014) Cutaneous involvement in canine leishmaniosis due to *Leishmania infantum* (syn. *L. chagasi*). Vet Dermatol 25(61–71):e22

Eosinophilic Pustulosis

84. Carlotti D, Prost C, Magnol J-P et al (1989) La maladie d'Ofugi (pustulose eosinophilique sterile). Prat Méd Chir Anim Comp 24: 131–138
85. Gross TL, Ihrke PJ, Walder EJ et al (2005) Sterile eosinophilic pustulosis. In: Skin diseases of the dog and cat: clinical and histopathological diagnosis, 2nd edn. Blackwell Science, Ames, pp 417–418
86. Gross TL, Ihrke PJ, Walder EJ et al (2005) Canine flea allergy dermatitis. In: Skin diseases of the dog and cat: clinical and histopathological diagnosis, 2nd edn. Blackwell Science, Ames, pp 208–211
87. Miller WH, Griffin CE, Campbell KL (2013) Canine fleabite dermatitis. In: Muller & Kirk's- small animal dermatology, 7th edn. Elsevier, St. Louis, pp 405–410
88. Scott DW (1987) Sterile eosinophilic pustulosis in dog and man: comparative aspects. J Am Acad Dermatol 16:1022–1026

Pustular Diseases in Cats

Pemphigus Foliaceus

89. Bizikova P, Burrows A (2019) Feline pemphigus foliaceus: original case series and a comprehensive literature review. BMC Vet Res 15(1):22
90. Miller WH, Griffin CE, Campbell KL (2013) Feline pemfigus foliaceus. In: Muller & Kirk's- small animal dermatology, 7th edn. Elsevier, St. Louis, pp 447–448
91. Preziosi DE, Goldschmidt MH, Greek JS et al (2003) Feline pemphigus foliaceus: a retrospective analysis of 57 cases. Vet Dermatol 14:313–321. Blackwell Publishing Ltd

Scaling Diseases in Dogs and Cat

Bacterial Overgrowth Syndrome (BOGS)

92. Jasmin P, Pin D, Carlotti DN (2001) Efficacy and interest of a systemic antibiotic treatment with cephalexin in dogs affected with bacterial overgrowth (BOG). In: Proceedings of the 7th FECAVA and 47th Annual Congress of the FK-DVG, Berlin, October, p 51
93. Pin D, Carlotti DN, Jasmin P et al (2006) Prospective study of bacterial overgrowth syndrome in eight dogs. Vet Rec 158(13): 437–441

Malassezia Overgrowth

94. Ahman S, Perrins N, Bond R (2007) Carriage of Malassezia spp. yeasts in healthy and seborrhoeic Devon Rex cats. Med Mycol 45(5):449–455
95. Bond R, Patterson-Kane JC, Lloyd DH (2004) Clinical, histopathological and immunological effects of exposure of canine skin to Malassezia pachydermatis. Med Mycol 42:165–175
96. Chen T, Hill PB (2005) The biology of *Malassezia* organisms and their ability to induce immune responses and skin disease. Vet Dermatol 16(1):4–26
97. Forster-Van Hijfte MA, Curtis CF, White RN (1997) Resolution of exfoliative dermatitis and *Malassezia pachydermatis* overgrowth

in a cat after surgical thymoma resection. J Small Anim Pract 38: 451–454
98. Godfrey DR (1998) A case of feline paraneoplastic alopecia with secondary Malassezia-associated dermatitis. J Small Anim Pract 39:394–396
99. Gross TL, Ihrke PJ, Walder EJ et al (2005) Malassezia dermatitis. In: Skin diseases of the dog and cat: clinical and histopathologic diagnosis. Blackwell Publishing, Oxford, pp 142–146
100. Guaguère E, Prélaud P (1996) Étude rétrospective de 54 cas de dermite a Malassezia pachydermatis chez le chien: resultats epidemiologiques, cliniques, cytologiques et histopathologiques. Prat Méd Chir Anim Comp 31:309–323
101. Linek M, Rüfenacht S, Brachelente C et al (2015) Nonthymoma-associated exfoliative dermatitis in 18 cats. Vet Dermatol 26(1): 40–45
102. Mauldin EA, Scott DW, Miller WH et al (1997) *Malassezia* dermatitis in the dog: a retrospective histopathological and immunopathological study of 86 cases (1990–95). Vet Dermatol 8:191–202
103. Mauldin EA, Morris DO, Goldschmidt MH (2002) Retrospective study: the presence of *Malassezia* in feline skin biopsies. A clinicopathological study. Vet Dermatol 13:7–13
104. Miller WH, Griffin CE, Campbell KL (2013) Feline *Malassezia* dermatitis. In: Muller & Kirk's- small animal dermatology, 7th edn. Elsevier, St. Louis, pp 243–249

Cutaneous Candidiasis

105. Carlotti D, Pinn D (1999) Candida pododermatitis in the dog: a report of 5 cases. In: Proceedings of the 15th AAVD/ACVD meeting, Maui, pp 55–56
106. Greene CE (2017) Algal and fungal diseases. In: Infectious diseases of the dog and cat, 5th edn. W.B. Saunders, Philadelphia
107. Gross TL, Ihrke PJ, Walder EJ et al (2005) Candidiasis. In: Skin diseases of the dog and cat: clinical and histopathologic diagnosis. Blackwell Publishing, Oxford, pp 9–11
108. Miller WH, Griffin CE, Campbell KL (2013) Feline candidiasis. In: Muller & Kirk's- small animal dermatology, 7th edn. Elsevier, St. Louis, pp 267–268
109. Mueller RS, Bettenay SV, Shipstone M (2002) Cutaneous candidiasis in a dog caused by *Candida guilliermondii*. Vet Rec 150(23): 728–730
110. Scott DW, Miller WH, Griffin CE (2001) Muller & Kirk's small animal dermatology, 6th edn. W.B. Saunders, Philadelphia, pp 361–363

Dermatophytosis

111. Carlotti DN, Bensignor E (1999) Dermatophytosis due to Microsporum persicolor (13 cases) or Microsporum gypseum (20 cases) in the dog. Vet Dermatol 10:17–27
112. Frymus T, Gruffydd-Jones T, Pennisi MG et al (2013) Dermatophytosis in cats: ABCD guidelines on prevention and management. J Feline Med Surg 15(7):598–604
113. Gross TL, Ihrke PJ, Walder EJ et al (2005) Feline and canine dermatophytosis. In: Skin diseases of the dog and cat: clinical and histopathologic diagnosis. Blackwell Publishing, Oxford, pp 410–415
114. Moriello K, Coyner K, Paterson S et al (2017) Diagnosis and treatment of dermatophytosis in dogs and cats: Clinical Consensus Guidelines of the World Association for Veterinary Dermatology. Vet Dermatol 28(3):266–e68

Demodicosis and Other Superficial Ectoparasites

115. Ferreira D, Sastre N, Ravera I et al (2015) Identification of a third feline Demodex species through partial sequencing of the 16S rDNA and frequency of Demodex species in 74 cats using a PCR assay. Vet Dermatol 26(4):239–e53
116. Gross TL, Ihrke PJ, Walder EJ et al (2005) Pustular and nodular diseases with adnexal destruction. In: Skin diseases of the dog and cat: clinical and histopathological diagnosis, 2nd edn. Blackwell Science, Ames, pp 442–449
117. Miller WH, Griffin CE, Campbell KL (2013) Demodicosis. In: Muller & Kirk's- small animal dermatology, 7th edn. Elsevier, St. Louis, pp 305–314
118. Mueller RS (2004) Treatment protocols for demodicosis: an evidence-based review. Vet Dermatol 15(2):75–89
119. Mueller RS, Bensignor E, Ferrer L et al (2012) Treatment of demodicosis in dogs: 2011 clinical practice guideline. Vet Dermatol 23(2):86–96, e20–e21
120. Sastre N, Ravera I, Villanueva S et al (2012) Phylogenetic relationships in three species of canine Demodex mite based on partial sequences of mitochondrial 16S rDNA. Vet Dermatol 23(6): 509–e101

Erosion in Dogs

121. Gross TL, Ihrke PJ, Walder EJ et al (2005) Pustular diseases of epidermis. In: Skin diseases of the dog and cat: clinical and histopathological diagnosis, 2nd edn. Blackwell Science, Ames, pp 4–9
122. Ihrke PJ (1996) Bacterial skin disease in the dog: a guide to canine pyoderma. Veterinary Learning Systems, Trenton, pp 1–97
123. Miller WH, Griffin CE, Campbell KL (2013) Superficial bacterial fulliculitis. In: Muller & Kirk's- small animal dermatology, 7th edn. Elsevier, St. Louis, pp 194–195

Ulcerative Diseases in Dogs

Deep Pyoderma

124. Bassett RJ, Burton GG, Robson DC (2004) Antibiotic responsive ulcerative dermatoses in German Shepherd dogs with mucocutaneous pyoderma. Aust Vet J 82:485–489
125. Gross TL, Ihrke PJ, Walder EJ et al (2005) Pyotraumatic dermatitis. In: Skin diseases of the dog and cat: clinical and histopathological diagnosis, 2nd edn. Blackwell Science, Ames, pp 116–118
126. Ihrke PJ (1996) Bacterial skin disease in the dog: a guide to canine pyoderma. Veterinary Learning Systems, Trenton, pp 1–97
127. Miller WH, Griffin CE, Campbell KL (2013) Bacterial skin diseases. In: Muller & Kirk's- small animal dermatology, 7th edn. Elsevier, St. Louis, pp 185–205
128. Olivry T, Rossi MA, Banovic F (2015) Mucocutaneous lupus erythematosus in dogs (21 cases). Vet Dermatol 26:256–e55
129. Rosser EJ Jr (2006) German shepherd dog pyoderma. Vet Clin North Am Small Anim Pract 36(1):203–211, viii. Review
130. Wiemelt SP, Goldschmidt MH, Greek JS et al (2004) A retrospective study comparing the histopathological features and response to treatment in two canine nasal dermatoses, discoid lupus erythematosus and mucocutaneous pyoderma. Vet Dermatol 15: 341–348

131. Wisselink MA, Koeman JP, Van den Ingh TS et al (1990) Investigations on the role of flea antigen in the pathogenesis of German shepherd dog pyoderma (GSP). Vet Q 12(1):21–28

Canine Leishmaniasis

132. Gross TL, Ihrke PJ, Walder EJ et al (2005) Cutaneous leishmaniosis. In: Skin diseases of the dog and cat: clinical and histopathologic diagnosis. Blackwell Publishing, Oxford, pp 312–316
133. Solano-Gallego L, Koutinas A, Miro G et al (2009) Directions for the diagnosis, clinical staging, treatment and prevention of canine leishmaniosis. Vet Parasitol 165:1–18

Phytiosis

134. Dykstra MJ, Sharp NJ, Olivry T et al (1999) A description of cutaneous-subcutaneous pythiosis in fifteen dogs. Med Mycol 37: 427–433
135. Oldenhoff W, Grooters A, Pinkerton ME (2013) Cutaneous pythiosis in two dogs from Wisconsin, USA. Vet Dermatol. First published: 28 December
136. Peano A, Molinar Min AR, Fondati A et al (2023) Cutaneous Pythiosis in 2 dogs, Italy. Emerg Infect Dis 29(7):1447–1450
137. Rodrigues Hoffmann A, Ramos MG, Walker RT, Stranahan LW (2023) Hyphae, pseudohyphae, yeasts, spherules, spores, and more: a review on the morphology and pathology of fungal and oomycete infections in the skin of domestic animals. Vet Pathol 60(6):812–828
138. Yolanda H, Krajaejun T (2022) Global distribution and clinical features of pythiosis in humans and animals. J Fungi (Basel) 8(2): 182

Ulcerative Diseases in Cats

Feline Indolent Ulcer

139. Buckley L, Nuttall T (2012) Feline eosinophilic granuloma complex(ities): some clinical clarification. J Feline Med Surg 14(7): 471–481
140. Colombini S, Hodgin EC, Foil CS et al (2001) Induction of feline flea allergy dermatitis and the incidence and histopathological characteristics of concurrent indolent lip ulcers. Vet Dermatol 12: 155–161
141. Gross TL, Ihrke PJ, Walder EJ et al (2005) Feline indolent ulcer. In: Skin diseases of the dog and cat: clinical and histopathological diagnosis, 2nd edn. Blackwell Science, Ames, pp 128–130

Eosinophilic Granuloma

142. Bardagi M, Fondati A, Fondevila D et al (2003) Ultrastructural study of cutaneous lesions in feline eosinophilic granuloma complex. Vet Dermatol 14:297–303
143. Buckley L, Nuttall T (2012) Feline eosinophilic granuloma complex(ities): some clinical clarification. J Feline Med Surg. 14(7): 471–481
144. Fondati A, Fondevila D, Ferrer L (2001) Histopathological study of feline eosinophilic dermatoses. Vet Dermatol 12(6):333–338
145. Gross TL, Ihrke PJ, Walder EJ et al (2005) Feline indolent ulcer. In: Skin diseases of the dog and cat: clinical and histopathological diagnosis, 2nd edn. Blackwell Science, Ames, pp 128–130
146. Miller WH, Griffin CE, Campbell KL (2013) Feline eosinophilic granuloma. In: Muller & Kirk's- small animal dermatology, 7th edn. Elsevier, St. Louis, pp 714–715

Herpesvirus Ulcerative Dermatitis

147. Flecknell PA, Orr CM, Wright AI et al (1979) Skin ulceration associated with herpesvirus infection in cats. Vet Rec 104:313–315
148. Gross TL, Ihrke PJ, Walder EJ et al (2005) Feline herpesvirus ulcerative dermatitis. In: Skin diseases of the dog and cat: clinical and histopathological diagnosis, 2nd edn. Blackwell Science, Ames, pp 124–127
149. Hargis AM, Ginn PE, Mansell JEKL et al (1999) Ulcerative facial and nasal dermatitis and stomatitis in cats associated with feline herpesvirus 1. Vet Dermatol 10:267–274
150. Lee M, Bosward KL, Norris JM (2010) Immunohistological evaluation of feline herpesvirus-1 infection in feline eosinophilic dermatoses or stomatitis. J Feline Med Surg 12:72–79
151. Sánchez MD, Goldschmidt MH, Mauldin EA (2012) Herpesvirus dermatitis in two cats without facial lesions. Vet Dermatol 23(2): 171–173, e35
152. Suchy A, Bauder B, Gelbmann W et al (2000) Diagnosis of feline herpesvirus infection by immunohistochemistry, polymerase chain reaction, and *in situ* hybridization. J Vet Diagn Invest 12:186–191

Sporotrichosis

153. Crothers SL, White SD, Ihrke PJ et al (2009) Sporotrichosis: a retrospective evaluation of 23 cases seen in northern California (1987–2007). Vet Dermatol 20(4):249–259
154. Dunstan RW, Reimann KA, Langham RF (1986) Feline sporotrichosis. J Am Vet Med Assoc 189:880–883
155. Fleury RN, Taborda PR, Gupta AK et al (2001) Zoonotic sporotrichosis. Transmission to humans by infected domestic cat scratching: report of four cases in Sao Paulo, Brazil. Int J Dermatol 40:318–322
156. Gross TL, Ihrke PJ, Walder EJ et al (2005) Sporotrichosis. In: Skin diseases of the dog and cat: clinical and histopathological diagnosis, 2nd edn. Blackwell Science, Ames, pp 298–301
157. Madrid IM, Mattei AS, Fernandes CG et al (2012) Epidemiological findings and laboratory evaluation of sporotrichosis: a description of 103 cases in cats and dogs in southern Brazil. Mycopathologia 173(4):265–273
158. Miller WH, Griffin CE, Campbell KL (2013) Sporotrichosis. In: Muller & Kirk's- small animal dermatology, 7th edn. Elsevier, St. Louis, pp 249–252
159. Silva JN, Passos RL, Menezes CR et al (2015) Diagnostic accuracy assessment of cytopathological examination of feline sporotrichosis. Med Mycol 53(8):880–884

Plaques and Nodules

Bacterial Nodular Diseases in Dogs and Cats

Nocardiosis and Actinomycosis

160. Greene CE (2017) Actinomycosis and Nocardiosis. In: Infectious diseases of the dog and cat, 5th edn. W.B. Saunders, Philadelphia
161. Gross TL, Ihrke PJ, Walder EJ et al (2005) Actinomycosis and nocardiosis. In: Skin diseases of the dog and cat: clinical and histopathological diagnosis, 2nd edn. Blackwell Science, Ames, pp 272–275
162. Kirpensteijn J, Fingland RB (1992) Cutaneous actinomycosis and nocardiosis in dogs: 48 cases (1980-1990). Department of Clinical Sciences, College of Veterinary Medicine, Kansas State University, Manhattan. J Am Vet Med Assoc 201(6):917–920
163. Malik R, Krockenberger MB, O'Brien CR et al (2006) Nocardia infections in cats: a retrospective multi-institutional study of 17 cases. Aust Vet J 84(7):235–245
164. Miller WH, Griffin CE, Campbell KL (2013) Bacterial diseases. In: Muller & Kirk's- small animal dermatology, 7th edn. Elsevier, St. Louis, pp 212–214
165. Siak MK, Burrows AK (2013) Cutaneous nocardiosis in two dogs receiving ciclosporin therapy for the management of canine atopic dermatitis. Vet Dermatol 24:453–e103
166. Zeni Condas LA, de Farias MR, Keller Siqueira A (2023) Molecular identification and antimicrobial resistance pattern of Nocardia isolated from 14 diseased dogs and cats. Braz J Microbiol 54(2): 1287–1294

Botryomycosis

167. Gross TL, Ihrke PJ, Walder EJ et al (2005) Bacterial pseudomycetoma. In: Skin diseases of the dog and cat: clinical and histopathological diagnosis, 2nd edn. Blackwell Science, Ames, pp 275–276
168. Miller WH, Griffin CE, Campbell KL (2013) Bacterial diseases. In: Muller & Kirk's- small animal dermatology, 7th edn. Elsevier, St. Louis, pp 212–214
169. Scott DW (2007) Bacterial pseudomycetoma (Botryomicosis) in dogs: two new case reports and review of the literature. Jpn J Vet Dermatol 13(3):135–140

Mycobacteriosis

170. Davies JL, Sibley JA, Myers S et al (2006) Histological and genotypical characterization of feline cutaneous mycobacteriosis: a retrospective study of formalin-fixed paraffin-embedded tissues. Vet Dermatol 17(3):155–162
171. Dedola C, Zobba R, Pinna Parpaglia ML (2014) First report of canine leprosy in Europe: molecular and clinical traits. Vet Rec 174(5):120
172. Greene CE (2017) Mycoplasmal and bacterial diseases. In: Infectious diseases of the dog and cat, 5th edn. W.B. Saunders, Philadelphia
173. Gross TL, Ihrke PJ, Walder EJ et al (2005) Feline leprosis syndrome. In: Skin diseases of the dog and cat: clinical and histopathological diagnosis, 2nd edn. Blackwell Science, Ames, pp 276–288
174. Gunn-Moore DA (2014) Feline mycobacterial infections. Vet J 201(2):230–238
175. Laprie C, Duboy J, Malik R et al (2013) Feline cutaneous mycobacteriosis: a review of clinical, pathological and molecular characterization of one case of *Mycobacterium microti* skin infection and nine cases of feline leprosy syndrome from France and New Caledonia. Vet Dermatol 24(6):561–569, e133–4 Epub 2013 Sep 2
176. Malik R, Wigney D, Dawson D et al (2000) Infection of the subcutis and skin of cats with rapidly growing mycobacteria: a review of microbiological and clinical findings. J Feline Med Surg 2(1):35–48
177. Malik R, Shaw SE, Griffin C et al (2004) Infections of the subcutis and skin of dogs caused by rapidly growing mycobacteria. J Small Anim Pract 45(10):485–494
178. Malik R, Smits B, Reppas G et al (2013) Ulcerated and nonulcerated nontuberculous cutaneous mycobacterial granulomas in cats and dogs. Vet Dermatol 24(146–153):e32–e33
179. Miller WH, Griffin CE, Campbell KL (2013) Mycobcterial infections. In: Muller & Kirk's- small animal dermatology, 7th edn. Elsevier, St. Louis, pp 207–212
180. Smits B, Willis R, Studdert V et al (2012) Case clusters of 'leproid granulomas' in foxhounds in New Zealand and Australia. Vet Dermatol 23:465–e88

Protozoa

Leishmaniasis

181. Abramo F, Albanese F, Gattuso S et al (2021) Skin lesions in feline leishmaniosis: a systematic review. Pathogens. 10(4):472
182. Gross TL, Ihrke PJ, Walder EJ et al (2005) Cutaneous leishmaniosis. In: Skin diseases of the dog and cat: clinical and histopathologic diagnosis. Blackwell Publishing, Oxford, pp 312–316
183. Pennisi MG, Hartmann K, Lloret A et al (2013) Leishmaniosis in cats: ABCD guidelines on prevention and management. J Feline Med Surg 15(7):638–642
184. Poli A, Abramo F, Barsotti P et al (2002) Feline leishmaniosis due to *Leishmania infantum* in Italy. Vet Parasitol 106:181–191
185. Solano-Gallego L, Koutinas A, Miro G et al (2009) Directions for the diagnosis, clinical staging, treatment and prevention of canine leishmaniosis. Vet Parasitol 165:1–18

Toxoplasma and *Neospora*

186. Anfray P, Bonetti C, Fabbrini F et al (2005) Feline cutaneous toxoplasmosis: a case report. Vet Dermatol 16:131–136
187. Clough B, Frickel EM (2017) The toxoplasma parasitophorous vacuole: an evolving host-parasite frontier. Trends Parasitol 33(6):473–488
188. Donahoe SL, Lindsay SA, Krockenberger M et al (2015) A review of neosporosis and pathologic findings of Neospora caninum infection in wildlife. Int J Parasitol Parasites Wildl 4(2):216–238
189. Dubey JP, Carpenter JL (1993) Histologically confirmed clinical toxoplasmosis in cats: 100 cases (1952–1990). J Am Vet Med Assoc 203:1556–1566
190. Dubey JP, Metzger FL Jr, Hattel AL et al (1995) Canine cutaneous neosporosis: clinical improvement with clindamycin. Vet Dermatol 6(1):37–43

191. Greene CE (2017) Protozoal diseases. In: Infectious diseases of the dog and cat, 5th edn. W.B. Saunders, Philadelphia
192. Hoffmann AR, Cadieu J, Kiupel M et al (2012) Cutaneous toxoplasmosis in two dogs. J Vet Diagn Invest 24(3):636–640
193. La Perle KM, Del Piero F, Carr RF et al (2001) Cutaneous neosporosis in two adult dogs on chronic immunosuppressive therapy. J Vet Diagn Invest 13(3):252–255
194. Legnani S, Pantchev N, Forlani A et al (2016) Emergence of cutaneous neosporosis in a dog receiving immunosuppressive therapy: molecular identification and management. Vet Dermatol 27(1):49–e14
195. Little L, Shokek A, Dubey JP et al (2005) Toxoplasma gondii-like organisms in skin aspirates from a cat with disseminated protozoal infection. Vet Clin Pathol 34:156–160
196. Lloret A, Fondevila D et al (2002) Cutaneous neosporosis during treatment of pemphigus foliaceus in a dog. J Am Anim Hosp Assoc 38(5):415–419
197. Mann TR, Cadore GC, Camillo G et al (2016) Canine cutaneous neosporosis in Brazil. Vet Dermatol 27:195–197
198. Moore A, Burrows AK, Malik R et al (2022) Fatal disseminated toxoplasmosis in a feline immunodeficiency virus-positive cat receiving oclacitinib for feline atopic skin syndrome. Vet Dermatol 3(5):435–439
199. Ordeix L, Lloret A, Fondevila D et al (2002) Cutaneous neosporosis during treatment of pemphigus foliaceus in a dog. J Am Anim Hosp Assoc 38(5):415–419
200. Park CH, Ikadai H, Yoshida E et al (2007) Cutaneous toxoplasmosis in a female Japanese cat. Vet Pathol 44:683–687
201. Perl S et al (1996) Pyogranulomatous dermatitis associated to *Neospora caninum* in a dog. World Small Anim Vet Assoc 21:417

Worms

Dirofilariasis

202. Albanese F, Abramo F, Braglia C et al (2013) Nodular lesions due to infestation by *Dirofilaria repens* in dogs from Italy. Vet Dermatol 24:255–e56
203. Giori L, Garbagnoli V, Venco L et al (2010) What is your diagnosis? Fine-needle aspirate from a subcutaneous mass in a dog. Mixed neutrophilic-eosinophilic inflammation with Dirofilaria fragments. Vet Clin Pathol 39(2):255–256
204. Hargis AM, Lewis TP, Duclos DD et al (1999) Dermatitis associated with microfilariae (Filarioidea) in 10 dogs. Vet Dermatol 10(2):95–107
205. Mackenzie CD, Behan-Braman A, Huptman J, Geary T (2017) Assessing the Viability and Degeneration of the Medically Important Filarial Nematodes -Chapter:6
206. Tarello W (2002) Cutaneous lesions in dogs with Dirofilaria (Nochtiella) repens infestation and concurrent tick-borne transmitted diseases. Vet Dermatol 13:267–274

Fungi

Dermatophytic Kerion

207. Albanese F, Caruso C (2007) Dermatophytic kerion: etiopathological, clinical, diagnostic and therapeutic aspects of 39 dogs. Veterinaria 21:9–18
208. Cornegliani L, Persico P, Colombo S (2009) Canine nodular dermatophytosis (kerion): 23 cases. Vet Dermatol 20(3):185–190
209. Gross TL, Ihrke PJ, Walder EJ et al (2005) Pustular and nodular diseases with adnexal destruction. In: Skin diseases of the dog and cat. Clinical and histopathological diagnosis, 2nd edn. Blackwell Science, Oxford, pp 420–459
210. Koutinas AF, Saridomichelakis M, Lekkas S et al (2003) Clinical and histopathological aspects of dermatophyte kerion in the dog: a retrospective study of 20 spontaneous cases. Vet Dermatol 14:243. (Abstract)
211. Miller WH, Griffin CE, Campbell KL (2013) Dermatophytosis. In: Muller & Kirk's- small animal dermatology, 7th edn. Elsevier, St. Louis, pp 231–241
212. Moriello K, Coyner K, Paterson S et al (2017) Diagnosis and treatment of dermatophytosis in dogs and cats: Clinical Consensus Guidelines of the World Association for Veterinary Dermatology. Vet Dermatol 28(3):266–e68

Dermatophytic Pseudomycetoma

213. Abramo F, Vercelli A, Mancianti F (2001) Two cases of dermatophytic pseudomycetoma in the dog: an immunohistochemical study. Vet Dermatol 12:203–207
214. Bianchi MV, Laisse CJM, Vargas TP et al (2017) Intra-abdominal fungal pseudomycetoma in two cats. Rev Iberoam Micol 34(2): 112–115
215. Chang SC, Liao JW, Shyu CL et al (2011) Dermatophytic pseudomycetomas in four cats. Vet Dermatol 22(2):181–187
216. Hobi S, Tam WYJ, Tse M et al (2024) Microsporum canis causes cutaneous and extracutaneous feline dermatophytic pseudomycetomas: molecular identification and clinicopathological characteristics. J Fungi (Basel) 10(8):576
217. Peano A, Gräser AF et al (2012) Use of microsatellite markers for typing of Microsporum canis isolates causing pseudomycetoma in cats. CABI Digital Library

Eumycetoma/True Mycetoma

218. Albanese F, Muscatello LV, Michelutti A et al (2022) Canine eumycetoma caused by Madurella pseudomycetomatis. Med Mycol Case Rep 35:51–53
219. Gross TL, Ihrke PJ, Walder EJ et al (2005) Infectious nodular and diffuse granulomatous and pyogranulomatous diseases of the dermis. In: Skin diseases of the dog and cat. Clinical and histopathological diagnosis, 2nd edn. Blackwell Science, Oxford, pp 301–303
220. Guillot J, Garcia-Hermoso D, Degorce F et al (2004) (2004) Eumycetoma caused by Cladophialophora bantiana in a dog. J Clin Microbiol 42:4901–4903
221. Hoffmann AR, Ramos MG, Walker RT et al (2023) Hyphae, pseudohyphae, yeasts, spherules, spores, and more: a review on the morphology and pathology of fungal and oomycete infections in the skin of domestic animals. Vet Pathol
222. Sun P-L, Peng P-C, Wu P-H et al (2013) Canine eumycetoma caused by Cladophialophora bantiana in a Maltese: case report and literature review. Mycoses. 56(3):376–381

Phaeohyphomycosis

223. Abramo F, Bastelli F, Nardoni S et al (2002) Feline cutaneous *phaeohyphomycosis* due to *Cladophyalophora bantiana*. J Feline Med Surg 4:157–163

224. Beccati M, Vercelli A, Peano A et al (2005) Phaeohyphomycosis by Phialophora verrucosa: first European case in a cat. Vet Rec 157(3):93–94
225. Fondati A, Gallo MG, Romano E et al (2001) A case of feline *phaeohyphomycosis* due to Fonsecaea pedrosoi. Vet Dermatol 12: 297–301
226. Greene CE (2017) Fungal and algal diseases. In: Infectious diseases of the dog and cat, 5th edn. W.B. saunders, Philadelphia
227. Gross TL, Ihrke PJ, Walder EJ et al (2005) Cutaneous infections of other opportunistic fungi. In: Skin diseases of the dog and cat: clinical and histopathologic diagnosis. Blackwell Publishing, Oxford, pp 301–312
228. Miller WH, Griffin CE, Campbell KL (2013) Chromomycosis (phaeohyphomycosis and chromoblastomycosis). In: Muller & Kirk's- small animal dermatology, 7th edn. Elsevier, St. Louis, pp 253–255
229. Rothenburg LS, Snider TA, Wilson A et al (2017) Disseminated phaeohyphomycosis in a dog. Med Mycol Case Rep. 15:28–32

Protothecosis

230. Carfora V et al (2017) Evidence of a Prototheca zopfii genotype 2 disseminated infection in a dog with cutaneous lesions. Mycopathologia 182:603–608
231. Falcaro C, Furlanello T, Binanti D et al (2021) Molecular characterization of Prototheca in 11 symptomatic dogs. J Vet Diagn Invest 33(1):156–161
232. Huth N, Wenkel RF, Roschanski N et al (2015) Prototheca zopfii genotype 2-induced nasal dermatitis in a cat. J Comp Pathol 152(4)
233. Masuda M, Jagielski T, Danesi P et al (2021) Protothecosis in dogs and cats-new research directions. Mycopathologia 186(1):143–152
234. Papadogiannakis EI et al (2016) Cutaneous disease as sole clinical manifestation of protothecosis in a Boxer dog. Case Report. Vet Med:1–4
235. Whipple KM, Wellehan JF, Jeon AB et al (2020) Cytologic, histologic, microbiologic, and electron microscopic characterization of a canine Prototheca wickerhamii infection. Vet Clin Pathol. 49(2):326–332

Blastomycosis

236. Arceneaux KA, Taboada J, Hosgood G (1998) Blastomycosis in dogs: 115 cases (1980-1995). J Am Vet Med Assoc 213(5): 658–664
237. Brömel C, Sykes JE (2005) Epidemiology, diagnosis, and treatment of blastomycosis in dogs and cats. Clin Tech Small Anim Pract 20(4):233–239
238. Garofalo J (2022) Disseminated blastomycosis in a 1-year-old, male golden retriever-poodle crossbred dog. Can Vet J 63(8): 855–858
239. Greene CE (2017) In: infectious diseases of the dog and cat, 5th edn. W.B. Saunders, Philadelphia, pp Section III

Histoplasmosis

240. Brömel C, Sykes JE (2005) Histoplasmosis in dogs and cats. Clin Tech Small Anim Pract 20(4):227–232
241. Heilmann RM, Nabity MB, Bryan LK et al (2021) Mucocutaneous nasal histoplasmosis in an immunocompetent dog. BMC Vet Res 17(1):192
242. Lin Blache J, Ryan K, Arceneaux K (2011) Histoplasmosis. Compend Contin Educ Vet 33(3)
243. Mavropoulou A, Grandi G, Calvi L et al (2010) Disseminated histoplasmosis in a cat in Europe. J Small Anim Pract 51(3): 176–180

Cryptococcosis

244. Greene CE (2017) Fungal and algal diseases. In: Infectious diseases of the dog and cat, 5th edn. W.B. saunders, Philadelphia
245. Gross TL, Ihrke PJ, Walder EJ et al (2005) Cutaneous cryptococcosis. In: Skin diseases of the dog and cat: clinical and histopathologic diagnosis. Blackwell Publishing, Oxford, pp 295–298
246. Malik R, Jacobs GJ, Love DN (2001) Cryptococcosis: new perspectives on etiology, pathogenesis, diagnosis, and clinical management. In: August JR (ed) Consultations in feline medicine. W.B. Saunders Co, Philadelphia, pp 39–50
247. Martins DB, Zanette RA, França RT et al (2011) Massive cryptococcal disseminated infection in an immunocompetent cat. Vet Derm 22(2):232–234
248. Miller WH, Griffin CE, Campbell KL (2013) Feline cryptococcosis (torulosis, european blastomycosis). In: Muller & Kirk's- small animal dermatology, 7th edn. Elsevier, St. Louis, pp 262–265
249. Pennisi MG, Hartmann K, Lloret A et al (2013) Cryptococcosis in cats: ABCD guidelines on prevention and management. J Feline Med Surg 15(7):611–618
250. Xiarong L, Heitman J (2006) The biology of the Cryptococcus neoformans species complex. Ann Rev Microbiol 60:69–105

Sterile Diseases

Cutaneous Mineralization

251. Bertazzolo W, Toscani L, Calcaterra S et al (2003) Clinicopathological findings in five cats with paw calcifications. J Feline Med Surg 5:11–17
252. Doerr KA, Outerbridge CA, White SD et al (2013) Calcinosis cutis in dogs: histopathological and clinical analysis of 46 cases. Vet Dermatol 24(3):355–361, e78–e79
253. Gross TL (1997) Calcinosis circumscripta and renal dysplasia in a dog. Vet Dermatol 8:27–32
254. Gross TL, Ihrke PJ, Walder EJ et al (2005) Calcinosis circumscripta. In: Skin diseases of the dog and cat: clinical and histopathological diagnosis, 2nd edn. Blackwell Science, Ames, pp 378–380
255. Scott DW, Buerger RG (1988) Idiopathic calcinosis circumscripta in the dog: a retrospective analysis of 130 cases. J Am Anim Hosp Assoc 24:651–658
256. Tafti AK, Hanna P, Bourque AC (2005) Calcinosis circumscripta in the dog: a retrospective pathological study. J Vet Med A Physiol Pathol Clin Med 52(1):13–17

Eosinophilic Granuloma and Eosinophilic Plaque

257. Albanese F, Tieghi C, De Rosa L et al (2009) Feline perforating dermatitis resembling human reactive perforating collagenosis: clinicopathological findings and outcome in four cases. Vet Dermatol 20(4):273–280

258. Bardagi M, Fondati A, Fondevila D et al (2003) Ultrastructural study of cutaneous lesions in feline eosinophilic granuloma complex. Vet Dermatol 14:297–303
259. Fondati A, Fondevila D, Ferrer L (2001) Histopathological study of feline eosinophilic dermatoses. Vet Dermatol 12(6):333–338
260. Gao Z, Lu SJ, Shan SJ (2023) Acquired perforating dermatosis: a clinicopathologic study, and the features of dermoscopy and reflective confocal microscopy of 37 cases. Skin Res Technol 29(7): e13416
261. Gross TL, Ihrke PJ, Walder EJ et al (2005) Feline eosinophilic plaque and granuloma. In: Skin diseases of the dog and cat: clinical and histopathological diagnosis, 2nd edn. Blackwell Science, Ames, pp 109–111
262. Joffe DJ, Allen AL (1995) Ulcerative eosinophilic stomatitis in three Cavalier King Charles spaniels. J Am Anim Hosp Assoc 31: 34–37
263. Madewell BR, Stannard AA, Pulley LT et al (1980) Oral eosinophilic granuloma in Siberian Husky dogs. J Am Vet Med Assoc 177:701–703
264. Miller WH, Griffin CE, Campbell KL (2013) Feline eosinophilic granuloma complex. In: Muller & Kirk's- small animal dermatology, 7th edn. Elsevier, St. Louis, pp 714–716
265. Porcellato I, Giontella A, Mechelli L et al (2014) Feline eosinophilic dermatoses: a retrospective immunohistochemical and ultrastructural study of extracellular matrix remodelling. Vet Dermatol 25(2):86–94, e26

Panniculitis

266. Boynosky NA, Stokking LB (2014) Potassium-bromide associated panniculitis. J Small Anim Pract 55:640–642
267. Contreary CL, Outerbridge CA, Affolter VK et al (2015) Canine sterile nodular panniculitis: a retrospective study of 39 dogs. Vet Dermatol 26:1–10
268. German AJ, Foster AP, Holden D et al (2003) Sterile nodular panniculitis and pansteatitis in three weimaraners. J Small Anim Pract 44:449–455
269. Gross TL, Ihrke PJ, Walder EJ et al (2005) Disease of the panniculus. In: Skin diseases of the dog and cat: clinical and histopathological diagnosis, 2nd edn. Blackwell Science, Ames, pp 538–560
270. Kim HJ, Kang MH, Kim J-H et al (2011) Sterile panniculitis in dogs: new diagnostic findings and alternative treatments. Vet Dermatol 22:352–359
271. Mellanby RJ, Stell A, Baines E et al (2003) Panniculitis associated with pancreatitis in a Cocker Spaniel. J Small Anim Pract 44:24–28
272. O'Kell AL, Inteeworn N, Diaz SF et al (2010) Canine sterile nodular panniculitis: a retrospective study of 14 cases. J Vet Intern Med 24:278–284
273. Scott DW, Anderson WI (1988) Panniculitis in dogs and cats: a retrospective analysis of 78 cases. J Am Anim Hosp Assoc 24:551–559
274. Yamagishi C, Momoi Y, Kobayashi T et al (2007) A retrospective study and gene analysis of canine sterile panniculitis. J Vet Med Sci 69:915–924

Sterile Granulomatous Dermatitis and Lymphadenitis (Juvenile Cellulitis)

275. Inga A, Griffeth GC, Drobatz KJ et al (2020) Sterile granulomatous dermatitis and lymphadenitis (juvenile cellulitis) in adult dogs: a retrospective analysis of 90 cases (2004-2018). Vet Dermatol 31(3):219–e47
276. Jeffers JG, Duclos DD, Goldschmidt MH (1995) A dermatosis resembling juvenile cellulitis in an adult dog. J Am Anim Hosp Assoc 31(3):204–208
277. Gross TL, Ihrke PJ, Walder EJ et al (2005) Juvenile sterile granulomatous dermatitis and lymphadenitis. In: Skin diseases of the dog and cat: clinical and histopathological diagnosis, 2nd edn. Blackwell Science, Ames, pp 327–330
278. Miller WH, Griffin CE, Campbell KL (2013) Juvenile cellulitis. In: Muller & Kirk's- small animal dermatology, 7th edn. Elsevier, St. Louis, pp 708–709
279. Reimann KA, Evans MG, Chalifoux LV et al (1989) Clinicopathologic characterization of canine juvenile cellulitis. Vet Pathol 26: 499–504
280. Scott DW, Miller WH (2007) Juvenile cellulitis in dogs: a retrospective study of 18 cases (1976–2005). Jpn J Vet Dermatol 13(2): 71–79

Plasma Cell Pododermatitis

281. Bensignor E, Merven F (2011) Nasal plasma cell dermatitis in cats. Letter to the Editor. Vet Dermatol 22:286
282. Biezus G, Grimade T, Cristo M et al (2020) Plasma cell pododermatitis associated with feline leukemia virus (FeLV) and concomitant feline immunodeficiency virus (FIV) infection in a cat. Top Companion Anim Med 41:100475
283. Brosseau G (2022) Feline plasma cell pododermatitis. Can Vet J 63(5):545–548
284. Declercq J, De Bosschere H (2010) Nasal swelling due to plasma cell infiltrate in a cat without plasma cell pododermatitis. Vet Dermatol 21(4):412–414
285. Gross TL, Ihrke PJ, Walder EJ et al (2005) Plasma cell pododermatitis. In: Skin diseases of the dog and cat: clinical and histopathological diagnosis, 2nd edn. Blackwell Science, Ames, pp 363–364
286. Gruffydd-Jones TJ, Orr CM, Lucke VM (1980) Foot pad swelling and ulceration in cats: a report of five cases. J Small Anim Pract 21: 381–389
287. Miller WH, Griffin CE, Campbell KL (2013) Plasma cell pododermatitis. In: Muller & Kirk's- small animal dermatology, 7th edn. Elsevier, St. Louis, pp 718–719
288. Pereira PD, Faustino AMR (2003) Feline plasma cell pododermatitis: a study of 8 cases. Vet Dermatol 14(6):333–337

7 The Cyto-architectures (Arrangements) in Skin Cytology

The term *cyto-architectures* refers to the arrangement that cells continue to maintain on cytological samples and represent the direct expression of their original anatomical organization. They can be defined as the point of connection between cytologists and pathologists [1–3].

In cytology, while the details of cells such as the chromatin model, the morphology and size of the nucleoli, and the cytoplasmic characteristics are best evaluated at high magnification, the architectural model is better evaluated when examining slides at low-medium magnification.

The evaluation of cyto-architectures is very useful not only to obtain the diagnosis of neoplasia but, in some cases, also to define which histotype that specific neoplasia belongs to. Finally, it should be emphasized that when a cyto-architecture can be recognized on the slides, it does not mean that we are observing a neoplasia with certainly; in fact, a recognizable arrangement of cells on cytological slides can also come from hyperplastic tissues, particularly in the course of epithelial lesions.

Knowledge of basis of the histopathology of skin tumors greatly enhances the ability of cytologists to interpret the cyto-architectures observed on cytological specimens.

Since cells of epithelial tissues are tightly connected to each other via intercellular adhesion molecules, the most common arrangement observed on cytological specimens is those belonging to epithelial structures such as the epidermis or follicular and glandular adnexal epithelia. Their recognition can be very useful in differentiating a glandular tissue from a basal cluster of follicular or epidermal cells when, for example, papillary cyto-architectures or palisade arrangements of cylindrical cells are clearly recognizable.

An arrangement of cells that resembles the typical organization of the original tissues can also be found in some cytological specimens from soft tissue mesenchymal sarcomas; in these cases, vessels or collagen acts as structures that allow cells to remain in groups, drawing an architecture. As regards histopathological preparations, the growth patterns are very numerous, but not all are recognizable on cytological preparations [3]; therefore, in this chapter only those observable with cytology will be considered.

The main cellular patterns that can be found during hyperplastic or neoplastic skin lesions and that are discussed in this chapter are reported below:

- *Pavement-like*
- *Honeycomb*
- *Acinar*
- *Palisade*
- *Trabecular*
- *Papillary/micropapillary*
- *Ductal*
- *Fasciculate*

7.1 Pavement-Like Arrangement

In this cytological pattern, cells are jointed between them and arranged in a single layer similar to floor tiles. This cells' disposition is typical of pluristratified-flattened epithelia in which sheets of epithelia detach from the tissue and are released on slides. To take on a morphology that resembles floor tiles, the cytoplasm must be large and with a clearly distinguishable polygonal profile, and the nuclei must be central and uniform in size; this arrangement is better recognizable in cells with a clear squamous differentiation as occurs for the epidermidis or epithelia of mucosa.

Pavement-like arrangements can be indeed observed in many epithelial lesions, both hyperplastic and neoplastic such as hyperplastic epidermal or mucosal epithelia, squamous cell carcinoma, follicular neoplasia such as the infundibular keratinizing acanthoma, and less frequently some cases of apocrine neoplasia (Figs. 7.1 and 7.2). In the latter neoplasia, the most common pattern is the honeycomb ones, but when cytoplasm is larger than usual, a pavement-like appearance can be imagined. In well-differentiated squamous tissues, cells are regularly organized, and their edges are

F. Albanese, *Canine and Feline Skin Cytology*, https://doi.org/10.1007/978-3-031-95713-0_7

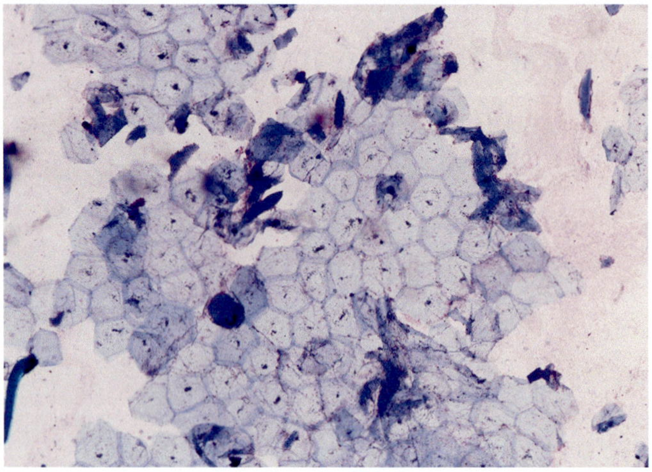

Fig. 7.1 Pavement-like arrangement: sheet of regular in size polygonal corneocytes collected with acetate adhesive tape from a scaly disorder in a dog

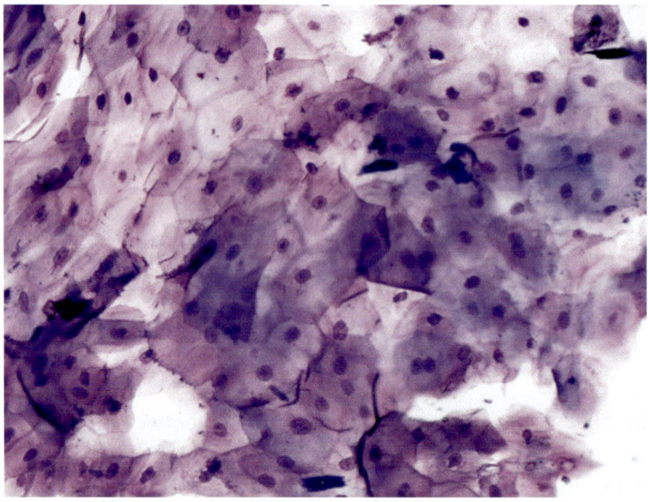

Fig. 7.2 Pavement-like arrangement: large sheet of squamous epithelium from a hyperplastic spinous layer

perfectly evident; in less-differentiated squamous cell neoplasia, the arrangement of cells can be less clear, and cytoplasm borders can be less regular and detailed; with a little experience, a cytological arrangement that resembles a pavement can be recognized by a cytologist in less well-differentiated and organized epithelial lesions also (Figs. 7.3, 7.4, and 7.5).

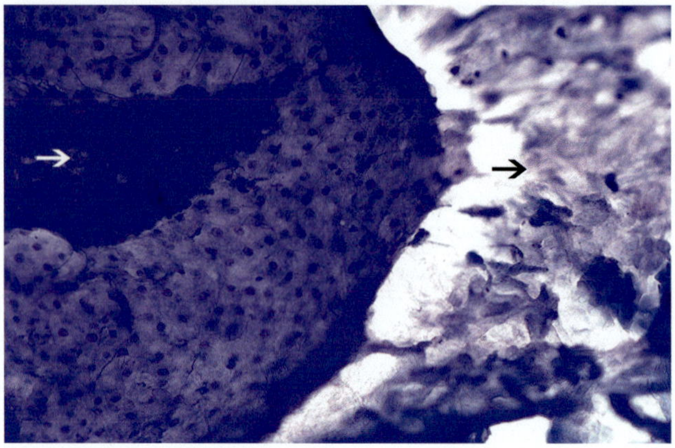

Fig. 7.3 Pavement-like arrangement: large sheet of squamous nucleated epithelium from an IKA. Note as basal keratinocytes (white arrow) and anucleate corneocytes (black arrow) are also present along squamous nucleated cells

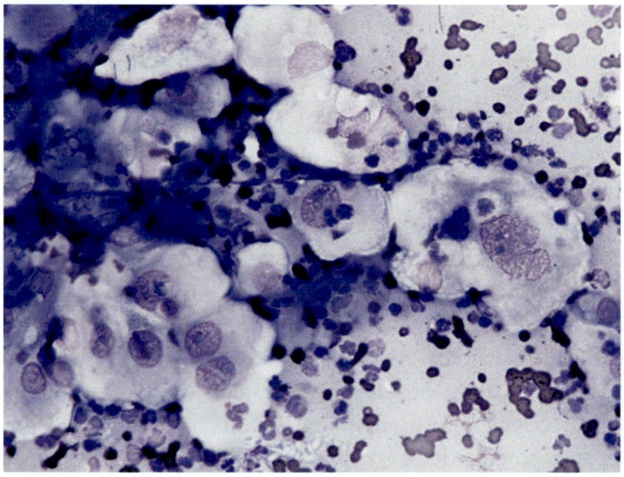

Fig. 7.4 Pavement-like arrangement: small cluster of malignant neoplastic squamous cells, different in size, but in which a pavement-like arrangement yet recognizable

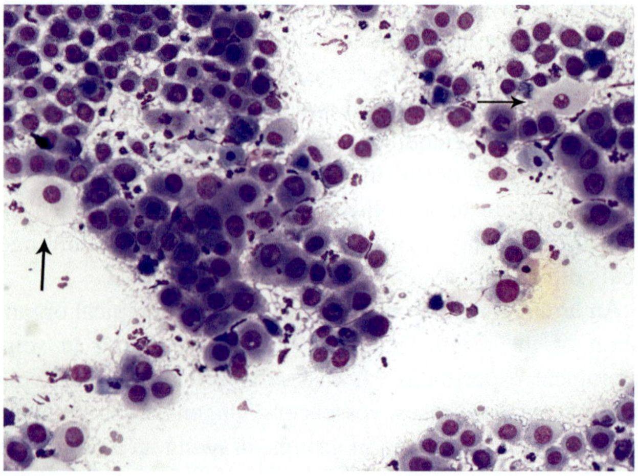

Fig. 7.5 Pavement-like arrangement: large cluster of moderately differentiated carcinomatous squamous cells where, despite a loss of cohesivity, a pavement-like pattern can be recognizable, yet two well-differentiated squamous cells are recognizable (arrows)

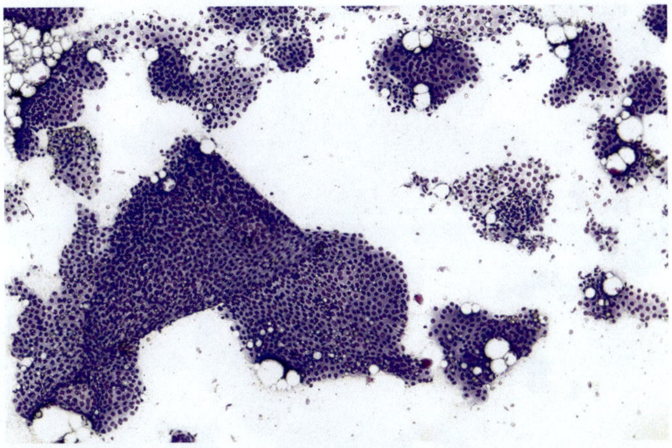

Fig. 7.6 Honeycomb arrangement: large sheets of nucleated epithelium from a sweat gland adenoma: the regular size of nuclei and small cytoplasm resemble the cells of a favum

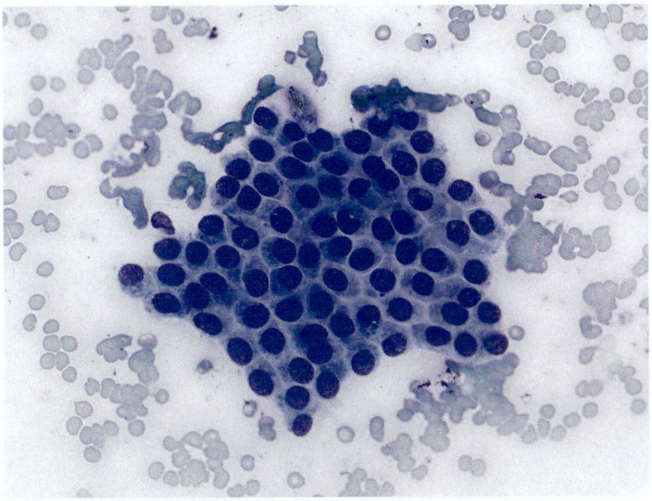

Fig. 7.7 Honeycomb arrangement: high magnification of an apocrine cluster; an orderly arrangement of cells of the same sizes resembling a favus is clearly evident

7.2 Honeycomb Arrangement

Some epithelial tissues may assume a morphology that, on cytological examination, resembles a *honeycomb*; in skin lesions, this cellular arrangement is evident especially in cells coming from the apocrine glands, while in internal organs, similar arrangements are commonly observed, for example, in prostatic benign lesions. As we will see later, in the palisade pattern, sweat glands are easily recognizable by their cuboidal and cylindrical morphology with polar nucleus and secretion by cytoplasmic decapitation, but in many cases of hyperplasia or well-differentiated adenomas, apocrine cells can be arranged on slides as flattened sheet of cells losing their cuboidal or cylindrical shape. This double morphological aspect can be explained since, when the pressure of sweat into the glandular lumens is high, the apocrine cells are squashed and, once placed on the slides, assume a flattened appearance; in practice, the cells are not observed laterally like palisades as occur for not flattened apocrine cells, but from top to bottom.

In this cyto-architecture, since cells are of the same size and hexagonal in shape, with central round nuclei, they assume an appearance very similar to the morphology of a honeycomb (Figs. 7.6 and 7.7).

When the cytoplasm of apocrine glands is larger than usual, these arrangements are cytologically not so dissimilar to those above described for the pavement-like arrangement, and the two epithelial patterns can be overlapped (Fig. 7.8).

In practice, despite these differences can seem of a few diagnostic utility, it can be concluded that the honeycomb structure of epithelial clusters coming from skin lesions reinforces the suspicion of a glandular tissue of apocrine origin.

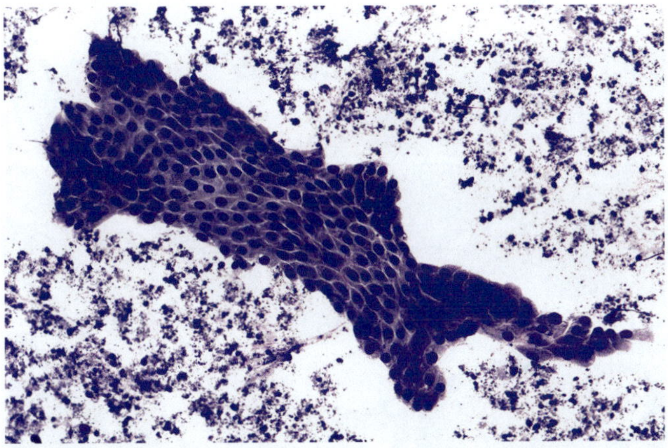

Fig. 7.8 Honeycomb arrangement: regular arrangement of a cluster coming from a ceruminous benign neoplasia. The larger cytoplasm can mimic a pavement-like arrangement

7.3 Acinar Arrangement

In Latin *acinus* means berry, so in pathology, the term "acinus" refers to a group of cells that resemble a multilobed "berry," such as a raspberry (Figs. 7.9 and 7.10). In cytology the three-dimensional aspect of glandular tissue is lost, and an acinus is observed as a group of cells peripherally arranged around an empty central area that represents the glandular lumen; in many cases, amorphous material representing glandular secretion is observed in the center of the lumen (Figs. 7.9 and 7.10).

Apart from mammary glandular tissue, acinar glands are not present in the glandular adnexa of the skin; in fact, both apocrine (sweat glands) and sebaceous glands are not

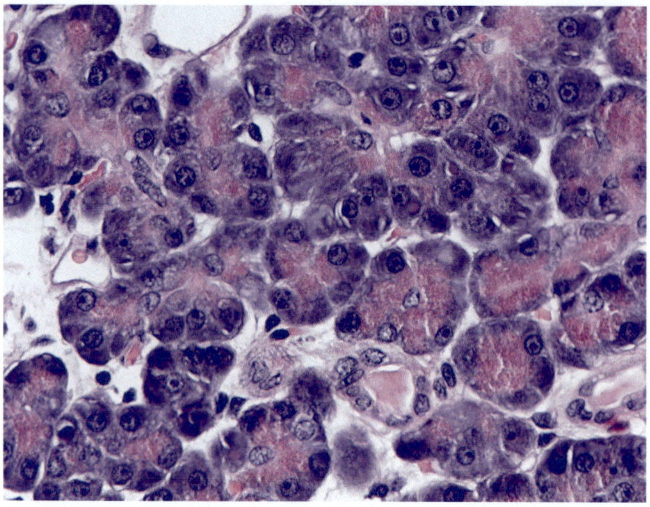

Fig. 7.9 Acinar arrangement: normal acini of a pancreatic lobule. The acinar cells that compose the acinus are polygonal with a broad base and a narrow apex arranged to delineate a narrow central lumen; the apical cytoplasm is markedly eosinophilic due to the presence of zymogen granules (courtesy of DVM L.V. Muscatello—Italy)

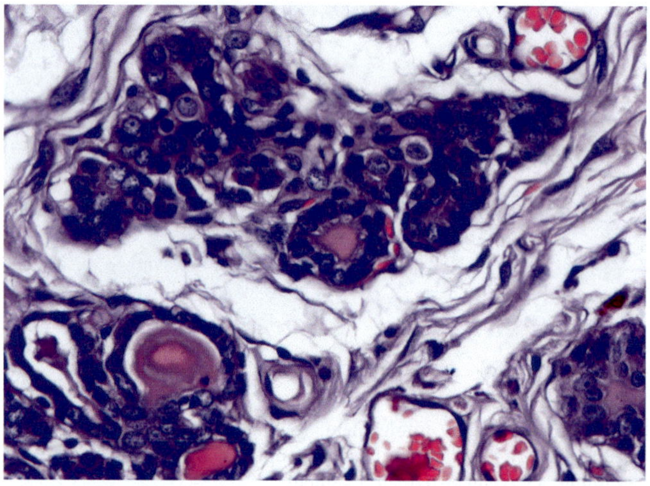

Fig. 7.10 Acinar arrangement: normal acini of a mammary lobule. The luminal epithelial cells are polygonal and surround a small central lumen containing a few amount of eosinophilic secretion (courtesy of DVM L.V. Muscatello—Italy)

anatomically acinar. In most cases, in neoplasms of the apocrine glands, both sweat and ceruminous, cells can take on an arrangement that resembles an acinus, when, as mentioned, cells are organized around an optically empty central area (Fig. 7.11).

Since apocrine glands are simple tubular glands and therefore not composed of acinar structures, the term acinar arrangement must be considered incorrect when describing a cytological sample obtained from apocrine glands; therefore, apart from the impossibility of forming acini by apocrine sweat glands, when arrangements that resemble the silhouette of acini are found, it would be more

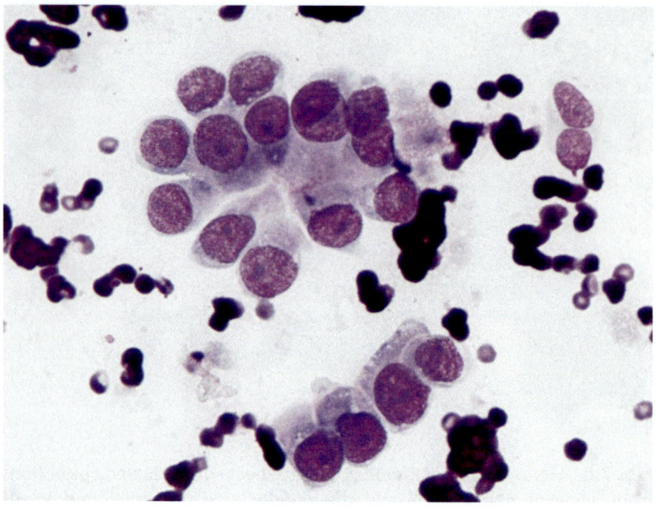

Fig. 7.11 Cytology of acinar arrangement: acinar and palisade arrangements collected from the digit of a cat with metastasis of a primary bronchial carcinoma (lung-digit syndrome)

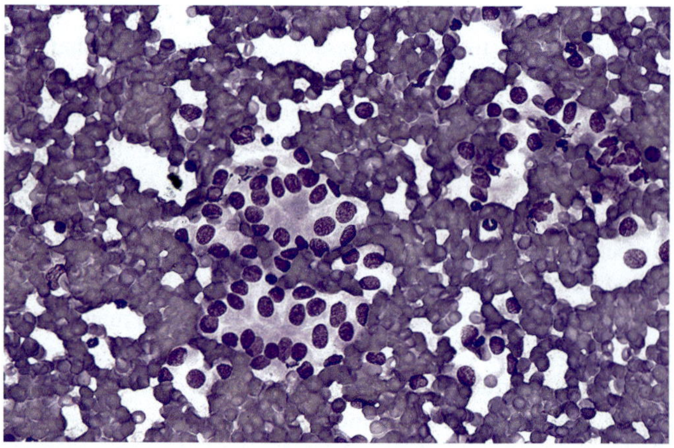

Fig. 7.12 Anal sac carcinoma: two rosettes are recognizable as cells arranged at the periphery of a virtual central lumen. Despite the cytological similarity, these formations are not acini

appropriate to use terms such as pseudo-acinar or acinar-like arrangements. From what has been said, real primary acinar cyto-architectures are not possible in the skin since the latter is devoid of acinar glandular tissue. If acini are found, assuming that they really are, a cutaneous metastasis must be considered from a neoplasm of the mammary gland or from a neoplastic internal organ composed of acinar glands.

Caution should be exercised when observing acini-like structures in cytological examination; in fact, in some histotypes of anal sac carcinoma, some cells may assume a disposition resembling an acinus; in this case, such cellular arrangements, named *rosette*, are not a real anatomical structure but only represents a disposition of single cells arranged around a virtual space that doesn't represent a real glandular lumen, as discussed in Chap. 8 (Fig. 7.12).

7.4 Palisade Arrangement

Some epithelial neoplasms are histologically organized as linear aggregates of basal cells, held together by intercellular junctions. This organization is well recognized in both histology and cytology and is represented by cells with a cuboidal to cylindrical shape, depending on the tissue of origin, which are arranged as rows of cells that resemble the arrangement of the square poles of a palisade (palisade arrangement). When these rows are represented by basal cells, composed of oval nuclei of regular size, often close together and in contact with each other, and with little or barely recognizable cytoplasm, a trichoblastoma must be suspected, a very common hair follicle neoplasm in both dogs and cats (Figs. 7.13, 7.14, 7.15, and 7.16).

Palisades also characterize apocrine gland tissue; in fact, the cytological architectures recall the anatomy of sweat gland in which, as tubular gland, a central lumen is bordered by a monolayer of cuboidal to cylindrical cells: the latter show a basal nucleus and an apical border that undergoes to a cytoplasmic decapitation originating the sweat. The sweat, in fact, is simply composed by detached pieces of the apical part of the cytoplasm of apocrine cells. In well-differentiated apocrine tumors as well as in course of hyperplastic apocrine lesions due to chronic stimulus on the skin, a number of variably length rows of cells arranged in palisades can be observed (Figs. 7.17, 7.18, 7.19, and 7.20).

When the latter are simultaneously present along with other clusters with honeycomb arrangements, an apocrine origin of the cells can be ascertained (Fig. 7.21).

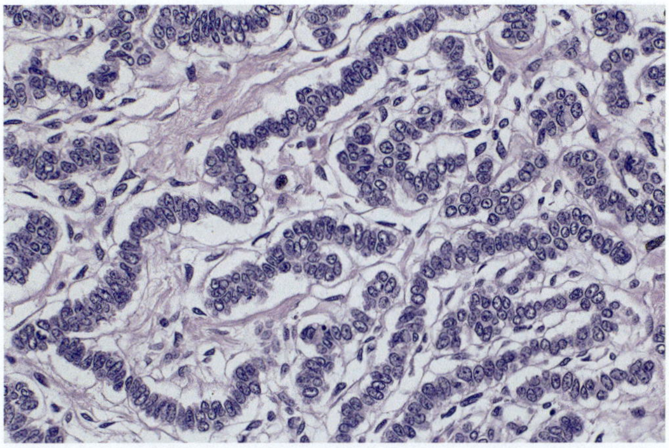

Fig. 7.14 Histology of trichoblastoma: linear rows of basal keratinocytes arranged in palisades

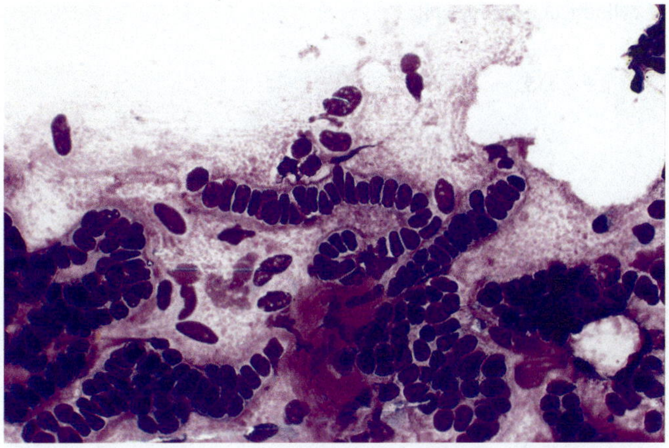

Fig. 7.15 Cytology of trichoblastoma: basal cells rows immersed in eosinophilic stroma

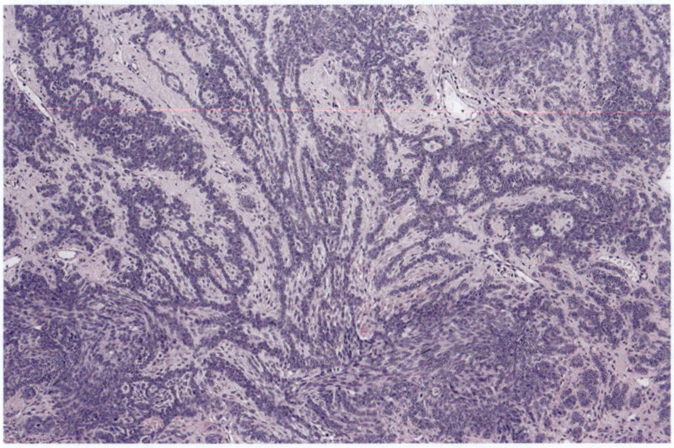

Fig. 7.13 Histology of trichoblastoma: basal keratinocytes arranged in rows composed of adjacent cells forming long palisades

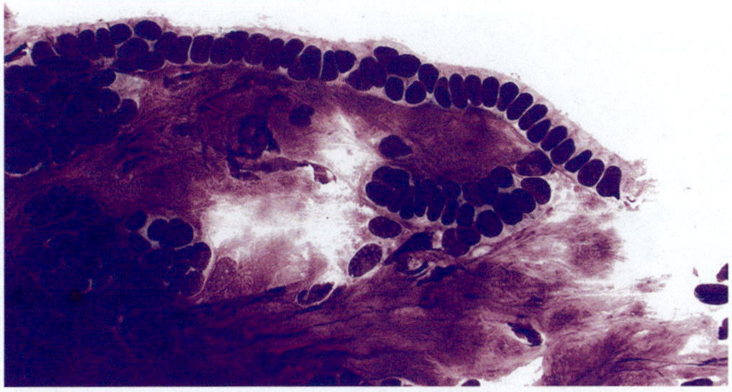

Fig. 7.16 Cytology of trichoblastoma: regular basal cells arranged in rows give to the cluster a palisade morphology

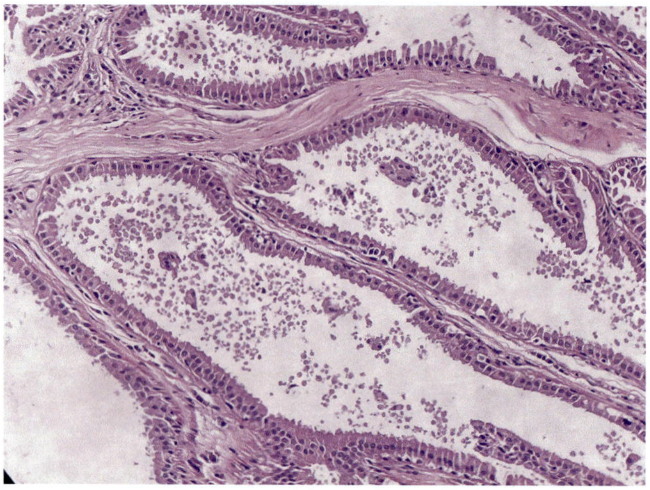

Fig. 7.17 Histology benign apocrine cystadenoma: multiple tubular areas containing amorphous blebs of sweat and lined by a monolayer of cylindrical apocrine cells

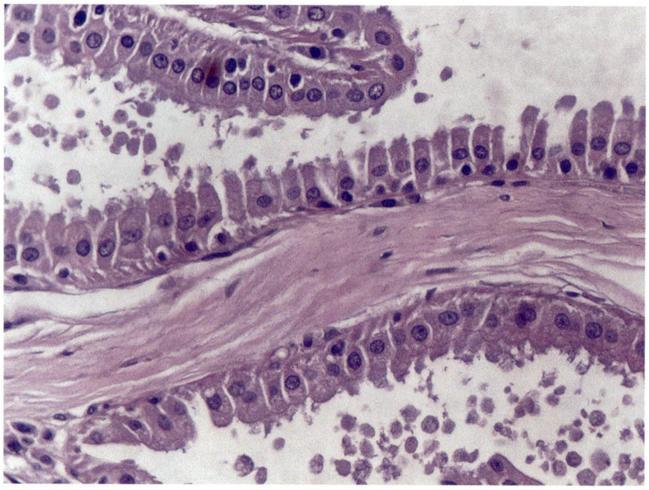

Fig. 7.18 Histology benign apocrine adenoma: close up of the same picture n. 7.17 in which palisading arrangement is clearly recognizable

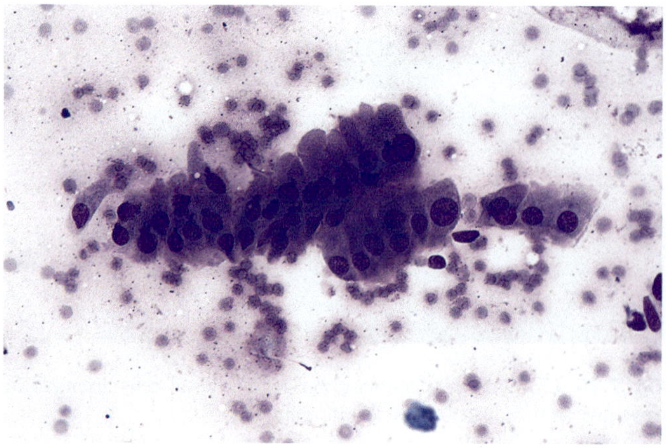

Fig. 7.19 Cytology sweat gland hyperplasia: rows of cylindrical apocrine cells

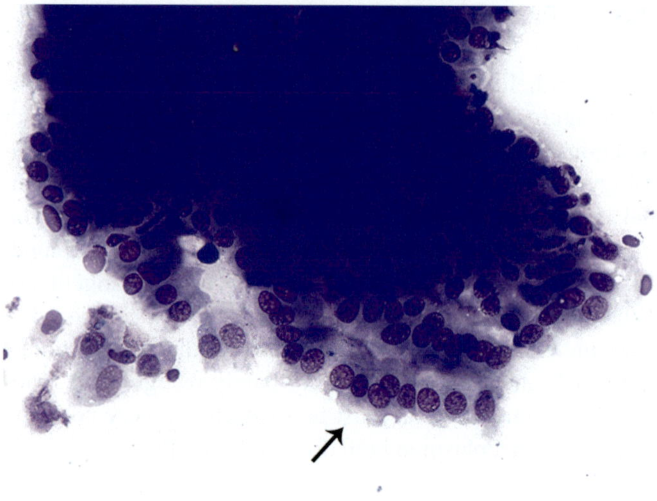

Fig. 7.20 Cytology sweat gland hyperplasia: simultaneous presence of honeycomb and palisade arrangements (arrow)

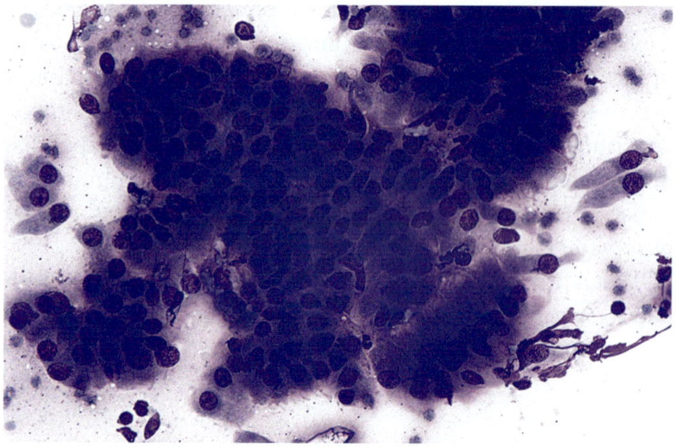

Fig. 7.21 Cytology sweat gland benign neoplasm: simultaneous presence of honeycomb and palisade arrangements indicates an apocrine origin of cells

7.5 Trabecular Arrangement

The word *trabecula* comes from the Latin word "trabes" meaning beam. In pathology, trabeculae represent anatomical structure that resembles, in form and function, a pillar or beam, and a typical example is the bones' trabeculae.

Trabeculae are also slangily used to define the morphology of some cluster of cells that can be observed in both healthy and pathologic skin epithelial lesions.

Trabecular arrangement is not indeed specific to a particular neoplasm, but is commonly observed in different epithelial neoplasia such as trichoblastoma and glandular tumors both apocrine (sweat and ceruminous glands) and of sebaceous origin such as sebaceous epithelioma and hepatoid glands. In cytology samples trabeculae are represented by large groups of cells, most commonly composed to the basal portion of epithelial of tissues, from roundish to irregular profile, of various sizes and very often branched.

Unlike other cyto-architectures, they do not have great diagnostic importance but, when observed, allow the tissue of origin to be classified as epithelial (Figs. 7.22, 7.23, 7.24, and 7.25).

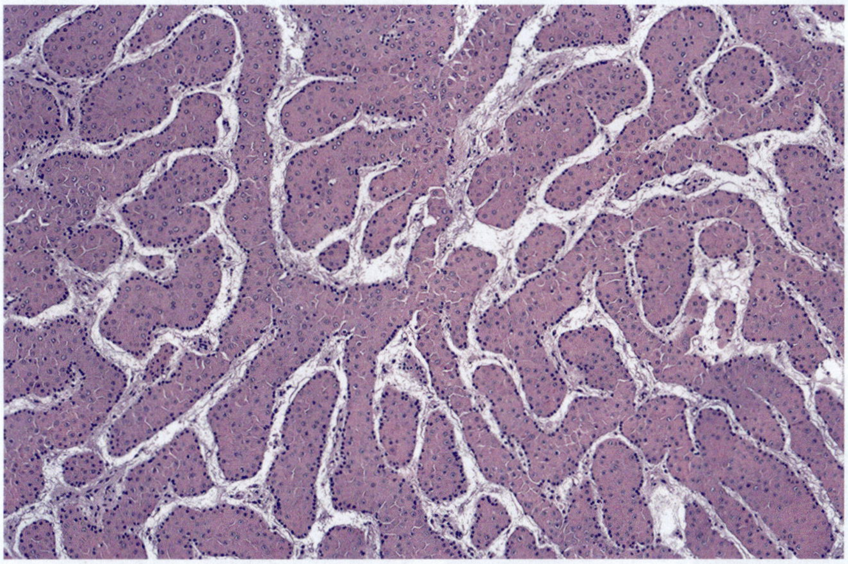

Fig. 7.22 Histology of hepatoid gland adenoma: many trabeculae of hepatoid gland tissue

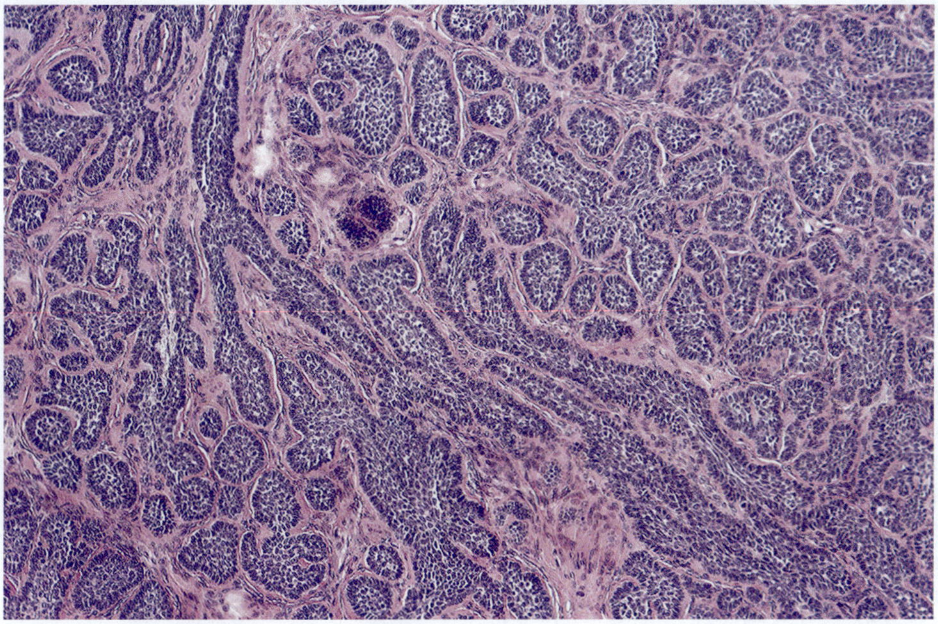

Fig. 7.23 Histology of trichoblastoma: trabeculae supported by a thin collagen stroma

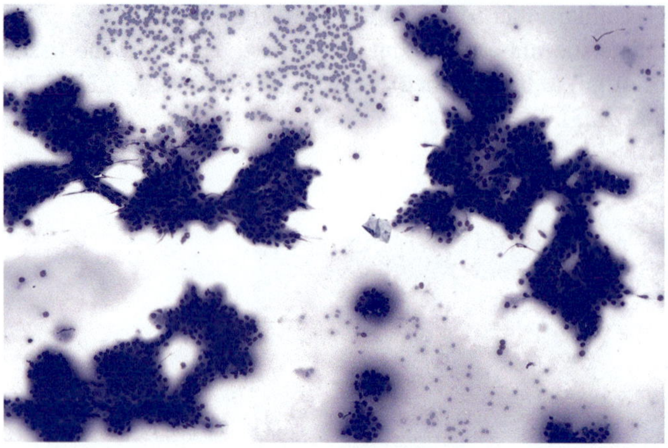

Fig. 7.24 Cytology of hepatoid glands: multiple glandular trabeculae; some of them show a branched arrangement

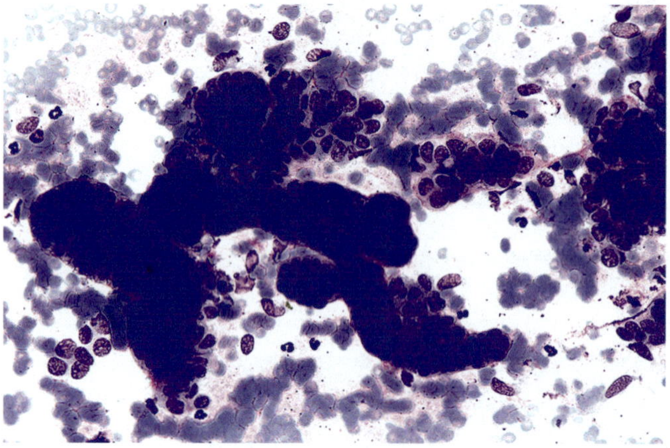

Fig. 7.25 Cytology of trichoblastoma: epithelial trabeculae with roundish profile

7.6 Papillary/Pseudopapillary Arrangement

In pathology, the term *papilla* refers to an epithelial projection similar to a nipple, supported by a fibro-vascular axis; when the epithelial projection does not present a fibro-vascular ventral core, the term *pseudopapillae* is used. Apart from the mammary gland, in the skin, true papillae can be observed only in some apocrine tumors of the sweat glands. Their morphological recognition, not always easy in cytology, allows not only to define their origin as apocrine but also to define the histotype (apocrine papillary adenoma or carcinoma) (Figs. 7.26 and 7.27). On cytological slide papillae are recognizable as elongated cluster of cells, which line peripherally a supporting fibrous or fibro-vascular stroma. In some cases, papillae can be branched assuming a frond-like appearance (Figs. 7.28 and 7.29).

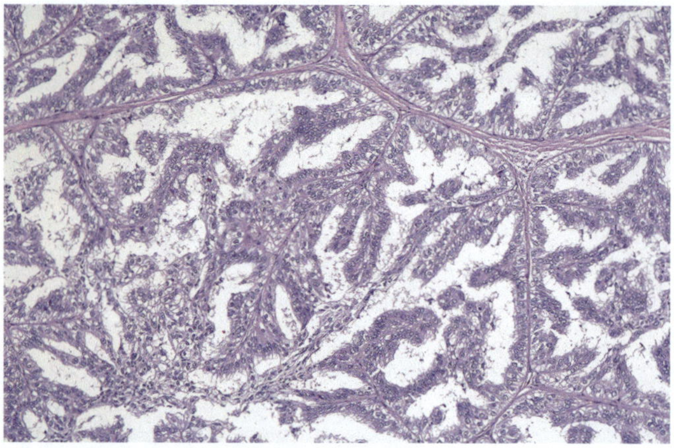

Fig. 7.26 Histology of papillary apocrine carcinoma: numerous papillae protrude into the glandular lumens

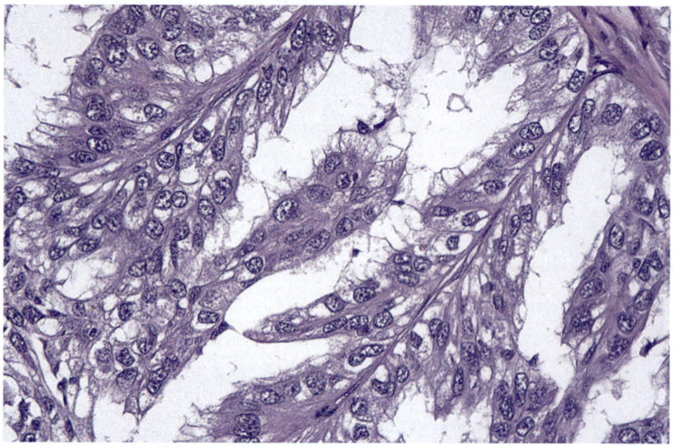

Fig. 7.27 Histology of papillary apocrine carcinoma: a fibro-vascular stroma composes the core of the papillae

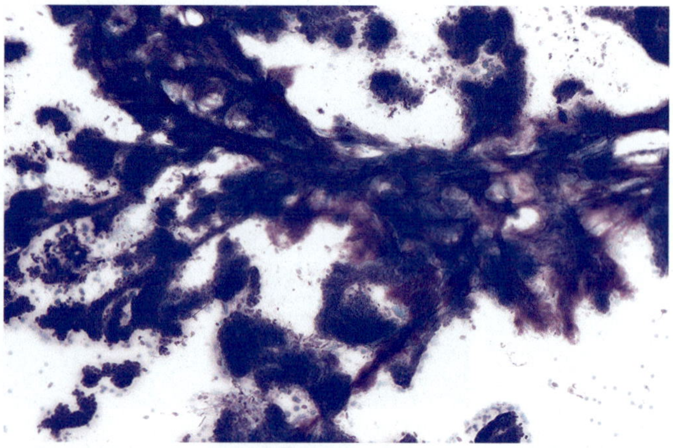

Fig. 7.28 Cytology of papillary apocrine carcinoma: branched papillae assuming a frond-like appearance. Note as cytological features resemble the histopathology of the previous two images

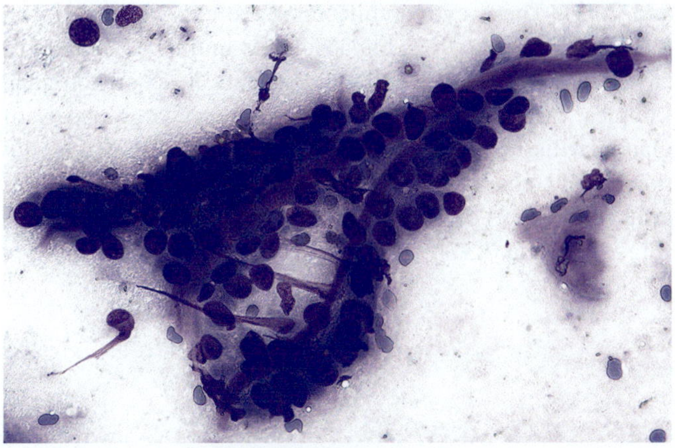

Fig. 7.29 Cytology of papillary apocrine carcinoma: a stromal axis is well recognizable in the center of papillae

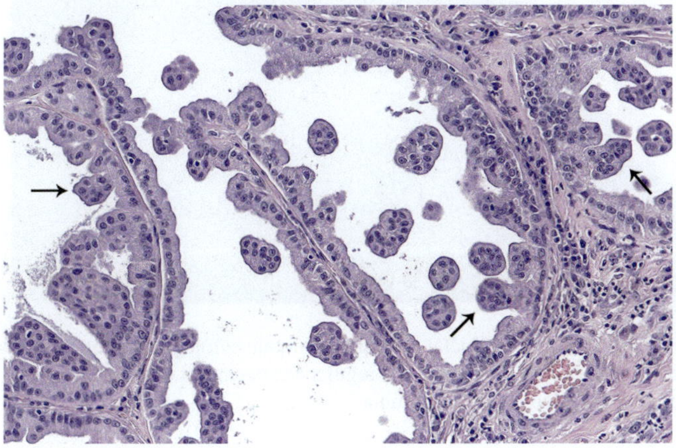

Fig. 7.30 Histology of papillary apocrine adenoma: papillae and pseudopapillae (arrows) are present on the same tumor

Finally, in some apocrine tumors, numerous pseudopapillae can be observed, alone or together with true papillae, from which they sometimes branch off. On cytological slides, pseudopapillae are not easy to recognize because they are represented by small groups of epithelial cells of a rounded or elongated shape, even if in some cases they can be larger; if no papillary arrangement is present on the same slide, pseudopapillae cannot be easily recognized because they can be similar to other epithelial aggregates (Figs. 7.30, 7.31, 7.32, and 7.33).

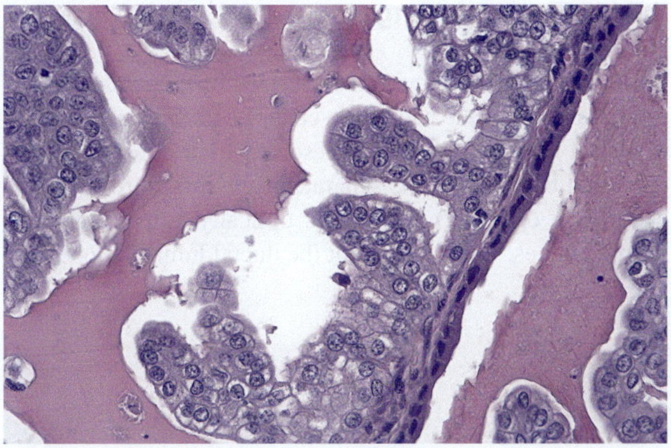

Fig. 7.31 Histology of papillary apocrine adenoma: pseudopapillae protrude into the lumen

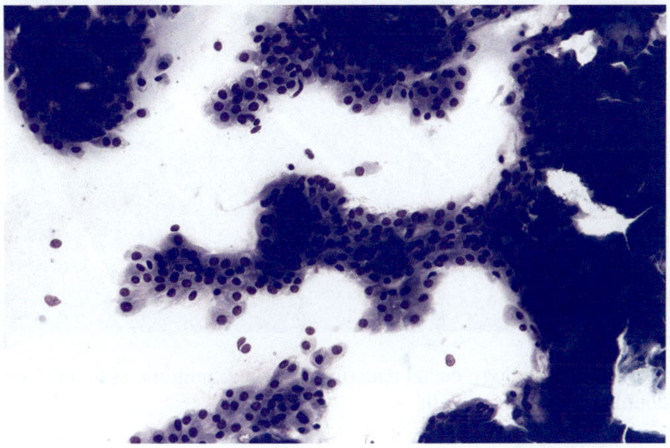

Fig. 7.32 Cytology of papillary apocrine adenoma: large pseudopapillae sampled from the same case of picture Fig. 7.31. Note as a fibrovascular axis is not present

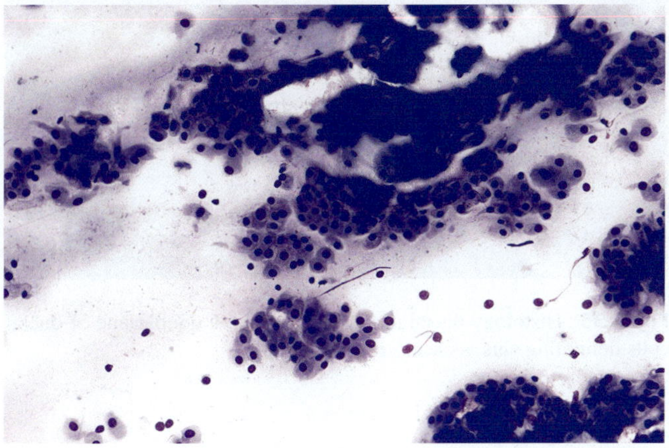

Fig. 7.33 Cytology of papillary apocrine adenoma: small pseudopapillae are not differentiable from not papillary neoplastic clusters

7.7 Ductal Arrangement

Some apocrine tumors are characterized by a ductal origin or differentiation of the neoplastic cells. The histological aspect of ductal differentiation is represented by a double layer of epithelial basophilic cells separated by a thin fibrous stroma and which often forms a central optically empty slits with sharped edges, which represent the ductal lumina (Figs. 7.34 and 7.35). These cyto-architectures are easy to identify on histological preparations but absolutely unlikely on cytological preparations. In very cellular and lucky cytological samples, however, it is possible to collect large sheets of glandular tissue in which clear aspects of ductal differentiation remain, characterized by an opposed bilayer of basal cell arranged in rows of variable length with a thin ductal lumina (Figs. 7.36 and 7.37).

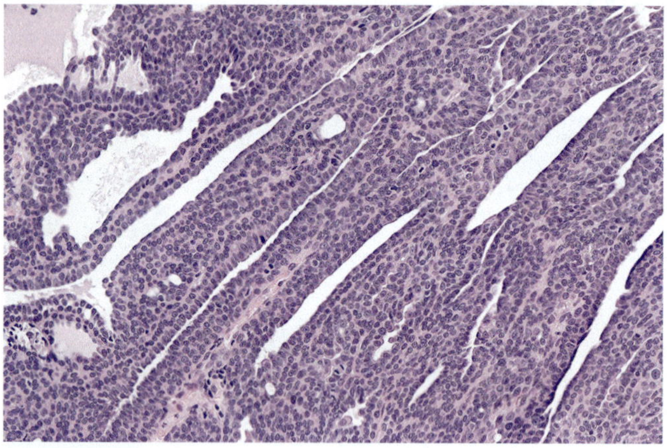

Fig. 7.34 Histology ductal apocrine adenoma: multiple slits lined by double rows of ductal cells

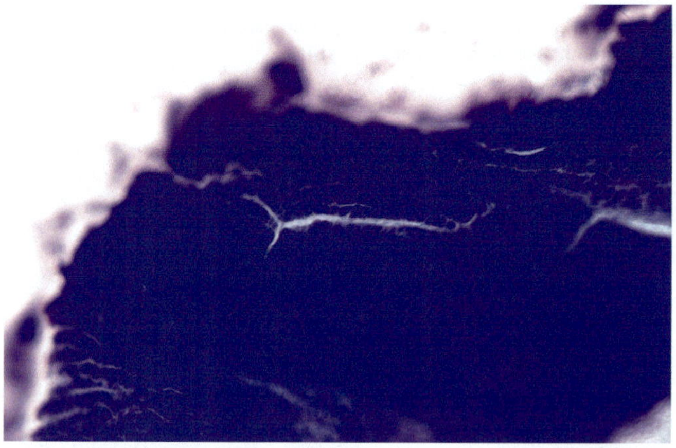

Fig. 7.36 Cytology ductal apocrine adenoma: cytology depicting of the same case of Fig. 7.34. Note the regular arrangements of the double line of cells that delimit ductal lumina

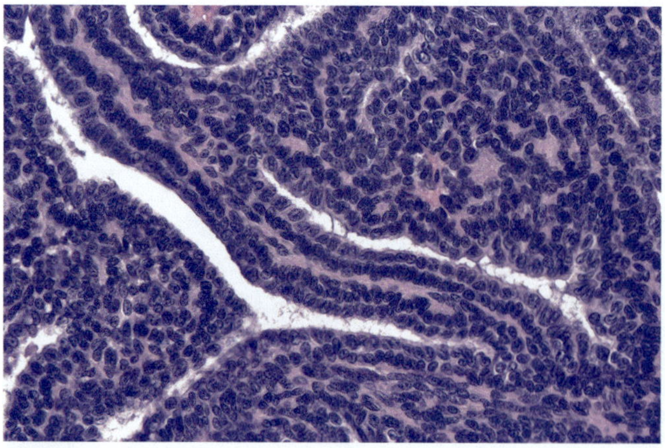

Fig. 7.35 Histology ductal apocrine adenoma: a double line of ductal cells delimiting slits is clearly recognizable

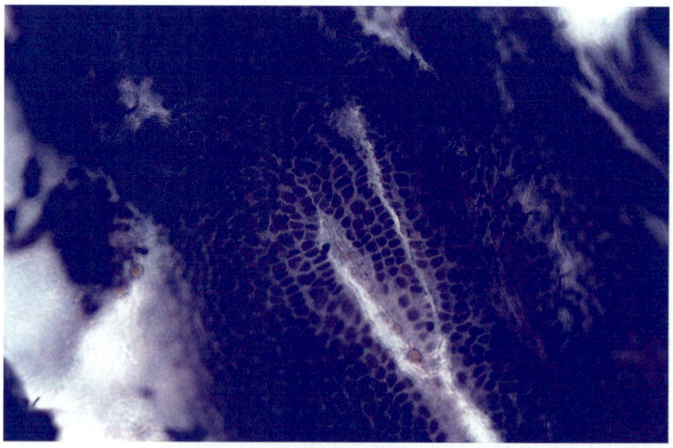

Fig. 7.37 Cytology ductal apocrine adenoma: at high magnification the double line of cells and the ductular lumen are clearly recognizable

7.8 Fasciculated Arrangement

The interpretation of histopathological patterns in mesenchymal spindle cell tumors of soft tissue is of fundamental importance to define the histotype of the tumor under examination or to differentiate between a neoplasm and a reactive fibroblastic process, such as granulation tissue or focal reactive fibroplasia (ex nodular fasciitis).

Since the common denominator of these tumors is the spindle morphology of the neoplastic cells and their arrangements in bundles, the different growth patterns are grouped under the name of *fasciculated* pattern. The fasciculated arrangements include many patterns which are not pathognomonic of a specific neoplasia, since they can be shared by different soft tissue neoplasms; anyway their recognition can be useful to orient the diagnosis.

7.8.1 Bundles and Fascicles Arrangement

This is a pattern common to most STSs in which cells are organized in differently sized bundles and fascicles, oriented in any direction such as *parallel* and *interwoven/interlacing* (Figs. 7.38 and 7.39). This pattern is not easily observed on STS cytology but in highly cellular specimens that release large groups of cells; the interlacing fasciculated bundle pattern can be recognized in low-medium magnification (Figs. 7.40 and 7.41). Fibrosarcoma and nerve sheath tumors (NST) are the most common neoplasia in which those arrangements can be observed.

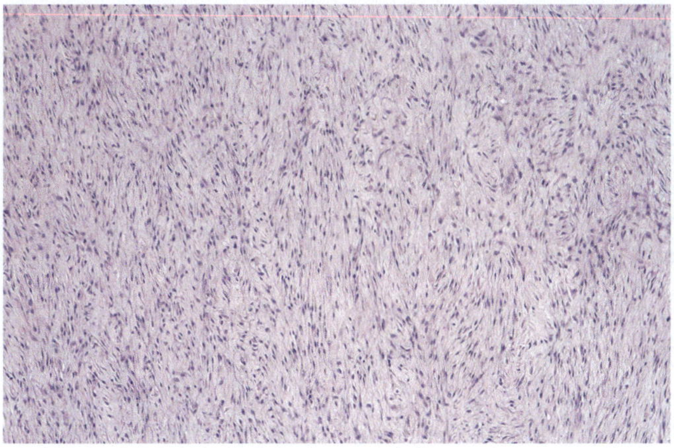

Fig. 7.38 Histology fasciculated pattern (parallel streams): mostly parallel bundles of neoplastic fibroblasts (fibrosarcoma)

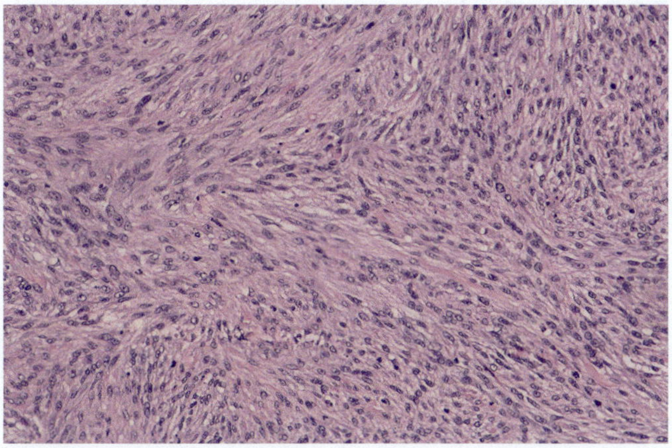

Fig. 7.39 Histology fasciculated pattern: interwoven bundles of neoplastic fibroblasts (fibrosarcoma)

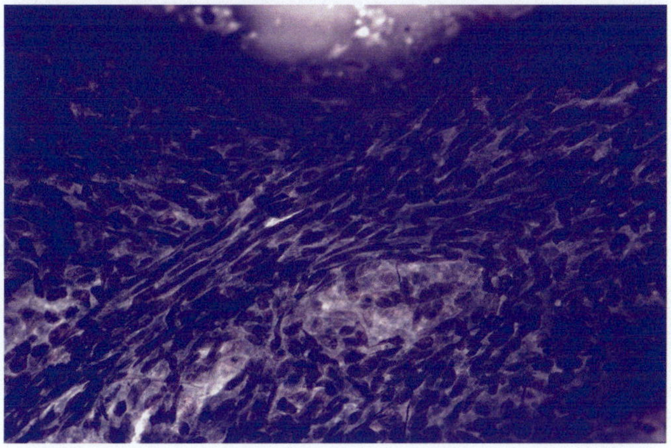

Fig. 7.40 Cytology fasciculated pattern (interwoven bundles): bundles of neoplastic fibroblasts show an interlacing/interwoven arrangement (fibrosarcoma)

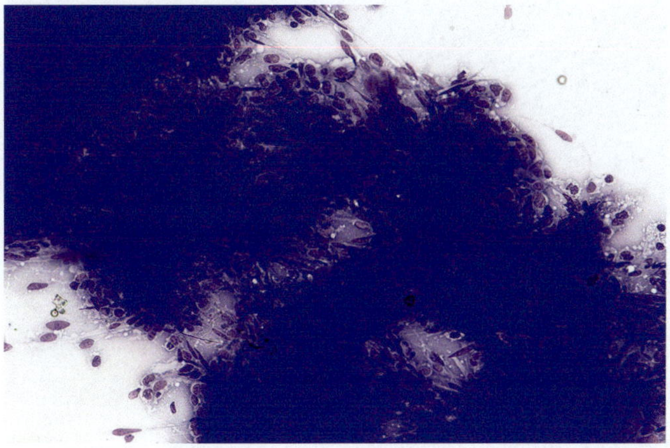

Fig. 7.41 Cytology fasciculated pattern (interwoven bundles): bundles of neoplastic fibroblasts show an interlacing/interwoven arrangement (fibrosarcoma)

7.8.2 Herringbone Arrangement

In fibrosarcoma, and less frequently in some NSTs, a cellular arrangement resembling a herringbone is frequently observed on a histopathological specimen. Starting from a central illusory or stromal axis, bundles of spindle cells intersect with one another at acute angles, imitating the spines of a fishbone and to which are given the name of *herringbone arrangement* (Figs. 7.42 and 7.43). Also this pattern, easily observed on histology, can be glimpsed with a little imagination on cytological specimens (Fig. 7.44).

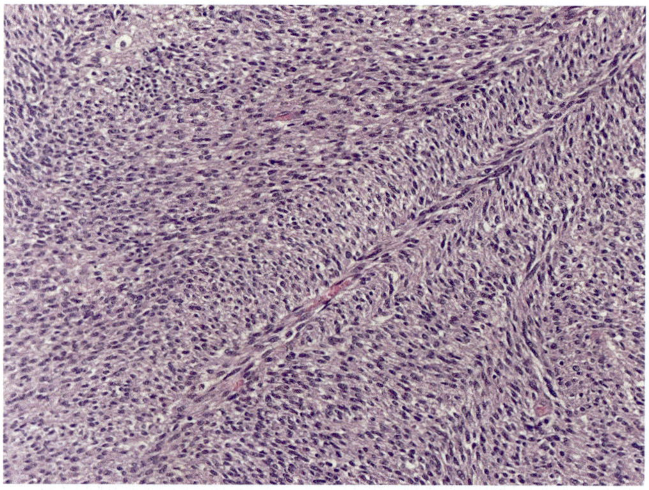

Fig. 7.42 Histology fasciculated pattern (herringbone arrangements): acute angle intersection of spindle neoplastic fibroblasts (fibrosarcoma)

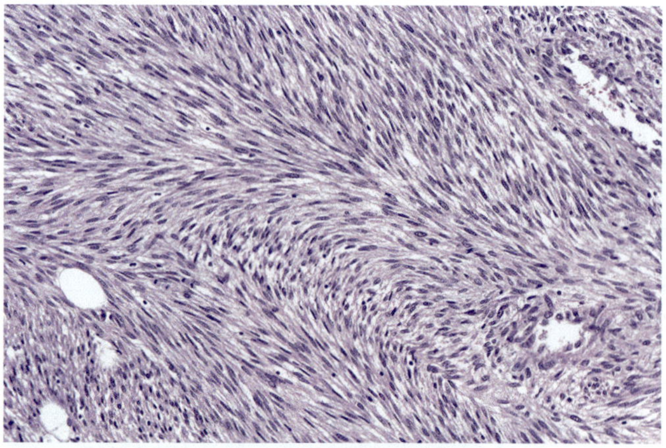

Fig. 7.43 Histology fasciculated pattern (herringbone arrangements): acute angle intersection of spindle neoplastic fibroblasts mimics a fish skeleton (fibrosarcoma)

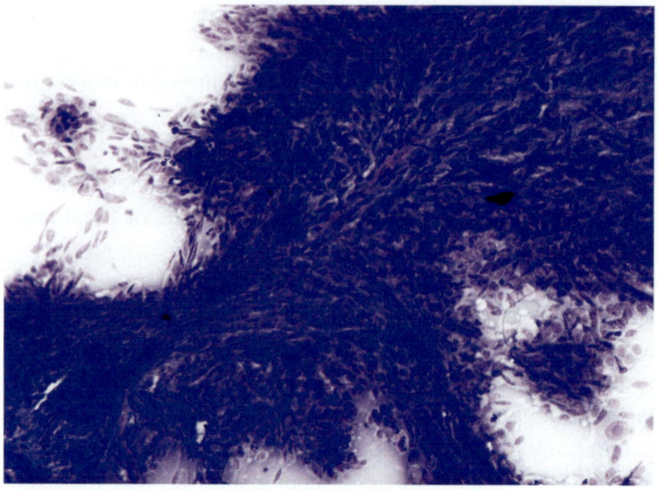

Fig. 7.44 Cytology fasciculated pattern (interwoven bundles): bundles of neoplastic fibroblasts show an interlacing/interwoven arrangement (fibrosarcoma)

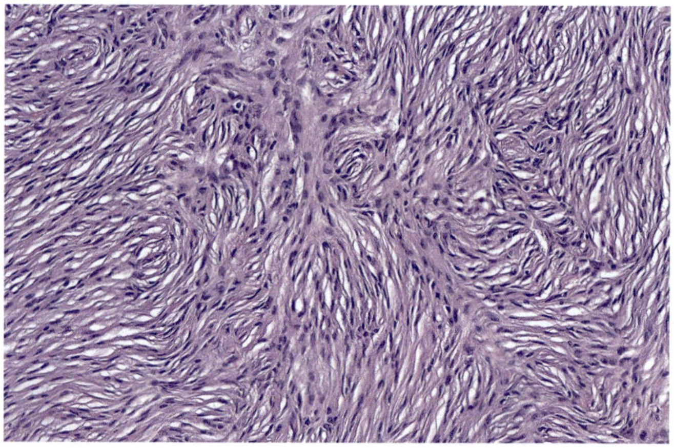

Fig. 7.45 Histology fasciculated pattern (fingerprint pattern): bundles of neoplastic fibroblasts show an irregular multifocal partially whorled interwoven pattern (PWT)

7.8.3 Storiform with Fingerprint Arrangements

This pattern, well recognizable in histology, is poorly appreciable on cytological preparations. Aspects of poorly defined vortices can give rise to the so-called fingerprint arrangements, very frequent in PWTs (Fig. 7.45).

7.8.4 Whorled Arrangement

On histopathological samples, in the whorled pattern, cells are arranged radially in spirals around a stem; the latter is usually represented by capillaries (PWT), by axons (NST), or by a collapsed illusory area; despite this pattern is mostly

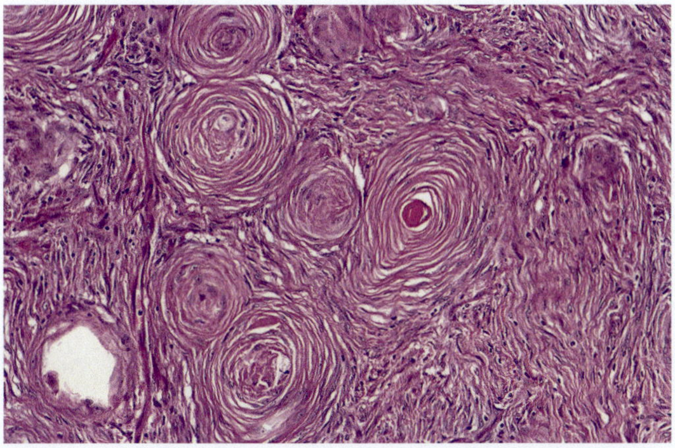

Fig. 7.46 Histology fasciculated pattern (whorled pattern): many whorls composed of spindle cells around an illusory stem. Immunohistochemistry allowed diagnosing a PWT

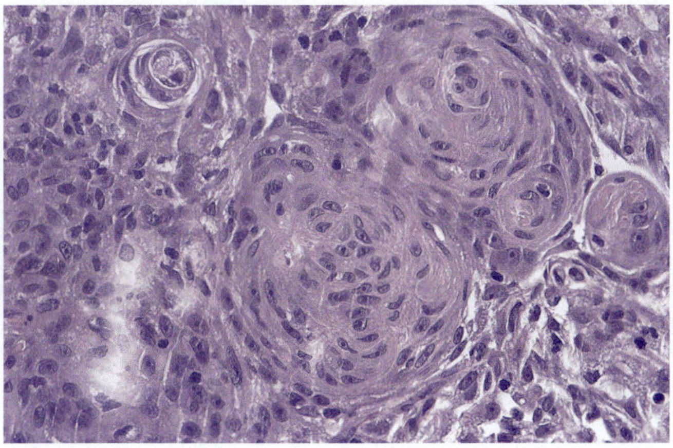

Fig. 7.47 Histology fasciculated pattern (whorled pattern): many whorls composed of spindle cells around an illusory, maybe collagen, stem. Immunohistochemistry allowed diagnosing a PWT

observed in PWTs and less frequently in NSTs, when the whorl is arranged around an illusory stem, a fibrosarcoma cannot be excluded from the differentials (Figs. 7.46 and 7.47).

In cytology, this pattern is very rarely observed, and, if present, no capillary or axons are recognizable in the center of the whorls. The neoplastic spindle cells are arranged in a spiral pattern; the cellular details are not clearly evident due to the thickness of the cell groups that give rise to a rounded three-dimensional structure composed of overlapping spindle cells but which allow a spiral arrangement to be recognized (Figs. 7.48 and 7.49). Since whorls are more frequently observed in PWTs than NSTs, and as abovementioned central vessels are never visible at the center of the whorls, other cytological aspects must be searched on cytological slides to orient the diagnosis. As discussed below, the cytological recognition of clear perivascular arrangements,

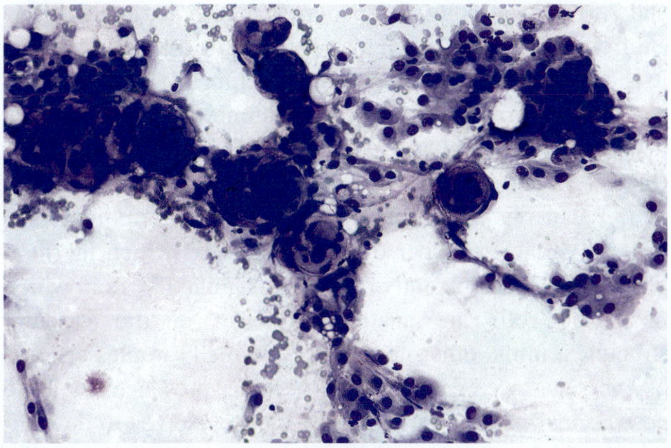

Fig. 7.48 Cytology of fasciculated pattern (whorled pattern): many whorls are recognizable as roundish structures composed of spindle cells arranged in a spiral pattern

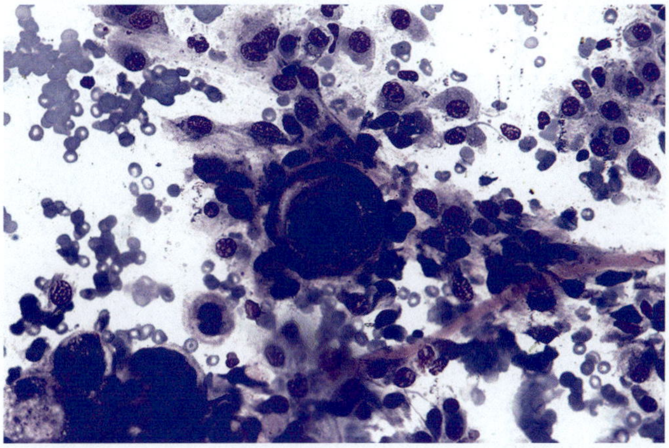

Fig. 7.49 Cytology of fasciculated pattern (whorled pattern): at high magnification and playing with focus, whorls are better recognized

morphologically completely different from the whorls in discussion, is strongly suggestive of PWT, and, when associated to whorls and to the so named "insect head" cells and "crown" cells, it becomes a hallmark of that neoplasia as it will be discussed in Chap. 8.

7.8.5 Perivascular Arrangement

The histopathological evidence of a *perivascular pattern* is diagnostic of tumors originating from the perivascular cells such as myopericytes, hemangiopericytes, etc. Due to their natural anatomical localization (around the vessels), the visualization of perivascular arrangements is comprehensible for this group of neoplasia. Perivascular pattern is easily recognizable when, on histological specimens, spindle cells arrange in a spiral way, around a central small vascular lumen representing a transversally cut capillary (Figs. 7.50

and 7.51). In cytology, this pattern is morphologically different from what observed on histopathological samples; in fact, it is characterized by long and sometimes branched small capillaries, surrounded by a large number of spindle cells that, from the capillary axis, depart in a centrifugal way (Figs. 7.52 and 7.53).

In cutaneous neoplastic cytology, other tumors may also show a variable number of vessels, around which the neoplastic cells are apparently organized. Unlike PWTs in which neoplastic cells are anatomically part of the vascular structures, in the other neoplasms in which samples are rich of vessels, we cannot define their presence as true perivascular pattern, but we must limit ourselves to describing the high presence of blood vessels associated to the neoplastic cells. Hepatoid glands neoplasia, lipoma, and liposarcoma are three classical examples in which capillaries can be very numerous, but the cytological definition of perivascular arrangements could be incorrect (Figs. 7.54 and 7.55).

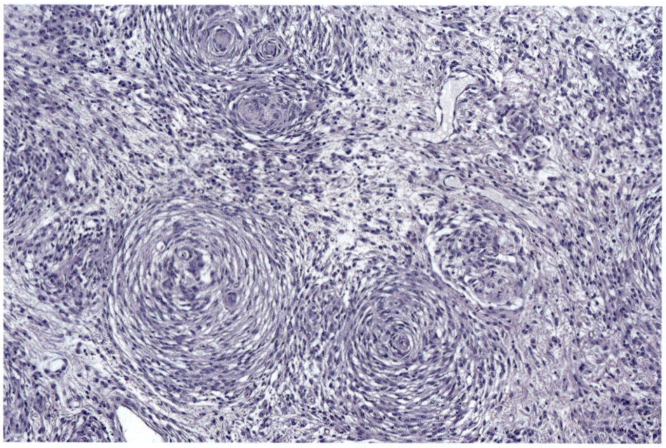

Fig. 7.50 Histology of fasciculated pattern (perivascular pattern): many whorls around small capillaries characterize the histology of PWTs

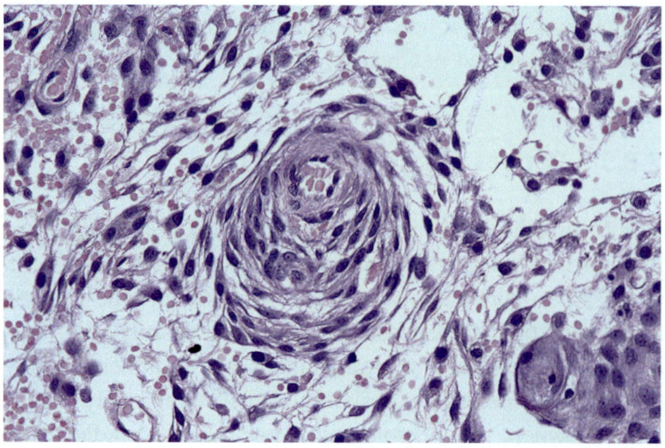

Fig. 7.51 Histology of fasciculated pattern (perivascular pattern): at high magnification, a central blood capillary is easily recognizable

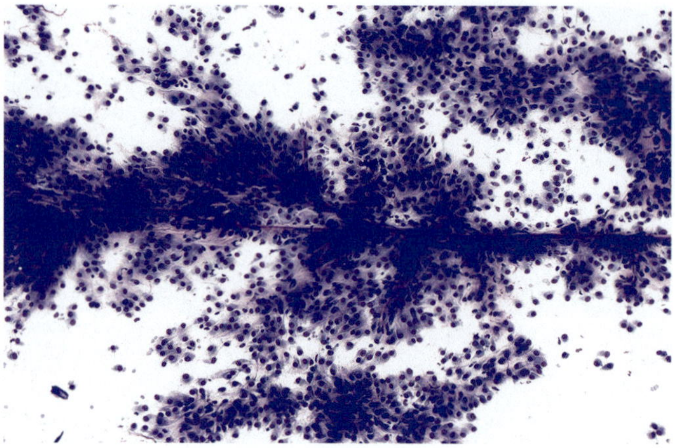

Fig. 7.52 Cytology of fasciculated pattern (perivascular pattern): many spindle neoplastic cells are arranged around central blood capillary

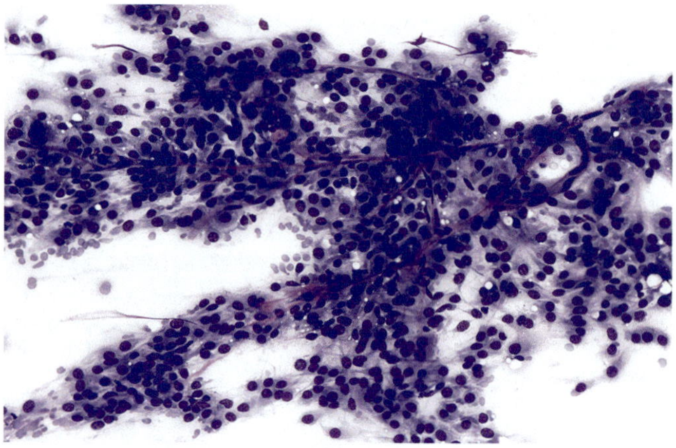

Fig. 7.53 Cytology of fasciculated pattern (perivascular pattern): a long capillary resembles the stem of a flower, while spindle cells the petals. Typical perivascular arrangement in course of a PWT

Fig. 7.54 Cytology of mimicking perivascular pattern: many capillaries along with hepatoid gland trabeculae in an adenomatous neoplasia

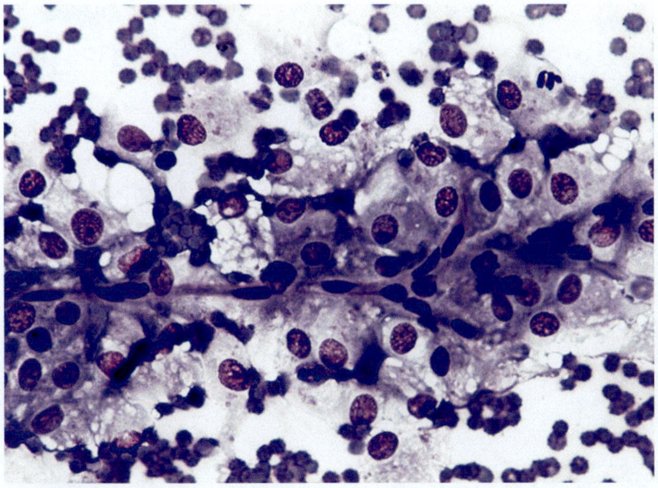

Fig. 7.55 Cytology of mimicking perivascular pattern: a branched capillary surrounded by neoplastic adipocyte is a common cytological feature in the case of liposarcoma

7.8.6 Nerve Sheath Growth Pattern

On histological samples of *peripheral nerve sheath tumors* (NST), a nervous pattern is well known and is composed of cellular arrangements that give rise to particular morphological aspects of growth.

Some NSTs are in fact characterized by denser cellular areas composed of bundles of spindle cells organized in clusters (Antoni A), by hypocellular areas with spindle cells that mimic a cobweb-like meshwork (Antoni B), and by the so-called Verocay bodies, a component of "Antoni A" pattern, that are morphologically composed by two opposing groups of spindle cells, arranged in parallel way resembling a pseudo-palisade; these two areas are interspersed with an central zone composed of acellular material (Figs. 7.56, 7.57, and 7.58).

In cytology, these aspects cannot be recognized, but, in some samples in which Antony A figures are very numerous, parts of Verocay bodies can be supposed to present when group of spindle cells, with elongated oval nuclei, are arranged in a parallel way and keep together by an eosinophil amorphous intercellular matrix (Figs. 7.59 and 7.60).

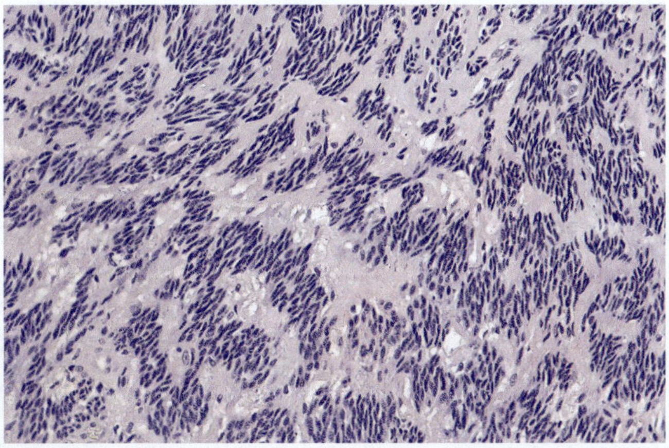

Fig. 7.56 Histology NST: nerve sheath growth pattern; many Verocay bodies are present

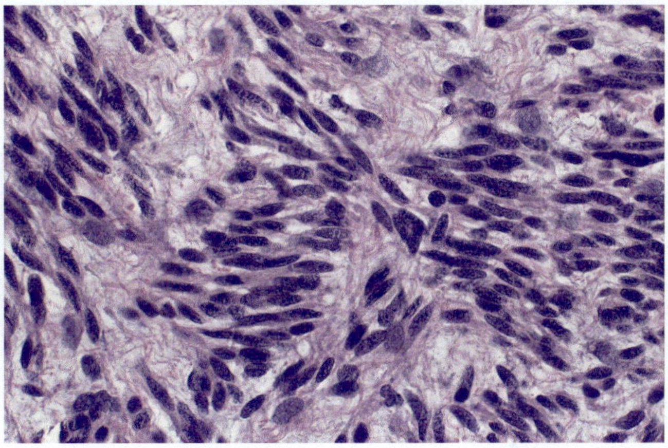

Fig. 7.57 Histology NST: spindle cells are arranged in groups parallel interspersed with acellular material

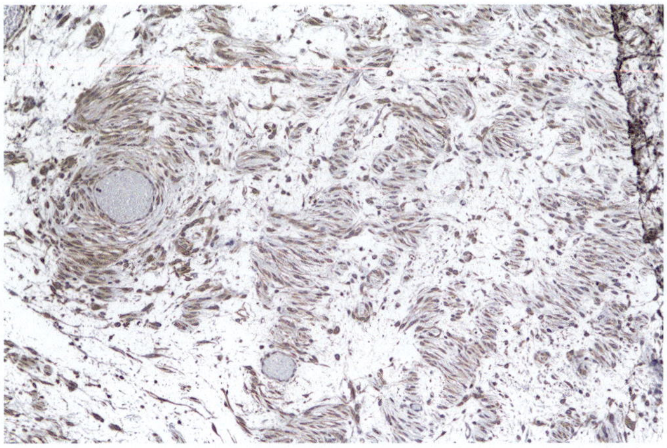

Fig. 7.58 Histology NST (glial fibrillary acidic protein (GFAP)): immunohistochemistry marker confirms the nervous origin of neoplastic cells. Same case of Fig. 7.56

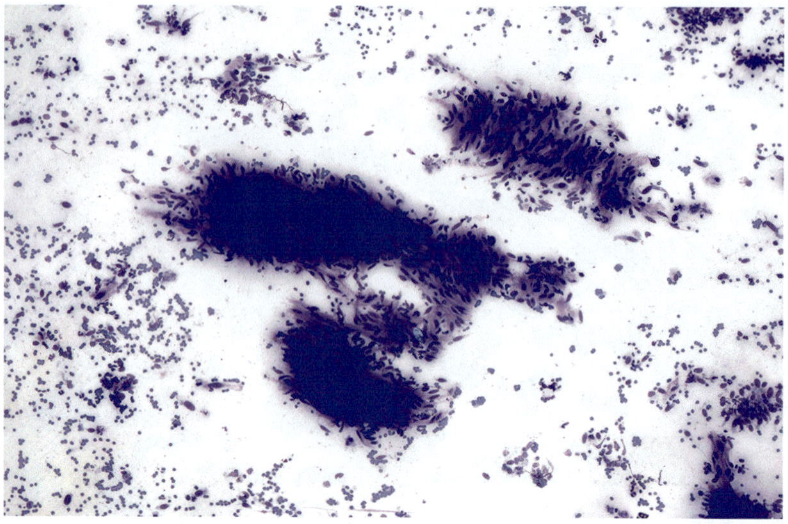

Fig. 7.59 Cytology NST: multiple groups of spindle cells with nuclei arranged in a parallel way representing part of Verocay bodies. Same case of Fig. 7.56

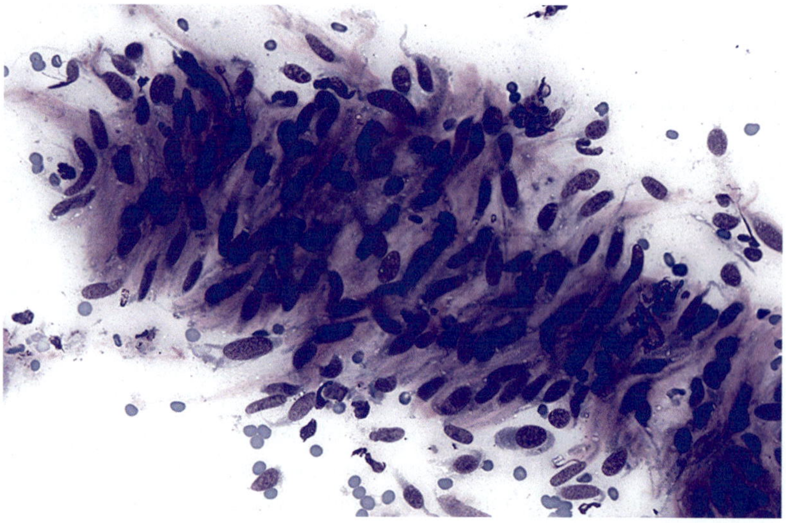

Fig. 7.60 Cytology NST (Verocay body): close up in which the parallel arrangement of nuclei embedded in an eosinophil matrix is better visible

References

1. Kiupel M, Goldschmidt MH, Klopfleisch R, Munday JS, Sruggs JL (2019) Surgical pathology of tumors of domestic animals, Epithelial tumors of the skin, vol 1, 3th edn. Davis-Thompson DVM Foundation
2. Masserdotti C (2006) Architectural patterns in cytology: correlation with histology. Vet Clin Pathol 35(4):388–396
3. Roccabianca P, Schulman FY, Avallone G, Foster RA, Scruggs JL, Dittmer K, Kiupel M (2020) Surgical pathology of tumors of domestic animals, Tumors of soft tissue, vol 3, 3th edn. Davis-Thompson DVM Foundation

8 Cytology of Skin Tumors

8.1 Introduction

In the last decades, many papers highlighted the importance of cytology in diagnosing skin tumors and its sensitivity and specificity when correlated with the histopathological results [1–4]. Skin tumors are very numerous as they can originate from any part of the epidermis, dermis, adnexa, and adipose tissue as well as from hematopoietic cells which can be localized in the skin or that can reach it via bloodstream. Skin metastases originating from internal malignancies or mammary tumors are not included in this wide group of neoplasia, as they will be covered in Chap. 9. With rare exceptions, such as in the case of ulcerative lesions associated to squamous cell carcinoma or in the case of exfoliative erythematous lesions that characterize some clinical variants of canine epitheliotropic lymphoma, almost all skin neoplasia are represented by nodules or plaques; for this reason, an approach based on clinical sign, as used for inflammatory skin diseases (Chap. 6), could be less useful.

In the previous "WHO Histological Classification of Tumors of Domestic Animals," published in 1998, skin tumors were divided into epithelial, melanocytic, mesenchymal, and unclassifiable; in addition to these four groups, metastatic tumors to the skin, cysts, hamartoma, and neoplastic-like lesions were also included. This classification is obsolete, as it didn't include many skin tumors, which are now well documented in many articles and books on histopathology and clinical oncology [5–8, 11, 12].

In recent years, the study of histopathology and the increased use of immunohistochemical markers allowed a broader and more detailed classification of tumors of the skin of pets. Recently, the Davis Thompson Foundation on behalf of very skilled worldwide veterinary pathologists is publishing several volumes, which represent for the scientific world as the new classification of all animal neoplasia, both cutaneous and not [7].

From a histological point of view, the new classification of skin tumors divides the neoplasia on the basis of the origin of the cells, in epithelial tumors, hematopoietic tumors, and mesenchymal tumors [7]. If we look at what has been published in the scientific literature about the *cytological classification* of skin tumors in dogs and cats, we find a "hybrid" classification; in fact, some tumors take the name on the basis of the arrangement of cells in the tissue (e.g., epithelial tumors), while other skin neoplasia are instead grouped exclusively on the basis of the their morphology (e.g., round cell tumors and spindle cells tumors). Neoplasms of melanocytic origin, such as melanocytoma and melanoma, further contribute to creating confusion in this cytological classification, as neoplastic melanocytes are characterized by a remarkable cytological polymorphism: they can, in fact, be round or clearly spindle-shaped or organized according to an epithelioid arrangement, and therefore, in theory, if the inclusion criteria is based exclusively on the cell silhouette, melanocytic neoplasia could be included in all three of the aforementioned groups [9, 10].

This type of classification is convenient for pathologists and clinicians who are interested in cytology and allows them to quickly orient the cytological suspicion through a simplified algorithm. However, it should be emphasized that in cytology there are morphological rules but that these are not always respected by the cells; in fact, for example, if by definition epithelial tumors are cytologically characterized by clusters of cells kept together through intercellular junctions, in some cases, neoplastic epithelial cells lose their cohesivity, and many of them are released as discrete cells mimicking a round cell tumor. These morphological characteristics that do not follow the "guideline" of the cytological classification will be discussed later when peculiar cytological features of each neoplasm will be treated.

In this book skin neoplasms are divided into the following four groups:

F. Albanese, *Canine and Feline Skin Cytology*, https://doi.org/10.1007/978-3-031-95713-0_8

- *Round cell tumors*
- *Epithelial tumors*
- *Spindle cell tumors (mesenchymal tumors)*
- *Melanocytic tumors*

8.2 Round Cell Tumors

Round cell tumors (RCT), although coming from different cell lines, origin almost all from hematopoietic tissue. RCTs share the silhouette of cells, which are roundish to oval, with well-defined margins and "discrete," the latter term means that cells are single and not arranged in groups of cells jointed together by intercellular junction as epithelial ones or kept together by extracellular matrix as occurs for mesenchymal cells; for this reason another cytological feature that characterizes this group of tumors is the high number of cells usually released during sampling. Although the above cytological features are observed in the majority of RCTs, some exceptions are well known, for example, as some histiocytic sarcoma, included in the RCTs group as belonging to histiocytic diseases, can be spindle in shape and can therefore be confused with mesenchymal soft tissue cells. In the opposite, some neoplasia classified in other groups can exhibit cells with an unequivocal round shape, as occurs in lipoma/liposarcoma, some melanomas and anaplastic carcinomas, or poorly differentiated sarcomas. From the above, it is deduced that this cytological classification is not free from risks of interpretation and should only be used as a diagnostic orientation; if diagnosis cannot be achieved through cytology, further histopathological and sometimes immunohistochemical investigations must be performed.

Tumors that are conventionally included in the RCT group are five: *mast cell tumors*, *lymphomas*, *plasma cell tumors*, *transmissible venereal tumors*, and the *histiocytic diseases/disorders*.

8.2.1 Mast Cell Tumor

Mast cell tumors (MCTs) are very common in dogs and common in cats [12, 29]. MCTs originate from dermal mast cells. In dogs, it is the most common form of skin cancer and can appear as nodules or plaques of varying sizes, ranging from a few millimeters up to large masses of many centimeters. The nodules are usually single, dermal, or subcutaneous in location, dome- or button-shaped, pedunculated, alopecic, erythematous or hyperpigmented, firm, soft or gelatinous in consistency, and sometimes ulcerated (Figs. 8.1, 8.2, 8.3 and 8.4). In Shar-Pei dogs, which

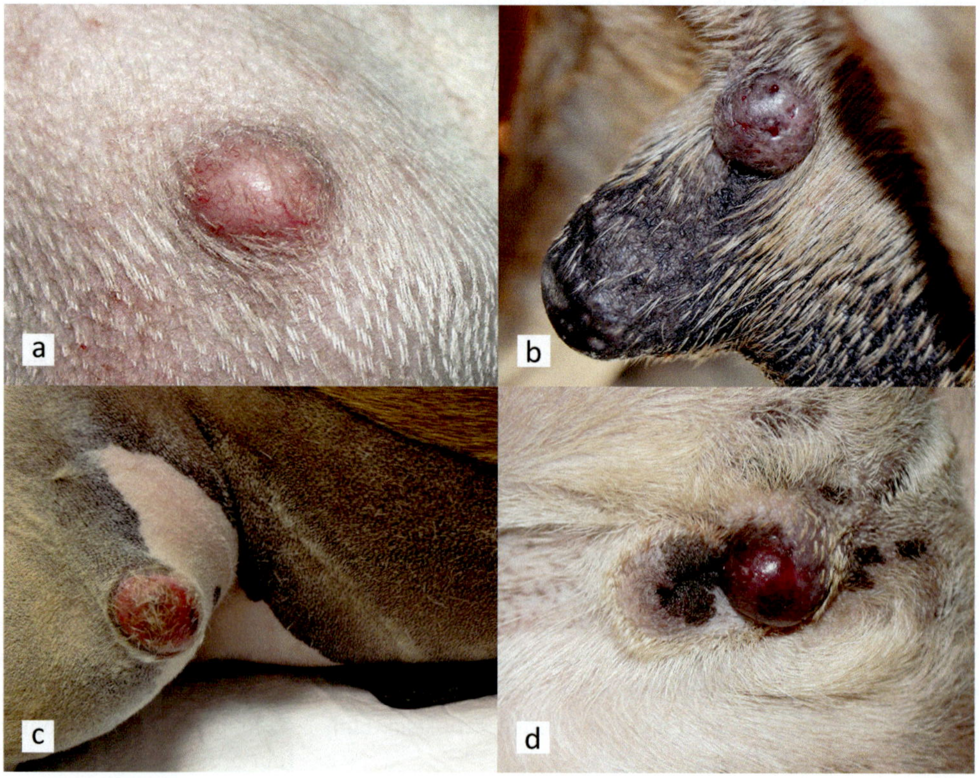

Fig. 8.1 Canine mast cell tumors (MCTs): single alopecic and erythematous nodules

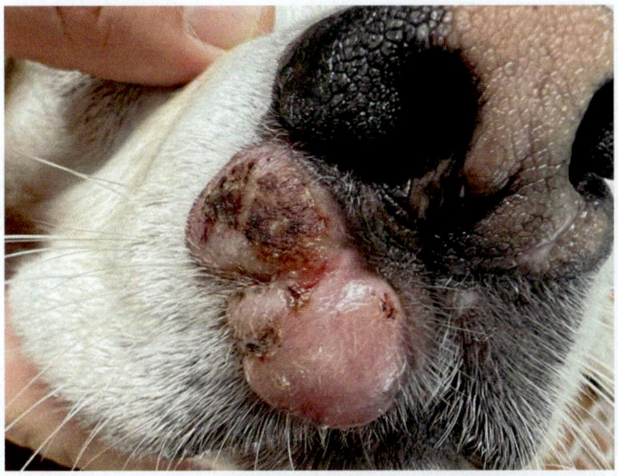

Fig. 8.2 Multiple MCTs confluent nodules on the muzzle of a mixed breed dog

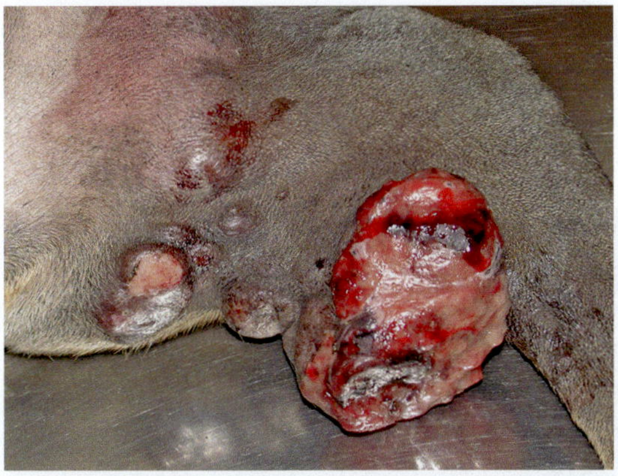

Fig. 8.3 Canine MCTs: multiple ulcerated masses on the leg of a Boxer

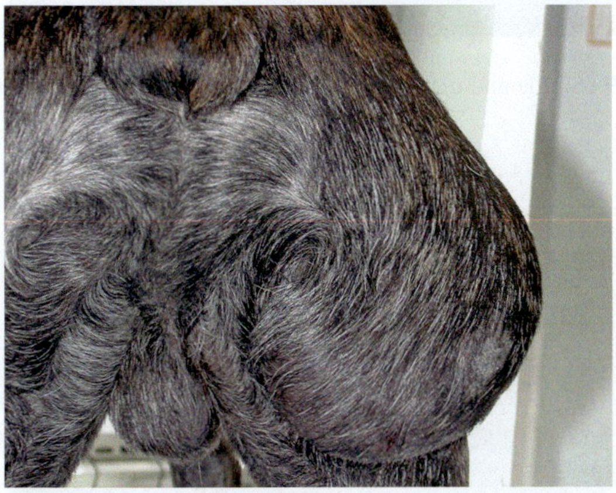

Fig. 8.4 Large soft and of pasty consistency mass, on right rump of a Boxer

physiologically have a high quantity of mucin in the dermis, the MCTs frequently present as swelling mass with undefined margins (Fig. 8.5) [14]. Rarely MCTs can be multiple, for example, two or three nodules, or can be very numerous and spread all over the body (Fig. 8.6). Adults and old dogs are

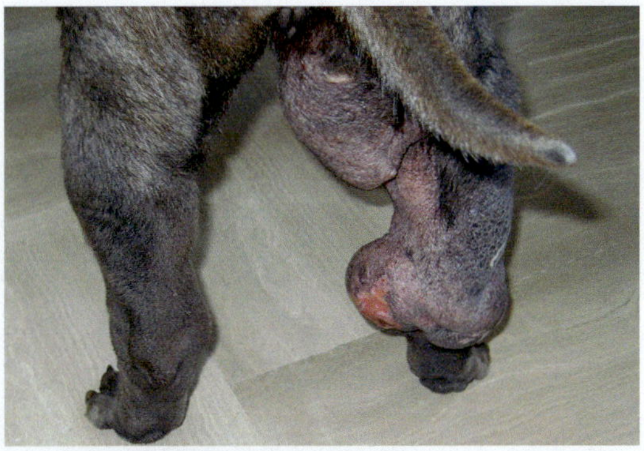

Fig. 8.5 Diffuse swelling with indistinct margins is a typical clinical presentation of MCTs in Shar-Pei dogs

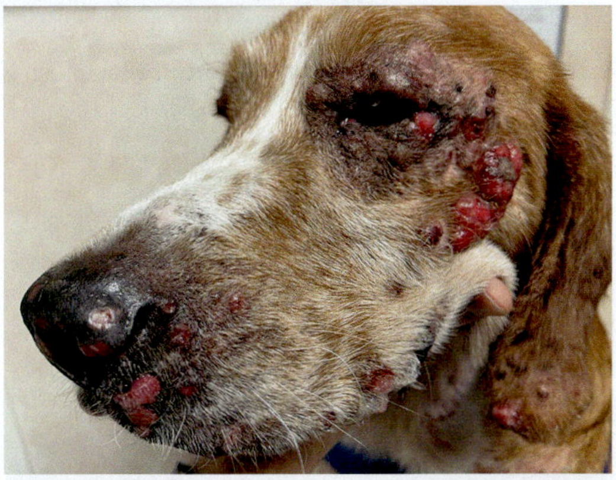

Fig. 8.6 Multiple ulcerated nodules on the face and pinna in a high-grade MCT

most commonly affected, but rarely, single or multiple MCTs can also be observed in very young animals. When multiple skin MCTs are observed in puppies or very young dogs (less than 1 year), the term of cutaneous mastocytosis is usually used, also if a great confusion and overlapping of clinical presentations, therapies, and outcome between mastocytosis and mast cell tumors is currently present on veterinary literature [30]. Many canine breeds, such as Boxer, Shar-Pei, Labrador, Golden Retriever, English Setter, Pug, French Bulldog, Boston Terrier, etc., are strongly predisposed to skin MCT.

A transitory change in the size occurring after a slight trauma, such as after a FNB has been performed, is commonly observed in many canine MCTs; this clinical behavior is unique to MCTs, in which nodules tend to increase in size and then, after a few hours, return to the original size. This phenomenon is due to the degranulation of mast cells, causing the release of intragranular vasoactive substances, which result in an intra- and peri-lesional edema. In some cases, as result of this degranulation, diffuse and marked erythema of the skin covering the tumor can be observed (Darier's sign) (Figs. 8.7 and 8.8).

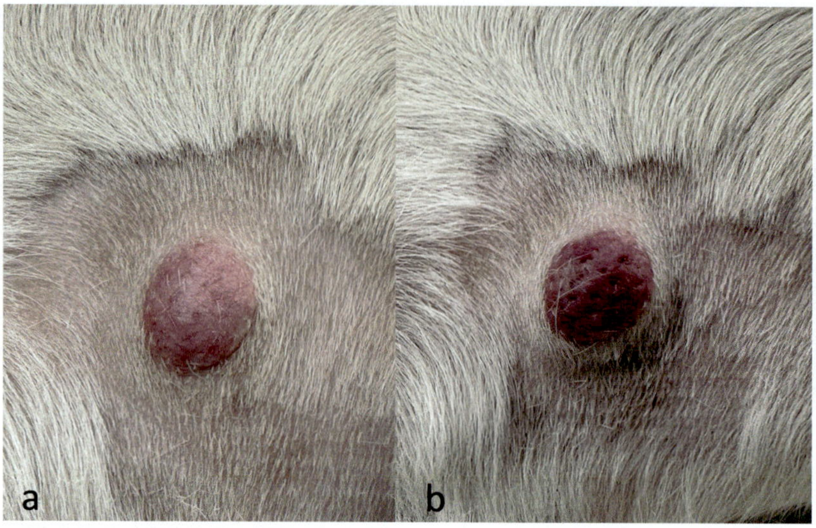

Fig. 8.7 Canine MCT.): (**a**) before and (**b**) after the sampling. Note how the nodule has increased in size, swollen, and changed the color (Darier's sign)

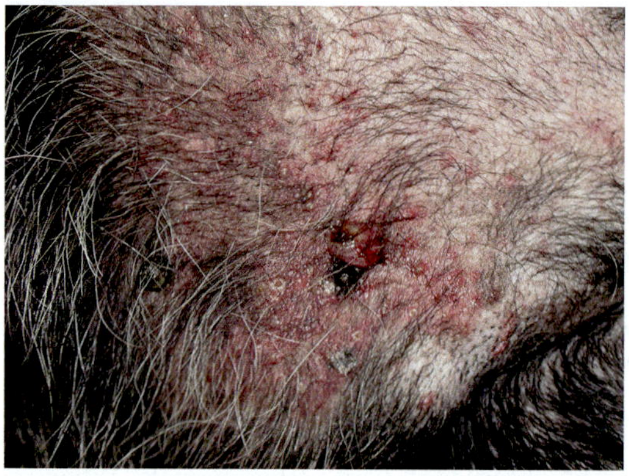

Fig. 8.8 Redness of the skin around an MCT characterizes Darier's sign

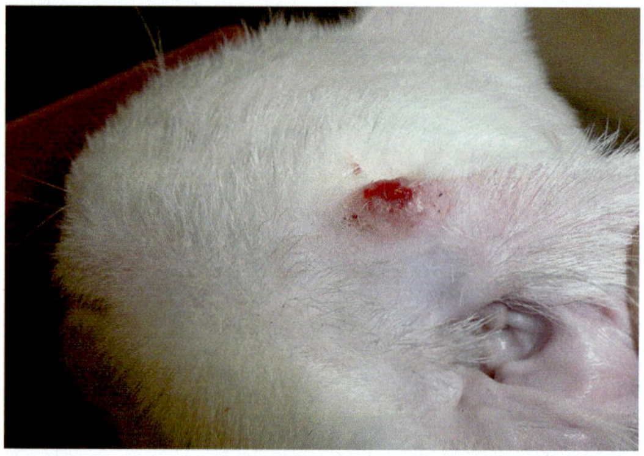

Fig. 8.9 Single and ulcerated MCT at the base of the pinna in a cat

Feline MCT usually occurs as a small, single, alopecic, pinkish, or erythematous nodule, mostly located on the head, eyelids, pinna, or extremities (Figs. 8.9 and 8.10). The size goes from very few millimeters to few centimeters. In many patients, multiple both intact and ulcerated nodules or plaques are present; the head and face are more frequently involved, but nodules can be present all over the body (Figs. 8.11, 8.12 and 8.13) [14, 18, 20]. The single tumors are more frequently confined to the skin, whereas the multicentric lesions can represent a metastasis from a primitive splenic or intestinal MCT. In any case, when a MCT is diagnosed, searching for a possible primary or metastatic visceral lesion must always be performed. A clinical variant of MCT affecting young Siamese cats, from a few months to 4 years of age, has occasionally been reported (Figs. 8.14 and 8.15). This subtype, called *atypical* or *poorly granulated*

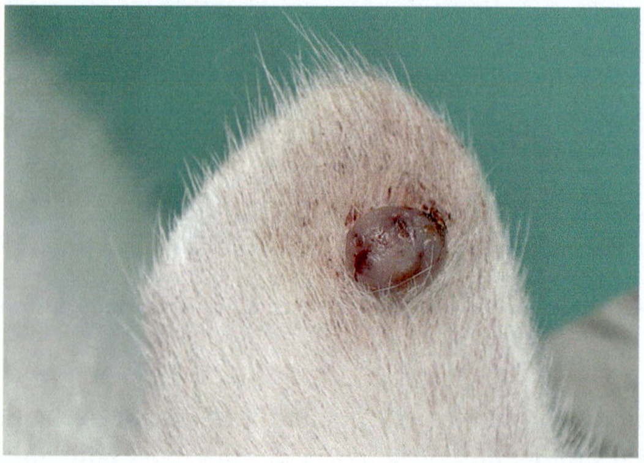

Fig. 8.10 A MCT in a cat: small and alopecic nodule on the pinna

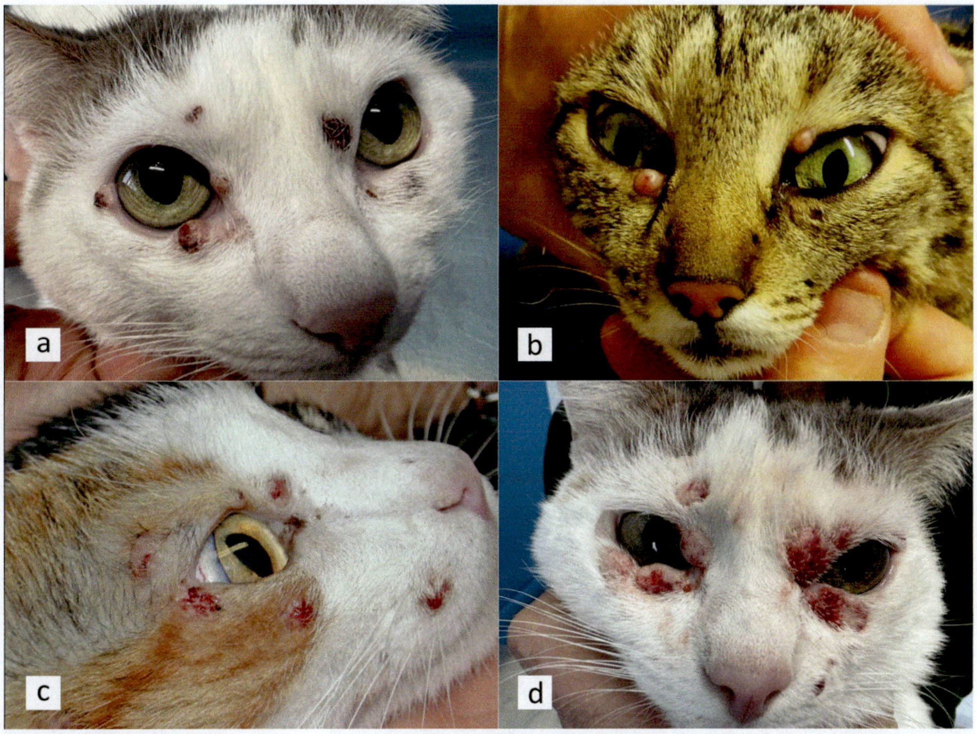

Fig. 8.11 Multiple nodules, some of them ulcerated, on the face of four cats with multicentric MCTs

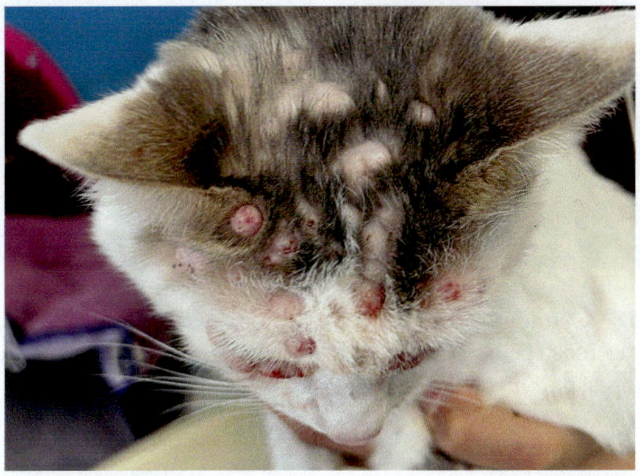

Fig. 8.12 Multiple metastatic MCTs on the head of a cat. The primary neoplasm was located in the spleen

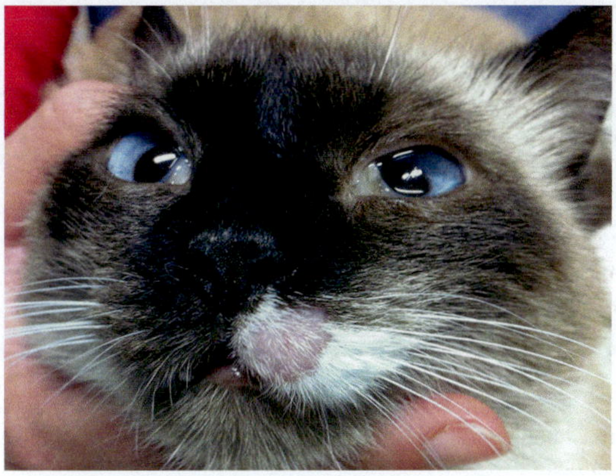

Fig. 8.14 Atypical MCT in a Siamese cat

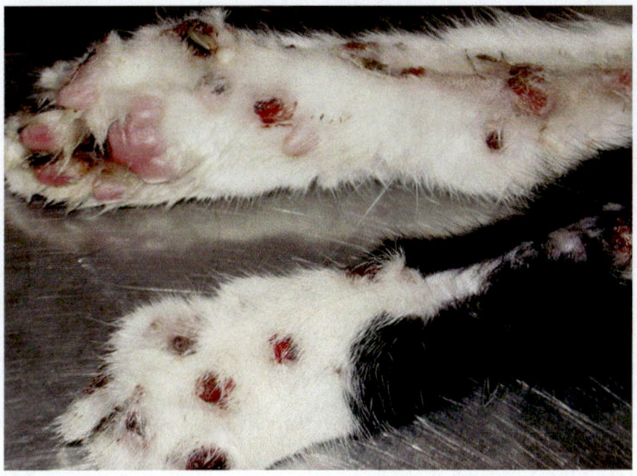

Fig. 8.13 Multiple metastatic MCTs on the extremities. The primary neoplasm was located in the spleen

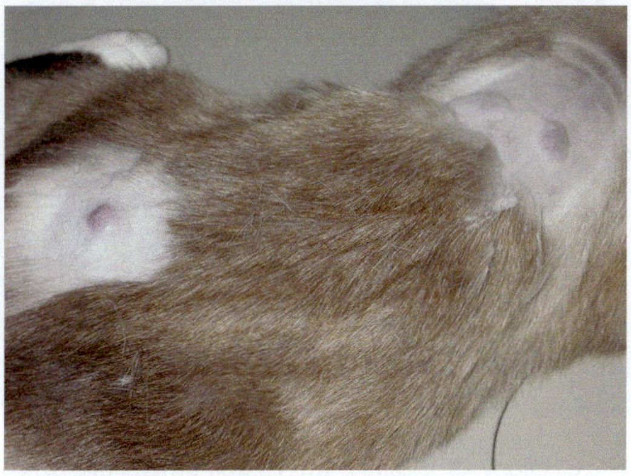

Fig. 8.15 Multiple MCTs on the same cat as in Fig. 8.14

MCTs, is clinically characterized by the spontaneous regression of the nodules. Since its rarity and the few and dated published reports, the real existence of this variant in young Siamese cats is debated. Atypical or poorly granulated MCTs, in which cells can mimic histiocytes, are observed in aged cats.

Cytological Findings

In well-differentiated MCTs, the cytological diagnosis is easy and immediate, especially when proper sampling is performed, and slides are highly cellular [24, 27]. In dogs, the MCTs are composed of round cells characterized by many intracytoplasmic granules, which assume a characteristic metachromatic *purple color* with the use of Romanowsky-type stain (Figs. 8.16 and 8.17). In well-differentiated MCTs, the granules absorb most of the dye, and the nuclei are barely noticeable in the center of the cell as a faded and pale azul round area, surrounded by intensely stained purple granules (Fig. 8.18).

In all the cytological textbooks and papers relating to canine MCTs, the possibility that Romanowsky-type dye is not capable of staining granules is constantly reported. The failure to visualize the granules could occur because the highly undifferentiated neoplastic cell has not more granules in the cytoplasm, and this is an understandable occurrence given that the undifferentiated neoplastic cell moves away from its normal morphology and composition; nevertheless, there are some MCTs in which the neoplastic cells are well-differentiated and have a round silhouette, with regular central nuclei, but the granules that are surely present in the cytoplasm are not highlighted by routine staining such as the Romanowsky-type (Fig. 8.19). In some cases, in the

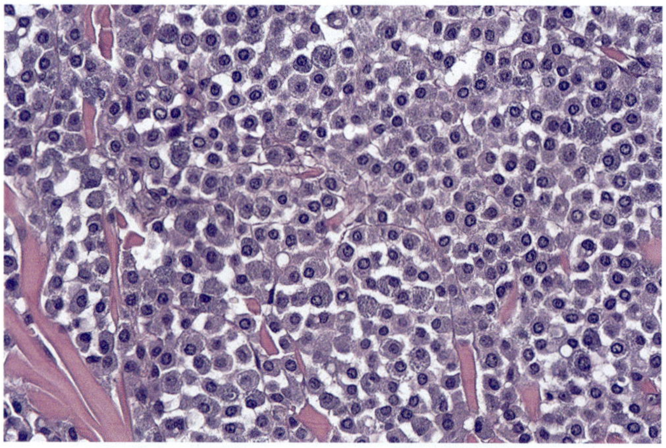

Fig. 8.16 Histopathology of a canine MCT: well-differentiated neoplastic mast cells with regular central round nuclei and cytoplasm filled with granules

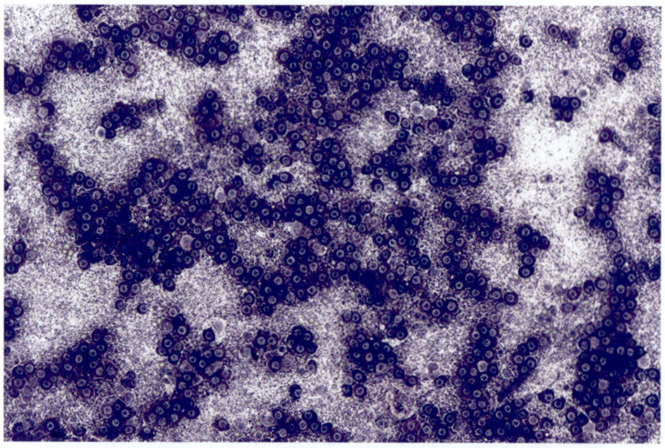

Fig. 8.17 Cytology of a canine MCT: well-differentiated neoplastic mast cells with regular central round nuclei and cytoplasm filled with granules

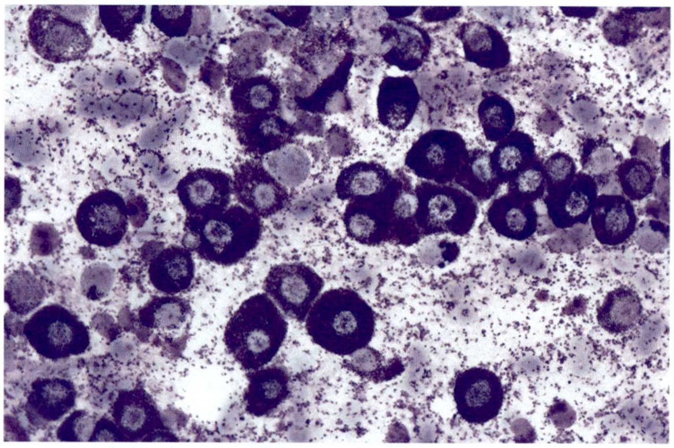

Fig. 8.18 Cytology of a canine MCT: close-up of well-differentiated neoplastic mast cells. Note as nuclei are pale, barely evident and the intracytoplasmic metachromatic granules are intensely purple

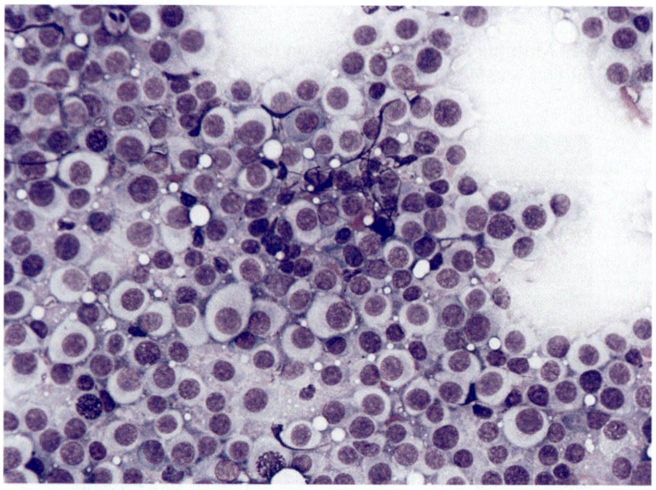

Fig. 8.19 Cytology of an apparently not granulated canine MCT: no metachromatic granules are evident into the cytoplasm of mast cells. In these situations it is not possible to rule out other RCTs with cytology

cytoplasm of these mast cells apparently devoid of granules, punctiform, rounded, achromatic, micro-vacuolar- like areas can be observed, which, once stained with Toluidine Blue or Giemsa dye, make the granules visible; this phenomenon is discussed in Chap. 2 (Fig. 8.20).

In most canine well-differentiated MCTs, a variable number of eosinophils, from scarce to very numerous, are usually observed; eosinophils are recruited by the eosinophil chemotactic intragranular substances and can be of great help in suspecting MCTs when an undifferentiated RCT, in which metachromatic granules are not present or not visible, is observed [17] (Figs. 8.21, 8.22 and 8.23).

Unfortunately, in poorly differentiated MCTs, it seems that the number of eosinophils is lower compared to that in well-differentiated ones and their presence is not a constant. Furthermore, eosinophils do not exclusively infiltrate MCTs since also some lymphoma, despite less frequently, can recall eosinophils in variable number. In some slides coming from well-differentiated MCTs, the number of eosinophils can be higher than the neoplastic mast cells; in these cases, when samples are not sufficiently cellular, cytologists must be aware of this possibility before formulating a diagnosis of eosinophilic inflammation, and especially if the clinical aspect is compatible with a neoplasia, a histopathology examination of the lesion must always be suggested (Fig. 8.24).

Many reactive fibroblasts are commonly detected especially in well-differentiated canine MCTs, probably as a result of the release of substances that stimulate fibrosis from mast cells. Fibroblasts can exhibit cytological atypia such as anisokaryosis and prominent nucleoli, and, in some samples, they can be so numerous that, especially for an inexperienced cytologist, a misdiagnosis of mesenchymal neoplasm could be made (Figs. 8.25, 8.26 and 8.27).

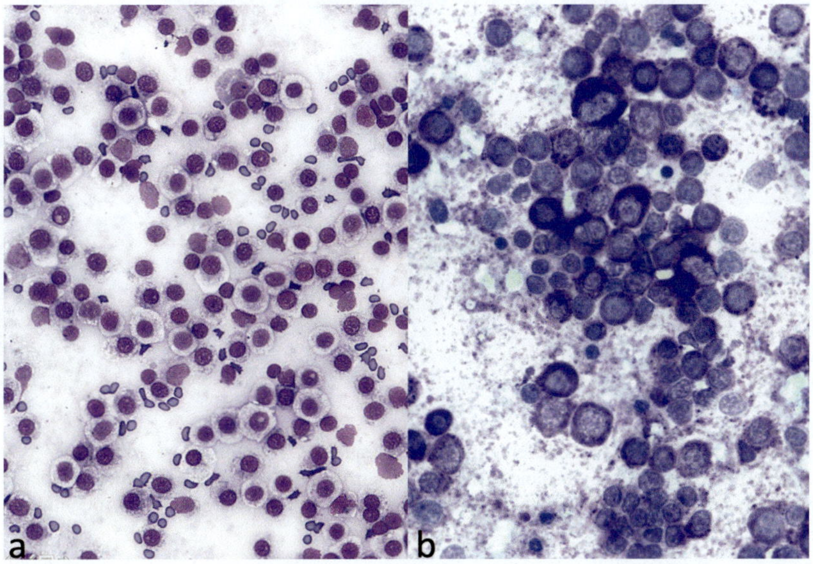

Fig. 8.20 (**a**) Many achromatic micro vacuole-like areas are recognizable in the cytoplasm of neoplastic mast cells; (**b**) the same specimen after stained with Toluidine blue dye: note as many metachromatic purple in color granules are highlighted

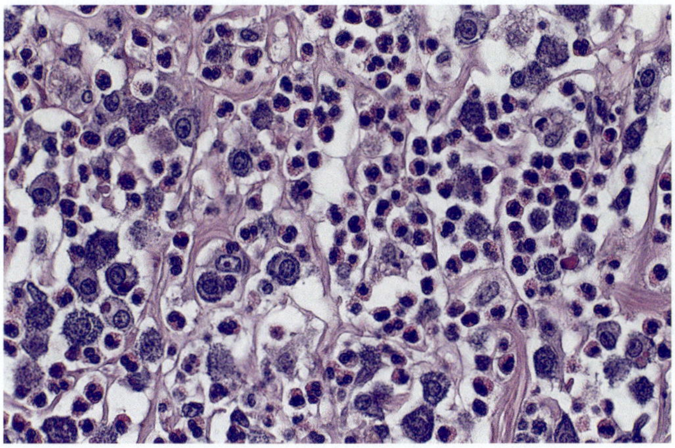

Fig. 8.21 Histopathology of a well-differentiated low-grade canine MCT: many eosinophils are present

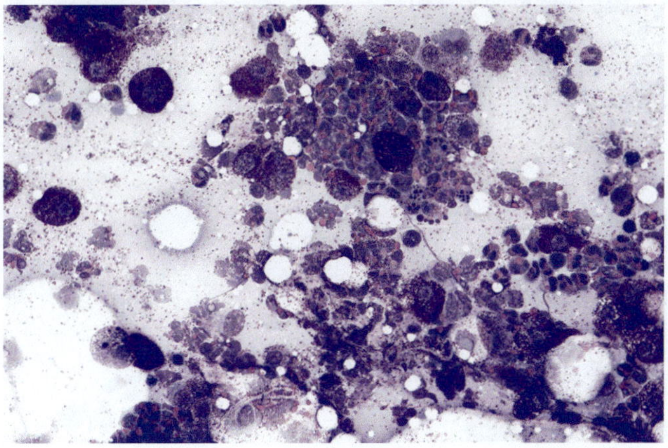

Fig. 8.22 Cytology of a well-differentiated canine MCT: many eosinophils along with granulated mast cells

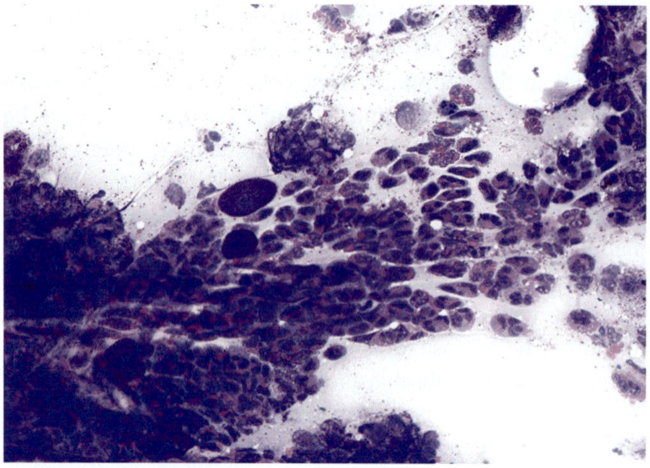

Fig. 8.23 Cytology of a well-differentiated canine MCT: note as eosinophils are more numerous than mast cells

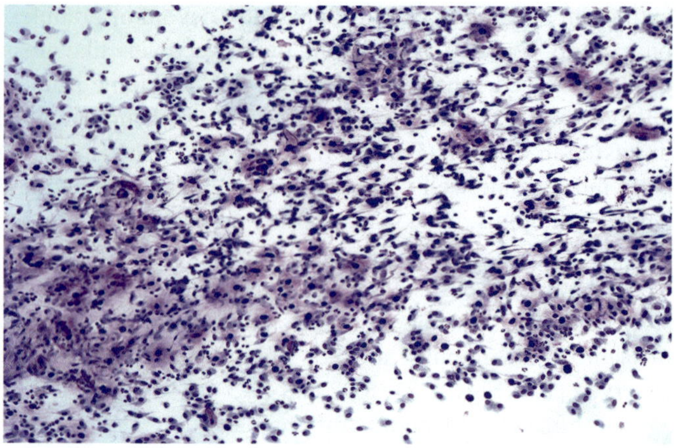

Fig. 8.26 Cytology of a canine MCT: countless reactive fibroblasts along with few mast cells

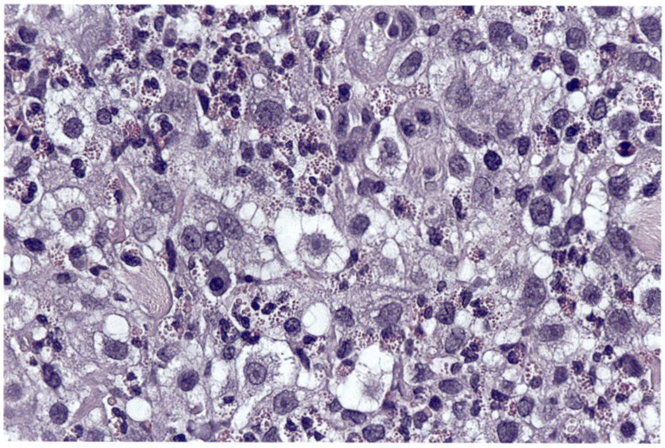

Fig. 8.24 Histology of undifferentiated high-grade MCT: infiltrating eosinophils increase the suspicion of MCT when no intracytoplasmic granules are evident in neoplastic cells

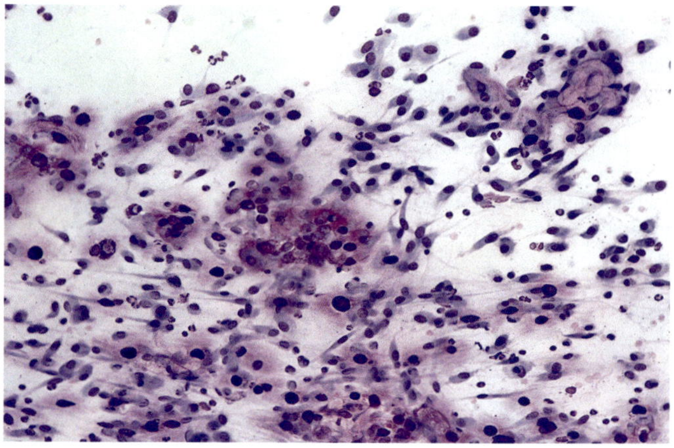

Fig. 8.27 Cytology of a canine MCT: when reactive fibroblasts are more numerous of mast cells, a spindle cell tumor can be misdiagnosed by beginners

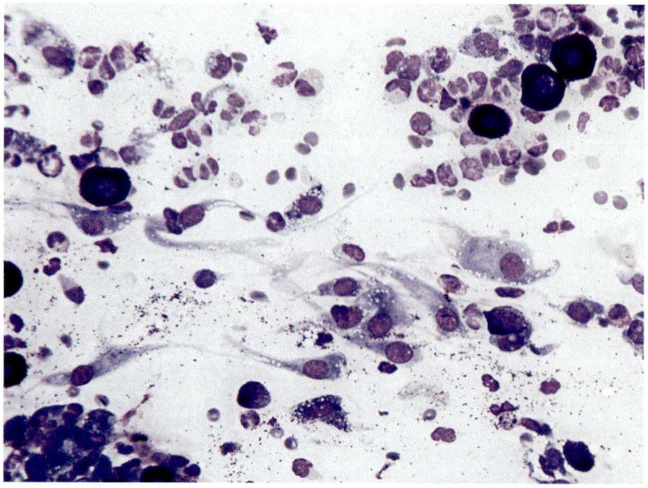

Fig. 8.25 Cytology of a canine MCT: reactive fibroblasts along with few mast cells and eosinophils

In most well-differentiated MCTs, a variable number of short to slender and long worm-like collagen fibers are frequently observed, probably because of the increase in number and in activity of fibroblasts which are stimulated by the substances released by mast cells; other hypotheses could be that the collagen fibers are simply mechanically separated by the high number of neoplastic cells that usually characterize MCTs or by the presence of intralesional edema that facilitates the separation between the collagen fibers and allows for more cellular sampling (Figs. 8.28, 8.29, 8.30 and 8.31). In many cases, the number of collagen fibers is very high and with a larger size than normal; these features give to the tumors a *keloid appearance* (Fig. 8.32). The coexisting of eosinophils, collagen fibers, and fibroblasts is a very common finding in cytological slides coming from well-differentiated MCTs.

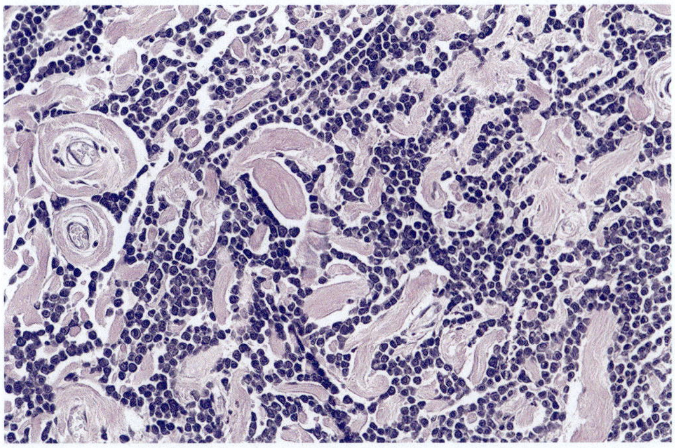

Fig. 8.28 Histology of a well-differentiated canine MCT: mast cell proliferation can separate the collagen fibers

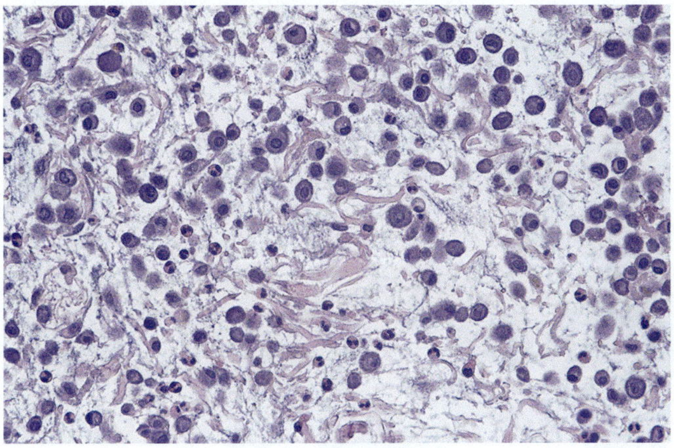

Fig. 8.29 Histology of a well-differentiated canine MCT: the edema of tissue contributes to separate the collagen fibers

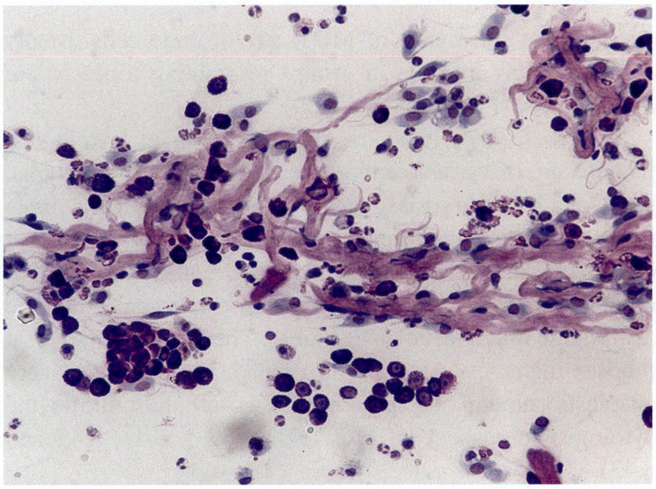

Fig. 8.30 Cytology of a low-grade canine MCT: well-differentiated mast cells, along with many slender and pink in color collagen fibers

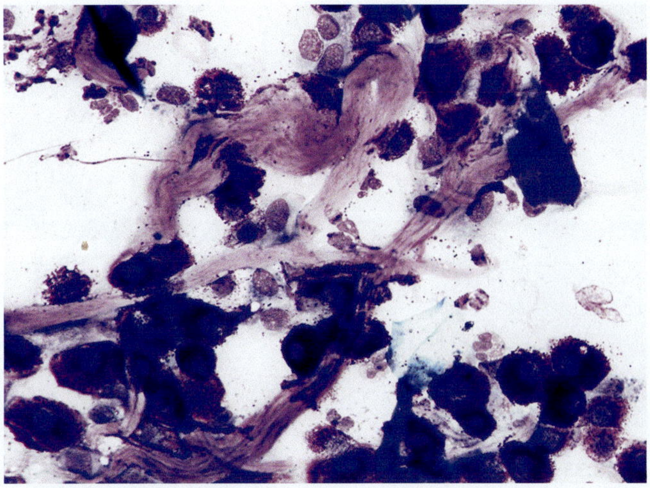

Fig. 8.31 Cytology of a canine MCT: at high magnification, the morphology of the collagen fibers is better recognizable

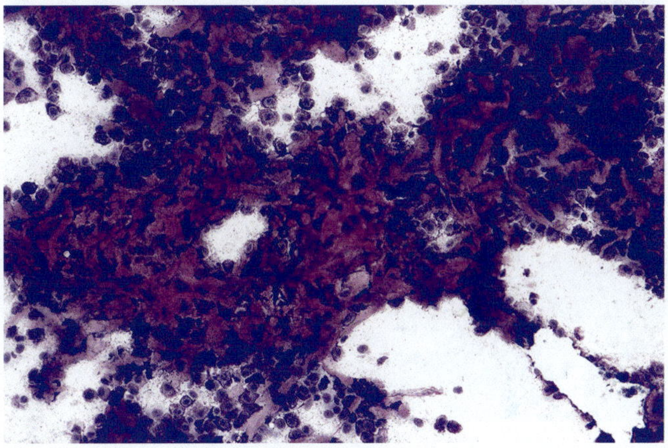

Fig. 8.32 Cytology of canine MCT: note the high amount of collagen fibers giving a keloidal aspect to the neoplasm

Other occasional minor cytological features can be observed in course of well-differentiated MCTs; since in histopathological samples of many well-differentiated MCTs a benign hyperplasia of apocrine sweat glands occurs, along with neoplastic mast cells, a variable number of clusters of basal apocrine cells arranged in honeycomb arrangements or in short palisades can also be observed on slides; their presence has not a prognostic value (Figs. 8.33, 8.34 and 8.35). In rare MCTs a myxoid background can be observed as background; mast cells are embedded in a dense, viscous, violet amorphous material that represents mucinous material. This is a constant feature in MCTs of Shar-Pei dog in which, as known, the dermis is physiologically rich in mucin (Figs. 8.36 and 8.37).

Finally, in very rare cases, neoplastic cells can be released in groups on a clear background, and, at low magnification, they can resemble large cluster of cells; this is a very curious

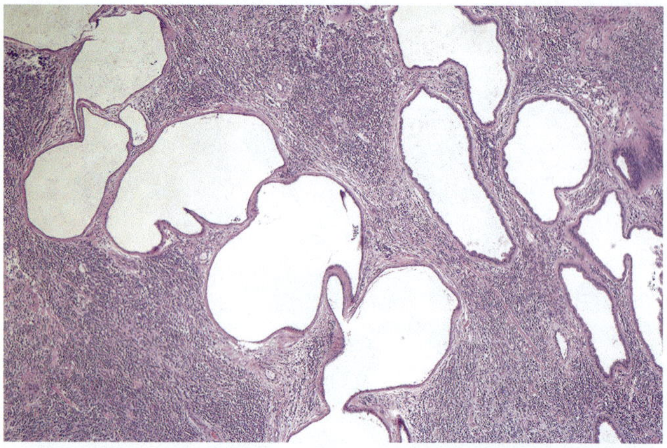

Fig. 8.33 Histology of canine MCT: many dilated tubules of apocrine sweat glands are recognizable at low magnification as achromatic areas bordered by a single line of columnar secreting cells

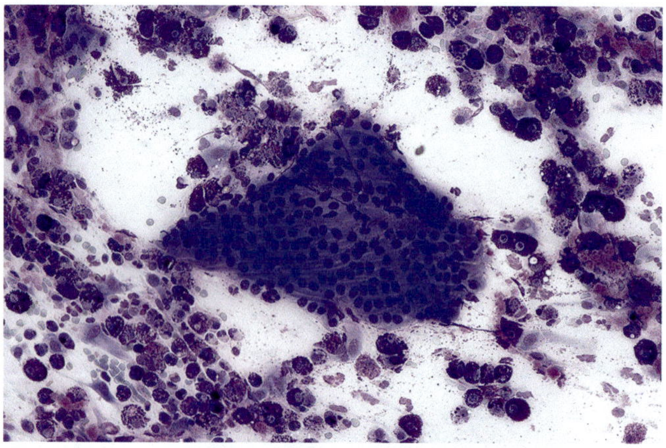

Fig. 8.34 Cytology of canine MCT: a large cluster of apocrine gland in the context of a well-differentiated MCT

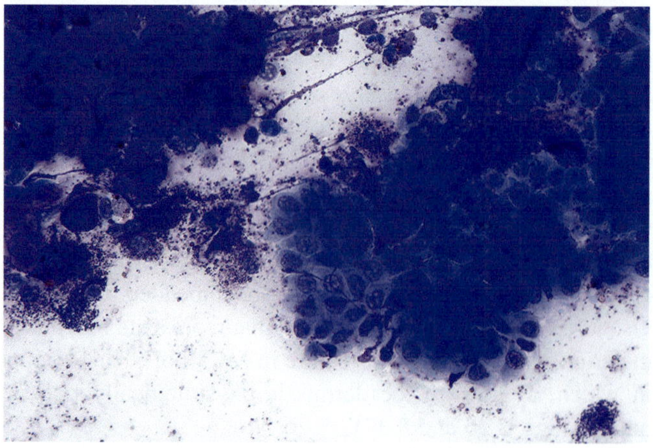

Fig. 8.35 Cytology of canine MCT: a large cluster of apocrine gland is easily recognizable in the lower-right part of the figure

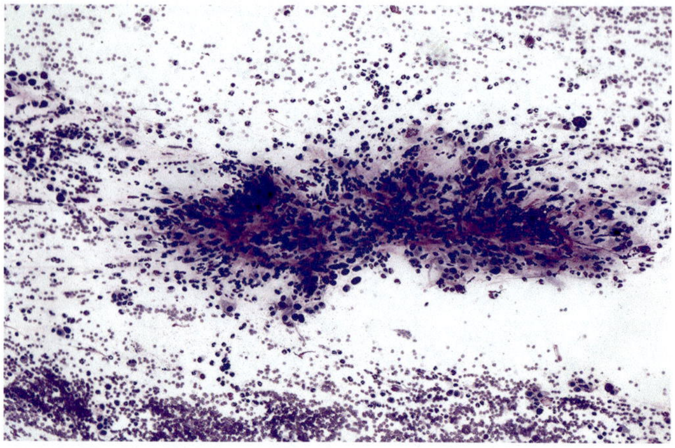

Fig. 8.36 Cytology of a canine MCT: mast cells are embedded in a myxoid dense background: note as cells are oriented in a linear disposition (windrowing appearance)

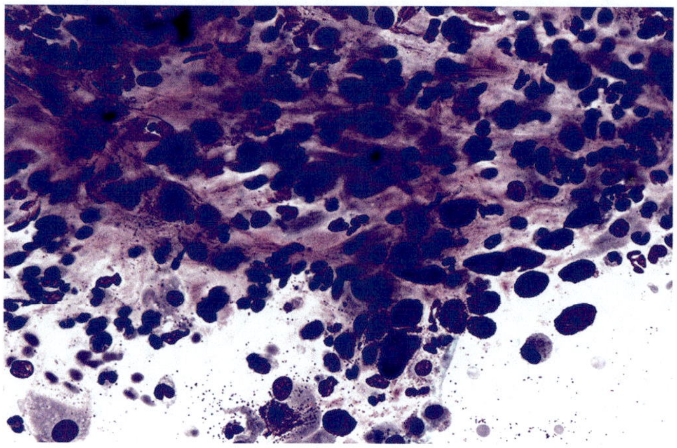

Fig. 8.37 Cytology of canine MCT: at high magnification the violet myxoid background is better recognizable between the neoplastic mast cells

arrangement since cells of MCTs are discrete cells usually released singly and not in groups resembling a *placentoid arrangement* (Figs. 8.38 and 8.39).

In *poorly differentiated MCTs*, malignant cytological features are of variable entity: the nuclei show more atypia, such as anisokaryosis, can be single and very large (macrokaryosis) or multiple, bizarre in shape, and with coarse chromatin and prominent nucleoli. Granules are few or completely absent, and the cytoplasm can contain multiple achromatic microvacuolar-like areas. The number of atypical mitoses increases. As afore-mentioned, sometimes, in very anaplastic tumors, the presence of eosinophils allows the cytologist to suspect an MCT (Figs. 8.40, 8.41, 8.42 and 8.43).

Although a cytological evaluation of cellular morphology may provide many indications regarding the degree of

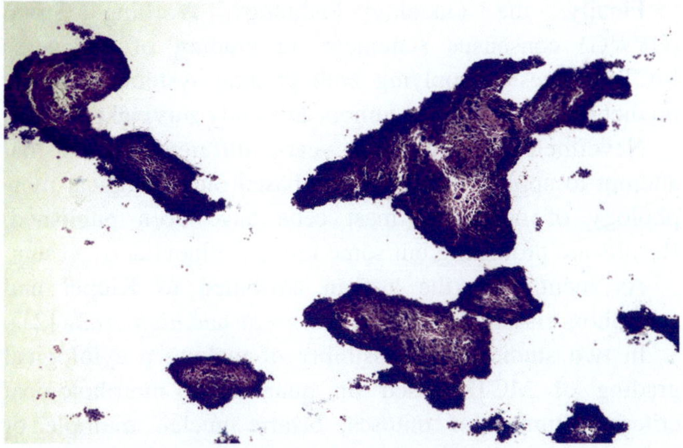

Fig. 8.38 Cytology of canine MCT: at low magnification, well-defined large groups of mast cells can be confused with epithelial cells

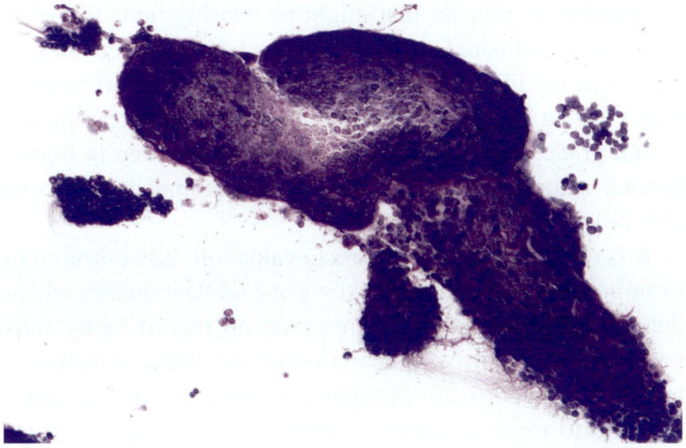

Fig. 8.39 Cytology of canine MCT: at high magnification the morphology of mast cells is easily recognizable also in the grouped cells

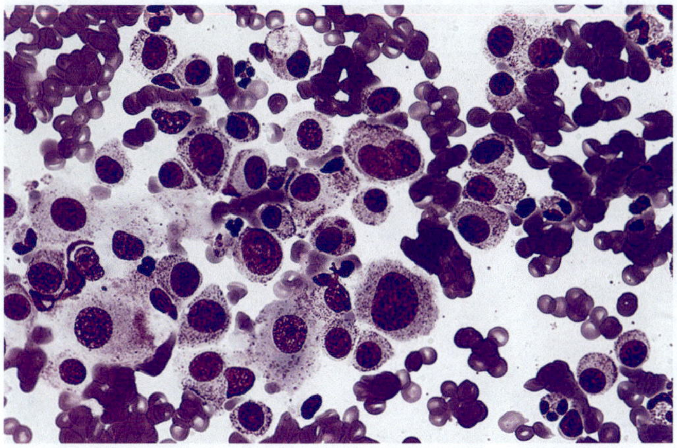

Fig. 8.40 Cytology of a poorly differentiated canine MCT: note the evident features of atypia, such as anisokaryosis and atypical silhouette of nuclei. Many metachromatic granules are evident yet

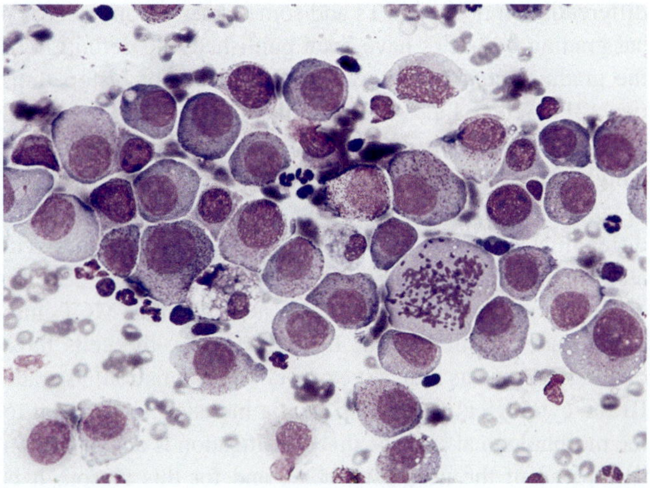

Fig. 8.41 Cytology of a poorly differentiated canine MCT: despite highly undifferentiated, the neoplastic mast cells show recognizable intracytoplasmic granulation

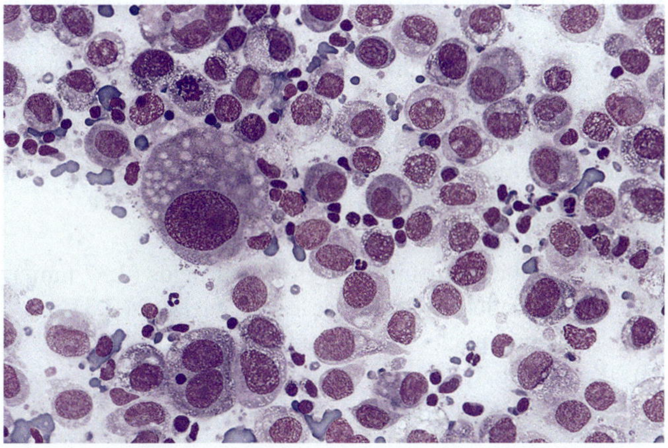

Fig. 8.42 Cytology of undifferentiated canine MCT: severe cytological atypia are recognizable; no granules are present to orient the origin of cells as mast cells

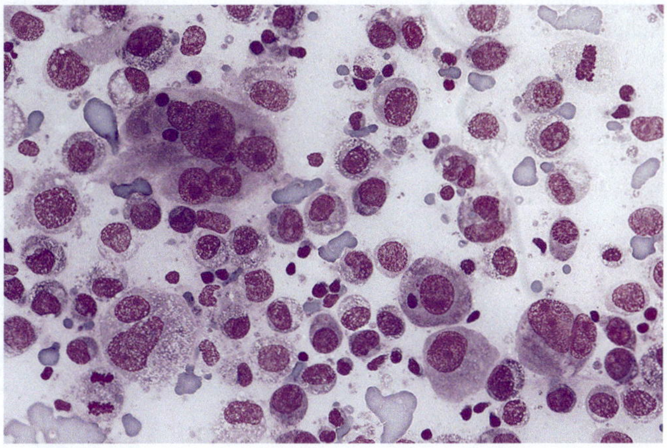

Fig. 8.43 Cytology of a poorly differentiated canine MCT: severe cytological atypia. Scattered purple granules are still detectable in the cytoplasm of some neoplastic cells

differentiation of the MCTs and some papers about cytological grading of MCTs have been published, the definition of the grading remains of exclusive pertinence of histopathology [21]. When slides show remarkable features of malignancy (anisokaryosis, atypical mitosis, multinuclearity etc.), a high-grade MCT can be diagnosed with cytology, whereas in cases in which MCTs are composed exclusively of well-differentiated mast cells, a low grade can be hypothesized, and histopathological confirmation is mandatory.

The Patnaik histopathological system classifies MCTs into three grades, from I to III. In the first two, the cells have cytological aspects indicative of well differentiation; in grade III the cytological features of atypia increase. In addition to the morphological aspects, this classification is also based on the spread of the neoplastic cells, and for this reason, it is clear how it is not cytologically possible to differentiate between grade I and II MCT. Since in different studies, significant differences in grading MCTs with the three-tier Patnaik system between different pathologists have been shown, a two-tier classification system was proposed by Kiupel and colleagues, classifying MCTs as *low grade* and *high grade*. With this histopathological method, a reduced interobserver difference in grading MCTs compared with the Patnaik system has been obtained. The morphological criteria to define a high-grade MCT include the following four cytomorphological characteristics:

(a) ≥7 mitotic figures in 10 high-power fields (2.37 mm^2)
(b) ≥3 multinucleated cells (at least 3 nuclei) in 2.37 mm^2
(c) ≥3 bizarre nuclei in 2.37 mm^2
(d) Karyomegaly (nuclear diameters vary by at least two-fold in ≥10% of cells) (Figs. 8.44 and 8.45)

Finally, the Oncology–Pathology Working Group (OPWG) consensus statement on grading of cutaneous MCTs suggested applying both grading systems to better predict behavior of some tumors and study survival rate [13].

Nevertheless, in the last years, different studies that attempt to apply grading criteria based on cytological morphology of neoplastic mast cells have been published. Results are promising but some limits are inevitably present.

As mentioned, the system attributed to Kiupel and co-authors classifies MCTs as *low grade* and *high grade* [21].

In two studies, the possibility of making a cytological grading of MCTs based on quantitative morphological criteria (number of mitoses, bizarre nuclei, multiple or karyomegalic nuclei), which form the basis of the Kiupel histopathological classification, has been assessed.

Both studies demonstrated a very high 94% correspondence between cytological and histological grading.

A risk to this cytological attempt of grading was found on the number of mitoses that might be missing on cytological specimens compared to histology, and since seven mitotic figures in ten high-power fields (2.37 mm^2) are sufficient to move from a *low* to a *high grade*, the risk of underestimating a high grade is always present with cytology even in highly cellular samples composed of well-differentiated mast cells [19, 28].

It is obvious that for a correct evaluation, it is essential to examine very cellular preparations of excellent quality where the cells must be perfectly preserved; in fact, if many cells break during incorrect preparation of the slides, it makes it impossible to correctly evaluate an improper slide preparation and to make a correct assessment of the nuclear shape (bizarre nuclei) and karyomegaly, all parameters included in the Kiupel classification.

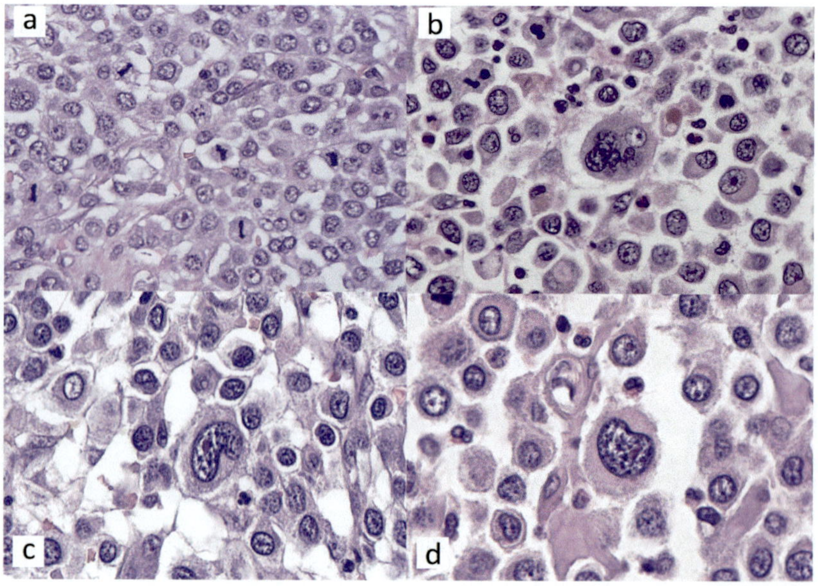

Fig. 8.44 Histopathology of high-grade MCT. Kiupel's cytological criteria of malignancy: (**a**) five mitotic figures; (**b**) multinucleated neoplastic mast cells (more than three nuclei); (**c**) bizarre morphology of nuclei; (**d**) karyomegaly (nuclear diameter more than twofold of other cells)

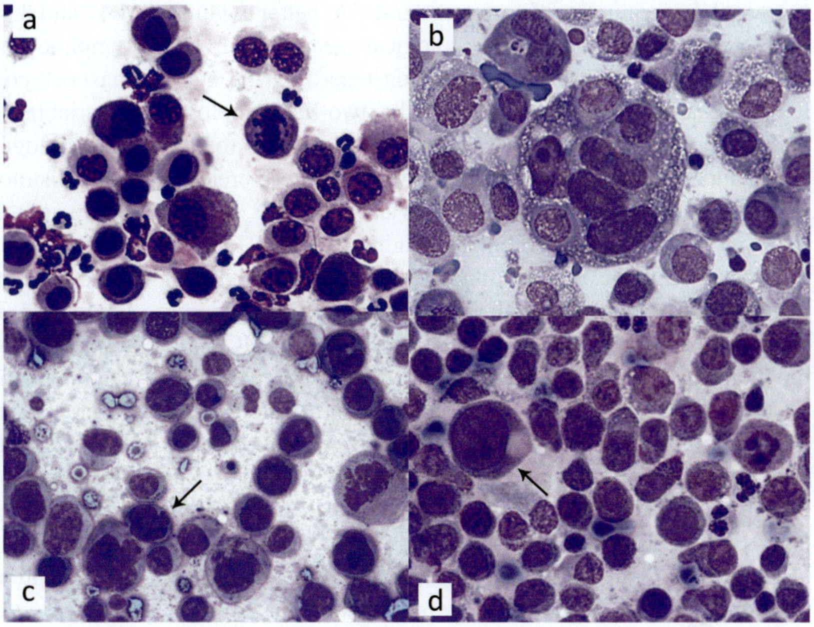

Fig. 8.45 Cytology of high-grade MCT. Kiupel' cytological criteria of malignancy: (**a**) mitotic figure (arrow); (**b**) multinucleated neoplastic mast cells (more than three nuclei); (**c**) bizarre morphology of nuclei (arrow); (**d**) karyomegaly (nuclear diameter more than twofold of other cells) (arrow)

To minimize these criticisms, Camus et al. developed a cytological classification system that did not apply to the Kiupel histological classification criteria alone but related it to the survival period of the affected patient. The number of cytoplasmic granules, divided into strongly, mixed, and poorly granulated, proved to be the most important data in predicting survival, followed, in order of importance, by anisokaryosis, multinucleation, binucleation, and mitotic figures. In this modified scheme, tumors are cytologically classified as high grade if they meet one or both of two criteria:

1. Poor granulation in most cells
2. The presence of two of the following features of atypia:
   - Anisokaryosis (karyomegaly—twice the size of a standard nucleus)
   - Binucleated or multinucleated cells
   - Nuclear pleomorphism (bizarre shape)
   - Mitotic figures

This study concluded that cytologic grade is a useful predictor for treatment planning and prognostication [15].

Since the degree of granulation is the most important criterion in this study, a criticism could be represented by the small quantity of granules that some mast cells may present but which is not linked to their actual reduced presence in the cytoplasm, but rather to the inability of the Romanowsky-type staining to highlight the granules normally present, as can occur in some cases. So the use of BT or Giemsa staining is mandatory to avoid to underestimate the amount of intracytoplasmic granules.

A recent study modified the cytological parameters of Camus' study, replacing the binucleated cells and nuclear pleomorphism with the evaluation of the number of fibroblasts and collagen fibers, which are commonly observed in cytological preparations, especially of low-grade mast cell tumors [23].

The result of this study demonstrated the high number of these two cytological parameters (fibroblasts and collagen fibers) is correlated with a better survival.

In all these studies, however, a number of cases emerged with discordant results between cytological and histological grades and in particular a higher number of cytological diagnosis of high-grade MCTs than that diagnosed with histology; however, this cytological overestimation is of lesser importance than the opposite, since histology remains the reference examination that must always be performed for the final diagnosis, for the definition of the grading, and for the evaluation of the margins.

The overall agreement between cytology and histology in these studies ranged from 77% to 94%. It should be considered that MCT is a tumor whose biological behavior does not always respect the grade; indeed, in a percentage of cases, despite a high-grade cytological diagnosis, later established as low grade by histology, the neoplasm resulted in the death of the patients in a much shorter period to what was expected based on the literature. For this reason, it has been speculated but not demonstrated that cytological grade may be predictive of outcome even in histologically low-grade tumors.

A recent paper reports the use of Ki-67 and Ki-67 × AgNOR markers in cytological specimens of 45 canine MCTs; the author demonstrated that cytological grading of canine MCTs increases prognostic tool from 75% to 80% when cytological slide evaluation was associated with the proliferative markers Ki-67 and Ki-67 × AgNOR in smears. A large number of cases are needed to better evaluate the utility of these markers on cytological specimens [21].

The last but necessary consideration regarding the cytological grading is that, until now, the two-tier histologic grading system is restricted to dermal MCTs, while for the subcutaneous one, a classification is not standardized creating a lot of confusion for both pathologists and clinicians. So, the attempt to make a cytological grading of MCTs that comprises only the dermal tumors will be re-evaluated in the next years when more studies will be published regarding both histopathology and cytology of subcutaneous MCTs; at the moment, a recent prospective study demonstrated that the two-tier histologic grading enabled the identification of all MCTs with aggressive biological behavior, regardless of their cutaneous or subcutaneous location, and this can speculate that in a future the two-tier grading system of Kiupel could also be extended to subcutaneous MCTs [27].

Feline MCTs are usually well differentiated, and cytological similarity with those of the dog is recognizable. As in the dog, the morphology of feline mast cells does not always allow for histological correlation; moreover, in cats, a true grading of MCTs has not yet been standardized. There are currently three recognized histological types: *mastocytic well-differentiated MCT*, *mastocytic pleomorphic MCT*, and the *atypical MCT*, the latter formerly known by the term *histiocytic MCT* [14, 18, 20, 28]. To avoid confusion with the nomenclature of feline histiocytic disorders, some authors have therefore suggested abolishing the definition of *histiocytic MCT* and renaming this atypical subtype using the term *poorly granular mast cell tumor* [22].

The most common subtype in cats is the *well-differentiated mastocytic*, more frequently observed as a single nodule. Cells are round to polygonal in shape and usually adhere strongly to each other, giving at low magnification a *pavement-like* appearance; the nuclei are usually central, and granules are finer and less intensely stained, compared with those of dogs, assuming a "dusty" appearance (Figs. 8.46, 8.47 and 8.48). In some mast cell MCTs, aggregates of small lymphocytes may be observed in the context of the neoplastic mast cell population; if the needle is inserted into these areas, a variable number of small lymphocytes may be present in the cytological preparations (Figs. 8.49 and 8.50).

A histological variant of feline MCT, considered a subgroup of the well-differentiated mastocytic subtype, is characterized by many *multinucleated mast cells*, in which the nuclei show a uniform size, and the cytoplasm is heavily granulated (Figs. 8.51 and 8.52) [22]. The multinuclearity in feline MCTs has not the same meaning as for canine neoplasia, in which, as mentioned, it is always one of the cytological criteria of malignancy. In cats, multinucleation must be evaluated at the light of cytological aspects of the other cells present on the slide: when multiple nuclei are associated to well-differentiated mast cells and if, in the same multinucleated cell, nuclei are uniform in size and shape, they must be interpreted as belonging to the abovementioned *well-differentiated mastocytic MCT* (Fig. 8.53). In case of

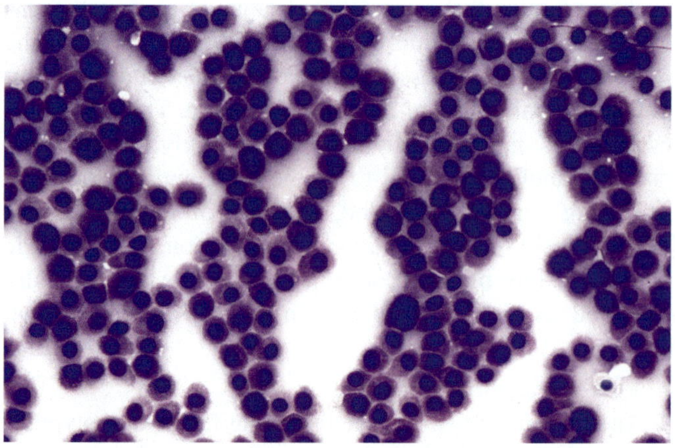

Fig. 8.47 Cytology of a feline mastocytic MCT: highly cellular sample. Note the characteristic polygonal shape of the feline mast cells

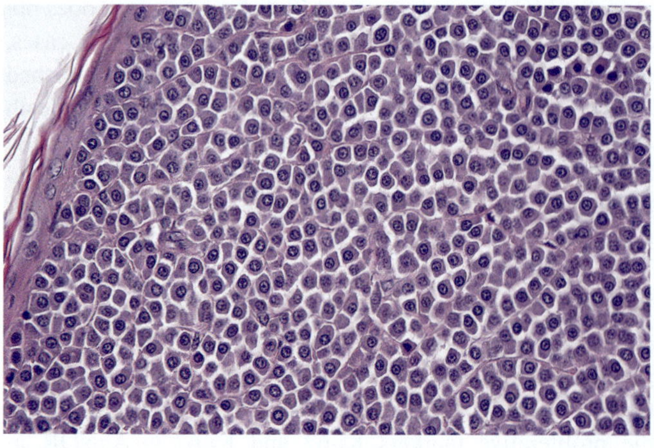

Fig. 8.46 Histology of a feline mastocytic MCT: round to polygonal mast cells in which intracytoplasmic granules are barely visible

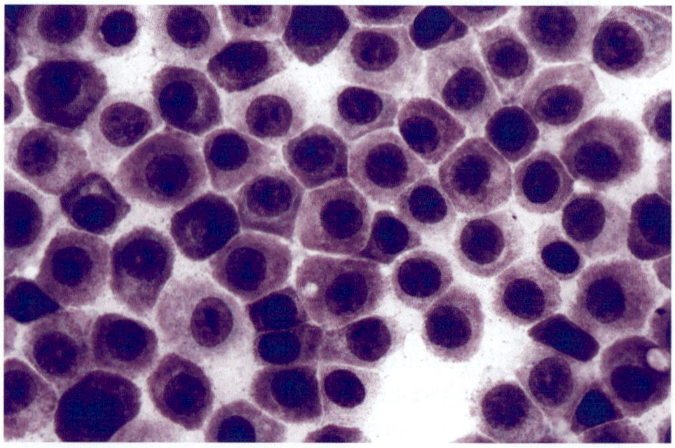

Fig. 8.48 Cytology of a feline mastocytic MCT: at high magnification, the fine purple intracytoplasmic granules are more evident ("dusty" appearance)

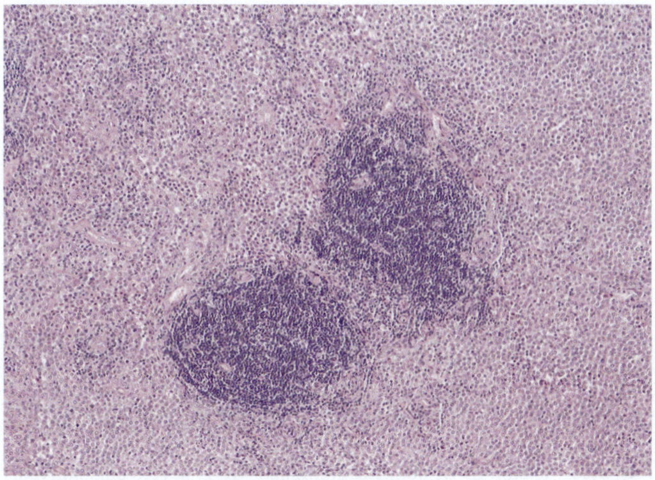

Fig. 8.49 Histology of a feline mastocytic MCT: two large aggregates of lymphocytes are present in a mastocytic MCT

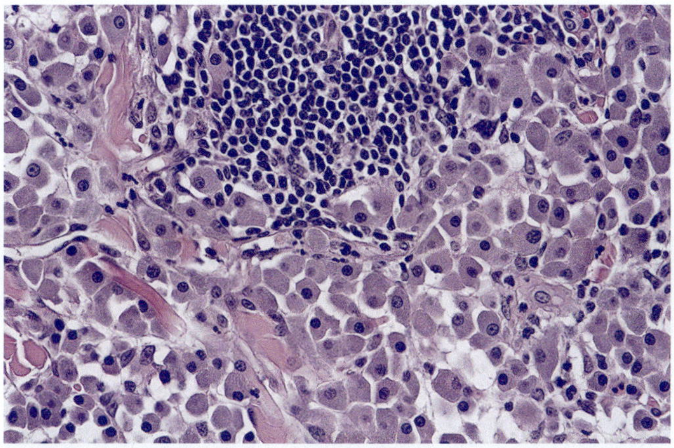

Fig. 8.50 Histology of a feline mastocytic MCT: at high magnification small mature lymphocytes are better recognizable

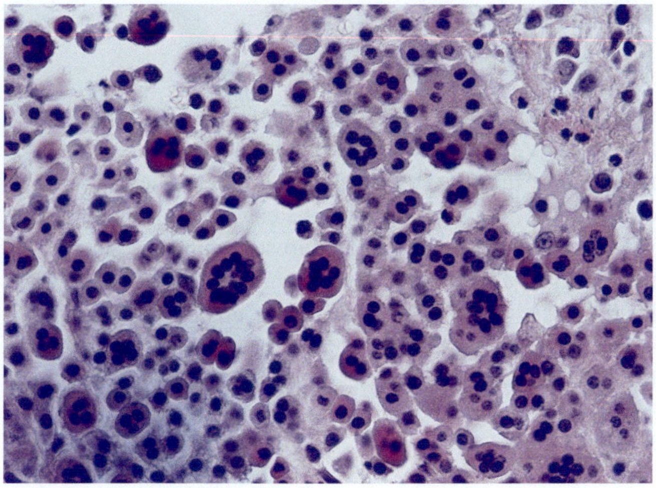

Fig. 8.51 Histology of a feline multinucleated mastocytic MCT: note the high number of multinucleated mast cells

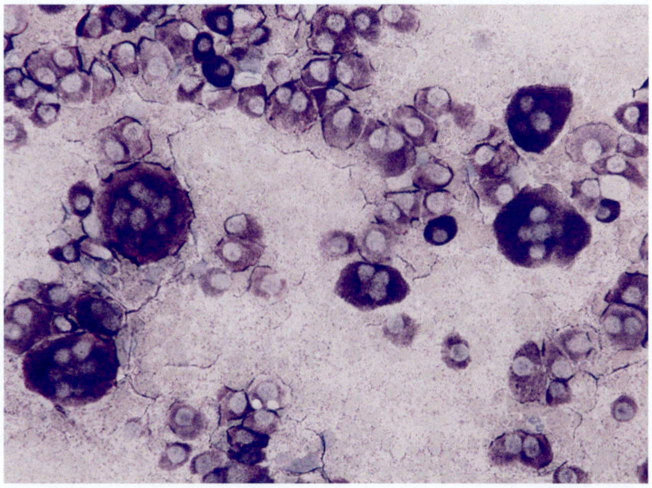

Fig. 8.52 Cytology of a feline multinucleated mastocytic MCT: many multinucleated mast cells are present along with other well-differentiated mast cells

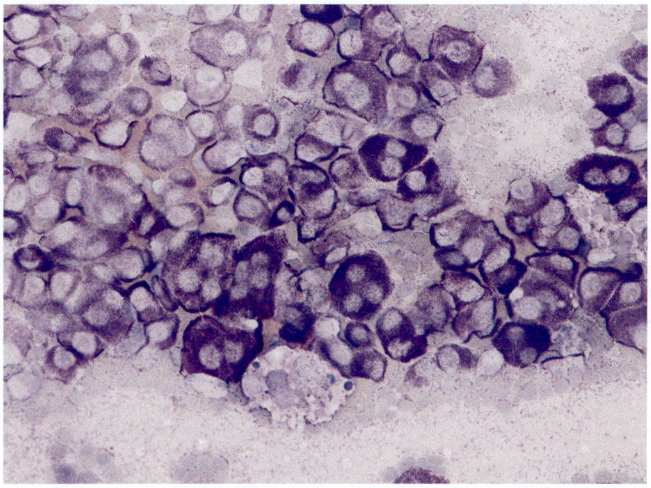

Fig. 8.53 Cytology of a feline multinucleated mastocytic MCT: many multinucleated mast cells. Note as nuclei are of the same size and shape

multinucleation as part of a poor-differentiated MCT, cells must be interpreted as malignant; in these cases, nuclei of the same cell are usually characterized by anisokaryosis (Fig. 8.54).

In *poorly differentiated MCTs*, the atypical malignant aspects of cells increase, and in the so named *pleomorphic MCT* variant, cell morphology can be very atypical, and, when granules are not visible, a MCT cannot be cytologically diagnosed, and other undifferentiated tumors must be included in differentials [17, 22, 26] (Figs. 8.55, 8.56 and 8.57).

In the cat, the *atypical* or *poorly granular* subtypes are characterized by cells that morphologically resemble histiocytes, with round or oval large nuclei, sometimes

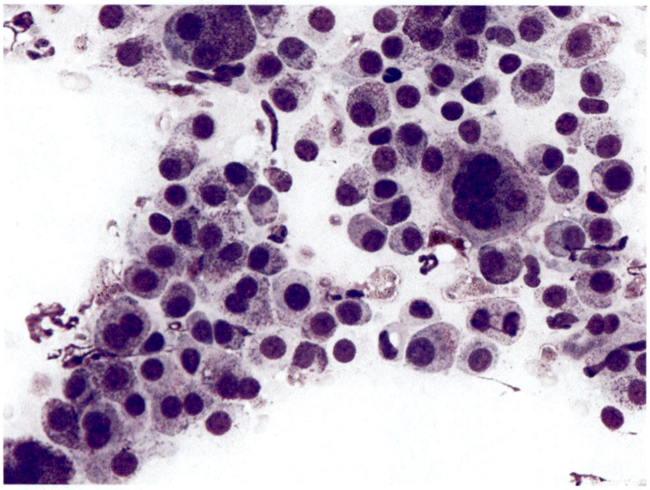

Fig. 8.54 Cytology of a feline high-grade MCT: many atypical mast cells. Note as in multinucleated neoplastic mast cells, nuclei are of different sizes and shapes

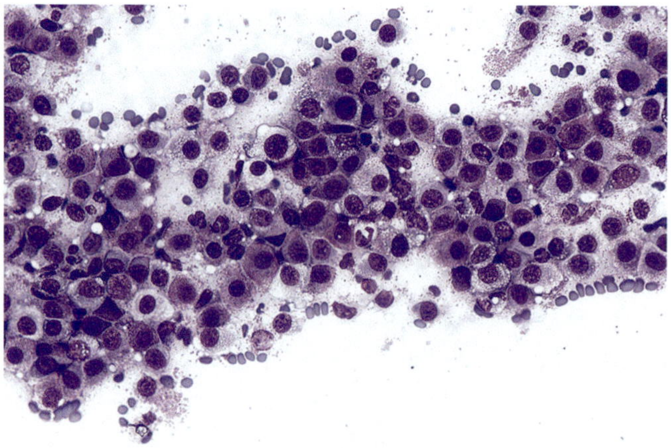

Fig. 8.55 Cytology of less-differentiated MCT: anisokaryosis, anisocytosis, and variable amount of metachromatic granules

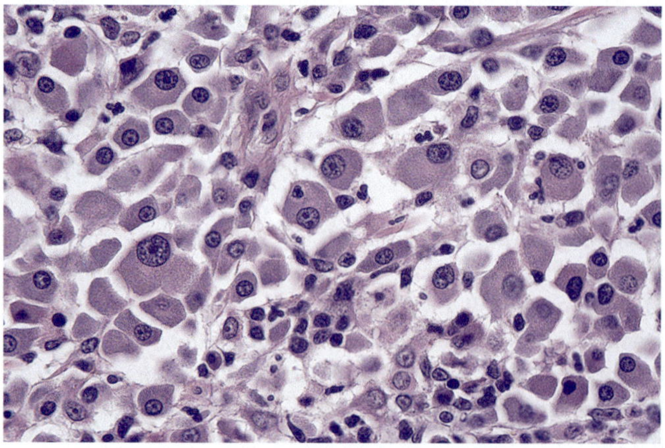

Fig. 8.56 Histology of poor-differentiated MCT: severe anisokaryosis and anisocytosis. Granules are barely perceptible

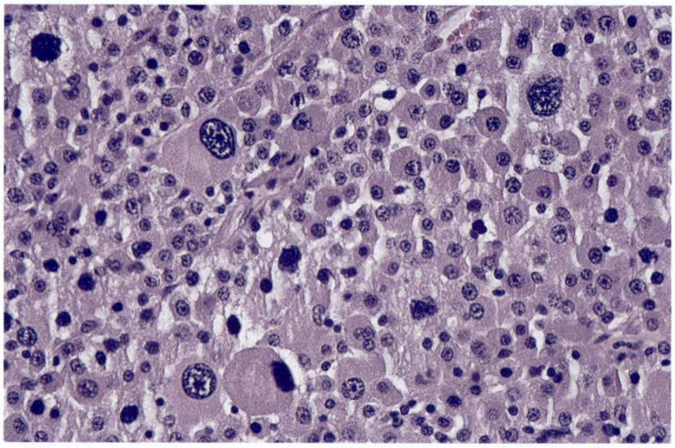

Fig. 8.57 Histology of poor-differentiated MCT: severe anisokaryosis and anisocytosis on neoplastic cells. Since granules are not visible, other undifferentiated RCTs can be suspected

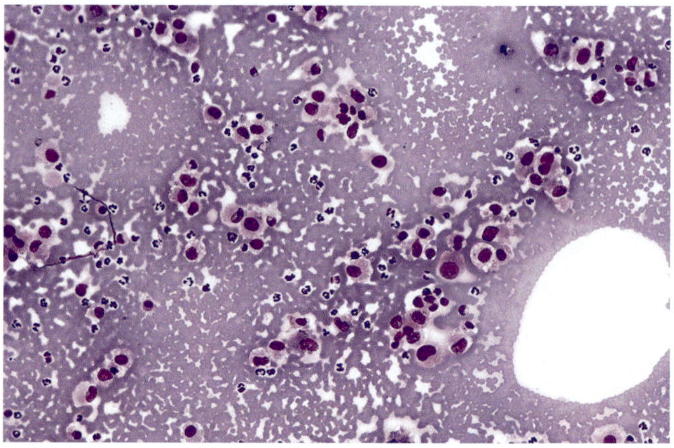

Fig. 8.58 Cytology of an atypical feline MCT: at high magnification many single cells in oval shape, with indented nuclei and large cytoplasm

slightly indented, and a larger cytoplasm that contains very few or no granules and sometimes small vacuole-like achromatic areas. In some cases, toluidine blue or Giemsa staining can highlight granules. Some cells can have a spindle shape, which can cause them to be confused with a mesenchymal neoplasm or a histiocytic disease such as feline progressive histiocytosis. Despite the number of eosinophils is usually scarce in most well-differentiated feline MCTs, it seems to be higher in the atypical subtype, and this could be very useful in suspecting a poorly granular MCT (Figs. 8.58, 8.59 and 8.60).

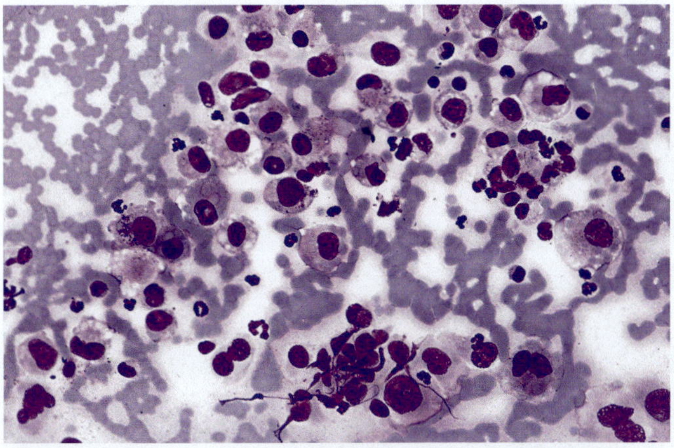

Fig. 8.59 Cytology of an atypical feline MCT: many single cells with morphology that mimics histiocytes. Few metachromatic purple granules allow defining cells as mast cells

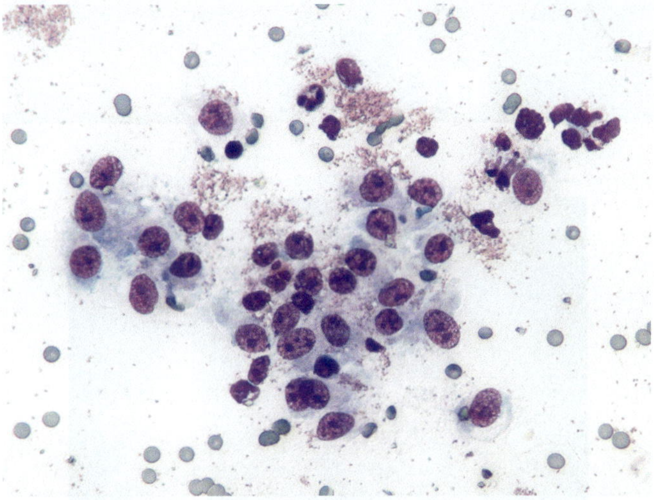

Fig. 8.60 Cytology of an atypical feline MCT in a young Siamese cat with multiple nodules that undergo spontaneous regression: many single cells oval to spindle in shape, with slightly indented nuclei. Note the high number of granules of eosinophils in the background

8.2.2 *Lymphoma* and Other Lymphoproliferative Disorders

Cutaneous lymphomas are uncommon tumors in dogs and rare in cats. Histologically, the cutaneous lymphomas are classified in *epitheliotropic lymphoma*, in which the neoplastic cells have a tropism for epidermis and adnexal epithelia, and *non-epitheliotropic lymphoma*, in which the neoplastic cells are localized in the dermis and can spread to the panniculus and the epithelia are not involved [40]. The epitheliotropic lymphoma (CEL) is always T phenotype, whereas the phenotype of non-epitheliotropic lymphoma can be both B and T. In some dogs with CEL, an immune-histochemical co-expression for CD4 and CD20, respectively, for T and B neoplastic cells, has been reported [34–36].

In dogs, the macroscopic aspects of cutaneous lymphoma differ greatly depending on the location of the neoplastic cells.

Epitheliotropic lymphoma (CEL) is a skin neoplasm of adult to old dogs in which four different clinical presentations are recognized:

(a) Multifocal to diffuse desquamation with variable degrees of erythema (exfoliative erythroderma)
(b) Plaques and nodules
(c) Depigmentation, erythema, erosions, and ulcers of the mucocutaneous junctions and on the nose (mainly on vertical part with spreading to upper haired lips)
(d) Erosions, ulcers, and plaques localized exclusively to the oral cavity

Such lesions may occur individually or simultaneously on the same animal. Given its clinical polymorphism, epitheliotropic lymphoma can be clinically defined as “the great imitator,” with numerous differential diagnoses depending on the different clinical presentation with which it manifests itself (Figs. 8.61, 8.62, 8.63, 8.64 and 8.65) [32, 36–38, 41, 43, 44].

In *non-epitheliotropic lymphoma (CNEL)*, multiple nodules and/or plaques, dermal or subcutaneous in location, of different sizes, alopecic, erythematous, and often ulcerated, characterize lesions [44]. The more superficial lesions often assume a characteristic bizarre morphology represented by donut-like and arciform-like shapes or by serpiginous in appearance plaques (Figs. 8.66, 8.67 and 8.68). This clinical presentation is strongly indicative, but not pathognomonic, of non-epitheliotropic lymphoma; in fact, it can also be observed in course of cutaneous reactive histiocytosis as discussed below.

Feline *epitheliotropic lymphoma* is extremely rare, and in the few published reports, it is grossly characterized by multifocal to generalized alopecia, erythema, and desquamation or by multiple, often ulcerated nodules and plaques of varying sizes. Single lesion can be also possible but is less common. Dermal or subcutaneous nodules and plaques, of different size and number, characterize instead the *non-epitheliotropic lymphoma* in cats [39, 47] (Fig. 8.69).

Canine and feline *cutaneous lymphocytosis* (CL) is a rare lymphoproliferative disorder of the skin, which etiology and pathogenesis are not completely understood yet. CL is seen more in cats than in dogs [33, 47]. CL shares clinical and histological similarities with the same disease recognized in humans, and for its definition different terms have been used including *cutaneous pseudolymphoma*, *lymphocytoma cutis*, *cutaneous lymphoid hyperplasia*, and *indolent lymphoma*. Adult/elderly patients are mainly affected, with skin lesions

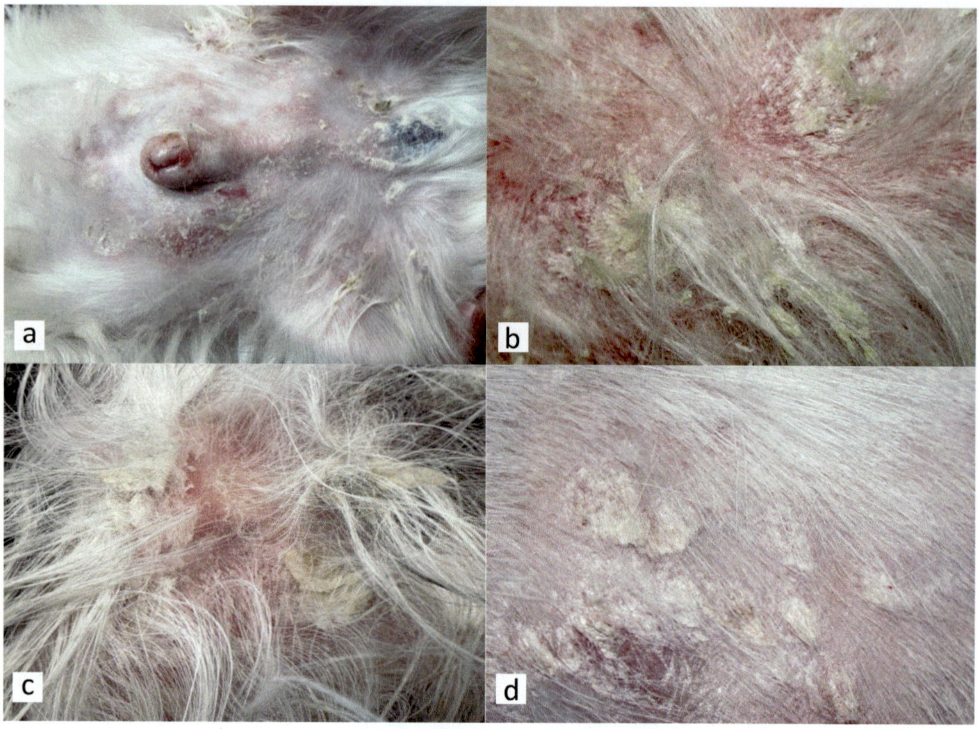

Fig. 8.61 Epitheliotropic lymphoma: (**a**) multifocal exfoliative dermatitis on the abdominal skin of a dog with CEL. Note as lesions mimic a superficial spreading pyoderma); (**b**) severe and diffuse skin erythema with large scales; (**c**) exfoliative dermatitis and erythema in a dog with CEL; (**d**) annular lesions covered by scales. Note as lesions mimic a superficial pyoderma (epidermal collarettes)

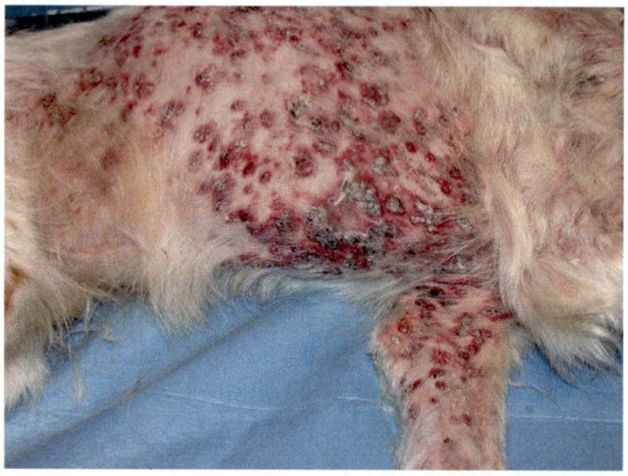

Fig. 8.62 Nodular CEL: multiple and generalized erythematous plaques and nodules on the skin

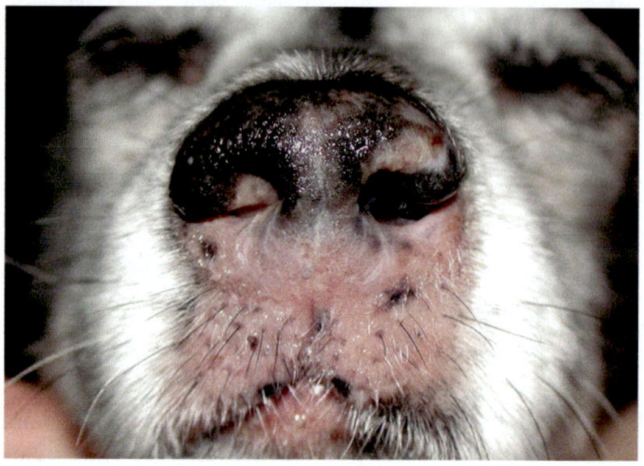

Fig. 8.64 Canine epitheliotropic lymphoma: inflammatory depigmentation of the vertical part of the nasal planum and upper lips

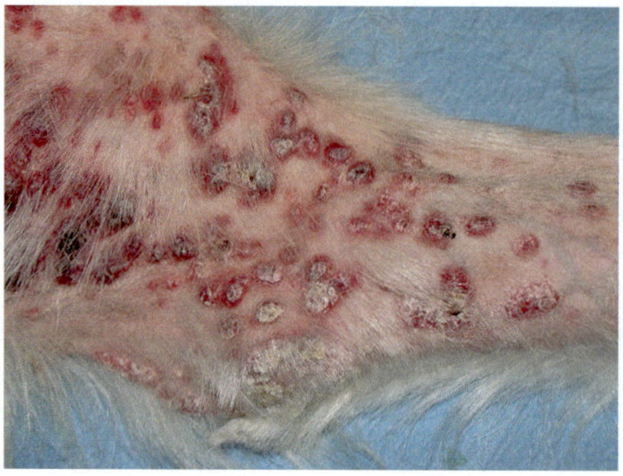

Fig. 8.63 Nodular CEL: close-up of the plaques and nodular lesions on the inner surface of the elbow

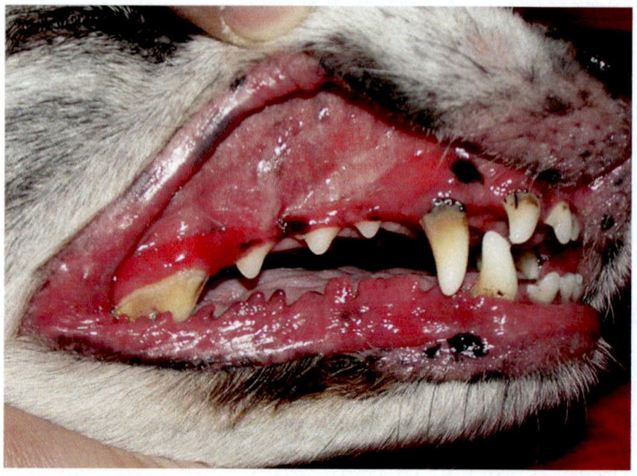

Fig. 8.65 Severe erythema and swelling of the oral mucosa and mucocutaneous junction depigmentation in a dog affected by CEL

that show a great clinical polymorphism, localized or diffuse, mostly represented by alopecia, erythema, macules, plaques, and exfoliative dermatitis with variable degrees of erythema. From a histological point of view, the lesions are characterized by a proliferation of mature lymphocytes that extended from the superficial to the deep dermis and up to the superficial panniculus (Figs. 8.70, 8.71 and 8.72).

In a study of 19 cats and 10 dogs affected by CL, scattered and focal epitheliotropism of mature lymphocytes is observed only in half of the cases and in none of them did they take on aspects of aggregation similar to "Pautrier's microabscesses" as usually occurs in epitheliotropic lymphoma [33]. Lymphocytic infiltrates are diffusely CD3+ and are associated, in most cases, with CD20+ nodular aggregates in the dermis. No clonal rearrangement was

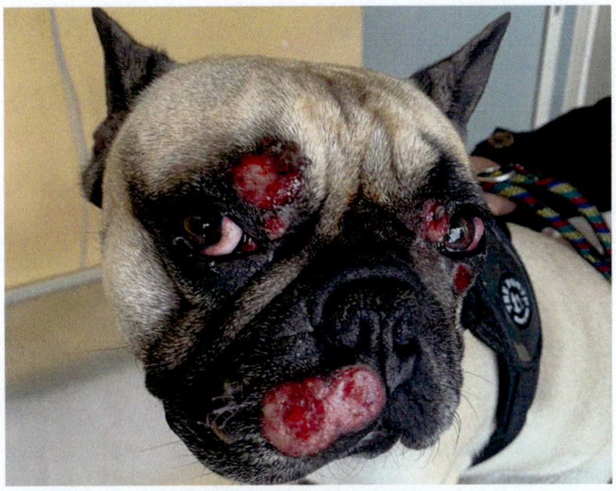

Fig. 8.68 Canine non-epitheliotropic lymphoma: ulcerated plaques on the face of a French Bulldog (courtesy of DVM M. Beccati, Italy)

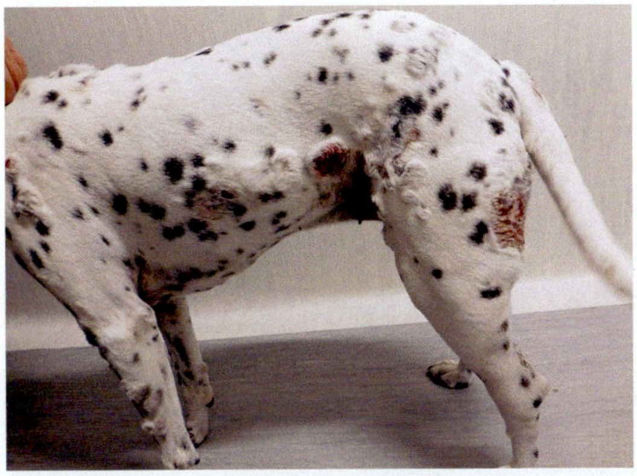

Fig. 8.66 Canine non-epitheliotropic lymphoma: multiple intact and ulcerated nodules and plaques spread all over the body

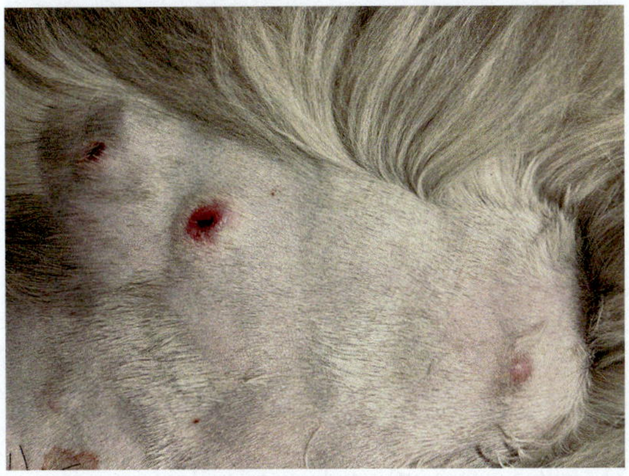

Fig. 8.69 Feline non-epitheliotropic lymphoma: multiple dermal–hypodermal nodules in a cat with non-epitheliotropic B cell lymphoma

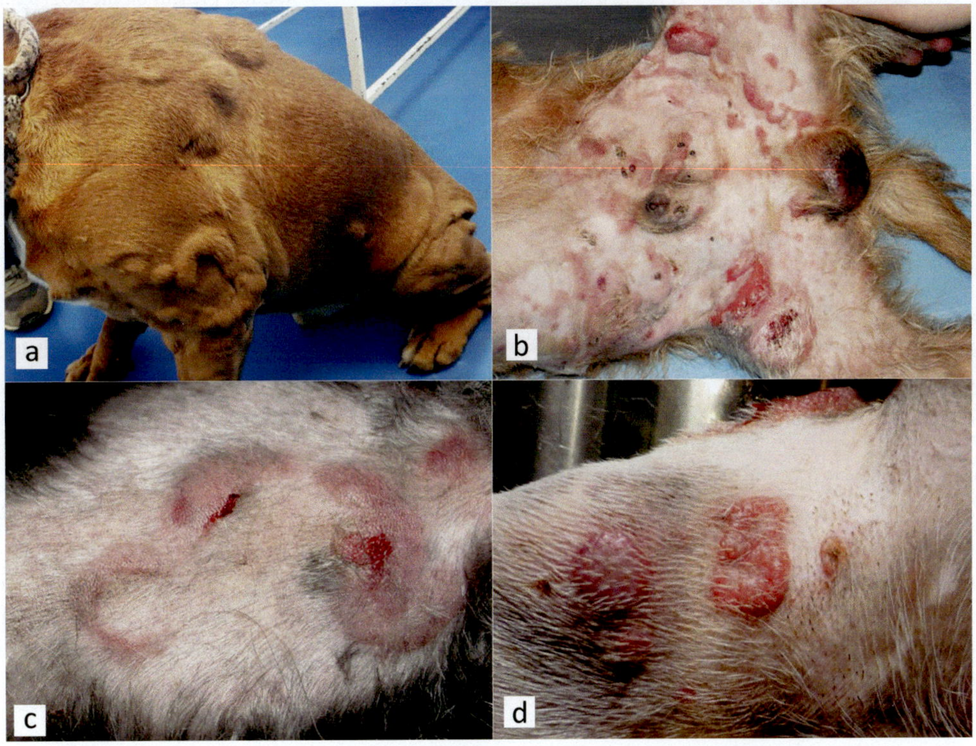

Fig. 8.67 Canine non-epitheliotropic lymphoma: (**a**) subcutaneous roundish to serpiginous nodules in a Boxer (courtesy of DVM E. Brogi-Italy); (**b**) nodules and serpiginous in shape abdominal plaques; (**c**) serpiginous and letter C-shaped erythematous plaques; (**d**) serpiginous plaques in a dog with CNEL

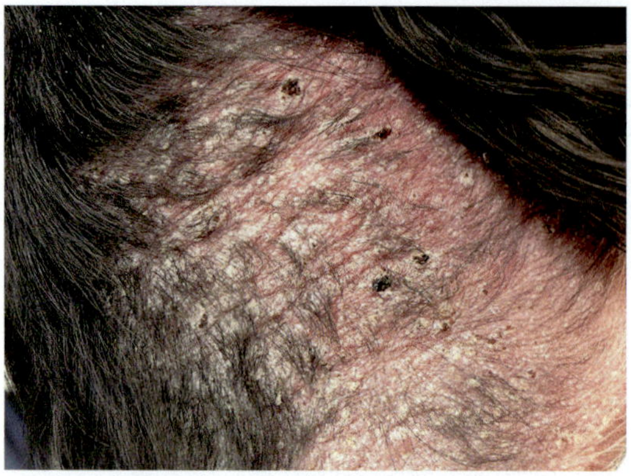

Fig. 8.70 Canine cutaneous lymphocytosis: alopecia, erythema, and scales on the neck of a dog

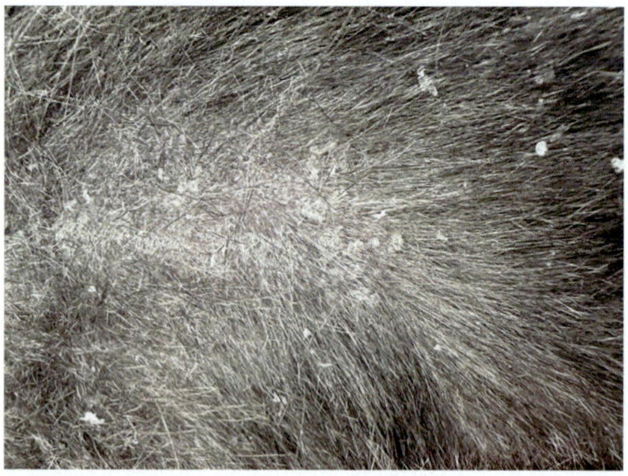

Fig. 8.71 Feline cutaneous lymphocytosis: alopecia, scales, and erosion on the trunk of a cat with diffuse exfoliative dermatitis

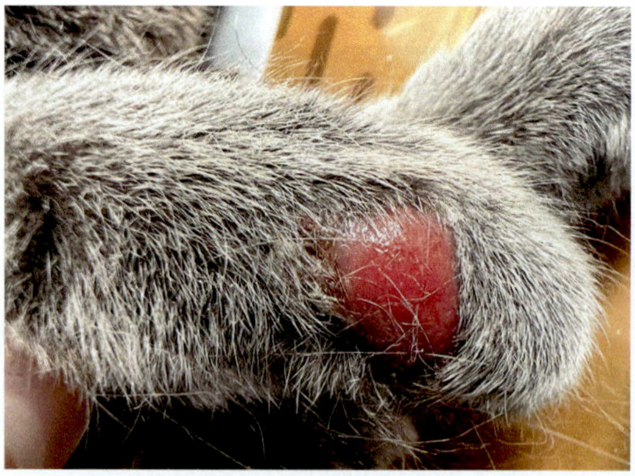

Fig. 8.72 Feline cutaneous lymphocytosis: erythematous nodule on a digit of the same cat of the previous image

identified in feline CL, while in dogs a clonal rearrangement of the BCR was found in four dogs (40%) and a clonal rearrangement of the TCR was found in one animal only (10%). One dog (10%) showed rearrangements in both the BCR and TCR gene loci. These results suggest that the absence or poor epitheliotropism associated with a diffuse cutaneous infiltration of small and well-differentiated lymphocytes may represent an important diagnostic element to differentiate between CL and CEL and that, while in cats, based on the clonal arrangements results, CL must be considered, to date, as a non-lymphomatous skin disorder, this discrimination between CL and CEL is more difficult in dogs. Further works will better define this very particular entity.

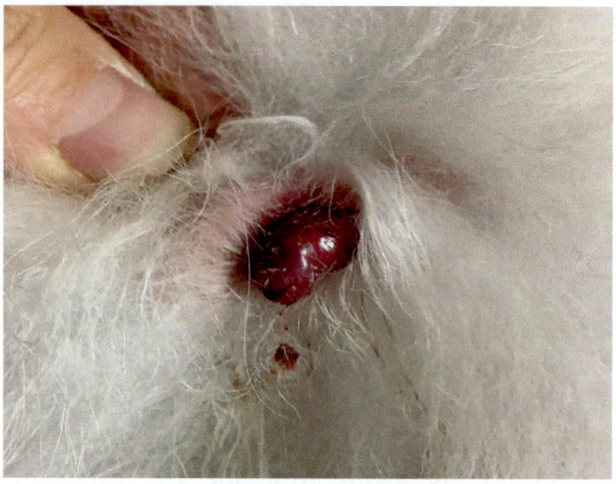

Fig. 8.73 B cell lymphoma with Mott cell differentiation: alopecic and erythematous nodule in a dog with concomitant gastric lymphoma

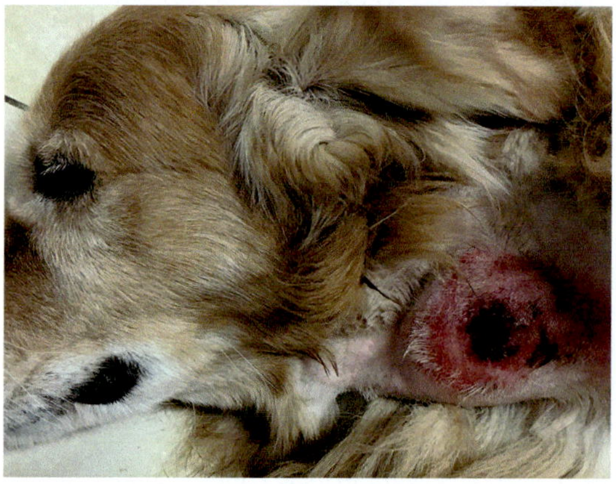

Fig. 8.74 Large granular lymphocyte lymphoma: ulcerated mass on the neck of a dog. The lesion is one of multiple cutaneous and subcutaneous nodules present on the dog' skin

Other rare to very rare variants of lymphoma with skin involvement are reported in dogs. *Canine inflamed non-epitheliotropic cutaneous T cell lymphoma*, *subcutaneous panniculitis-like T cell lymphoma*, *B cell lymphoma with Mott cell differentiation*, and *large granular lymphocyte lymphoma* are all rare forms of lymphoma reported in dogs [31, 42, 45, 46, 48] The first two lymphomas are primary cutaneous/subcutaneous in location, and the last two are subtypes of lymphoma which usually affect the internal organs and that exceptionally can involve the skin. A cutaneous lymphoma at injection sites has been diagnosed in 17 cats [46b] (Figs. 8.73 and 8.74).

Cytological Findings

As occurs in other round cell tumors, apart from those cases in which the CEL does not occur as nodules (exfoliative erythroderma, swelling with depigmentation of the nose, and mucocutaneous junctions), cytopathological samples are usually rich in cells.

Epitheliotropic lymphoma can show different histopathological patterns, which determine the grossly appearance of clinical lesions and, consequently, the quantity of cells that can be released of the lesion. In the case of the histopathological variant of CEL, named Pagetoid reticulosis, the neoplastic lymphoid cells are exclusively localized in the epidermis and in the wall of the follicular adnexa; in these cases the only way to obtain representative cytological samples is by imprint technique after having lifted the thick scales that, once removed, can expose the first layers of epidermis in which neoplastic cells are located. Since in most exfoliative clinical presentation of CEL, neoplastic lymphocytes are also present into the dermis, the removal of the scales can also allow to remove all the epidermal layers and reach the superficial sub-epidermal dermis; also in this case the imprint technique permits to collect a variable number of neoplastic lymphocytes (Figs. 8.75, 8.76 and 8.77).

Slides prepared from exfoliative lesions usually have scarce to moderate number of lymphoid cells, and in ulcerated lesions where the superficial dermis is reached, a variable number of inflammatory cells, mainly neutrophils and some histiocytes, are often associated with the neoplastic cells. In these cases, because in many CEL the cells do not show marked atypia, the coexistence with neutrophils can create mistaken interpretation in favor of an inflammatory chronic skin disease. In most cases of exfoliative CEL, neoplastic cells are round, with slightly not uniform in size nuclei, and chromatin is more often fine and regular with not evident nucleoli; cytoplasm can be scarce to medium in size and pale blue in color as to create some doubts with histiocytic cells. Severe atypia are not usually evident in CEL (Figs. 8.78, 8.79 and 8.80).

As discussed in Chap. 1 regarding the sampling techniques, FNB can be also used for collecting cells from flattened lesions. In the case of CEL in which neoplastic cells are also present in the sub-epidermal dermis, it is possible collect a large number of cells by an intradermal fine needle biopsy (Figs. 8.81, 8.82 and 8.83).

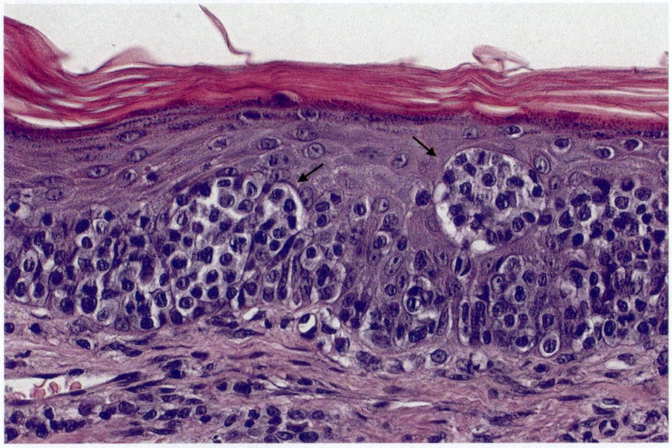

Fig. 8.76 Histology of epitheliotropic lymphoma: intra-epidermal aggregates of neoplastic lymphocytes characterized the so named "Pautrier' microabscesses

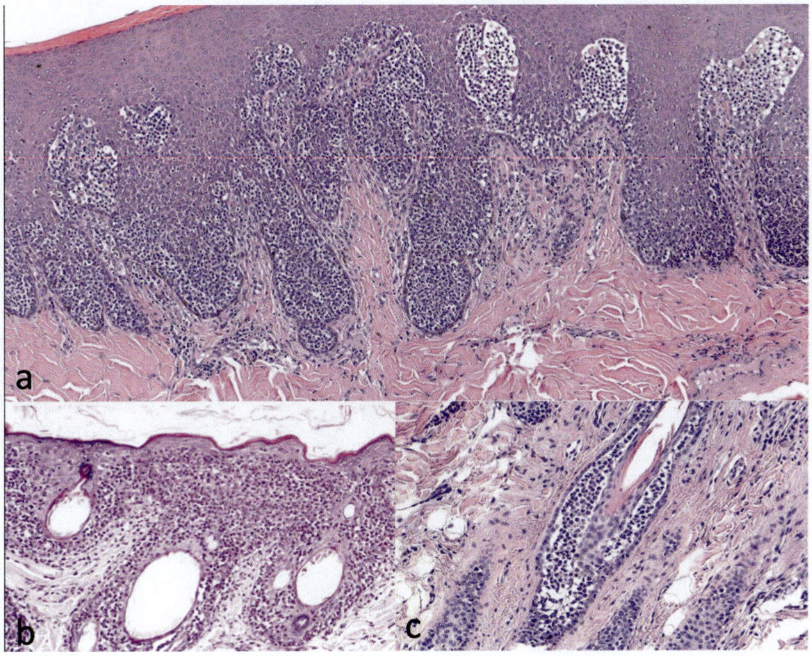

Fig. 8.75 Histology of epitheliotropic lymphoma: (**a**) marked thickening of the epidermis diffusely infiltrated by neoplastic lymphocytes; (**b**) neoplastic lymphoid cells are recognizable in the contest of the epidermis that appears "swollen"; (**c**) Pagetoid reticulosis variant of CEL: note as all the follicular wall is selected infiltrated by neoplastic lymphocytes

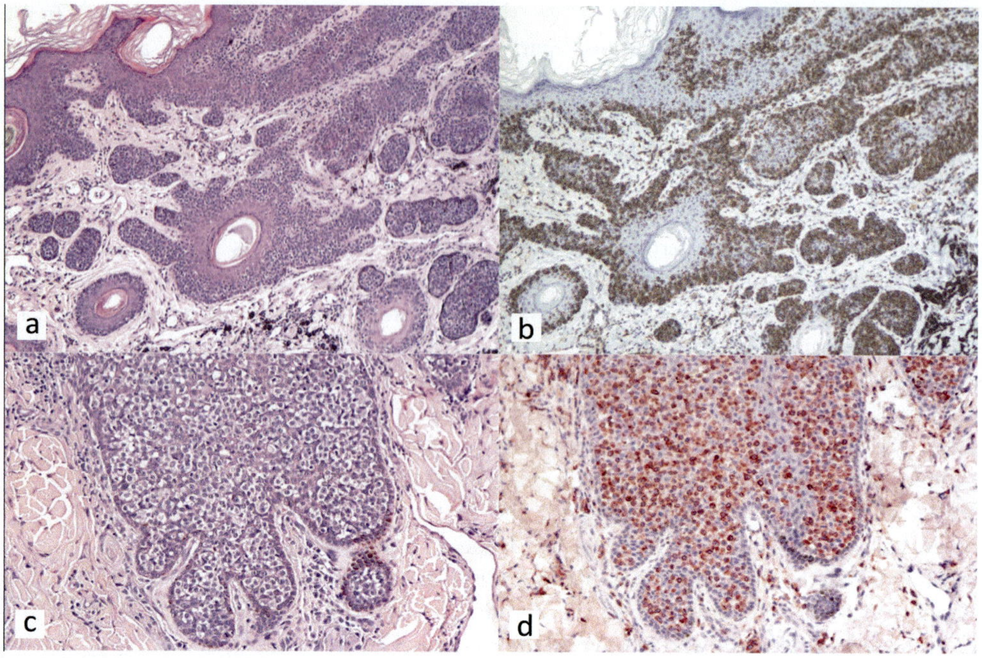

Fig. 8.77 Histology of epitheliotropic lymphoma: (**a**) Pagetoid reticulosis variant of CEL, only the epidermis is involved by neoplastic lymphocytic infiltration; (**b**) immunohistochemistry (CD3), neoplastic cells are positive to T lymphocyte marker; (**c**) higher magnification shows multiple small white spots in the epidermis which represent the pale cytoplasm of neoplastic infiltrating lymphocytes; (**d**) the latter can be clearly marked with immunohistochemistry

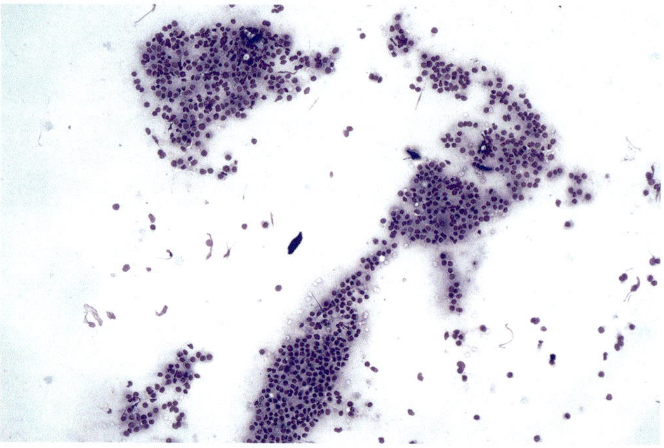

Fig. 8.78 Cytology of epitheliotropic lymphoma: imprint technique. Many neoplastic lymphocytes are collected from exfoliative lesions

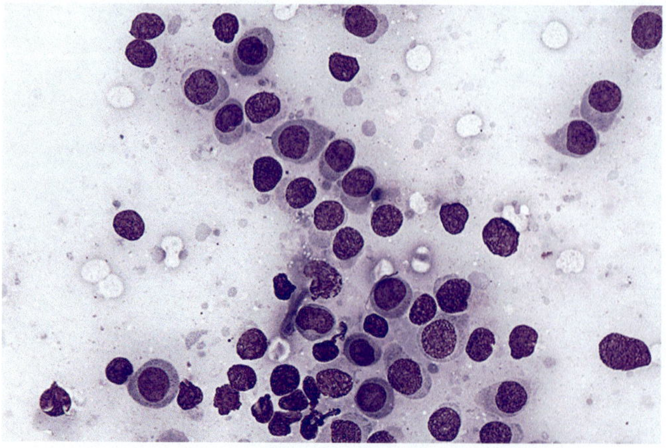

Fig. 8.79 Cytology of epitheliotropic lymphoma: imprint technique. Neoplastic lymphocytes have regular round nuclei, fine chromatin pattern, and pale medium in size cytoplasm

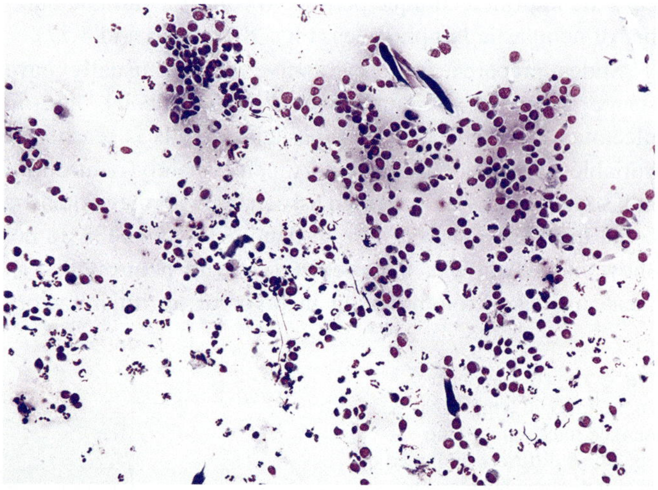

Fig. 8.80 Cytology of epitheliotropic lymphoma: imprint technique. Neoplastic lymphocytes along with neutrophils are commonly sampled from ulcerated lesions with secondary pyoderma

This diagnostic method is very interesting; in fact, since many of these dogs affected with exfoliative CEL are pruritic, clinical signs may be confused with other infectious or allergic diseases (e.g., dermatophytosis, leishmaniasis, atopic dermatitis). The cytological detection of many lymphoid cells via imprint or intradermal FNB can orient the clinicians to perform a skin biopsy without lost time.

In the presence of a CEL in the tumoral stage or with plaque/nodular lesions, the fine needle biopsy is the technique of choice for sampling large quantities of cells. Despite the above discussed morphology of cells in course of CEL, in large nodular or plaque-like lesions with diffuse dermal involvement, neoplastic cells can show more atypical

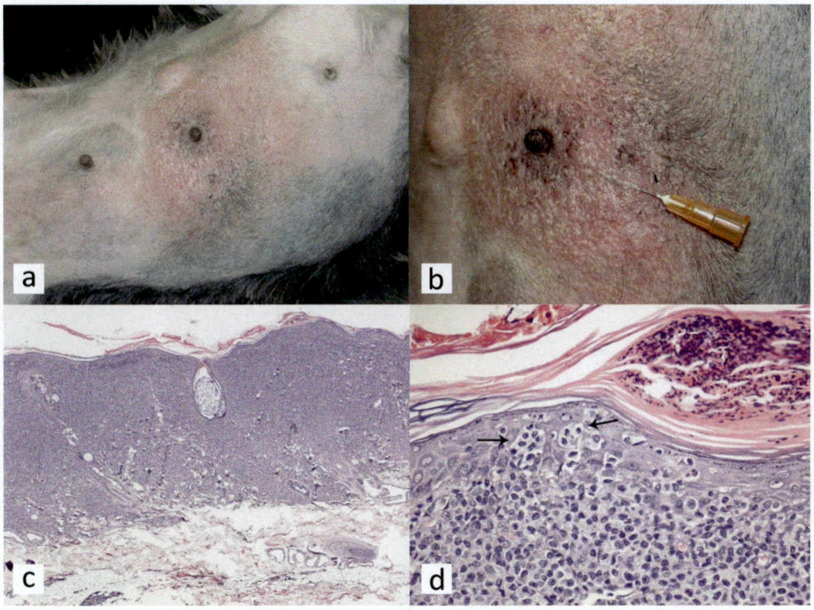

Fig. 8.81 (**a**) CEL: Erythema and exfoliation on the skin of a dog with CEL; (**b**) fine needle biopsy in the same lesion of previous picture; (**c**) histology of epitheliotropic lymphoma diffuse infiltration of neoplastic lymphocytes into the upper dermis; (**d**) at low magnification some little nests of lymphocytes (Pautrier's microabscesses) are easily recognizable as indicator of epitheliotropism of the neoplastic cells (arrows)

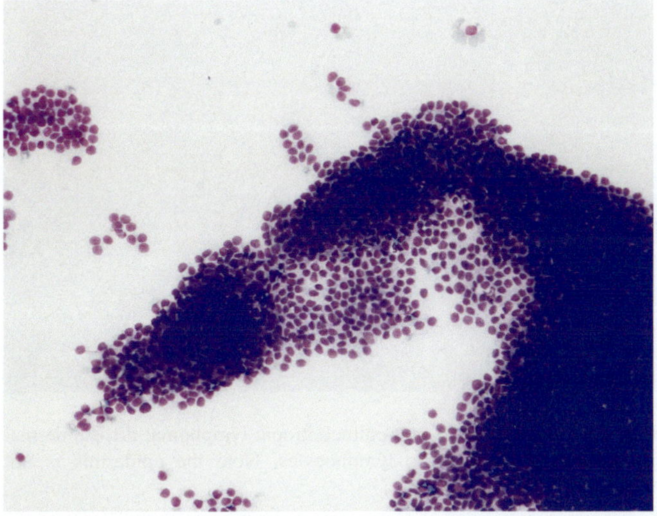

Fig. 8.82 Cytology of epitheliotropic lymphoma: FNB. Many neoplastic lymphocytes are collected from the same lesion of Fig. 8.81

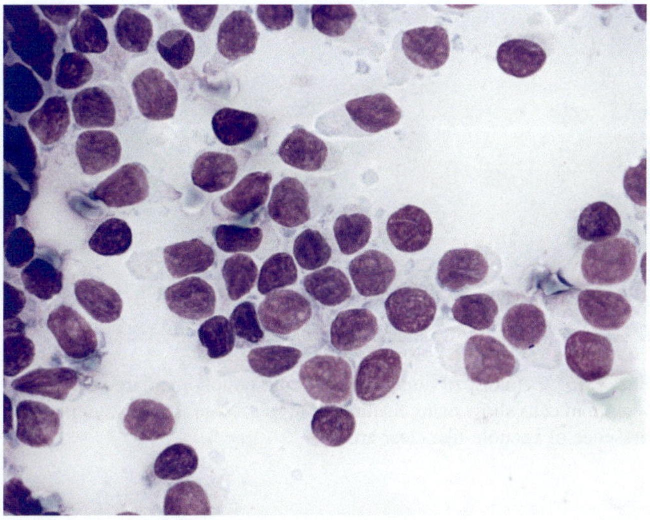

Fig. 8.83 Cytology of epitheliotropic lymphoma: at high magnification the typical morphology of neoplastic lymphoid cells can be appreciated

features; cells are small to medium in size; nuclei can show more relevant alterations, with single or multiple nuclear folds, stippled to coarse chromatin, high nucleus/cytoplasm (N/C) ratio, and cytoplasm that shows varying tonality of blue.

In the case of *non-epitheliotropic lymphomas*, histology is characterized by the presence of a diffuse neoplastic lymphoid infiltrating the dermis and the panniculus; cells are usually larger and more pleomorphic. Anisocytosis, anisokaryosis, multiple or macro nucleoli, intracytoplasmic vacuolar-like areas, and atypical mitosis can be frequently detected (Figs. 8.84, 8.85 and 8.86). As most lymphomas originate from T lymphocytes, it is sometimes possible to observe nuclei with multiple and deep folds that give the nuclei a *cerebriform* or *petal-like* appearance; this morphology is not pathognomonic of lymphoid cells since in rare case of histiocytic diseases nuclei can assume this appearance (Fig. 8.87). These cytological features can only suggest the T origin of neoplastic lymphocytes, but do not allow the differentiation between epitheliotropic and non-epitheliotropic lymphoma, for which histopathological and immunohistochemical examinations are necessary.

When cells are spread in the subcutis, many lipid droplets can be collected with cytology, together with a variable number of neoplastic cells. The latter often consist mostly of *bare* nuclei (without the cytoplasm), and only a small number of cells have the cytoplasm still intact. In these cases, it must be careful to not make a misdiagnosis of

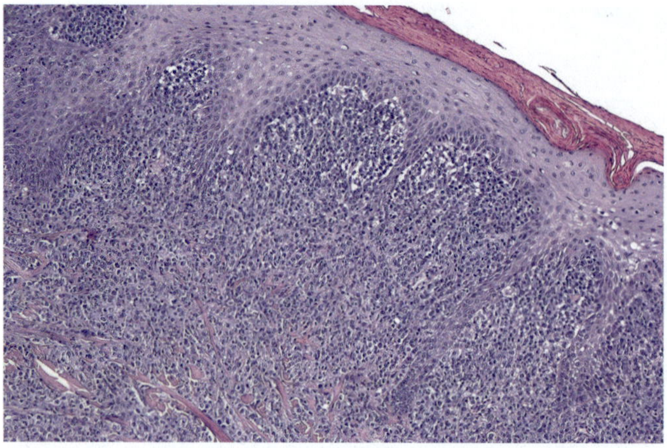

Fig. 8.84 Histology of non-epitheliotropic lymphoma: diffuse dermal proliferation of neoplastic lymphocytes. Note the epidermis is not involved

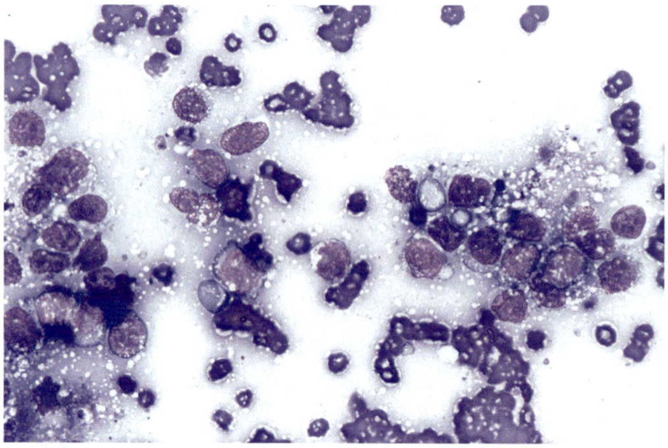

Fig. 8.85 Cytology of non-epitheliotropic lymphoma: large neoplastic lymphoid cells show many features of atypia. Note the intracytoplasmic presence of vacuole-like clear areas

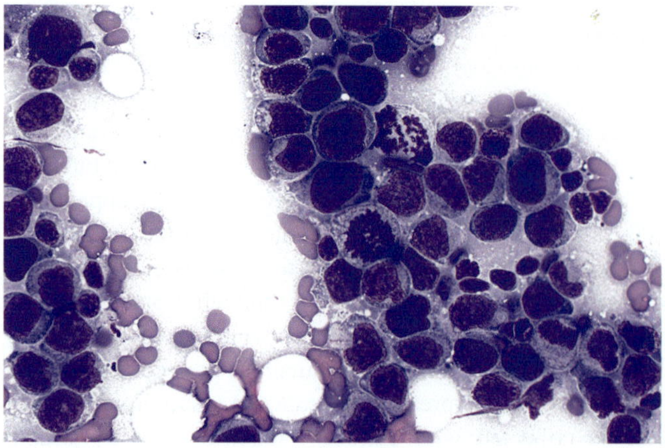

Fig. 8.86 Cytology of non-epitheliotropic lymphoma: very large neoplastic lymphocytes, with folded and bizarre in shape nuclei. Note two mitotic neoplastic cells

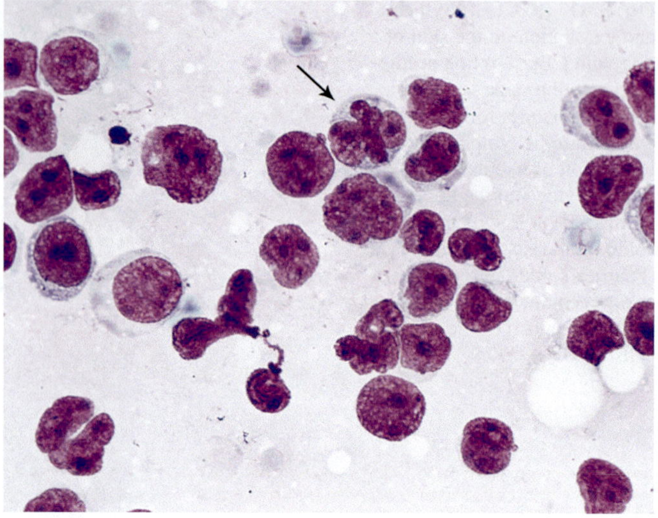

Fig. 8.87 Cytology of non-epitheliotropic lymphoma: the *cerebriform* or *petal* appearance of nuclei suggests their T origin (arrow)

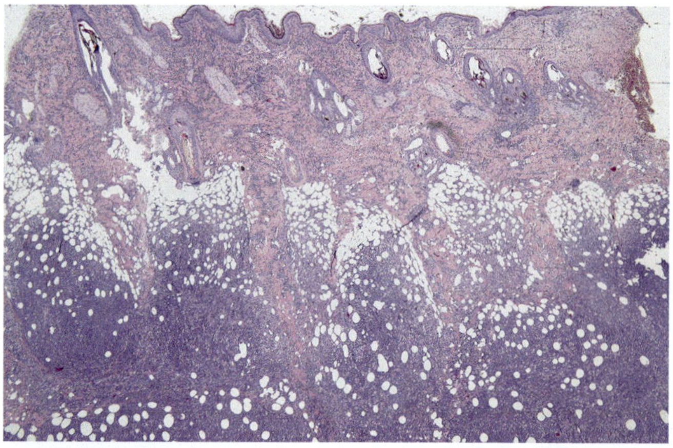

Fig. 8.88 Histology of non-epitheliotropic lymphoma: diffuse dermal proliferation of neoplastic lymphocytes. Note the epidermis is not involved

panniculitis. The usually high number of lympho-glandular bodies spread on the background helps to interpret the bare nuclei as being of lymphoid origin.

This cytological finding can occur also in case of the so named *panniculitis-like T cell lymphoma*. In this histopathological variant of cutaneous lymphoma localized in the adipose panniculus, neoplastic cells do not show cytological features different from other non-epitheliotropic lymphomas, but the localization of the cells to subcutis can create some interpretative problem in the case of scarcely cellular samples; in fact, in the same cases, only few round lymphoid cells are collected along with many uninterpretable naked nuclei. Since a variable amount of histiocytes is also commonly present along with the neoplastic lymphocytes, and the slides are constantly characterized by a lipid background, a misdiagnosing of granulomatous panniculitis can be done with cytology (Figs. 8.88, 8.89 and 8.90).

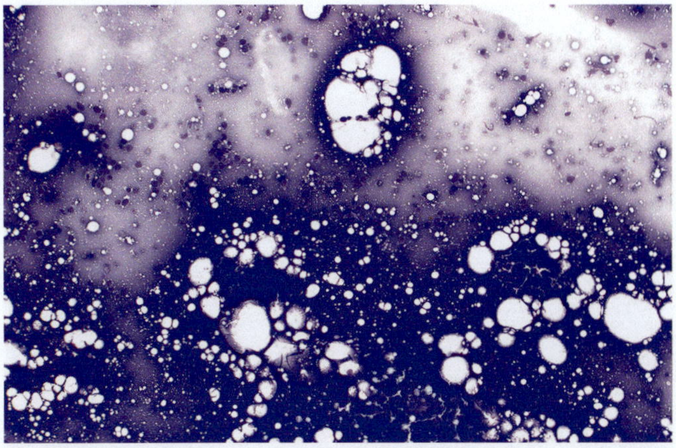

Fig. 8.89 Cytology of non-epitheliotropic lymphoma: amorphous and lipid background on which many round cells are present

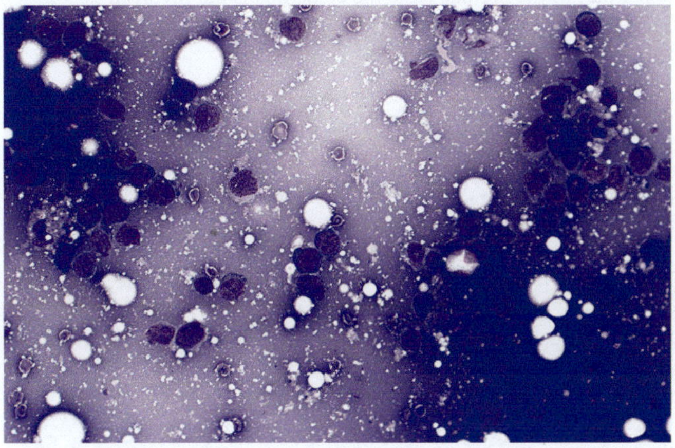

Fig. 8.90 Histology of non-epitheliotropic lymphoma: at high magnification naked nuclei and some lymphoid neoplastic intact cells are recognizable

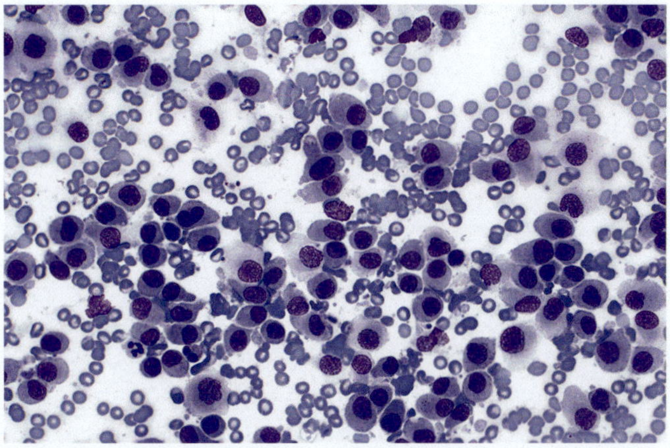

Fig. 8.91 Many large neoplastic cells with voluminous and pale cytoplasm. In this case, the large cytoplasm and folded nuclei can generate doubts regarding the lymphoid origin of cells

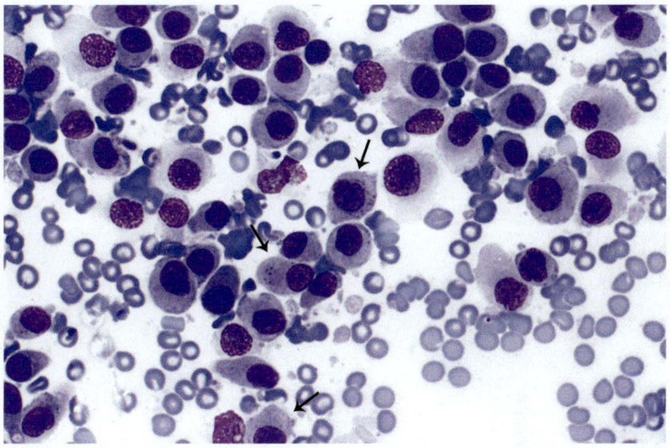

Fig. 8.92 Cytology of LGL lymphoma: at high magnification, intracytoplasmic granules are recognizable (arrows)

A so named *inflamed non-epitheliotropic cutaneous T cell lymphoma* has been described in dogs. In this uncommon variant of lymphoma, the infiltrating cells are mixed with a high amount of histiocytes, plasma cells, and neutrophils, and in some cases an important infiltrate of eosinophil has also been found. Other population of cells include small, uniformly sized lymphocytes, and a variable number of intermediate to large, atypical lymphocytes; the latter have large vesicular or hyperchromatic nuclei with prominent nucleoli. Different histological pattern are recognized, and in some of them, the localization of the infiltrate along with the prominent histiocytic presence creates difficulties in differentiating the lymphoma with a cutaneous reactive histiocytosis. In this lymphoma, cytology is usually not diagnostic, and in cases a mixed population of inflammatory cells is intermingled with large atypical round cells, an inflamed non-epitheliotropic cutaneous T cell lymphoma should be suspected.

Large granular lymphocyte lymphoma (LGLL) is a very rare T cell lymphoma in dogs, and the skin is not the primary localization; in fact, it is rarely involved in animal with this typology of lymphoma. Cytology of LGLL is characterized by medium- to large-sized round cells with granules of variable size, from fine to large azurophilic granules, sometimes barely visible. The chromatin is irregular and the nucleolus is sometimes visible (Figs. 8.91, 8.92 and 8.93).

B cell lymphoma with Mott cell differentiation is a very rare lymphoma in which the skin is not the primary organ; in fact, its report on the skin is occasional and must be interpreted as part of a lymphoma involving other internal organs. Cytology is very characteristic since many small- to medium-sized neoplastic lymphoid cells are accompanied by other cells that show a clear differentiation in Mott cells. With Romanowsky-type stains, the atypical Mott cells contain many pale blue intracytoplasmic inclusions consistent with Russell bodies (Figs. 8.94 and 8.95).

Cutaneous lymphocytosis (CL) is an uncommon lymphoproliferative disorders which origin is to date, not yet clear. Since the wide polymorphism of clinical lesions, cells can be collected with different sampling techniques

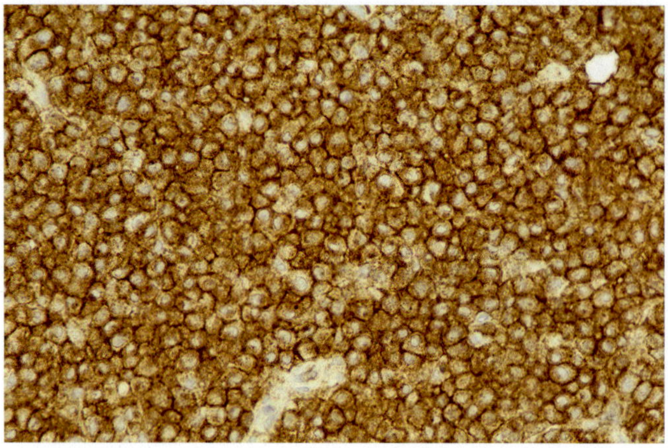

Fig. 8.93 Histology of LGL lymphoma coming from the same cells presented in the two previous cytological pictures (immunohistochemistry): strong and diffuse positivity to CD3 marker that confirms the T lymphocytic origin of neoplastic cells

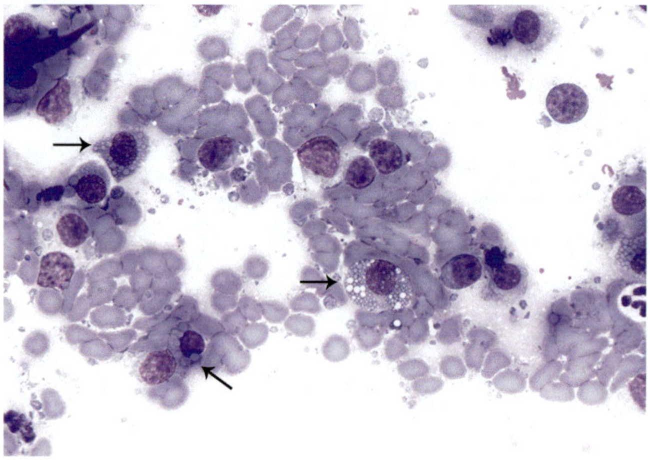

Fig. 8.94 Cytology of B cell lymphoma with Mott cell differentiation: many neoplastic lymphoid cells some of them showing clear Mott cell differentiation (arrows)

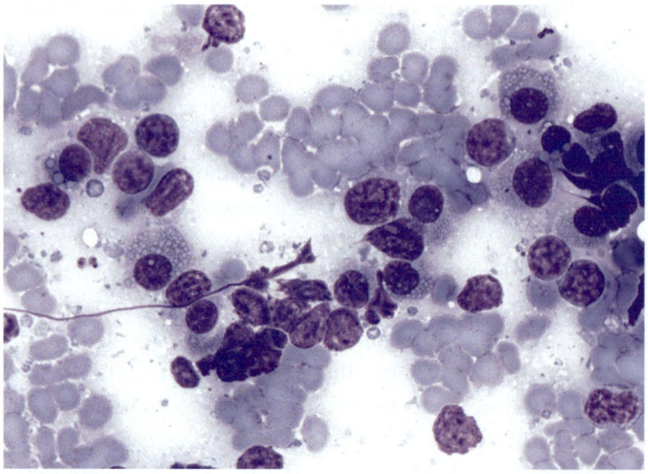

Fig. 8.95 Cytology of B cell lymphoma with Mott cell differentiation: at high magnification, intracytoplasmic roundish and overlapped Russell bodies confirm the differentiation of neoplastic lymphocytes in Mott cells

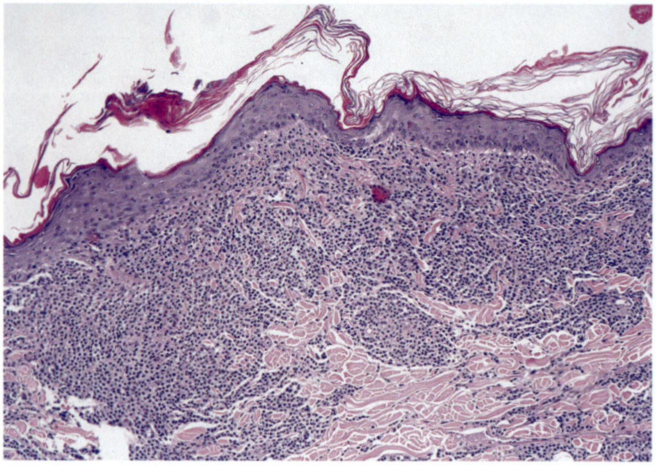

Fig. 8.96 Histology of canine CL: many quite well-differentiated lymphocytes with a diffuse band-like arrangement, located in the upper part of the dermis. Note the epidermis is not involved

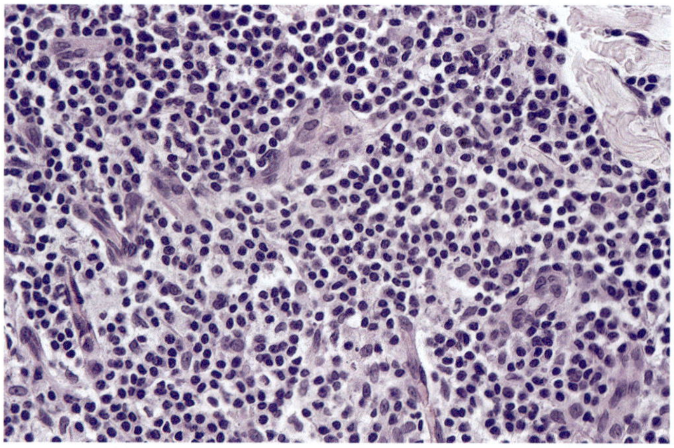

Fig. 8.97 Histology of canine CL: at high magnification, the lacking of atypia of lymphoid cells is better evaluable

depending on the typology of lesions. Plaques and nodules can be collected through FNB, while for flattened and scaly lesions, imprint technique or the intradermal FNB/FNAB technique can be performed, as above discussed for CEL. Cytology from CL shows a large number of mature small lymphocytes, without morphological atypia which allows cytologists to orient the suspicion of disease; in some cases, along with lymphoid cells, a small number of histiocytes, mast cells, and eosinophils can be sampled, especially in cats as reported in literature [33]. The final diagnosis of CL requires histopathology, immunohistochemistry, and clonal rearrangement test (Figs. 8.96, 8.97, 8.98 and 8.99).

In the practice, single or multiple nodules and plaques, cytologically diagnosed as cutaneous lymphomas, have spontaneously recovered.

For this reason, especially when single nodules are observed, the presence of a cytologically atypical lymphoid

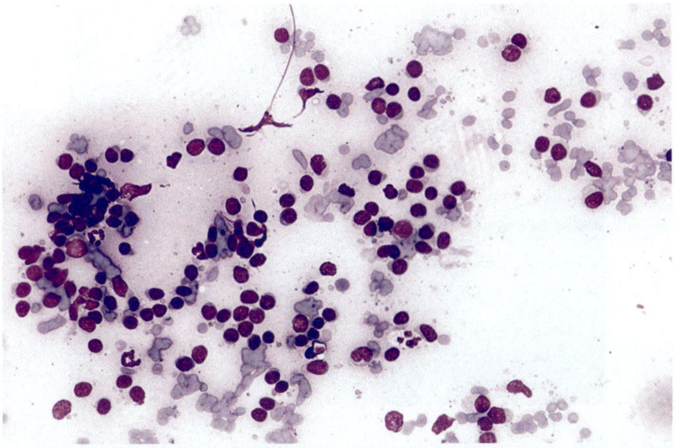

Fig. 8.98 Cytology of feline CL: small- and medium-sized lymphocytes collected with FNB from a cat with an exfoliative form of cutaneous lymphocytosis

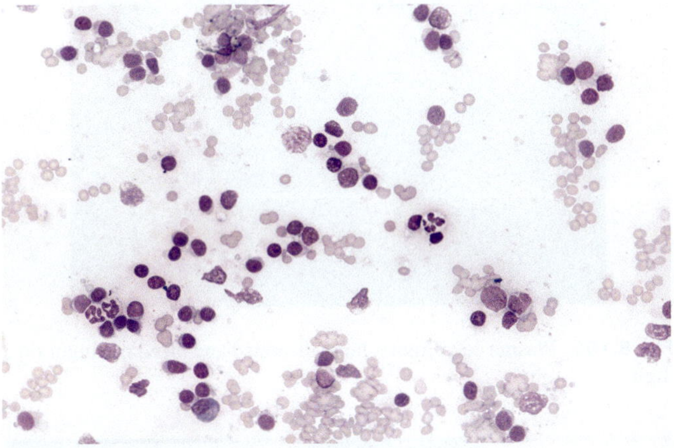

Fig. 8.99 Cytology of feline CL: hyperchromatic nuclei with scarce cytoplasm and lacking of morphological atypia characterize the cytology of cutaneous lymphocytosis

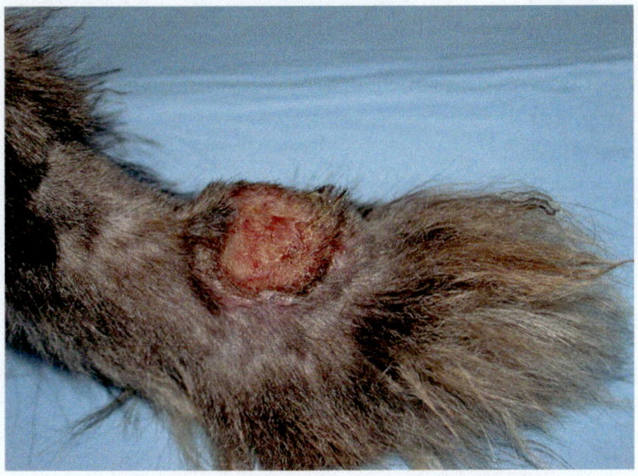

Fig. 8.100 Large and ulcerated pruritic plaque on the fore extremity of an old dog. The lesion started since 3 months and spontaneously regressed

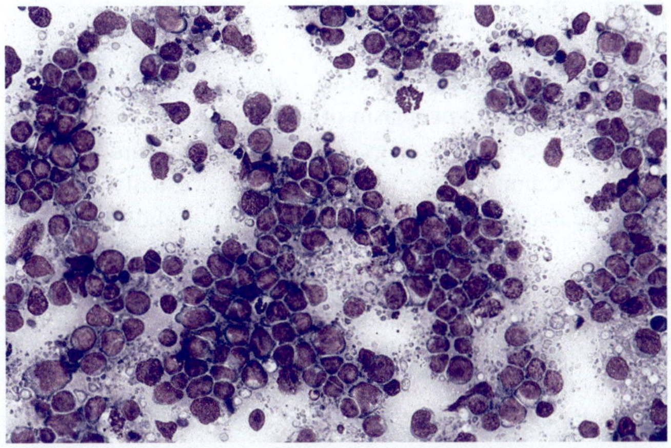

Fig. 8.101 Cytology of the lesion in Fig. 8.100: note the high number of medium-large lymphoid cells. Many lympho-glandular bodies are spread on the background

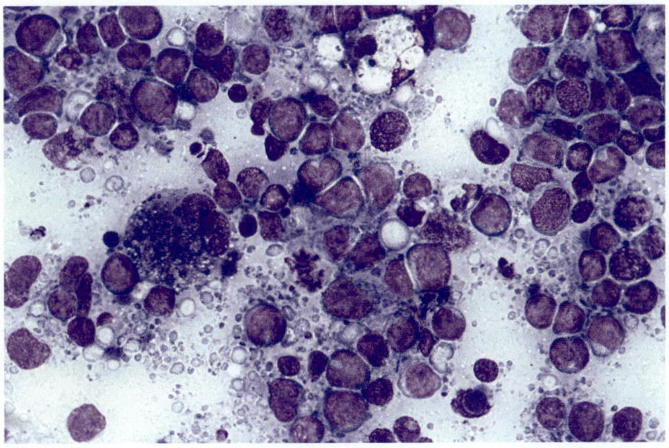

Fig. 8.102 Cytology of the lesion in Fig. 8.100: at high magnification the cytological features mimicking a lymphomatous lesions are better evaluable

population of cells must be interpreted with great caution, before making diagnosis of lymphoma. In dogs, these lymphoid lesions that mimic a neoplasia and undergo spontaneous remission are not well characterized, but their clinical behavior has allowed such lesions to be defined as non-neoplastic lymphoid inflammatory reactions of unknown cause that could be considered in the group of lesions defined in human medicine as pseudo-lymphomatous lesions. From the above, based exclusively on cell morphology, many single nodules diagnosed as lymphoma only using histopathology could have been non-neoplastic lymphomatous proliferations, since, once removed, it's not demonstrable that they would have regressed on their own. In doubtful cases, to obtain the definitive diagnosis of lymphoma, the clonality of the cells, which should be monoclonal in neoplasia and polyclonal in reactive lesions, must be assessed. Therefore, the cytological diagnosis of cutaneous lymphoma in dogs and cats should always be made with caution, especially when only a single nodule is present (Figs. 8.100, 8.101 and 8.102).

8.2.3 Plasma Cells Tumor

Extramedullary or cutaneous *plasma cell tumor* or *plasmacytoma* is a neoplasm of plasma cells, not so uncommonly observed in dogs and extremely rare in cats [50, 53]. Canine cutaneous plasmacytoma usually develops as single nodule that is often alopecic and brightly reddish in color with a cherry-like appearance. Lesions are of varying sizes, from few mm to some cm, usually round/oval in shapes and rarely with irregular and ulcerated surface mainly located on the toes, pads, interdigital areas, pinna, nose, and lips [50, 51] (Figs. 8.103, 8.104 and 8.105). In rare cases, nodules are multiple and can be spread all over the body; in these cases, when a bone marrow multiple myeloma is not present, the term of plasmacytosis is used. Golden Retriever, Labrador, Cocker Spaniel, and WHWT must be included in the most frequently breeds predisposed to this syndrome [49] (Figs. 8.106 and 8.107).

In cats, they are very rare neoplasms and are characterized by a single nodule, mostly reported on the legs, face, and tail [53].

In humans, another lesion that originates from plasma cells and is called *primary localized amyloidosis* or *nodular amyloidosis* or *amyloidoma* is recognized. This type of lesion has been rarely reported in dogs, but its cause and, for some researcher, its real existence in canine patients are still debated. As in humans, it is hypothesized that a local traumatic event may be the cause of the onset of amyloidoma [58]. Lesion is usually characterized by a single nodule [57] (Fig. 8.108).

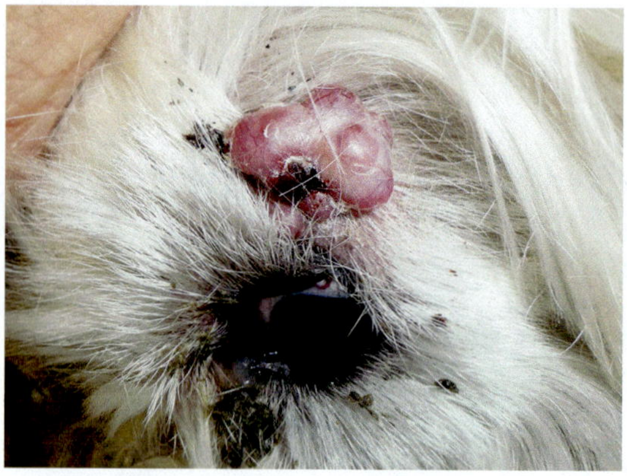

Fig. 8.104 Extramedullary plasma cell tumors with an irregular shape on the eyelid of a West Highland White Terrier

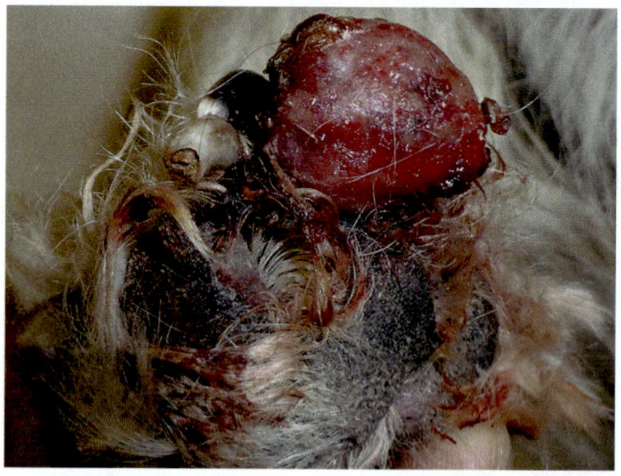

Fig. 8.105 Plasma cell tumors: large ulcerated and bleeding tumor on a digit

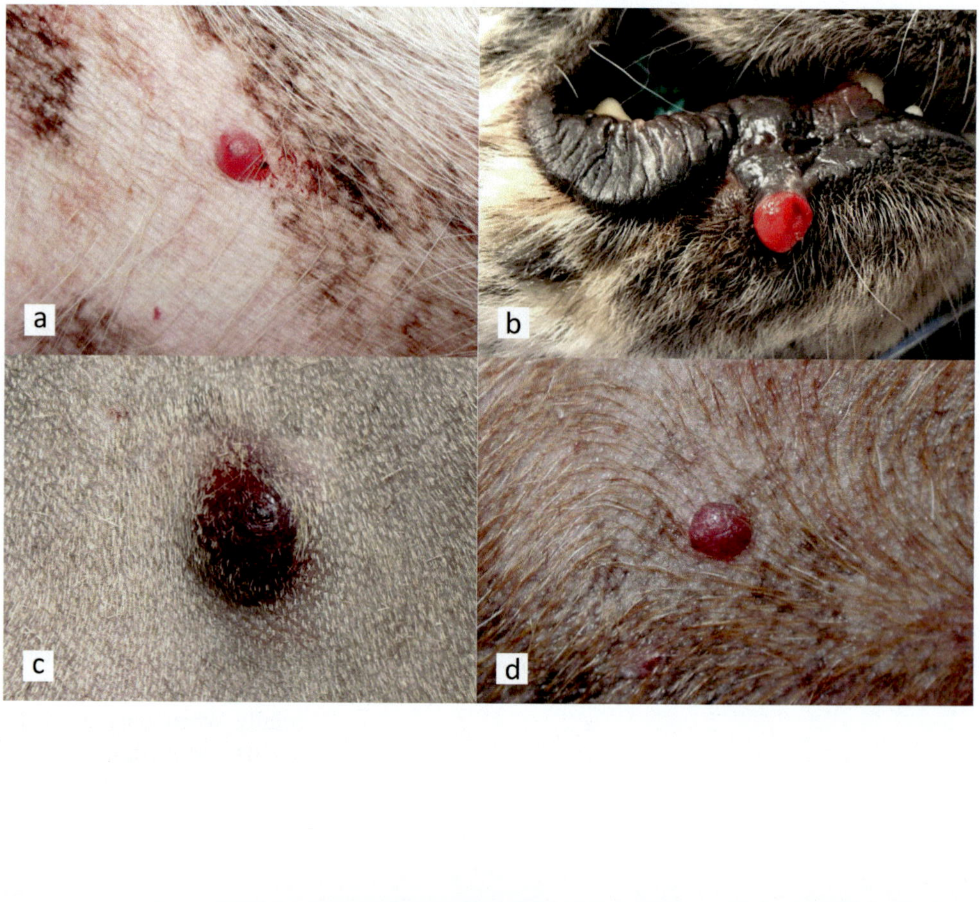

Fig. 8.103 Single small erythematous cutaneous plasma cell tumors

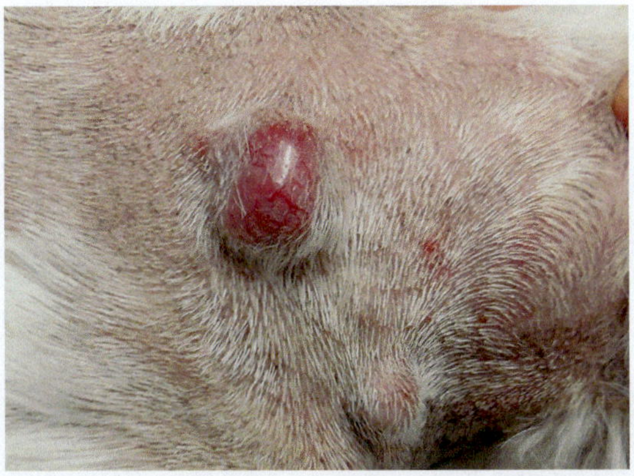

Fig. 8.106 Plasma cell tumors: two cutaneous plasma cell tumors on the skin of a dog with plasmacytosis

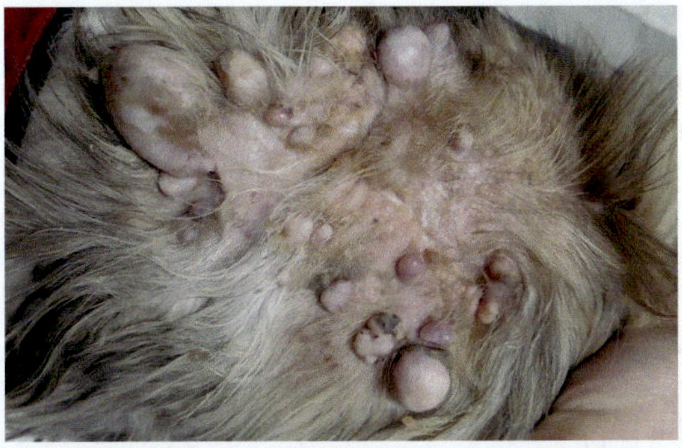

Fig. 8.107 Plasmacytosis: multiple skin nodules and masses on the skin of a Cocker Spaniel with plasmacytosis

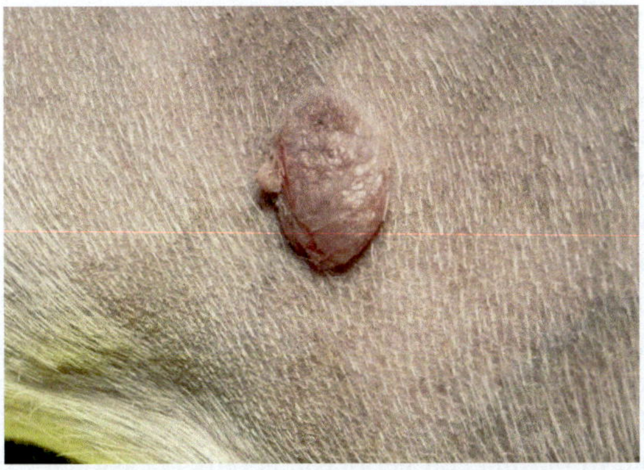

Fig. 8.108 Amyloidoma: single alopecic erythematous nodule on the skin of a dog (courtesy of DVM A. Negro-Italy)

Cytological Findings

Canine *cutaneous plasmacytoma* is classified into five histological subtypes, and therefore, the polymorphism of this tumor is remarkable. The five histotypes are *hyaline*, *mature*, *cleaved*, *asynchronous*, and *polymorphous* [50, 51, 54]. This detailed classification based on the cytological morphology

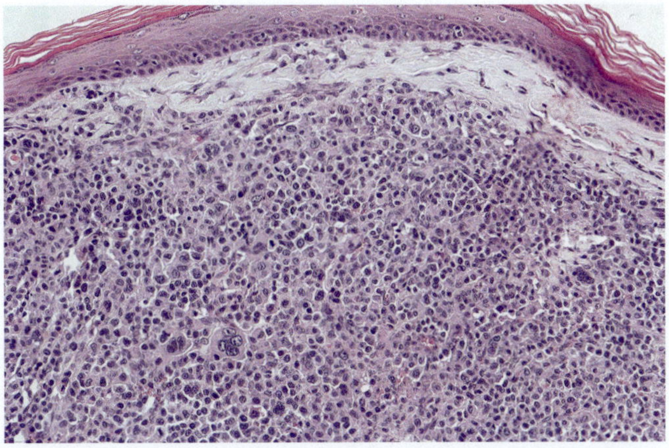

Fig. 8.109 Histopathology of well-differentiated plasmacytoma: round neoplastic plasma cells, some of them multinucleated, are separated by the epidermis by a Grenz zone

does not seem to have any implications regarding the biological behavior of the neoplasm. In cats, there appears to be four subtypes [53]. Cytologically, it is not always possible to recognize to which histotype the cells belong, also because in some cases, different histological types can coexist in the same tumor. Therefore, from a cytological point of view, it is only possible to classify the grade of differentiation based on cell morphology and define whether the neoplasia is *well or poorly differentiated* or *pleomorphic*; the former are characterized by a monomorphic population of small cells, with roundish or oval eccentric or central nuclei and regular and sometimes stippled chromatin, and, in some cells, an achromatic area between the nucleus and the cytoplasm, representing the *Golgi complex*, is still recognizable. Scattered macrokaryosis, binucleation, and multinucleated cells, the latter with a variable number of overlapped nuclei, are common findings also in well-differentiated plasma cell tumors and represent important cytological findings, which help to differentiate a plasmacytoma from a histiocytoma (Figs. 8.109, 8.110, 8.111 and 8.112). The plasma cell tumor is the prototypical neoplasm that escapes to rules of malignant cytological criteria because, despite the clear morphological atypia of the cells (anisocytosis, macrokaryosis, multinucleation), it is not usually associated with malignant behavior.

These above cytological features, which guide cytologists to the suspicion of a plasma cell tumor, are usually not recognizable in *pleomorphic plasmacytomas*, in which cells show varying degrees of anisocytosis, anisokaryosis, bizarre and kidney-shaped nuclei, and a large number of multinucleated cells. In this histotype the neoplastic cells can be confused with those of a histiocytic sarcoma or an amelanotic melanoma. Immunohistochemistry is mandatory in differentiating these neoplasms (Figs. 8.113, 8.114, 8.115 and 8.116).

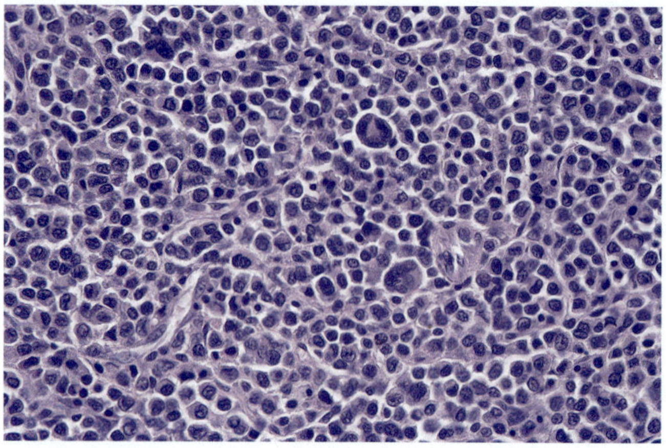

Fig. 8.110 Histopathology of well-differentiated plasmacytoma: at high magnification scattered cell with megalic nucleus and some multinucleated cells are better recognizable

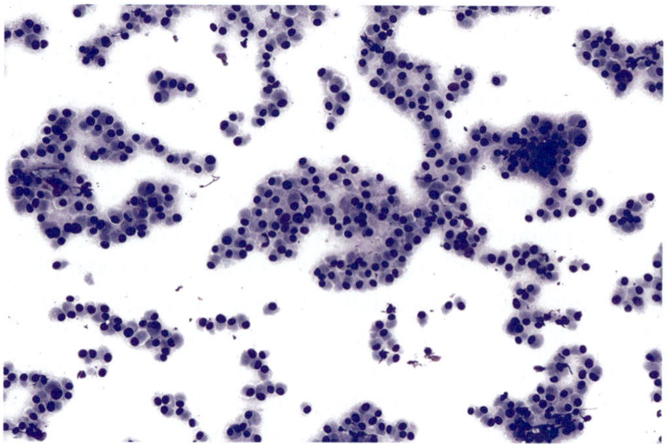

Fig. 8.111 Cytology of well-differentiated cutaneous plasma cell tumors: oval, uniform-sized neoplastic plasma cells with eccentric nuclei. Note cells preserve a recognizable morphology

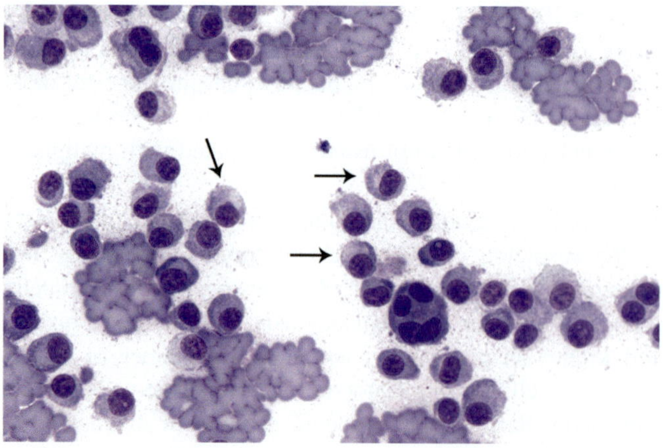

Fig. 8.112 Cytology of well-differentiated cutaneous plasma cell tumors: at high magnification the Golgi apparatus is yet recognizable in some cells (arrows). Note the multinucleated neoplastic cell

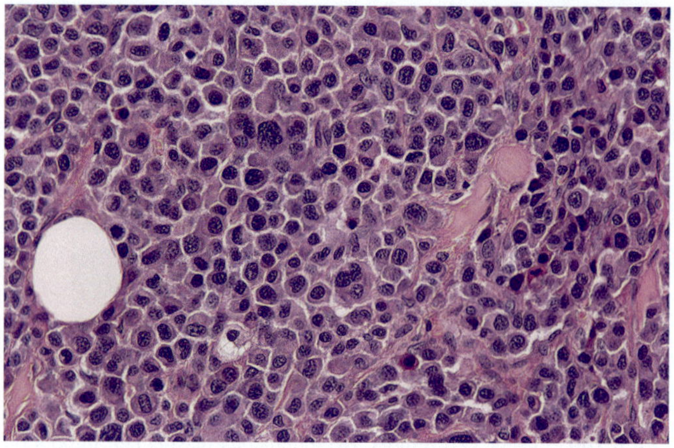

Fig. 8.113 Histopathology of poor-differentiated plasma cell tumor: round and discrete neoplastic plasma cells with evident atypias represented with anisokaryosis and irregular and overlapped multiple nuclei

In histological specimens of a few percentages of plasma cell tumors, it is possible to detect an amorphous dense and eosinophilic material interposed between and in the cytoplasm of the neoplastic cells. This acellular material represents the *amyloid* produced by the same neoplastic cells [55, 56].

The most common forms of amyloid in human beings are (a) amyloid AL (derived from the light chains of immunoglobulin kappa and lambda and secondary to plasma cell and lymphoid disorders), (b) amyloid AA (derived from serum amyloid A, which occurs in a context of chronic inflammation, infections, or neoplasms other than lymphoid or plasma cell neoplasms), and c) amyloid beta-2-microglobulin, which develops in patients on long-term hemodialysis.

Most cutaneous amyloidosis in dogs consists of deposition of monoclonal light-chain immunoglobulins (amyloid L) and are reported in cases of plasma cell disorders such as localized plasma cell proliferation, plasma cell dyscrasia, or extramedullary cutaneous plasmacytoma. Immunoglobulin secreting cells, B lymphocytes, and plasma cells are associated with the deposition of AL-amyloid. As mentioned above, amyloidosis occurs in a small percentage of extramedullary plasmacytomas that can process immunoglobulin (AL)-associated amyloid; therefore, the finding of amyloid in plasma cell neoplasia is only occasional.

Immunohistochemistry has demonstrated amyloid derived from the immunoglobulin lambda light chain (A lambda) in amyloid-containing extramedullary plasmacytomas in dogs and cats [54, 55]. This dominant expression of lambda light chain is probably due to the preponderance of lambda light chain immunoglobulin in plasma cells of these species [54].

In plasmacytomas, the amount of amyloid is variable, and, in some cases, it is so minimal that it is necessary to use Congo red stain to highlight it as described in Chap. 2; in rare cases, the amount is so large that a variable number of plasma

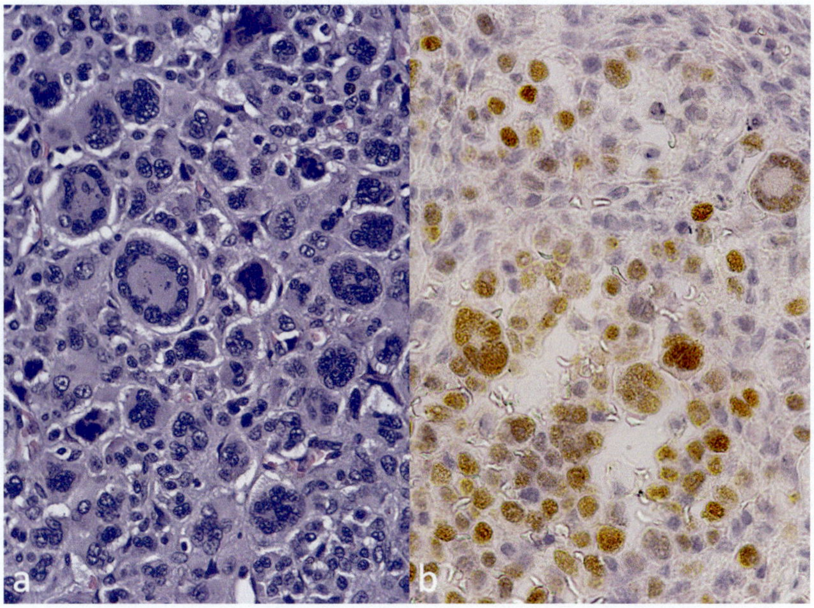

Fig. 8.114 Pleomorphic plasma cell tumor: (**a**) histopathology shows severe atypias are evident; morphological features resembling plasma cells are completely lost; (**b**) immunohistochemistry—clear nuclear positivity to MUM-1 marker, confirm the plasmacellular origin of the neoplastic cells

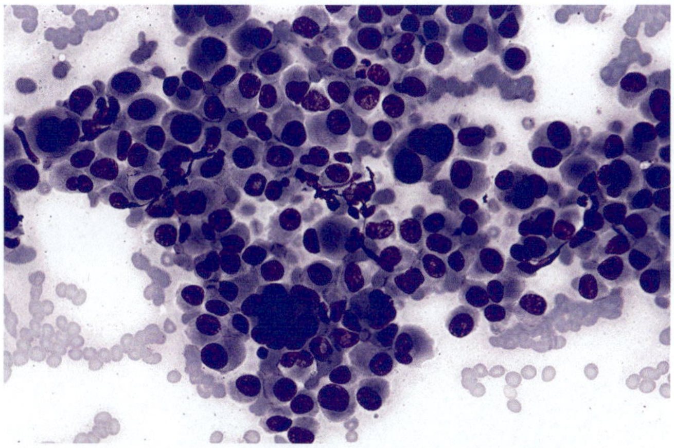

Fig. 8.115 Cytology of pleomorphic cutaneous plasma cell tumors: pleomorphic neoplastic cells with marked anisokaryosis, macrokaryosis, and multiple nuclei

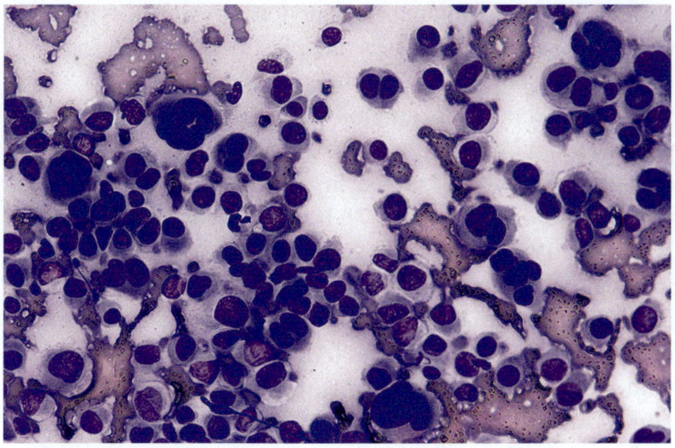

Fig. 8.116 Cytology of pleomorphic cutaneous plasma cell tumors: pleomorphic neoplastic cells with marked anisokaryosis, macrokaryosis, and multiple nuclei

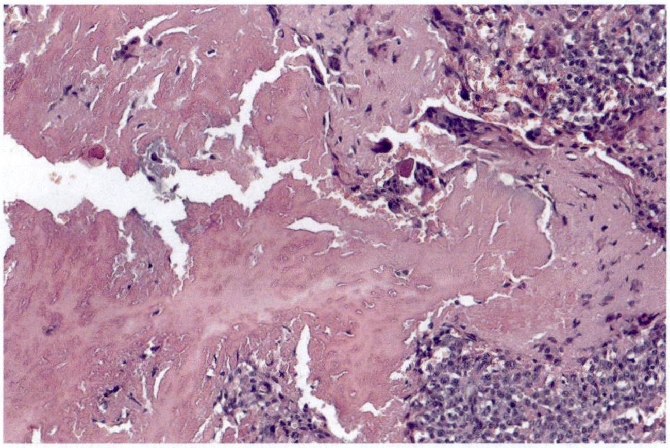

Fig. 8.117 Histology of a plasma cell tumor with amyloid production: large amount of amyloid surrounded by neoplastic plasma cells

cells are spread between or around the amyloid (Figs. 8.117 and 8.118).

On cytological samples stained with Romanowsky-type stain, when amyloid is present in scarce amount, it is recognizable as brightly pinkish amorphous material between plasma cells and in their cytoplasm; instead, when amyloid is produced in large amount, a very thick material which takes a deep blue color, is observed; in these cases, the more classical pinkish color of amyloid is barely recognizable at the periphery of these thick areas, where its thickness is reduced or the material is fragmented. In these cases, the use of Congo red dye is mandatory to highlight amyloid, which is fundamental in diagnosing a plasmacytoma when typical morphological aspects of plasma cells are lacking (Figs. 8.119, 8.120, 8.121 and 8.122).

Since amyloid is a foreign body, it stimulates a granulomatous inflammation, and numerous histiocytic giant cell

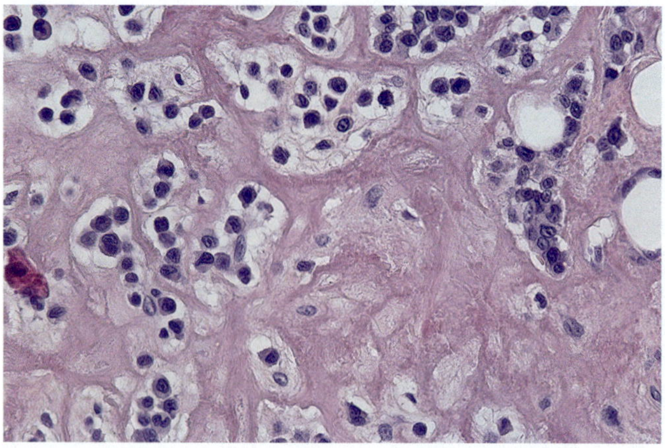

Fig. 8.118 Histology of a plasma cell tumor with amyloid production: dense amorphous eosinophilic material represent amyloid. Many neoplastic cells are embedded in it

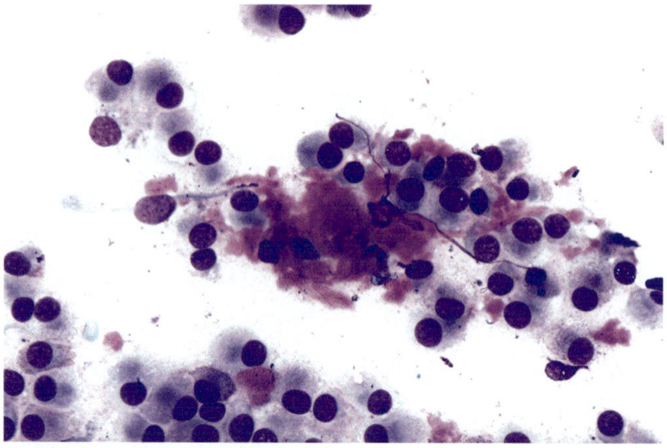

Fig. 8.119 Cytology of plasma cell tumors: neoplastic plasma cells with brightly pinkish amyloid

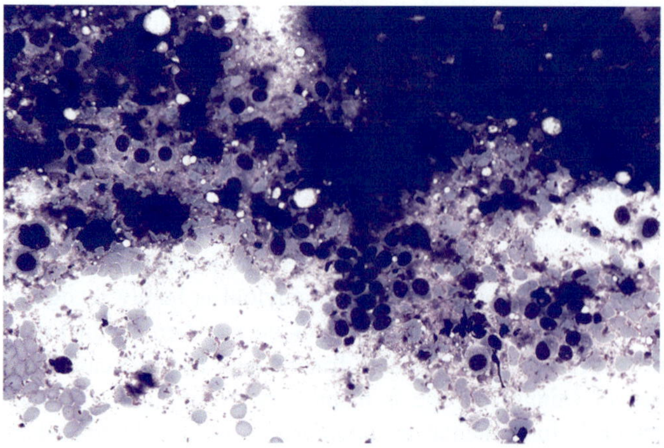

Fig. 8.120 Cytology of amyloid: very thick dark blue material represents large amount of amyloid. Note pinkish color can be only observed at its periphery as small pieces of fragmented material

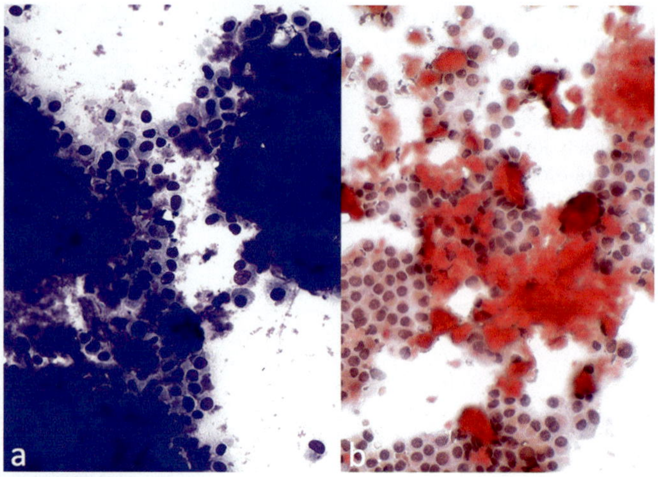

Fig. 8.121 Cytology of amyloid: (**a**) Romanowsky-type stain, with this staining, thick amyloid material take a deep blue color; (**b**) Congo red stain, the deep blue material of the previous case is strongly positive to Congo red confirming its amyloid nature

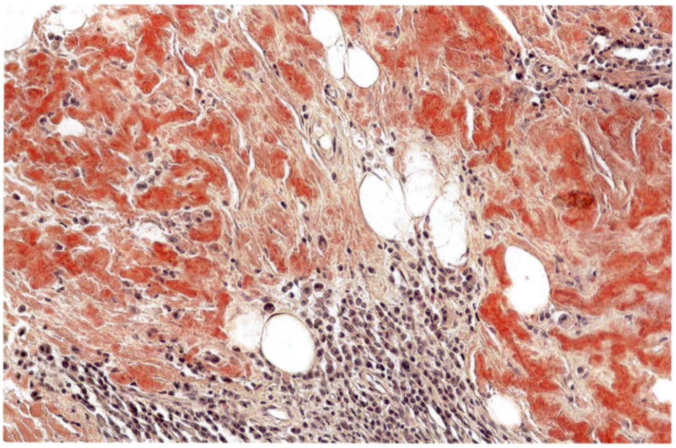

Fig. 8.122 Histology Congo red stain: the large amount of Congo red-positive material represents amyloid

macrophages infiltrate between plasma cell tumors in order to surround the large amount of amyloid; beginner cytologists can confuse these cells as neoplastic multinucleate cells (Figs. 8.123 and 8.124).

The detection of amyloid is not pathognomonic of plasmacytoma; indeed, as above discussed, a large amount of amyloid can be observed in *nodular primary cutaneous amyloidosis/amyloidoma* [57]. It is believed that AL amyloid is produced by local plasma cells and whose cause, in human medicine, is attributed to post-traumatic phenomena; indeed, in humans, *amyloidomas* can develop either in the context of lymphoproliferative disorders (AL amyloid) or in association with local trauma, infections, surgery, or chronic inflammation (AA amyloid).

In veterinary medicine, the reported cases are very few, and therefore this lesion is not clearly codified, so much so

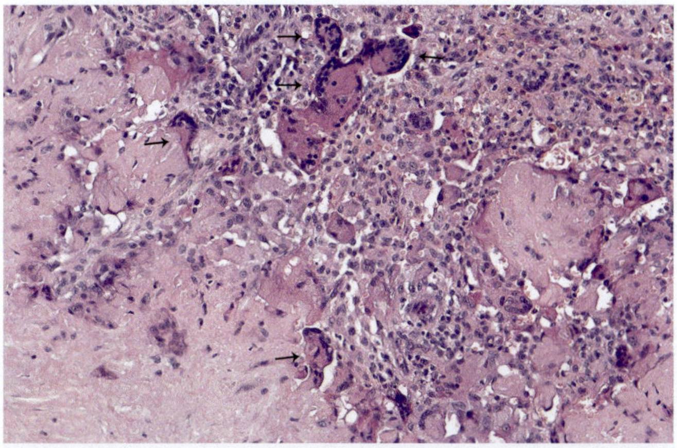

Fig. 8.123 Histology of plasma cell tumor with amyloid: many multinucleated macrophages in attempting to phagocyte amyloid (arrows)

that its existence is even debated and some authors believe that they may simply represent plasmacytomas with abundant production of amyloid [56, 57].

Histologically, *cutaneous amyloidoma* consists of abundant deposition of amyloid within the dermis, which may extend to the subcutaneous tissue. Slides from these lesions show a large amount of amorphous eosinophilic material representing *amyloid*, along with a few numbers of well-differentiated to slightly atypical plasma cells.

This amyloid is always produced by plasma cells which, however, are usually well differentiated to occasionally atypical and are arranged in groups around the amyloid foci or scattered among it. Amyloid may be so abundant that it is difficult to demonstrate its association with the few plasma cells present. Lymphocytes, macrophages, and giant cells, the latter attempting to remove amyloid, complete the cytological pattern (Figs. 8.125 and 8.126).

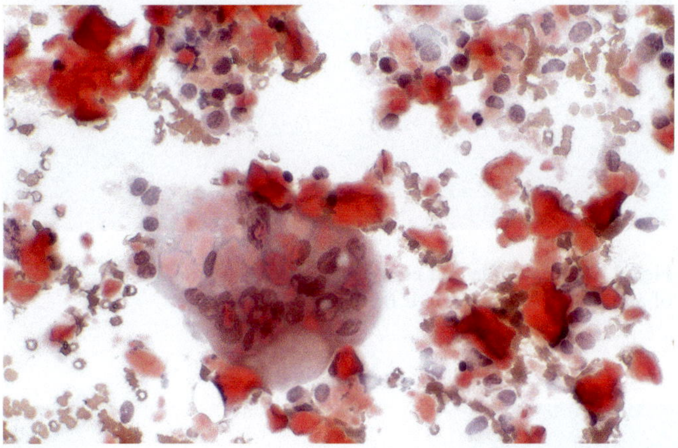

Fig. 8.124 Cytology of plasma cell tumor with amyloid: a multinucleated macrophage phagocytizing amyloid, along with many discrete neoplastic plasma cells

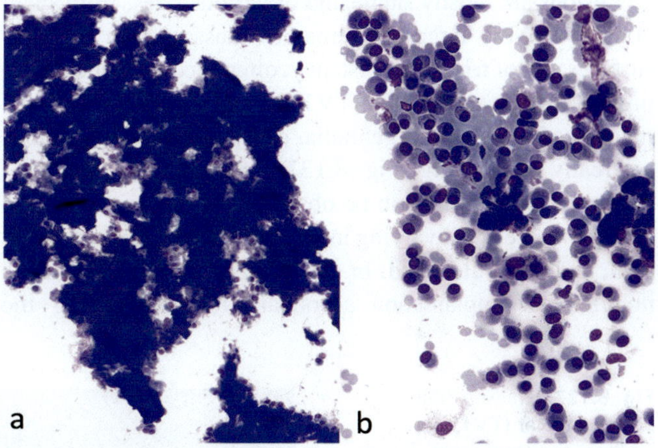

Fig. 8.126 Cytology of amyloidoma: (**a**) large amount of amorphous amyloid along with few plasma cells; (**b**) note all plasma cells are well differentiated and no anisocytosis, macronuclei, or multiple nuclei are present

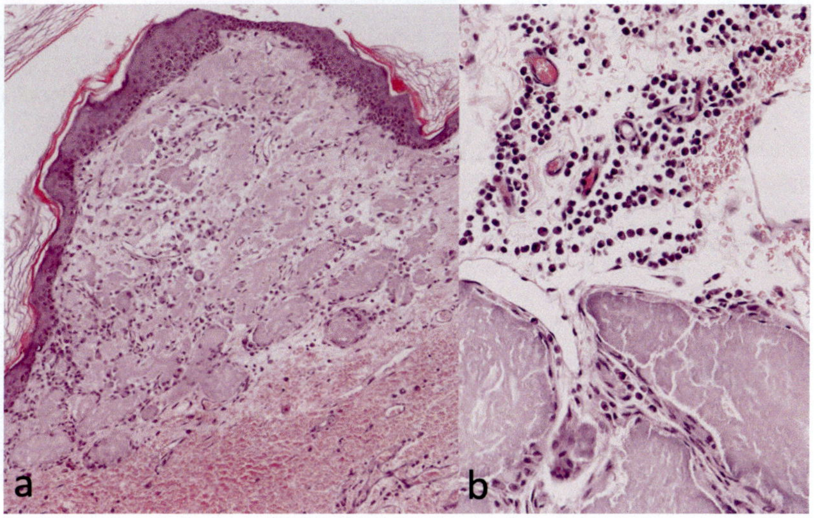

Fig. 8.125 Histology of amyloidoma: (**a**) small sub-epidermal nodule, mostly composed of amyloid along with quite well-differentiated plasma cells; (**b**) at medium magnification well-differentiated plasma cells are better recognizable

8.2.4 Transmissible Venereal Tumor

The canine *transmissible venereal tumor* (TVT) is a rare neoplasm of canidae, mostly reported in sub-tropical geographic areas [62, 63, 67]. TVT cells have unique characteristics, showing a karyotype composed of only 59 chromosomes, very different from the 78 chromosomes of normal canine diploid cells. This tumor has the peculiarity of being transmitted from one dog to another, via the direct implantation of viable cells. It has been demonstrated that the ancestral TVT cell has been propagated over time, passing from animal to animal until the present day; in practice, the neoplastic cell behaves like a *parasite* [61, 68, 69].

Social and sexual behavior and fighting between stray dogs are the basis of the location of the lesions, which, as mentioned above, result from direct implant of neoplastic cells. The skin lesions are mostly seen in immune-compromised or in stray dogs living in a poor state of health.

Lesions are mostly single and of various sizes, from small nodules to very large erythematous masses, with irregular surfaces and of friable consistency, often ulcerated and bleeding. The classical lesions of TVT are mainly observed on the mucosa of the external genitalia, and in many cases lesions are really devastating (Fig. 8.127). In some dogs single or multiple skin nodules can be observed, in association or not with genital lesions, varying in size, located on any part of the skin, and often ulcerated. In stray dogs with only cutaneous nodules, an implantation secondary to bites during the fighting has been hypothesized. In dogs with multiple lesions, TVT cells are spread via the bloodstream or via lymphatic, and these two ways of transmission justify the metastases reported in different internal organs [70]. In one case report, TVT cells were detected in the peripheral blood in a severely immune-compromised dog that had developed hundreds of TVT nodules spread all over the body [59] (Figs. 8.128, 8.129 and 8.130).

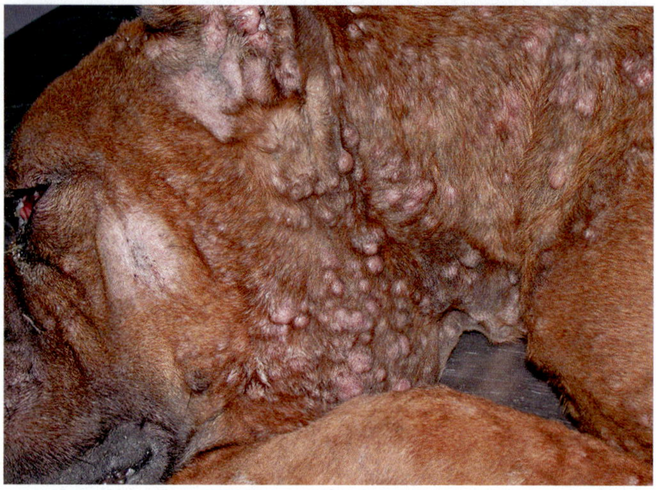

Fig. 8.128 Multiple and confluent TVT nodules on the face and neck of an immune-compromised Boxer

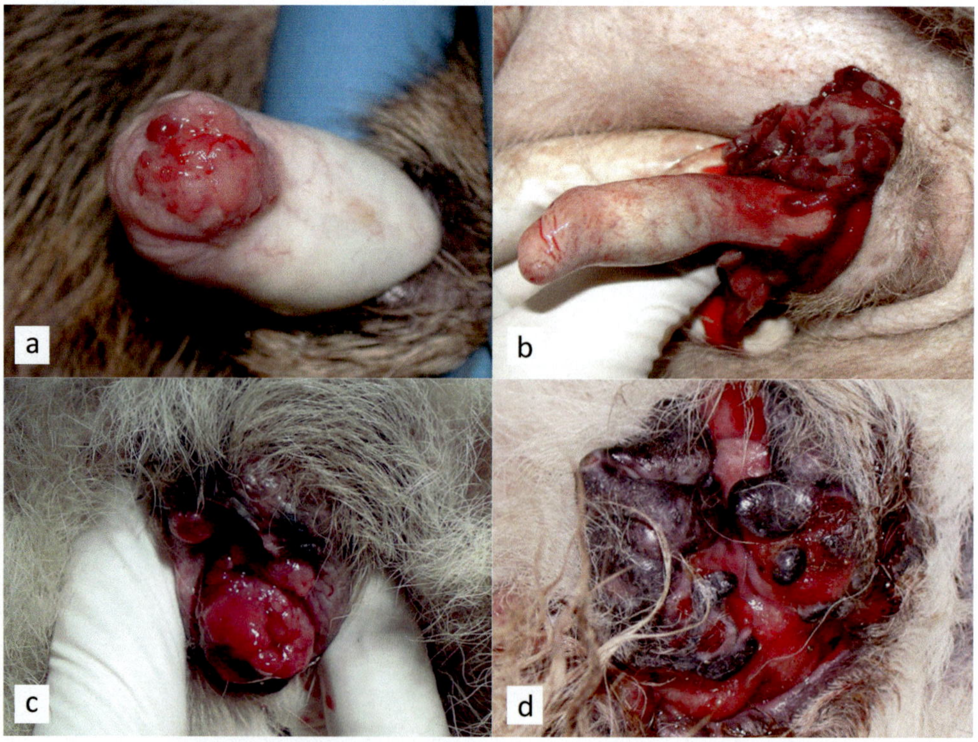

Fig. 8.127 Transmissible venereal tumor (TVT): (**a**) erythematous neoformation with irregular surface on the penis; (**b**) fleshy, bleeding, and irregular in shape mass located at the basis of the penis; (**c**) neoformation protruding from the vulvar mucosa in a bitch; (**d**) deforming mass on the vulva of a stray bitch

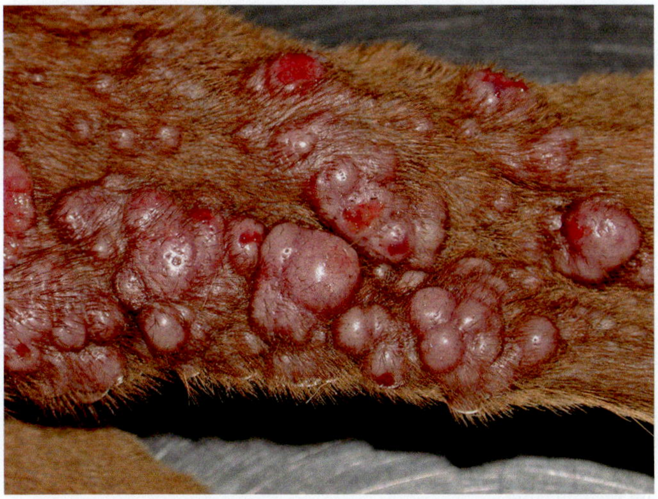

Fig. 8.129 Multiple merging nodules on leg of the same dogs in Fig. 8.128

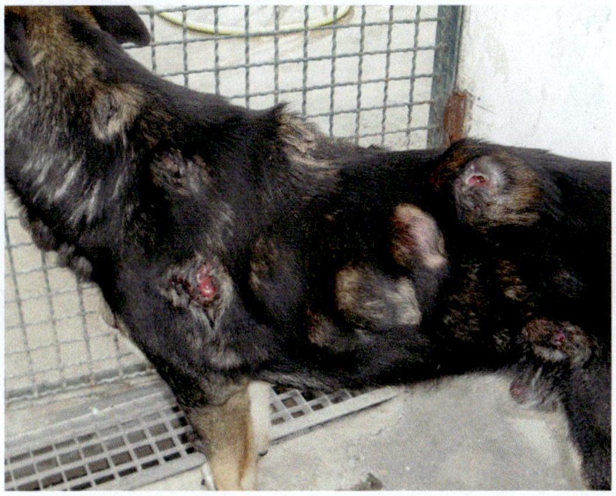

Fig. 8.130 Multiple and masses on the thorax of a stray dog with multiple TVTs

Cytological Findings

The nature of TVT cells is subject of discussion from many years, and despite some papers seem to agree on their histiocytic nature, based on the demonstration of their phagocytic activity and on the positivity of several immunohistochemical markers for histiocytes, other reports or anecdotal experiences in which IHC markers for histiocytes were constantly negative (IBA-1, Cd11c, CD1) do not agree with the histiocytic origin in favor of a subpopulation of lymphoid nature (CD45RA, CD49d) [60, 64–67, 70].

The samples collected from TVT lesions are characterized by very high cellularity, with densely packed cells, which are sometimes difficult to identify as discrete round cells at low magnifications (Figs. 8.131 and 8.132).

The nuclei are round, central or eccentric, with a regular chromatin pattern, sometimes stippled, and a single nucleolus, when visible; anisokaryosis can be evident. Cytoplasm is large and pale and characterized by the presence of a variable number, usually high, of intracytoplasmic and non-overlapping vacuoles, clearly rounds, achromatic, of uniform size and with well-defined borders (Figs. 8.133 and 8.134). A great advantage from cytology compared with histology is that, in the latter, vacuoles are not easily recognizable (Fig. 8.135). In some TVTs the number of vacuoles is scarce, and, especially if lesions do not involve the genitals, other RCTs must be considered as differentials (Fig. 8.136). The anisocytosis is more obvious than in the other well-differentiated RCTs. Another constant and very characteristic feature of the TVTs is the presence of a large number of *mitotic cells*, which must not be misinterpreted as an indicator of malignant behavior, as the TVT usually has a good prognosis due to its excellent sensitivity to chemotherapy (Figs. 8.137 and 8.138). TVT, although very rarely, can spontaneously regress following an attack of cytotoxic T lymphocytes; therefore, the detection of a high amount of infiltrating mature lymphocytes could be interpreted as a positive prognostic index.

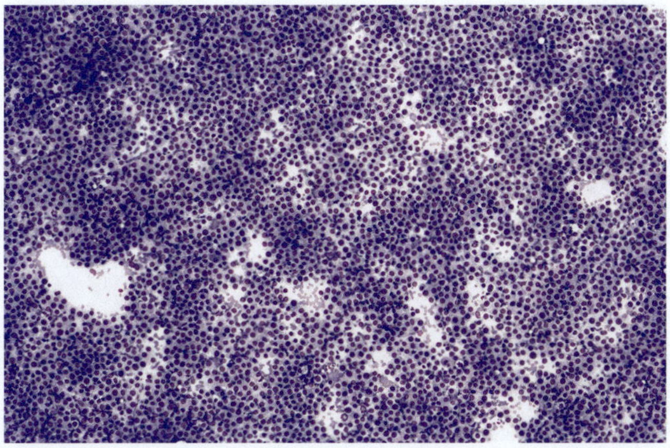

Fig. 8.131 Cytology of TVT: at low magnification, the discrete arrangement of TVT cells is not recognizable

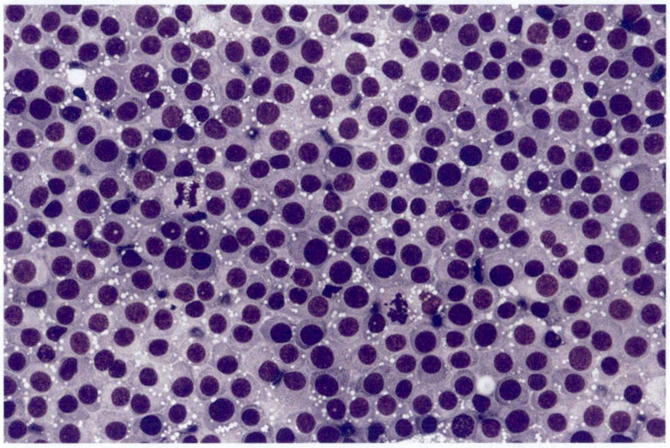

Fig. 8.132 Cytology of TVT: at high magnification, although the cells are tightly packed, the cytoplasmic margins are quite recognizable

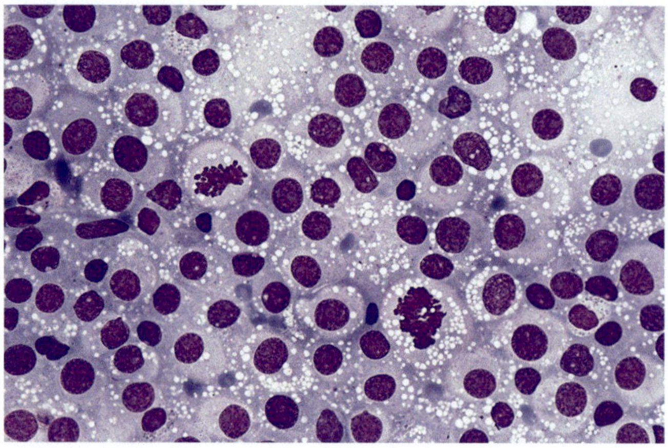

Fig. 8.133 Cytology of TVT: cytoplasm of TVTs is filled with round, non-overlapped, with clear margin vacuoles

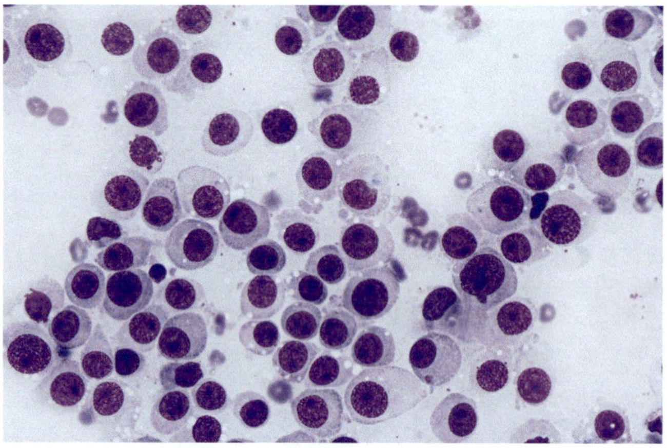

Fig. 8.134 Cytology of TVT: The cytoplasm of neoplastic cells is lacking the typical vacuoles observed in this neoplasm; furthermore, note that there are no cells in mitosis

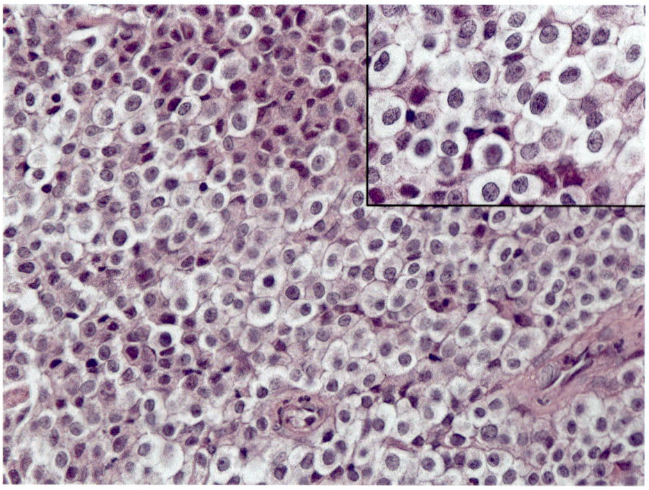

Fig. 8.135 Histology of TVT: discrete round cells with large pale cytoplasm. Note that with histology, the typical intracytoplasmic vacuoles are not as clearly evident as in cytological specimens (inset)

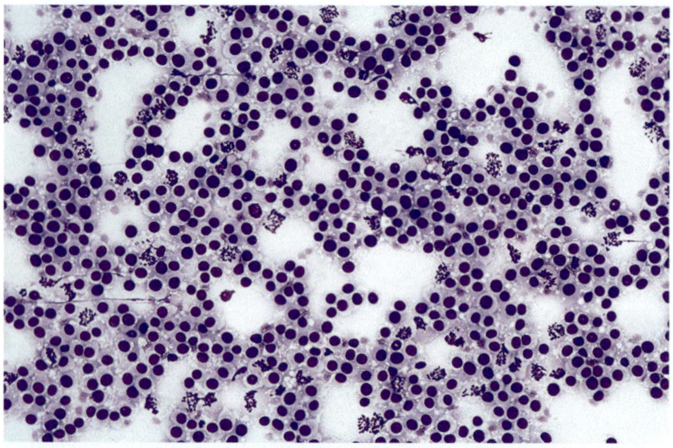

Fig. 8.136 Cytology of TVT: at low magnifications, many mitotic cells are recognizable

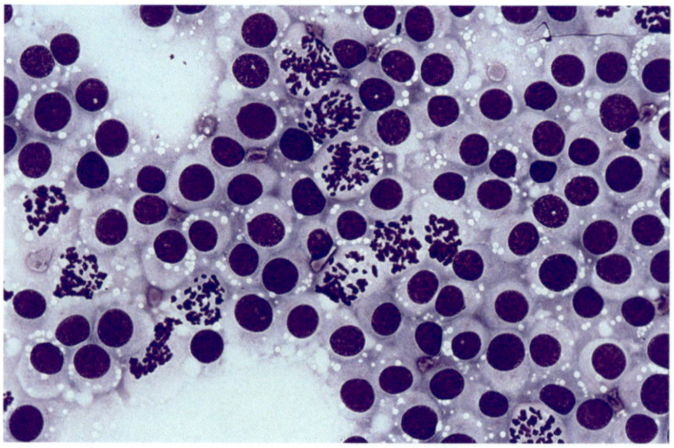

Fig. 8.137 Cytology of TVT: round discrete cells, strongly vacuolated with many mitotic figures, characterize the TVT cells

8.2.5 Histiocytic Diseases

Canine and feline *histiocytic diseases* encompass a complex group of reactive/neoplastic disorders, originating from dendritic cells (DC), also named *antigen-presenting cells* (APCs), residing in the epidermis, dermis, and subcutis.

The term *histiocytes* is used to generically describe both *dendritic cells* (DC) and those of the *macrophage lineage*. In the group of diseases named "histiocytic diseases," the majority involves proliferations of cells of various DC lineages. Thanks to the development of canine-specific markers for differentiation molecules of macrophages and dendritic cells, the identification of the cell lineages involved in canine histiocytic diseases has been possible.

This group includes the *benign cutaneous histiocytoma*, *Langerhans cell histiocytosis*, *reactive histiocytosis (cutaneous and systemic)*, and the *histiocytic sarcoma complex (localized and disseminated)* [82, 83].

Some of the latter are not considered true neoplasms: *histiocytoma* is in fact a Langerhans cell proliferation that regresses spontaneously, and *reactive cutaneous histiocytosis* is considered to be a reactive inflammatory response of interstitial DC cells, based on the reported spontaneous regression of some cases and on the good response to different immune-modulatory drugs; for this reason, many authors prefer to group all these forms under the umbrella of *histiocytic disorders/diseases* and not as histiocytic neoplasia [82, 83, 86].

Although some of these disorders have some anamnestic, clinical, and cytological peculiarities, differentiating them from other neoplastic and non-neoplastic diseases is not always easy.

A typical example is cutaneous reactive histiocytosis, which can be cytologically confused with other histiocytic diseases of macrophagic lineage, such as sterile granuloma syndrome or some nodular forms of leishmaniasis; in addition, the spindle variant of the histiocytic sarcoma cannot be cytologically differentiated from some anaplastic sarcomas of soft tissue with giant cells. In these cases, a definitive diagnosis can be only obtained with the use of immunohistochemical investigations.

In cats, only the so-called feline dendritic progressive histiocytosis has been well documented as a true cutaneous histiocytic disorder; the few reported cases of histiocytic sarcoma in cats are mostly hemophagocytic (macrophagic lineage) and visceral in location [73, 77].

8.2.5.1 Benign Cutaneous Histiocytoma

The canine cutaneous *histiocytoma* is a benign proliferation of Langerhans cells, the APCs mostly residing in the epidermis [82, 83]. It is common in young dogs (less than 3 years of age), although it can be observed in dogs of all ages. It usually appears as a single nodule, localized mainly on the limbs, face, muzzle, eyelids, and pinna, but it can grow in any part of the body. The canine histiocytoma usually has a dome- or button-like shape; it can be alopecic, erythematous, with a crusty surface and grows rapidly (Fig. 8.138).

Although cutaneous *histiocytoma* displays a clinical behavior that alarms many owners, such as the rapid growth and tendency to ulcerate, it is characterized by a benign outcome resulting in spontaneous regression within a few weeks (Fig. 8.139). In rare cases, single histiocytomas do not spontaneously regress and can persist for a long time (*persistent cutaneous histiocytomas)* [79] (Fig. 8.140). Nevertheless, Langerhans cells, as dendritic presenting cells, normally migrate to the lymph nodes, a secondary enlargement of the latter occurs very rarely; usually, these forms also tend to regress spontaneously. Cases of multiple histiocytoma have been reported in literature, but in recent

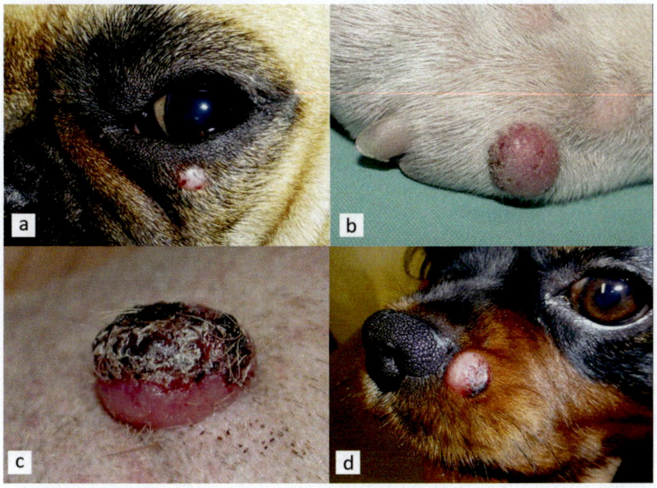

Fig. 8.138 Canine benign histiocytoma: (**a**) small alopecic papular–nodular lesion on the eyelid of a French Bulldog; (**b**) alopecic and pinkish nodule on the digit of a young dog; (**c**) typical benign canine histiocytoma represented with a dome-shaped alopecic, erythematous, and crusty nodule; (**d**) alopecic nodule on the muzzle of a young CKCS

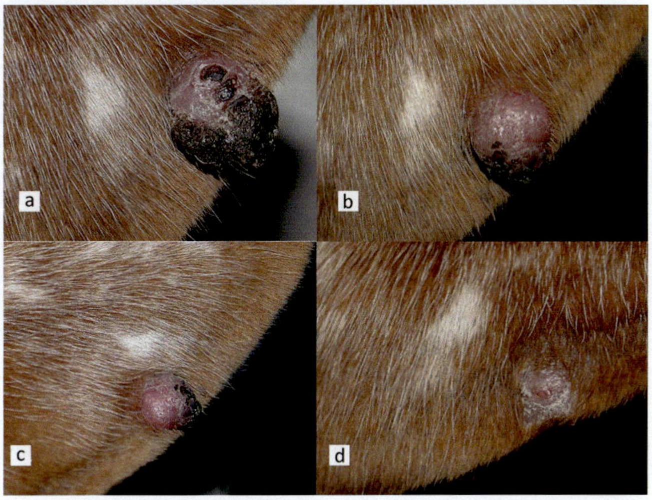

Fig. 8.139 Benign cutaneous histiocytoma. Complete regression of the nodule during over 2 months: (**a**) day 0; (**b**) day 40; (**c**) day 60; (**d**) day 80

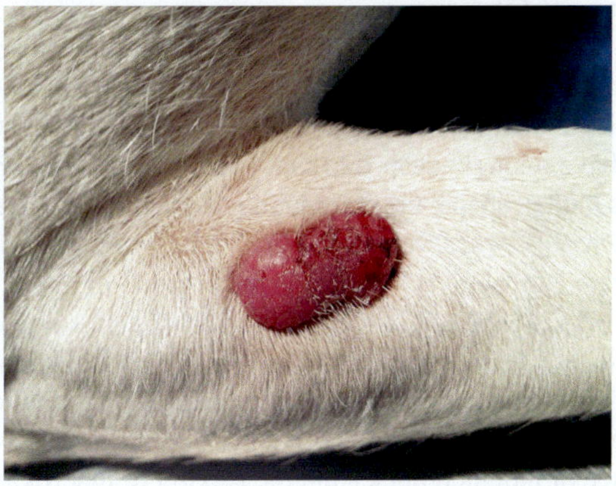

Fig. 8.140 Persistent cutaneous histiocytoma. The nodule was removed after 8 months from the cytological diagnosis of canine benign histiocytoma; the diagnosis was confirmed with histopathology and immunohistochemistry

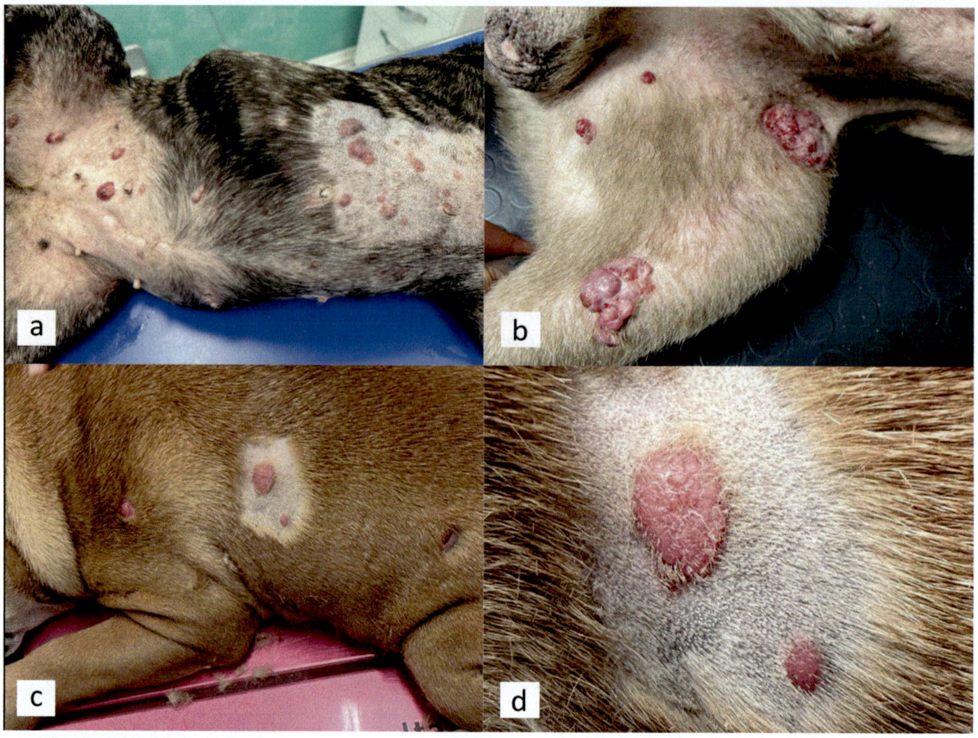

Fig. 8.141 Canine Langerhans cell histiocytosis (LCH): (**a**) multiple erythematous nodules and plaques on the skin of a French Bulldog; (**b**) multiple, erythematous, alopecic, and with irregular surface nodules, in a Shar-Pei (courtesy of DVM V. Rinaldi-Italy; (**c**) multiple nodules and plaques on the thorax of an English Bulldog (courtesy of DVM L. Zumbo-Italy); (**d**) close-up of the same dog in Fig. 8.141c. (Courtesy of DVM L. Zumbo-Italy)

years, with new clinical–pathological knowledge, there has been a greater tendency to think that these cases may belong to the so-called Langerhans cell histiocytosis (LCH) well documented also in human medicine [83]. Dogs can develop up to hundreds of skin lesions represented by nodules and skin plaques, often erythematous and which may ulcerate. More rarely, lesions may be located in the mouth, at the mucocutaneous junctions, and affect the lymph nodes (migration of APCs) and internal organs, especially lungs, with possible fatal outcome, as occurs in the multi-organs form observed in humans. Shar-Pei dogs seem predisposed to LCH [79, 83, 84] (Fig. 8.141).

Cytological Findings

Slides from *histiocytoma* are usually highly cellular and are characterized by round discrete cells with slight anisocytosis and with a central or eccentric, round, or oval nucleus, which often have a single deep fold that gives the nuclei a characteristic kidney shape; the chromatin is regular or stippled, and the nucleolus, when visible, is small. The cytoplasm is from moderate to abundant and slightly azul or pale grey (Figs. 8.142, 8.143, 8.144, 8.145 and 8.146). In some cases, the cells are less characteristic and smaller, with scarce and more blue in color cytoplasm, creating difficulties in differentiating between with neoplastic plasma cells; the presence of indented nuclei and the lacking of macrokaryosis and of multinucleated cells orient the diagnosis to a histiocytoma (Fig. 8.147).

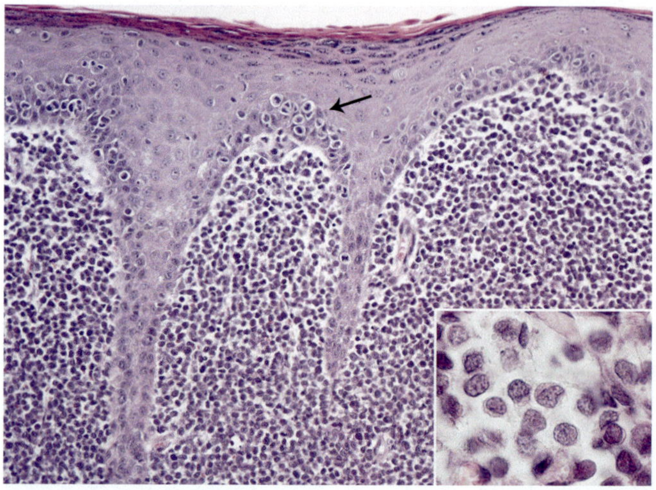

Fig. 8.142 Histopathology of cutaneous histiocytoma. Diffuse dermal proliferation of round and discrete cells. Note the epitheliotropism (arrow) and the deep fold of some nuclei giving them a kidney-shaped appearance (inset)

Occasionally the cytoplasm may contain a low number of vacuole-like areas even if, rarely, they can be numerous; in these cases cells can mimic those observed in TVT (Figs. 8.148 and 8.149). It is not uncommonly observed in slides of benign histiocytoma, few cells with spindle silhouette along with the more classical round cells; only occasionally spindle cells are in large number (Fig. 8.150). When specimens were collected in the regression phase, they are

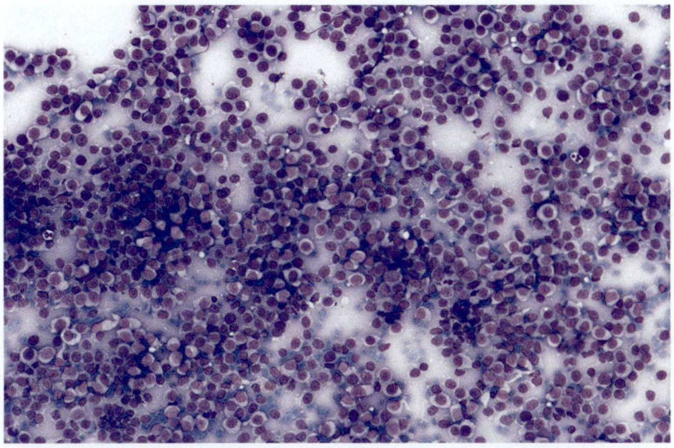

Fig. 8.143 Cytology of cutaneous histiocytoma. Highly cellular specimen of neoplastic round cells

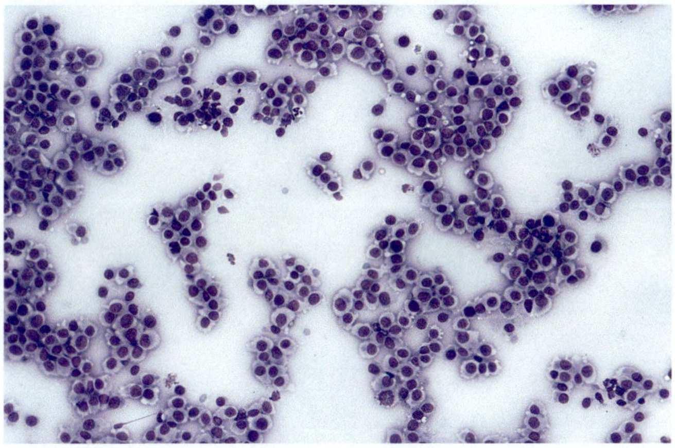

Fig. 8.144 Cytology of cutaneous histiocytoma. At medium magnification round silhouette and discrete appearance of cells are better visible

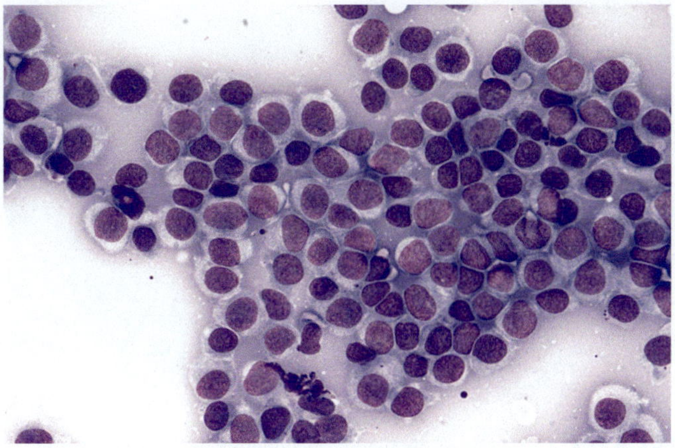

Fig. 8.145 Cytology of cutaneous histiocytoma. Round cells with large and azul pale cytoplasm

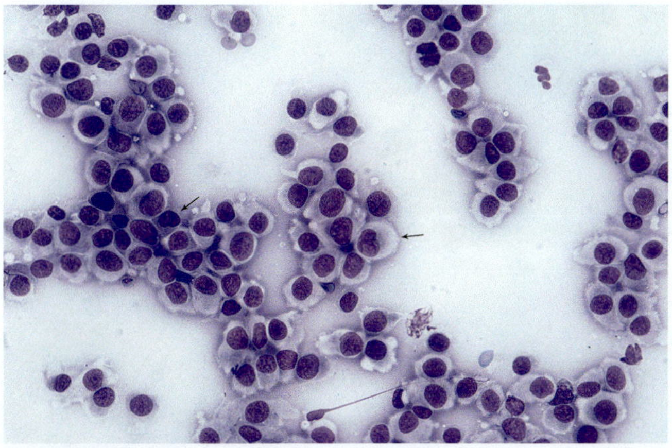

Fig. 8.146 Cytology of cutaneous histiocytoma. Round cells with large and azul pale cytoplasm; note the kidney shape of two nuclei (arrows)

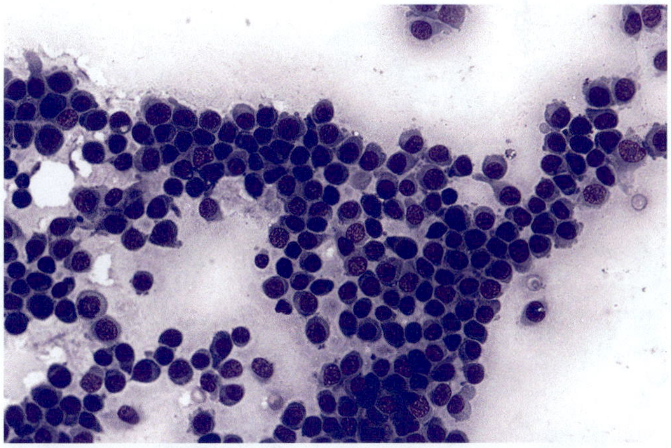

Fig. 8.147 Cytology of cutaneous histiocytoma. When the cytoplasm is small and deep blue in color, histiocytes can be confused with neoplastic plasma cells

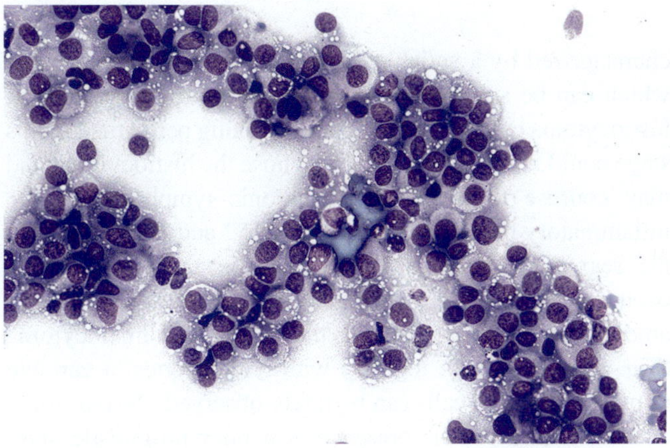

Fig. 8.148 Cytology of cutaneous histiocytoma. Many similar to vacuole areas in the cytoplasm of histiocytes

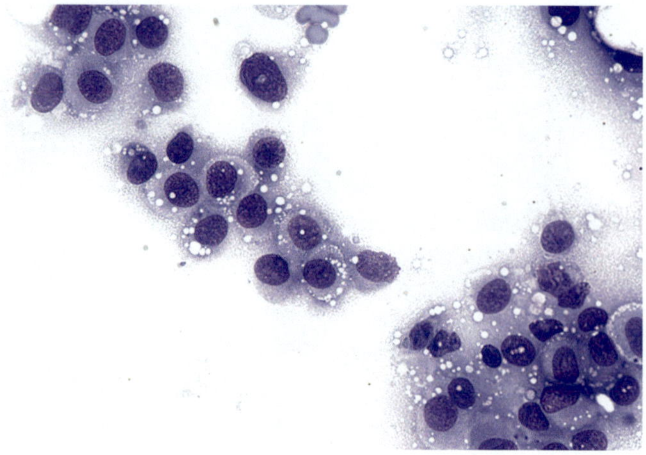

Fig. 8.149 Cytology of cutaneous histiocytoma. Large vacuoles-like, non-overlapped, achromatic areas with clear margins are recognizable in the cytoplasm. Note as these cells mimic those of TVT

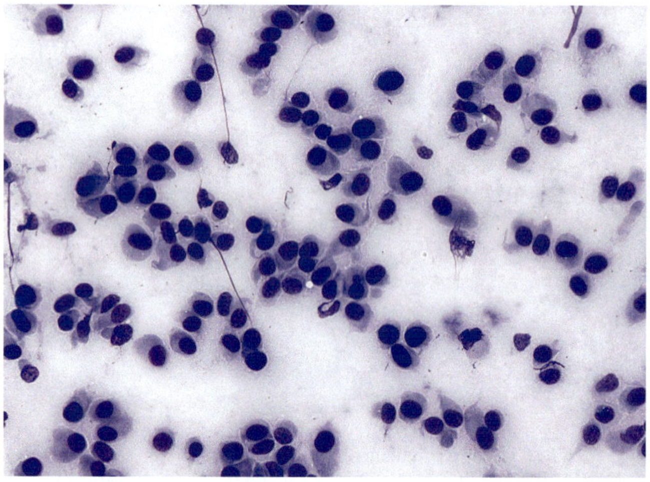

Fig. 8.150 Cytology of cutaneous histiocytoma. When the cytoplasm is small and deep blue in color, histiocytes can be confused with neoplastic plasma cells

characterized by a variable number of small T lymphocytes, which can be so numerous that they exceed the number of histiocytoma cells; the cytological sampling performed at this stage could make it difficult to diagnose the histiocytoma and may confuse the lesion for a chronic lympho-histiocytic inflammatory process (Figs. 8.151, 8.152 and 8.153). During the regression phase, intraneoplastic necrosis phenomena occur; cytology shows acellular, granular, and dark in color amorphous material along the cells of histiocytoma (Fig. 8.154). Finally, together with lymphocytes, a variable number of plasma cells can be rarely observed; despite some authors consider their presence is a poor prognostic sign, which is observed in persistent histiocytomas, this hypothesis not seems to be based on scientific evidence.

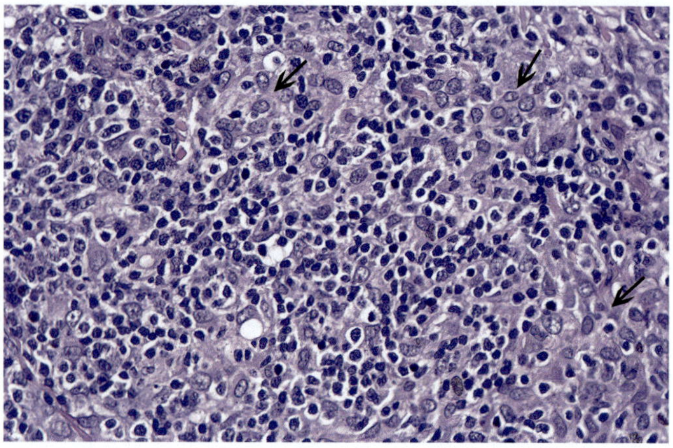

Fig. 8.151 Histology of cutaneous histiocytoma in regression phase: the smaller and hyperchromatic nuclei represent infiltrating lymphocytes. Larger cells representing histiocytes are yet recognizable (arrows)

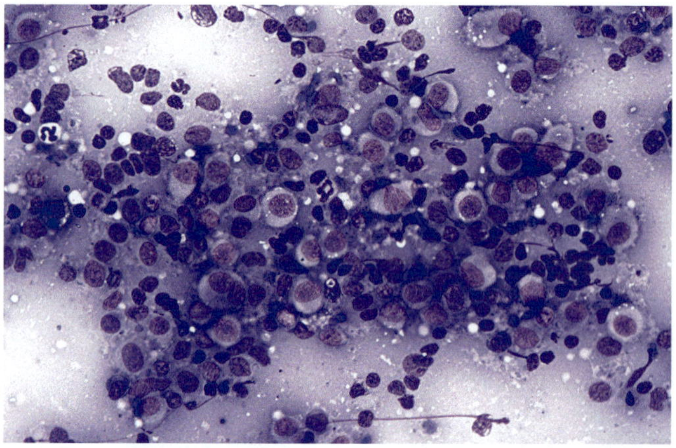

Fig. 8.152 Cytology of cutaneous histiocytoma in regression phase: mixed population of cells composed of histiocytes (larger with pale cytoplasm) and lymphoid cells (smaller, scarce deep blue cytoplasm, bare nuclei)

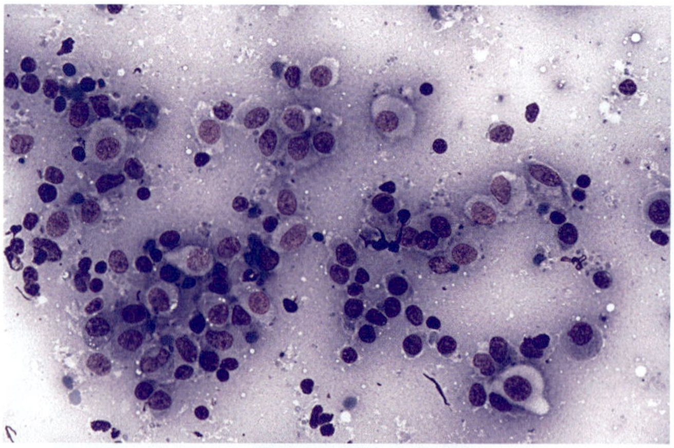

Fig. 8.153 Cytology of cutaneous histiocytoma in regression phase: mixed population of cells composed of histiocytes (larger with pale cytoplasm) and lymphoid cells (smaller, scarce deep blue cytoplasm, bare nuclei)

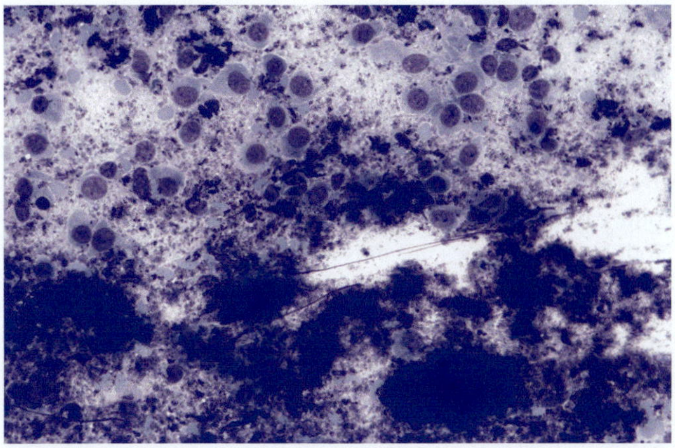

Fig. 8.154 Cytology of cutaneous histiocytoma in regression phase: violet granular amorphous acellular background representing necrosis

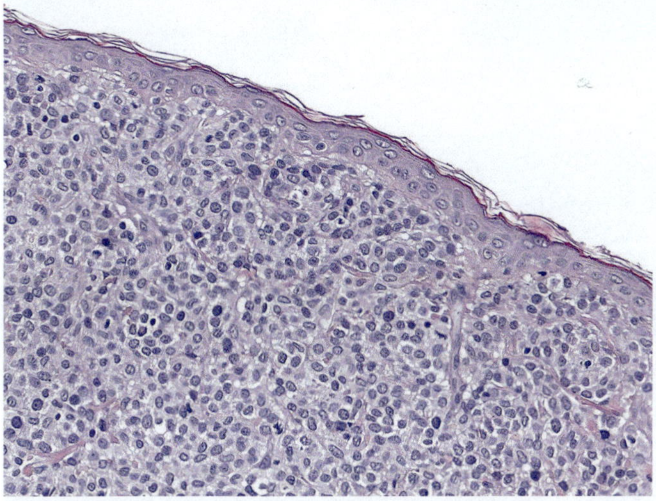

Fig. 8.155 Histology Langerhans cell histiocytosis: dermal proliferation of histiocytic cells. Note morphology is the same of cutaneous benign histiocytoma

The cytological aspects of LCH are in most cases similar to those described for benign canine histiocytoma, but in some cases, especially more aggressive ones, cells can exhibit more degrees of atypia such as anisocytosis, anisokaryosis, and multinuclearity, which are never observed in benign histiocytoma (Figs. 8.155 and 8.156).

In practice, since these two forms of Langerhans cell disorder also share the immune phenotype of histiocytes, a clinical differentiation between them is not possible, unless clear cytological atypia or diffusion into internal organs is observed, which obviously directs the diagnosis toward LCH (Fig. 8.157).

In case of cytological doubts in the differentiation with other histiocytic diseases, immunohistochemistry can be used to mark the Langerhans cells. It must be emphasized however that although E-cadherin is an important functional marker expressed by Langerhans cells and not expressed by other APCs, it is not a specific marker for LCs, as it can also result positive in other RCTs such as lymphoma and plasma cell tumors; furthermore, to increase the diagnostic difficulties, not all cases of histiocytoma and LCH are positive for E-cadherin. From what has been said, a positive E-cadherin sample allows excluding other histiocytic disorders, but in cases in which one wants to differentiate the diseases that originate from LCs from another RCT, the results must be interpreted carefully in light of the cytological and clinical–anamnestic data [80].

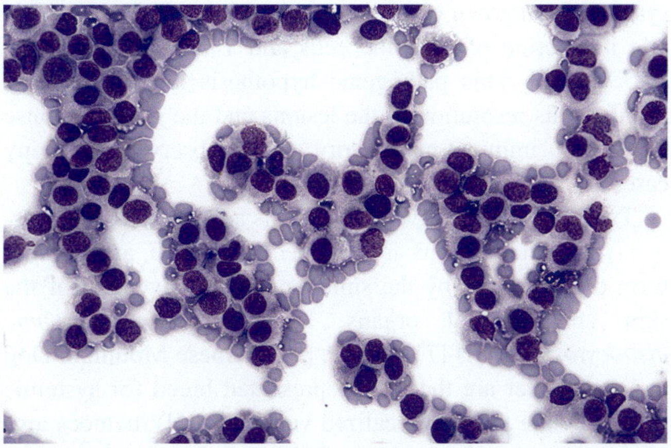

Fig. 8.156 Cytology of Langerhans cell histiocytosis. Round and uniform in size cells, with roundish to kidney-shaped nuclei

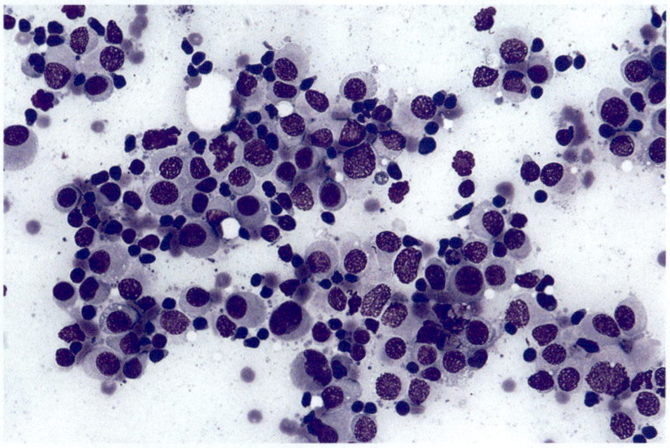

Fig. 8.157 Cytology of Langerhans cell histiocytosis (lymph nodes): many atypical histiocytic cells along with lymphoid cells. Note some cells show cytological criteria of atypia such as anisocytosis and multinuclearity

8.2.5.2 Cutaneous and Systemic Reactive Histiocytosis

Cutaneous reactive histiocytosis (CRH) is an inflammatory lympho-histiocytic proliferation of the dendritic APCs residing in the dermis; a probable deregulation of the immune system of unknown origin, probably arising from the defective interaction of dendritic cells and T cells, is suspected [82, 83, 86]. This pathogenic hypothesis is based on the spontaneous resolution of the lesions and the good response to several immune-modulatory drugs observed in many cases.

Two forms of reactive histiocytosis are reported: the *cutaneous*, in which lesions are confined to the skin, and another form characterized by the simultaneous involvement of the skin and internal organs, named *reactive systemic histiocytosis* (SRH) [71, 81, 82, 83]. Bernese Mountain Dog and Rottweiler are the most represented breed for systemic variant, while the skin localized variant (CRH) includes also many other breeds such as Golden Retriever, Labrador, Cocker Spaniel, Giant Schnauzer, etc. Not all cases of SRH, however, have skin lesions. Skin lesions are represented by multiple nodules or plaques, dermal and subcutaneous in location, of variable size and appearance, sometimes showing a donut or an arc shape, which look very similar to the lesions observed in dogs with non-epitheliotropic lymphoma; in some cases, nodules can merge giving origin to large lesions. Nodules and plaques are mostly localized on the head, face, trunk, scrotum, nose, and extremities but can be spread all over the body (Figs. 8.158 and 8.159). Sporadic cases exclusively localized to the nose are reported. Many dogs simultaneously present involvement of the conjunctiva and nasal mucosa. In the latter case, the nose can assume a so-called *clown nose* appearance [76].

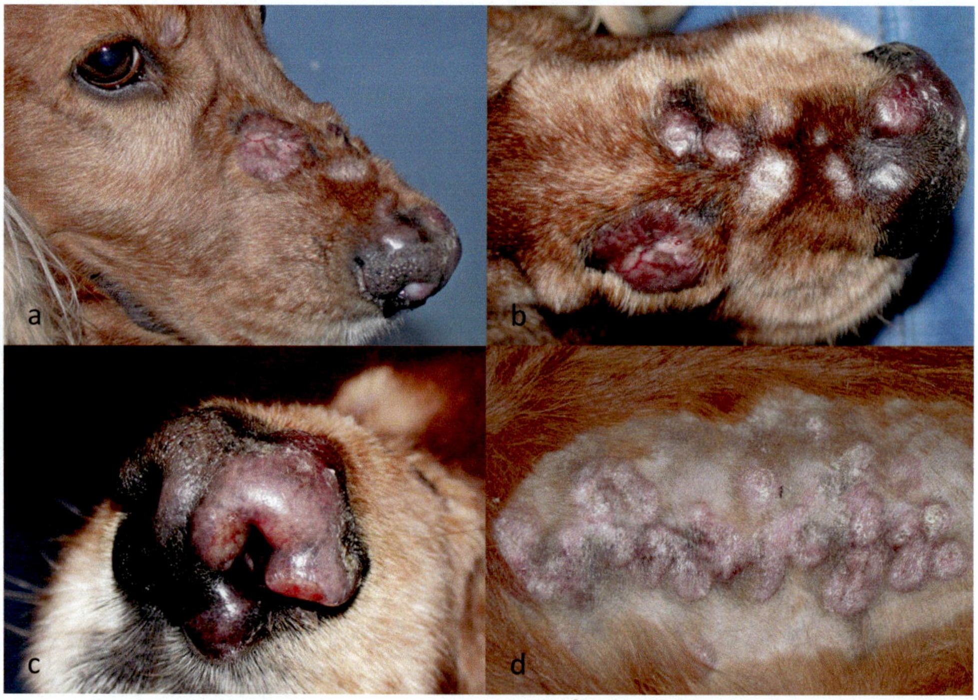

Fig. 8.158 Nodules and donut-shaped plaques on the face, bridge of the nose, nose (clown nose appearance), and confluent plaques on the trunk, in a Cocker with reactive cutaneous histiocytosis

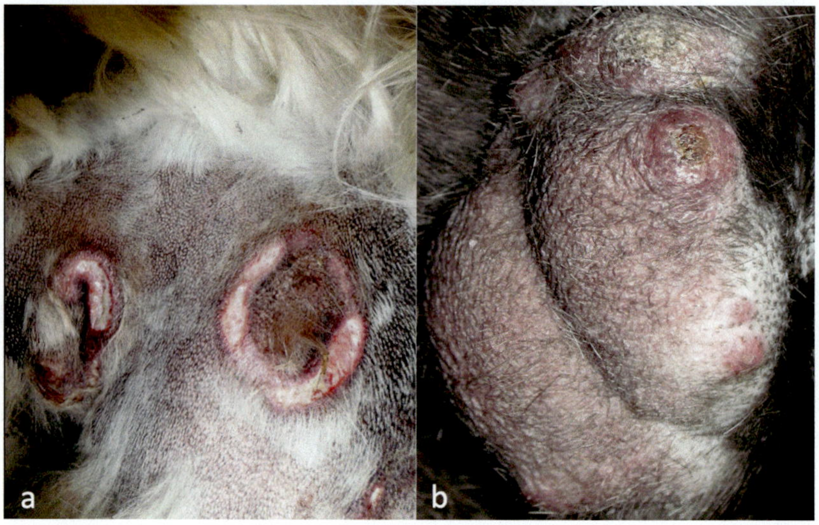

Fig. 8.159 Reactive cutaneous histiocytosis. (**a**) Large annular and ulcerated plaques on the sternal area of a Golden Retriever; (**b**) multiple nodules on the scrotum

Cytological Findings

The cytological features depend on the stage of the disease. Histologically, CRH is characterized by both peri-adnexal and free in the dermis lympho-histiocytic granulomas, sometimes angiocentric with secondary lympho-histiocytic vasculitis, which involve also the deep dermis (bottom heavy) and infiltrating the panniculus [76, 82, 83] (Figs. 8.160, 8.161 and 8.162). In early stages, via FNB, the possibility of the needle penetrating the granulomas is very low; thus, samples usually contain few cells. In the advanced stages of the disease, granulomas can merge, and the infiltrate changes from nodular to diffuse, allowing more successful collection of cells. In many cases, cells are released in scattered groups on slides, mainly as bare nuclei immersed in an amorphous blue-gray background that represents cytoplasmic material coming from the broken histiocytes; it is rare to observe a high amount of intact dendritic cells with a recognizable silhouette. More representative specimens are characterized by roundish to oval histiocytes with oval, folded, or kidney-shaped nuclei, regular chromatin, inconspicuous nucleoli, and large and pale azul cytoplasm, sometimes with poorly defined margins and occasionally containing small achromatic areas similar to vacuoles. Many histiocytes are represented by bare nuclei with cytoplasm spread on the background. In association with histiocytic dendritic cells, a variable but usually high number of lymphocytes are constantly present; occasional few plasma cells and neutrophils are also detected (Figs. 8.163, 8.164 and 8.165). In most cases, cytological features are not differentiable from those observed in sterile granuloma syndrome or sarcoid reaction (cutaneous sarcoidosis), as the APCs can be cytologically similar to non-activated histiocytes of the macrophage lineage. The definitive diagnosis, in cases where the clinical

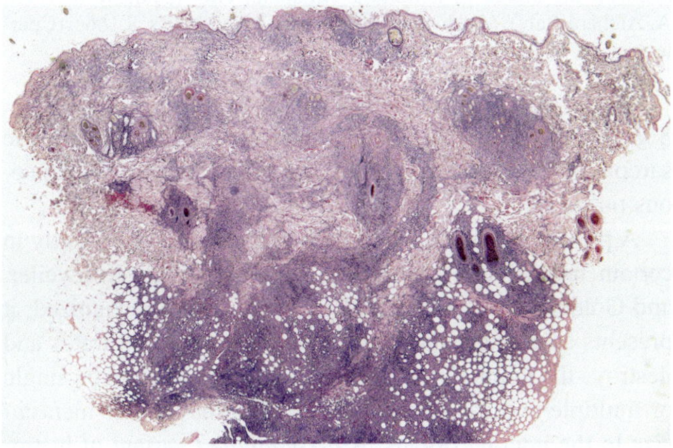

Fig. 8.160 Histopathology of reactive cutaneous histiocytosis: multiple deep and confluent dermal peri-adnexal nodules and deep dermal and adipose panniculus infiltration

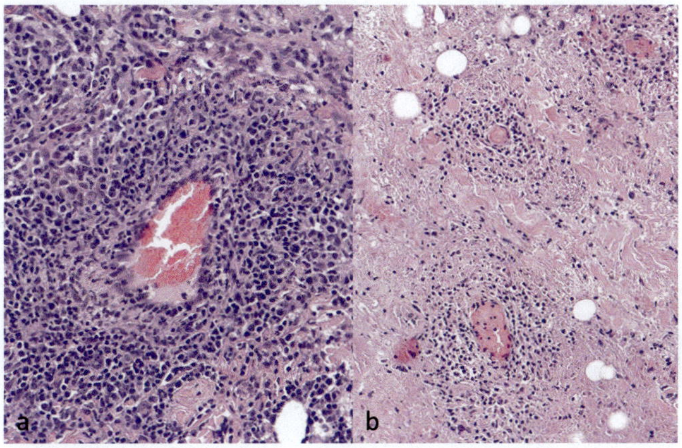

Fig. 8.162 Histopathology of reactive cutaneous histiocytosis: (**a**) angioinvasive arrangement of infiltrating dendritic histiocytes and lymphocytes; (**b**) lympho-histiocytic vasculitis surrounded by necrotic tissue

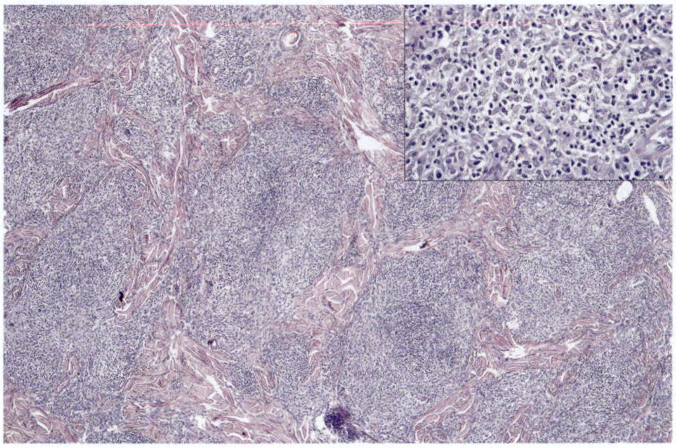

Fig. 8.161 Histopathology of reactive cutaneous histiocytosis: multiple elongated dermal nodules—at high magnifications, large histiocytes and small lymphocytes are recognizable (inset)

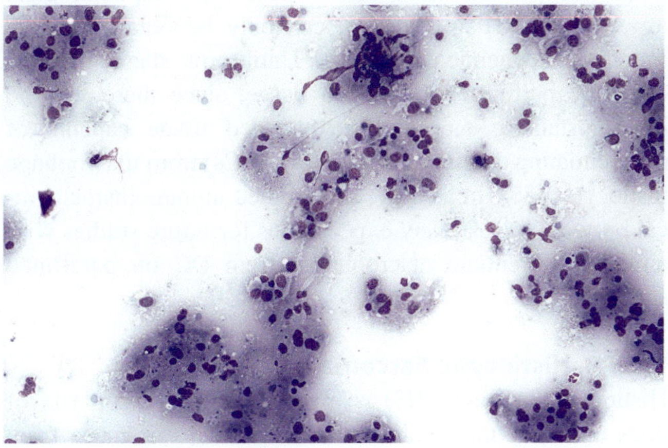

Fig. 8.163 Cytology of reactive cutaneous histiocytosis: on an opaque purple background (dissolved cytoplasms), many naked nuclei and some histiocytes and lymphocytes are recognizable on the slide

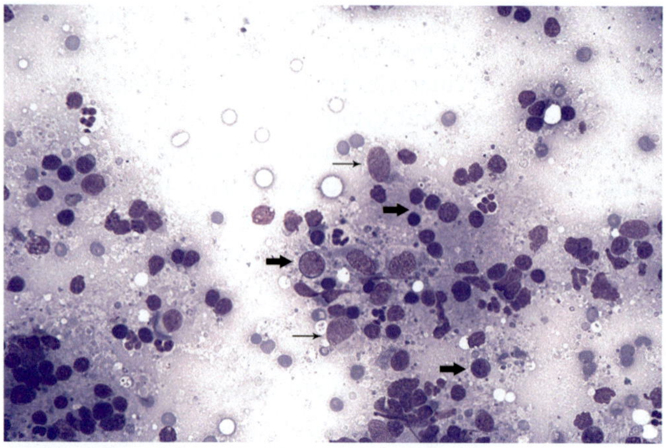

Fig. 8.164 Cytology of reactive cutaneous histiocytosis: many lymphoid cells (thick arrows) and dendritic histiocytes (thin arrows)

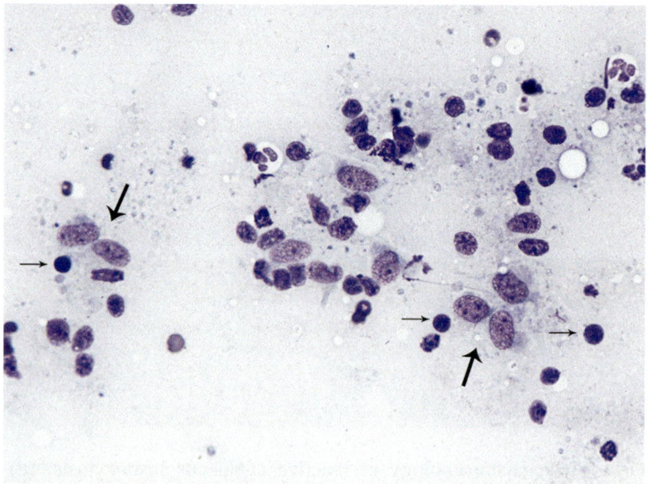

Fig. 8.165 Cytology of reactive cutaneous histiocytosis: lymphoid cells (thin arrows) and dendritic histiocytes (thick arrows)

presentation is doubtful and no peculiar histopathological aspects of CRH are present, can only be obtained using immunohistochemical markers. Until now these markers could be performed on frozen tissue, since there was no immunostaining on paraffin-embedded tissue capable of differentiating dendritic histiocytes (APCs) from macrophage histiocytes [83]. In a study, a combined immunohistochemical protocol opens many expectations for future studies with aim to differentiate macrophages from DC on paraffined samples [78].

8.2.5.3 Histiocytic Sarcoma

Histiocytic sarcoma *(HS)* is a rare malignant neoplasia of dogs and very rarely observed in cats that originates from interstitial dendritic cells present in many organs [72, 81, 82, 83]. In the skin, the APCs from which the histiocytic sarcoma originates are present in the dermis and subcutaneous tissue.

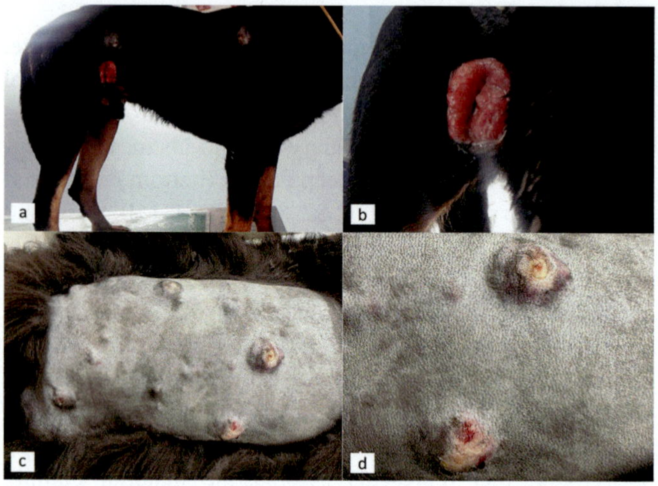

Fig. 8.166 Histiocytic sarcoma: (**a**) multiple nodules in a Rottweiler (courtesy of DVM V.Fraldi-Italy); (**b**) close-up of an ulcerated mass (courtesy of DVM V. Fraldi-Italy); (**c**) multiple nodules of different sizes on the thorax of a Bernese Mountain Dog (courtesy of DVM A. Archetti-Italy); (**d**) close-up of the same lesion in Fig. 8.166c. (Courtesy of DVM A. Archetti-Italy)

A primarily periarticular HS, observed more frequently in certain breeds such as Bernese Mountain Dog, Rottweiler, and Golden and Flat-Coated Retriever, is well recognized; it presents as a single mass of variable size that infiltrates and destroys the joints [74, 81–83, 85]. Skin lesions can be single or multiple and can remain localized to the skin or metastasize to the internal organs. A *disseminated* variant of histiocytic sarcoma that starts with lesions in multiple visceral organs and that can involve even the skin is well recognized. All these forms with multicentric lesions are defined by the term *disseminated histiocytic sarcoma*, supplanting the old term *malignant histiocytosis* (Fig. 8.166).

Cytological Findings

The cytological pattern of histiocytic sarcoma is strongly pleomorphic, and as all the discrete cell tumors, slides are always of high cellularity. It is possible to find two different morphological variants: the first, most common, is prevalently composed with round cells, and another form, less frequent, is mainly composed of spindle-shaped cells. In any case, the malignant cytological features of the neoplastic cells are usually remarkable. The neoplastic cells are characterized by marked anisocytosis, with very large, round or oval, often indented or kidney-shaped nuclei, with one or more bizarre nucleoli and large cytoplasm, slightly basophilic and often vacuolated (Figs. 8.167, 8.168, 8.169,

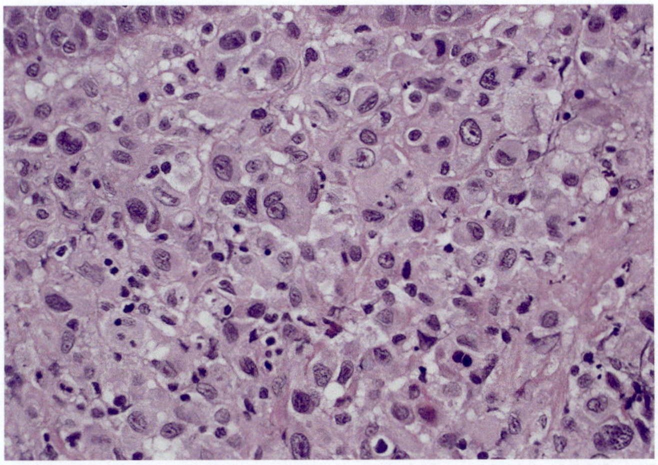

Fig. 8.167 Histology of histiocytic sarcoma: round malignant neoplastic dendritic cells, characterized by severe anisocytosis, anisokaryosis, and vacuolated cytoplasm

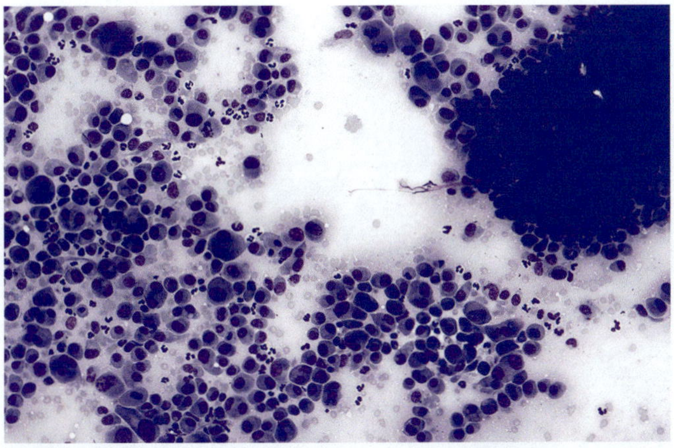

Fig. 8.168 Cytology of histiocytic sarcoma: highly cellular specimens with many neoplastic roundish dendritic cells

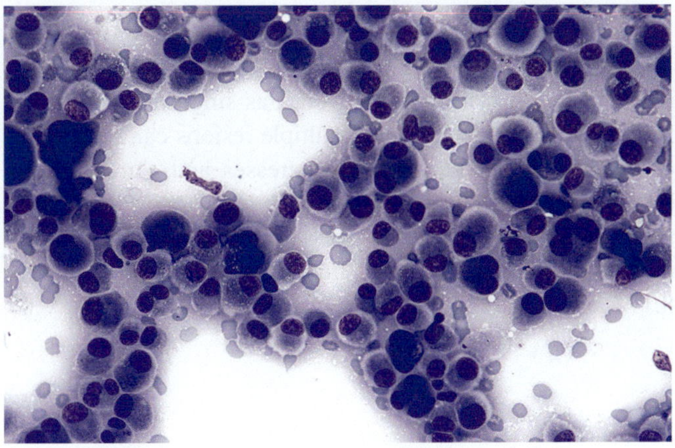

Fig. 8.169 Cytology of histiocytic sarcoma: roundish to oval neoplastic DCs with double or multiple nuclei and vacuolated cytoplasm

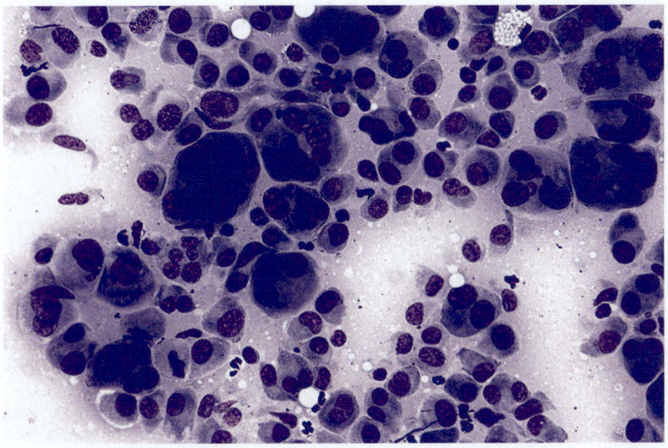

Fig. 8.170 Cytology of histiocytic sarcoma: round pleomorphic neoplastic dendritic histiocytes with anisocytosis, anisokaryosis, multiple nuclei, and vacuolated cytoplasm

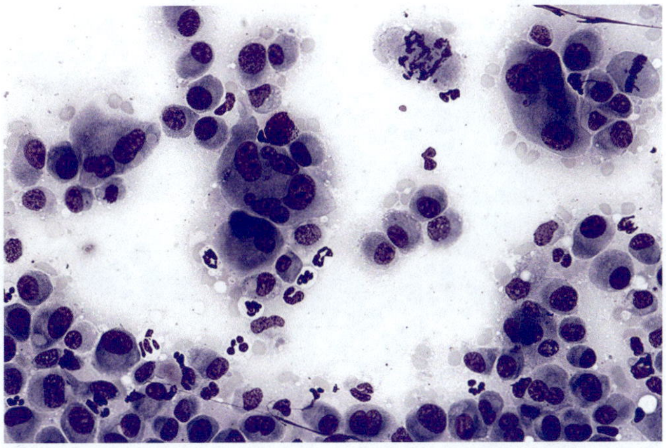

Fig. 8.171 Cytology of histiocytic sarcoma: note the atypical mitotic figures and the clear anisokaryosis in the same multinucleated cell

8.170 and 8.171). When vacuolizations are not so evident, the morphology of neoplastic DCs can be similar to those observed in pleomorphic plasmacytomas and in undifferentiated amelanotic melanoma; in these cases, cytology cannot differentiate them, and clinical presentation, breed predisposition, and histology with immunohistochemistry are fundamental for the diagnosis. The spindle cells have round to oval, often indented, and distorted nuclei; the presence of numerous multinucleated cells can create diagnostic difficulties as the cytological features are very similar to those observed in some anaplastic soft tissue sarcomas with a high percentage of giant cells, formerly named malignant fibrous histiocytoma and recently renamed *pleomorphic sarcoma.* In some samples it is also possible to find a variable number, often high, of infiltrating leukocytes (Figs. 8.172,

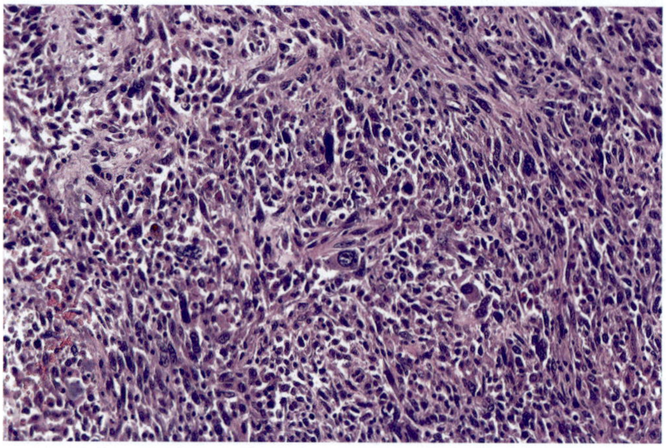

Fig. 8.172 Histology of histiocytic sarcoma: spindle-shaped neoplastic dendritic cells. Marked atypia and multinuclearity characterize the cells. Note they cannot be morphologically differentiated from a spindle-shaped undifferentiated soft tissue sarcoma

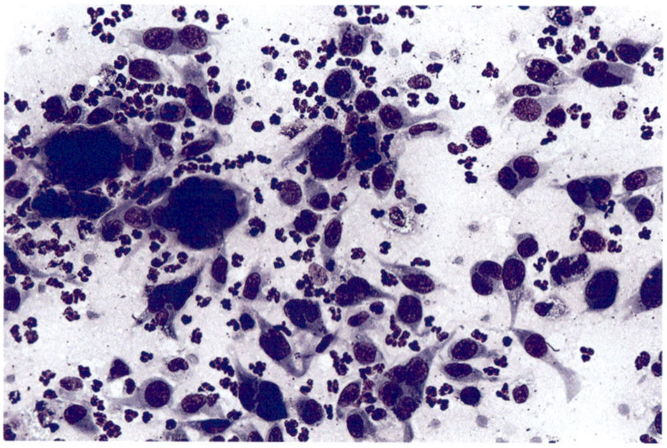

Fig. 8.173 Cytology of histiocytic sarcoma: spindle-shaped neoplastic dendritic cells. Marked atypia and multinuclearity characterize the cells; neutrophils don't take part of cytological pattern of HS but are present secondarily to ulceration on the tumor

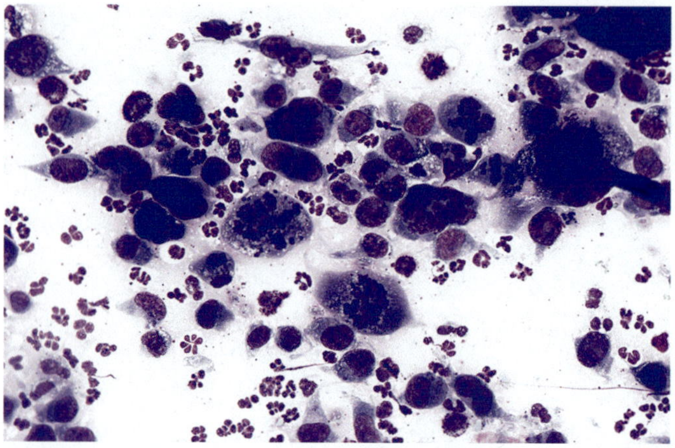

Fig. 8.174 Cytology of histiocytic sarcoma: spindle-shaped neoplastic dendritic cells. Marked malignant features and atypical mitotic figures. It is impossible to differentiate with cytology a STS

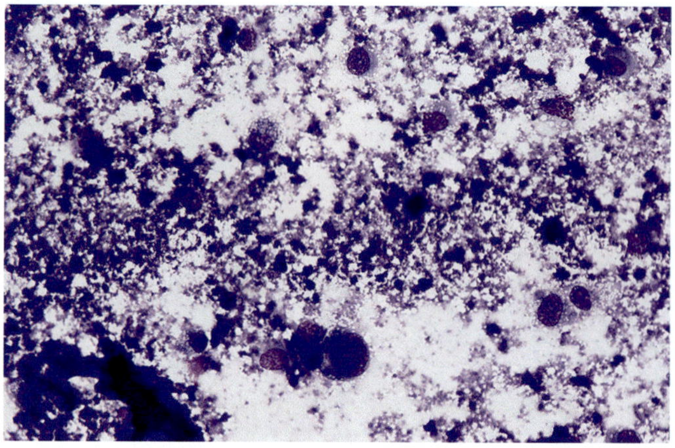

Fig. 8.175 Cytology of histiocytic sarcoma: amorphous and granular necrotic background; some neoplastic round malignant DCs are evident

8.173 and 8.174). Intratumoral necrosis is a frequent evolution of histiocytic sarcomas (Fig. 8.175).

8.2.5.4 Feline Progressive Dendritic Histiocytosis

Feline progressive histiocytosis is a histiocytic disorder with involvement of the skin, observed in adult to old cats and originating from interstitial APCs [82, 83, 87].. In a paper, in most cases of FPH, neoplastic cells were positive to E-cadherin marker; so, an origin from Langerhans cells and not from interstitial APCs has been supposed [75]. In most reported cases, lesions usually start as single nodule on the limb or head, and with time, multiple lesions can develop to other skin sites. The disease progresses and can spread to internal organs. The cutaneous nodules vary in size, from small (papular nodules) to very large plaque and masses, mostly located on the head, limbs, and trunk (Figs. 8.176, 8.177 and 8.178).

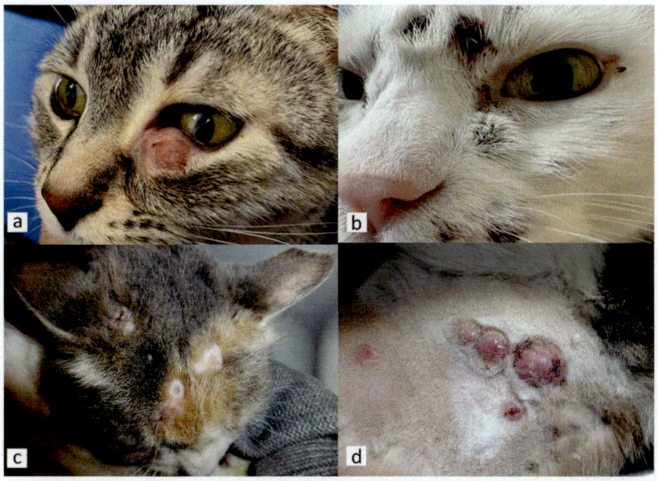

Fig. 8.176 Feline progressive histiocytosis. (**a**) Nodule on the eyelid in a cat with multifocal lesions; (**b**) multiple small nodule on the face. (Courtesy of DVM M. Beccati-Italy); (**c**) multiple alopecic nodules on the head; (**d**) multiple alopecic nodules on the trunk of the same cat in Fig. 8.176c

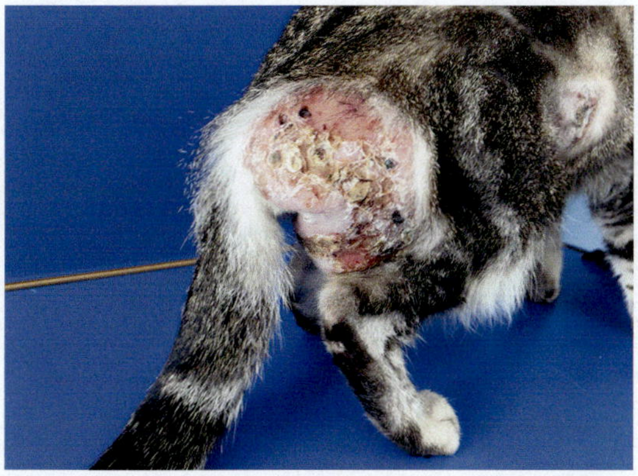

Fig. 8.177 Feline progressive histiocytosis: large mass involving the back and the base of the tail

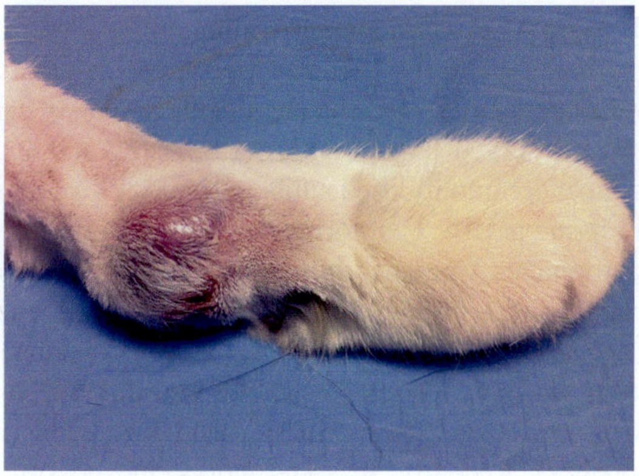

Fig. 8.178 Feline progressive histiocytosis. Single, alopecic, and erythematous nodule on the left foreleg

Cytological Findings

Slides from *feline progressive histiocytosis* can show different cytological morphology depending not only from the stage of the disease but also from the well-known variable morphology of histiocytes. In the initial stages, histiocytes can be well differentiated and with scarce atypias; cells vary from round to polygonal or spindloid, with round to folded kidney-shaped nuclei, sometimes double with scant cytoplasm stained pale blue (Figs. 8.179, 8.180 and 8.181). In some cases, morphological atypia is more evident, and even the shape of the cells can take on a clear spindle-shaped silhouette; when all cells are spindle in shape, a mesenchymal soft tissue sarcoma or a spindle amelanotic melanoma cannot

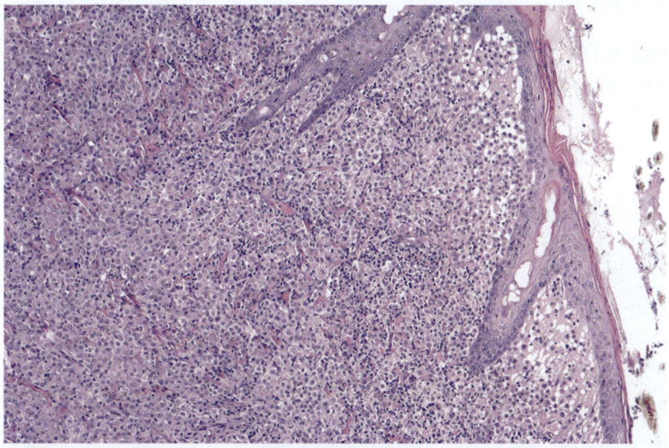

Fig. 8.179 Histology of feline progressive histiocytosis. Massive dermal proliferation of, from round to polygonal, discrete DCs, with scarce atypias, replacing all the skin adnexa. The smaller and dark cells are infiltrating lymphocytes

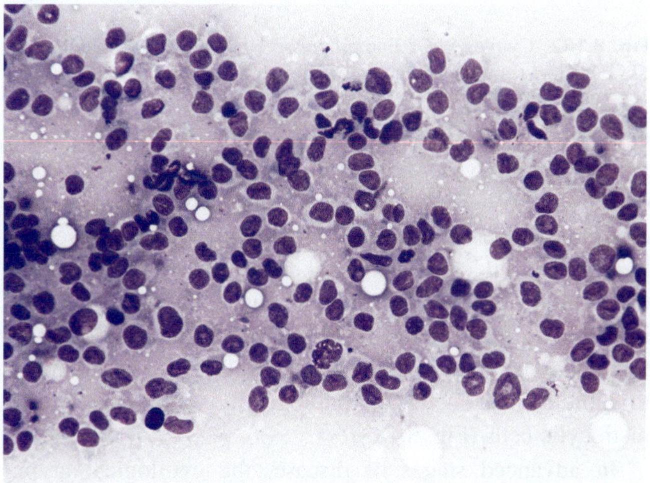

Fig. 8.180 Cytology of feline progressive histiocytosis. Numerous dendritic cells with a round profile and no obvious morphological atypia. The nuclei are round or reniform and medium in size; the cytoplasm is scant to medium and pale blue in color

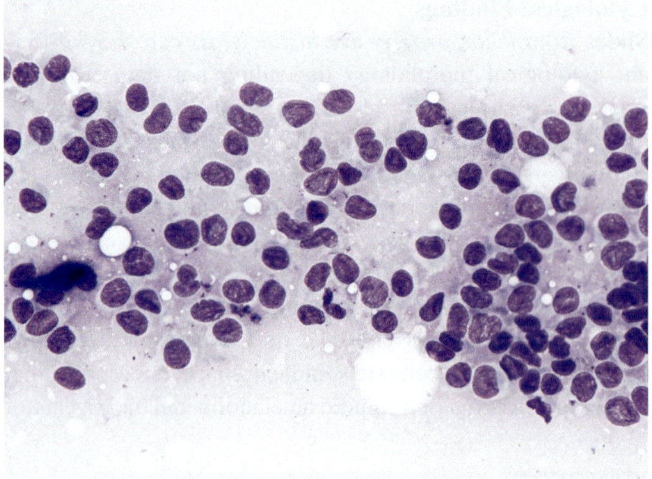

Fig. 8.181 Cytology of feline progressive histiocytosis. Round cells with numerous reniform nuclei suggesting a histiocytic origin of the cells

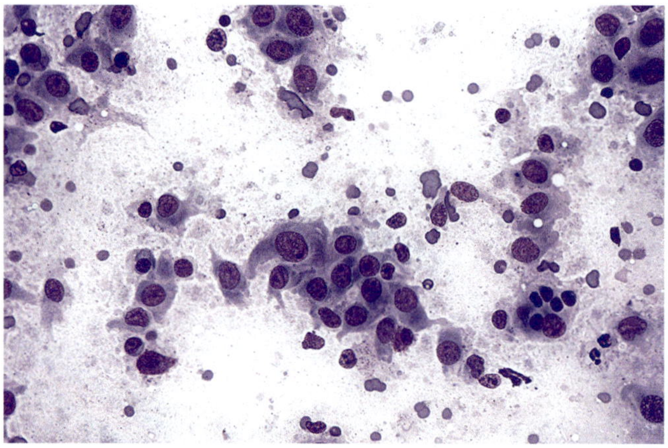

Fig. 8.182 Cytology of feline progressive histiocytosis. Spindle dendritic histiocytes with evident anisocytosis and anisokaryosis

cytologically be ruled out (Fig. 8.182). In some cases, a variable quantity of infiltrating lymphocytes and/or mast cells can be observed. In the latter case, when the mast cells are numerous and metachromatic granules are dispersed on the background, care must be taken not to confuse the histiocytic cells with poorly granulated mast cells; this possibility increases in the so named histiocytic or atypical variant of feline mast cell tumor, in which mast cells can show roundish to spindloid silhouette, morphologically similar to those of histiocytic cells (Fig. 8.183).

In advanced stages of disease, the cytological atypia becomes much more evident, and cells are similar to those observed in histiocytic sarcoma (anisokaryosis, anisocytosis, bizarre nuclei, large and multiple macronucleoli, vacuolization of cytoplasm); for this behavior in modifying cytological morphology with the chronicity of disease and

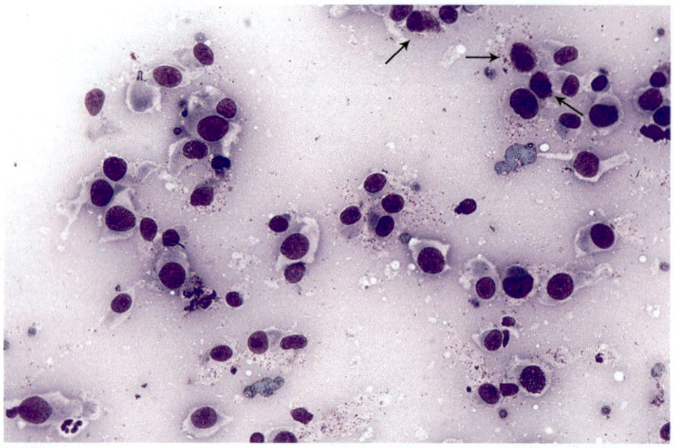

Fig. 8.183 Cytology of feline progressive histiocytosis. Morphologically mixed histiocytic population, predominantly composed of spindle cells. Note how well-differentiated infiltrating mast cells are present (arrows) and many metachromatic granules dispersed on the background. In this case, a diagnosis of histiocytic feline mast cell tumor can be wrongly done

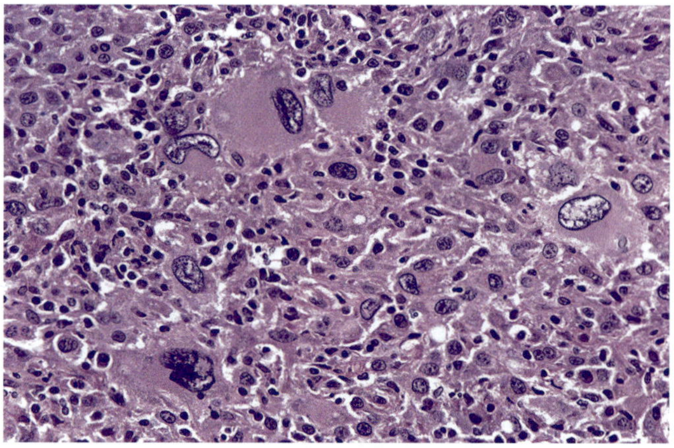

Fig. 8.184 Histology of feline progressive histiocytosis. Severe atypias represented with anisokaryosis, anisocytosis, mega and bizarre in shape nuclei, large and multiple macronucleoli

since FPH and HS share a similar immunophenotype, some authors believe that FPH can represent a multifocal low-grade histiocytic sarcoma. Leucophagocytosis can frequently be observed (Figs. 8.184, 8.185, 8.186 and 8.187).

The most commonly used antibody to label histiocytes on paraffin tissue is IBA-1; unfortunately, it does not differentiate macrophages from dendritic cells, and, in the case of differential with a granulomatous lesion, the IBA-1 marker is not decisive. To date, there are no immunohistochemical markers that work on paraffin samples and that differentiate the two histiocyte populations (dendritic cells versus macrophages). Recently, in some works, the possibility of using immunocytochemistry on frozen cytological samples has been suggested; the use of specific markers for the cytological identification of various populations of dendritic

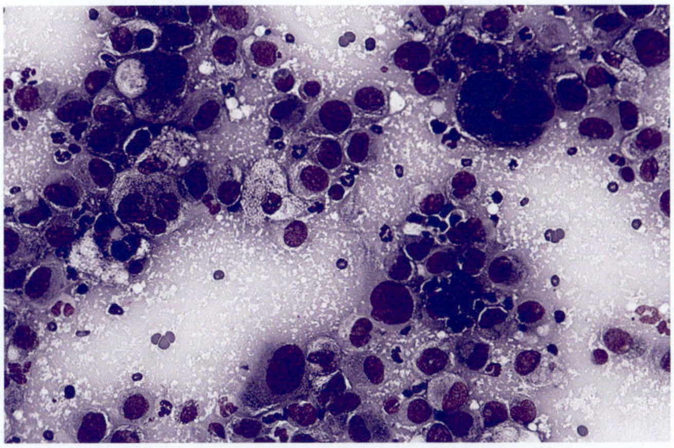

Fig. 8.185 Cytology of feline progressive histiocytosis. Malignant and roundish dendritic cells, with severe anisokaryosis, anisocytosis, and multiple nuclei with different shapes. Note cells are similar to those of canine HS

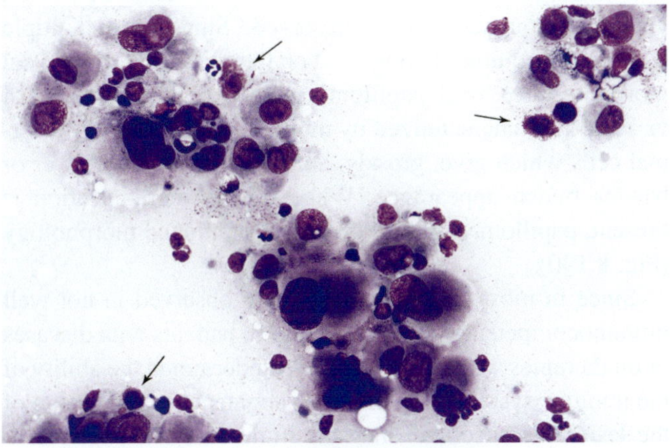

Fig. 8.186 Cytology of feline progressive histiocytosis. Malignant dendritic cells with severe anisokaryosis and anisocytosis. Note the infiltrating well-differentiated mast cells

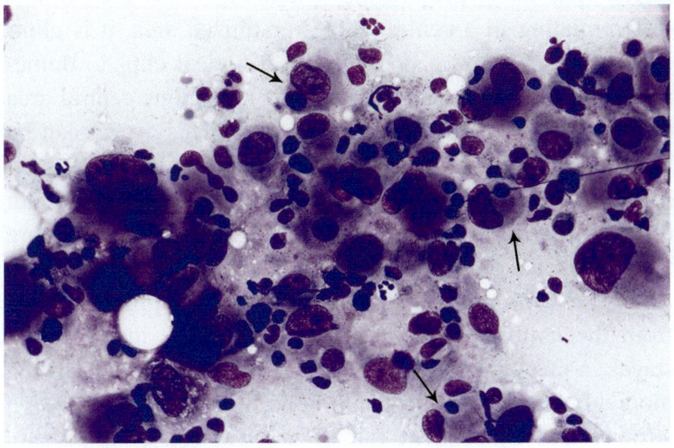

Fig. 8.187 Cytology of feline progressive histiocytosis. Malignant dendritic cells with severe anisokaryosis and anisocytosis. Note the phagocytosis of lymphocytes by neoplastic cells

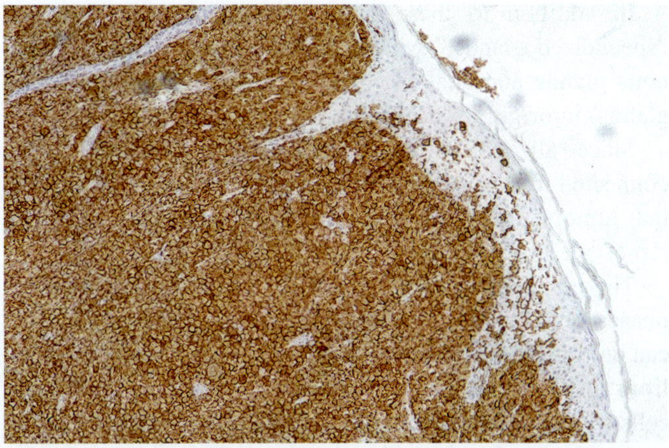

Fig. 8.188 Immunohistochemistry of feline progressive histiocytosis. Strong and diffuse positivity to IBA-1 marker confirms the histiocytic nature of cells, but not differentiates dendritic cells from macrophages

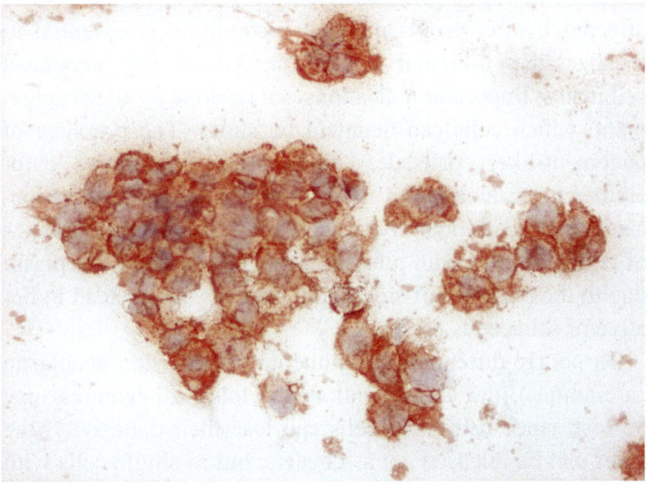

Fig. 8.189 Immunocytochemistry of feline progressive histiocytosis. Positivity to CD11c antibody confirms the dendritic nature of histiocytic cells

histiocytes allows them to differentiate not only from other dendritic cells but also from macrophages [87] (Figs. 8.188 and 8.189).

8.3 Epithelial Tumors

Epithelial tumors encompass a wide number of neoplasia which origin from the epithelia of the epidermis and skin adnexa.

Epithelial cells present in the skin, from where neoplasia can originate, are the keratinocytes from both *superficial epidermis* and *follicular wall*; furthermore, neoplasms can origin from *sebaceous* and *apocrine* glandular epithelia.

In addition to these tumors, neoplasia of so named "specialized glands," which can be both apocrine (ceruminous glands and anal sac glands) and sebaceous (hepatoid glands) in origin, is also discussed.

Since all the epithelial cells are, by nature, tightly connected to each other by intercellular junctions, all epithelial tumors share the cytological characteristic of being represented by cohesive clusters of cells.

The epithelial tumors are composed of cells in different phases of differentiation, and for these characteristics, they can be very different in shape and size; indeed, they can range from small basal cells with scarce cytoplasm to medium/large cells with abundant cytoplasm and round to ovoid nuclei with chromatin ranging from fine to coarse. Since some of these neoplasias, both follicular and epidermal, can be keratinized, cells from partial to full keratinized corneocytes can be observed along with basal cells. As will be discussed in the section regarding follicular tumors, keratin can also take on different morphologies based on the portion of the follicle affected by the neoplasm; their recognition is essential to localize the follicular area from which the neoplasm originates. Important indications can be done by the arrangement, which cells can maintain on slides. The presence of pavement-like, palisade, ductal, and papillary cyto-architectures, along with typical features of apical secretion, can cytologically identify the apocrine tumors, while lobules of cohesive cells with a foamy appearance of the cytoplasm due to the presence of small lipid droplets, are instead indicative of sebaceous origin.

In poorly differentiated epithelial tumors (i.e., anaplastic carcinomas), the abovementioned cytological features may be lost; since neoplastic cells can lose their cohesivity, the latter can be released not as clusters, but as single cells with variable severity of cytological atypia. In these cases, when only single cells are observed on slides, it is difficult to include the neoplasia in the group of epithelial tumors, and some cytological features such as keratinization or secretory activity aspects can orient cytologists. A classic example of epithelial neoplasm in which the cells often appear singly (discrete) is the *squamous cell carcinoma*, one of the most common epithelial tumors in cats. In this skin neoplasia, the clear squamous aspect of cells permits to make a diagnosis of epithelial tumor, also if clustered cells are not visible, while, in other situation, as in *inflammatory mammary carcinoma*, cells cannot show specific aspects of secretory differentiation and since many of these tumor are acantholytic, a cytological differentiation from a round cell tumor is not always so immediate.

The epithelial neoplasms of the skin in dogs and cats are numerous, and only the most important, or those for which the cytological examination is of great diagnostic aid, are described in detail in this chapter.

8.3.1 Papilloma

The role of *papillomavirus* on the development of skin neoplasia has been well documented both in dogs and cats [88, 89–91, 92, 93, 95]. Regarding the role of papillomavirus in skin neoplasia in cats, it will be discussed in squamous cell carcinoma since, differently to dogs, *papilloma* is not a typical tumor of cats. In dogs, two types of cutaneous papilloma are recognized: a *viral papilloma*, caused by various canine papillomavirus and a *squamous papilloma*, in which no viral pathogenesis has been documented. *Viral papilloma* is a very common epithelial tumor of dogs, and it is more commonly observed in the oral cavity, e.g., muco-cutaneous junction of the lips and nostrils of young puppies. Lesions can be single or multiple, and in immunodepressed dogs, very large masses can be observed. Single and multiple nodules are commonly observed on the skin also. The typical morphology of viral papilloma is an exophytic nodule, of variable size, characterized by multiple projections of epidermal cells which give, grossly, an artichoke or cauliflower or banana bunch appearance. When marked keratinization is present, papilloma can lose the abovementioned morphology (Fig. 8.190).

Since in most cases the lesions are observed in not well immunocompetent young puppies or in patients with diseases or on therapies with drugs that have undermined the ability of the immune system, very often the spontaneous regression of the lesions is observed once the immune system returns to function properly.

A morphological variant of canine papilloma, so named *inverted papilloma*, is well known in dogs. Since histologically, inverted papilloma is characterized by epidermal endophytic papillary projections extending into the dermis differentiating in a central fully keratinized area, it is clinically represented as a nodule with a typical cup- to dome-shaped appearance with a depressed keratinous central area that gives it an umbilicated appearance. This aspect can be lost when keratinous material protruding from the center is of high amount and gives origin to an exophytic horn of keratin [90, 93]. Inverted papilloma can be single or multiple and are usually located on the hairless skin of the thighs and abdomen (Fig. 8.191). A clinical variant of inverted papilloma arises from the nailbed epithelium [94]. Clinically, in the initial stages, a swelling of the phalanx is observed which, in the more advanced stages, leads to bone lysis from compressive atrophy and consequent fall of the nail (onychomadesis).

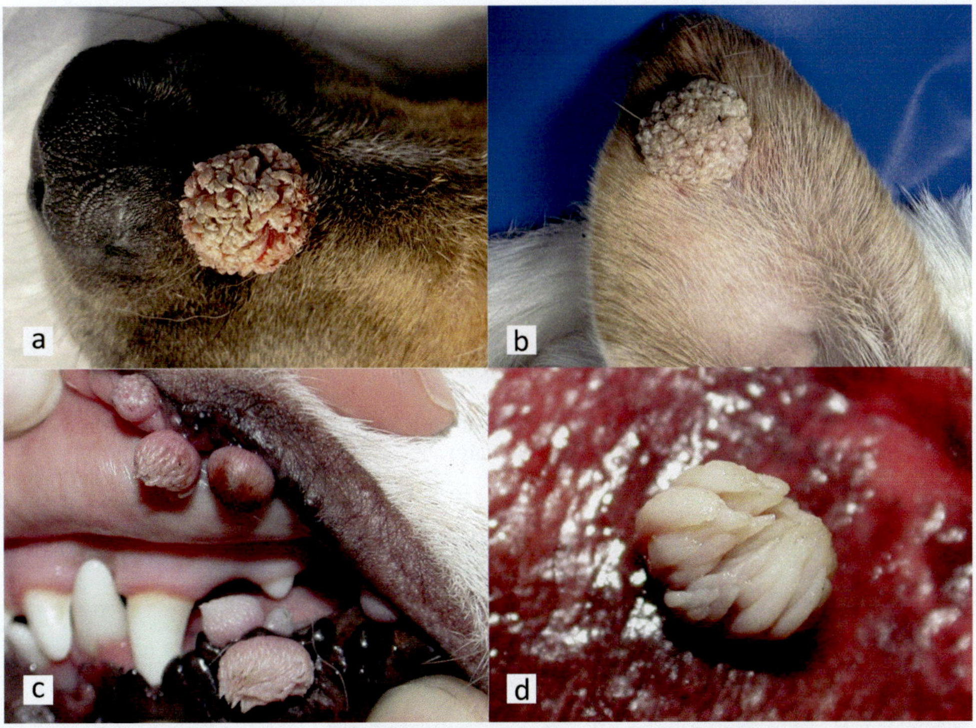

Fig. 8.190 Canine conventional papilloma: irregular with artichoke appearance nodules on the oral mucosa and skin of four young dogs

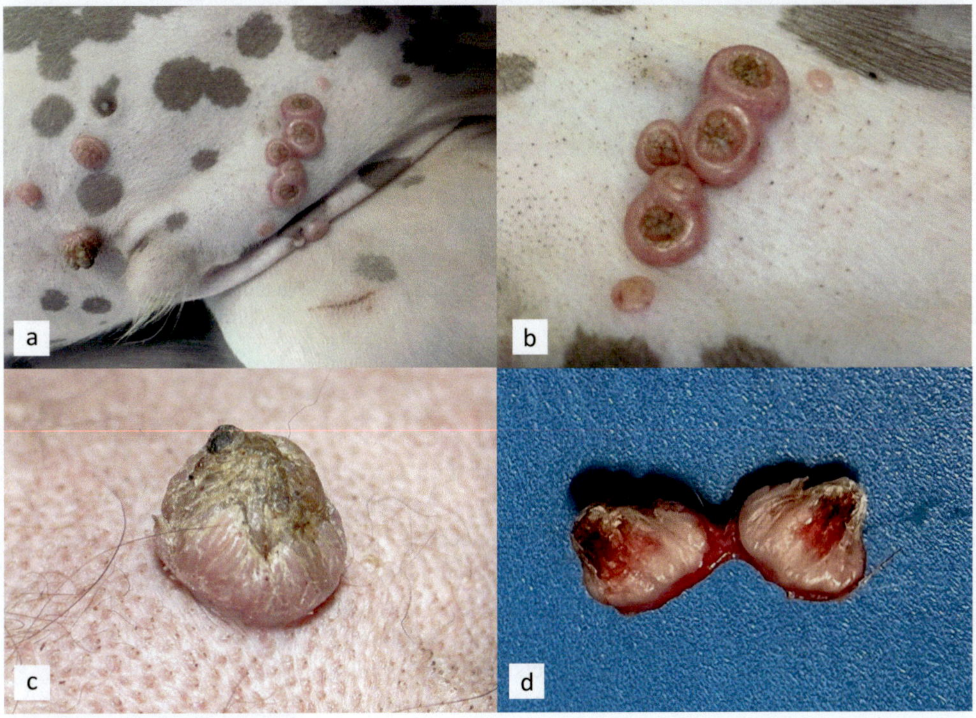

Fig. 8.191 Canine inverted papilloma: (**a**) multiple dome-shaped nodule with keratinous center (courtesy of DVM F. Necci-Italy; (**b**) close-up of the same lesions of the previous figure—note the typical umbilicated shape; (**c**) inverted papilloma with marked keratin material protruding from the center; (**d**) section of the papilloma showed in the previous image

From a clinical point of view, the differential diagnoses to consider are certainly squamous cell carcinoma and subungual keratoacanthoma. Since bone lysis and nail loss also occur in the other two epithelial neoplasms of the nail bed, the clinical and radiological aspects are not decisive, and the definitive diagnosis requires histology (Figs. 8.192 and 8.193). Squamous papilloma not due to papillomavirus is more commonly observed as single nodule on the skin of adult-old dogs. Since their pathogenesis is not linked to a virus, but it is thought be a benign reaction of epidermis originated by an unknown underlying stimulus, the spontaneous recover must not be expected as for viral papilloma occurs; also this type of papilloma can assume an artichoke or cauliflower-like appearance [88].

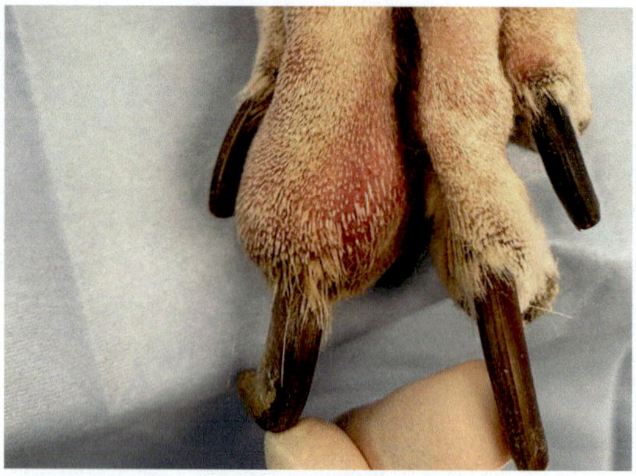

Fig. 8.192 Inverted papilloma of nailbed: swelling of third phalanx with onychogryphosis. (Courtesy of DVM A. Prolo-Italy)

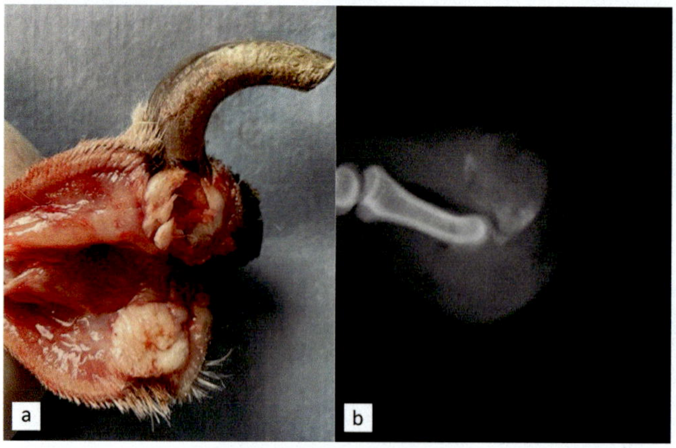

Fig. 8.193 Inverted papilloma of nailbed: (**a**) after digit amputation, it is possible to evaluate the localization of the papilloma and its cup-shaped morphology (courtesy of DVM A. Prolo-Italy); (**b**) radiography of the same digit—note the complete bone lysis of the third phalanx

Cytological Findings

In histology, the conventional *viral papilloma* is composed with exophytic epidermal projections covered with a thick ortho- or parakeratotic keratinized corneum layer. Nucleated keratinocytes show different cytopathic viral effects, from few to very evident, especially in spinous layer, in which koilocytes, cells characterized by shrunken pyknotic nuclei surrounded by a clear achromatic halo, are frequently observed. Other viral cytopathic effects are ballooning degeneration of keratinocytes, intranuclear eosinophilic inclusions, and hypergranulosis. In the rare histological variant (Le Net type), many brightly eosinophilic intracytoplasmic inclusions are clearly evident [88] (Figs. 8.194, 8.195 and 8.196).

The above-described viral cytopathic effects, which can be easily observed in histopathology of both exophytic and inverted papilloma, are extremely hard to find on cytological

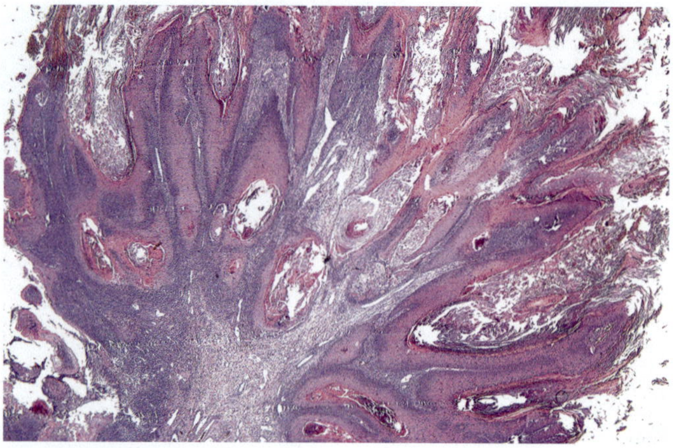

Fig. 8.194 Histology of conventional viral papilloma: exophytic epidermal digitation covered with thick parakeratotic stratum corneum

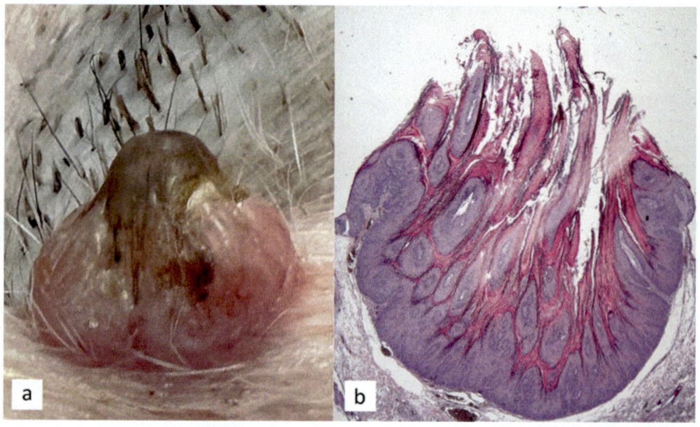

Fig. 8.195 Histology of inverted viral papilloma: clinical and histological representation of inverted papilloma: note, compared with conventional one, as epidermal endophytic papillary projections extend into the dermis causing the typical dome shape of this papilloma

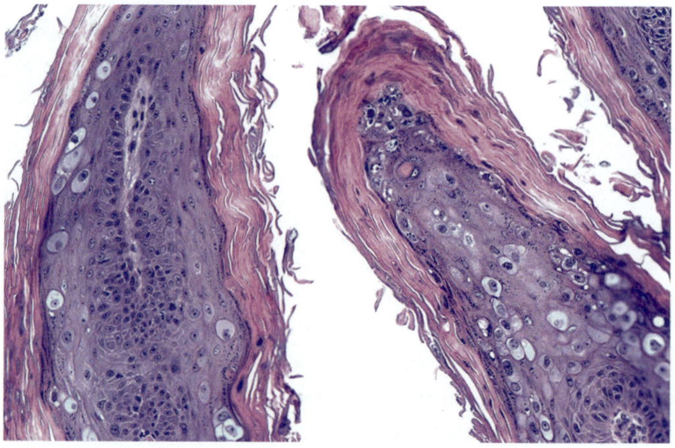

Fig. 8.196 Histology of viral papilloma: cytopathic effects represented by hypergranulosis, ballooning degeneration, and koilocytes are clearly evident at high magnification

specimens. With FNB technique, it is quite impossible to put the needle into the more superficial spinous or granulosum layer and collect, for example, koilocytes and cells from the very thick stratum corneum are usually collected. Only with scraping technique, performed on the cut surface of the tumor after its surgical removed, it is possible to collect a higher number of nucleated cells and have a better chance of finding cells with viral cytopathic effects (i.e., koilocytes or with large keratohyalin granules). Furthermore, the number of koilocytes is usually not so high in papillomas, and they are mainly located at the more superficial area; so the possibility to collect them is, anyway, very few. Finally, since the cytological morphology of koilocytes is not standardized with Romanowsky-type stain, it is not clear how it can be cytologically surely recognizable. The most common cytological feature we can find in the cells infected by canine papillomavirus is the presence of intracytoplasmic violet fibrillar stripes, as commonly observed in histopathological sections; small roundish vacuole-like achromatic areas, maybe associated with cytoplasmic degeneration secondary to virus, are also recognizable (Figs. 8.197, 8.198 and 8.199).

In comparison with the above-discussed inverted papillomas of the skin, in those located at the nail bed, viral cytopathic changes are usually absent and therefore, if present, are extremely rare to observe on cytological preparations. The cells sampled from nailbed inverted squamous papilloma are represented by a large number of squamous epithelial cells, with large cytoplasm both with fully keratinized cytoplasm and with the nucleus still present and characterized by low N/C ratio (Figs. 8.200, 8.201 and 8.202). Unfortunately, the cytological findings can be misinterpreted as belonging to nailbed squamous cell carcinoma (SCC), especially when the latter is well differentiated and with minimal atypia.

Furthermore, to increase confusion, a similar cytology can be observed also in another benign tumor of the nail bed, the

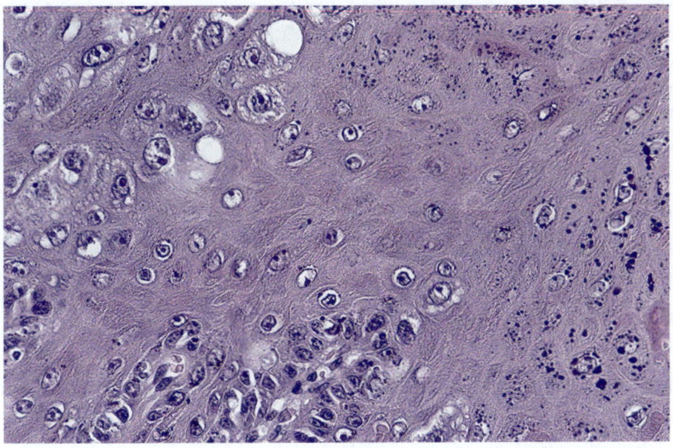

Fig. 8.197 Histology of viral papilloma: cytopathic effects represented by intracytoplasmic fibrillar stripes are clearly evident in almost all the infected keratinocytes

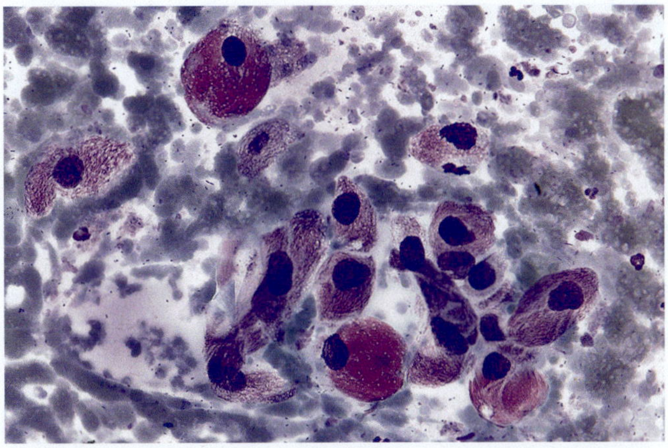

Fig. 8.198 Cytology of viral papilloma: cytopathic effects represented by intracytoplasmic purple fibrillar stripes are better evident with cytology. Many small vacuoles are also recognizable

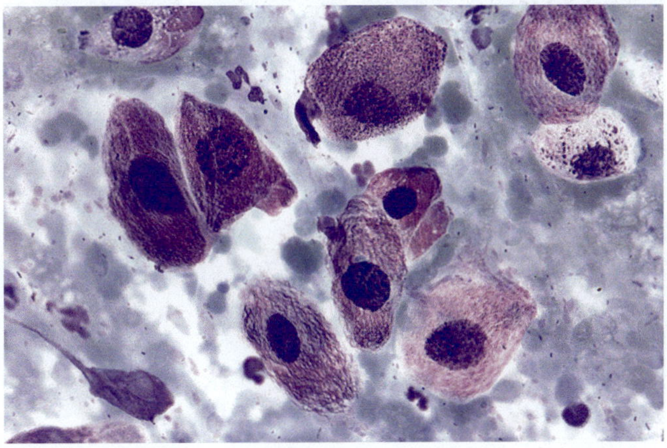

Fig. 8.199 Cytology of viral papilloma: the infected cells show cytoplasm filled with purple fibrillar stripes indicative of the cytopathic viral activity

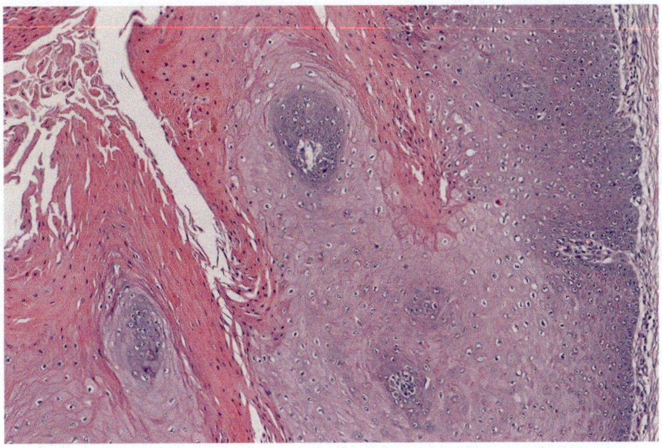

Fig. 8.200 Histology of nailbed inverted papilloma: proliferation of benign squamous nucleated keratinocytes, which undergo incomplete keratinization (parakeratotic hyperkeratosis)

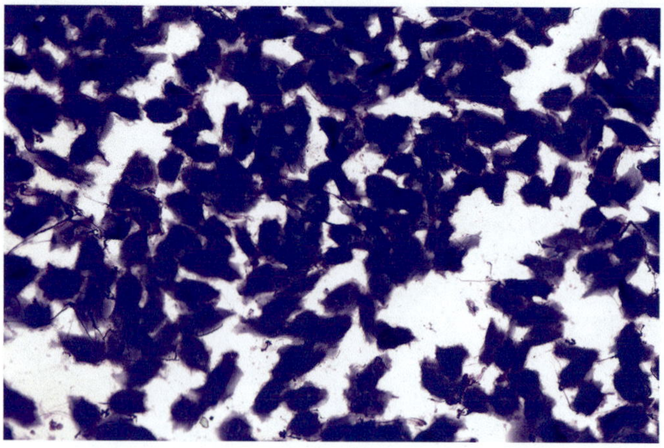

Fig. 8.201 Cytology of nailbed inverted papilloma: many polygonal squamous keratinocytes both anucleate and with nuclei: note viral cytopathic effects are not evident

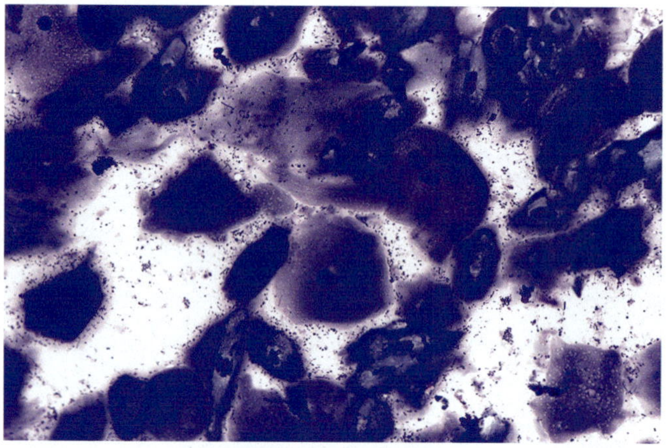

Fig. 8.202 Cytology of nailbed inverted papilloma: in this image, moderate cytological atypia are present and can cause a misdiagnosis of SCC

subungual keratoacanthoma (SK), an uncommon, benign neoplasm that origins from the epithelium of the nail bed and that must not be confused with *infundibular keratinizing acanthoma* (IKA) that, instead, origins from the upper parts of the follicle and that will be discussed with follicular neoplasms (Figs. 8.203, 8.204, 8.205 and 8.206).

Since once the finger is surgically removed and they not always are subjected to histological examination, the prevalence of subungual keratoacanthoma is underestimated in favor of the nailbed squamous cell carcinoma, the latter diagnosis based on a misinterpreted cytology. Summarizing, inverted papilloma, subungual keratoacanthoma, and well-differentiated nailbed SCC are all epithelial neoplasms arising from the nailbed epithelium and which share clinical

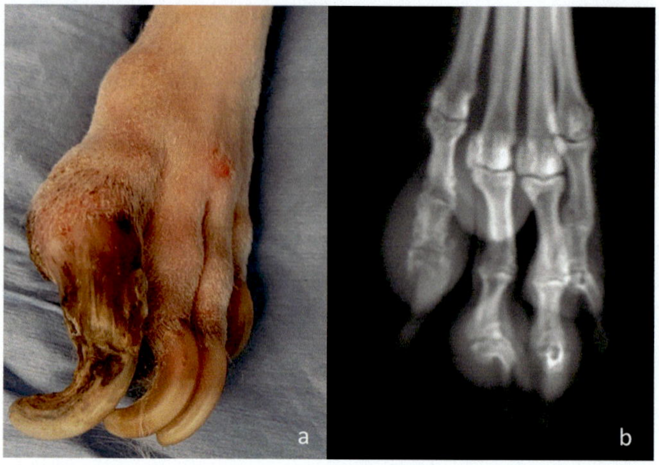

Fig. 8.203 Nailbed keratoacanthoma. (**a**) Swelling of the phalanx with marked hyperkeratosis and onychogryphosis in a Maltese dog (courtesy of DVM G. Bartoletta-Italy); (**b**) note osteolysis of the third phalanx is radiographically evident

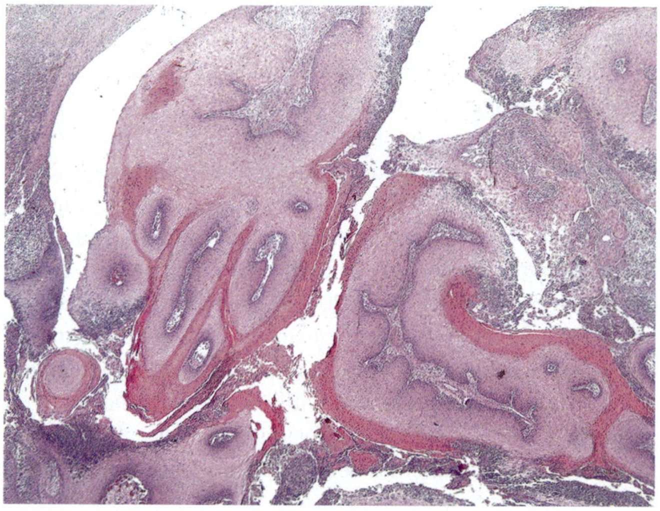

Fig. 8.204 Histology of nailbed keratoacanthoma: proliferation of benign squamous nucleated keratinocytes, which undergo incomplete keratinization (parakeratotic hyperkeratosis)—note the histological pattern is similar to those of nailbed inverted papilloma from whom is not easy to differentiate also through histology

(swollen of the digit, onychogryphosis, and/or onychomadesis), radiological (bone lysis), and cytological similitudes (mixed nucleated squamous keratinocytes and keratin lamellae); for these reasons, histopathology is mandatory for diagnosis.

Cytology of non-viral papilloma can show large keratinocytes, both nucleated and not, with angular cytoplasmic edges. No cytoplasmic fibrillar structures or other viral signs of cells infection must be expected.

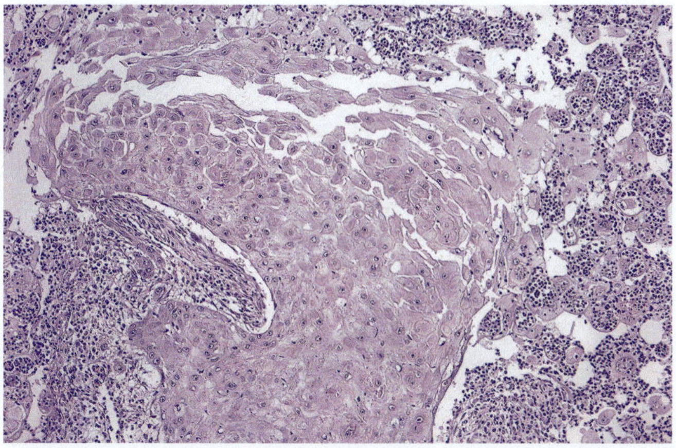

Fig. 8.205 Histology of nailbed keratoacanthoma: at high magnification, squamous nucleated cells can show quite "dysplastic" features but not show clear cytological criteria of malignancy

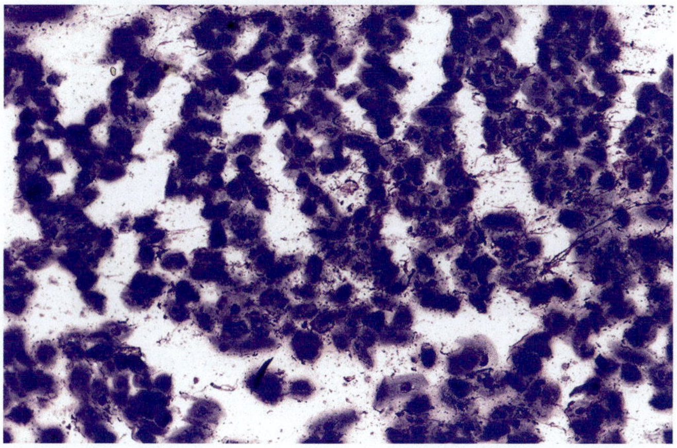

Fig. 8.206 Cytology of nailbed keratoacanthoma: many nucleated squamous cells, showing scarce aspects of atypia maybe also due to secondary inflammation

8.3.2 Squamous Cell Carcinoma

Squamous cell carcinoma (SCC) is the most common epithelial tumor in cats, while it is less common in dogs. The neoplasia originates mainly from the cells of the spinous layer of the epidermis [98, 99, 104, 109].

Different histotypes of SCC are recognized: *conventional*, *acantholytic*, *spindle cell*, *verrucous*, *papillary*, and *clear cell* [88, 98].

The main cause of SCC in pets is linked to chronic sun exposure, and therefore SCCs are very frequent in animals with white skin, especially on glabrous or poorly haired areas and with thinner skin.

The role of papillomavirus in the development of SCC has been widely documented in dogs, and the ongoing discovery of new viral variants is a constant confirmation [88, 91, 92, 95]. In the last years, many papers have been published regarding the presence and the role of papillomavirus in *Bowenoid* in situ *squamous cell carcinoma* (BISC) that is mainly associated with *Felis catus papillomavirus-2* (FcaPV-2) [97, 100, 103, 105–108, 110].

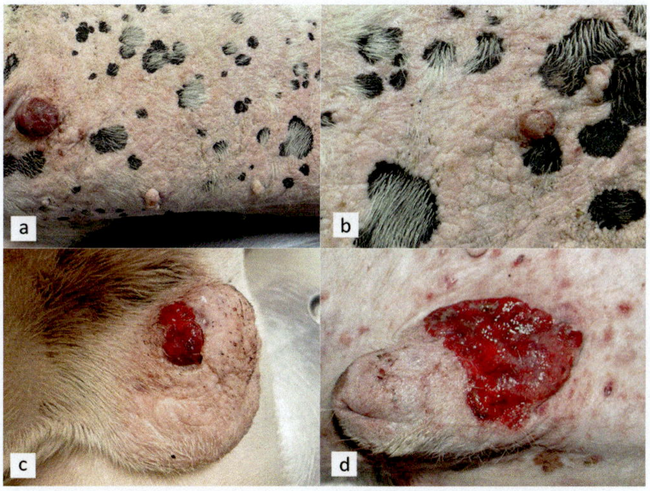

Fig. 8.207 Solar-induced SCC in dogs: (**a**) exophytic and erythematous SCC on the non-pigmented skin of the abdominal area in a Dogo Argentino; (**b**) same dog of the previous image with lesions localized only in the areas in which pigment is not present. Note hairs are only present in the non-injured pigmented areas; (**c**) ulcerative and erythematous SCC on the scrotum; (**d**) ulcerated plaque-like SCC on the penis of a dog with white skin

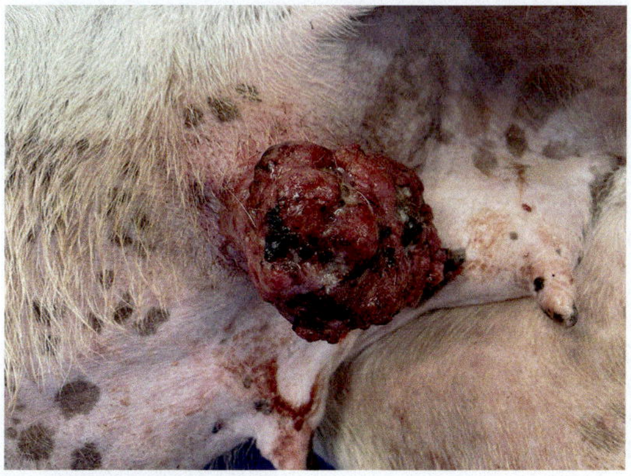

Fig. 8.208 Exophytic and ulcerated large nodular SCC on the abdominal area of a dog with non-pigmented skin

In dogs, the SCCs appear as nodules, plaques, or exophytic masses with not well-defined margins and often with an ulcerated and bleeding surface. These tumors are mainly located on hairless and thin skin, such as the groin, abdomen, and scrotum, especially in dogs with white skin, which is less protected from sunlight (Figs. 8.207, 8.208 and 8.209). In many dogs, lesions are observed in association with actinic keratosis, which represents an in situ carcinoma that precedes the development of an infiltrating SCC. Ulcerative SCCs are also observed on the nose, particularly in

Labradors and Golden Retrievers [101] (Fig. 8.210). Another area where SCC develops in dogs, considered highly malignant, is the nail bed; this location is most commonly observed in black-haired giant breeds such as in giant Schnauzers, Rottweilers, and Labradors, where multiple toes of more than one leg can be simultaneously affected [96, 102] (Figs. 8.211 and 8.212). It must be stressed, as in the early stages of development of nailbed SCC, lesions are characterized by the swelling of the digit and periungual area and cannot be clinically differentiated from an inflammatory paronychia. As abovementioned, other nailbed neoplasia, such as inverted papilloma and nailbed keratoacanthoma, can be presented with the same clinical

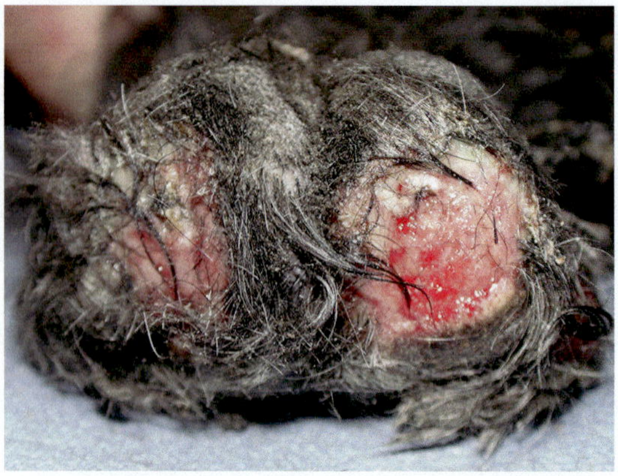

Fig. 8.211 Multidigital SCC of the nail bed in a Giant Schnauzer

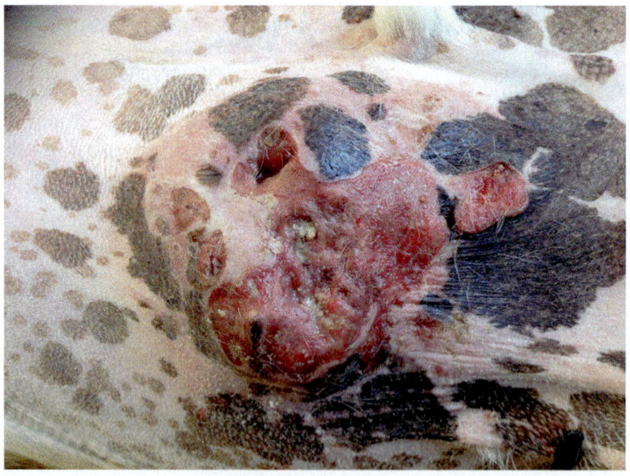

Fig. 8.209 Large ulcerated mass on the abdominal skin of a Dogo Argentino with solar-induced SCC

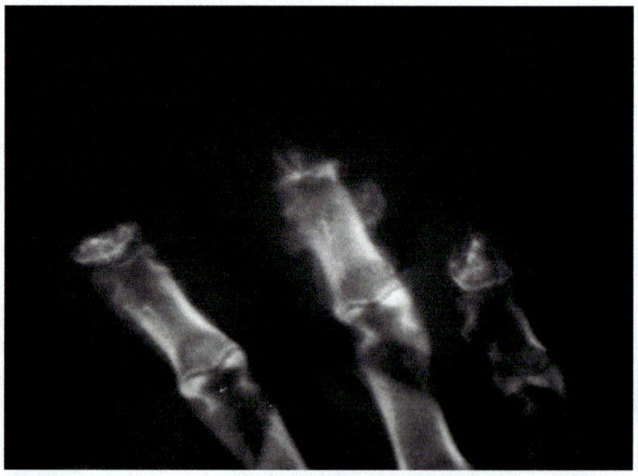

Fig. 8.212 Severe osteolysis of three adjacent digits of the same dog as in Fig. 8.211

Fig. 8.210 Nasal SCC in dogs: (**a**) swelling and nose profile alteration in an old Springer Spaniel; (**b**) same dog of the previous image; note as the anatomical silhouette of the nose is severely modified; (**c**) swelling and ulceration of the ventral right nostril of a Labrador Retriever. (Courtesy of DVM F. Necci-Italy); (**d**) diffuse ulceration of the nose of a Golden Retriever with SCC. (Courtesy of DVM M. Annoni-Italy)

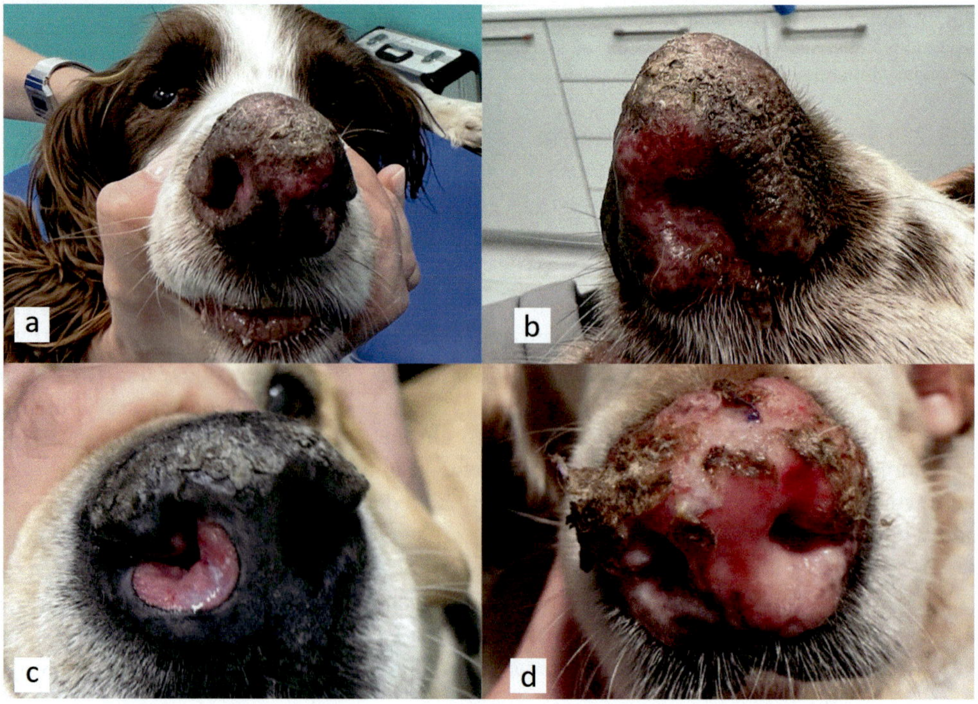

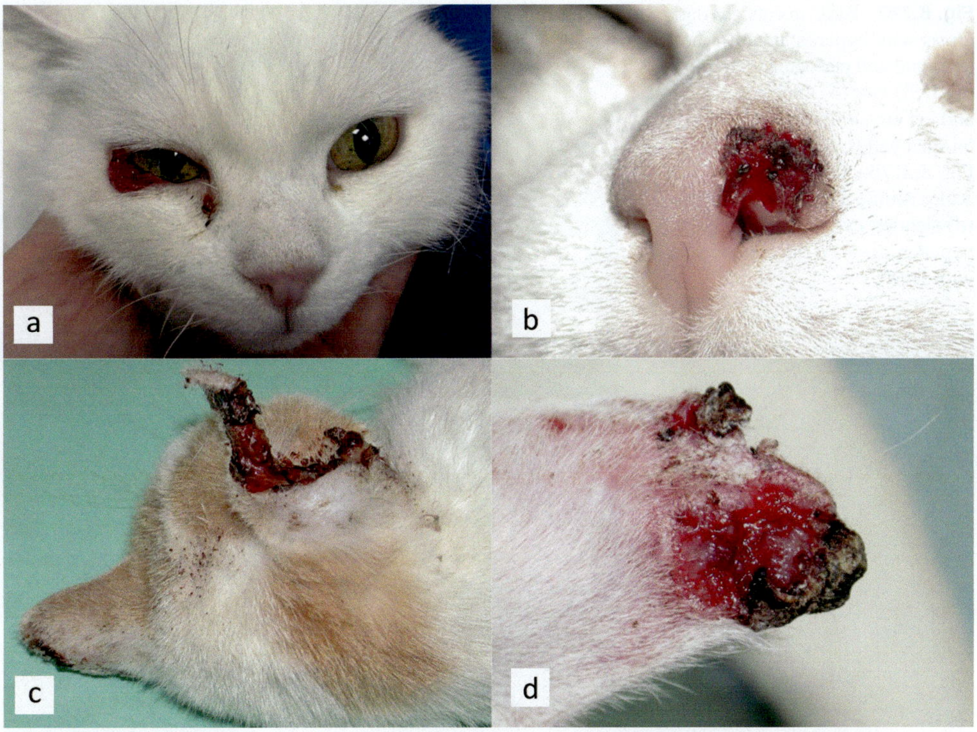

Fig. 8.213 Solar-induced SCC in cats. Ulcerative lesions on the (**a**) eyelid; (**b**) nose; (**c**) and (**d**) pinna

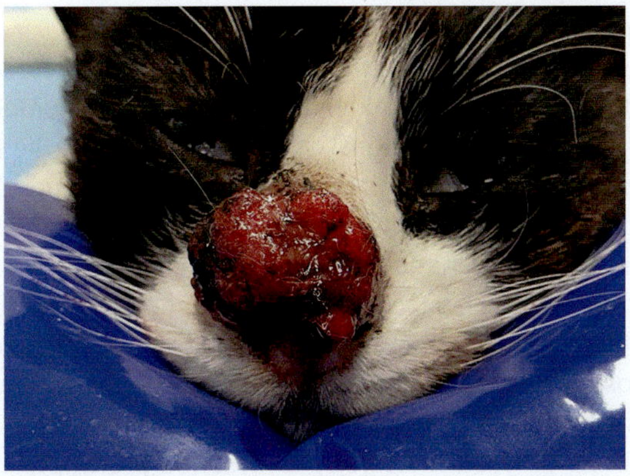

Fig. 8.214 Exophytic and ulcerated SCC of the nose (Courtesy of DVM B. Basta - Italy)

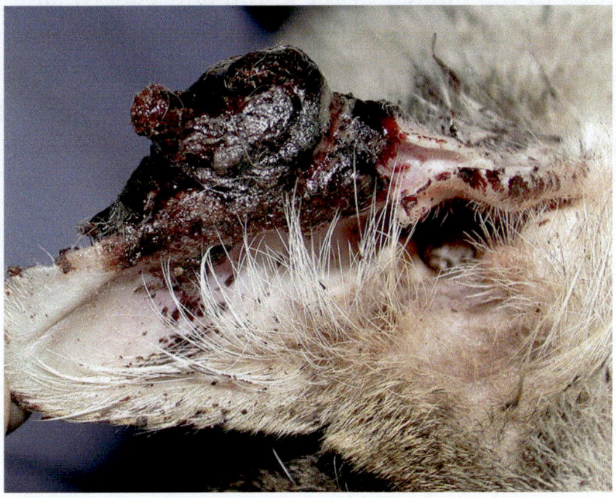

Fig. 8.215 Nodular exophytic SCC covered by hematic crusts on the pinna of a cat

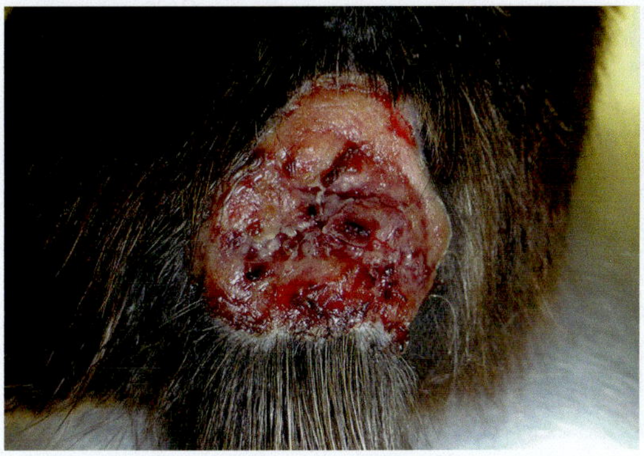

Fig. 8.216 Large ulcerated mass on the sacrum area in an old cat with SCC

appearance, and osteolysis of the third phalanx is also a common finding to all these tumors.

In cats, SCC is frequently observed on the nose, eyelids, and pinna. Even in this species, the white coat is the main predisposing factor. In this species, SCC occurs more frequently with ulcerative lesions that, in advanced stages, cause the loss of large portions of tissue disfiguring the affected areas (Fig. 8.213). In cats, exophytic lesions are much less common than in dogs, but sometimes can be observed (Figs. 8.214, 8.215 and 8.216).

Cases of non-sun-induced SCCs are observed in cats with pigmented skin and are very often secondary to papillomavirus infection and sometimes develop from viral plaque transformation.!In cats with BISC, skin lesions can be single or multiple and are usually characterized by lesions with

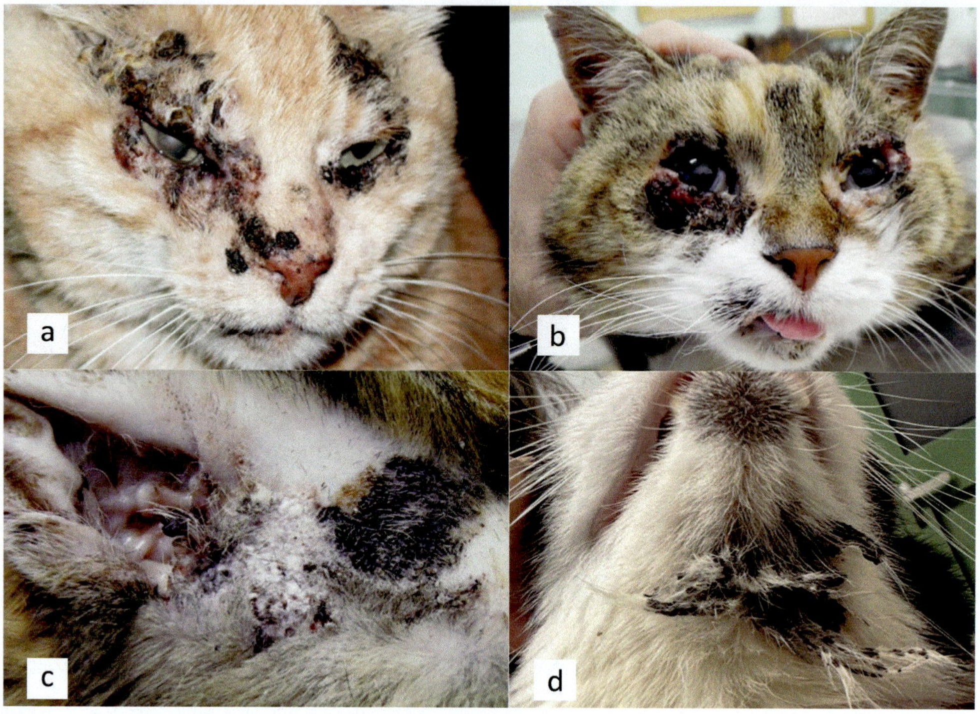

Fig. 8.217 BISC in cats: (**a**) dark crusts with hyperkeratosis in the temporal and periocular skin of an old DSH cat; (**b**) thick ulcerated and covered with hematic crusts on the eyelids; (**c**) same cat in Fig. 8.217b—note the whitish scales mimic a dermatophytosis; (**d**) slender, exophytic, and dark hyperkeratosis on the chin of a cat with BISC

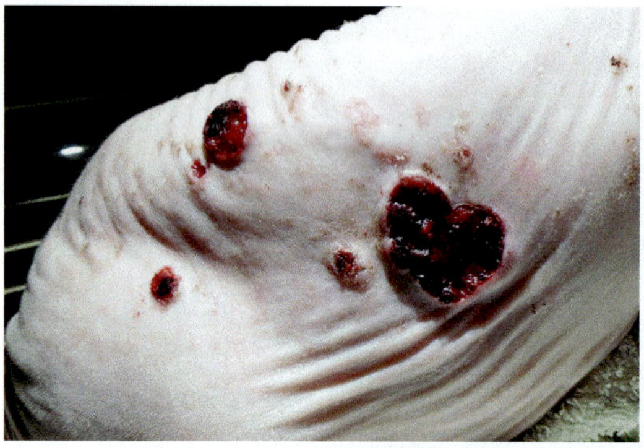

Fig. 8.218 Basal cell carcinoma and squamous cell carcinoma secondary to BISC in a Sphinx cat

different clinical aspects: gray to yellowish scales, multiple plaques, often covered with thick exophytic dark hyperkeratosis and crusts, and thick lesions with corrugate surface; lesions are mainly located on the face (temporal area), muzzle, and pinna but can be observed in any part of the body. For these clinical aspects that mimic a dermatitis more than a neoplastic lesion, they are often confused with inflammatory lesions by most veterinarians (Figs. 8.217 and 8.218).

Some BISC can transform in *basal cell carcinoma*; in these cases the infiltrating growth of the neoplastic epithelial cells in the deep dermis can give origin to the plaque or nodular lesions observed clinically.

Cytological Findings

The knowledge of the morphology of keratinocytes is very important since their cytological identification provides very useful diagnostic indications, as occurs in cases of the follicular or epidermal keratinizing neoplasia. Furthermore, their recognition is very useful in the context of inflammatory lesions, in which epithelial tissues can show morphological alteration that can be confused with neoplastic atypia. To describe morphological changes of keratinocytes that resemble those observed in neoplastic cells, the terms "dysplasia" or "dysplastic cell" are commonly used by cytologists. Since "dysplasia" is used in biology to describe preneoplastic not reversible changes in epithelial tissues (dysplastic tissue), it should not be used to describe the cytological change seen in individual epithelial cells; in fact, the morphological alterations of the nuclei, of the chromatin pattern, and of the cytoplasm which allow a cell to be defined as "dysplastic" are not standardized for single cells (i.e., keratinocytes). So, a term as "atypical cell" maybe would be more indicated to describe a cell that shows cytomorphological features that mimic a neoplasia.

The cytological features of SCC are linked to its grade of differentiation, and different aspects can coexist within the same neoplasia. SCC can be well, mildly, or poor differentiated, and these differences can determine a great variability of cytological features on slides. In most cases, neoplastic cells are clearly differentiated, and for this reason, they are easily interpreted as squamous keratinocytes (Fig. 8.219). Depending on the degree of differentiation, variable malignant features characterize the histopathological samples (Fig. 8.220). In most instances, SCCs are well recognizable as are cytologically mainly characterized by epithelial cells with squamous differentiation; the latter are usually released as discrete cells and with a variable number of small cluster of less differentiated cells. The latter are numerous in poorly differentiated tumors. Although the SCC is an epithelial neoplasia, it is the prototypal epithelial neoplasia that escapes to the cytological rules that want the cells of the epithelia neoplasias must be released in clusters; the squamous cells generally tend to lose cohesivity with others and are released as single cells. The main cytological characteristic of SCC, already detectable when observing slides at low magnifications, is the typical *asynchronous maturation* of squamous cells. The nuclei of the same cells are of different size (anisokaryosis), and marked differences are also evident regarding the ratio between the nucleus and the cytoplasm in the context of cell belonging to the same neoplasia; indeed, a reduction of the size of the nucleus does not always follow an adequate squamous differentiation of cytoplasm. Squamous differentiation is cytologically recognizable when cells show large pale blue (azul) cytoplasm, polygonal to roundish in shape and with glassy appearance; many cells can simultaneously show spindled or irregular cytoplasmic silhouette. The number of clusters of less differentiated epithelial neoplastic cells is scarce and not always present (Figs. 8.221 and 8.222).

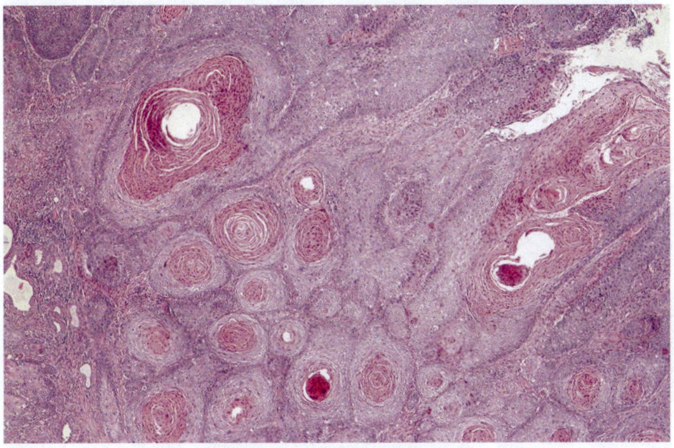

Fig. 8.219 Histology of SCC: multiple nests and trabeculae of well-differentiated neoplastic epithelial squamous cells; many parakeratotic horny pearls are recognizable

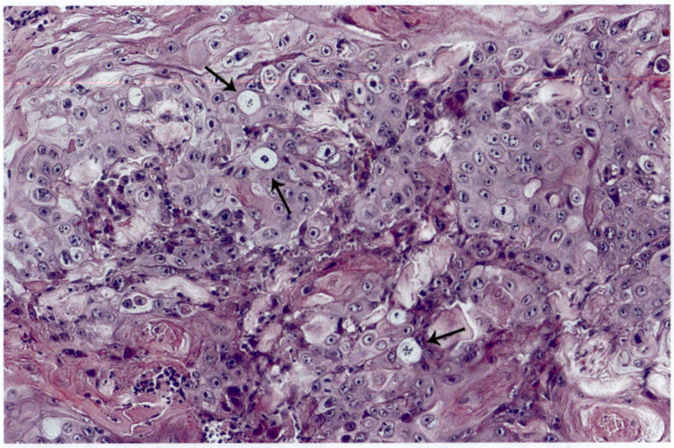

Fig. 8.220 Histology of SCC: numerous malignant cytological features represented by anisocytosis, anisokaryosis, macronuclei with large nucleoli, atypical mitoses (arrows) and dyskeratinization

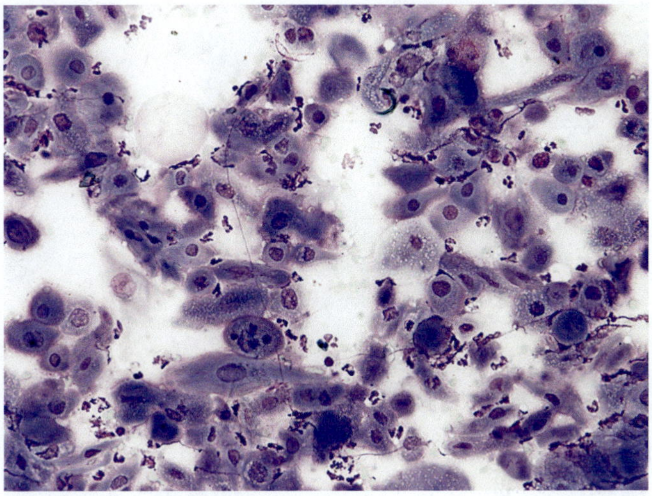

Fig. 8.221 Cytology of SCC. Discrete squamous neoplastic keratinocyte showing marked asynchronous maturation

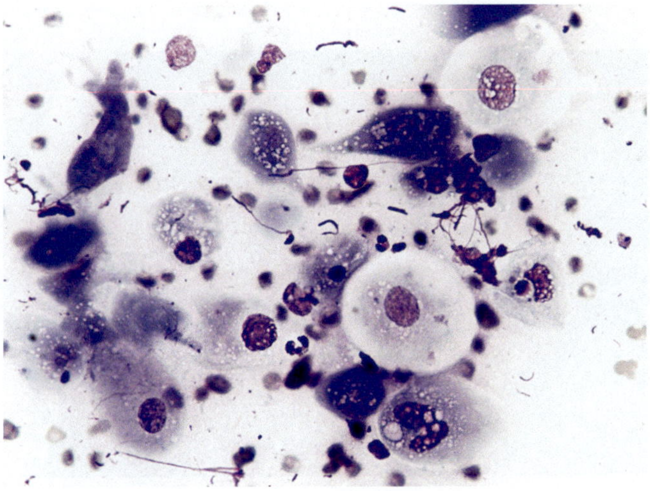

Fig. 8.222 Cytology of SCC. Asynchronous keratinization and atypia in neoplastic keratinocytes

In *mildly* differentiated SCCs, the diagnosis can be less immediate, but usually, along with less differentiated cluster of keratinocytes, a variable number of scattered squamous cells are usually detected (Fig. 8.223). More difficult is the cytological diagnosis of *poor-differentiated* SCC as very few cells exhibit squamous differentiation and slides may present clusters of neoplastic keratinocytes with a high N/C ratio without any obvious squamous differentiation (Figs. 8.224 and 8.225). However, in highly cellular slides, scattered cells with squamous features can be detected (Fig. 8.226). In case no squamous differentiation is detected, the clinical presentation and the site of the lesion can help the cytologists to orient the suspicion. Finally, it must be highlighted as in rare case also other non-epidermal neoplasia (adnexal) can show clear squamous differentiation as occurs in some apocrine tumors and more frequently in infundibular keratinizing acanthoma (IKA), a benign follicular neoplasia, in which, as will be discussed below, some cytological aspects can mimic some hyperkeratotic SCCs.

The cytological features of SCC are represented by a variable entity of anisocytosis and anisokaryosis. Nuclei are from small to very large, sometimes double or multiple, with coarse chromatin and large, multiple, and bizarre nucleoli; mitoses can be numerous. Small rounded nuclear and perinuclear vacuolar-like roundish achromatic areas of unknown origin and features of cannibalism are also commonly observed (Figs. 8.227, 8.228, 8.229 and 8.230).

A rare histological variant, the *spindle cell SCC*, is characterized by a large amount of spindle-shaped neoplastic keratinocytes; when the cytoplasmic tail is single, cells can

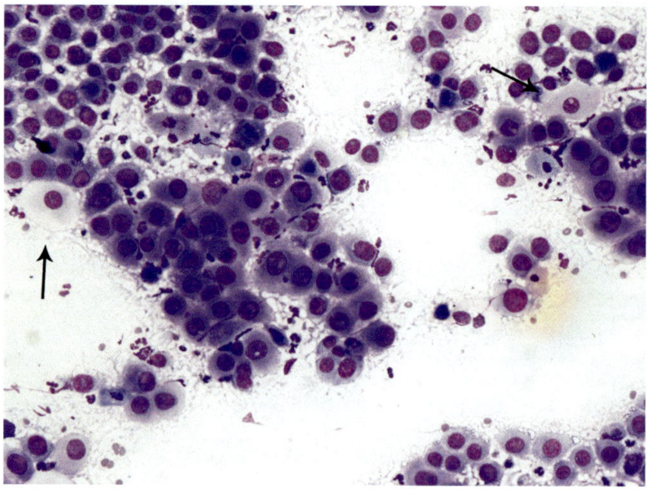

Fig. 8.223 Cytology of SCC. Mildly differentiated neoplastic keratinocytes. Note squamous differentiation is recognizable only in few cells (arrows)

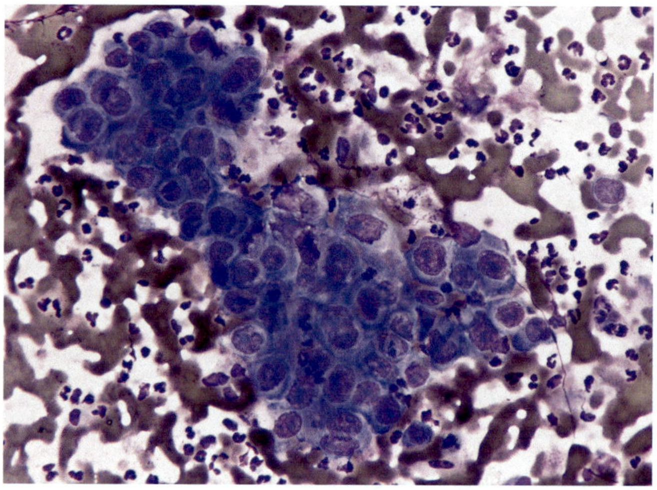

Fig. 8.225 Cytology of SCC. Undifferentiated neoplastic keratinocytes; note no features of squamous differentiation are present

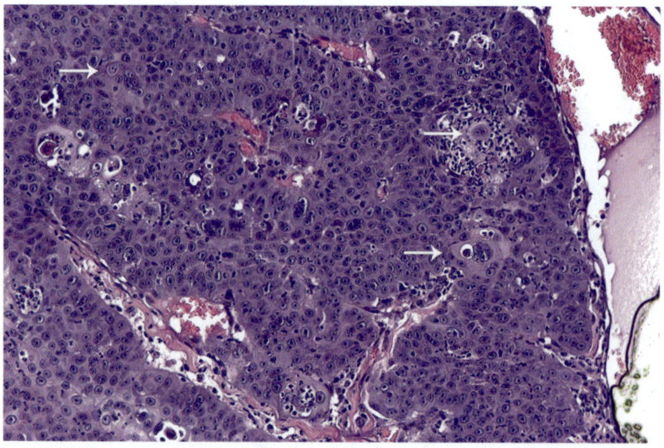

Fig. 8.224 Histology of poorly differentiated SCC. Malignant epithelial proliferation of poorly differentiated keratinocytes, which show many criteria of cytological malignancy such as anisokaryosis, atypical mitoses, chaotic organization, and multinucleation

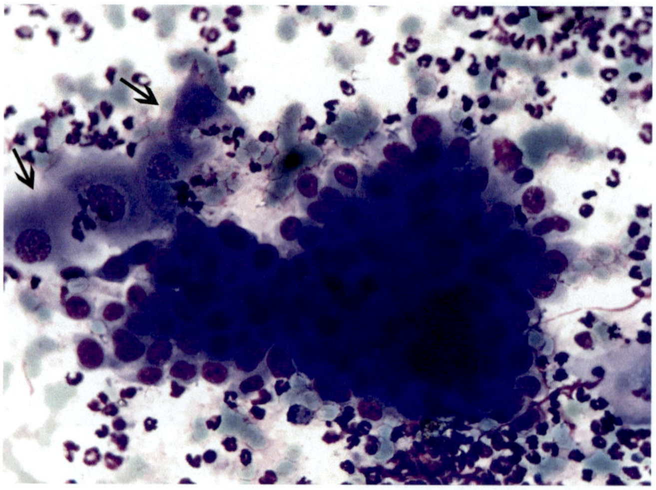

Fig. 8.226 Cytology of SCC. Large cluster of basaloid undifferentiated neoplastic keratinocytes along with few cells with squamous differentiation (arrows)

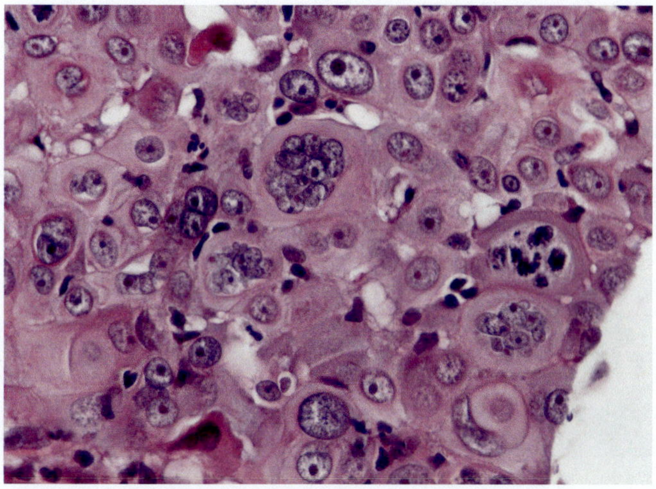

Fig. 8.227 Histology of SCC. Severe aspects of cytological malignancy are simultaneously present. Note the high number of malignant multinucleated cells

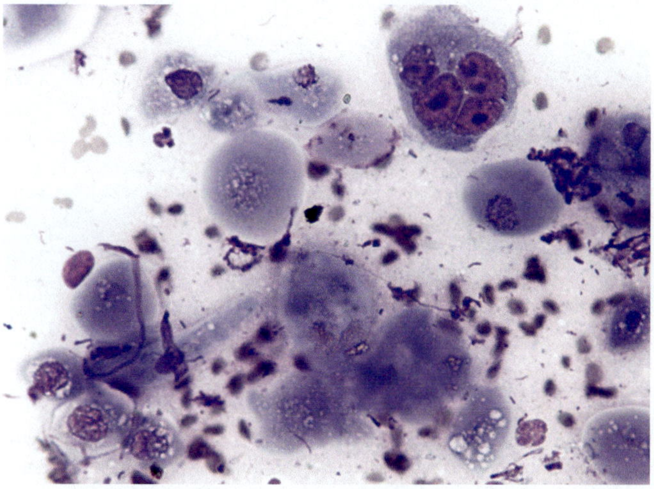

Fig. 8.228 Cytology of SCC. A malignant multinucleated keratinocyte is recognizable

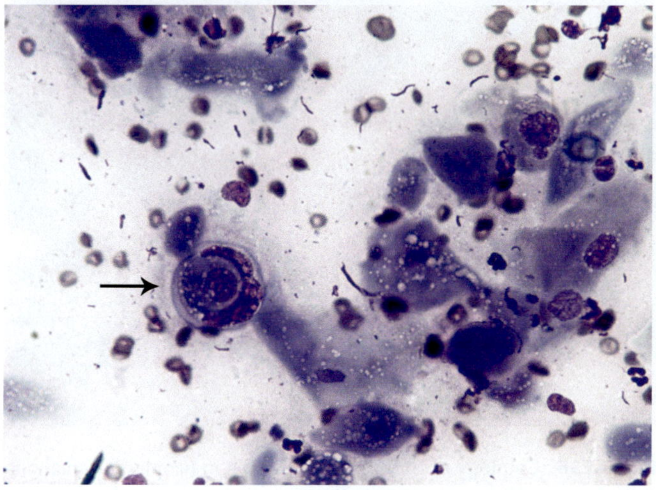

Fig. 8.229 Cytology of SCC (cannibalism). A neoplastic keratinocyte internalized into another neoplastic keratinocyte (arrow)

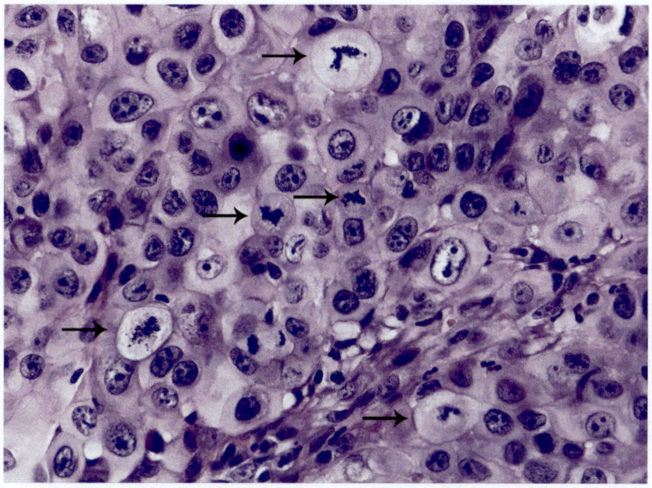

Fig. 8.230 Histology of SCC. Many atypical mitoses (arrows)

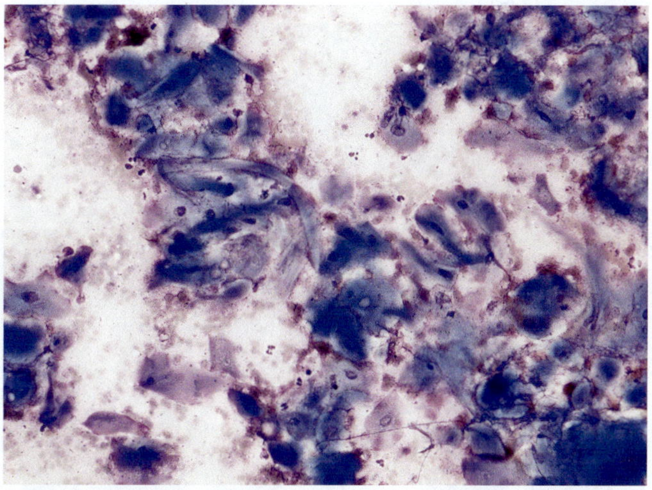

Fig. 8.231 Cytology of spindle SCC. Many atypical mitoses (arrows)

assume a so-called *tadpole* appearance. In this spindle subtype of SCC, the high number of fusiform cells is the main cytological feature, and the latter should not be confused with fibroblasts/myofibroblasts, especially because the stroma in which many SCC growth is very often desmoplastic (Figs. 8.231 and 8.232).

Since in most SCCs many areas of complete squamous differentiation are present, represented by mature lamellar keratin, and as the latter is highly antigenic and irritating, cytopathological samples from SCCs are often accompanied by a secondary predominantly neutrophilic or pyogranulomatous inflammation. This phenomenon must be carefully considered when examining specimens from inflammatory skin diseases where keratinocytes may exhibit moderate to severe cytomorphological alterations, mimicking the atypia observed in SCCs. In these cases, the knowledge of the sampled lesion and its clinical aspect is fundamental to orient the cytological suspicion (Figs. 8.233 and 8.234).

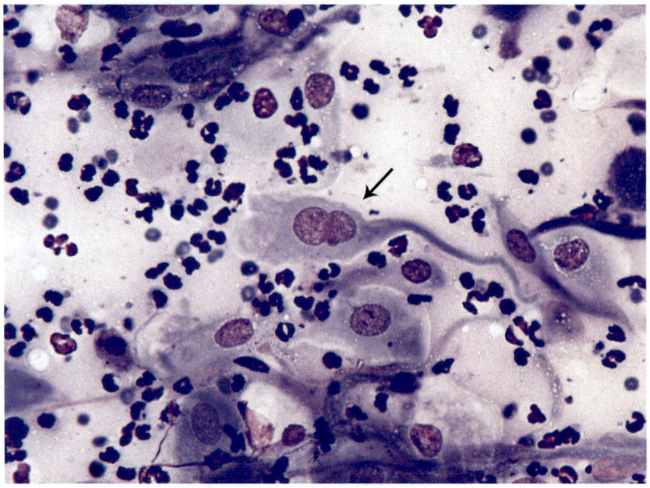

Fig. 8.232 Cytology of spindle SCC. Multiple spindle squamous neoplastic cells. The spindle or *tadpole* appearance of some neoplastic squamous cells should not be confused with mesenchymal cells

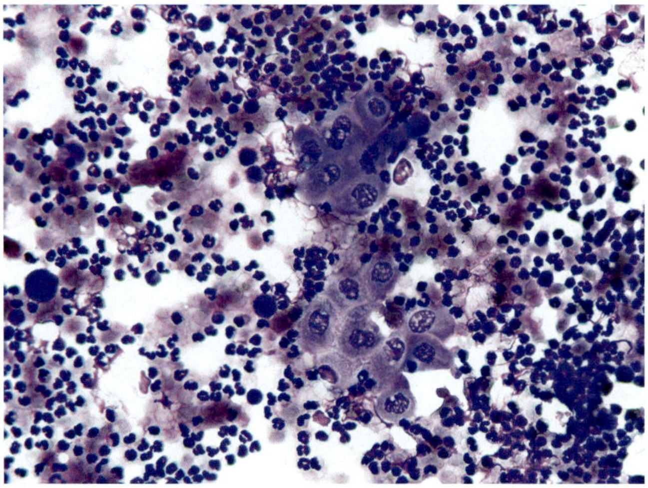

Fig. 8.233 Cytology non-neoplastic squamous cells: irregular chromatin pattern and altered silhouette of nuclei are due to the inflammatory environment in which squamous keratinocytes were located

As aforementioned, some histological variants of SCC such as *verrucous* are not characterized by clear aspects of cytological malignancy and cannot be cytologically differentiated by other benign follicular keratinizing tumors such as IKA or, as discussed for nailbed neoplasias, the nailbed inverted papilloma and nailbed keratoacanthoma (Figs. 8.235 and 8.236).

A common variant of feline SCC secondary to the oncogenic action of papillomavirus, called *Bowenoid* in situ *carcinoma (BISC)*, is well recognized in cats. In this variant of BISC, because the lesions are flattened or slightly raised and covered with thick dark hyperkeratotic crusts, cytological examination is usually not performed. The possibility of collecting cells suggestive of BISC by FNB, in the early stage of neoplasia, is in fact very low because the neoplastic

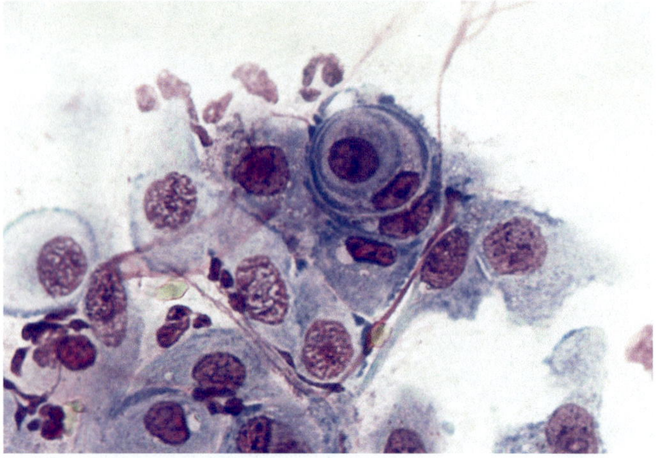

Fig. 8.234 Cytology non-neoplastic squamous cells: a double internalization of non-neoplastic keratinocytes (cannibalism) collected from a chronic deep pyoderma

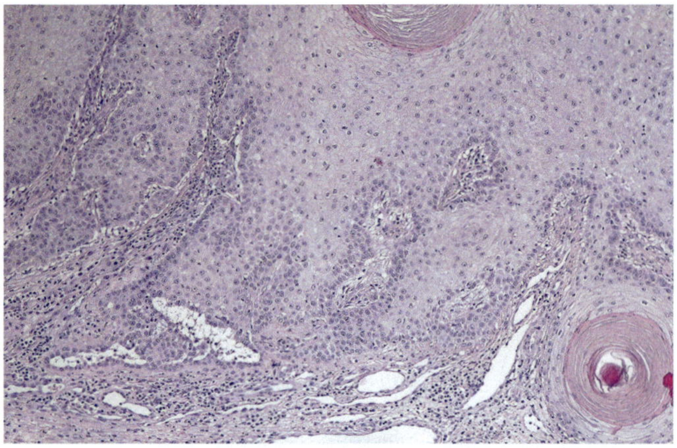

Fig. 8.235 Histology of SCC. Close-up of a SCC with minimal cytological alteration. Cytology collected from these lesions is not differentiable from other benign follicular and epidermal neoplasia

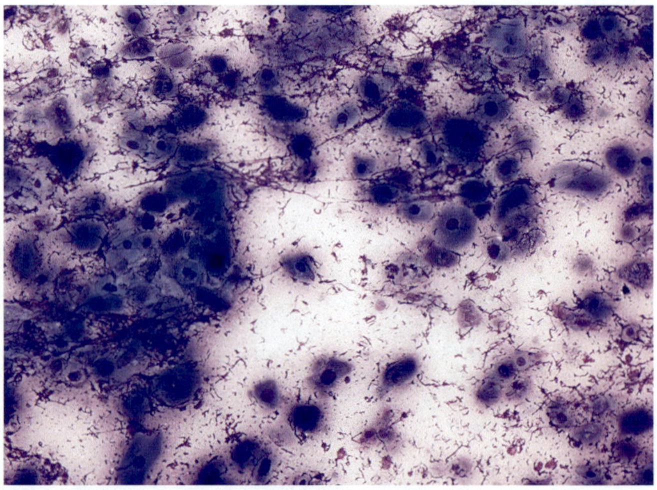

Fig. 8.236 Cytology of SCC. Minimal atypia in neoplastic squamous cells. In some IKAs and some nailbed benign neoplasms, the morphology of cells can be worst than that showed in this image

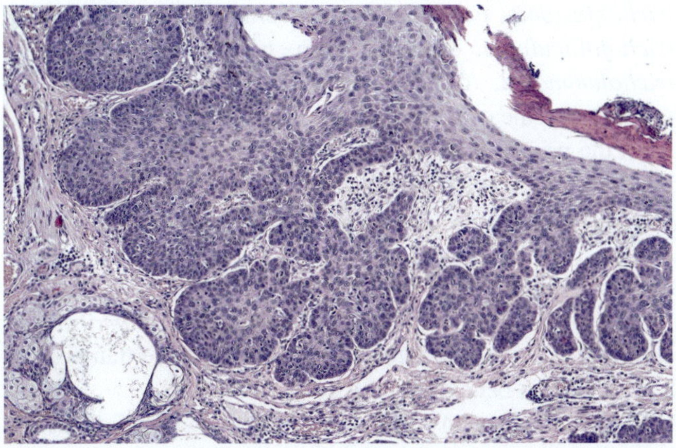

Fig. 8.237 Histology of BISC. Thick epidermis with basal aspects involving the follicular wall and forming irregular trabeculae with round profile bulging to the dermis. Note as atypia of neoplastic keratinocytes are not so marked do not show a squamous differentiation

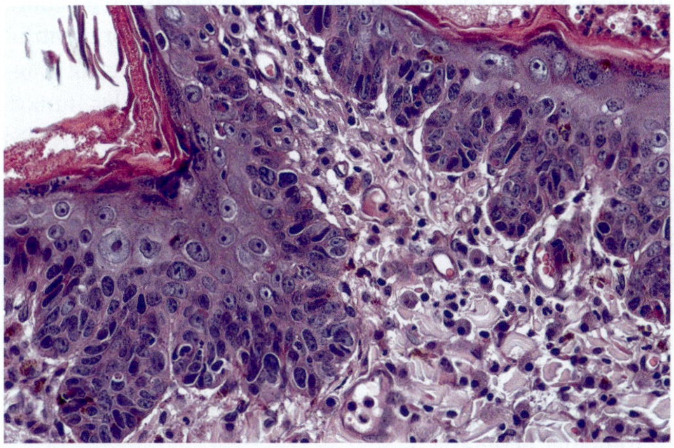

Fig. 8.238 Histology of BISC. Malignant cytological features are represented by loss of basal polarity, crowding of cells, chaotic organization, anisocytosis, anisokaryosis, and evident nucleoli

cells remain confined for a long time in the epidermis without passing through the basement membrane (in situ) (Figs. 8.237 and 8.238). Based on this anatomical feature of BISC, the only possible cytological sampling technique is that of superficial skin scraping that allows to tear from the epidermis, groups of neoplastic keratinocytes. Since in the early stages of BISC the morphological aspects of neoplastic keratinocytes are not so atypical, the basaloid-appearing cells collected must be carefully evaluated since they may not be distinguishable from basal epidermal cells originating from an inflammatory lesion that has determined epidermal hyperplasia; the anamnesis and clinical presentation become decisive in these cases, and, in any case, the definitive diagnosis requires histopathological examination. In advanced stages of BISC, but it is not a constant rule, the cells tend to assume more atypical morphological alterations with cytological

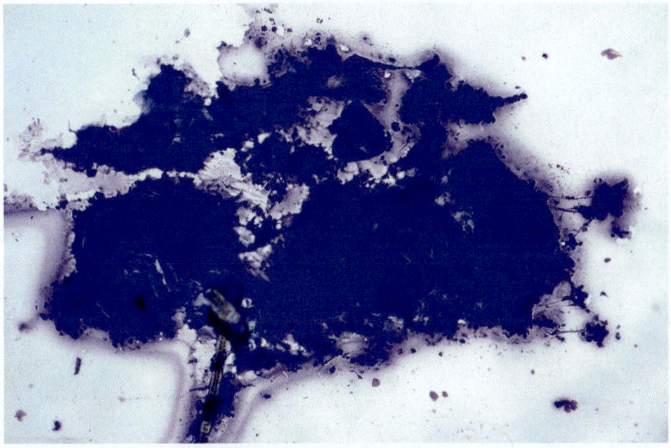

Fig. 8.239 Cytology of BISC. Sampling technique: skin scraping. Large cluster of basaloid neoplastic cells in which no evident malignant cytological aspects are clearly evident

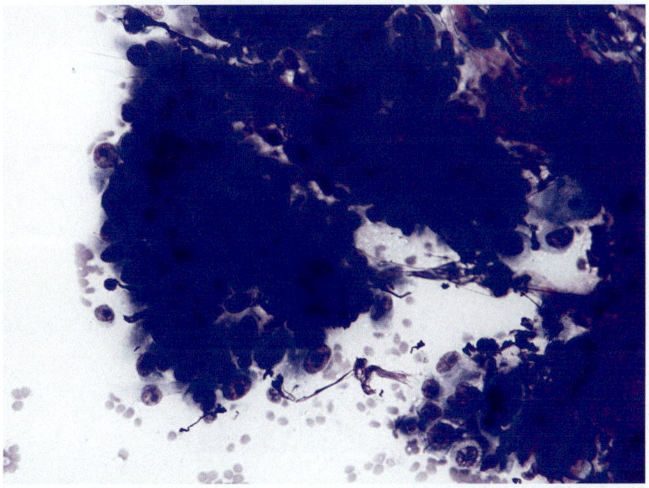

Fig. 8.240 Cytology of BISC. Sampling technique: skin scraping. In this image, crowding of cells with chaotic organization, anisokaryosis, and prominent nucleoli represents malignancy cytological criteria

malignancy features such as anisocytosis anisokaryosis, prominent nucleoli, and high mitotic index (Figs. 8.239, 8.240 and 8.241). Despite being of epidermal origin, a squamous differentiation of neoplastic keratinocytes does not occurs as rule, and focal areas of squamous differentiation can be only occasionally observed in BISC.

In some cases of BISC, epithelial proliferation goes deeper to form nodular lesions or plaques, with denser cellularity and which give rise to histological aspects of basal cell carcinoma; in these cases, being a nodule, it is possible to sample the cells by FNB (Fig. 8.242).

It must be emphasized, however, that even in these chronic cases of BISC with more nodular formations, it is not given that a greater number of cytomorphological features of malignancy are appreciated on the cytological preparations and, if we evaluate the situation only from a

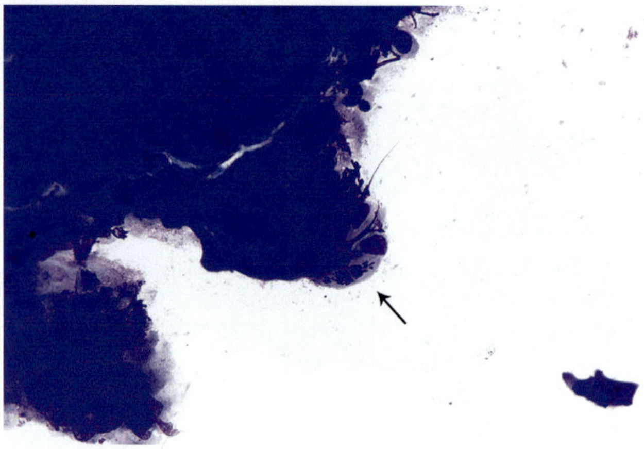

Fig. 8.241 Cytology of BISC. Sampling technique: skin scraping. Cluster of crowding basaloid neoplastic keratinocytes with chaotic organization; note the nucleus in mitosis (arrow)

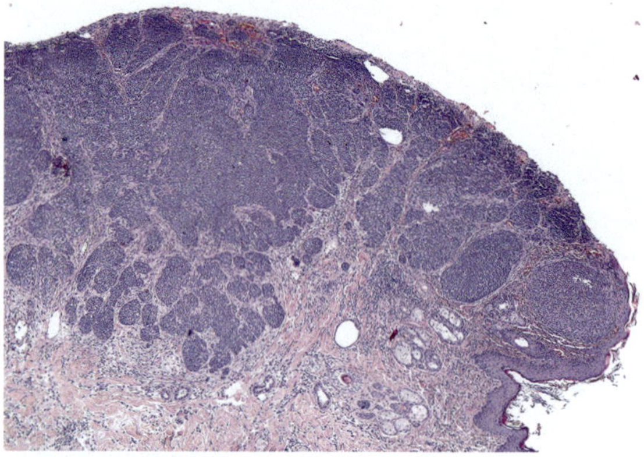

Fig. 8.242 Histology of BISC. Dense basaloid proliferation that bulges to the deep dermis forming a nodule. Other multifocal characteristic lesions of BISC were present on the skin of the affected cat

cytological point of view, could be not cytologically easy to differentiate a nodular BISC from another basal cell neoplasm such as a trichoblastoma or an apocrine tumor. The overall evaluation, which includes the anamnesis and the simultaneous presence of any other neoplastic lesions suggestive of BISC, will guide the diagnosis toward a basal epidermal neoplasia arising on a Bowenoid carcinoma.

8.3.3 Follicular Cysts and Follicular Tumors

The group of follicular lesions is very numerous and includes both cysts and neoplasia [116, 117, 126].

Follicular cysts are classified into *infundibular*, *isthmic*, *matrical*, and *hybrid*. Neoplastic lesions are very numerous and comprise *infundibular keratinizing acanthoma* (IKA), which originates from the infundibulum and isthmus; *trichoblastoma*, derived from the follicular primitive germ; *trichofolliculoma*, derived from trichogenic epithelium; *tricholemmoma*, derived from the isthmic or inferior segments of the external root sheath; *pilomatricoma*, derived from the primitive basal cells of the follicle that differentiate into hair matrix cells; and *trichoepithelioma*, derived from all three parts of the follicle. Other rare cutaneous benign cystic lesions include *dilated pore of Winer* in cats and *Warty dyskeratoma in dogs* [112, 117, 122, 126].

Except for some rare cases of *malignant pilomatricoma* and *malignant trichoepithelioma*, tumors of the hair follicle have a benign behavior and a good prognosis.

Before discussing the cytology of the various follicular tumors, it is essential to briefly refresh some anatomical concepts that concern the structure of the hair follicle and the different types of keratinization associated to the different portions of the follicle itself. The morphological knowledge of follicular keratins is in fact of considerable importance in recognizing the segment of origin of the follicular neoplasia. In most cases, in fact, when veterinary cytologists read slides collected from a follicular cyst or a follicular tumor plentiful of keratin, they usually formulate a generic cytological diagnosis of inclusion cyst or keratinizing follicular tumor. Knowledge of the different morphologies of keratin increases the cytologist's ability to identify, in most cases, from which type of follicular segment the keratin and possibly the associated keratinocytes were taken. Furthermore, it must be aware, because many follicular tumor are heterogeneous as composition and can present simultaneously both solid and cystic areas, cytology can provide highly variable findings based on the area in which the needle is inserted; for this reason the results can be not constant also for the same group of follicular neoplasia.

The hair follicle is a very complex structure that can be anatomically divided into three segments: the *infundibulum*, the *isthmus*, and the *bulb*. These three different parts are composed of keratinocytes, which undergo very different keratinization processes. The *infundibulum* is the upper portion of the follicle: it extends from the external opening (follicular ostium) to the outlet of the sebaceous gland, and it is characterized by the same type of keratinization present on external epidermis (*epidermal keratinization*), in which the production of lamellar keratin (scales) passes through the stratum granulosum rich in keratohyalin granules. The *isthmus*, which is delimited at the top by the sebaceous gland duct and at the bottom by the insertion of the arrector pili muscle, represents the middle portion. In this area the granular layer is missing, and there is therefore a so-called trichilemmal keratinization; the keratin produced is amorphous and eosinophilic on routine histological stains. The third and deeper follicular portion, which begins at the bottom of the insertion point of the arrector pili muscle, is the *bulb*. In this follicular portion, the cells that reside in the

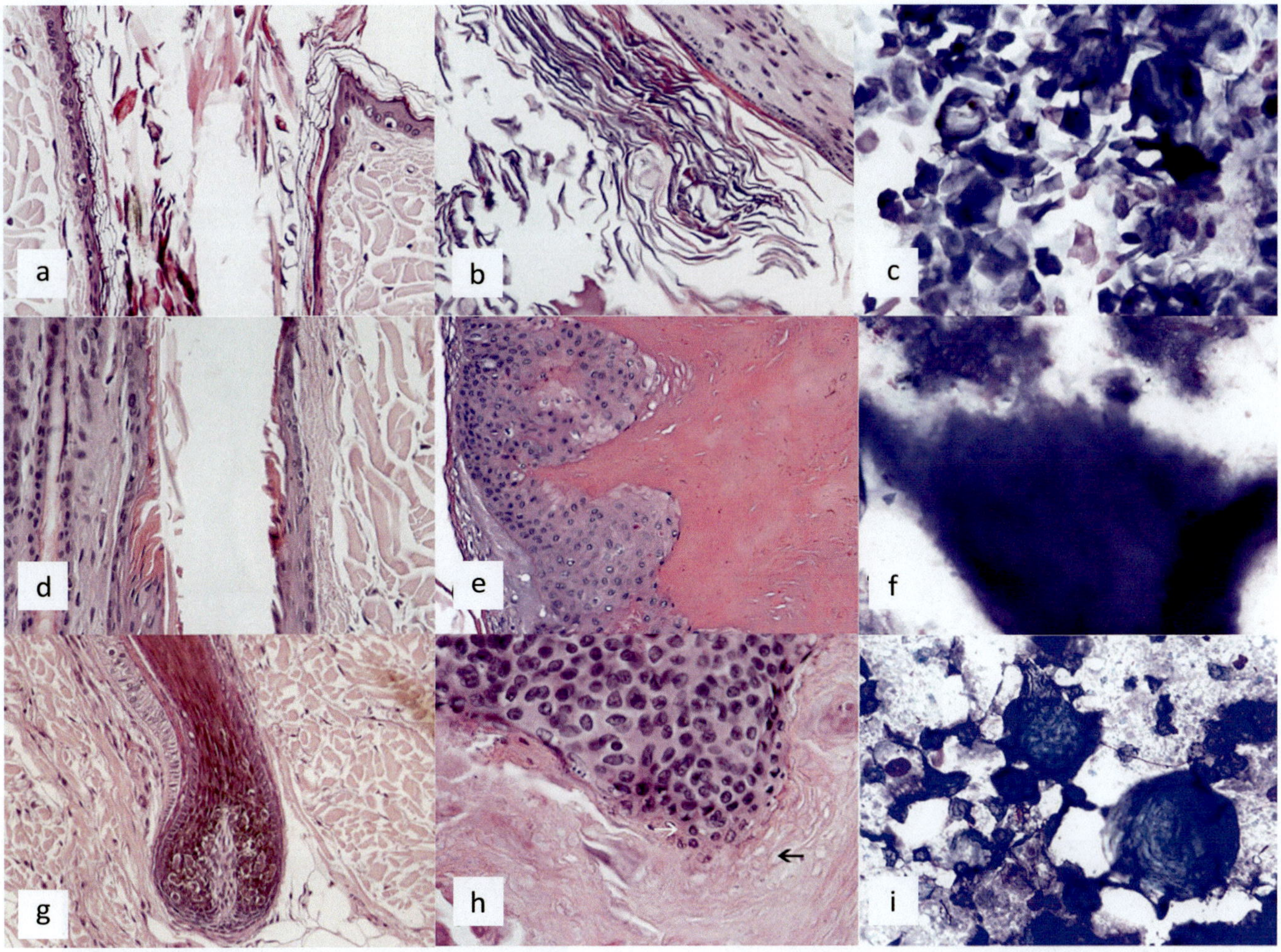

Fig. 8.243 Anatomy and different typology of keratinization of the hair follicle: (**a**) Histology: infundibulum. (**b**) Histology: note as keratinization, passing through the stratum granulosum, origins lamella keratin (infundibular cyst). (**c**) Cytology: lamellar corneocytes are the only present cells (infundibular cyst). (**d**) Histology: isthmus. (**e**) Histology: trichilemmal keratin. Note the epidermis is lacking of granulosum layer (trichilemmal cyst). (**f**) Histology: dense deep blue, amorphous keratin material (trichilemmal cyst). (**g**) Histology: hair bulb in anagen phase. (**h**) Histology: abrupt keratinization of matrical basal cells (white arrow) in ghost keratin cells (black arrow) (matrical cyst). (**i**) Cytology: ghost keratin cells both in cluster and singularly dispersed on the bottom (matrical cyst)

matrix produce both the keratin of the hair and the cells of the root sheath. The type of keratinization that occurs in this follicular area is called *matrical keratinization* and is characterized by the production of keratinized cells that conserve the cellular outline of nuclei forming an optically empty central area; because of this cytological aspect, these cells are defined as *ghost cells* (Fig. 8.243).

For an exhaustive description of the cytological aspects of follicular lesions, the latter are discussed singularly, highlighting their main microscopic features.

8.3.4 Follicular Cysts

In most cases, benign follicular cysts are infundibular and represented as dermal single nodular lesions mainly localized to the dorsal trunk and tail, but they can be observed in any part of the body (Fig. 8.244). Less commonly, they can be multiple; breeds such as German Shepherd, Shih Tzu, and Pekingese are predisposed to multiple infundibular cysts. Also in cats, multiple infundibular cysts can be observed, especially in Sphinx cats [114, 121, 124, 125] (Figs. 8.245 and 8.246). A clinical variant presents since the birth and mainly localized on the head area; these lesions, named *dermoid cysts*, are anatomically characterized by multiple and confluent infundibular cysts with a clinical aspect of plaque with irregular surface. Tufts of hairs can emerge from superficial openings.

In "hairless" breeds, such as Chinese crested dog, small multiple infundibular cysts are normally observed since the genetically researched follicular dysplasia of this breed; comedones are normally present on the skin of these cats, and in

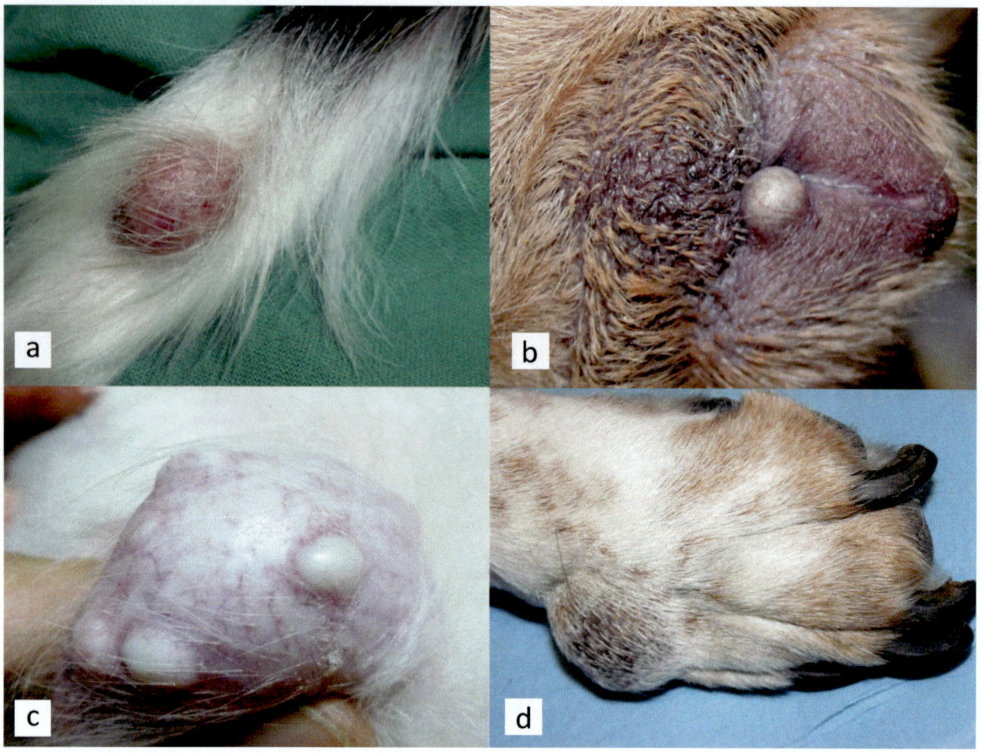

Fig. 8.244 Follicular cysts: (**a**) alopecic and erythematous infundibular cyst on the tail; (**b**) single small infundibular cyst on the vulva of a bitch; (**c**) small whitish infundibular cysts on the skin of the scrotum; (**d**) hybrid cyst on the extremity of a Basset Hound

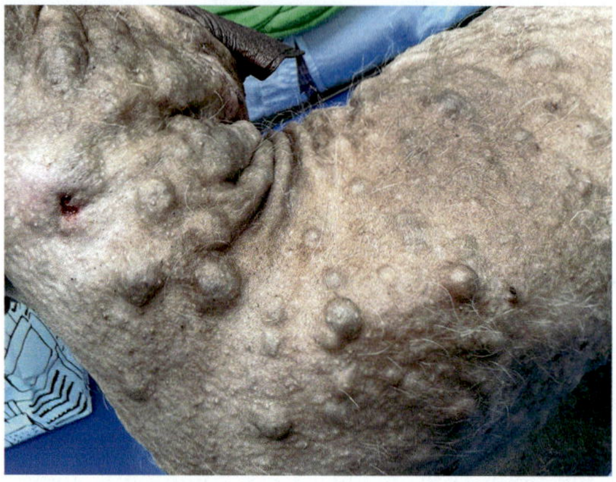

Fig. 8.245 Multiple follicular infundibular cysts in a Sphynx cat

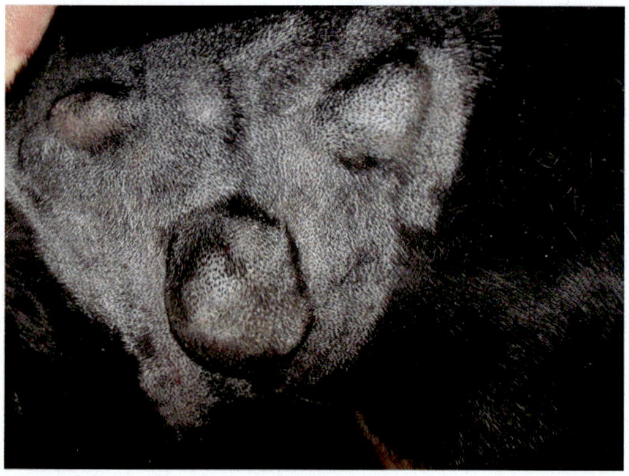

Fig. 8.246 Multiple large follicular infundibular cysts in a Dobermann

some cases, the follicular lumen filled of keratin enlarges so much to give origin to small infundibular cysts (milia).

A similar clinical situation, less severe and more localized, can occur in dogs suffering from hypercortisolism; in these dogs, the atrophy of the epidermis and of follicular wall caused by the chronic action of cortisol is the cause of large comedone development and their secondary transformation in infundibular follicular cysts, especially in glabrous and more thin skin area, such as the axilla and ventral abdomen.

The skin which covers follicular cysts is usually intact, alopecic, or normally haired and sometimes hyperpigmented. The breaking of the cystic wall causes the spreading of keratin into the dermis, resulting in a severe pyogranulomatous reaction and discharging of dense, yellowish, grainy, of a creamy appearance, keratinous material mixed with blood and inflammatory cells (Fig. 8.247).

These macroscopic aspects are also commonly observed in keratinizing follicular neoplasms; therefore, its observation can only indicate the presence of a nodule containing keratin material but cannot differentiate between a cyst and a keratinizing follicular neoplasia. In other cases, the contents of the cyst may be dark and have a more fluid consistency.

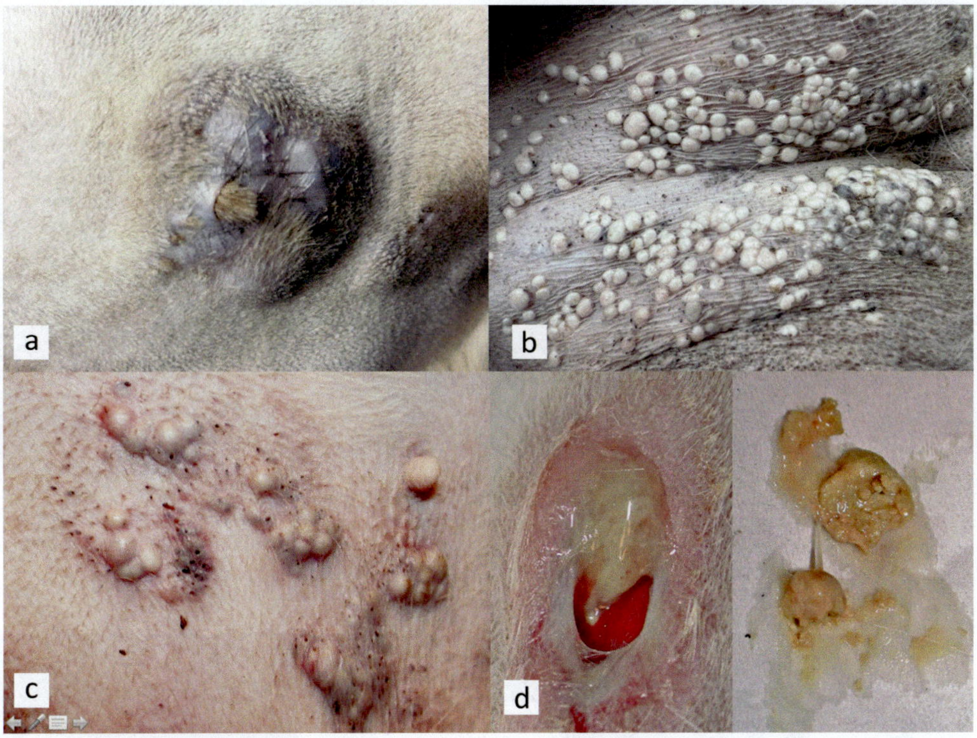

Fig. 8.247 Infundibular cysts (**a**) dermoid cysts on the head of a young dog: note tufts of hairs go out from surface openings. (**b**) countless small infundibular cyst on the neck of a crested Chinese dog; (**c**) small whitish infundibular cysts (milia) and comedones on the axilla of a dog with spontaneous hypercortisolism; (**d**) follicular infundibular cyst with broken wall —keratin material has dense and grainy appearance

Another characteristic, very indicative of a keratinizing lesion, is the difficulty encountered in transferring the collected material on the slide; in fact, since the content is very dense in consistency, the lumen of the needle is usually obstructed, and a strong pressure on the plunger of the syringe is necessary to deposit the dense, grainy, and greasy in appearance material on the slide.

Even follicular neoplasms are clinically indistinguishable and are presented as individual nodules of variable size, from a few to up 10 centimeters, sometimes ulcerated and secondarily inflamed due to the spreading of the keratin in the dermis.

Cytological Findings

- *Infundibular Cysts*

Histology of infundibular cyst is characterized by a dermal cyst filled with keratin and lined with a squamous epithelium comprising the granular layer (Figs. 8.248 and 8.249). Cytological specimens are usually highly cellular and composed by anucleate lamellar corneocytes with a polygonal profile and by a varying blue/azul tonality of colors with Romanowsky-type dye; this different affinity with the stain is maybe linked both to the level of keratinization and to the presence of lipid material that normally surrounds the cells (lipid envelope) and that, if not properly dissolved by the alcoholic fixative, can cause inadequate penetration of the stains into the cells (Fig. 8.250). In pigmented animals also the corneocytes that fill the cyst can be variably pigmented

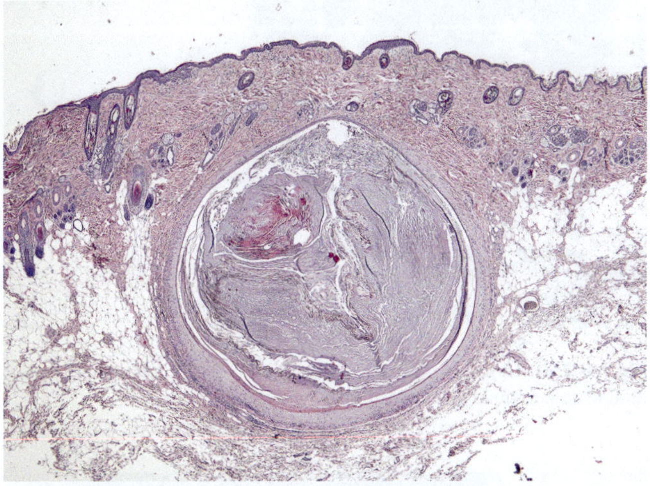

Fig. 8.248 Histopathology of an infundibular cyst. Large cyst filled with lamellar corneocytes

(Fig. 8.251). In dermoid cyst, along with typical corneocytes, fragments of hair shafts can be collected since in these congenital follicular lesions the cystic cavity is plentiful of hairs (Fig. 8.252). The alcoholic fixative can cause the dissolution of the cholesterol present in the lipid envelope, with formation of crystals represented by transparent and geometric formations of variable size (rectangular or polygonal negative image). Cysts that contain keratin a more fluid content may cytologically show an amorphous and intensely blue background in which, usually, many cholesterol crystals are associated, the latter also with amorphous, grainy not

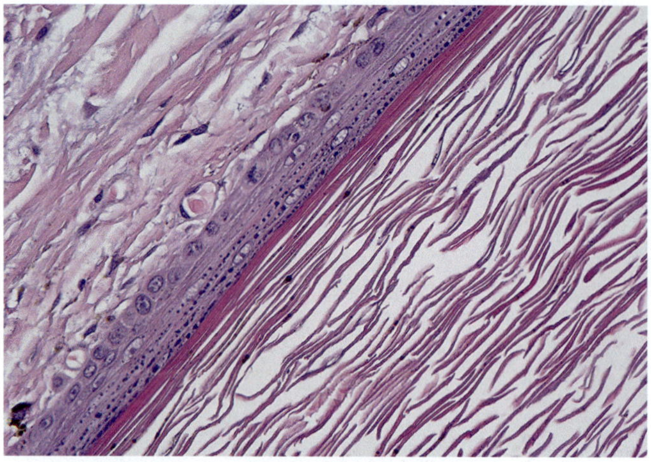

Fig. 8.249 Histopathology of an infundibular cyst. Close-up of the epidermal wall of the infundibular cyst where the typical way of squamous keratinization is clearly appreciable. Note that the keratin is produced via the granular layer

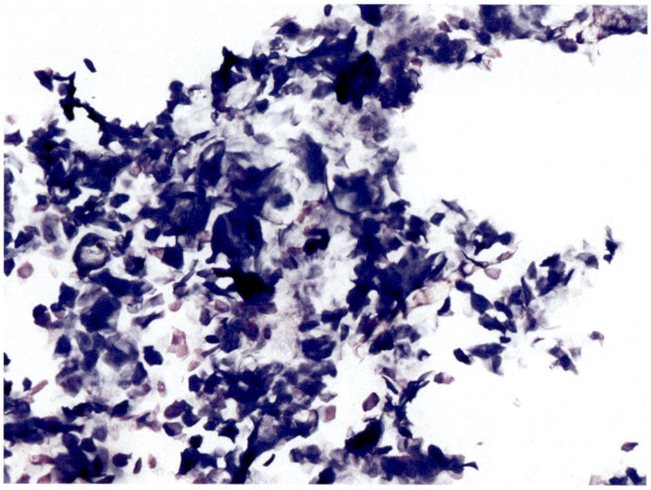

Fig. 8.250 Cytology of an infundibular cyst. Many polygonal anucleate keratinocytes stained blue, violet, and red

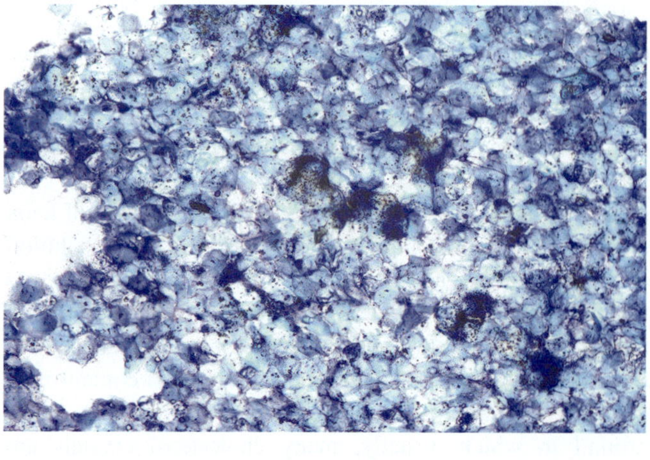

Fig. 8.251 Cytology of an infundibular cyst. Many polygonal anucleate keratinocytes stained pale blue and with variable amount of melanin

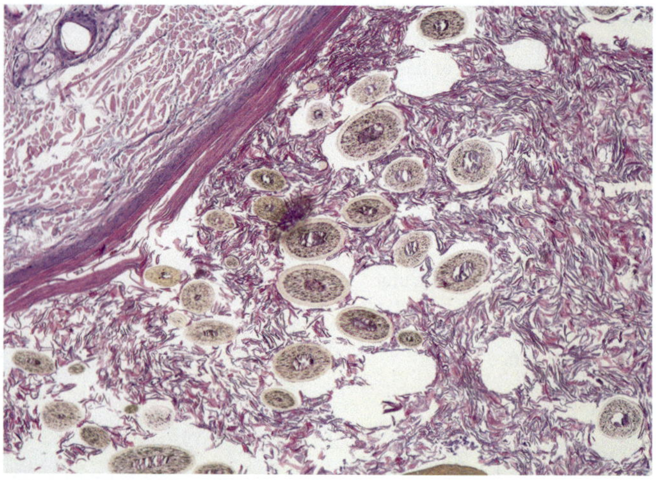

Fig. 8.252 Histopathology of a dermoid cyst. Many hair shafts are present in the lumen of follicular cyst

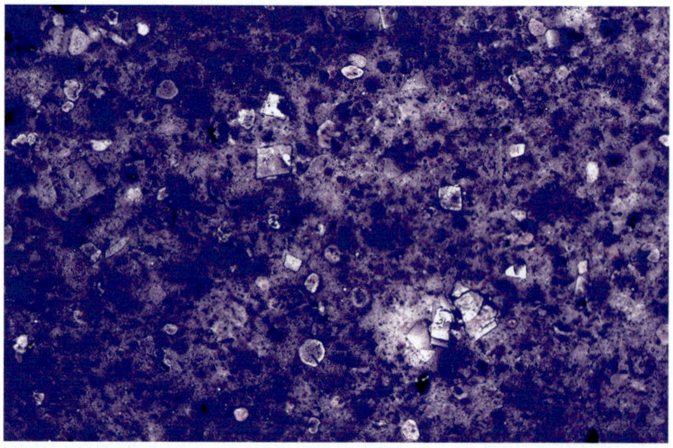

Fig. 8.253 Cytology of an infundibular cyst. Deep blue material on background; no keratinocytes are clearly recognizable. Many unstained negative images, with geometrical silhouette, represent cholesterol crystals

geometrical appearance (Figs. 8.253 and 8.254). It must be stressed that when the content is very fluid and no many corneocytes are present, it is very hard to differentiate a fluid infundibular cysts from the content of an apocrine lesion (sweat gland), in which the secretion can show a similar cytological appearance; furthermore, the presence of cholesterol crystals doesn't rule out apocrine lesions since crystals can originate from the ductal wall. Finally, due to excessive growth of the cyst, the wall can break, and part of the keratin material can spread into the dermis; since keratin is present only in hairs, scales, and nails, and all these structures are external, keratin is not in contact with the skin's immune system. When keratin exits the cyst and is in contact with the dermis, it is very irritating and antigenic, evoking a strong pyogranulomatous inflammation composed of neutrophils, macrophages, epithelioid macrophages, and multinucleated macrophagic cells (Fig. 8.255). These cytological findings

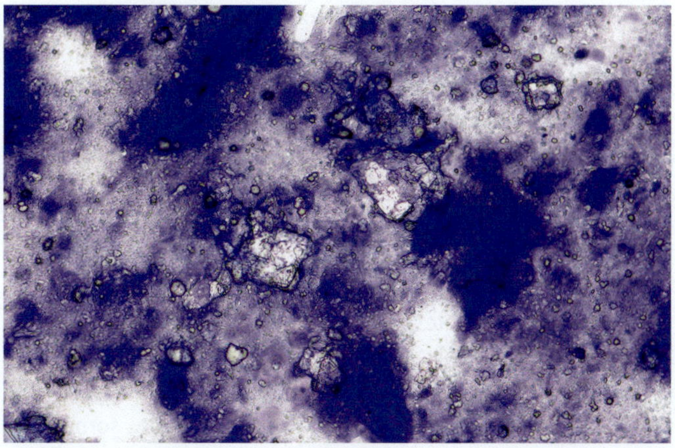

Fig. 8.254 Cytology of an infundibular cyst. More liquid keratin material in which corneocytes are barely visible and in which both geometrical and amorphous cholesterol crystals are recognizable, the latter maybe secondary to the dissolution of geometrical ones

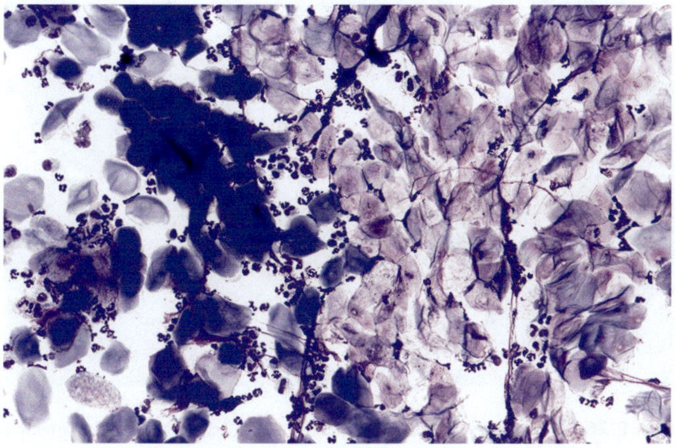

Fig. 8.255 Cytology of an inflamed infundibular cyst. Many polygonal keratinocytes along with neutrophils and macrophages

are commonly seen in all epidermal and follicular keratinizing lesions where keratin comes in contact with the dermis.

Isthmic Cysts

Isthmic cysts originate from the outer root sheath of the follicle, and for this reason, they are characterized by trichilemmal keratinization. At isthmic level indeed, the granular layer is not present, so the keratin produced ex abrupt from nucleated epithelium is not lamellar but amorphous, pale, and more homogeneous. The pure isthmic-trichilemmal cyst is very rare, and it is mainly observed in Siberian Husky (Figs. 8.256 and 8.257). Cytology shows amorphous dense keratin material that stains deep blue (Fig. 8.258).

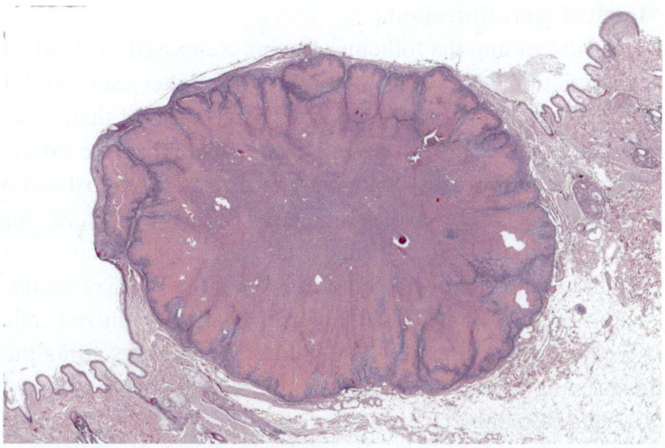

Fig. 8.256 Histopathology of an isthmic cyst. Dermal cysts filled with brightly pink trichilemmal keratin

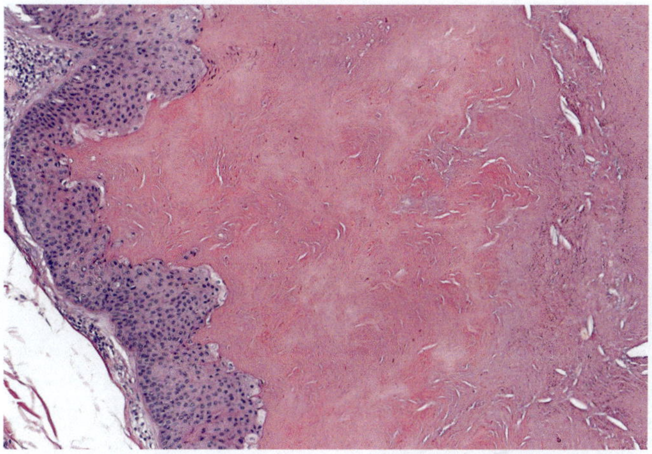

Fig. 8.257 Histopathology of an isthmic cyst. At high magnification amorphous trichilemmal keratin abruptly produced from the follicular wall is recognizable. Note that the granular layer is not present

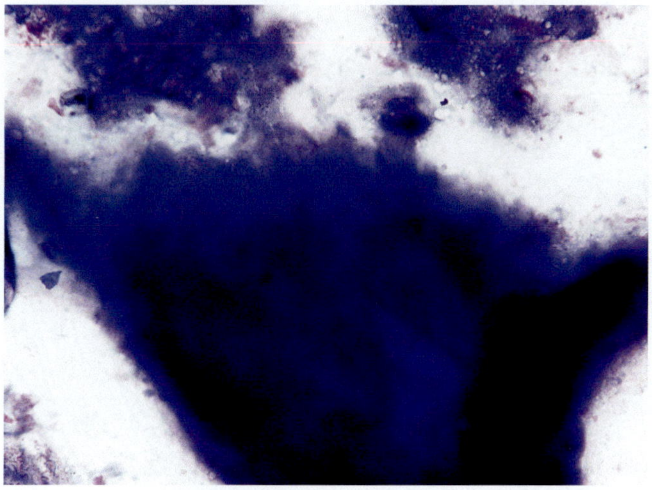

Fig. 8.258 Cytology of an isthmic cyst. Trichilemmal keratinization is characterized by amorphous basophilic keratinous material

Matrical Keratinization

Before discussing the follicular lesions composed of matrical differentiation, the *ghost cells*, since the latter can exhibit different cytological aspects, a brief description of their morphology is necessary. Knowing how to recognize the different morphologies with which hair matrix keratin (ghost cells) can manifest itself is of great help to the cytologist for diagnosing the different follicular tumors. The most common presentation of matrical keratin is as a variable in size roundish to oval or elongated cluster of achromatic keratinized cells in which a round central well-delimited area, representing the pre-existing nuclear membrane, is observed; for this central area that is barely viewable and that represents the nucleus that is no more present, the keratinized cell is defined as *ghost cell* (Figs. 8.259, 8.260, 8.261 and 8.262). When the latter are not so cohesive or clusters are broken during slide preparation, ghost cells can be released singularly appearing as small polygonal cells with sharp angles and sometimes with a spindle to starry appearance (Figs. 8.263 and 8.264). In some cases, the outline of the area where the nuclei were present is highlighted, and this happens even when the ghost cells are very pigmented; furthermore, since the ghost cells are usually organized in a regular order, the arrangement of the cells resembles the spots of the fur of a large feline (*leopard skin* appearance). This curious cytological aspect is of great help in interpreting the matrical keratin when the silhouette of ghost cells is not clearly recognizable (Figs. 8.265 and 8.266). Finally, matrical ghost cells can assume a linear to trabecular and slender shape, sometimes with sharped edges, both achromatic or hyperpigmented; in these cases, especially when they are strongly pigmented,

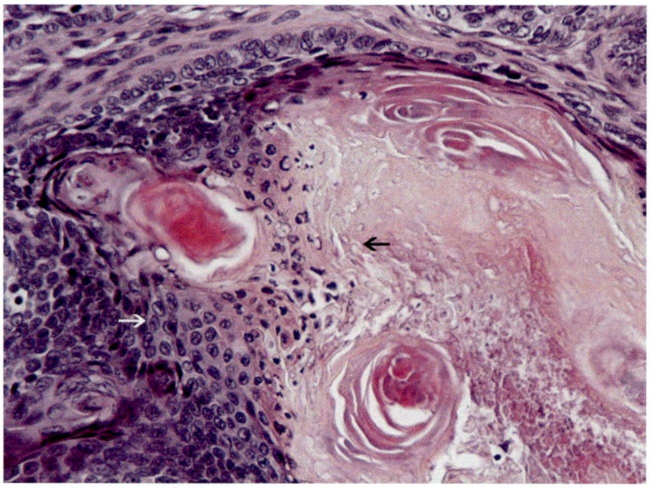

Fig. 8.259 Histology of a matrical cyst: abrupt keratinization of matrical basal cells (white arrow) in ghost keratin cells (black arrow)

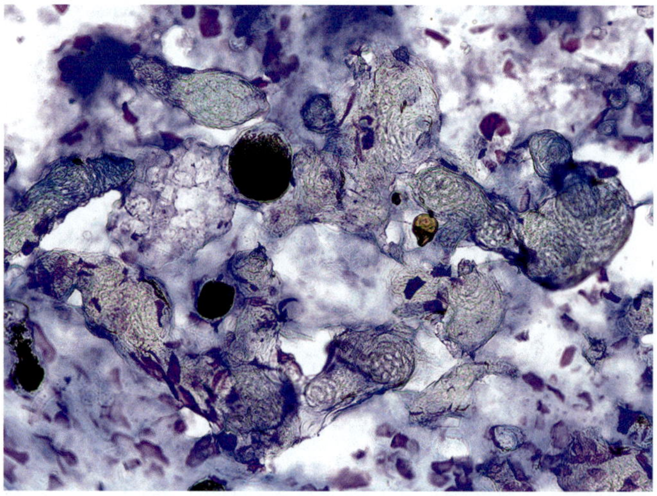

Fig. 8.261 Cytology of ghost cells: the roundish to trabecular groups of apparently amorphous achromatic material represent groups of matrical ghost cells

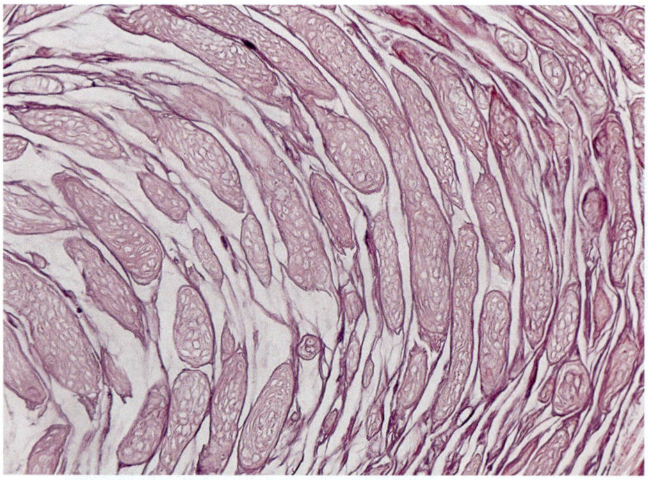

Fig. 8.260 Histology of ghost cells: matrical ghost cells arranged in oval or elongated groups

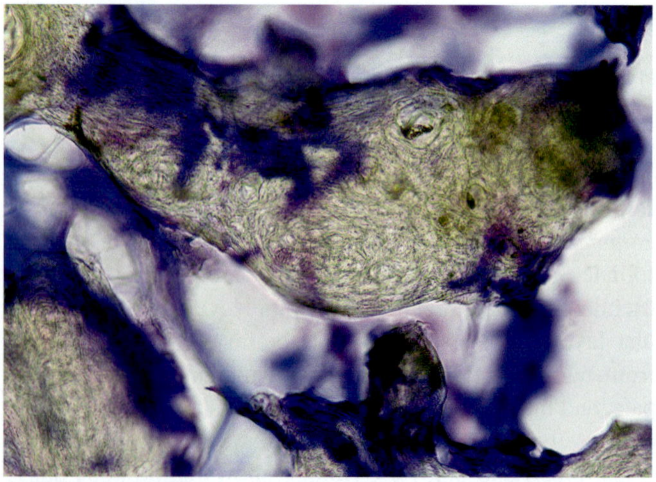

Fig. 8.262 Cytology of ghost cells: at high magnification the silhouette of matrical ghost cells is recognizable

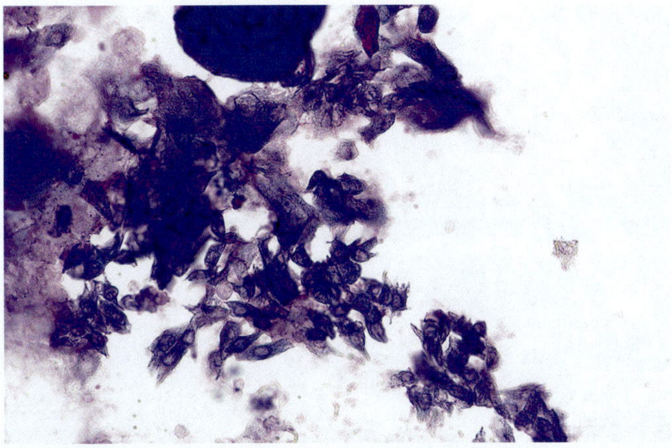

Fig. 8.263 Cytology of ghost cells: single ghost cells after the breakage of clusters of matrical keratin during slide preparation

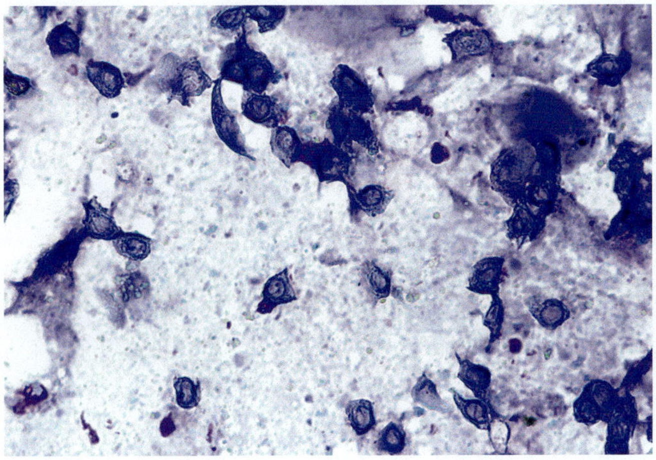

Fig. 8.264 Cytology of ghost cells: single ghost cells show a typical polygonal to spindle shape

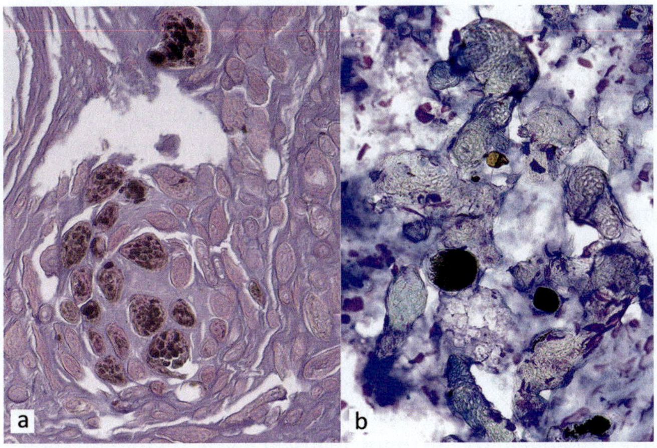

Fig. 8.265 Both achromatic and pigmented groups of ghost cells: (**a**) histology—note pigmented cluster as a leopard-skin appearance; (**b**) cytology—achromatic and pigmented groups of ghost cells

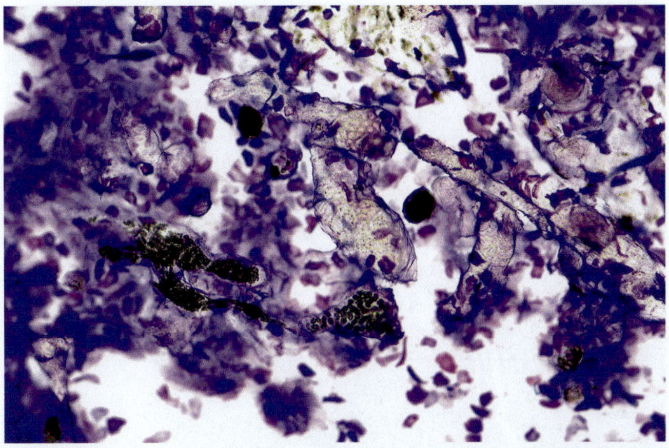

Fig. 8.266 Cytology of ghost cells. Both achromatic and pigmented groups of ghost cells along with lamellar keratin

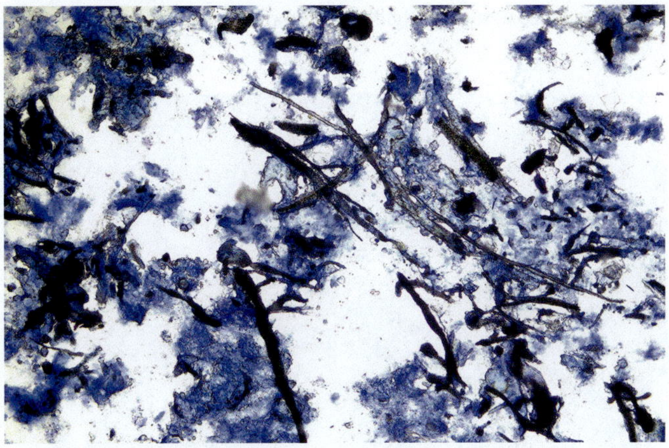

Fig. 8.267 Cytology of ghost cells. Thin and long, slender linear structures representing keratin in matrical differentiation

ghost cells are not always easily recognizable on cytology, but looking them at high magnification and playing with the focus of the microscope, their morphology can be highlighted. These structures could represent abortive trichogenesis (Figs. 8.267, 8.268 and 8.269).

Ghost cells are not pathognomonic of a particular follicular lesion, since they can be detected on slides coming from *matrical cysts*, *hybrid cysts*, *trichoepitheliomas*, and especially *pilomatricomas*; all these lesions are in fact characterized by matrical keratinization and can exhibit a variable amount of matrical ghost cells in cytological specimens. In practice, they can be present in all the follicular diseases in which the matrical basal epithelium is involved. Fortunately, with a little bit experience, other cytological features, as discussed below, can help cytologists to differentiate between follicular lesions.

The matrical ghost cells are only a few microns in size, and care must be taken not to confuse them with corneocytes that are observed during orthokeratotic lamellar

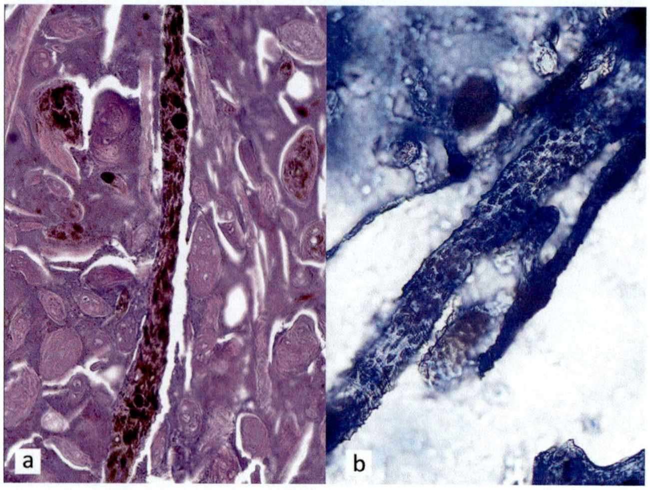

Fig. 8.268 Linear matrical keratin: (**a**) histology in which both achromatic and pigmented ghost cells are present; (**b**) cytology—the leopard-skin appearance allows to highlight the former nuclear areas

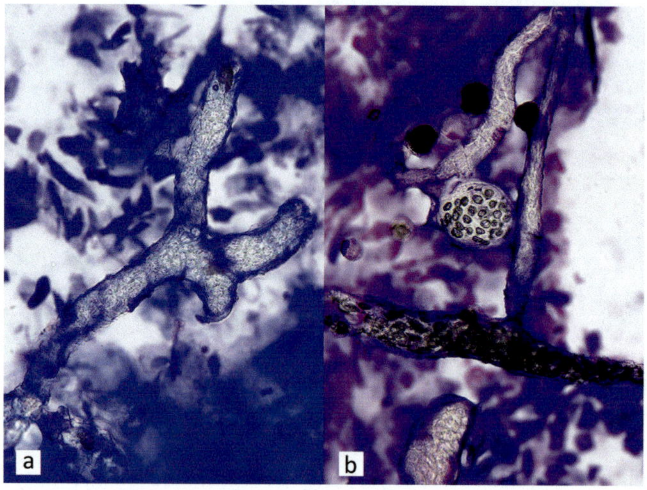

Fig. 8.269 Cytology of ghost cells: (**a**) trabecular branching of achromatic ghost cells; (**b**) thin round to oval to linear matrix keratin material

hyperkeratosis associated with marked hyperplasia of the nucleated epidermis (acanthosis). In these cases, the fully mature keratinocytes (corneocytes), which in a normal keratinization process are represented by anucleate lamellae, although completely keratinized, can retain a rounded central area delimited by a line that represents the former nuclear area and which gives the keratinocyte a typical appearance that imitates the ghost cells of the matrical keratin (Figs. 8.270 and 8.271).

Unlike matrical ghost cells, these superficial cornified lamellae are larger, with angular edges, and for this reason the author prefers to use the term "superficial" or "false ghost" cells to describe them. In practice, these false ghosts are nothing more than fully keratinized corneocytes, much more voluminous than true matrical ghost cells, and that

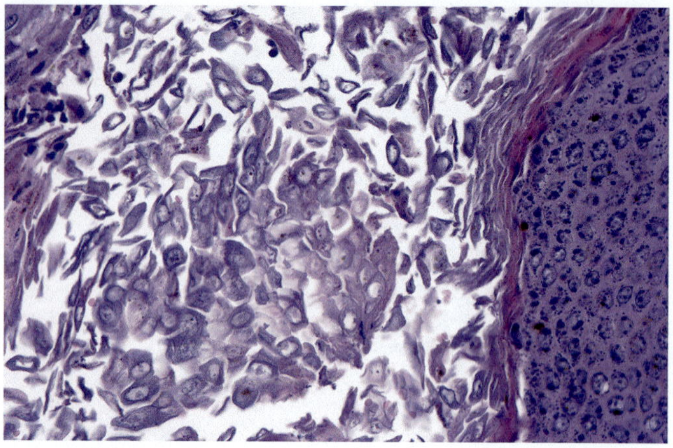

Fig. 8.270 Cytology of superficial ghost cells: false ghost cells in a trichofolliculoma

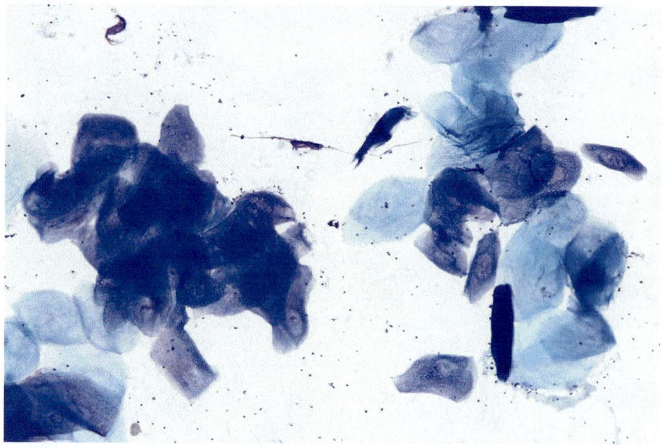

Fig. 8.271 Cytology of superficial ghost cells: the remnant nuclei are evident in the center of the fully keratinized cells

probably represent the result of a final keratinization process that occurs very quickly (Fig. 8.272).

A part for epidermal hyperplastic lesions, *false ghost cells* are also seen in other keratinizing follicular tumors such as trichofolliculoma, IKA, or other benign neoplasms with an altered and tumultuous hyperkeratosis occurring; so, it is necessary to know how to recognize and interpret them in order to avoid diagnosing a tumor that involves the hair matrix incorrectly.

Matrical Cysts

Matrical cysts originate from the bulbar portion of the anagen follicle and are histologically composed of a single large cyst lined with basaloid matrical keratinocytes that abruptly transform into "ghost cells," which represent *incomplete or abortive trichogenesis.* In pure matrical cysts, cytology shows exclusively ghost cells without other typology of keratin (Fig. 8.273).

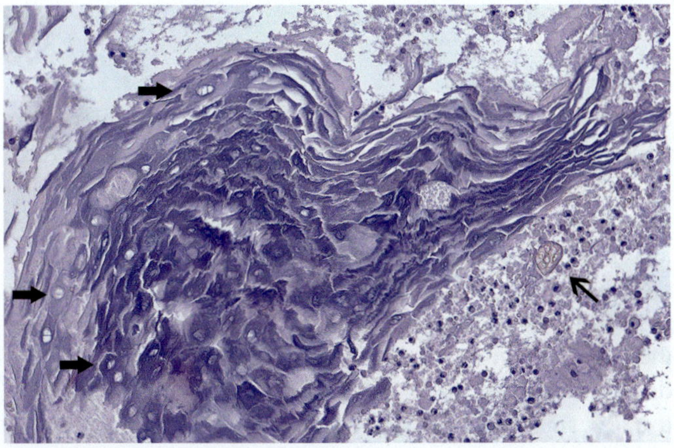

Fig. 8.272 Cytology of ghost cells: note the difference in size between the matrical ghost cells (thin arrow) and the superficial "false" ghost cells (large arrows)

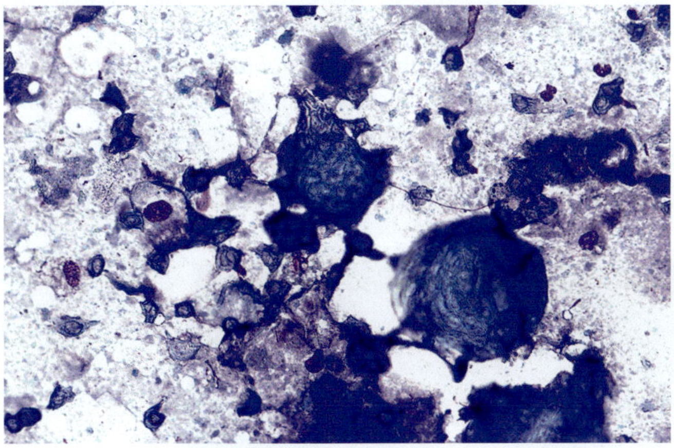

Fig. 8.273 Cytology of ghost cells: matrical ghost cells arranged in round groups and as single polygonal cells dispersed on the bottom of slide

Hybrid Cysts

The *hybrid cysts* involve two or all three portions of the follicle and therefore are cytologically characterized by the simultaneous presence of different types of keratins. A combination of lamellar corneocytes, matrical ghost cells, and, occasionally, trichilemmal amorphous keratin can be observed on the same slide (Figs. 8.274 and 8.275). It must be highlighted as, when both lamellar corneocytes and ghost cells are detected, care must be taken in formulating a diagnosis of mixed infundibular–matrical cyst (hybrid cyst), because similar cytological features can be observed in trichoepithelioma.

Although in trichoepithelioma nucleated basal cells are a typical and important finding for making a diagnosis, in the histological variant of cystic trichoepithelioma, the tumor is mainly constituted by a large cyst lined by a not too thick basal epithelium; therefore, if the needle is inserted in the center of the cyst and no basal cells are sampled, the slides will be composed only of lamellar and matrical keratin, as occurs in hybrid cysts, from which it is possible to differentiate only histologically.

As abovementioned, in all follicular keratinizing lesions, an inflammation can occur secondarily to the breakage of the cysts' wall; in these case, many inflammatory cells, comprising multinucleated giant macrophages, try to eliminate the foreign body (keratin). It must be highlight that cytologically, the presence of many giant cells can be interpreted as epithelial basal cells, and a pilomatricoma can be confused with hybrid cysts (Fig. 8.276).

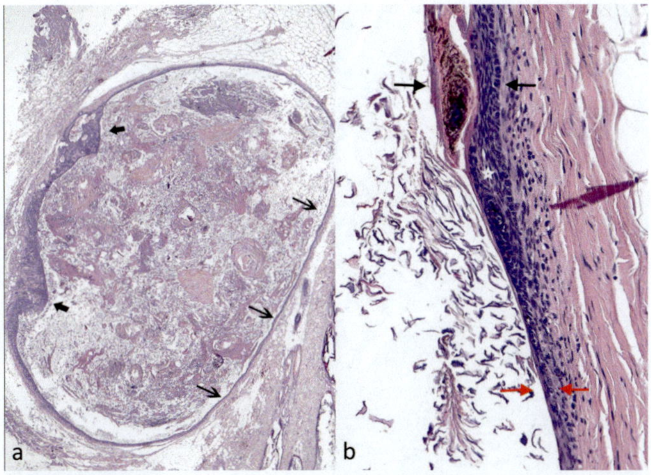

Fig. 8.274 Histology of a hybrid cyst. (**a**) Large cyst filled with both lamellar and matrical (ghost) keratin. Note the different thick and differentiation of the opposite cyst's wall: epidermoidal on the top that origins lamellar corneocytes (thin arrow) and matrical at the base that origins ghost cells (large arrow); (**b**) at high magnification the matrical epithelium differentiated in ghost keratin (black arrows) and keratin lamellae produced through the granulosum stratum (red arrows) are clearly recognizable. The passage of different keratinizing epithelia is evident (white star)

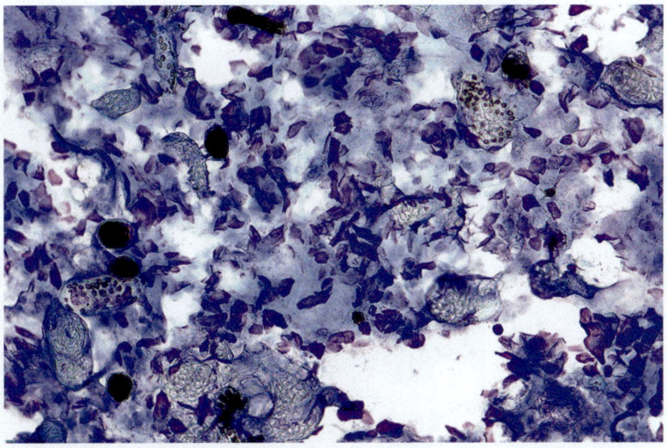

Fig. 8.275 Cytology of a hybrid cyst. Mix of lamellar corneocytes together with multiple clusters of ghost cells. No epithelial basal cells are present

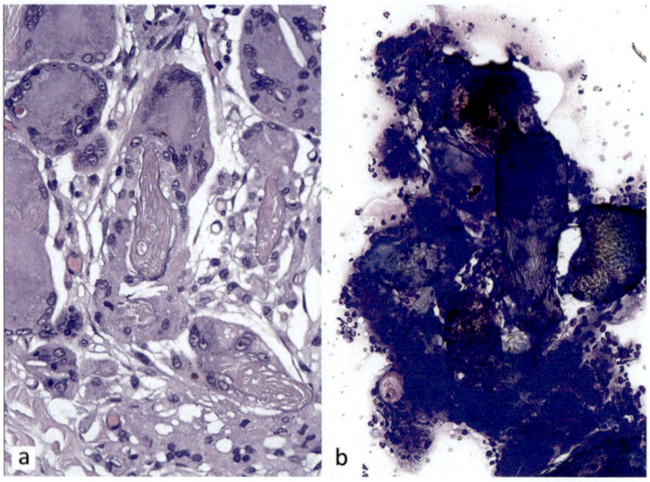

Fig. 8.276 Inflamed hybrid cyst. (**a**) Histology: many large groups of ghost cells are surrounded and phagocytized by multinucleated macrophages. (**b**) Cytology: the nucleated cells surrounding matrical keratin are not easily recognizable as multinucleated macrophagic cells

8.3.4.1 Follicular Neoplasms

(a) *Trichoblastoma*

Trichoblastoma, once defined as *basal cell* tumors, is a neoplasm arising from the cells of the hair germ, from which the follicle originates; it is one of the most common follicular neoplasia in dogs and cats. Trichoblastoma can be observed as regular round exophytic nodular lesions, usually alopecic and with a smooth surface, sometimes pigmented, of varying size, from few to five centimeters. Trichoblastoma is one of the most common skin tumors in cats, most frequently seen on the head and neck; in this species, trichoblastoma is usually small in size and frequently hyperpigmented. In dogs, it occurs in any breed, but it is more frequently detected in the miniature Poodle, mainly on the neck and at the base of the pinna. Many follicular both cysts and neoplasms are pigmented also in dogs (Fig. 8.277).

Cytological Findings

The cytological aspect that all *trichoblastomas* have in common is the presence of clusters of highly cohesive basal cells, composed of uniform small roundish cells, with scarcely perceptible cytoplasm and central round nuclei of uniform size.

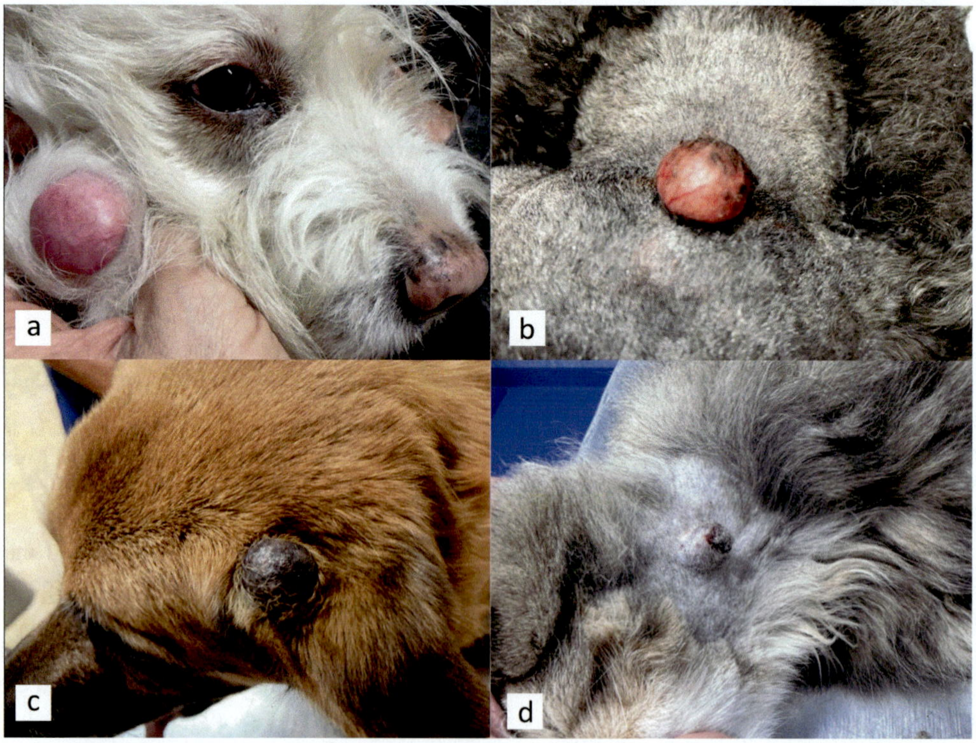

Fig. 8.277 Trichoblastoma. (**a**) Alopecic and with smooth surface nodule on the face of a mixed breed dog; (**b**) alopecic and erythematous with narrowed base on the neck of a Poodle; (**c**) hyperpigmented and exophytic nodule on the pinna; (**d**) ulcerated nodule on the neck of a cat

Different histotypes of trichoblastoma are recognized and classified based on the histological pattern. In the current histological classification, trichoblastoma patterns are classified as *ribbon*, *medusoid*, *trabecular*, *granular*, *spindle cells*, and *solid cystic* [116].

Since some trichoblastomas can show a mixed histological pattern (i.e., ribbon along with granular or ribbon along with spindle), prior to this latter classification, some subtypes were thought to be variants of the same pattern and not classified as separate histotypes; however, given the benign behavior of all trichoblastoma, any attempt at cytological classification has no prognostic value.

Epithelial basal cells are tightly packed together, so a variable number of clusters are collected by FNB. Due to this characteristic, the clusters can maintain on cytological samples the original dispositions they have in histology; therefore, with a minimum knowledge of the histopathological patterns of trichoblastoma, the most expert cytologists not only can make diagnosis of trichoblastoma with cytology but also are able to identify its histotype.

The knowledge of the typical cellular architectures of trichoblastoma allows cytologists to formulate a certain diagnosis of trichoblastoma, and other basal cell tumors such as of apocrine (sweat glands) or sebaceous nature (sebaceous epithelioma without evident sebaceous differentiated cells) can be ruled out. With the usually benign behavior of the basal cell neoplasia, the value of this cytological differentiation could be useless, but some basal cell carcinomas or some sebaceous epitheliomas can occasionally have malignant behavior; for the sebaceous neoplasms, as it will be highlighted below, the problem could occur when basaloid clusters are poorly differentiated, and sebaceous differentiation is not evident; fortunately, the sebaceous epithelioma, in which sebocytic differentiation is not evident, is extremely rare.

Usually cytological samples are highly cellular and cells are released in clusters with well-defined edges. Clusters can be of different shape: roundish, trabecular, and organized in long rows composed of both single and double basal cells, which can show a centrifugal pattern of growth starting from a central cluster. In some clusters, a tridimensional arrangement is cytologically recognizable. In ribbon and medusoid histotypes, long palisade of basal cells can be observed; nuclei tend to press against each other so much so as to deform their profile assuming a typical vertically oriented arrangement. Basal cells are uniform in size, with roundish to oval nuclei with small nucleoli and with scarce amount of cytoplasm; some differences can be observed based on the various histotypes. In particular cases, as we will see later, the basal cells can separate from each other and be released on the slides as single cells (Figs. 8.278, 8.279, 8.280, 8.281, 8.282, 8.282, 8.283, 8.284, 8.285, 8.286, 8.287 and 8.288) [123].

Since epithelial aggregates are normally embedded in a variable amount of stromal collagen, it is common to find, along with extracellular eosinophilic material, few stromal spindle cells and pinkish extracellular matrix. In some cases, the stroma is strongly myxoid, and, on slides, it is easily recognized as dense to pinkish to purple amorphous material on the background in which a variable number of basaloid clusters are imbedded.

Occasionally, trichoblastoma with abundant plump stromal cells is reported in dogs [120] (Figs. 8.289 and 8.290).

A *spindle cell trichoblastoma*, characterized cytologically by clusters of densely packed spindle-shaped basaloid cells, is well recognized in cats and occasionally dogs (Figs. 8.291 and 8.292). Because mixed patterns are often present in trichoblastomas within the same tumor, especially in dogs, it is not uncommon to see spindle-shaped and round basal cells on the same smears. In few cellular specimens or in those composed predominantly or exclusively of spindle cells (spindle trichoblastoma), because the basal cells often tend to break away individually from the edges of the aggregates, care must be taken not to confuse them with the spindle-shaped mesenchymal cells of soft tissue sarcomas; this is a mistake often made by inexperienced cytologists or those

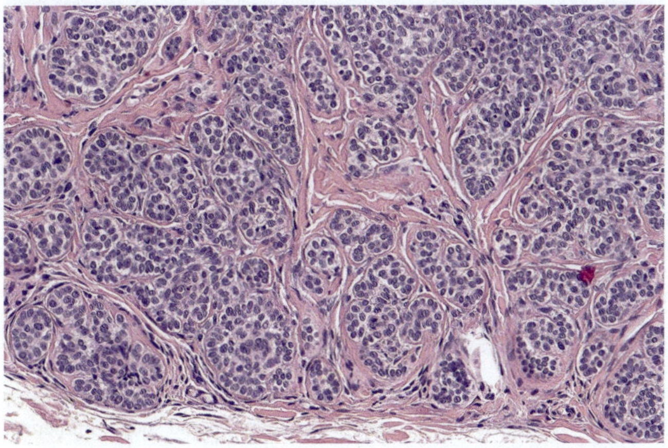

Fig. 8.278 Histopathology of a trichoblastoma: multiple nests of basaloid keratinocytes

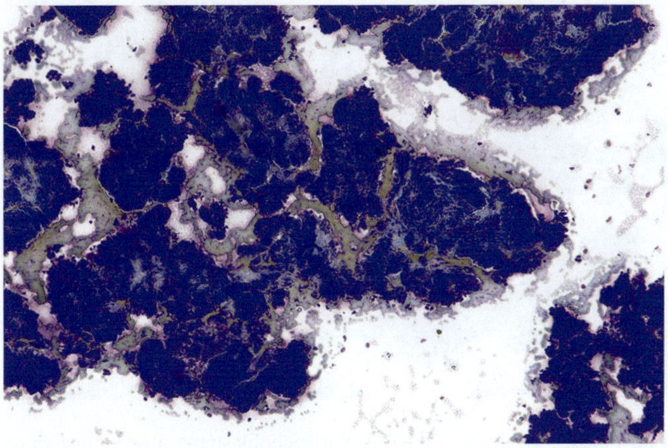

Fig. 8.279 Cytology of a trichoblastoma: roundish nests of basaloid cells with scarce cytoplasm

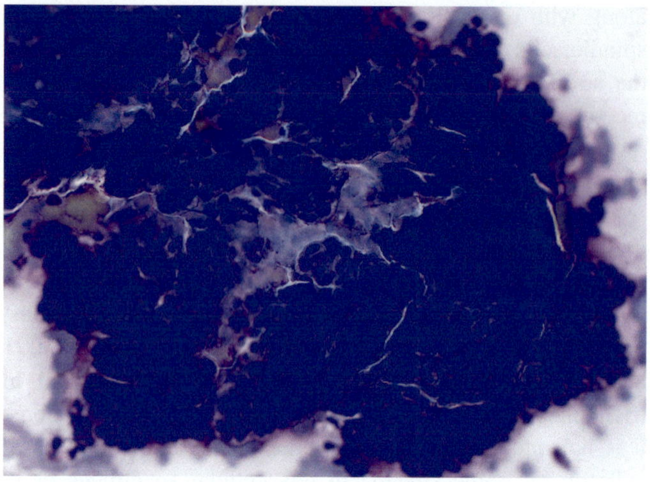

Fig. 8.280 Cytology of a trichoblastoma: roundish and overlapped nests of uniform in size basaloid cells with scarce cytoplasm. Note clustered cells maintain their organization also on cytological slides

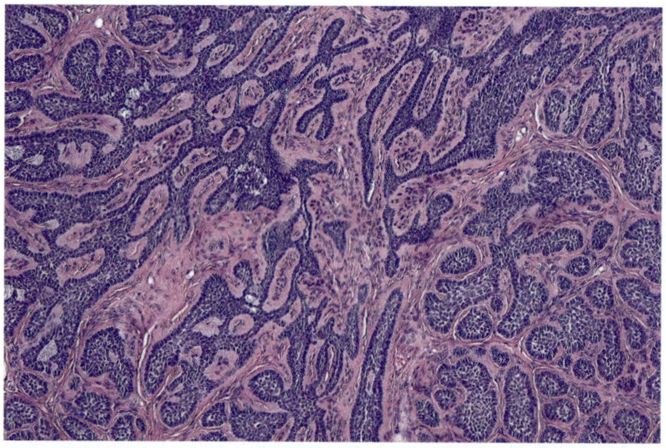

Fig. 8.281 Histopathology of a trabecular trichoblastoma: clusters of basal cells arranged in trabeculae of variable size and shape, immersed in a highly cellular stroma

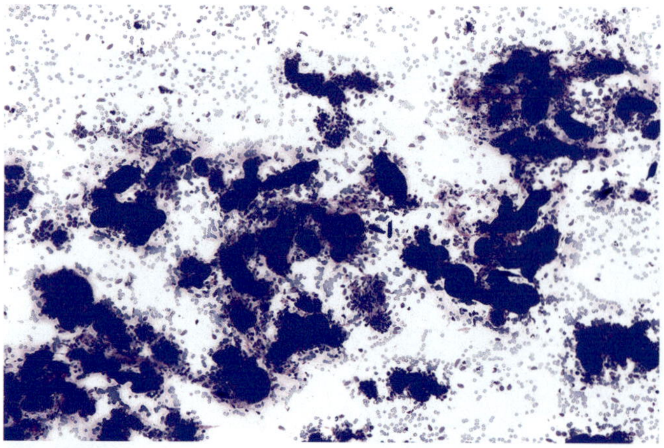

Fig. 8.282 Cytology of a trabecular trichoblastoma: many clusters of basal cells arranged in trabeculae

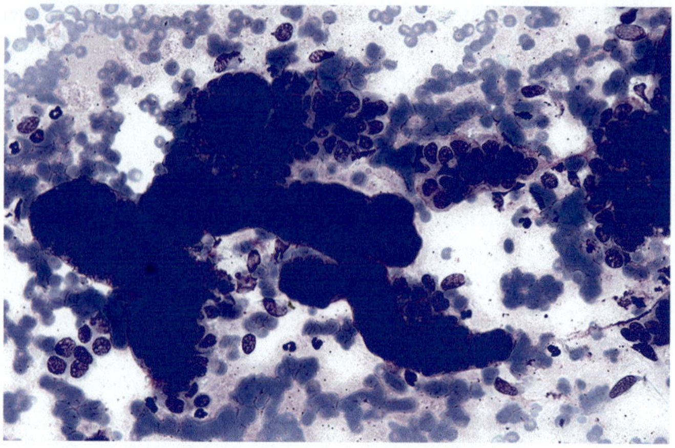

Fig. 8.283 Cytology of a trabecular trichoblastoma: at high magnification, many clusters of basal cells arranged in trabeculae with a tridimensional appearance

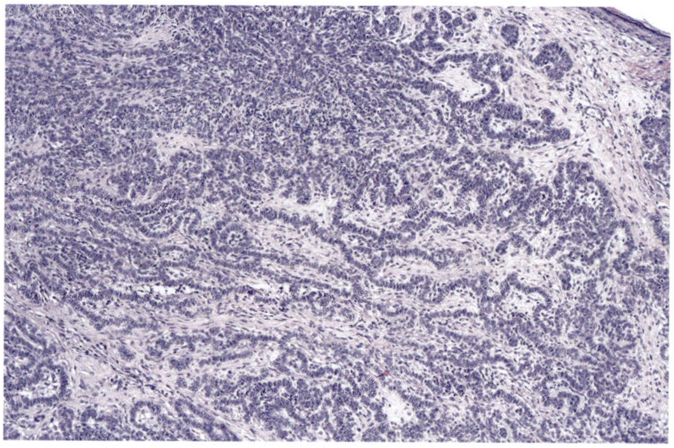

Fig. 8.284 Histopathology of a ribbon-type trichoblastoma: singularly layered basal cells arranged in long rows

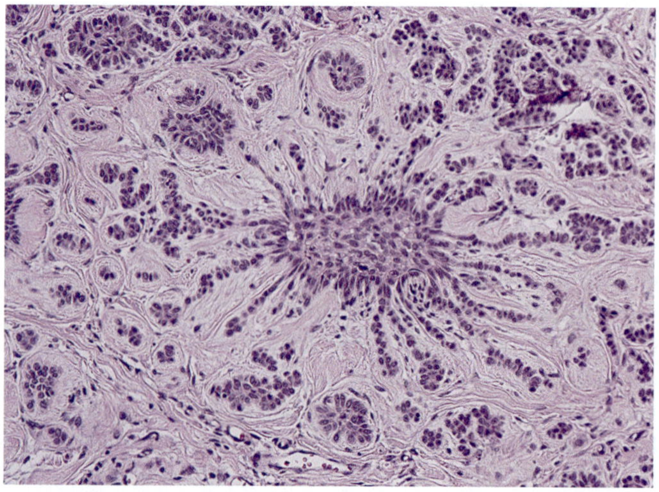

Fig. 8.285 Histopathology of a ribbon-type trichoblastoma (medusa head variant): note rows of basal cells start from the center with a centrifugal direction

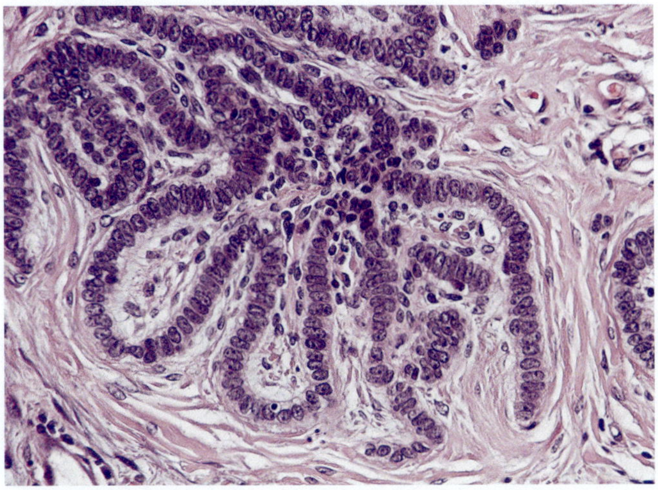

Fig. 8.286 Histopathology of a ribbon-type trichoblastoma: serpiginous rows of single or double basaloid keratinocytes; note the vertically arranged palisade of nuclei

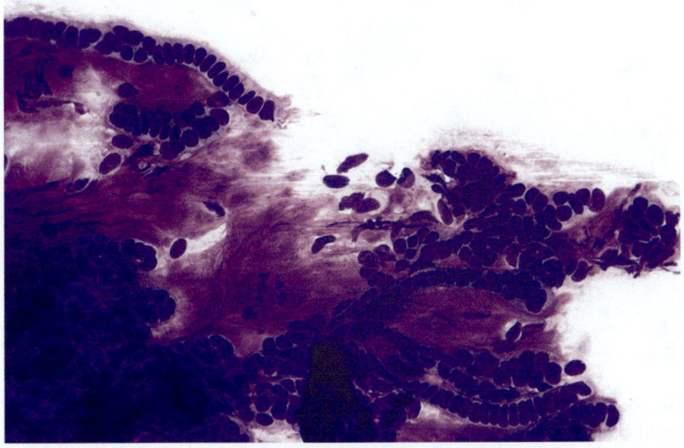

Fig. 8.287 Cytology of a trichoblastoma: typical rows of basaloid keratinocytes embedded in a pinkish abundant extracellular stromal matrix

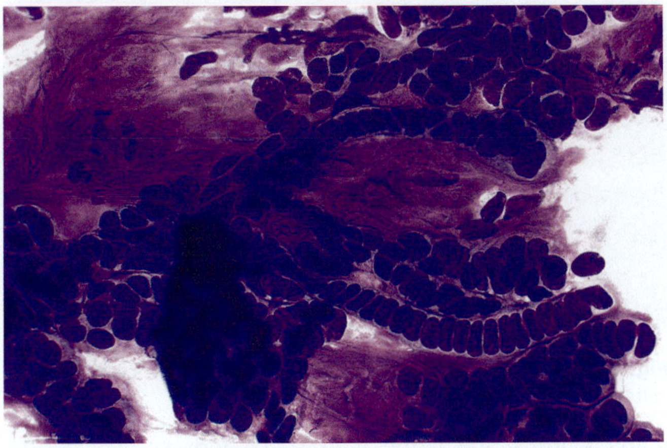

Fig. 8.288 Cytology of a trichoblastoma: at high magnification the typical arrangement of the rows of basal keratinocytes is clearly recognizable

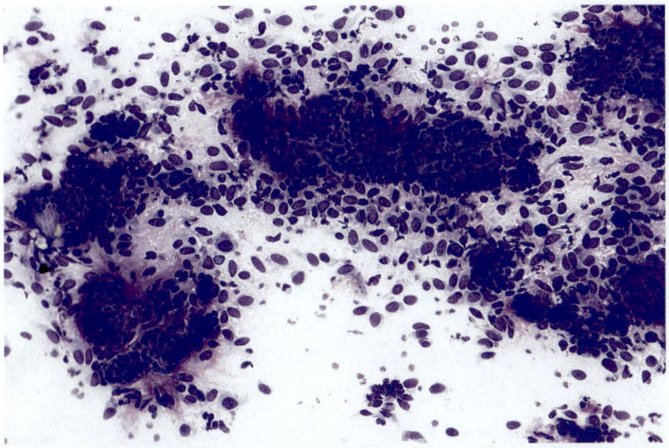

Fig. 8.289 Cytology of trichoblastoma: clusters of small basaloid cells along with many, plump, spindle cells representing stromal fibroblasts/ myofibroblasts

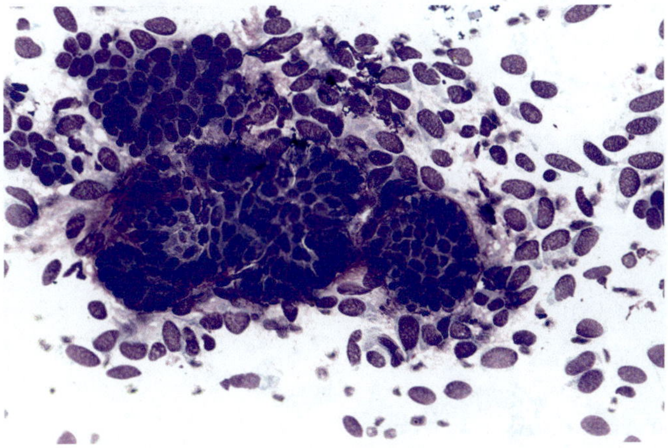

Fig. 8.290 Cytology of trichoblastoma: clusters of small basaloid cells along with many, plump, spindle cells representing stromal fibroblasts/ myofibroblasts

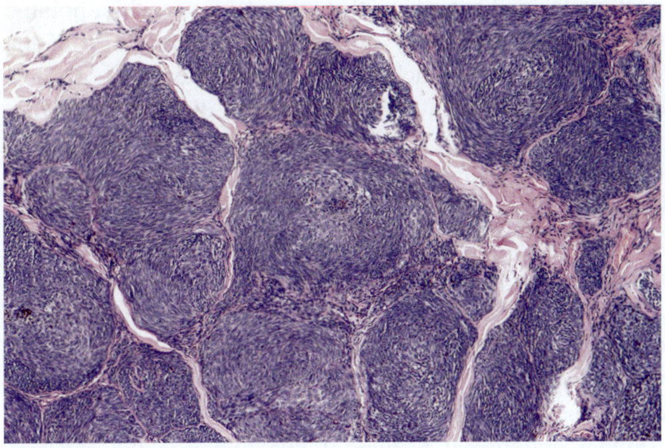

Fig. 8.291 Histology of spindle cell trichoblastoma: multiple nests of basal cells with clear spindle silhouette

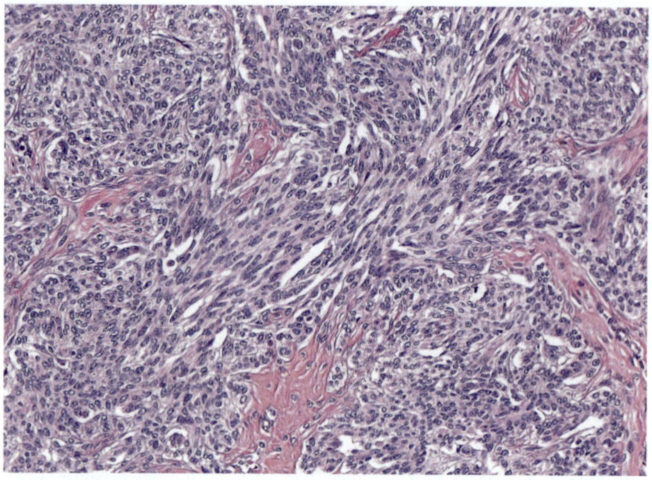

Fig. 8.292 Histology of spindle cell trichoblastoma: at high magnification the spindle shape of basal follicular cells is better visible

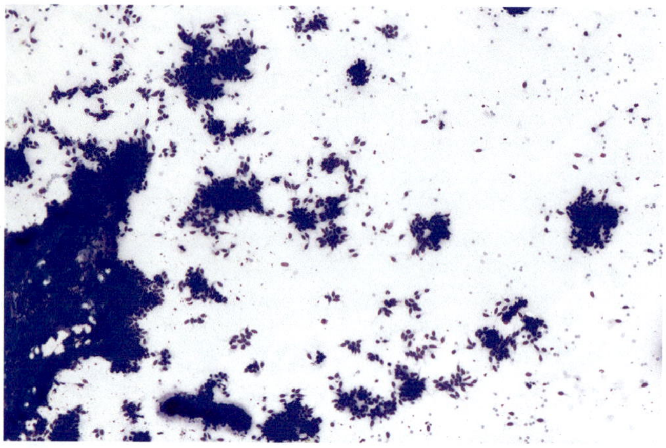

Fig. 8.293 Cytology of spindle cell trichoblastoma: large clusters of cells along with singularly released spindle cells

unfamiliar with skin pathology. In fact, the basal keratinocytes of spindle cell trichoblastoma have elongated, oval nuclei, and cytoplasm that may show very thin, pointed tails, making them very difficult to differentiate from the spindle cells of mesenchymal tumors. Fortunately, in most cases, together with the individually dispersed spindle-shaped keratinocytes, some groups of cells can be recognized which allow us to orient the suspicion toward an epithelial neoplasm (Figs. 8.293, 8.294 and 8.295) [116, 117].

The *granular cell* variant of trichoblastoma is uncommon in dogs and very rare in cats.

The cells are roundish or oval and most frequently have an eccentric nucleus. The distinctive morphological sign of this variant, not present in other trichoblastomas, is the presence of reddish to purple dotted "granules" in the cytoplasm. Another cytological feature of this tumor is the constant presence, on cytological preparations, of single neoplastic basal cells; this occurs because the cells tend to detach easily from clusters compared to other trichoblastomas (Figs. 8.296, 8.297, 8.298, 8.299, 8.300, 8.301 and 8.302). For this reason, smears are characterized by the simultaneous presence of cohesive aggregates together with many single cells. This particular cytological feature may cause misdiagnosis,

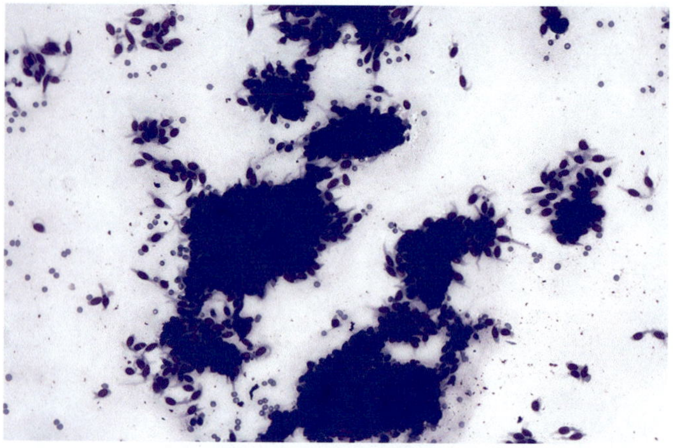

Fig. 8.294 Cytology of spindle cell trichoblastoma: clusters of basal cells—note at the periphery of the clusters, spindle cells are detaching

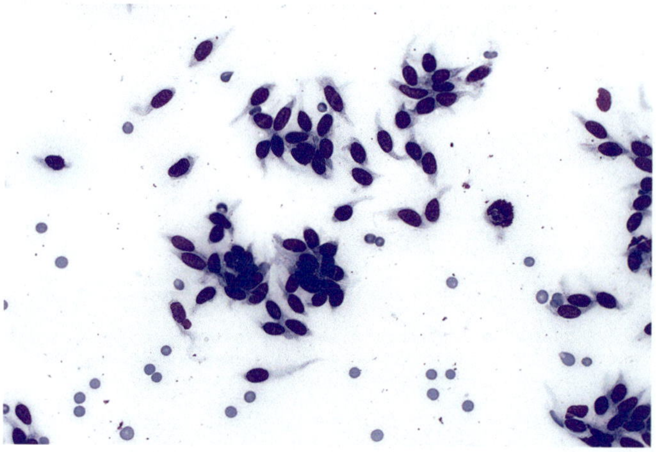

Fig. 8.295 Cytology of spindle cell trichoblastoma: spindle follicular basal cells. It is very complex to define as epithelial these cells

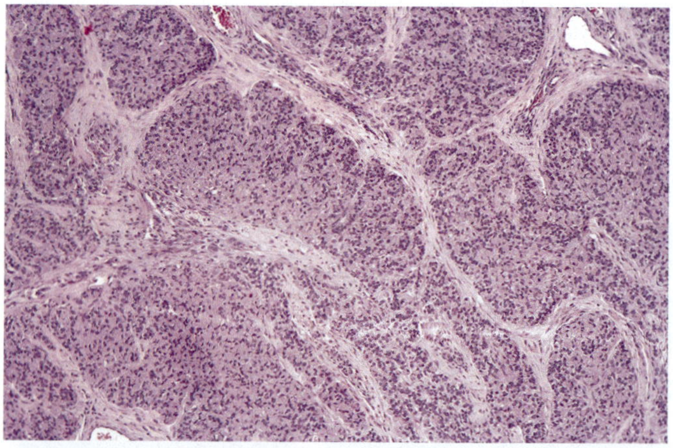

Fig. 8.296 Histology of granular cell trichoblastoma: nests of basal cells with clear cytoplasm and eccentric nuclei

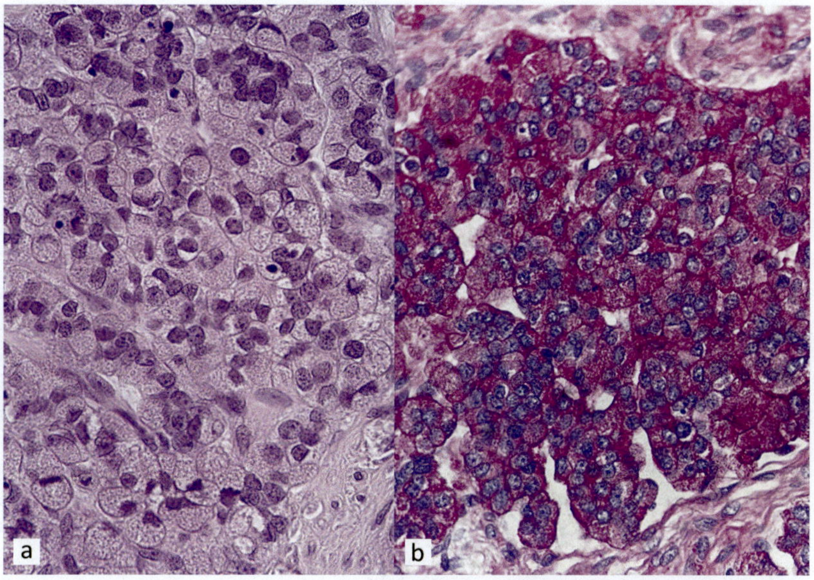

Fig. 8.297 Histology of granular cell trichoblastoma: (**a**) intracytoplasmic granules are barely evident with H-E stain; (**b**) PAS staining highlights the intracytoplasmic granules

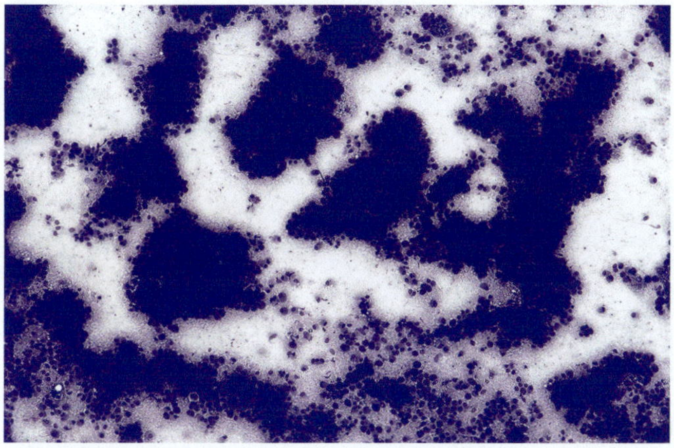

Fig. 8.298 Cytology of granular cell trichoblastoma: along with many clusters, single basal cells detach from the cluster and are released singularly on the bottom of the slide

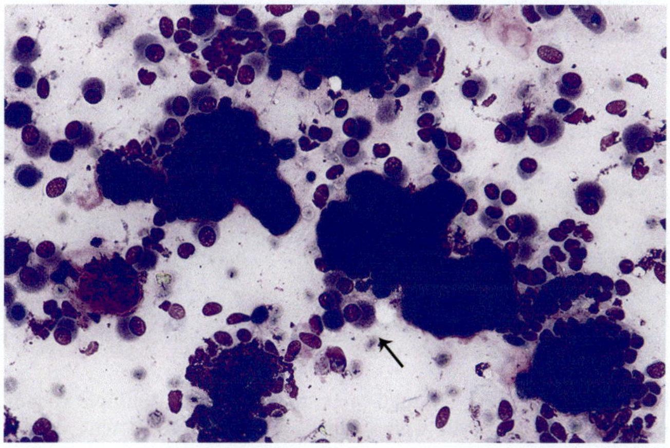

Fig. 8.300 Cytology of granular cell trichoblastoma: clusters of basal cells along with many singularly arranged round to oval in shape cells. Note in some cells intracytoplasmic granules are recognizable (arrow)

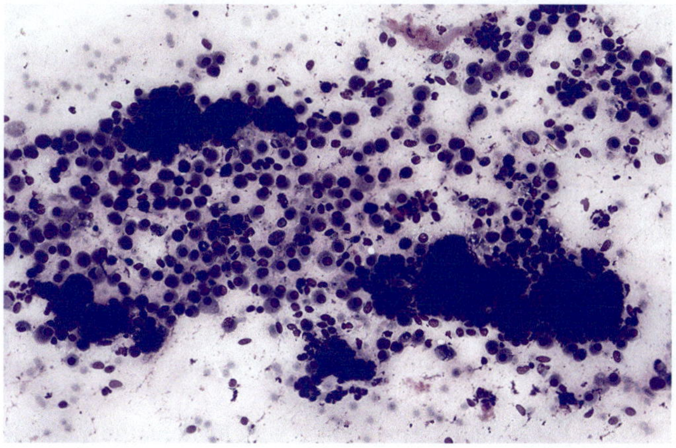

Fig. 8.299 Cytology of granular cell trichoblastoma: clusters of basal cells along with many singularly arranged round to oval cells

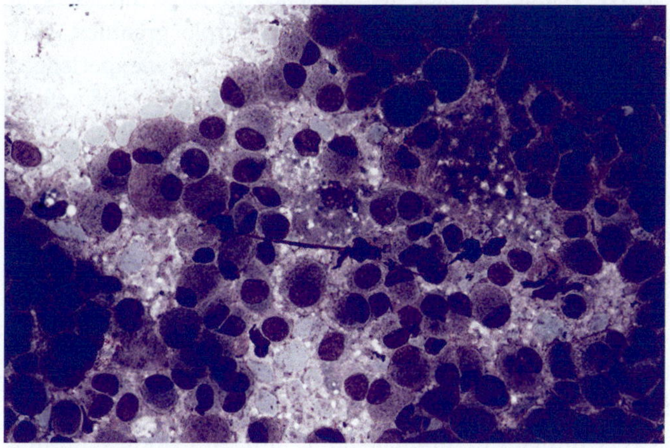

Fig. 8.301 Cytology of granular cell trichoblastoma: at high magnification, reddish granules are better evaluable

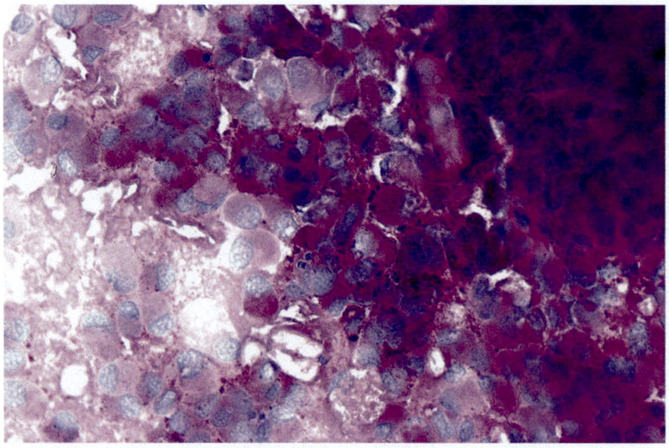

Fig. 8.302 Cytology of granular cell trichoblastoma: intracytoplasmic granules are clearly PAS positive

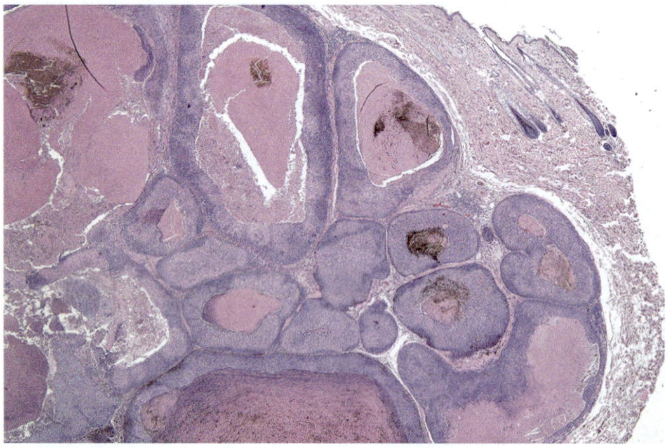

Fig. 8.303 Histology of solid-cystic trichoblastoma: multiple islands of neoplastic basal cells with a central cystic area filled with debris, melanin, and melanophages

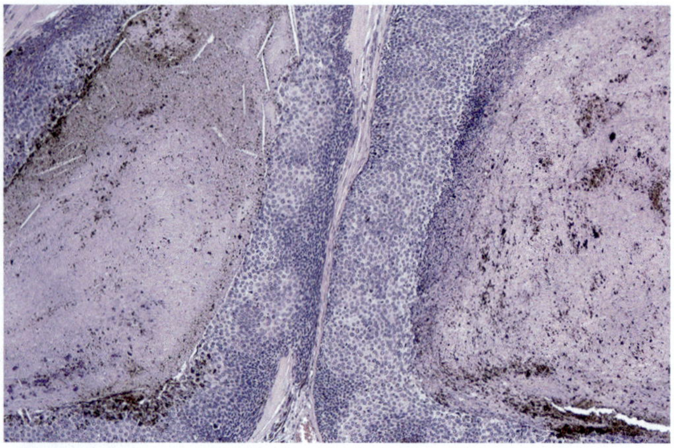

Fig. 8.304 Histology of solid-cystic trichoblastoma: at high magnification basal cells appear regular in size and shape. Note the necrotic and melanized intracystic material

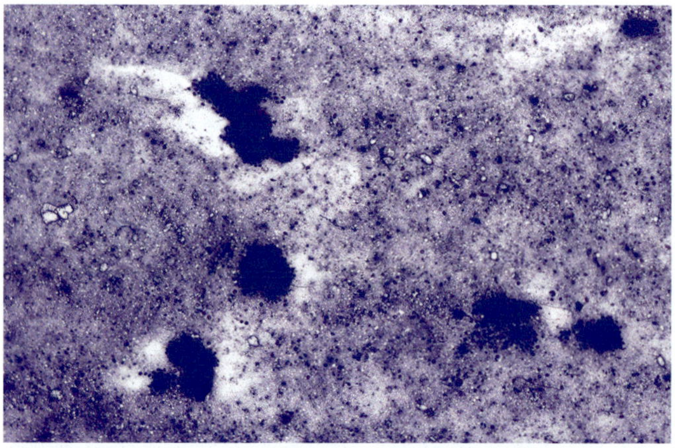

Fig. 8.305 Cytology of solid-cystic trichoblastoma: large cluster of benign basal cells immersed in a necrotic amorphous blue in color background

especially in less cellular samples in which the clusters that would allow the cells to be recognized as epithelial are not clearly visible; in fact, the presence of large numbers of discrete cells, with intracytoplasmic purple granules, may be misdiagnosed as mast cell tumor by inexperienced cytologists. The intracytoplasmic granules of these keratinocytes are PAS positive, and this dye affinity can also be exploited for cytological samples allowing better highlighting of the granules. As with other trichoblastomas, different histotypes can coexist, and in those in which the granular differentiation of keratinocytes is poor, the cytological diagnosis is not easy.

In the presence of the *solid/cystic* histological variant of trichoblastoma, the diagnostic difficulties increase, and in fact, in addition to defining it as a basal epithelial tumor, it is very difficult to differentiate it, through cytology, from a solid/cystic apocrine adenoma (sweat glands); the latter is very common in cats. In some cases, these two tumors are difficult to differentiate even histologically, and only the presence of tubular structures or abortive ductal formation allows the diagnosis to be oriented toward a glandular neoplasm (Figs. 8.303 and 8.304).

The slides are composed of an amorphous, detrital, deep blue background that represents the contents of the "cystic" areas that form in the center of the solid areas of the neoplastic basal cells.

The basal cells, organized in regular groups of monomorphic cells of uniform size, are present on the necrotic background described above in which scattered naked nuclei are also recognizable.

Often, the basaloid cells of trichoblastoma, especially of the solid/cystic histotypes, are strongly pigmented (Figs. 8.305 and 8.306).

When the amount of pigment is very marked, the material sampled with the needle and deposited on the slide may macroscopically suggest a melanocytic neoplasm (Fig. 8.307). At low magnification, the marked amount of

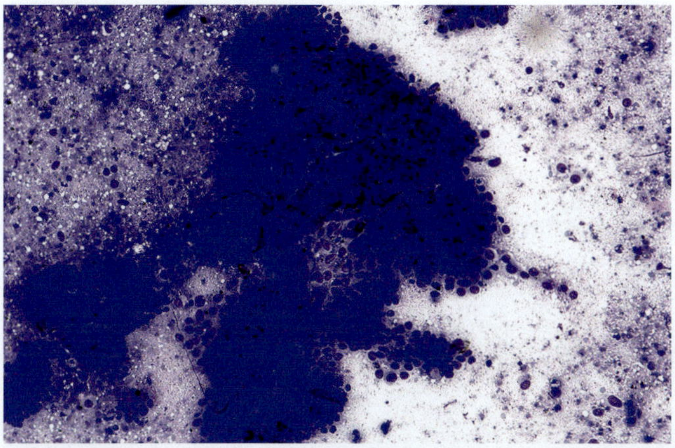

Fig. 8.306 Cytology of solid-cystic trichoblastoma: multiple clusters of benign basal cells immersed in a necrotic amorphous blue in color background

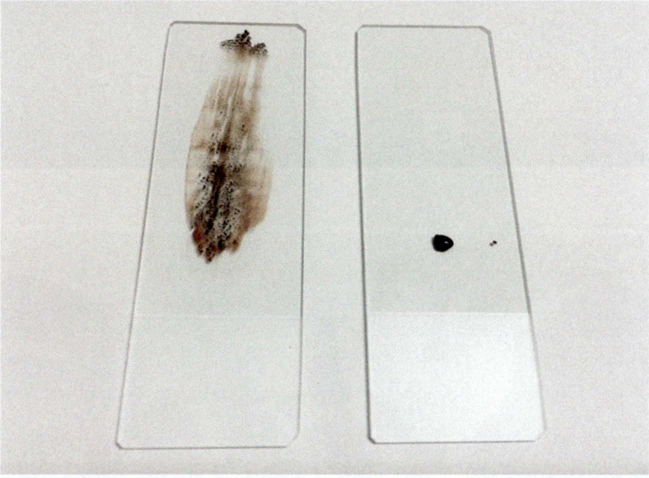

Fig. 8.307 Dark material collected via FNB from a highly pigmented trichoblastoma. The macroscopic appearance of the color always raises the suspicion of a melanocytic neoplasm

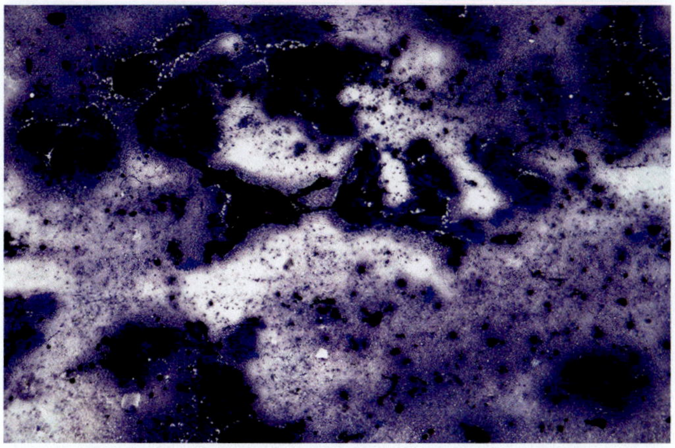

Fig. 8.308 Cytology of hyperpigmented trichoblastoma: the large amount of melanin obscures the silhouette of neoplastic basal cells

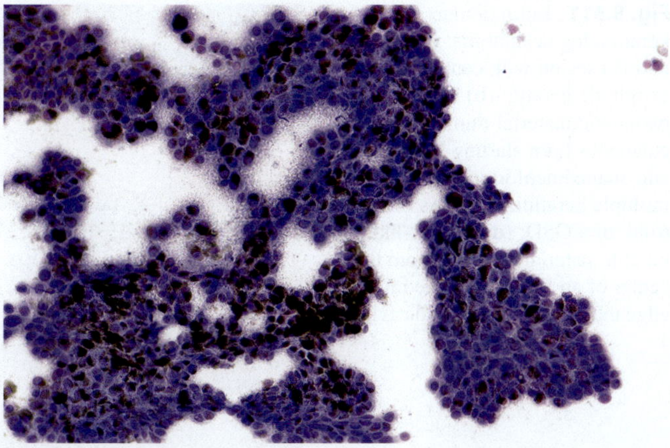

Fig. 8.309 Cytology of hyperpigmented trichoblastoma: at medium magnification the epithelial arrangement is easily detectable

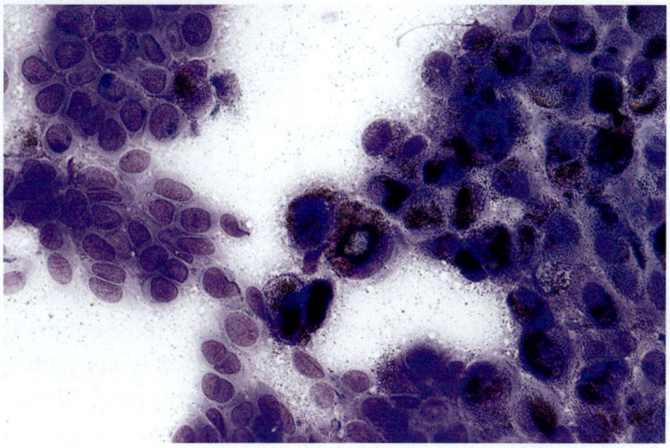

Fig. 8.310 Cytology of hyperpigmented trichoblastoma: pigmented basal cells

melanin may obscure the cellular details creating the suspicion that the cells may be neoplastic melanocytes (Fig. 8.308). By observing carefully at a higher magnification, or when the degree of pigmentation is not so marked, it is possible to recognize the epithelial origin of the pigmented cells (pigmented basaloid aggregates) (Figs. 8.309 and 8.310). Since a variable number of macrophages are normally present in the cystic areas, they can phagocytize both the melanin (melanophages) and the dark blue amorphous material that represents the necrotic material contained in the intratumoral cystic areas.

(b) *Infundibular Keratinizing Acanthoma*

The *infundibular keratinizing acanthoma* (IKA) is a benign follicular tumor that arises from both infundibular and isthmic segment of the follicle [116]. Sometimes the presence of some macroscopic characteristics can direct the clinician to suspected follicular neoplasia, such as when a small central crater from which keratin material protrudes,

Fig. 8.311 Infundibular keratinizing acanthoma: (**a**) small nodular lesion with central exophytic keratin; (**b**) keratin exophytic material mimics a cutaneous horn starting from a cup-shaped neoformation; (**c**) multiple keratinizing IKAs on the trunk of a GSD; (**d**) creamy-like keratin material goes out from the center of an IKA—note the raised edge that delimit the central cavity

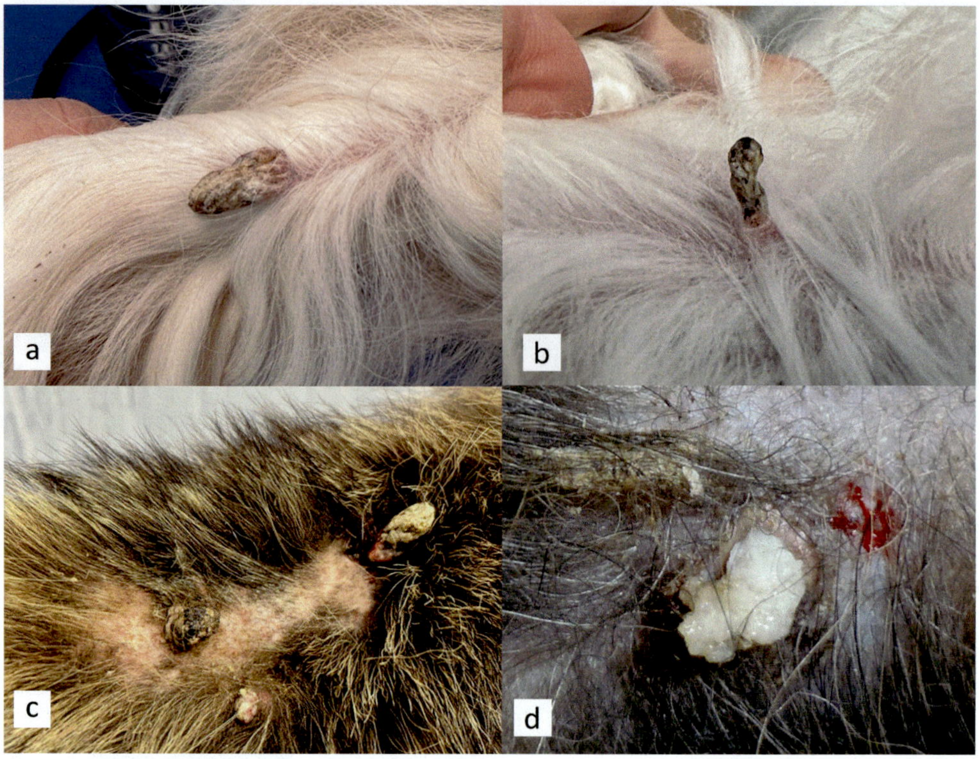

sometimes as long exophytic horn, suggests the presence of an IKA. In small IKAs the central pore filled with keratin cannot be recognizable. The tail is one of the most common involved body areas. This clinical appearance is not pathognomonic of IKA since also other very rare benign keratinizing lesions such as warty dyskeratoma in dogs and dilated pore of Winer in cats can share a similar appearance and unfortunately also they can be observed on the tail. In some predisposed breeds, such as German Shepherd dogs and Shih Tzu, multiple IKA can be observed, usually spread on the trunk (Fig. 8.311).

Cytological Findings

IKAs are histologically characterized by a proliferation of epithelial cells forming a thick wall of solid nests and trabeculae of keratinocytes composed of the same layers of the infundibular epidermidis and therefore show an epidermoid type of keratinization due to the presence of the granulosum layer. In the conventional IKA, epithelial keratinocytes delimit a central cavity filled with lamellar keratin that has an opening on the skin surface through which the keratin protrudes. When the keratin material is abundant, a thick exophytic elongated keratin horn can emerge from the center of the nodule. Multiple keratinizing pearls are present in the thickness of the wall, composed of a variable amount of concentric lamellar keratin. In the initial stages of IKA, lesions can be very small, and the central cystic area filled with lamellar anucleate keratin is not present yet; in these cases, the amount of lamellar keratin can be few or absent on cytology, and only both nucleated and fully keratinized keratinocytes are collected (Figs. 8.312 and 8.313).

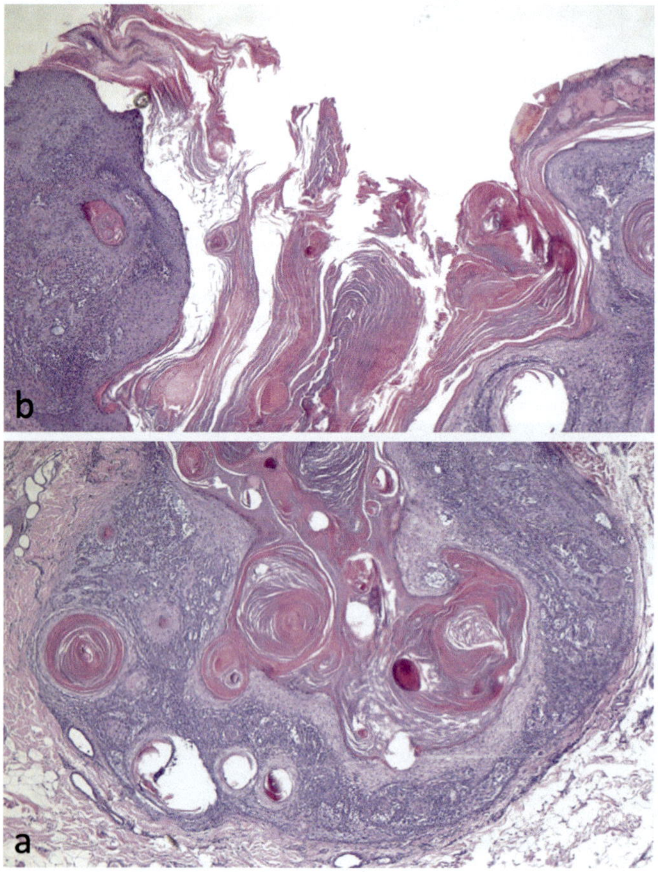

Fig. 8.312 Histopathology of an IKA: (**a**) base of IKA composed of a thick wall of squamous keratinocytes delimiting a large cyst filled with lamellar keratin; (**b**) external opening of the central cyst from which lamellar keratin material protrudes

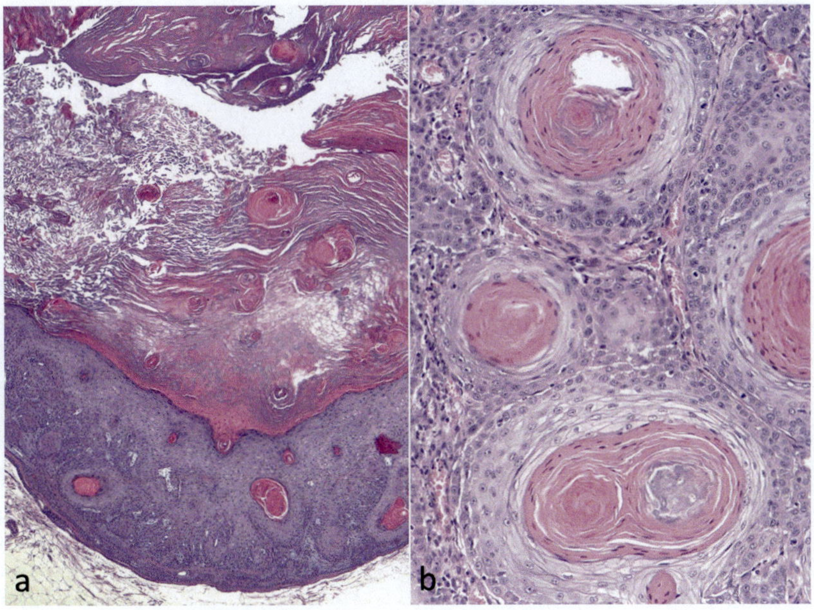

Fig. 8.313 Histopathology of an IKA: (**a**) base of IKA composed of a thick wall of squamous keratinocytes delimiting a large cyst filled with lamellar keratin and containing many keratin pearls: (**b**) close-up of the keratinizing pearls composed of concentric lamellar keratin surrounded by squamous nucleated epithelium

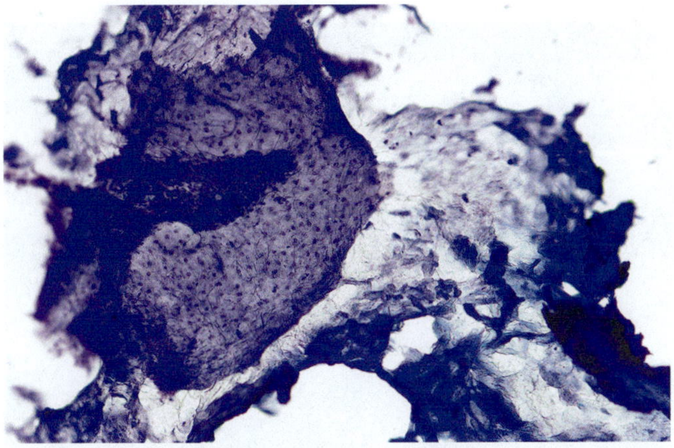

Fig. 8.314 Cytology of an IKA: the coexistence of basal keratinocytes, superficial squamous cells, and corneocytes is strongly suggestive of IKA

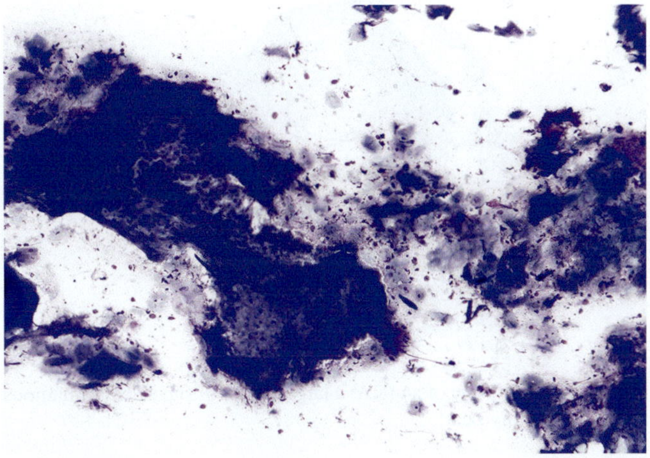

Fig. 8.315 Cytology of an IKA: large sheet of superficial squamous cells along with many corneocytes

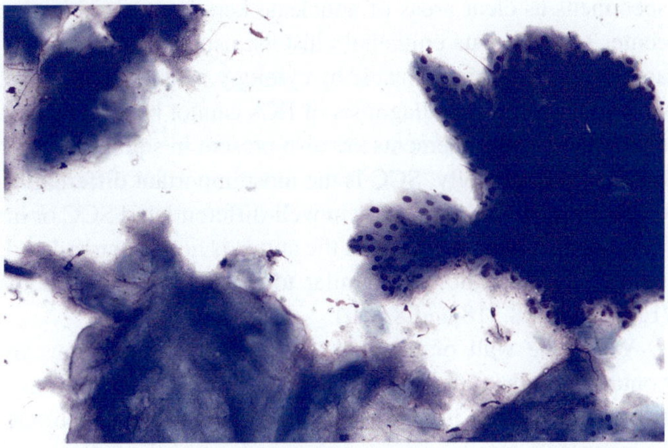

Fig. 8.316 Cytology of an IKA: large cluster of superficial squamous cells along with many corneocytes

If only the center of the IKAs is sampled, lamellar corneocytes and sometimes cholesterol crystals characterize the cytological specimens. Unfortunately, these cytological features are the same observed in the infundibular cysts, and therefore in this case the two entities cannot be cytologically differentiated. Despite this possibility, since the wall of the IKA is usually thick and composed of multiple layers of nucleated keratinocytes, it is common, along with lamellar keratin, to collect a variable number of nucleated keratinocytes; so, clusters at different stages of maturation, both basal, spinous, and granulosum keratinocytes, are simultaneously observed in the same samples. In most cases, the basal component is scarce, and large sheets of squamous spinous keratinocytes and corneocytes are collected (Figs. 8.314, 8.315 and 8.316). Since IKA is composed of a

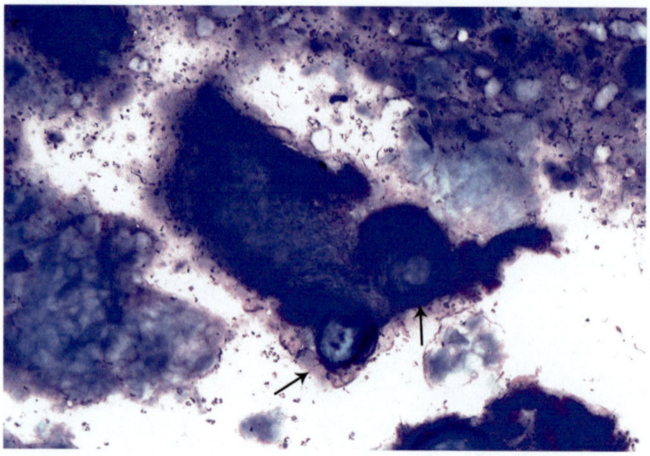

Fig. 8.317 Cytology of an IKA: large cluster of superficial squamous cells in which two keratinizing pearls are recognizable (arrows)

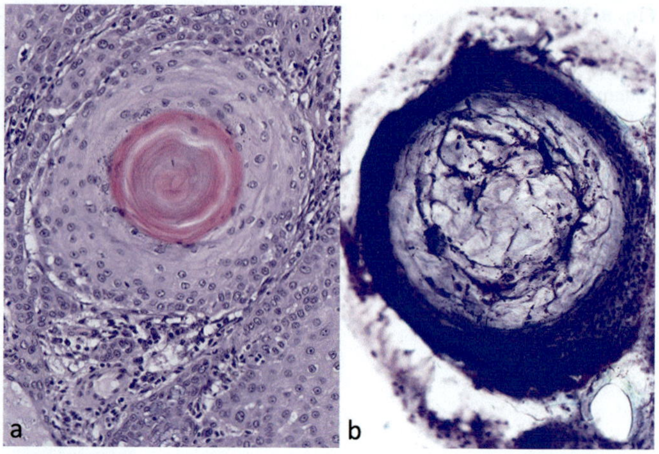

Fig. 8.319 Cytology of an IKA: (a) histology—keratinizing pearls. Thick layer of squamous keratinocytes that delimit an eosinophilic cystic wall containing keratin organized in concentric lamellae. (b) Cytology—note how the cytological finding is incredibly similar to the histological counterpart

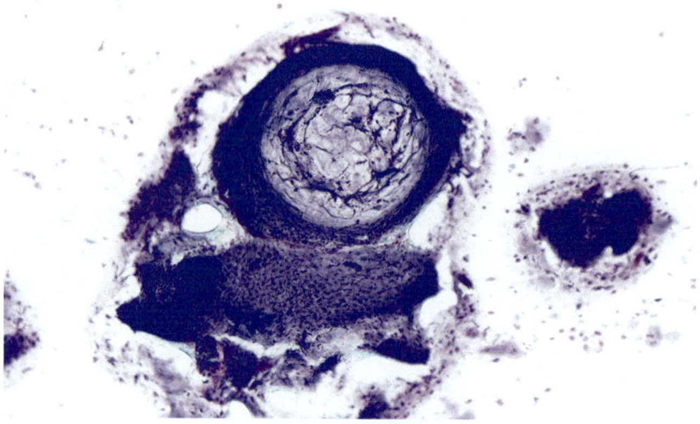

Fig. 8.318 Cytology of an IKA: a large cluster of superficial squamous and a large horny pearl

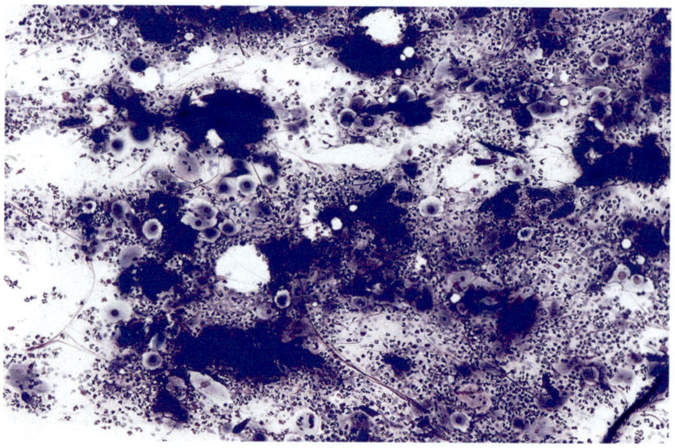

Fig. 8.320 Cytology of an IKA: many single nucleated keratinocytes immersed in an inflammatory background. Cytological atypia are few but cannot be differentiated from a well-differentiated SCC

squamous epithelial wall containing multiple concentric lamellar keratin cystic foci (horny or keratinizing pearls), the latter can exceptionally be observed in cytological specimens as clear areas of anucleate keratin material in the center of squamous epithelial clusters; usually, horny pearls are not so easily recognizable by cytology, and when they are detected, a definitive diagnosis of IKA cannot be made since these cellular arrangements are also present in squamous cell carcinoma. Generally, SCC is the most important differential diagnosis for IKA, especially in well-differentiated SCC or in its verrucous variant, in which the atypia is not so marked and the cytology can be very similar to that of follicular tumor (Figs. 8.317, 8.318 and 8.319).

When the wall of the IKA brakes, and keratin goes in contact with the dermis, a severe secondary pyogranulomatous inflammation occurs; nucleated squamous keratinocytes can be released singularly, and the inflammatory environment can be the cause of the development of morphological nuclear atypia of keratinocytes. Anisokaryosis and anisocytosis can be relevant, and cells can be misinterpreted as neoplastic squamous cells belong to a SCC (Figs. 8.320 and 8.321).

The thick stratum corneum produced in IKA may show impaired keratinization, and, together with normal corneocytes, a variable amount of superficial ghost cells (false ghost cells) may be present. As discussed above, the size and polygonal silhouette of these corneocytes, which have a central round halo reminiscent of the former nuclear area and are part of the superficial epidermoid-type keratinization, are different from the true ghost cells produced by the matrix epithelium and should not be confused with them. Since the coexistence of lamellar keratin, basal epithelium,

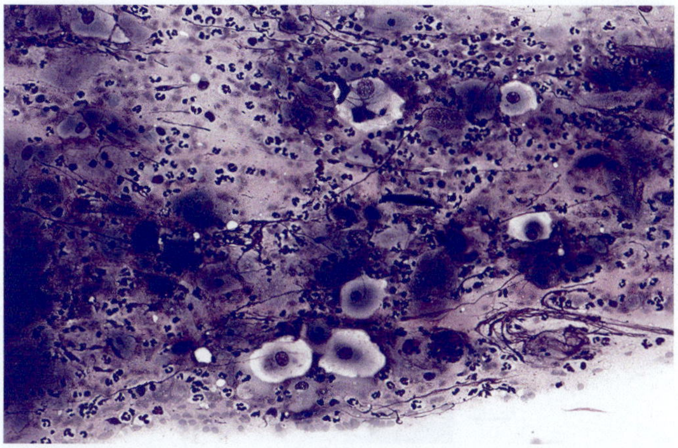

Fig. 8.321 Cytology of an IKA: at high magnification the morphology of squamous cells with evident asynchrony keratinization can be better valuable

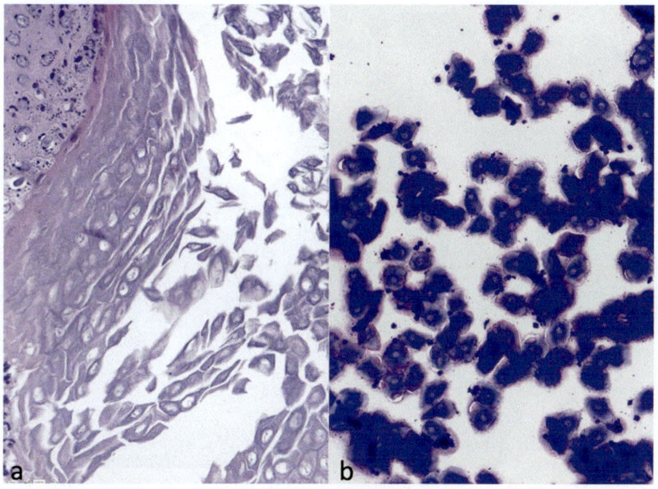

Fig. 8.322 Superficial ghost cells: (**a**) histology—inner part of the wall of an IKA in which granulosum layer is evident with hyperkeratosis that shows a large number of superficial (false) ghost cells; (**b**) cytology—cytological evidence of large, polygonal corneocytes with ghost cell appearance

and ghost cells on the same preparation is a cytological finding observed in trichoepithelioma, it is important to recognize and differentiate the larger superficial ghost keratinocytes (false ghost cells) from the smaller matrix ones (real matrix ghosts), to avoid diagnosing trichoepithelioma instead of IKA (Fig. 8.322).

In some IKAs, a strong mucinous stroma can be present, and, when marked, it is clearly recognizable on the background of slides along with the epithelial clusters and lamellar corneocytes.

(c) *Trichoepithelioma*

Trichoepithelioma is a common follicular neoplasia arising from all three segments of the follicle. Clinically it can be

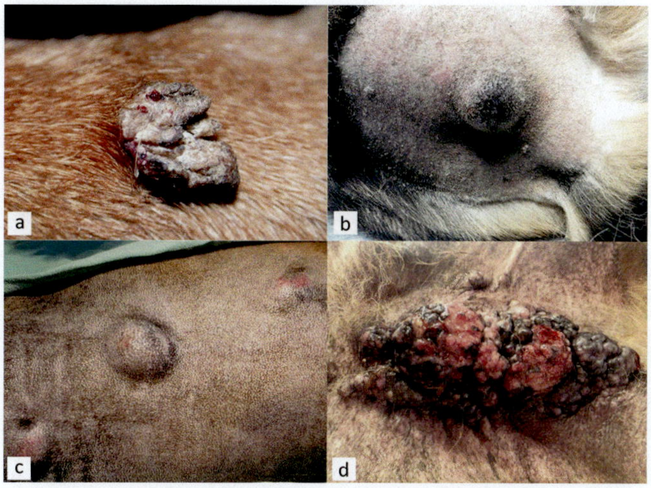

Fig. 8.323 Trichoepithelioma: (**a**) single hyperkeratotic exophytic nodule. Note morphology mimics a papilloma; (**b**) pigmented nodule on the rump of a GSD; (**c**) multiple trichoepitheliomas in a Setter Gordon; (**d**) large and with irregular surface mass on the abdominal skin of a dog with malignant trichoepithelioma. (Courtesy of DVM F. Clemente-Italy)

observed as a single skin nodule, of variable size and with no peculiar clinical appearance. In some predisposed breeds such as Setter Gordon or Golden Retrievers, lesions can be multiple. Less frequently, trichoepithelioma is represented with a nodule or plaque with irregular and ulcerated surface, especially in its less common malignant variant (malignant trichoepithelioma) (Fig. 8.323) [116].

Cytological Findings

Trichoepithelioma originates from all three segments of the follicle and is therefore histologically characterized by multiple cystic nodules containing different types of keratins delimited by nucleated epithelia; the latter are composed of both basaloid matrical epithelium and stratified squamous epithelium, the latter originating from the infundibulum of the follicle. This double epithelial composition determines the simultaneous presence of matrical keratin (ghost cells) and anucleate fully keratinized corneocytes (lamellar keratin) (Figs. 8.324 and 8.325). The isthmic portion is poorly represented in this neoplasm; therefore trichilemmal keratin is not predominant either.

As the cells that compose the wall of the cysts come from all segments of the follicle, different types of keratins fill the lumens. In the area of matrical cells, abrupt keratinization forming ghost cells is evident, whereas in follicular areas where epidermal keratinization is present (infundibulum), lamellar keratin is produced. As the walls of cysts are composed of multiple layers of nucleated cells, several clusters of small basal matrical keratinocytes along with more squamous keratinocytes of infundibular wall are usually sampled by FNB; so, in cytological specimens, the simultaneous

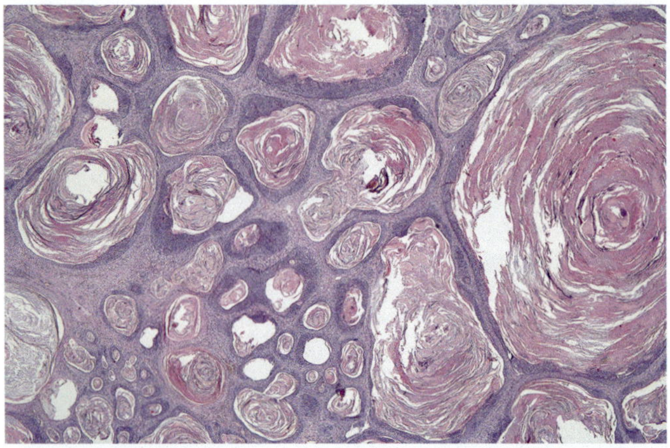

Fig. 8.324 Histology of trichoepithelioma: multiple dermal cysts of variable in size and with lumen filled with keratin

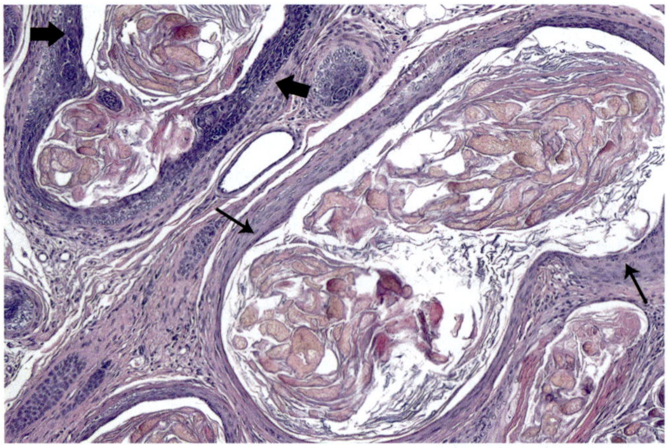

Fig. 8.325 Histology of trichoepithelioma: at high magnification the double type of keratinization of the wall is identifiable—in the same tumor, there is epidermoid keratinization with production of lamellar keratin (corneocytes) (long arrows) and matrical keratinization with production of ghost cells (short arrows)

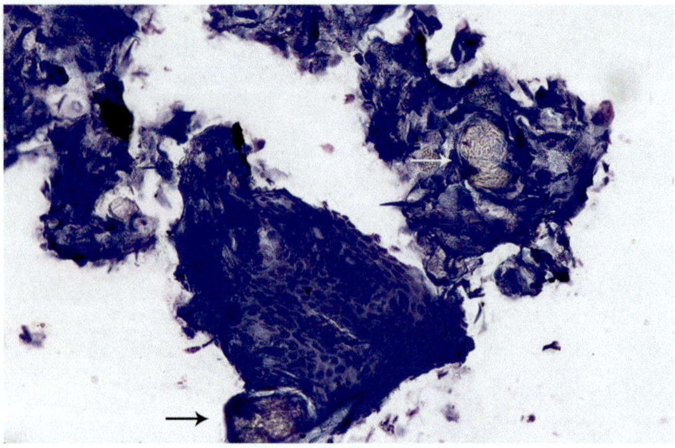

Fig. 8.326 Cytology of trichoepithelioma: a large cluster of basal cells, with regular and uniform nuclei, along with both lamellar and matrical keratinization; the latter highlighted with white and black arrows

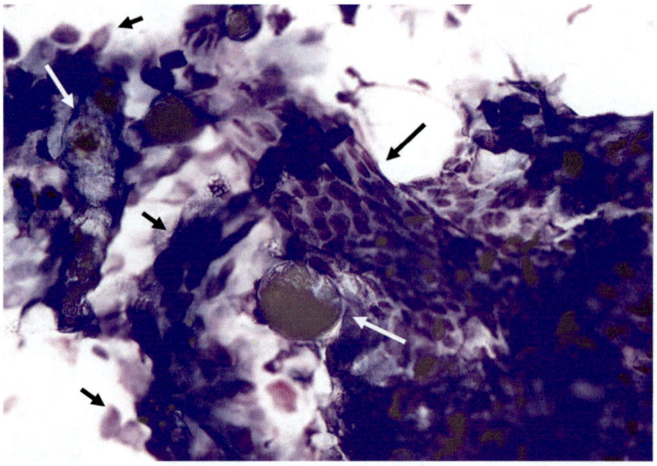

Fig. 8.327 Cytology of trichoepithelioma: a large cluster of pigmented basal cells, with regular and uniform nuclei (black long arrows), along with both lamellar (short black arrows) and matrical keratinization (white arrows)

presence of these two types of nucleated keratinocytes along with the two different keratins (ghost and lamellae) is the hallmark of this follicular neoplasia (Figs. 8.326 and 8.327).

In *cystic trichoepithelioma*, small satellite cystic nests of basal cells surround or are displaced laterally by a large main cyst; in this histological variant, the cytology may be different depending on where the needle is inserted. If the needle enters the large cyst in which the wall is composed of a few flattened layers of basal keratinocytes, a mix of lamellar corneocytes and matrix ghost cells without nucleated basal cells may be obtained; while if the sampled area of the tumor is composed of smaller cysts with a thick basal epithelial wall, in addition to the two types of keratin, a variable quantity of nucleated basal cells can be sampled; in the first case, the cytology is similar to that of hybrid cyst, from which it cannot be cytologically differentiated; in the second case, the presence of basal aggregates allows for a cytological diagnosis of trichoepithelioma (Figs. 8.328 and 8.329).

Fortunately, almost all trichoepithelioma have a benign behavior, and a false negative cannot create serious outcome for the patient.

The percentage amount of the different keratins (matrical and lamellar) depends on the part of the follicle involved and can be different between the various cystic formations of the same tumor; usually, however, in trichoepithelioma, the number of ghost cells is lower than or equal to that of the lamellar keratin. The latter is an important cytological aspect to differentiate trichoepithelioma from pilomatricoma; in fact, since the latter is a pure neoplasm of the matric keratinocytes, the keratin is represented almost exclusively by ghost cells, and it is therefore not possible to visualize lamellar keratin, or, if

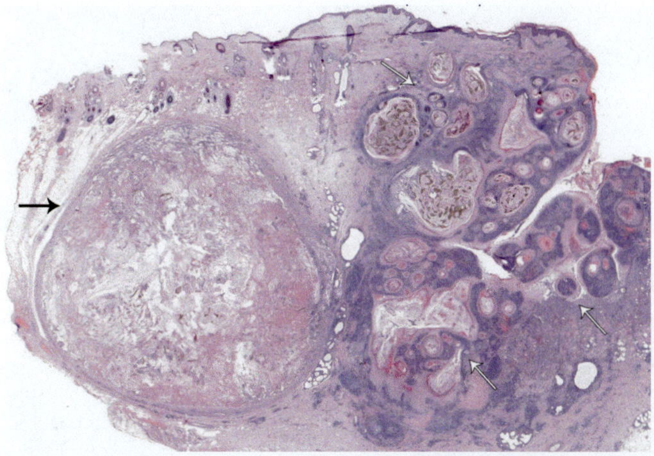

Fig. 8.328 Histology of cystic trichoepithelioma: heterogeneous composition—large cyst lined by a thin wall (black arrow) and many smaller satellite cysts with a thick basal epithelial wall (white arrows). Cytology can be different depending on where the needle is inserted

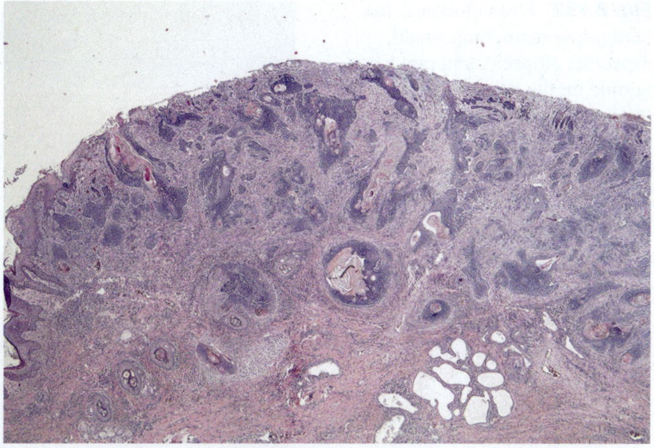

Fig. 8.330 Histology of malignant trichoepithelioma: irregular in shape nests and trabeculae of basal neoplastic keratinocytes with infiltrating pattern. Note as only few cysts show keratinization, and this is a histological feature of poor differentiation of the tumor

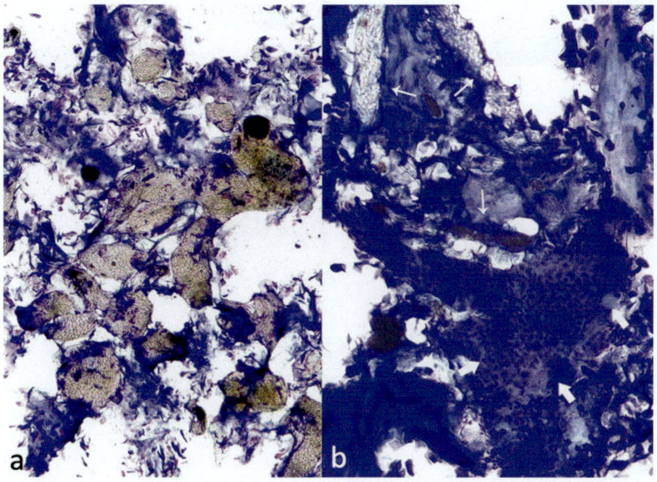

Fig. 8.329 Cytology of cystic trichoepithelioma: (**a**) lamellar and matrical ghost cells; no basal epithelial cells are evident; in these case, cytology cannot differentiate a cystic trichoepithelioma from a hybrid cyst; (**b**) large cluster of basal cells, with regular and uniform nuclei (short arrows), along with both lamellar and matrical keratinization; both achromatic and pigmented ghost cells (long arrows). In this case a cytological diagnosis of trichoepithelioma can be done

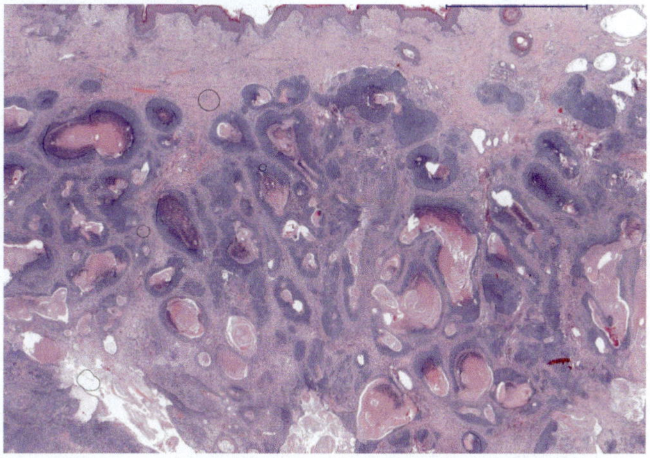

Fig. 8.331 Histology of malignant trichoepithelioma: irregular in shape cystic nests and trabeculae of basal neoplastic keratinocytes with infiltrating pattern. In these tumor aspects of multiple keratinization are present

sometimes present, it is in very small quantities compared to trichoepithelioma. As for all the keratinizing follicular tumor, the breakage of cystic lesions and spreading of keratin in the dermis evoke a strong pyogranulomatous with giant cell secondary inflammation.

Malignant trichoepithelioma is less frequently observed compared with the benign counterpart [116, 117]. Cytology features depend on the degree of differentiation and if necrosis and degeneration of the cyst content occur. Basal cells have scant cytoplasm and can show varying criteria of atypia and lost cohesivity. The presence of irregular clusters of basal cells, with increased anisokaryosis and a detectable number of mitotic cells, usually few in benign trichoepitelioma, can increase the suspicion of malignant neoplasia. Other diagnostic features such as infiltration of subcutaneous tissue and lymphatic vessels need a histopathological examination (Figs. 8.330 and 8.331).

(d) *Pilomatricoma*

Pilomatricoma is a common follicular neoplasm mainly observed in canine breeds such as Basset Hounds, Beagle, and many other hunting dogs; no typical clinical appearances are associated to this follicular neoplasm (Fig. 8.332) [116].

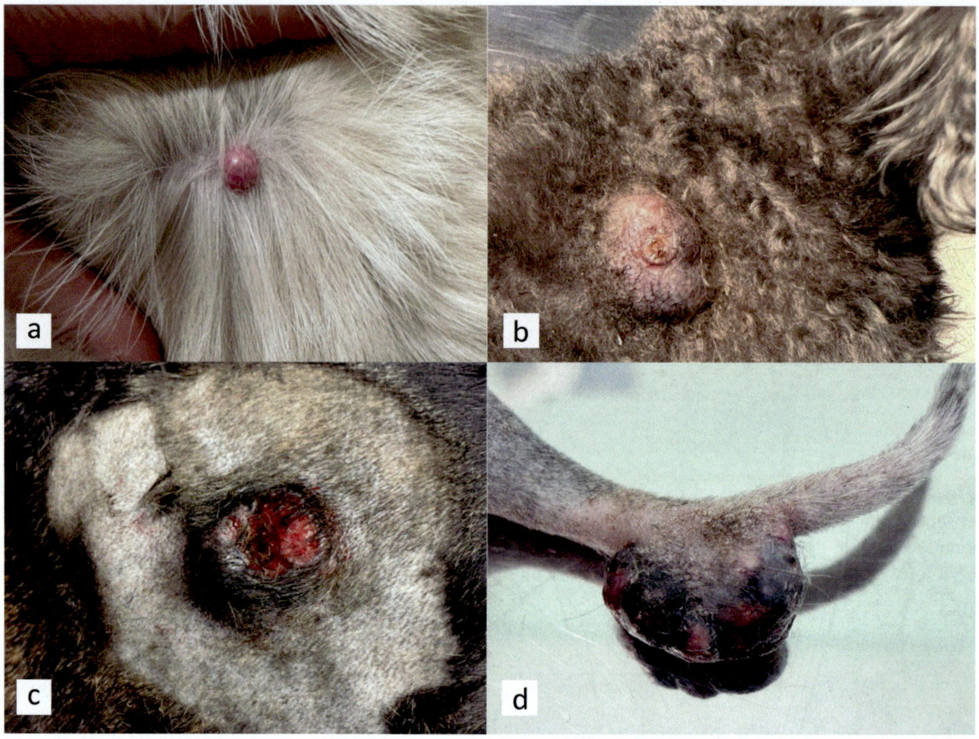

Fig. 8.332 Pilomatricoma: (**a**) benign pilomatricoma, small, aspecific, alopecic, and pinkish nodule on the skin of a Golden Retriever; (**b**) large erythematous and ulcerated malignant pilomatricoma on the shoulder of a dog; (**c**) ulcerated malignant pilomatricoma; (**d**) large malignant pilomatricoma on the tail of a dog. (Courtesy of DVM G. Di Bella-Italy)

Cytological Findings

Pilomatricoma is a follicular tumor derived from the primitive basal germ cells of the follicle that differentiate into *hair matrix cells*; as trichoepithelioma, it is characterized by multiple cysts, but in pilomatricoma, the wall is delimited only by basaloid cells with large and ovoid nuclei that dissolve originating keratinized ghost cells, as described above (Figs. 8.333, 8.334 and 8.335).

Classical cytology of benign pilomatricoma is characterized by a variable amount of clusters of basaloid matrical cells associated to ghost cells, which are the only type of keratin observed; even if, as previously mentioned, it is possible to observe a very small amount of lamellar keratin, the latter is so few that is not relevant on slides. Basaloid matrical cells are uniform in size and arranged in clusters with regular and well-defined profile. Cytoplasm is so scarce that it is not easily detected. In lucky cases, the ghost cells are numerous and incorporated or strictly connected to basal matrical keratinocytes demonstrating what occurs in this follicular neoplasia in which an aberrant *ex abrupt keratinization* starts directly from basal cells (Figs. 8.336, 8.337 and 8.338) [119].

Matrical ghost cells can be heavily pigmented, and in some neoplasia, the ghost cells undergo calcification or osseous metaplasia, and therefore, in cytological samples, it is possible to observe mineralization and a variable number of secondary inflammatory cells, especially giant histiocytic cells; for this characteristic, this tumor is also known in human medicine as *necrotizing* and *calcifying epithelioma*

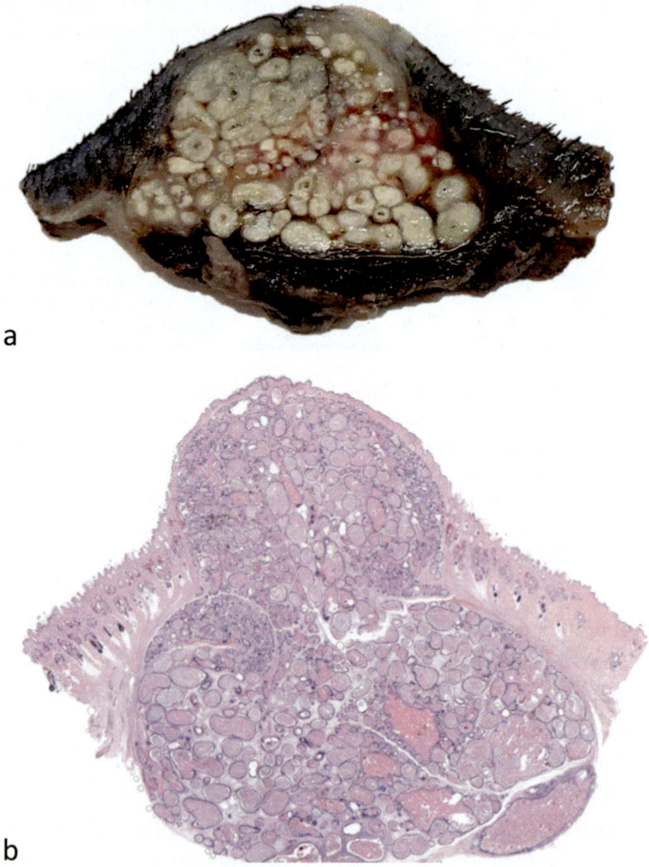

Fig. 8.333 Benign pilomatricoma: (**a**) macroscopic neoplasm section shows many small nodules with a cystic central area; (**b**) histopathology—multiple and small keratinizing nodules composed of basal cells

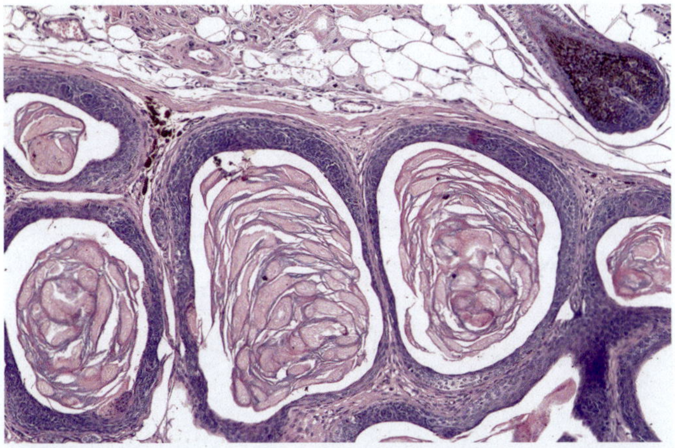

Fig. 8.334 Histology of *benign* pilomatricoma: nodular cystic lesions lined by matrical keratinocytes and filled of ghost cells

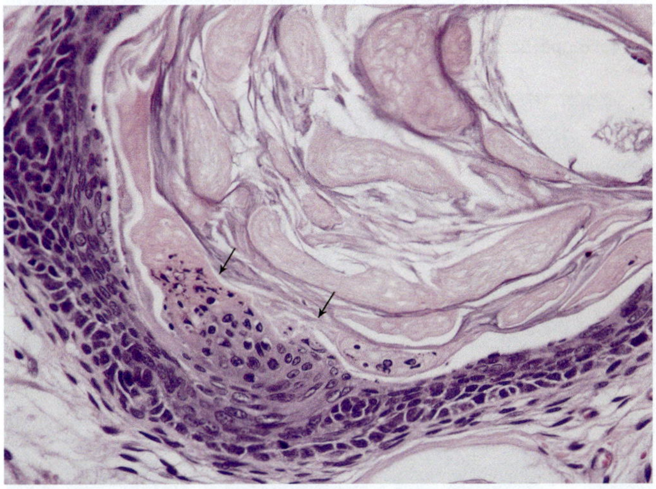

Fig. 8.335 Histology of benign pilomatricoma: at high magnification, the ex abrupt matrical keratinization is recognizable as basal cells which directly transform themselves in ghost cells (arrows)

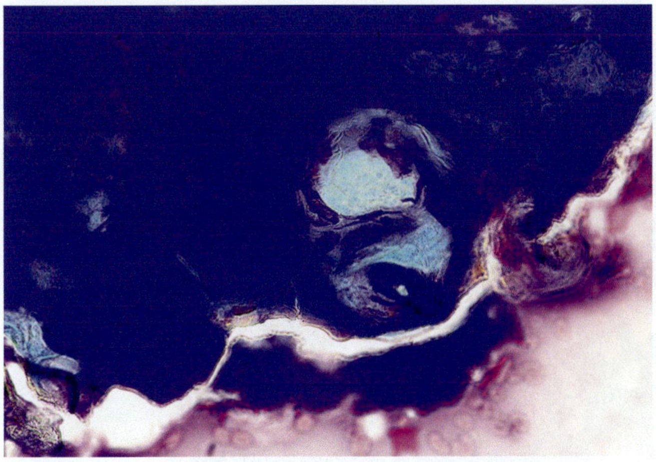

Fig. 8.336 Cytology of benign pilomatricoma: a large cluster of uniform in size basal cells in which two clear azul areas composed of ghost cells are incorporated

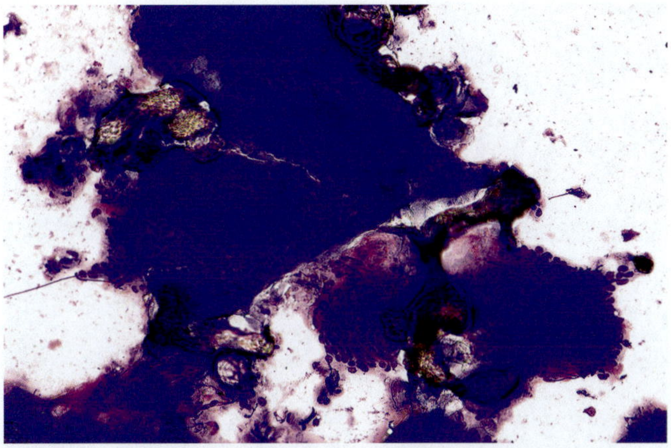

Fig. 8.337 Cytology of benign pilomatricoma: numerous clusters of benign basal matrix cells in which numerous ghost cells are adhered, indicating the *ex abrupt* keratinization process they undergo

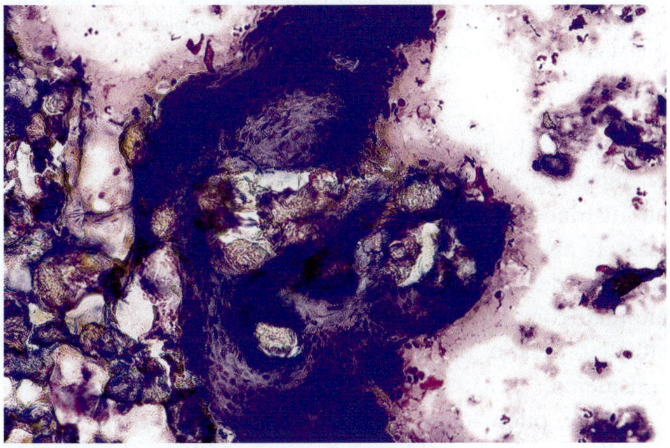

Fig. 8.338 Cytology of benign pilomatricoma: numerous clusters of benign basal matrix cells in which numerous ghost cells are adhered, indicating the *ex abrupt* keratinization process they undergo

of Malherbe. Pyogranulomatous inflammation is also seen in ulcerated tumors or when the cyst wall ruptures and ghost cells spread into the dermis. When multinucleated macrophage cells (giant cells) are numerous and epithelioid macrophages are overlapped and crowded around the matrical keratin in an attempt to phagocytize it, their morphology is not as clearly recognizable as single cells and may be confused with large clusters of malignant basal cells. Therefore, care must be taken before diagnosing a malignant pilomatricoma with cytology if the cellular morphology is not clear, as we could be faced with a benign pilomatricoma or even a banal inflamed matrical cyst (Fig. 8.339).

The real differential diagnosis is the matrical cyst. It should be emphasized that the presence of large amounts of ghost cells on slide as unique population of cells is a cytological finding strongly suggestive of matrical cyst. In these

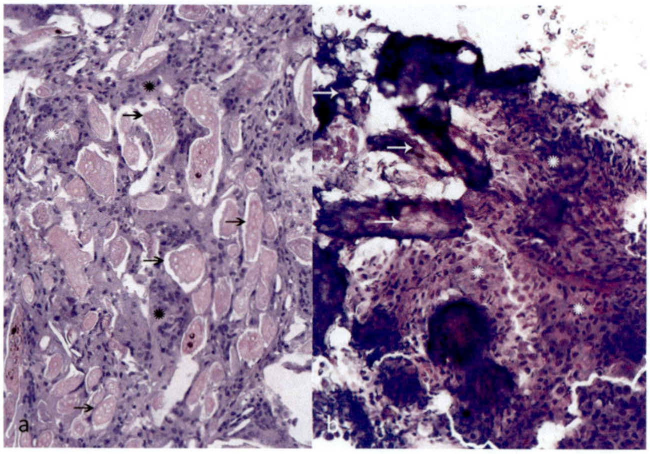

Fig. 8.339 Inflamed benign pilomatricoma. Histological and cytological comparative aspects of severe inflammation due to ghost cells spreading in the dermis. Stars and arrows highlight, respectively, inflammatory cells and ghost cells (black color for histology and white for cytology). In this case, inflammatory cells must not be confused with epithelial neoplastic cells; furthermore, despite no basal cells are evident, these images come from an inflamed benign pilomatricoma

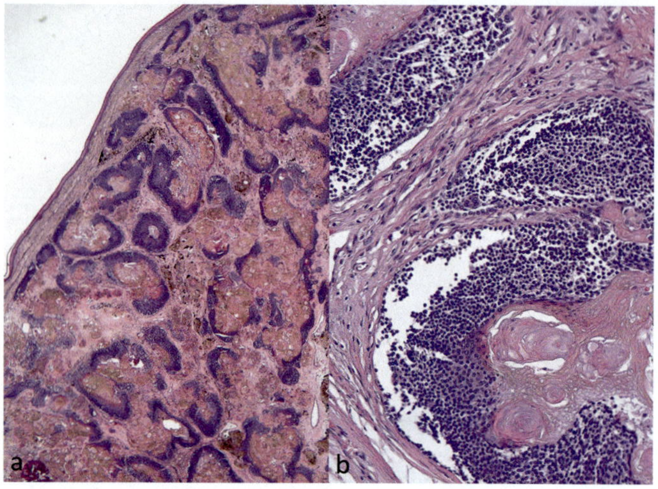

Fig. 8.340 Histology of malignant pilomatricoma: (**a**) many irregular and incomplete cystic lesions, filled with matrical keratin; (**b**) loss of cohesivity of neoplastic matrical basal cells is clearly recognizable. Note the eosinophilic groups of ghost cells

lesions in fact, the cyst's wall is very thin, and the possibility of sampling a significant amount of basaloid cells as in pilomatricoma that is composed of many small hypercellular cysts is absolutely exceptional; therefore, when only ghost cells are detected, a matrical cyst is more probable than a pilomatricoma. Regarding other differentials, in case few lamellar keratin is present, a trichoepithelioma could be considered.

Rarely, pilomatricoma may be malignant and can exceptionally metastasize, although for some authors, the percentage of the malignant variant is underestimated [115, 118]. Cytology of the malignant pilomatricoma is characterized by usually highly cellular slides, with many neoplastic basal cells, arranged in clusters with irregular shape and with not well-defined borders; cytoplasm is so scarce that apparently seems absent. Cells tend to be less cohesive and can be released as discrete cells, very often as naked nuclei, especially at the periphery of the clusters; this is a common cytological finding in malignant pilomatricoma. Variable degrees of anisocytosis and anisokaryosis are present. Mitotic figures can be from low to numerous. Since malignant pilomatricoma can be poorly differentiated, the number of ghost cells can be very low, and, without their visualization, the matrical origin of basaloid neoplastic cells cannot be defined. In many cases the number of ghost cells is enough to not create doubts about the matrical origin of the neoplastic basal cells. As discussed for malignant trichoepithelioma, the invasiveness' features of the neoplasia can be only highlighted with histology (Figs. 8.340, 8.341, 8.342 and 8.343).

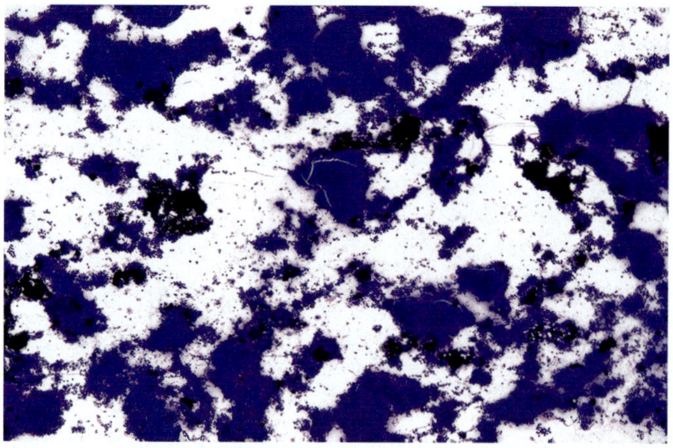

Fig. 8.341 Cytology of malignant pilomatricoma: highly cellular sample composed of epithelial clusters. The irregular apparently amorphous areas are groups of ghost cells

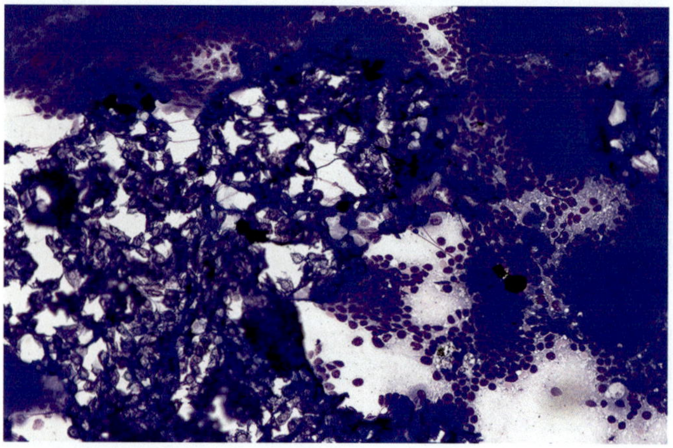

Fig. 8.342 Cytology of malignant pilomatricoma: at medium magnification scarcely cohesive clusters of neoplastic basal cells with irregular and not well-defined margins are present. Note the high number of ghost cells represented from polygonal to spindloid appearance

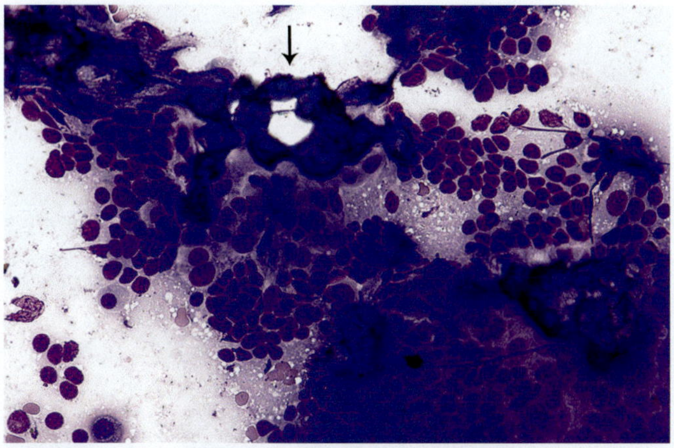

Fig. 8.343 Cytology of malignant pilomatricoma: at medium magnification scarcely cohesive clusters of neoplastic basal cells with irregular and not well-defined margins are present. Note the high number of ghost cells represented from polygonal to spindloid appearance

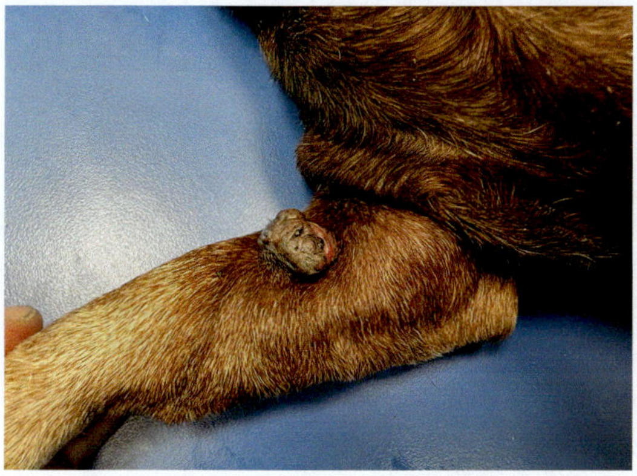

Fig. 8.344 Trichofolliculoma: pigmented nodule on the fore right arm of a Dachshund

(e) *Trichofolliculoma*

Trichofolliculoma is a rare tumor of the follicle, of the trichogenic epithelium, characterized histologically by the presence of multiple cystic area delimited by basal epithelium which keratinizes, for the presence of the granulosum layer, into anucleate horny lamellae (corneocytes). Histologically multiple primary and secondary follicles are associated and sometimes may include sebaceous glands (Fig. 8.344) [116].

Cytological Findings

The main components of trichofolliculoma are keratin lamellae, which fill cysts, and in some cases they are the only collected cells. Few clusters of basaloid epithelium can be sometimes collected. Cytological features depend on if the needle is inserted into the large cysts; in that case the basal nucleated part can be missed, and cytology is composed of lamellar anucleate keratin, and infundibular cysts or IKAs must be considered as differentials (Figs. 8.345 and 8.346).

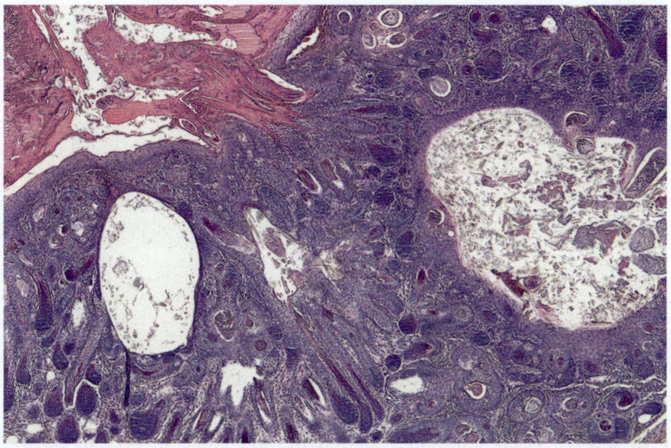

Fig. 8.345 Histology of trichofolliculoma: multiple cystic areas delimited by basal epithelium, which keratinizes in lamellar corneocytes. Note multiple primary and secondary follicles at the periphery of the epithelium and the central large cystic area filled with lamellar keratin (upper left part of the image)

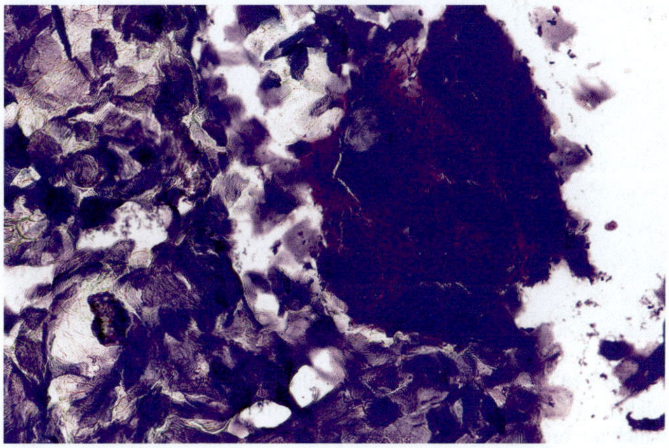

Fig. 8.346 Cytology of trichofolliculoma: many anucleate keratinocytes along with a large cluster of nucleated basaloid keratinocytes, without cytological aspects of atypia

(f) *Tricholemmoma*

Tricholemmoma is an uncommon benign follicular tumor with differentiation to the isthmic segment of the external root sheath. A malignant counterpart can exceptionally occur. Large nests and cords of epithelial cells characterize this neoplasia for which two histotypes are recognized, *isthmic* and *inferior*, depending on which portion of the follicle is the neoplasia origin [116]. Grossly, neoplasms can be presented as not-specific dermal nodules sometimes with an exophytic growth (Fig. 8.347).

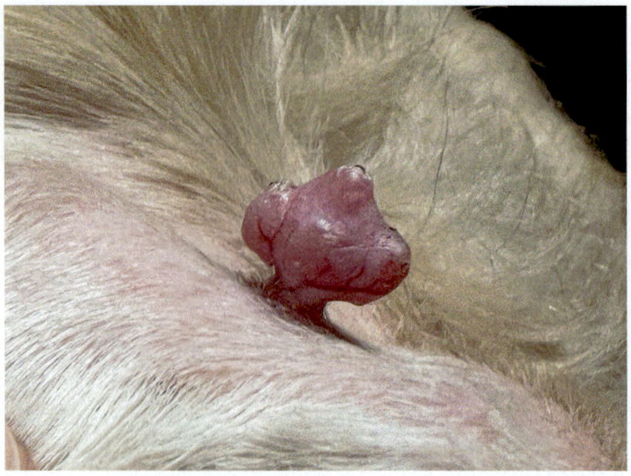

Fig. 8.347 Tricholemmoma: exophytic and alopecic nodule on the skin of dogs

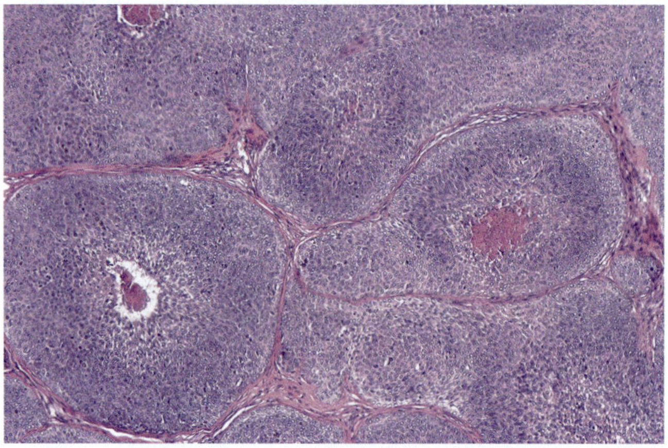

Fig. 8.348 Histology of benign isthmic tricholemmoma: multiple islands of regular shape composed of epithelial cells, of uniform size, which are organized around a central area in which a trichilemmal-type keratinization is observed

Cytological Findings

Histologically tricholemmoma are composed of epithelial islands of cells with a central area in which a trichilemmal-type keratin, eosinophilic in color, is produced (Fig. 8.348). Cytologically, a variable number of clustered epithelial cells, with regular nuclei and small to medium pale blue cytoplasm, characterize benign tricholemmomas. The cells do not show squamous differentiation, and, since the isthmic variant is the most commonly observed histotype, an amorphous material can be detected along with the nucleated epithelial clusters, which represents the trichilemmal keratin produced by the neoplastic cells themselves. In some cases, a detachment of single cells from clusters is present (Figs. 8.349 and 8.350). Malignant tricholemmoma is absolutely rare in dogs, and the severity of cytological atypia depends on the degree of differentiation (Figs. 8.351, 8.352 and 8.353).

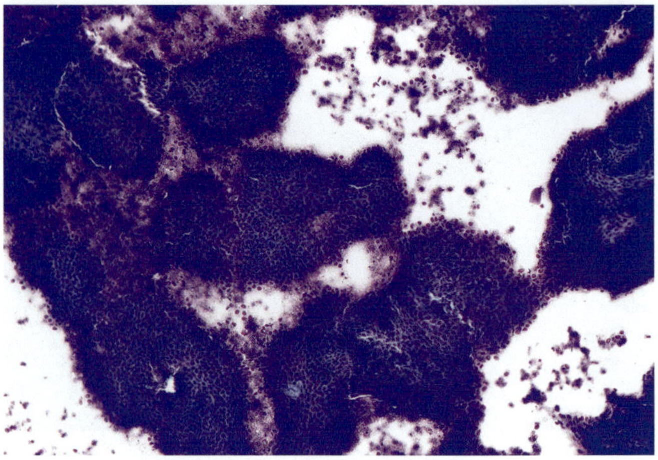

Fig. 8.349 Cytology of benign isthmic tricholemmoma: large clusters of epithelial cells, uniform in size and morphology

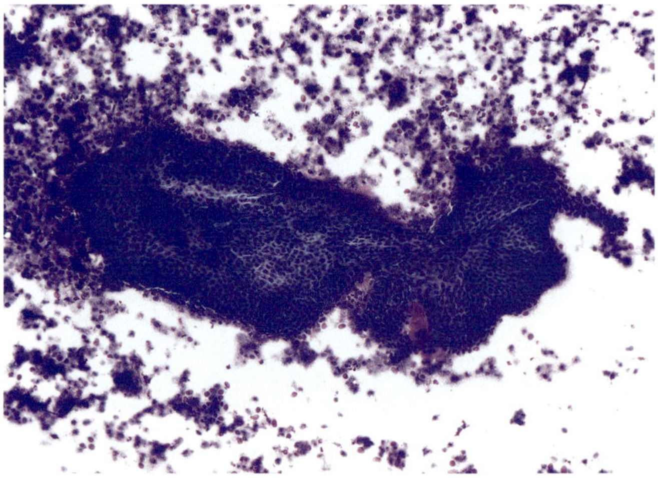

Fig. 8.350 Cytology of benign isthmic tricholemmoma: large cluster of uniform in size epithelial neoplastic cells. Atypia are few

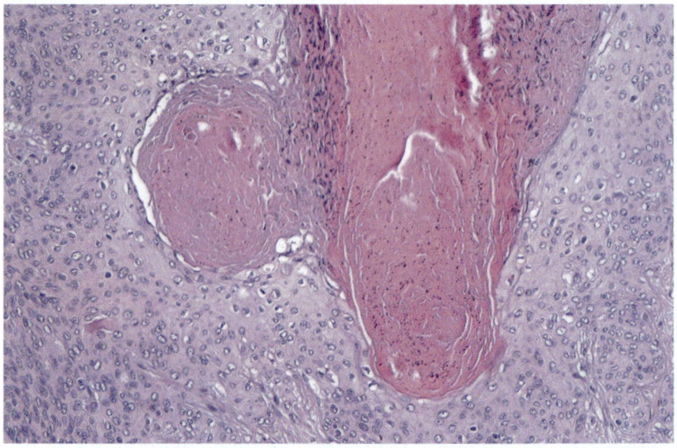

Fig. 8.351 Histology of malignant isthmic tricholemmoma: neoplastic follicular epithelial cells, with moderate degree of atypia. Note the trichilemmal-type keratinization in the center of the neoplastic island

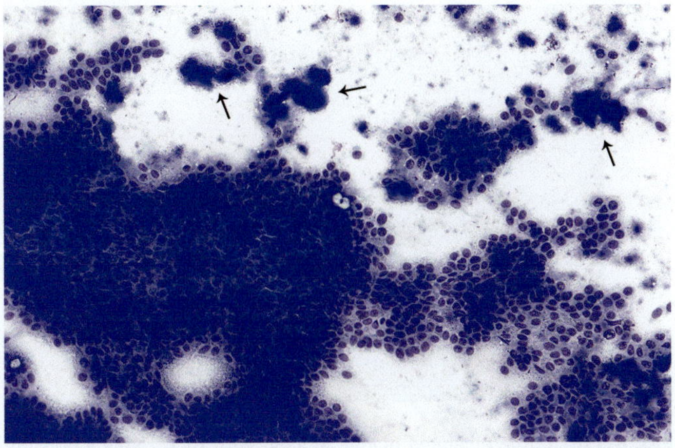

Fig. 8.352 Cytology of malignant isthmic tricholemmoma: large cluster of neoplastic external root sheath cells, with irregular profile; cells are starting to lose cohesivity. Note the blue in color trichilemmal keratin (arrows)

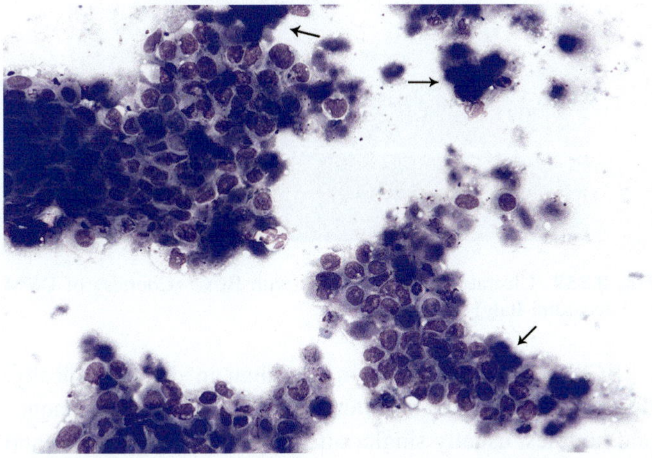

Fig. 8.353 Cytology of malignant isthmic tricholemmoma: evident atypia represented by cyto-architectural disorganization, anisokaryosis, and loss of cohesivity. Note the blue color in trichilemmal keratin (arrows)

(g) *Warty Dyskeratoma*

Warty dyskeratoma is a very rare benign epidermal skin neoplasm of dog. Usually it appears as a cup-shaped nodule with a central pore from which keratin protrudes, as a result of an dilated follicle-like invagination [116]. This clinical aspect is quite similar to that observed in IKAs. In some cases, no pore is evident and a morphological nonspecific nodule can be observed (Fig. 8.354).

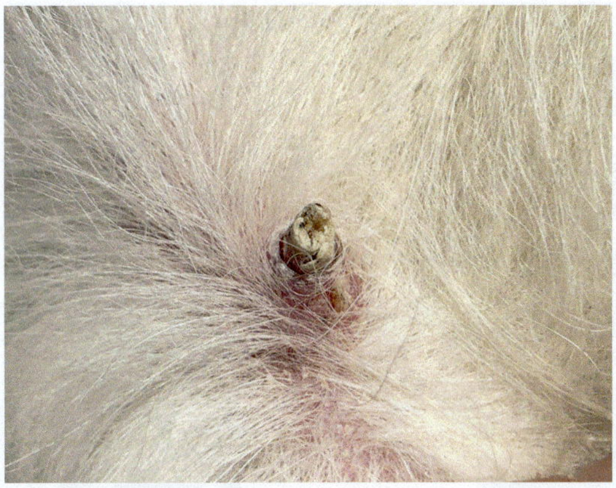

Fig. 8.354 Warty dyskeratoma: small and exophytic alopecic nodule on the tail of a dog. Note the cup-shaped and the keratin material that protrudes from the center. The nodule perfectly mimics an IKA

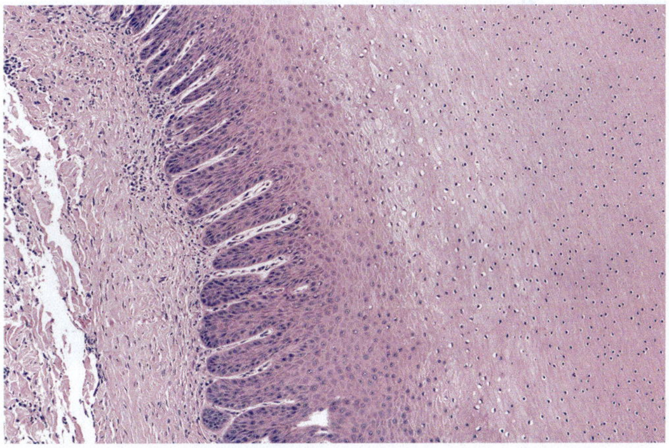

Fig. 8.355 Histology of warty dyskeratoma: note the typical rete pegs of the basal part of the peripheral epithelium, which extend into the dermis and line a central cyst filled with keratin

Cytological Findings

Since the internal part of this benign neoplasm shows dyskeratinization, acantholysis of keratinocytes can be observed, which are collected as the only typology of cells when the needle penetrates the intraluminal part. The cytological evidence of many lamellar squamous keratinocytes, both nucleated and completely keratinized, must always place in the differential diagnosis other squamous keratinizing neoplasms such as papilloma, IKA, and some SCC characterized by poor aspects of cytological atypia. The final diagnosis requires histopathological examination (Figs. 8.355, 8.356 and 8.357).

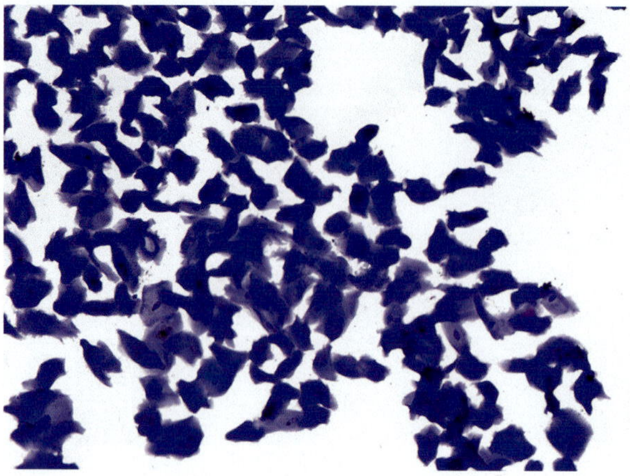

Fig. 8.356 Cytology of warty dyskeratoma: squamous both nucleate and anucleate keratinocytes

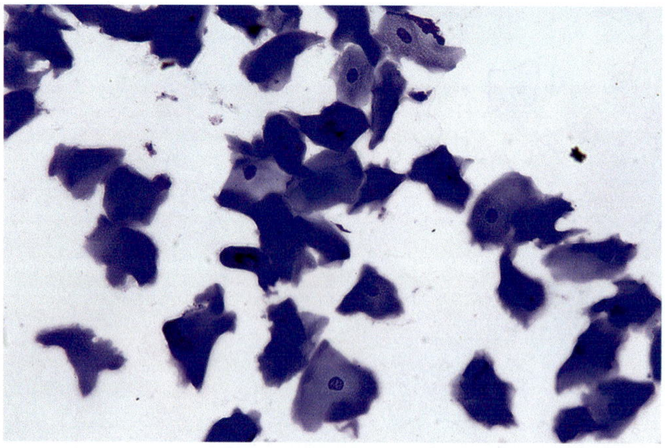

Fig. 8.357 Cytology of warty dyskeratoma: squamous both nucleate and anucleate keratinocytes

8.3.5 Basal Tumor of Epidermidis

In the past, "basal" neoplasms included all tumors composed of cells with basal appearance, creating great confusion about their differentiation and classification, also because, for some of them, histopathological differentiation can be difficult; in fact, some basal tumors share the same pattern of growth, as occurs for cystic trichoblastomas, solid-cystic apocrine adenomas, and some epidermal basal tumors that show solid-cystic features. With the increase in morphological knowledge and immunohistochemical discoveries regarding epidermal and adnexal basal neoplasms, many tumors that were once grouped under the umbrella of basal tumors have been reclassified as trichoblastomas and ductal neoplasms of the apocrine glands or basal cells of sebaceous glands; for this reason, currently, in the new WHO classification, only tumors originating from the basal cells of the epidermis, namely, basal cell tumor (BCT) and basal cell carcinoma (BCC), have remained as basal cell tumor [116, 117].

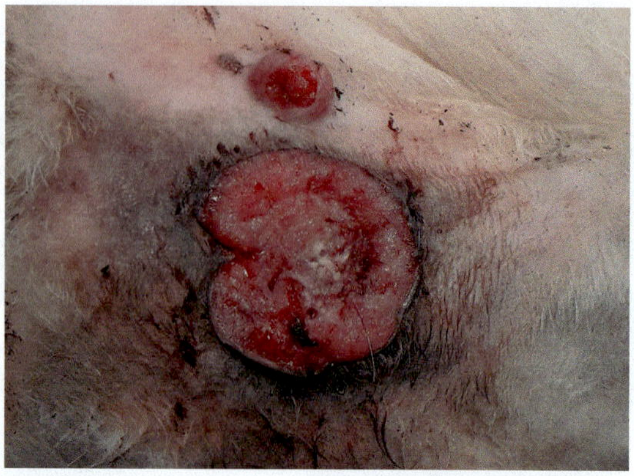

Fig. 8.358 Large ulcerated plaque in a cat with BCC. Note a dorsally second smaller lesion. (Courtesy of DVM R. Benedetti-Italy)

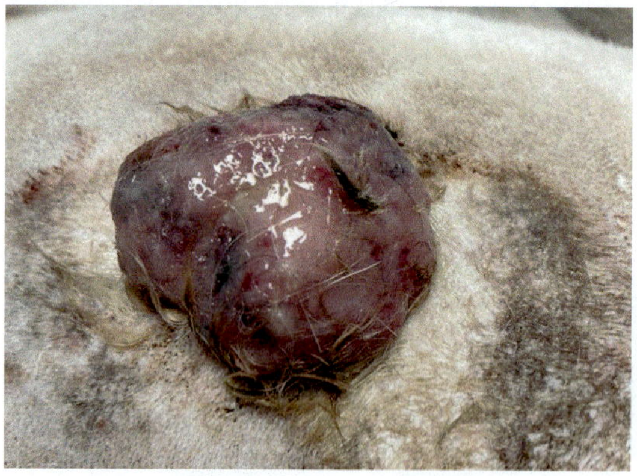

Fig. 8.359 Ulcerated nodule in a cat with BCC. (Courtesy of DVM L. Luccardini-Italy)

BCT is observed more in cats than in dogs. Clinically, basal cell tumors are characterized by the presence of plaques and nodules, usually single, often ulcerated (Figs. 8.358 and 8.359). In cats, similar lesion can represent an evolution of BISC and can be multiple.

Histologically, BCTs are composed of a neoplastic proliferation of basal cells with a horizontally oriented pattern of growth, composed of nests and trabeculae of uniform basal epithelial cells that tend to go down into the superficial dermis; a line of palisade-arranged basal cells is usually recognizable at the periphery of the epithelial islands. Over the time, especially in carcinomas, cells infiltrate the deep tissues (Figs. 8.360 and 8.361).

In cytological specimens from benign BCTs, the cells show a uniform round to polygonal silhouette with barely recognizable cytoplasm and are released in cohesive clusters without atypia; in these cases, if the history of patient and the morphology of the lesions are unknown, it is really complex to differentiate cytologically a basal tumor from a trichoblastoma or from other basal adnexal gland benign neoplasms (Figs. 8.362 and 8.363).

In BCCs instead, epithelial basal clusters have less defined margins, and cells are less organized and overlapped;

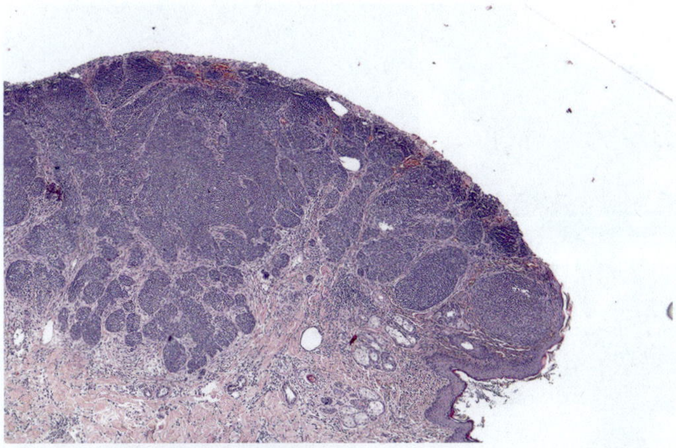

Fig. 8.360 Histology of BCT: multiple islands and trabeculae of uniform in size benign basal cells. The overlying epidermis is diffusely ulcerated

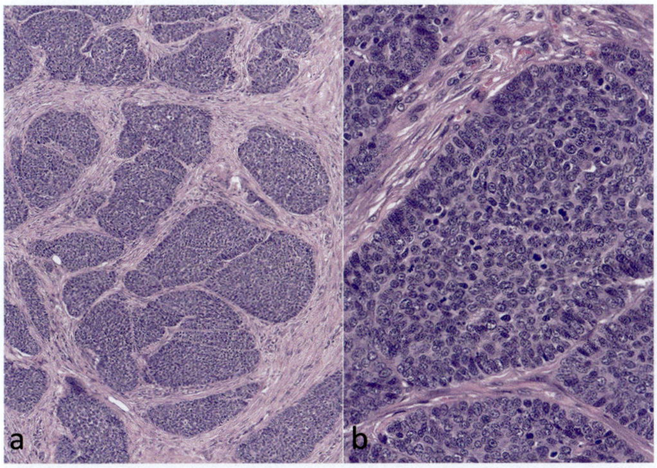

Fig. 8.361 Histology of BCT: (**a**) regular in size multiple islands of benign basal cells. (**b**) At high magnification the regular size and shape of basal cells are clearly recognizable. Note the palisade arrangements of the basal peripheral cells

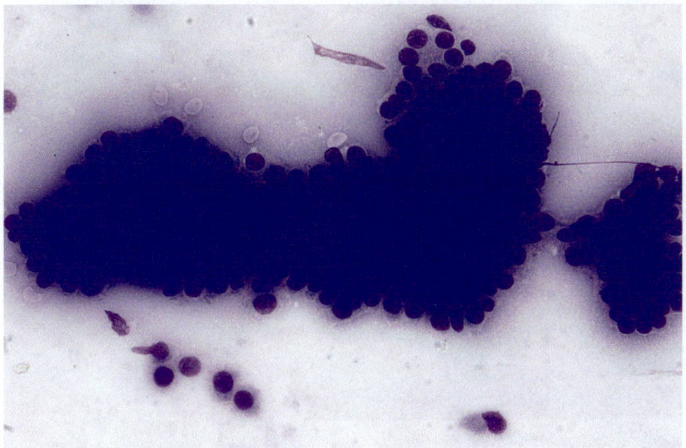

Fig. 8.362 Cytology of BCT: large cluster of benign basal cells. Note as cells are uniform in size

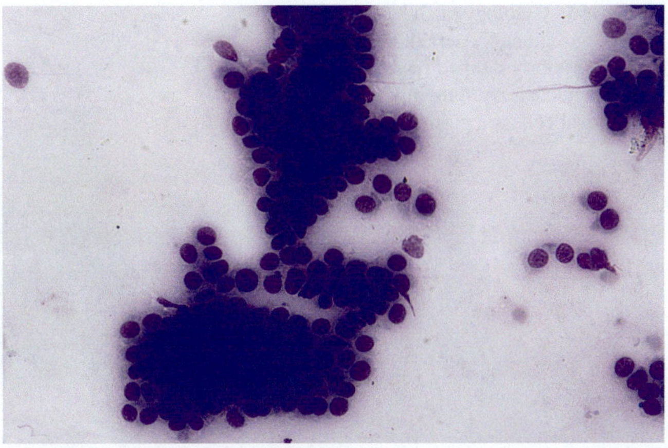

Fig. 8.363 Cytology of BCT: clusters of benign basal cells. Note the palisade arrangements of peripheral cells

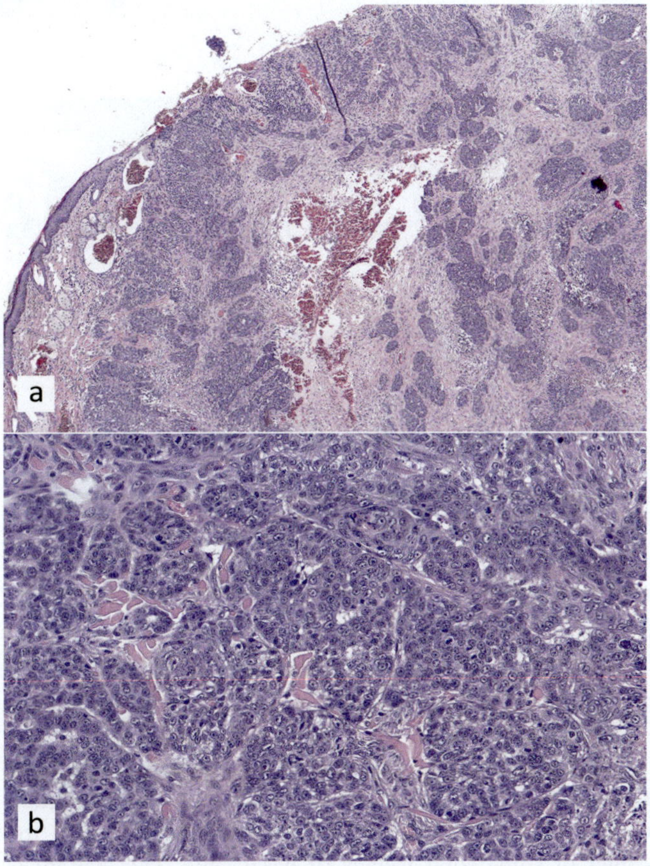

Fig. 8.364 Histology of BCC: (**a**) multiple irregular islands and trabeculae of basal cells; (**b**) at high magnification many islands of basal cells, with not well-defined margins and chaotic organization, are recognizable. Anisokaryosis and prominent nucleoli are evident

anisocytosis and anisokaryosis are more evident and the number of mitoses increases. The infiltrating aspect of BCC can only be assessed by histological examination. In BCCs characterized with scarce atypia, the cytological features cannot be differentiated from the benign counterpart (Figs. 8.364, 8.365 and 8.366).

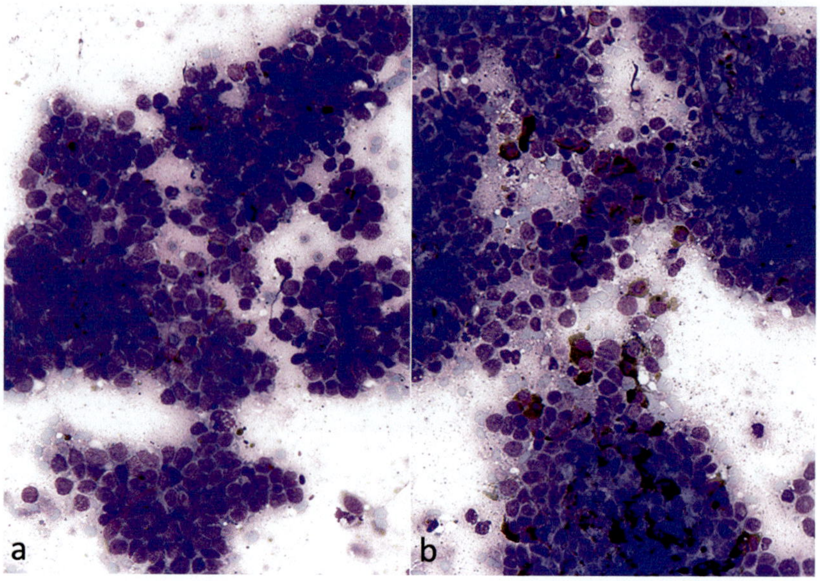

Fig. 8.365 Cytology of BCC: (**a**) multiple irregular clusters of basal cells with poorly defined margins. Note the nuclear overlapping; (**b**) many basal cells are pigmented

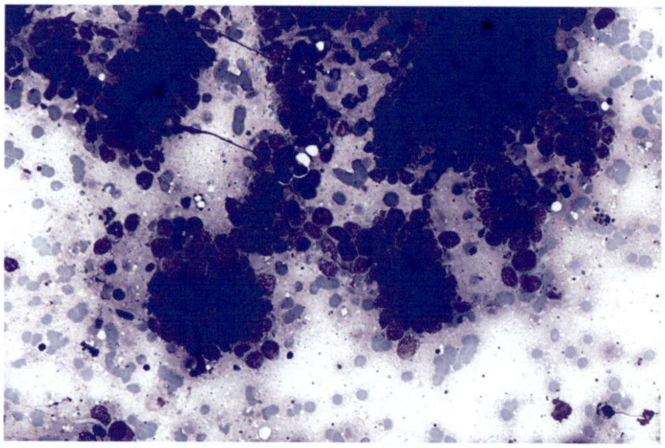

Fig. 8.366 Cytology of BCC: at high magnification, irregular shape of cluster, overlapped nuclei, and mild anisokaryosis are better recognizable

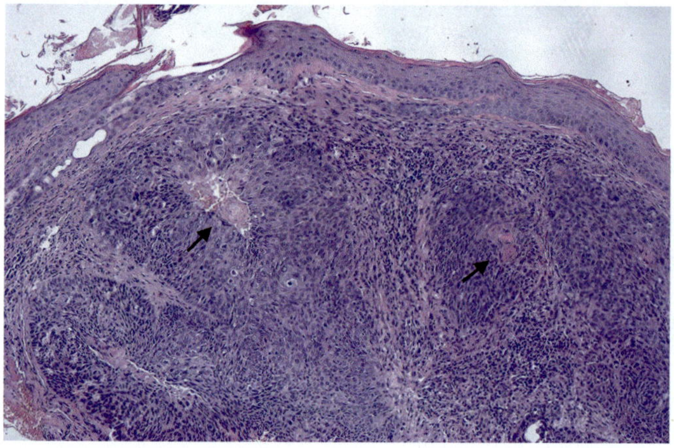

Fig. 8.367 Histology of squamous BCC: islands of basal cells with central keratinized area

Usually, basal epidermal neoplasms are not keratinizing tumors, but in rare cases, a squamous differentiation can be observed with formation of squamous foci in the center of some solid islands of basal cells. This variant is separated from the basal tumors in WHO classification, maybe since in humans it is a well-defined distinct entity. In any cases histopathology is very similar to a basal cell carcinoma, but the presence of abrupt squamous keratinization is evident, and a variable amount of keratin can be collected along with the basal cells (Figs. 8.367, 8.368 and 8.369).

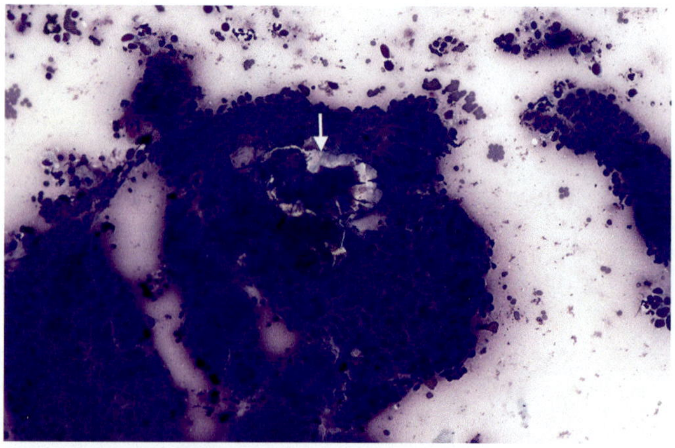

Fig. 8.368 Cytology of squamous BCC: clusters of basal cells with central keratinized area (white arrow)

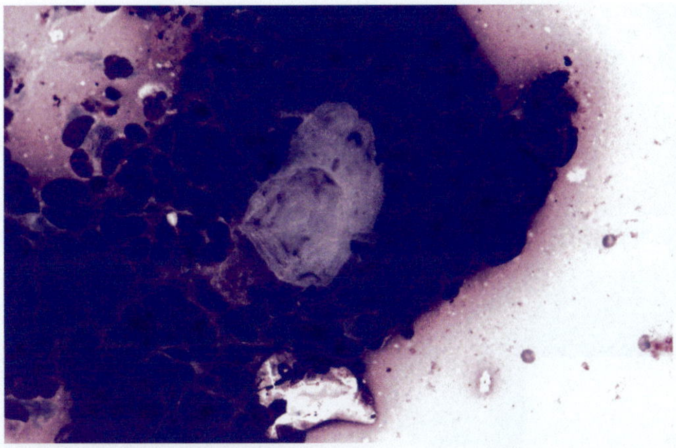

Fig. 8.369 Cytology of squamous BCC: clusters of basal cells with central area in which fully keratinized corneocytes are evident

8.3.6 Sebaceous Gland Tumors and "Modified" Sebaceous Gland Tumors

Tumors of the *sebaceous glands* can originate from both mature sebocytes and immature cells that line peripherally the differentiated sebocytes and are known as *basal*, *germinative*, or *reserve cells*. The reserve cells differentiate into lipidized sebocytes. The new histological classification of sebaceous glands includes, as neoplasms of reserve cells, only the *sebaceous epithelioma*, whereas tumors arising from mature sebocytes comprise *sebaceous hyperplasia*, *sebaceous adenoma*, and the *sebaceous ductal adenoma* [129, 130]. The malignant counterparts of both basal and mature sebocytic tumors, formerly, respectively, named as *sebaceous epitheliomatous carcinoma* and *sebaceous sebocytic carcinoma*, are not more present on the new histopathological classification of sebaceous neoplasia [116]. Since a malignant tumor with clearly sebaceous differentiation is rare and most of the malignant sebaceous neoplasia origin from the basal cells, the malignant sebaceous neoplasms are now included under the name of *sebaceous adenocarcinoma* [129]. The sebaceous tumors of the eyelids are named *meibomian gland*, in honor of the German researcher Heinrich Meibom; these neoplasms comprise both adenoma and epithelioma.

Other types of modified glands of sebaceous origin are present in the skin; these "specialized" glands, named *perianal/circumanal glands* or *hepatoid glands*, so named since of their histomorphological similarity to hepatocytes, are well recognized in dogs.

Clinical Findings

Benign hyperplasia of the sebaceous glands is the most frequent type of sebaceous tumors, especially in older dogs such as English Cocker Spaniel and Miniature Poodle, whereas it is rarer in cats [129, 130]. Grossly, lesions are characterized by nodules of various sizes, usually of 0.5–1 cm in size, single or multiple, with a *cauliflower* or *blackberry-like* appearance, usually pink in color or black/brown in animal with pigmented skin. In many cases, a yellow–orange film of waxy sebum covers their surface (Figs. 8.370 and 8.371).

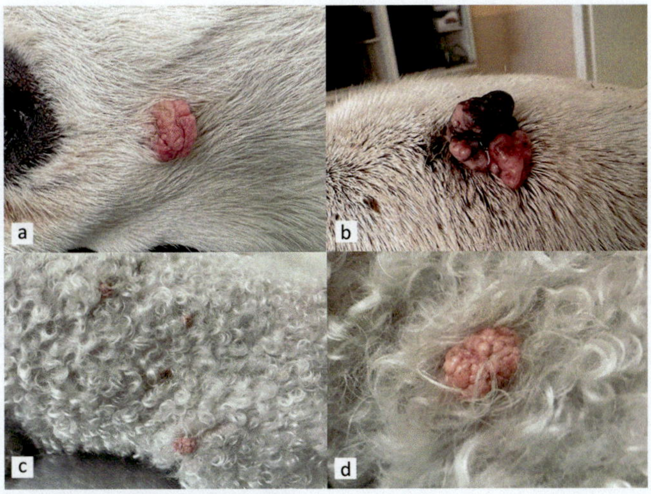

Fig. 8.370 Sebaceous neoplasms: (**a**) sebaceous hyperplasia—alopecic, exophytic small nodule; (**b**) irregular in shape alopecic sebaceous adenoma; (**c**) multiple alopecic and pinkish nodule in a Poodle with multifocal sebaceous hyperplasia; (**d**) at high magnification the typical white blackberry appearance is well recognizable

Sebaceous "sebocytic" adenoma is clinically similar to the hyperplastic lesions, while *sebaceous epithelioma* does not have a typical macroscopic aspect and usually occurs as single lesions, variable in size, of irregular shape, alopecic, erythematous, and often ulcerated.

The *sebaceous epithelioma*, even if in most cases has a benign behavior and surgery is curative, can rarely metastasize [127]. Since also in author experience some sebaceous epitheliomas have metastasized, a more careful approach to these "supposed" benign lesions must be kept, since their biological behavior is not predictable. Anyway, the meibomian sebaceous epithelioma seems to be benign since no metastatic lesions have been reported for this neoplasia when localized on eyelids. In animals with multiple lesions, the simultaneous presence of hyperplasia, adenomas, and epitheliomas can be observed; furthermore, different sebaceous histotypes can be present in the same nodule such as adenoma along with epithelioma.

Sebaceous adenocarcinomas of the sebaceous glands are rare and are usually characterized by a single nodule that is of varying size and often ulcerated. The malignant transformation involves mainly the basal cells (Fig. 8.372).

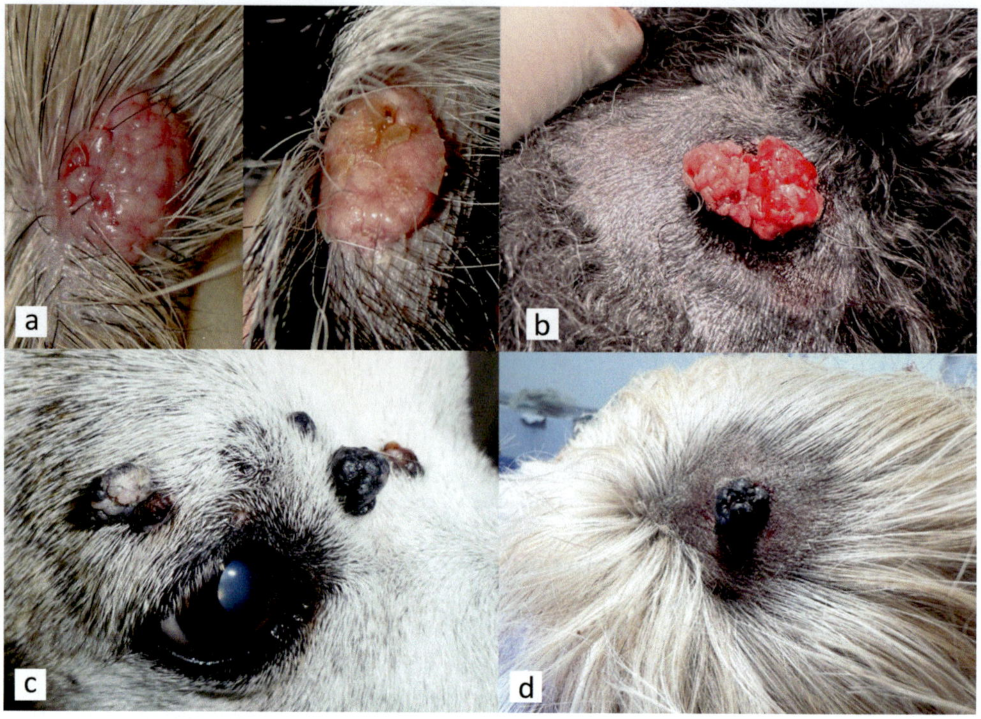

Fig. 8.371 Sebaceous neoplasms: (**a**) sebaceous hyperplasia (left image) and adenoma (right image). Note the greasy in appearance material covering the surface of the adenoma; (**b**) irregular, erythematous, and papilloma-like sebaceous adenoma; (**c**) multiple alopecic pigmented nodules of benign sebaceous lesions; (**d**) heavily pigmented sebaceous adenoma. Note as it is clinically not differentiable from melanocytic tumors or from a pigmented follicular neoplasia

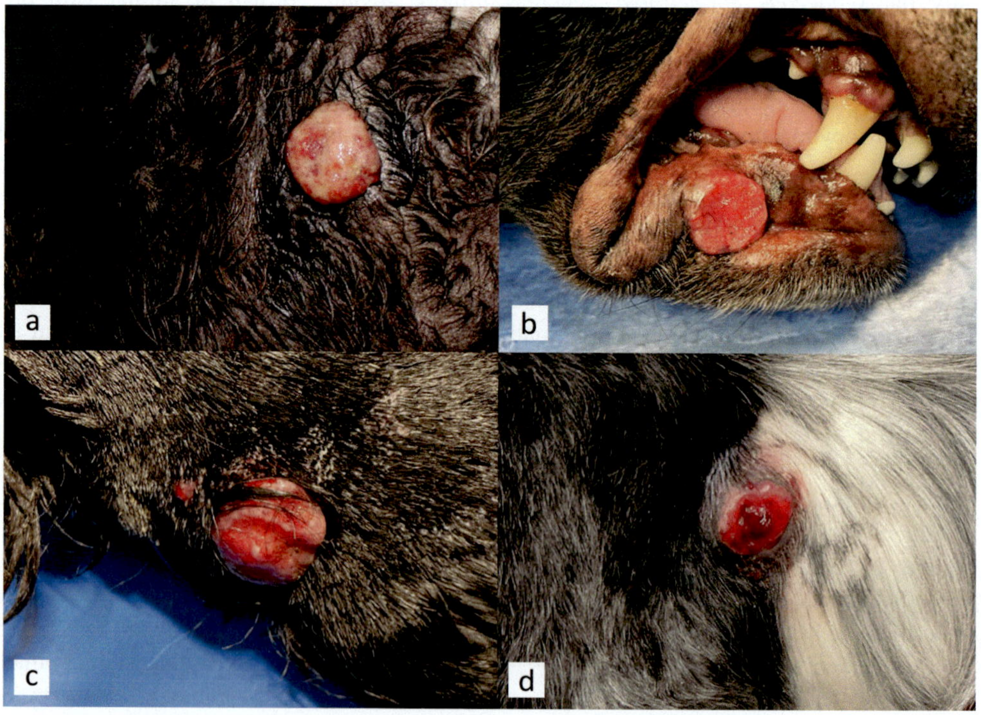

Fig. 8.372 Sebaceous neoplasms: (**a**) sebaceous epithelioma—alopecic, erythematous, and bleeding nodule on the face of a dog; (**b**) ulcerated sebaceous epithelioma on the lower right lip; (**c**) erythematous and ulcerated sebaceous epithelioma; (**d**) sebaceous carcinoma on the neck of a dog: note as it is impossible to clinically differentiate from a sebaceous epithelioma

Cytological Findings

The cells of *sebaceous hyperplasia* and *sebaceous adenoma* do not differ morphologically from the normal sebocytes, nor can they be differentiated from each other using cytology. In these benign lesions, sebocytes are organized in cohesive clusters, with usually round, uniform in size nuclei, mainly located in the center of the cells, and with a cytoplasm filled with small overlapping lipid droplets that give cells a characteristic *foamy* appearance. The lipid droplets are different in size in dogs, whereas they are more uniform and regular in cats. Often, at the periphery of the aggregates of sebocytes, arranged in a single row, reserve cells characterized by small, round, hyperchromatic nuclei and scant cytoplasm are usually observed; as mentioned, these cells are immature cells that differentiate into sebocytes (Figs. 8.373, 8.374, 8.375 and 8.376).

Sebum is cytologically characterized by an achromatic greasy material that does not take dye, and in the context of which only a few corneocytes and a variable amount of intact or destroyed sebocytes, indicative of their *holocrine* way of secretion, are recognizable. In practice, sebum is nothing other than the cytoplasmic content released by the dissolution of mature sebocytes (Figs. 8.377, 8.378 and 8.379).

In some cases, because the sebaceous glands are anatomically arranged around a large central dilated duct lined with a keratinized stratified epithelium, it is possible to sample large amounts of sebum mixed with a variable number of corneocytes (Fig. 8.380). In rare cases, intact cyto-architectures can be observed on slides (Fig. 8.381). Large amount of sebum, in the author's experience, are more commonly observed in meibomian gland benign

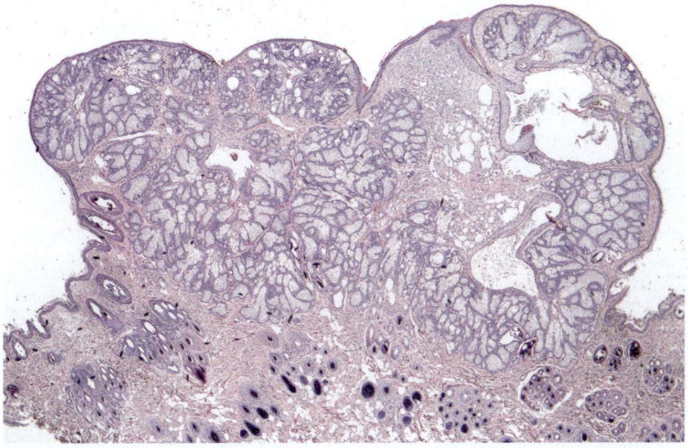

Fig. 8.373 Histopathology of sebaceous adenoma: roundish to elongated lobules of uniform-sized sebocytes lined with a single layer of basaloid cells

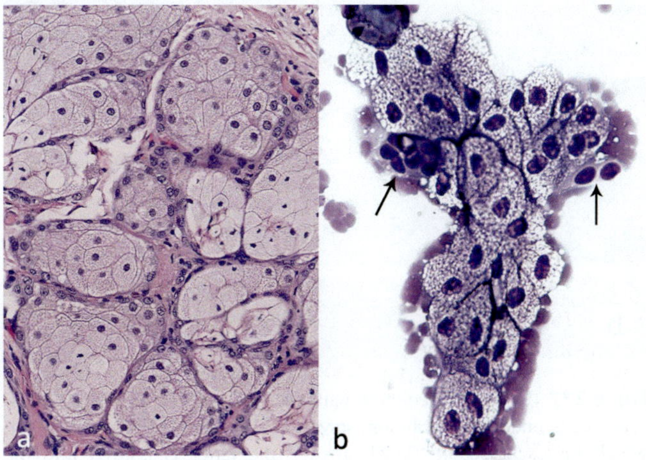

Fig. 8.375 Histopathology of a sebaceous lesion in cat: (**a**) sebaceous cells show small and regular in size lipid droplet; (**b**) the intracytoplasmic lipid droplets show a more uniform size compared with those observed in dogs. The arrows indicate the reserve cells

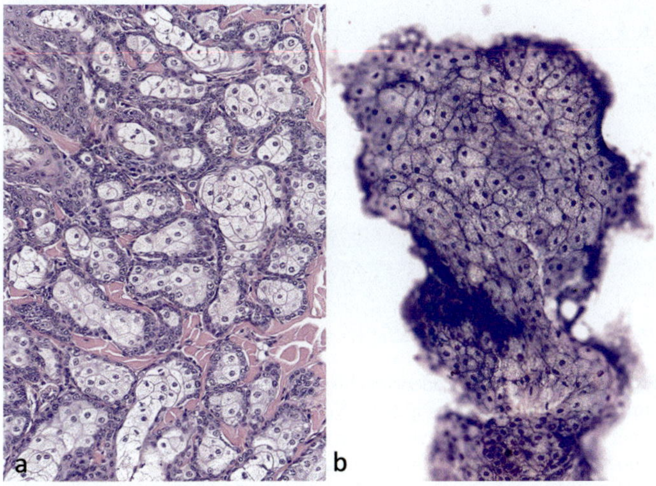

Fig. 8.374 Histopathology of a sebaceous lesion in dog: (**a**) histology—lobules of sebaceous cells lined by a single line of reserve immature cells; (**b**) typical "foamy appearance" of sebocytes with cytoplasm filled with lipid droplets

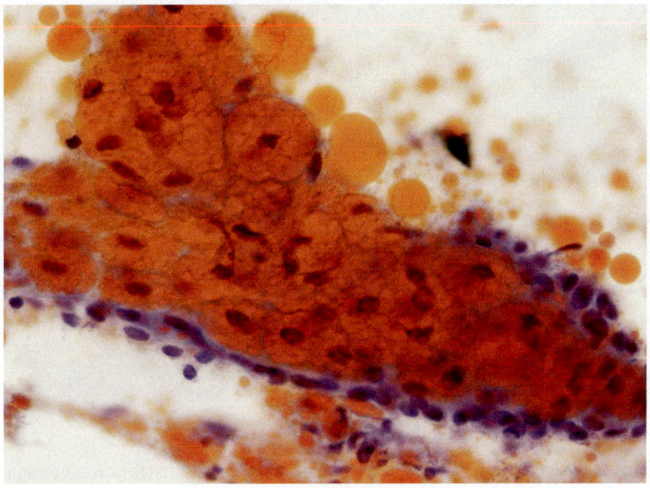

Fig. 8.376 Cytology of a sebaceous cluster: the oil red O dye highlights in brightly red the lipid cytoplasmic content; note the single basal germinative cells at the periphery of the mature sebocytes

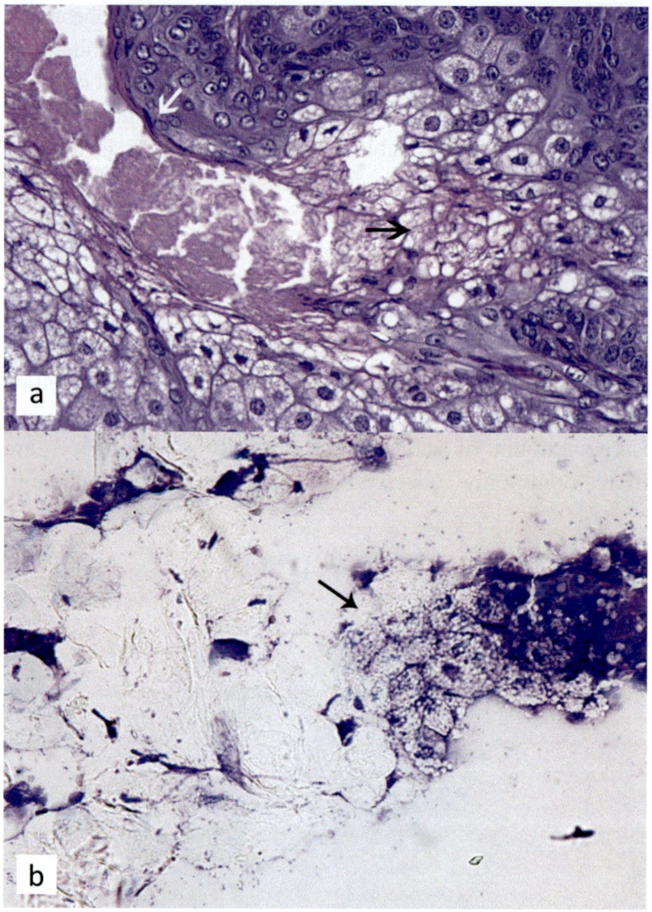

Fig. 8.377 Holocrine secretion pathway: (**a**) histology—mature sebaceous cells undergo dissolution, spreading all their cytoplasmic content which represent sebum (black arrow); the latter is poured into the follicular lumen through a duct lined with keratinizing squamous epithelium (white arrow); (**b**) cytology—the black arrow indicates the moment of cellular dissolution which gives rise to sebum which mixes with ductal-derived horny lamellae

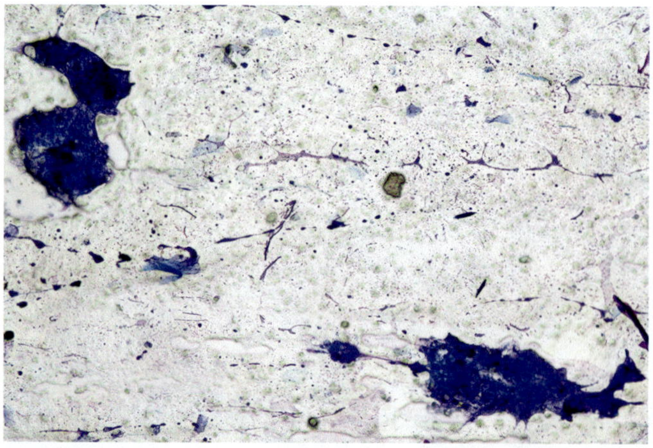

Fig. 8.378 Sebum cytology: two large sebaceous clusters on a greasy, dense, from achromatic to yellowish background that represents sebum

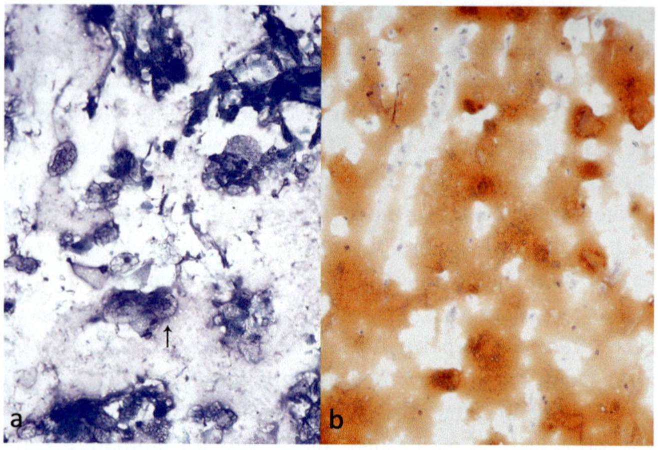

Fig. 8.379 Holocrine secretion: (**a**) sebum—few intact sebocytes (arrow), a lot of dissolved lipid droplets (sebum) and some lamellar keratin coming from the sebaceous duct; (**b**) oil red O stains red the lipids that compose the sebum

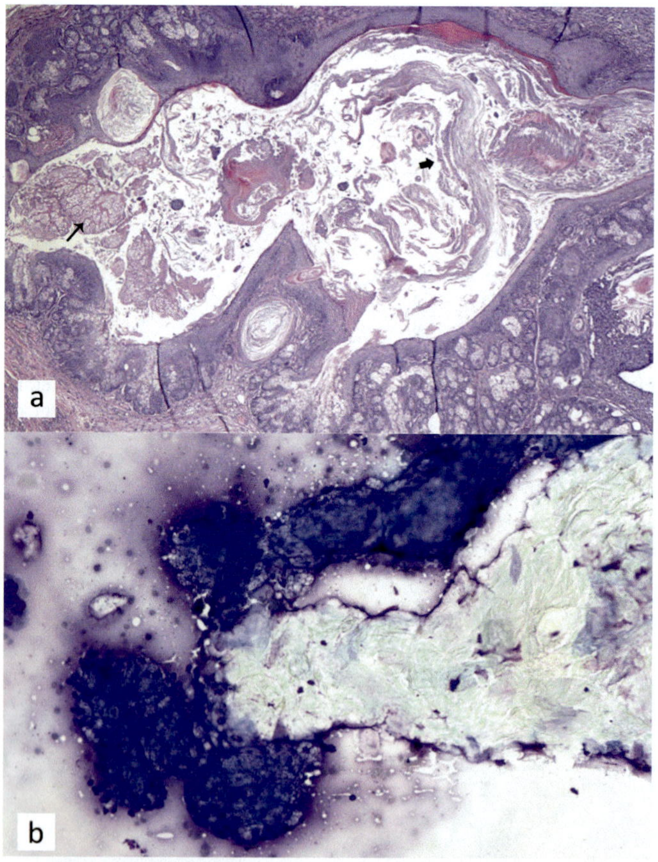

Fig. 8.380 Sebaceous adenoma histology: (**a**) large central ectatic duct, surrounded by sebaceous glands. Note as the content is a mix of sebum (long arrow) and corneocytes (short arrow); (**b**) a large cluster of mature sebocytes that produce sebum through a holocrine pathway of secretion. Note as many corneocytes are also present, the latter coming from the dilated duct filled with lamellar keratin

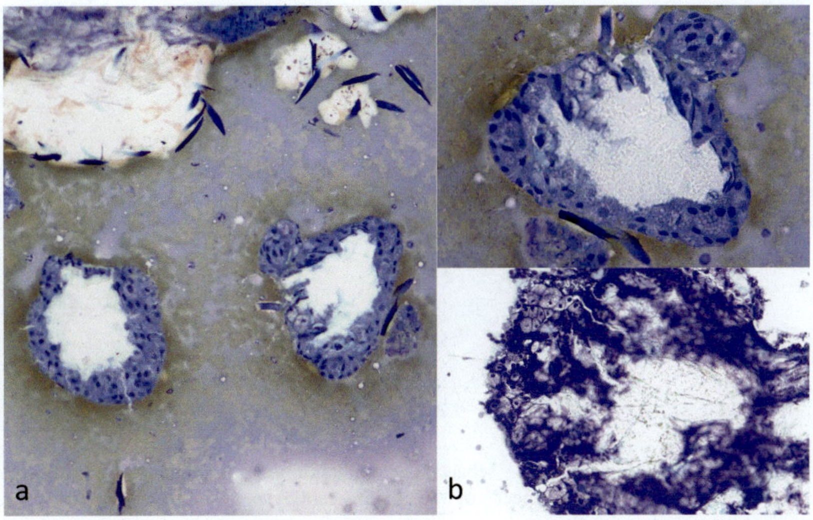

Fig. 8.381 Cytology sebaceous adenoma: (**a**) two clusters of sebocytes arranged around a clear acellular area; (**b**) close-up of two aggregates with a central area filled with sebum and corneocytes

adenomas, often associated with secondary granulomatous inflammation, mainly represented by lipophagocytic macrophages. The cytological differentiation between *benign sebaceous hyperplasia* and *adenoma*, as for two other benign sebaceous lesions, *sebaceous hamartoma* and *sebaceous nevus*, is not possible. It has been speculated that, if the specimens are composed of a higher number of reserve cells, with larger aggregates of less uniform sebocytes, an adenomatous lesion should be suspected more than hyperplasia or a hamartoma. However, this differentiation has no prognostic significance because of the benign behavior of all these sebaceous lesions. Furthermore, many well-differentiated sebaceous clusters can be the main population of cells collected from non-neoplastic reactive hyperplastic benign lesions such as *acral lick granuloma* or *fibroadnexal hamartoma*. In these cases, the simultaneous presence on slides of regular cluster of apocrine sweat glands along with the clinical presentation of the lesion can help to rule out a primary sebaceous lesion.

Very rarely, both in hyperplasia and adenoma, a few elements with hepatoid differentiation can be observed especially in anatomic areas where the latter are more numerous.

Sebaceous epithelioma is a very common neoplasia that originates from the reserve cells of the sebaceous glands (Fig. 8.382). Cytologically, large aggregates of uniform basaloid cells, regular in shape and size, in the context of which a variable number of mature sebocytes, are observed. The mature sebocytes, single or in small groups, are easily detectable since the presence of multiple and very small lipid droplets, which fill their cytoplasm (foamy appearance) (Figs. 8.383, 8.384 and 8.385).

Although, in almost all cases, few scattered mature sebocytes are present, it may happen that only many large clusters of immature basal cells are present on the smears; in

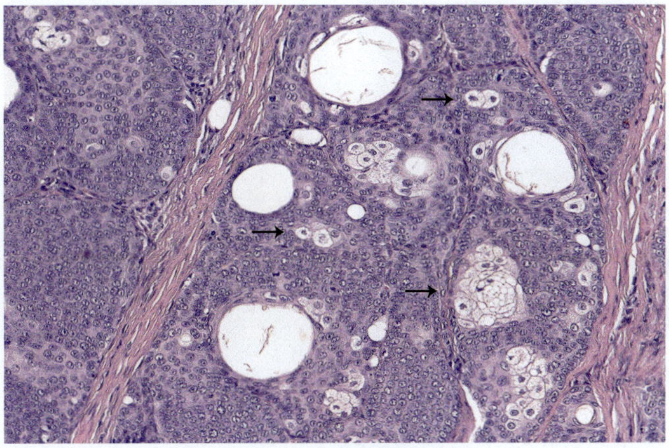

Fig. 8.382 Histopathology of a sebaceous epithelioma: large nests of basaloid germinative cells in which both single and small clusters of mature sebocytes are recognizable (*arrows*)

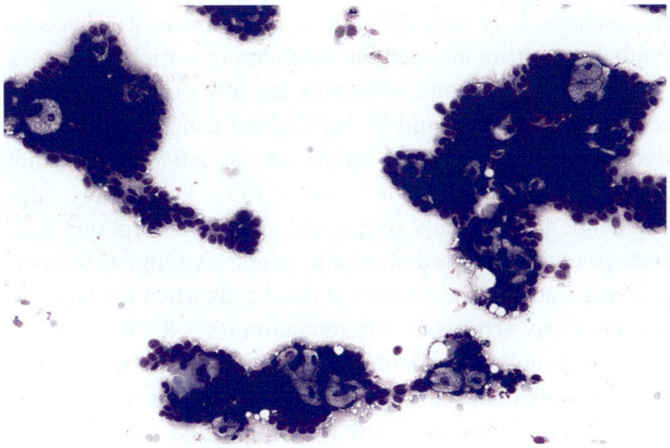

Fig. 8.383 Cytology of a sebaceous epithelioma: large sheets of basaloid cells in which many mature sebocytes are recognizable

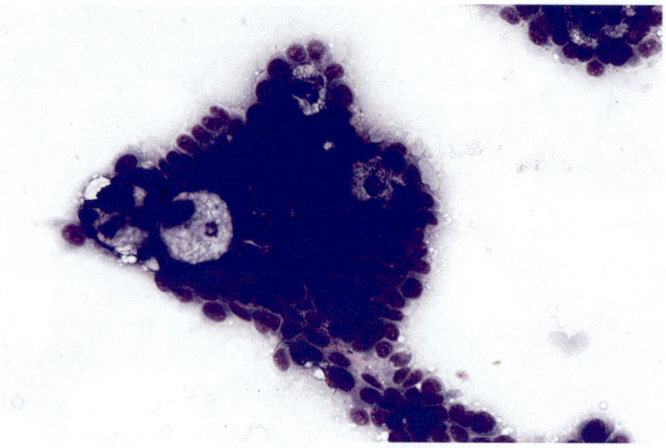

Fig. 8.384 Cytology of a sebaceous epithelioma: at high magnifications the "foamy aspect" of the mature sebocytes is more evident between basal cells

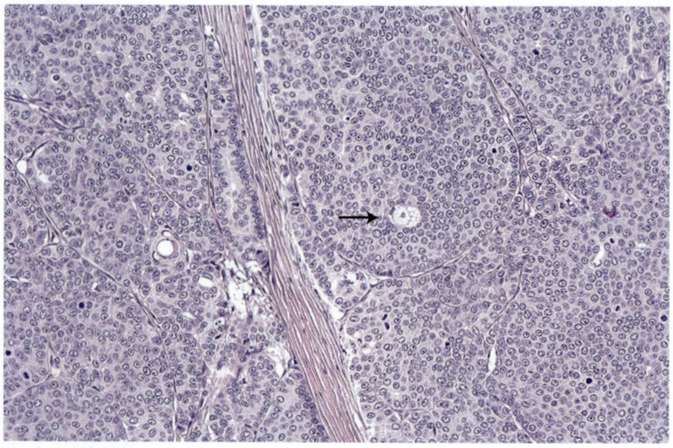

Fig. 8.386 Histology of a sebaceous epithelioma: solid nodules composed of regular in shape basal reserve cells in which only two adjacent mature sebocytes are recognizable (arrow)

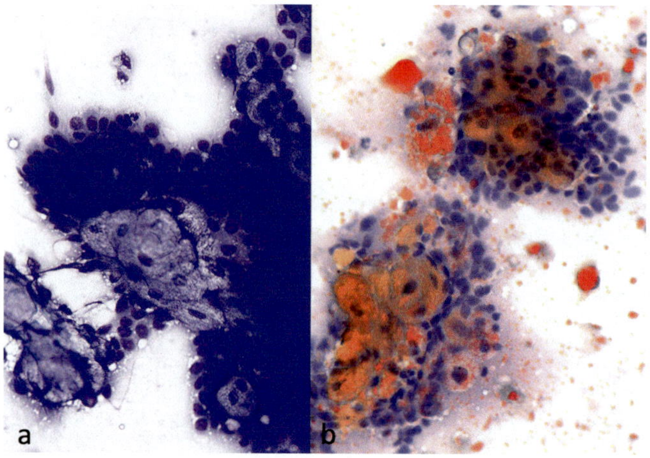

Fig. 8.385 Cytology of a sebaceous epithelioma: (**a**) a cluster of basal cells with many mature sebocytes; (**b**) lipid droplet are highlighted with oil red O staining

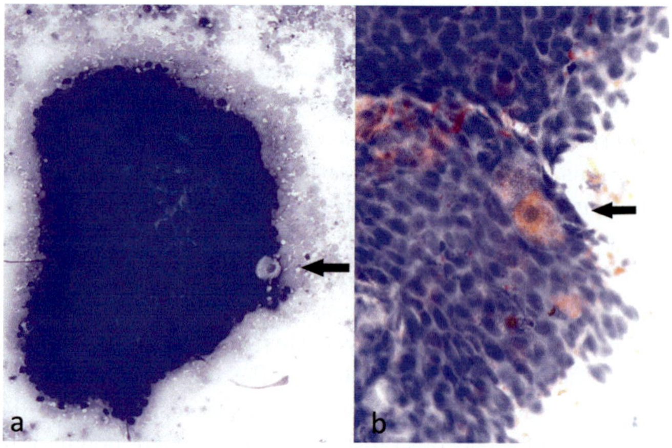

Fig. 8.387 Cytology of a sebaceous epithelioma: (**a**) single mature sebocytes are present in this basal cell cluster (arrow); (**b**) sample from the same nodule in which the oil red O stain highlights the lipid content of the cytoplasm of a mature sebocyte (arrow)

these cases, it is very difficult to differentiate a sebaceous epithelioma from an apocrine adenoma or a trichoblastoma. In cases where mature sebocytes are few or not so clearly recognizable, they could be highlighted using the oil red O dye, which stains the lipids on a red–orange color (Figs. 8.386, 8.387, 8.388 and 8.389). Fortunately, this evenience is absolutely occasional, and an amorphous lipid background, composed of sebum, orients cytologists to interpret as sebaceous the clusters of basal cells when the latter do not show any sebaceous differentiation (Fig. 8.390).

In some nodules, a double differentiation can be observed with the simultaneous presence of both immature basal cells (reserve or germinative cells) and lipidized cells (mature sebocytes); in these cases cytological features can be different depending on in which part of the tumor the needle is inserted (Fig. 8.391).

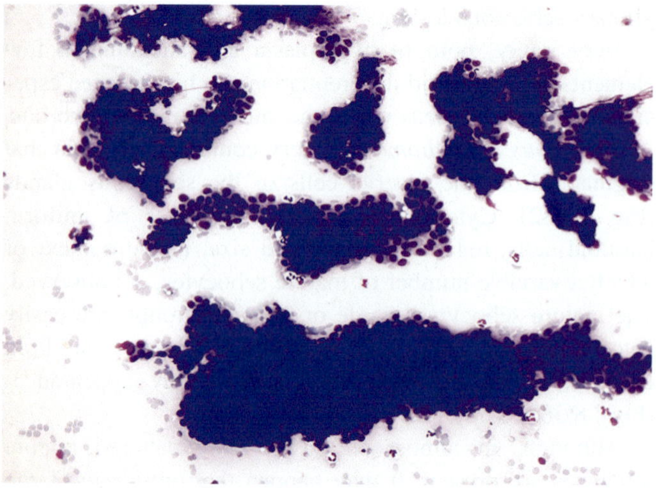

Fig. 8.388 Cytology of a sebaceous epithelioma: multiple trabecular in shape clusters of sebaceous basal cells. Note as sebaceous differentiation is not evident

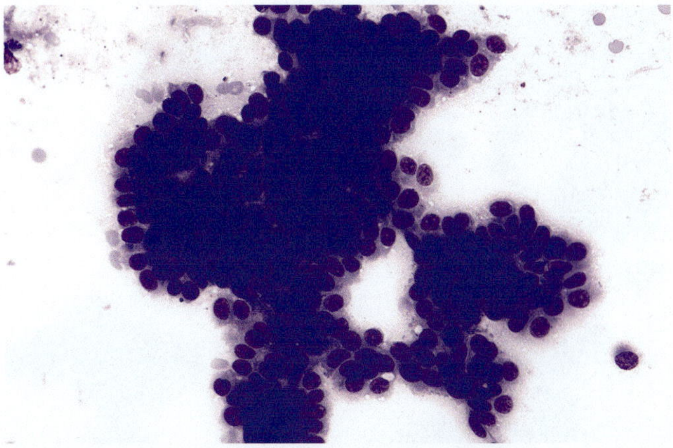

Fig. 8.389 Cytology of a sebaceous epithelioma: cluster of sebaceous basal cells without sebaceous differentiation

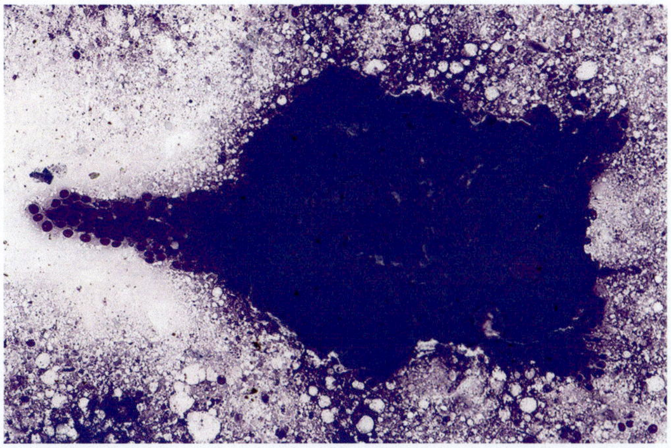

Fig. 8.390 Cytology of a sebaceous epithelioma: no sebaceous differentiation is evident between basal cells. The lipid background (sebum) orients the suspicion to a sebaceous nature of basal cells

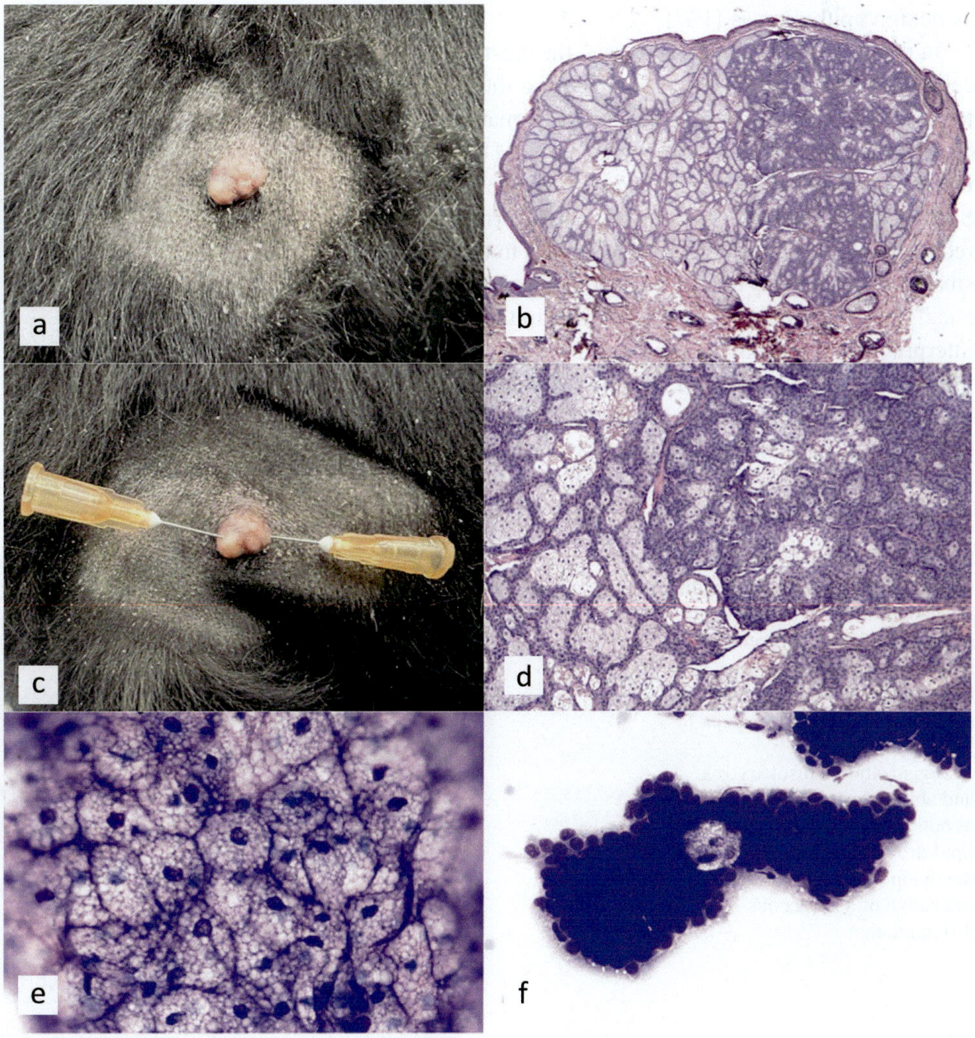

Fig. 8.391 Mixed composition of a sebaceous nodule: (**a**) small and alopecic sebaceous nodule; (**b**) histology of the same nodule: note the heterogeneity of the cells, mature on the left and immature on the right; (**c**) needles inserted in opposite parts of the nodule; (**d**) histology—high magnification in which the difference between the left (mature) and right (immature) sebaceous areas of the same nodule is better recognized; (**e**) the mature sebocytes collected by sampling only the left area of the tumor would have led to a cytological diagnosis of sebaceous hyperplasia/adenoma; (**f**) the large cluster of basal cells with a single mature sebocyte collected by sampling only the right area of the tumor would have led to a cytological diagnosis of sebaceous epithelioma

Sebaceous adenocarcinoma shows cytological features that differ according to the grade of differentiation of the same glands. In case the neoplasia origin from the reserve basal cells, in some cases, especially in well-differentiated cases in which atypia are very few, it is impossible to differentiate a sebaceous epithelioma from a basal epitheliomatous carcinoma; in these cases, histopathology is mandatory to evaluate the infiltrative growth pattern of the tumor and presence and morphology of mitotic figures. An interpretation problem also occurs when severely undifferentiated malignant carcinomas show no sebaceous differentiation; in these cases, no other carcinoma, such as apocrine carcinoma, can be excluded. It should be remembered that the biological behavior of sebaceous epithelioma, a benign lesion by etymological definition, does not always follow the rules that associate biological behavior with cellular morphology, as cases of metastasis of lesions that morphologically did not manifest malignant aspects in cytological and histological preparations have been reported and have been diagnosed as benign epitheliomas [127].

In the case of malignant neoplasia, the clusters of basaloid cells have less regular profile compared to the benign counterpart; loss of cohesivity, larger nuclei in mature sebaceous cells, and larger in size intracytoplasmic lipid droplets are all features indicative on malignancy (Figs. 8.392, 8.393 and 8.394). In other cases, the cells are characterized by a more recognizable sebaceous differentiation due to the presence of smaller lipid droplets that give the so defined "foamy appearance" to the cytoplasm; in these situations, it is easier to interpret the neoplastic cells as coming from the sebaceous tissue. In poorly differentiated carcinomas, the poor cohesion of the neoplastic cells combined with the marked anisocytosis, anisokaryosis, and the presence of a high

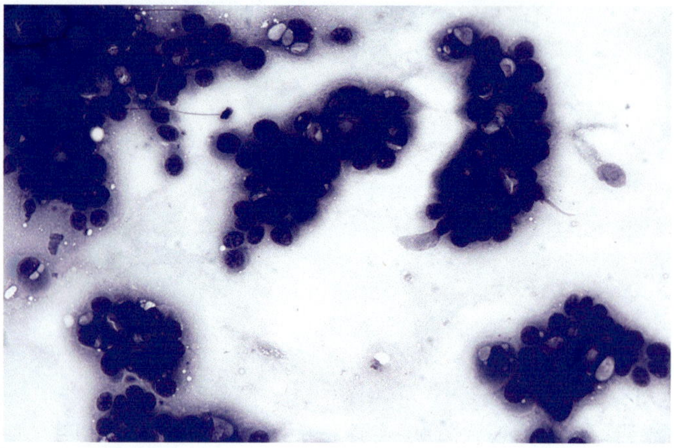

Fig. 8.393 Cytology of a sebaceous carcinoma: clusters of poorly differentiated cells in which large intracytoplasmic lipid droplets are evident

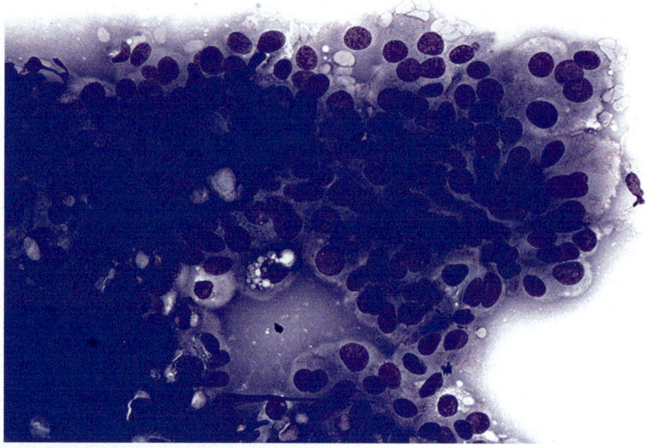

Fig. 8.394 Cytology of a sebaceous carcinoma: irregular cluster with overlapped cells in which both large and overlapped small lipid droplets are recognizable

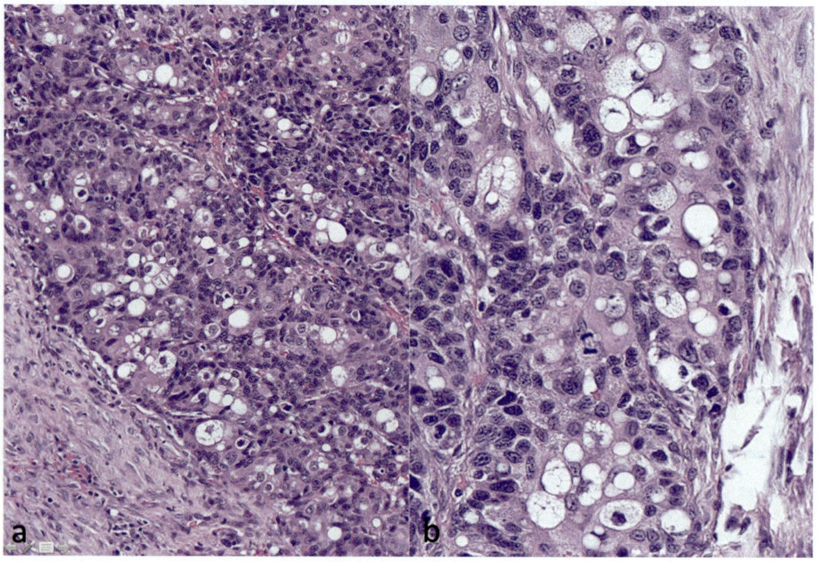

Fig. 8.392 Histology of a sebaceous carcinoma: (**a**) irregular in shape lobules of atypical and scarcely differentiated cells in many of which a sebaceous differentiation is recognizable; (**b**) malignant cytological features such as architectural disorganization, anisokaryosis, and anisocytosis are evident. Note as both large and overlapped small lipid droplets are recognizable in the cytoplasm of atypical cells indicative of a sebaceous differentiation

number of atypical mitotic figures can allow the cytologist to define the diagnosis (Figs. 8.395 and 8.396).

Modified sebaceous gland tumors include neoplasms of the canine *perianal/circumanal* or, as commonly defined, *hepatoid gland cells* [129, 130].

The *hyperplasia* and *adenomas* of the *hepatoid glands* are very common in intact male dogs and develop mainly in the perianal area, very rich in these glands. Other body areas such as the tail and the prepuce are frequently affected. Rarely hepatoid tumors have been reported on other body areas such as the back, the thighs, and the neck. The clinical appearance is variable. Nodules vary in size, from few mm to up 3 cm, and can be single or multiple, erythematous, and ulcerated. Hepatoid glands are also physiologically present in the so-called supracaudal organ or tail gland, a mix of sebaceous and hepatoid glands located in the dorsal portion of the tails of dogs, a few centimeters from the sacrum. As for the sebaceous glands, even from hepatoid glands, hyperplastic lesions, adenomas, epitheliomas, and carcinomas may develop (Figs. 8.397 and 8.398).

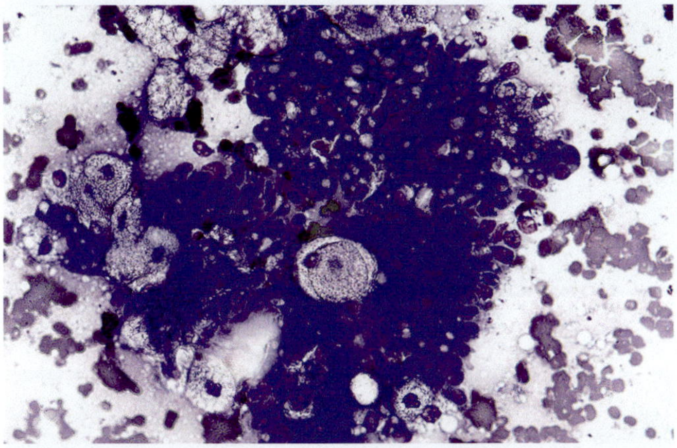

Fig. 8.395 Cytology of a sebaceous carcinoma: irregular cluster of malignant cells in which many mature cells are recognizable

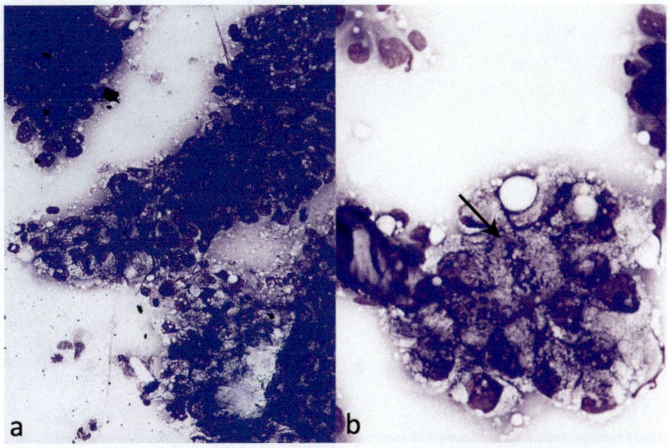

Fig. 8.396 Cytology of sebaceous carcinoma: (**a**) severely atypical and malignant features in sebaceous clusters; the sebaceous origin of the cells cannot be confused; (**b**) malignant clusters of sebaceous mature cells; note as mitosis involves a mature cell a clear sign of malignancy (arrow)

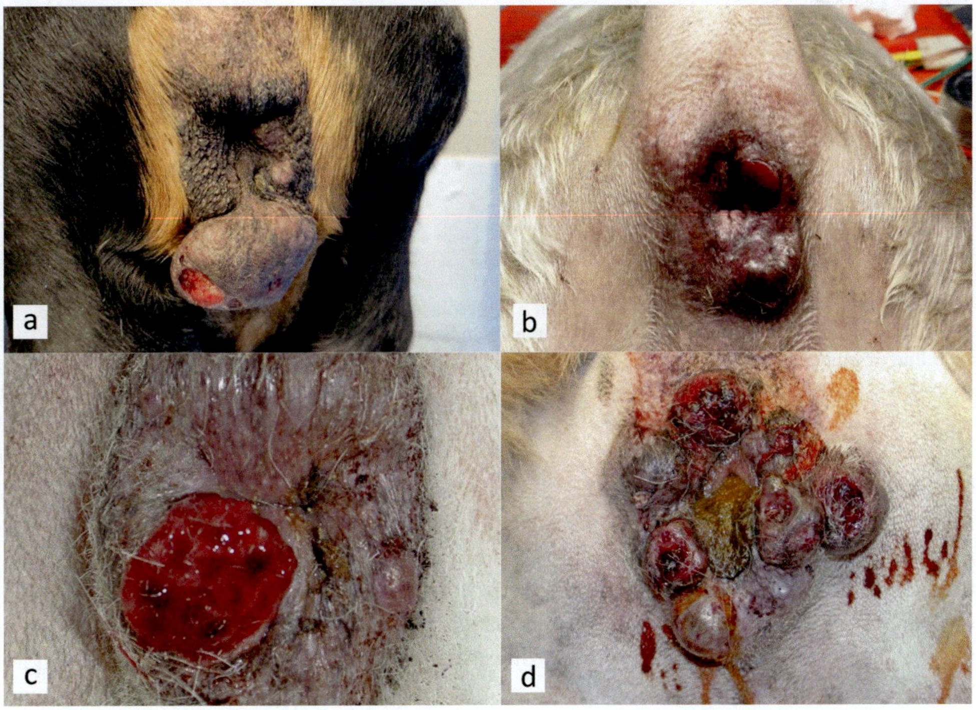

Fig. 8.397 Hepatoid gland neoplasms: (**a**) partially ulcerated hepatoid gland adenoma on the ventral perianal area; (**b**) two nodules on the anus of a dog with hepatoid gland adenomas; (**c**) large ulcerated hepatoid gland adenoma along with a smaller intact nodule; (**d**) many confluent and ulcerated hepatoid gland carcinomas

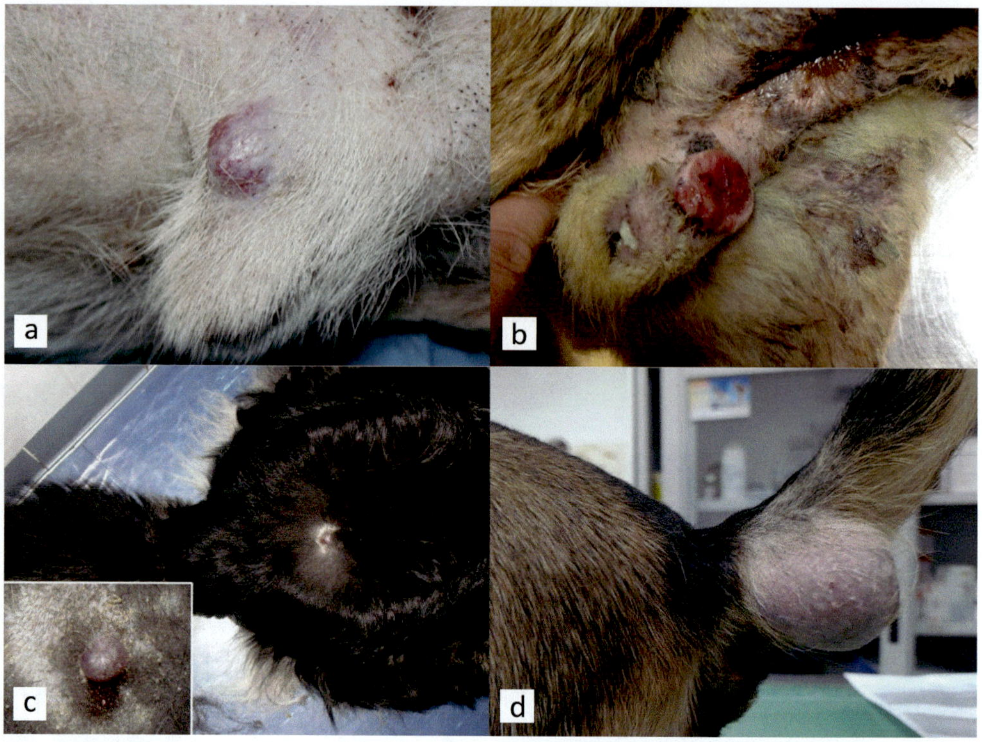

Fig. 8.398 Hepatoid gland neoplasms: (**a**) hepatoid gland adenoma on the prepuce of a dog; (**b**) ulcerated plaque of hepatoid gland carcinoma on the preputial skin; (**c**) small alopecic nodule on the back—cytology and histology confirmed a hepatoid gland adenoma; (**d**) large intact hepatoid gland carcinoma on the tail (Courtesy of DVM L. Zumbo-Italy)

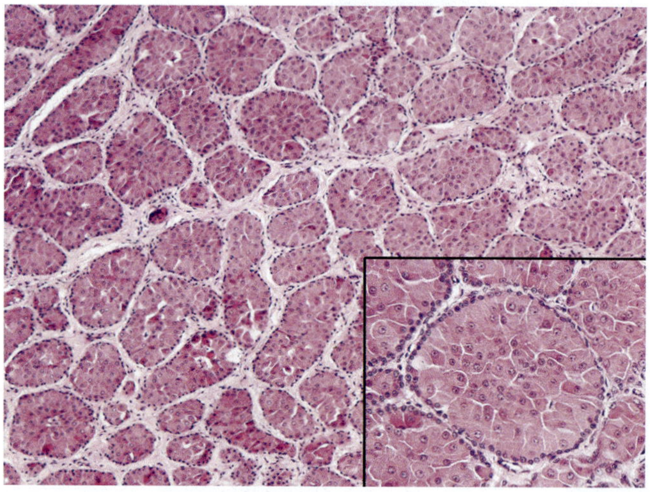

Fig. 8.399 Histology of hepatoid gland hyperplasia: well-differentiated roundish islands of mature hepatoid glands surrounded by a line of immature reserve basal cells, better evaluable in the inset. Lobules are separated by abundant stroma

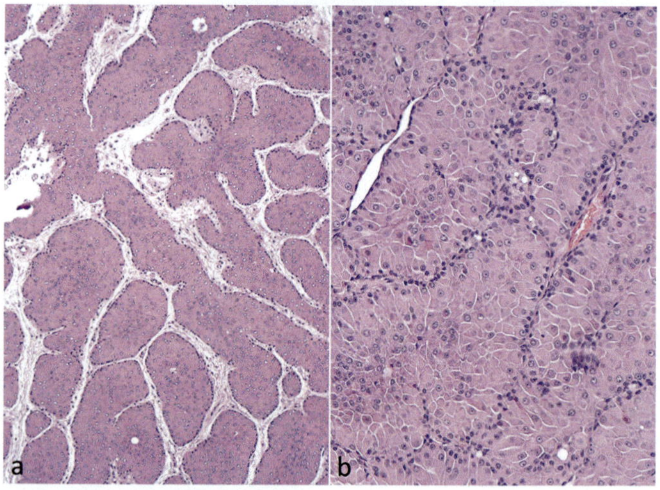

Fig. 8.400 Histology of hepatoid gland adenoma: (**a**) well-differentiated islands and trabeculae of mature hepatoid glands surrounded by a line of immature reserve basal cells. Lobules are separated by abundant stroma; (**b**) islands of well-differentiated benign neoplastic hepatoid glands separated by an imperceptible stroma

Cytological Findings

Hepatoid glands are so named since their morphology resembles the liver cells. Histologically they are composed of islands and trabeculae of modified sebaceous glands, separated by vascularized collagen stroma; cells are polygonal in shape with large cytoplasm and with a line of basal reserve cells (Figs. 8.399 and 8.400).

The cytology of *hyperplasia* and *adenoma* of *hepatoid glands* reflects what abovementioned for histology; slides usually show numerous and large from roundish to trabecular in shape clusters of polygonal to round cells with round nuclei, centrally located, usually with one or two clearly appreciable nucleoli and a large cytoplasm, pinkish to violet in color and with a grainy aspect. At the periphery of the clusters, reserve cells usually arranged in a single to multiple layers are well recognizable in some tumors. The reserve cells are often represented as small hyperchromatic nuclei since cytoplasm is scarce and usually not visible. These cells are those from which the hepatoid gland epithelioma and carcinoma can arise (Figs. 8.401, 8.402, 8.403, 8.404, 8.405 and 8.406). In many specimens, a variable number of bare nuclei due to the breakage of the cytoplasm during the preparation of slides are also frequently observed (Fig. 8.407).

Cytological differentiation between hyperplasia and adenoma is not simple, and based on histopathological knowledge, when irregular trabeculae of hepatoid cells are associated with numerous blood vessels, an adenoma should be suspected, while, in the case of hyperplasia, the cells show a more regular aggregation, with smaller clusters with a rounded profile, well-defined margins, and usually absent blood vessels (Fig. 8.408). It should be kept in mind, however, that since most adenomas are histologically surrounded

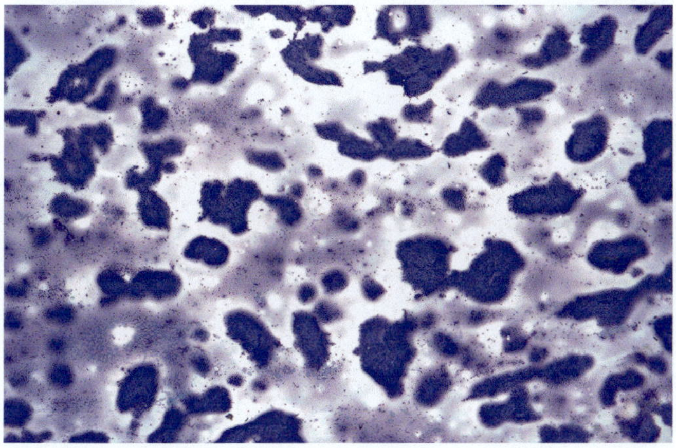

Fig. 8.401 Cytology of hepatoid gland adenoma: many clusters of hepatoid glands. Samples are very often highly cellular

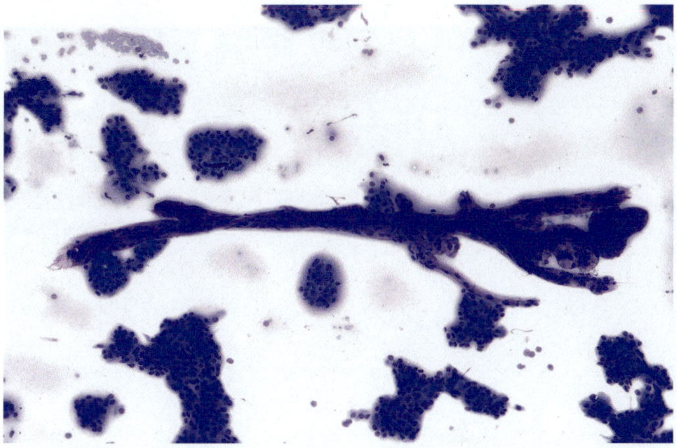

Fig. 8.402 Cytology of hepatoid gland adenoma: roundish and trabecular in shape clusters of benign hepatoid glands. Note the long and branched hepatic vessel in the center of the picture

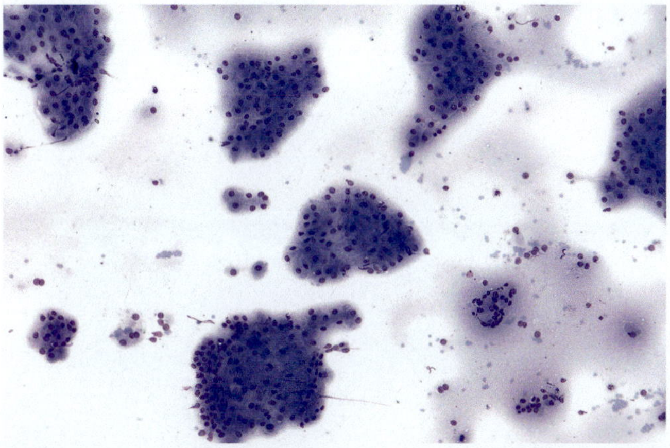

Fig. 8.403 Cytology of hepatoid gland adenoma: roundish clusters of benign hepatoid glands

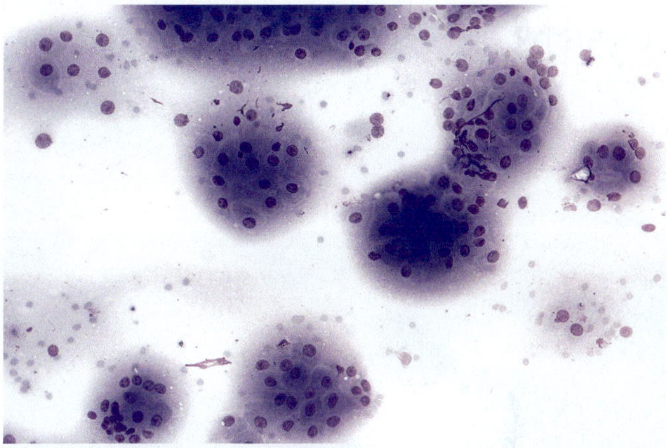

Fig. 8.404 Cytology of hepatoid gland adenoma: clusters of polygonal hepatoid cells with violet cytoplasm

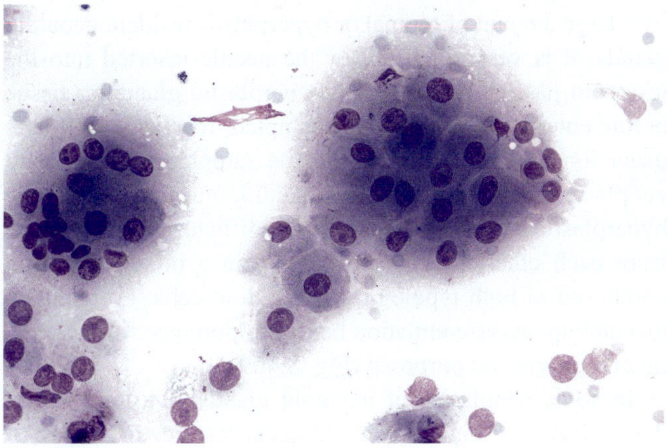

Fig. 8.405 Cytology of hepatoid gland adenoma: at high magnification the central round nucleus with a single nucleolus and the grainy aspect of the cytoplasm are better recognizable

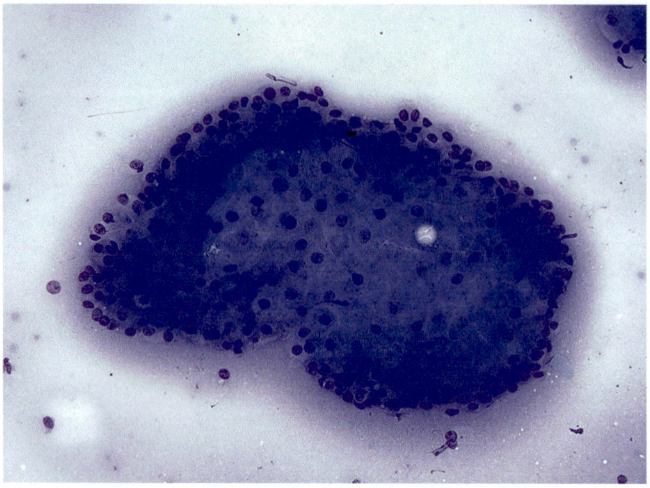

Fig. 8.406 Cytology of hepatoid gland adenoma: large cluster of mature hepatoid gland with multilayers of peripheral basal reserve cells

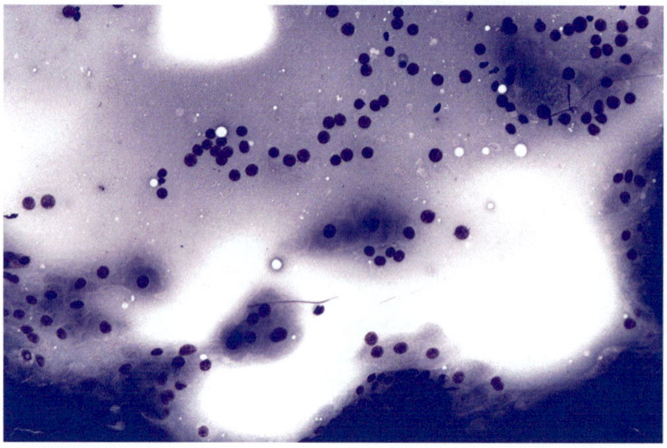

Fig. 8.407 Cytology of hepatoid gland adenoma: on a background composed of violet material mostly represented by broken cytoplasms, many naked nuclei are detectable

Fig. 8.408 Cytology of hepatoid gland adenoma: many trabecular aggregates of hepatoid glands along with hematic vessels are a cytological feature of adenoma more than hepatoid gland hyperplasia

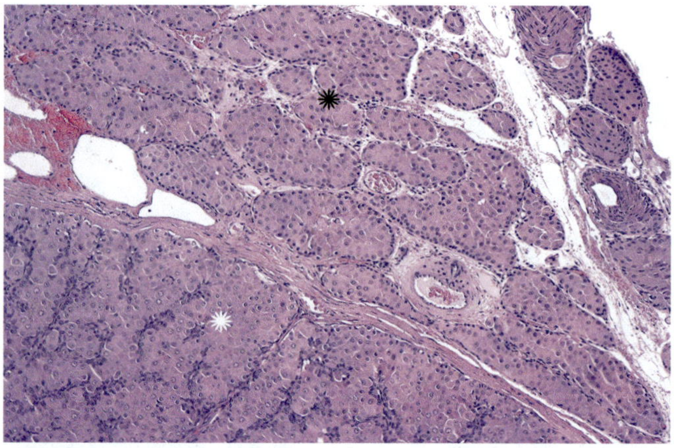

Fig. 8.409 Histology of hepatoid gland hyperplasia and adenoma: hepatoid neoformations are very often heterogeneous; the upper part of the lesion is composed of hyperplastic lobules (black flower), while the lower part is composed of an adenoma (white flower). A fibrous stroma separates the two benign proliferations

by a large amount of normal or hyperplastic resident hepatoid glands, it is very common for the needle inserted into the lesion to pass through normal/hyperplastic glandular tissue before entering the lesion to be sampled; for this reason, it is usual for non-neoplastic cells to be sampled together with neoplastic cells; summarizing, in addition to the fact that the hyperplastic and neoplastic cells are difficult to differentiate from each other, it is normal to obtain a mixed sampling composed of both typologies of hepatoid cells. Fortunately, this attempt at differentiation has no importance for diagnostic and prognostic purposes (Fig. 8.409).

In some neoplasms of hepatoid glands, cytology shows their sebaceous origin, and cytological features indicative of sebaceous differentiation can be observed; these findings are represented by intracytoplasmic overlapped lipid droplets, typical of sebocytes (Figs. 8.410 and 8.411).

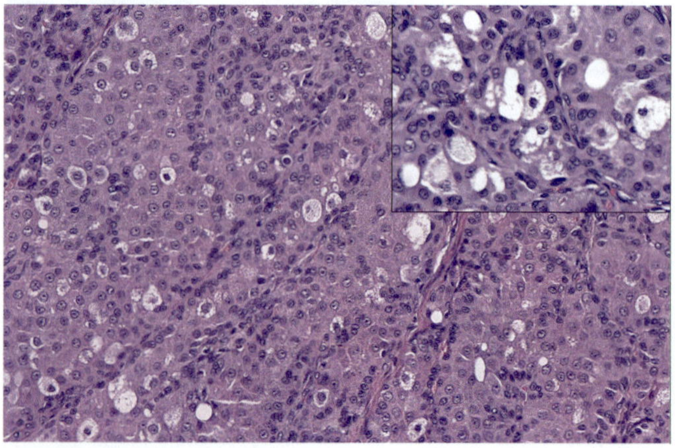

Fig. 8.410 Histology of hepatoid gland adenoma with sebaceous differentiation: sebocytic differentiation easily recognizable as "foamy" lipid cells, better visible in the insert

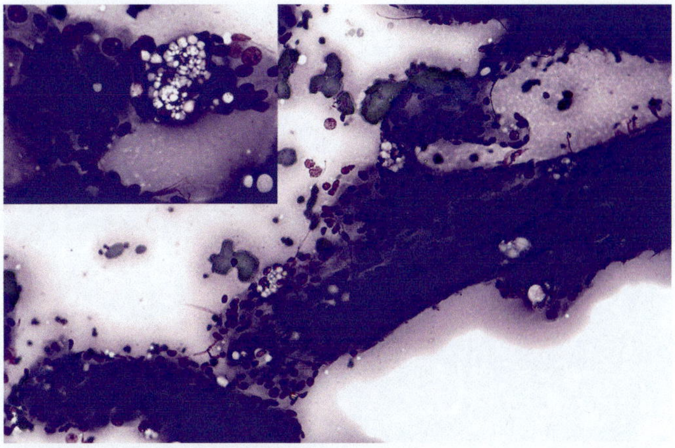

Fig. 8.411 Cytology of hepatoid gland adenoma with sebaceous differentiation: hepatoid aggregates showing sebocyte differentiation easily recognizable as "foamy" lipid cells, better visible in the insert

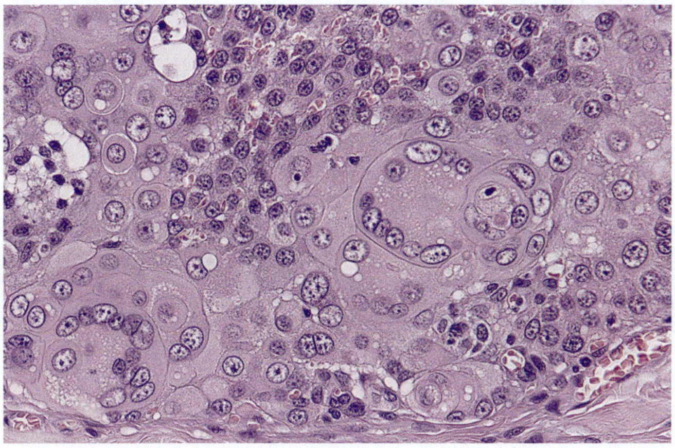

Fig. 8.412 Histology of hepatoid gland neoplasm: many multinucleated hepatoid cells along with classical hepatoid cells

A variant of benign hepatoid adenoma with *multinucleated cells* is rarely reported. A variable number of cells with more than a nucleus can be admixed to normal hepatoid cells. This cytological aspect is not indicative of the behavior of the neoplasia. No data about the malignant counterpart is reported (Figs. 8.412, 8.413 and 8.414).

When most cells are represented by reserve basal cells, with regular morphology, an *epithelioma* of hepatoid glands must be suspected (Figs. 8.415, 8.416 and 8.417).

Regarding *hepatoid gland carcinoma*, it is clearly recognizable only when cell atypia is obvious, since in the case of well-differentiated carcinoma, cells could not show clear cytological criteria of malignancy, so,a definitive diagnosis can only be obtained by histopathology.

Many papers are published in the last years regarding this topic in the attempt to evaluate both the correlation between cytology and histology and also if the cytological features of

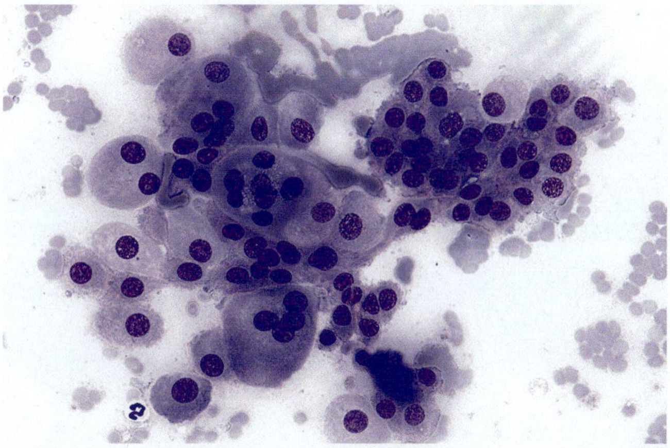

Fig. 8.413 Cytology of hepatoid gland adenoma: a cluster of typical mono-nucleated cells along with some multinucleated hepatoid cells. Note as the nuclei are uniform in size, with fine chromatin pattern and no features of atypia are recognizable

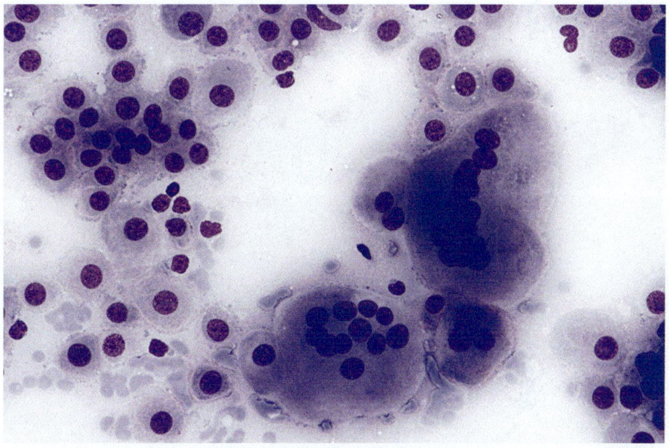

Fig. 8.414 Cytology of hepatoid gland adenoma: many multinucleated hepatoid cells are recognizable. This feature is only rarely observed

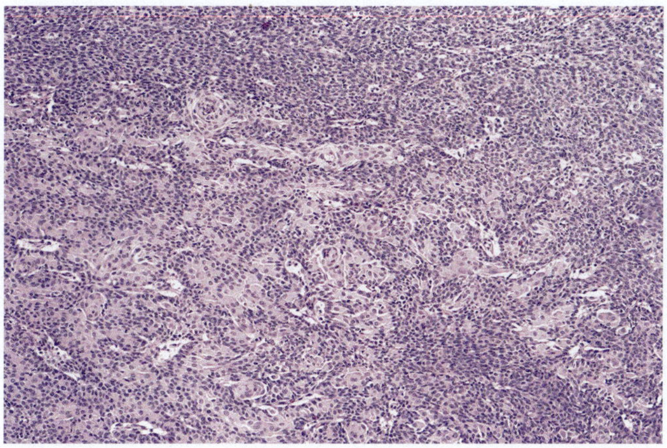

Fig. 8.415 Histology of hepatoid gland epithelioma: massive presence of basal reserve cells—in the lower part of the pictures, few differentiated hepatoid cells are recognizable

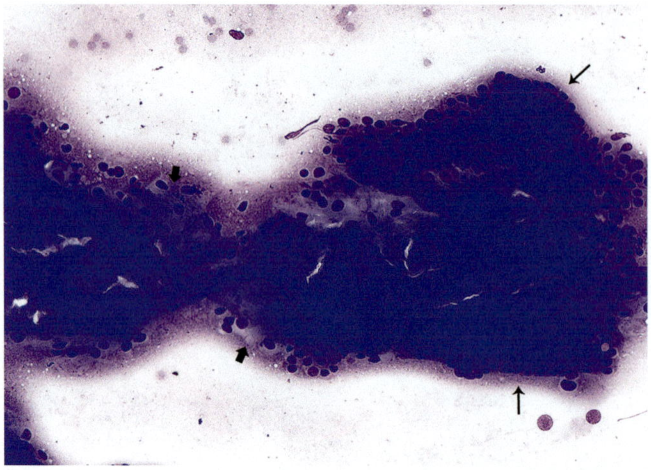

Fig. 8.416 Cytology of epithelioma of the hepatoid glands: large epithelial cluster composed mainly of immature germ cells (long arrows) among which only a few mature hepatoid cells are present (short arrows)

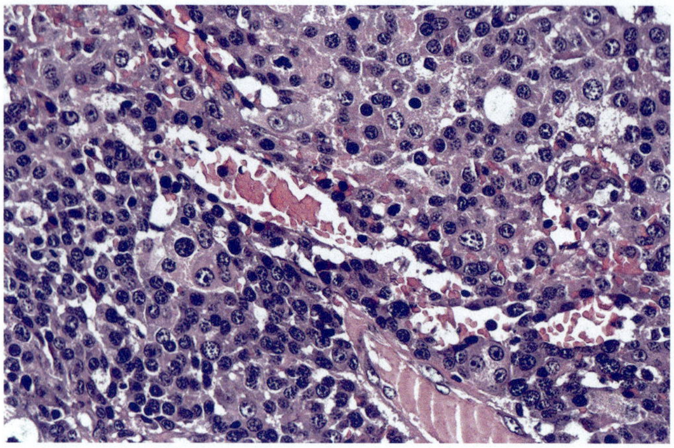

Fig. 8.418 Histology of hepatoid gland carcinoma: severe malignant atypia such as anisocytosis, macrokaryosis, anisokaryosis, and architectural disorganization

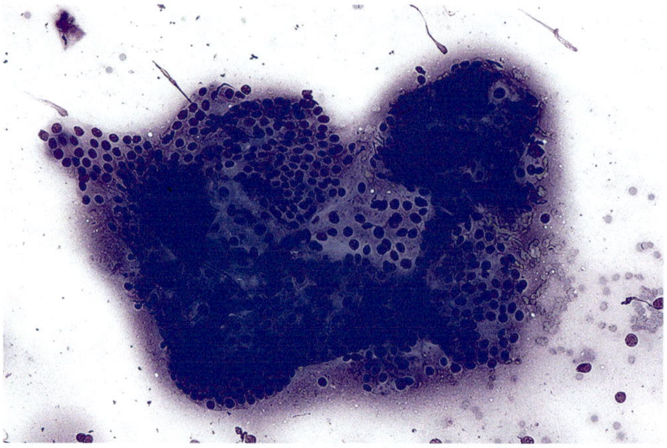

Fig. 8.417 Cytology of epithelioma of the hepatoid glands: large epithelial cluster composed mainly of immature germ cells arranged in honeycomb pattern among which only a few mature hepatoid cells are recognizable at its central area

Fig. 8.419 Cytology of hepatoid gland carcinoma: malignant features are cytologically represented by anisocytosis, macrokaryosis, anisokaryosis, and architectural disorganization. The neoplastic malignant cells show sebaceous differentiation

hepatoid gland neoplasia could be predictive of their outcome.

In a study, 84.37% of the samples from 32 dogs demonstrated a matching between cytological and histopathological diagnoses; only five cases were misdiagnosed [134].

Another study on 83 dogs demonstrated that the proportion of the basaloid reserve cells is not an indication of potential malignancy for the tumor [131]. Finally, in a preliminary study on 77 hepatoid gland tumors from 56 dogs, the authors elaborate a cytological algorithm which evaluates some parameters such as (a) the disposition of basal cells in the hepatoid clusters, (b) the dissection of hepatoid clusters by the basal reserve cells, (c) anisocytosis/anisokaryosis and nuclear activation, and (d) vacuolization of cytoplasm [133]. In benign tumor the basal cells were usually detectable as a rim of monolayer hyperchromatic nuclei with scant cytoplasm usually non-visible, located at the periphery of mature hepatoid cells. Despite also in benign tumors, reserve cells can be irregularly admixed with hepatoid cells, this is a cytological feature commonly observed in malignant hepatoid gland tumors, in which a dissection of mature cells from basal cells occurs. According to the authors, the simultaneous presence of reserve cells with anisocytosis/anisokaryosis and nuclear activation is a strong suspect for hepatoid carcinoma. In practice a cytological diagnosis of carcinomas of hepatoid glands must be made only when cellular atypia is evident. As some well-differentiated carcinomas are cytologically indistinguishable from adenoma, histopathological examination is necessary for a final decision (Figs. 8.418, 8.419 and 8.420).

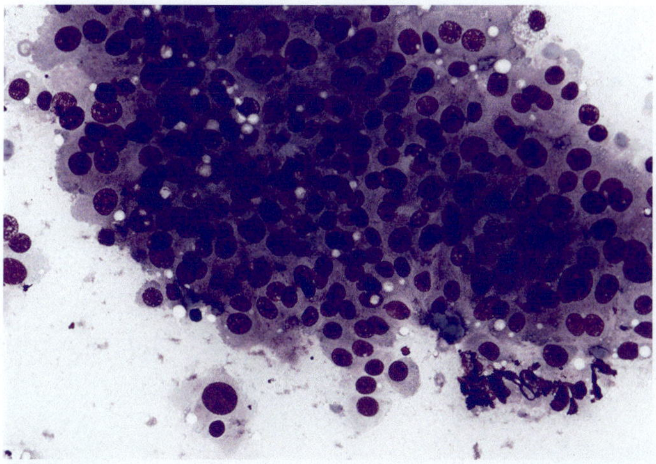

Fig. 8.420 Cytology of hepatoid gland carcinoma: chaotic organization of the cluster. Nuclei are overlapped and with evident anisokaryosis

As abovementioned for *hepatoid adenoma*, also in the case of *carcinomas*, the malignant proliferation can be surrounded by large lobules of both hyperplastic and adenomatous benign neoplasia; it represents the major limitation for the cytology of hepatoid glands, since the possibility of not inserting the needle into the malignant area can increase the percentage of false-negative result. For this reason, it is the author's opinion that cytology of hepatoid glands should be performed in any case, only for confirmation of their origin, but the definitive differentiation between the benign and malignant histotypes is always based on histopathological results.

Finally, metastasis to intra-abdominal lymph nodes starting from benign hepatoid gland neoplasia is occasionally reported in literature; this situation creates further confusion in cytological defining between the benign and malignant hepatoid cell tumors [132].

8.3.7 Apocrine Gland Tumors (Sweat Glands) and Modified Apocrine Gland Tumors

Sweat gland tumors are very common in dogs and cats and almost all originate from the *epitrichial glands*, which are distributed throughout the body. *Atrichial* glands, which empty the sweat directly on the skin surface, are *eccrine* or *merocrine* and are only present in the skin of the paw pads. Tumors that originate from the eccrine glands are extremely rare in dogs and cats [137, 138].

Two other glandular tissues originating from "modified apocrine glands" are present on the skin: the *ceruminous glands* and the *anal sac glands*, the latter covering the inner wall of the anal sacs.

Apocrine neoplasms have not a peculiar clinical appearance and are represented by nodular lesions of varying sizes, from small to very big masses, solid, solid-cystic, or very often mainly cystic in composition.

The mainly cystic lesions, such as cystadenomas, can be covered by an atrophic to very thin and alopecic skin because of the compression exerted by the liquid on their walls; in these cases, the surface is very soft to palpation. During FNAB, the sweat collected is usually colorless or rarely, in chronic cases, dark. Once emptied of its contents, the thin atrophic wall is deflated and assumes the appearance of a vesicle, flabby in consistency; in some cases, the remaining solid tissue can be palpated at its base.

Feline *apocrine cystomatosis* is a term used to describe a peculiar clinical lesion of the apocrine glands, represented by multiple small, dark bluish cystic lesions, found on the face, eyelids, and head and/or the inner surface of the pinna up to the external auditory canal. Rarely, the lesion may be single. Unlike the cystadenomas described above in dogs, these cysts are not neoplastic in origin, but simply represent dilated benign tubules of the apocrine glands; Persian and Himalayan cats appear to be predisposed to this condition, especially with facial and peripalpebral involvement, while DSH cats develop predominantly external ear canal cystomatosis [136, 140].

Adenomas and carcinomas of the apocrine glands do not have peculiar macroscopic aspects and can occur mainly on the head as nodules of variable size, alopecic, and often ulcerated. One clinical feature that can be observed in more superficial and alopecic lesions and that may suggest the apocrine origin of the tumor is the bluish color of the external skin, usually due to the dark bluish content of the cysts present in the context of many apocrine tumors. Apocrine tumors in cats are located mostly on the head and neck as small, pigmented skin nodules, which represent solid/cystic adenomas, but can be very large in size. Multiple lesions are not usual in neoplastic apocrine lesions a part for metastatic malignant neoplasia, which can spread on other skin areas (Figs. 8.421, 8.422, 8.423 and 8.424).

Cytological Findings

Cytological features of *apocrine lesions* depend on the typology of the investigated lesion. Anatomically, *sweat gland tumors* are simple tubular apocrine glands in which the

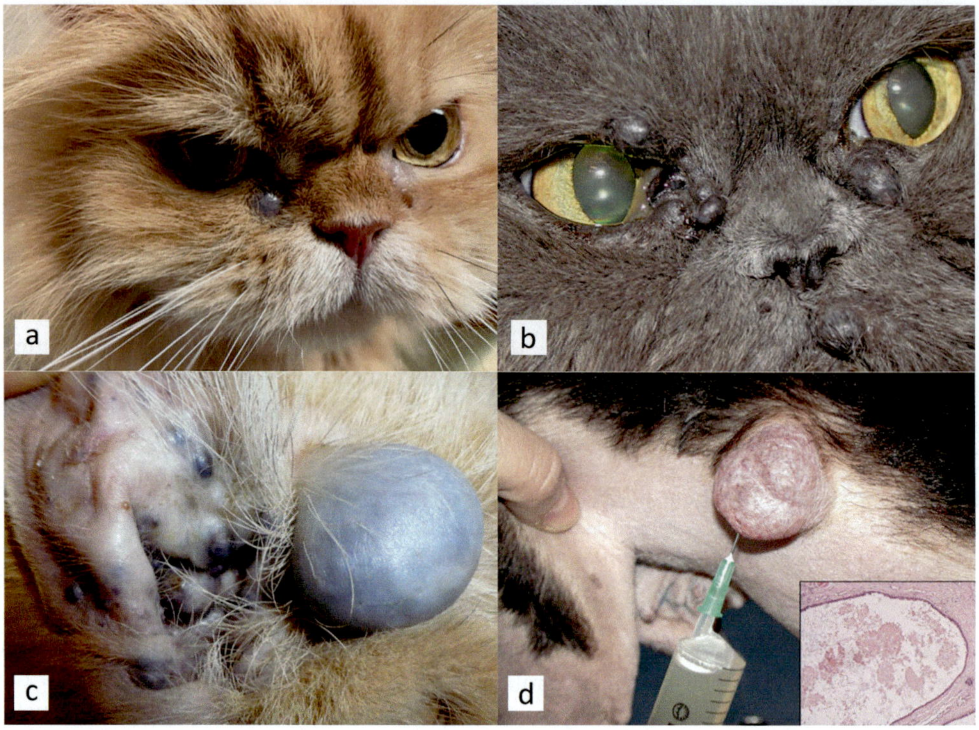

Fig. 8.421 Apocrine benign neoplasia: (**a**) dark blue single nodule of cystomatosis on the face of a cat; (**b**) multiple dark blue confluent nodules in a cat with facial cystomatosis; (**c**) a simultaneous presence of both a large and many smaller apocrine cysts on the ear and skin of the peri-auricular area in a cat with apocrine cystomatosis; (**d**) large cystadenoma in a dog. Note the very thin skin pressed by the high amount of intratubular sweat. Many dilated cystic area are histologically recognizable (inset)

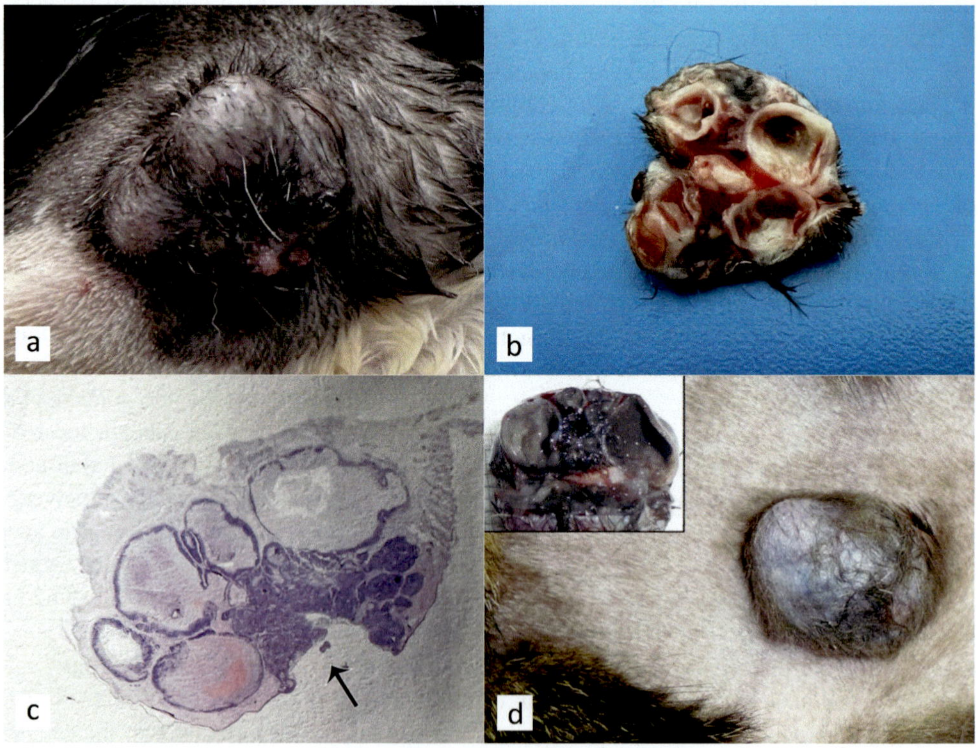

Fig. 8.422 Apocrine benign ductal solid-cystic adenoma: (**a**) multilobated partially alopecic nodule in a cat; (**b**) same nodule after surgical removal; note the multiple cystic areas; (**c**) histological appearance of the tumor on slide stained with H-E stain—note as the neoplasia is composed of multiple cystic lesions and as a solid part at their basis is also recognizable (arrow); (**d**) solid-cystic apocrine neoplasia in a cat—note the multilocular cystic areas (inset)

secretory portion is lined by a single layer of cuboidal to columnar epithelial cells, which secrete sweat through cytoplasmic decapitation of the cell itself. In practice, *sweat* is composed of fragments of the apical part of the cytoplasm through blebbing formation (cytoplasmic decapitation) (Fig. 8.425). Each apocrine gland is associated with a follicle, into whose lumen, through a short duct lined by squamous epithelium, the secreted sweat is released. *Atrichial* glands,

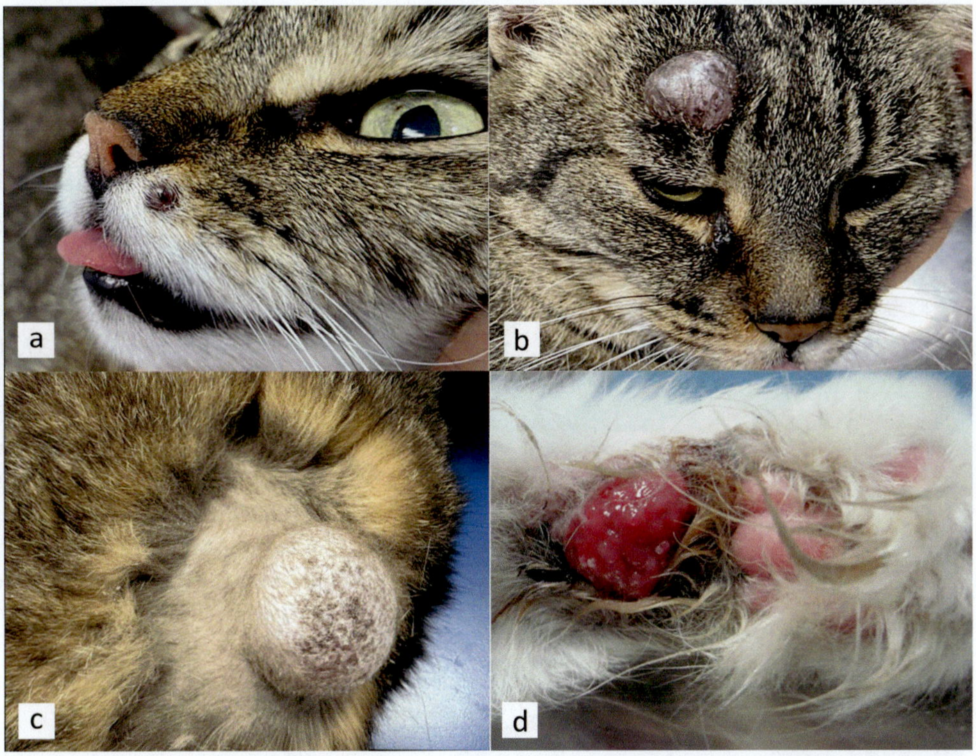

Fig. 8.423 Feline apocrine gland neoplasms: (**a**) small dark apocrine ductal adenoma on the muzzle; (**b**) dark alopecic ductal adenoma involving the apocrine glands; (**c**) large, soft, predominantly cystic ductal adenoma involving the apocrine glands; the nodule completely deflated after removal of the secretion, leaving only a very small solid portion; (**d**) ulcerated apocrine carcinoma on the ventral part of the foot of an elderly cat. (Courtesy of DVM C. Penna-Italy)

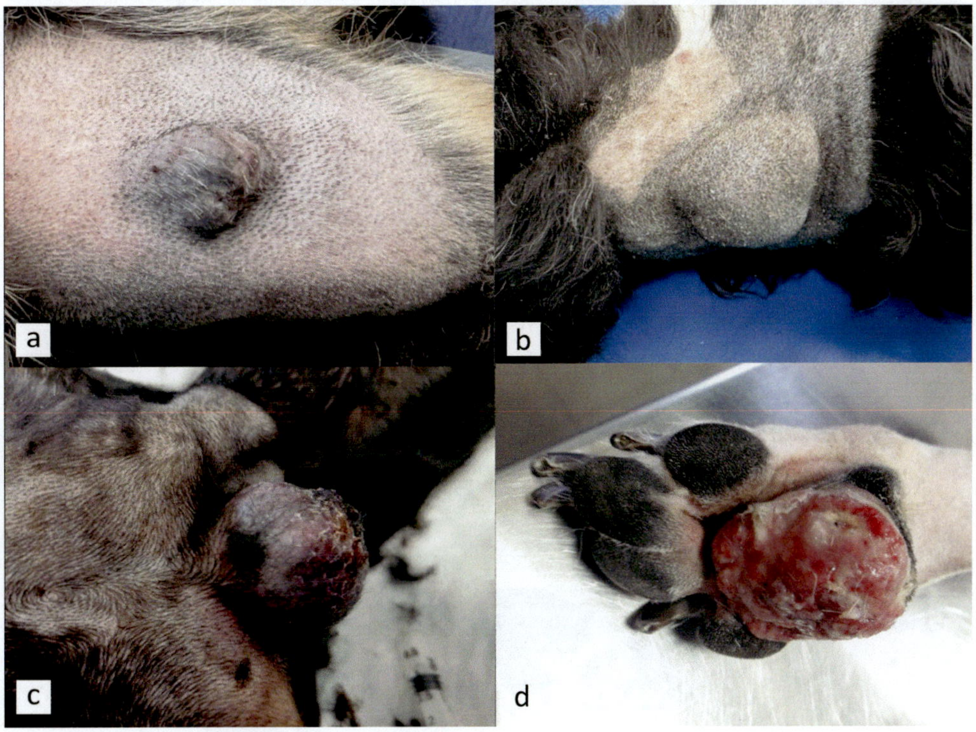

Fig. 8.424 Canine apocrine gland neoplasms: (**a**) apocrine cystadenoma with irregular and bluish in color surface; (**b**) large apocrine benign neoplasm on the neck; (**c**) apocrine adenocarcinoma on the neck of a dog. (Courtesy of DVM R. Caccia-Italy); (**d**) ulcerated mass on the pad of a dog with eccrine carcinoma. (Courtesy of DVM C. Penna-Italy)

which empty the sweat directly on the skin surface, are *eccrine* or *merocrine* and are only present in the skin of the paw pads. From a cytological point of view, these two types of sweat glands are non-differentiable.

The histological classification of the *sweat gland tumors* has undergone considerable changes, as many tumors once diagnosed as *basal cell tumors* are, with the advent of immunohistochemical stains, reclassified as *apocrine*. Neoplasms

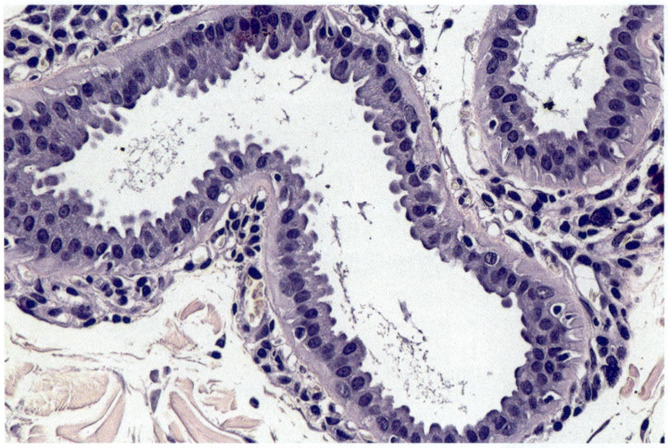

Fig. 8.425 Histology of sweat glands: normal tubule in which apical decapitation of the cytoplasm of apocrine cells is clearly recognizable and representing sweat

that originate from sweat glands include many both non-neoplastic and neoplastic lesions. Among the former, *single or multiple apocrine cysts* and *apocrine cystomatosis* are included, whereas the neoplasms include many variants characterized by different histomorphological features.

Apocrine tumors include the *apocrine adenoma/cystadenoma*, with its *simple*, *ductal*, *papillary, solid-cystic*, and *complex* and *mixed* histological variants, and malignant *adenocarcinoma* which is in turn divided into *simple*, *complex* and *mixed*, *cystic, ductal*, *solid-cystic*, and the *clear cell* histological variants [137, 138, 139, 145].

In cats with *apocrine cystomatosis*, many macrophages, with cytoplasm filled with coarse deep blue material representing apocrine gland secretion, are collected. The intracytoplasmic amorphous material should not be confused with the melanin pigment and consequently make an incorrect diagnosis of melanocytic neoplasm. In many cases, cholesterol crystals are also associated to macrophages (Fig. 8.426).

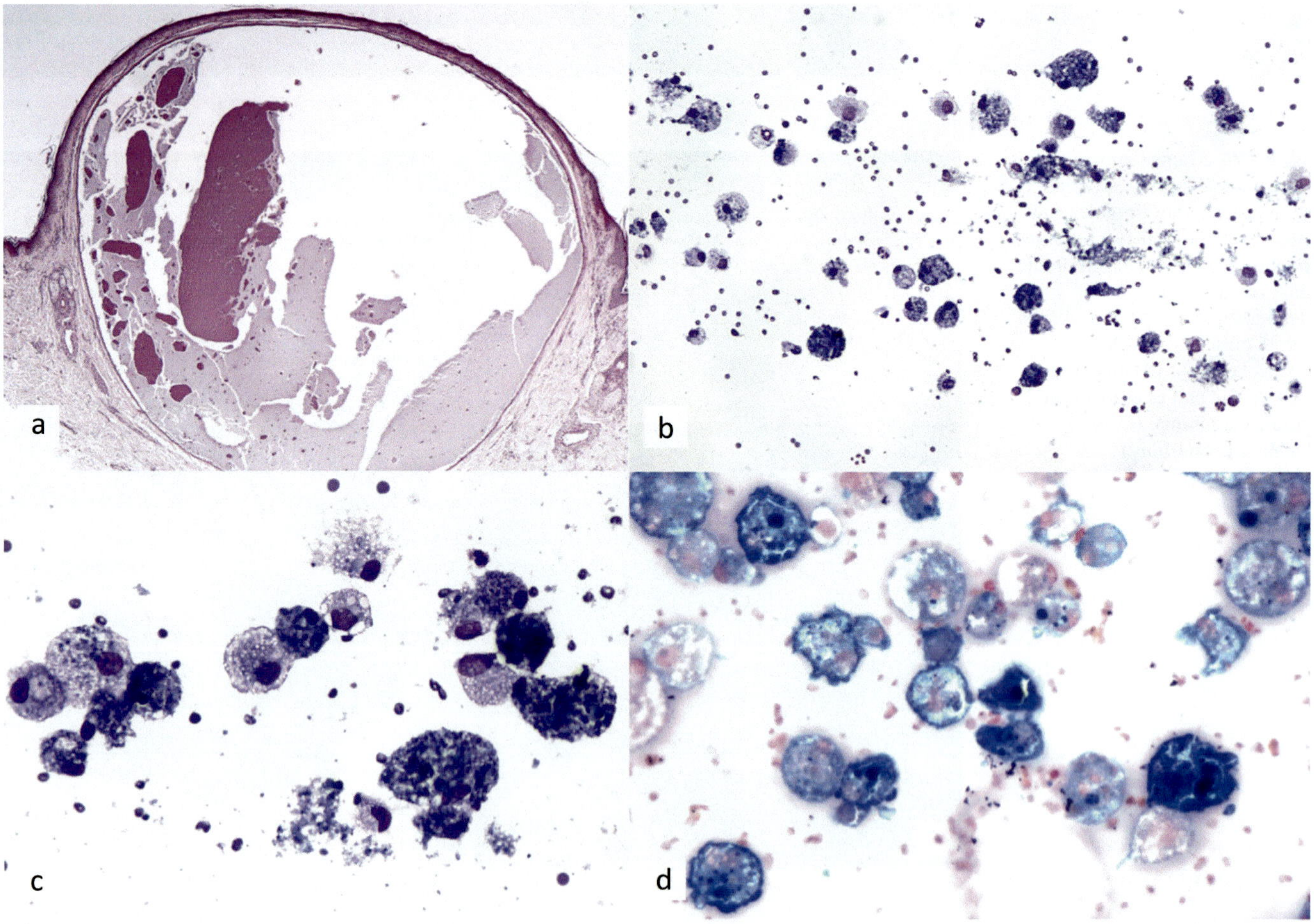

Fig. 8.426 Apocrine cystomatosis: (**a**) Histology—large apocrine cyst filled with sweat. (**b**) Cytology—many foamy macrophages. (**c**) Cytology—at high magnification, macrophages filled with amorphous deep blue secretion. (**d**) Cytology of Perls Prussian blue staining: phagocytized intracytoplasmic secretion can be highlighted with the cytochemical Perls Prussian blue staining for its high iron content

The most common benign tumor of the sweat glands is the *apocrine cystadenoma*; the neoplasm is composed of multiple ectatic cystic apocrine tubules, histologically supported by a fibrovascular connective stroma and lined by a single row of cylindrical/cuboidal or flattened apocrine cells, the latter following the pressure of sweat exerted on the wall (Fig. 8.427). Slides obtained from apocrine cystadenomas, since these lesions are mainly composed of multiple cysts, frequently contain only a clear usually acellular liquid or variable amounts of amorphous deep blue material of smooth appearance representing sweat secretion as discussed in Chap. 5; few macrophages and crystals of cholesterol may be also observed on slides (Fig. 8.428). If no apocrine cells are detectable on slides, a cytological examination of the sediment, after centrifugation of the fluid, increases the possibility of visualizing some epithelial clusters (Fig. 8.429).

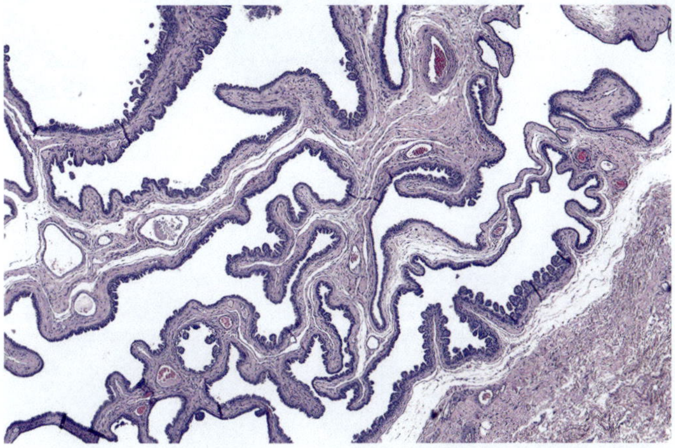

Fig. 8.427 Histology of cystadenoma: multiple ectatic glands lined with columnar cells with pseudo-papillary arrangement

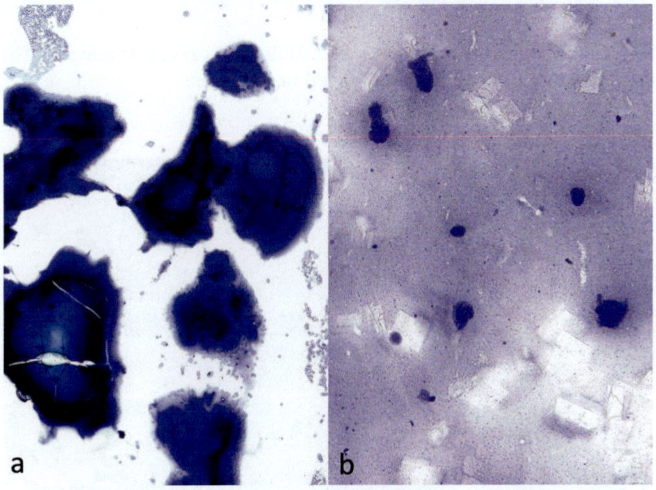

Fig. 8.428 Cytology of cystadenoma: (**a**) clods of apocrine secretion (sweat). There are no epithelial cells in this cytological specimen; (**b**) on a purple dot-like background, macrophages filled with secretion and numerous cholesterol crystals represented by achromatic geometric figures are recognizable

Cytology of well-differentiated *simple apocrine tumors* is characterized by the presence of multiple clusters of basaloid cells, organized in trabecular or honeycomb arrangements. For a beginner cytologist, regular well-differentiated cluster of basal cells are not easily differentiable from those observed in trichoblastoma or other basal tumors, especially if there are no other peculiar cytological features indicating a sweat gland origin.

Cytology of well-differentiated *apocrine adenomas* usually shows very cohesive clusters of uniform-sized basaloid cells with well-defined margins, round to polygonal in shape, with round uniform in size nuclei and scant slightly blue cytoplasm. The cells are usually yielded in sheets, with a *honeycomb* or a *pavement-like* arrangement; in some apocrine tumors, a pathognomonic arrangement of cells is the *palisade* one, in which secretory cells maintain a disposition in rows, a clear columnar/cylindrical shape, basal nuclei, and the presence of apical cytoplasmic blebs, indicative of secretion by cytoplasmic decapitation. The presence of linear palisades should not be confused with the basal rows discussed above for trichoblastoma; in fact, in the latter, the palisades are composed of basal epithelial cells with very scarce cytoplasm and no aspects of secretory activity (Figs. 8.430, 8.431, 8.432, 8.433 and 8.434).

Two concepts regarding the cytology of apocrine sweat glands must be pointed. The first is the incorrect way, used by many cytologists, of using the term *acinar arrangements*, whenever a group of epithelial cells arrange themselves in a circular way around a central area, which recalls an acinar anatomical structure. This definition would presuppose that the sweat glands were anatomically composed of acini; but,

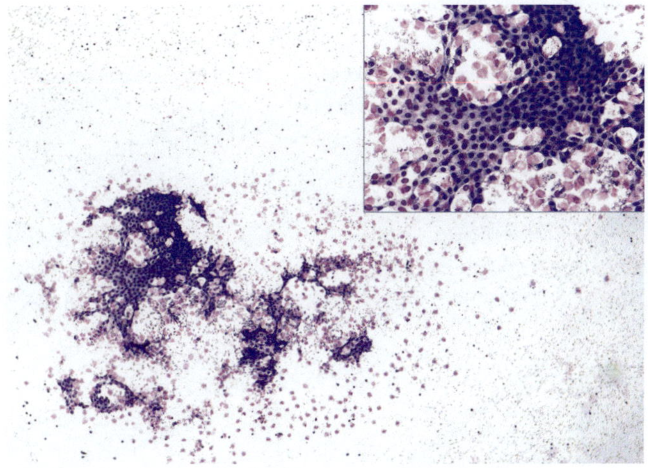

Fig. 8.429 Cytology of cystadenoma. A single large cluster of apocrine cells was sampled only after centrifuging the collected fluid with the needle. Note the cells partially damaged by the cell concentration technique (inset)

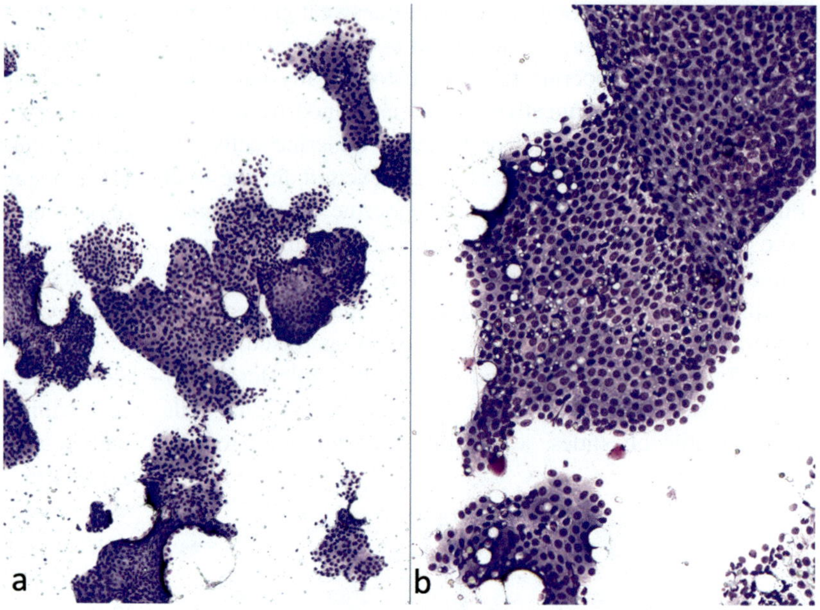

Fig. 8.430 Cytology of a benign apocrine tumor (apocrine adenoma): (**a**) large groups of uniform apocrine cells with a honeycomb arrangement; (**b**) at high magnification the regular morphology size and uniform size of cells are better evaluable

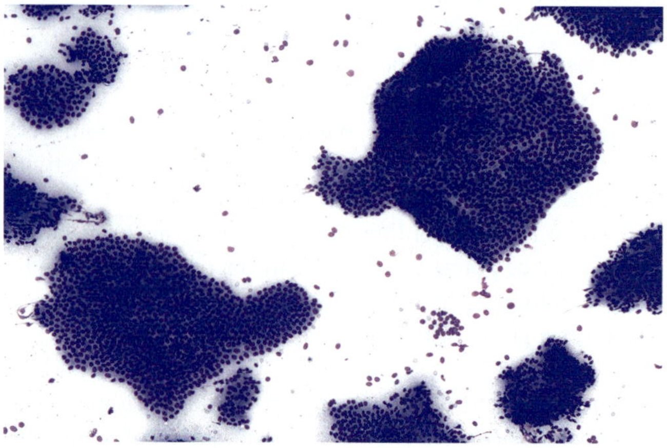

Fig. 8.431 Cytology of an apocrine adenoma: large groups of uniform in size and shape apocrine cells

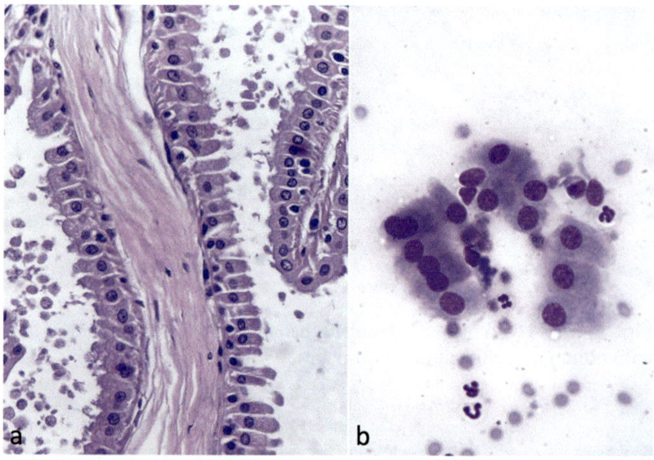

Fig. 8.433 Palisade arrangements: histological (**a**) and cytological (**b**) comparison of apocrine cells arranged in palisades

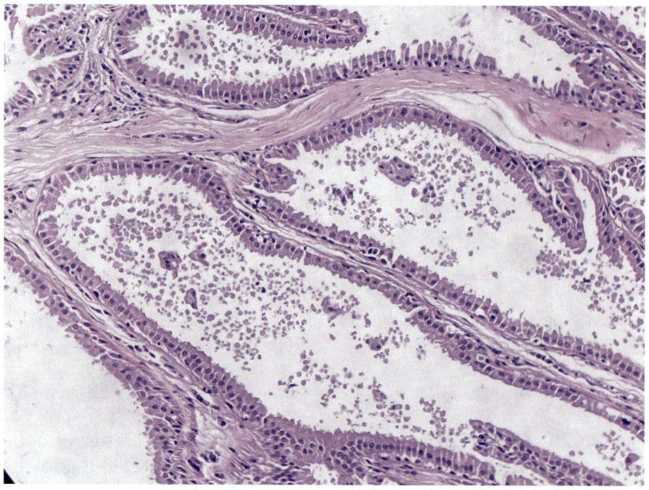

Fig. 8.432 Histology of benign sweat gland tumor. Note the palisade arrangement of cells which line the tubules

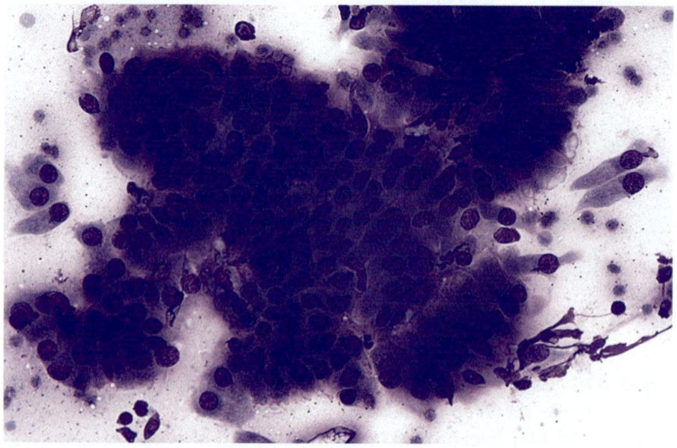

Fig. 8.434 Cytology of apocrine adenoma. Well-differentiated apocrine cells: note the palisade arrangements at periphery

since the sweat glands, as already discussed, are simple tubular glands, it is not possible that they can form acini, and therefore all the arrangements that the cells assume on the slides with aspects that recall the anatomy of an acinus are not true acini. It would be more correct, when these dispositions originating from the sweat glands is detected, to describe them as *pseudo-acinar* or *acinar-like* arrangements, even if this description is totally useless. From the above, if true acinar structures are present on the slides, a cutaneous metastasis of a mammary tumor or coming from another glandular tissue composed of acini, such as from internal abdominal organs (e.g., pancreas) or lung tissue (e.g., bronchi), should be suspected.

The second point is that the presence on the slides of only apocrine epithelial aggregates, organized either in clusters with honeycomb appearance or in well-differentiated palisades, must not allow to make the diagnosis of apocrine neoplasia; in fact, there are reactive skin situations that determine a hyperplasia of the apocrine glandular tissues and that, when sampled, release numerous epithelial cells of a sweat nature that can be absolutely confused with well-differentiated neoplastic cells. A clear example is the acral lick granuloma, an inflammatory lesion frequently observed on the extremities, especially of the carpus, which, following continuous licking, determines a marked hyperplasia of the apocrine glands that can be collected in high number with FNB; therefore the evaluation of the clinical lesion from where cells are collected is mandatory in differentiating a neoplasia from a reactive hyperplastic apocrine lesion (Fig. 8.435). Finally, another important cytological finding that indicates the apocrine nature of the epithelial cells is the presence of dark blue dotted granular material in the cytoplasm, which is never present in other basal follicular or epidermal neoplasms (Fig. 8.436).

To define the histotype of the apocrine tumors, histology is necessary, also if, in cases of well-differentiated tumors and in smears of great quality and high cellularity, the arrangement of the cells can perfectly report the histopathological pattern of growth, as discussed in Chap. 7.

Since most apocrine neoplasms are composed of cystic tubular structures producing sweat, it is very common to find, along with the apocrine epithelial clusters, a variable amount of deep blue, dense, smooth, amorphous, and acellular material in the background, which represents sweat; its finding on slides is of great help in confirming the diagnosis of apocrine lesion, especially when the basal cell clusters do not show other cytological characteristics of apocrine differentiation (Figs. 8.437 and 8.438).

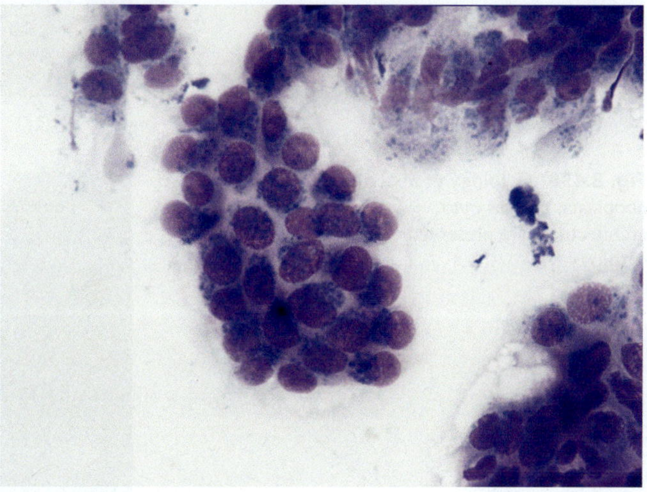

Fig. 8.436 Well-differentiated apocrine cells: presence of intracytoplasmic punctate and deep blue in color material suggesting the apocrine natures of clusters

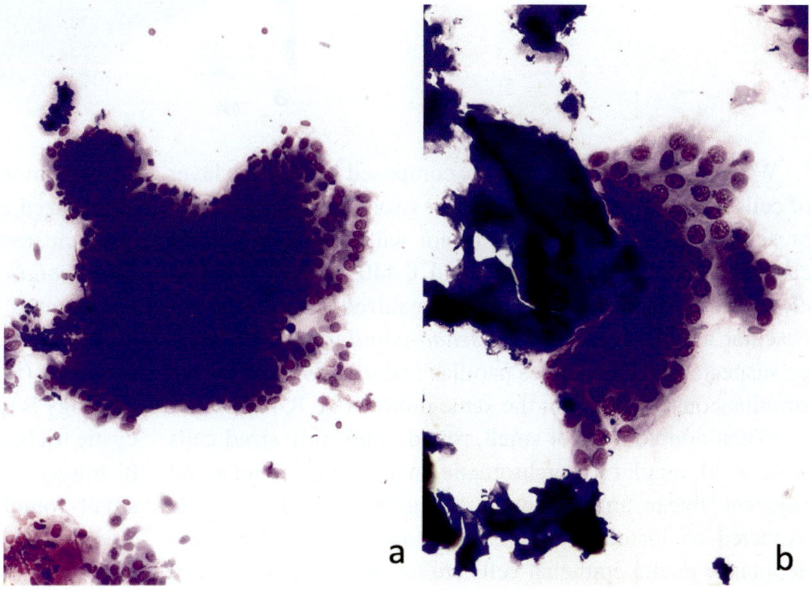

Fig. 8.435 Cytology of apocrine cells. In very well-differentiated samples, it is impossible to differentiate, with cellular morphology alone, (**a**) hyperplastic apocrine cells from (**b**) benign neoplastic cells

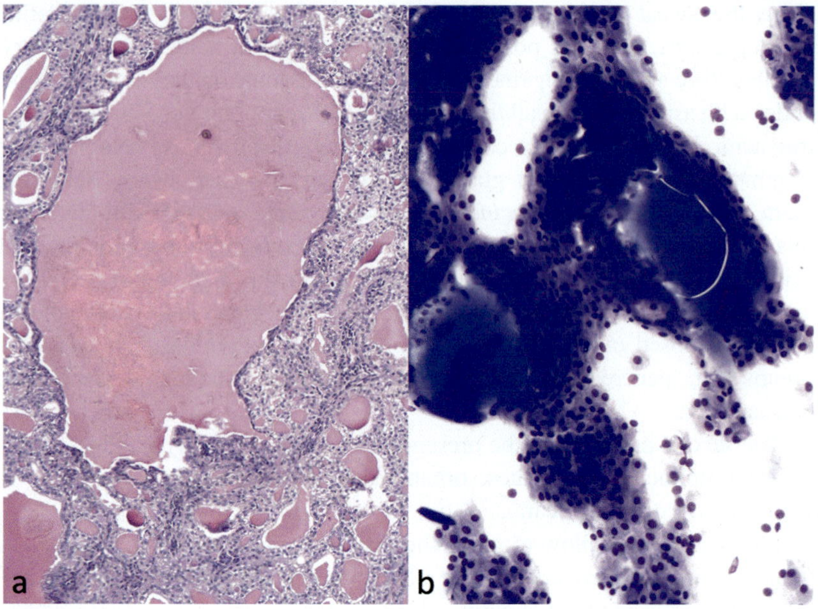

Fig. 8.437 Apocrine adenoma: (**a**) histology—multiple cystic tubules of different size, filled with eosinophilic secretion; (**b**) cytological representation of the same lesion showed in Fig. 8.437a; tubular cystic structures maintain their anatomical arrangement. The secretion stains deep blue with Romanowsky-type dye

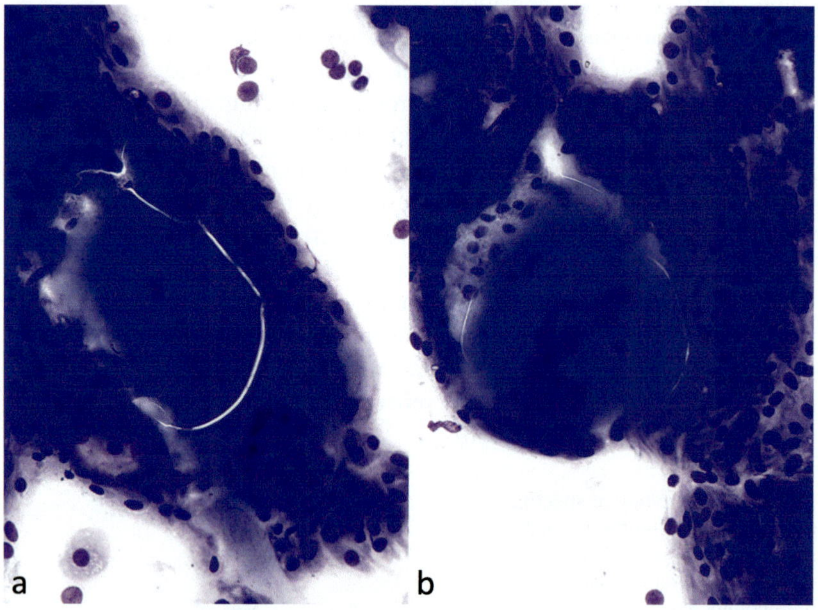

Fig. 8.438 Cytology of apocrine neoplasia: note as cyto-architectures are preserved on cytology

When *papillary* arrangements, composed by a single layer of cells oriented around a vascular or stromal axis, are clearly present on slides, an apocrine tumor with *papillary pattern* can be diagnosed (Figs. 8.439 and 8.440). When small to elongated trabeculae of cells, not organized around a fibro-vascular axis, are detected, a *pseudo-papillary* growth must be suspected; in many cases papillae and *pseudo-papillae* are simultaneously present in the same tumor (Fig. 8.441).

When double rows of small, round, uniformly sized cells with small, regular hyperchromatic nuclei compose long and adjacent linear arrangements, a *ductal pattern* is to be expected on histology (Figs. 8.442 and 8.443). When the neoplastic ductal epithelial cells are not basal but arise from the squamous differentiated cells, both nucleated polygonal keratinized cells and fully keratinized cells, along with apocrine clusters, are detectable on slides.

In *solid-cystic* variant, an amorphous acellular necrotic background with debris, bare nuclei, and variably sized clusters composed of regular in size basal apocrine cells is observed (Figs. 8.444, 8.445, 8.446, 8.447 and 8.448). This cytology is quite impossible to be differentiated from that of a cystic trichoblastoma, and this is also very difficult to do with histology if no apocrine features in the solid areas, such as small roundish achromatic areas, which should represent abortive intracellular ductules, are recognizable.

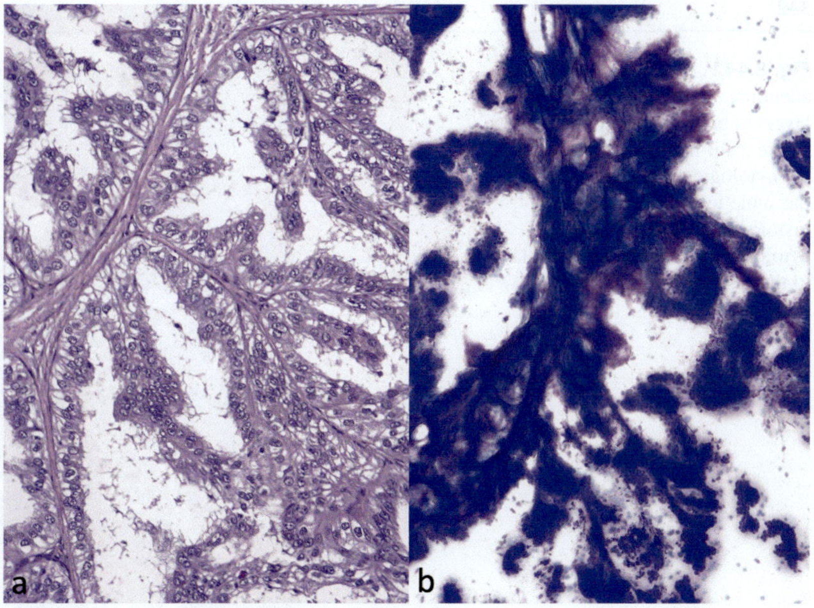

Fig. 8.439 Cytology of an apocrine neoplasia: (**a**) histology—many papillae with fibrovascular axis are recognizable; (**b**) cytological representation of the same lesions showed in histological image, in which long papillary arrangements are recognizable

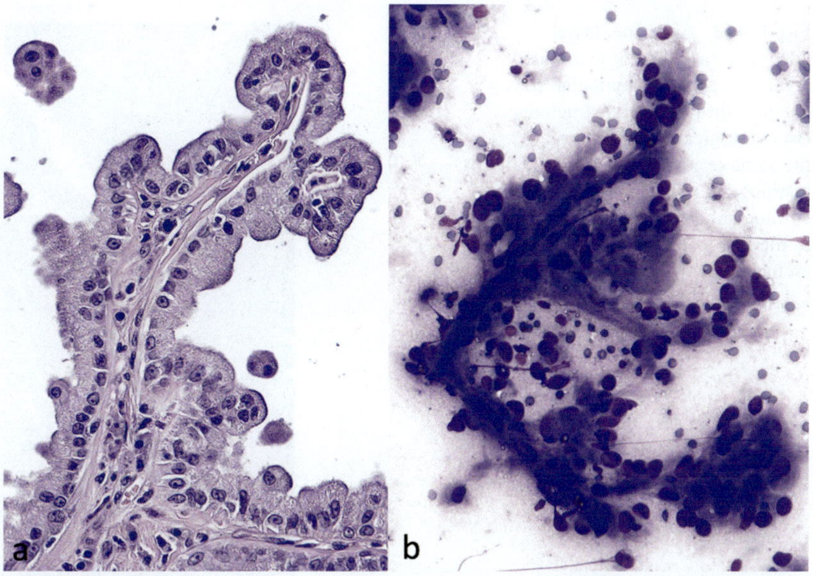

Fig. 8.440 Cytology of an apocrine neoplasia: (**a**) histology—papillae with pseudo-papillae; note as epithelial cells are arranged around a fibrovascular axis; (**b**) cytological representation of a papillary arrangement

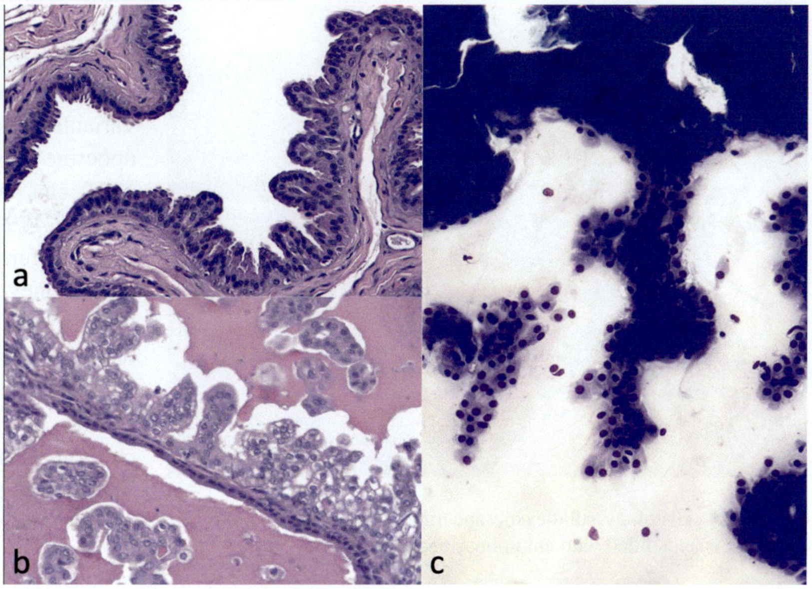

Fig. 8.441 Pseudo-papillary arrangement: (**a**) and (**b**) small papillae protruding into the lumen of neoplastic apocrine tubules. Note how the central axis is not present; (**c**) cytological representation of differently sized pseudo-papillae, collected from an apocrine neoplasm

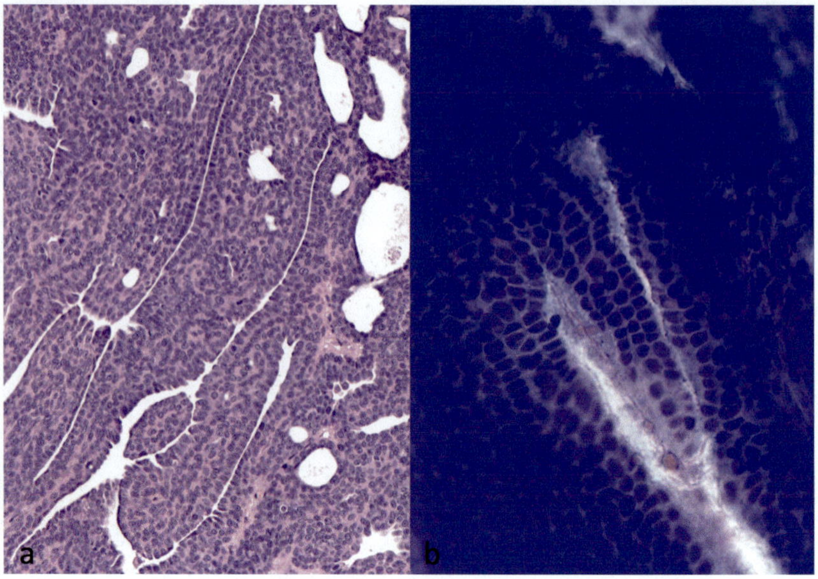

Fig. 8.442 Apocrine ductal adenoma. (**a**) Histology: double layers of basophilic epithelial cells create a slit-like ductal structures; (**b**) cytological representation of the same lesions showed in histological image. Note as cytological arrangement is similar to the histopathological architecture

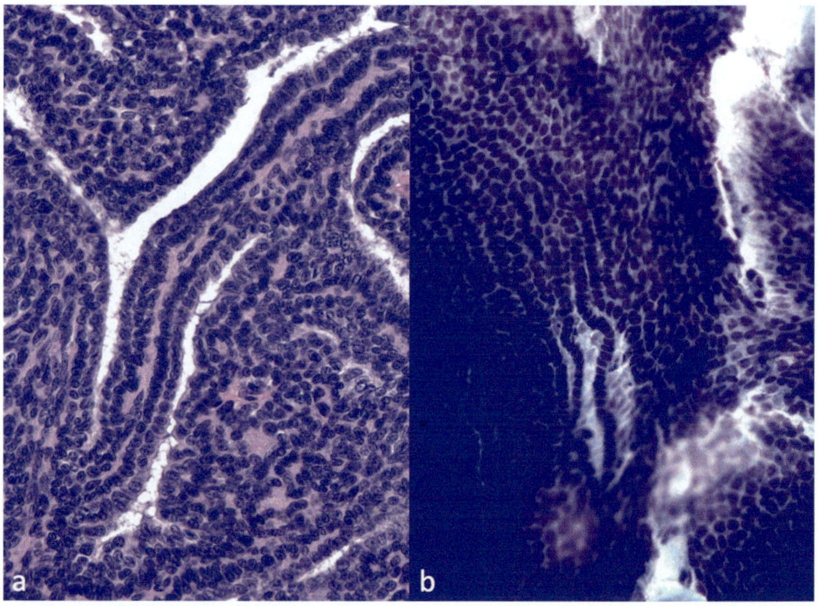

Fig. 8.443 Apocrine ductal adenoma. (**a**) Histology: double layers of basophilic epithelial cells create a slit-like ductal structures; (**b**) cytological representation of the same lesions showed in histological image. Note as cytological arrangement is similar to the histopathological architecture

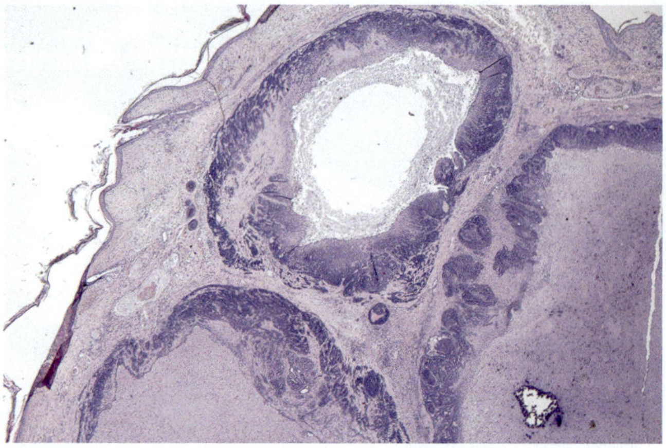

Fig. 8.444 Histology solid-cystic apocrine adenoma: multiple cystic epithelial islands filled with amorphous necrotic material

In exceptional cases, sweat glands apocrine tumors can exhibit a *sebaceous differentiation*, as occurs in some variants of mammary tumors; in these cases the presence of apocrine clusters in which some cells transform in sebaceous cells can be observed in cytological specimens [141] (Figs. 8.449 and 8.450). Cytology of these neoplasms is not easily differentiable from that of a sebaceous epithelioma; the clear honeycomb arrangement, the lacking of a sebaceous background, and the presence of sweaty secretion adjacent to basal cells, can orient the diagnosis to an apocrine neoplasia.

These are all examples, in which the cytological research of the cyto-architectures of apocrine neoplasms, allows cytologists not only to make a correct diagnosis but also to suspect the pattern of the tumor. These cytological aspects unfortunately are not so frequently detectable and are evident

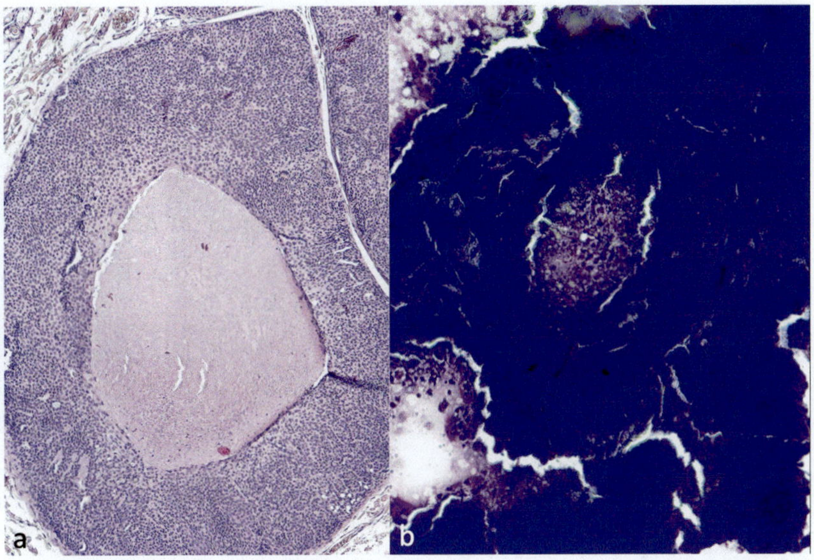

Fig. 8.445 Solid-cystic apocrine adenoma: (**a**) histology—close-up of an epithelial island of basal cell with a central cystic area; (**b**) cytological representation of the same lesions showed in histological image. Note as cytological arrangement is similar to the histopathological architecture

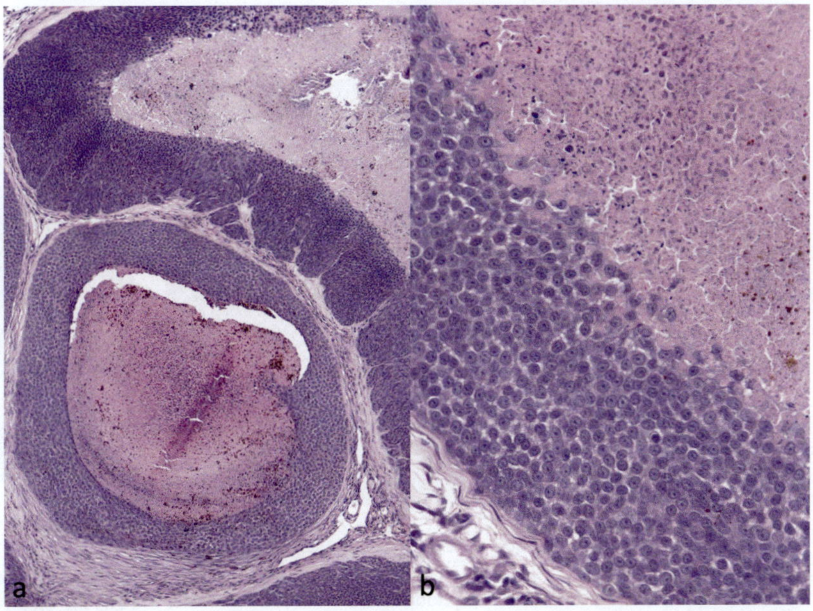

Fig. 8.446 Histology of solid-cystic apocrine adenoma: (**a**) multiple cystic epithelial islands filled with amorphous necrotic material; (**b**) close-up of intracystic material. Note also the uniform size and shape of the basal epithelial apocrine cells

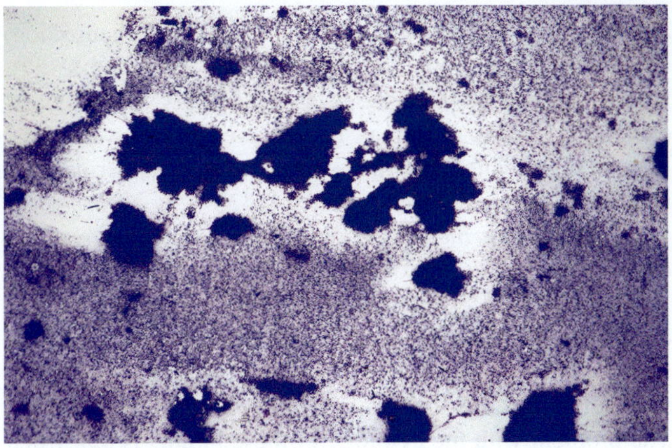

Fig. 8.447 Cytology of solid-cystic apocrine adenoma: on a detrital and amorphous background, many large clusters of basal cells are detectable at low magnification

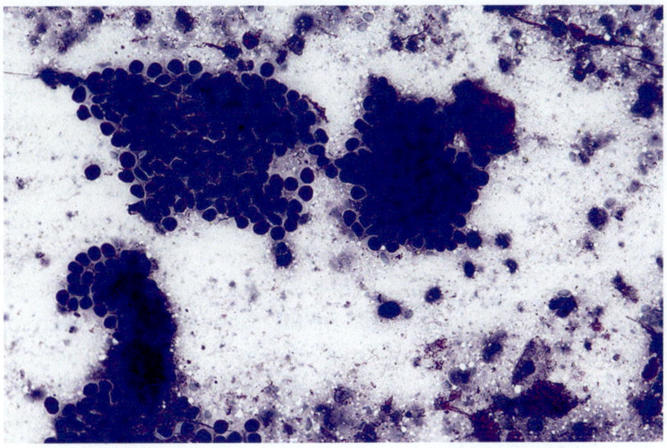

Fig. 8.448 Cytology of solid-cystic apocrine adenoma: at high magnification, uniform in size basal apocrine cells are better recognizable

Fig. 8.449 Cytology of an apocrine neoplasm with sebaceous differentiation: large mass of basal cells with scattered sebaceous differentiation (cells with apparently foamy cytoplasms) (black arrows). Note the large clod of secreted sweat, of intense blue color, of smooth appearance, which supports the apocrine origin of the neoplastic cells (white arrow)

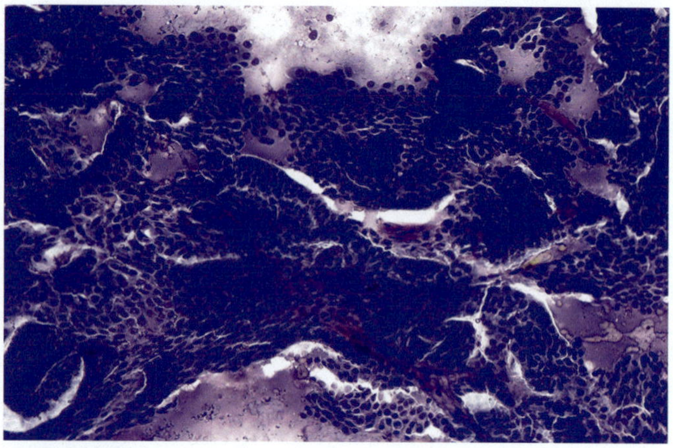

Fig. 8.451 Cytology of an apocrine carcinoma: large groups of well-differentiated malignant apocrine cells. Note as the morphology does not show clear cytological atypia

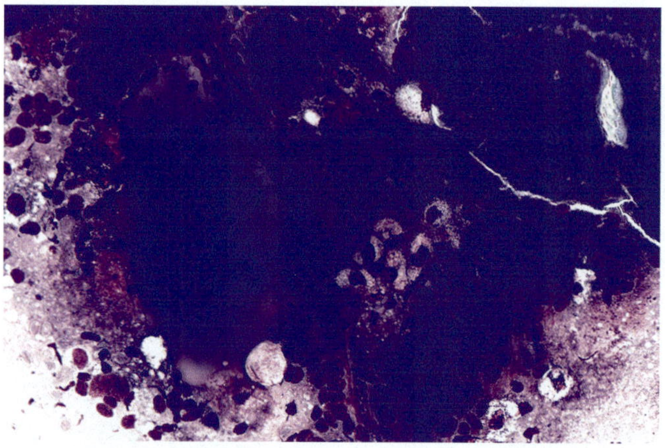

Fig. 8.450 Cytology of an apocrine neoplasm with sebaceous differentiation: close-up of Fig. 8.449 in which apocrine cells with sebaceous differentiation and the deep blue amorphous sweat secretion are better identifiable

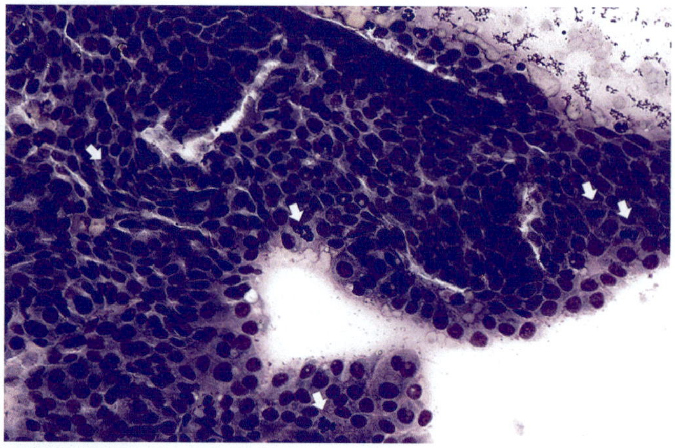

Fig. 8.452 Cytology of well-differentiated apocrine carcinoma: large cluster of well-differentiated but malignant apocrine cells. Note as the morphology does not show clear cytological atypia but many mitotic cells are clearly evident (arrows)

only in well-differentiated neoplasia while they are lost in malignant ones.

Apocrine gland carcinomas show malignant cytological features of varying severity, but in the case of some well-differentiated apocrine gland carcinomas, cytological examination does not allow differentiation from adenomas, and histopathological examination is necessary to evaluate the growth pattern, degree of infiltration, and other diagnostic parameters (Figs. 8.451 and 8.452); this is a very important point if we emphasize that in the same tumor, different degrees of differentiation may be present simultaneously and sampling may not be representative of the entire tumor (Figs. 8.453 and 8.454). In less-differentiated carcinomas, the cyto-architectural features that distinguish epithelial cells as apocrine are lost in favor of morphologic atypia such as anisocytosis, anisokaryosis, architectural disorganization, and loss of cellular cohesion (Figs. 8.455, 8.456 and 8.457). Rarely, cutaneous metastases from primary sweat gland carcinomas of the skin may be observed in dogs and cats. Metastases to internal organs from primary apocrine skin neoplasms have also been reported [142].

Modified apocrine gland tumors include neoplasms of the *ceruminous glands* and of the *glands of anal sacs* [137].

Neoplasms of *ceruminous glands* are found in the external ear canal and can be of variable size, alopecic, erythematous, with irregular and friable surface and often ulcerated. In cats, more than half are malignant as morphology (Fig. 8.458). The problem with the cytology of these tumors is that when the tumor is located in the upper part of the vertical external auditory canal, it can be easily sampled with a needle,

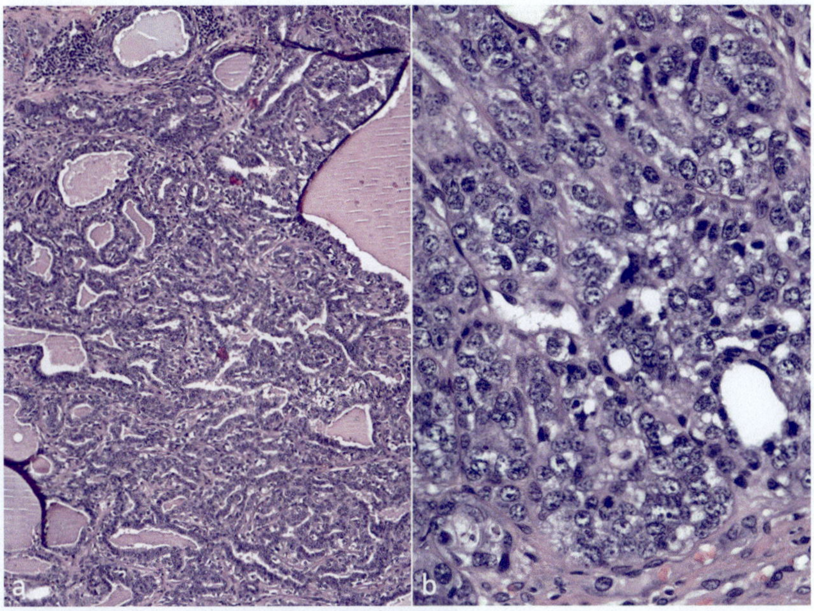

Fig. 8.453 Histology or an apocrine carcinoma: (**a**) in this area on neoplasm, no marked malignant cytological aspects are clearly evident; (**b**) in other areas of the same tumor, malignant features represented as chaotic cell organization and anisokaryosis are evident

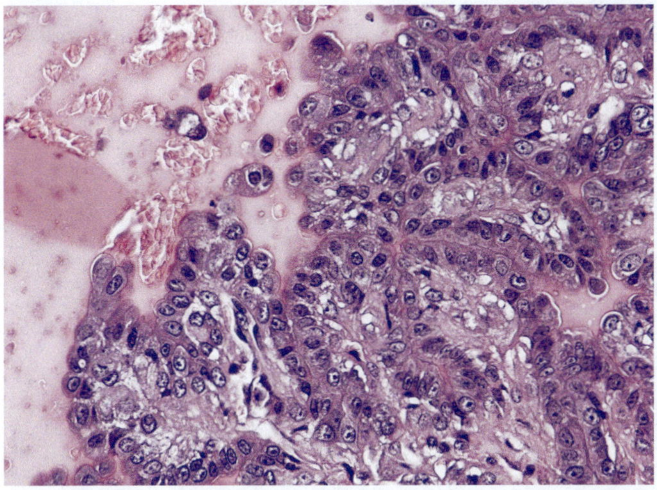

Fig. 8.454 Histology or an apocrine carcinoma: atypia of neoplastic cells are evident

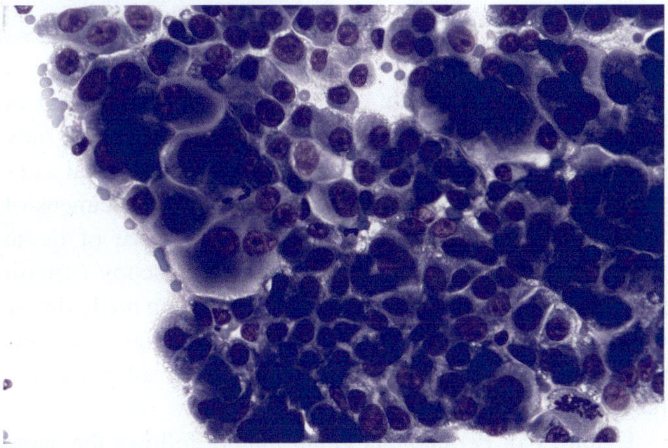

Fig. 8.456 Cytology of apocrine carcinoma: anisocytosis, anisokaryosis, and loss of cellular cohesiveness are all cytological features indicative of malignancy of epithelial tissues

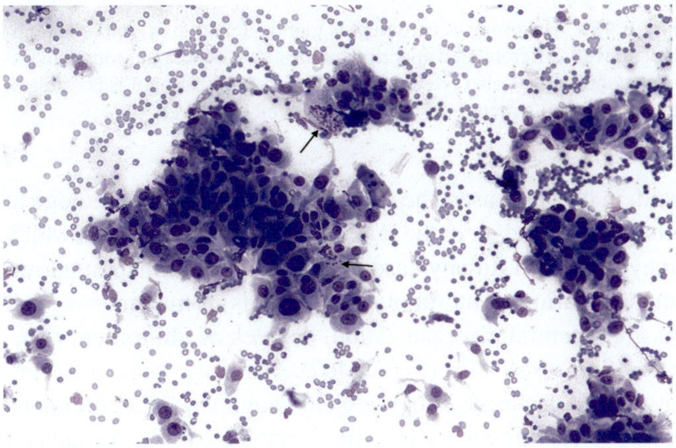

Fig. 8.455 Cytology of an apocrine carcinoma: many malignant cytological features are present. Note the two atypical mitotic cells (arrows)

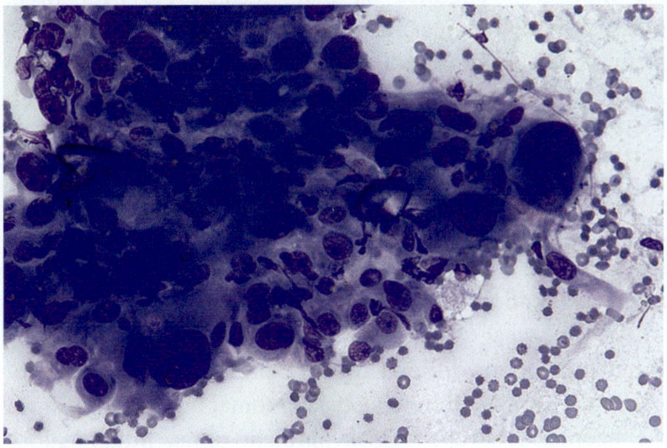

Fig. 8.457 Cytology of apocrine carcinoma: marked anisocytosis and anisokaryosis are malignant cytological features clearly evident at higher magnifications

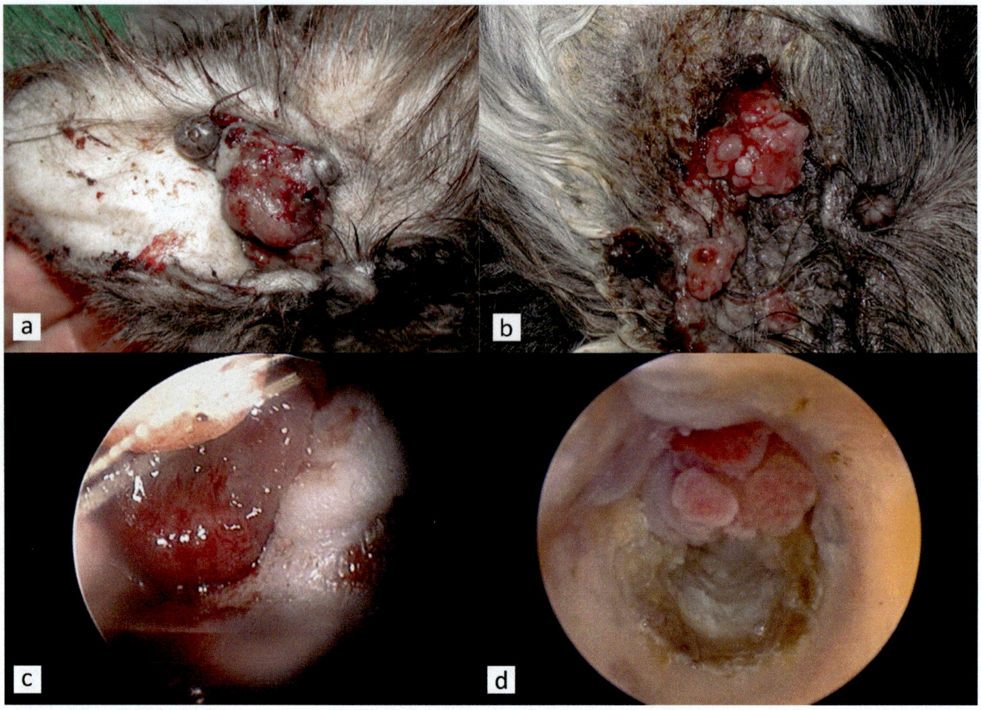

Fig. 8.458 Ceruminous apocrine gland neoplasia: (**a**) multiple neoplasms in a cat, the largest of which protrudes from the external ear canal; (**b**) multiple neoplasms occluding the ear canal in a dog; (**c**) erythematous and with smooth surface ceruminous gland carcinoma in the vertical external canal of a cat; (**d**) parietal multilobed ceruminous carcinoma in a cat with irregular surface (Courtesy of DVM F. Necci-Italy)

whereas, when the tumor is located deep in the vertical or horizontal segment of the external auditory canal, it can only be reached by otoscopy and a good quality cytological sampling is not always obtained. Most cytological specimens of these tumors are obtained by squashing a piece of tissue sampled with a biopsy forceps via video-otoscopy (squash prep). In this regard, the squash prep preparation method may create difficulties in evaluating some morphological aspects of the epithelial cells which, with this method, suffer inevitable morphological damage.

Cytology of *ceruminous gland tumors* follows the same rules as described for other apocrine neoplasms, and similarly, when present, the detection of large amounts of large, dark blue intracytoplasmic granules is sufficient to define their origin; the same material is present at the bottom of the slide or is engulfed by macrophages. As with sweat glands, the cytological features depend on the histological picture of the lesion examined; large cellular sheets, papillae and/or pseudo-papillae or palisades, may be collected with FNB. In simple benign ceruminous tumors, large sheets of uniform, regularly sized epithelial cells are observed; honeycomb or pavement-like arrangements are easily detectable. Since many tubes are normally filled with amorphous blue secretion, the latter may be found at the bottom (Figs. 8.459, 8.460 and 8.461).

As for sweat gland tumors, also for most ceruminous gland tumors are difficult to differentiate between an adenoma and a well-differentiated carcinoma; furthermore, since the neoplasia always causes purulent otitis, inflammatory cells are sampled together with neoplastic epithelial cells, and, in chronic cases, the former may undergo morphological changes secondary to the inflammatory environment in which they are found. In any case, histological examination allows the evaluation of morphological aspects of malignancy such as architectural disorganization, degrees of infiltration, number of mitoses, etc. (Fig. 8.462). Anyway, only if severe atypia is detected, a cytological diagnosis of ceruminous gland carcinoma can be cytologically formulated (Figs. 8.463, 8.464 and 8.465).

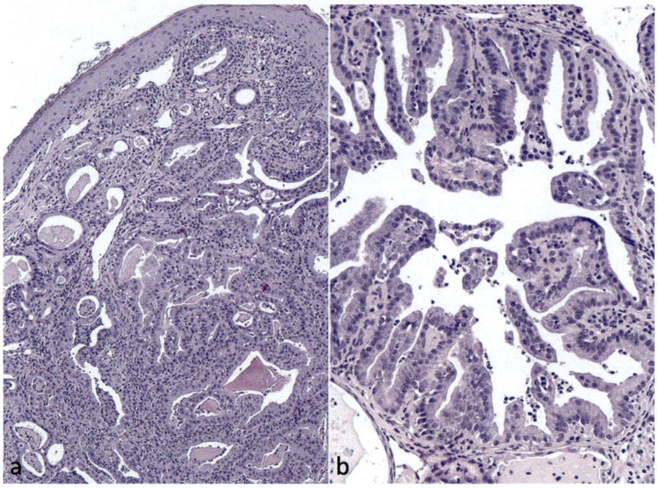

Fig. 8.459 Histology of ceruminous gland adenoma: (**a**) multiple regular and well-differentiated apocrine tubules. A secondary chronic inflammatory infiltrate is also present; (**b**) at high magnification it is possible to evaluate as the structure of the glands is well preserved

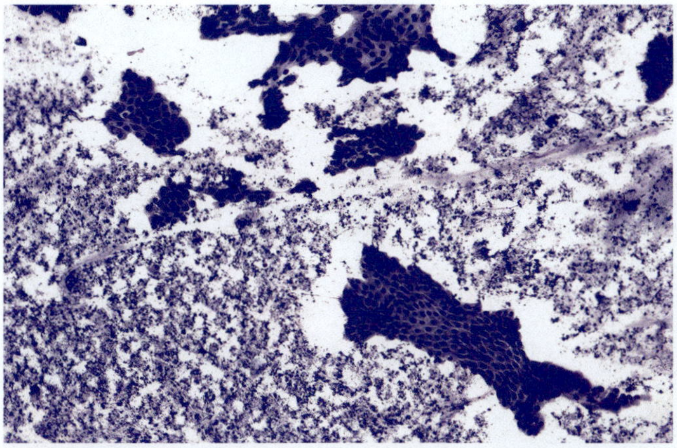

Fig. 8.460 Cytology of ceruminous gland adenoma: on a detrital and amorphous background, many large clusters of epithelial neoplastic apocrine cells with a honeycomb to pavement-like arrangement are recognizable

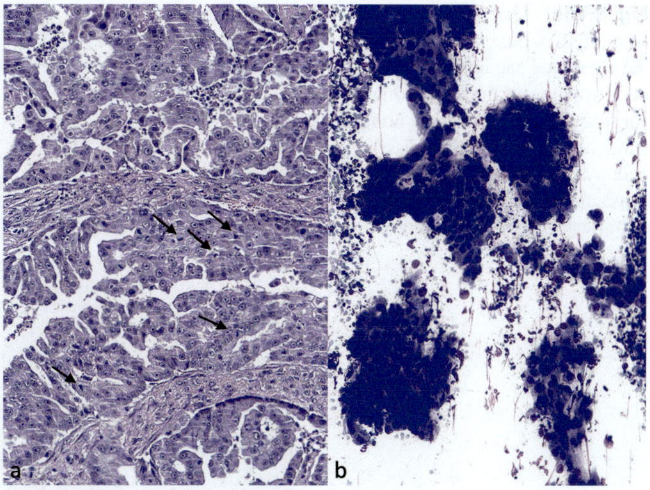

Fig. 8.463 Ceruminous gland carcinoma: (**a**) Histology—well-differentiated carcinoma; note the high number of mitotic cells. (**b**) Cytology: clusters with poorly defined edges. A secondary inflammation is also present

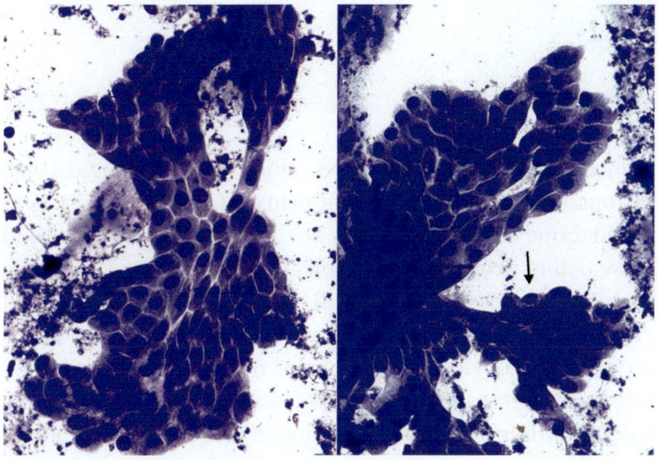

Fig. 8.461 Cytology of ceruminous gland adenoma: many large clusters of epithelial neoplastic apocrine cells with pavement-like arrangements are recognizable. Note as no malignant atypia is visible. In the right photo, a pseudo-papillary structure is recognizable (arrow)

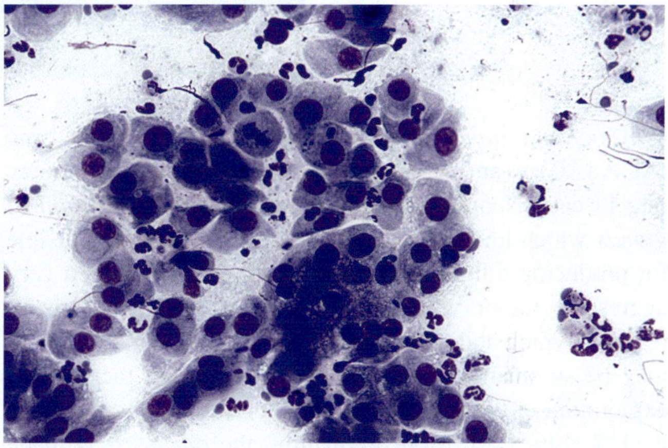

Fig. 8.464 Cytology of gland carcinoma: anisocytosis, anisokaryosis, loss of cohesiveness, and mitotic cells are all cytological features of malignancy

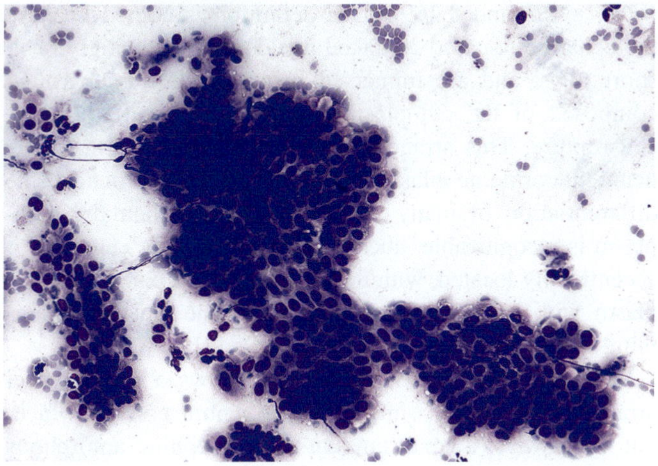

Fig. 8.462 Cytology of well-differentiated ceruminous gland carcinoma: note as only slight features of malignancy are recognizable at the periphery of this cluster (anisokaryosis, loss of cohesiveness)

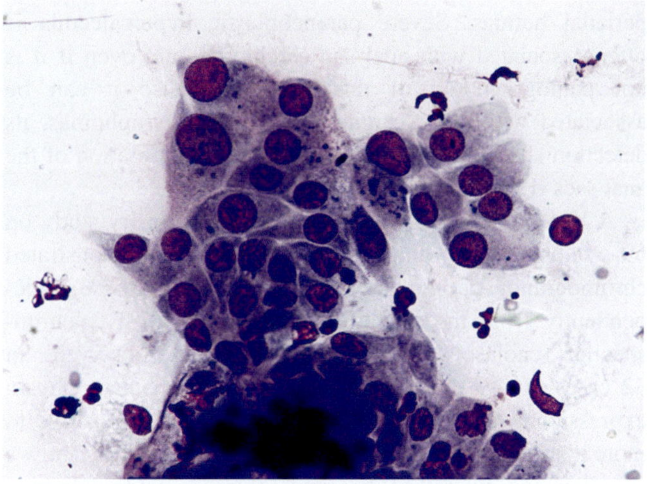

Fig. 8.465 Cytology of glandular carcinoma: despite malignancy, pavement-like and palisade arrangements are still recognizable. Intracytoplasmic dark blue granular material is indicative of apocrine cells

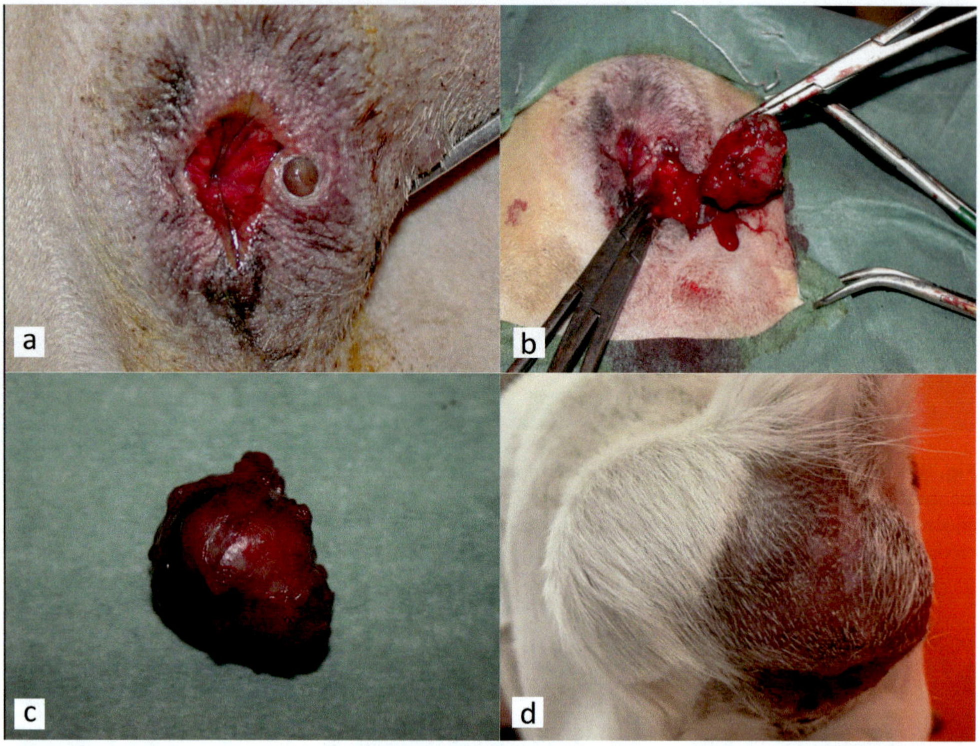

Fig. 8.466 Anal sac carcinoma: (**a**) a small nodule protruding from the right anal sac opening; (**b**) neoformation of the sac during surgical removal; (**c**) same nodule surgically removed; (**d**) mass occupying the entire paranal area originating from the anal sac. (Courtesy of DVM M. Beccati-Italy)

Apocrine gland adenocarcinoma of the *anal sac* (AGASAC) is an uncommon tumor in dogs and extremely rare in cats. Neoplasia origins from the *apocrine modified glands* which line the anal sac wall and that are responsible for producing a fluid used for marking territory. It is a very aggressive tumor with a high potential for metastasis to regional lymph nodes or distant organs. Clinically, AGASAC may be so small that it cannot be detected; in fact, digital exploration of the anal sacs must be performed to appreciate it [137, 146–148]. More frequently, the diagnosis is made because a large mass is detected lateral to the anus; since the tumor grows internally, compressing the rectum, it frequently causes difficulty in defecation. The masses can sometimes, due to their size and position, be grossly confused with perianal hernias. Severe paraneoplastic hypercalcemia is often associated with anal sac carcinoma, and even if it is not pathognomonic of this tumor, because it can be associated with other carcinomas or some lymphomas, its detection should always suggest a careful exploration of the anal sacs (Fig. 8.466).

A histopathological and immunohistochemistry study on 69 samples of anal sac carcinomas demonstrated chromogranin A immunoreactivity in 28 cases (40.6%), independently from the histotype. Furthermore NSE (neuron-specific enolase) immunoreactivity was observed in 32 cases (46.4%), while in 15.9% of cases, synaptophysin expression was present. Many neoplasms were positive to more markers. These data highlight how some tumors show a simultaneous positivity to more than a neuroendocrine marker, and this demonstrated, along with cytological and histopathological findings, that many AGASAC have a neuroendocrine differentiation or may be derived from pre-existing neuroendocrine cells [147] (Fig. 8.467). Despite its very aggressive biological behavior, the cytology of AGASAC usually does not show obvious atypia, and this is a typical cytological feature of this neoplasm. Different histological patterns are reported, classified as *rosette*, *tubular*, *solid*, *papillary*, and *mixed*, although frequently a combination of more patterns is present in a single tumor.

The cytological features observed on smears can be different depending on the pattern, but consistently, the specimens are highly cellular and composed of large groups of variably cohesive epithelial cells associated with a so-called neuroendocrine appearance of exfoliated cells (Figs. 8.468 and 8.469). This definition is coined because the cells are released as naked nuclei, arranged singly or in short rows, and are immersed in a pale blue background composed of the cytoplasmic material released following cell rupture. This arrangement is similar to that shown by neuroendocrine neoplasms and confirms the neuroendocrine differentiation of many AGASAC [147]. When the cytoplasm is recognizable, nuclei are round or oval, centrally or eccentrically located, with fine chromatin and pale blue cytoplasm with poorly defined margins. Some clusters resembling tubular structures may be observed (Fig. 8.470).

In the *Rosette pattern*, histology shows a spoke-wheel arrangement of cells surrounding a central core in which, but not always, a small amount of eosinophilic amorphous

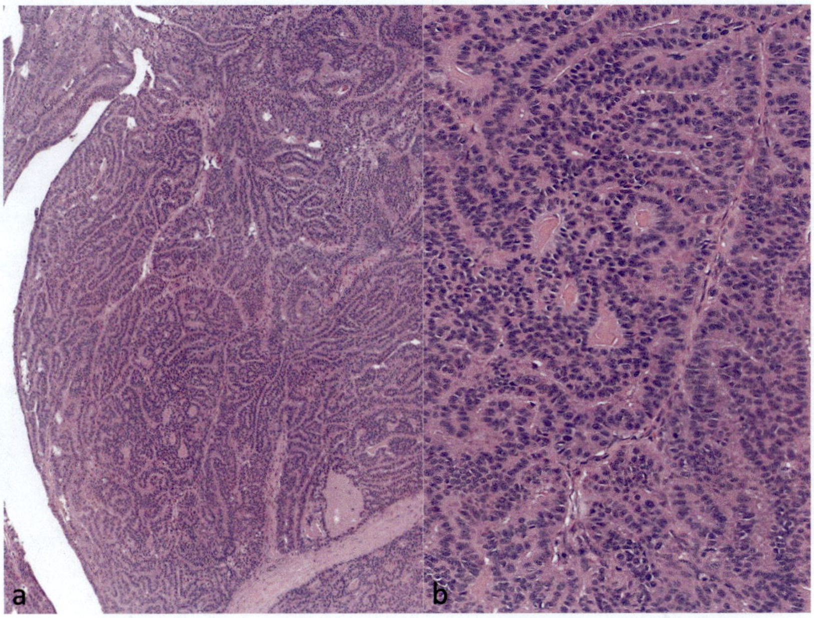

Fig. 8.467 Histology of anal sac adenocarcinoma: (**a**) many lobules of anal sac glands with a regular organization; (**b**) at high magnification some Rosettes are recognizable

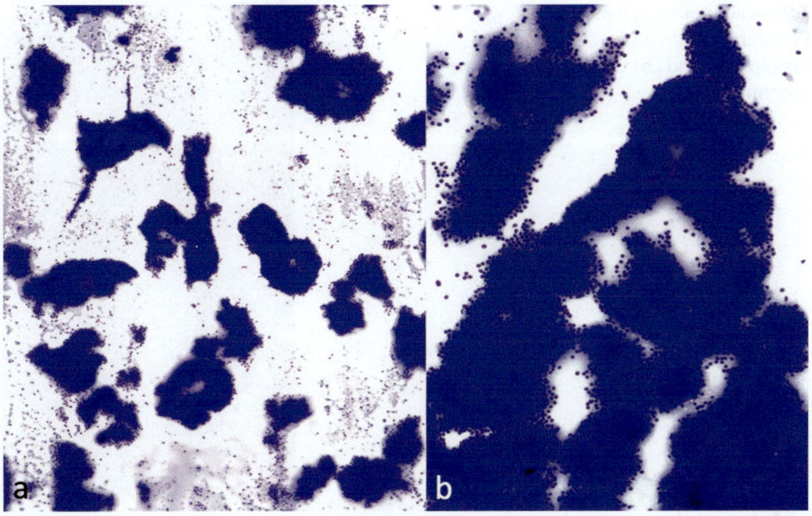

Fig. 8.468 Cytology of anal sac adenocarcinoma: (**a**) many large from roundish to trabecular in shape clusters of apocrine glands; (**b**) at high magnification, the uniform in size neoplastic cells are better recognizable

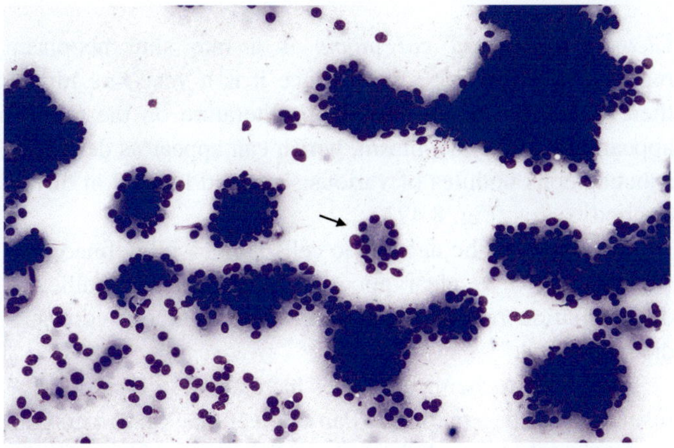

Fig. 8.469 Cytology of anal sac adenocarcinoma: irregular clusters of apocrine glands with few atypia. Note the Rosette arrangement (arrow)

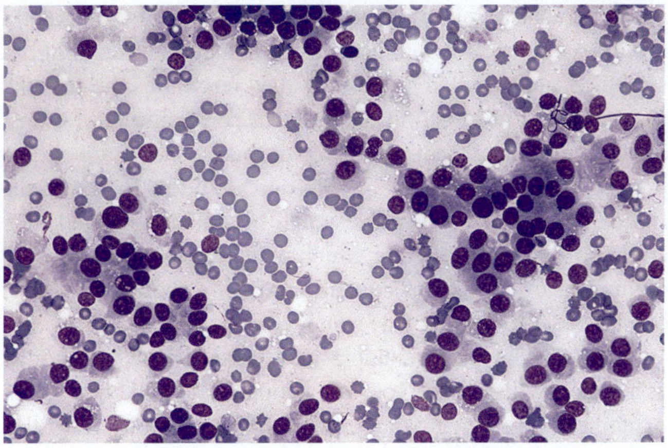

Fig. 8.470 Cytology of anal sac adenocarcinoma: when the cytoplasm is present, it is of medium size and slightly blue-gray. Rosettes and nuclei with palisade arrangements are also evident

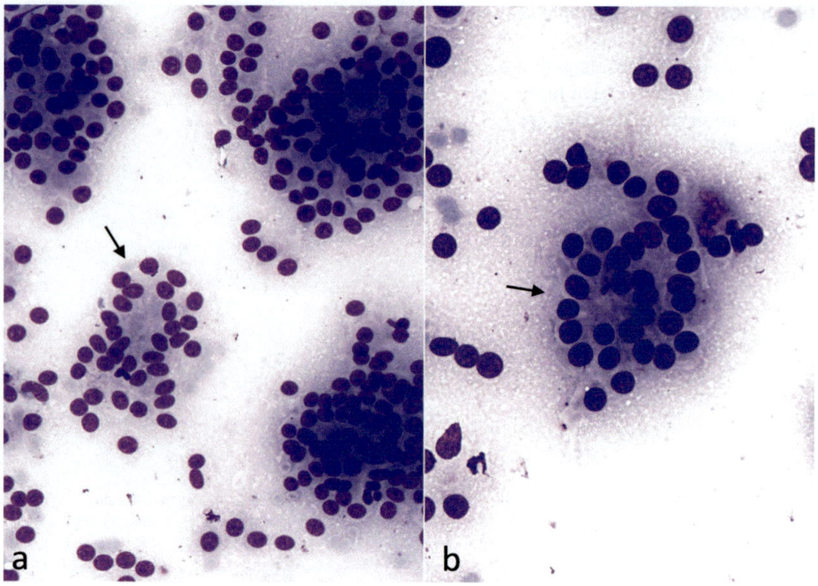

Fig. 8.471 Cytology of anal sac adenocarcinoma: rosettes and bare dispersed nuclei are recognizable

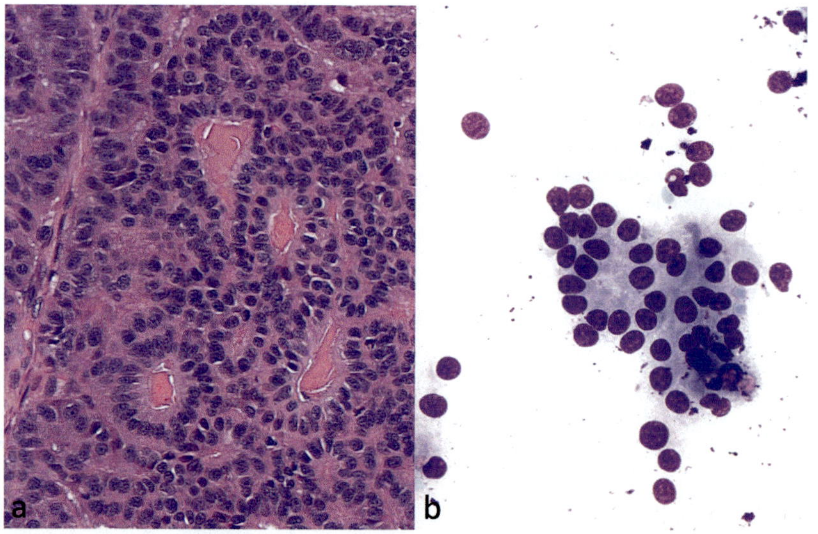

Fig. 8.472 Cytology of anal sac adenocarcinoma: comparison between Rosette on (**a**) histological and (**b**) cytological specimens

material is observed; this arrangement, which resembles an acinar structure, represents only a disposition of single cells around a central area and not a glandular lumen (anatomical structure) and is called a "Rosette" as it resembles the cellular arrangements observed in histological preparations of many central nervous system neoplasms in humans and which are precisely called Rosettes. In cytological slides, Rosettes are frequently detected and represented by nuclei arranged in a circle with a central area in which pale blue material, mostly composed of cytoplasm content spread on the background, can be observed (Figs. 8.471 and 8.472).

Finally, even the paraneoplastic hypercalcemia commonly associated with this type of neoplasia could be better explained by neuroendocrine differentiation of the AGASAC.

8.3.8 Clear Cell Adnexal Carcinoma

Clear cell adnexal carcinoma is a rare skin neoplasm reported in dogs [150, 151]. Since it is a very rare tumor, there is no significant data in the literature on the clinical appearance of this neoplasm, which can appear as dermal or subcutaneous nodules of various sizes and located in different body areas (Fig. 8.473).

The origin of the neoplastic cells is not well defined, and the absence of a clear apocrine, sebaceous, or follicular differentiation makes the classification of these neoplasms difficult [155].

Immunohistochemical studies have always given positive results for the markers Pan-cytokeratin and vimentin confirming the epithelial nature of the tumor. The positivity to melanocytic markers observed in some tumors suggested a

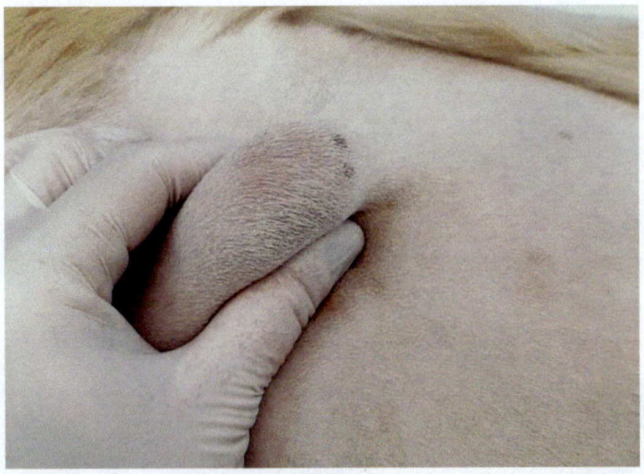

Fig. 8.473 Clear cell adnexal carcinoma: dermal-subcutaneous mass on the chest of a dog (courtesy of DVM F. Gallizia)

possible origin from the cells of the hair matrix, but the positivity was only focal and aspects of keratinization were never observed in all the neoplasms investigated. The negativity to the histochemical staining oil red O makes a sebaceous origin unlikely, and the positivity to the marker CAM-5.2 has oriented the suspicion toward an apocrine nature [155].

Another attempt to define the origin of the clear cell adnexal carcinoma is the constant presence of PAS-positive cytoplasmic granules that was abolished by pretreatment with diastase, indicating the presence of glycogen. Since glycogen is present in the outer root sheath cells as in other follicular neoplasms such as granular trichoblastoma, a follicular origin has been speculated; as for the outer root sheath cells, these are positive for CK16 and KL1 but resulted both negative in one study; however, apocrine cells also contain glycogen in the cytoplasm and therefore PAS positivity a sweat gland origin cannot be excluded. All these inconsistent results therefore leave uncertainty as to which epithelium the tumor cells originate from.

The predominant histological appearance in this adnexal tumor is precisely the presence of epithelial cells with large, clear, and finely vacuolated cytoplasm (clear cells), organized in nests separated by a thin stroma [150] (Figs. 8.474 and 8.475).

The cytological hallmark of these tumors, however, is the characteristic *tigroid background* on which the cells are dispersed, knowledge of which is essential to suspect a clear cell adnexal carcinoma. As discussed in Chap. 5, this amorphous material is from reticulated to lacy or frayed in appearance, is pink or purple in color, and forms streaks that are thought to resemble the stripes on a tiger's coat, hence the name *tigroid* [154] (Fig. 8.476).

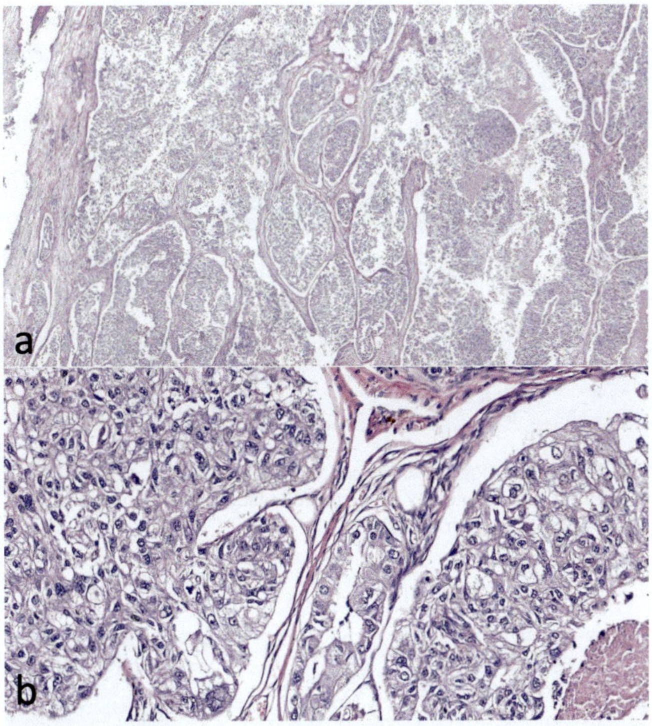

Fig. 8.474 Histology of clear cell adnexal carcinoma: (**a**) multiple islands on neoplastic cells separated by thin stroma; (**b**) at high magnification, large atypical cells with large clear cytoplasm are evident

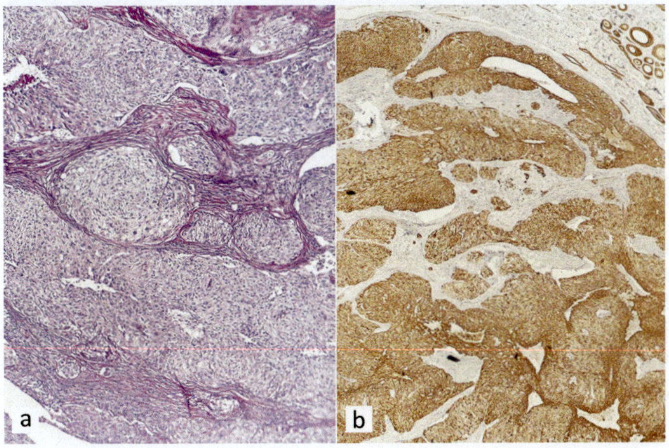

Fig. 8.475 Histology of clear cell adnexal carcinoma: (**a**) multiple islands on neoplastic cells separated by fibrous stroma; (**b**) immunohistochemistry—the strong positivity to Pan-cytokeratin marker confirms the epithelial origin of neoplastic cells

This background material is simply the cytoplasmic content of the neoplastic cells, which is spread on the bottom of the slides during preparation. The *tigroid background* is only recognizable on cytological specimens and exclusively in

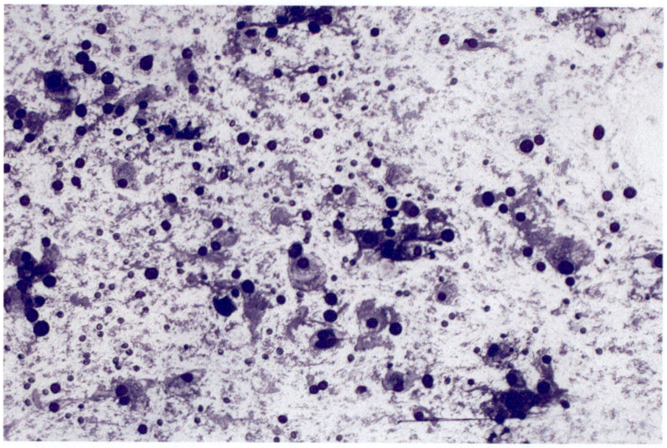

Fig. 8.476 Cytology of clear cell adnexal carcinoma: typical tigroid background represented by a reticulated to lacy or frayed in appearance, pink or purple in color material

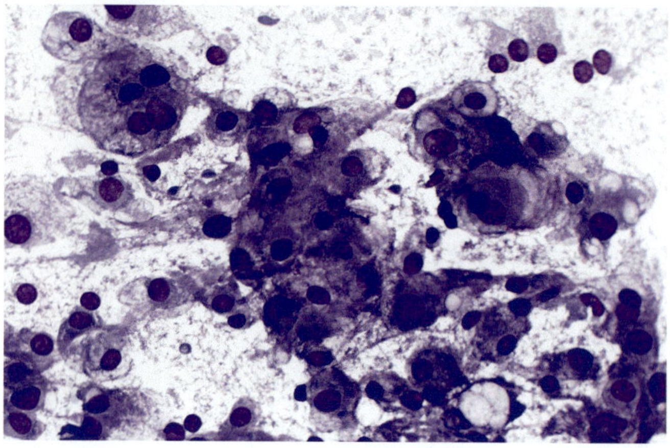

Fig. 8.478 Cytology of clear cell adnexal carcinoma: neoplastic cells show different morphology and variable degrees of malignant cytological features

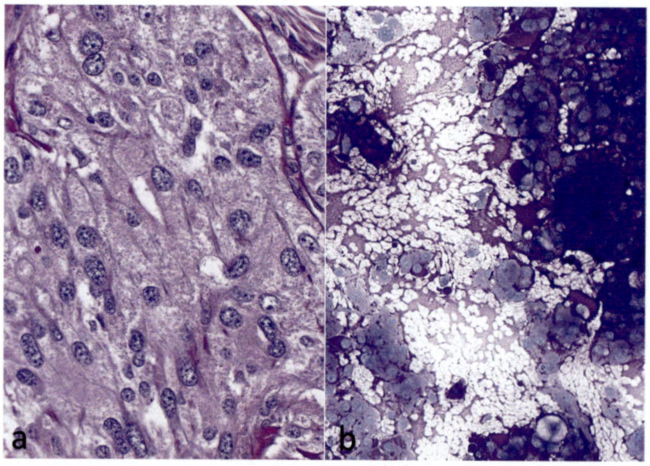

Fig. 8.477 Clear cell adnexal carcinoma (PAS staining): (**a**) histology—large amount of PAS-positive punctate material is highlighted into the cytoplasm of neoplastic clear cells; (**b**) on cytological slides, PAS-positive material is evident both into the cytoplasm and spread on the bottom of the slide, supporting the high amount of glycogen contained in the cytoplasm

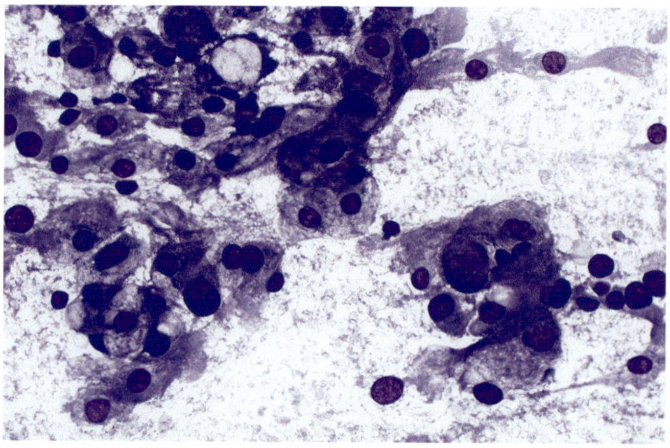

Fig. 8.479 Cytology of clear cell adnexal carcinoma: the filamentous fragments of broken cytoplasmic material give rise to the so-called tigroid background

those stained with Romanowsky-type stains and therefore cannot be seen on histopathological preparations; since the cytoplasm of neoplastic cells is composed of large quantities of glycogen, it is possible to highlight it using PAS stain on histological preparations. Since the tigroid background is also observed in the course of other non-cutaneous neoplasms, in which the cytoplasm is rich in glycogen, such as in seminomas, it is believed that it is precisely glycogen that determines the microscopic characteristics of this background: the PAS staining can be also used on cytological samples in order to mark the background [152] (Fig. 8.477).

Clear cell adnexal carcinoma cells show marked pleomorphism and are usually characterized by obvious cytological atypia. Cell morphology ranges from round to distinctly spindle shape with cytoplasm that tends to break up, taking on a frayed appearance and whose details are lost in the background, resulting in the presence of numerous naked nuclei on slides. The nuclei of clear adnexal cell cells show many atypical cytological features that are commonly seen on slides, such as anisocytosis, anisokaryosis, bizarre shapes, and large nucleoli. Multiple nuclei are often seen (Figs. 8.478 and 8.479). Because in many cases the epithelial cells show poor cohesion and tend to be released as single cells on the slides, and since a large number of these cells have a clear spindle-shaped silhouette, it is very common for those unfamiliar with this neoplasm to confuse it with a soft tissue sarcoma [155] (Fig. 8.480).

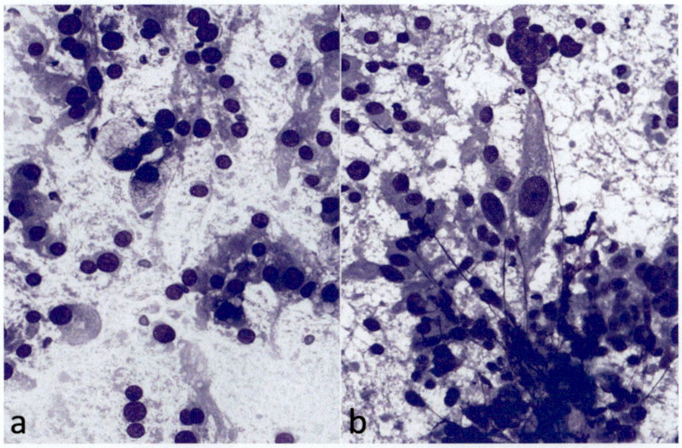

Fig. 8.480 Cytology of clear cell adnexal carcinoma: (**a**) the different morphology of neoplastic cells create many problems regarding their interpretation; (**b**) note the large spindle cell with large nucleus; these pleomorphic aspects can be misdiagnosed as belonging to a soft tissue sarcoma

8.3.9 Fibroadnexal Hamartoma (Fibroadnexal Dysplasia)

The term *fibroadnexal hamartoma* (FH) refers to a benign hamartomatous malformation of the skin appendages that can involve the connective tissue, the follicles, and the sebaceous and apocrine glands [156, 157].

Because different skin structures are involved, names such as *fibroadnexal dysplasia*, *adnexal nevus*, *focal adnexal dysplasia*, and *folliculo-sebaceous hamartoma* are used to interchangeably describe this benign lesion. FH is the most frequent hamartomatous lesion in dogs, and although the hamartomatous lesions indicate a congenital disorder, this type of lesion is acquired, and, even if the pathogenesis remains controversial, it is thought that they might originate in areas where reactive fibroplasia has occurred, probably secondary to a traumatic event. Nevertheless, a primary defect of the pilo-sebaceous unit cannot be excluded in some cases [158, 159].

FHs are characterized by a single alopecic, raised, nodular, or plaque-like lesion, covered by pinkish or pigmented skin, and also if they are mainly located on the limbs and digits, they can be observed everywhere. In many cases, when the lesion is not completely alopecic and some hairs are scattered present on the surface, lesions assume a characteristic *pincushion* appearance. In some cases, FH can be represented as multiple merging papular-nodular lesions, which form large plaques with irregular surface (Figs. 8.481, 8.482, 8.483 and 8.484).

Cytological Findings

The histology of FH shows the presence of dysmorphism of follicles, often with dilated and filled with keratin lumen, associated with hyperplasia of both apocrine and sebaceous

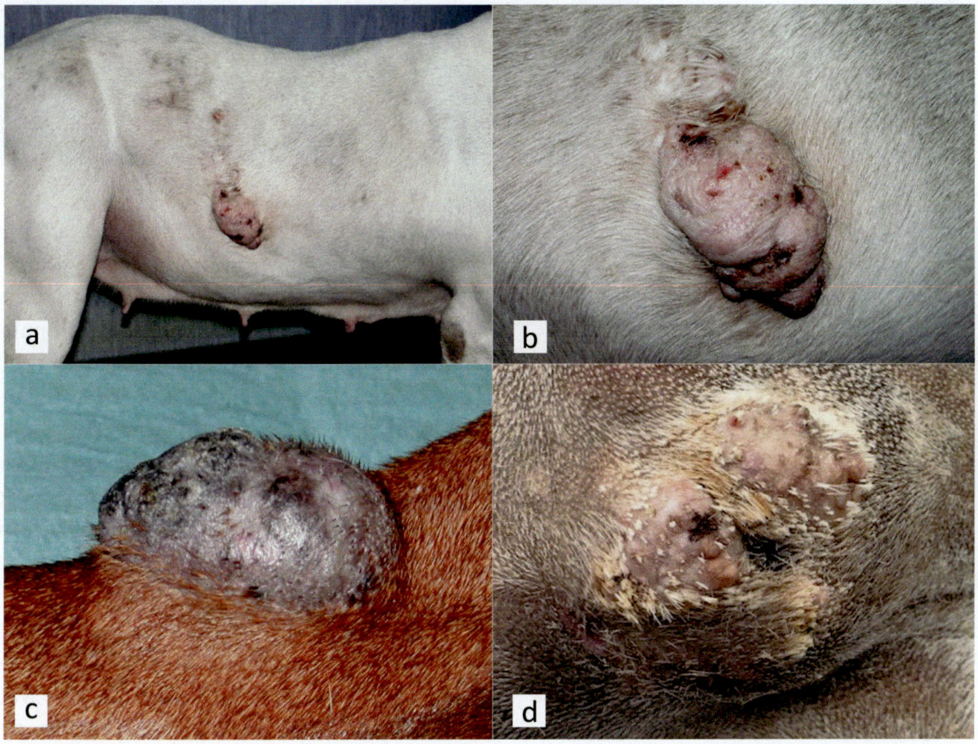

Fig. 8.481 Fibroadnexal hamartoma: (**a**) large alopecic mass on the thorax of a dog; (**b**) close-up of the previous image; (**c**) pigmented and with alopecic surface nodule; (**d**) irregular in shape nodule with a pincushion appearance

Fig. 8.482 Fibroadnexal hamartoma: (**a**) alopecic exophytic nodule; (**b**) pigmented nodule on the lip

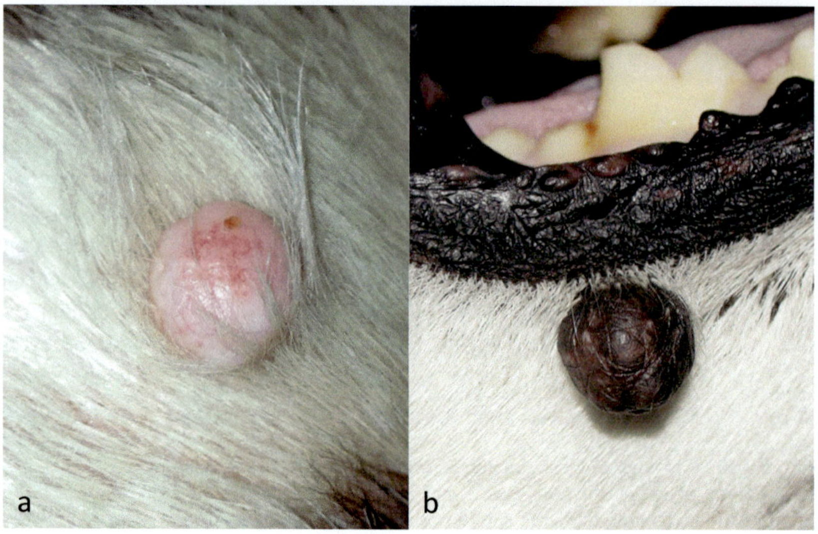

Fig. 8.483 Fibroadnexal hamartoma: interdigital lesions are secondary to chronic pododermatitis

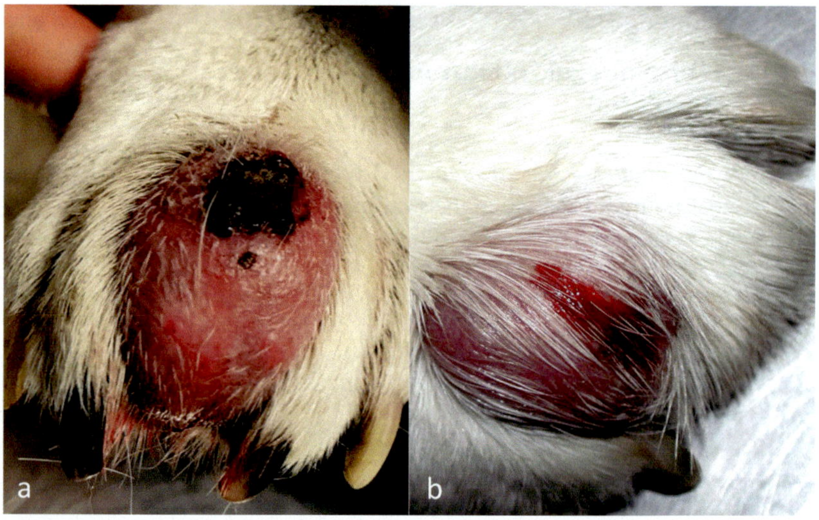

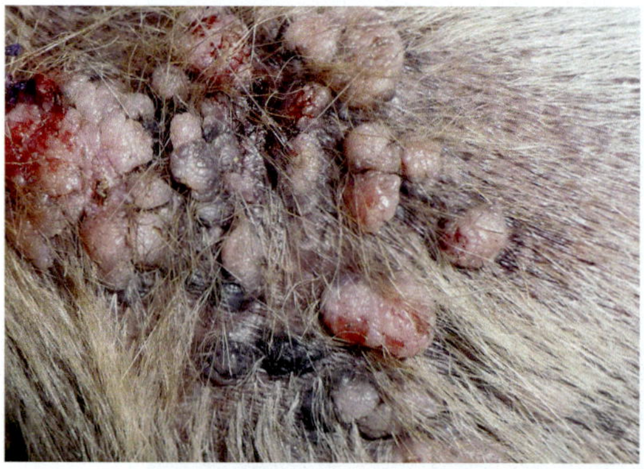

Fig. 8.484 Fibroadnexal hamartoma: large plaque composed of multiple confluent small nodules

glands; all these hyperplastic adnexa are immersed in abundant fibrous collagen stroma (Fig. 8.485). For this reason, cytology of FHs is variable and depends on which adnexa are mainly represented. Usually, the two populations most commonly represented on slides are both sebaceous and apocrine glands, present in variable numbers and without morphological atypia [159].

When only sebaceous or apocrine clusters are collected, the lesions can be confused with a sebaceous hyperplasia/adenoma or an apocrine neoplasia, respectively; when both glands are collected, a FH can be strongly suspected (Figs. 8.486 and 8.487). In some cases, along with the above-described clusters, reactive fibroblasts, collagen bundles, and keratin lamellae can be visualized (Fig. 8.488). Following the breakage of the follicular wall and the spread of corneocytes into the dermis, a severe pyogranulomatous inflammation is commonly observed on slides and can completely change the typical FH cytology.

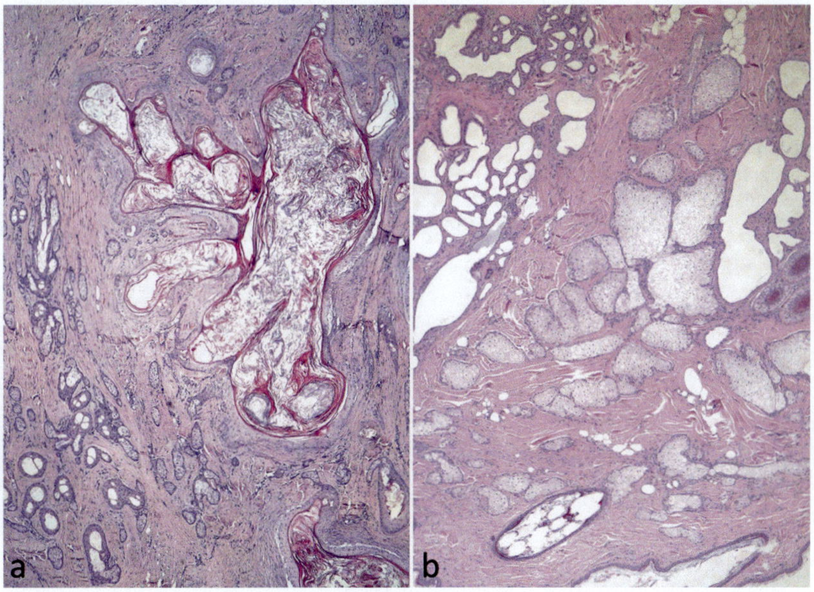

Fig. 8.485 Histology of fibroadnexal hamartoma: (**a**) large and irregular in shape follicle and many sebaceous gland immersed in a fibrous stroma; (**b**) many hyperplastic both sebaceous and apocrine glands

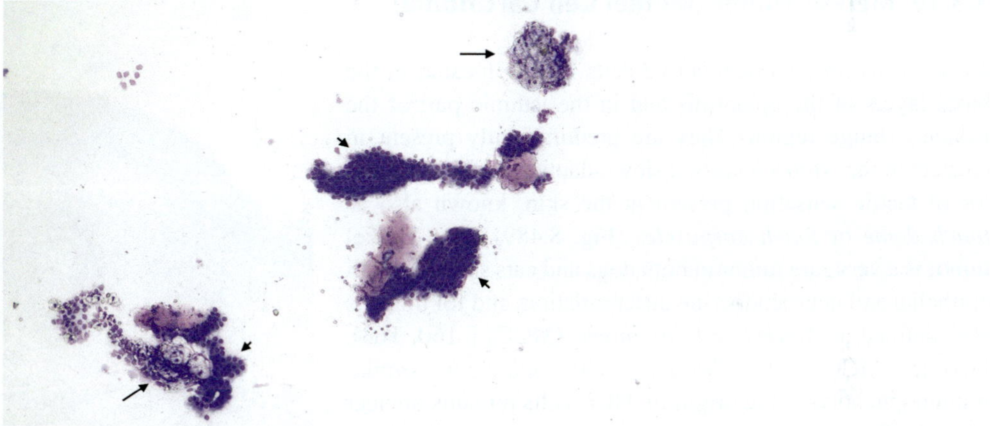

Fig. 8.486 Cytology of fibroadnexal hamartoma: simultaneous presence of cluster of both apocrine (short arrows) and sebaceous glands (long arrows). No atypia is present in both glandular cells

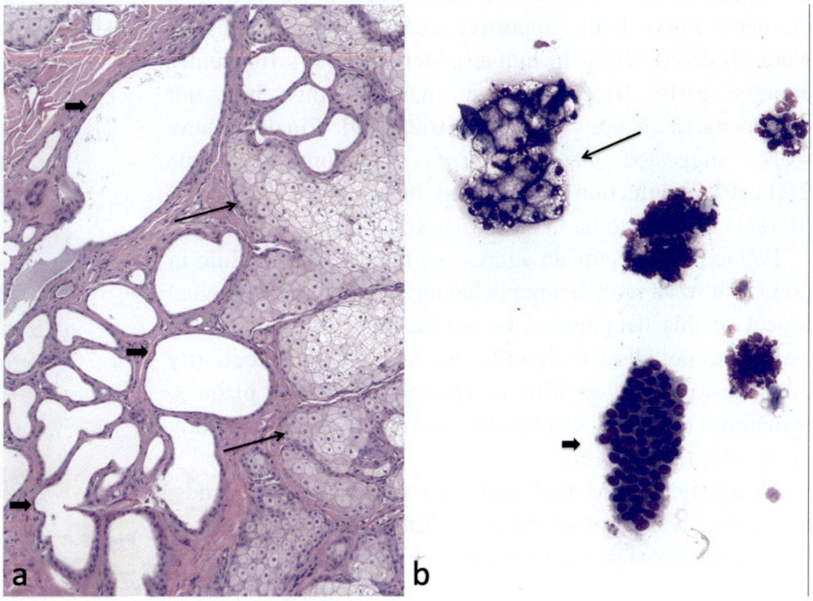

Fig. 8.487 Fibroadnexal hamartoma: comparison between (**a**) histology and (**b**) cytology. The long arrows indicate sebaceous glands, while the short arrows the apocrine glands

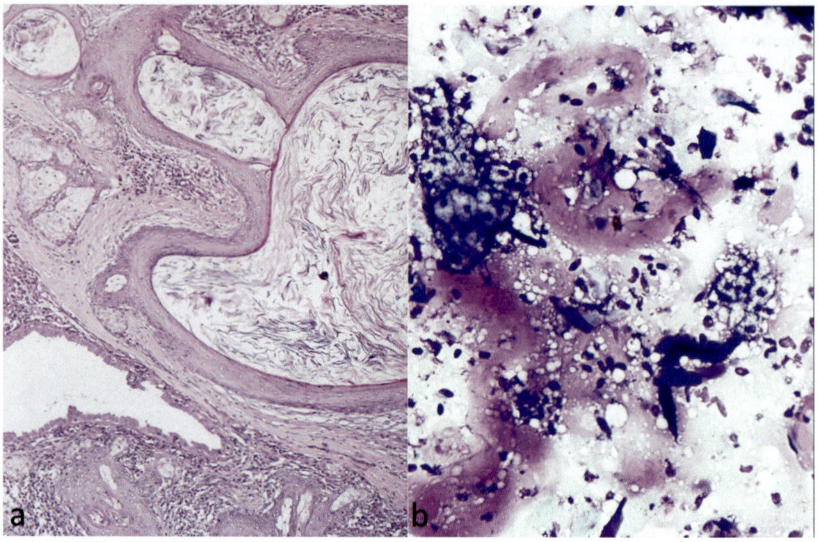

Fig. 8.488 Fibroadnexal hamartoma: (**a**) histology shows an inflamed and fibrous stroma in which hyperplastic adnexa are immersed; (**b**) cytology in which fibroblasts, collagen fibers, and clusters of sebaceous glands are recognizable

8.3.10 Merkel Tumor (Merkel Cell Carcinoma)

Merkel cells are neuroendocrine cells mainly located in the basal layers of the epidermis and in the isthmic part of the follicles (bulge region); they are predominantly present in clusters in the *tylotrich pads*, a slow-adapting mechanoreceptor of tactile sensation present in the skin, known also as *touch dome* or *touch corpuscles* (Fig. 8.489). The Merkel tumor is a very rare tumor in both dogs and cats showing both epithelial and neuroendocrine differentiation, and for this it is also defined as *Merkel cell carcinoma* (MCC) [160, 163]. Despite MCCs and Merkel cells share a similar immunophenotype, the origin of MCC cells remains unclear [162, 169].

Different types of cells have been proposed as the cells which could give origin to MCCs, including epidermal/dermal epithelial stem cells and dermal fibroblasts; nevertheless, in many works both supportive and contradictory results were observed [168]. In humans Merkel tumors frequently express early B-cell lineage markers; this has not demonstrated in dogs and cats [162, 169]. Finally, many works suggested that *Felis catus* papillomavirus type 2 (FcaPV2) infection is involved in the development of Merkel cell carcinoma in cats [161, 164, 165].

Feline MCC seems an aggressive tumor in cats, while in dogs it shows a more benign behavior. Regarding the clinical aspect of this neoplasms, the scarce number of published cases does not allow to describe any macroscopic peculiarity of the lesions, and on what is reported, nodules or plaques, sometimes ulcerated, can be observed in any part of the body [166, 168] (Fig. 8.490).

Histologically, Merkel cells are organized as islands, trabeculae, and rows of polygonal/rounded cells, separated by thin stroma. The nuclei are round and with a variable in

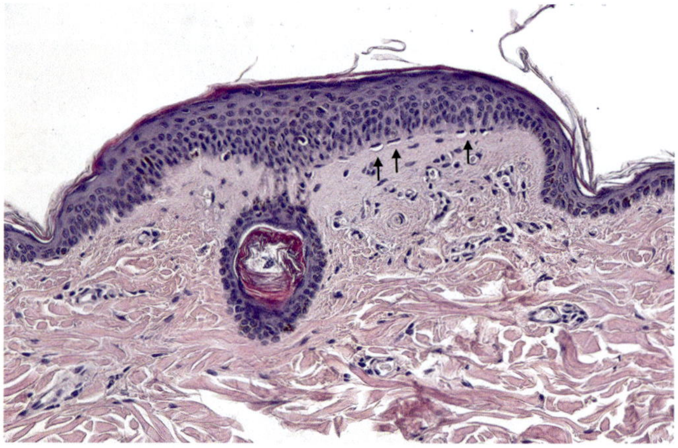

Fig. 8.489 Histology of tylotrich pad: Merkel cells are located at the basis of epidermis, morphologically recognizable as clear cells with the nucleus on the top of the cell and with their long axis parallel to the epidermis (arrows)

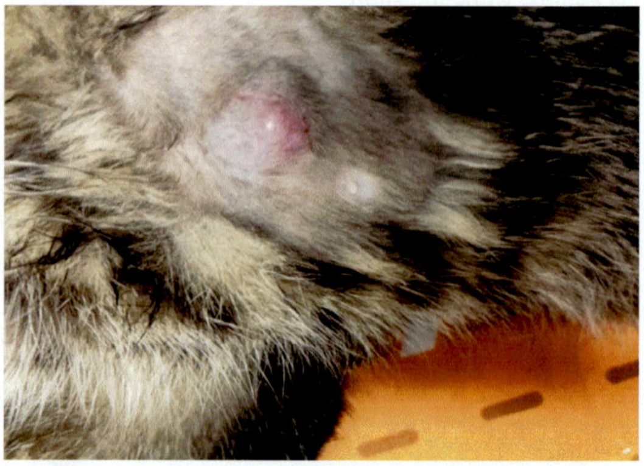

Fig. 8.490 Alopecic and erythematous Merkel tumor on the skin of a cat. (Courtesy of DVM F. Bonadeo-Italy)

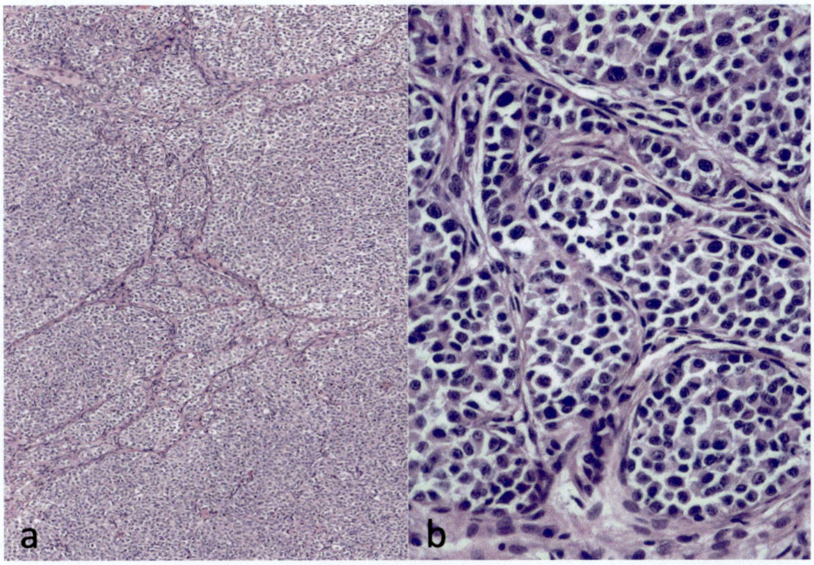

Fig. 8.491 Histology of Merkel cell carcinoma: (**a**) multiple islands of neoplastic cells separated by thin stroma; (**b**) at high magnification, cells look roundish with slight anisokaryosis and partial loss of cohesiveness

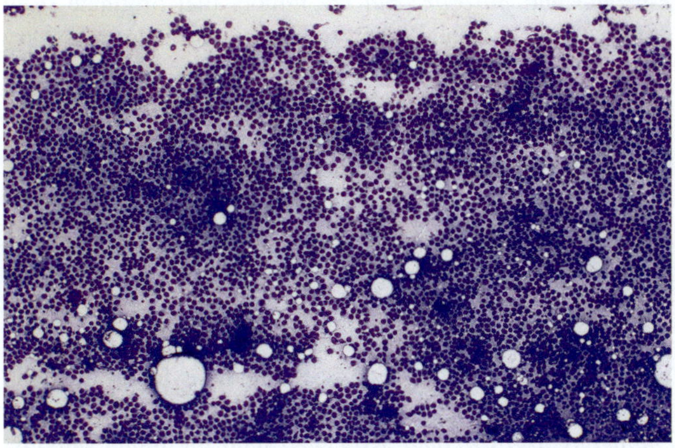

Fig. 8.492 Cytology of Merkel cell carcinoma: high cellular sample of small in size and discrete neoplastic cells

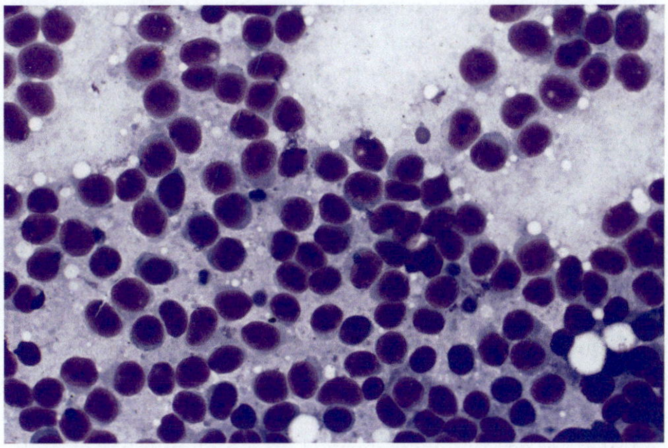

Fig. 8.493 Cytology of Merkel cell carcinoma: at high magnification neoplastic cells show round shape, high N:C ratio, eccentric nuclei, and scarce blue cytoplasm. Note as cells look like of lymphoid origin

size nucleolus [160, 163]. Immunohistochemical studies showed a constant positivity to pan-cytokeratin, marker of epithelial tissues, and mixed variable positivity to neuron-specific enolase (NSE), synaptophysin and less frequently chromogranin, markers of neuroendocrine tissues (Fig. 8.491).

In cytology, as for other neuroendocrine neoplasms, the aggregated cells tend to separate between themselves and to be released on slides as discrete cells. The cytological characteristics are not particular, and Merkel cells are often confused with lymphoid cells due to their rounded shape, eccentric nuclei, with dotted chromatin, which occupies almost the entire cell and with a small, marginal, and intense blue cytoplasm. If small clusters of cells are recognizable along with the discrete ones, an epithelial tumor can be suspected; in case no clusters are detected, Merkel tumors are usually confused with a round cell tumor, especially with lymphoma [163] (Figs. 8.492, 8.493).

8.4 Mesenchymal Cell Tumors (Spindle Cell Tumors)

The term *mesenchymal tumors* is a generic term used to refer to a group of neoplasms that originate from both cutaneous and subcutaneous *soft tissues* and that are included under the umbrella of *soft tissue neoplasms* (STN); when they are morphologically malignant, they are defined as *soft tissue sarcomas* (STS).

In STN group, all the neoplasia originating from the *connective tissue* of the dermis and subcutis (fibroblasts/fibrocytes), the *blood* and *lymphatic vessels* (endothelial cells, smooth muscle, pericytes), the *peripheral nerves* (Schwann cells, perineural cells), the *muscles* of the hairs (erector smooth muscle) or of the fascia, and *adipose tissue* (adipocytes) are included.

Connective tissue, which is composed of resident cells (fibroblasts) along with an amorphous extracellular matrix composed of fibrils and gel-like substances, has the role of supporting and connecting many tissues and organs.

In general, *soft tissue mesenchymal tumors* cytologically share the fusiform morphology of cells, due to the presence of single or multiple cytoplasmic tails; for this reason, to define them, many cytologists use the term *spindle cell tumors*. However, the fusiform shape of the cells cannot be considered a dogma, since in cytology some exceptions are recognized; in fact, tumors such as lipomas and liposarcomas that belong to STNs have a round shape on cytological preparations. On the contrary, fusiform cells do not always indicate a mesenchymal neoplasm of soft tissues; indeed, as mentioned, some melanomas, trichoblastomas, squamous cell carcinomas, and histiocytic sarcomas can show a clear fusiform silhouette, even if they are not included in the group of STNs.

The cytoplasmic borders of mesenchymal cells are usually poorly defined with regard to those of the epithelial and round cells, and nuclei can range from round to oval or elongated (elliptical) in shape.

There is a tendency to say that mesenchymal tumors release few cells on slides, but this is a common misconception, as the number of yielded cells depends on the type of neoplasm. Among mesenchymal spindle cell tumors, there are benign poorly cellular proliferations of collagen that obviously do not produce highly cellular specimens, and consequently the slides show only scattered spindle cells or collagen fibers. In highly cellular tumors such as perivascular tumors, the most common cutaneous sarcoma in dogs, samples are usually characterized by a high number of cells.

Therefore, the cellularity of samples taken from these tumors is highly variable and not strictly linked to their benign or malignant behavior.

The presence of the *extracellular matrix*, an amorphous eosinophilic intercellular protein substance that keeps the cells together, is detected in some spindle cell tumors; in other STNs are pre-existing stromal structures as the vessels to keep together the neoplastic cells (perivascular architecture).

Regarding the cytology of SCNs, if compared with other groups of neoplasms, the morphological appearance of the neoplastic cells allows cytologists to define their mesenchymal tissue origin and to describe their cytological features of malignancy, but, in most cases, it does not allow to define from which mesenchymal soft tissue the cells originate. With experience and thanks to the better knowledge of the pathology of these skin tumors, some peculiar cytological features can orient or, in rare cases, permit to obtain the diagnosis of a specific tumor already with the cytological examination.

Finally, it should be emphasized that the evaluation of the cytological features of spindle cell neoplasms is not free from interpretative doubts; indeed, while some well-differentiated low-grade sarcomas show only a few cytological aspects of atypia, some cases of focal reactive fibroplasia can be characterized by fibroblasts and myofibroblasts which show features of atypia. In these cases, the clinical presentation, the history, and, if present, the coexistence of inflammatory cells are fundamental for orienting the diagnosis, which often requires a histological investigation and/or immunohistochemistry markers.

Mesenchymal neoplasms are very numerous, and only the most important or those for which cytology appears most useful are discussed in this chapter.

8.4.1 Fibroma and Fibrosarcoma

Fibroma and *fibrosarcoma* are tumors that originate from dermal and subcutaneous fibroblasts. Morphology is very variable, usually characterized by nodules of different sizes. In dogs, both tumors develop mainly in adult/old patients, usually presenting with a single nodule, ranging in size, located mainly on the limbs, head, and trunk and may exceptionally present as multiple nodules. In cats the head and pinna are frequently involved and in chronic cases lesions can be ulcerative. Fibrosarcoma is uncommon in dogs compared with other STNs and are common in cats, while fibroma is more common in dogs than in cats [170–174, 189] (Figs. 8.494, 8.495 and 8.496). In young feline leukemia virus (FelV)-positive cats, a very rare form of *multicentric cutaneous fibrosarcoma* associated with the feline sarcoma virus (FeSV) has been reported [176].

In cats, the *fibrosarcoma* that arouses most interest is certainly the *feline injection-site sarcoma* (FISS), once defined as *vaccine-associated sarcoma*, for which various studies have shown that other injected substances, not linked to vaccination, such as long-acting glucocorticoids or antibiotics, or the implant of microchip or the persistence of not-absorbable suture threads may all determine a state of chronic inflammation that predisposes the neoplastic transformation of primitive mesenchymal cells [177–180, 182–184]. Because in cats injections are usually administered in the interscapular area, the development of post-injection sarcomas is frequently observed at this body site and is characterized by a single subcutaneous nodule of irregular shape and ranging in size from a few centimeters to a large, infiltrating, and sometimes ulcerated mass (Fig. 8.497). In the past, it was believed that all interscapular sarcomas originate from fibroblasts and therefore were classified as *fibrosarcoma*. Despite fibrosarcoma is the most common recognized

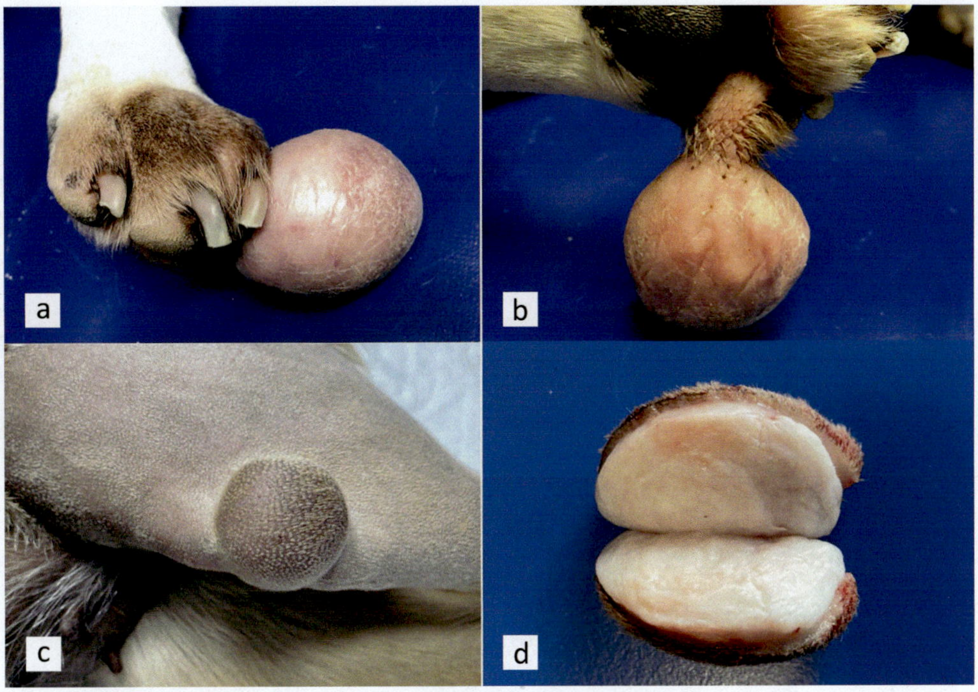

Fig. 8.494 Canine fibroma: (**a**) large alopecic and pedunculated nodule on a pad of a dog (courtesy of DVM M. Perazzetta-Italy); (**b**) note as the fibroma is connected to the skin with a very few tissue; (**c**) well-defined nodule on the knee of a dog; (**d**) fibrous appearance and whitish color of the surface once the nodule has been sectioned

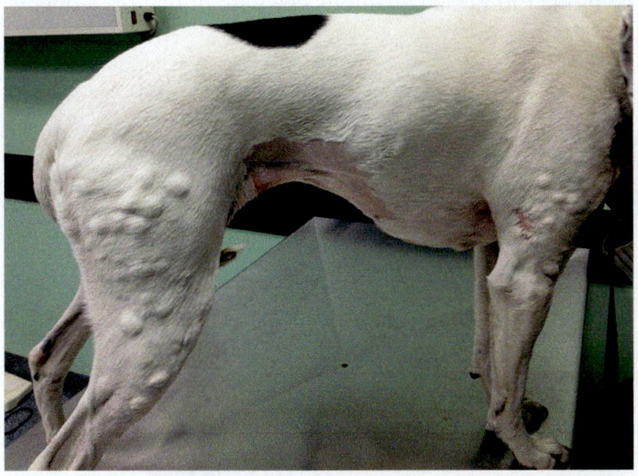

Fig. 8.495 Multiple fibromas are spread on the extremities and on the trunk of a mixed-breed dog

histotype in FISS, immunohistochemical and molecular studies have shown that these sarcomas can have a fibroblastic, myofibroblastic, myoblastic, osteoblastic, chrondroblastic, or other undifferentiated origin. Since in many cases, it is not possible to identify the exact histotypes without immunohistochemistry, and since it is possible that two different histotypes can coexist in the same tumor, the generic term *feline injection-site sarcoma* is currently used.

Cytological Findings

The cytology of conventional fibroma is usually characterized by a low to moderate number of spindle-shaped cells, occasionally released singularly or in small groups kept together by extracellular matrix or collagen fibers. The amount of released cells depends on the cellularity of the neoplasia. The nuclei of spindle cells are oval or elongated and uniform in size and shape, sometimes with a scarcely evident nucleolus and with tailed cytoplasm that usually has indistinct and slightly basophilic borders (Figs. 8.498 and 8.499).

Without the knowledge of the clinical presentation of the lesion from where cells are sampled, cells of fibroma cannot be differentiated from fibroblasts coming from other benign collagen lesions such as skin tag (fibroepithelial polyp), collagen nevus, or dermatofibrosis [190].

A subtype of fibroma, named *keloidal fibroma*, in which the neoplastic spindle-shaped cells are intermingled with a large number of very thick, eosinophilic, hyaline, or glassy collagen fibers, is reported in dogs [175, 181, 186]. The term "keloidal" comes since the aspect of these collagen fibers resembles that observed in human keloids. In almost all cases, the nodules are subcutaneous, from 1 to around 2 cm in size, discoidal to oval in shape and located on the lateral area of trunk, back, or rump. These clinical aspect and localization are very similar to that observed in case of focal reactive fibroplasia (nodular fasciitis), in which a local trauma is hypothesized as trigger. A *keloidal fibrosarcoma* is also well known in dogs. Cytological features are not so different from the above-described benign counterpart, and its definition as fibrosarcoma seems linked to the higher cellularity and more evident morphological atypia compared

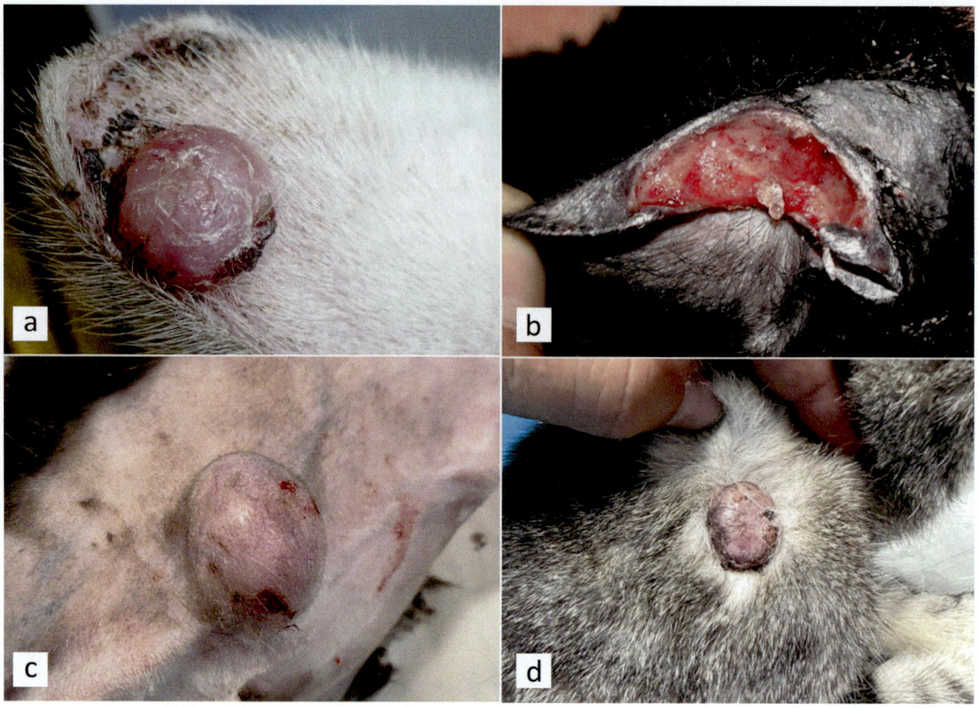

Fig. 8.496 Feline fibrosarcoma: (**a**) nodule on the pinna; (**b**) ulcerated lesion on the pinna mimics a squamous cell carcinoma; (**c**) alopecic nodule on the abdomen of a cat; (**d**) alopecic nodule on the thorax of the cat

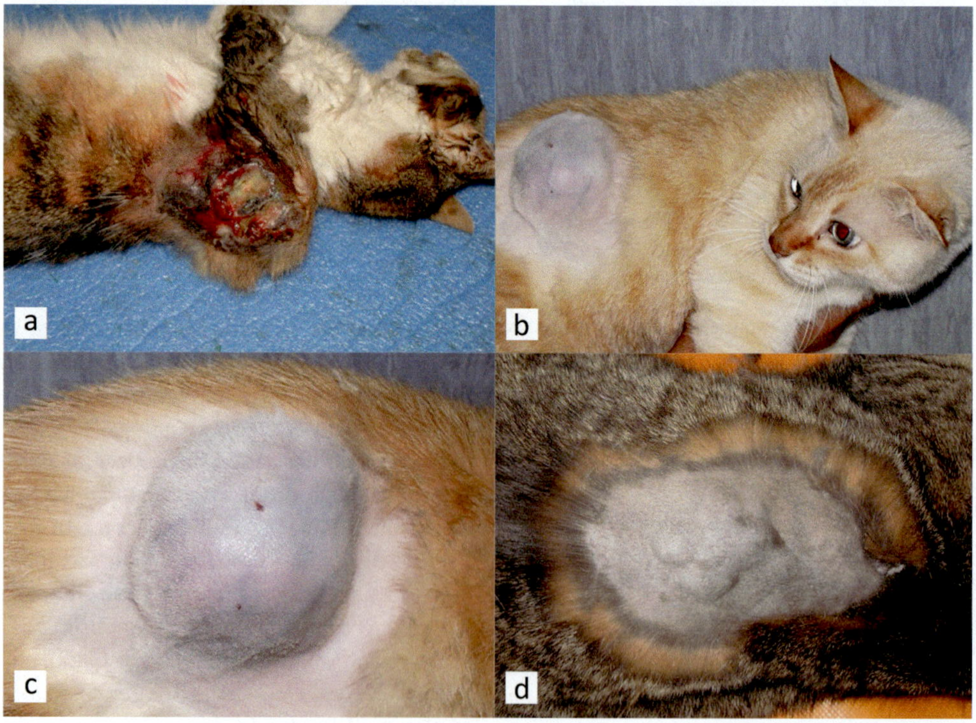

Fig. 8.497 Feline FISS: (**a**) interscapular ulcerated mass between in an old cat; (**b**) large mass on the thorax; (**c**) close-up of the previous lesion; (**d**) multilobed mass on the dorsal thorax of a cat

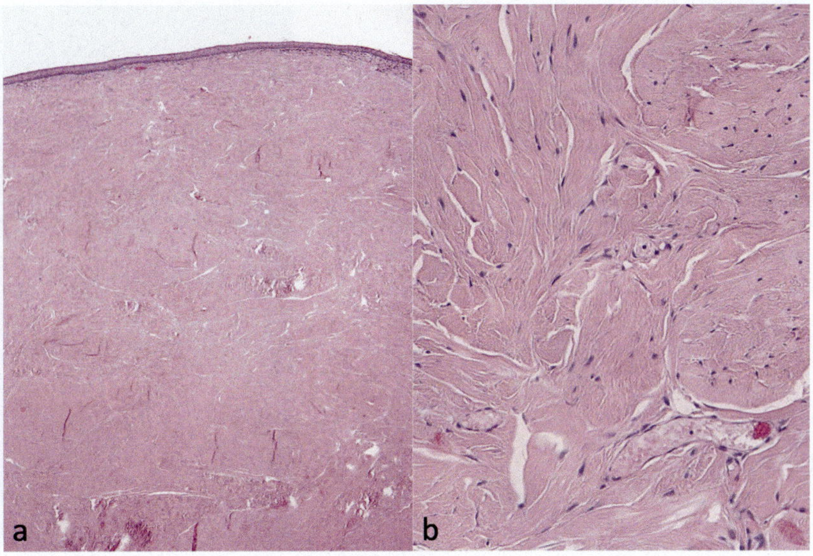

Fig. 8.498 Histology of fibroma: (**a**) subcutaneous proliferation of fibroblasts arranged in regular bundles; (**b**) at high magnification few spindle cells immersed in a collagen stroma: no atypia is recognizable

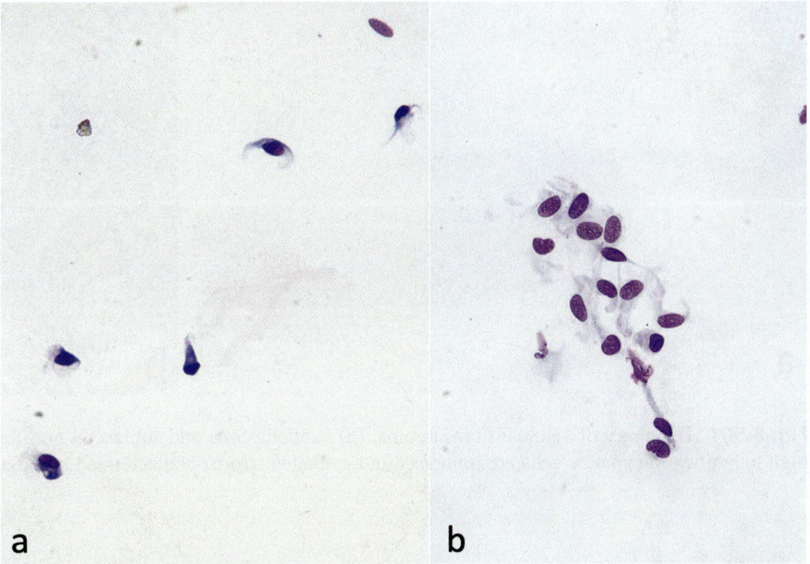

Fig. 8.499 Cytology of fibroma: (**a**) spindle and regular in shape benign fibroblasts are released singularly; (**b**) small group of benign fibroblasts. No atypia is present

to those observed in keloidal fibroma. The author has followed numerous cases of keloidal fibroma/fibrosarcoma and believes that the higher number of spindle cells, on which the main differentiation between a fibroma and a fibrosarcoma is based, is a not-sufficient cytological parameter to describe the existence of a benign and malignant variant. Furthermore, since in both cases specimens are highly cellular and prognosis is good, a differentiation between the two variants could be useless also histologically [173, 181] (Figs. 8.500, 8.501, 8.502, 8.503 and 8.504).

Histology of *conventional fibrosarcoma* is composed of bundles of spindle cells embedded in a collagenous matrix and arranged in a fasciculate pattern that includes storiform, parallel or interlacing bundles, herringbone, and, less frequently, whorled-like arrangements [189] (Fig. 8.505). These cellular architectures are not commonly detectable in cytology since most cells are released singularly, but in more lucky samples, in which a high number of cells are kept together in large groups, some cellular arrangements can be recognizable, as discussed in Chap. 7. We must be aware that some of aforementioned arrangements are not pathognomonic of fibrosarcoma since are also present in other soft tissue sarcomas such as NSTs and perivascular wall tumors (PWTs). In an animal with clinical presentation suggestive of fibrosarcoma, the detection of these cellular arrangements associated to many collagen fibers is strongly indicative of fibrosarcoma.

Fig. 8.500 Keloidal fibrosarcoma: note how since to be highlighted, the nodules must be kept between the fingers, testifying to their deep location. (**a**) Ellipsoidal subcutaneous nodule in a Dobermann (courtesy of Animal Care Clinic-Italy); (**b**) subcutaneous nodule in a Pitt bull (courtesy of DVM S. Bruno-Italy)

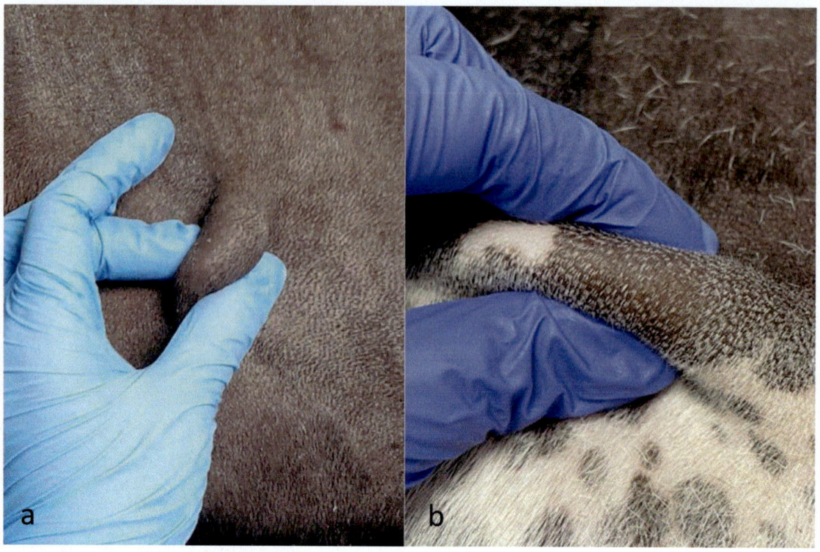

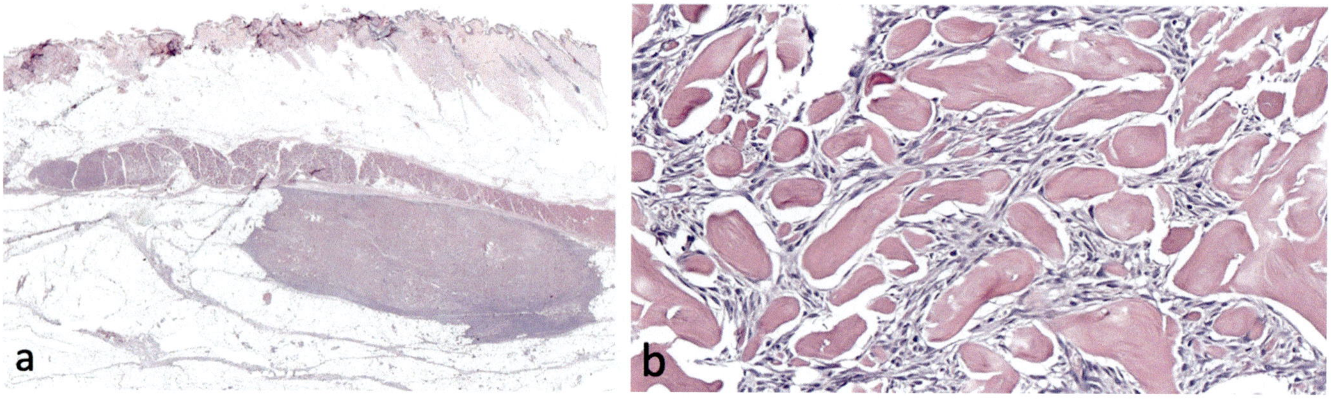

Fig. 8.501 Histology of keloidal fibrosarcoma: (**a**) subcutaneous and subfascial proliferation of neoplastic fibroblasts with an oval shape; (**b**) at high magnification spindle cells are immersed in a collagen stroma characterized by large, eosinophilic, and very thick collagen fibers

Fig. 8.502 Cytology of keloidal fibroma: (**a**) large amount of thick glassy collagen fibers; (**b**) at high magnification few spindle cells are recognizable at the periphery of the collagen fibers

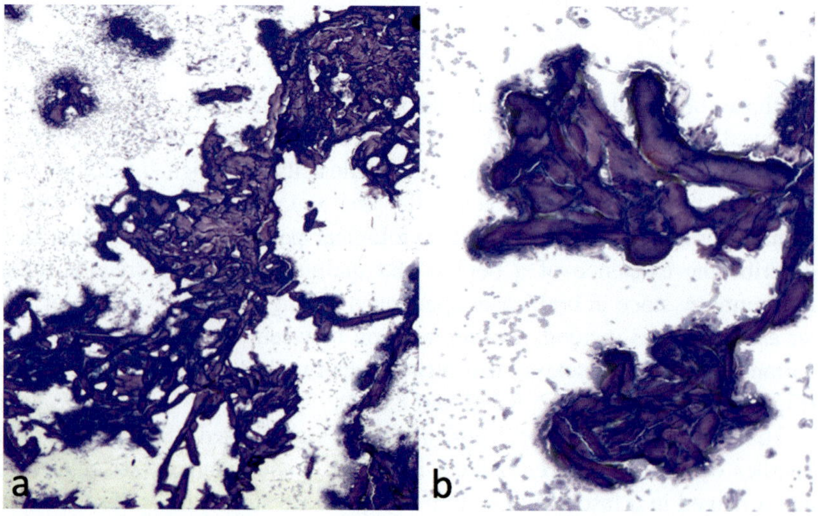

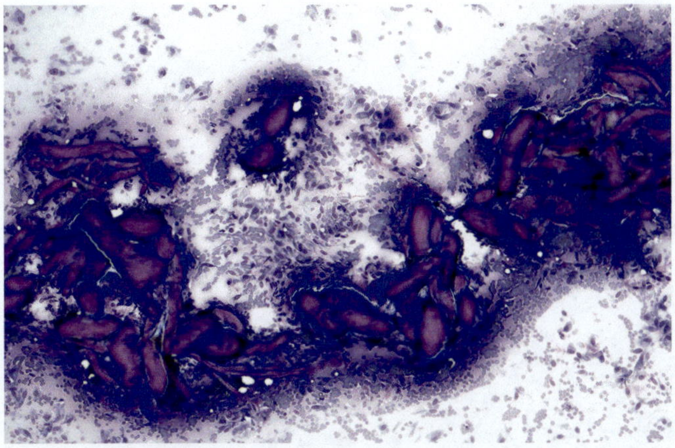

Fig. 8.503 Cytology of keloidal fibrosarcoma: many neoplastic fibroblasts are observed along large and thick collagen fibers

The *conventional fibrosarcoma* is cytologically characterized by the increased cellularity compared with fibroma, even though the number of cells observed, as in many mesenchymal neoplasms, is linked to the amount of stromal collagen produced by neoplastic cells. For this reason, the neoplastic cells can be released singly or arranged in large groups, often kept together by an extracellular matrix or collagen fibers. In large groups, cells can maintain the same arrangement that have in the histological tissue. The high presence of collagen fibers is a strong indicator of the fibroblastic origin of cells (Figs. 8.506, 8.507 and 8.508).

Cytological features are variable and depend on the degree of differentiation; it ranges, in low-grade neoplasms, from uniform spindle cells with few aspects of atypia, which are not differentiable from some fibroplasia or nodular fasciitis, to markedly atypical cells with severe malignant cytological

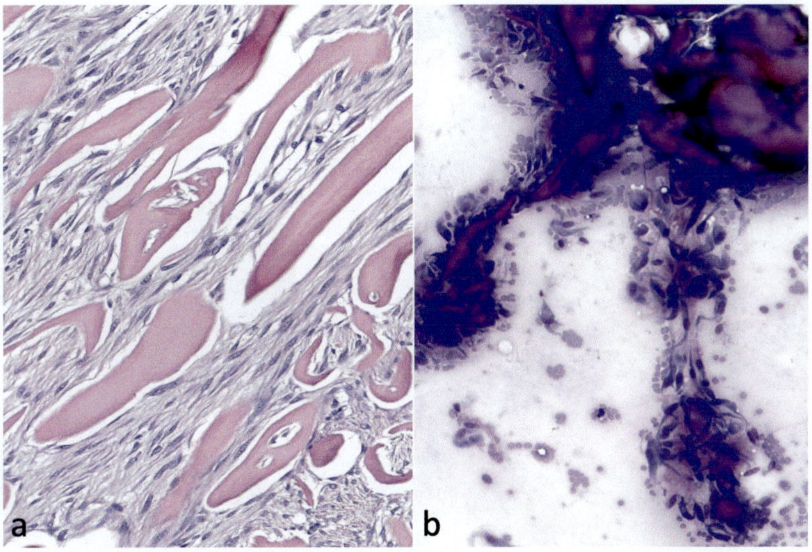

Fig. 8.504 Comparison between histology (**a**) and cytology (**b**) of keloidal fibrosarcoma: many neoplastic fibroblasts are observed along large and thick collagen fibers

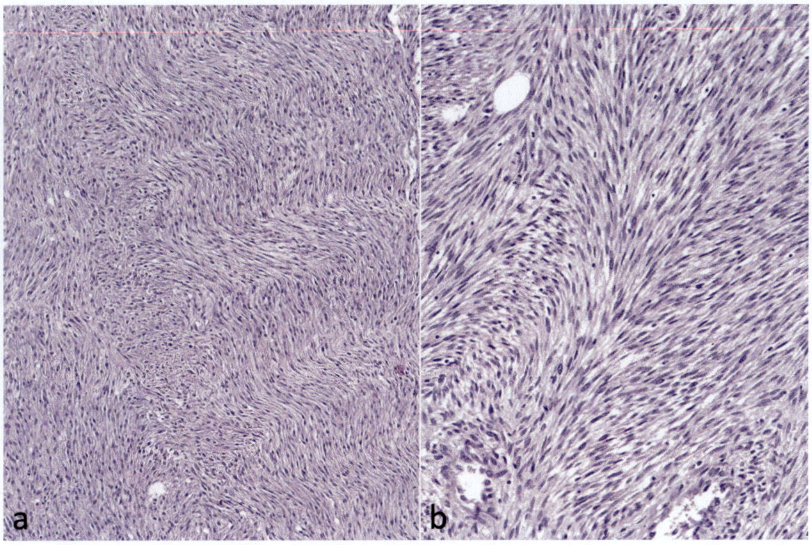

Fig. 8.505 Histology of conventional fibrosarcoma: (**a**) bundles of spindle cells arranged in interlacing bundles, and (**b**) so named "herringbone" arrangement

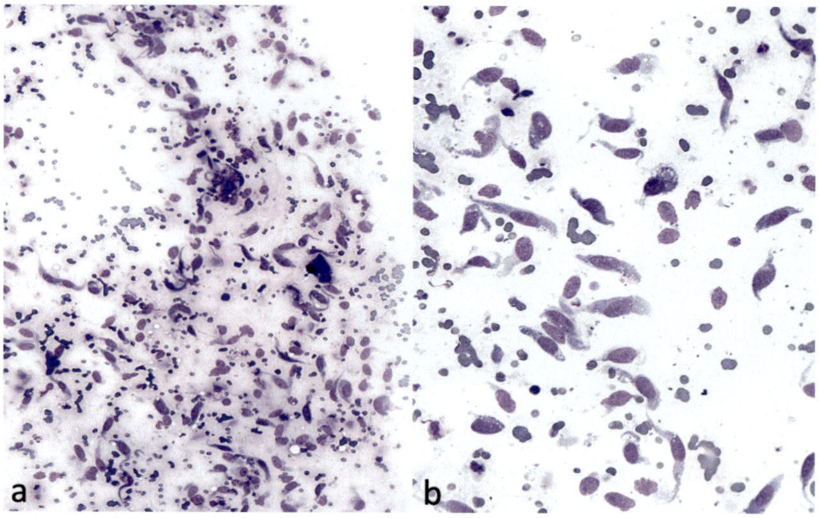

Fig. 8.506 Cytology of fibrosarcoma: (**a**) high number of neoplastic spindle fibroblasts released as single cells; (**b**) at high magnification cells do not show marked atypia. Histology diagnosed a low-grade (grade I) fibrosarcoma

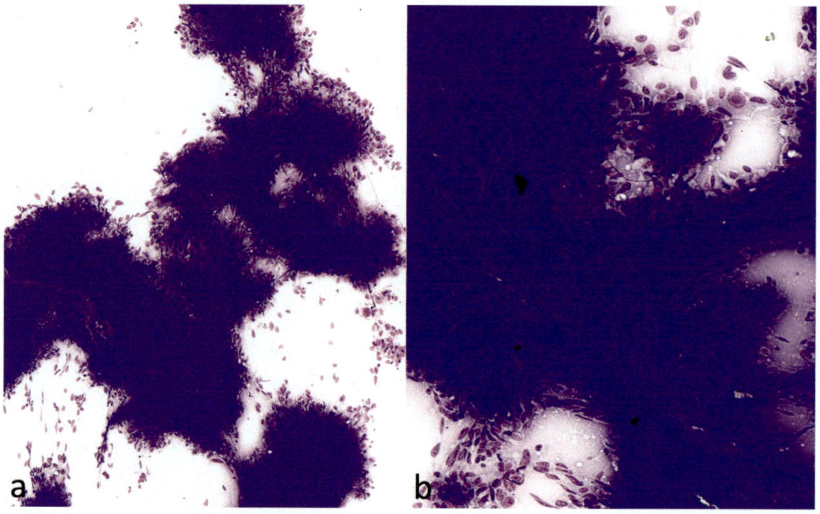

Fig. 8.507 Cytology of fibrosarcoma: (**a**) neoplastic spindle fibroblasts released in large cohesive bundles; (**b**) at higher magnification, a herringbone-like organization of neoplastic cells can be ipotized

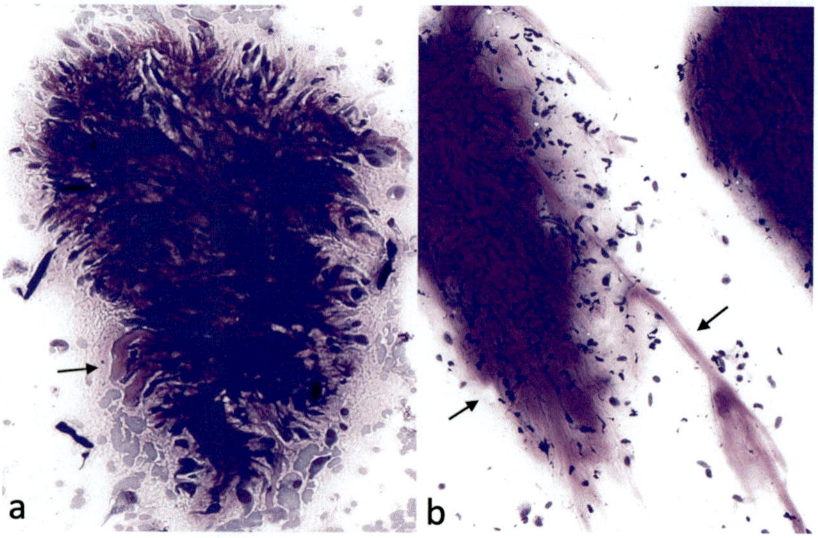

Fig. 8.508 Cytology of fibrosarcoma: (**a**) neoplastic fibroblasts are kept together by both extracellular matrix and collagen fibers (arrow); (**b**) large amount of collagen fibers (arrows)—note as the latter are more numerous than neoplastic cells

features in undifferentiated neoplasia. It should be emphasized that for STSs, the only cytological interpretation does not allow, in the presence of little or moderate atypia, to define the neoplasia as low grade; in fact, according to the literature, the grading system of STSs is based, in addition to the recognizable cellular morphology, also on parameters such as the number of mitotic cells and the percentage of necrosis. The latter parameters must be taken into account during the cytological evaluation of all soft tissue sarcomas. However, all these grading parameters can only be definitively assessed through histopathology [170].

Cells of fibrosarcoma can show spindle to plump shape with large round to oval, single or multiple nuclei, often multiple and prominent nucleoli, and caudate cytoplasm with indistinct borders, deep blue in color, and sometimes with small achromatic vacuolar-like areas. Some cells can show double or multiple tails that can give cells a star-like appearance (Figs. 8.509, 8.510 and 8.511). In poor-differentiated tumors, the severe malignant cytological criteria can completely change the morphology of cells observed in conventional fibrosarcoma; cells can lose the spindle shape and can become roundish to oval. Atypical mitotic cells are present in variable numbers. These tumors usually contain a limited number of multinucleated cells, but the latter can be very numerous in more undifferentiated tumors; feline fibrosarcoma usually shows more giant cells than the canine ones.

A historical problem regarding the cytological interpretation of spindle cells, reported in all textbooks of cytology, is the difficulty that can occur in differentiating, with cytology alone, a STN from a reactive fibroblastic process (granulation tissue). In some cases, in fact, especially when the cytological atypia is not so marked as to be certainly associated with an STS, it is not easy to define whether the spindle cells are neoplastic or reactive; in a study performed on 25 dogs, it was reported that the cytological characteristics that direct the suspicion toward fibroplasia rather than a sarcoma seem to be the presence of low cellularity, the presence of spindle cells arranged individually, and an increased ratio between inflammatory cells and spindle cells [185]. Unfortunately,

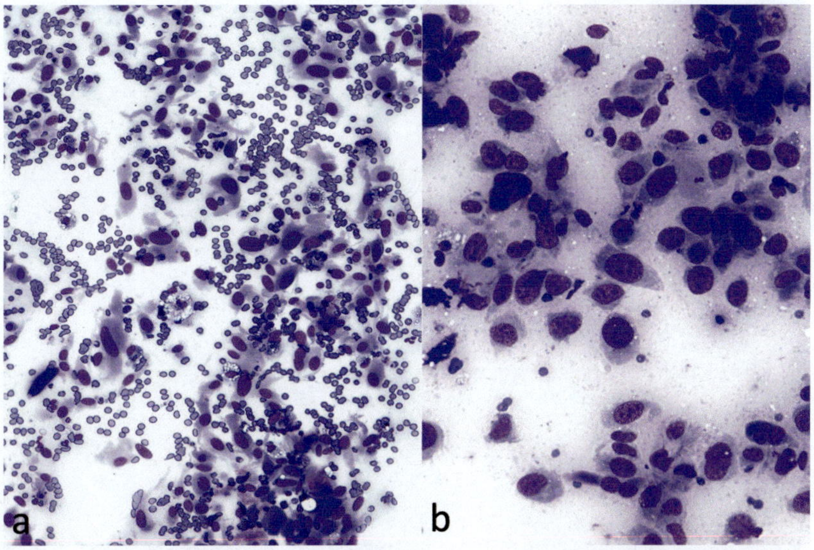

Fig. 8.509 Cytology of canine fibrosarcoma: (**a**) more atypical neoplastic spindle fibroblasts are released singularly; (**b**) singularly released spindle neoplastic cells in which moderate features of atypia are recognizable

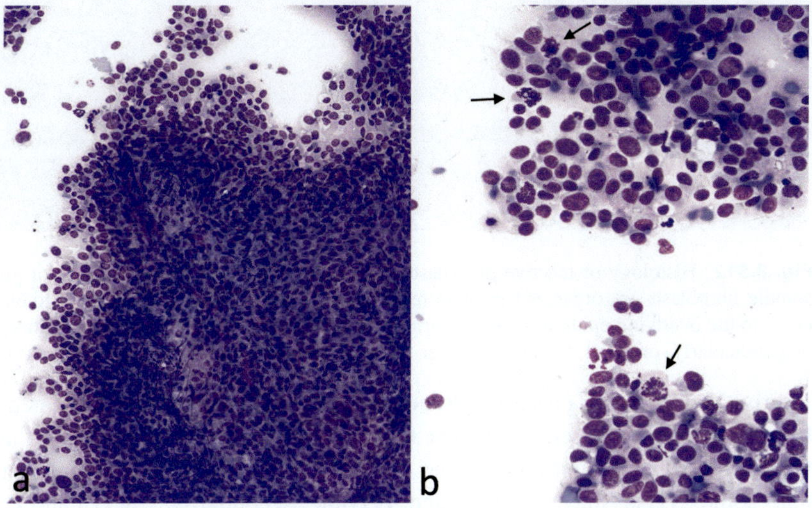

Fig. 8.510 Cytology of feline fibrosarcoma: (**a**) large clusters of both oval and slender fibroblasts: note that the cells do not show the conventional appearances that would allow us to define with certainty cytologically an STS; (**b**) at higher magnification, marked atypia and numerous mitotic cells are recognizable (arrows). Histology and immunohistochemistry were necessary to diagnose a fibrosarcoma

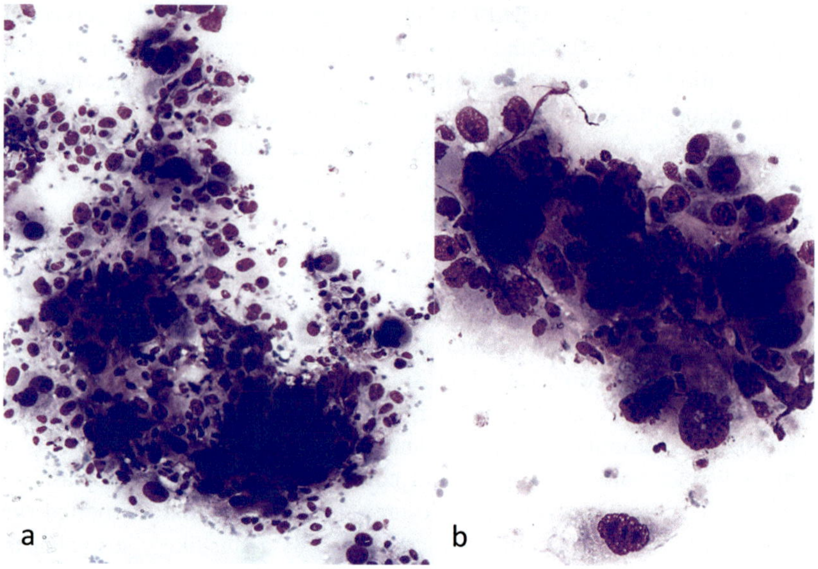

Fig. 8.511 Cytology of feline fibrosarcoma: (**a**) severely malignant neoplastic cells showing anisokariosis and macrokariosis; (**b**) at higher magnification the malignant features are clearly evident. A morphological diagnosis cannot be done, and final diagnosis of undifferentiated fibrosarcoma was achieved with immunohistochemistry

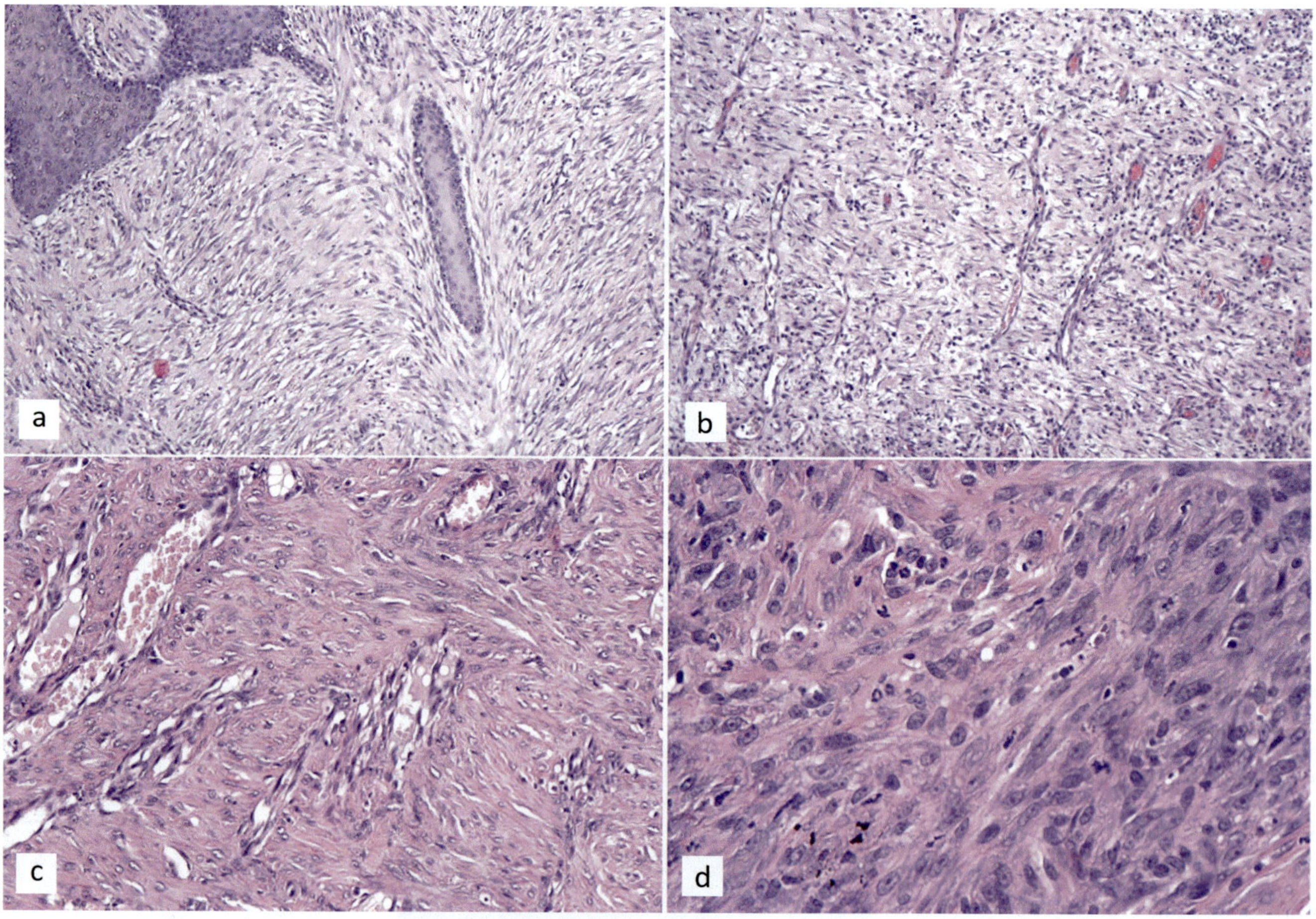

Fig. 8.512 Histology of reactive fibroblasts in granulation tissue: (**a**) spindle fibroblasts are organized in quite regular bundles parallel oriented to the overlying epidermis; (**b**) uniform in size spindle fibroblasts perpendicularly oriented to small caliber vessels, histopathological hallmark of granulation tissue; (**c**) reactive fibroblasts oriented perpendicularly to blood vessels with reactive endothelium; (**d**) at high magnification, morphological atypia are better evaluated and can be misinterpreted with neoplastic fibroblasts

reactive and repairing processes do not always show a number of inflammatory cells sufficient to orient the suspicion toward a reactive lesion and, on the contrary, it is possible that some STNs may ulcerate and become secondarily complicated with inflammatory cells; in these cases, the anamnesis and the clinical findings play a fundamental role in orienting the cytological diagnostic suspicion (Figs. 8.512, 8.513 and 8.514).

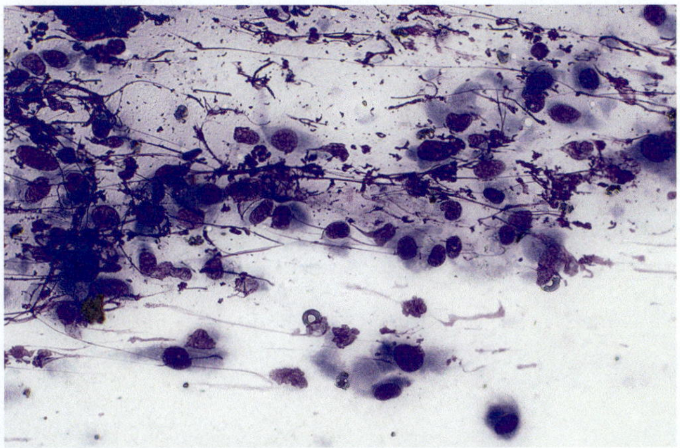

Fig. 8.513 Cytology of reactive fibroblasts: many spindle reactive fibroblasts along with a neutrophilic inflammation. Note as mild anisokaryosis and anisocytosis are present; the latter must be interpreted with caution

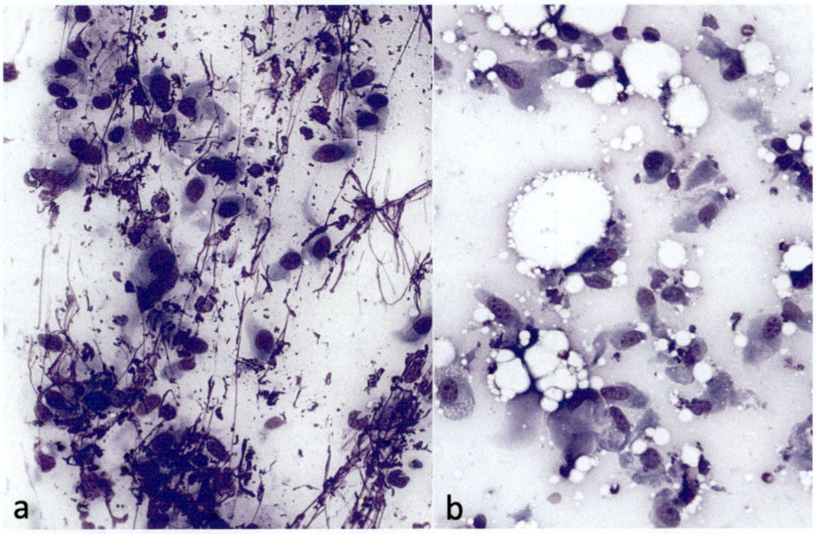

Fig. 8.514 Comparative cytology: (**a**) reactive fibroblasts are not so morphologically different from some (**b**) soft tissue neoplasms

An example of a localized reactive spindle cells lesion often devoid of inflammatory cells is the so-called nodular fasciitis. *Nodular fasciitis*, renamed *focal reactive fibroplasia*, is a subcutaneous benign proliferation of both fibroblasts and myofibroblasts attached or infiltrating the muscular fascia more rarely, only circumscribed in the deep dermis. Histology is characterized by bundles of fibrocyte, fibroblasts, and myofibroblasts admixed to few inflammatory cells, especially lymphocytes, and cytological specimens show spindle cells with features similar to a fibroma although, in some cases, a variable degree of anisokaryosis, with a prominent nucleoli and deep blue cytoplasm which mimic sarcomatous cells, is evident. In many cases a slightly myxoid background is recognizable (Figs. 8.515, 8.516, 8.517 and 8.518). When these morphological atypias are more evident and few or no inflammatory cells are

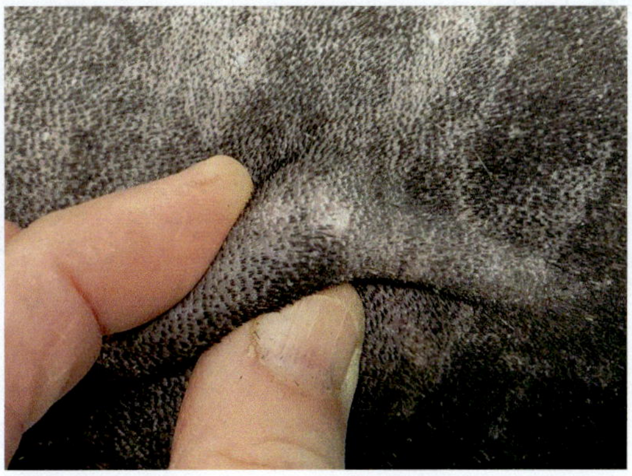

Fig. 8.515 Focal reactive fibroplasia (nodular fasciitis): subcutaneous nodule in an AmStaff. Note as the lesion is similar to those observed in the case of keloidal fibroma/fibrosarcoma (courtesy of DVM L. Luccardini-Italy)

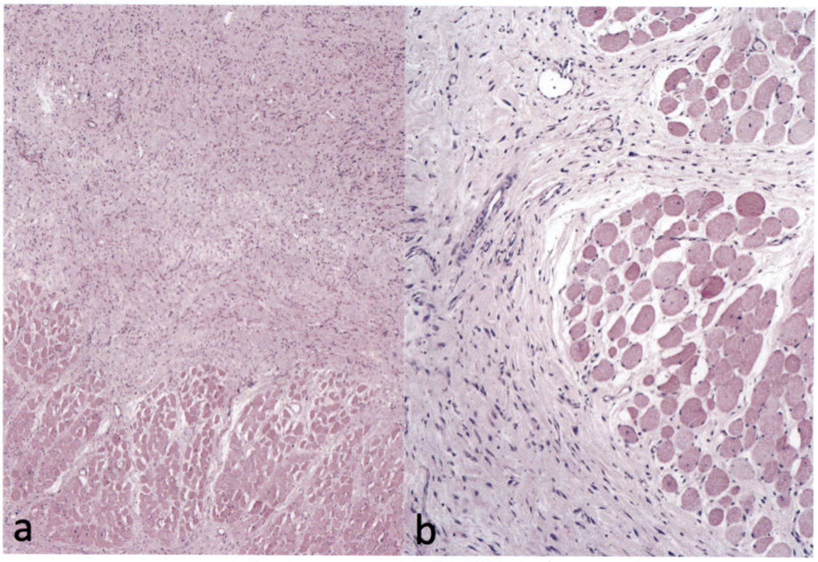

Fig. 8.516 Histology of focal reactive fibroplasia (nodular fasciitis): (**a**) spindle-shaped cells immersed in a collagenous stroma that infiltrates the muscular fascia with spiked margins; (**b**) at high magnification, infiltration of the underlying fascia is recognized

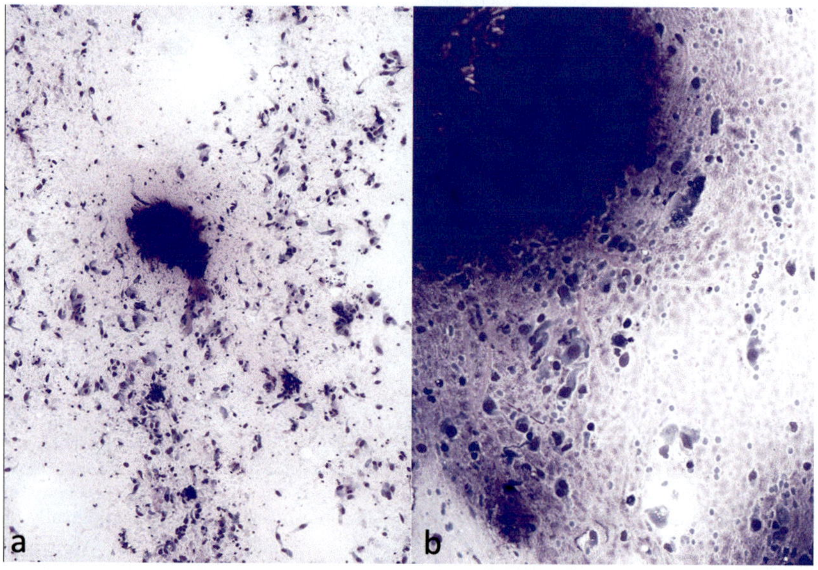

Fig. 8.517 Cytology of focal reactive fibroplasia (nodular fasciitis): (**a**) many spindle cells singularly released. Cells can be a mixed between fibro- and myofibroblasts; (**b**) reactive spindle cells with few atypical cytological features are recognizable on a myxoid background

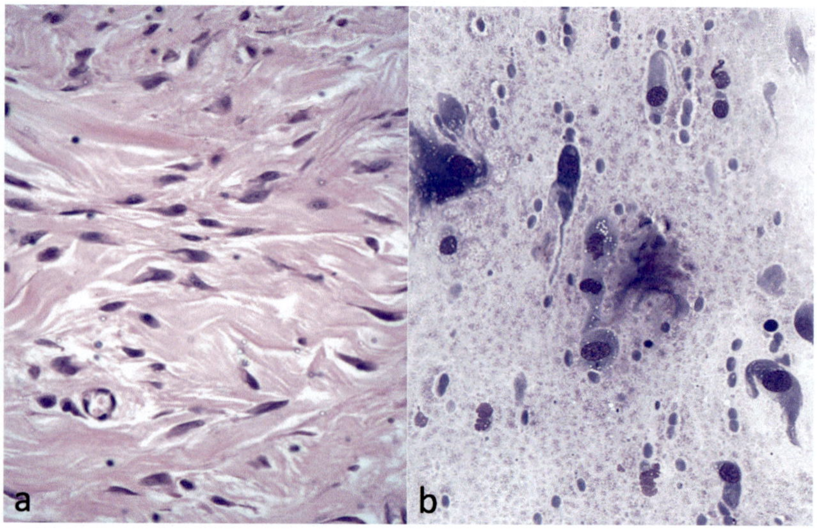

Fig. 8.518 Focal reactive fibroplasia (nodular fasciitis): (**a**) spindle reactive cells show poor aspects of atypia; (**b**) spindle and starry reactive cells can mimic neoplastic fibroblasts/myofibroblasts

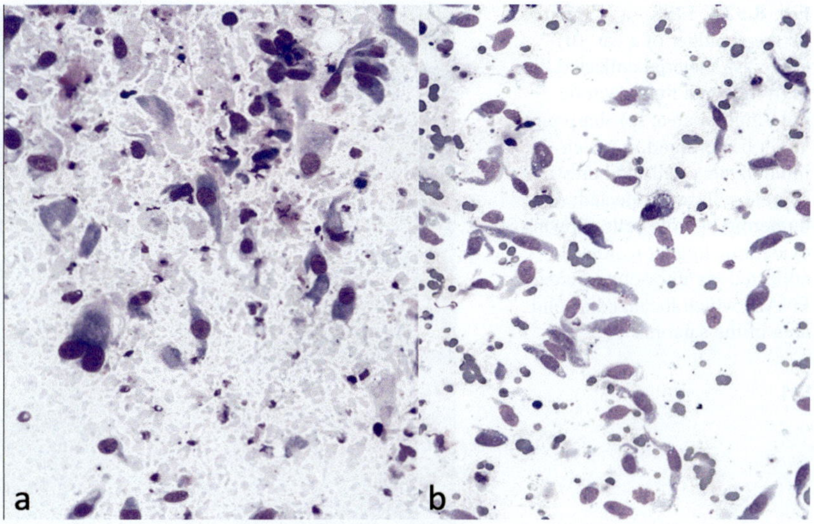

Fig. 8.519 Comparison between reactive and neoplastic spindle cells: (**a**) spindle reactive cells in nodular fasciitis; (**b**) spindle neoplastic cells in a low-grade fibrosarcoma—note as cytological morphology is more suggestive for neoplasia in nodular fasciitis compared with fibrosarcoma, indicating, as in some cases, a cytological diagnosis cannot be done

simultaneously present, a wrong cytological diagnosis of low-grade fibrosarcoma is often formulated [189] (Fig. 8.519).

Furthermore, since focal reactive fibroplasia (nodular fasciitis) and keloidal both fibroma and fibrosarcoma share the same clinical aspect, the body site localization, the shape and size of lesions, and the fact that almost all the lesions are observed in short-haired dogs, it is speculated by the author that all these lesions could represent local reactions secondary to trauma, with some of them characterized by a strong keloidal reaction of collagen, which inhibits to the lesion to spontaneously regress as occurs for focal reactive fibroplasia.

When FISS origins from neoplastic fibroblasts, the cells exhibit the same morphological aspects as described for other fibrosarcomas or other STNs that are not induced by injections. In poor-differentiated cases, the malignant cytological criteria can be very marked as well as the number of multinucleated cells. The cytological diagnosis of FISS is not easy, and in most cases it is not possible, but it is strongly suspected when slides collected from interscapular neoformations show the coexistence of spindle neoplastic cells along with small lymphocytes and necrotic debris, neutrophils, and macrophages, the latter containing intracytoplasmic amorphous material representing the vaccine adjuvant or other injectable products (Fig. 8.520).

When FISS is represented with a large mass, it can contain multiple cavities filled with fluid or viscous necrotic liquid. This histological finding is very important since it is not uncommon to put the needle in these necrotic areas and collect only fluid, along with few spindle cells and many necrotic bare nuclei on a protein background; in these cases, with the guide of ultrasound, it is possible to localize solid areas of the mass in which to perform a FNB and collects a more significant number of neoplastic cells (Fig. 8.521).

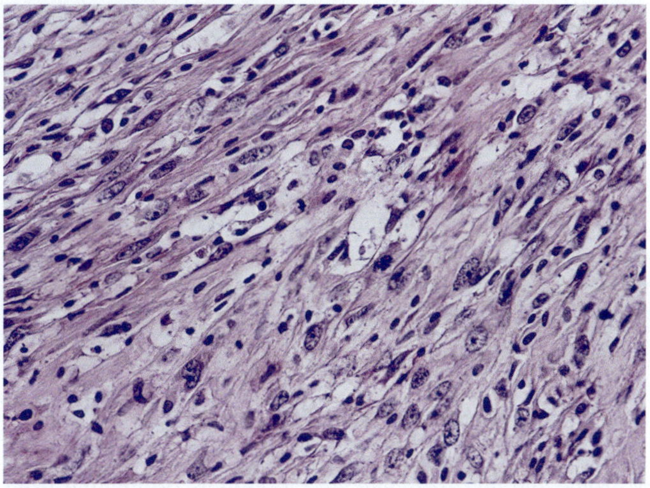

Fig. 8.520 Histology of FISS: bundles of neoplastic spindle cells with malignant atypia represented by large nuclei, anisokaryosis, and mitotic cells. Note as an inflammatory population of cells, mainly lymphocytic, is present

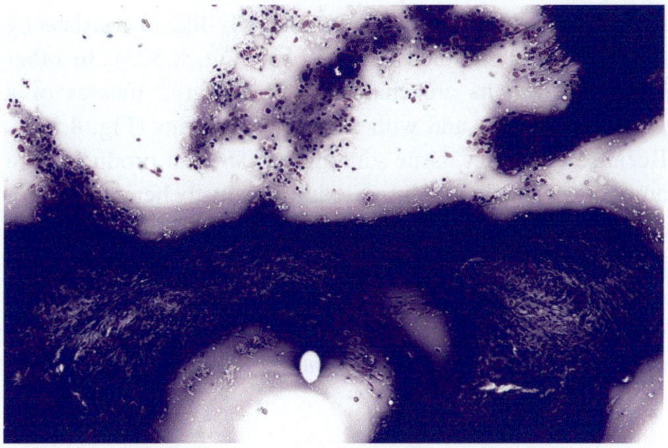

Fig. 8.521 Cytology of FISS: large bundles of neoplastic spindle cells along with many lymphocytes and macrophages

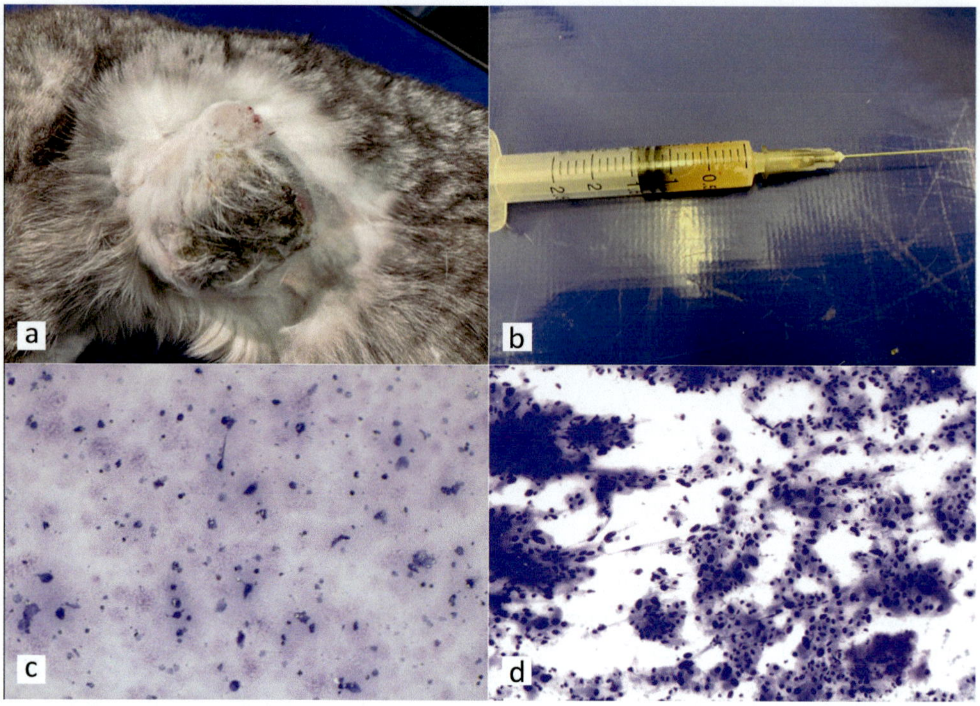

Fig. 8.522 FISS: (**a**) large mass on the shoulder of a cat: (**b**) yellowish material collected by FNAB from a fluid necrotic intratumoral area; (**c**) sharp pink-violet background in which inflammatory cells and only a few spindle cells are collected; (**d**) numerous spindle cells showing evident malignant features collected via ultrasound-guided FNAB, which localized a solid area of the sarcoma

The detection of inflammatory cells does not authorize a diagnosis of FISS as a not post-injection sarcoma can ulcerate and secondarily inflamed; therefore, histopathological confirmation is needed (Fig. 8.522).

8.4.2 Myxoma and Myxosarcoma

Myxoma and *myxosarcoma* are rare tumors originating from dermal or subcutaneous fibroblasts and rarely from other primitive mesenchymal cells that produce large amounts of mucin, mucopolysaccharides rich in hyaluronic acid [191–193]. Macroscopically, they are characterized by nodules of various sizes and mainly localized to the limbs and thorax, which can yield a dense and viscous, jelly-like in consistency material when sampled with FNAB (Fig. 8.523). In other cases, the lesions are characterized by large masses of a pasty consistency and with ill-defined margins (Fig. 8.524). Because other soft tissue sarcomas (STS) can produce large amounts of mucin and because immunohistochemistry is not always performed on these tumors, it is possible that many of the tumors previously diagnosed as myxoma and myxosarcoma were spindle cell lipomas and PWTs with myxoid differentiation, which are incidentally the most common STS in dogs, especially when the tumors were located on the extremities.

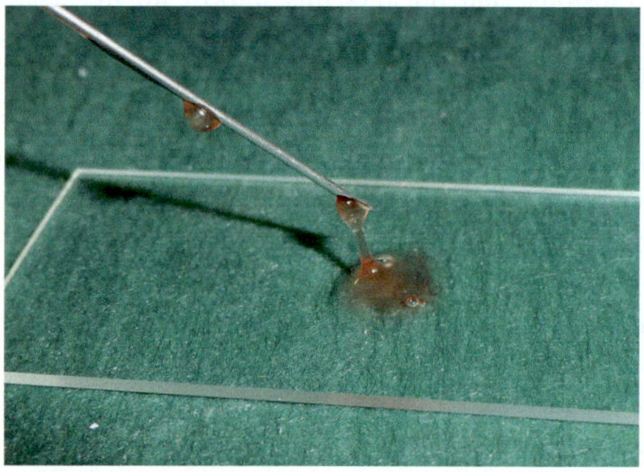

Fig. 8.523 Dense and viscous, jelly-like material collected from a myxosarcoma

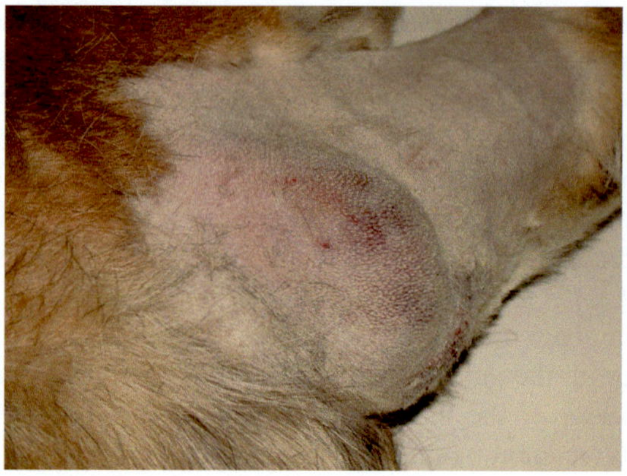

Fig. 8.524 Large myxosarcoma on the elbow of a German Shepherd

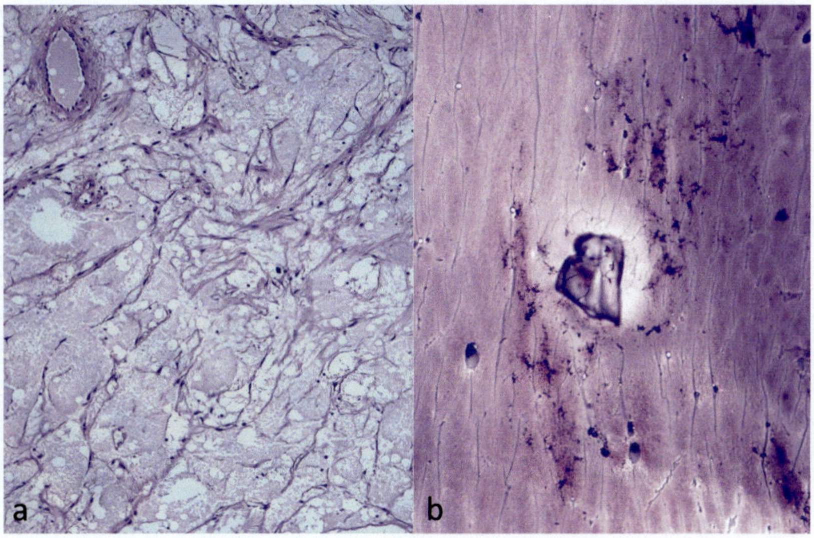

Fig. 8.525 Myxoma. (**a**) Scattered benign fibroblasts in a grainy slightly azul material representing myxoid substance; (**b**) cytology—large amount of pinkish, grainy to fibrillar extracellular myxoid material in which few and not atypical spindle cells are present

Cytological Findings

The cytology of these tumors follows their histological characteristics, revealing a background composed of abundant intercellular matrix in the form of amorphous and finely granular or fibrillar pink material. *Myxoma* cells have oval or elongated nuclei with small nucleoli; the cytoplasm is typically spindle-shaped or stellate, with barely visible edges, which are often lost within the intercellular matrix they themselves produce. A small amount of myxoid material can be observed in the cytoplasm (Fig. 8.525). One of the main differential diagnoses with fibroma is *spindle cell lipoma*; in this skin neoplasm, the cells are typically spindle-shaped and immersed in a myxoid background as described for myxoma. Histopathology or histochemical stains for lipids on fresh tissue are mandatory for the final diagnosis, as discussed below.

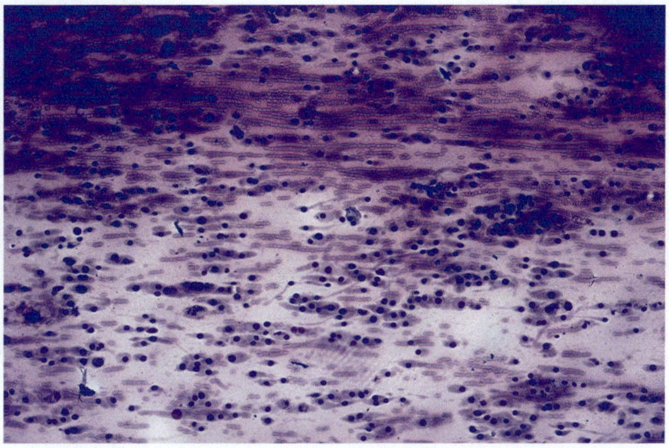

Fig. 8.526 Cytology of a myxosarcoma. Many malignant spindle fibroblasts are embedded in a dense myxoid material. Note as cells are oriented in a linear row disposition (windrowing appearance)

Well-differentiated *myxosarcomas* are cytologically indistinguishable from myxomas, especially in specimens with few cells. Malignant myxosarcomas are characterized by many atypical and pleomorphic cells, ranging from round to spindle-shaped or stellate, with large, sometimes multiple nuclei, and by numerous and atypical mitoses. The cytoplasm is large and contains an eosinophilic material (mucin) that is also evident against the extracellular background. In any mesenchymal neoplasm where there is a large amount of mucinous background, the dense consistency of the mucopolysaccharide material tends to distribute the neoplastic cells into long linear rows, giving them an arrangement on the slides called *windrowing appearance* (Figs. 8.526 and 8.527). Myxoid material can be highlighted using Alcian blue stains; this dye belongs to the copper phthalocyanine group and binds to the polyanions of acid mucopolysaccharides. By treating the slides with Alcian blue, a polyvalent basic stain, the acidic mucins are stained, forming the "Monastral blue" complex with the dye; in histology it is used to stain mucoid/myxoid extracellular matrix components, such as glycosaminoglycans and proteoglycans. With Alcian blue stain, mucins stain blue while nuclei stain red (Fig. 8.528). This stain is not commonly used in cytology, since it is not specific of a tumor; in fact, other STNs such as some PWTs or the rarer spindle cell lipoma can produce mucinous material; for this reason its presence on slides does not allow a cytological diagnosis of myxoma/myxosarcoma.

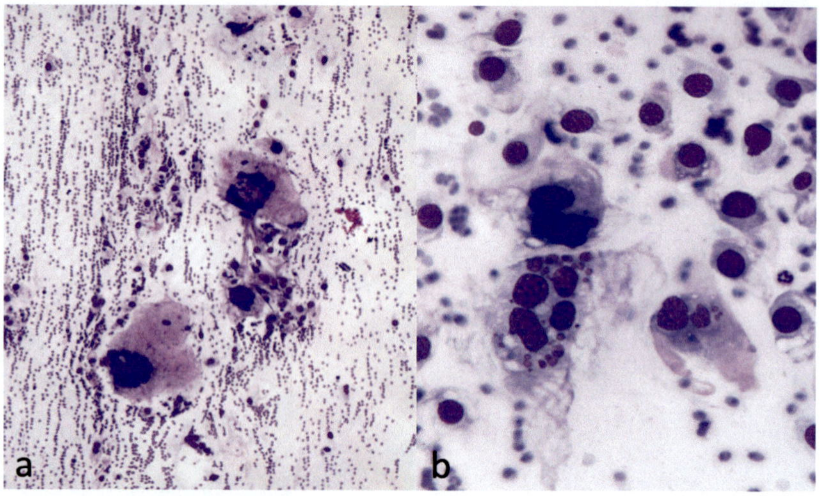

Fig. 8.527 Cytology of a myxosarcoma: (**a**) myxoid background, windrowing appearance, and two large malignant multinucleated spindle cells with cytoplasm filled with pinkish myxoid substance; (**b**) severe cytological atypia in a myxosarcoma; mucin is represented by the pinkish intracytoplasmic material

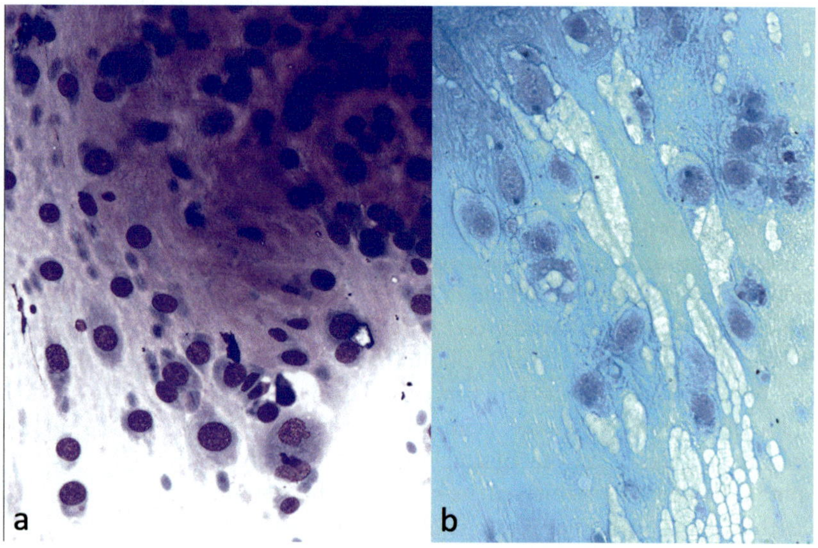

Fig. 8.528 Cytology of a myxosarcoma: (**a**) large amount of myxoid material in which spindle neoplastic fibroblasts are embedded; (**b**) Alcian blue dye: myxoid background is highlighted as azul in color material. Many spindle neoplastic cells are recognizable

8.4.3 Hematic and Lymphatic Vessel Tumors

8.4.3.1 Hemangioma and Hemangiosarcoma

Hemangioma (HA) and *hemangiosarcoma* (HSA) are, respectively, benign and malignant tumors that originate from endothelial cells of blood vessels of the dermis and subcutis [200, 201, 207].

A single nodule that can be round to oval and dark blue or slately blue in color can represent a dermal HA; when more deep in location and if covered by glabrous skin, the surface can show a dark bluish appearance, while no peculiar aspects characterize a subcutaneous HA with haired skin (Fig. 8.529). The most common histotype of HA is the *cavernous* HA, while other histological variant of HA such as *capillary*, *inflammatory lobular capillary*, *mixed capillary-cavernous*, *spindle cell*, *epithelioid*, and *angiokeratoma*, is less commonly observed [207]. The cutaneous HSA is usually primitive, and a single nodule of variable sizes, usually deeply red in color and sometimes of a fleshy appearance, clinically characterizes it [199, 202–205, 208, 209]. In dogs, single or multiple lesions can rarely represent a metastatic spread from a primary splenic or heart auricle HSA, as discussed in Chap. 9. In any case, when a cutaneous HSA has been diagnosed, staging of the neoplasia is mandatory. Both HA and HSA can be sun-induced. The UVA-induced HSAs, which are commonly observed on the glabrous abdominal skin of short-haired and with white skin dogs, such as Argentine Dogo, American Staffordshire, Pitbull, Dalmatian, Beagle etc. have usually a benign behavior. The groin, the parapreputial area, and the abdominal skin are the most frequently affected areas from solar-induced HSA. Very often these vascular neoplasia are associated to solar dermatitis or sun-induced actinic keratosis and squamous cell carcinoma [200, 207]. In cats, cutaneous HSA are less frequently observed, and the nose and pinna seem the most commonly involved skin areas (Fig. 8.530).

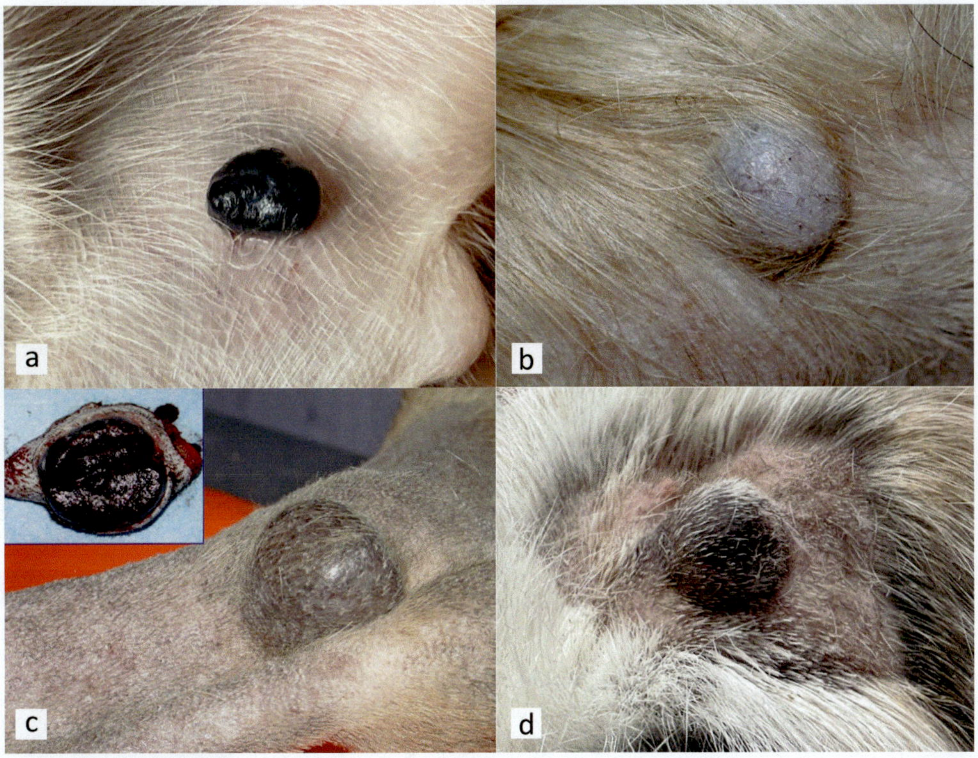

Fig. 8.529 Hemangioma: (**a**) exophytic apparently black nodule on the groin of a dog; (**b**) slately blue in color nodular HA; (**c**) dark nodule in a dog; (**d**) HA with pigmented skin

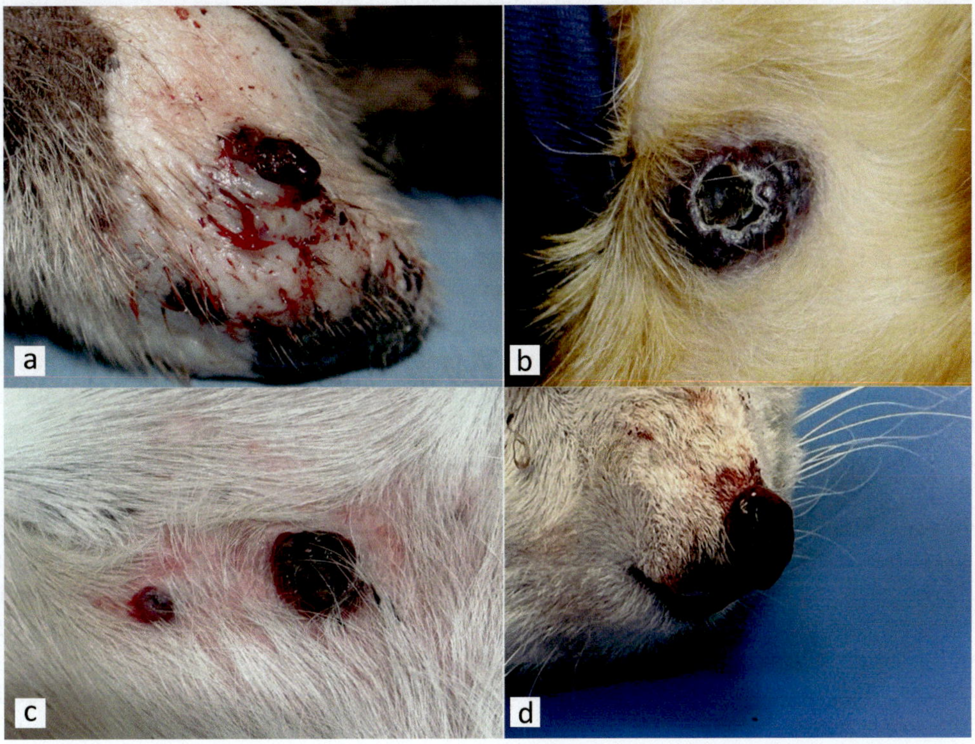

Fig. 8.530 Hemangiosarcoma: (**a**) small bloody HSA on the penis of a Pitt bull; (**b**) HSA with necrotic crusty central area; (**c**) multiple sun-induced HSAs on the abdomen of a dog with white skin; (**d**) hemorrhagic HSA on the nose of a cat

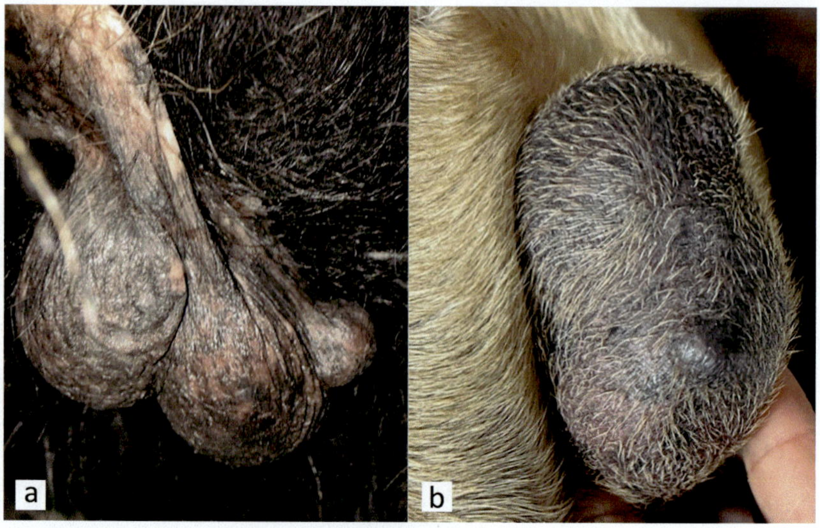

Fig. 8.531 Scrotal vascular hamartoma: (**a**) nodule on the scrotum of 1.5 cm in size; (**b**) small nodule of few mm on the scrotum of a dog

The spectrum of vascular skin neoplasia is vast, and other benign blood vessel proliferations are well recognized in dogs and cats.

A typical *vascular hamartoma* of the *scrotum* is well known on dogs. The lesion, represented as dermal nodule, from 0,5 to 2 cm in size, is usually single, and only rarely multifocal lesions are detected. The cause of why this lesion commonly occurs on the scrotum is not known, and the sun exposition is not involved since the skin of the scrotum of the affected dogs is usually pigmented [200, 206] (Fig. 8.531).

Another uncommon vascular lesion is the *hemangioendothelioma*, described in three different histological types: *epithelioid*, *retiform*, and *kaposiform*. In this group of blood vessel neoplasias, single or multiple lesions are reported, and their biological behavior is not well defined; in fact, these lesions are classified as *intermediate* vascular lesions in the new classification of skin tumors [207].

In cats, and less frequently in dogs, a so named *progressive angiomatosis* is reported. This vascular benign lesion is characterized by dermal/subcutaneous vascular proliferation that mainly affects the extremities and digital skin but that can be observed in other body sites as trunk and face. Lesions are represented by flattened or raised erythematous lesions which can merge and spread into the deep tissue, but if are deeply onset, they can be represented as swollen lesion with not well-defined margins [195, 196, 206] (Fig. 8.532).

A clinical finding observed in almost all vascular lesions, but especially in cavernous hemangioma, is the rapid filling of the needle cone with dark blood as soon as the needle is inserted into the neoformation; if we think to the anatomy of this tumor that is composed of multiple and large vascular lacunae, this behavior is understandable (Fig. 8.533).

Cytological Findings

As mentioned above, *hemangioma* is a benign vascular neoplasm, and its most common histotype is the *cavernous*

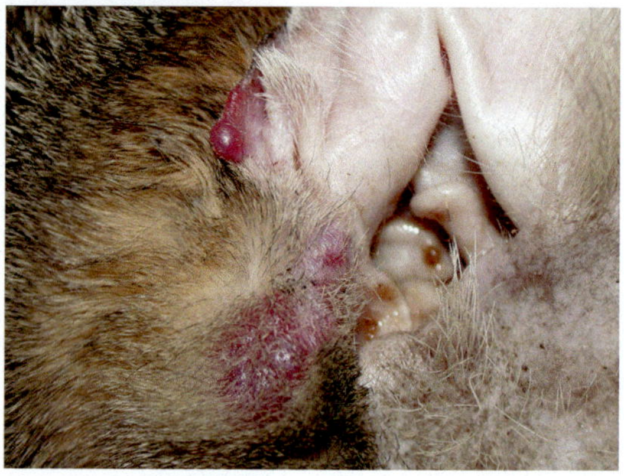

Fig. 8.532 Feline cutaneous angiomatosis: benign vascular proliferation composed of multiple lesions, with flattened and not well-defined margins

hemangioma; it is composed of multiple regular to irregular vascular lacunae filled with blood and lined by a single layer of flattened endothelial cells that are separated by a variable amount of variably cellular stroma (Fig. 8.534). Since the possibility of collecting endothelial cells with FNB is quite impossible, slides are composed only by red and white blood cells; so, when a few spindle-shaped cells are occasionally found in the blood background, they most likely represent stromal spindle cells (Fig. 8.535). In the case of cutaneous HA, the clinical presentation, associated with cytological specimen composed only of blood, is highly suggestive of a benign vascular neoplasm, and the suspicion of a blood-contaminated specimen collected from a nonvascular lesion can be ruled out, especially if many and repeated specimens yield only hematic slides. However, doubts may arise in the presence of a subcutaneous HA, especially if the clinical appearance is not typical of a vascular lesion; also in these cases, if only blood cells are released, a vascular neoplasm should be suspected. Anyway, when instead large

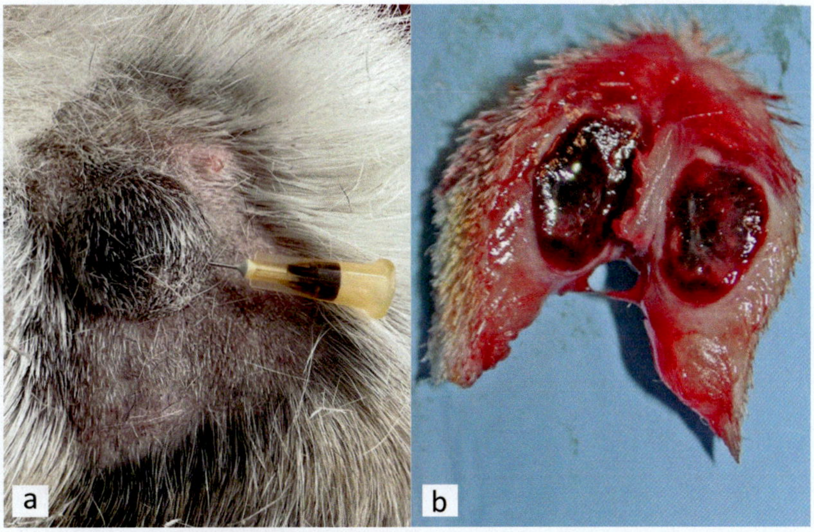

Fig. 8.533 Cavernous hemangioma: (**a**) note as the needle cone is filled with blood; (**b**) the removed and cut nodule shows a well-defined and blood-filled lesion

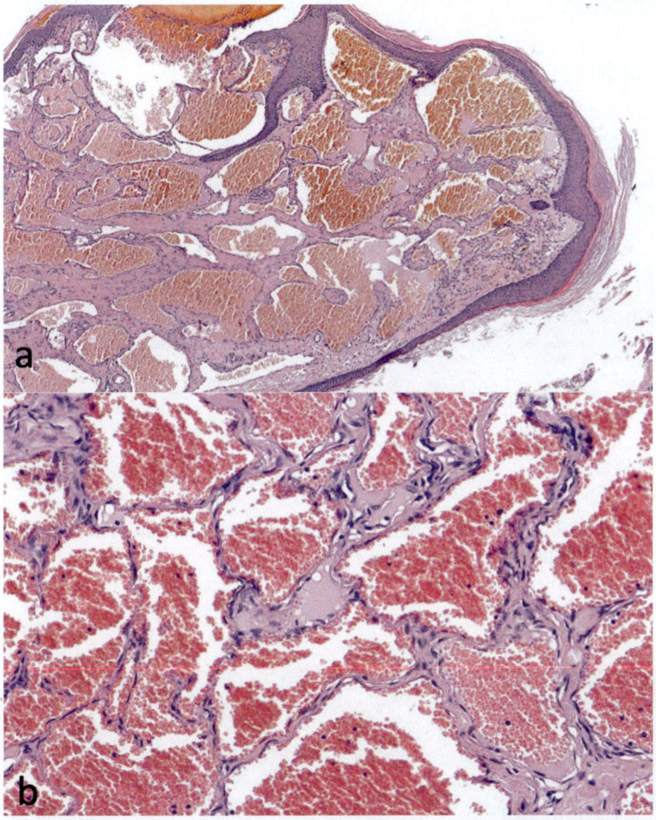

Fig. 8.534 Histology of cavernous hemangioma: (**a**) multiple dermal vascular lacunae filled with blood; (**b**) at higher magnification the flattened endothelial cells are recognizable along with the few stromal cells

subcutaneous nodules are sampled and a large amount of blood is collected, vascular neoplastic lesions must be differentiated from a hematoma; the lack of a traumatic history, the presence of platelets, and the lack of erythrophagocytic macrophages containing hemoglobin derivate such as *hemosiderin* and *hematoidin pigment* allow cytologists to differentiate an HA from a hematoma.

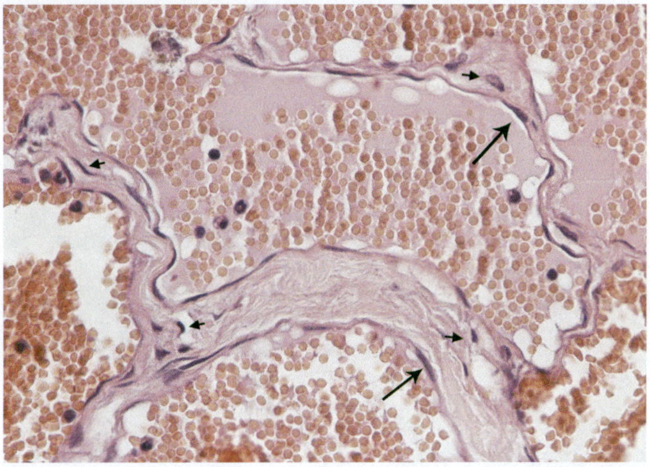

Fig. 8.535 Histology of cavernous hemangioma. Note as endothelial cells which line the vascular lacunae are flattened (long arrows). Spindle stromal fibroblasts are present around the vascular lacunae (heads of arrow). The possibility to sample endotheliocytes is quite impossible with cytology

HSA presents different histopathological patterns since it can be composed with many areas of well-differentiated endothelial cells, forming irregular vascular channels, and/or solid areas composed of highly anaplastic endothelial cells, sometimes without clearly recognizable vascular channels; in the latter scenario, it is difficult to differentiate HSA from other STSs without immunohistochemical investigation. An epithelioid histological variant is recognized (Figs. 8.536 and 8.537). An abundant hematic background and a variable number of pleomorphic cells (spindle, star-like or oval in shape), showing atypical features that are related to the degree of differentiation of the tumor, characterize the cytological slides of HSA. In highly cellular specimens and if more solid areas are collected, large groups of overlapped neoplastic endotheliocytes are visible (Figs. 8.538, 8.539 and 8.540). The cytoplasm is of variable intensity of blue and

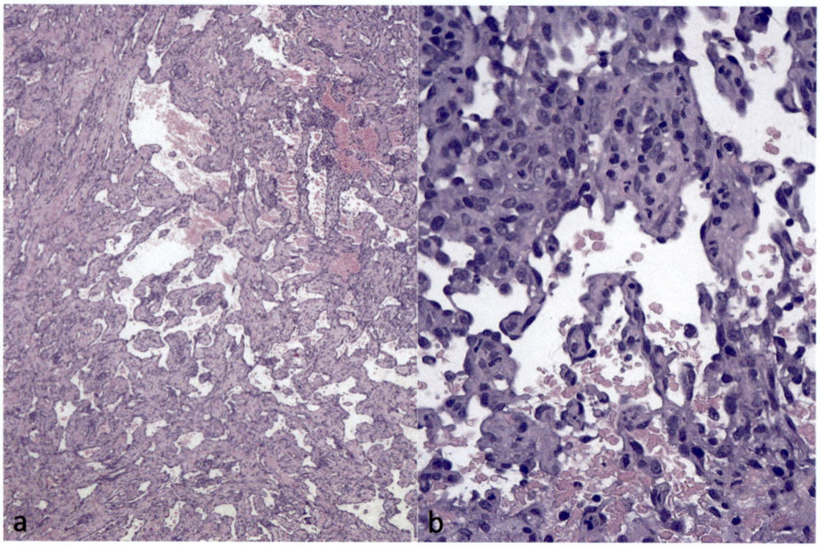

Fig. 8.536 Histology of hemangiosarcoma: (**a**) in conventional HSA, multiple irregular and incomplete vascular channels lined by neoplastic endotheliocytes; (**b**) at high magnification, it is better recognizable as vascular lacunae are lined by neoplastic cells showing moderate degree of atypias

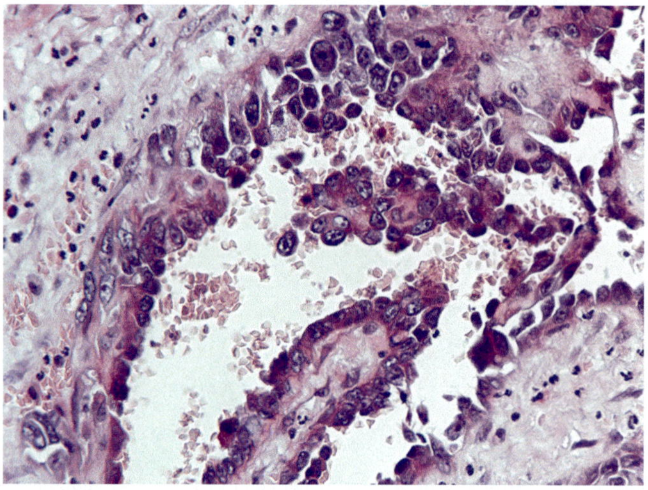

Fig. 8.537 Histology of hemangiosarcoma: marked malignant cytological features of the endothelial cells represented by anisokaryosis, anisocytosis, large nucleoli, and atypical mitosis

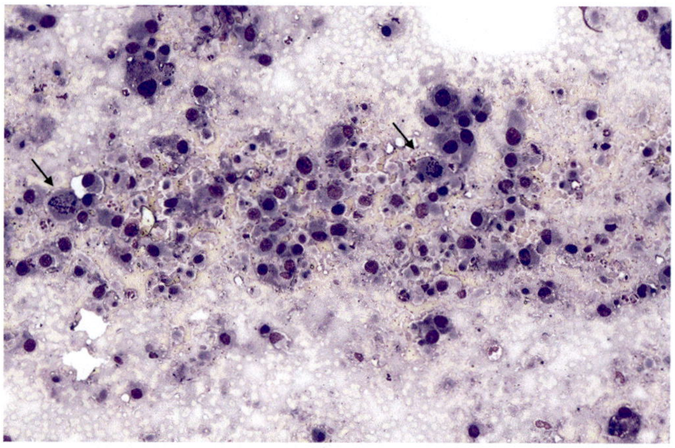

Fig. 8.538 Cytology of hemangiosarcoma: on a hematic background, many spindle malignant cells are present

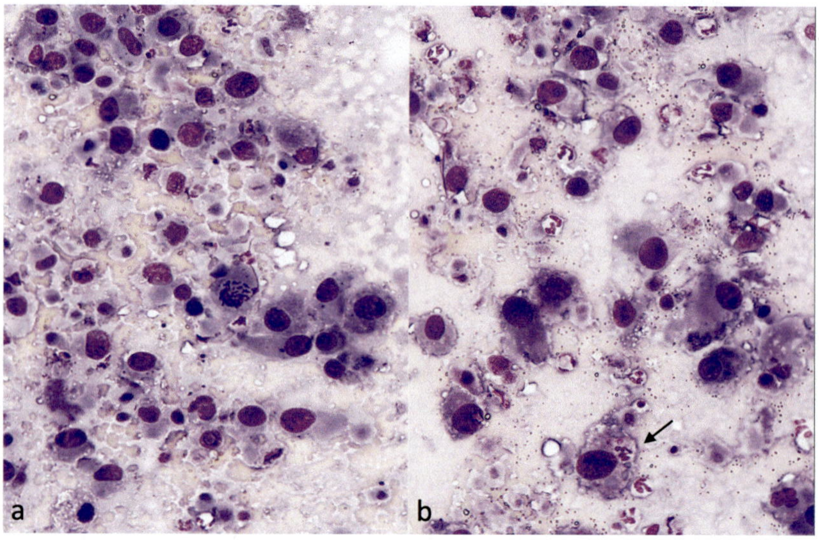

Fig. 8.539 Cytology of hemangiosarcoma: (**a**) marked cytological features of malignancy of endothelial cells represented by anisokaryosis, anisocytosis, and large nucleoli. In the center of the image, a mitotic cell is recognizable; (**b**) malignant endothelial cells—note that the neoplastic cells are internalizing a neutrophil and some red blood cells (arrow)

may have indistinct borders, and small punctate simil-vacuoles are frequently observed (Fig. 8.541) [197, 202].

The cytology of HSA, as for most mesenchymal neoplasms, does not permit identification of their origin, for which histopathology is mandatory. A strong suspicion of HSA can be generated when angioformative features represented by small and irregular groups of neoplastic cells, sometimes with a simil-whorl arrangement, set around a small central area containing a single or few red blood cells (RBCs), are observed. This cytological finding, however, creates many interpretative doubts, since some cytologists want to see, in these arrangements, an attempt to form blood vessel outlines by the neoplastic cells, aspects that can also be observed in histological samples of some HSAs; however, more frequently, the presence of RBCs in the cytoplasm of neoplastic cells could be justified as internalization or anomalous phagocytic behavior that neoplastic endotheliocytes can undergo (Figs. 8.542 and 8.543). Since HSA can show intralesional necrosis, intravascular thrombi, along with neoplastic endothelial cells, debris, necrotic nuclei of both inflammatory and neoplastic cells, and fragmented erythrocytes, can be detected on slides giving a "dirty" aspect of the background (Fig. 8.544). Another uncommon cytological aspect rarely observed in some cutaneous HSAs is the presence of hematopoietic precursors such as megakaryocytes and rubrocytes.

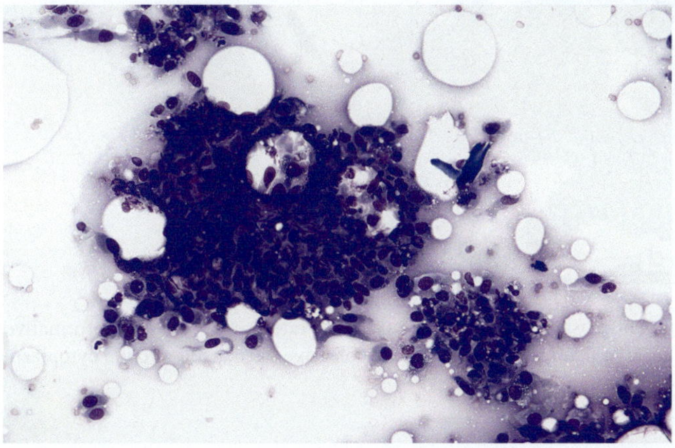

Fig. 8.540 Cytology of hemangiosarcoma: neoplastic cells are released both in groups and singularly; this finding depends on the pattern of growth of the neoplasia

The high amount of blood in the background is not a key parameter unique to HSA, since also in other STNs and especially in perivascular tumors (PWT), due to their anatomical location, the blood background is a common cytological finding; in contrast to PWT, in HSA the atypical cytological aspects tend to be more marked, and the samples are usually less cellular. Furthermore, PWT very often presents pathognomonic cytological features as discussed below.

In *scrotal-type vascular hamartoma* and in *hemangioendothelioma*, slides are constantly hematic. In lucky samples, a variable amount of spindle cells can be observed. Since in these vascular proliferative neoplasms, many stromal fibroblasts, myofibroblasts, and pericytes are present around the vessels, it is not possible differentiate, with cytology, if the collected spindle cells are of endothelial or, more probably, of stromal origin (Figs. 8.545 and 8.546).

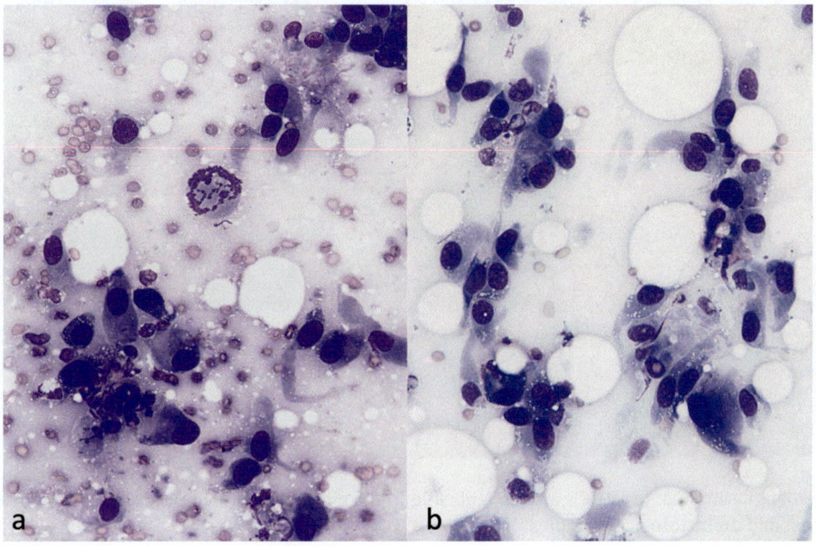

Fig. 8.541 Cytology of hemangiosarcoma: (**a**) spindle neoplastic endotheliocytes with intracytoplasmic vacuole-like areas; (**b**) many small punctate simil-vacuoles are recognizable in the cytoplasm on neoplastic cells

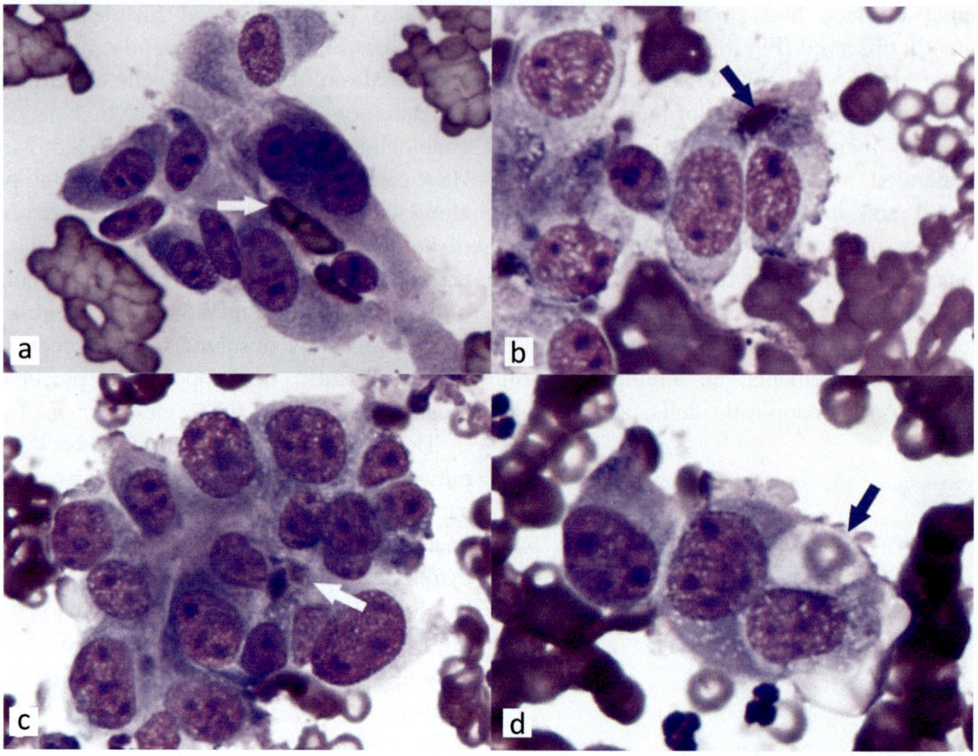

Fig. 8.542 Cytology of a hemangiosarcoma. (**a**, **b**) single or a few RBCs are contained between neoplastic cells (arrows); (**c**) a single RBC surrounded by aggregates of endotheliocytes with epithelioid arrangement (arrow). These cytological features are observed in many hemangiosarcomas and could be interpreted as angioformative arrangements; (**d**) a single RBC is clearly internalized into the cytoplasm of a neoplastic cell (arrow)

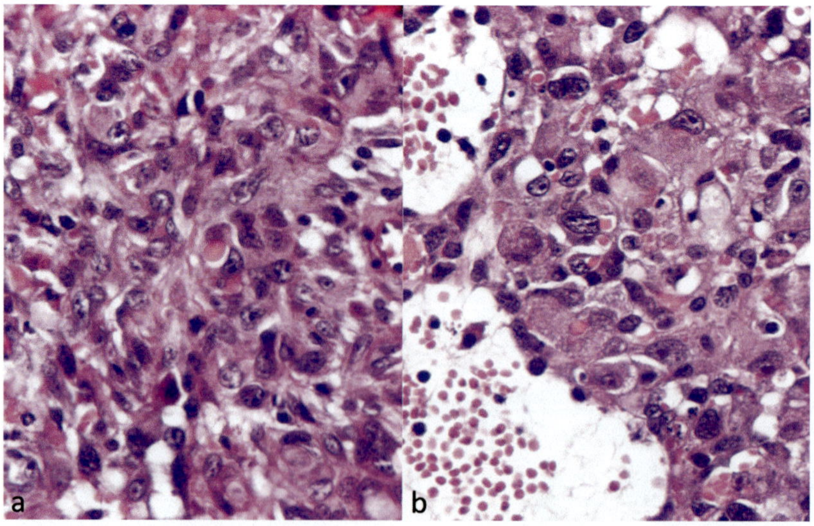

Fig. 8.543 Histology of a hemangiosarcoma: (**a**) RBCs are recognizable in the cytoplasm of neoplastic endotheliocytes: (**b**) both intracellular and extracellular RBCs are recognizable on histological samples

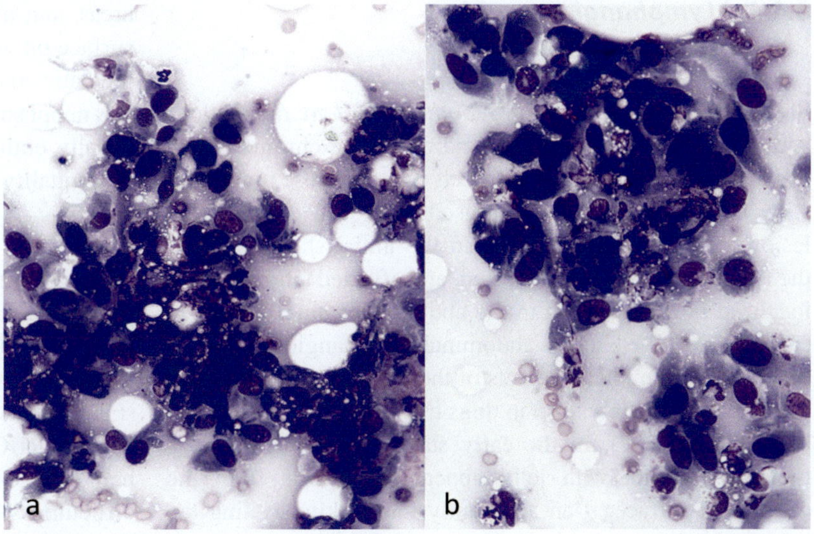

Fig. 8.544 Cytology of a hemangiosarcoma. Debris material composed of necrotic nuclear material and RBC fragments are recognizable between neoplastic cells

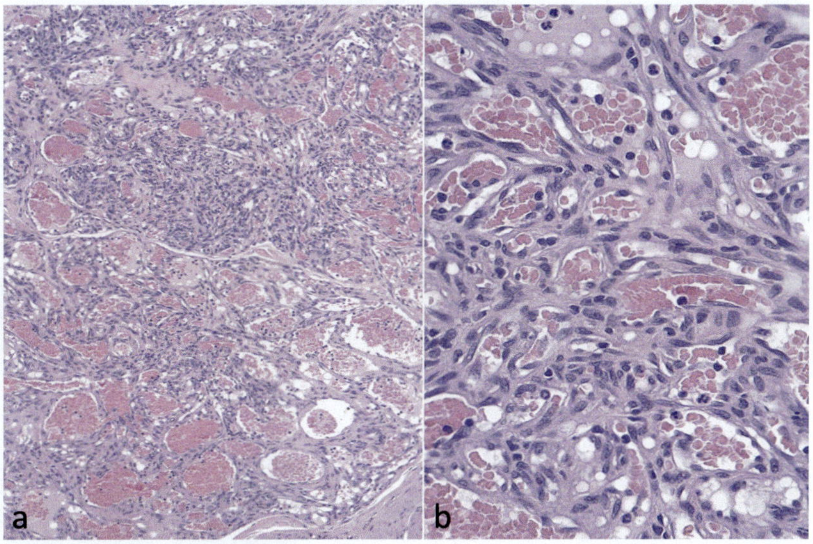

Fig. 8.545 Histology scrotal-type vascular ham*artoma*: (**a**) multiple capillary lined by a single line of non-atypical endotheliocytes. Many spindle cells represented by pericytes, fibroblasts, and myofibroblasts are visible around the vessels; (**b**) at high magnification spindle cells are recognizable in the inter-vascular stroma

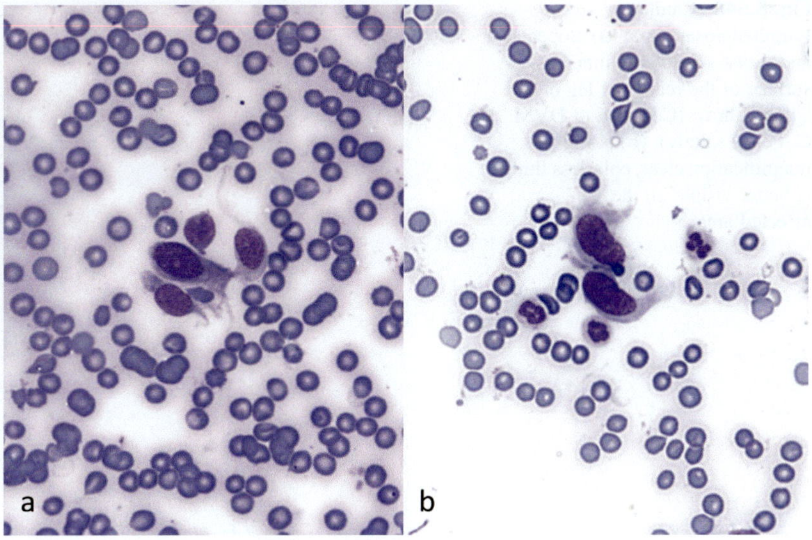

Fig. 8.546 Cytology scrotal-type vascular hamartoma: few spindle cells on a hematic background: with higher possibility, those cells are not endothelial but of stromal origin (fibroblasts and myofibroblasts)

8.4.4 *Lymphangioma* and *Lymphangiosarcoma*

Skin neoplasms of *lymphatic endothelial cell* are rare in dogs and cats. The lesions vary depending on the chronicity and can have very subtle clinical aspects that, for those unfamiliar with this neoplasm, are confused with an inflammatory lesion, delaying the diagnosis. In the initial stages, in fact, the lesions are poorly demarcated, and affected animals show localized or diffuse areas in the abdomen or inner thighs. In cats, the so-called feline abdominal lymphangiosarcoma, a tumor of the lymphatic vessels of the abdominal skin region, is well known; rarely, also in dogs this type on neoplasia can be observed [207]. In the early stage of this neoplasia, a transparent fluid, with jelly appearance, oozes from the abdominal surface of an apparently intact skin, wetting the haircoat. Edema, alopecia, diffuse erythema, hemorrhagic areas, and ulceration develop with time; in these stages, the surface on which the cat rests is usually dirty by a yellowish-thematic material (Figs. 8.547 and 8.548). The outcome of this neoplasm is fatal, metastases are exceptional, but cats are usually euthanized due to the spreading of the lesions and poor quality of life [198].

Cytological Findings

Histology of *lymphangiosarcoma* is characterized by multiple irregular and with slit-like appearance channels, lined from one or more layers of spindle neoplastic cells, very often plump, with variable degrees of malignancy (Fig. 8.549). Usually, the lumens of the channels do not contain RBCs, but in some case, especially if intralesional hemorrhages occur, they can be present RBCs, creating difficulties in differentiation between a lymphatic to a hematic origin of the neoplasia (abdominal HSA).

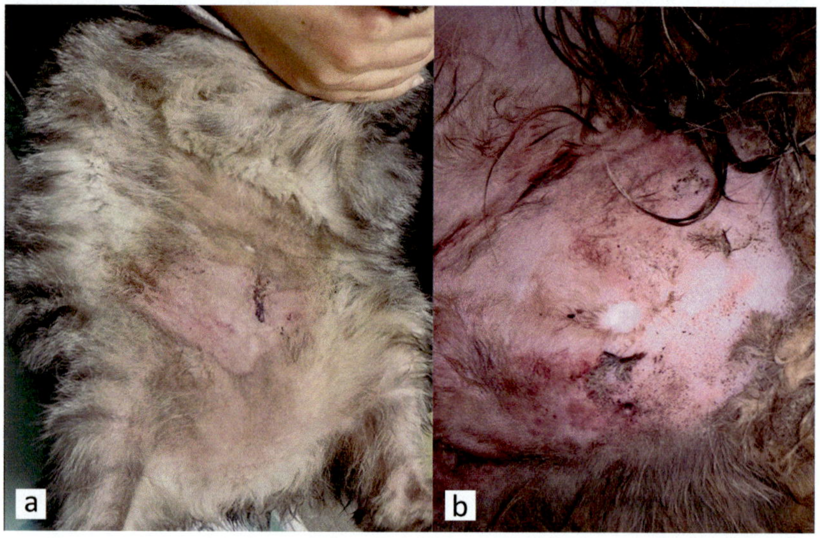

Fig. 8.547 Feline abdominal lymphangiosarcoma: (**a**) alopecia and moist skin on the abdominal area; (**b**) mild swelling, erythema, and wet hair on the abdominal skin of a cat

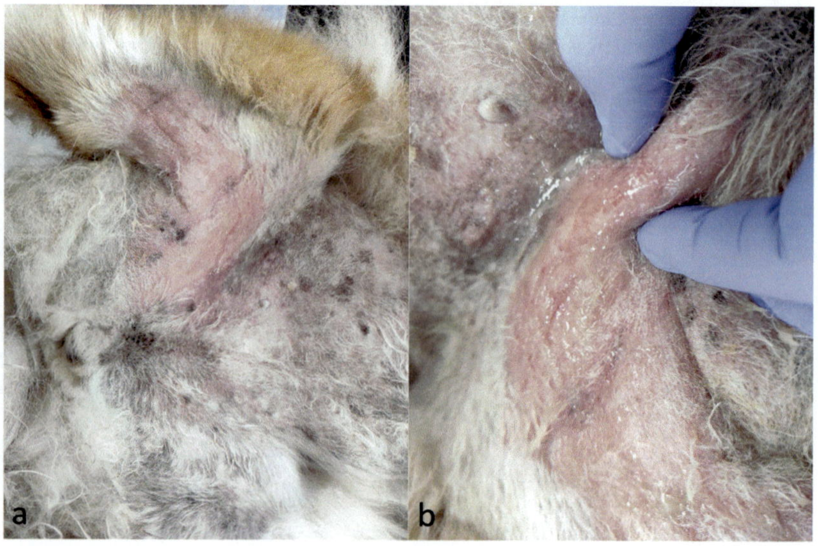

Fig. 8.548 Canine lymphangiosarcoma: (**a**) alopecia and moist skin on the inner surface of the rear right leg of a Chow-Chow. (Courtesy of DVM C. Penna—Italy); (**b**) at high magnification clear, colorless fluid is better visible on the affected area

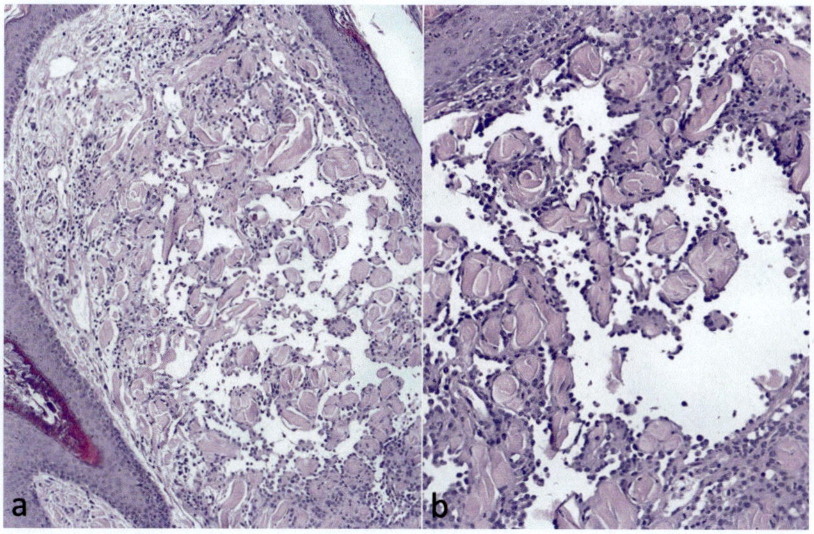

Fig. 8.549 Histology of lymphangiosarcoma: (**a**) the light areas are neoplastic lymph vessels infiltrating the superficial dermis; (**b**) at higher magnification the neoplastic cells, from spindle to plump in shape, lining the irregularly shaped lymph vessels, are clearly visible

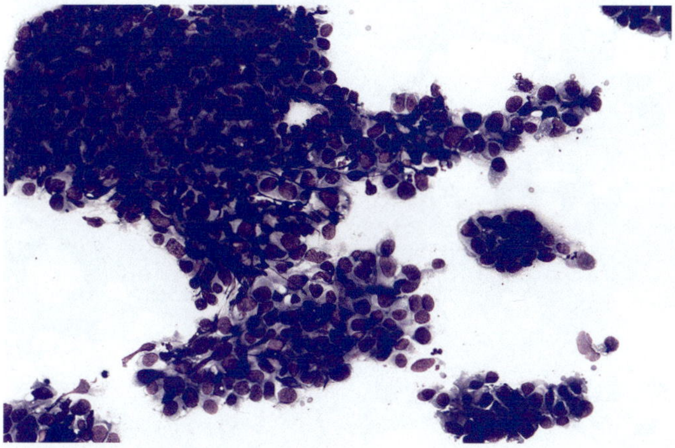

Fig. 8.550 Cytology of lymphangiosarcoma: large group of oval to spindle-shaped neoplastic cells on a clear background

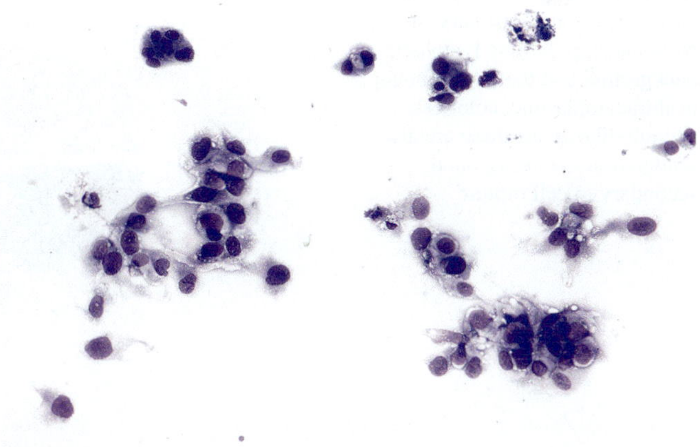

Fig. 8.551 Cytology of lymphangiosarcoma: spindle-shaped neoplastic cells on a clear background

Lymphangiosarcoma cytology is characterized by a usually clear background, at most characterized by light blue material representing the lymph, in which the neoplastic cells are spindle to round in shape, released either in clusters or as single cells; the cells show variable degrees of anisocytosis and anisokaryosis, and the cytoplasm may contain multiple vacuole-like, achromatic, and punctate. The neoplastic cells cannot be differentiated from those belonging to other STNs, and but when spindle-shaped neoplastic cells are released on a clear background, and if there is a very suggestive clinical presentation, a lymphatic neoplasm is strongly suspected (Figs. 8.550, 8.551, 8.552 and 8.553).

8.4.5 Perivascular Wall Tumors

Perivascular wall tumors (PWTs) are mesenchymal neoplasia belonging to a heterogeneous group of neoplasms that originate from the vascular smooth muscle cells (vSMC)-pericyte continuum. These cell lines surround the endothelial cells of the blood vessel walls, including *pericytes*, *myopericytes*, *adventitial cells*, and *myofibroblasts*. In veterinary medicine, a detailed description of these tumors is relatively recent, and in the last decade, many papers were published on this group of neoplasia [211, 212, 214, 216, 217].

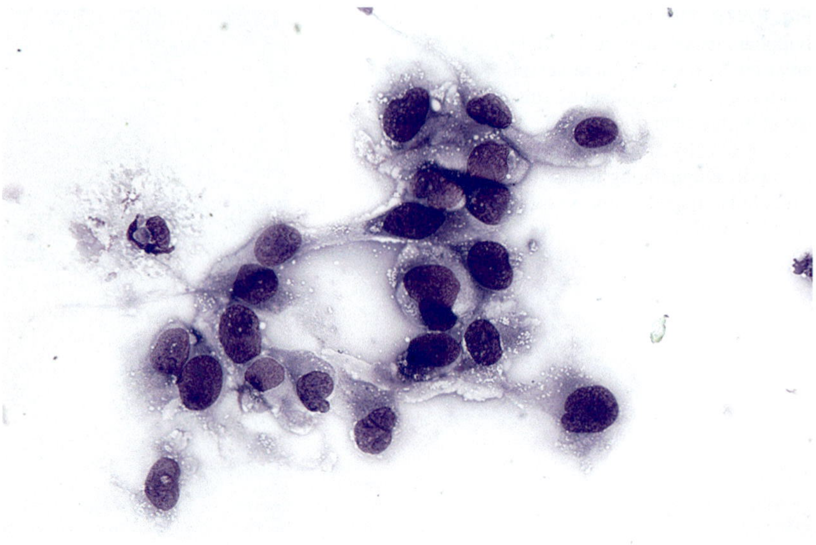

Fig. 8.552 Cytology of lymphangiosarcoma: at higher magnification multiple vacuole-like, achromatic, and punctate intracytoplasmic areas are recognizable

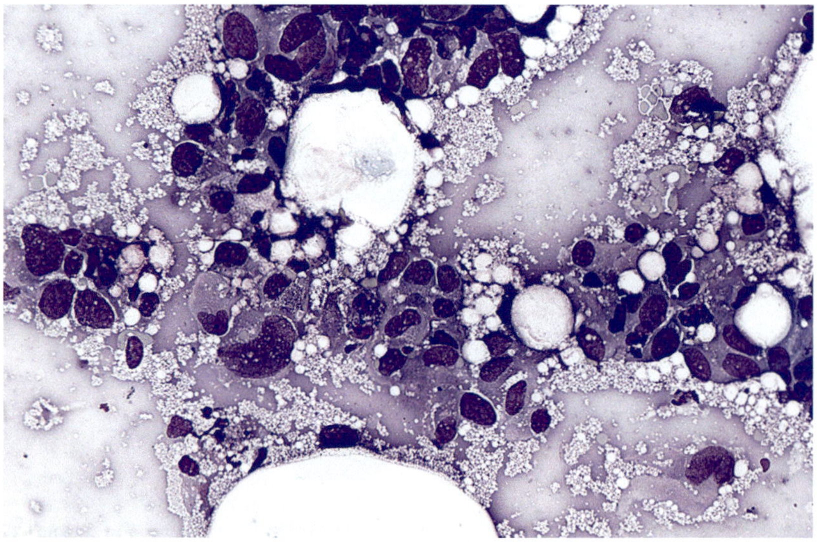

Fig. 8.553 Lymphangiosarcoma cytology: atypical lymphatic endothelial cells showing clear anisocytosis and anisokaryosis. Note the slightly blue lymphatic background and the large amount of intracytoplasmic, colorless, vacuole-like areas; these are also scattered in the background secondary to cell rupture

For a long period, PWTs were diagnosed without ascertaining the exact line of the neoplastic cell, causing great confusion in their classification. New knowledge in veterinary pathology has shown that in the past, although they were histologically indistinguishable from other PWTs, many perivascular neoplasms were misdiagnosed as *hemangiopericytomas* (HPCs).

On the basis of the combination of the histological pattern and positivity to different immunohistochemical antibodies, different types of PWTs, originating from different lines of perivascular cells, have been demonstrated. As a result of these researches, many PWTs formerly diagnosed as HPCs, which originate from *pericytes*, an undifferentiated mesenchymal cell with contractile function that partially surrounds the sub-endothelial cells (basal lamina), were instead *myopericytomas* (MPCs), whose neoplastic cells are *myopericytes*, cells that surround the endothelial cells of medium-sized vessels (arterioles and venules). True HPCs exist in dogs, but they are much rarer than previously reported and, unlike the MPCs, seem to have a more aggressive biological behavior [210, 211]. Their differentiation, impossible with cytology, is not clear if it could have a relevant prognostic value, and to date, the knowledge are few and not definitive.

Other PWTs, including the *angioleiomyoma/angioleiomyosarcoma*, which originate from smooth muscle of the vascular wall; the *angiofibroma*, originating from cells of the vascular adventitia; and the *angiomyofibroblastoma* and *angiomyxoma*, originating from perivascular myofibroblasts, are rarely reported in veterinary medicine [211].

Grossly, PWTs are present, with single, cutaneous, or subcutaneous nodules, mainly localized on the limbs, especially on the elbows and hocks and less frequently on the

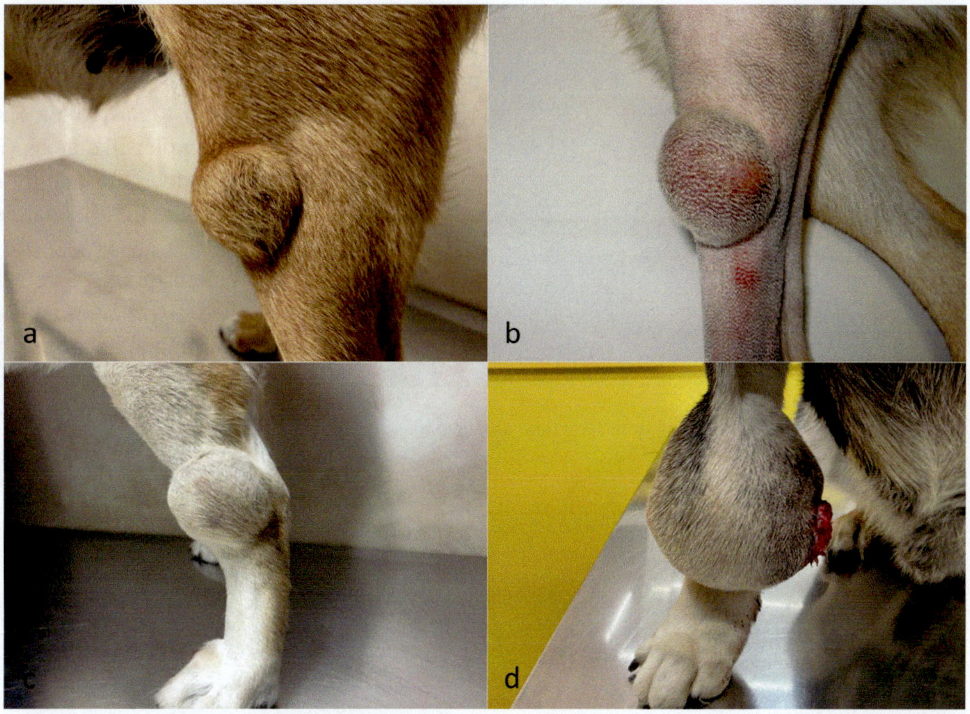

Fig. 8.554 PWT clinical presentation: single nodule/mass on the leg is the most common clinical presentation of canine perivascular wall tumor

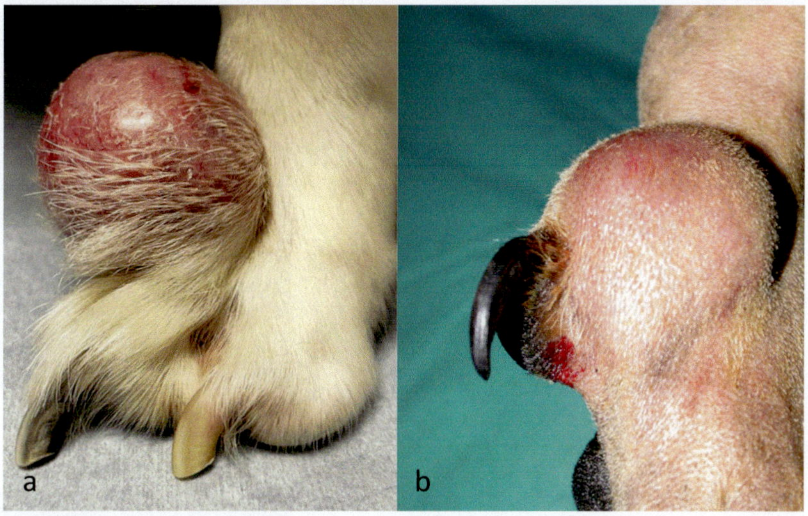

Fig. 8.555 PWT clinical presentation: (**a**) exophytic nodule on a digit. (Courtesy of DVM A. Prolo –Italy); (**b**) PWT involving the digit I of the right fore paw

abdomen, chest, or other body areas (Figs. 8.554, 8.555 and 8.556). The size ranges from a few centimeters up to large masses measuring more than 15–20 cm. The PWTs tend to very rarely metastasize but have a high index of local recurrence in that, because of their peculiar growth along the small vessels, the presence of satellite nodules, and the frequent localization in areas with scarce subcutaneous tissue, surgical resection is often incomplete [218, 219].

Cytological Findings

The morphology of PWTs has been well documented, and when all the cytological features are present on the same specimen, they allow cytologists to confirm the perivascular nature of the neoplastic cells [213].

Histologically four main patterns are recognized: *perivascular whorls, placentoid pattern, staghorn (antler-like) vessels, and bundles*, with the first two being the most frequent patterns of PWTs [217]. In very cellular slides, a suggestive organization of cells (cyto-architectures) can be recognizable (Figs. 8.557, 8.558 and 8.559).

The slides range from poorly to highly cellular, and in the latter, cells seem organized in large clusters, and the typical spindle silhouette of PWT cells can be clearly recognized only to medium–high magnification (Fig. 8.560). With regard to its perivascular origin, the samples can be variably

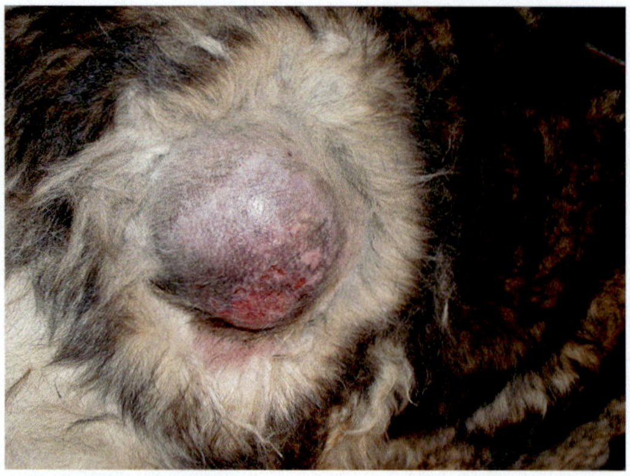

Fig. 8.556 PWT clinical presentation: large and alopecic mass on the lateral thigh of a dog

hemodiluted; also in many cases the amount of blood is very few. Slides from conventional PWTs are characterized by spindle cells, usually with a wispy, tailed, and slightly blue in color cytoplasm, sometimes rich of roundish and achromatic vacuolar-like areas (Fig. 8.561). In well-differentiated tumors, nuclei are mainly roundish if compared with other sarcomas or slightly oval, with finely regular chromatin, and the nucleoli are not so obvious; nuclei can often be double and paired in an eccentric arrangement to give cells a so-called *insect head* appearance or can be multiple (five or more) arranged haphazardly or at the periphery of the cytoplasm giving cells a typical aspect defined as *crown cells* [213]. These two morphologies are not always present on slides, but when present, they are indicative of a PWT (Figs. 8.562, 8.563 and 8.564).

Fig. 8.557 PWT histology: (**a**) the perivascular arrangement is the histological hallmark of PWT; (**b**) at high magnification a central hematic vessel is recognizable

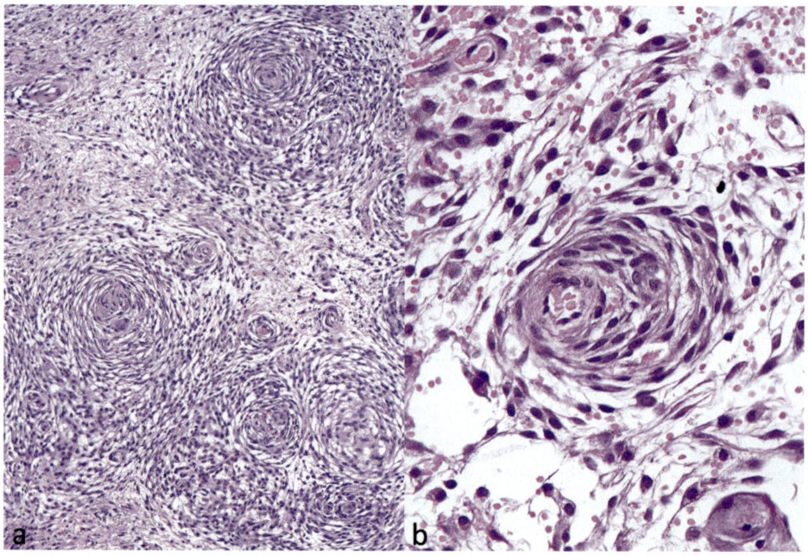

Fig. 8.558 PWT histology: (**a**) many vortex arrangements not involving blood capillaries; (**b**) at high magnification collagen is recognizable in the center of whorls. In these cases it is not possible to differentiate a PWT from a NST

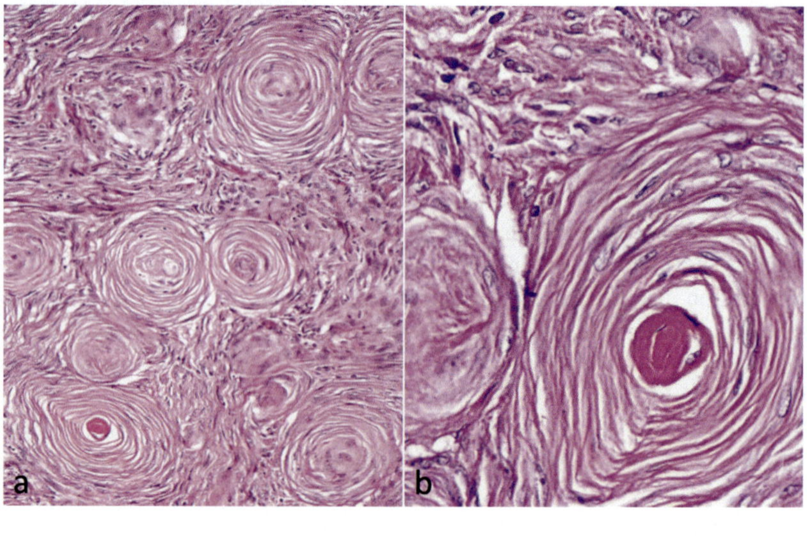

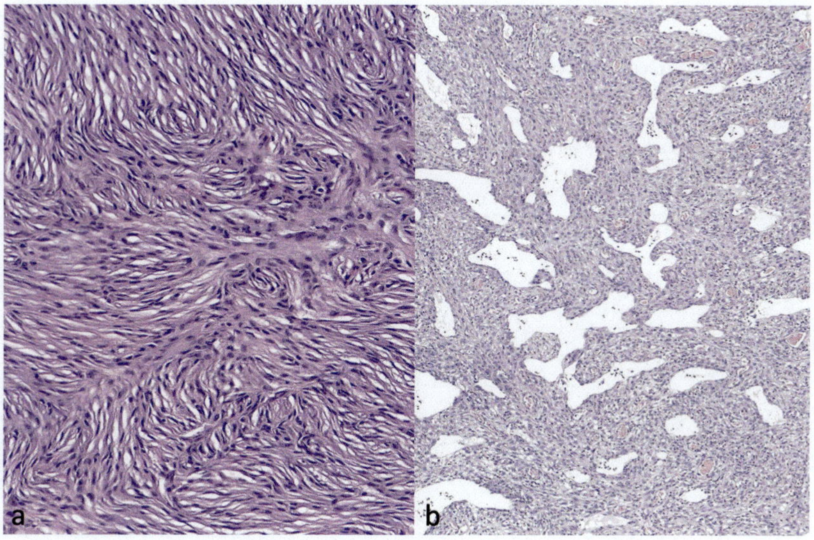

Fig. 8.559 PWT histology: (**a**) fasciculate pattern with a storiform fingerprint arrangements; (**b**) staghorn pattern

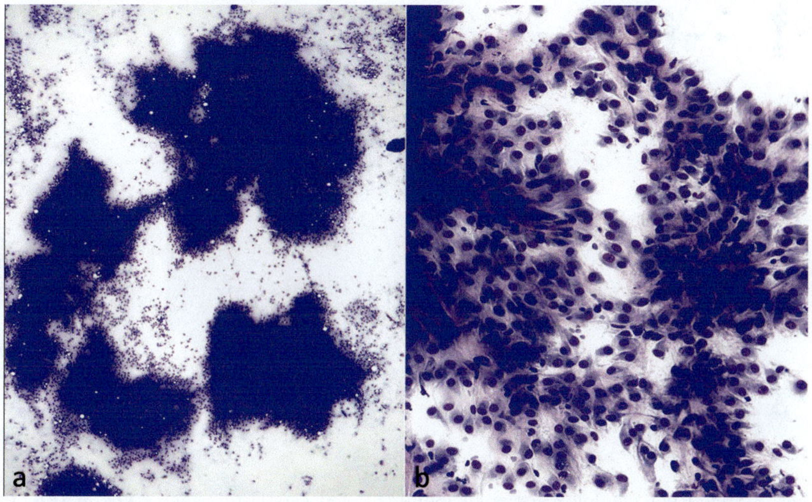

Fig. 8.560 PWT cytology: (**a**) large group of perivascular wall tumor cells; (**b**) at high magnification the typical neoplastic spindle cells of PWT are clearly evident

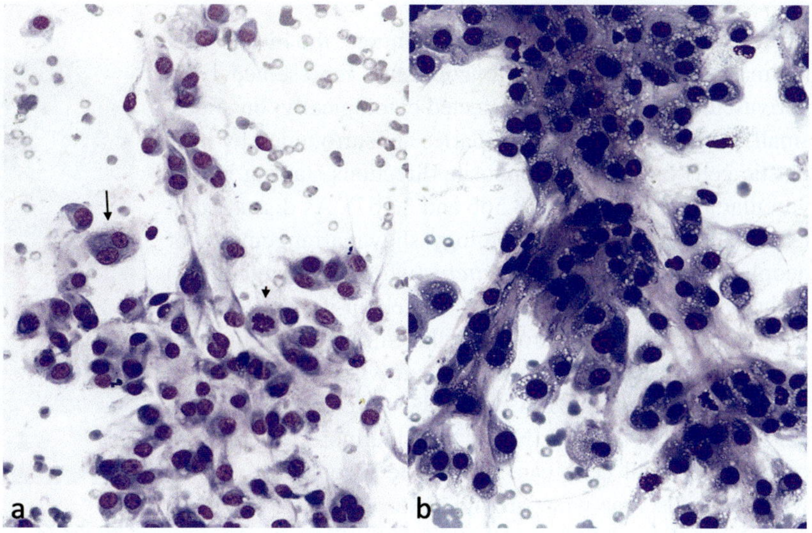

Fig. 8.561 PWT cytology of conventional tumor: (**a**) single spindle cells with roundish nuclei and wispy, tailed cytoplasm. Note the insect head cell (long arrow) and the mitotic cell (head of arrow); (**b**) perivascular neoplastic cells show many roundish and achromatic vacuole-like intracytoplasmic areas

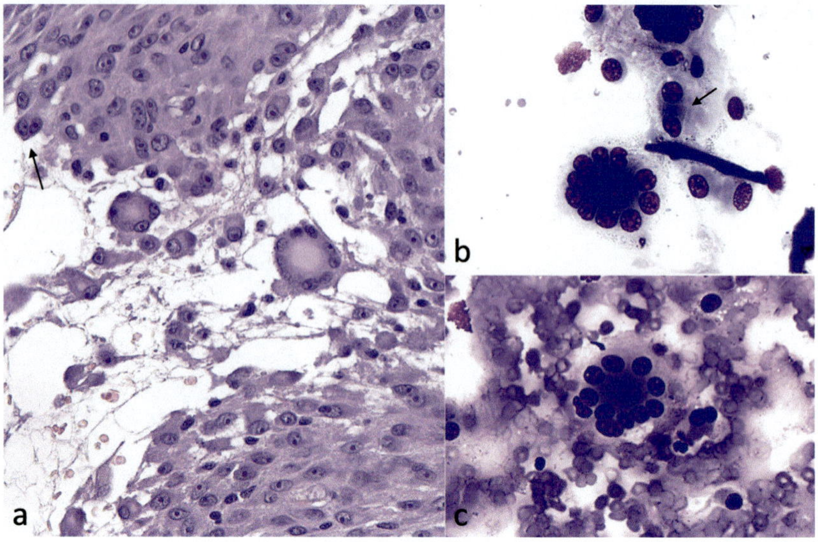

Fig. 8.562 PWT: conventional tumor. (**a**) Histology: two *crown cells* and one *insect head cell* (arrow) along with spindle neoplastic cells. (**b**) Cytology: two *crown cells* and an *insect head cell* (arrow). (**c**) Cytology: *insect head cells*

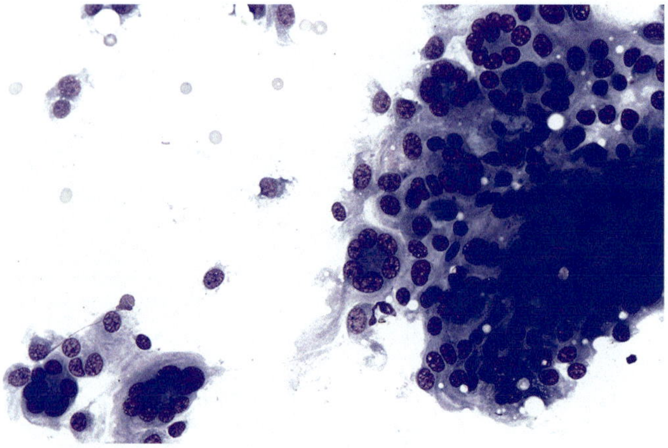

Fig. 8.563 PWT cytology: numerous multinucleated neoplastic cells. Some with nuclei arranged in an haphazardly way and some with the characteristic crown-like morphology

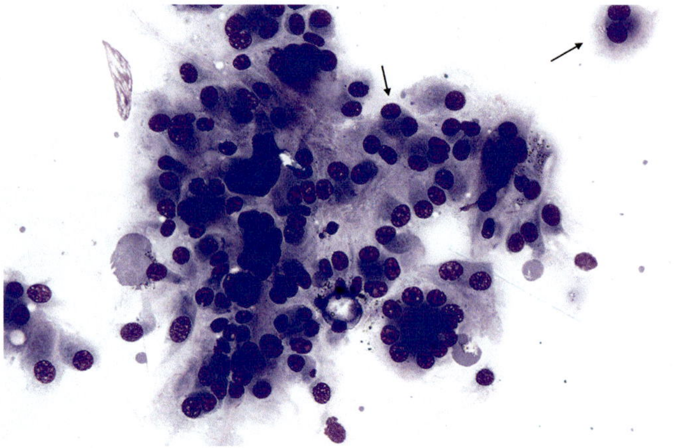

Fig. 8.564 PWT cytology: *crown cell* and *insect head cells* (arrows) are simultaneously present in this picture

In conventional perivascular wall tumors, the histological hallmark of these groups of neoplasms, represented by a perivascular pattern, can be detected cytologically; on slides, small, long, and branched capillaries are surrounded by neoplastic cells branching in opposite directions starting by the vascular axis (Figs. 8.565, 8.566 and 8.567). As discussed in Chap. 7, in some PWTs, histology shows arrangements of neoplastic cells in a *vortex pattern* (whorled arrangements) without showing the presence of vessels in its center; these whorls can be arranged around collagen fibers, and in these cases no perivascular arrangements are visible on slides. In this situation the neoplasia cannot be differentiated from a nerve sheath tumor (NST). Furthermore, neoplastic perivascular wall cells can also create incomplete whorls, with the cells that wrap around themselves in whorl-like formations (Figs. 8.568, 8.569 and 8.570). Since to date, no

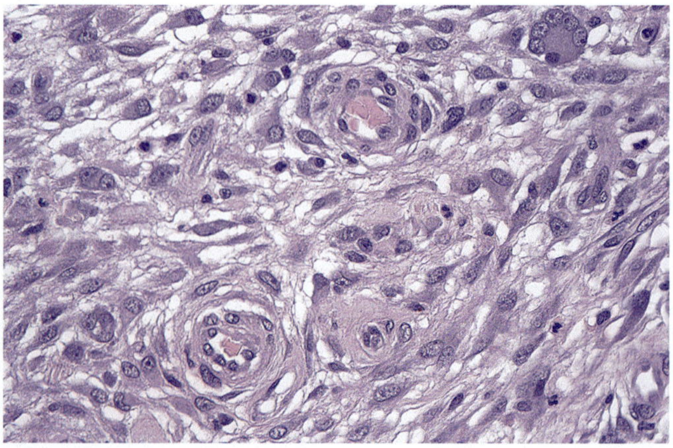

Fig. 8.565 PWT histology: two perivascular arrangements are clearly recognizable

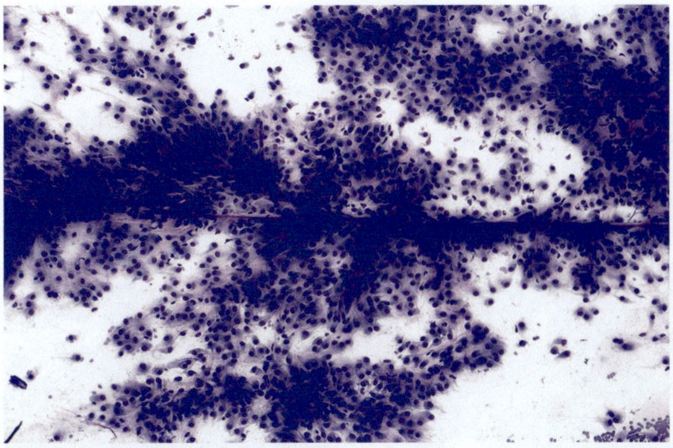

Fig. 8.566 PWT cytology: a long vessel surrounded with a lot of spindle PWT cells characterizes the perivascular arrangement in cytology

other type of mesenchymal tumors shows all the cytological features mentioned, and their presence has a very high predictive value for PWTs.

Lastly, it is not so uncommon to see a myxoid background in PWTs, recognizable as violet dense background material in which cells are embedded and that orient them in a linear disposition (windrowing appearance). The myxoid material can be highlighted with Alcian blue staining. In these cases, if no other peculiar cytological aspects of PWT are present, such as insect-head cells, crown cells, and perivascular arrangements, it is not possible differentiate a myxoid PWT from a myxosarcoma or another STN with myxoid differentiation (Figs. 8.571, 8.572 and 8.573). In some PWTs, a particular arrangement can be observed that resembles that reported in the so-called storiform *cartwheel pattern*; the cells appear to grow in a centrifugal direction starting from a central area (Figs. 8.574 and 8.575). In poor-differentiated

Fig. 8.567 PWT cytology: at high magnification from a central vascular axis, peripheral neoplastic spindle cells branch in opposite directions

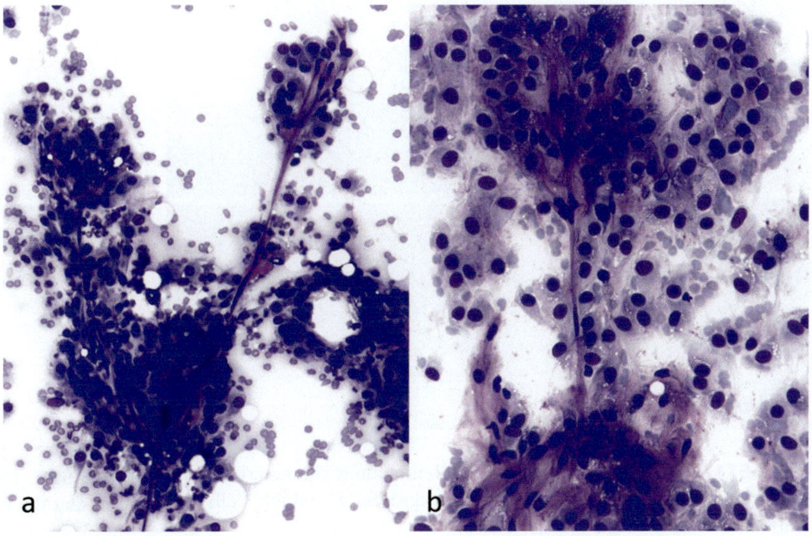

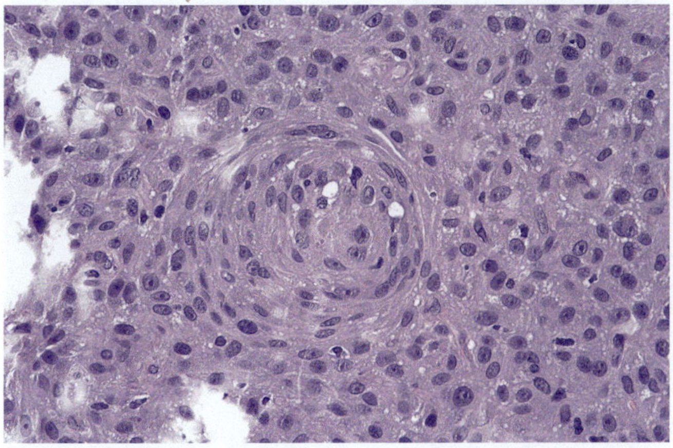

Fig. 8.568 PWT histology: whorl arrangement. Note as a hematic capillary is not present in the center

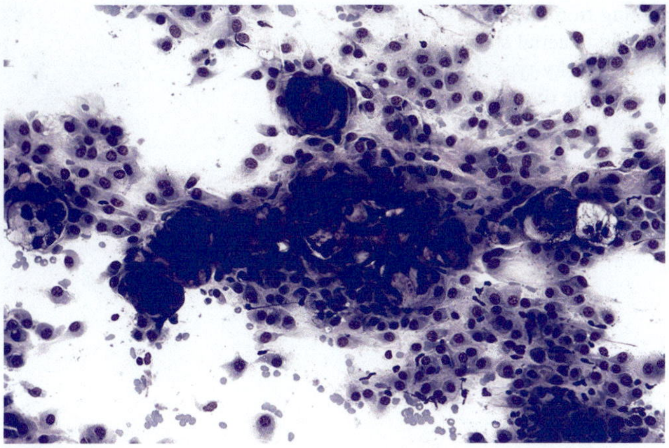

Fig. 8.569 PWT cytology: whorl arrangement. Many whorls are recognizable with cytology also

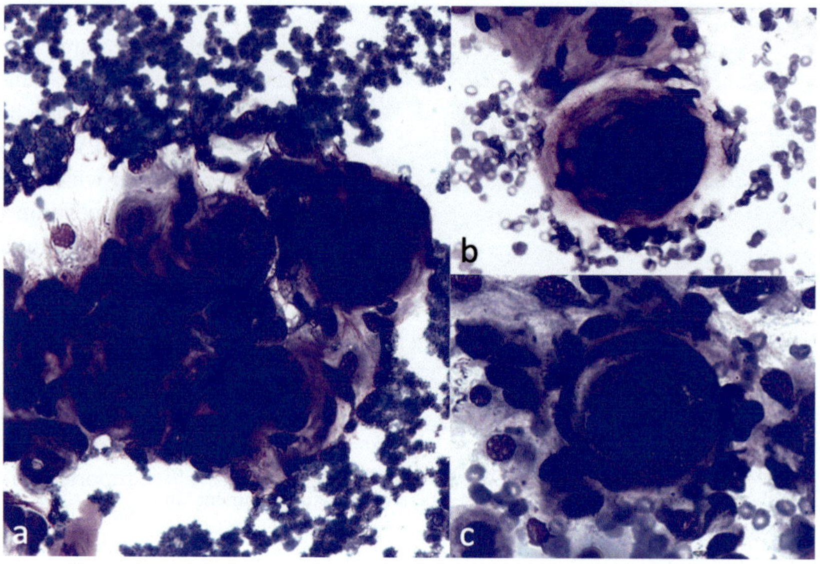

Fig. 8.570 PWT cytology: whorl arrangement. (**a**) at higher magnification whorls are recognizable even if cell details are not so clear; (**b**) and (**c**) close-up of cytological whorling arrangement

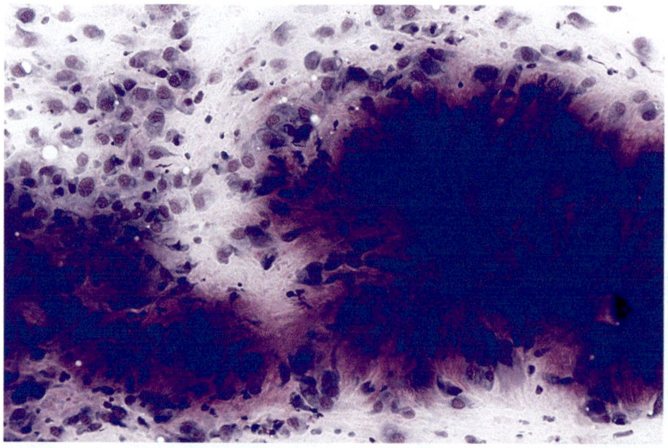

Fig. 8.571 PWT cytology: myxoid differentiation is characterized by a pinkish amorphous material on the background in which cells are embedded

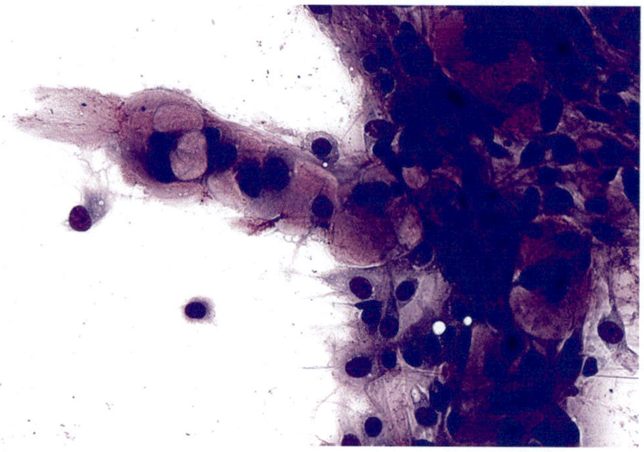

Fig. 8.572 PWT cytology: large intracytoplasmic globular formations represent the mucin produced by neoplastic cells

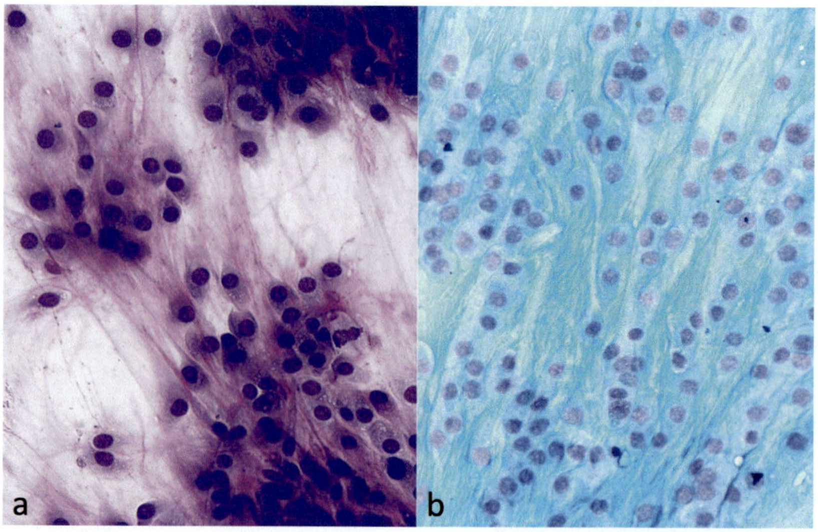

Fig. 8.573 PWT cytology: cells coming from the same sample—myxoid material stained with (**a**) Romanowsky-type dye and (**b**) Alcian blue dye marks pale azul the mucin

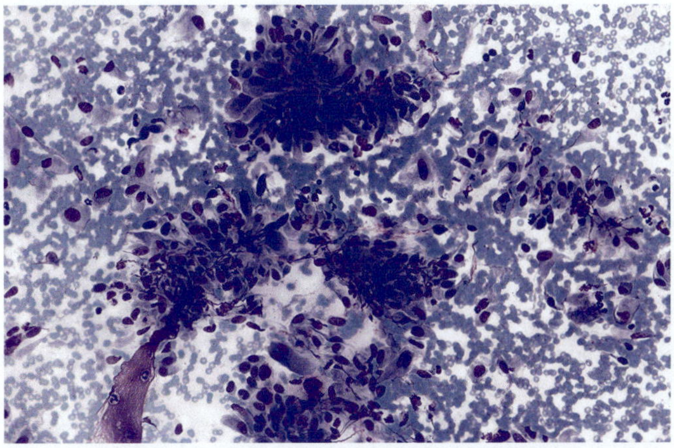

Fig. 8.574 PWT cytology. Cartwheel arrangements of poorly differentiated PWT cells

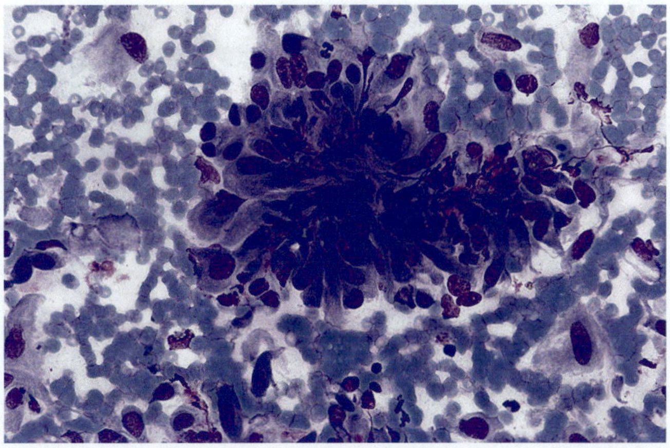

Fig. 8.575 PWT cytology. Cartwheel arrangements: the cells appear to grow in a centrifugal direction starting from a central area

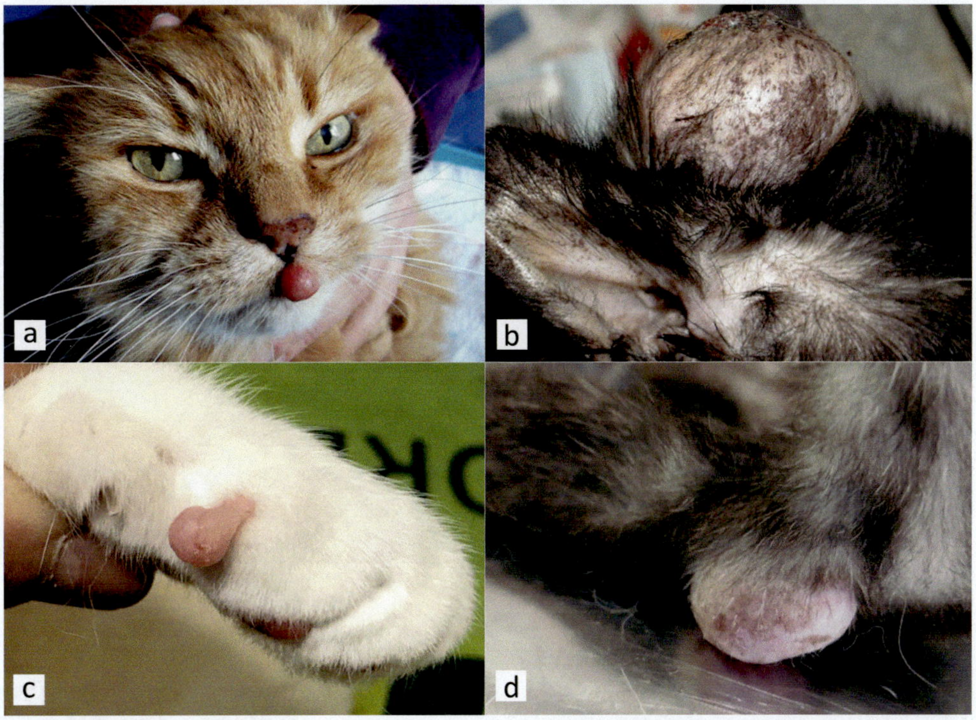

Fig. 8.576 Clinical presentation of NST in cats: (**a**) small alopecic nodule on the upper lip; (**b**) large alopecic and pedunculated nodule on the pinna of a cat; (**c**) small alopecic and pedunculated nodule on the digit; (**d**) alopecic nodule on the ventral surface of the tarsus

PWTs, the morphology can be so far from the abovementioned features that neoplasia cannot be cytologically differentiated from other undifferentiated STSs.

8.4.6 Nerve Sheath Tumors

Peripheral nerve sheath tumors (PNSTs) of the skin, previously known as *Schwannomas*, in the new WHO classification are named only *nerve sheath tumor* (NST) because the term peripheral is thought useless since only peripheral nerves are present in the skin [225, 226]. They are canine and feline tumors that originate from *Schwann cells* (nerve sheath) and *perineural cells* localized in the dermis and subcutis. They are classified into benign (NSTs) and malignant (MNSTs) [223–225, 226, 229]. The tumor is very rare in dogs and more frequent in cats; single nodules can arise anywhere, although they are more frequently observed on the head, pinna, neck, limbs, and digits [221, 222, 224, 227] (Figs. 8.576 and 8.577).

Cytological Findings

The NSTs are spindle cell tumors that may have histopathological characteristics in common with other mesenchymal neoplasms and in particular with PWTs; with which can share, for example, histological pattern such as *whorled* and

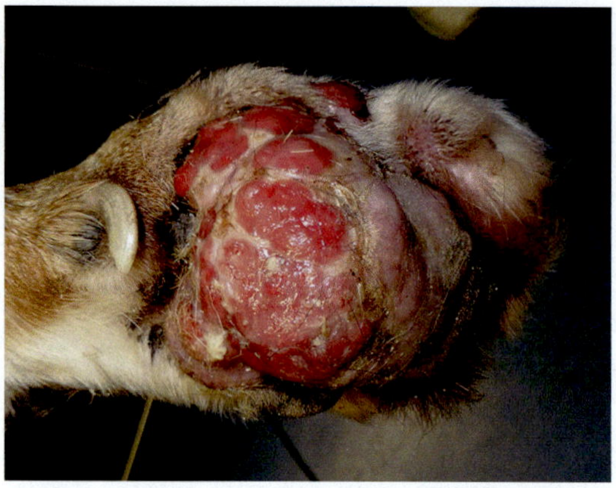

Fig. 8.577 Large and ulcerated malignant NST on the paw of a dog

finger print (Fig. 8.578). In NSTs, the whorls may be organized around a virtual area, or collagenous tissue is recognizable at their center; as discussed above, also in PWTs the whorls may share the same histopathological features. For this reason, to differentiate these two groups of neoplasms, a broad panel of immunohistochemical markers is mandatory, and the final diagnosis may be missed even after such immunolabelling [220]. In a work, the authors suggest that NGFR and Olig2 are the most useful immunohistochemical markers to distinguish NSTs and PWTs, since, in the latter, both markers seem to be frequently negative [229] (Fig. 8.579). In some NSTs, the *Antoni A* and *Antoni B* patterns, or the presence of *Verocay bodies*, orient the diagnosis to a NST [225, 226, 228] (Fig. 8.580).

In well-differentiated NSTs, cytology samples are from scarcely to highly cellular, and cells are spindle-shaped with

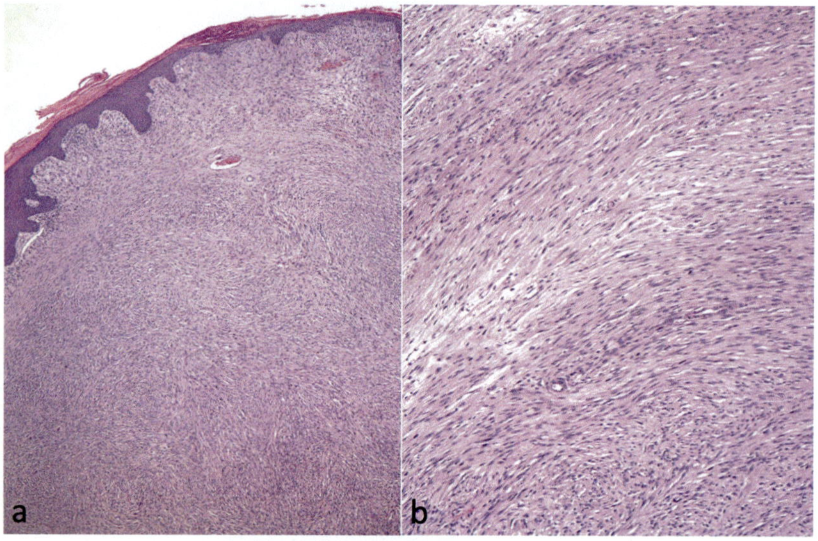

Fig. 8.578 Histology of NST: (**a**) dermal proliferation of spindle cells arranged in regular interwoven bundles; (**b**) parallel bundles are recognizable to higher magnification

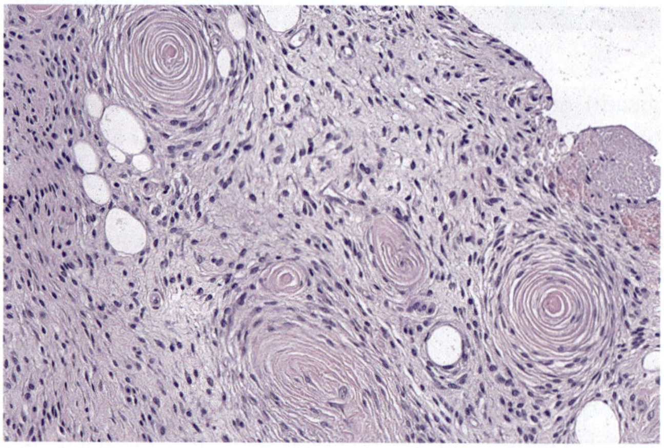

Fig. 8.579 Histology of NST: many whorls around collagen; this pattern can be also observed in other soft tissue neoplasia

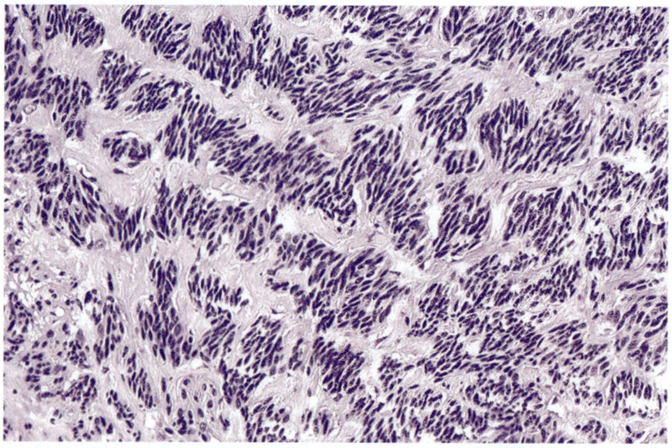

Fig. 8.580 Histology of NST: Antoni A pattern with many Verocay bodies characterize some nerve sheath tumors

oval or elongated nuclei and tailed cytoplasm. When cells are few on the slides, their morphology does not allow differentiation from cells belonging to another STN or from reactive fibroblasts.

As with other STNs, in samples from NSTs containing large groups of cells, it is possible to recognize bundles of spindle-shaped cells arranged in interwoven or parallel pattern; when cells are tightly packed, the spindle-shaped morphology of nerve cells can be better recognized only at the periphery of these groups (Figs. 8.581, 8.582 and 8.583). In many cases a large production of myxoid material is observed, creating difficulty to differentiate the NST with a myxoma and spindle cell lipoma.

In samples coming from NST histologically characterized of *Antoni A pattern*, the presence of spindle cells organized in pseudo-palisade resembling the *Verocay bodies* is strongly

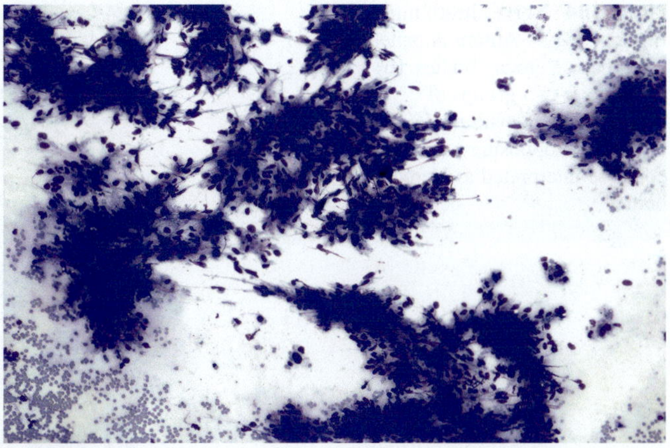

Fig. 8.583 Cytology of NST: storiform arrangement of neoplastic nerve sheath cells

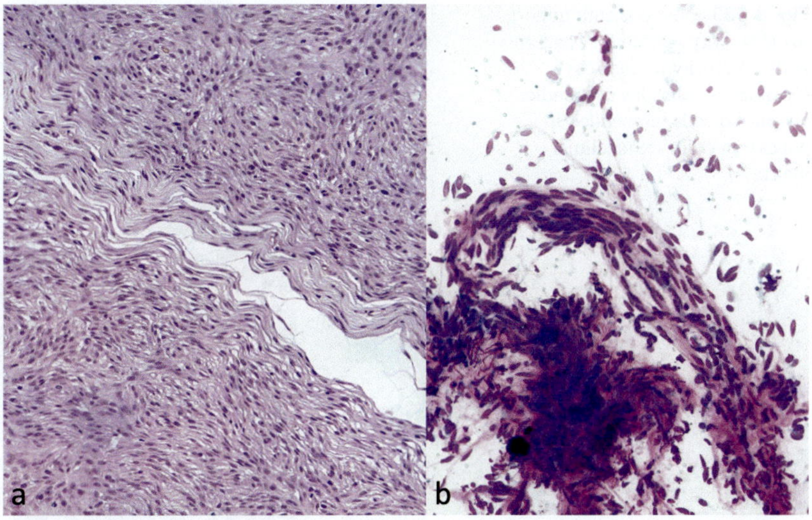

Fig. 8.581 Nerve sheath tumor: (**a**) Histology—parallel and undulated bundles of slender spindle neoplastic cells. (**b**) Cytology—a bundle of slender of spindle neoplastic cells is recognizable

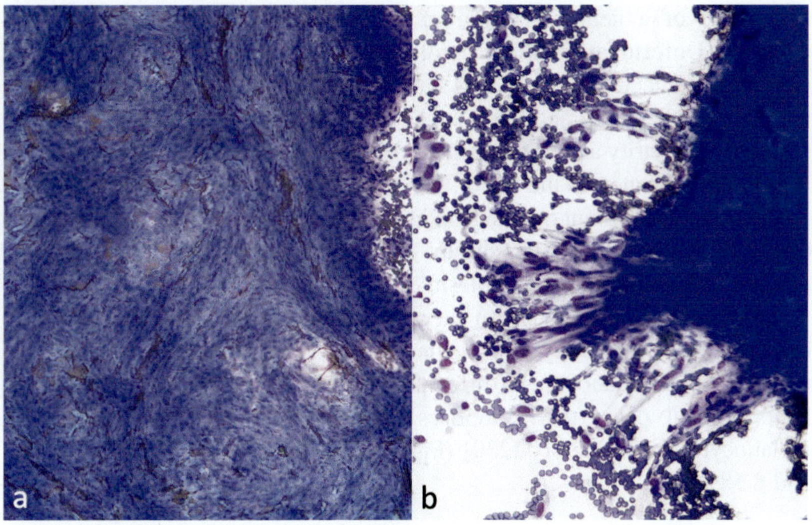

Fig. 8.582 Cytology of NST: (**a**) large group of neoplastic nerve sheath cells. Note as interwoven bundles of cells are barely recognizable; (**b**) at higher magnification spindle-shaped cells are clearly recognizable

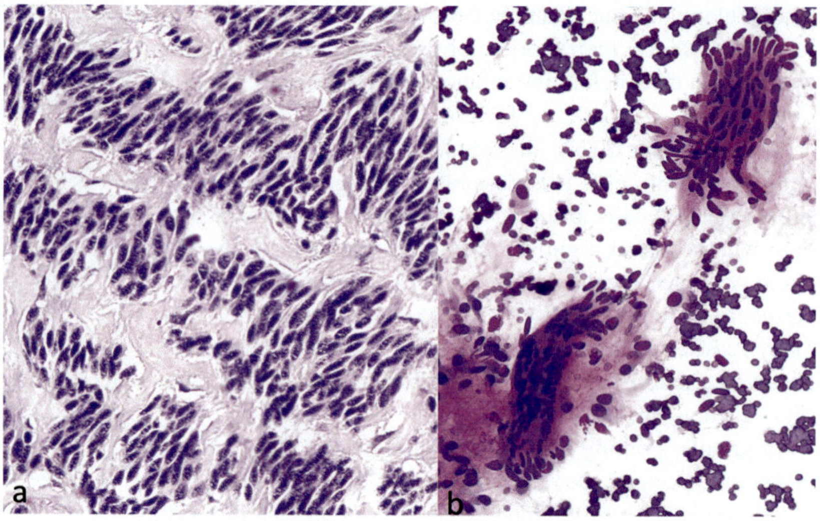

Fig. 8.584 Nerve sheath tumor: (**a**) Histology—Antoni A pattern with many Verocay bodies. (**b**) Cytology—two groups of neoplastic cells, with nuclei arranged in opposing palisades, could be interpreted as Verocay bodies

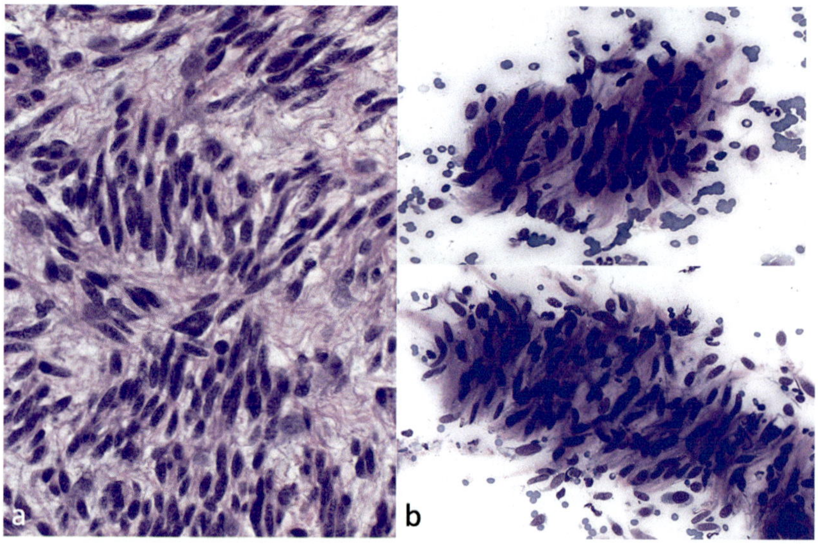

Fig. 8.585 Nerve sheath tumor: (**a**) Histology—Antoni A pattern with a Verocay bodies. (**b**) Cytology—palisades of spindle neoplastic cells, embedded in a pinkish in color extracellular matrix

suggestive of a neuronal origin of cells, as described in Chap. 7. Unfortunately, these cytological findings are not commonly detected in dogs and cats compared to human medicine (Figs. 8.584 and 8.585).

Hypercellularity and severe degrees of cellular atypia confirm the malignancy of the tumor (MNST).

A pigmented variant of this neoplasia has been reported in dogs, in which neoplastic sheath cells show a variable amount of melanin in the cytoplasm.

Since melanocytic cells can be spindle shape, all these cases are cytologically misdiagnosed as spindle cell melanocitoma/melanoma. To differentiate them, an immunohistochemical panel that comprises multiple markers for melanocytes is mandatory [230] (Figs. 8.586, 8.587, 8.588 and 8.589).

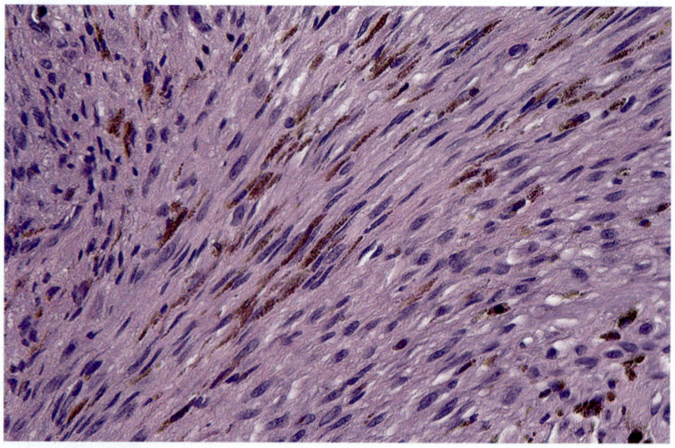

Fig. 8.586 Histology of pigmented NST: parallel bundles of pigmented nerve sheath cells. Note on the basis of morphology, cells cannot be differentiated from neoplastic spindle melanocytes

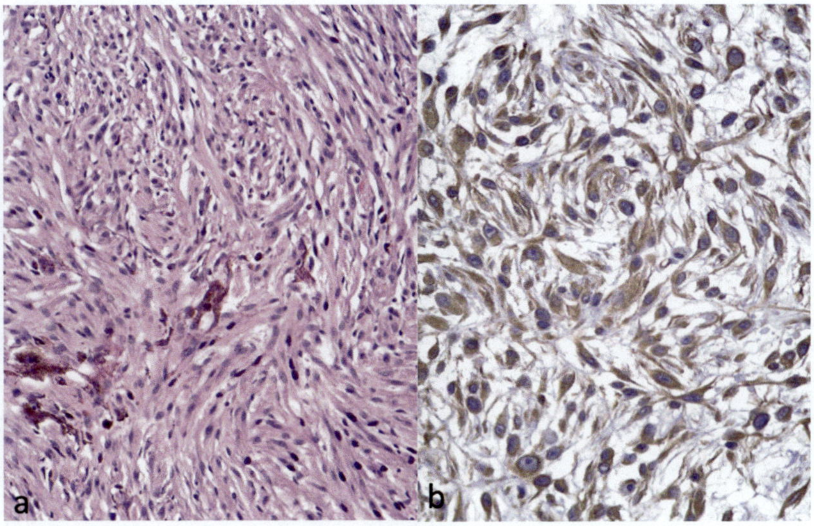

Fig. 8.587 Histology of pigmented NST: (**a**) interwoven bundles of nerve sheath neoplastic cells in some of which melanin pigment is clearly recognizable; (**b**) strong cytoplasmic immunoreactivity to GFAP confirms the nerve nature of the cells

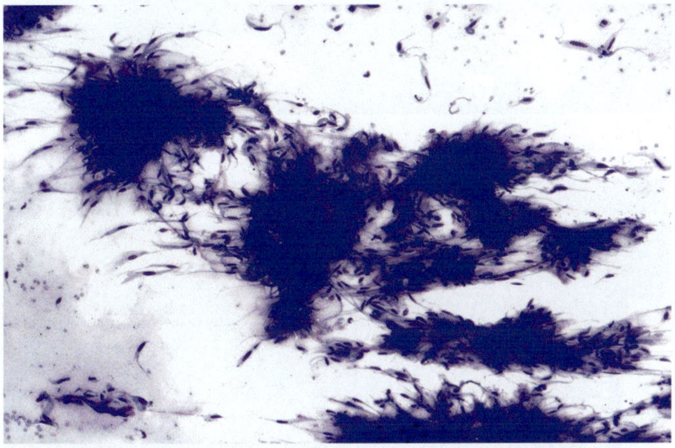

Fig. 8.588 Cytology of pigmented NST: spindle and slender nerve sheath cells with a storiform arrangement. At this magnification intracytoplasmic melanin pigment cannot be evaluable

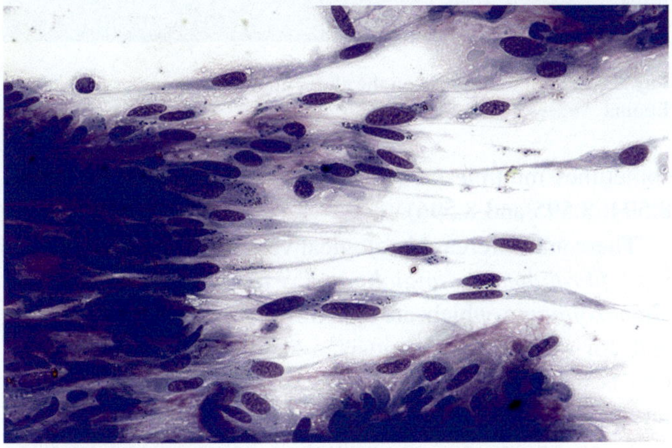

Fig. 8.589 Cytology of pigmented NST: at higher magnification intracytoplasmic melanin is appreciable. It is impossible with cytology to define that these cells are not melanocyte but nerve sheath cells

8.4.7 Lipoma, Hibernoma, and Liposarcoma

Lipoma is a benign tumor of the adipose tissue and is very common in dogs and rare in cats. Lipoma occurs with single or multiple nodules, ranging in size from a few centimeters up to voluminous masses, soft elastic in consistency, usually with well-defined and palpable margins, except for the *infiltrative lipoma*, in which the fat infiltrates the underlying striated muscles [235, 241]. Lipoma is mainly located on the trunk and on the proximal part of the limbs, especially in adult/elderly bitches and obese subjects (Fig. 8.590). *Hibernoma* is a rare benign tumor of brown adipose tissue; brown fat is responsible for producing heat and which takes its name from the fact that adipose cells have a greater number of mitochondria and greater vascularization. Of the few cases reported in veterinary literature, the most frequently reported sites are the retrobulbar and palpebral fat, comprising the third eyelid and conjunctiva [240, 242]. Occasional reports are reported in the subcutis [231]. *Liposarcomas* are less frequent than lipomas and are not derived from malignant transformation of the latter, but arise directly from malignant subcutaneous lipoblasts [233, 235, 241] (Fig. 8.591). The liposarcomas can be single or multiple and usually more firm on palpation than lipomas.

Cytological Findings

A first clue to the possible presence of a lipoma is provided by the macroscopic observation of colorless lipid droplets that remain on the slide after sampling; these droplets do not tend to dry as they dissolve once the slides are immersed in the alcohol fixative (Fig. 8.592). This feature should not lead to a macroscopic diagnosis of lipoma since normal fat can show the same macroscopic appearance. It should be emphasized that although a high number of blood vessels and the presence of perivascular cyto-architectures is

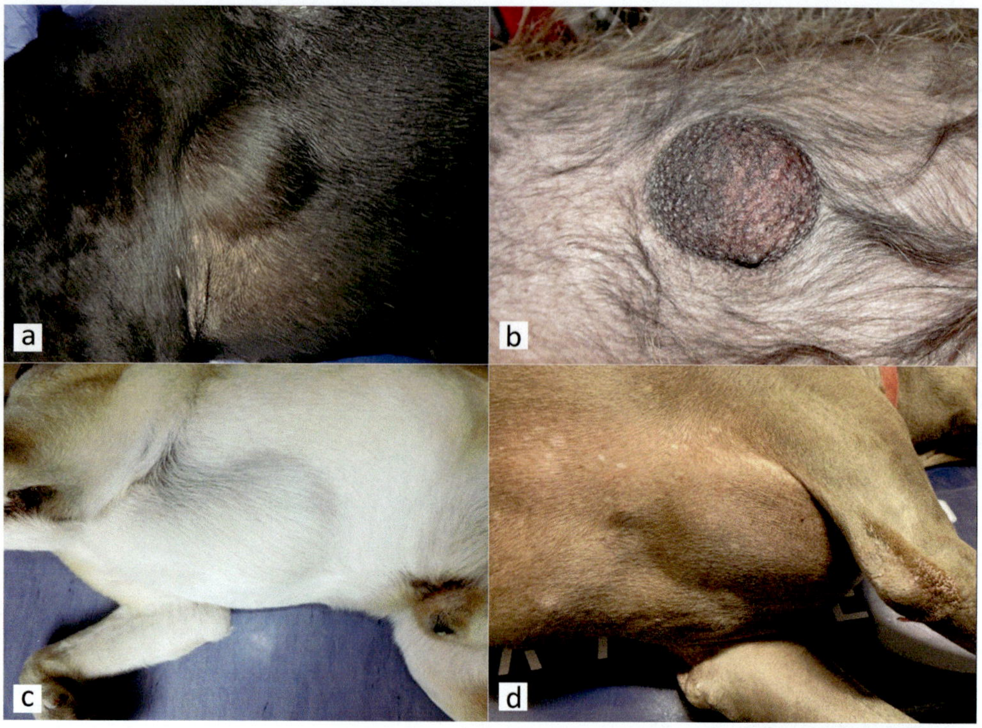

Fig. 8.590 Lipoma: (**a**) subcutaneous mass on the thorax of a dog; (**b**) large lipoma on the abdominal area in a dog; (**c**) large lipomatous mass on the sternal area; (**d**) large lipoma on the sternal area

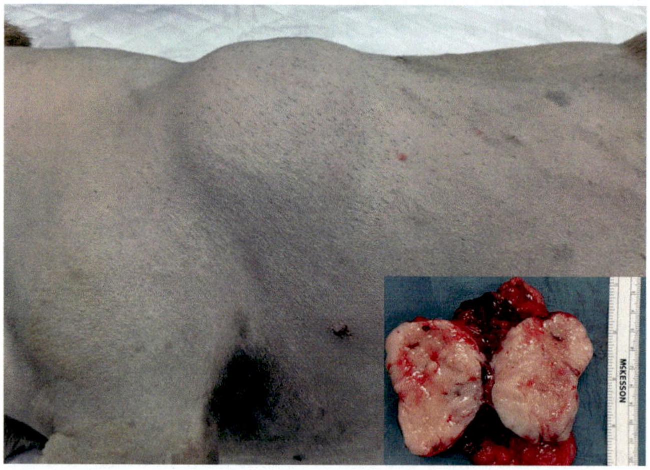

Fig. 8.591 Liposarcoma: large subcutaneous mass on the thorax of a dog. Typical macroscopic aspect of adipose tissue neoplasms is recognizable in the insert. (Courtesy of DVM M. Annoni—Italy)

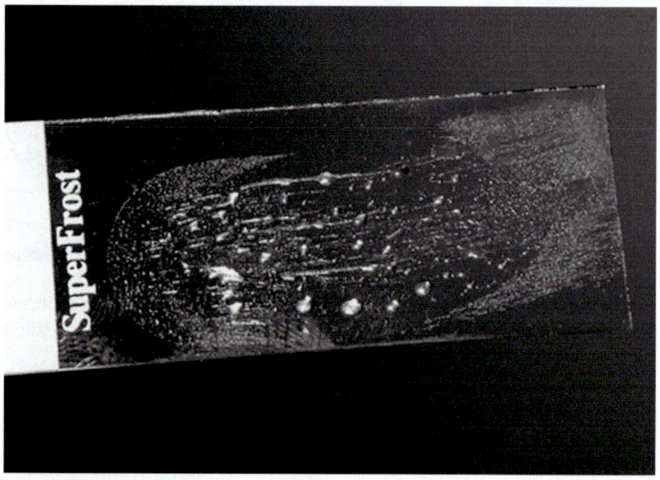

Fig. 8.592 Lipid droplets on the slides after squashing the content of a lipoma

commonly observed in lipomas, it is not possible to formulate a cytological diagnosis of lipoma since it is not easy to differentiate normal adipocytes from neoplastic lipomatous tissue.

In any case, the cytology of lipoma is characterized by large fat-filled cells, characterized by a marked anisocytosis; the cells are arranged singly or more often in groups, with pyknotic and hyperchromatic nuclei, from round to oval and localized at the periphery of the cells. Since the intracellular fat does not stain with the Romanowsky-type dye, the cytoplasm takes on an optically empty appearance, and sometimes multiple fat drops are recognizable (Figs. 8.593, 8.594, 8.595 and 8.596).

There are different histological variants of lipoma, including *fibrolipoma*, *angiolipoma*, *angiofibrolipoma*, and *chondrolipoma* which are not differentiable with cytology, and, for a definitive diagnosis, a histopathological examination is mandatory [241]. A rare *spindle cell lipoma* is also reported in dogs [232] (Fig. 8.597).

Some lipoma has an infiltrating behavior and tends to spread between the muscle fibers. In these cases, along with

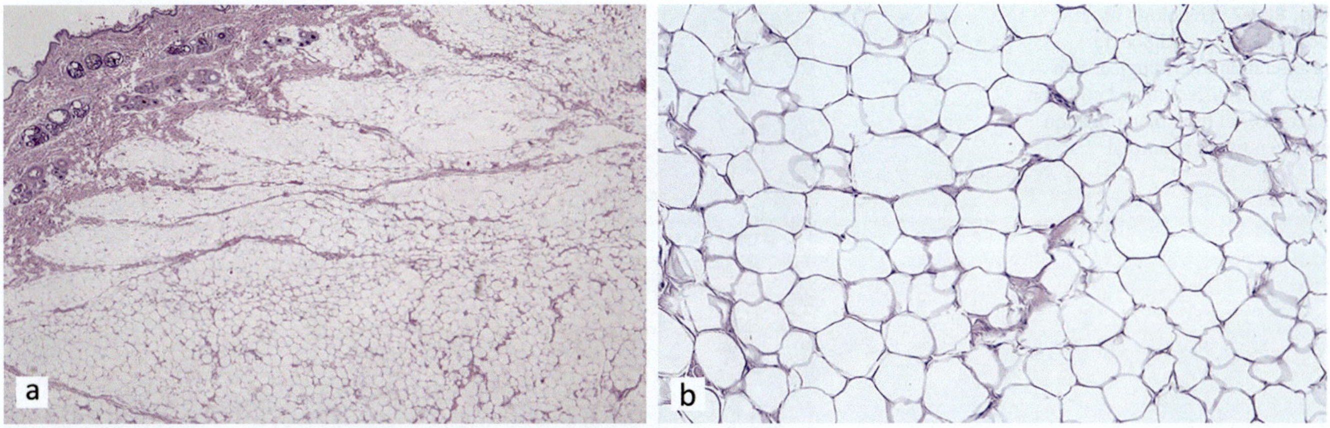

Fig. 8.593 Histology of lipoma: (**a**) groups of neoplastic benign adipocytes separated by thin fibrous stroma; (**b**) at high magnification, optically empty cells characterize lipoma. Nuclei are not always detectable since they are very small

Fig. 8.594 Cytology of lipoma: (**a**) large aggregate of adipocytes. Cells are large with small nuclei and optically empty cytoplasm; (**b**) at higher magnification the small, oval, and usually marginalized nuclei are recognizable

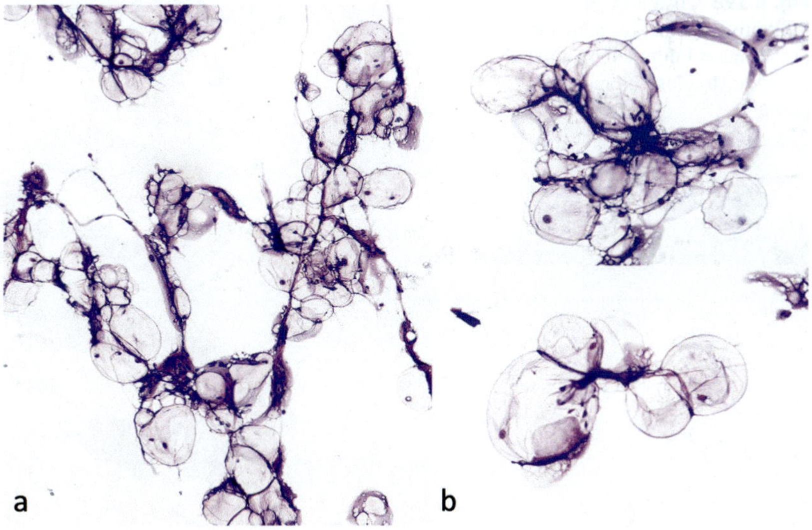

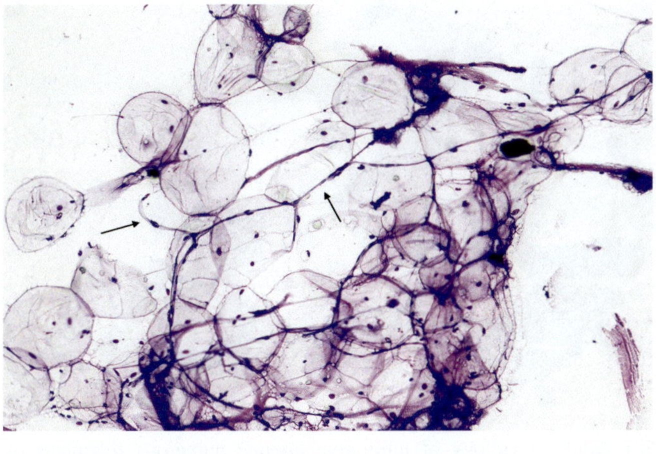

Fig. 8.595 Cytology of lipoma: large aggregate of adipocytes. Note the presence of hematic vessels (arrow)

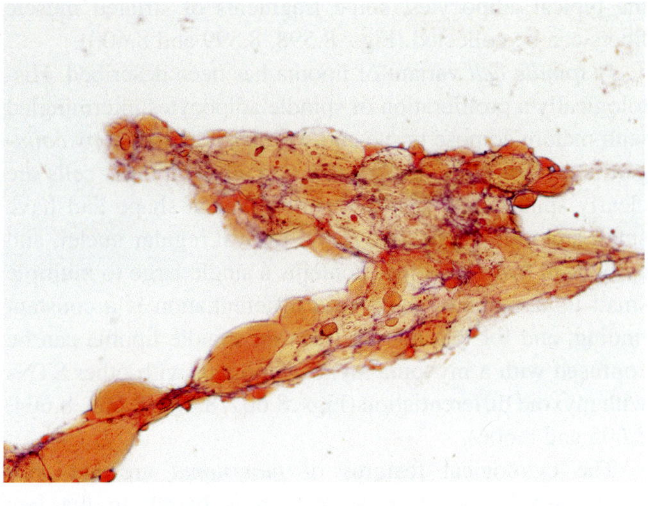

Fig. 8.596 Cytology of lipoma: with oil red O, the lipid content of adipocytes stains orange-red

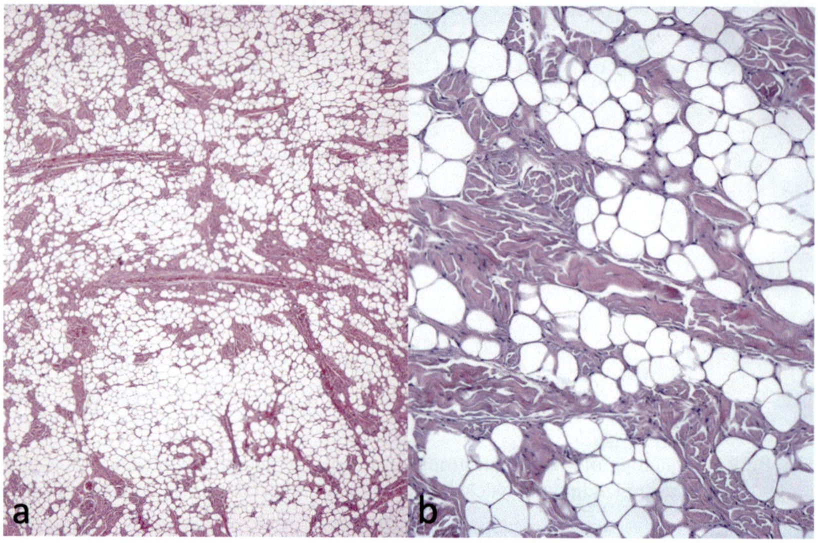

Fig. 8.597 Histology of fibrolipoma: (**a**) adipocyte proliferation intermingled with many well-differentiated collagen fibers; (**b**) at high magnification collagen tissue is normal

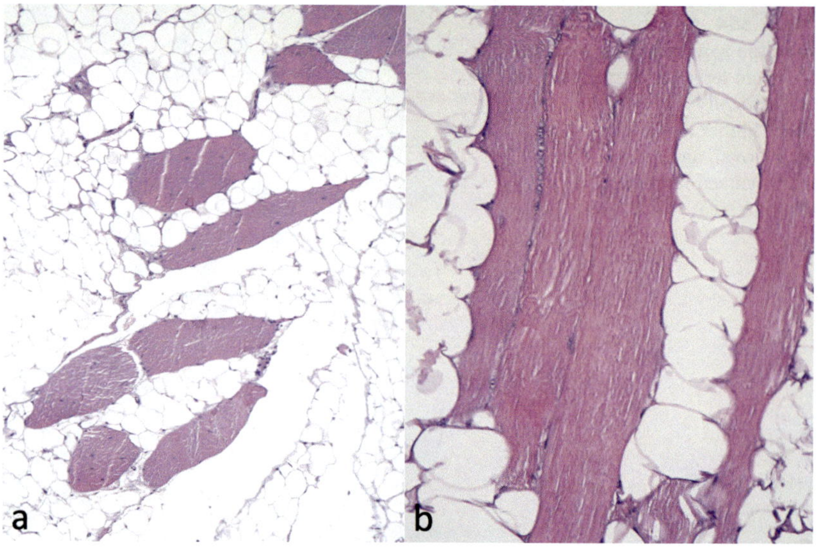

Fig. 8.598 Histology of infiltrative lipoma: (**a**) adipocyte proliferation infiltrated by striated muscles; (**b**) close-up muscular tissue

the typical adipocytes, some fragments of striated muscle fibers can be collected (Figs. 8.598, 8.599 and 8.600).

A *spindle cell* variant of lipoma has been described. Histologically a proliferation of spindle adipocytes intermingled with mature adipose tissue and with a so defined *ropy collagen* is recognizable. In this rare variant of lipoma, cells are clearly spindle and less frequently oval in shape and have small and uniform in size, round to oval regular nuclei, and cytoplasm that sometimes contains a single large to multiple small lipid droplets. A myxoid differentiation is a constant finding, and for this reason, this rare spindle lipoma can be confused with a myxoma/myxosarcoma or with other STNs with myxoid differentiation (Figs. 8.601, 8.602, 8.603, 8.604, 8.605 and 8.606).

The cytological features of *hibernoma* are not well characterized, and their description is based on the few published clinical cases [239]. Histology shows large adipose

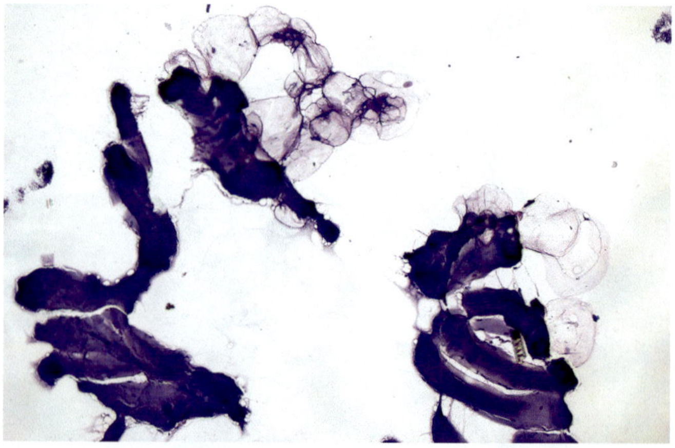

Fig. 8.599 Cytology of infiltrative lipoma: numerous fragments of striated muscle tissue with small, oval, and uniform in size nuclei along with mature adipocytes

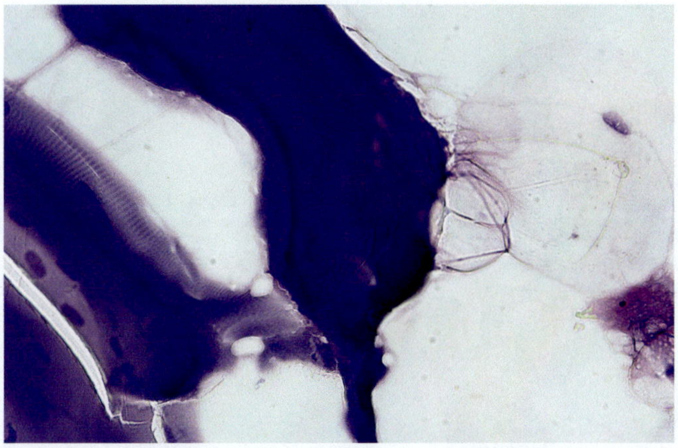

Fig. 8.600 Cytology of infiltrative lipoma: at high magnification the striations of the muscle fibers and the adipocyte with an oval nucleus displaced at the cell periphery are recognizable

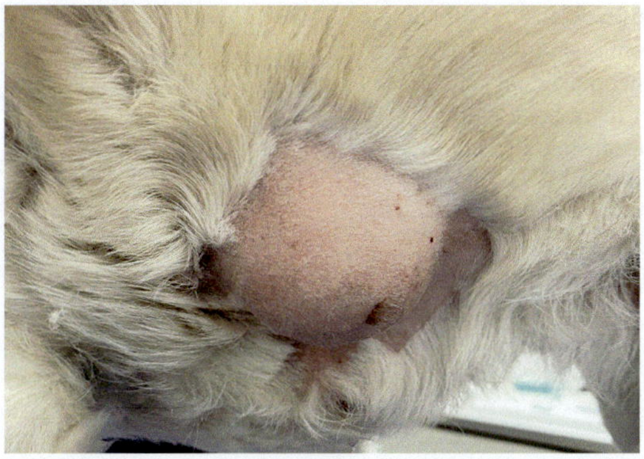

Fig. 8.601 Large mass on the lateral thorax of a Golden Retriever with a spindle cell lipoma

Fig. 8.602 Histology of spindle cell lipoma: (**a**) H-E stain—spindle and uniform in size cell adipocyte on a myxoid background. (**b**) Alcian blue stain—the myxoid background assumes a pale azul color. The collagen fibers are haphazardly arranged assuming a morphological aspect defined as "Ropey collagen"

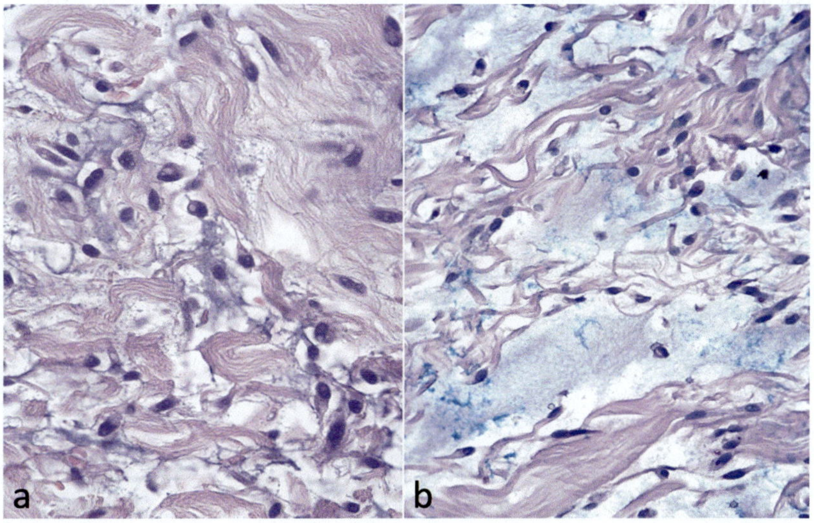

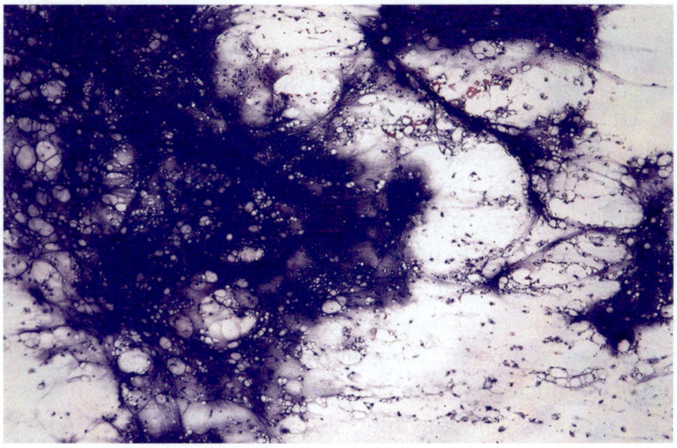

Fig. 8.603 Cytology of spindle cell lipoma: at low magnification, many small spindle nuclei are barely perceptible. Note the characteristic both myxoid and lipidic background

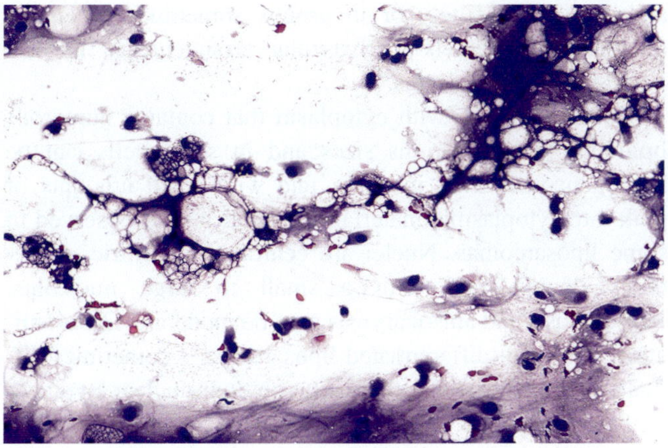

Fig. 8.604 Cytology of spindle cell lipoma: along with lipid droplets, two intact adipocytes, and a violet myxoid background, many spindle and less frequently plump cells are recognizable

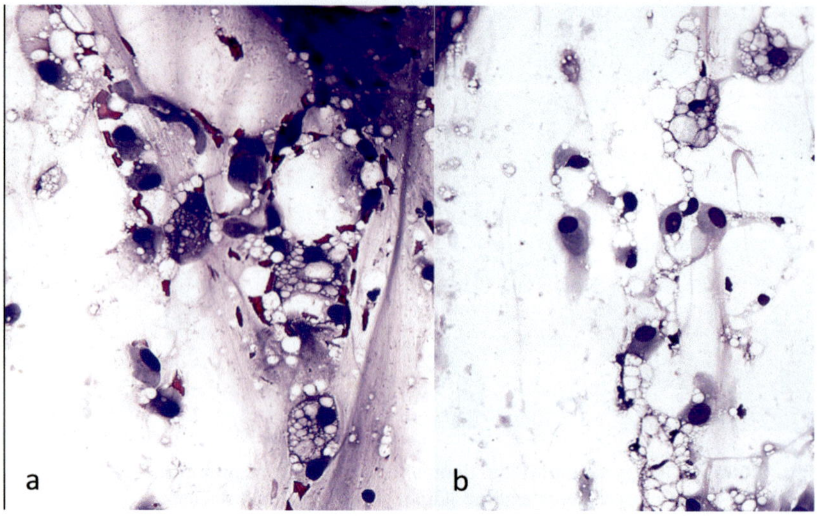

Fig. 8.605 Cytology of spindle cell lipoma: spindle to oval and plump neoplastic fat cells. Note as some of them have many intracytoplasmic lipid droplets

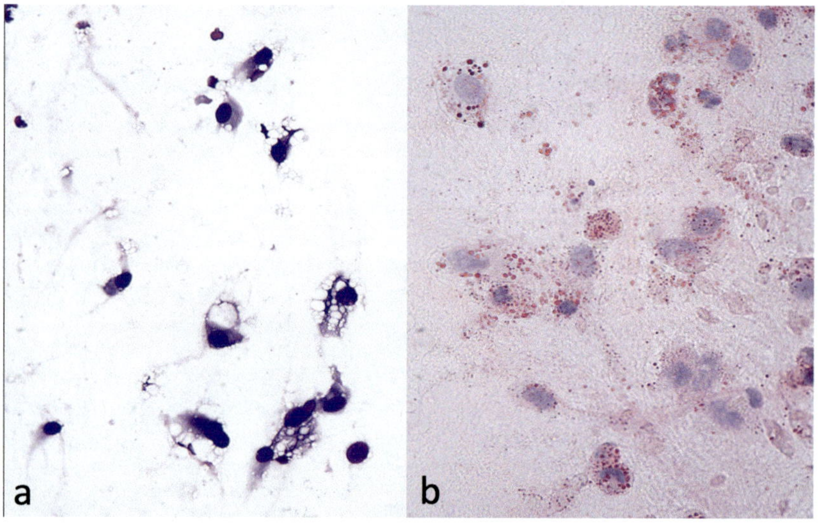

Fig. 8.606 Cytology of spindle cell lipoma: (**a**) spindle to oval and plump neoplastic fat cells with clear intracytoplasmic lipid droplets; (**b**) lipid material can be demonstrated using oil red O stain. Fat is stained orange and it is clearly intracytoplasmic

cells with many lipid droplets. The cytological slides are highly cellular, composed of voluminous round cells arranged either singly or in groups sometimes organized around blood vessels (perivascular arrangement), like all adipose tissue neoplasms.

Cells are round, with cytoplasm that contains numerous lipid droplets of various sizes and in some cells can be homogeneous, violet in color, and with small amounts of pink intracytoplasmic material; the latter is also observed in some liposarcomas. Nuclei are central to eccentric, sometimes double, and with a small to large nucleolus. Anisocytosis and anisokaryosis may be moderate. Differentiation from well-differentiated liposarcoma is sometimes difficult, but in hibernomas there is no obvious nuclear atypia, or if there is, it is occasional; furthermore, other cytological features that characterize liposarcoma should not be present, i.e., nuclei pressed against the cell wall and deformed by the presence of abundant intracytoplasmic lipid material. Unfortunately no specific immunohistochemistry markers are available for their differentiation also on histopathological specimens; in fact, the brown fat marker *Uncoupling Protein 1* (UCP1) is also positive for liposarcoma (Figs. 8.607, 8.608, 8.609, 8.610, 8.611 and 8.612).

The cytology of liposarcoma can be very polymorphic depending on the histological variant and on the degree of differentiation of the neoplastic cells. Histological variants include *well-differentiated*, *dedifferentiated*, *myxoid*, and *pleomorphic* liposarcoma (Figs. 8.613 and 8.614).

In well-differentiated liposarcoma, the background is constantly lipidic and characterized by overlapped droplets of various sizes. In some cases the nuclei are scarcely visible or only bare nuclei can be detected; care must be done in these cases, since beginner cytologists can erroneously diagnose a panniculitis. Neoplastic cells are roundish in shape, and, in most cases of well-differentiated liposarcoma, they are released individually; in highly cellular specimens, cells are

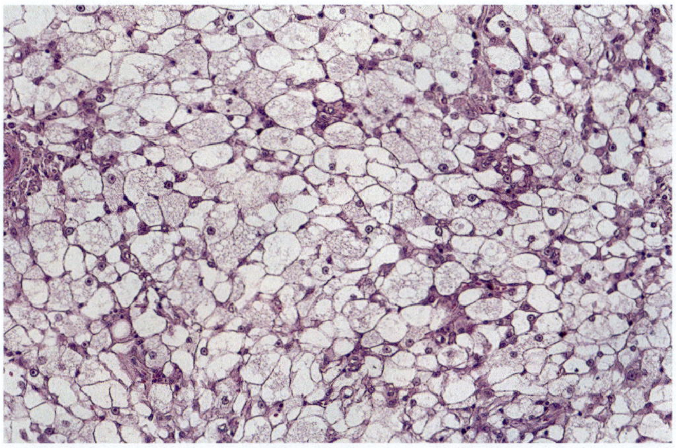

Fig. 8.607 Histology of hibernoma: adipose cells, with large cytoplasm filled with lipid droplets and with central nuclei

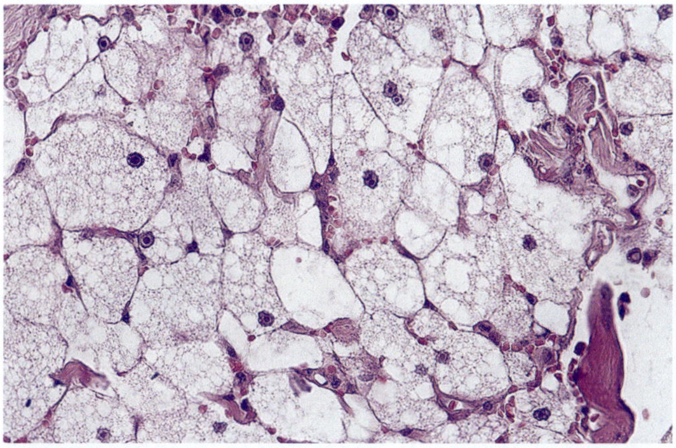

Fig. 8.608 Histology of hibernoma: at higher magnification, cells show central nuclei with single and often evident nucleolus. Note as no cytological features of lateralization and pressure of the nuclei on the cell wall are present

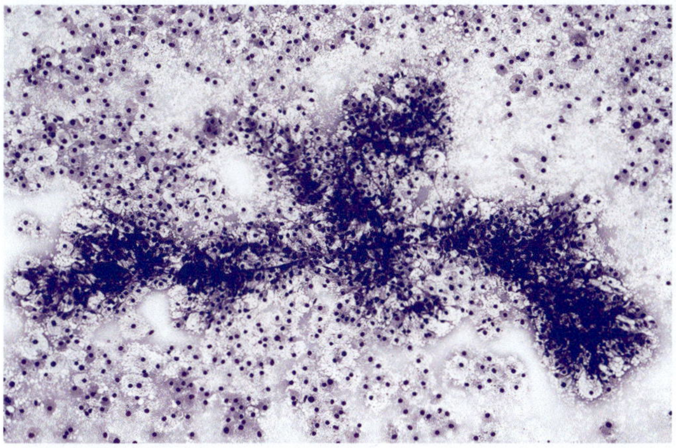

Fig. 8.609 Cytology of hibernoma: a perivascular arrangement of many adipose cells

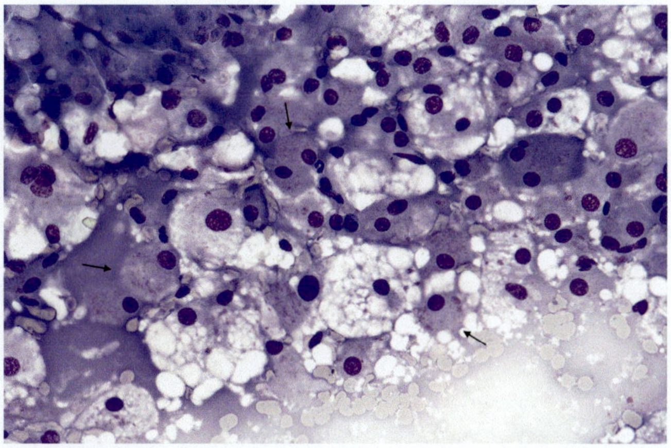

Fig. 8.610 Cytology of hibernoma: adipose cells both with intracytoplasmic lipid droplets and with no lipogenic and homogeneous cytoplasm in which small violet material is recognizable (arrows); with this cytology it is very hard to differentiate a hibermoma from a well-differentiated liposarcoma

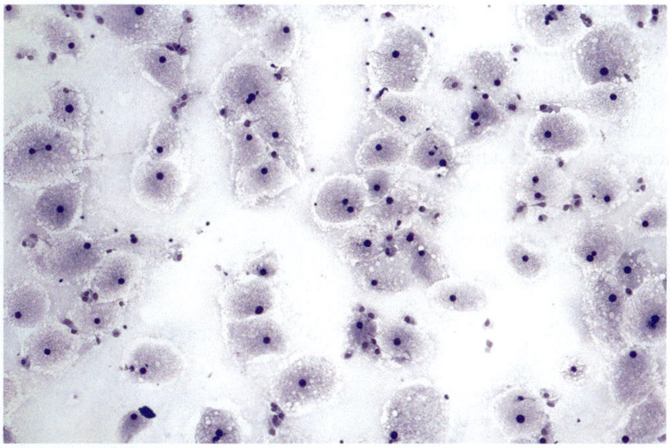

Fig. 8.611 Cytology of hibernoma: adipose cells with large cytoplasm and roundish, regular in size, centrally located nuclei

released in overlapped groups with round discrete cells that can be observed only at their periphery. Perivascular arrangement is frequently observed. The nuclei are round or oval, with variable degrees of anisokaryosis, while the nucleoli are variably perceptible. The cytoplasm is medium-large and rich in achromatic lipid droplets of various sizes. In not fully lipidized cells, cytoplasm can be homogeneous and violet in color and can show amorphous pinkish material (Figs. 8.615 and 8.616).

An aspect that cytologically must always be searched in order not to misdiagnose a liposarcoma with a neoplasm of non-lipogenic nature that has simply infiltrated the subcutaneous tissue is the presence of lipid material in the cytoplasm of the neoplastic cells in confirming their adipocytic nature and, more importantly, the features of peripheral marginalization of the nuclei and changes in the nuclear silhouette due to the pressure of the intracytoplasmic fat; when the latter

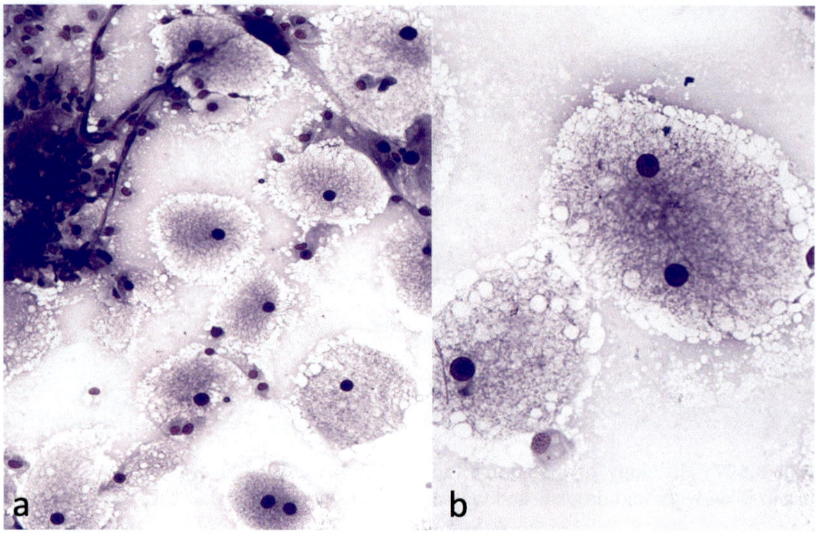

Fig. 8.612 Cytology of hibernoma: (**a**) adipose cells with large cytoplasm and roundish, regular in size, centrally located nuclei. Perivascular arrangement is recognizable; (**b**) at high magnification multiple fat droplets are recognizable into the cytoplasm

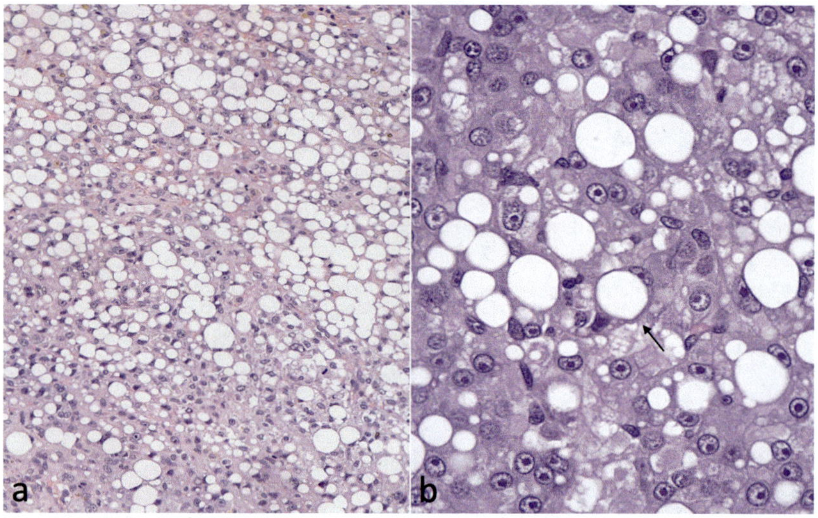

Fig. 8.613 Histology of well-differentiated liposarcoma: (**a**) neoplastic adipocytes with clear adipose differentiation; (**b**) at high magnification the adipose nature of the neoplastic cells is clearly recognizable. Note the signet ring appearance of the adipocyte due to the lipid content pressing the nucleus onto the cell wall. This feature is a hallmark of liposarcoma

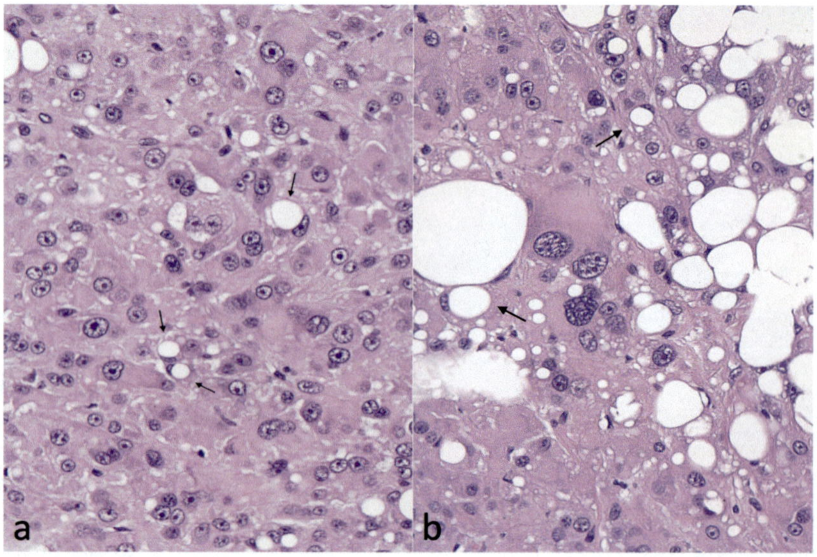

Fig. 8.614 Histology of liposarcoma: (**a**) dedifferentiated adipose neoplasm in which some typical characteristics of the fatty origin of the cells are evident (arrows); (**b**) pleomorphic liposarcoma—severe cytological atypia such as anisokaryosis and macrokaryosis. Some characteristics of fatty differentiation are still recognizable (arrows)

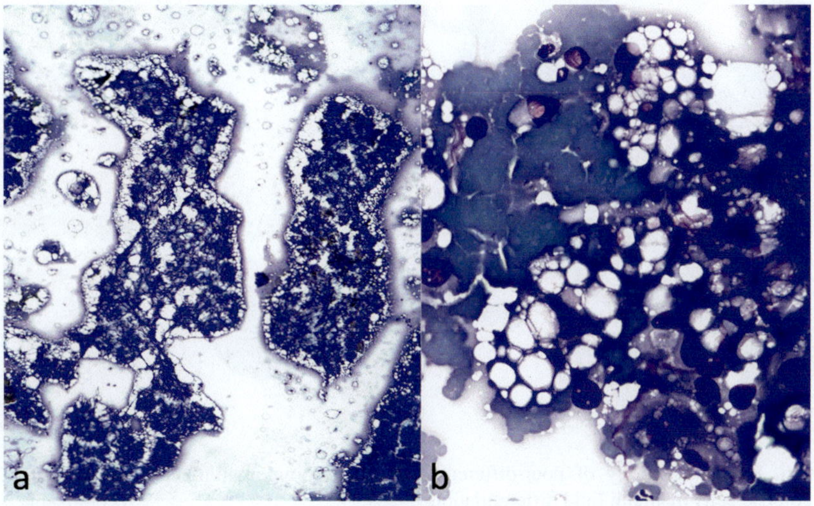

Fig. 8.615 Cytology of well-differentiated liposarcoma: (**a**) at low magnification large groups of cells are observed on a lipid background; (**b**) at higher magnification multiple overlapped droplets of various sizes are recognizable as indicator of an adipose origin of cells

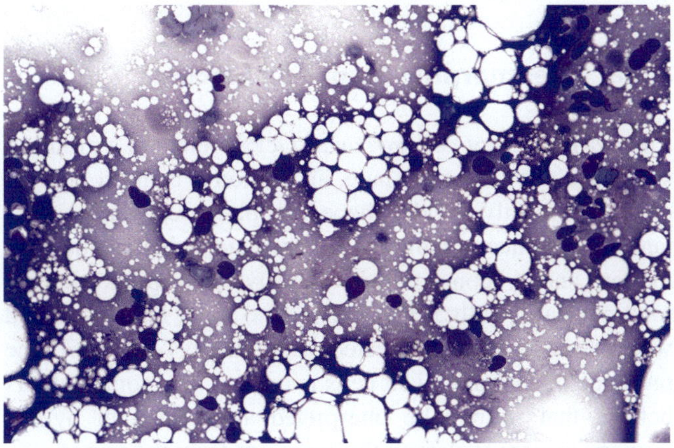

Fig. 8.616 Cytology of well-differentiated liposarcoma: on a protein and lipid background, some naked nuclei are present. In these cases, care must be taken to misdiagnose panniculitis

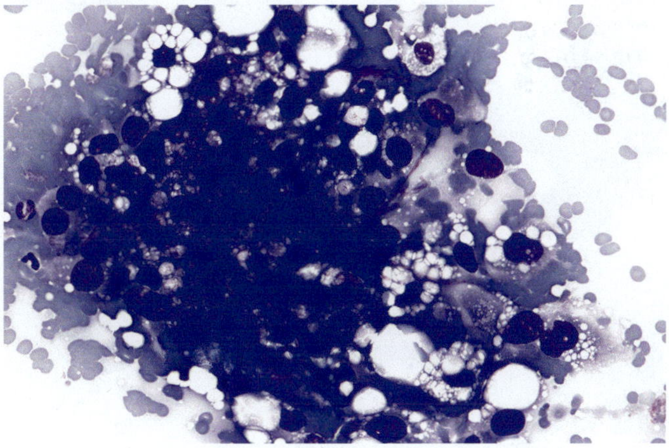

Fig. 8.617 Cytology of well-differentiated liposarcoma: at higher magnification, multiple overlapped and of various sizes lipid droplets are recognizable as indicator of adipose origin of neoplastic cells

feature occurs, cells assume a "signet-ring" appearance (Figs. 8.617 and 8.618).

In samples from dedifferentiated liposarcomas, the morphological atypia of cells depends on the grade of malignancy; fortunately also in less differentiated neoplasia, and especially if the slides are highly cellular, some cells can maintain an adipocytic differentiation (Figs. 8.619 and 8.620).

In the *pleomorphic* variant of liposarcoma, marked atypia and many multinucleated cells can be observed. In these cases, if not intralipid droplets are clearly evident, it is not possible to differentiate a liposarcoma from other anaplastic soft tissue sarcomas with giant cells or melanoma. In general, if the lipid content of cytoplasmic would be confirmed, the special stain oil red O can be used; with this dye, lipids are highlighted with a red–orange color [237]. This staining may be superfluous if lipids are clearly recognizable in the cytoplasm but could be of great diagnostic utility in the case of non-lipogenic liposarcomas (Figs. 8.621 and 8.622).

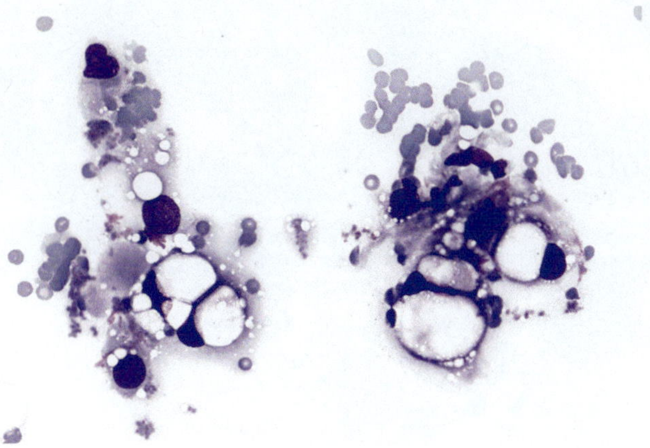

Fig. 8.618 Cytology of well-differentiated liposarcoma: a single large lipid droplet can cause the nucleus to press against the cell wall and create a signet-ring morphology

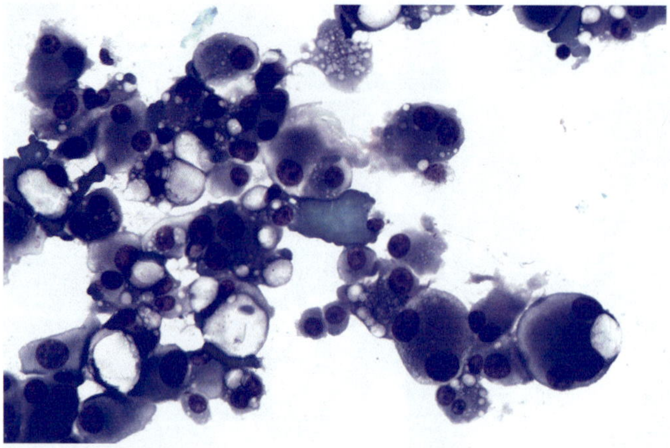

Fig. 8.619 Cytology of poor-differentiated liposarcoma: malignant adipose cells maintain lipid differentiation

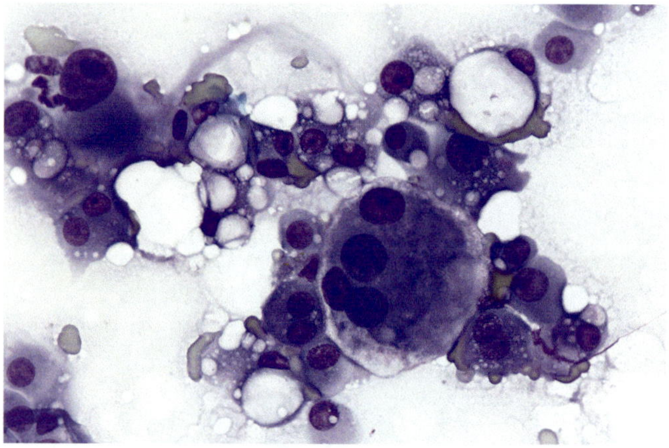

Fig. 8.620 Cytology of pleomorphic liposarcoma: malignant adipose cells maintain lipid differentiation

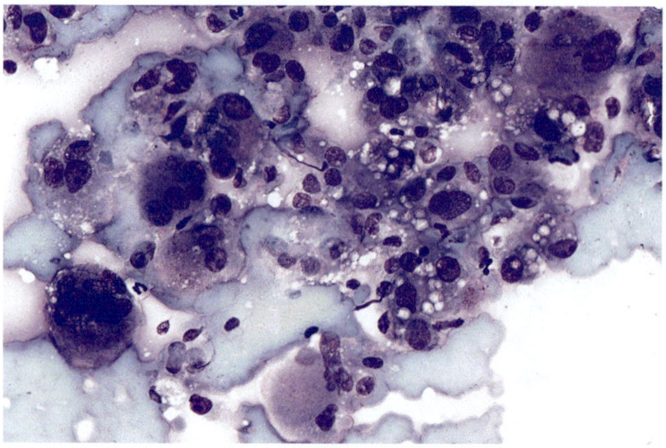

Fig. 8.621 Cytology of pleomorphic liposarcoma: only few lipid droplets are present in the cytoplasm. In these cases, the differentiation with other "vacuolated" sarcomas or melanoma cannot be done with cytology

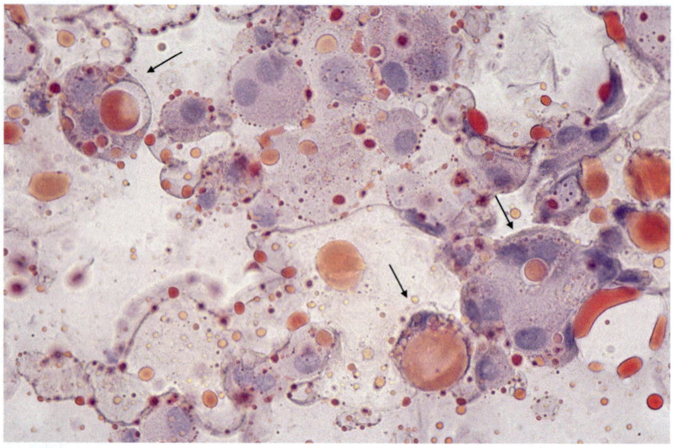

Fig. 8.622 Cytology of pleomorphic liposarcoma: lipid material is recognizable into the cytoplasm of some neoplastic adipocytes (arrows)

8.4.8 Pleomorphic Giant Cell Sarcoma

Pleomorphic giant cell sarcoma (GcPS) refers to a group of soft tissue sarcomas that share, in addition to being composed of predominantly spindle-shaped neoplastic cells, the simultaneous presence of a mononuclear roundish histiocytic population and a constant and usually high proportion of multinucleated cells (giant cells), also them histiocytic and which represent multinucleated macrophages [244–246]. The presence of so many giant cells is unclear, and some authors believe that it may be interpreted as an attempt by the immune system to defend the organism from the tumor.

Previously classified as *malignant fibrous histiocytoma* or *anaplastic giant cell sarcoma*, this type of mesenchymal STSs has recently been grouped under the umbrella of *pleomorphic giant cell sarcomas*. As above, the tumors share the fact that they are composed of cells characterized by very different morphologies, justifying the term "pleomorphic."

The cell origin of these tumors is unknown, and even if they may originate from many different cell lineages, according to immunohistochemical studies conducted in cats, the species most affected by these tumors, spindle cells appear to belong to fibroblasts/myofibroblasts [244]. Furthermore, since in cats this tumor is observed predominantly in the interscapular space, flank, and chest, basically the areas where vaccinations are performed, it has been thought that these tumors may be, at least in most cases, a histological variant of FISS.

These neoplasms, uncommon in dogs and more frequent in cats, are clinically characterized by single and occasionally multiple nodules, of variable size and irregular in shape, mainly located on the trunk and limbs, but can be rarely observed also in other body areas (Figs. 8.623 and 8.624).

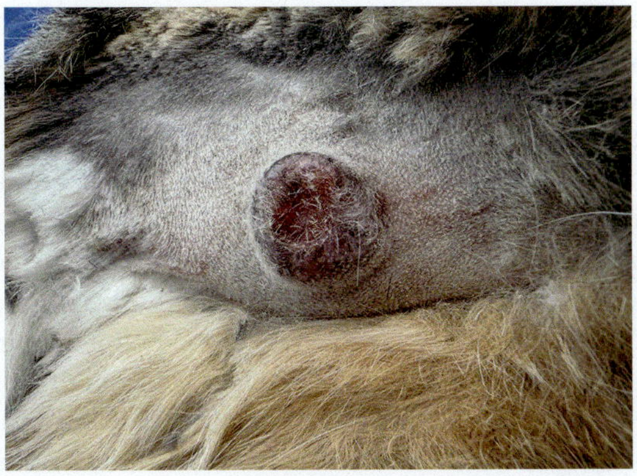

Fig. 8.623 Giant cell pleomorphic sarcoma: ulcerated sarcoma on the sternal region of a cat

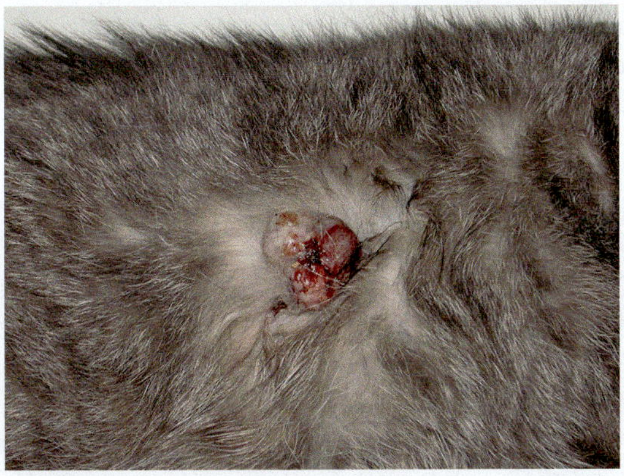

Fig. 8.624 Giant cell pleomorphic sarcoma: ulcerated sarcoma on the thorax of a cat

Cytological Findings

As previously mentioned, the cytology of this group of neoplasms is usually characterized by highly cellularized slides, whose cells are characterized by a marked polymorphism and pleomorphism, and composed of a prevalent number of spindle cells, a variable number of round/oval macrophage histiocytes, and many multinucleated giant cells, also of macrophage origin (Figs. 8.625 and 8.626). In these pleomorphic sarcomas, in fact, the multinucleated cells are not of neoplastic origin, but are histiocytic (macrophage giant cells), and according to some researchers, they would indicate an immune-mediated reaction by the immune system against the tumor. From a morphological point of view, the multinucleated cells are characterized by a very high number of nuclei and above all by the fact that the size and shape of the same are uniform and regular and do not show atypia (Figs. 8.627 and 8.628).

It must be emphasized that in STSs of any origin (fibroblasts, muscle or nerve cells, myxosarcomas, etc.), it is not uncommon to observe multinucleated malignant neoplastic cells. In those cases, however, the characteristics of

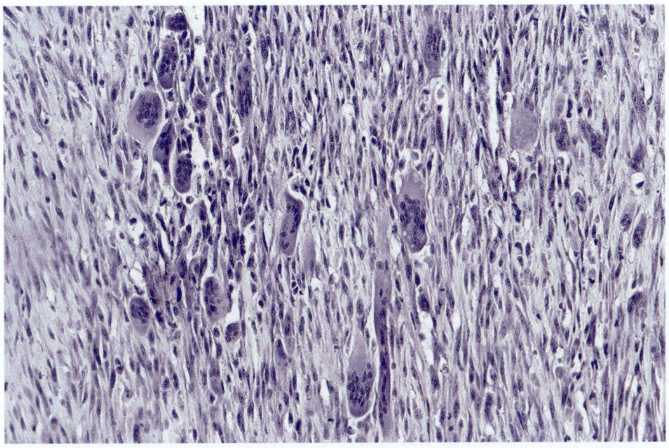

Fig. 8.625 Histology of GcPS: mixed population of cells with most spindle cells and many multinucleated cells

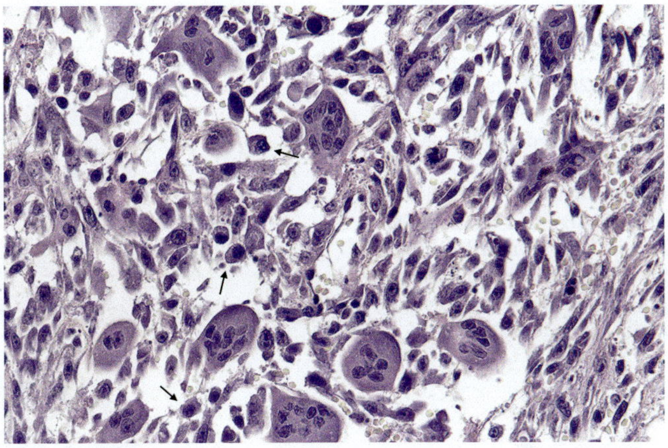

Fig. 8.626 Histology of GcPS: along with spindle and multinucleated cells, roundish and plump cells can be also recognizable (arrows). Note as all the nuclei into the giant cells have the same size and shape

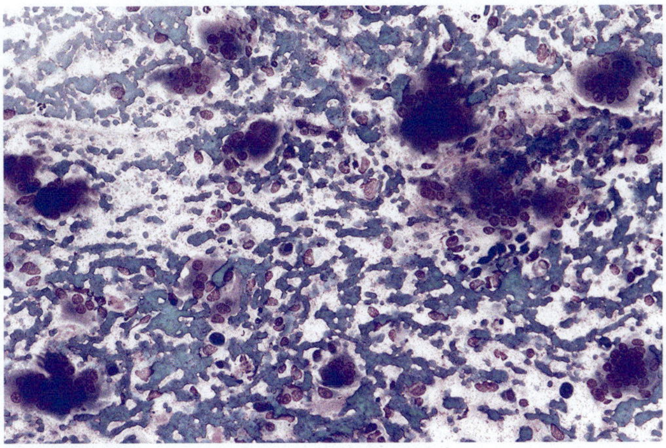

Fig. 8.627 Cytology of GcPS: on a slightly proteinaceous background, a mix of spindle, roundish, and many multinucleated cells are present

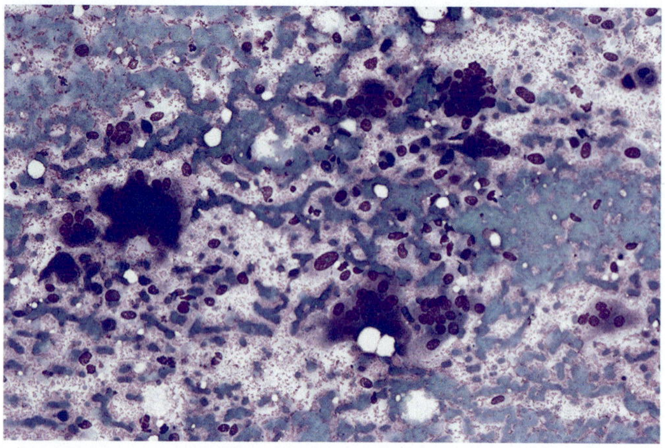

Fig. 8.628 Cytology of GcPS: note as the size and shape of nuclei present in the same multinucleated cell are regular and do not manifest morphological atypia

nuclear malignancy are usually marked, and an evident anisokaryosis is observed in the context of the same multinucleated cell; in practice, multinucleated cells are cells of neoplastic nature.

To differentiate multinucleated malignant cells from reactive multinucleated histiocytic cells (macrophages), the immunohistochemical maker IBA-1, specific for histiocytic cells, can be used. With this marker, spindle-shaped neoplastic cells, of certain non-histiocytic nature, but predominantly fibroblasts/myofibroblasts, are negative, while both mononuclear and multinucleated macrophages are clearly IBA-1 positive (Figs. 8.629 and 8.630). To create further confusion, is the fact that many histiocytic giant cells can show a clear spindle morphology to be suspected belong to the neoplastic spindle population of cells. Spindle-shaped cells of GcPS show variable degrees of atypia, represented by anisocytosis and anisokaryosis, irregular chromatin, multiple large and bizarrely shaped nucleoli, and atypical mitoses (Figs. 8.631 and 8.632).

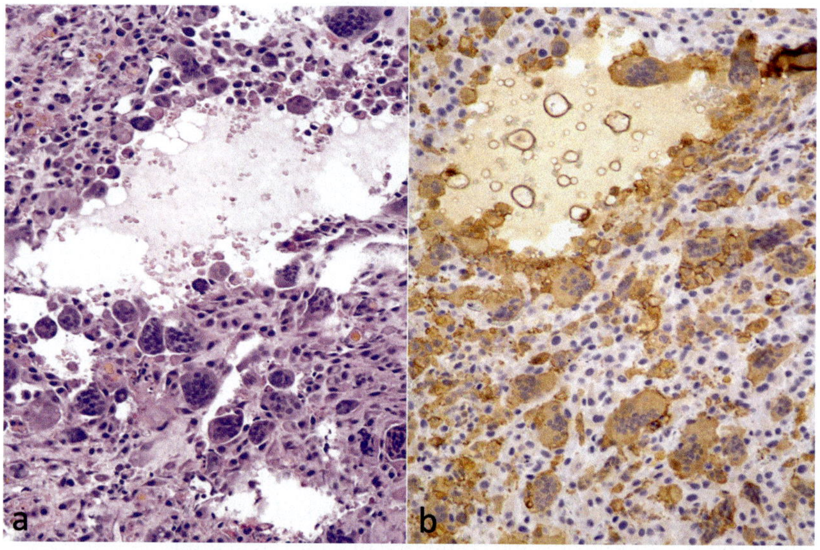

Fig. 8.629 Pleomorphic giant cell sarcoma: (**a**) histology H-E stain—a mixed population of spindle and roundish and multinucleated cells characterize the histology of PGcS. (**b**) Same lesion stained with IBA-1: note as all the multinucleated cells are positive, while the spindle ones are negative to the marker

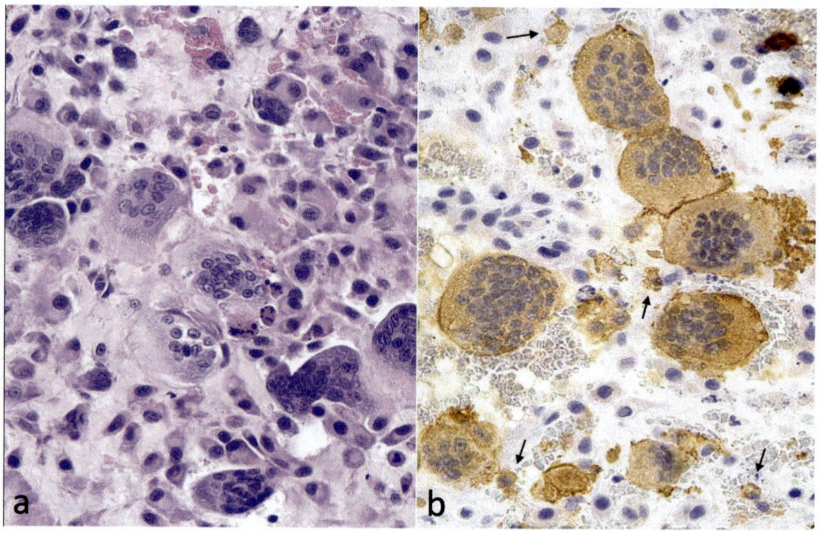

Fig. 8.630 Pleomorphic giant cell sarcoma: (**a**) histology H-E stain—at higher magnification the morphologically mixed cell population is better recognizable. Note as, in multinucleated cell, the size of nuclei is uniform; (**b**) IBA-1 marker—note as only multinucleated cells and some mononucleated cells (histiocytes) are positive to the marker (arrows). Spindle neoplastic cells are negative

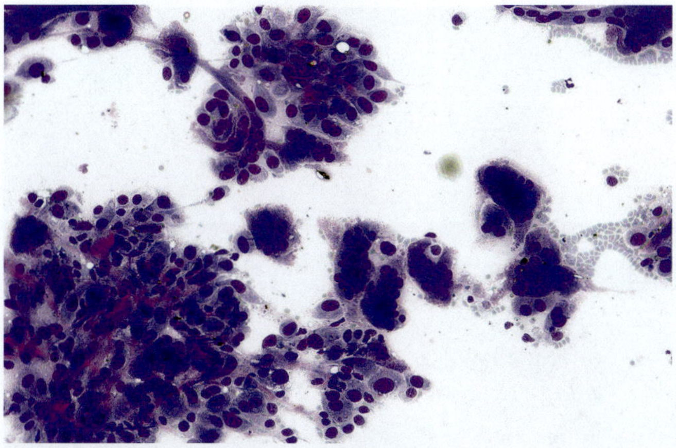

Fig. 8.631 Cytology of GcPS: Atypical spindle neoplastic cells mixed with many giant inflammatory cells—note the different shapes and morphologies of the two typology of cells

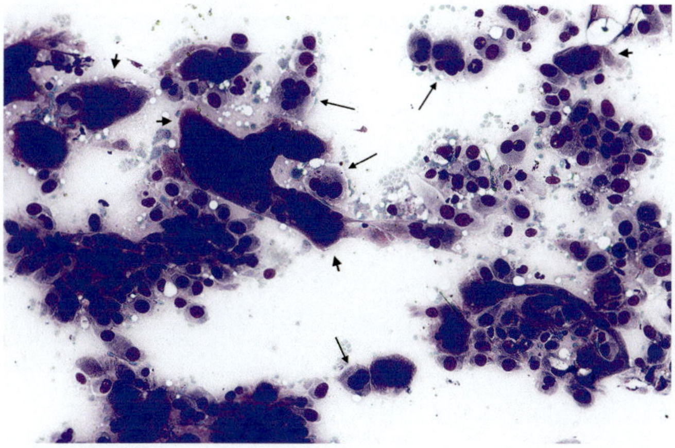

Fig. 8.632 Cytology of GcPS: note as the reactive giant cells can be spindle in shape. Few binucleated and multinucleated with few nuclei can represent malignant cells (long arrows); the morphology of their nuclei is different from whose of giant histiocytic cells (short arrows)

8.4.9 Dermatofibrosis

The term *nodular dermatofibrosis* (ND) refers to a skin benign neoplasm in which multiple nodular dermal lesions are observed, composed of redundant collagen tissue associated with a scarce number of fibrocytes. ND is predominantly seen in the German Shepherd dogs (GSD) and less frequently in Boxers and Golden Retrievers; rarely it can be observed in other pure breeds or in mixed breed dogs [249, 250].

The disease is very similar to Birt-Hogg-Dubé syndrome (BHD) in humans, a rare autosomal dominant disease, linked to a mutation in the BHD gene, an ortholog of the canine Folliculin gene (FLCN), which causes the development of renal carcinomas. On the basis of the human disease, for years, also in dogs, skin lesions have been associated to internal organ neoplasms such as renal cystadenocarcinomas or uterine leiomyomas; for this reason, it was believed to be a cutaneous paraneoplastic syndrome linked to an internal neoplasia [247]. Since dermatofibrosis is often not associated with any neoplasm in dogs, it cannot be classified as a paraneoplastic syndrome, but instead the skin lesions and tumors could be two faces of the same genetic mutation in the gene encoding FLCN, present on chromosome 5 [248, 251].

Clinical lesions are characterized by firm dermal nodules, located mainly on the head and extremities. The lesions are both nodular and plaque-like with ill-defined margins that are macroscopically appreciated as skin elevations. Nodules can ulcerate (Figs. 8.633, 8.634).

Cytological Findings

The histopathology of ND is characterized by a benign proliferation of pre-existing dermal collagen in which a usually small number of uniformly sized spindle cells, representing

Fig. 8.633 Nodular dermatofibrosis: (**a**) multiple dermal nodules and plaques on both fore limbs of a GSD; (**b**) three confluent nodules with not well-defined margins in GSD; (**c**) multiple nodules, some of them ulcerated in a GSD with ND. The leg appears swollen and edematous, but this appearance is instead due to the multiple dermal adjacent fibrous benign neoplasms

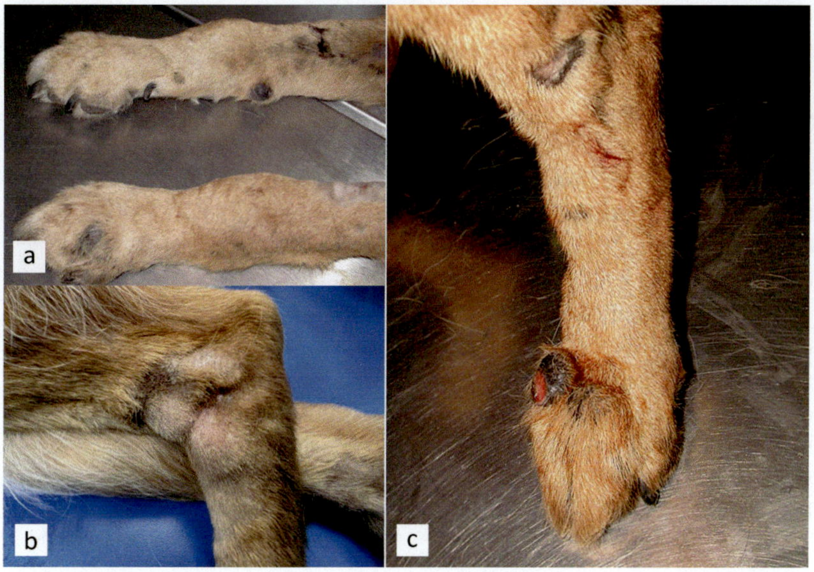

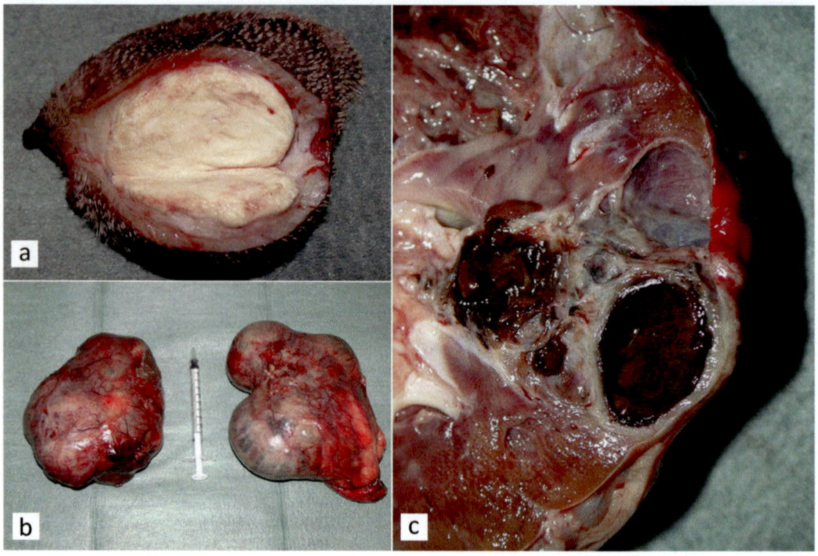

Fig. 8.634 Nodular dermatofibrosis: (**a**) nodule in the section shows a fibrous appearance and consistency; (**b**) macroscopic appearance at autopsy of two neoplastic kidneys of a GSD affected by dermatofibrosis; (**c**) cystoadenocarcinoma—macroscopic aspect of the neoplastic kidney. Note the large cystic areas that displace and replace the renal parenchyma

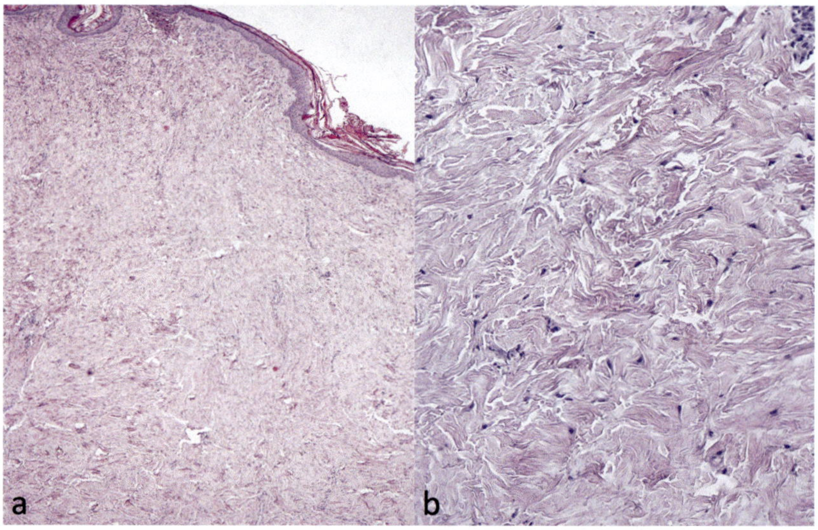

Fig. 8.635 Histology of nodular dermatofibrosis: (**a**) proliferation of mature collagen in the dermis displacing the adnexa; (**b**) at high magnification, a normal number of fibrocytes/fibroblasts are evident between the collagen bundles

fibrocytes, can be recognized (Fig. 8.635). Any attempt to collect cells via FNB is disappointing, and therefore in order to try to obtain some cells, a FNAB, and therefore a sampling with aspiration, must be performed. In the most fortunate cases, it is possible to sample a variable, but always a small number of fibrocytes/fibroblasts, characterized by oval nuclei of uniform size, free from atypia and with cytoplasm showing thin tails whose details can be lost in the background; in some cases fragments of collagen fibers can be collected (Figs. 8.636 and 8.637).

As can be easily understood, the small number of cells composed exclusively of fibrocytes does not allow a cytological diagnosis, also because the result is similar to that obtainable from the sampling of a fibroma, a dermatofibroma, or a skin tag (fibroepithelial polyp); therefore, the clinical aspect will guide the final diagnosis, which however remains histopathological.

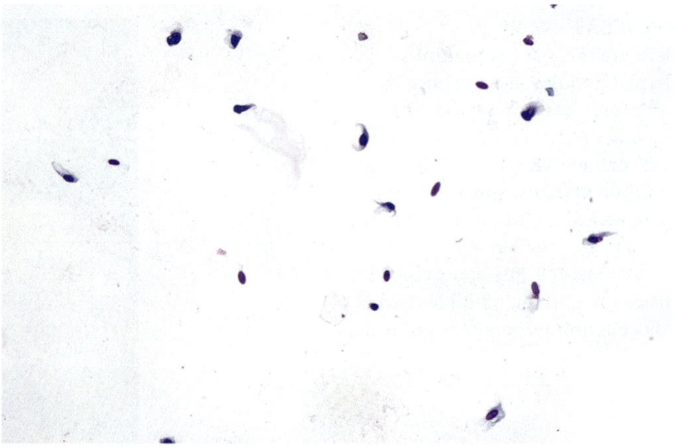

Fig. 8.636 Cytology of nodular dermatofibrosis: in very fortunate cases, it is possible to sample a significant number of small fibrocytes/fibroblasts with a clear spindle-shaped silhouette, on an acellular background

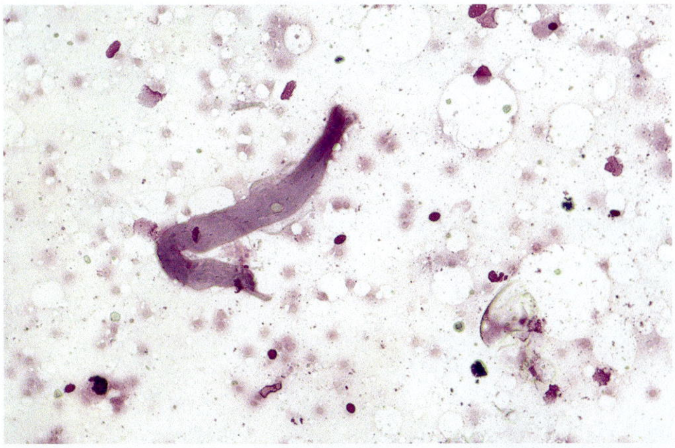

Fig. 8.637 Cytology of nodular dermatofibrosis: few spindle fibrocytes/fibroblasts along with a pinkish in color collagen fibers; the latter can be collected only if forced aspiration is used (FNAB)

8.4.10 Fibroepithelial Polyp (Skin Tag)

Fibroepithelial polyps are benign collagen lesions also known as *skin tags*, *acrochordon*, *fibromatous polyp*, and *fibromatous plaque*. They are very common lesions in dogs and rare in cats, alopecic, pink in color, but sometimes pigmented, of variable size, from round to filiform and pedunculated. Small and multiple lesions are often located in the sternal area, and it is hypothesized that they may be the consequence of collagen hyperplasia secondary to chronic friction (Figs. 8.638 and 8.639).

Cytological Findings

Histology of skin tag is characterized by a proliferation of paucicellular mature collagen fibers, in some cases with an increased vascularization. The cellular component is scarce but sometimes can be a little bit higher (Figs. 8.640 and 8.641). Cytological slides coming from these lesions are usually acellular if collected only via FNB. When instead a

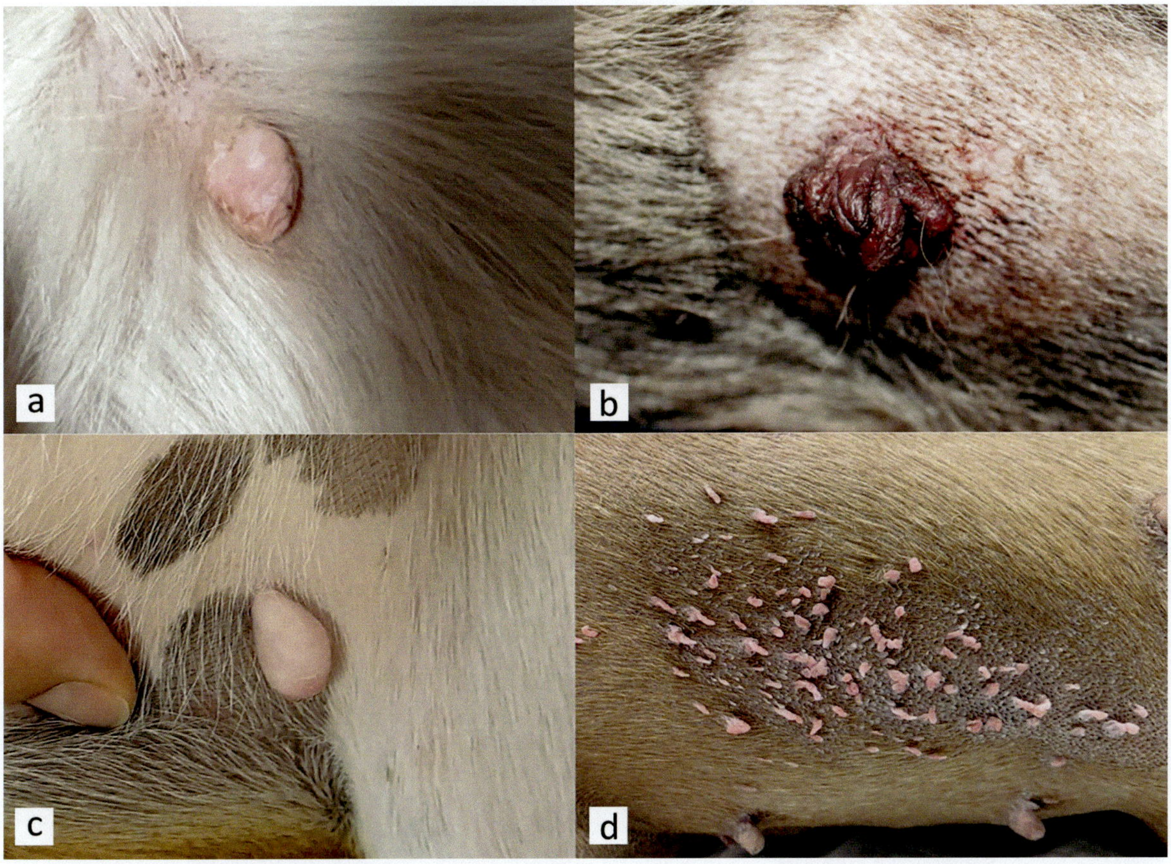

Fig. 8.638 Skin tag: (**a**) small and alopecic, pinkish in color skin tag; (**b**) pigmented and alopecic nodule in a dog; (**c**) alopecic and pedunculated 1 cm nodule, with narrow base; (**d**) multiple filiform fibroepithelial polyps on the sternal skin in a dog. (Courtesy of DVM M. Beccati—Italy)

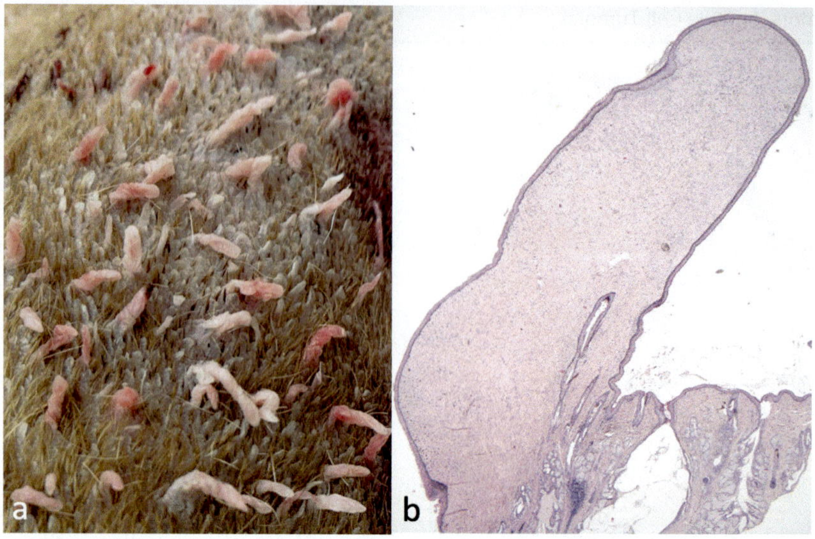

Fig. 8.639 Skin tag: (**a**) multiple filiform fibroepithelial polyps (courtesy of DVM M. Beccati—Italy); (**b**) histological representation of the previous lesions—mature collagen forming an exophytic filiform lesion

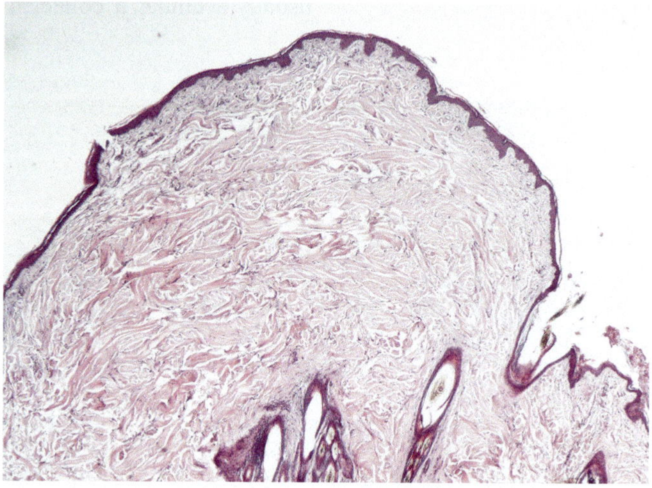

Fig. 8.640 Histology of skin tag: dermal nodule composed of mature collagen. Note that the nucleated cells are so few that they are not recognizable at this magnification

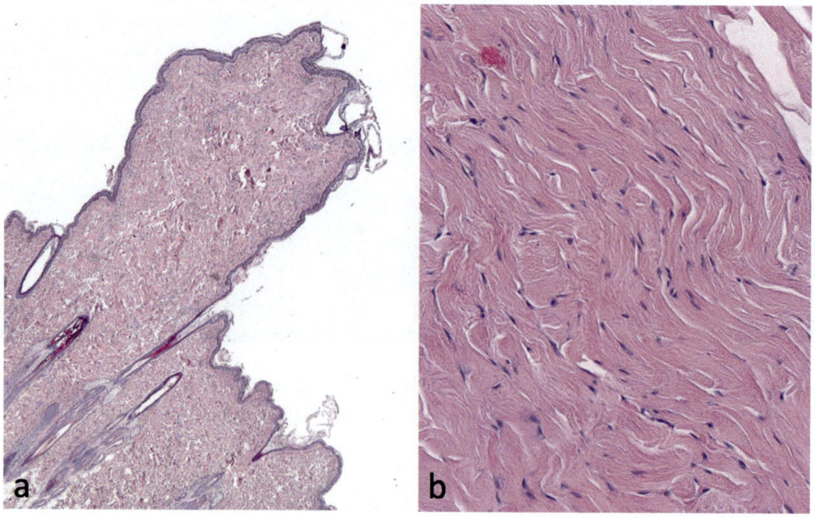

Fig. 8.641 Histology of skin tag: (**a**) exophytic proliferation of mature collagen; (**b**) dermal nodule composed of mature collagen. Note that the nucleated cells are so few that they are not recognizable at this magnification

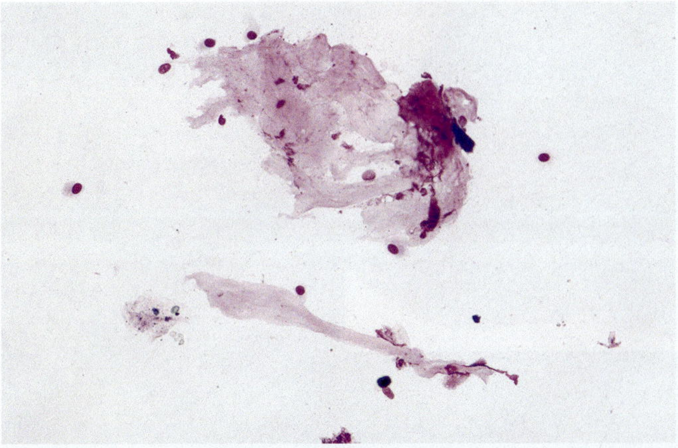

Fig. 8.642 Cytology of skin tag: scattered bare nuclei and spindle fibrocytes along with collagen fibers

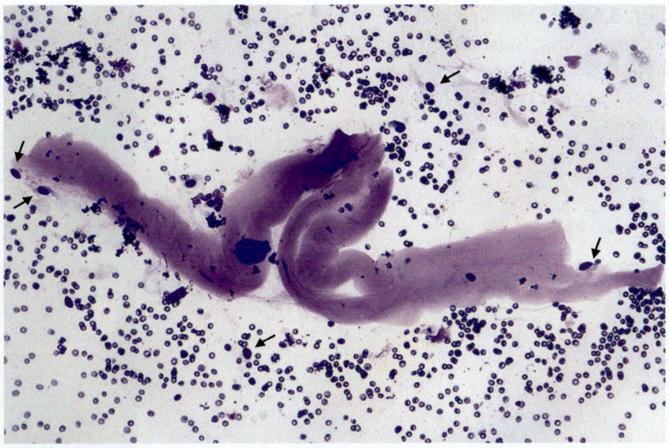

Fig. 8.643 Cytology of skin tag: collagen fibers on a hematic background due to the traumatic technique of FNAB. Some spindle fibrocytes and their bare nuclei are recognizable (arrows)

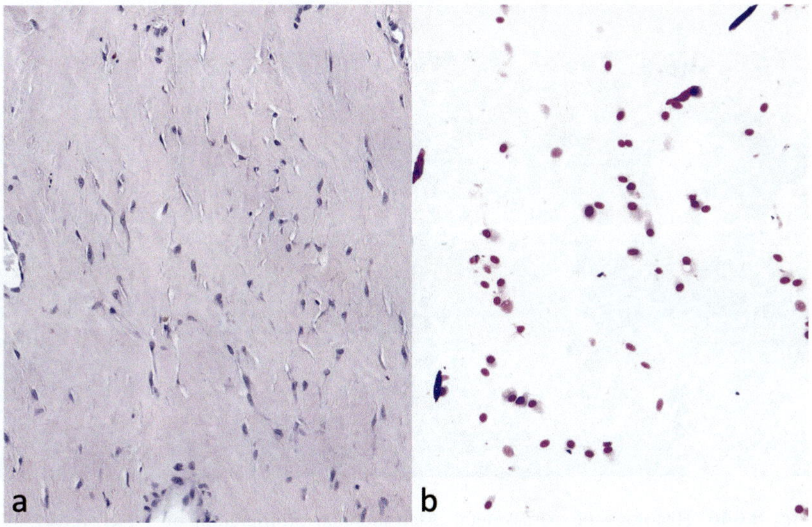

Fig. 8.644 Skin tag: (**a**) Histology—dense collagen fibers with increased numbers of fibrocytes. (**b**) Cytology: many non-atypical fibrocytes collected from the skin tag of previous histology images. Such cellular preparations are rare in the course of fibroepithelial polyps, and when they do occur, it is possible that the needle has collected cells from a more cellular site

forced aspiration is performed, a mix of collagen fibers and a variably amount of fibrocytes can be collected and observed (Figs. 8.642, 8.643 and 8.644). These cytological features are similar to those observed in course of dermatofibrosis, collagen nevi, or hamartomas, and in some cases also the histopathology can be not easy to differentiate between these benign proliferations of collagen.

8.4.11 Extraskeletal Osteosarcoma

Extraskeletal osteosarcomas (EOS) are very rare tumors that can arise in various soft tissues including the skin. It is a very aggressive tumor and is fatal in a short time. The few cases reported in the literature involving the skin do not allow us to define its specific clinical aspects. The reported cases report that lesions are characterized by nodule or mass in the subcutaneous tissue, without primary bone involvement (Fig. 8.645).

Cytological Findings
Histology of EOS shows a proliferation of malignant osteoblasts, mesenchymal cells that produce osteoid. Several histological variants of extraskeletal osteosarcoma are reported: chondroblastic, osteoblastic, fibroblastic, and giant cell type [254, 256] (Fig. 8.646).

Cytological slides usually have good cellularity with neoplastic cells that can assume a round, oval, or spindle shape; anisocytosis and anisokaryosis depend on the degree of differentiation and the histological variant. Nuclei are central or eccentric, and in some cases, they seem protrude outside the cytoplasmic profile; when most neoplastic cells assume this appearance, care must be done to not confuse osteoblasts with neoplastic plasma cells or melanocytes. In some cases, small optically empty areas, similar to vacuoles, may be

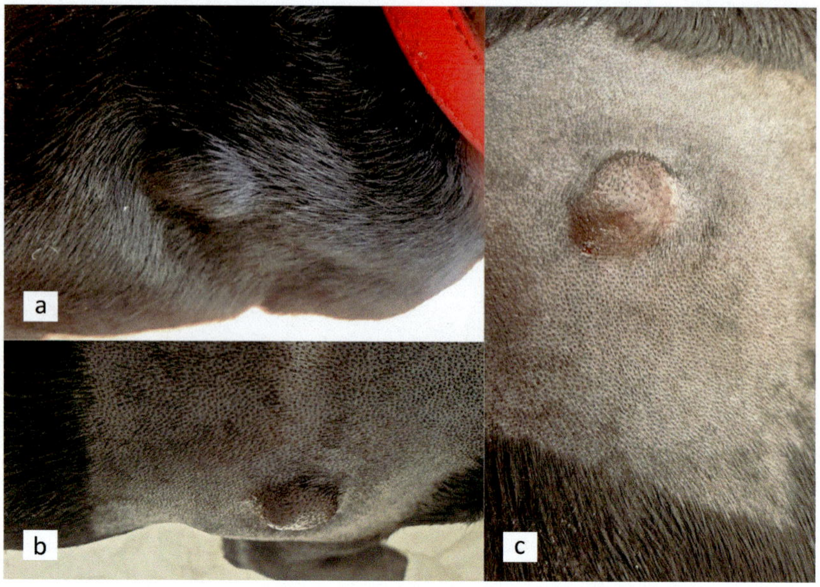

Fig. 8.645 Extraskeletal osteosarcoma: Subcutaneous nodule on the thorax of a dog. (Courtesy of DVM E. Tapinassi—Italy)

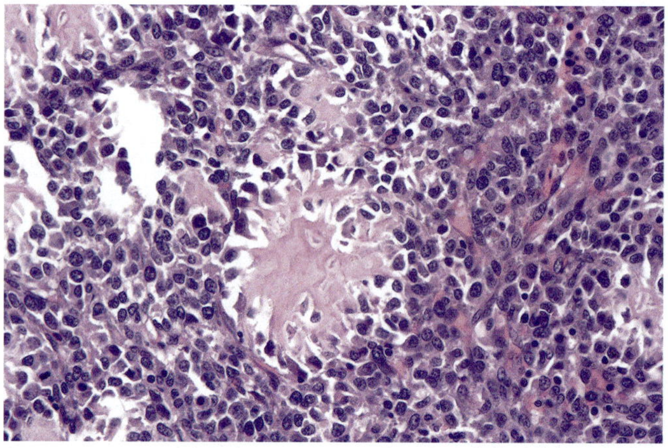

Fig. 8.646 Histology of extraskeletal osteosarcoma: proliferation of neoplastic osteoblasts, with different morphology and with clear bone matrix production

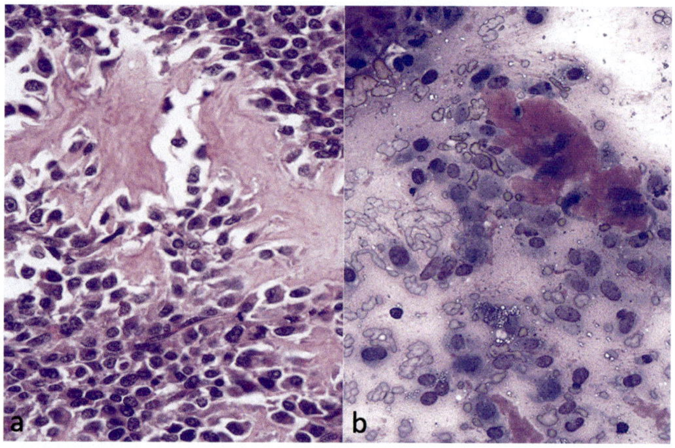

Fig. 8.647 Extraskeletal osteosarcoma: (**a**) histology—osteoid material is represented as brightly pink amorphous material between neoplastic osteoblasts. (**b**) Cytology: also on cytological slides, bone matrix is easily recognizable as fibrillar, pinkish amorphous material

found in the cytoplasm. An important cytological feature is the presence of osteoid material, represented as bright pink amorphous and fibrillar material, among the osteoblastic cells (Figs. 8.647 and 8.648).

When these cytological features are evident, it is easy to suspect an osteosarcoma, despite its rarity in the skin [253]. If the morphological features are not suggestive of osteosarcoma and there is no osteoid material, it is more difficult to make a cytological diagnosis, given the similarities with other soft tissue sarcomas or amelanotic melanomas. In these cases, the substrate (BCIP/NBT) can be used on cytological slides. 5-Bromochromogen, 4-chlorogen, 3-indolyl phosphate/nitroblue tetrazium (BCIP/NBT) is a very sensitive substrate for the detection of alkaline phosphatase in cytoplasm of osteoblasts; alkaline phosphatase hydrolyzes BCIP, which is oxidized by atmospheric oxygen, producing

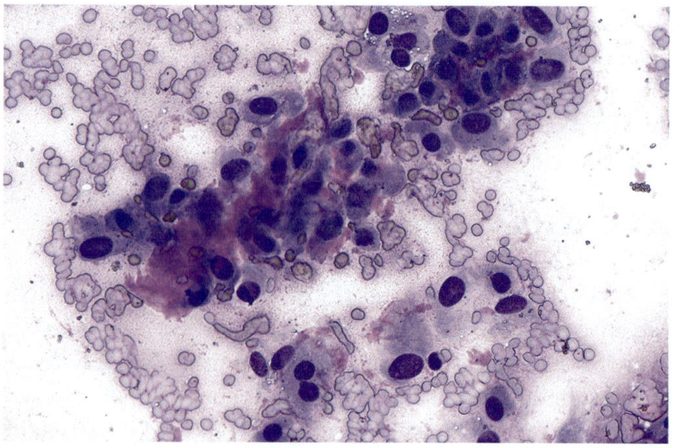

Fig. 8.648 Cytology of extraskeletal osteosarcoma: pink and amorphous bone matrix dispersed among osteoblasts

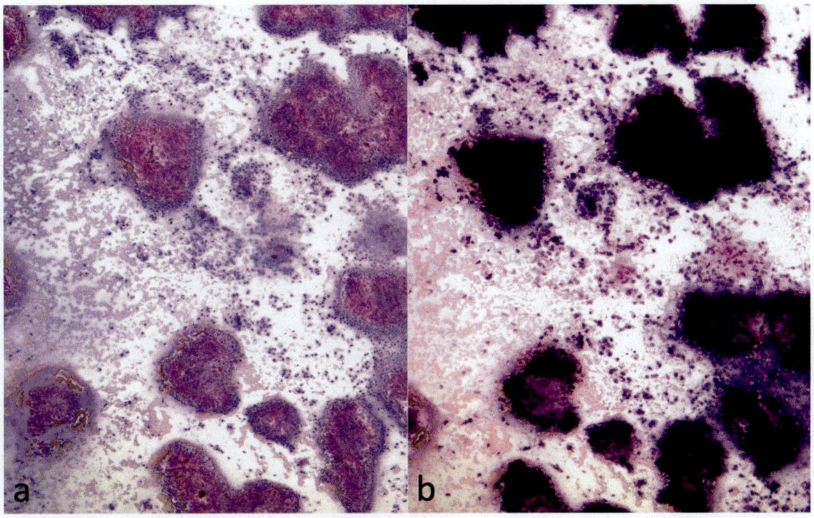

Fig. 8.649 Cytology of extraskeletal osteosarcoma: (**a**) large groups of malignant osteoblasts stained with Romanowsky-type dye; (**b**) same part of the slide after used substrate BCIP/NBT—note as the cytoplasm assumes a dark color, indicative of the presence of alkaline phosphatase

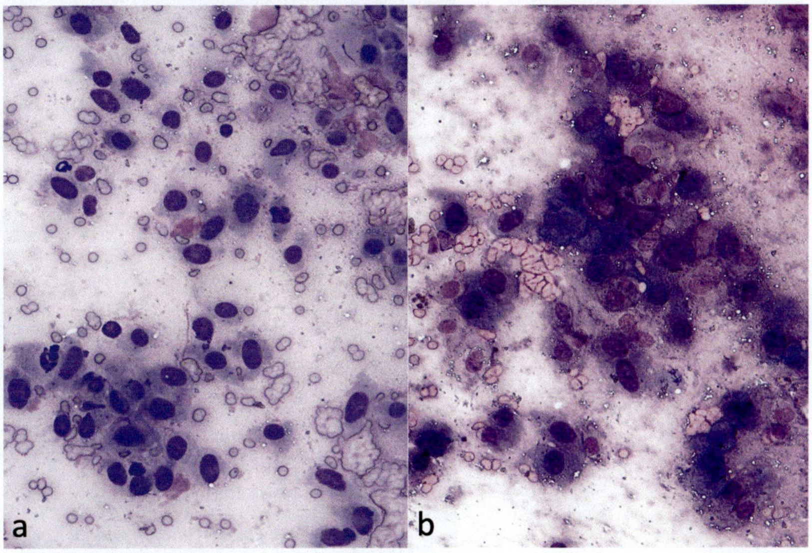

Fig. 8.650 Cytology of extraskeletal osteosarcoma: (**a**) from oval to spindloid neoplastic osteoblasts; few amount of osteoid is recognizable; (**b**) after used substrate BCIP/NBT, the cytoplasm of osteoblasts becomes dark and of grainy appearance

an insoluble product and a color change (blue) at the reaction site (cytoplasmic expression); in the presence of NBT (P-nitroblue tetrazolium chloride), after a reduction reaction, an insoluble cytoplasmic precipitate is formed, dark blue in color and with a finely granular appearance [252, 255]. This reaction can only be done on slides previously stained with Romanowsky-type stain, and the cytoplasm changes color to dark gray-brown (Figs. 8.649 and 8.650).

8.5 Melanocytic Tumors

Melanocytic skin tumors are quite common in dogs and rare in cats [260, 263, 265, 272, 273].

Melanocytic neoplasms can arise from epidermal, dermal, and follicular melanocytes and are classified into two main groups that, in the past, were differentiated on the basis of histological features alone: *melanocytoma* (benign melanocytic neoplasia) and *melanoma* (malignant melanocytic neoplasia). *Epitheliod*, *sp*indle, and *round cells* are the most common histological variants of canine and feline skin melanocytic neoplasia although, in most cases, mixed patterns are present in the same neoplasm. Other less frequent histopathological variants are the *signet-ring cell melanoma* and the *balloon cell melanoma*.

The degree of pigmentation can be variable, and both heavily pigmented and completely amelanotic neoplasias are possible.

Single skin nodules, of variable size, from very small to large, alopecic and usually highly pigmented, characterize the clinical presentation of melanocytic neoplasia, which are mainly located on the head (lips), limbs (digit and nail-bed), and trunk, but they can be observed in any skin area. Black haired dog breeds such as Rottweiler, Dobermann, etc.

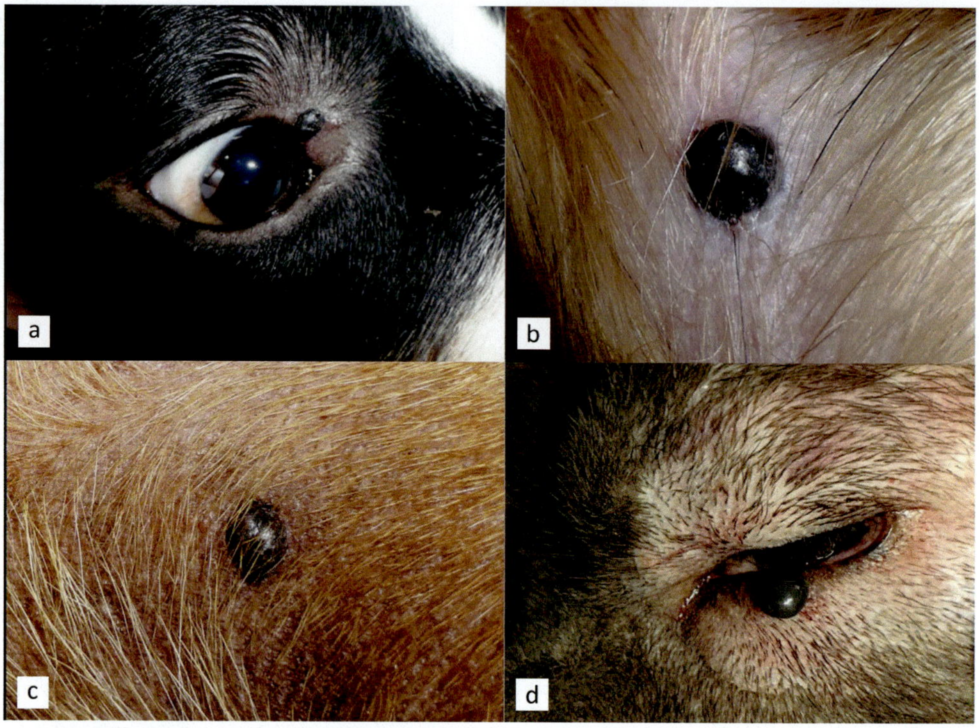

Fig. 8.651 Melanocytic tumors: (**a**) small pigmented melanocytoma on the eyelid; (**b**) pigmented alopecic melanocytoma in a dog; (**c**) small pigmented nodule; (**d**) melanoma on the eyelid of a Shar-Pei

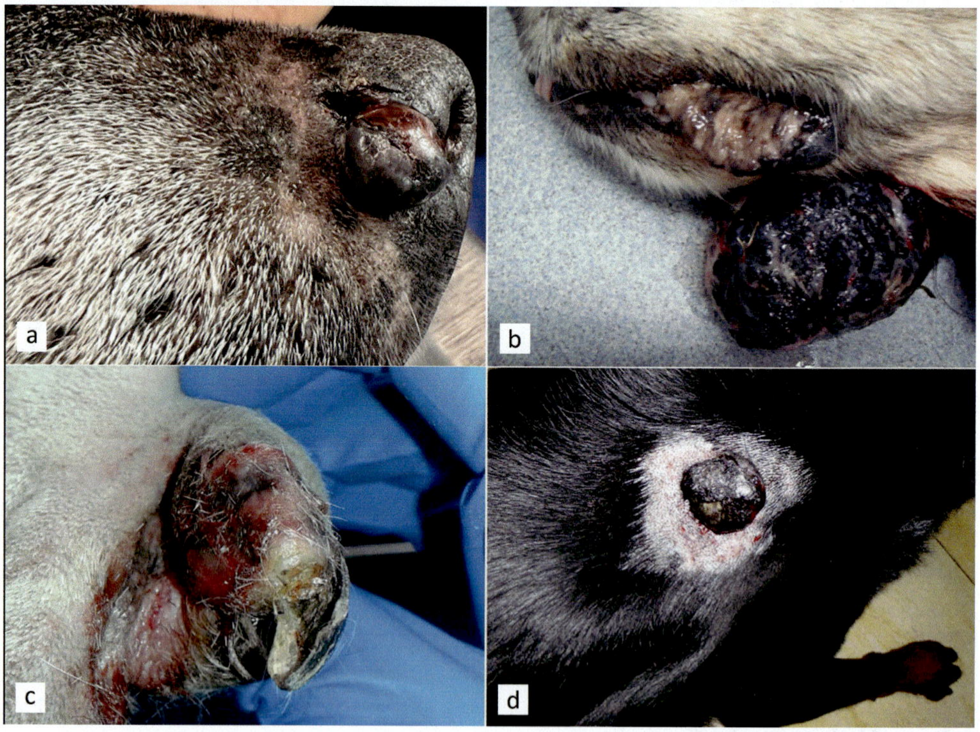

Fig. 8.652 Melanocytic tumors: (**a**) melanoma on the nose of a GSD; (**b**) large melanoma on the intermandibular area of an English setter; (**c**) nail-bed melanoma; (**d**) large pigmented mass on the shoulder of a Rottweiler

seem to be predisposed. In cats, cutaneous melanocytic tumors are rarer than dogs (Figs. 8.651, 8.652, 8.653 and 8.654).

The body site localization is often thought indicative of the behavior of melanocytic neoplasia; for example, melanocytic tumors that occur on the eyelids (mucocutaneous) have usually a benign behavior, while nailbed melanomas are more aggressive and are more frequently observed on large dogs with black haircoat [267]. Over the years, many works have reported that oral and lip melanomas

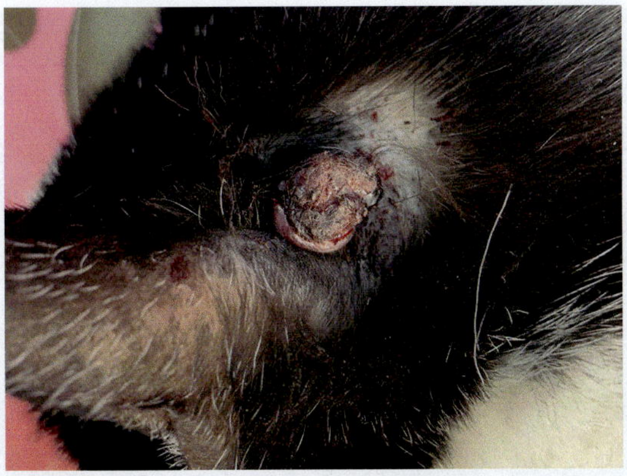

Fig. 8.653 Signet-ring melanoma in cat: single and ulcerated nodule on the head. (Courtesy of DVM L. Navas-Italy)

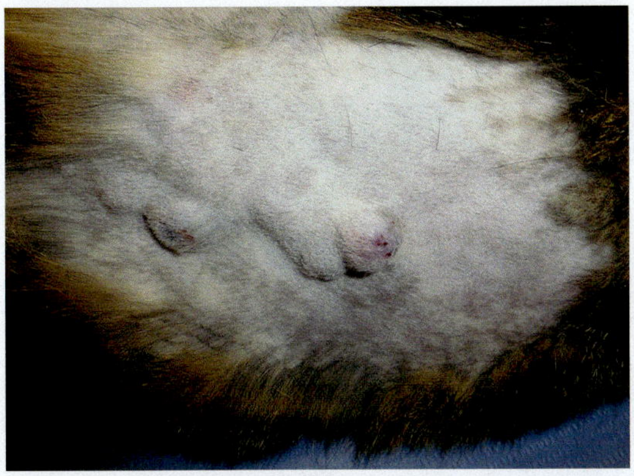

Fig. 8.654 Signet-ring melanoma in cat: multiple skin nodules on the trunk of an old cat. (Courtesy of DVM C. Damiani-Italy)

have a higher rate of malignancy compared with those arising at other body sites; some studies have shown that, although melanoma growth in this area is more aggressive, the percentage of good prognoses is higher than previously reported [268].

However, the site of the lesion is not sufficient to formulate a certain prognosis; in fact, the parameters to differentiate melanomas from melanocytomas and to evaluate the prognosis are continually being studied and are based not only on localization but also on histological features such as the thickness of the neoplasm, the degree of pigmentation, the number and morphology of mitoses, as well as on evaluation of cell proliferation index (ki67) [257, 266, 267].

Cytological Findings

Since many histological variants on melanocytic tumors are recognized, the cytological slides are highly polymorphic, and melanocytes can be defined as "the great imitators," as they are the only cells that can assume all possible morphologies; for this reason, when melanin is not present in the cytoplasm, and the morphological aspects of cells do not allow a diagnosis to be formulated, melanoma must always be included in the differential diagnoses. Melanocytes may show a perfectly round silhouette, a clear spindle cell morphology, some neoplasms may show plasmacytoid aspects, and in other cases, neoplastic cells may be arranged in groups resembling epithelial cells (epithelioid melanoma); these multiple morphological aspects can create difficulties regarding the group of tumors in which melanocytic tumors should be included. Normally these neoplasms are classified separately, but, if we wanted to evaluate them exclusively on the basis of cellular morphology and not on the basis of their origin, melanomas could be classified among RCTs, fusiform tumors, or, less frequently, epithelial tumors.

Despite these differences in shape, the presence of a variable amount of melanin pigment allows to interpret them as of melanocytic origin. Usually most of the benign melanocytic tumors of the skin are pigmented allowing a fast diagnosis, while, when the melanin pigment is completely absent (amelanotic melanoma), the conclusive diagnosis is very hard even for the most skilled cytologists.

The neoplastic melanocytes collected from a *melanotic melanocytoma* are usually well differentiated and characterized by poor cytological polymorphism. Neoplastic cells are mostly roundish and less frequently spindloid, with round to oval nuclei and moderate to large cytoplasm, filled with uniform rod- to round-shaped black or dark green melanin granules; in heavily pigmented neoplasms, the granules may be so numerous that they completely obscure the nuclei (Figs. 8.655 and 8.656). In the slides of these heavily pigmented tumors, when observed at low magnification, a dark black melanin background, arranged both in large clumps and in multiple linear streaks, is commonly observed along with a variable number of pigmented melanocytes; nuclei, when evident, don't show evident morphological atypia. Mitosis are usually not present, also if they could be underestimated due to the high amount of melanin (Figs. 8.657, 8.658 and 8.659). In the absence of clear cytological aspects of malignancy, the cytological distinction between melanocytoma and well-differentiated melanoma can be complex, for which reason in this cases it is necessary to perform confirmatory histopathology.

Malignant melanocytic tumors can be heavily pigmented also; morphological atypia such as anisocytosis, anisokaryosis, macronucleoli, and mitoses are microscopic findings that orient the diagnosis. The presence of mitotic cells is a very important cytological clue for differentiating melanocytoma from melanoma (Fig. 8.660).

As mentioned, neoplastic melanocytes can exhibit a *round/oval*, *spindle*, *plasmacytoid*, signet ring, and *balloon* shape or can be arranged in an *epithelioid* appearance; these different cytological morphologies are often present together in the same tumor.

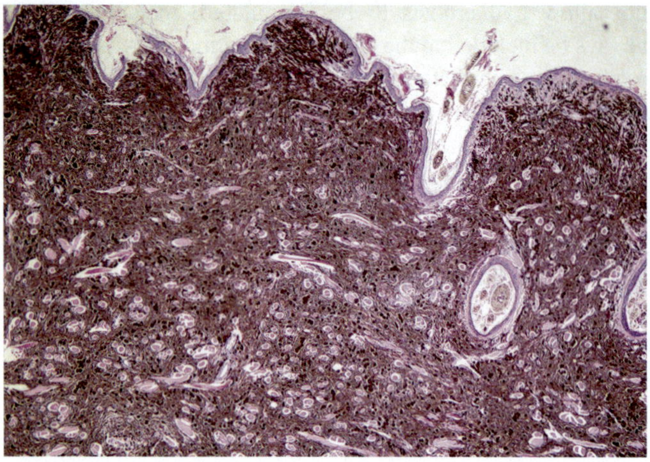

Fig. 8.655 Histology of melanocytoma: dermal proliferation of heavily pigmented benign melanocytes

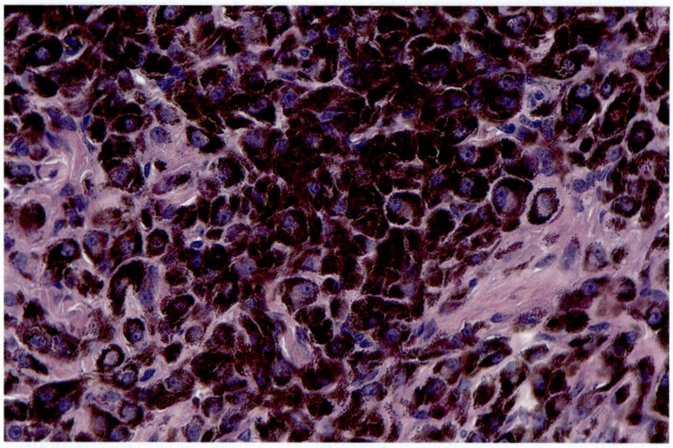

Fig. 8.656 Histology of melanocytoma: heavily pigmented melanocytes in which the nuclei are partially recognizable

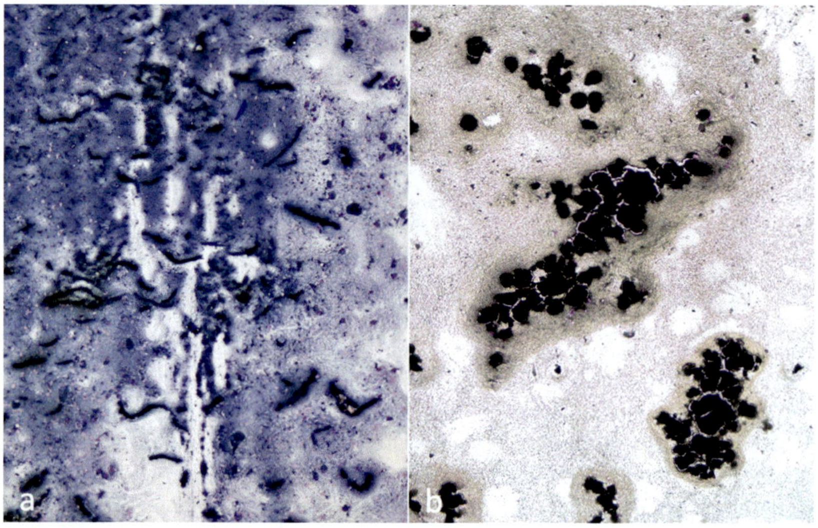

Fig. 8.657 Cytology of melanocytoma: (**a**) dark linear streaks of melanin are the typical background observed in heavily pigmented melanocytoma; (**b**) multiple large clods of melanin—at this magnification, some pinkish spot representing nuclei is barely viewable among melanin

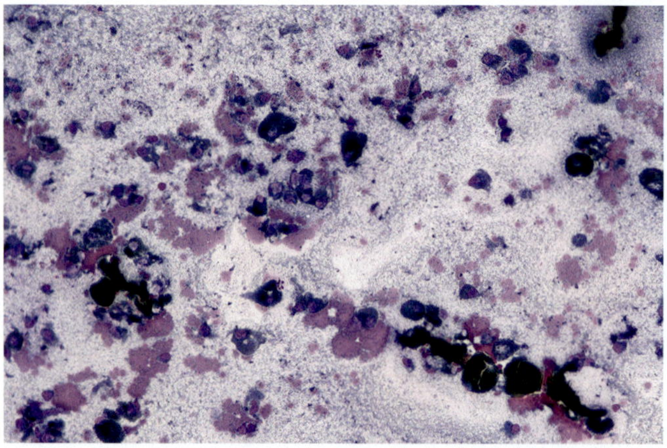

Fig. 8.658 Cytology of melanocytoma: many melanocytes, regular in shape and size, with a high amount of intracytoplasmic melanin pigment—despite the high amount of latter, nuclei are visible and interpretable

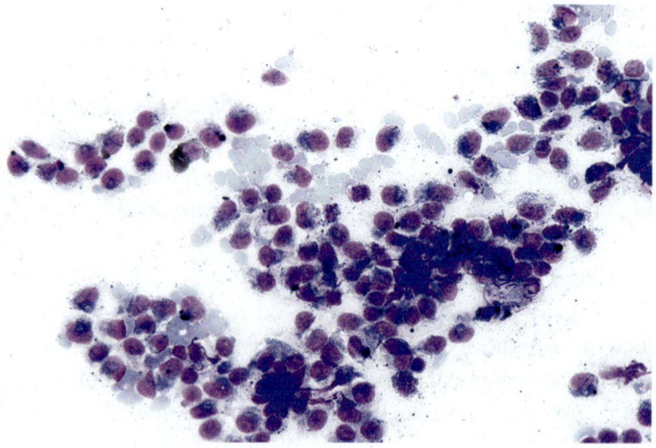

Fig. 8.659 Cytology of melanocytoma: since the moderate amount on melanin, oval to indented nuclei are well recognizable

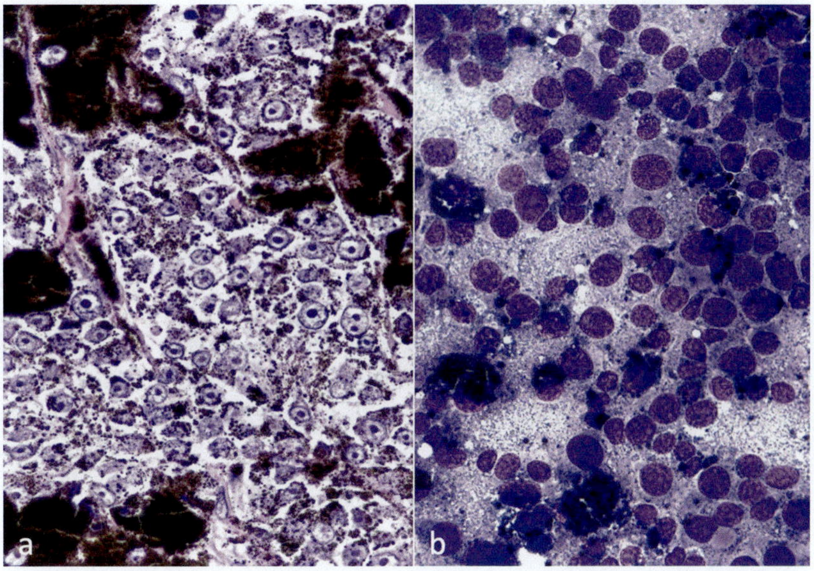

Fig. 8.660 Melanoma: (**a**) histology of pigmented melanoma—anisocytosis, anisokariosys, and large nucleoli are recognizable. (**b**) Cytology of pigmented melanoma—in cytology, the nucleoli are usually less evident than in histology

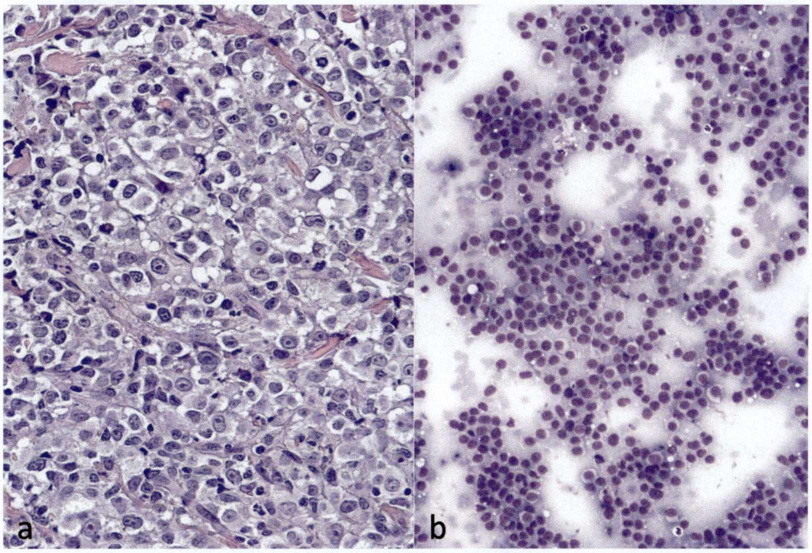

Fig. 8.661 Round cell amelanotic melanoma: (**a**) histology—round cells are recognizable. (**b**) Cytology of round cell amelanotic melanoma: many round melanocytes

In *round cell melanoma*, the neoplastic melanocytes are released as discrete cells, show round silhouette, with large nuclei, roundish and rarely folded, with large pale blue cytoplasm. Nucleolus can be inconspicuous. When intracytoplasmic pigment is not visible, cells can be confused with histiocytes, and in the case of single lesion, a benign cutaneous histiocytoma can be erroneously diagnosed with cytology. Fortunately, in most cases, few pigmented cells are recognizable (Figs. 8.661, 8.662 and 8.663). Also in cats, when melanoma is amelanotic, the morphology of neoplastic cells can be confused with a histiocytic dendritic neoplasia such as feline progressive histiocytosis, especially if some more spindloid melanocytes are present along with the roundish ones (Figs. 8.664 and 8.665).

In a human being, a *plasmacitoid* histological variant of melanoma is well known; only few reports are published on

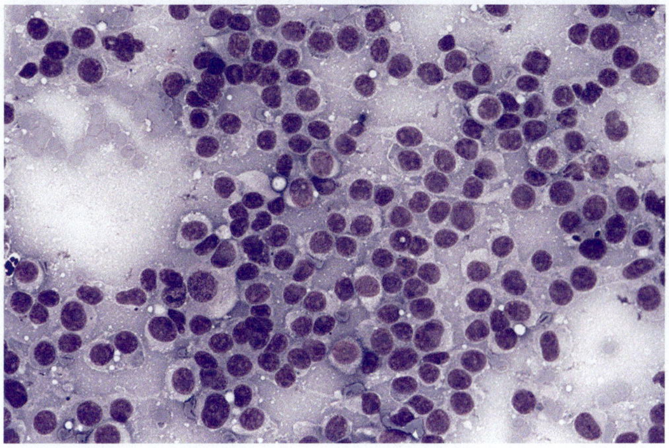

Fig. 8.662 Round cell amelanotic melanoma: at medium magnification the round shape of neoplastic melanocytes is clearly recognizable. Note as cells resemble histiocytes

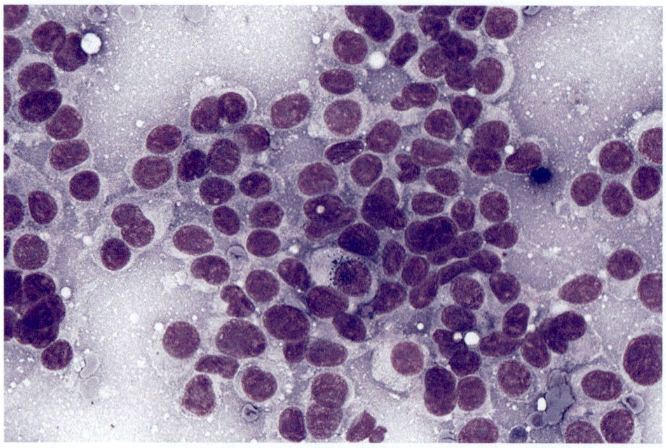

Fig. 8.663 Round cell amelanotic melanoma: at high magnification cells can be confused with histiocytes—a single pigmented cell allows to define cells as melanocytes

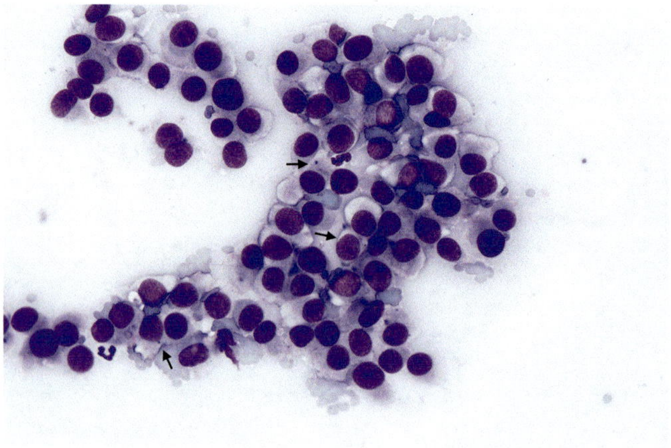

Fig. 8.664 Round cell amelanotic melanoma in a cat: discrete round cells with large cytoplasm. Few spot of melanin is recognizable in some cells (arrows). Note as cells are similar to histiocytes

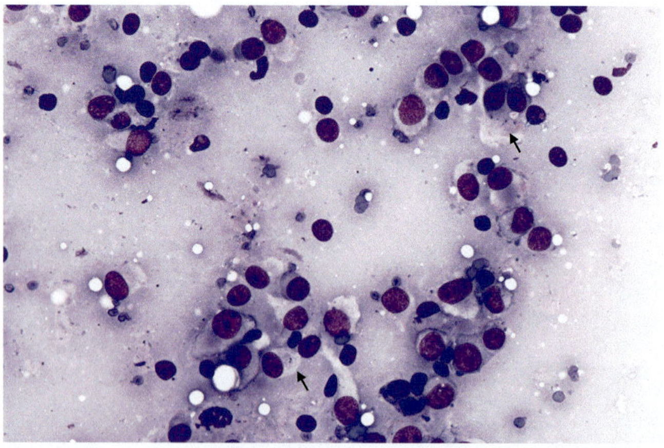

Fig. 8.665 Round cell amelanotic melanoma in a cat: discrete round to spindle cells with large cytoplasm. Very few melanin is recognizable in some cells (arrows). Note as cells are similar to dendritic histiocytes and a misdiagnosis of feline progressive histiocytosis can be done

veterinary literature regarding this morphology of neoplastic melanocytes. In case cells are completely without pigment, the cytological differentiation with a plasma cell tumor can be difficult, and, also with histology, the diagnosis cannot be ensured until immunohistochemistry is performed [264] (Fig. 8.666).

In *spindle cell variant*, cells can manifest a clear spindle shape with a variable degree of pigmentation; histologically the pattern of growth can mimic a soft tissue sarcoma, and if no melanin is detectable on specimens, immunohistochemistry is mandatory for obtaining the diagnosis [272] (Figs. 8.667, 8.668 and 8.669). It must be stressed, as previously discussed for NSTs, rare nerve sheath tumors can be slightly pigmented, and both with cytology and histopathology, the differentiation with a spindle melanoma is not possible without IHC markers. When melanoma is completely

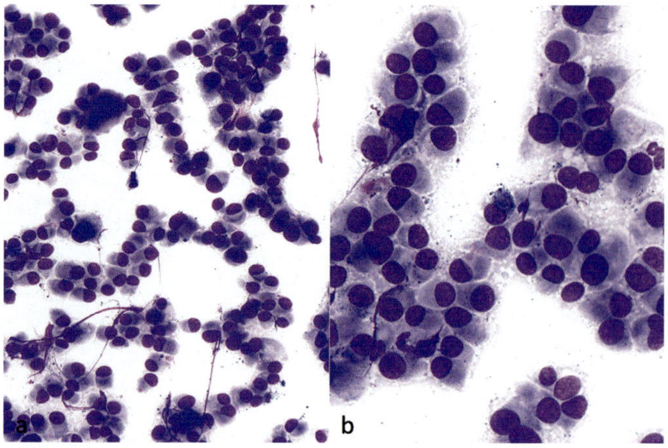

Fig. 8.666 Plasmacytoid melanoma: (**a**) melanocytes show a morphology very similar to neoplastic plasma cells; (**b**) some cells have intracytoplasmic melanin that allows the cytological diagnosis of melanoma to be formulated

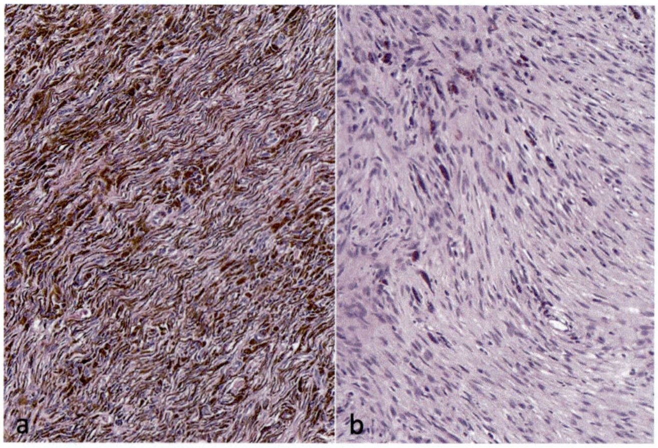

Fig. 8.667 Histology of spindle cell melanoma: (**a**) parallel and wavy bundles of heavily pigmented spindle melanocytes; (**b**) ordered and interwoven bundles with little intracytoplasmic melanin

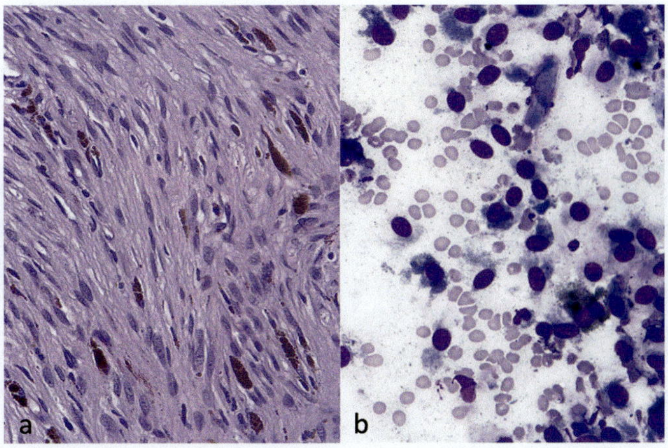

Fig. 8.668 Spindle cell melanoma: (**a**) histology—at high magnification the intracytoplasmic melanin is clearly evident in the spindle cells; (**b**) cytology—spindle-shaped pigmented melanocytes. The cytoplasm is barely perceptible thanks to the pigment that outlines its profile

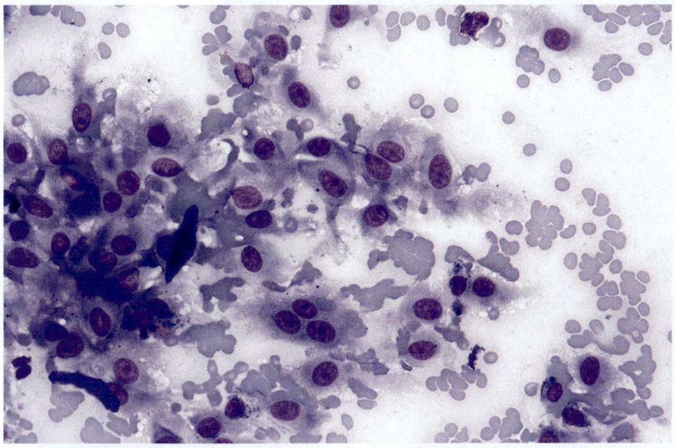

Fig. 8.669 Cytology of spindle cell melanoma: spindle cells are diagnosed as melanocytes due to the intracytoplasmic pigment that is recognizable in some of them

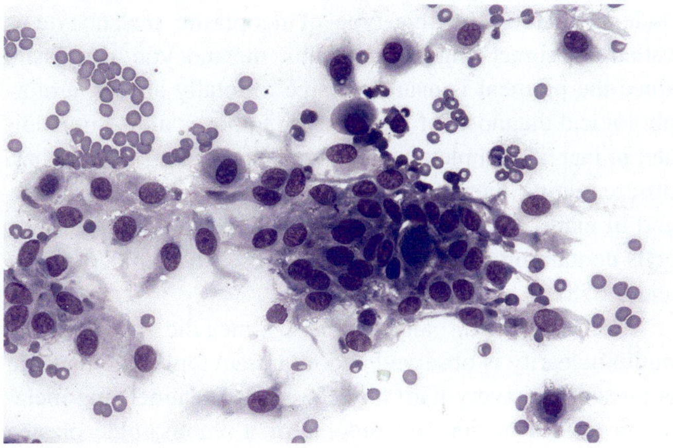

Fig. 8.670 Cytology of spindle cell melanoma: spindle cells do not have intracytoplasmic melanin. In these cases it is impossible to differentiate a soft tissue sarcoma

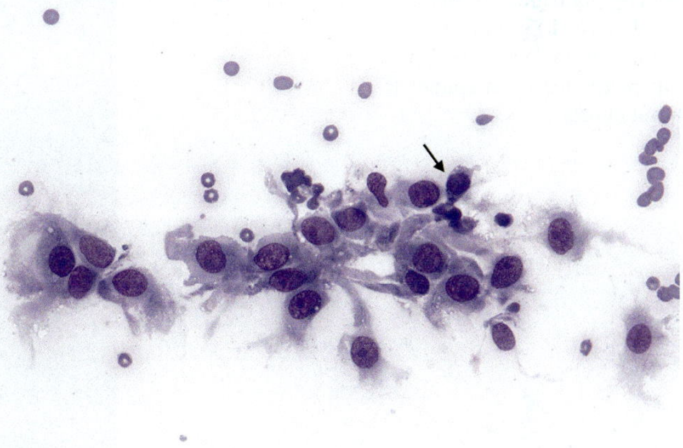

Fig. 8.671 Cytology of spindle cell melanoma: if it were not for a single cell containing intracytoplasmic melanin, it would not be possible to cytologically differentiate a soft tissue sarcoma (arrow)

amelanotic, the differentiation from a soft tissue mesenchymal neoplasia is not possible; in the same way, in the presence of severe malignant cytological atypia in which multinucleated cells are also present, a spindle cell amelanotic melanoma cannot be cytologically differentiated from a pleomorphic soft tissue sarcoma or a spindle cell histiocytic sarcoma (Figs. 8.670 and 8.671).

Cytological arrangements of cells that mimic epithelial cells can be observed in the case of melanoma with *epithelioid* aspects; these arrangements are clearly recognizable on histology while are not so clear evident with cytology. Epithelioid dispositions are not observed as alone cytological pattern but along with singularly dispersed melanocytes, both roundish and spindle in shape. On slides, some groups of neoplastic melanocytes maintain an arrangement in pseudo-aggregates which can mimic cluster of epithelial cells with

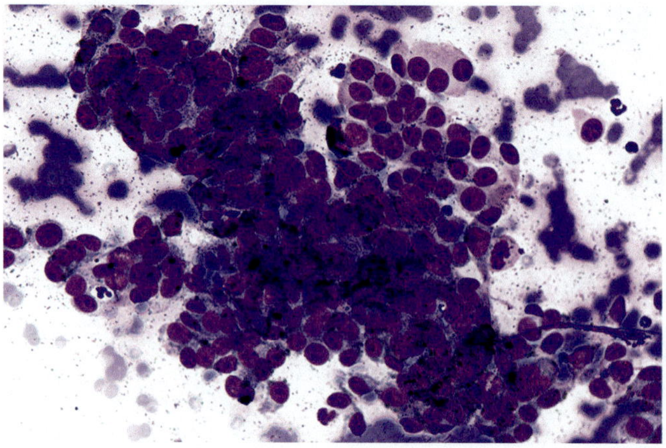

Fig. 8.672 Cytology of melanoma: large group of pigmented melanocytes resembles an epithelioid arrangement

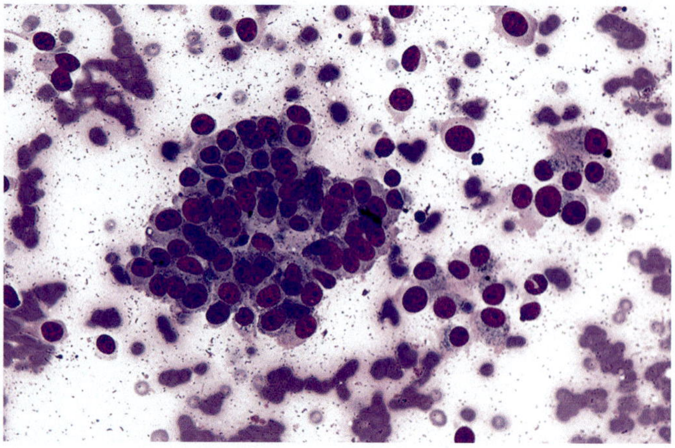

Fig. 8.673 Cytology of melanoma: clear epithelioid arrangement of malignant melanocytes

Fig. 8.674 Cytology: (**a**) pigmented trichoblastoma—heavily pigmented clusters of basal cells; (**b**) group of pigmented melanocytes can be misinterpreted as an epithelial basal cluster

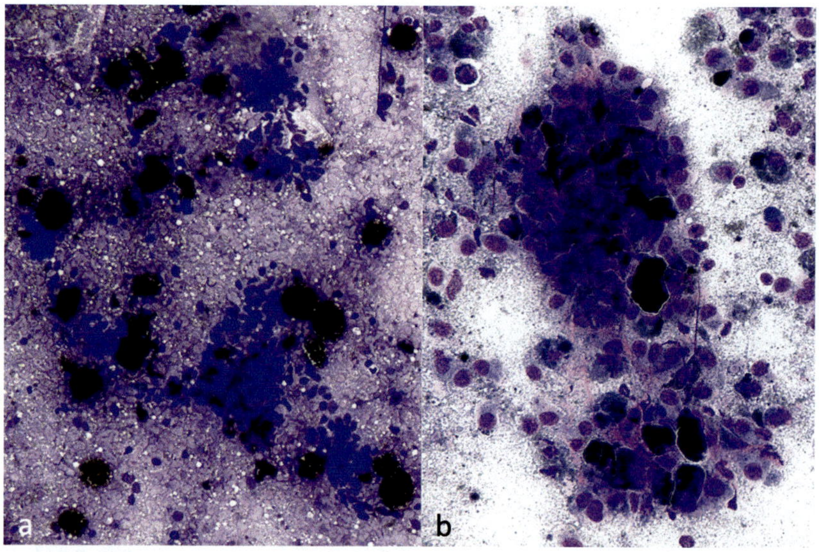

which beginner cytologists must not confuse; fortunately, also these grouped cells are usually variably pigmented (Figs. 8.672 and 8.673). Beginners must be aware of this cytological pitfall.

In some *pigmented melanocytic neoplasms* and characterized of few atypia, in which cells can be released in heavily pigmented cohesive groups, the cytological slides can be confused with epithelial basal melanized tumors, such as basal cell tumors or trichoblastomas (Fig. 8.674).

A histological variant of melanoma, named *signet-ring cell melanoma*, is well recognized in dogs and cats [258, 261, 272]. In this subtype, neoplastic melanocytic cells not only have a typical oval shape that resembles a plasma cell, with a round to crescent-shape eccentrically located nucleus, but cytoplasm contains a characteristic slightly darker large globular area delineated with a peripheral weakly line that is located between the nucleus and the cytoplasm; with a few of imagination, especially when nuclei have a crescent-shape, the latter are compared to a signet and the darker large globular area to the hole of a ring, it is understandable why this type of neoplastic melanocyte is called as signet-ring cell. In this melanocytic neoplasm, since the pigment is usually scarce to totally absent, a morphological diagnosis of melanocytic tumor is not so immediate; in fact, in completely amelanotic signet-ring melanomas, differentiating them from a plasmacytoma is very difficult, and in many cases histopathology is requested to detect few cells containing melanin pigment (Figs. 8.675, 8.676, 8.677 and 8.678).

When marked anysocytosis, megalic nuclei, or multinuclearity is observed and no intracytoplasmic melanin is present, it is very hard to differentiate a signet-ring melanoma from a histiocytic sarcoma or a pleomorphic plasma cell tumor. In these cases, immunohistochemical markers on histopathological specimens are mandatory to achieve diagnosis. Fortunately, in many cases, scattered pigmented cells

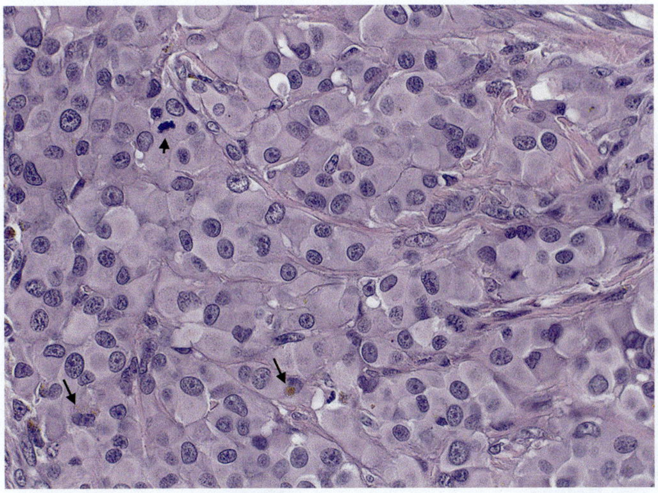

Fig. 8.675 Histology of signet-ring cell melanoma: melanocytic neoplastic cells show oval shape with eccentric nuclei—note as few cells contain melanin (arrows)

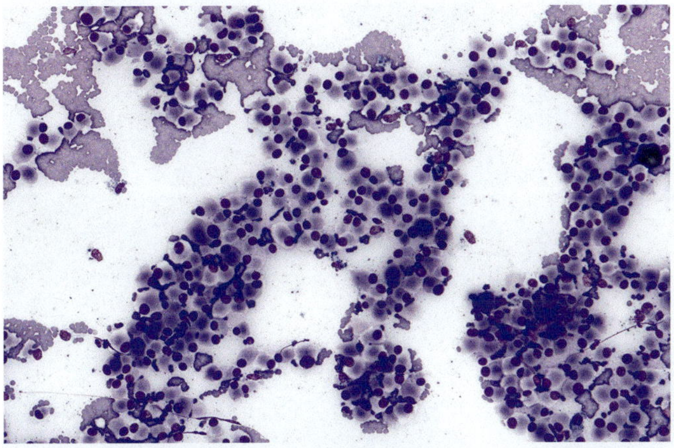

Fig. 8.676 Cytology of signet-ring cell melanoma: many melanocytic neoplastic cells with oval shape and eccentric nuclei—a darker round intracytoplasmic area can be recognizable at low magnification

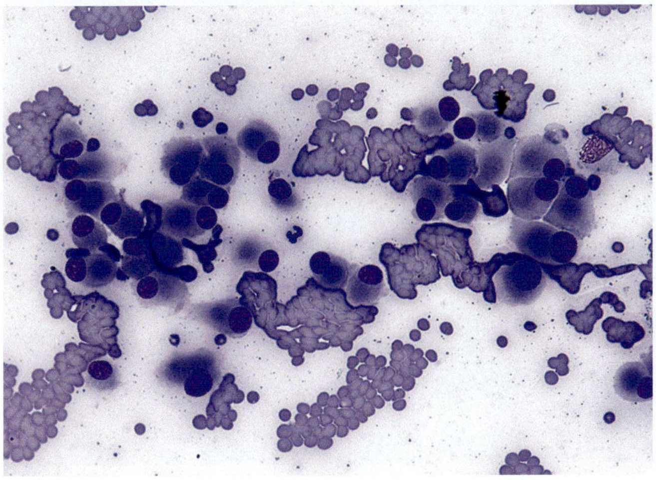

Fig. 8.677 Cytology of signet-ring cell melanoma: the signet-ring appearance of neoplastic melanocytes characterizes this variant of melanoma

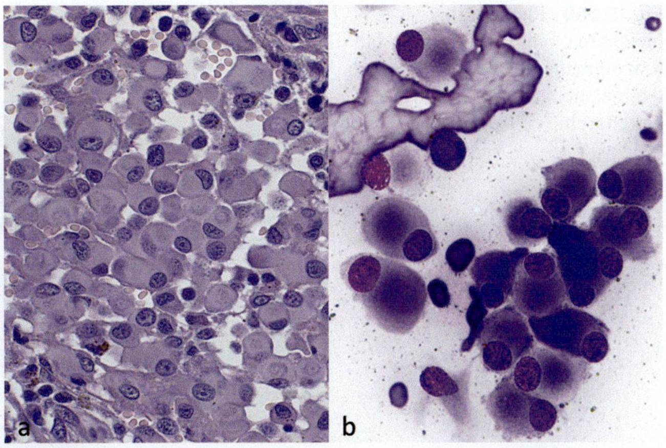

Fig. 8.678 Signet-ring cell melanoma: comparison between (**a**) histology and (**b**) cytology of a signet-ring cell melanoma in which the intracytoplasmic globular area that lateralizes the nucleus is clearly evident

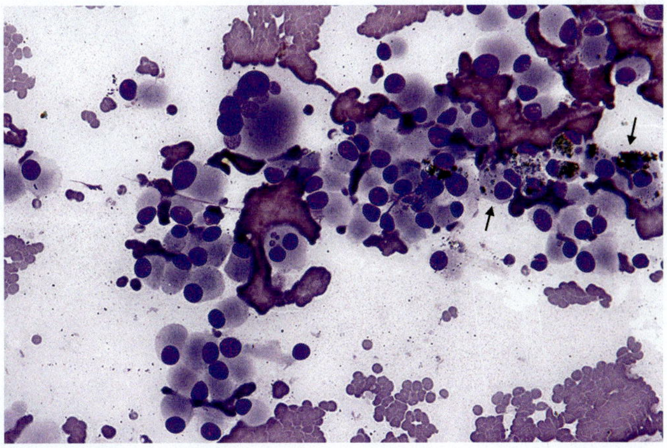

Fig. 8.679 Cytology of signet-ring cell melanoma: pleomorphic melanocytes with multiple atypical nuclei. The melanin present in some cells confirms the melanocytic nature of the signet-ring shaped cells

or few melanin pigment dispersed on the bottom of slides can help to orient the diagnosis (Fig. 8.679).

Balloon cell melanoma is a very rare variant of melanoma, cells are large with wide clear cytoplasm, and nuclei are smaller than usual, giving this aspect to cells [272, 274]. Since usually amelanotic, cells are not easily differentiable from other round cell tumors; also in these cases, if melanin is not evident in some areas of the tumor on histopathological specimens, immunohistochemical markers are needed for diagnosis. Between differentials, liposarcoma, hibernoma, and clear cell carcinomas must be considered.

In general, in poorly differentiated melanoma, marked aspects of malignancy are observed; nuclei with large and often multiple and with bizarre-shaped macronucleoli are commonly detected. In very anaplastic amelanotic

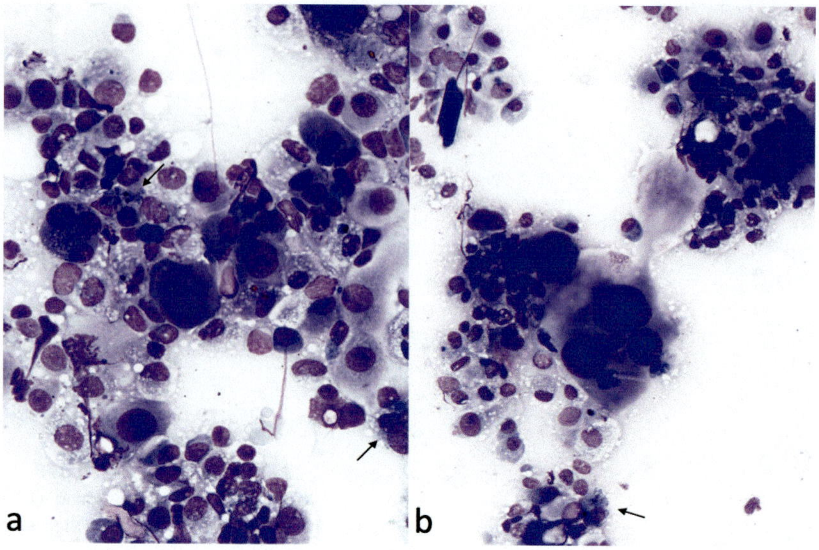

Fig. 8.680 Pleomorphic melanoma: only the detection of few pigmented cells can suggest the melanocytic origin of these pleomorphic cells (arrows)

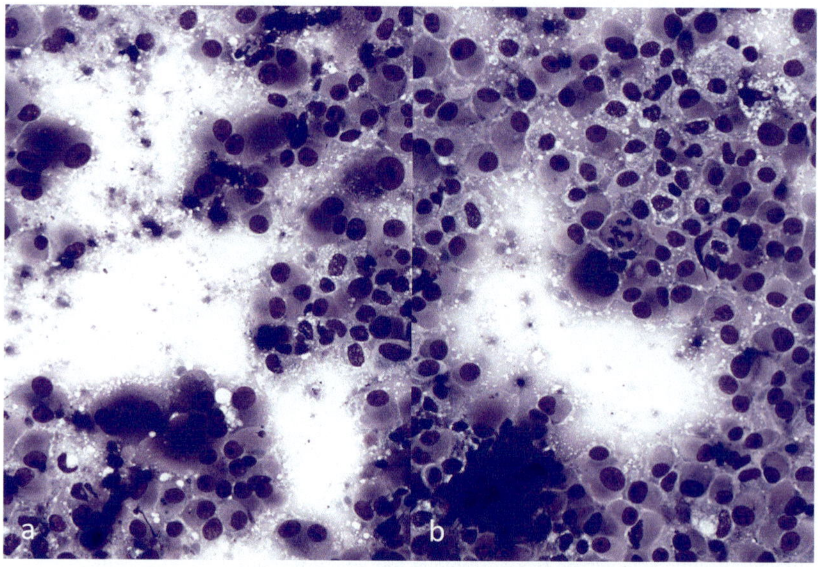

Fig. 8.681 Pleomorphic amelanotic melanoma: no intracytoplasmic melanin and severe atypical cells characterize these two images. In these cases it is impossible to differentiate a melanoma from a histiocytic sarcoma or a pleomorphic plasma cell tumor

melanomas, especially when characterized by spindle and multinucleated cells, a cytological diagnosis is not possible. Similar cytological findings can be indeed observed in specimens coming from anaplastic soft tissue sarcoma or dendritic histiocytic sarcomas (Figs. 8.680 and 8.681).

In amelanotic melanomas, the detection of melanophages intermingled to not pigmented cells can increase the suspicion of a melanocytic neoplasia. The melanophages may be recognizable and therefore differentiated from melanocytes, by the appearance of nuclear shape and chromatin disposition, the presence of cytoplasmic vacuoles filled with a coarser melanin pigment. This cytological differentiation is not always easy, particularly when the amount of pigment obscures the nucleus (Fig. 8.682).

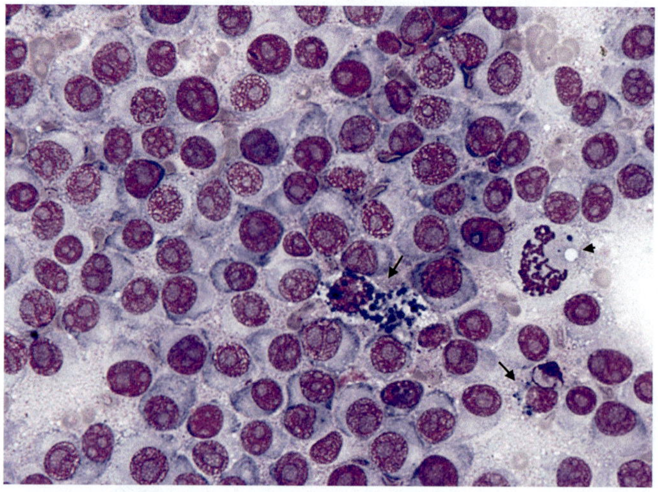

Fig. 8.682 Amelanotic melanoma: two macrophages filled with melanin pigment (melanophages) can alert to the suspicion of an amelanotic melanoma

Bibliography

1. Chalita MC, Matera JM, Alves MT et al (2001) Nonaspiration fine needle cytology and its histologic correlation in canine skin and soft tissue tumors. Anal Quant Cytol Histol 23:395–399
2. Cohen M, Bohling MW, Wright JC et al (2003) Evaluation of sensitivity and specificity of cytologic examination: 269 cases (1999–2000). J Am Vet Med Assoc 222:964–967
3. Ghisleni G, Roccabianca P, Ceruti R et al (2006) Correlation between fine-needle aspiration cytology and histopathology in the evaluation of cutaneous and subcutaneous masses from dogs and cats. Vet Clin Pathol 35:24–30
4. Griffiths GL, Lumsden JH, Valli VEO (1984) Fine needle aspiration cytology and histologic correlation in canine tumors. Vet Clin Pathol 13:13–17
5. Grant Mazie M (2015) Jubb, Kennedy, and palmer's: pathology of domestic animals, vol 1, 6th edn. Elsevier
6. Gross TL, Ihrke PJ, Walder EJ et al (2005) Skin diseases of the dog and cat: clinical and histopathologic diagnosis, 2nd edn. Blackwell Science, Oxford
7. Kiupel M (2022) Surgical Pathologyof tumors of domestic animals. Davis-Thompson Foundation-Gurnee, Illinois
8. Meuten DJ (2016) Tumors in domestic animals. Wiley Blackwell
9. Raskin RE, Meyer DJ (2022) Canine and feline cytology. In: A color atlas and interpretation guide, 4th edn. WB Saunders Co, Philadelphia
10. Valenciano AC, Cowell RL (2019) Cowell and Tyler's diagnostic cytology and hematology of the dog and cat, 5th edn. Elsevier/ Masson, St. Louis
11. Withrow SJ, Vail DM, Page R (2019) Withrow and MacEwen's small animal clinical oncology, 6th edn. Elsevier-Saunders, St. Louis
12. Yager JA, Wilcock BP (1994) Dermatopathology and skin tumors, vol 1, Color atlas and text of surgical pathology of the dog and cat. Mosby Inc, London

Round Cell Tumors Mast Cell Tumors

13. Berlato D, Bulman-Fleming J, Clifford CA et al (2021) Value, limitations, and recommendations for grading of canine cutaneous mast cell tumors: a consensus of the Oncology-Pathology Working Group. Vet Pathol 58(5):858–863
14. Blackwood L, Murphy S, Buracco P et al (2012) European consensus document on mast cell tumours in dogs and cats. Vet Comp Oncol 10:e1–e29
15. Camus MS, Priest HL, Koehler JW, Pierezan F, Costa MP, Lavalle GE (2016) Cytologic criteria for mast cell tumor grading in dogs with evaluation of clinical outcome. Vet Pathol 53(6):1117–1123
16. Goldschmidt MH, Hendrick MJ (2002) Tumors of the skin and soft tissues. In: Meuten DJ (ed) Tumors in domestic animals, 4th edn. Iowa State Press/Blackwell, Ames, pp 45–117
17. Gross TL, Ihrke PJ, Walder EJ et al (2005) Skin diseases of the dog and cat: clinical and histopathologic diagnosis, 2nd edn. Blackwell Science, Oxford, pp 853–865
18. Henry C, Herrera C (2013) Mast cell tumors in cats: clinical update and possible new treatment avenues. J Feline Med Surg 15:41
19. Hergt F, von Bomhard W, Kent MS, Hirschberger J (2016) Use of a 2-tier histologic grading system for canine cutaneous mast cell tumors on cytology specimens. Vet Clin Pathol 45(3):477–483
20. Johnson TO, Schulman FY, Lipscomb TP et al (2002) Histopathology and biologic behavior of pleomorphic cutaneous mast cell tumors in fifteen cats. Vet Pathol 39:452–457
21. Kiupel M, Webster JD, Bailey KL et al (2011) Proposal of a 2-tier histologic grading system for canine cutaneous mast cell tumours to more accurately predict biological behaviour. Vet Pathol 48: 147–155; 21b. Marouda C, Anagnostou T, Brunetti B et al (2024) Cutaneous canine mast cell tumor: the use of proliferative markers (Ki-67 and Ki-67 × AgNOR) in cytological samples for diagnosis and prognosis. Vet Sci. 11(1):23
22. Melville K, Smith KC, Dobromylskyj MJ (2014) Feline cutaneous mast cell tumours: a UK-based study comparing signalment and histological features with long-term outcomes. J Feline Med Surg 17(6):486–493
23. Paes PRO, Horta RS, Luza LC et al (2022) Inclusion of fibroblasts and collagen fibrils in the cytologic grading of canine cutaneous mast cell tumors. Vet Clin Pathol 51(3):339–348
24. Patnaik AK, Ehler WJ, MacEwen EG (1984) Canine cutaneous mast cell tumor: morphologic grading and survival time in 83 dogs. Vet Pathol 21:469–474
25. Pedraza F, Grandi F, Rocha NS (2011) The need for cytologic/ histologic correlation studies to establish a cytologic grading system for canine mast cell tumors in veterinary medicine. Vet Clin Pathol 40:280–281
26. Sabattini S, Bettini G (2010) Prognostic value of histologic and immunohistochemical features in feline cutaneous mast cell tumors. Vet Pathol 47:643–653
27. Sabattini S, Brocanelli A, Zaccone R et al (2024) The 2-tier grading system identifies canine cutaneous and/or subcutaneous mast cell tumors with aggressive biological behavior regardless of growth model. Vet Pathol 61(6):874–881
28. Scarpa F, Sabattini S, Bettini G (2014) Cytological grading of canine cutaneous mast cell tumours. Vet Comp Oncol 14(3): 245–251
29. Welle MM, Rohrer Bley C, Howard J et al (2008) Canine mast cell tumours: a review of the pathogenesis, clinical features, pathology and treatment. Vet Dermatol 19:321–339
30. Yang C, Bradley CW, Preziosi D et al (2023) Cutaneous mastocytosis in 8 young dogs and review of literature. Vet Pathol; 30b. Wyatt EK, Affolter V, Borio S et al (2024) Mastocytosis in the skin in dogs: a multicentric case series. Vet Comp Oncol 22(1): 136–148

Cutaneous Lymphoma

31. Adachi M, Igarashi H, Okamoto M et al (2022) Large granular lymphocyte lymphoma in the skin and urinary bladder of a dog. J Vet Med Sci 84(2):296–301
32. Affolter VK, Gross TL, Moore PF (2009) Indolent cutaneous T-cell lymphoma presenting as cutaneous lymphocytosis in dogs. Vet Dermatol 20(5–6):577–585
33. Albanese F, Abramo F, Marino M et al (2022) Feline and canine cutaneous lymphocytosis: reactive process or indolent neoplastic disease? Vet Sci 9(1):26
34. Azuma K, Ohmi A, Goto-Koshino Y et al (2022) Outcomes and prognostic factors in canine epitheliotropic and nonepitheliotropic cutaneous T-cell lymphomas. Vet Comp Oncol 20(1):118–126
35. Brachelente C, Affolter VK, Fondati A et al (2015) CD3 and CD20 coexpression in a case of canine cutaneous epitheliotropic T-cell lymphoma (mycosis fungoides). Vet Pathol 53(3):563–566
36. Dettwiler M, Mauldin EA, Jastrebski S et al (2023) Prognostic clinical and histopathological features of canine cutaneous epitheliotropic T-cell lymphoma. Vet Pathol 60(2):162–171
37. Fontaine J, Bovens C, Bettenay S et al (2009) Canine cutaneous epitheliotropic T-cell lymphoma: a review. Vet Comp Oncol 7:1–14

38. Fontaine J, Heimann M, Day MJ (2010) Canine cutaneous epitheliotropic T-cell lymphoma: a review of 30 cases. Vet Dermatol 21(3):267–275
39. Fontaine J, Heimann M, Day MJ (2011) Cutaneous epitheliotropic T-cell lymphoma in the cat: a review of the literature and five new cases. Vet Dermatol
40. Gross TL, Ihrke PJ, Walder EJ et al (2005) Skin diseases of the dog and cat—clinical and histopathologic diagnosis, 2nd edn. Wiley-Blackwell, Oxford, pp 876–888
41. Keller SM, Moore PF (2012) A novel clonality assay for the assessment of canine T cell proliferations. Vet Immunol Immunopathol 145(1–2):410–419
42. Kodama A, Sakai H, Kobayashi K et al (2008) B-cell intestinal lymphoma with Mott cell differentiation in a 1-year-old miniature dachshund. Vet Clin Pathol 37(4):409–415
43. Miller WH, Griffin CE, Campbell KL (2013) Tumors of lymphoid origin. In: Muller & Kirk's small animal dermatology, 7th edn. Elsevier, St. Louis, pp 810–816
44. Moore PF, Affolter VK, Graham PS et al (2009) Canine epitheliotropic cutaneous T-cell lymphoma: an investigation of T-cell receptor immunophenotype, lesion topography and molecular clonality. Vet Dermatol 20(5–6):569–576
45. Moore PF, Affolter VK, Keller SM (2013) Canine inflamed nonepitheliotropic cutaneous T-cell lymphoma: a diagnostic conundrum. Vet Dermatol 24(1):204–11.e44-5
46. Noland EL, Keller SM, Kiupel M (2018) Subcutaneous panniculitis-like T-cell lymphoma in dogs: morphologic and Immunohistochemical classification. Vet Pathol 55(6):802–808
46b. Roccabianca P, Avallone G, Rodriguez A et al (2016) Cutaneous lymphoma at injection sites: pathological, immunophenotypical, and molecular characterization in 17 Cats. Vet Pathol 53 (4):823–832
47. Rook KA (2019) Canine and feline cutaneous epitheliotropic lymphoma and cutaneous lymphocytosis. Vet Clin North Am Small Anim Pract 49(1):67–81. https://doi.org/10.1016/j.cvsm.2018.08.007. Epub 2018 Sep 24
48. Stacy NI, Nabity MB, Hackendahl N et al (2009) B-cell lymphoma with Mott cell differentiation in two young adult dogs. Vet Clin Pathol 38(1):113–120

Plasma Cell Tumors

49. Boostrom BO, Moore AS, DeRegis CJ et al (2017) Canine cutaneous plasmacytosis: 21 cases (2005–2015). J Vet Intern Med 31(4): 1074–1080
50. Cangul IT, Wijnen M, Van Garderen E et al (2002) Clinico-pathological aspects of canine cutaneous and mucocutaneous plasmacytomas. J Vet Med A Physiol Pathol Clin Med 49(6): 307–312
51. Gross TL, Ihrke PJ, Walder EJ et al (2005) Skin diseases of the dog and cat—clinical and histopathologic diagnosis, 2nd edn. Wiley-Blackwell, Oxford, pp 866–872
52. Ehrensing G, Craig LE (2018) Intravascular neoplastic cells in canine cutaneous plasmacytomas. J Vet Diagn Invest 30(2): 329–332
53. Majzoub M, Breuer W, Platz SJ et al (2003) Histopathologic and immunophenotypic characterization of extramedullary plasmacytomas in nine cats. Vet Pathol 40:249–253
54. Platz SJ, Breuer W, Pfleghaar S et al (1999) Prognostic value of histopathological grading in canine extramedullary plasmacytomas. Vet Pathol 36:23–27
55. Rowland PH, Valentine BA, Stebbins KE et al (1991) Cutaneous plasmacytomas with amyloid in six dogs. Vet Pathol 28:125–130
56. Woldemeskel M (2012) A concise review of amyloidosis in animals. Vet Med Int
57. Woldemeskel M (2007) Primary localized nodular cutaneous amyloidosis in a male neutered Golden retriever. Dtsch Tierarztl Wochenschr 114(12):473–475
58. Zanelli M, Palicelli A, Sanguedolce F et al (2022) Cutaneous involvement in diseases with plasma cell differentiation: diagnostic approach. Curr Oncol 29(5):3026–3043

Transmissible Venereal Tumor

59. Albanese F, Marconato L, Salerni F (2006) Extragenital transmissible venereal tumour associated with circulating neoplastic cells in an immunologically compromised dog. Vet Comp Oncologia 4(1): 57–62
60. Albanese F, Poli A, Millanta F et al (2002) Primary cutaneous extragenital canine transmissible venereal tumor with Leishmania-laden macrophages: a further suggestion of histiocytic origin? Vet Dermatol 13:243–246
61. Cohen D (1985) The canine transmissible venereal tumor: a unique result of tumor progression. Adv Cancer Res 43:75–112
62. Das U, Das AK (2000) Review of canine transmissible venereal sarcoma. Vet Res Commun 24:545–556
63. Faro TAS, de Oliveira EHC (2023) Canine transmissible venereal tumor – from general to molecular characteristics: a review. Anim Genet 54(1):82–89
64. Gonzalez CM, Griffey SM, Naydan DK et al (2000) Canine transmissible venereal tumour: a morphological and immunohistochemical study of 11 tumours in growth phase and during regression after chemotherapy. J Comp Pathol 122(4):241–248
65. Marchal T, Chabanne L, Kaplanski C et al (1997) Immunophenotype of the canine transmissible venereal tumour. Vet Immunol Immunopathol 57(1–2):1–11
66. Mozos E, Mendez A, Gomez-Villamandos JC et al (1996) Immunohistochemical characterization of canine transmissible venereal tumor. Vet Pathol 33:257–263
67. Mukaratirwa S, Gruys E (2003) Canine transmissible venereal tumour: cytogenetic origin, immu- nophenotype, and immunobiology. A review. Vet Q 25(3):101–111
68. Murchison EP, Wedge DC, Alexandrov LB et al (2014) Transmissible dog cancer genome reveals the origin and history of an ancient cell lineage. Science 343(6169):437–440
69. Murgia C, Pritchard JK, Kim SY et al (2006) Clonal origin and evolution of a transmissible cancer. Cell 126(3):477–487
70. Souto EPF, Rissi DR, Oliveira AM et al (2024) Transmissible venereal tumour with encephalic metastasis in dogs. J Comp Pathol 216:25–32

Histiocytic Diseases

71. Affolter VK, Moore PF (2000) Canine cutaneous and systemic histiocytosis: reactive histiocytosis of dermal dendritic cells. Am J Dermatopathol 22(1):40–48
72. Affolter VK, Moore PF (2002) Localized and disseminated histiocytic sarcoma of dendritic cell origin in dogs. Vet Pathol 39(1): 74–83
73. Affolter VK, Moore PF (2006) Feline progressive histiocytosis. Vet Pathol 43:646–655
74. Constantino-Casas F, Mayhew D, Hoather TM et al (2011) The clinical presentation and histopathologic-immunohistochemical classification of histiocytic sarcomas in the Flat Coated Retriever. Vet Pathol 48(3):764–771

75. Coste M, Prata D, Castiglioni V et al (2019) Feline progressive histiocytosis: a retrospective investigation of 26 cases and preliminary study of Ki67 as a prognostic marker. J Vet Diagn Invest 31(6):801–808
76. Gross TL, Ihrke PJ, Walder EJ, Affolter VK (2005) Reactive histiocytosis. In: Skin diseases of the dog and cat: clinical and histopathological diagnosis, 2nd edn. Blackwell Science, Ames, pp 323–327
77. Kraje AC et al (2001) Malignant histiocytosis in 3 cats. J Vet Intern Med 15:252–256
78. Lazzarini G, Abramo F, Albanese F et al (2023) Combined immunohistochemical protocols to differentiate macrophages within the mononuclear-phagocyte system. Ann Anat 249:152107
79. Maina E, Colombo S, Stefanello D (2014) Multiple cutaneous histiocytomas treated with lomustine in a dog. Vet Dermatol 25: 559–562, e98–e99
80. Moore PF (1984) Systemic histiocytosis of Bernese mountain dogs. Vet Pathol 21(6):554–563
81. Moore PF, Rosin A (1986) Malignant histiocytosis of Bernese mountain dogs. Vet Pathol 23(1):1–10
82. Moore PF (2014) A review of histiocytic diseases of dogs and cats. Vet Pathol 51(1):167–184
83. Moore PF (2023) Histiocytic diseases. Vet Clin North Am Small Anim Pract 53(1):121–140
84. Nagata M, Hirata M, Ishida T et al (2000) Progressive Langerhans' cell histiocytosis in a puppy. Vet Dermatol 11(4):241–246
85. Hirabayashi M, Chambers JK, Kishimoto TE et al (2022) Establishment and characterisation of cell lines and xenograft mouse models of canine systemic histiocytosis and disseminated histiocytic sarcoma. Vet Comp Oncol 20(2):465–475
86. Palmeiro BS, Morris DO, Goldschmidt MH et al (2007) Cutaneous reactive histiocytosis in dogs: a retrospective evaluation of 32 cases. Vet Dermatol 18(5):332–340
87. Pinto da Cunha N, Ghisleni G, Scarampella F et al (2014) Cytologic and immunocytochemical characterization of feline progressive histiocytosis. Vet Clin Pathol 43(3):428–436

Epithelial Tumors

Papilloma

88. Goldschmidt et al (2021) In: Kiupel M (ed) Epithelial tumors of the skin, vol 1. Davis-Thompson DVM Foundation, pp 25–33
89. Lange CE, Jennings SH, Diallo A et al (2019) Canine papillomavirus types 1 and 2 in classical papillomas: high abundance, different morphological associations and frequent co-infections. Vet J 250:1–5
90. Lange CE, Tobler K, Brandes K et al (2010) Canine inverted papillomas associated with DNA of four different papillomaviruses. Vet Dermatol 21:287–291
91. Munday JS, Knight CG, Luff JA (2022) Papillomaviral skin diseases of humans, dogs, cats and horses: a comparative review. Part 1: papillomavirus biology and hyperplastic lesions. Vet J 288: 105897
92. Munday JS, Knight CG, Luff JA (2022) Papillomaviral skin diseases of humans, dogs, cats and horses: a comparative review. Part 2: pre-neoplastic and neoplastic diseases. Vet J 288:105898
93. Orlandi M, Mazzei M, Vascellari M et al (2021) Localization and genotyping of canine papillomavirus in canine inverted papillomas. J Vet Diagn Invest 33:1069–1078
94. Resendes AR, Trainor KE, Bera M et al (2023) Claw bed inverted squamous papilloma associated with canine papillomavirus type 2 in a dog. Vet Dermatol
95. Zaugg N, Nespeca G, Hauser B et al (2005) Detection of novel papillomaviruses in canine mucosal, cutaneous and in situ squamous cell carcinomas. Vet Dermatol 16:290–298

Squamous Cell Carcinoma

96. Aupperle-Lellbach H, Heidrich D, Conrad D et al (2023) Comparative study of digital squamous cell carcinoma in Giant, standard, and miniature schnauzers. Animals (Basel) 13(12):1990
97. Favrot C, Welle M, Heimann M et al (2009) Clinical, histologic, and immunohistochemical analyses of feline squamous cell carcinoma in situ. Vet Pathol 46(1):25–33
98. Goldschmidt et al (2021) In: Kiupel M (ed) Epithelial tumors of the skin, vol 1. Davis-Thompson DVM Foundation, pp 43–63
99. Gross TL, Ihrke PJ, Walder EJ, Affolter VK (2005) Epidermal tumors. In: Skin diseases of the dog and cat: clinical and histopathological diagnosis, 2nd edn. Blackwell Science, Ames, pp 581–589
100. Kok MK, Yamashita-Kawanishi N, Chambers JK et al (2019) Pathologic characterization of *Felis catus* papillomavirus type 5 (FcaPV-5)-associated viral plaques and Bowenoid in situ carcinoma in a Domestic Shorthair cat. J Vet Med Sci 81(5):660–666
101. Lascelles BD, Parry AT, Stidworthy MF et al (2000) Squamous cell carcinoma of the nasal planum in 17 dogs. Vet Rec 147(17): 473–476
102. Marconato L, Murgia D, Finotello R et al (2021) Clinical features and outcome of 79 dogs with digital squamous cell carcinoma undergoing treatment: a SIONCOV Observational Study. Front Vet Sci 8:645982
103. Mazzei M, Forzan M, Carlucci V et al (2018) A study of multiple Felis catus papillomavirus types (1, 2, 3, 4) in cat skin lesions in Italy by quantitative PCR. J Feline Med Surg 20(8):772–779
104. Murphy S (2013) Cutaneous squamous cell carcinoma in the cat current understanding and treatment approaches. J Feline Med Surg 15(5):401–407
105. Munday JS, Fairley R, Atkinson K (2016) The detection of Felis catus papillomavirus 3 DNA in a feline bowenoid in situ carcinoma with novel histologic features and benign clinical behavior. J Vet Diagn Invest 28(5):612–615
106. Munday JS, Sharp CR, Beatty JA (2019) Novel viruses: update on the significance of papillomavirus infections in cats. J Feline Med Surg 21:409–418
107. Nespeca G, Grest P, Rosenkrantz WS et al (2006) Detection of novel papillomavirus-like sequences in paraffin-embedded specimens of invasive and in situ squamous cell carcinomas from cats. Am J Vet Res 67(12):2036–2041
108. Vascellari M, Mazzei M, Zanardello C et al (2019) *Felis catus papillomavirus* types 1, 2, 3, 4, and 5 in feline Bowenoid in situ carcinoma: an in situ hybridization study. Vet Pathol 56(6): 818–825
109. Webb JL, Burns RE, Brown HM et al (2009) Squamous cell carcinoma. Compend Contin Educ Vet 31(3):E9
110. Wilhelm S, Degorce-Rubiales F, Godson D et al (2006) Clinical, histological and immunohistochemical study of feline viral plaques and bowenoid in situ carcinomas. Vet Dermatol 17(6):424–431
111. Withrow SJ, Vail DM, Page R (2019) Withrow and MacEwen's small animal clinical oncology, 6th edn. Elsevier-Saunders, St. Louis

Follicular Cysts and Follicular Neoplasia

112. Abramo F, Pratesi F, Cantile C et al (1999) Survey of canine and feline follicular tumours and tumour-like lesions in Central Italy. J Small Anim Pract 40:479–481
113. Adedeji AO, Affolter VK, Christopher MM (2017) Cytologic features of cutaneous follicular tumors and cysts in dogs. Vet Clin Pathol 46(1):143–150
114. Campos AG, Cogliati B, Guerra JM et al (2014) Multiple trichoblastomas in a dog. Vet Dermatol 25(1):48–e19
115. Carroll EE, Fossey SL, Mangus LM et al (2010) Malignant pilomatricoma in 3 dogs. Vet Pathol 47:937–943
116. Goldschmidt et al (2021) In: Kiupel M (ed) Epithelial tumors of the skin, vol 1. Davis-Thompson DVM Foundation
117. Gross TL, Ihrke PJ, Walder EJ, Affolter VK (2005) Follicular tumors. In: Skin diseases of the dog and cat: clinical and histopathological diagnosis, 2nd edn. Blackwell Science, Ames, pp 608–639
118. Martano M et al (2013) Malignant pilomatricoma with multiple bone metastases in a dog: histological and immunohistochemical study. Exp Ther Med 5:1005–1008
119. Masserdotti C, Ubbiali FA (2002) Fine needle aspiration cytology of pilomatricoma in three dogs. Vet Clin Pathol 31(1):22–25
120. Mineshige T et al (2014) Trichoblastoma with abundant plump stromal cells in a dog. J Vet Med Sci 76:735–739
121. Park JK, Hong IH, Ki MR et al (2010) Multiple perianal infundibular follicular cysts in a dog. Vet Dermatol 21(3):303–306
122. Scott DW, Anderson WI (1991) Canine hair follicle neoplasms: a retrospective analysis of 80 cases (1986–1987). Vet Dermatol 2: 143–150
123. Stockhaus C, Teske E, Rudolph R et al (2001) Assessment of cytological criteria for diagnosing basal cell tumours in the dog and cat. J Small Anim Pract 42(12):582–586
124. Toma S, Noli C (2005) Isotretinoin in the treatment of multiple benign pilomatrixomas in a mixed-breed dog. Vet Dermatol 16(5): 346–350
125. White A, Stern A, Campbell K et al (2013) Multiple (disseminated) follicular cysts in five dogs and one cat. Vet Rec 173(11):269
126. Wiener DJ (2021) Histologic features of hair follicle neoplasms and cysts in dogs and cats: a diagnostic guide. J Vet Diagn Invest 33(3):479–497

Sebaceous and Hepatoid Glands Neoplasia

127. Bettini G, Morini M, Mandrioli L et al (2009) CNS and lung metastasis of sebaceous epithelioma in a dog. Vet Dermatol 20: 289–294
128. Chantawong P, Mamom T, Tangtrongsup S et al (2022) First study on the immunohistochemical expression of cyclooxygenase-2 and clinicopathological association in canine hepatoid gland neoplasms. Vet World 15(10):2432–2441
129. Gross TL, Ihrke PJ, Walder EJ, Affolter VK (2005) Sebaceous tumors. In: Skin diseases of the dog and cat: clinical and histopathological diagnosis, 2nd edn. Blackwell Science, Ames, pp 641–664
130. Goldschmidt et al (2021) In: Kiupel M (ed) Epithelial tumors of the skin, vol 1. Davis-Thompson DVM Foundation
131. Evans SJM, Connolly SL, Schaffer PA et al (2018) Basal cell enumeration does not predict malignancy in canine perianal gland tumor cytology. Vet Clin Pathol 47(4):634–637
132. McCourt MR, Levine GM, Breshears MA et al (2018) Metastatic disease in a dog with a well-differentiated perianal gland tumor. Vet Clin Pathol 47(4):649–653
133. Sabattini S, Renzi A, Rigillo A et al (2019) Cytological differentiation between benign and malignant perianal gland proliferative lesions in dogs: a preliminary study. JSAP
134. Yumusak N, Caliskan M, Kutsal O (2016) Fine needle aspiration cytology (FNAC) in the diagnosis of canine hepatoid gland tumors-a comparative study with histopathology. Ankara Univ Vet Fak Der 63:259–266

Sweat Gland and Anal Sac Neoplasia

135. Baharak A, Reza K, Shahriar D et al (2012) Metastatic apocrine sweat gland adenocarcinoma in a terrier dog. Asian Pac J Trop Biomed 2(8):670–672
136. Chaitman J, van der Woerdt A, Bartick TE (1999) Multiple eyelid cysts resembling apocrine hidrocystomas in three Persian cats and one Himalayan cat. Vet Pathol 36:474–476
137. Goldschmidt et al (2021) In: Kiupel M (ed) Epithelial tumors of the skin, vol 1. Davis-Thompson DVM Foundation
138. Gross TL, Ihrke PJ, Walder EJ, Affolter VK (2005) Sweat gland tumors. In: Skin diseases of the dog and cat: clinical and histopathological diagnosis, 2nd edn. Blackwell Science, Ames, pp 665–689
139. Kalaher KM, Anderson WI, Scott DW (1990) Neoplasms of the apocrine sweat glands in 44 dogs and 10 cats. Vet Rec 127:400–403
140. Marignac G, Barlerin L, Meunier V et al (2002) Apocrine cystadenomatosis in three related Persian cats. In: Proceedings of the ISVD meeting, Nice, p 30
141. Michishita M, Yasui J, Nakahira R et al (2013) Apocrine sweat gland ductal adenoma with sebaceous differentiation in a dog. Case Rep Vet Med 2013:1–4
142. Moon CH, Min HA, Lee HB et al (2022) Retroperitoneal metastatic apocrine gland ductal adenocarcinoma in a beagle dog. Vet Sci 9(5):234
143. Repasy AB, Selmic LE, Kisseberth WC (2022) Canine apocrine gland anal sac adenocarcinoma: a review. Top Companion Anim Med 50:100682. https://doi.org/10.1016/j.tcam.2022
144. Sakai H, Murakami M, Mishima H et al (2012) Cytologically atypical anal sac adenocarcinoma in a dog. Vet Clin Pathol 41(2): 291–294
145. Simko E, Wilcock BP, Yager JA (2003) A retrospective study of 44 canine apocrine sweat gland adenocarcinomas. Can Vet J 44: 38–42
146. Shoieb AM, Hanshaw DM (2009) Anal sac gland carcinoma in 64 cats in the United Kingdom (1995–2007). Vet Pathol 46(4): 677–683
147. Suzuki K, Morita R, Hojo Y et al (2013) Immunohistochemical characterization of neuroendocrine differentiation of canine anal sac glandular tumours. J Comp Pathol 149(2–3):199–207
148. Williams LE, Gliatto JM, Dodge RK et al (2003) Carcinoma of the apocrine glands of the anal sac in dogs: 113 cases (1985–1995). J Am Vet Med Assoc 223(6):825–831
149. Wong H, Byrne S, Rasotto R et al (2021) A retrospective study of clinical and histopathological features of 81 cases of canine apocrine gland adenocarcinoma of the anal sac: independent clinical and histopathological risk factors associated with outcome. Animals (Basel) 11(11):3327

Clear Cells Adnexal Carcinoma

150. Goldschmidt et al (2021) In: Kiupel M (ed) Epithelial tumors of the skin, vol 1. Davis-Thompson DVM Foundation

151. Kari K, Miller MA, Ramos-Vara JA (2013) Clear cell differentiation in cutaneous tumors of dogs and cats. Vet Pathol 50(6): 1050–1057
152. Jiménez-Heffernan JA, Rodríguez-García AM, Cima L (2022) A comprehensive review of the "tigroid" background cytological concept: what, when, where and why? Pathologica 114(2): 121–127
153. Yasuno K, Nishiyama S, Suetsugu F et al (2009) Cutaneous clear cell adnexal carcinoma in a dog: special reference to cytokeratin expression. J Vet Med Sci 71(11):1513–1517
154. Piviani M, Sánchez MD, Patel RT (2012) Cytologic features of clear cell adnexal carcinoma in 3 dogs. Vet Clin Pathol 41(3): 405–411
155. Schulman FY, Lipscomb TP, Atkin TJ (2005) Canine cutaneous clear cell adnexal carcinoma: histopathology, immunohistochemistry, and biologic behavior of 26 cases. J Vet Diagn Invest 17(5):403–411

Fibroadnexal Hamartoma (Fibroadnexal Dysplasia)

156. Abramo F, Guido P, Davide L et al (2003) Canine fibroadnexal hamartoma (focal adnexal dysplasia): an epidemiological survey. In: Proceedings of the 9th ESVD meeting, Tenerife, p 154. (poster)
157. Goldschmidt et al (2021) In: Kiupel M (ed) Epithelial tumors of the skin, vol 1. Davis-Thompson DVM Foundation
158. Lores FH, Conceicao LG (2009) Nevi and cutaneous hamartomas in dogs: retrospective clinical and epidemiologic study of 81 cases. Ciência Rural, Santa Maria 39(9)
159. Lores FH, Conceicao LG, Acha LMP (2019) Fibroadnexal hamartoma in the dog: retrospective epidemiological and histopathological study of 102 cases. Cienc Rural 49(9):1–7

Merkel Tumor (Merkel Cells Carcinoma)

160. Bagnasco G, Properzi R, Porto R et al (2003) Feline cutaneous neuroendocrine carcinoma (Merkel cell tumour): clinical and pathological findings. Vet Dermatol 14(2):111–115
161. Chambers JK, Ito S, Uchida K (2023) Feline papillomavirus-associated Merkel cell carcinoma: a comparative review with human Merkel cell carcinoma. J Vet Med Sci 85(11):1195–1209
162. Dohata A, Chambers JK, Uchida K et al (2015) Clinical and pathologic study of feline Merkel cell carcinoma with Immunohistochemical characterization of Normal and neoplastic Merkel cells. Vet Pathol 52(6):1012–1018
163. Goldschmidt et al (2021) In: Kiupel M (ed) Epithelial tumors of the skin, vol 1. Davis-Thompson DVM Foundation
164. Ito S, Chambers JK, Sumi A et al (2022) Involvement of Felis catus papillomavirus type 2 in the tumorigenesis of feline Merkel cell carcinoma. Vet Pathol 59(1):63–74
165. Ito S, Chambers JK, Sumi A et al (2023) Genomic integration and expression of *Felis catus* papillomavirus type 2 oncogenes in feline Merkel cell carcinoma. Vet Pathol 60:21–34
166. Joiner KS, Smith AN, Henderson RA et al (2010) Multicentric cutaneous neuroendocrine (Merkel cell) carcinoma in a dog. Vet Pathol 47(6):1090–1094
167. Patnaik AK, Post GS, Erlandson RA (2001) Clinicopathologic and electron microscopic study of cutaneous neuroendocrine (Merkel cell) carcinoma in a cat with comparisons to human and canine tumors. Vet Pathol 38(5):553–556
168. Sumi A, Chambers JK, Doi M et al (2018) Clinical features and outcomes of Merkel cell carcinoma in 20 cats. Vet Comp Oncol 16(4):554–561
169. van der Steen FEMM, Grinwis GCM, Weerts EAWS, Teske E (2020) Feline and canine Merkel cell carcinoma: a case series and discussion on cellular origin. Vet Comp Oncol 19(2):393–398

Mesenchymal (Spindle Cell) Neoplasia

Fibroma and Fibrosarcoma

170. Dennis MM, McSporran KD, Bacon NJ et al (2011) Prognostic factors for cutaneous and subcutaneous soft tissue sarcomas in dogs. Vet Pathol 48(1):73–84
171. Dobromylskyj MJ, Richards V, Smith KC (2021) Prognostic factors and proposed grading system for cutaneous and subcutaneous soft tissue sarcomas in cats, based on a retrospective study. J Feline Med Surg 23(2):168–174
172. Dobromylskyj MJ (2022) Feline soft tissue sarcomas: a review of the classification and histological grading, with comparison to human and canine. Animals (Basel) 12(20):2736
173. Evans SJM, Frank CB, Avery PR, Moore AR (2018) What is your diagnosis? Subcutaneous mass on a dog. Vet Clin Pathol 47(1): 160–161
174. Gross TL, Ihrke PJ, Walder EJ, Affolter VK (2005) Fibroma and Fibrosarcoma. In: Skin diseases of the dog and cat: clinical and histopathological diagnosis, 2nd edn. Blackwell Science, Ames, pp 719–727
175. Gumber S, Wakamatsu N (2015) Vaccine-associated fibrosarcoma with keloidal differentiation in a cat. J Vet Diagn Invest 23(5): 1061–1064
176. Harasen GL (1984) Multicentric fibrosarcoma in a cat and a review of the literature. Can Vet J 25(5):207–210
177. Hartman K, Day MJ, Thiry E et al (2015) Feline injection-site sarcoma ABCD guidelines on prevention and management. J Feline Med Surg 17:606–613
178. Hauck M (2003) Feline injection site sarcomas. Vet Clin North Am Small Anim Pract 33(3):553–557
179. Hendrick M (1998) Historical review and current knowledge of risk factors involved in feline vaccine-associated sarcomas. J Am Vet Med Assoc 213(10):1422–1423
180. Hendrick MJ, Shofer FS, Goldschmidt MH et al (1994) Comparison of fibrosarcomas that developed at vaccination sites and at nonvaccination sites in cats: 239 cases (1991–1992). J Am Vet Med Assoc 205(10):1425–1429
181. Little LK, Goldschmidt M (2007) Cytologic appearance of a keloidal fibrosarcoma in a dog. Vet Clin Pathol 36(4):364–367
182. Macy DW, Hendrick MJ (1996) The potential role of inflammation in the development of postvaccinal sarcomas in cats. In: Rosenthal RC, Jeglum KA (eds) Controversies in clinical oncology, vol 26. W.B. Saunders, Philadelphia, pp 103–109
183. Madewell BR, Griffey SM, McEntee MC et al (2001) Feline vaccine-associated fibrosarcoma: an ultrastructural study of 20 tumors (1996–1999). Vet Pathol 38(2):196–202
184. Martano M, Morello E, Buracco P (2011) Feline injection-sarcoma: past, present and perspectives. Vet J 188:136–141. 488 4 Cytology of Skin Tumours
185. Masserdotti C, Drigo M (2022) Cytologic features of reactive fibroplasia in cutaneous and subcutaneous lesions of dogs: a retrospective study. Vet Clin Pathol 51(2):244–251
186. Mikaelian I, Gross TL (2002) Keloidal fibromas and fibrosarcomas in the dog. Vet Pathol 39(1):149–153

187. Millanta F, Asproni P, Aquino G et al (2020) Cytologic grading of canine and feline spindle-cell sarcomas of soft tissues and its correlation with histologic grading. Top Companion Anim Med 41:100458
188. Palić J, Saey V, van der Lugt J, Baumgärtner W, Hülskötter K (2023) Cytologic, histopathologic, and immunohistochemical features of keloidal fibroma in a dog. Vet Clin Pathol 52(4): 676–680
189. Roccabianca P, Schulman Y, Avallone G et al (2020) Surgical pathology of tumors of domestic animals 3: tumor like lesions. In: Tumors of soft tissues, 4th edn. Davis Thomson Foundation, Gurnee
190. Sanchez-Redondo S, Hare CHZ, Constantino-Casas F et al (2021) Correlation between cytologic features and histologic grades in cutaneous and subcutaneous soft tissue sarcomas in dogs-A pilot study. Vet Clin Pathol 50(2):236–239

Myxoma and Myxosarcoma

191. Graadt van Roggen JF, Hogendoorn PC, Fletcher CD (1999) Myxoid tumours of soft tissue. Histopathology 35:291–312
192. Gross TL, Ihrke PJ, Walder EJ et al (2005) Skin diseases of the dog and cat: clinical and histopathologic diagnosis, 2nd edn. Blackwell Science, Oxford, pp 727–730
193. Roccabianca P, Schulman Y, Avallone G et al (2020) Surgical pathology of tumors of domestic animals 3: tumor like lesions. In: Tumors of soft tissues, 4th edn. Davis Thomson Foundation, Gurnee
194. Willems SM, Schrage YM, Baelde JJ et al (2008) Myxoid tumours of soft tissue: the so-called myxoid extracellular matrix is heterogeneous in composition. Histopathology 52(4):465–474

Hematic and Lymphatic Vessel Tumors

195. Abramo F, Vascellari M, Colzi G et al (2022) Identification of histopathological criteria for the diagnosis of canine cutaneous progressive angiomatosis. Vet Sci 9(7):340
196. Baron CP, Puntel FC, Fukushima FB et al (2020) Progressive cutaneous angiomatosis in the metatarsal region of a cat. J Am Vet Med Assoc 256(2)
197. Bertazzolo W, Dell'Orco M, Bonfanti U et al (2005) Canine angiosarcoma: cytologic, histologic, and immunohistochemical correlations. Vet Clin Pathol 34(1):28–34
198. Bellamy E, Larsen Moberg H, Suárez-Bonnet A et al (2024) Feline ventral abdominal wall angiosarcoma: haemangiosarcoma or lymphangiosarcoma? Clinical and pathological characteristics in nine cases. J Feline Med Surg 26(1)
199. Brown NO, Patnaik AK, MacEwen EG (1985) Canine hemangiosarcoma: retrospective analysis of 104 cases. J Am Vet Med Assoc 186:56–58
200. De Nardi AB, de Oliveira Massoco Salles Gomes C, Fonseca-Alves CE et al (2023) Diagnosis, prognosis, and treatment of canine Hemangiosarcoma: a review based on a consensus organized by the Brazilian Association of Veterinary Oncology, ABROVET. Cancers (Basel) 15(7):2025
201. Gross TL, Ihrke PJ, Walder EJ, Affolter VK (2005) Hemangioma and Hemangiosarcoma. In: Skin diseases of the dog and cat: clinical and histopathological diagnosis, 2nd edn. Blackwell Science, Ames, pp 741–753
202. Hargis AM, Ihrke PJ, Spangler WL et al (1992) A retrospective clinicopathologic study of 212 dogs with cutaneous hemangiomas and hemangiosarcomas. Vet Pathol 29:316–328
203. Johannes CM, Henry CJ, Turnquist SE et al (2007) Hemangiosarcoma in cats: 53 cases (1992–2002). J Am Vet Med Assoc 231(12):1851–1856
204. McAbee KP, Ludwig LL, Bergman PJ et al (2005) Feline cutaneous hemangiosarcoma: a retrospective study of 18 cases (1998–2003). J Am Anim Hosp Assoc 41(2):110–116
205. Miller MA, Ramos JA, Kreeger JM (1992) Cutaneous vascular neoplasia in 15 cats: clinical, morphological and immunohistochemical studies. Vet Pathol 29:329–336
206. Olivieri L, Nardini G, Pengo G et al (2010) Cutaneous progressive angiomatosis on the muzzle of a dog, treated by laser photocoagulation therapy. Vet Dermatol 21:517–521
207. Roccabianca P, Schulman Y, Avallone G et al (2020) Surgical pathology of tumors of domestic animals 3: tumor like lesions. In: Tumors of soft tissues, 4th edn. Davis Thomson Foundation, Gurnee, pp 149–199
208. Szivek A, Burns RE, Gericota B et al (2012) Clinical outcome in 94 cases of dermal haemangiosarcoma in dogs treated with surgical excision: 1993–2007*. Vet Comp Oncol 10(1):65–73
209. Ward H, Fox LE, Calderwood-Mays MB et al (1994) Cutaneous hemangiosarcoma in 25 dogs: a retrospective study. J Vet Intern Med 8:345–348

Perivascular Wall Tumors

210. Avallone G, Boracchi P, Stefanello D et al (2014) Canine perivascular wall tumors: high prognostic impact of site, depth, and completeness of margins. Vet Pathol 51(4):713–721
211. Avallone G, Helmbold P, Caniatti M et al (2007) The spectrum of canine cutaneous perivascular wall tumors: morphologic, phenotypic and clinical characterization. Vet Pathol 44(5):607–620
212. Avallone G, Stefanello D, Ferrari R, Roccabianca P (2020) The controversial histologic classification of canine subcutaneous whorling tumors: the path to perivascular wall tumors. Vet Comp Oncol 18(1):3–8
213. Caniatti M, Ghisleni G, Ceruti R et al (2001) Cytological features of canine haemangiopericytoma in fine needle aspiration biopsy. Vet Rec 149:242–244
214. Dennis MM, McSporran KD, Bacon NJ et al (2011) Prognostic factors for cutaneous and subcutaneous soft tissue sarcomas in dogs. Vet Pathol 48(1):73–84
215. Mentzel T, Dei Tos AP, Sapi Z et al (2006) Myopericytoma of skin and soft tissues: clinicopathologic and immunohistochemical study of 54 cases. Am J Surg Pathol 30(1):104–113
216. Palmieri C, Avallone G, Cimini M et al (2013) Use of electron microscopy to classify canine perivascular wall tumors. Vet Pathol 50(2):226–233
217. Roccabianca P, Schulman Y, Avallone G et al (2020) Surgical pathology of tumors of domestic animals 3: tumor like lesions. In: Tumors of soft tissues, 4th edn. Davis Thomson Foundation, Gurnee
218. Stefanello D, Avallone G, Ferrari R (2011) Canine cutaneous perivascular wall tumors at first presentation: clinical behavior and prognostic factors in 55 cases. J Vet Intern Med 25(6): 1398–1405
219. Suzuki S, Uchida K, Nakayama H (2014) The effects of tumor location on diagnostic criteria for canine malignant peripheral nerve sheath tumors (MPNSTs) and the markers for distinction between canine MPNSTs and canine perivascular wall tumors. Vet Pathol 51(4):722–736

Nerve Sheath Tumors (NST)

220. Chijiwa K, Uchida K, Tateyama S (2004) Immunohistochemical evaluation of canine peripheral nerve sheath tumors and other soft tissue sarcomas. Vet Pathol 41:307–318
221. Cooper-Khan RS, Frankovich AN, Thompson CA, Thomovsky SA, Lewis MJ (2024) Clinical findings and outcome in 30 dogs with presumptive or confirmed nerve sheath tumors. Vet Sci 11(5): 192
222. Gaitero L, Añor S, Fondevila D et al (2008) Canine cutaneous spindle cell tumours with features of peripheral nerve sheath tumours: a histopathological and immunohistochemical study. J Comp Pathol 139:16–23
223. Gross TL, Ihrke PJ, Walder EJ, Affolter VK (2005) Peripheral nerve sheath tumor. In: Skin diseases of the dog and cat: clinical and histopathological diagnosis, 2nd edn. Blackwell Science, Ames, pp 789–797
224. Koestner A, Higgins RJ (2002) Tumours of the nervous system. In: Meuten DJ (ed) Tumours in domestic animals, 4th edn. Iowa State Press, Ames, p 697e738
225. Roccabianca P, Schulman Y, Avallone G et al (2020) Surgical pathology of tumors of domestic animals 3: tumor like lesions. In: Tumors of soft tissues, 4th edn. Davis Thomson Foundation, Gurnee
226. Schulman FY, Johnson TO, Facemire PR et al (2009) Feline peripheral nerve sheath tumors: histologic, immunohistochemical, and clinicopathologic correlation (59 tumors in 53 cats). Vet Pathol 46:1166–1180
227. Soto S, Blatter S, Hobi S et al (2024) Feline facial spindle cell tumors in 29 cats: Histomorphological and Immunohistochemical characterization. Animals (Basel) 14(7):1103
228. Suzuki S, Uchida K, Nakayama H (2014) The effects of tumor location on diagnostic criteria for canine malignant peripheral nerve sheath tumors (MPNSTs) and the markers for distinction between canine MPNSTs and canine perivascular wall tumors. Vet Pathol 51(4):722–736
229. Tekavec K, Švara T, Knific T et al (2022) Histopathological and Immunohistochemical evaluation of canine nerve sheath tumors and proposal for an updated classification. Vet Sci 9(5):204
230. Warren AL, Miller AD, de Lahunta A et al (2020) Four cases of the Melanotic variant of malignant nerve sheath tumour: a rare, aggressive neoplasm in young dogs with a predilection for the spinal cord. J Comp Pathol 178:1–8

Lipoma and Liposarcoma and Ibernoma

231. Amorim I, Faria F, Taulescu M et al (2021) Nipple Hibernoma in a dog: a case report with literature review. Front Vet Sci 8:627288
232. Avallone G, Pellegrino V, Muscatello LV et al (2017) Spindle cell lipoma in dogs. Vet Pathol 54(5):792–794
233. Baez JL, Hendrick MJ, Shofer FS et al (2004) Liposarcomas in dogs: 56 cases (1989–2000). J Am Vet Med Assoc 224:887–891
234. Dzimira S, Kapusniak V, Madej JA (2015) Immunohistochemical diagnostic of hibernoma in dog. Pol J Vet Sci 18:233–236
235. Gross TL, Ihrke PJ, Walder EJ, Affolter VK (2005) Lipoma and Liposarcoma. In: Skin diseases of the dog and cat: clinical and histopathological diagnosis, 2nd edn. Blackwell Science, Ames, pp 766–778
236. Liggett AD, Frazier KS, Styer EL (2002) Angiolipomatous tumors in dogs and a cat. Vet Pathol 39:286–289
237. Masserdotti C, Bonfanti U, De Lorenzi D et al (2006) Use of Oil Red O stain in the cytologic diagnosis of canine liposarcoma. Vet Clin Pathol 35(1):37–41
238. Messick JB, Radin MJ (1989) Cytologic, histologic, and ultrastructural characteristics of a canine myxoid liposarcoma. Vet Pathol 26:520–522
239. Piccione J, Dial SM (2020) Cytologic appearance of hibernoma in two dogs. Vet Clin Pathol 49(1):125–129. https://doi.org/10.1111/vcp.12836. Epub Mar 22
240. Ravi M, Schobert CS, Kiupel M et al (2014) Clinical, morphologic, and immunohistochemical features of canine orbital hibernomas. Vet Pathol 51(3):563–568
241. Roccabianca P, Schulman Y, Avallone G et al (2020) Surgical pathology of tumors of domestic animals 3: tumor like lesions. In: Tumors of soft tissues, 4th edn. Davis Thomson Foundation, Gurnee
242. Stuckey JA, Rankin AJ, Romkes G et al (2015) Subconjunctival hibernoma in a dog. Vet Ophthalmol 18(1):78–82

Pleomorphic Sarcoma with Giant Cells

243. Couto SS, Griffey SM, Duarte PC et al (2002) Feline vaccine-associated fibrosarcoma: morphologic distinctions. Vet Pathol 39: 33–41
244. de Cecco BS, Argenta FF, Bianchi RM et al (2021) Feline giant-cell pleomorphic sarcoma: cytologic, histologic and immunohistochemical characterization. J Feline Med Surg 23(8):738–744
245. Dobromylskyj M (2022) Feline soft tissue sarcomas: a review of the classification and histological grading, with comparison to human and canine. Animals (Basel) 12(20):2736
246. Roccabianca P, Schulman Y, Avallone G et al (2020) Surgical pathology of tumors of domestic animals 3: tumor like lesions. In: Tumors of soft tissues, 4th edn. Davis Thomson Foundation, Gurnee

Dermatofibrosis

247. Ciccarelli S, Di Bello A, Valastro C et al (2019) Unilateral renal cystadenocarcinoma and nodular dermatofibrosis in a mixed-breed dog carrying a FLCN gene mutation. Vet Dermatol. https://doi.org/10.1111/vde.12719
248. Lingaas F, Comstock KE, Kirkness EF et al (2003) A mutation in the canine BHD gene is associated with hereditary multifocal renal cystadenocarcinoma and nodular dermatofibrosis in the German shepherd dog. Hum Mol Genet 12(23):3043–3053
249. Lium B, Moe L (1985) Hereditary multifocal renal cystadenocarcinomas and nodular dermatofibrosis in the German shepherd dog: macroscopic and histopathologic changes. Vet Pathol 22(5):447–455
250. Moe L, Lium B (1997) Hereditary multifocal renal cystadenocarcinomas and nodular dermatofibrosis in 51 German shepherd dogs. J Small Anim Pract 38(11):498–505
251. Zanatta M, Bettini G, Scarpa F et al (2013) Nodular dermatofibrosis in a dog without a renal tumour or a mutation in the folliculin gene. J Comp Pathol 148(2–3):248–251

Extraskeletal Osteosarcoma

252. Barger A, Graca R, Bailey K et al (2005) Use of alkaline phosphatase staining to differentiate canine osteosarcoma from other vimentin-positive tumors. Vet Pathol 42:161–165
253. Langenbach A, Anderson MA, Dambach DM, Sorenmo KU, Shofer FD (1998) Extraskeletal osteosarcomas in dogs: a

retrospective study of 169 cases (1986–1996). J Am Anim Hosp Assoc 34(2):113–120
254. Patnaik AK (1990) Canine extraskeletal osteosarcoma and chondrosarcoma: a clinicopathologic study of 14 cases. Vet Pathol 27(1):46–55
255. Ryseff JK, Bohn AA (2012) Detection of alkaline phosphatase in canine cells previously stained with Wright–Giemsa and its utility in differentiating osteosarcoma from other mesenchymal tumors. Vet Clin Pathol 41:1–5
256. Roccabianca P, Schulman Y, Avallone G et al (2020) Surgical pathology of tumors of domestic animals 3: tumor like lesions. In: Tumors of soft tissues, 4th edn. Davis Thomson Foundation, Gurnee

Melanocytic Tumors

257. Bergin IL, Smedley RC, Esplin DG et al (2011) Prognostic evaluation of Ki67 threshold value in canine Oral melanoma. Vet Pathol 48:41–53
258. Cangul IT, van Garderen E, van der Linde-Sipman JS et al (2001) Canine balloon and signet-ring cell melanomas: a histological and immunohistochemical characterization. J Comp Pathol 125(2–3): 166–173
259. Choi C, Kusewitt DF (2003) Comparison of tyrosinase-related protein-2, S-100, and Melan A immunoreactivity in canine amelanotic melanomas. Vet Pathol 40(6):713–718
260. Gross TL, Ihrke PJ, Walder EJ, Affolter VK (2005) Melanocytic tumors. In: Skin diseases of the dog and cat: clinical and histopathological diagnosis, 2nd edn. Blackwell Science, Ames, pp 608–639
261. Hirz M, Herden C (2016) Cutaneous amelanotic signet-ring cell malignant melanoma with interspersed myofibroblastic differentiation in a young cat. J Vet Diagn Invest 28(4):429–435
262. Jeon MD, Leeper HJ, Cook MR, McMillan SK, Bennett T, Murray CA, Tripp CD, Curran KM (2022) Multi-institutional retrospective study of canine foot pad malignant melanomas: 20 cases. Vet Comp Oncol 20(4):854–861
263. Luna LD, Higginbotham ML, Henry CJ (2000) Feline non-ocular melanoma: a retrospective study of 23 cases (1991–1999). J Feline Med Surg 2(4):173–181
264. Mendes RE, Tolbert RH, Thorn CE, Rissi DR (2024) Dermal melanoma with plasmacytoid differentiation in a dog. J Vet Diagn Invest 36(6):941–944
265. Polton G, Borrego JF, Clemente-Vicario F et al (2024) Melanoma of the dog and cat: consensus and guidelines. Front Vet Sci 11: 1359426
266. Sabattini S, Renzi A, Albanese F et al (2018) Evaluation of Ki-67 expression in feline non-ocular melanocytic tumours. BMC Vet Res 14(1):309
267. Schultheiss PC (2006) Histologic features and clinical outcomes of melanomas of lip, haired skin, and nail bed locations of dogs. J Vet Diagn Invest 18(4):422–425
268. Smedley RC, Spangler WL, Esplin DG et al (2011) Prognostic markers for canine melanocytic neoplasms: a comparative review of the literature and goals for future investigation. Vet Pathol 48(1): 54–72
269. Smedley RC, Bongiovanni L, Bacmeister C et al (2022) Diagnosis and histopathologic prognostication of canine melanocytic neoplasms: a consensus of the Oncology-Pathology Working Group. Vet Comp Oncol 20(4):739–751
270. Spangler WL, Kass PH (2006) The histologic and epidemiologic basis for prognostic considerations in canine melanocytic neoplasia. Vet Pathol 43:136–142
271. Porcellato I, Orlandi M, Lo Giudice A (2023) Canine melanocytes: immunohistochemical expression of melanocytic markers in different somatic areas. Vet Dermatol 34(4):284–297
272. van der Linde-Sipman JS, de Wit MM, van Garderen E et al (1997) Cutaneous malignant melanomas in 57 cats: identification of (amelanotic) signet-ring and balloon cell types and verification of their origin by immunohistochemistry, electron microscopy, and in situ hybridization. Vet Pathol 34(1):31–38
273. Withrow SJ, Vail DM, Page R (2019) Withrow and MacEwen's small animal clinical oncology, 6th edn. Elsevier-Saunders, St. Louis
274. Wilkerson MJ, Dolce K, DeBey BM et al (2003) Metastatic balloon cell melanoma in a dog. Vet Clin Pathol 32(1):31

9 Cutaneous Metastasis from Non-primary Skin Tumors

9.1 Introduction

Skin metastases from non-primary skin neoplasms originating from internal organs are rarely seen in canine and feline dermatology. Sporadic cases of metastases from *gastric*, *intestinal*, *prostatic*, *renal*, and *bronchial carcinomas*, as well as *testicular seminomas* and some sarcomas such as *hemangiosarcomas* and *osteosarcomas*, have been reported. The spread of neoplastic cells occurs mainly via the bloodstream, and, therefore, metastases can develop anywhere in the skin. The most frequently observed skin metastases are those related to bronchogenic carcinoma in cats, which characterize the so-called lung-digit syndrome. In this neoplastic disease, metastases are multicentric and mostly localized to the digits, and, less frequently, other skin areas may be involved. Although mammary tumors are outside the scope of this book, in dogs, as in humans, a clinical variant of mammary neoplasia known as *inflammatory mammary carcinoma* is well documented. This neoplasia is characterized by the spread of neoplastic cells via the lymphatic system, which causes, in dogs, skin lesions that are sometimes confused with dermatitis; for this reason, this variant of mammary neoplasia is discussed in this chapter.

9.2 Metastatic Bronchial Carcinoma in Cats (Lung–Digit Syndrome)

The term *lung–digit syndrome* (LDS) is used to describe a clinical syndrome characterized by metastases secondary to the diffusion, via the bloodstream, of a primary bronchogenic carcinoma [1, 3, 4, 9, 10]. Old cats are more predisposed, although, in rare cases, LDS has been reported in adult patients. In most cases LDS is represented by the development of multiple lesions, mainly located on several digits of different feet; the digits appear swollen, ulcerated, discharging purulent exudate mixed with blood, painful on palpation and with deviation of the axis of the nail and/or onychomadesis (Fig. 9.1). The neoplastic cells cause osteolysis and cats develop lameness (Fig. 9.2).

In some cats, extra-digital cutaneous localization, both as a single lesion and in combination with the digital lesions, is reported [2, 8–10]. Some cases of extra-skin metastases, predominantly affecting skeletal muscles and bones, have also been reported [6]. Thoracic radiographs usually show a single and well-circumscribed primary lung cancer; in rare cases, it is possible to find multiple lesions, whereas in other patients, the cancer is not radiographically detectable, especially when neoplasia is located at the diaphragmatic lobe (Figs. 9.3, 9.4, 9.5, and 9.6). Most cats do not show respiratory symptoms such as coughing, pleural effusion, and dyspnea, as we can expect in presence of a primary pulmonary neoplasm (Fig. 9.7) [7]. The angioinvasive power of this type of cancer and the fact that cats have a high digital blood flow that permits better heat dissipation are speculated to be the cause of the multiple digital metastases. Another hypothesis is that, on the digital area, a particular vascular perfusion present composed of tortuous or narrow vascular nets favors to the neoplastic emboli to localize and give origin to nodular lesions [7].

F. Albanese, *Canine and Feline Skin Cytology*, https://doi.org/10.1007/978-3-031-95713-0_9

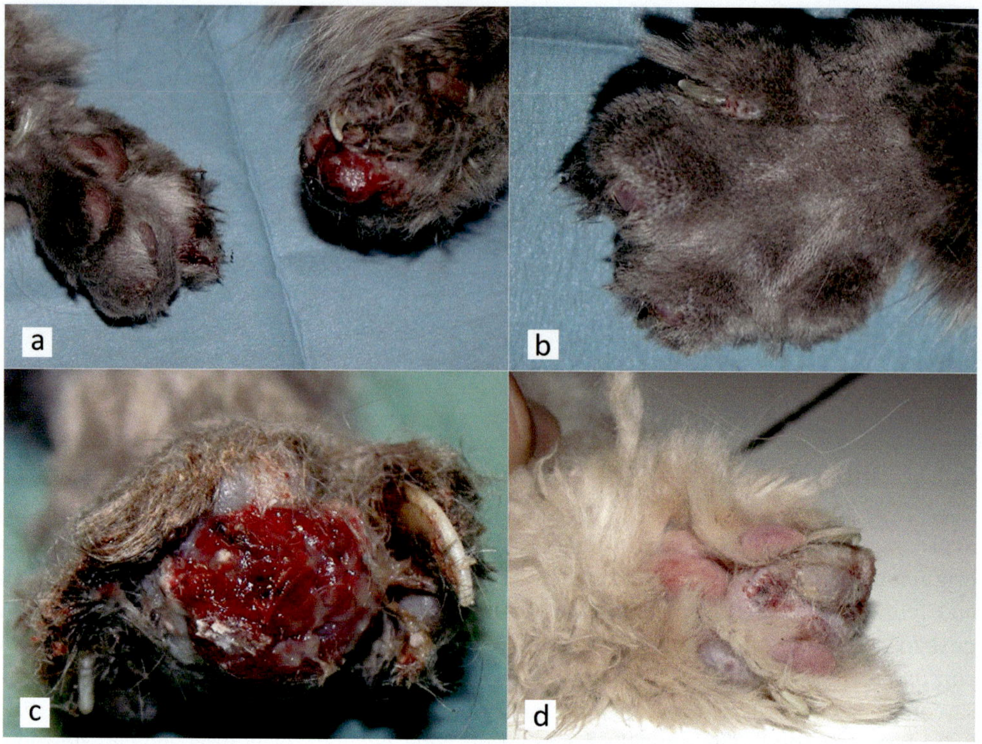

Fig. 9.1 Lung–digit syndrome: (**a**) multiple digital swelling in a Persian cat; (**b**) swelling of all fingers with onychomadesis in the same cat of the previous image; (**c**) close-up view of a metastatic digit associated to severe ulcerative lesions with secondary onychomadesis; (**d**) severe alteration of a digit in a cat with LDS

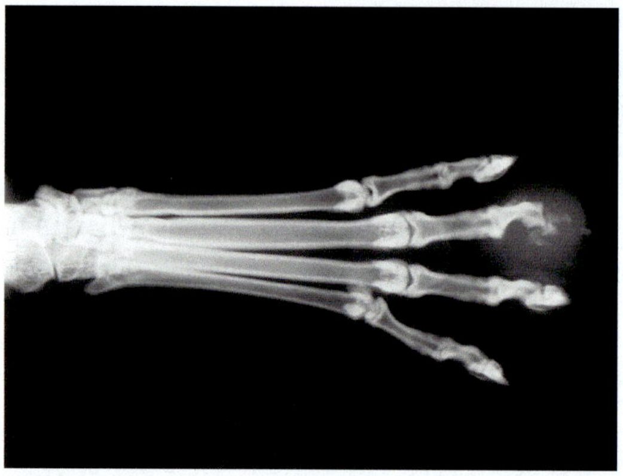

Fig. 9.2 Osteolysis of the third phalange due to a metastatic bronchial carcinoma. (Courtesy of DVM F. Leone-Italy)

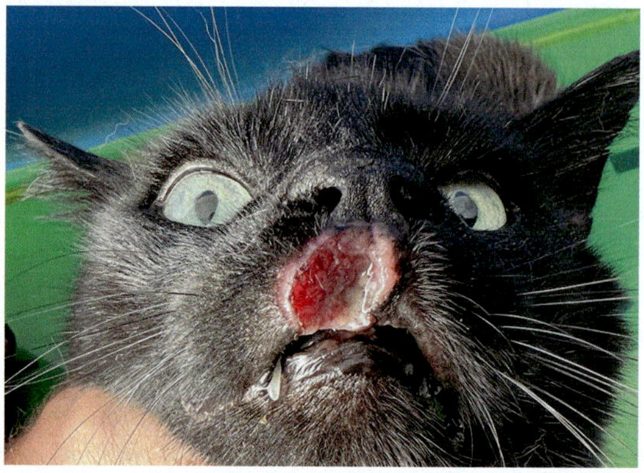

Fig. 9.4 Large ulcerated plaque on the upper lip: note as this uncommon localization of metastatic lung cells can mimic an indolent ulcer. Multiple digital lesions were simultaneously present

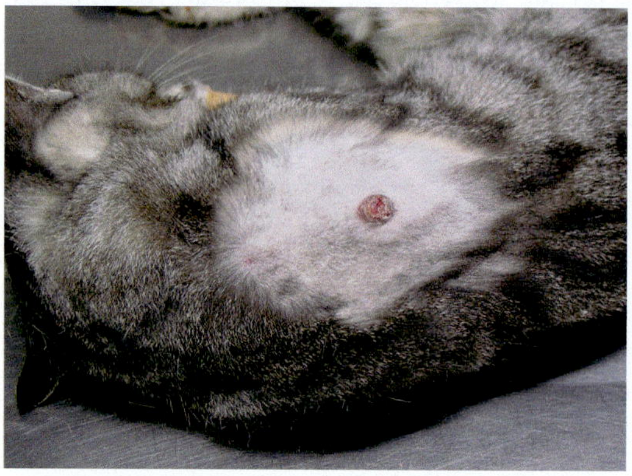

Fig. 9.3 Small ulcerated metastatic nodule on the skin of the chest of a cat. The primary neoplasm was confirmed as a bronchogenic carcinoma. (Courtesy of DVM W. Bertazzolo-Italy)

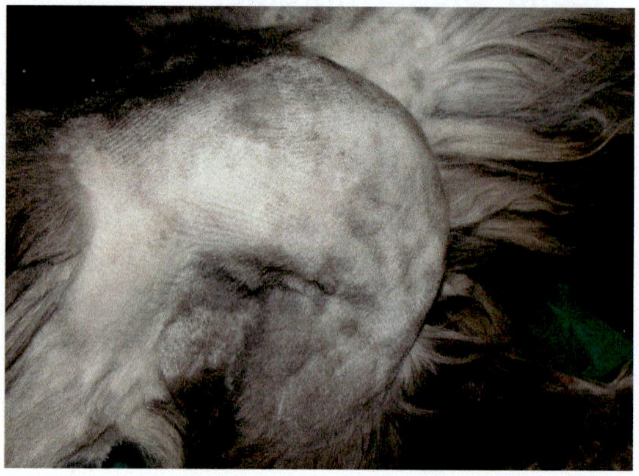

Fig. 9.5 Enlargement of the scapula bones due to metastasis of bronchial carcinoma. Multiple digital lesions were simultaneously present

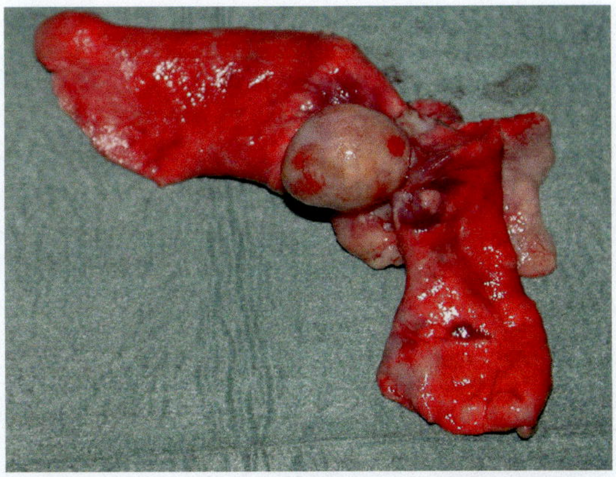

Fig. 9.6 Primary bronchial carcinoma. The small neoplasia was not evident at the radiographic investigation since located in the mediastinal area

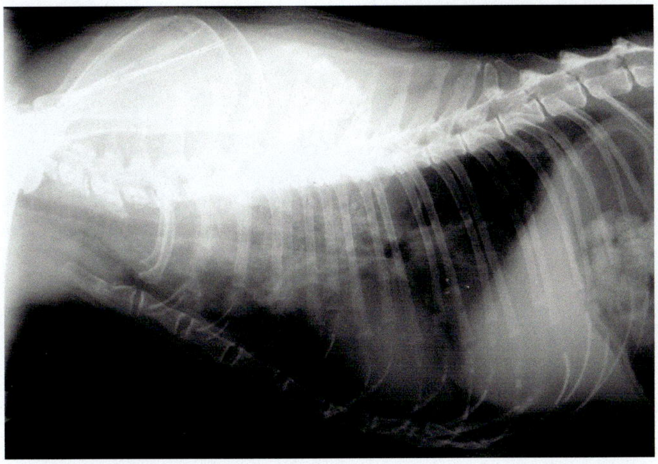

Fig. 9.7 Primary bronchial carcinoma. Radiographic evidence of a large mass into the thorax. (Courtesy of DVM F. Leone-Italy)

Cytological Findings

Three histological subtypes of bronchial carcinoma are recognized as causes of lung–digit syndrome: *adenosquamous* (mucoepidermoid) carcinoma, *squamous cell* carcinoma, and *adenocarcinoma* (Fig. 9.8) [5].

Due to the existence of different histotypes of bronchial carcinoma, it is easy to understand as cytological samples can show different features. The slides are usually characterized by highly cellular clusters of epithelial cells, more or less cohesive, from round/polygonal to cuboidal/columnar in shape, which show varying features of atypia such as anisocytosis, anisokaryosis, pleomorphic nuclei, and prominent nucleoli. The presence of acinar-like or palisade arrangements allows to confirm the glandular nature of the neoplasia, but if typical features of the respiratory epithelium are not present, a diagnosis of DLS cannot be cytologically done (Figs. 9.9, 9.10, and 9.11).

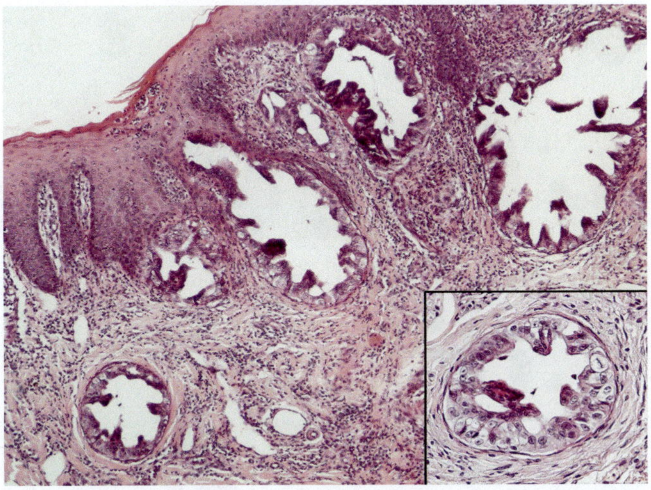

Fig. 9.8 Histopathology of the skin metastasis of a bronchogenic carcinoma: multiple intravascular neoplastic glandular cells, composed by polygonal to columnar epithelial malignant cells. Neoplastic intravascular thrombus (*inset*)

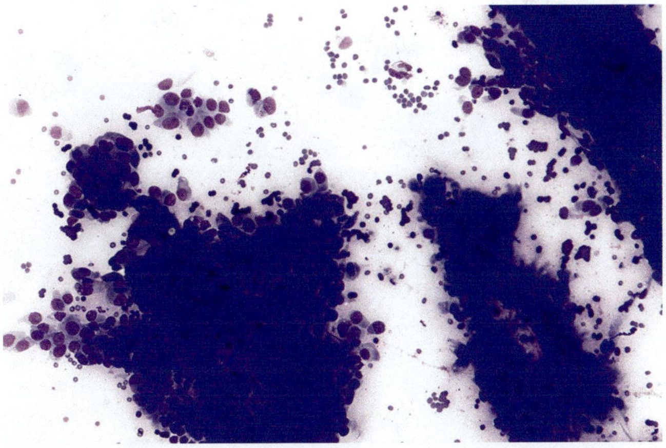

Fig. 9.9 Cytology of bronchogenic carcinoma: large clusters of neoplastic epithelial cells

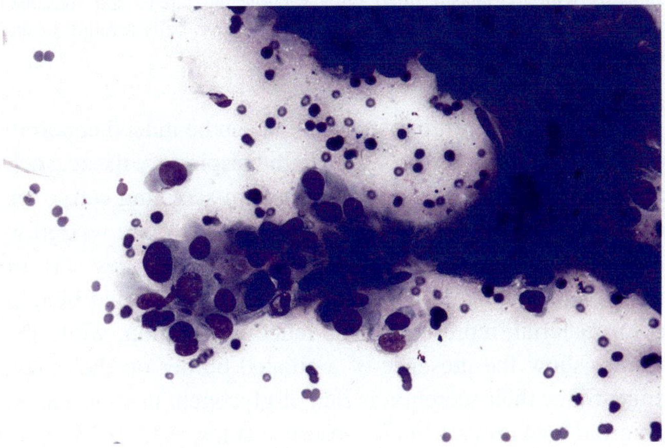

Fig. 9.10 Cytology of bronchogenic carcinoma: at high magnification many criteria of malignancy are evident, but no cytological features of respiratory origin of cells are present

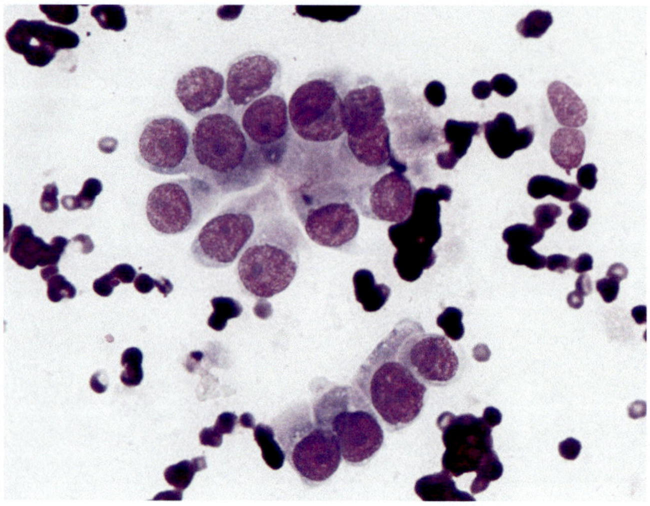

Fig. 9.11 Cytology of bronchogenic carcinoma: clusters of cuboidal epithelial cells showing both acinar and palisading arrangements

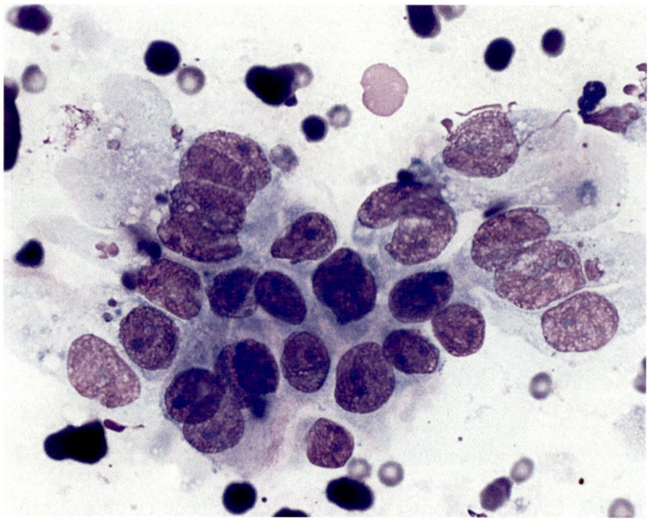

Fig. 9.12 Cytology of bronchogenic carcinoma: palisades of cylindrical cells with cytoplasm filled with vacuole-like areas and mucous material (goblet cells). In the center of the picture, cells arrange in an acinar-like cyto-architecture

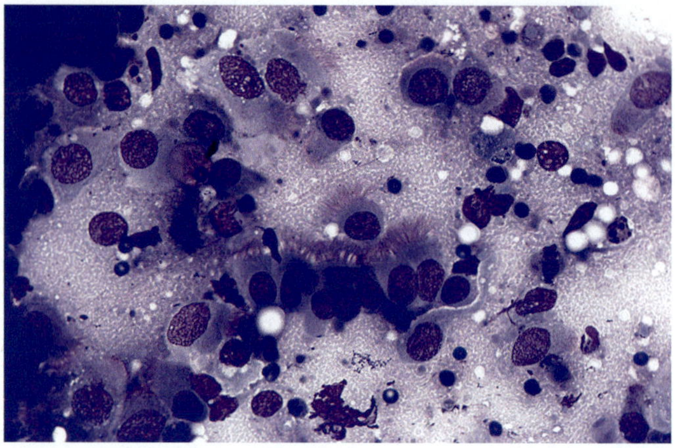

Fig. 9.13 Cytology of bronchogenic carcinoma: nice palisade of ciliated cells on a tigroid background

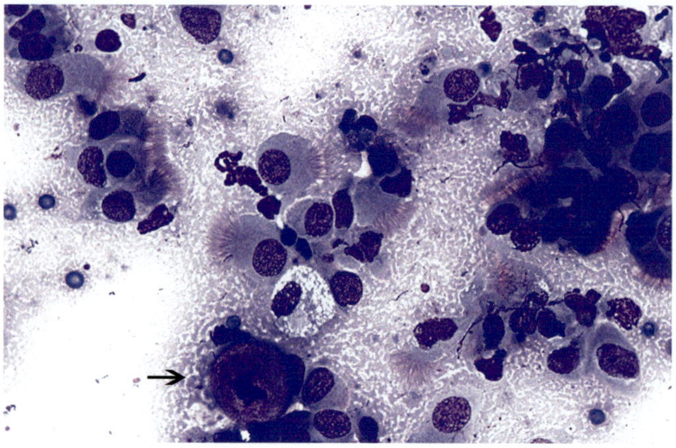

Fig. 9.14 Cytology of bronchogenic carcinoma: many single or in short palisades arrangement ciliated cells. Note the mucous material that fills the cytoplasm of goblet cells (arrow) and that creates a tigroid background when dispersed on the bottom of slide

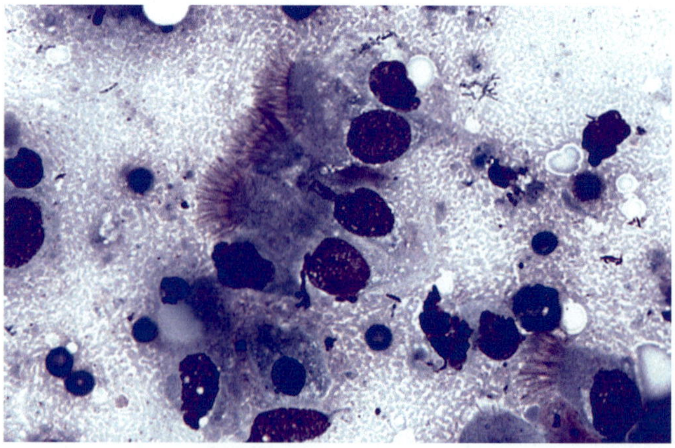

Fig. 9.15 Cytology of bronchogenic carcinoma: at high magnification the cilia are clearly recognizable

The bronchial origin of carcinoma can be indeed certainly confirmed when cells belonging to the respiratory tissue, such as ciliated and/or goblet cells (the mucin-secreting cells), are observed on cytological specimens. The latter are recognizable by the presence of vacuolar-like spaces or of intracytoplasmic amorphous and coarse granular and blue in color material, representing the mucous secretion, while the former show the presence of a ciliated border on the apical edge. Since their secretion is rich of glycogen, in some cases, a tigroid background can be observed (Figs. 9.12, 9.13, 9.14, and 9.15). These two cell types are more common in the adenocarcinoma subtype and, to a lesser extent, in the adenosquamous carcinoma variant. As the adenocarcinoma

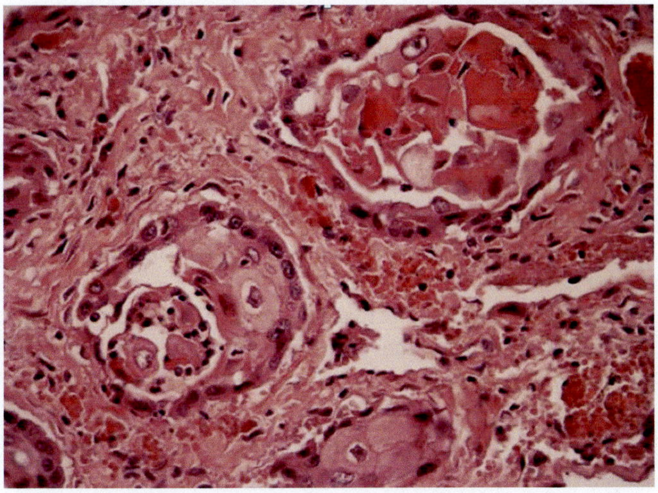

Fig. 9.16 Histology of squamous cells bronchogenic carcinoma: multiple neoplastic emboli with squamous differentiation

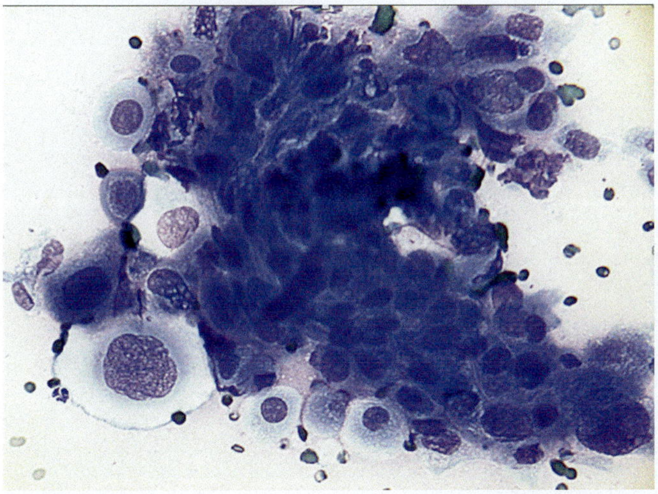

Fig. 9.17 Some squamous neoplastic cells are recognizable along undifferentiated malignant epithelial cells arranged in cluster

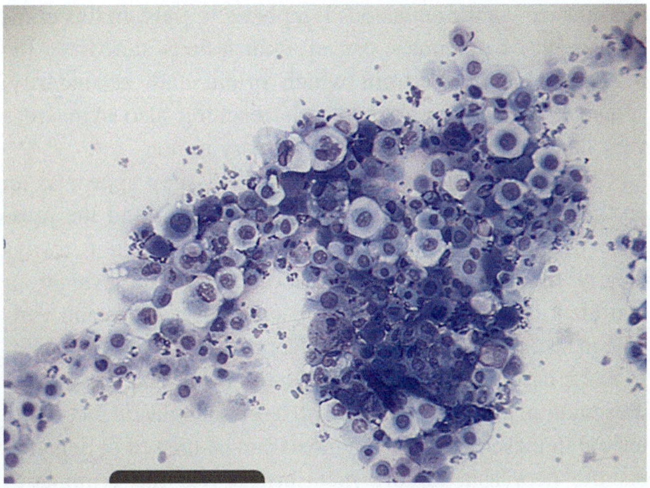

Fig. 9.18 Cytology of squamous cells bronchial carcinoma: clearly squamous differentiation of cells. In these cases, the differential with a primary SCC of the nail bed is not cytologically possible

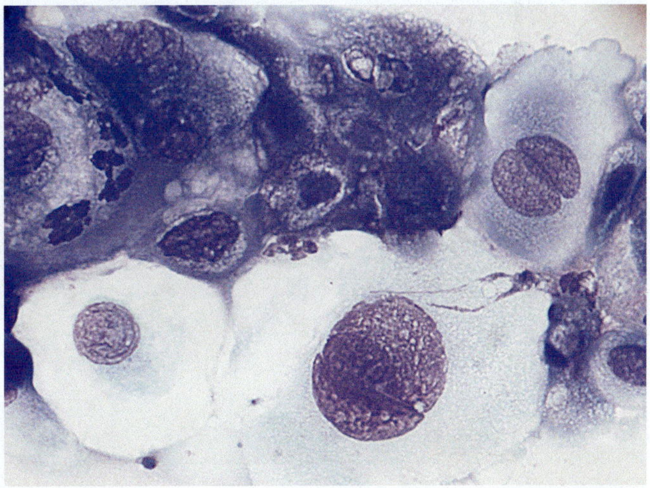

Fig. 9.19 High magnification in which squamous cells are easily recognizable for their large cytoplasm, clear, pale azul and of vitreous aspect

is the less frequent subtype of LDS, the detection of ciliated cells is not a constant finding; therefore, when they are not visible on slides, the cytological confirmation of metastatic bronchial carcinoma is not possible, but anyway, it cannot be ruled out and can be strongly suspected if the clinical presentation is very suggestive of LDS.

In specimens collected from the adenosquamous subtype, it is possible to detect some epithelial cells showing squamous differentiation; the latter are the predominant cytological population in the histological SCC subtype (Figs. 9.16, 9.17, 9.18, and 9.19). For these reasons, cytological differentiation between a primary SCC of the nail bed and a digital metastasis from a bronchogenic SCC is very difficult and often not possible. When multiple digits are involved, a primary SCC of nail bed is clinically ruled out, but many doubts can be present when a single digit is involved or only a digit is apparently altered.

9.3 Cutaneous Metastasis of Inflammatory Mammary Carcinoma

Inflammatory mammary carcinoma (IMC) is the most malignant, aggressive, and lethal variant of mammary cancer in dogs, and it is characterized by a clinical presentation that is completely different from that of other mammary carcinomas [11, 12, 14, 15]. It is a typical cancer of bitches but it can exceptionally be observed in male dogs [16]. In human beings, studies on inflammatory breast cancer showed the high angiogenic and angioinvasive potential of the tumor, which contributes to its aggressive power [13].

Although mammary carcinomas in dogs can metastasize to internal organs and, via bloodstream, to any area of the skin, the peculiarity of IMC is that neoplastic cells metastasize via the cutaneous lymphatic vessels. In this clinical variant of mammary cancer, skin lesions start from the mammary neoplasm from which origin and, secondarily, spread through lymphatic vessels reaching also skin areas very distant from the primary mammary tumor.

Clinically, in the early stage, the IMC may appear as an erythematous swelling of the mammary area, and the presence of an underlying mammary neoplasia is not always clearly recognizable with palpation; with time, single or confluent nodular–papular lesions or small nodules and variable in size plaques, very often ulcerated and bleeding, occur. In some cases, no raised lesions are evident, and the grossly presentation can be confused with a non-neoplastic ulcerative lesion; in these cases, the diagnosis can be missed or delayed. As the neoplasia rapidly metastasizes via the lymphatics vessels, in many dogs, the lesions can be observed far from the mammary area, such as the inner surface of the thighs and axilla and on the lateral area of the chest and shoulders (Figs. 9.20, 9.21, and 9.22).

In some cases, lesions of IMC are not grossly present when the mammary neoplasms are surgically removed since only micrometastasis is present in the dermal lymphatic vessels; within a few days from surgery, erythematous and nodular lesions rapidly develop and spread from the sutured area (Fig. 9.23).

It must be pointed out that the clinical aspects of the skin lesions explain why the adjective *inflammatory* has been coined to define this variant of mammary cancer. The term "inflammatory" is indeed related to the clinical appearance of the lesion, and it is absolutely not associated to the microscopic presence of inflammatory cells in the context of the neoplastic proliferation.

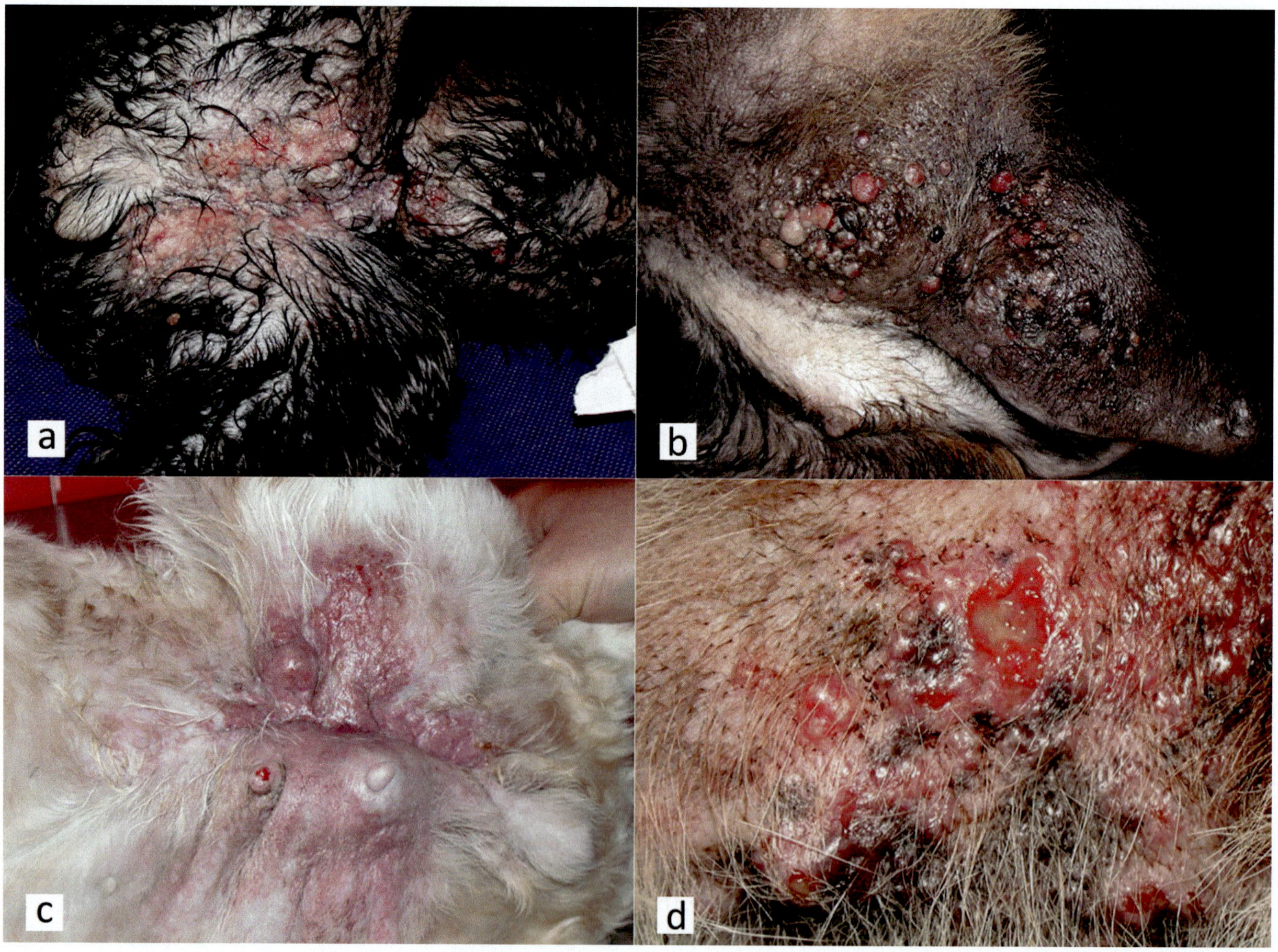

Fig. 9.20 Inflammatory mammary carcinoma: (**a**) multiple, small, erythematous, and ulcerated nodular papules and plaques, on the abdomen and inner thigh of an old Cocker spaniel; (**b**) multiple neoplastic nodular papules and nodules on the abdomen and leg of a Rottweiler; (**c**) swelling and erythema of the skin covering a large mammary neoplasia. Note that the lesions are spread over the left thigh forming a large erythematous plaque; (**d**) close-up of intact and ulcerated papular–nodular lesions on the mammary area of a bitch

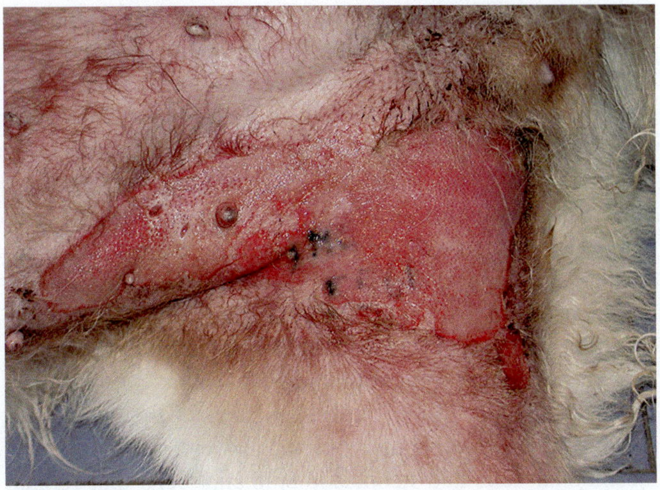

Fig. 9.21 Large erythematous skin ulceration on the right legs of a mixed-breed dog: the unusual presentation of an IMC can be confused with a dermatitis

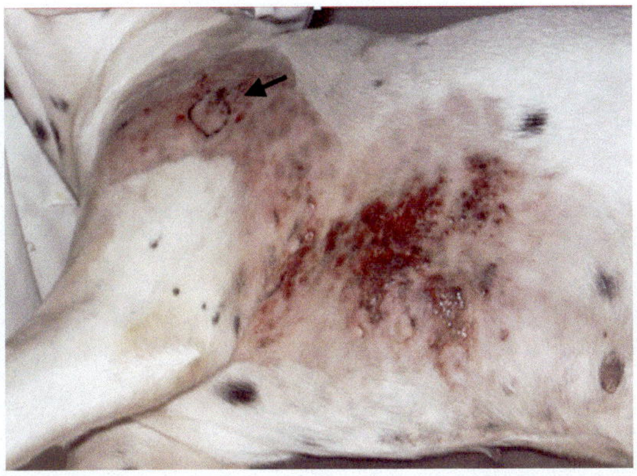

Fig. 9.22 Multiple ulcerative papular lesions that spread from the first left mammary gland to the skin of the scapular area. Note as the lesion spread: the area marked with black marker is where biopsies were performed (arrow)

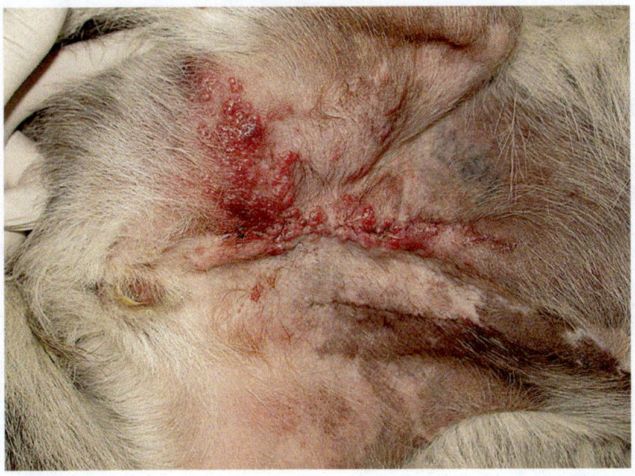

Fig. 9.23 Multiple erythematous papules, nodules, and large plaques adjacent to the sutured area. The lesions spread rapidly over the ventral thigh within a few days of surgery

Cytological Findings

The histopathological features of IMC are characterized by the spread of neoplastic cells through the lymphatic vessels (Figs. 9.24 and 9.25). This metastatic behavior justifies why the neoplasia is also defined as *carcinomatous lymphangiectasia*. The cytological samples collected from intact papular–nodular lesions show highly cellular specimens usually characterized by a clear background, sometimes slightly hemocontaminated, with many clusters of malignant epithelial cells showing various features of malignancy as occurs in other tumors of the mammary glands; anisocytosis, anisokaryosis, voluminous nuclei of bizarre shape, macronucleoli, nuclear molding, cannibalism, and vacuolated-like cytoplasm that can contain granular secretion can be commonly observed (Figs. 9.26, 9.27, and 9.28).

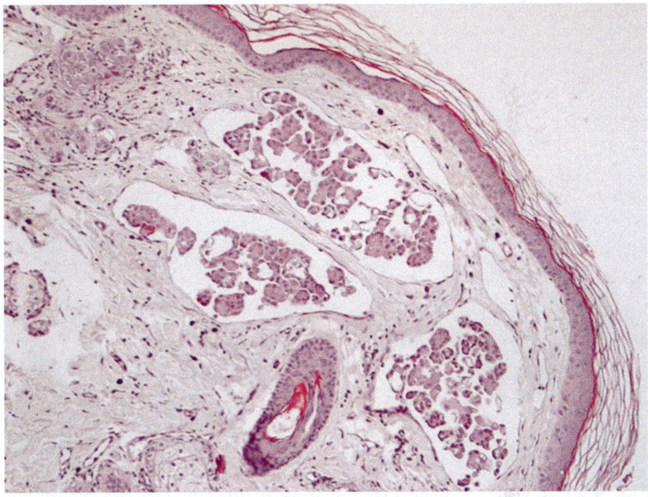

Fig. 9.24 Histopathology of IMC: multiple neoplastic thrombi in lymphatic vessels

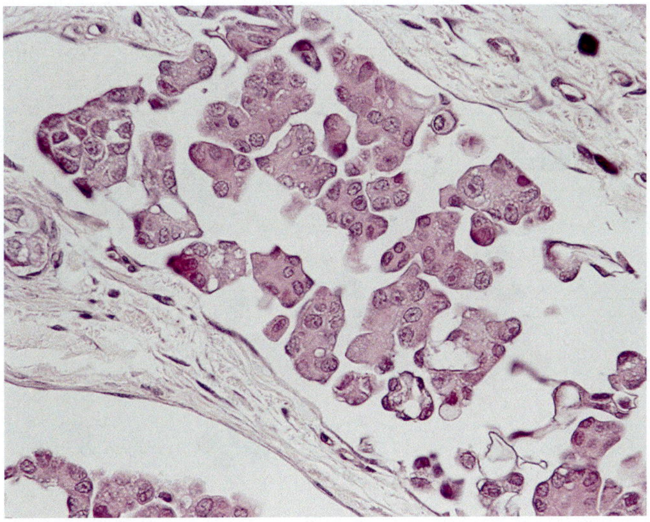

Fig. 9.25 Histopathology of IMC: at high magnification of the same picture n. 9.16

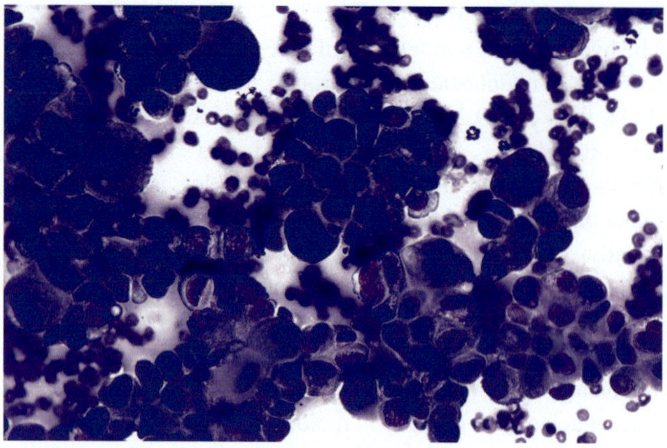

Fig. 9.26 Cytology of an IMC: large clusters of carcinomatous mammary

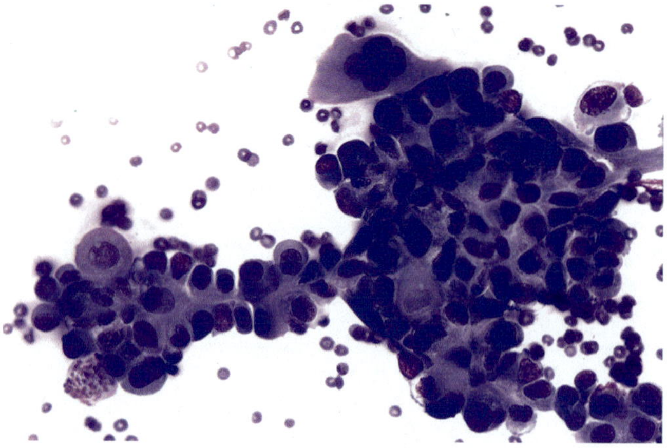

Fig. 9.27 Cytology of an IMC: at high magnifications the malignant criteria of neoplastic cells are better highlighted (anisocytosis, anisokaryosis, multinucleation, mitosis, chaotic organization)

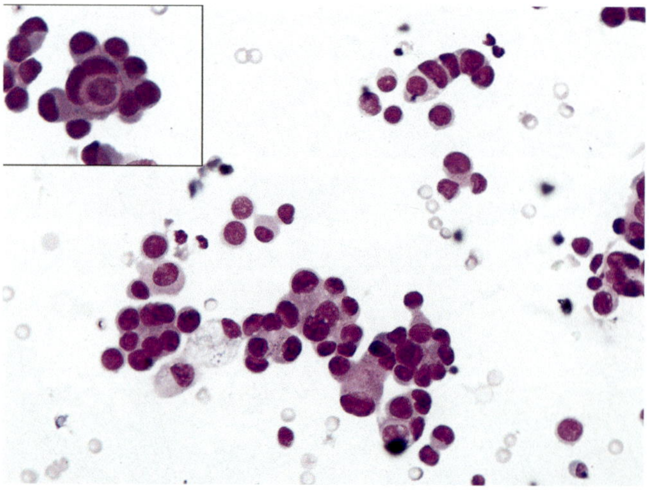

Fig. 9.28 Cytology of an IMC: many acinar-like arrangements and a small row of neoplastic cells are evident (arrow). Note the cannibalized neoplastic cell (inset)

One cytological finding observed in many IMC is the loss of cohesion of neoplastic cells which, when separated from each other, appear as discrete cells (acantholysis). When the needle does not take cells from solid areas of neoplasia, these discrete cells can be very numerous on the slides, so that they can be the only population present; in these cases, cytology may really be confused with a round cell neoplasia. Fortunately, in most cases, along with these discrete cells, small clusters of cells arranged in small rows or acinar-like arrangements are recognizable, allowing to interpret them as epithelial. Neoplastic epithelial cells during IMC may also show another cytomorphological feature, the presence of a "corona" or "fringe" at the periphery of the cytoplasm; since similar aspects are observed on cytoplasm of both mesothelial cells and neoplastic epithelial cells that remain in the cavitary fluids such as ascites or pleural effusions, it is probable that the same result can be linked to the long time standing of mammary neoplastic emboli into the lymphatic vessels and so in contact with lympha (Figs. 9.29, 9.30, 9.31,

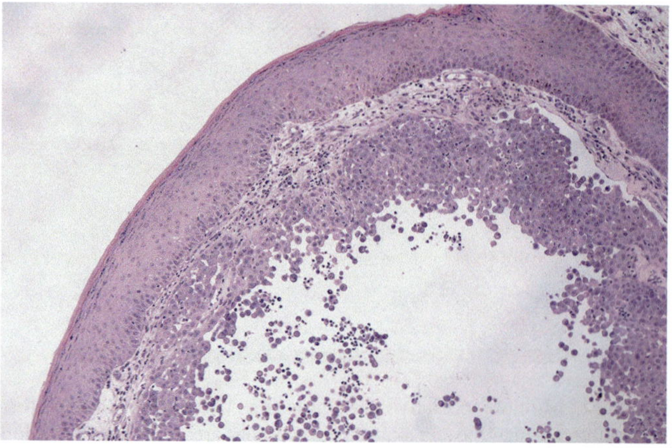

Fig. 9.29 Histology of an IMC: large intralymphatic neoplastic embolo. Note as single roundish cells detach from the solid neoplasia (acantholysis)

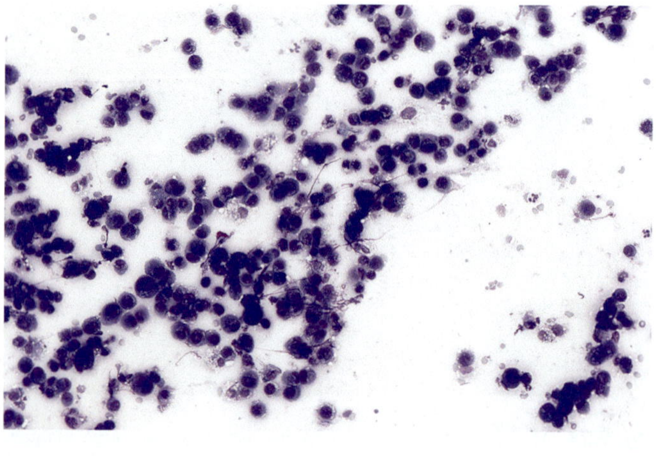

Fig. 9.30 Cytology of an IMC: many epithelial glandular cells released as discrete cells with roundish profile that mimic a RCT

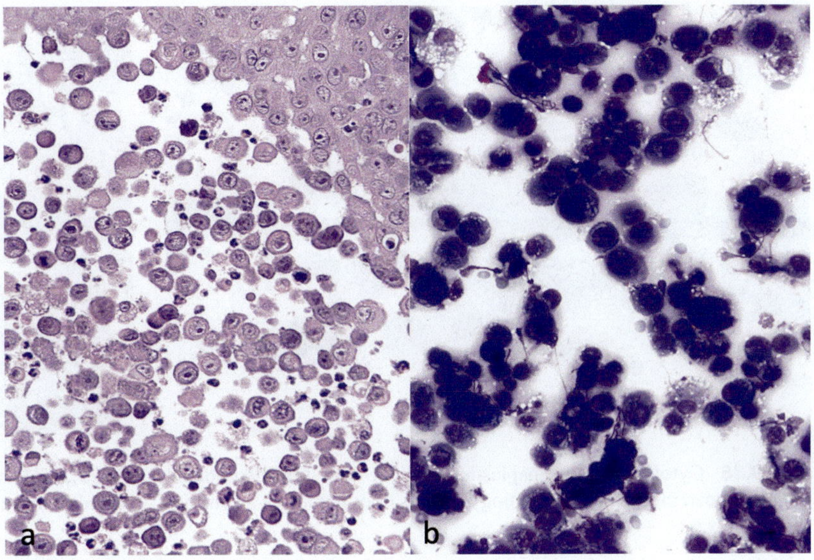

Fig. 9.31 Inflammatory mammary gland carcinoma: (**a**) histology: close-up of the malignant acantholytic epithelial mammary gland cells. Discrete round cells detached from the solid neoplasia; (**b**) cytological representation of the previous image. Note how this presentation can be confused with a RCT

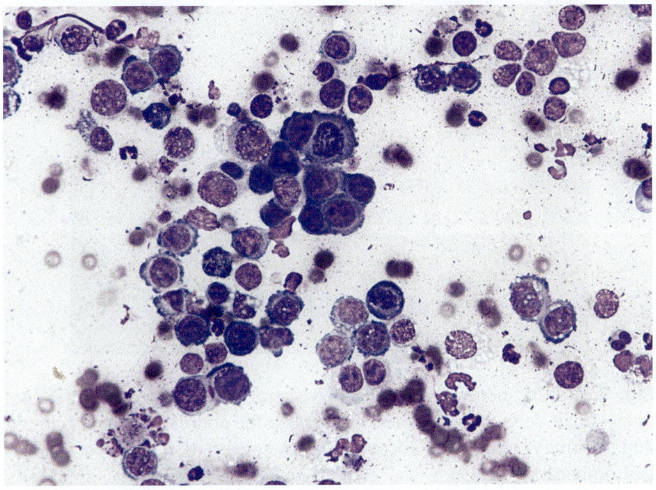

Fig. 9.32 Cytology of an IMC: many discrete round neoplastic cells can mimic a RCT. Note the fringe-like aspect of the periphery of the cytoplasm

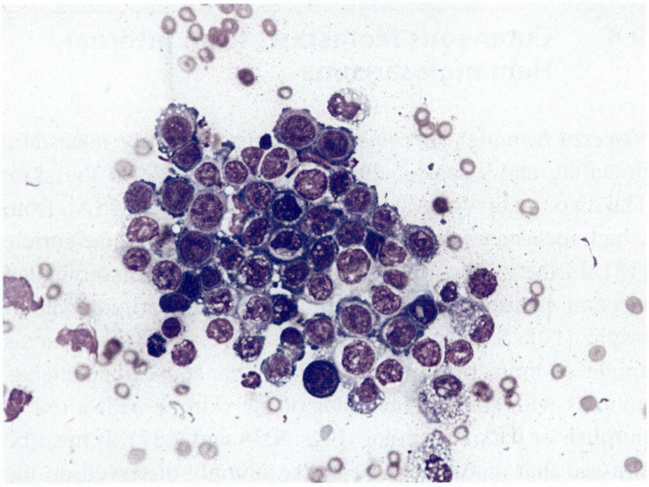

Fig. 9.33 Cytology of an IMC: a cluster of neoplastic glandular cells that are losing their arrangement in cluster. Note the fringe-like peripheral appearance of the cytoplasm

9.32, and 9.33). Single cells containing large secretory intracytoplasmic vacuoles that marginalize the nucleus, the so-called signet ring cells, can also be detected. In more anaplastic carcinomas, cells are poorly differentiated, and the cells' pleomorphism can create many doubts regarding their epithelial origin; in these cases, differential cytological diagnosis must include signet ring melanoma, histiocytic sarcoma, and pleomorphic plasma cell tumor. In case the clinical presentation of IMC is typical, the above differentials can be ruled out (Figs. 9.34 and 9.35).

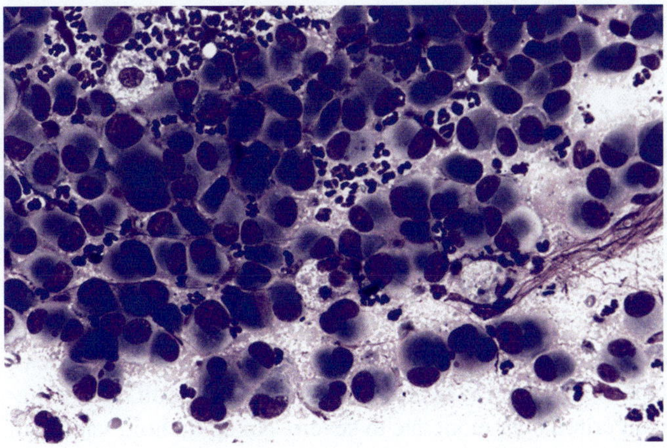

Fig. 9.34 Cytology of an anaplastic IMC: many discrete pleomorphic mammary gland cells. Note as the morphology and the disposition in single cells are not typical of the epithelial neoplasias

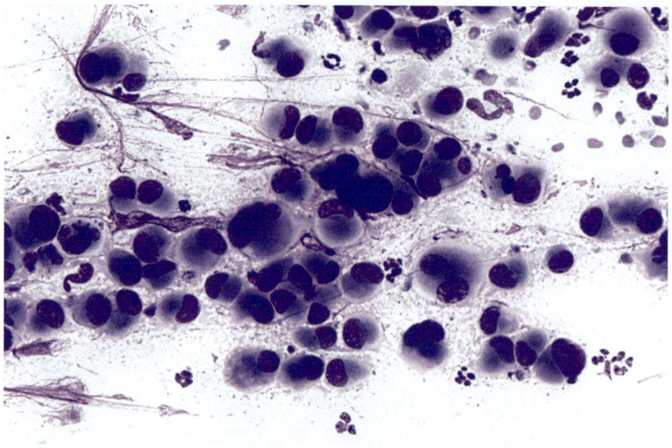

Fig. 9.35 Cytology of an anaplastic IMC: in very undifferentiated mammary carcinoma, the morphology of cells does not allow a cytological diagnosis

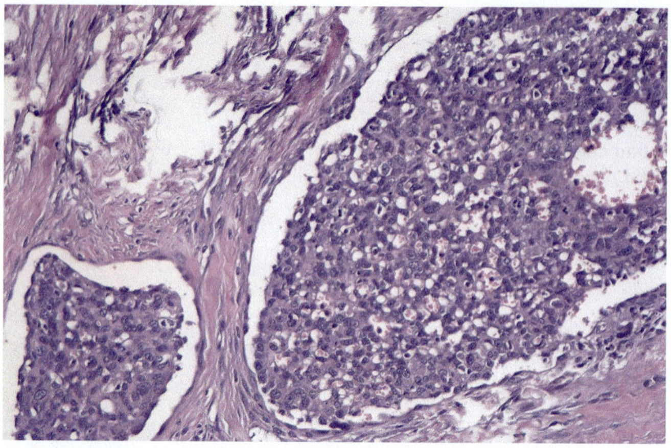

Fig. 9.36 Histology cutaneous hemangiosarcoma (HSA): large intravascular thrombi of neoplastic endothelial cells are clearly recognizable

9.4 Cutaneous Metastasis from Internal Hemangiosarcoma

Visceral hemangiosarcoma (vHSA) has a highly metastatic potential, and, occasionally, it can metastasize to the skin. The two main visceral locations of primary vHSA, from which metastases can start, are the spleen and cardiac auricle [18]. In the case of multiple skin tumors and disseminated visceral metastases, it is difficult to define the primary site of origin [19]. Skin metastases from HSA usually appear as single or more frequently multiple cutaneous/subcutaneous nodules with slately blue color of the skin or with a red to purplish or dark blue color (Figs. 9.36 and 9.37). It must be stressed that multiple HSAs are commonly observed on the non-pigmented skin of dogs, especially on the ventral abdominal area (Chap. 8). These HSAs are secondary to the chronic exposure to sun ray (UVA) and show a benign behavior; so, when multiple cutaneous HSA are observed, especially in predisposed breeds, they must not be interpreted as metastasis from visceral HSA.

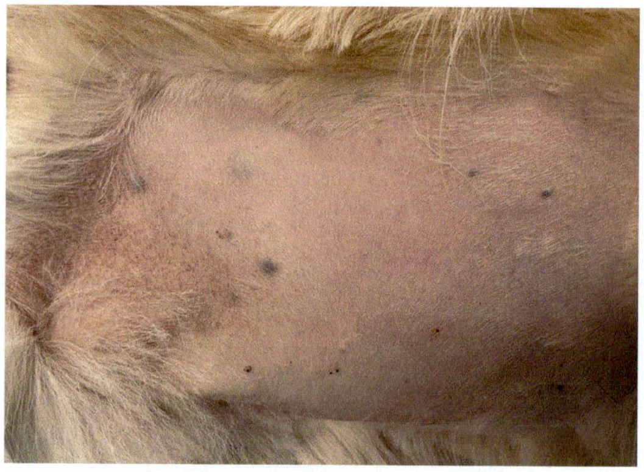

Fig. 9.37 Cutaneous metastasis of visceral hemangiosarcoma (HSA): many small deep violet colored nodules on the abdominal skin of an old dog. Primary HSA was located at the spleen. (Courtesy of DVM-L. Lombardini—Italy)

Cytological Findings

Cytological features are similar to those observed in HSA located at other body sites and are exhaustively described in Chap. 8; samples are always highly hematic and with moderate to low cellularity. Neoplastic cells may be spindle or pleomorphic, often arranged in loose pseudo-aggregates, with moderate to abundant basophilic cytoplasm, and often containing small, punctate, and colorless vacuole-like areas. Nuclei are round to oval, with prominent nucleoli, and malignant criteria are moderate to severe. An epithelioid variant of HSA has been described; in this uncommon histotype, round to polygonal cells are arranged in tight clusters that may resemble an epithelial tissue (Figs. 9.38 and 9.39) [17]. The

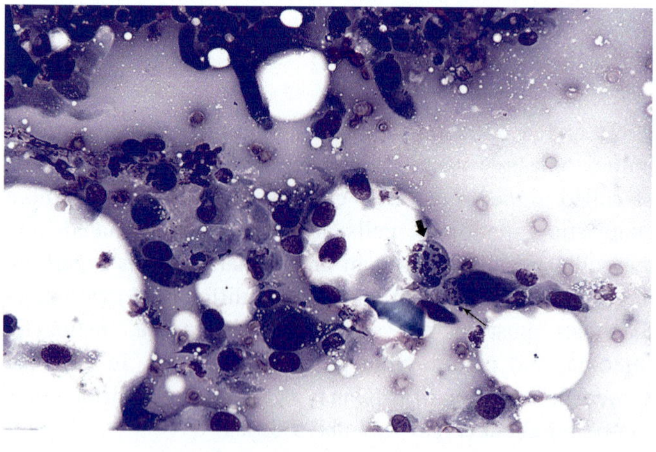

Fig. 9.38 Cytology of an HSA: a cluster of pleomorphic neoplastic cells with intracytoplasmic punctate vacuole-like areas. Note the dirty background, and nuclei in mitosis (short arrow) and karyorrhexis (thin arrows)

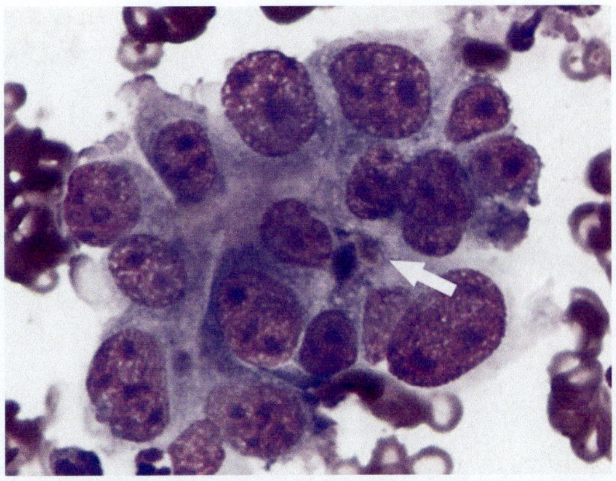

Fig. 9.39 Cytology of an HSA: epithelial-like appearance of endothelial neoplastic cells. The single RBC recognizable in the clustered cells "could" represent a vasoformative activity of endothelial neoplastic cells (arrow)

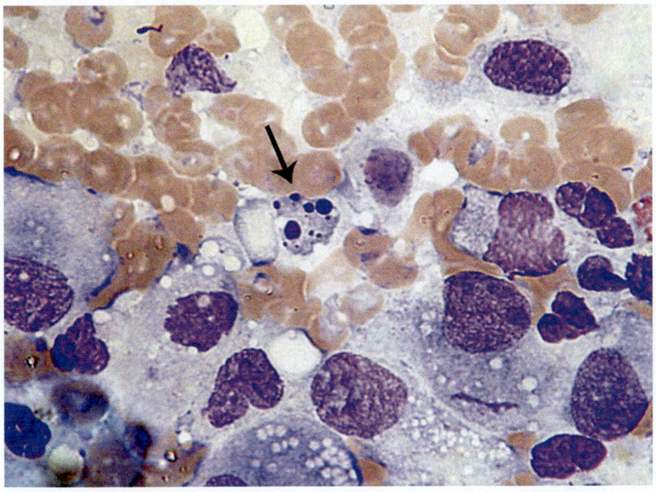

Fig. 9.40 Cytology of an HSA: a single apoptotic cell represented by a multiple fragmented roundish portion of the nucleus (arrow). (Courtesy of DVM W. Bertazzolo- Italy)

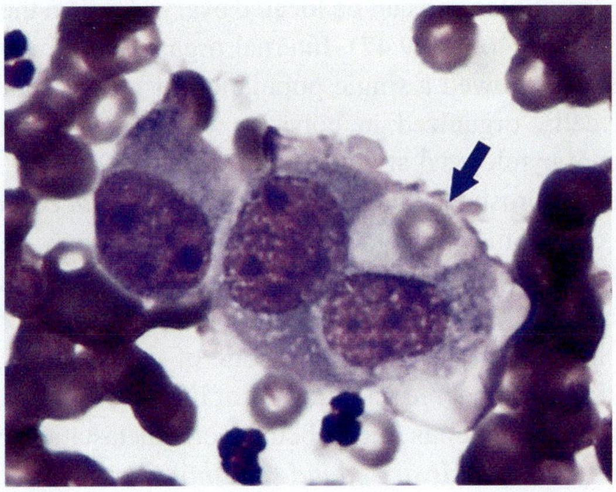

Fig. 9.41 Cytology of an HSA: malignant spindle endotheliocytes showing a probable vasoformative arrangement. A RBC is recognizable into the cytoplasm and can be a probable attempt by cells to form vessels from neoplastic endoteliocytes but can also represent an aberrant phagocytic activity of the malignant cell (arrow)

diagnosis of HSA is achieved with histology and, in doubt cases, by demonstrating, through immunohistochemistry, positivity to endothelial markers (CD31 and Factor VIII). An additional, less frequent cytological feature observed in HSA is the presence of apoptotic cells (most likely leukocytes), hemosiderophages, hematopoietic precursors and eosinophils, macrophages, and debris, scattered among the neoplastic cells (Fig. 9.40). The cytological finding of megakaryocytes and rubricytes in samples obtained from a hematic soft-tissue neoformation should always be suspected as coming from a malignant vascular tumor, even if neoplastic cells are not present in the cytological sample. Finally, another feature that could help in supporting a cytological diagnosis of HSA is the presence of vasoformative features: cells may be arranged in pseudoacinar aggregates, in swirling structures surrounding erythrocytes, or occasionally, single cells may produce an intracytoplasmic vascular channel containing erythrocytes [19, 20]. The latter features are absolutely hard to demonstrate with certainty on cytological samples and, in many cases, the fact that they are really vasoformative formations is doubious or and could be only a fantasious interpretation of cytologists (Fig. 9.41).

9.5 Other Rare Skin Metastasis

Other neoplasms have been reported to be responsible for skin metastases in dogs: a few case reports have been documented describing cases of skin metastases from *testicular seminoma* [26, 30, 31], *appendicular osteosarcoma* [23, 28], *duodenal carcinoma* [24], *transitional cell carcinoma* [29], *colonic adenocarcinoma* [32], *renal carcinoma* [33], *nasal neuro-endocrine tumor* [25], *prostatic adenocarcinoma* [21], disseminated *alveolar rhabdomyosarcoma* [27], and *gastric mucinous carcinoma* (Figs. 9.42, 9.43, and 9.44). [22]

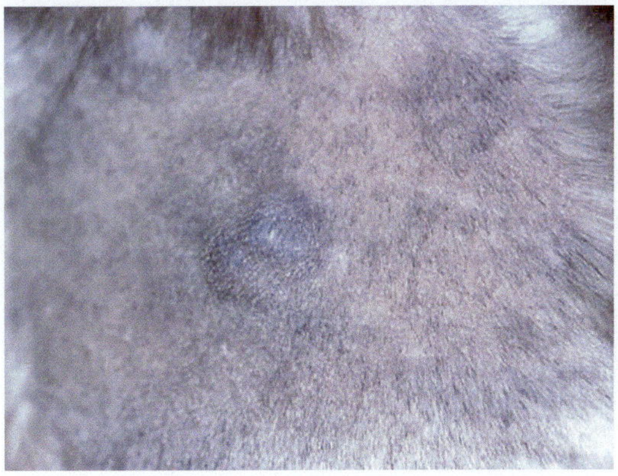

Fig. 9.42 Metastatic gastric adenocarcinoma presented as a small bluish cutaneous nodule. (Courtesy of DVM W. Bertazzolo—Italy)

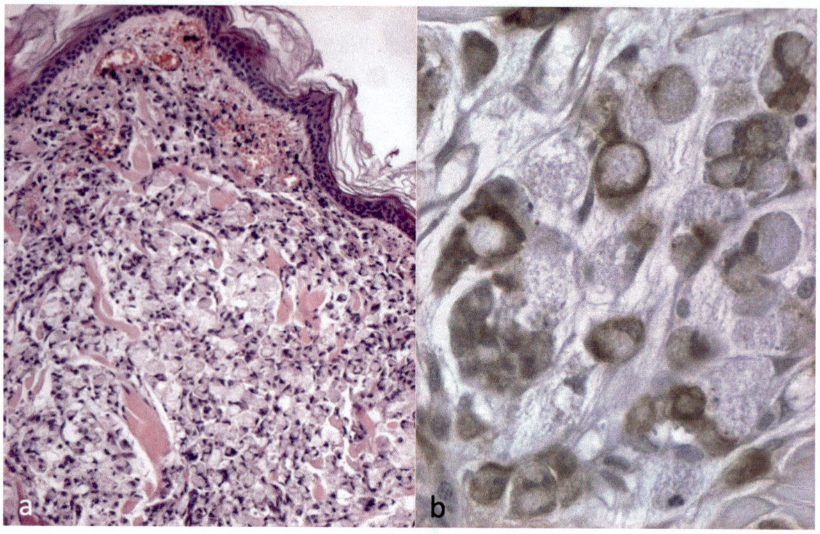

Fig. 9.43 Metastatic gastric adenocarcinoma: (**a**) histology: neoplastic cells infiltrating the dermis. Note the eccentric nuclei and large cytoplasm; (**b**) immunohistochemistry Pan-cytokeratin: neoplastic cells show strong positivity. (Courtesy of DVM W. Bertazzolo—Italy)

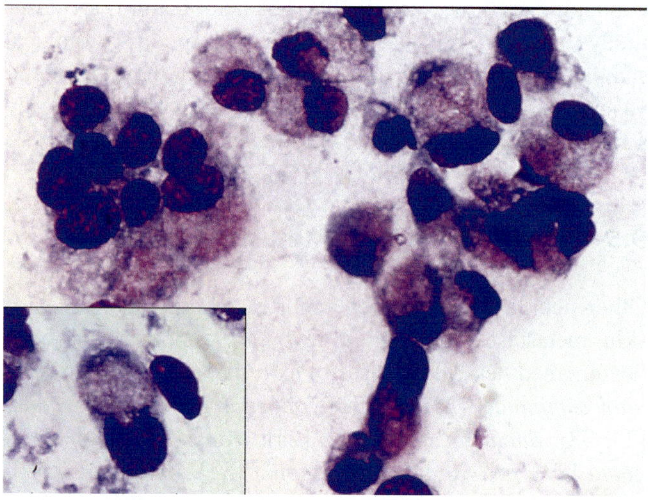

Fig. 9.44 Cytology of a gastric adenocarcinoma: neoplastic cells are arranged singly or as small clusters; the cytoplasm is large, eosinophilic, and of granular appearance. Occasionally, cells look like goblet cells (*inset*). (Courtesy of DVM W. Bertazzolo—Italy)

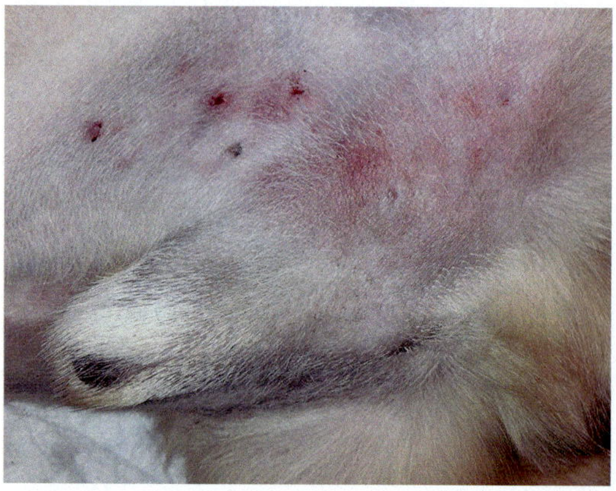

Fig. 9.45 Swelling and multiple small nodules on the abdomen of a dog. Histopathology allowed diagnosis of metastasis from a prostatic carcinoma. (Courtesy of DVM M. Annoni—Italy)

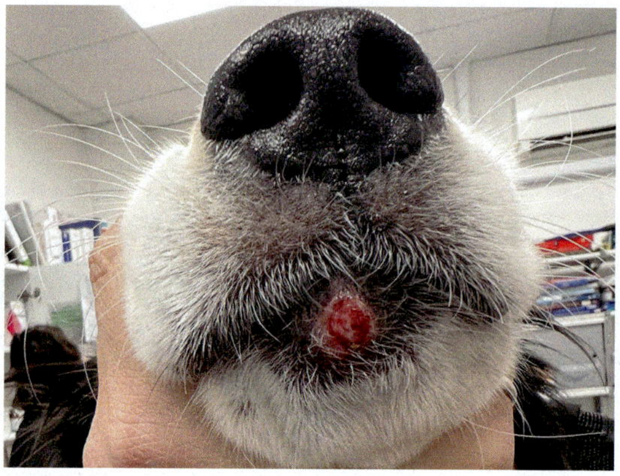

Fig. 9.46 Ulcerated nodule on the chin in a dog with cutaneous metastasis of a prostatic carcinoma

Rare cases of cutaneous metastases of prostatic carcinomas have been reported [17]. Nodules are more frequently multiples and can be located everywhere on the body (Figs. 9.45, 9.46 and 9.47). Internal organs can be involved also. Slides showed a single population of pleomorphic epithelial cells, organized in cohesive clusters, with variable malignant cytological atypias (Figs. 9.48 and 9.49).

In some prostatic cancers, such as *transitional cell carcinoma*, carcinomatous spindle-shaped cells that may be erroneously attributed to a mesenchymal origin can be detected. It is assumed that a de-differentiation toward a mesenchymal type, including the co-expression of cytokeratins and vimentin, is considered an indicator of increased malignancy.

Multiple cutaneous and subcutaneous metastatic nodules from a primary *salivary gland carcinoma* can also occur. Cytology is characterized by a large amount of amorphous eosinophilic matrix embedding pleomorphic and poorly cohesive atypical neoplastic cells (Figs. 9.50 and 9.51).

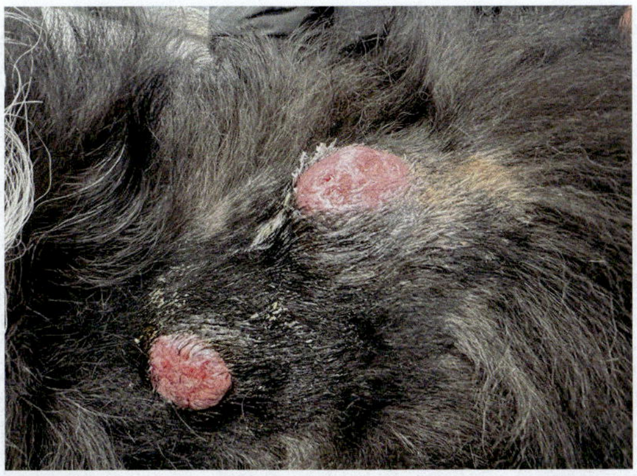

Fig. 9.47 Two ulcerated nodules on the thorax of a dog with metastatic prostatic carcinoma

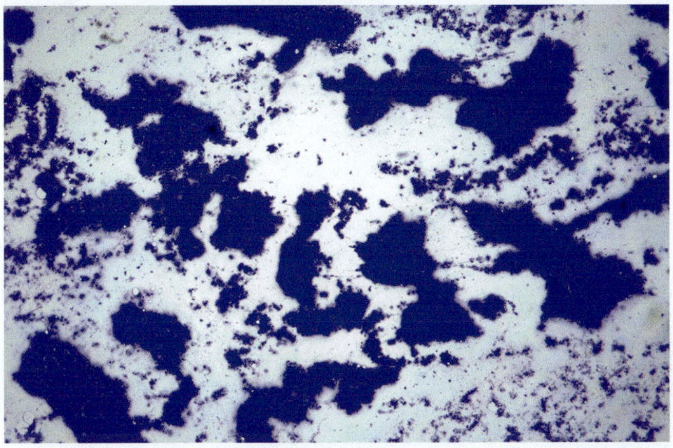

Fig. 9.48 Cytology of skin metastasis from a prostatic carcinoma: many large cluster of epithelial neoplastic cells

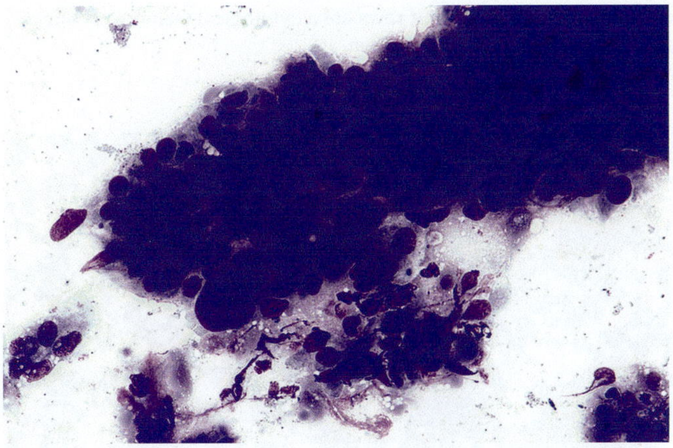

Fig. 9.49 Cytology of a prostatic carcinoma: at high magnification severe cytological atypia such as anisokaryosis, macronuclei, and multinucleation are easily recognizable

A cutaneous metastatic *Pheochromocytoma* was cytologically suspected by the author in a 9-year-old mixed-breed dog. A single subcutaneous metastatic nodule was on the right side of lumbar area, whereas the primary

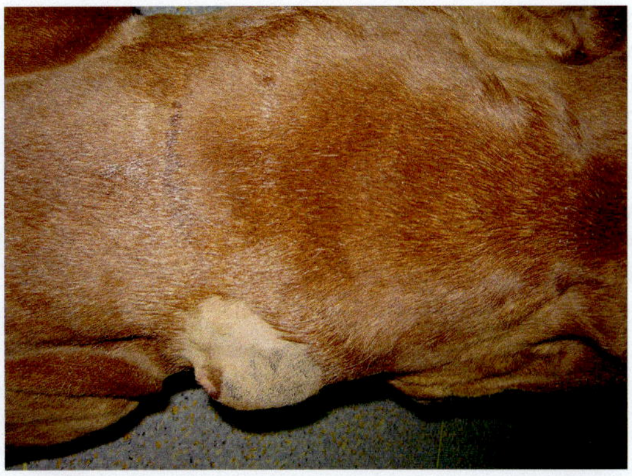

Fig. 9.50 Skin metastasis of a primary salivary carcinoma: large nodule on the right thorax of a Boxer. (Courtesy of DVM W. Bertazzolo—Italy)

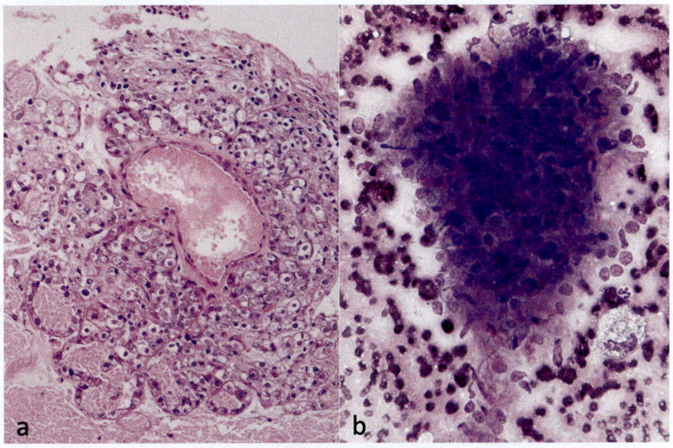

Fig. 9.51 Skin metastasis of a primary salivary carcinoma: histology—malignant neoplastic salivary cells show a disorganized arrangement but adenomera are still well recognizable; (**b**) cytology: loosely cohesive pleomorphic cells are embedded in an eosinophilic matrix. Cells show indistinct borders, moderate, slightly basophilic cytoplasm, monomorphic oval nuclei with coarse chromatin, and indistinct to prominent nucleoli. (Courtesy of DVM W. Bertazzolo—Italy)

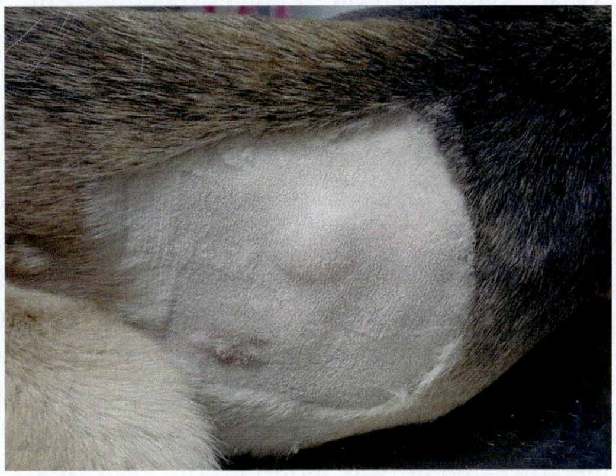

Fig. 9.52 Metastatic subcutaneous nodule from an adrenal pheochromocytoma. (Courtesy of DVM C. Damiani—Italy)

pheochromocytoma was present on the left adrenal gland (Fig. 9.52). The confirmation of the neuroendocrine origin of the tumor was achieved with histopathology and with

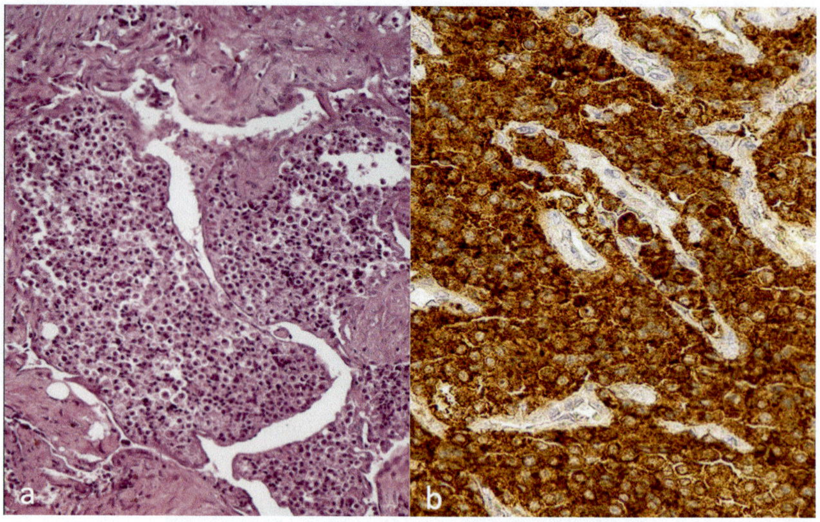

Fig. 9.53 Histology of skin metastasis from a pheochromocytoma: neoplastic cells are arranged in tight nests embedded in a fibrous, connective, stromal tissue; (**b**) immunohistochemistry—neoplastic cells show strong positivity to synaptophysin immunostaining

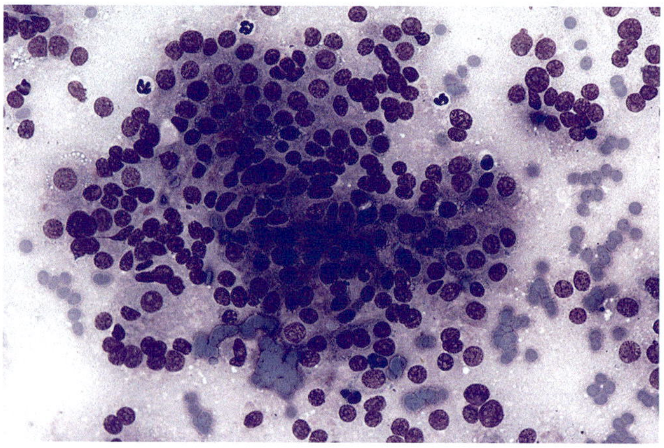

Fig. 9.54 Cytology of a metastatic pheochromocytoma: monomorphic population with round, bare nuclei, often arranged in rows and pseudo-acinar structures

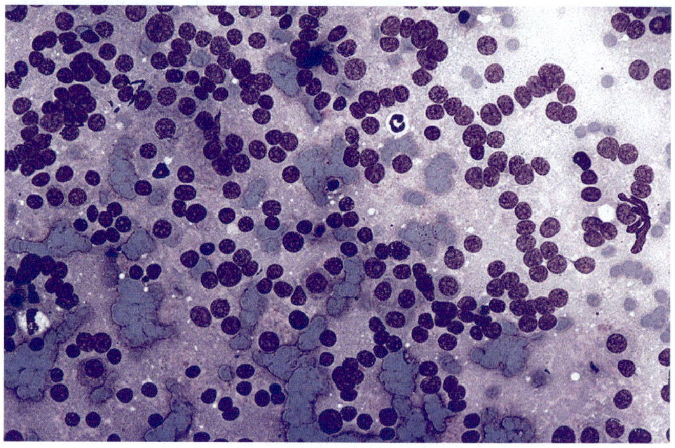

Fig. 9.55 Cytology of a metastatic pheochromocytoma: bare nuclei sometimes arranged in short rows are a cytological feature very suggestive of a neuroendocrine neoplasia

positivity to synaptophysin immune-staining (Fig. 9.53). Cytological findings were consistent with a neuroendocrine tumor, exhibiting many different sized bare nuclei, with both rows and acinar-like arrangement. When present, cytoplasm ranged from scarce to moderate and was slightly basophilic Some cytological features of malignancies were also observed (Figs. 9.53, 9.54 and 9.55).

Skin metastases from renal neoplasia are very rare [33]. A single metastatic nodule on the skin has been observed in a mixed breed dog of 9 years old suffering of renal adenocarcinoma with spreading to vena cava. A previous history of renal adenocarcinoma was present which was treated with chemotherapy. Nodule was alopecic, of 3 cm in size, and located on the left chest. Cytology of both kidney and skin showed the same population of neoplastic cells, represented by clusters of neoplastic epithelial cells showing marked criteria of malignancy such as anisokaryosis, anisocytosis, and macro-nuclei (Figs. 9.56 and 9.57).

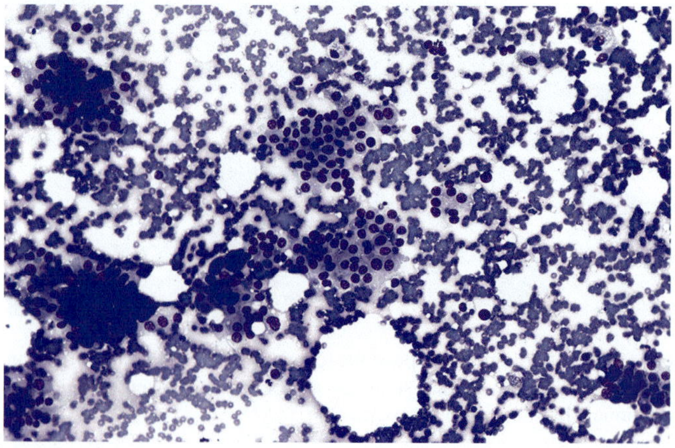

Fig. 9.56 Cytology of skin metastasis from a renal carcinoma: neoplastic cells are arranged in clusters on a hematic background

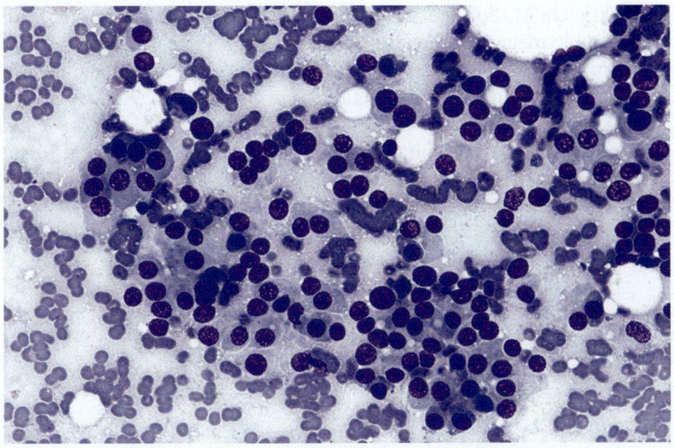

Fig. 9.57 Cytology of skin metastasis from a renal carcinoma: poorly cohesive neoplastic cells, with round nuclei, large cytoplasm, and few naked nuclei

References

Lung-Digit Syndrome

1. Estrada M, Lagadic M (1992) Métastases digitales d'un carcinome pulmonaire asymptomatique chez le chat: Etude d'une série de 11 cas. Prat Med Chir Anim Comp 27:791–795
2. Favrot C, Degorce-Rubiales F (2005) Cutaneous metastases of a bronchial adenocarcinoma in a cat. Vet Dermatol 16(3):183–186
3. Goldfinch N, Argyle DJ (2012) Feline lung-digit syndrome: unusual metastatic patterns of primary lung tumours in cats. J Feline Med Surg 14(3):202–208
4. Gottfried SD, Popovitch CA, Goldschmidt MH et al (2000) Metastatic digital carcinoma in the cat: a retrospective study of 36 cats (1992–1998). J Am Anim Hosp Assoc 36(6):501–509
5. Gross TL, Ihrke PJ, Walder EJ, Affolter VK (2005) Metastatic pulmonary carcinoma in cats. In: Skin diseases of the dog and cat: clinical and histopathological diagnosis, 2nd edn. Blackwell Science, Oxford, pp 703–706
6. Langlais LM, Gibson J, Taylor JA et al (2006) Pulmonary adenocarcinoma with metastasis to skeletal muscle in a cat. Can Vet J 47: 1122–1123
7. Moore AS, Middleton DJ (1982) Pulmonary adenocarcinoma in three cats with non-respiratory signs only. J Small Anim Pract 23: 501–509
8. Petterino C, Guazzi P, Ferro S et al (2005) Bronchogenic adenocarcinoma in a cat: an unusual case of metastasis to the skin. Vet Clin Pathol 34:401–404
9. Thrift E, Greenwell C, Turner AL et al (2017) Metastatic pulmonary carcinomas in cats ('feline lung-digit syndrome'): further variations on a theme. JFMS Open Rep 3(1):2055116917691069
10. van der Linde-Sipman JS, van den Ingh TS (2000) Primary and metastatic carcinomas in the digits of cats. Vet Q 22:141–145

Inflammatory Mammary Carcinoma

11. Clemente M, Pérez-Alenza MD, Peña L (2010) Metastasis of canine inflammatory versus non- inflammatory mammary tumours. J Comp Pathol 143(2–3):157–163
12. de M Souza CH, Toledo-Piza E, Amorin R, Barboza A, Tobias KM (2009) Inflammatory mammary carcinoma in 12 dogs: clinical features, cyclooxygenase-2 expression, and response to piroxicam treatment. Can Vet J 50(5):506–510
13. Kleer CG, van Golen KL, Merajver SD (2000) Molecular biology of breast cancer metastasis. Inflammatory breast cancer: clinical syndrome and molecular determinants. Breast Cancer Res 2(6): 423–429
14. Marconato L, Romanelli G, Stefanello D et al (2009) Prognostic factors for dogs with mammary inflammatory carcinoma: 43 cases (2003–2008). J Am Vet Med Assoc 235(8):967–972
15. Perez Alenza MD, Tabanera E, Peña L (2001) Inflammatory mammary carcinoma in dogs: 33 cases (1995–1999). J Am Vet Med Assoc 219:1110–1114
16. Silva DMD, Kluthcovsky LC, Jensen de Morais H et al (2019) Inflammatory mammary carcinoma in a male dog-case report. Top Companion Anim Med 37:100357

Cutaneous Metastasis from Visceral Hemangiosarcoma

17. Bertazzolo W, Dell'Orco M, Bonfanti U et al (2005) Canine angiosarcoma: cytologic, histologic, and immunohistochemical correlations. Vet Clin Pathol 34(1):28–34
18. Guaguère E, Vanderkerckhove M, Mialot M (1994) Cutaneous metastases of a visceral haemangiosarcoma in a Brittany Spaniel. Prat Med Chir Anim Comp 29(5):479–480
19. Hargis AM, Ihrke PJ, Spangler WL et al (1992) Retrospective clinicopathologic study of 212 dogs with cutaneous hemangiomas and hemangiosarcomas. Vet Pathol 29(4):316–328
20. Wilkerson MJ, Chard-Bergstrom C, Andrews G et al (2002) Subcutaneous mass aspirate from a dog. Vet Clin Pathol 31(2):65–68

Other Rare Skin Metastasis

21. Di Maria FM, Annoni M, Roccabianca P, Antoniazzi E, Bertazzolo W (2020) Cutaneous metastases of prostatic adenocarcinoma in two dogs. Vet Clin Pathol 49(3):459–464
22. Dell'Orco M, Bertazzolo W, Vergine M et al (2005) Gastric mucinous adenocarcinoma with cutaneous metastases in a dog: diagnosis by fine-needle aspiration cytology. J Small Anim Pract 46(9): 449–453
23. Gorman E, Barger AM, Wypij JM et al (2006) Cutaneous metastasis of primary appendicular osteosarcoma in a dog. Vet Clin Pathol 35(3):358–361
24. Juopperi TA, Cesta M, Tomlinson L et al (2003) Extensive cutaneous metastases in a dog with duodenal adenocarcinoma. Vet Clin Pathol 32(2):88–91
25. Koehler JW, Weiss RC, Aubry OA et al (2011) Nasal tumor with widespread cutaneous metastases in a Golden Retriever. Vet Pathol 49(5):870–875
26. Lucas X, Rodenas C, Cuello C et al (2012) Unusual systemic metastases of malignant seminoma in a dog. Reprod Domest Anim 47(4):e59–e61
27. Otrocka-Domagala I, Pazdzior-Czapula K, Gesek M et al (2015) Aggressive, solid variant of alveolar rhabdomyosarcoma with cutaneous involvement in a juvenile labrador retriever. J Comp Pathol 152(2–3):177–178
28. Parachini-Winter C, Curran KM, Pellin M, Laver T, Hanot C, Vernier TH, Séguin B (2019) Cutaneous and subcutaneous metastasis of

appendicular osteosarcoma in dogs: 20 cases. J Vet Intern Med 33(5):2200–2208
29. Reed LT, Knapp DW, Miller MA (2013) Cutaneous metastasis of transitional cell carcinoma in 12 dogs. Vet Pathol 50(4):676–681
30. Spugnini EP, Bartolazzi A, Ruslander D (2000) Seminoma with cutaneous metastases in a dog. J Am Anim Hosp Assoc 36(3): 253–256
31. Takiguchi M, Iida T, Kudo T et al (2001) Malignant seminoma with systemic metastases in a dog. J Small Anim Pract 42(7):360–362
32. Tomlinson MJ, McKeever PJ, Nordine RA (1982) Colonic adenocarcinoma with cutaneous metastasis in a dog. J Am Vet Med Assoc 180(11):1344–1345
33. Conceição LG, Mesquita BM, Shmitz Duarte T et al (2011) Renal cell carcinoma with cutaneous metastasis in a dog. Braz J Vet Pathol 4(2):132–137

Index

F. Albanese, *Canine and Feline Skin Cytology*, https://doi.org/10.1007/978-3-031-95713-0

GPSR Compliance

The European Union's (EU) General Product Safety Regulation (GPSR) is a set of rules that requires consumer products to be safe and our obligations to ensure this.

If you have any concerns about our products, you can contact us on ProductSafety@springernature.com

In case Publisher is established outside the EU, the EU authorized representative is:

Springer Nature Customer Service Center GmbH
Europaplatz 3
69115 Heidelberg, Germany

Batch number: 10730948

Printed by Printforce, the Netherlands